# Gene Expression

*Volume 3*

# Gene Expression

*Volume 3*

*Plasmids and Phages*

Benjamin Lewin

Editor, *Cell*

*A Wiley–Interscience Publication*

JOHN WILEY & SONS

New York London Sydney Toronto

**Library of Congress Cataloging in Publication Data:**

Lewin, Benjamin M
Gene expression.

"A Wiley-Interscience publication."
Bibliography: p.

CONTENTS: v. 1. Bacterial genomes.—v. 2. Eucaryotic chromosomes.—v. 3. Plasmids and phages.
1. Gene expression. 2. Molecular genetics. [DNLM: 1. Genetics, Biochemical. 2. Genetics, Microbial. 3. Molecular biology. 4. Proteins—Biosynthesis. QH450 G326]

QH450.L48 575.2'1 73-14382

ISBN 0-471-53170-7 (v. 3)
ISBN 0-471-02715-4 pbk

Printed in the United States of America

10 9 8 7 6 5 4 3 2

*In spite of Jonathan*

# PREFACE

*Gene Expression* is intended to provide a critical analysis of the organization and expression of the genetic apparatus at the molecular level. In this series I hope to demonstrate the design that becomes apparent when the genetic operations of procaryotic and eucaryotic cells are analyzed from the perspective of the gene. The first volume, *Bacterial Genomes,* considers procaryotic nucleic acid and protein synthesis; the second volume, *Eucaryotic Chromosomes,* discusses the structure and function of the eucaryotic genetic apparatus. The present volume, *Plasmids and Phages,* attempts to define the events involved in the expression and reproduction of these small genetic elements. This volume is therefore closely related to the first: there I have considered replication, transcription and translation, and the systems responsible for their control in bacteria; here I consider the uses to which these systems are put, what may be thought of as the genetic strategies displayed by plasmids and phages.

Two themes constitute this book: the nature of the systems by which genetic information is transferred between bacterial cells; and the means by which the small elements that constitute the plasmids and phages are perpetuated. The transfer of genetic information between bacterial genomes is accomplished by transformation with DNA as well as through the conjugation promoted by plasmids and the transduction sponsored by some phages. Although not a plasmid or phage function, transformation is included so that discussion of genetic transfer is not limited to the achievements of plasmid and phage vectors. As a vector, the sex factor is the best characterized plasmid; and its counterpart among the lysogenic phages is, of course, phage lambda. Discussion of genetic transfer by vectors therefore focuses on conjugation and sexduction and on specialized transduction. As self-perpetuating elements, the sex factor and the drug factors related to it, and the smaller colicinogenic factors, are the best characterized; these are the focus of this discussion. Reproductive mechanisms are well characterized for many phages and this book therefore is able to provide fairly detailed accounts of their life cycles. This includes the lytic cycle of lambda as well as the virulent infection that is the only option for survival open to nonlysogenic phages. Obviously it is scarcely possible to consider in a single volume all the systems that now have been investigated and discussion therefore concerns the best characterized: the large phage T4, the smaller phages T3 and T7; the single stranded DNA phages; and the RNA phages.

In this book as in the earlier volumes, I have been concerned not simply to summarize what is thought to be the status quo, but to consider critically the original experiments and to establish the limits of the conclusions that can be drawn from them. Sometimes, of course, my conclusions may differ from those of the authors reporting the data. And obviously it is as important to ask questions as to provide answers: any attempt to define the conclusions supported by current data can hardly be divorced from consideration of their further implications. I have tried therefore to point to the questions that may now be asked and to consider the lines along which their resolution may be possible.

I have followed my usual policy on references; I have attempted to provide a complete citation of those papers that comprise the mainstream of research (some 2000 in this volume) but without necessarily referring to every confirmatory report. Virtually all the references cited represent papers published in the major research journals; only in those rare instances where no other source is to be found have I relied upon contributions to symposia and other such volumes.

Finally it is a pleasure to thank colleagues who have provided material included in this book and especially Drs. E. P. Geiduschek and W. Szybalski, who read some of the chapters and made many helpful comments. Although any expression of gratitude here can only be inadequate, I should like to conclude by thanking my wife, Ann, without whose encouragement this book would never have been started nor completed.

BENJAMIN LEWIN

*Cambridge, Massachusetts*
*January 1977*

# CONTENTS

# Gene Expression

*Volume 3*

CHAPTER 1

# Transformation

## Genetic Transfer by DNA Molecules

### *Discovery of Transformation and the Transforming Principle*

Transformation was the first type of transfer of genetic information to be discovered in bacteria; and it was the subsequent identification of the "transforming principle" with deoxyribonucleic acid that led to the development of the concept that genetic information is stored in the form of nucleotide sequences. Transformation was first observed in the study of pneumococcal infection of mice reported by Griffith in 1928. Pneumococcus causes pneumonia in man, but the bacterium is pathogenic for mice and so on infection may prove fatal. Virulence in pneumococci depends upon a component of the surface of the bacterium, the capsular polysaccharide, which confers protection against phagocytosis in the animal and gives colonies of virulent bacteria a smooth (S) appearance. Several "types" of pneumococcus differ from each other in the constitution of their capsular polysaccharide and thus in the antigenic response which can be provoked against them. Mutants may be derived (from any type) which fail to synthesize the capsular polysaccharide; these R bacteria produce colonies with a rough surface and are avirulent when injected into mice.

An appeal of this system for experimental analysis is that it is not difficult to distinguish R and S bacteria and to type the S strains according to their antigenic properties. As Griffith described the system, "virulence and type characters are closely related in the pneumococcus. When pneumococci are grown in homologous immune serum, some descendents become attentuated in virulence and these can be recognized by their formation on solid media of a distinctive variety of colony known as the R form of pneumococcus. The virulent or S form of pneumococcus produces in fluid media, and still more abundantly in the peritoneal cavity of the mouse, a soluble substance which, although not itself antigenic, gives a copious precipitate with the appropriate antiserum. Each type of pneumococcus forms a special soluble substance which has no affinity for an antiserum prepared against any other pneumococcal type and it is to this property that the remarkably clear definition of the serological races of pneumococcus is due. . . . As a result of its change to the R form, the pneumococcus generally loses this power of producing soluble substance.''

The critical observation which Griffith made in this system was that injection into mice of either avirulent mutant R bacteria or of heat killed virulent S bacteria did not harm the animal; but injection of a mixture of the two bacterial preparations killed some animals and from them he could recover live virulent S bacteria. When live R bacteria derived from type II were injected together with dead S bacteria of type I, the live virulent (S) bacteria recovered post mortem were of type I; thus some property of the dead type I bacteria allows them to confer upon living R bacteria the ability to make the capsular polysaccharide characteristic of type I. This change induced in the properties of the R bacteria was called *transformation*.

These experiments were confirmed and extended to other types of pneumococcus by Dawson at the Rockefeller laboratories in 1930, but, like Griffith, he was at first unable to demonstrate transformation in vitro. In 1931, however, Dawson and Sia converted R bacteria incubated in vitro with heat killed S bacteria to give live S bacteria; and Alloway (1932, 1933) was then able to use cell free and filtered extracts of dead S bacteria to transform R bacteria into the specific type of S bacteria from which the cell free preparations were derived.

The significance of Griffith's original observations was not realized at the time and the idea that the changes induced by transformation are hereditary was proposed only later. One early idea, for example, was that some autocatalytic utilization by the live R bacteria of the capsular polysaccharide from dead S bacteria might be implicated. The nature of the system was seen clearly, however, by Avery and his colleagues in their studies at the Rockefeller laboratories of the chemical nature of the transforming principle. In describing the pneumococcal system, McCarty, Taylor and Avery (1946) wrote: ''The phenomenon of transformation of pneumococcal types provides an outstanding example of the induction of specific and heritable modifications in microorganisms. Basically the phenomenon represents the transformation of a nonencapsulated (R) variant derived from one specific type of pneumococcus into encapsulated (S) cells of heterologous specific type. By the techniques employed at present, this is accomplished by growing the nonencapsulated cells in a special serum broth to which has been added the active fraction extracted from encapsulated pneumococci of a heterologous type. The production of a new polysaccharide capsule is induced in the R cells so that they acquire the type specificity of the organisms from which the extract was obtained. The property of forming the new capsule is transmitted indefinitely to subsequent generations, and, in addition the substance responsible for the induction of transformation is itself reduplicated in the transformed cells. It is thus apparent that one is dealing with hereditary bacterial modifications, which are predictable and subject to direct experimental control."

The importance of identifying the *transforming principle,* the name given to the active component of the cell free system, was obvious, and in introducing their classic results, Avery, Macleod and McCarty (1944) noted that "biologists have long attempted by chemical means to induce in higher organisms predictable and specific changes which thereafter could be transmitted in series as hereditary characters. Among microorganisms the most striking example of inheritable and specific alterations in cell structure and function that can be experimentally induced and are reproducible under well defined and adequately controlled conditions is the transformation of specific types of pneumococci. . . . The present paper is concerned with a more detailed analysis of the phenomenon of transformation of specific types of pneumococcus. The major interest has centered in attempts to isolate the active principle from crude bacterial extracts and to identify if possible its chemical nature or at least to characterize it sufficiently to place it in a general group of known chemical substances."

Chemical analysis of the transforming principle suggested that it comprised deoxyribonucleic acid. That it might be protein or ribonucleic acid was excluded by the failure of trypsin, chymotrypsin or ribonuclease to impair its activity. Confirmation that the transforming principle was DNA was provided by the experiments of McCarty and Avery (1946a), in which deoxyribonuclease was purified and it was shown that it irreversibly destroyed the transforming principle. Demonstration of the general nature of this conclusion came from the improved purification of transforming DNA which McCarty and Avery (1946b) used to achieve transformation of pneumococcus types II and VI as well as the type III previously used.

The significance of the identification of genetic material with DNA was immediately realized by Avery, who in 1943 described the isolation of the transforming principle in a letter (quoted by Hotchkiss, 1966). "When alcohol reaches a concentration of about 9/10 volume there separates out a fibrous substance which on stirring the mixture wraps itself about the glass rod like thread on a spool and the other impurities stay behind as a granular precipitate. The fibrous material is redissolved and the process repeated several times. In short, this substance is highly reactive and on elementary analysis conforms *very* closely to the theoretical values of pure desoxyribose nucleic acid (thymus) type (who could have guessed it). This type of nucleic acid has not to my knowledge been recognised in pneumococcus before, though it has been found in other bacteria. . . . If we are right, and of course that is not yet proven, then it means that nucleic acids are not merely structurally important but functionally active substances in determining the biochemical activities and specific characteristics of cells and that by means of a known chemical substance it is possible to induce predictable and hereditary changes in cells."

The implications of the conclusion that the transforming principle and the cellular component which it influences possess different chemical natures were well expressed by Avery, Macleod and McCarty. "The inducing substance, on the basis of its chemical and physical properties, appears to be a highly polymerized and viscous form of sodium desoxyribonucleate. On the other hand, the type III capsular substance, the synthesis of which is evoked by this transforming agent, consists chiefly of a nonnitrogenous polysaccharide constituted of glucose-glucuronic acid units linked in glycosidic union. The presence of the newly formed capsule containing this type specific polysaccharide confers on the transformed cells all the distinguishing characteristics of pneumococcus type III. Thus, it is evident that the inducing substance and the substance produced in turn are chemically distinct and biologically specific in their action and that both are requisite in determining the type specificity of the cell of which they form a part." By the early fifties, when studies had started on the nature of the cellular events involved in transformation, the distinction between genetic material and the products of its expression was implicit.

The implication of DNA in some genetic capacity was clear to the participants in these and subsequent experiments, although how such functions might be exercised remained a matter for imagination. In concluding their classic paper, Avery, Macleod and McCarty thus expressed the caution that "if the results of the present study on the nature of the transforming principle are confirmed, then nucleic acids must be regarded as possessing biological specificity the chemical basis of which is as yet undetermined." Extensive scepticism of the idea that DNA is the transforming principle prevailed among others, however, largely due to misapprehension of its structure. Because DNA was thought to have a simple, tetrameric repeating structure, objections were raised to the idea that it might have any biological specificity—its role was thought to be solely structural. Indeed, although its presence was recognized in eucaryotic chromosomes, it had not even been established as a component of all procaryotic cells. One frequent criticism was that a small amount of contaminating protein might be present to provide the active component in transforming DNA preparations, but this, of course, was inconsistent with the responses of transforming principle to deoxyribonucleases and proteinases. And increased purification of transforming DNA by Hotchkiss reduced to ridiculously small proportions the amount of contaminating protein that might be present (see Hotchkiss, 1966). Two important advances helped to establish the role of DNA as genetic material. One, of course, was the demonstration by Hershey and Chase in 1952 that DNA carries the genetic information of a bacteriophage (see Chapter 4). And another was the growing realization that the structure of DNA was more complex than hitherto thought, this leading into the elucidation of the double helix structure and culminating in the breaking of the genetic code.

In implicating DNA as the active agent in the transforming principle, it was of obvious importance to demonstrate that bacteria other than pneumococcus can exchange genetic information by transformation and to show that markers other than the capsular polysaccharide can be carried by transforming DNA. In 1951, Alexander and Leidy used R and S forms of Hemophilus influenzae to show that DNA preparations from type specific S cells can change nontypeable R cells into the specific type from which the DNA was derived. And working with pneumococcus, Ephrussi-Taylor (1951) showed that other capsular phenotypes in addition to the original R and S forms can be transferred by DNA preparations derived from appropriate bacteria. Also in 1951, Hotchkiss reported that resistance to penicillin can be transferred between pneumococci. The use of medium containing penicillin provided an easy selective technique to isolate mutants resistant to different levels of the antibiotic and then to identify transformants of sensitive cells (compared with the more cumbersome agglutination technique used to identify transformants of antigenic type). These experiments clearly excluded any interpretation that the role of DNA might be connected in some physiological (rather than hereditary) manner with the synthesis of capsular polysaccharide.

Identification of the genetic material with DNA was therefore by this time clear, as Hotchkiss noted. "The sudden appearance of penicillin resistant pneumococci after transformation so far removed in time from any exposure to penicillin, is a strong indication that the genetic mechanism which has preserved the resistance property in the donor strain has been transferred to the receptor strain. It is therefore of considerable interest that this property is apparently carried in the desoxyribonucleate fraction prepared from these cells." And in another early paper on transformation, noting that several characters in addition to the original R and S phenotypes could be transformed, Ephrussi-Taylor (1951) summarized the situation. "Transformation consists of the replacement of the nucleic acid of an autoreproducing cell element by a nucleic acid of an extraneous origin having a closely related structure and function."

### *Kinetics of Transformation*

Transformation has now been achieved in many species of bacteria, although it remains best characterized in pneumococcus and H.influenzae, with which the first experiments were performed. With these and other bacteria, transformation can be demonstrated for any characteristic whose acquisition be measured in the recipient bacteria. Quantitative studies to measure the frequency of transformation require markers which can be easily selected and counted in the recipient population, a condition not fulfilled by the original mutants of capsular polysaccharides. Resistance to antibiotics was used to provide such markers in early studies of transformation of pneumococcus and H.influenzae; B.subtilis later came into use as a system for studying transformation and the many auxotrophic mutants available have extended the range of markers that can be utilized. We shall concentrate here largely on these three bacterial species.

Transformation involves two successive types of event: the transfer of genetic material from donor to recipient; and the utilization of donor genetic material within the recipient. Because transformation is achieved by the addition of donor DNA to recipient bacteria, this method of genetic transfer differs from others in making it possible to establish by direct experiment the structural parameters that are demanded for genetic activity in transforming DNA. Questions such as whether the DNA must be duplex or single stranded, whether there are any necessary minimum or maximum lengths, can be answered by subjecting the donor DNA molecules to appropriate treatments before transformation. By using preparations of radioactively labeled transforming DNA, it is possible also to follow the fate of the nucleic acid after its entry into a recipient cell. Such experiments show that transforming DNA enters recipient cells most effectively when provided in the duplex state, but that only one of the two donor strands is utilized for providing genetic information after uptake; recombination with the host genome is necessary to establish donor DNA in the recipient cell as a source of genetic information, and following such integration the donor sequence is replicated as part of the recipient and its descendent genomes.

A characteristic of all transformation systems is that an increase in the concentration of the transforming DNA preparation causes a proportional increase in the number of transformants, up to a plateau level. Figure 1.1

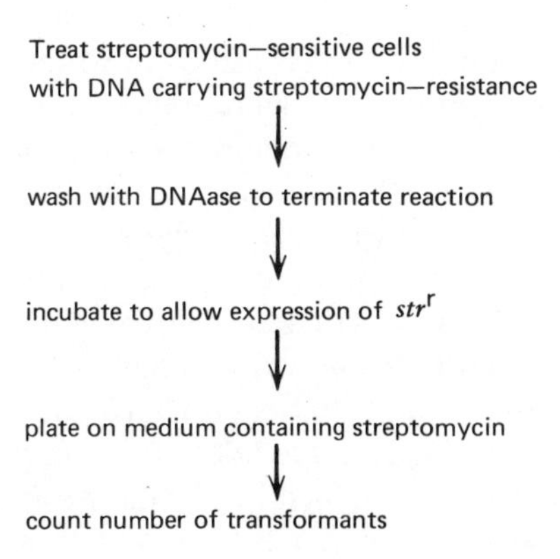


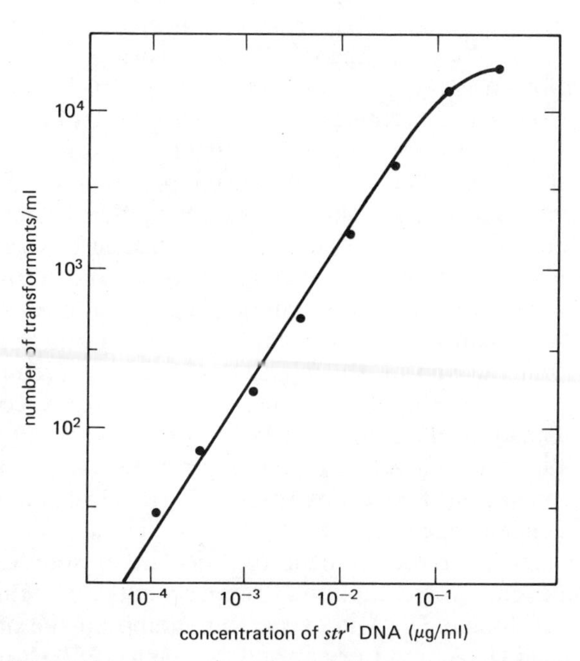


**Figure 1.1:** relationship between number of transformants and concentration of transforming DNA. The protocol used in this experiment is given above the results, which show that the number of transformants increases in proportion to the concentration of transforming DNA up to a plateau level at which recipient bacteria are saturated with transforming DNA. Data of Hotchkiss (1957).

shows an experiment of Hotchkiss (1957) in which streptomycin-sensitive recipient cells of pneumococcus were exposed to varying concentrations of DNA extracted from streptomycin-resistant bacteria. After 5 minutes of exposure to the donor DNA preparation, cells were incubated to allow the donor *str*$^r$ gene to be expressed in the recipient cells and then plated on medium containing streptomycin. The number of survivor colonies gives the number of transformants. Similar experiments have been reported with other markers and with other systems (see Hayes, 1968). These one-hit kinetics imply that each transformation event results from interaction of a single DNA molecule with a recipient bacterium.

That incorporation of transforming DNA into a recipient bacterium is necessary for acquisition of a new character is shown by the use of labeled donor DNA. By using $^{32}$P labeled DNA to transform pneumococci for streptomycin-resistance, Lerman and Tolmach (1957) showed that the incorporation of label into recipient cells is proportional to the frequency of transformation. It is therefore the amount of donor DNA taken up by recipient bacteria that determines the probability of transformation.

The plateau reached when addition of further transforming DNA does not increase the number of transformants represents saturation of the ability of recipient bacteria to take up the donor DNA. Experiments with both Hemophilus and pneumococcus have confirmed this conclusion by showing that addition of nontransforming DNA (extracted from bacteria of the recipient type) to the transforming preparation does not alter the number of transformants when the total DNA concentration is below the plateau level; but the nontransforming DNA becomes a competitive inhibitor of the transforming DNA when the total DNA concentration exceeds the plateau level (Hotchkiss, 1954; Alexander, Leidy and Hahn, 1954). Below the plateau level, addition of more DNA thus results simply in increased uptake, but above this level DNA molecules must compete for entry to recipient bacteria.

There appears to be no requirement for specificity of source for DNA to enter pneumococci. In competition experiments, Hotchkiss (1954) showed that calf thymus DNA in excess can inhibit uptake of transforming pneumococcal DNA; and Lerman and Tolmach (1957) showed that $^{32}$P labeled E.coli DNA is taken up as efficiently by pneumococci as the homologous transforming pneumococcal DNA. Both Streptococcus and B.subtilis show a similar lack of demand for specificity in uptake, but Schaeffer et al. (1960) reported that uptake in Hemophilus is restricted to DNA preparations of related species (see also Socca et al., 1974). Heterologous DNA taken up by recipients probably is treated in the same way as homologous DNA, except that upon failure to recombine with the host genome it becomes degraded (see Piechowska et al., 1975). An exception to the lack of specificity in uptake by B.subtilis and streptococcus is seen with glucosylated T4 DNA, which is taken up only very poorly; since the phage DNA in nonglucosylated form is taken up as efficiently as homologous DNA, the presence of glucose residues on the DNA must reduce its affinity for the cellular receptors (Soltyk et al., 1975; Ceglowski et al., 1975).

No formal upper limit has been set on the size of transforming DNA, but breakage during preparation of DNA of pneumococci or H.influenzae during preparation usually means that the size is of the order of $10^7$ daltons (about 10 genes). (Further breakage may take place during uptake so that the final length of donor DNA within the cell is less than this.) Only about 0.5% of the genome is therefore carried on each molecule of transforming DNA, so that there is only this low probability that any individual donor molecule will carry a particular genetic marker. Thus Hayes (1968) calculated that when the saturation plateau is reached in a typical Hemophilus or pneumococcal transformation, between 100 and 200 donor molecules have been taken up by the entire recipient population for each transformed bacterium.

Early experiments suggested that a minimum length of DNA might be necessary for transformation; Litt et al. (1958) showed that DNA of less than $10^6$ daltons did not appear to be taken up by pneumococci. Subsequent experiments, however, suggested that transformation can take place with lengths of DNA smaller than this. Cato and Guild (1968) fractionated pneumococcal DNA and measured the transformants per unit weight of DNA as a function of its size. The activity fell from $8 \times 10^5$ transformants per $\mu$g of DNA at a size of $1.2 \times 10^6$ daltons to $8 \times 10^3$ transformants per $\mu$g of DNA at a size of $2.9 \times 10^5$ daltons. Morrison and Guild (1972a) reported similar experiments with B.subtilis; transforming activity of a DNA preparation declines with size, but no minimum critical size was noted. The effect of length is probably concerned with events following uptake, when donor DNA must be integrated into the recipient genome; the probability of this occurrence may depend on donor length.

Transformation depends upon the incorporation into the recipient genome of donor DNA carrying new genetic information. In experiments using $^{32}P$ labeled transforming DNA carrying a marker for streptomycin resistance, Fox and Hotchkiss (1960) followed the transforming activity of this donor DNA as a function of radioactivity. Streptomycin-sensitive bacteria were exposed to the transforming DNA for about 30 minutes, washed with DNAase to remove any DNA that had not been taken up, and then the DNA was reextracted and tested for transforming activity by addition to a fresh culture of streptomycin-sensitive recipients. The transforming activity/$^{32}P$ unit at first has an activity comparable to that of the original donor DNA and with subsequent growth of the bacteria the ratio increases in parallel with the total bacterial number. The marker carried by the original donor DNA has therefore become part of the recipient genome and is replicated in each division cycle.

Five minutes after entry into a recipient, a length of donor DNA displays its original ability to transform a genetic marker; and this activity is multiplied with its subsequent replication. Before its original properties are regained at 5 minutes, however, the transforming DNA passes through an *eclipse phase* during which only a small part of its activity can be recovered. Following the protocol illustrated in Figure 1.2, Fox (1960) exposed p-nitrobenzoic acid-resistant pneumococci to transforming DNA carrying the streptomycin-resistance marker for 3 minutes, washed for 1 minute with DNAase to remove extraneous DNA molecules, and then reextracted DNA after varying times. To provide a control by which the activity of the donor (streptomycin-resistant) marker can be assessed, the ability of the reextracted DNA to transform for the recipient marker, p-nitrobenzoic acid resistance, was measured as well as its ability to transform for streptomycin-resistance. Immediately after exposure to DNA, the ratio of streptomycin-resistance to p-nitrobenzoic acid resistance activity in the reextracted DNA was very low (about 5% of its ultimate level), but increased with longer periods of incubation before extraction to a maximum value between 5 and 10 minutes. Further incubation then fails to promote any increase in the ratio. A donor transforming DNA therefore loses its transforming activity immediately upon entering a recipient, but regains it during the 5 minute eclipse period; at the end of this period, the donor marker has become part of the recipient chromosome and so its activity relative to a recipient marker has reached a plateau. Venema et al. (1965a,b) reported a similar eclipse phase in B.subtilis. However, no eclipse was detected in H.influenzae in a study by Voll and Goodgal (1965).

Changes in the state of transforming DNA after entry into a recipient suggest that only one strand of the donor DNA is utilized by the recipient. Lacks (1962) treated pneumococci with $^{32}P$ labeled transforming DNA for 5 minutes and then incubated the cells for 0, 3, 10 or 40 minutes before extracting the DNA from the recipients and examining the state of the label. The reextracted DNA was separated by chromatography on MAK (methylated albumin kieselguhr) columns into four classes: degraded fragments (oligonucleotides or nucleotides), sequences eluting together with RNA, single strands of DNA and duplex strands. Table 1.1 shows that immediately after entry, almost none of the radioactivity is present as duplex DNA; about half is in the form of degraded fragments, a small amount separates with RNA (due simply to the technical conditions of

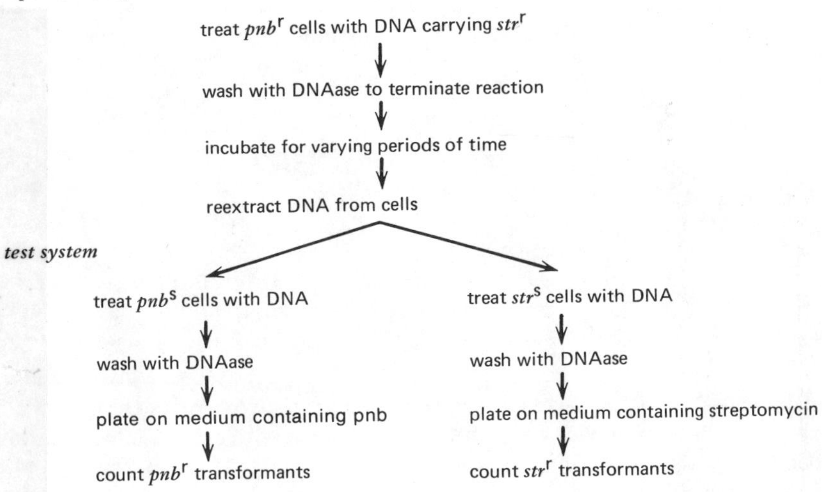


*results*

number of $str^r$ transformants/number of $pnb^r$ transformants in test system represents activity of original $str^r$ donor marker in DNA reextracted in experimental system

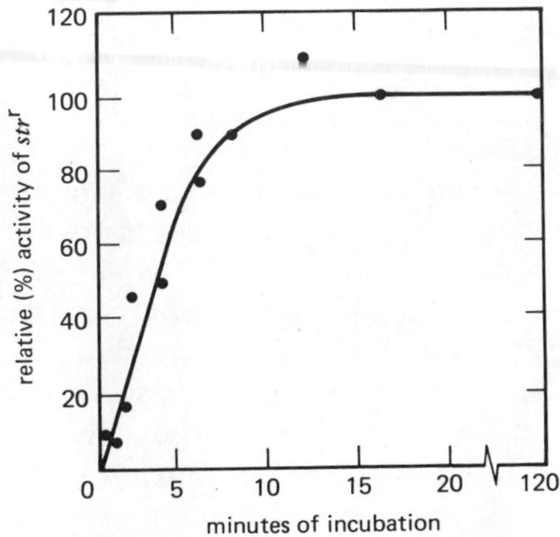


**Figure 1.2:** use of reextraction experiments to follow fate of transforming DNA. In the experimental system, recipient cells carrying the marker $pnb^r$ are treated with donor DNA carrying the marker $str^r$. DNA is extracted from these cells after varying periods of time and then the test system is used to determine its activity for transforming the recipient and donor markers. After short periods of incubation, the donor marker activity is low relative to the control of the recipient marker activity; during this eclipse phase the donor marker activity increases until after 10–15 minutes it reaches its maximum level. Data of Fox (1960).

**Table 1.1:** fate of transforming DNA in recipient cell

| | minutes of incubation after addition of $^{32}P$ DNA | | | |
|---|---|---|---|---|
| | 0 | 3 | 10 | 40 |
| percent of counts in: | | | | |
| degraded fragments | 50 | 47 | 32 | 20 |
| RNA-associated | 0 | 6 | 10 | 15 |
| single stranded DNA | 47 | 18 | 3 | 3 |
| native DNA | 3 | 29 | 55 | 62 |
| percent of maximum recovery of streptomycin resistance marker | 5 | 50 | 100 | 100 |

Immediately after $^{32}P$ labeled DNA has been incubated with recipient cells for 5 minutes, half of the radioactivity is found in the form of degraded fragments and almost all the remainder in the form of single strands. Upon incubation, the proportion of single stranded DNA declines and the proportion of counts in native DNA increases. A small amount of DNA separates on these columns with RNA; the decline with time of the counts in the degraded fraction is presumably due to reutilization. The recovery of a streptomycin resistance marker is shown relative to the final level achieved and correlates well in time with the decline in single stranded and increase in native DNA. Data of Lacks (1962).

operating such columns), and about half appears in the form of single strands. After 3 minutes, some of the $^{32}P$ is present as duplex DNA and this proportion increases with time. As can be seen from the table, the time of loss of label from single strands and appearance in double strands corresponds well with the time of recovery (assayed by genetic means) of a streptomycin-resistance marker. In similar experiments, Lacks, Greenberg and Carlsson (1967) followed the state of DNA at earlier times after transformation. The results again showed a good correlation between decline of label in single strands, appearance in double strands, and recovery of a streptomycin-resistance marker. These results suggest that one strand of the duplex DNA is degraded immediately upon entry into the recipient, causing loss of its transforming activity; the eclipse period represents the time that it takes for the single strand to reenter the duplex state, regaining its ability to transform. By 5–10 minutes after uptake, the single strand has been integrated into recipient DNA (presumably by replacing the corresponding recipient strand) and is subsequently replicated as part of the recipient genome. That pneumococcal DNA is in a single stranded state during the eclipse period is shown also by the results of an experiment in which Ghei and Lacks (1967) reextracted DNA shortly after uptake, annealed with single strands, and then determined its transforming activity; this treatment results in the recovery of a donor marker from its state of eclipse, presumably by effecting its conversion from the single stranded to duplex state. The degradation of one strand of donor DNA is not specific for sequences homologous to the recipient chromosome, for Lacks, Greenberg and Carlsson (1967) found that a similar fate is suffered by E.coli DNA taken up by pneumococci, although, of course, no subsequent integration occurs.

A similar conversion of duplex donor DNA into single stranded molecules that represent the intermediate structures preceding the formation of recombinants has been observed in the transformation of B.subtilis (Piechowska and Fox, 1971; Davidoff-Abelson and Dubnau, 1973a,b). In H.influenzae there appears to be only a more limited extent of conversion of donor duplexes into single strands (Sedgwick and Setlow, 1976; and see below).

The mechanism of genetic recombination in transformation is therefore the replacement of a single strand sequence in the recipient chromosome by the corresponding sequence of the donor single strand. The physical association of donor DNA with the recipient genome is paralleled by the appearance of recombinant DNA that carries both donor and recipient genetic markers (see Ghei and Lacks, 1967; Voll and Goodgal, 1961). After transformation, there is a lag before the donor gene product is synthesized, reported at 6 minutes for pneumococcus by Lacks and Hotchkiss (1960b) and apparently somewhat longer in other systems. After this a constant rate of expression is attained. The delay between uptake of donor DNA and expression of the markers that it carries presumably reflects the need for its insertion into the recipient genome to regain the duplex state before it can be expressed. This sequence of events is illustrated in Figure 1.3.

As might be predicted from this sequence of events, uptake of transforming DNA by a recipient cell is not sufficient to ensure transformation. After entry, the donor DNA must be integrated into the host genome before it can be reproduced as part of the recipient DNA and can function as a template for genetic expression. Integration presumably does not occur in every cell which takes up donor DNA. Using transforming DNA carrying information allowing pneumococcal mutants lacking amylomaltase to synthesize the enzyme, Lacks and Hotchkiss (1960) found that cells in different mutant cultures may take up the same amount of wild type DNA, but may show different extents of genetic transformation. Steps subsequent to the uptake of DNA, culminating in the integration of

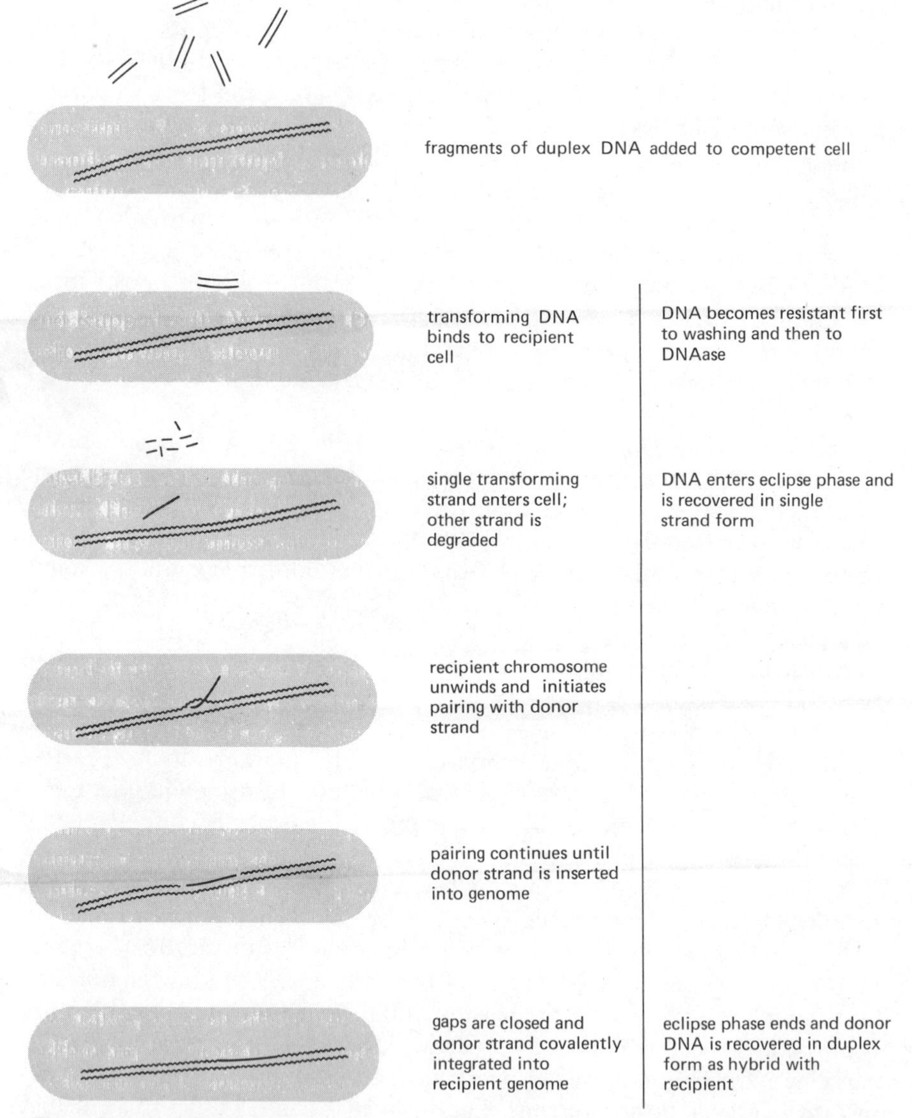


**Figure 1.3:** model for transformation of pneumococcus. The steps through which transforming DNA passes are shown but their sequence is not entirely established; for example, it is not clear whether one strand of the transforming DNA is degraded during or following entry into the cell. Phenotypic expression of a donor marker takes place only after the final stage shown here of covalent integration into the recipient genome.

a donor sequence as part of the recipient genome, are therefore important in establishing the extent of transformation; this may depend on the individual mutation and these marker effects are discussed later.

A variation on transformation with bacterial DNA is to use as donor material the DNA of a phage for which the recipient bacterium can act as host. Following uptake, the phage DNA enters the normal life cycle, which with a lytic phage leads to lysis of the cell and the release of progeny phage particles. In this case, there is no need for the donor DNA to be integrated in the resident genome. This is known as *transfection* and is assayed by the formation of plaques when infected bacteria burst to release progeny particles. The efficiency of transfection, however, is generally only about $10^{-5}$–$10^{-6}$ of that seen when the DNA is introduced by injection following adsorption of the phage particle (for review see Notani and Setlow, 1974). This protocol may be useful for investigating the phage cycle, especially with mutants that are defective in the stages prior to injection of DNA.

## Entry and Integration of Donor DNA

### *Competence of Bacterial Populations*

During the early development of transformation systems it was found that only bacteria cultured under certain conditions could support transformation; and this led to the use of transformation "rituals" to obtain bacteria in a receptive condition. As early as 1946, McCarty, Taylor and Avery observed that physiological conditions influence frequencies of transformation and suggested that changes in the cell membrane might be responsible for controlling ability to take up DNA. It soon became apparent that the *competence* of bacteria to be transformed is not a permanent feature, even once acquired under suitable growth conditions, but represents a transient phase in the life of a population (Hotchkiss, 1954; Thomas, 1955). The time of occurrence of competence and its duration is characteristic for each bacterial species and varies with the conditions of growth.

A common observation has been that competence develops towards the end of logarithmic growth, some time before a population enters the stationary phase. The development of competence can be followed by determining the number of transformants in a population exposed to transforming DNA at various times after the initial inoculation into the medium. Ephrussi-Taylor and Freed (1964) reported that under their conditions competence developed 90–100 minutes after initiating a culture of pneumococcus and lasted for 7–15 minutes. Although competence develops synchronously among the cells of a culture, this is not correlated with cell division; the timing appears to be a consequence of establishment of a critical cell concentration rather than of a change in the growth phase.

Typical data on the development of competence in pneumococcal cultures were reported by Tomasz and Hotchkiss (1964), who found that noncompetent bacteria grow in a suitable medium for a time which is inversely proportional to the starting number. Towards the end of the growth phase, a small number of competent cells appears. As Figure 1.4 shows, competence then develops very rapidly, the number of competent cells doubling every 3–5 minutes (much faster than the rate of cell growth). Although development of competence is a function of cell concentration, occurring when a critical concentration is reached, it is relatively independent of the growth rate and phase of growth. The duration of competence is limited, less than two hours in the culture of pneumococcus shown in Figure 1.4. The competent period is not invariable but may depend on growth conditions; for example, Lawson and Gooder (1970) reported that its duration in Streptococci of strain Challis is 6 hours in medium supplemented with serum but only 1 hour in synthetic medium.

Data on the development of competence reveal only the time at which competence develops in the population and do not reveal what proportion of its cells acquire competence. Nor do they show whether the duration of competence is the same in all cells: does competence develop and disappear synchronously in the population or is its duration shorter than the competent period, with the time of acquisition varying so that at any moment during the competent phase only some cells are in the competent state? The proportion of competent cells can be calculated by genetic analysis, by comparing the frequency observed for double transformation of two unlinked markers with the frequency predicted from the product of the single transformation frequencies (see below); in general this is high in pneumococcus, often apparently 100%, high also in H.influenzae, but much lower (10–20%) in B.subtilis. An attempt to measure the proportion of competent cells by autoradiography, reported by Javor and Tomasz (1968), is consistent with these calculations; 100% of the pneumococcal cells can take up labeled DNA, but only 15% of a competent population of B.subtilis can do so. These values imply that in pneumococcus and H.influenzae, when all cells may be competent during the competent phase, competence develops synchronously in the population; differences in the physiological state of competent and noncompetent B.subtilis cells (see below) suggest that this is true also of the Bacillus.

An important characteristic of the development of competence in

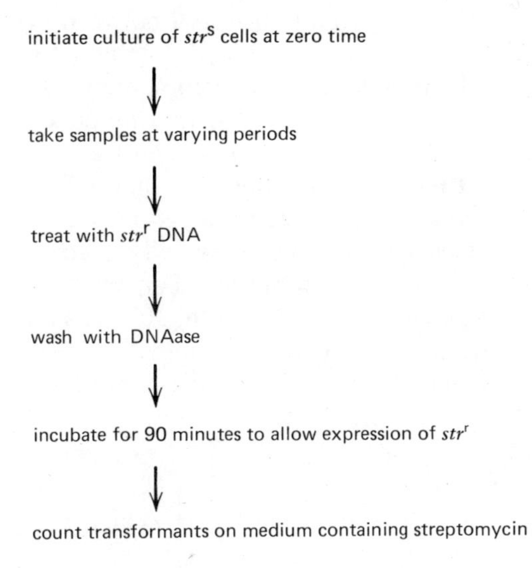


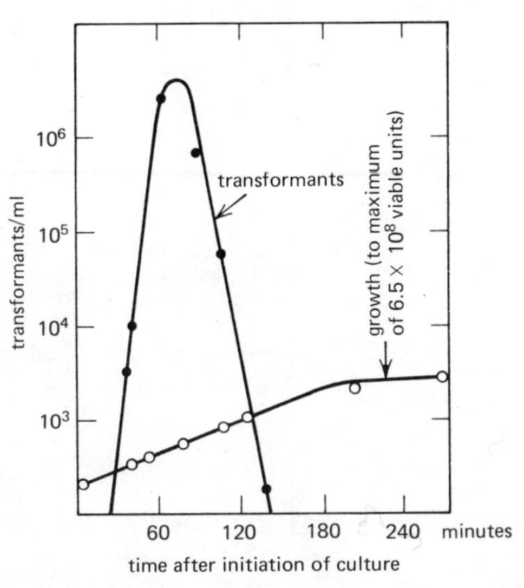


**Figure 1.4:** development of competence in pneumococcus. The protocol used in this experiment is illustrated above the results, which show that competence develops rapidly in the culture and is only of limited duration. Data of Tomasz (1965).

pneumococci is that it can be transferred between cells. Once induced by the presence of a few competent cells, competence therefore spreads rapidly through the population. Since competence spreads at a rate much faster than the growth rate, it is clear that its development is not due to any parameter such as exhaustion of culture medium. As Figure 1.5 shows, mixing experiments demonstrate that a small number of competent cells may cause competence to develop in a much larger number of noncompetent cells. Using two cultures which could be distinguished by their resistance to different antibiotics, Tomasz (1965) mixed $10^6$ cells/ml of a highly competent culture (*a*) with $10^7$ cells/ml of a noncompetent culture (*b*). Some 80–90% of the noncompetent cells became competent, although the original competent cells constituted only 10% of the population.

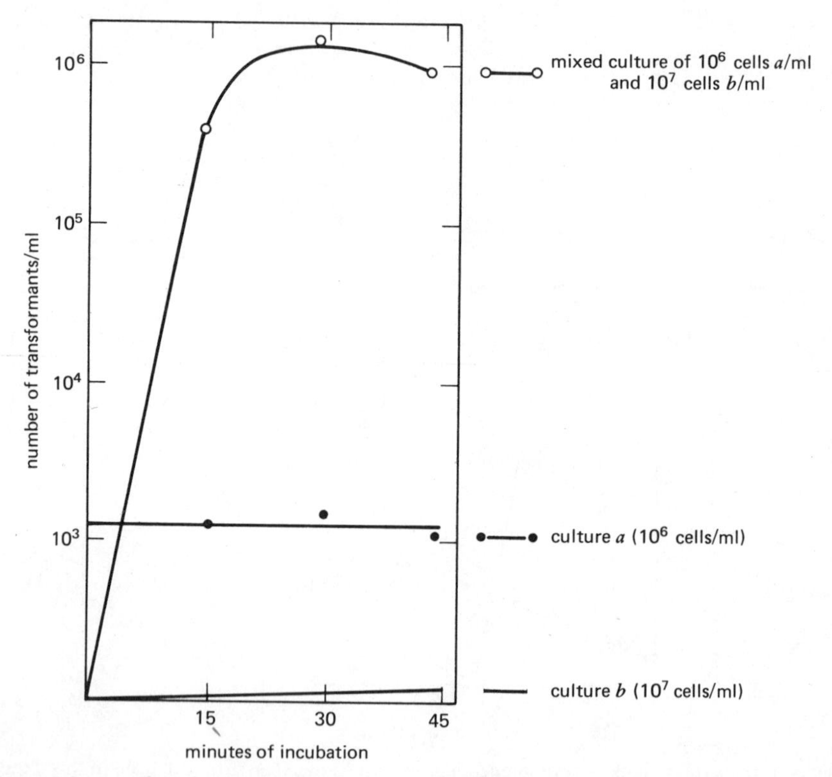


**Figure 1.5:** transfer of competence between cells. When a competent culture (*a*) is mixed with a much larger number of cells of a noncompetent culture (*b*), the number of transformants in the mixed culture is much greater than that in the competent culture. Competence has been transferred from the competent cells to the noncompetent cells in the mixed culture. Data of Tomasz (1965).

Transfer of competence does not involve any physical contact of cells since it can still occur when the two cultures are separated by a millipore filter; but a dialysis membrane prevents transfer of competence, implying that some macromolecule may be implicated. The development of competence in a culture is closely correlated with its ability to transfer competence when incubated with another culture, implying that this transfer is the mechanism responsible for the usual spread of competence through a population. Isolation of the factor inducing competence (the CF, competence factor) was first reported for Streptococcus strain Challis by Pakula and Walczak (1963). The factor can be isolated only from competent cells and provokes competence when it is inoculated into a noncompetent culture (comprising cells grown in poor culture conditions); it is able also to provoke competence in cells of another Streptococcus strain, Wicky, that usually never develop competence when grown in isolation and can be induced to do so only by interaction with strain Challis. This factor was purified by Leonard and Cole (1972); it is a small protein, less than 5000 daltons in size, and is very basic, with an isoelectric point of about pH 11.0. The competence factor of pneumococcus was isolated by Tomasz (1965) in the form of a preparation whose activity was sensitive to trypsin; Tomasz and Mosser (1966) reported that the factor can be extracted only from competent cells and is a protein of about 10,000 daltons. The protein does not interact with DNA in vitro and its activity in vivo is probably mediated at the cell surface.

Two classes of theory have been advanced to account for the development of competence. One model is to suppose that the structure of the cell wall is critical and that it is in a state permitting uptake of DNA only during the restricted period which defines the competent phase. In the form proposed by Spizizen et al. (1966), this model ascribes a more or less passive role to the cell wall, postulating that its permeability to macromolecules may change with growth conditions. An alternative model, favored by more recent results, is to suppose that competence results from the synthesis of specific receptor sites at the cell surface; this view is supported by the kinetics of interaction of competent pneumococci with DNA (which appear in general to follow those of classical enzyme-substrate reactions) and by the demonstration that new proteins must be synthesized for the acquisition of competence. Of course, these models are not mutually exclusive; it is possible, for example, to postulate that both general structural changes and the establishment of specific receptor sites are necessary for the development of competence. But although changes in cell surface properties to allow uptake of DNA are prerequisite for development of competence, they are not sufficient; for it is possible to persuade bacteria to take up DNA by mechanisms independent of the usual system, so that both competent and noncompetent bacteria gain the DNA; but only the competent bacteria are able to utilize its genetic information. In addition to whatever changes are necessary to permit uptake of DNA, the establishment of competence must thus include development of systems concerned with the stages of transformation succeeding uptake.

Competent cells of pneumococcus possess an antigen (detected by reaction with rabbit serum) which is absent from noncompetent cells. Tomasz and Beiser (1965) observed that production of this antigen is provoked when competence factor is added to noncompetent cells. The reaction against the antigen inhibits the development of competence both in cells which develop it in culture and in cells in which its production is stimulated by addition of extraneous competence factor. This is consistent with the idea that the competence factor acts on cells to induce competence in normal cultures, one step in the response of the cells being the synthesis of a new protein. A mutant strain of pneumococcus that does not depend upon the competence factor has been isolated by Lacks and Greenberg (1973), by selecting for cells able to be transformed in the presence of trypsin (which destroys the competence factor). These cells clearly must differ from wild type in the mechanism by which they acquire competence.

Inhibition of protein or RNA synthesis during treatment with competence factor prevents acquisition of competence, suggesting that synthesis of new proteins is necessary for its development. Tomasz (1970) followed the effects of inhibition of various synthetic processes in cells at different stages of transformation: being treated with competence factor; in the competent cells before addition of DNA; and during DNA uptake. Inhibition of DNA synthesis has no effect on any of these processes. Inhibition of protein or RNA synthesis inhibits transformation when added during acquisition of competence or before addition of DNA to competent cells; it is not clear what effect it may have at the time of DNA uptake. Establishment of the competent state thus requires protein synthesis, both in order to synthesize competence factor and then to synthesize the proteins whose production it stimulates. The same situation prevails in H.influenzae and in Streptococcus, in which Ranhand and Lichstein (1969) and Leonard (1973), respectively, found that inhibition of protein synthesis prevents the onset of competence.

The importance of the cell wall in the development of competence is made clear by the results reported by Tomasz, Zanati and Ziegler (1971). Choline residues of the techoic acid component of the pneumococcal cell envelope are needed for several functions; the absence of choline results in defective cell division because daughter cells are unable to separate.

Choline deprivation prevents cells from responding to the competence factor by binding DNA and achieving transformation; and the ability of cells to respond to competence factor disappears rapidly after transfer to a growth medium in which ethanolamine is substituted for choline.

Because the pneumococcal autolytic enzyme degrades cell walls that contain choline, but fails to attack regions in which ethanolamine has been substituted, it is possible to use it to assay the incorporation of choline into the walls of bacteria previously grown for several generations in ethanolamine. Reappearance of sensitivity to the enzyme correlates well with the recurrence of ability to respond to competence factor. The incorporation of new choline molecules into the cell surface takes place at a growing zone located in the equatorial region of the bacterium; and it is possible that it is this growth zone of the envelope that possesses sites where DNA molecules enter during transformation. That competence is correlated with changes in the cell surface is implied also by the observation of Javor and Tomasz (1968) that competent cells of pneumococcus show a tendency to aggregate not typical of noncompetent cells. Tomasz and Zanati (1971) reported that agglutination results from the deposition of a trypsin-sensitive component on the pneumococcal spheroplast membrane after competence factor has bound to its cellular receptors.

Changes in membrane permeability that may reflect alterations induced by the competence factor have been observed by Seto and Tomasz (1975) in experiments in which pneumococci were treated with activator under the conditions needed to induce competence and then were transferred to certain postincubation media. Depending on the nature of the medium, cells either became leaky or were converted to protoplasts (cells lacking the rigid outer cell wall but retaining the membrane) by the action of cellular autolysin. An attractive interpretation of these results is to suppose that these changes in the outer cell structure mimic those that usually occur when cells become competent, but are exaggerated in magnitude by the suspension in postincubation medium. This suggests a model in which competence factor acts upon the cell to cause a limited release of autolysin into the periplasmic space (the region between the membrane and the cell wall). Then the autolysin might weaken the cell wall sufficiently to make receptors for DNA accessible to donor molecules in the medium.

Competence in B.subtilis differs in several respects from the state shown by pneumococcus or H.influenzae. Although competence develops at a comparable time, towards the end of the growth phase, only a minority of the cells becomes competent and the duration of the competent phase is longer, usually several hours. Goldsmith et al. (1970) reported that treatment with pronase has no effect upon the development of competence and observed also that mixing experiments show no transfer of competence between cultures. This suggests that any component comparable to the competence factor of pneumococcus must be tightly bound to the cell wall and so is unable to pass between cells. This leaves the implication that it is solely the physiological state of the cell that determines competence, in which case all those cells that become competent must independently enter this state during the same period. Although a critical cell concentration again appears to be necessary for acquisition of competence, the rate of its development is independent of concentration. Perhaps the smaller proportion of competent cells and the increased duration of competence in B.subtilis reside in differences from other bacterial systems in the mechanism of acquisition and loss of competence.

Competent cells of B.subtilis display clear differences in physiological state from noncompetent cells. Nester and Stocker (1963) observed that transformants of B.subtilis are resistant to killing by penicillin for some time following transformation; and Nester (1964) observed that this resistance is displayed while B.subtilis cells are in the competent phase and is lost when competence disappears. Since most (about 90%) of the cells of a recipient culture are killed by penicillin in such experiments, presumably competence has developed only in the surviving 10%. Penicillin acts only on growing bacteria, implying that competence is a property displayed only by nongrowing cells.

Another experiment supporting this conclusion was performed by McCarthy and Nester (1967), who used a tritium suicide technique to investigate macromolecular synthesis in competent B.subtilis. Cells are killed by the incorporation of highly tritiated precursors only if they are engaged in synthesis of the appropriate macromolecule. According to this criterion, protein synthesis is normal in competent cells but RNA synthesis is somewhat reduced; DNA synthesis shows the greatest reduction. A further demonstration that there are gross differences between competent and noncompetent cells was the direct observation of Dooley, Hadden and Nester (1971) that they can be separated on a density gradient: competent cells appear to be smaller and less dense.

Experiments suggesting that competent cells of B.subtilis are not engaged in replication were reported by Bodmer (1965), who made use of a thymine-requiring strain that suffers thymineless death if incubated in medium lacking thymine while it is engaged in DNA synthesis. (This phenomenon is discussed in Volume 1 of Gene Expression.) Acquisition of competence, however, is not prevented by lack of thymine, implying that competent cells are not replicating their DNA. And in another experiment, just before donor DNA was added, competent cells were transferred to a medium containing the heavy precursor analogue, bromodeoxyuridine (BUdR), whose incorporation increases the density of DNA. After transformation, the activity of the donor DNA remained largely in the light part of the gradient; since it was not incorporated into regions containing BUdR, the competent cells could not be synthesizing DNA. In noncompetent cells, by contrast, preexisting recipient DNA enters the hybrid state as a result of replication. When cells are incubated with BUdR during and following transformation, the donor DNA continues to display little association with the hybrid label; newly transformed cells also therefore fail to engage in DNA synthesis. These experiments thus imply that competence is achieved only by nonreplicating cells and that this inactivity continues for some time after the entry of transforming DNA. That cessation of cell division is necessary for acquisition of competence has been reported by Spencer and Herriott (1965) for H.influenzae also.

Following entry of transforming DNA into cells of pneumococcus, integration takes place fairly rapidly and the transforming gene functions may be replicated as part of the host genome and expressed within a period measured in minutes. A much greater lag between the integration of donor DNA and expression of the transformed phenotype is shown in B.subtilis, however, in which transformants may become detectable only after some hours. In reextraction experiments when donor DNA carried a *try*$^+$ marker and recipient bacteria carried a *his*$^+$ gene, Stocker (1963) and Nester and Stocker (1963) found that *try*$^+$ *his*$^+$ recombinant DNA was formed within 30 minutes after uptake. Yet the transformed phenotype is not expressed (as measured by the synthesis of tryptophan synthetase) until some four hours after uptake. When the temperature is increased so that the cells grow more rapidly, the lag time is reduced, so that it seems to be related to the growth rate of the culture. Transformants remain resistant to penicillin killing for 3–5 hours after uptake of DNA; and no increase in the number of transformants is observed during this period. These results therefore suggest that failure to multiply explains the lag before the number of transformants rises and a consequence of this physiological state must be the delay in expression of donor genes. In other experiments, Chasin and Magasanik (1970) found a delay of only about 45 minutes before expression of the histidase gene following transformation of B.subtilis. No way to reconcile the difference between these sets of data is immediately apparent, but it remains clear that the lag period in B.subtilis is greater than in pneumococcus or H.influenzae.

Early attempts to achieve transformation with E.coli all were unsuccessful. However, Mandel and Higa (1970) found that treatment with $CaCl_2$ allowed E.coli cells to take up noninfective phage DNA; Cohen,

Chang and Hsu (1972) found that under these conditions E.coli cells become effective recipients for plasmid DNA. The $CaCl_2$ probably causes changes in the structure of the cell wall that are necessary for uptake of DNA. In both these sets of experiments, the donor DNA molecules were autonomous replicating units that then were maintained in this form; they did not undergo recombination with the bacterial genome. Using the $CaCl_2$ treatment, Cosley and Oishi (1973) were able to transform E.coli cells for several nutritional markers; they found that it is necessary to use recipients that are defective in the *recBC* enzyme system. These genes code for a nuclease that is part of a recombination-repair pathway; presumably in the *recBC*$^+$ cells the transforming DNA is degraded before it can recombine with the resident genome. By applying the $CaCl_2$ treatment to *recBC*$^-$ strains, it may therefore be possible to investigate transformation in E.coli. This may make it possible to investigate which genes are necessary for integration of transforming DNA and how they are related to the genes involved in other forms of genetic exchange that have been characterized in E.coli.

We may ask what significance the acquisition of competence and the transformation with donor DNA that it makes possible has for the normal life cycle of bacteria. An indication that such transfer of genetic information may occur with pneumococcus in the wild is provided by the observation of Ottolenghi and Hotchkiss (1962) that pneumococcal cultures release DNA into the medium, thus making material available to act as donor molecules. On the other hand, Steinhart and Herriot (1968) observed that cells of H.influenzae do not release DNA; transforming activity equivalent to no more than 0.1% of the cellular DNA could be found in the medium. The physiological significance of transformation therefore remains to be established.

### *Uptake of DNA*

Transformation is initiated by random collisions between bacteria and donor DNA, the reaction depending upon the concentrations of competent cells and added DNA. Donor DNA is first bound reversibly and then is converted into an irreversibly bound state. The initial interaction probably occurs by an ionic binding; most of the DNA initially bound can be released if the subsequent irreversible binding is inhibited (for review see Tomasz, 1969). The donor DNA enters the cell as a linear molecule; this is converted to a single strand which is utilized in transformation and the subsequent stages of the reaction are those concerned with its integration.

The usual protocol for transformation is to add duplex DNA to competent cells, allow a suitable period of time for uptake, and then to end the reaction by addition of DNAase; this degrades any DNA still in solution or that is only loosely attached to the surface of the recipient cells, while leaving undamaged the DNA that has become "irreversibly" associated with the recipients. The acquisition by donor DNA of resistance to DNAase has often been taken as the criterion to define the point at which a donor molecule is committed to entering the recipient cell. However, other treatments of recipient cells shortly after addition of donor DNA may prevent transformation, even though the time has passed when the donor molecules become DNAase resistant. This means that although donor DNA has become inaccessible to DNAase, this resistance cannot be equated with its penetration into the cell, but must represent a transient stage of entry, following which the DNA penetrates the cell sufficiently far to be resistant to these other treatments as well as to DNAase.

The kinetics of transformation thus depend upon the treatment used to terminate the reaction. In the protocol used by Levine and Straus (1965), B.subtilis cells either were simply washed or were treated with DNAase following uptake of DNA; then the cells were transferred to new medium and grown up to be examined for transformants. When washing alone is used to terminate the reaction, transformants appear in the population almost immmediately after the addition of DNA. But when DNAase is used, no transformants appear until the cells have been incubated with DNA for at least 1–2 minutes before addition of the enzyme. This implies that donor DNA remains susceptible to DNAase during the immediate entry process, although it is not removed by washing during this period.

Acquisition of resistance to DNAase by donor DNA is an energy-dependent process that may be inhibited by compounds which interfere with energy metabolism. This appears to be the reason why cyanide ions inhibit transformation. Straus (1970) observed that cyanide can prevent transformation of B.subtilis for 6–8 minutes after resistance to DNAase has been attained. This means that although the donor DNA has become resistant to enzymatic attack, it has not entered the cell during this period. Another treatment that inhibits uptake of duplex DNA is exposure to EDTA. Morrison (1971) reported that with B.subtilis addition of EDTA may prevent transformation immediately after donor DNA has become resistant to DNAase. On the other hand, Seto and Tomasz (1974) observed that with pneumococci EDTA (and other chelators of divalent cations) inhibit entry to the DNAase-resistant state, although the initial binding of DNA to the cell is not prevented. Probably the difference in the stage at which EDTA appears to act relative to the acquisition of DNAase resistance reflects the different structures of the cell surface in B.subtilis and D.pneumoniae. The important general conclusion suggested by these experiments, then, is that there are several stages in the uptake of donor DNA; and these may be distinguished by their different susceptibilities to various inhibitors of transformation.

Early experiments suggested that transforming DNA must be duplex. But although preparations of single stranded DNA are less effective, they can be used for transformation, apparently in all bacterial systems. Miao and Guild (1970), for example, found that competent cells of pneumococcus are transformed by single stranded DNA to a level about 0.5% of that shown by duplex DNA; the reduction in transformation efficiency is probably due largely or entirely to the greater difficulty of uptake of single strands. Conditions required to promote the entry into recipient cells of single stranded DNA, however, may differ from those established for duplex DNA, presumably because different uptake systems are implicated. The treatments necessary before single stranded DNA becomes effective as a transforming agent for B.subtilis or H.influenzae, for example, would be deleterious to the cells if prolonged; and so it is reasonable to suppose that it is the uptake of duplex DNA which represents the "physiological" mechanism.

Duplex DNA is taken up by competent cells of H.influenzae at neutral pH, with an optimum at pH 6.8; no uptake occurs below a pH of 5.5. Postel and Goodgal (1966) found that at low pH, with an optimum of 4.8, cells can take up single stranded DNA as efficiently as they take up duplex DNA at neutral pH. Cells which take up single stranded DNA at pH 4.8 show few transformants when plated directly. But after a period of incubation at pH 7.4, transformation can be observed, although only at the low level of 0.2% of that seen with the same amount of duplex DNA at pH 7.4. Transformation by single stranded DNA shows the usual dependence on DNA concentration, cell concentration, and competence of the cells. Exposure of H.influenzae cells to EDTA after, but not before, uptake of single stranded DNA increases the transformant number about tenfold, although this treatment has no effect upon transformation by duplex DNA. This emphasizes that there are differences, at least in the stages of penetration, between transformation by duplex and by single stranded DNA. Although the uptake of single stranded DNA at low pH does not require the cells to be competent, transformation results only in competent cells; this implies that development of competence is necessary for stages in transformation beyond those concerned with the irreversible binding of duplex DNA. The steps of transformation by single stranded DNA subsequent to its entry into the cell are presumably the same as those applying to duplex DNA, a conclusion supported by the observation of Postel and Goodgal (1967) that the same genetic linkage is observed between markers whether duplex or single stranded DNA is used. The expression and replication of transformants after uptake is similar for duplex and single stranded DNA, although the kinetics are not identical.

EDTA also is effective in allowing single stranded DNA to transform cells of B.subtilis, although its action appears to differ from that shown with H.influenzae, since it is effective with the Bacillus only if added before the uptake of DNA. Chilton and Hall (1968) observed that 6% transformation, relative to that shown by duplex DNA, can be achieved by single strands in the presence of EDTA. Tevethia and Mandel (1971) noted another similarity with the H.influenzae system, in observing that reduction of the pH to 6.1 enhances uptake of single stranded DNA by B.subtilis. Tevethia and Caudill (1971) showed that single strand transformation is proportional to the extent of transformation by duplex DNA in the population, so that it must depend upon the competence of the cells.

### *Conversion of Duplex DNA to Single Strands*

Labeled transforming DNA is converted to single strands upon entering recipient pneumococci; and the reconversion of these single strands to the duplex form accompanies the rise in the number of transformants during incubation following uptake. After the eclipse period, the number of transformants per unit of transforming DNA increases in parallel with the replication of the transformed cells. This suggests a model for pneumococcal transformation in which donor DNA enters the cell and is converted to single strands which are then integrated into the recipient genome. The same events are involved in transformation of B.subtilis and H.influenzae, but their precise order may not necessarily be the same as with pneumococcus.

There has been some debate about the stage of transformation at which the conversion of duplex transforming DNA to single strands takes place. Lacks (1962) suggested that hydrolysis of one strand might take place as donor material enters the cell; one model might be to suppose that the degradation is related to uptake, so that an enzyme located close to the cell surface helps "pull in" the donor DNA as it degrades one of the two strands. On the other hand, Fox (1966) proposed that the single strands may represent only those donor molecules that succeed in associating with the recipient genome to yield transformants, so that creation of single strands might be a consequence of, rather than a prerequisite for, transformation. While it is possible to recover donor DNA after its entry into the recipient cell but before its association with the host chromosome, it is difficult to know whether changes in its state are causally related to transformation; for example, additional nucleases may be present whose degradation of DNA is not related to its incorporation. To dissect transformation and determine the step at which the generation of single strands becomes necessary may therefore require the isolation of mutants defective at this stage.

Utilization of only a single donor strand is common to transformation in pneumococcus, H.influenzae, and B.subtilis. But the timing of events in these three bacterial species is different. In pneumococcus, a short eclipse period separates the entry of donor DNA from the recovery of its ability to transform, which is more or less simultaneous with the expression of the transformed phenotype; during the eclipse period the donor DNA can be retrieved in the single stranded state. An eclipse period also has been reported in B.subtilis, but competent cells are in a different physiological state from noncompetent cells and the transformed phenotype is not expressed until quite some time after uptake; the state of the donor DNA following entry has not been followed with the same clarity as in pneumococcus, but it is possible that conversion to single strands may occur during this time. No eclipse period has been detected in Hemophilus, and until recently single strands could not be detected immediately following uptake, but only later as part of the recipient genome; now short single stranded regions have been detected in donor DNA, but it remains true that no general conversion of donor material into this state can be detected until after recombination with the recipient genome. The generation of single stranded DNA may therefore precede association with the recipient genome in pneumococcus and Bacillus but may accompany it in Hemophilus.

Upon binding to pneumococcal cells, donor DNA is degraded. The initial binding reaction can be distinguished from the subsequent stages of transformation by adding the donor DNA to competent recipients in the presence of EDTA. In this case the DNA adsorbs to the bacteria but no further reaction occurs. Seto and Tomasz (1974) found that when divalent cations then are added to release the block, two events occur: donor DNA enters the cells in a DNAase-resistant state, concomitant with a rise in the number of transformants; and DNA is released into the medium in the form of small (acid soluble) oligonucleotides. Similar conclusions were reported previously by Morrison and Guild (1972b, 1973), who found that DNA bound to cells (in the DNAase-sensitive state) can be eluted with guanidium hydrochloride; the eluted DNA is smaller than the original DNA preparation due to the introduction of single strand breaks. These results suggest that cleavage of the DNA takes place during its uptake. The effect of EDTA may be to inhibit nucleases that depend on divalent cations; the inhibition of transformation caused by the EDTA then would leave the implication that the action of these nucleases is essential. However, more direct evidence than this is needed to show whether this degradation is necessary for transformation or simply represents a hazard to which donor DNA is subject. Even if required for transformation, the degradation of donor DNA may not necessarily be associated with its uptake; Seto et al. (1975) reported that when competent pneumococci are incubated with a preparation of DNA, all of the DNA is degraded during a 30 minute period in which less than 5% of it becomes irreversibly associated with the cells. Only competent cells have the ability to support this reaction; and it is necessary for the DNA to be in contact with the cells. This suggests that the nucleases responsible for attacking donor DNA are present at the cell surface and that DNA may be degraded and then released without being taken up.

Which enzymes are responsible for attacking the DNA and what is the effect upon transformation of mutations in their genes? At first it appeared that neither of the major DNAases of pneumococcus is implicated in transformation, for Lacks (1970) found that the double mutant *end⁻ exo⁻*, which contains only some 3% of the level of wild type endonucleolytic and exonucleolytic activity, is normal in the uptake and integration of donor DNA. Three residual DNAase activities were identified in the *end⁻ exo⁻* mutant cells by Lacks, Greenberg and Neuberger (1974). One is associated with a DNA polymerase, probably analogous to the polymerase-exonuclease activity of the corresponding enzyme in E.coli. Another is an exonuclease that produces mononucleotides. And there is an endonuclease which generates oligonucleotides similar in distribution to those released to the medium during DNA uptake.

This endonuclease activity was implicated in transformation by the characteristics of the *noz* mutants isolated from pneumococci by a selection procedure that assays for failure in a DNAase assay. Mutants of the *end⁻ noz⁻* genotype are transformed at a reduced level, generally 0.1–1.0% of normal; uptake of DNA is reduced in parallel with the decline in transformation. No endonuclease activity can be detected in these cells. Although at first it was thought that the *noz* mutation identified an enzyme separate from that coded by the *end* gene, Lacks, Greenberg and Neuberger (1975) found that the *noz* and *end* mutations are closely linked and may in fact map at one locus. And the enzyme remaining in the *end⁻* cells is similar in properties to the major endonuclease. These results suggest that the *end⁻* mutation may leave the cells able to synthesize a mutant enzyme whose activity is sufficient to support transformation; and by introducing a second mutation in its gene, the *noz* mutation completely abolishes the enzyme activity. This implies that only a small proportion of the amount of endonuclease activity usually present in the wild type cell is needed to support transformation. (This assumes that the in vitro assay reflects activity in vivo.) That the *end* and *noz* mutations lie in the same gene might be confirmed by a test for genetic complementation; or that the residual endonuclease of *end*$^-$ mutants is related to the major enzyme of wild type cells might be confirmed by comparing the physicochemical properties of the two proteins.

Cells altogether lacking the endonuclease activity remain able to bind DNA in a form susceptible to DNAase, and presumably located on the outside of the cell since it can be removed by agitation. This implies that the degradation of DNA occurs as a step subsequent to, and independent of, the initial binding reaction. (A further class of mutants, described as *ntr,* were isolated in the same selection protocols as the *noz* mutants; these cells do not bind or take up DNA and appear to be transformed not at all. No change in the activity of any DNAase is found in *ntr* cells. We may speculate that the *ntr* mutation identifies a gene whose product is necessary for the initial binding of DNA so that in its absence the DNA never subsequently becomes accessible to the endonuclease; alternatively, the mutation may prevent acquisition of competence with a pleiotropic effect upon DNA binding activity.)

By examining fractions of cell extracts for endonuclease activity, Lacks and Neuberger (1975) showed that the major endonuclease is associated with the cell membrane. What is the action of this enzyme upon donor DNA? Cells that possess the enzyme subject donor DNA to double strand breakage. Its involvement in this cleavage can be tested by comparing the fate of DNA bound to cells deficient in the endonuclease. Lacks and Greenberg (1976) found that when DNA is bound to *end*$^-$ *noz*$^-$ cells in the presence of EDTA (to prevent entry into the cell), upon elution its size is unaffected when examined in the duplex form. But upon denaturation it releases shorter single strands. This suggests that the initial binding of DNA to the cell is accompanied by nicking of single strands, with a nick made roughly every $2 \times 10^6$ daltons (6000 bases) of single stranded DNA. This action is independent of the major endonuclease. In cells only partially blocked in endonuclease activity, the initial binding and nicking is succeeded by the introduction of double strand breaks, presumably made by the endonuclease as entry to the cell is initiated. It is not clear whether the duplex breaks are made independently of the previous single strand breaks or whether they represent conversion of the nicks into double strand breaks. Nor is the relationship between the introduction of double strand breaks and the degradation of one strand of donor DNA clear yet. One speculation that has been made is that the endonuclease is itself responsible for converting duplex DNA to the single stranded state in an action that succeeds the generation of duplex breaks. On the other hand, it remains possible that the endonuclease must initiate this process but that some other enzyme accomplishes the removal of the unwanted strand. Presumably the oligonucleotide fragments released into the medium are derived from the strand that is degraded, while the other enters the cell.

Cells of B.subtilis are able to degrade exogenous DNA both endonucleolytically and exonucleolytically. Dubnau and Cirigliano (1972a) noted that duplex DNA added to B.subtilis cells suffers endonucleolytic breaks during uptake, so that the final length of integrated donor DNA is somewhat less than the size of the molecules originally added to the cells. Joenje and Venema (1975) found that when competent and noncompetent cells are separated by their different buoyant densities, transforming DNA may be inactivated by exposure to either culture, more effectively by the competent bacteria. The weight of the duplex DNA preparation decreased on exposure to either set of cells; that is both can make endonucleolytic breaks. In addition the competent cells degraded some DNA exonucleolytically, a capacity not possessed by the noncompetent cells. Exposure to EDTA inhibited both transformation and exonucleolytic attack, suggesting that the breakdown may be part of the transformation process. The behavior of B.subtilis cells towards exogenous DNA thus appears to be broadly similar to, although not ncccssarily identical with, that of pneumococcal cells.

Before transforming DNA can become part of the recipient genome it must be transported from the membrane where it enters the cell to the nucleoid body that represents the compactly folded bacterial chromosome. Vermeulan and Venema (1974a,b) followed the uptake of $^3H$ labeled DNA in B.subtilis by incubating the transforming preparation with competent cells for a short time and then localizing the label by electron microscope autoradiography after various intervals. Over a one hour period DNA migrates from the cell periphery to the nucleoid bodies, association being completed within 15–60 minutes. The transforming DNA appears to be preferentially associated with the mesosomes during its movement. Mesosomes are convoluted invaginations of the cell membrane and two types can be distinguished in these cells: some are connected only to the plasma membrane; but others are connected also to the nuclear bodies. The number of nuclear mesosomes is increased in competent cells, consistent with the idea that they may provide the mechanism by which DNA is transported to the nucleoids. Another observation that may be relevant to this concept was noted by Joenje, Konings and Venema (1975): isolated membrane vesicles can bind DNA.

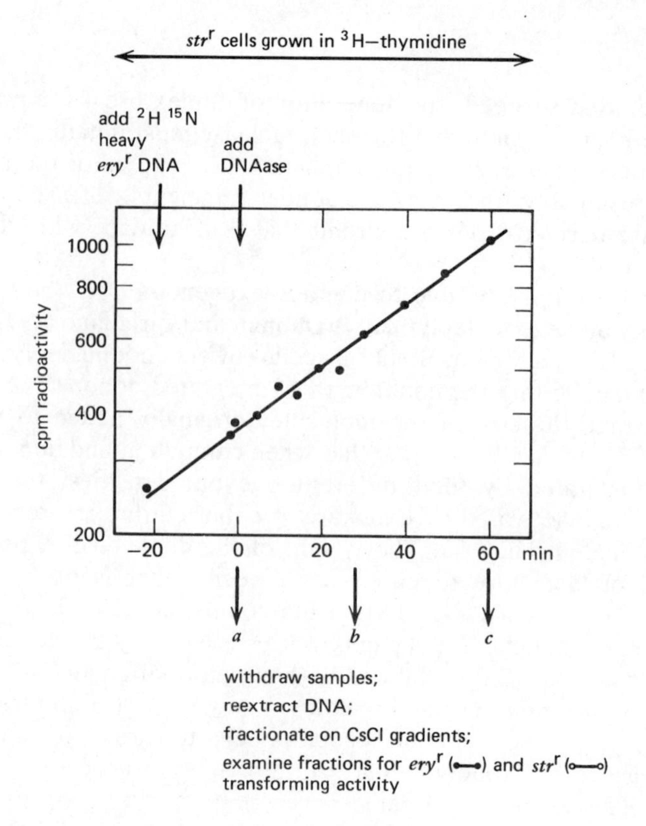


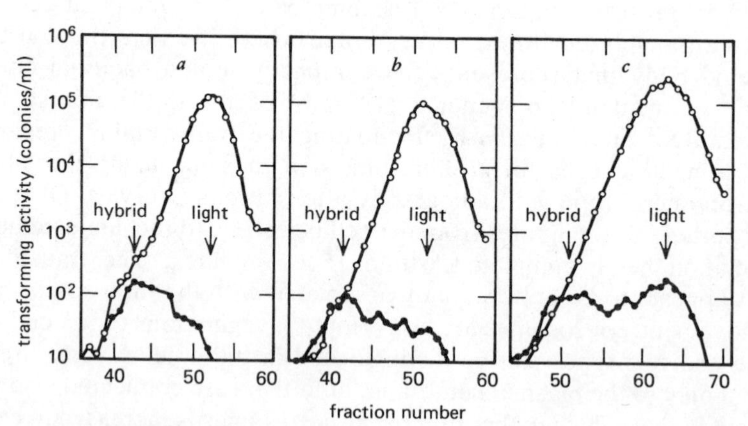


**Figure 1.6:** integration of single donor strand into recipient genome of pneumococcus. Streptomycin-resistant cells grown in $^3$H-thymidine were treated with heavy DNA carrying the marker erythromycin resistance and the reaction terminated with DNAase as shown above. The culture was then allowed to multiply and cell division was followed by the increase in radioactivity. Samples were taken after 0, 30, 60 minutes; the transforming activity for the donor ($ery^r$) marker is found in DNA of hybrid density at 0 minutes,

### *Integration of Donor DNA*

Using transforming pneumococcal DNA carrying a density label, Fox and Allen (1964) showed that upon transformation the donor molecules enter a density position close to that of recipient DNA, presumably because they have been integrated into much longer regions of the light recipient DNA. However, sonication to reduce the size of the recipient-donor recombinant molecules releases fragments of hybrid density and denaturation shows that these consist of one light strand (recipient) and one heavy strand (donor). In reviewing early events in pneumococcal transformation, Fox (1966) noted that the donor DNA is close to the recipient density when the recombinant DNA is analyzed at a fragment size of $20 \times 10^6$ daltons, but is revealed at hybrid density at a fragment size of $1–2 \times 10^6$ daltons. In subsequent experiments, Gurney and Fox (1968) infected competent pneumococci with donor DNA carrying a heavy density label, reextracted the DNA immediately after transformation, and then sheared the preparation to various molecular weights; as the molecular size of the preparation declines, the density of the donor marker increases towards the fully hybrid position. From these data it is possible to calculate that the general spectrum of integrated lengths is the range from $10^6 - 5 \times 10^6$ daltons (1500–8000 bases).

If transformation involves integration of only single donor strands into a recipient genome, host markers should be present on the light strand of the hybrid duplex; and when this DNA replicates the host markers should leave the hybrid region and enter a light duplex. Gurney and Fox (1968) tested this prediction by treating recipient cells carrying a marker for streptomycin resistance with donor DNA carrying the closely linked marker for erythromycin resistance. At various times after the addition of donor material, DNA was extracted and separated by its density into fractions that were examined for the presence of the recipient (*str*$^r$) and donor (*ery*$^r$) markers. As Figure 1.6 shows, immediately following transformation a peak of streptomycin resistance can be found at hybrid density; but at later times, when replication has intervened, all the *str*$^r$ activity returns to light density. Replication also sees the transfer of the donor transforming activity for erythromycin resistance from solely hybrid density to peaks at both hybrid and light density.

---

corresponding to the insertion of a single (heavy) donor strand into the (light) recipient genome. At 30 and 60 minutes, increasing amounts of donor marker activity are found at the light position, corresponding to replication of the donor genetic information. Most recipient marker (*str*$^r$) transforming activity is found at the light position; but the shoulder at the hybrid position (most prominent at 0 minutes) demonstrates that in some cells the strand complementary to that carrying the recipient information has been replaced by donor material. Data of Gurney and Fox (1968).

The integration of only one strand of a duplex transforming DNA, and the ability of single strands to achieve transformation (albeit under unusual conditions), imply that the genetic information of only one strand of a duplex transforming DNA is utilized in the steps succeeding uptake. By renaturing the denatured DNA of pneumococcus resistant to streptomycin in the presence of excess denatured wild type DNA, Marmur and Lane (1960) obtained a preparation in which all the sequences of DNA from the drug-resistant bacteria were present in the form of hybrid DNA molecules. Yet these molecules were able to transform cells to streptomycin resistance. Thus the strand of the hybrid duplex derived from the resistant cells is sufficient to cause transformation. We should note, however, that the discovery of correction systems that act on mispaired DNA (made since this experiment was performed) make the interpretation of this and similar experiments less than direct, since hybrid DNA sequences in transforming or in integrated DNA may be subject to such correction.

Recovery from eclipse in pneumococcus is temperature dependent; Ghei and Lacks (1967) found that a reduction in temperature from 30°C to 23°C slowed the rate of recovery 2–3 times while an increase in temperature from 30°C to 37°C roughly doubled the rate of recovery. Similarly a reduction in the ability of donor DNA to integrate in H.influenzae was noted by Sedgwick and Setlow (1976) when the temperature was reduced. In a study of the effect of temperature on the eclipse period in pneumococcus conducted by Shoemaker and Guild (1972), donor DNA was added to competent cells incubated at various temperatures and then reextracted and tested for transforming activity. At 37°C there is an immediate rise in transforming activity from uptake until 10 minutes later, when the eclipse period ends and maximum transforming activity is recovered. At 25°C, however, there is a lag of 1–2 minutes before any activity is recovered and it then takes longer to achieve the maximum extent of recovery from eclipse. At still lower temperatures the lag period is increased further and the maximum recovery is further decreased. When the rate of recovery is expressed relative to the temperature of incubation in the form of an Arrhenius plot, two reactions appear to be involved, with different activation energies. The biological significance of these observations is not entirely clear, but one possible speculation is that the two reactions inferred from these data may represent successive stages of integration, for example perhaps the unwinding of recipient DNA and its pairing with donor single strands.

Experiments using heavy donor DNA of B.subtilis showed that uptake is succeeded by covalent association with recipient DNA; using this approach, Bodmer and Ganesan (1964) obtained an average length for the integrated material of about 2.5 × $10^6$ daltons (8000 bases). A similar figure was obtained by Dubnau and Cirigliano (1972a). The status of the donor DNA immediately after uptake in B.subtilis has appeared to be different from that seen in pneumococcus, perhaps because of physiological differences between these bacteria, especially where the cell wall and mechanism of uptake are concerned. Venema et al. (1965a,b) suggested that the DNA preparations containing both donor and recipient markers that can be recovered immediately following transformation may in fact be due not to recombination but instead may represent an intermediate form of association. When reextracting donor B.subtilis DNA soon after transformation, Dubnau and Abelson (1971) also suggested the presence of a complex containing a single donor strand associated with the recipient duplex, although disagreeing with the conclusion of Venema et al. that the recovery of donor activity may precede the integration step. They suggested that the association of single donor strands with the recipient duplex precedes the formation of covalent linkage which establishes recombination.

The cause of the earlier confusion about the eclipse phase and the single stranded nature of donor DNA following uptake in B.subtilis was clarified by the experiments of Piechowska and Fox (1971). They found that at 30 seconds after the uptake of heavy DNA, reextracted DNA does not appear to be associated with recipient DNA when analyzed on density gradients; the donor DNA can be recovered with cell components partially sensitive to pronase digestion and is released to the position characteristic of free DNA by heating in high salt concentrations. But reconstruction experiments showed that these characteristics can be acquired by denatured DNA added to cells at the time of lysis. After uptake some of the donor DNA appears to be in the form of single strands which then appear in such complexes. On incubation the denatured donor strands in these structures disappear and about half appears at the density characteristic of recipient DNA, presumably due to its integration. Shearing the recipient DNA to a size of 4 × $10^6$ daltons yields fragments whose density suggests that integration takes place in lengths of about $10^6$ daltons (3000 bases). These results therefore suggest that the fate of donor DNA in B.subtilis may be similar to that in pneumococcus, but this is obscured by its association with cellular structures during isolation.

Following its uptake by recipient cells of H.influenzae, donor DNA does not pass through a single stranded intermediate state; it is taken up in the duplex form and appears to be converted into single strands only concomitant with the integration of donor material into the recipient genome. This is consistent with the absence of an eclipse phase. To examine the fate of Hemophilus donor DNA, Notani and Goodgal (1966) used a protocol in which the recipient genome was labeled with $^3$H-thymidine while the donor DNA carried a radioactive $^{32}$P label and a heavy density label. After addition of donor DNA to competent recipients, its association with the host genome can be followed by assaying the activity of the transforming marker in the different fractions of DNA separated on a density gradient; the presence of recipient DNA in these fractions is marked by $^3$H-thymidine and the donor material can be assayed by its $^{32}$P label. (We should note that it is only the immediate product of integration that is significant in this protocol. Its replication should have the same consequences whether duplex or single strands of donor DNA are integrated: all the transforming activity should be transferred from heavy to hybrid density after one round of replication of an integrated donor duplex; but since only the heavy strand of the hybrid duplex generated by insertion of a single donor strand carries the transforming sequence, in the first round of replication only the hybrid and not the light products have the donor marker. In any case, this conclusion assumes that there is no correction of mismatched hybrid DNA; any such activity alters the transforming ability of hybrid DNA and so emphasizes the necessity to confine experiments to the immediate product of integration.)

The donor activity at first remains at its original heavy density, not associated with recipient DNA; thus it has entered the cell in its original duplex form. When it becomes associated with recipient DNA it enters fractions which display the hybrid density that would be expected from the insertion of a single donor strand into the recipient genome. (The reduction in donor density would be less if it were due to the insertion of heavy duplex segments. Of course, the shift to hybrid density is ameliorated by the covalent linkage of heavy donor material to the light recipient DNA and it is therefore necessary to examine recombinant molecules that are not too much longer than the size of the inserted donor material.) Denaturation of the reextracted transforming DNA confirms that the integrity of the single strands is maintained during integration, since almost entirely heavy single strands can be extracted. The size of the integrated donor DNA appears to be about $12 \times 10^6$ duplex daltons (about 18,000 bases), but its distribution is heterogeneous, corresponding to some variation in the lengths of integrated donor material. These results therefore suggest (although they do not conclusively prove) that only a single strand of donor DNA is integrated in the recipient genome, but that it is generated only when integration occurs and not previously.

That either strand of a transforming DNA preparation of H.influenzae can transform has been confirmed by the experiments of Goodgal and Notani (1968). In these experiments competent cells were transformed with heavy donor DNA and then hybrid density DNA with transforming activity was reextracted after varying times. This DNA was then denatured into its single strands and annealed with wild type (light density) DNA lacking the transforming marker. The duplexes generated by this treatment were tested for their possession of the transforming marker. Since one strand of the reconstituted duplex does not carry the transforming marker, its activity in transformation measures the presence of this marker on the strand that was annealed with it, that is the strand derived from the DNA extracted from transformed cells at different times. Thus any transforming activity in DNA of hybrid density is due to an original heavy donor strand; any activity in DNA of light density is due to a complement that must have been synthesized after integration of the donor material. According to this assay, at 10 minutes after addition of donor DNA almost all the transforming activity resides in the heavy strand; but at 40 minutes or 70 minutes, after one or two rounds of replication respectively, 50% of the activity was in the heavy and 50% in the light (newly synthesized) strands. The implication is that only one strand of transforming DNA need carry the donor information; and this may be either strand.

A search for single stranded regions in donor DNA at various times after uptake by H.influenzae recipients was made by Sedgwick and Setlow (1976), taking advantage of the S1 nuclease; this enzyme is extremely specific for single stranded DNA, removing any such regions from duplex DNA. In these experiments competent cells were incubated with labeled transforming DNA and uptake then terminated by addition of DNAase. Immediately following this protocol, about 13% of the donor DNA taken up by the cells was degraded by S1 nuclease; and over the next 40 minutes this value declined to about 4%. When the *rec2* or *KB*1 mutants of H.influenzae which appear to be deficient in the ability to integrate donor DNA are used as recipients, sensitivity to S1 nuclease does not appear in the donor DNA. This suggests that the generation of single stranded regions in the donor material is necessary for transformation. What is the location of the single stranded regions in the donor molecules? Because the size of the donor DNA appears to be unchanged by treatment with S1 nuclease, as judged by its rate of sedimentation through sucrose gradients, the single stranded regions may lie at the ends of the donor molecules (rather than within where their cleavage would reduce the size).

During acquisition of competence in H.influenzae single stranded regions appear in the recipient genome. LeClerc and Setlow (1975) found that when the DNA of competent cells is sedimented through alkaline sucrose gradients, a pulse label given after the acquisition of competence is found in a fraction of smaller size than a label previously incorporated. This suggests that DNA synthesized during the competent phase either represents short single strands or is associated with breaks in the integrity of the strands of the resident genome. The nature of these single strands is not clear: they might represent some intermediate necessary for transformation or might be an artifact of the protocols used to obtain competent cells. A suggestion that they are related to transformation is supported by the observation that a *rec2* mutant defective in integration of donor DNA does not generate this material (in addition to failing to generate the single stranded ends of donor DNA; on the other hand, the mutant *KB1* is deficient in production of donor single strand regions but remains able to generate single strands in the recipient genome). One possible speculation is that the single stranded regions in donor and recipient DNA both are needed to initiate integration. Perhaps we should add that the production of single stranded regions does not in itself make any implication about whether the bulk of donor DNA is integrated in single stranded or duplex form: it would be equally consistent with the integration of single strands or with the integration of duplex material through the creation of single stranded ends. The recipient DNA that is displaced by the donor material is presumably degraded, a supposition that is supported by the observation of Steinhart and Herriot (1968) that some degradation of donor DNA, corresponding to about 2% of the genome, occurs during transformation.

### *Mutants Defective in Transformation*

Little is known about the mechanism of genetic recombination during transformation. Since only a single strand of the donor DNA is implicated, the reaction constitutes the replacement of one recipient strand rather than an exchange between two duplex molecules. But several similarities may be expected between this process and the later stages of conventional genetic recombination between duplex DNA molecules, since the mechanism of assimilation of the donor single strand may well be similar to the formation of hybrid DNA. (Genetic recombination in bacterial and other systems is discussed in Volume 1 of Gene Expression.)

Integration of donor DNA requires the transforming single strand (in pneumococcus or B.subtilis) or single stranded end (in H.influenzae) to find the appropriate duplex region of the recipient into which it is to substitute. How such recognition may take place is not at all clear; and, indeed, synapsis between duplex DNA segments is no better understood. Given the large length of bacterial DNA compared with the short length of a donor transforming molecule, it is difficult to see how any sort of random diffusion process could explain synapsis. The mechanism that may be responsible remains entirely unidentified. Presumably, however, one reason for the low efficiency of transformation may be that those donor strands not rapidly achieving synapsis with an appropriate recipient segment are likely to suffer degradation in the host cell.

Integration may be initiated by base pairing between one end of the transforming donor single strand and its complement in the (presumably) unwound host DNA; pairing between donor and recipient single strands might then extend along the recipient genome, displacing the resident strand. The displaced recipient strand presumably must be degraded instead of indulging in the pairing with a complement that occurs in recombination between duplex molecules. Whether the insertion of donor single strands is terminated by some specific mechanism or continues simply until the end of the donor fragment is reached cannot be said. If transformation takes place by this sort of process, as illustrated in Figure 1.3, it is reasonable to suppose that the hybrid DNA may be the subject of attention by enzyme systems similar or identical to those involved in duplex recombination. The relationship between the events involved in donor strand assimilation during transformation and those that take place during genetic recombination and in repair is best revealed by the deficiencies in these processes apparent in mutants isolated in each of them.

Only a few mutants have been isolated that are defective in transformation; because no series of mutations affecting the various stages of transformation has been assembled in any one bacterial system, no complete genetic analysis of transformation is yet possible. However, those mutants that have been obtained so far in general show properties which are consistent with the models for transformation suggested by the analysis of the fate of donor DNA following uptake.

Mutants of H.influenzae selected for their poor transformation abilities were isolated by Caster, Postel and Goodgal (1970); streptomycin-sensitive cells were exposed to the mutagen nitrosoguanidine and then mutants were selected which yielded less than 0.1% transformation to streptomycin resistance upon treatment with *str*$^r$ DNA (the usual level of transformation under these conditions is about 1%). The mutants fell into four classes.

Class 1 mutants transform as well as wild type with single stranded DNA but fail to bind significant amounts of duplex DNA. These mutants may lack binding sites specific for duplex DNA; and their existence supports the concept that uptake mechanisms differ for single stranded and double stranded DNA. It is clear also from the properties of these mutants that the acquisition of competence involves more than simply the development of DNA binding sites, for they are susceptible to transformation by single stranded DNA only when in the competent state.

Class 4 mutants bind single stranded but not double stranded DNA, but fail to yield corresponding numbers of transformants; these may be similar to or may be variants of the class 1 mutants. Mutants of class 2 bind neither single stranded nor double stranded DNA and may therefore have some more general change in the cell surface that affects the uptake of DNA.

Mutants of class 3 bind both types of DNA, but yield only low numbers of transformants; these may therefore be defective in the stages succeeding uptake. Postel and Goodgal (1972a,b) found that the defect cannot be attributed to the degradation of DNA in the recipient cell (at least, not insofar as can be measured by entry into acid soluble fragments; simple endonucleolytic cleavage cannot be excluded). In wild type cells a single strand of the donor DNA becomes associated with the recipient genome following uptake. But in the mutant cells of class 3, the extent of association varies widely, from wild type levels to zero. When association does take place at normal levels it is stable to alkali, that is involves covalent linkage between donor and recipient material; this makes the nature of the defect less than obvious and these mutants have been termed *dab*⁻ (donor association biologically defective). The class 3 mutants in which donor DNA fails to associate or associates only rather poorly with the recipient genome have been termed *dad*⁻ (donor association defective). Some of the class 3 mutants show increased susceptibility to ultraviolet irradiation, although others remain in the wild type range and some are more resistant than wild type to ultraviolet. Although no detailed conclusions about the genes involved in transformation can be drawn from these results in lieu of data on the mapping and characterization of the mutations, they do suggest that some, although not necessarily all, of the functions involved in transformation may be involved also in repair systems.

An overlap between recombination and integration activities has been observed by Notani (1972), who noted that certain mutants deficient in recombination are transformed only at low frequencies, although the uptake of DNA by competent cells is as great as in the wild type. And Kooistra and Venema (1974) reported that a mutant sensitive to mitomycin is defective in the integration of donor DNA. Such a connection also is suggested by the properties of the *rec2* mutant, isolated for its deficiency in recombination, and which cannot act as a recipient for transforming DNA, apparently due to its inability to create single stranded regions in both the donor and resident DNA (LeClerc and Setlow, 1975; Sedgwick and Setlow, 1976). However, not all transformation functions are necessarily involved in recombination or repair activities. Beattie and Setlow (1971) isolated H.influenzae mutants defective in transformation by taking advantage of their resistance to the lethality of heterospecific transformation with donor DNA derived from H.parainfluenzae. The uptake of H.parainfluenzae DNA by wild type H.influenzae cells is lethal to the recipients (for review see Notani and Setlow, 1974); and mutant recipients unable to support transformation may therefore be selected by their survival after treatment with heterologous DNA. Most of the mutant isolates did not show increased sensitivity to ultraviolet or X-irradiation (although this implies only that the mutant functions are not involved in certain types of repair function). One of the mutants *(KB1)* fails to form single stranded regions in donor DNA, although it remains able to generate single stranded segments in the DNA of the resident genome of competent bacteria (Sedgwick and Setlow, 1976).

An overlap between the systems utilized in repair and recombination in B.subtilis and those implicated in transformation-mediated genetic exchange is suggested by the observations of Abelson and Dubnau (1971). In $rec^-$ cells the transformation frequency is reduced to 10–25% of that seen with $rec^+$ recipients, although the parameters of transformation—the eclipse period, recovery of donor marker activity, etc.—appear to be similar in both the wild type and mutant cells. This makes it likely that the *rec* mutation results in a defect at one of the later stages of transformation concerned with the integration of donor DNA. Another parallel with conventional recombination may be drawn from the observation of Dubnau et al. (1973) that mutants sensitive to mitomycin C and methylmethanesulfate are defective in both general recombination and the later stages of transformation.

A mutant in the development of competence has been isolated in Bacillus stearothermophilus by Streips and Welker (1971a), who used an assay in which cells are infected with the DNA of phage TP-1C. A $com^-$ mutant unable to support this transfection can do so when competence factor is added; this suggests that the mutation may prevent synthesis of the factor.

## Genetic Mapping by Transformation

### *Frequencies of Transformation*

Transformation results from the interaction between single cells and DNA molecules, a linear relationship prevailing between the number of transformants and the concentration of DNA (up to the plateau level beyond which the number of transformants does not increase). Double transformants may arise in either of two ways. For unlinked markers, two DNA molecules each carrying a different donor genetic marker may interact with a single cell; this cell may then become transformed for both characteristics, due to two independent integration events. In this case, the kinetics of double transformation display two hit kinetics; and when transformation frequencies are expressed relative to the number of competent cells (that is, rather than to the total number of viable cells), the frequency of double transformation is the product of the two single transformation frequencies. By a reversal of this argument, in fact the more usual sequence in which the calculation is performed, it is possible to calculate the proportion of competent cells by comparing the frequency of the double transformation event with the product of the single transformation frequencies (now expressed relative to the total cell count). For linked markers, both may be carried on a single molecule of DNA, in which case the recipient may be transformed for both characteristics in a single event. This results, of course, in an increased frequency of double transformation, dependent upon single hit kinetics.

As we have just implied, linkage between two markers cannot be assumed because the frequency of double transformation is greater than the product of the relevant single transformation frequencies; for spurious linkage will result whenever the proportion of competent cells is less than 100%. As Goodgal and Herriot (1961) pointed out, comparison of double with single transformation frequencies can be valid only when expressed relative to the number of competent cells. Thus for two markers transformed independently (assuming that both donor markers are present in the transforming DNA preparation in the same concentration, as will be the case whenever they were originally located on one genome, but too far apart to be linked, or if two different DNA preparations are used in equal amounts):

$$\frac{N_a}{f} \times \frac{N_b}{f} = \frac{N_{ab}}{f}$$

so that

$$N_{ab} = \frac{N_a \times N_b}{f}$$

or

$$f = \frac{N_a \times N_b}{N_{ab}}$$

where $N_a$, $N_b$, $N_{ab}$ are the frequencies of transformation for marker *a* alone, for marker *b* alone, and for joint transformation of markers *a* and *b*; and f is the proportion of competent cells.

Table 1.2 shows the frequencies of transformation obtained in two cultures of H.influenzae, one that had developed a high degree of competence and one which was only poorly competent, for two markers, streptomycin resistance and erythromycin resistance. In the high competency culture, the single transformation frequencies of $5.2 \times 10^{-3}$ and $7.5 \times 10^{-3}$ give a product of $3.9 \times 10^{-5}$; but the observed frequency of double transformation is greater than this, $5.2 \times 10^{-5}$. The discrepancy between the product of single transformation frequencies and the double transfor-

**Table 1.2:** frequencies of single and double transformation in H.influenzae

| | high competency culture | | low competency culture | |
|---|---|---|---|---|
| | cell number | transformation frequency | cell number | transformation frequency |
| number of viable cells | $2.4 \times 10^8$ | | $1.2 \times 10^8$ | |
| $str^r$ transformants | $1.25 \times 10^6$ | $5.2 \times 10^{-3}$ | $4.0 \times 10^4$ | $3.3 \times 10^{-4}$ |
| $ery^r$ transformants | $1.8 \times 10^6$ | $7.5 \times 10^{-3}$ | $4.7 \times 10^4$ | $3.9 \times 10^{-4}$ |
| $str^r$ $ery^r$ transformants | $1.25 \times 10^4$ | $5.2 \times 10^{-5}$ | $6.7 \times 10^2$ | $5.7 \times 10^{-6}$ |
| product of single $str^r$ and $ery^r$ transformation frequencies | | $3.9 \times 10^{-5}$ | | $1.3 \times 10^{-7}$ |
| proportion of competent cells $= \frac{str^r \times ery^r}{str^r\text{-}ery^r}$ | $0.75 = 75\%$ | | $0.023 = 2.3\%$ | |

Frequencies of transformation for streptomycin-resistance, erythromycin-resistance, and joint resistance to both markers, were measured in two cultures of H.influenzae, one in conditions of high competence and one in conditions of low competence. The transformation frequency is calculated as the number of observed transformants divided by the number of viable cells counted. The expected frequency of double transformation (if all the cells were competent) is given by the product of the observed frequencies of transformation for the single markers; but this is less than the observed frequency of double transformation. The proportion of competent cells is given by the ratio of the expected to observed frequency of double transformation for unlinked markers. Data of Goodgal and Herriot (1961).

mation frequency is more marked in the low competency culture, where the product of the single transformation frequencies of $3.3 \times 10^{-4}$ and $3.9 \times 10^{-4}$ takes the value of $1.3 \times 10^{-7}$; this is appreciably lower than the observed frequency of double transformation, $5.7 \times 10^{-6}$. In a conventional genetic analysis, the apparent increase in the frequency of double transformation compared with the product of the single transformation frequencies would imply linkage between the markers. In the transformation system, however, this is a spurious linkage resulting because these frequencies were expressed relative to the total number of cells, which is greater than the number of competent cells, thus giving estimates of transformation frequencies that are too low, their products therefore being much lower than the observed double transformation frequency. Dividing the product of the single transformation frequencies by the observed frequency of double transformation, that is, applying the formula $f = N_a \times N_b / N_{ab}$, gives the proportion of competent cells, 75% in the highly competent culture and 2.3% in the poorly competent culture. Looking at these figures in reverse, we can see that as the proportion of competent cells rises, the product of the single transformation frequencies more closely approximates the observed frequency of double transformation.

Calculation of competence from spurious linkage assumes that transformable cells are uninucleate. If they are binucleate, the proportion of nuclei transformed for each character will be between 0.5 and 1.0 of the proportion measured per bacterium. Thus the size of the competent fraction will be overestimated unless compensation is made when counting the number of cells. Another possible error may be introduced when two transforming markers are integrated onto different strands of the bacterial DNA and thus are segregated by replication. This reduces the frequency of double transformation caused by independent transforming events (see Notani and Setlow, 1974).

Genuine linkage between markers can result, of course, when they lie close together and so may be carried on a single molecule of transforming DNA. Goodgal (1961) observed that when a preparation of donor DNA from H.influenzae carries markers for streptomycin resistance and cathamycin resistance, the number of double transformants relative to DNA concentration is always a constant fraction of the number of single transformants. And the kinetics of transformation are single hit, that is, the transformants accumulate linearly with time. This contrasts with the situation observed when the $str^r$ and $cth^r$ markers are carried on different DNA molecules (that is, they are derived from two different bacterial populations); in this case the number of double transformants varies with the product of the single transformation frequencies (taking into account the proportion of competent cells) and displays two hit kinetics, that is, double transformants accumulate with time as expected from the need for two independent DNA cell interactions (Kent and Hotchkiss, 1964).

Since an increase in the number of double transformants over the product of the single transformation frequencies (relative to total cell number) cannot be taken as evidence for linkage, these other criteria must be used to detect linked markers: single hit kinetics for the double transformant; and a constant ratio of double to single transformants at varying concentrations of donor DNA and recipient cell number. Thus Voll and Goodgal (1961) observed that 20% of the transformants for either streptomycin-resistance or cathamycin-resistance carry both markers when the transforming DNA is extracted from $str^r$ $cth^r$ cells; this value provides a measure of linkage.

Finally we should note the other methods that have been used to calculate the proportion of competent cells. The most direct is autoradiography of cells following uptake of labeled DNA. Another is to rely on physiological differences between competent and noncompetent cells, although this has the disadvantage that the physiology varies with the bacterial species. A last method is to use transfection with a suitable phage, determining the proportion of cells that are killed by its development. The results obtained by these methods have been reviewed by Notani and Setlow (1974).

### *Efficiency of Integration*

The efficiency with which markers in donor DNA are integrated into the recipient genome may vary widely. Integration efficiencies cannot be determined in absolute terms but usually are described by comparing the number of transformants obtained with a donor marker with the number of transformants generated by a reference marker. An extensive series of investigations of the *amiA* locus of pneumococcus was started by Ephrussi-Taylor, Sicard and Kamen (1965); mutants in *amiA* are resistant to aminopterin and sensitive to inbalance in the molar concentrations of the branched chain amino acids, leucine, isoleucine, valine. Reciprocal crosses were performed with the DNA and cells:

$amiA^r$ $str^r$ donor DNA × $amiA^s$ $str^s$ recipient cells
$amiA^s$ $str^r$ donor DNA × $amiA^r$ $str^s$ recipient cells

By determining the number of transformants for the unlinked *amiA* and $str^r$ donor markers, it is possible to measure the efficiency of integration of any *amiA* mutation relative to the $str^r$ reference marker. Mutations fell into two nonoverlapping classes, those with an efficiency of integration close to that of the $str^r$ reference allele and those with an efficiency of integration some tenfold less. High efficiency (HE) mutations thus integrate with relative frequencies of 0.7–1.06; and low efficiency (LE) mutations integrate only with frequencies of 0.07–0.24.

In these one point crosses (so called because the two parents differ only at one *amiA* mutant site), all mutant sites except one showed the same efficiency of integration in whichever direction the cross was performed. This suggests that the parameter controlling the efficiency of integration is the difference between the sequences of the donor and recipient DNA molecules. Similar results were obtained by Lacks (1966) in measurements of the efficiencies of integration of a series of *mal*$^-$ mutants of pneumococcus that lack the enzyme amylomaltase, with the difference that four efficiency groups were found; taking into account the difference in integration efficiency of the reference markers used in these two studies, two of these groups correspond to those found at the *amiA* locus and the two additional groups possess intermediate efficiencies and very high efficiencies of integration. In these one point crosses, again only the relationship between the donor and recipient DNAs and not the orientation of the markers appeared to control the efficiency of integration (for review see Hotchkiss and Gabor, 1970).

An orientation effect is observed, however, when two point crosses are made, that is, the donor and recipient parents differ at two sites. By crossing two mutants and scoring the wild type recombinants, Lacks and Hotchkiss (1960a) found that the presence of a low efficiency site in the recipient much reduces the frequency of integration compared with the cross in which the low efficiency site is present only in the donor DNA. In a detailed study of transformation at the *mal* locus, Lacks (1966) found that this effect depends upon the distance between the two mutant sites; the ability of a low efficiency site in the recipient genome to prevent integration at a second site declines with the distance between them, an effect reported also in the *amiA* system by Gray and Ephrussi-Taylor (1967). The result of this effect is to make it impossible to construct a genetic map by conventional determination of recombination frequencies (for review see Ephrussi-Taylor and Gray, 1966); for genetic mapping relies upon the concept that only the distance between the two sites, and not their molecular nature, determines the probability that a recombination event will occur between them. Lacks therefore proposed the use for mapping by transformation of a "recombination index," in which the recombination frequency between two sites is divided by the efficiency of integration of the less efficient recipient marker. Although this was criticized by Ephrussi-Taylor et al. (1965), who found that it did not eliminate differences in reverse polarity in all their crosses with the *amiA* system, it has since been useful in constructing genetic maps of the *amiA* and *mal* loci (for review see Hotchkiss and Gabor, 1970).

To overcome the difficulties caused by the varying efficiencies of integration found with different recipient markers, Sicard and Ephrussi-Taylor (1965) mapped the *amiA* locus by using several series of two point transformation crosses. With any particular recipient, the efficiency of integration should remain the same for all donor DNA molecules, whatever donor mutant they carry, so long as the mutant site in the donor DNA is reasonably close to the recipient site of mutation. A series of two point crosses performed with one recipient and several donor DNAs therefore allows the donor mutations to be ordered on the basis of the number of recombinants that each forms with the recipient. Of course, since these are only two point crosses, they yield information only about the relative distance of each donor mutation from the recipient mutation; that is, they do not reveal whether donor mutant sites lie to the right of left of the recipient site. But by using several recipient marker strains, each of which is transformed by a series of donor mutant DNAs, it is possible to construct a genetic map of the gene. An example of two such series of crosses is shown in Table 1.3. These crosses utilized two recipient mutants, *r1* and *r22*, which lie at the two ends of the series of the mutant sites involved in these crosses. Each recipient was transformed by a series of eleven mutant DNAs and scored for wild type recombinants. The table shows the order relative to each recipient that can be deduced from these results. It is clear that each series leaves some ambiguity in order, but when the two series are combined each recipient marker provides good definition of the order of mutant sites close to it, leaving ambiguity only in the central region. This of course can be resolved by further crosses, for example, using *r6*, *r26*, *r7* as recipients. By making such crosses, a map was constructed for the gene which orders all the mutants unambiguously, although this transformation mapping does not establish the distances that are involved with any degree of precision.

What is the nature of the low efficiency and high efficiency mutations? Ephrussi-Taylor et al. (1965) observed that there was no correlation between the map position of a mutation and its efficiency of integration; the effect is simply site-specific. Reverse mutations can take place spontaneously in both the low efficiency and high efficiency class, which implies that both must contain point mutations, thus excluding any possible effect of the length of a mutation. (And one deletion that covered two sites of low efficiency point mutations was itself integrated with high efficiency.) There appeared to be some conflict between the number of mutant classes that could be categorized for integration efficiency, since

**Table 1.3:** mapping the *amiA* locus of pneumococcus by two point transformation crosses

| | relative frequency of recombination | |
|---|---|---|
| mutant in donor DNA | r1 recipient | r22 recipient |
| r1 | 0 | 0.89 |
| r6 | 0.41 | 0.52 |
| r7 | 0.17 | 0.78 |
| r9 | 0.40 | 0.08 |
| r11 | 0.14 | 0.69 |
| r20 | 0.40 | 0.14 |
| r22 | 0.50 | 0 |
| r24 | 0.33 | 0.15 |
| r25 | 0.07 | 0.74 |
| r26 | 0.27 | 0.42 |
| r27 | 0.03 | 0.79 |

order relative to r1 recipient:

r1———r27———r25———r11 = = = r7———r26———
0 0.03 0.07 0.14 0.17 0.27

r24———r20= = = r9= = =r6———r22
0.33 0.40 0.40 0.41 0.50

order relative to r22 recipient:

r22———r9————r20= = = r24———r26———r6————
0 0.08 0.14 0.15 0.42 0.52

r11———r25———r27———r7———r1
0.69 0.74 0.79 0.78 0.89

final genetic map:

r22——r9——r20——r24——r6= = =r26——r7——r11——r25——r27——r1

The relative frequency of recombination is the number of wild type recombinants from the two point donor mutant × recipient mutant cross, as a proportion of the number of transformants obtained for a reference *str*ʳ donor marker. The two recipient markers, r1 and r22, lie at the two ends of this genetic segment; a map order of donor mutant sites relative to each recipient can be constructed from the recombination frequencies as shown. Mutant sites that are not resolved from each other in these crosses are connected by equality signs. By taking the cross with the r1 recipient to define the sites of mutations close to r1, and by using the crosses with the r22 recipient to define the sites of mutations located close to r22, it is possible to produce a genetic map in which the only ambiguity is in the central region, concerning the positions of r6 and r26. Data of Sicard and Ephrussi-Taylor (1965).

Gray and Ephrussi-Taylor (1967) reported that with mutagens inducing point mutations they could identify only the two (high and low) classes in the *amiA* gene, whereas Lacks (1966) was able to identify four in the *mal*$^-$ gene. However, Tiraby and Sicard (1973b) found that the appearance of only two classes may result simply from failure to obtain a sufficient number of mutations; in the *amiA* locus, the HE and LE classes previously identified are predominant when large numbers of mutants are examined, but intermediate efficiency and very high efficiency classes can be found in smaller amounts. That the efficiency depends upon the type of mutation is implied by the observation of Tiraby and Fox (1974) that hydroxylamine and nitrous acid induce LE mutations almost exclusively.

The variations in efficiency that are characteristic of these mutant classes appear to result from the bias of repair systems that recognize mismatched bases in the DNA duplex. Since only one strand of donor DNA is integrated into the resident genome, the structure produced is a hybrid DNA possessing one strand with recipient information and one strand with donor information. This means that mismatching will occur at any mutant site; the nature of the mismatching will depend upon the type of mutational event. As Figure 1.7 illustrates, several types of event may ensue. If no repair system acts on the hybrid DNA, the two strands carrying recipient and donor sequences will separate at replication, each acting as template for synthesis of a complement, so that the two daughter cells into which they segregate will have different genotypes: one will carry the donor information, that is, it will be transformed; whereas the other will carry the recipient information, that is, it will not be transformed. This situation is revealed experimentally by the identification of clones that are mixed in genotype for the transforming character.

Mixed clones resulting from transformation were observed by Guerrini and Fox (1968a) when they analyzed the colonies derived from single viable cells that had been exposed to transforming DNA carrying the marker *suf-d* that confers sulfanilamide resistance in pneumococcus. Almost all the colonies that displayed the donor phenotype also possessed cells retaining the recipient marker. This demonstration that the transformed cells are genetically heterozygous implies that only one donor strand is involved in transformation and thus provides a genetic correlation to the biochemical demonstration that one donor strand is inserted into the resident DNA. The involvement of repair systems was implied by the observation of Guerrini and Fox (1968b) that when cells are treated with mitomycin C before they are mixed with transforming DNA, the proportion of pure transformed clones is much increased; mitomycin C introduces structural changes in DNA that are the subject of repair systems and this result therefore suggests that in repairing the damage

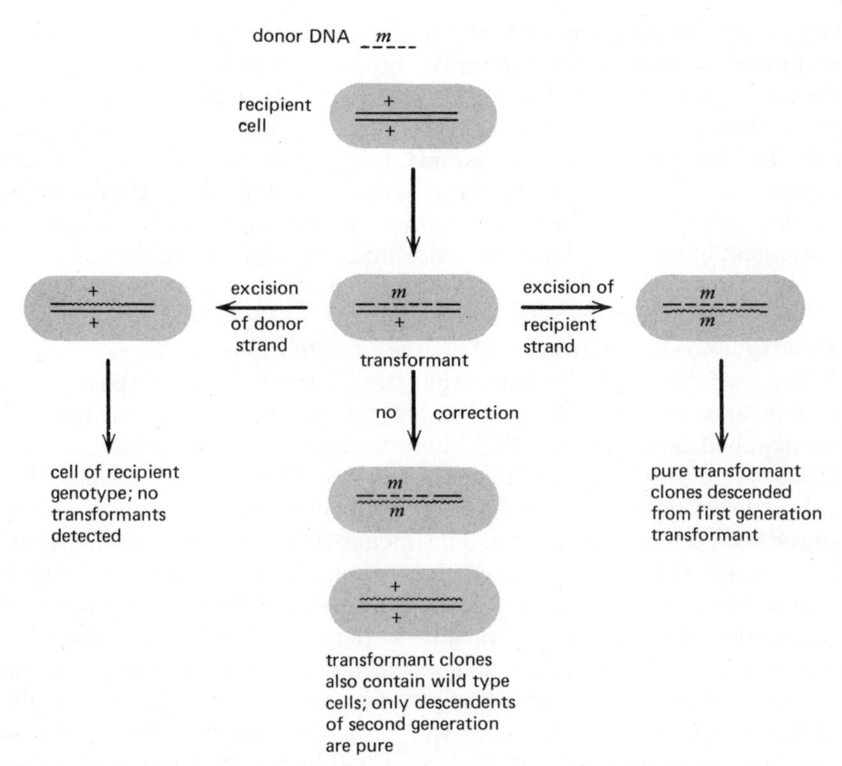


**Figure 1.7:** excision-repair of transformed cells. The insertion of donor DNA carrying a mutation (m) into the DNA of the wild type recipient (+ on each strand) generates a hybrid stretch of DNA, carrying mutant information on one strand and wild type information on the other. If this mismatch is not corrected (center), the transformed cell gives rise to progeny of two genotypes: homozygous for the mutant and homozygous for the wild type. Thus clones derived from the original transformant are mixed; only clones derived from cells of the next generation are pure. If excision of the donor strand takes place (left), the cell regains the recipient genotype and thus no transformation can be detected. If the recipient strand is excised (right), the cell becomes homozygous for the mutation and gives rise to pure clones, that is, it does so one division cycle before the descendents of transformants that were not corrected.

caused by mitomycin C the repair system also corrects the mismatched sequence resulting from insertion of donor DNA.

When an excision-repair system acts upon mismatched DNA, one strand of the DNA is removed and a replacement is synthesized using the remaining strand as template. This produces a perfectly matched duplex of DNA, representing the information of whichever of the two mismatched strands was not excised. In E.coli, where the processes of excision-repair have been best characterized, the enzymes responsible for the correction event appear to recognize distortions in the structure of the duplex, for the same excision-repair system acts upon damage induced by several different agents, such as ultraviolet irradiation or treatment with mitomycin C. The characterization of excision-repair enzymes in E.coli is discussed in Volume 1 of Gene Expression. The presence of such enzymes appears to be a general phenomenon in many types of cell and the analysis of the products of genetic recombination, especially in the ascomycetes fungi, suggests that the joining of duplex DNA molecules is accomplished by formation of hybrid DNA which also may be the subject of attention by such enzyme systems. In fact, recombination between sites located close together appears usually to be accomplished by the formation and correction of hybrid DNA rather than by the exchange of duplex segments of DNA (see Volume 1 of Gene Expression).

When correction of hybrid DNA takes place after integration of donor transforming DNA, only pure clones should be produced by individual cells. Figure 1.7 illustrates the consequences of correction in either direction. If the recipient strand is excised and replaced with a strand complementary to the inserted donor sequence, the cell bears the transformed genotype and all its progeny should be transformants. On the other hand, if the donor strand is excised and material complementary to the recipient strand is inserted, then the cell regains the recipient genotype and there is no evidence for transformation in its progeny; this situation cannot be recognized directly since recipient cells do not change in phenotype and there is therefore no trace of the entry of the transformed DNA.

The efficiency of integration of markers in donor DNA depends upon the activity of repair system(s) on the hybrid DNA that is involved. Ephrussi-Taylor and Gray (1966) proposed that the difference between low efficiency and high efficiency sites might be explained if the low efficiency sites usually are corrected by a repair system whereas the high efficiency sites usually are not corrected. To explain the low efficiency of integration of the LE marker sites in both orientations, it is necessary to postulate that the repair system most frequently excises the donor strand at these sites (in 90% of the cells gaining the transformed DNA) but sometimes excises the recipient strand (in the 10% of the cells that give rise to transformants, relative to the HE markers). One prediction of the model that the LE sites always are corrected whereas the HE sites are not corrected is that homozygous transformants for LE characters should appear one generation before homozygous transformants for HE characters (see Figure 1.7). This prediction was tested by Ephrussi-Taylor (1966) in experiments in which sister cells of pneumococcus were split apart when they were formed at successive divisions in synchronous cultures;

these cells were then grown into clones and were considered to carry the donor gene if transformed cells were present in the clone. The marker *opt*$^{r}$ which is transformed with low efficiency always is transmitted to both progeny at the second division after uptake of DNA; but the marker *str*$^{r}$ which is transformed with high efficiency is transmitted to both progeny only after the third division. These markers provide references against which other mutants can be assayed; it is possible to determine at which division any marker became homozygous by comparing the relative proportion of transformed cells with the proportion transformed for these two markers after 1, 2, and 3 divisions. All the LE sites of *amiA* that were tested became homozygous after two divisions, whereas the HE sites took a further division cycle before becoming homozygous.

The predilection of the repair system for excising the donor strand at LE sites in one point crosses, for excising neither strand at HE sites, and perhaps for excising the recipient strand at very high efficiency sites, offers an immediate explanation for three classes of mutation. Presumably it is the type of difference between the mutant sequence and the wild type sequence which dictates the response of the repair system; it is reasonable to suppose that excision-repair enzymes may recognize different types of mispairing in DNA with differing efficiencies. Tiraby and Fox (1974) proposed that the specificity of hydroxylamine and nitrous acid in inducing LE mutations suggests that the mispairs involved in this class are of the types A-C and G-T.

The effect of a low efficiency marker in reducing the efficiency of integration of a high efficiency marker located near to it can be explained by the supposition that the length of the sequence excised extends from the LE site to past the HE site; this accounts for the observation that the effect declines with distance between the two mutant sites (see Bresler, Kreneva and Kusher, 1971). The polarity observed in two point crosses, that it is LE sites in the recipient DNA that reduce integration of HE sites, remains to be explained. Perhaps the reason lies with the structure of the intermediates involved in the integration process. Although points such as this remain to be resolved so that a complete sequence of events in integration cannot yet be defined, the classification of sites of different efficiencies of integration and their properties show clearly that excision-repair systems are involved in transformation.

A gene function necessary for discriminating between LE and HE sites was first identified by Lacks (1970), when he isolated some pneumococcal mutants which yield high numbers of transformants for all donor markers, whether HE or LE in type. These *hex*$^{-}$ mutants appear to have lost the ability to excise the mispairing involved at LE sites. Tiraby, Fox and Bernheimer (1975) noted that all pneumococcal strains that they tested appear to possess the wild type *hex*$^+$ function since they are able to discriminate between LE and HE markers. Tiraby and Sicard (1973a) showed that in *hex*$^+$ cells, bacteria transformed for LE markers increase in number about one generation sooner than bacteria transformed for HE markers, an observation which suggests that the *hex*$^-$ function involves a repair event. In a study of the effect of the *hex*$^-$ mutation upon the efficiency of integration of mutants in all four classes—that is, including intermediate efficiency and very high efficiency as well as LE and HE—Tiraby and Sicard (1973b) found that no discrimination is observed between them; this implies that it is the single system identified by the *hex* locus in pneumococcus which is responsible for discriminating between types of mispairing in DNA. One prediction made by this model for the role of the *hex* locus is that repair of certain mismatched bases in *hex*$^-$ strains should be less efficient than in *hex*$^+$ strains; this was confirmed by Tiraby and Fox (1973) by the observation that the rate of spontaneous mutation is increased (from 4–30-fold at various loci) in the *hex*$^-$ mutant strain, presumably because of its defective function.

CHAPTER 2

# Conjugation

## Nature of Bacterial Variation

### *Identification of Mutants*

For the first part of this century, bacteria were regarded as a unique life form because of two unusual properties: they did not appear to possess a complex nuclear apparatus analogous to that of eucaryotic cells; and they possess an evident ability to adjust to new environments with changes that are inheritable. But although the bacterial genome does not take the form of the deoxyribonucleoprotein complex typical of eucaryotes and is not surrounded by a nuclear membrane, it is now known to be directly comparable in structure and function to the eucaryotic chromosome and to suffer heritable change only by mutation. The first experiments to imply that the genetic apparatus of bacteria is organized on lines familiar from eucaryotic genetics were those of Luria and Delbruck (1943), who devised the *fluctuation test* that allows adaptation to be distinguished from spontaneous mutation. Before the development of this test, the nature of heritable variation in bacteria had been a long standing controversy; for experiments which showed that only a small proportion of the bacteria in a culture may survive when transferred to new, selective conditions could be interpreted in either of two ways. Mutants might previously exist that are able to survive after the population is shifted to new growth conditions; or adaptation might be induced by exposure to the new conditions. To demonstrate that spontaneous mutation is the cause of variation, it was therefore necessary to show that the mutant bacteria are present in the population before exposure to the new conditions; the statistical analysis of the fluctuation test allows prior existence and subsequent adaptation to be resolved.

The system with which Luria and Delbruck worked was the acquisition of resistance to bacteriophage T1 by E.coli. When a culture of cells sensitive to the phage is plated in the presence of excess phage particles, virtually all the bacteria are infected and thus killed; but a few survive to yield colonies. The cells of these colonies are resistant because they have lost the ability to adsorb the phage, a characteristic which is inherited by their progeny. Since this result is obtained also with cultures started from single (sensitive) bacteria, the development of the resistant cells must have occurred by a change in the properties of the original cell type. The critical question that the fluctuation test resolves is whether the resistant cells result from an adaptation induced by exposure to the phage particles or whether they are caused by spontaneous mutations, occurring by chance to some cells and independent of exposure to the phage.

If resistance to phage (or, indeed, any other form of inherited characteristic) results from adaptation, at any given time each bacterium in the population must have a small, constant, random probability of acquiring resistance in response to exposure to phage. Thus the number of resistant colonies will depend only on the number of bacteria plated in the presence of the phage and does not depend on any occurrence prior to this exposure. But if resistance results from spontaneous mutation, each bacterium has a small, constant, random probability in each generation of suffering mutation to resistance; since all descendants of a resistant mutant inherit the characteristic, the time at which the mutation has occurred (prior to exposure to the phage) is a critical parameter. If the mutation occurred soon after initiation of a culture, exponential growth of the progeny of the resistant mutants will have generated many resistant cells by the time of exposure to the phage; if the mutation occurred at a later time, the population might contain only a few resistant cells derived from the mutant.

The principle of the fluctuation test is therefore to compare the numbers of resistant cells found in many independent cultures, all started from a few cells under identical conditions at the same time. After time for growth, each of these cultures is plated on medium containing phage and the number of resistant colonies is determined. If resistance develops by adaptation, all cultures should have the same number of resistant cells, since all cells have the same probability of adaptation in response to the phage and events prior to this time are irrelevant. If resistance depends on prior mutation, very different numbers of resistant cells may be found in each culture, varying from those in which a mutation occurred soon after establishment (generating many resistant cells by the time of plating) to those in which mutation occurred shortly before plating (leaving time to generate only a few resistant descendants). A fluctuation between the resistant numbers in each colony thus demonstrates that resistance to phage is caused by spontaneous mutation. The significance of an observed fluctuation can be assessed experimentally by comparing the fluctuation between independent cultures with that of a control in which samples of a single colony are examined. One important principle which the fluctuation test illustrates, and which runs through all of bacterial genetics, is that the subject of analysis is not the individual progeny typical of eucaryotic genetics but constitutes populations of bacteria.

Other methods of demonstrating that bacterial variation is caused by spontaneous mutation have been developed since the fluctuation test was devised. In the test of Newcombe (1949) plates are spread with (for example) bacteria sensitive to a virulent phage. After incubation to allow each bacterium to multiply into a colony, half the plates are rubbed with a spreader to redistribute the bacteria. Both the rubbed and nonrubbed plates are then treated with phage and the number of resistant colonies is counted. If adaptation is responsible for resistance, both types of plate should contain the same number of surviving colonies, for redistribution prior to exposure to the phage cannot influence their response. If spontaneous mutation is responsible, the rubbed plates should contain many more resistant colonies than the nonrubbed plates; for on the undisturbed plates each resistant mutant will have given rise to a single colony, remaining in one position, whereas on the respread plates each of the colonies will have been separated so that each of its members can grow into a separate resistant colony after plating with phage. Any character for which selection can be imposed without disturbing the bacteria (phages can be sprayed on, for example) can be tested; the results show that phage resistance and other characters are produced by spontaneous mutation.

Examining bacteria for hereditary characteristics demanded lengthy protocols until the technique of replica plating was introduced by Lederberg and Lederberg (1952). Previously, each colony had to be analyzed individually for every characteristic by transferring samples to a series of plates containing appropriate selective media. Analyses of many colonies for recombinants involving the inheritance of several characters could thus become limited by the very large number of individual actions required. Replica plating allows the pattern of growth on a "master" plate to be transferrred as a whole to a series of replica plates. The master plate is pressed gently onto a square of sterile velvet, which acts as an orderly array of many short needles; bacteria adhere to the velvet so that it carries an exact representation of the bacterial growth pattern of the master plate. Replica plates, each consisting of an appropriate medium, are then pressed onto the velvet pad to pick up from it a sample of the bacterial population; several such replicas can be printed from each pad. Up to one third of the bacteria of the master plate can be transferred to the pad and up to one third of these are transferred to the replicas. By keeping the orientation of master and replica plates the same, each colony can be examined for any characteristic by replica plating onto appropriate medium. This technique, now of wide application in examining recombinants, also was used to show that bacteria resistant to a phage arise in clones which pass on this characteristic in inheritance even if they are never exposed to the phage; for by exposing replica plates to a phage but retaining a series of master plates that are not exposed, it is possible to follow the development on master plates of resistant colonies.

### *Estimates for Mutation Rates*

Mutation rates are usually calculated as the probability of mutation per bacterium per generation, in other words as the probability that an event will occur during a single reproduction of the bacterial chromosome. The rate of spontaneous mutation usually is about $10^{-6}$ per locus. Because experimental conditions in general utilize exponentially growing (and unsynchronized) populations, calculation of the mutation rate requires a statistical analysis of the experimental data. The parameters that can be observed are the initial and final numbers of bacteria, the time required for doubling of the population, and the number of mutants; the parameters from which the mutation rate is calculated are the average number of bacteria and the number of bacterial divisions.

Because the number of bacteria in a growing population is continually increasing, the average number of bacteria during the generation time is not the initial number of cells but is the initial number divided by the exponential function, $\log_e 2 = 0.69$. A factor of $\log_e 2$ thus appears in the mutation rate calculation. Since there is a probability that mutation will occur in each bacterial division, the value required for the number of generations is the total number of individual bacterial divisions, which, of course, is much larger than the number of doubling times during the experiment since every bacterium divides in every doubling period. In exponential growth, much larger numbers are involved towards the end of the period than at the beginning; and a summation of the number of bacterial divisions shows that it may be represented by the expression $(N_f - N_i)/\log_e 2$, where $N_f$ is the final number and $N_i$ is the initial number of cells. The rationale that underlies this argument is reviewed in detail by Hayes (1968). When the final number of cells is very large compared with the initial number (a condition effectively fulfilled whenever five or more doubling times have elapsed), it is therefore practical to take the number of divisions as $N_f/\log_e 2$.

To calculate the rate of mutation, two series of plates are established. At appropriate intervals, one plate is used to provide a sample for estimating the total number of viable bacteria and the other to discover the number of mutants present. The mutation rate can then be represented by the expression

$$m = \frac{M_f - M_i}{N_f - N_i} \cdot \log_e 2$$

where $M_f$ and $M_i$ are the number of mutants detected on the final and initial plates. These data are readily obtained from the control (undisturbed) plates of the Newcombe test for variation. A drawback of this direct technique is that it can be applied only to mutant characteristics which allow selection to be imposed on the plates without disturbing the colonies on them.

Another approach to calculate mutation rates derives from the Luria-Delbruck fluctuation test. The Poisson distribution can be used to calculate the average number of mutations per culture (usually from the proportion of cultures that shows no mutations). The mutation rate is then given by

$$m = \frac{M_{average}}{N_f} \cdot \log_e 2$$

Other, more sophisticated methods of analyzing the data have been developed from the concept that in an exponentially growing culture the proportion of mutants increases with time (although these approaches usually assume that the growth rates of the mutant and wild type bacteria are the same, often not a valid condition). Since old mutants divide in harmony with the population and new ones are continually added, the mutant proportion increases (up to a plateau when the mutant number is so great that reversion and other factors establish an equilibrium). These methods of analysis have been reviewed by Hayes (1968).

A complicating factor in analyzing bacterial mutations is that there is often a delay in the appearance of the mutant phenotype. One cause for such delay is the *segregation lag*. Under conditions of rapid growth, each bacterial cell may possess more than one haploid nucleus (see Volume 1 of Gene Expression). When a mutation is dominant, it may be expressed immediately by the cell, but as Figure 2.1 shows, the number of mutant bacteria will not increase until sufficient divisions have taken place to yield a bacterium all of whose nuclei are mutant. If a mutation is recessive, the segregation lag remains the same, but the mutant phenotype is not expressed until after segregation has yielded wholly mutant cells; this delay in expression is the *phenotypic lag*. In calculating rates of mutation, it is therefore necessary to take into account the existence of multinu-

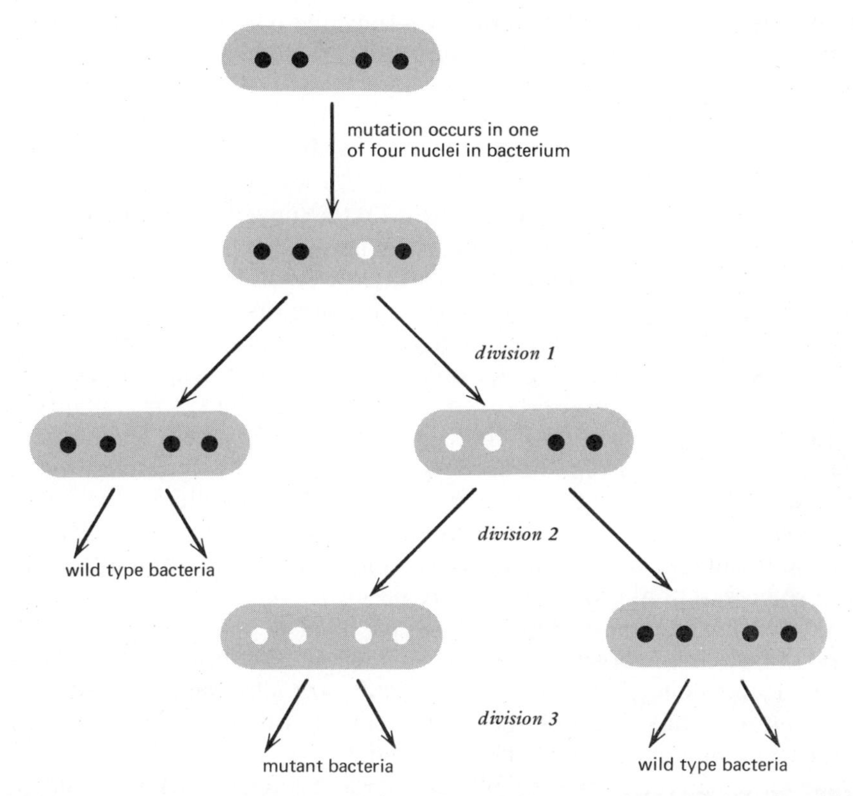


**Figure 2.1:** segregation and phenotypic lags. If a mutation occurs in one of several nuclei in a bacterium, there is a lag before entirely mutant bacteria are segregated; with four nuclei, as shown above, the first entirely mutant bacterium is produced by the second division and the number of mutants begins to increase only with the third division. If the mutation is dominant, it may be expressed immediately; but if it is recessive, there is a phenotypic lag until entirely mutant bacteria are produced, in the above example as a result of the second division.

cleate bacteria if the rate is to be expressed per nucleus (appropriate since the nucleus represents the unit subject to mutation) and to remember also that the segregation and phenotypic lags may cause the mutation rate to be underestimated.

The delay before a new characteristic is expressed also may vary with the nature of the activity of the gene product. An enzyme activity may be displayed immediately upon synthesis of the protein; but a mutation whose expression requires structural changes to be effected may take longer to become visible. Phage resistance, for example, is recessive to sensitivity since it involves the failure of the cell wall to adsorb phage.

Thus when a resistant allele replaces a sensitive one, it may take several generations before the preexisting receptors are diluted out and resistance is acquired.

### *Selection of Mutants*

Induction of mutations in bacteria was first achieved by irradiation with X-rays or ultraviolet light; many chemical agents, such as nitrosoguanidine or analogues of nucleotide precursors, have since been found to cause mutations with high efficiency. Mutants that are resistant to phages or to drugs can be isolated simply by plating large populations of sensitive cells on nutrient agar containing the phage or drug. The range of available mutants can be greatly extended by using the ability or inability to metabolize growth factors. E.coli is particularly well suited for this genetic analysis, for wild type bacteria synthesize all the factors required for growth and so can multiply in a synthetic medium in which glucose and ammonium ions are the sole sources of carbon and nitrogen. Mutants unable to survive on such minimal medium because they cannot synthesize some essential growth factor are described as *auxotrophs;* an auxotrophic strain can thus survive only on medium supplemented with all the growth factors in whose synthesis it is deficient. The term *prototroph* was introduced to describe a cell able to synthesize all essential growth factors, although this term does not imply that it is wild type since it may, of course, be mutant for other characteristics.

Given an auxotrophic strain of bacteria, prototrophs may easily be isolated and quantitated simply by selection on minimal medium lacking the necessary growth factor(s). Auxotrophic mutants, however, cannot be isolated in this way since their characteristic is the failure to grow on minimal medium. Replica plating offers a way to compare the abilities of bacteria to grow on different media; colonies which grow on the complete medium of the master plate, but which fail to grow on a replica plate provided with deficient medium, can be isolated and tested to see whether they are genetically deficient in the ability to synthesize whatever growth factors are absent from the deficient plate. This is not a very productive method, however, since it applies no selective pressure for survival of the desired mutant and since only some of the colonies that fail to grow ultimately prove to be stable auxotrophic mutants.

A selective method that allows auxotrophs to survive but kills prototrophs was made possible by the observation that penicillin kills growing but not nongrowing bacteria. As illustrated in Figure 2.2, by growing a mutagenized culture on medium deficient in some growth factor but containing penicillin, it is possible to kill selectively the prototrophs

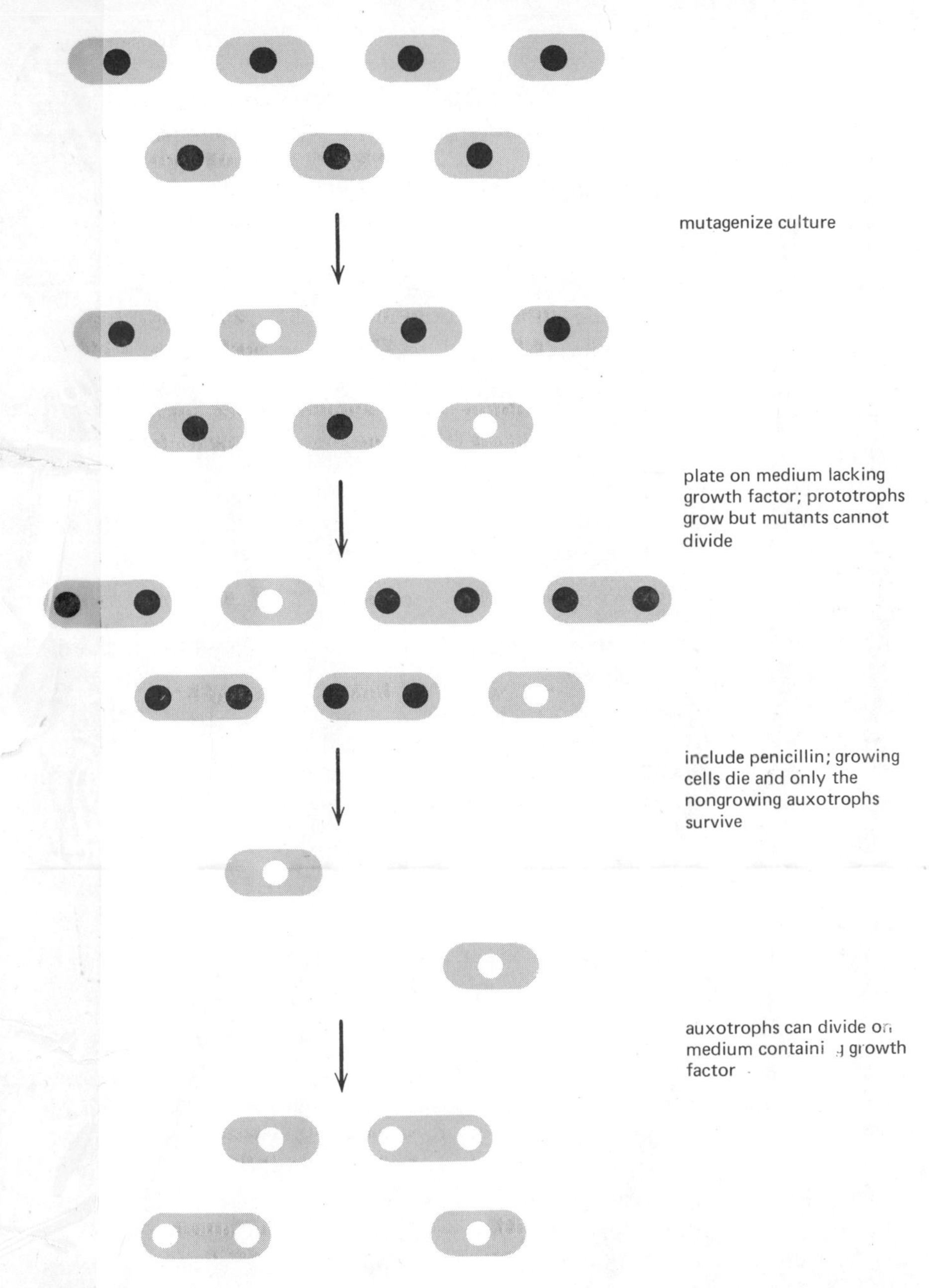


**Figure 2.2:** selection of auxotrophic mutants. After a culture has been mutagenized, it is plated on medium lacking a growth factor so that mutants unable to synthesize this factor cannot grow; inclusion of penicillin kills all growing cells but not the auxotrophs.

which grow normally, but leave untouched the auxotrophic cells that are unable to grow. Of course, although this method allows the application of selective pressure to obtain an improved yield of the desired mutants, it cannot easily be quantitated since it kills the cells that lack the mutation.

## Genetic Transfer Between Bacteria

### *Discovery of Unidirectional Genetic Transfer*

Derivatives of two lines, 58–161 and W677, originally isolated from E.coli K12 by Tatum (1945), were used for many of the early crosses made between bacteria; they proved especially productive for their relatively high yields of recombinants and for the ease with which other markers could be linked to the auxotrophic loci. Strain 58–161 cannot synthesize methionine and biotin, and so may be described as $met^-bio^-$, and strain W677 cannot manufacture threonine, leucine and thiamine (vitamin B1), and so is $thr^-leu^-thi^-$. (Nomenclature for genetic markers follows the suggestions of Demerec et al., 1966; genotypes are described by the forms $abc^-$ and $abc^+$ and the corresponding phenotypes by the forms $\text{Abc}^-$ and $\text{Abc}^+$.) Genetic transfer in E.coli was first reported by Lederberg and Tatum (1946a,b) in two brief communications in which they observed that when these two auxotrophic strains are grown together, prototrophs can be recovered. Thus mixing bacteria of the two strains:

$$bio^- \ met^- \ thr^+ \ leu^+$$
$$bio^+ \ met^+ \ thr^- \ leu^-$$

gives rise to prototrophs of the genetic constitution

$$bio^+ \ met^+ \ thr^+ \ leu^+$$

at a frequency of about $10^{-5}$ to $10^{-6}$ relative to the original cell number. (This frequency is much greater than the mutation rate, which is about $10^{-6}$ per locus and would therefore be only some $10^{-12}$ for reversion of a double mutant.)

That the formation of prototrophs is due to some form of genetic exchange was suggested by observations of the distribution of unselected markers in the prototrophs. The *selected markers* in crosses between auxotrophic strains, of course, are those which must be present for survival of the progeny. In the cross between the two double auxotrophs $bio^-met^-$, and $thr^-leu^-$, the two parental strains are grown upon media supplemented with biotin and methionine or threonine and leucine, re-

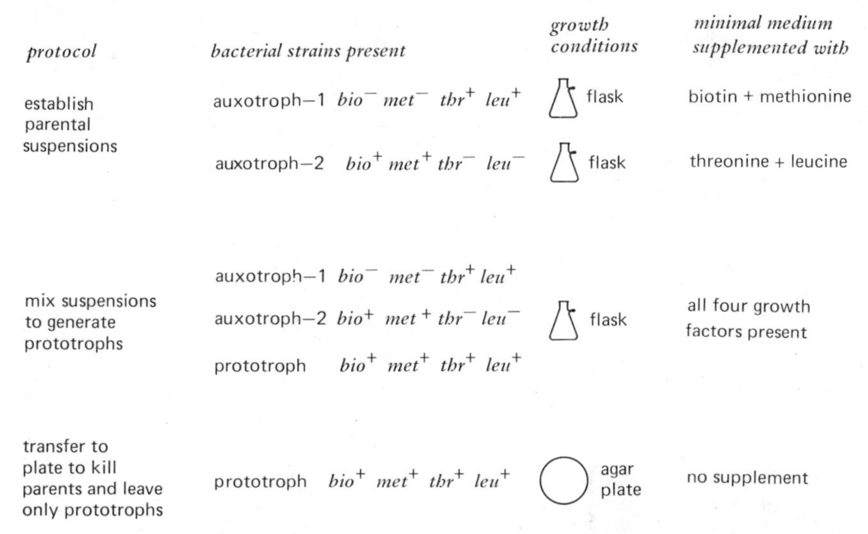


**Figure 2.3:** isolation of prototrophs from auxotrophic parents. The auxotrophic parents are each maintained by growth in the presence of the necessary factors that they are unable to synthesize; after mixing these suspensions, plating on medium lacking the growth factors allows only recombinant prototrophs to survive.

spectively. As illustrated in Figure 2.3, the bacteria are mixed and then transferred to plates with minimal medium, which lack all four of the growth factors; the parental genotypes cannot survive under these conditions, which select for prototrophic bacteria able to synthesize all four growth factors, that is possessing the four selected markers $bio^+$, $met^+$, $thr^+$ and $leu^+$.

But the two parental strains may differ in other markers that are not the subject of selection. Two *unselected markers* used in these early crosses, for example, govern the response to infection with phage T1 ($ton^r$ bacteria are resistant to killing by this phage whereas cells with the $ton^s$ allele are sensitive) and the ability to metabolize β-galactosides ($lac^+$ cells can do so whereas $lac^-$ cells cannot). The distribution of these unselected markers in the prototrophs can be determined readily by plating on medium to which phage T1 is added (when $ton^s$ cells are killed and a count of the surviving colonies gives the $ton^r$ number) and by growth on indicator medium (when $lac^+$ colonies become colored by metabolizing an appropriate indicator, whereas $lac^-$ colonies do not). This protocol is illustrated in Figure 2.4.

The distribution of unselected markers in the prototrophs was first observed to be nonrandom by Lederberg (1947). That is, there is a

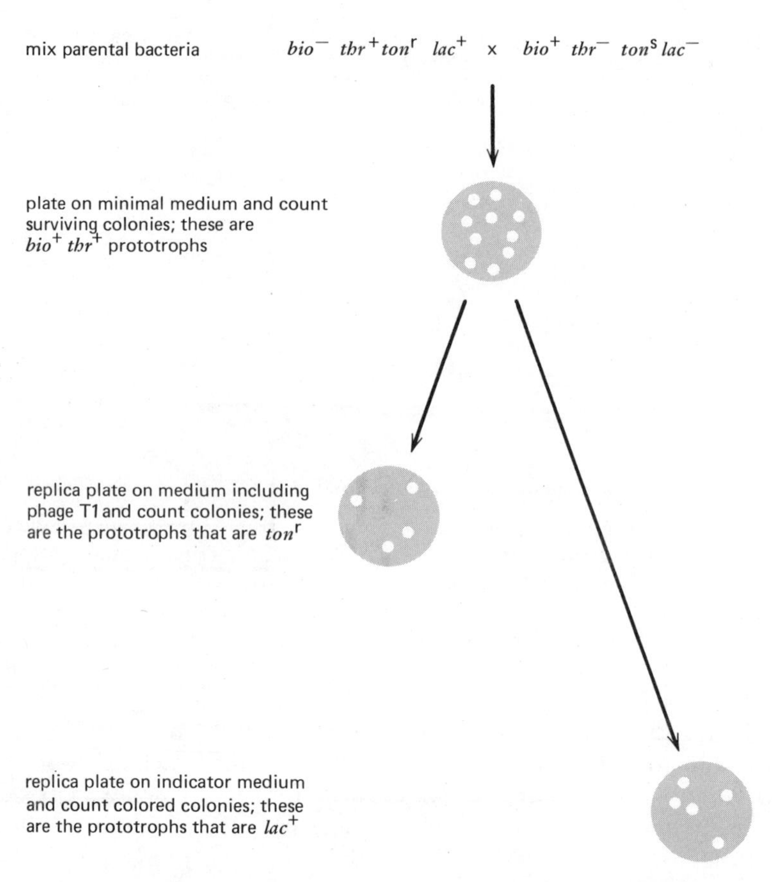


**Figure 2.4:** determination of unselected markers in prototrophs. Two auxotrophic strains which differ in unselected markers as well as in the auxotrophic markers are crossed and the prototrophs isolated; by plating on further media, these may be examined to see which unselected parental markers are present.

tendency for certain unselected markers to remain with the markers selected from their parental genotype. In the example shown in Table 2.1, the unselected *ton* allele of the first parent tends to remain associated with the selected markers, $thr^+$, $leu^+$, $thi^+$. This is, of course, the classical condition for defining a linkage group; the frequency with which any unselected allele remains associated with a selected allele in principle provides a measure of their distance apart in a linkage group (although the mechanisms of bacterial genetic transfer do not permit analysis in terms identical to those established for eucaryotes). All seven of the loci followed in these early experiments fell into one linkage group. Since segre-

**Table 2.1:** linkage between selected and unselected markers

| cross | distribution of *ton* alleles in the *thr*$^+$ *leu*$^+$ *thi*$^+$ *met*$^+$ *bio*$^+$ prototrophs | |
|---|---|---|
| | percent *ton*$^r$ | *ton*$^s$ |
| *thr*$^+$ *leu*$^+$ *thi*$^+$ *met*$^-$ *bio*$^-$ *ton*$^r$ × *thr*$^-$ *leu*$^-$ *thi*$^-$ *met*$^+$ *bio*$^+$ *ton*$^s$ | 75 | 25 |
| *thr*$^+$ *leu*$^+$ *thi*$^+$ *met*$^-$ *bio*$^-$ *ton*$^s$ × *thr*$^-$ *leu*$^-$ *thi*$^-$ *met*$^+$ *bio*$^+$ *ton*$^r$ | 36 | 64 |

The two parental strains were mixed and then plated on medium on which only the prototrophic recombinants can survive. These colonies were then analyzed to see whether they were resistant or sensitive to killing by phage T1. The two matings represent reciprocal crosses since in one the first parent carries the allele *ton*$^r$ and in the second it has the allele *ton*$^s$. The proportion of *ton*$^r$ recombinants is reversed in the two crosses, following the reversal of its orientation; the unselected *ton* allele of the first parent shows about 70% linkage to the unselected markers *thr*$^+$ *leu*$^+$ *thi*$^+$.

gation of parental markers was observed in the first generation clones, which represent recombinants of uniform genotype, the bacterial genome must be haploid; otherwise heterozygotes would be present at this stage and would segregate subsequently. Tatum and Lederberg (1947) thus suggested that the formation of prototrophs observed when two E.coli auxotrophs are grown together is due to genetic recombination between the selected markers. (For a review of these early results, see Hayes, 1953a.)

The definition of different classes of bacteria according to their sexual compatibilities in conjugation showed that there are two different types of cell. Lederberg, Cavalli and Lederberg (1952) observed that it was impossible to obtain prototrophs by crossing two of the auxotrophic bacterial stocks; certain derivatives of the 58-161 *bio*$^-$*met*$^-$ strain could not form recombinants with the standard W677 *thr*$^-$*leu*$^-$*thi*$^-$ strain. But although infertile when crossed with each other, these strains were able to recombine with other stocks. The mutually infertile strains were designated $F^-$ and strains that can recombine with them termed $F^+$. Compatibility can then be defined in terms of the crosses:

$F^- \times F^-$ sterile
$F^+ \times F^-$ fertile
$F^+ \times F^+$ apparently fertile

Thus an $F^+$ strain is defined experimentally as one that can support conjugation, with an $F^-$ strain or with another $F^+$ strain; an $F^-$ strain cannot support recombination when crossed with another $F^-$ strain. The only cross that is in fact fertile is $F^+ \times F^-$, since we now know that the apparent fertility of crosses between two $F^+$ populations results from the presence in them of some cells that behave as $F^-$. Conjugation between two E.coli cells thus demands one of each mating type. The standard strain W677 used to introduce the markers *thr*$^-$, *leu*$^-$ is $F^-$, but was not found to be sterile in early crosses because the other strains, including 58-161, largely are $F^+$; it was the isolation of an $F^-$ derivative from strain 58-161 that led to the definition of sexual compatibility.

That the two parental populations play different roles in genetic transfer was suggested also by the results of Hayes (1952a), who found that treatment of strain 58-161 with streptomycin, which is lethal to the cells, stimulates the formation of prototrophs; but treatment of cells of strain W677 with streptomycin prevents the formation of recombinants. Another distinction between the two strains can be drawn from the effect of irradiation with ultraviolet light. Hayes (1952b) observed that when strain 58-161 is irradiated, the formation of recombinants is increased 5–20 times. On the other hand, irradiation of the W677 parental cell population produces a fall in the number of recombinants roughly proportional to the drop in viable count.

These results suggest that strain 58-161 may be a donor of genetic material; donation must be able to continue during cell death induced by streptomycin treatment, probably because the killing effect is slow enough to allow at least some donor cells to sponsor transfer, and is enhanced also by ultraviolet irradiation. Strain W677, however, must provide recipient cells, which obtain genetic information from the donor; and the death of the recipient cells, induced by either streptomycin treatment or ultraviolet irradiation, thus prevents formation of recombinants. Genetic transfer is therefore unidirectional; and once the donor (or male) cells have transferred their genetic information, their participation is complete so that the further fate of the donors becomes irrelevant to the formation of recombinants. It is the recipient (or female) cells in which the formation of recombinants takes place by interaction between the resident genome and donor genetic material; and since recombination takes place within the recipient, the death of these cells prevents formation of the prototrophs.

By examining the sexual compatibilities of the 58-161 (donor) and W677 (recipient) strains, Hayes (1953b) demonstrated that the first can be characterized as $F^+$ and the second as $F^-$. This suggests that sexual compatibility reflects the ability of a strain to act as a donor of genetic material. An $F^+$ strain is therefore able only to donate genetic material; an $F^-$ strain is able only to receive it. When the same cross is performed between a male donor and either a male or female recipient, the cross $F^+ \times F^+$ gives some twentyfold fewer prototrophs than the cross $F^+ \times F^-$. This is consistent with the idea that crosses take place only between an $F^+$ donor and $F^-$ recipient; but although $F^+$ cells cannot act as recipients, the presence in the $F^+$ population of some cells that are $F^-$ allows conjugation to proceed, although at a rate that is reduced in accordance with the low concentration of true recipients.

Two consequences ensue when an $F^+$ culture is mixed with $F^-$ cells. One is the transfer of genetic information from donor to recipient with the subsequent formation of recombinants. A second is the transfer of the $F^+$ genotype. This was first noted when Lederberg et al. (1952) tested the compatibilities of the progeny of $F^+ \times F^-$ crosses and found that all were compatible, that is $F^+$. That $F^+$ is transmitted infectiously, independently of any genetic transfer, was revealed by an experiment in which three strains were mixed: an $F^+$*met*$^-$*thr*$^-$*leu*$^-$, an $F^-$*met*$^+$, and an $F^-$*thr*$^+$*leu*$^+$. Although the two $F^-$ strains cannot yield prototrophs when mixed together, they are able to do so when mixed in the presence of the $F^+$ strain. Since the $F^+$ strain does not contribute positive markers to help form prototrophs, its action must be to convert cells of one or both $F^-$ strains to the $F^+$ mating type, so that they in turn can conjugate with the auxotroph. Subsequent experiments showed that most or all of the cells of an $F^-$ culture may be converted to the $F^+$ state upon inoculation with some $F^+$ cells; the efficiency with which mating type is transferred is thus very much greater (about $10^6$-fold) than the frequency of genetic recombination. The sexual characteristic of E.coli is thus infectious. Another feature of $F^+$ cells is that when cultures are grown to high concentrations, the cells may behave as though $F^-$ although their genetic constitution remains unaltered—*phenocopies* have been produced.

### *Linkage between Unselected and Selected $F^+$ Markers*

Following transfer of donor genetic material, the recipient cell passes through a transient diploid stage, during which genetic recombination may take place between donor and recipient DNA. Although usually it is not possible to isolate cells in the diploid condition, Lederberg (1949) found that after crosses in which one or both parents possess a mutant allele termed *het*, the recombinant prototrophs behaved as heterozygous diploids; after a cross of *bio*$^-$*met*$^-$ *lac*$^+$ $\times$ *thr*$^-$ *leu*$^-$ *lac*$^-$, for example, the *bio*$^+$ *met*$^+$ *thr*$^+$ *leu*$^+$ prototrophs plated on the indicator medium EMB-lactose give variegated colonies instead of dark *lac*$^+$ or light *lac*$^-$. Unlike the stable recombinant prototrophs usually produced by such crosses, these cells were extremely unstable, continually giving rise, however, to stable haploid genotypes. By using various derivatives of the parental strains which differed in several markers, Lederberg showed that the segregants produced by the unstable prototrophs were always stable for all these markers. Since these experiments involved bacterial populations, there was no proof that the diploid condition existed in individual cells; this prompted the experiments of Zelle and Lederberg (1951) in which single cells were isolated and followed through subsequent generations. This made it clear that the persistent diploid condition applies to individual products of the genetic cross; and when these unusual diploids give stable segregants, segregation involves the separation of groups of markers derived from each parent.

Analysis of the progeny of the 58-161 $F^+ \times$ W677 $F^-$ crosses performed by Hayes (1953b) showed that most of the prototrophs possess unselected markers derived from the recipient W677 cells. There is thus a bias in the progeny of the cross towards the presence of recipient markers. Two possible interpretations were considered for this result. Conjugation between $F^+$ and $F^-$ might represent the transfer of an entire chromosome from donor to recipient, followed by the elimination of segments of the $F^+$ chromosome to leave only some donor markers available for recombination with the recipient genome. This type of model was favored by Lederberg et al. (1951). Or only small regions of the chromosome might be transferred from the $F^+$ cell to the $F^-$ cell; the genotype of the progeny would thus remain essentially that of the $F^-$ parent, with changes possible only at the limited number of genes introduced in the form of donor DNA. This model was favored by Hayes (1953b).

The behavior of the unusual persistent diploid progeny is most easily explained by the second model, for Lederberg, Cavalli and Lederberg (1952) found that these prototrophs exclude the markers contributed by the $F^+$ parent when they form stable segregants. This suggests that the usual mechanism of conjugation is the transfer of a small amount of donor genetic material from $F^+$ to $F^-$ cell; following transfer, donor genes may be utilized if they are inserted into the recipient genome by genetic recombination, but are lost if this does not happen. In the unusual partial heterozygotes, however, the diploid condition is revealed by its persistence and represents the continued presence of the donor genes introduced by the $F^+$ factor; loss of the donor genes en masse results in the formation of stable segregants. (Of course, recombination can take place between the donor chromosome fragment and the recipient genome, so that new genotypes also arise and are segregated during subsequent divisions.)

Conjugation by mating between $F^+$ and $F^-$ populations differs from conventional genetic analysis since only the recombinants that are selected can be analyzed; all other recombinant classes pass undetected. Thus only the prototrophs are examined; reciprocal recombinants are not available. And a point of theoretical rather than practical importance is that it is not the immediate products of conjugation that are analyzed but rather their progeny in the form of colonies. It is necessary when constructing linkage maps from crosses performed by conjugation to take into account the unidirectional and partial nature of genetic transfer. In the early genetic analysis reviewed by Lederberg et al. (1951), for example, it seemed that certain anomalies in the mapping data could be explained only by the construction of a branched linkage map. A linear linkage map for the E. coli chromosome could be drawn up, however, when it was realized that a restricted amount of genetic material passes only from the donor $F^+$ to the recipient $F^-$ population. For example, in the cross

$$F^+\ lac^+\ met^+\ thr^- \times F^-\ lac^-\ met^-\ thr^+$$

the prototrophic progeny have the genetic constitution *met*$^+$ *thr*$^+$ and thus represent selection for the acquisition by the recipient of the *met*$^+$ allele from the $F^+$ parent. The bias towards recipient markers caused by partial transfer of donor genetic material means that the prototrophs will be *lac*$^-$ unless a *lac*$^+$ gene was introduced together with the selected *met*$^+$ gene. Examining these prototrophs for the presence of the unselected *lac*$^+$ allele is thus equivalent to asking the question: how closely is the *lac* locus linked to the *met* locus? The closer the linkage, the greater the number of progeny that will be *lac*$^+$ because the donor DNA contained the unselected marker as well as the selected marker.

But if a cross with the same markers is performed in opposite orientation as

$$F^+\ lac^-\ met^-\ thr^+ \times F^-\ lac^+\ met^+\ thr^-$$

the selected marker is different, for the *met*$^+$ *thr*$^+$ prototrophs must have gained the *thr*$^+$ gene from the $F^+$ parent. The bias towards the presence of recipient markers means that the prototrophs will have the *lac*$^+$ marker unless an unselected *lac*$^-$ marker was introduced together with the selected *thr*$^+$ marker. The presence of the *lac*$^+$ allele in the prototrophs thus is not the significant measure; the proportion carrying the *lac*$^-$ allele, however, depends upon the linkage between *lac* and *thr*, since the closer are these two loci, the more frequently will both be introduced into the recipient together.

Crosses of reversed sexual polarity therefore give very different distributions of unselected markers in the prototrophs, with a strong bias towards the markers of the $F^-$ recipient (for reviews of early work see Hayes, 1953a; Lederberg, 1952).This explains the apparently aberrant results obtained when prototrophic recombinants are backcrossed to their parents (the prototrophs are $F^+$ although most of their genetic markers are those of the $F^-$ parent, so that a reversal in polarity occurs). To use $F^+ \times F^-$ crosses for genetic mapping, it is necessary to conduct analysis in terms of the relationship between unselected and selected donor markers; the presence of recipient markers in the progeny cannot be used to establish linkage. In the first cross described above, the selected marker is *met*$^+$ and the unselected marker which should be followed is *lac*$^+$; in the second cross, the selected marker is *thr*$^+$ and the unselected marker which can be tested for linkage to it is *lac*$^-$. The success of this analysis supports the idea that only part of the donor genome is transferred in $F^+ \times F^-$ crosses, since if the entire genome were transferred and thus available for genetic recombination, reversal of polarity would not change the distribution of unselected markers in the prototrophs.

Only a few markers were available for these early experiments and they appeared to fall into a small number of groups when linkage was examined. This led to suggestions such as that of Watson and Hayes (1953) that the bacterial genome might consist of three linkage groups, presumably occupying three chromosomes. But interactions can be demonstrated between the markers ascribed to the different groups and in an extensive analysis of data from early $F^+$ x $F^-$ crosses, Clowes and Rowley (1954) showed that the distribution of unselected markers was consistent with the construction of a single linear genome if this is broken into fragments during genetic transfer. Different zygotes thus gain chromosome fragments of different lengths and genetic constitution. This model reconciles the incomplete nature of genetic transfer with the existence of a single linkage group. Of course, the construction of a single linkage map has since been supported by the mapping of many more loci than were originally available.

The distribution of unselected markers after crosses between two $F^+$ populations supports the idea inferred from the low fertility that some of the cells of each $F^+$ population behave as $F^-$ recipients. Indirect evidence is that the prototrophic progeny of $F^+$ x $F^-$ crosses display no bias towards the unselected markers of either parent but both are equally well represented. This suggests that either parent may generate the $F^-$ recipient cell. And Hayes (1953b) reported that treatment of one $F^+$ parent with streptomycin restricts its behavior to the donor type by preventing it from providing recipient cells, an effect seen in the reduction of the representation of its unselected markers in the progeny. Irradiation of either $F^+$ parent with ultraviolet also increases its propensity to act as donor.

An obvious hypothesis to account for the ability of $F^+$ cells to transmit mating type very efficiently and to act as genetic donors less efficiently is to suppose that they possess an agent, the *sex factor* or *F factor*, which is itself infectious at high frequency and also has the ability on occasion to promote transfer of genetic markers. The presence of an F factor in a cell also has the effect of preventing it from acting as a recipient. Yet all attempts to prepare cell free extracts with the ability to transfer mating type or genetic markers proved unsuccessful. The F factor thus resembles temperate phages in the stability of its relationship with the host cell and in its efficiency of transmission to sensitive cells, but differs in its absence from filtrates and in the exclusion of lysis from its mechanism of action. From this evidence it is thus reasonable to suppose that the presence of the F factor is all that is needed to establish the $F^+$ genotype, but that contact between cells is needed both for its transmission and for its transfer of host genetic markers.

### *High Frequency Transfer by Hfr Donors*

The low frequencies with which recombinants occur in crosses of $F^+$ and $F^-$ populations, and the partial and random nature of genetic transfer, mean that it would be very difficult to use such matings to analyze bacterial conjugation and to construct a complete map of the bacterial chromosome. Genetic analysis of conjugation became possible with the discovery of a new class of donor, the *Hfr* mating type, which was isolated by Lederberg, Cavalli, and Lederberg (1952) and Hayes (1953b) by its ability to promote a high frequency of recombination. This Hfr strain, termed HfrH, was derived from the 58-161 $F^+$ and carried the markers *met*$^-$ *bio*$^-$; when it was crossed with the standard W677 $F^-$ *thr*$^-$ *leu*$^-$ *thi*$^-$ strain, about 1000 times more prototrophs were produced than in the comparable $F^+ \times F^-$ cross.

In compatibility tests the Hfr type resembles the $F^+$ in being able only to take the role of a donor. But the Hfr type is not infectious and usually is not transferred from Hfr to $F^-$ in the same way that $F^+$ is transferred when $F^+$ and $F^-$ populations are mixed; the progeny of an Hfr $\times$ $F^-$ are thus $F^-$ (although with some rare exceptions). The Hfr strain can lose its ability to promote efficient genetic transfer, always then reverting to the $F^+$ condition from which it originated. This suggests that Hfr cells are derived from $F^+$ cells by some change in the state of the F factor and may revert by reversing the change.

The behavior of the HfrH strain depends on the markers that are selected. Only some markers show the very high frequencies of recombination for which the strain was named, other markers being transferred at characteristic frequencies varying from high efficiency to the low efficiency shown in $F^+ \times F^-$ crosses. With HfrH as donor, selection for the donor markers $thr^+$, $leu^+$ or $lac^+$ results in high frequencies of recombination, but selection for $met^+$ yields a recombinant frequency very similar to that produced by $F^+ \times F^-$ crosses. Such observations suggest that Hfr $\times$ $F^-$ crosses share with $F^+ \times F^-$ crosses the characteristics of unidirectional and partial transfer; but whereas markers are transferred at random and with low frequencies by $F^+$ donors, transfer from the Hfr donor is nonrandom so that certain groups of linked markers are transferred at high frequency.

The procedure used for quantitative studies of Hfr $\times$ $F^-$ crosses is to mix exponentially growing cultures of the two parent strains in liquid medium; the population density of the $F^-$ recipient is commonly in a twentyfold excess to ensure that all donors suffer collision with recipients. The frequency of progeny formation is then measured as a proportion of the minority (Hfr) parent number. Samples of the mixture are removed after intervals, diluted to prevent further donor-recipient contacts, and plated on selective minimal medium to isolate recombinants. Since genetic markers are transferred from Hfr to $F^-$, the recipient strain always carries minus (auxotrophic) alleles; thus the $F^-$ parental cells die on selective medium and only the recombinants that have gained appropriate positive alleles from the donors can survive. When the Hfr strain is auxotrophic for a positive allele carried by the $F^-$ strain, the donor parent also is killed on selective medium (and the cross is then performed in the same way as the crosses of auxotrophic $F^+$ and $F^-$ populations). If the Hfr strain itself is prototrophic, however, it is able to grow on the minimal agar and so it is necessary to prevent its continued growth among the progeny. This is achieved by selection against some character possessed by the Hfr cells and absent from the $F^-$. One method is to use a $str^s$ HfrH parent and a $str^r$ $F^-$ recipient; since $str^s$ is almost never transferred by an HfrH donor, inclusion of streptomycin in the selective medium kills the HfrH parents just as the auxotrophic $F^-$ parents find the lack of growth factors to be lethal. An alternative is to use an Hfr parent sensitive to a virulent phage (with an $F^-$ recipient resistant to the phage) and to kill the donor cells by treatment with the phage before plating (see Hayes, 1957); it is again necessary, of course, to use a genetic locus that is not often transferred in the Hfr $\times$ $F^-$ cross. The recombinants that have been isolated by this protocol are then arranged in colonies in the form of a grid on another plate of selective medium and may then be transferred by replica plating onto other plates containing media that allow unselected markers to be examined.

The general approach used in Hfr × $F^-$ conjugation has been to follow the transfer of prototrophic alleles from the donor to a multiple auxotrophic recipient; the early work with Hfr strains which led to the elucidation of the mechanism of transfer was reviewed extensively by Wollman, Jacob and Hayes (1956). Two types of information can be gained by crossing a prototrophic Hfr donor with an auxotrophic $F^-$ recipient. The frequency of recombination can be determined for any selected marker by measuring the number of recombinants obtained on appropriate selective medium. And the genetic constitution of these recombinants can be determined, that is, the distribution in them of unselected Hfr markers can be followed.

### *Linear Chromosome Transfer by HfrH Donors*

Sexual mating in bacteria falls into two stages. *Conjugation* represents the sequence of events from the first donor-recipient contact to the completion of genetic transfer. This may presumably be divided into the steps: random collision; effective contact as specific attachment forms between donor and recipient; transfer of some of the genome from donor to recipient. These interactions result in the formation of a zygote; since the zygote contains the entire recipient chromosome but only part of the donor genome, it constitutes a partial diploid which may be heterozygotic for the donor genes. The second stage in the formation of new genotypes is *recombination*. Donor genetic material must be integrated into the resident genome by a process of breakage and reunion in which recipient DNA is replaced by corresponding donor sequences. And then the segregation of haploid recombinant cells from the zygote is succeeded by phenotypic expression of the donor genes. The zygote usually does not divide before recombination has occurred, since each recombinant colony (that is the selected progeny of a single zygote) contains cells of only one genotype.

Contact between bacteria is necessary for conjugation. When two parental populations are mixed together and then plated out on selective medium to isolate recombinants, it is not possible to follow the kinetics of genetic transfer, because conjugation may continue in mating pairs formed before the transfer to selective medium. The technique of *interrupted mating* was therefore introduced by Wollman and Jacob (1955, 1957). At various times after mixing the parental populations the culture is agitated violently in a blender before it is diluted and plated out on

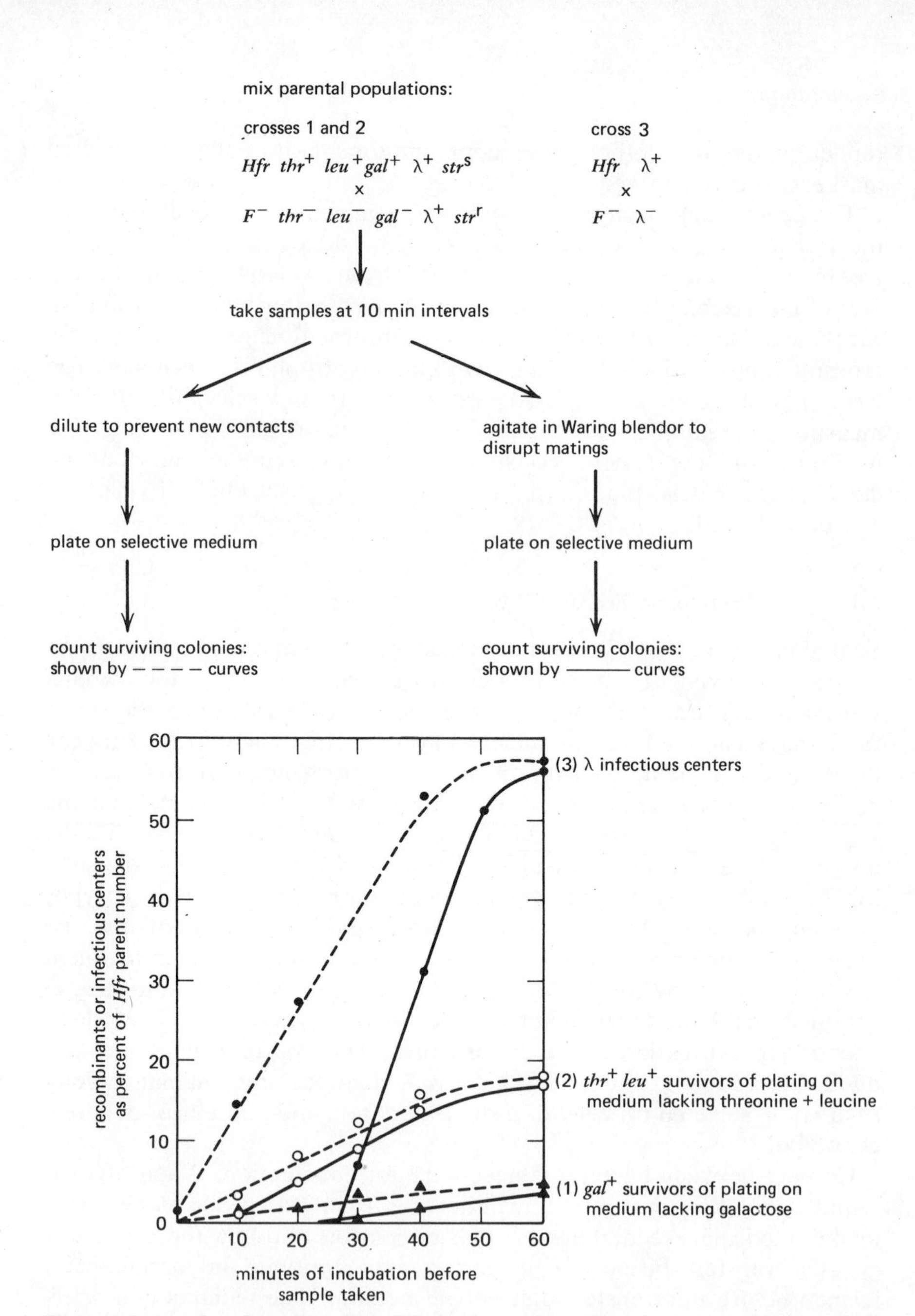


**Figure 2.5:** transfer of markers from Hfr to F⁻ parents. After the two strains have been mixed, either an uninterrupted (left) or interrupted (right) mating protocol can be followed. For crosses 1 and 2, plating on an appropriate selective medium allows cells to survive only if they can synthesize the omitted growth factors; by including streptomycin in the medium, the Hfr parent cells are killed, so that survivors must be F⁻ cells that have gained

selective medium. The agitation does not damage the cells but separates mating pairs, thus bringing conjugation to an immediate end.

Figure 2.5 compares the results obtained when mating populations are plated directly at various times after mixing and when agitation is introduced as an additional step before plating. The curves from the untreated samples (when only a dilution step separates mixing and selection) all rise from the origin to a plateau at about 50 minutes; but in the treated samples (both agitation and dilution intervene between mixing and selection) recombinants begin to appear only after a delay. The appearance of recombinants in the untreated samples can be taken to represent the formation of effective contacts since the dilution that terminates mixing prevents only formation of new contacts, while leaving existing mating pairs able to continue conjugation. The appearance of recombinants in the treated samples depends upon the formation of zygotes, since agitation separates mating pairs and thus prevents any further genetic transfer. Effective contacts thus appear to be formed immediately upon mixing, whereas 8 minutes are required before zygotes are produced that contain the $thr^+$ and $leu^+$ markers.

Different donor markers first appear in the progeny of interrrupted matings only after different periods have been allowed before mating is brought to a halt; $thr^+$ and $leu^+$ appear after about 8 and 9 minutes, but $gal^+$ and λ infective centers (see below) appear only after some 25 or 26 minutes. Since closely linked markers appear at the same time whereas distantly linked markers appear at different times, these kinetics suggest that there is a relationship between the time at which a given marker is transferred from Hfr to $F^-$ cell and its map location on the donor chromosome.

When HfrH $thr^+$ $leu^+$ is crossed with $F^-$ $thr^-$ $leu^-$ in an uninterrupted mating, the frequency with which any unselected donor marker is represented in the $thr^+$ $leu^+$ recombinant progeny is a fixed characteristic. But when the mating is interrupted, the frequency distribution of unselected markers depends upon the time allowed for mating before treatment in the blender. The results in Figure 2.6 show an analysis of the frequencies with which the unselected markers $gal^+$, $lac^+$ $ton^r$ and $azi^r$ are represented in the $thr^+$ $leu^+$ prototrophs. The two selected markers are among the earliest to be transferred from donor to recipient and so the presence of

---

the necessary positive markers from the Hfr. In cross 3, infectious centers are counted instead of surviving colonies (this is not specified in the figure); no selection is necessary as all $F^-$ cells that gain a $\lambda^+$ locus are killed. In uninterrupted mating, recombinants are found to increase in number from zero time; in interrupted matings, each marker appears in recombinants only after a characteristic delay. The same plateau level for any marker is reached in both uninterrupted and interrupted matings. Data of Wollman and Jacob (1957).

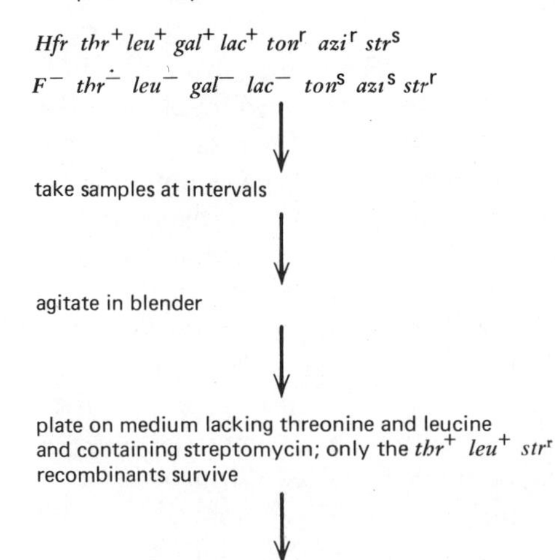


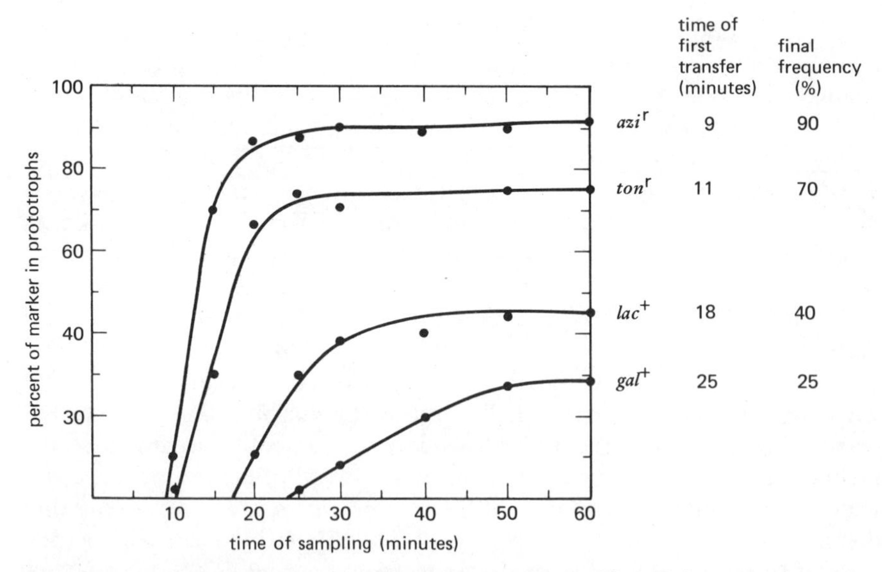


**Figure 2.6:** transfer of unselected markers to prototrophs in Hfr × $F^-$ interrupted matings. Prototrophic recombinants are selected and then examined for the presence of unselected markers by plating on appropriate media. Each unselected marker first appears in the prototrophs only after a characteristic time and reaches a final frequency that is inversely related to this delay. Data of Wollman and Jacob (1957).

the unselected markers in effect depends upon how long transfer has continued after its initiation. That is, these frequencies are expressed relative to the number of cells in which conjugation has proceeded as far as transfer of *thr*$^+$ and *leu*$^+$. Each unselected marker begins to appear in these prototrophs only after mating has been allowed to proceed for a characteristic time; after this time, the frequency with which it is represented in the prototrophs increases up to a maximum value that is the same as that achieved in uninterrupted matings. This plateau is achieved within 50 minutes after mixing.

The frequency finally attained by any marker is related to the time after mixing at which it is first transferred. Thus *azi*$^r$ first appears in the prototrophs at 9 minutes after mixing and attains a maximum frequency of 90%; *ton*$^r$ is first transferred at 11 minutes and reaches a plateau of 70% representation in the prototrophs; *lac*$^+$ is not transferred until 18 minutes after mixing and has a plateau frequency of 40%; and *gal*$^+$ is transferred only after 25 minutes and correspondingly has a low plateau frequency of 25%. These results suggest that the Hfr chromosome penetrates the F$^-$ recipient in an orientated manner. The *origin for transfer* is defined as the first locus to enter the recipient; it is followed 8–9 minutes later by the closely linked markers *thr*$^+$, *leu*$^+$ and *azi*$^r$, by *ton*$^r$ at 11 minutes, *lac*$^+$ at 18 minutes, *gal*$^+$ at 25 minutes, and λ at 26 minutes. By abruptly separating mating pairs, agitation in the blender terminates transfer of genetic material; and so only those markers lying in the region of the chromosome transferred during the period preceding the agitation can be expressed in the recombinants after mating has been interrupted at any time. The time at which any donor marker first appears in the recipient thus provides an estimate of its distance from the origin of transfer.

The correlation between time of first appearance and ultimate frequency of transfer suggests that matings suffer spontaneous separation of donor-recipient pairs after conjugation has begun. Breakage of the donor chromosome then has the same effect as that achieved in interrupted mating experiments; only markers previously transferred are available in the recipient. If the probability that a spontaneous break will occur depends simply on the period of conjugation, the likelihood of a break must be greater the farther a marker lies from the origin of transfer, so that the frequency with which it enters recipients is correspondingly reduced. (This simple proportionality would predict an exponential decline in the frequency of transfer with distance from the origin.) The frequency with which any unselected marker is represented in the recombinants thus provides another measure of its distance from the origin of transfer. Since all markers reach a plateau of transfer frequency at the same time, by 50 minutes after mixing all the Hfr bacteria capable of transferring genetic markers must have done so. Of course, in uninterrupted matings the frequency of spontaneous breakage depends upon the experimental conditions, so that factors such as the extent of shaking of cultures, or the procedures used in dilution, may influence the absolute frequencies of recombination.

The conclusion that agitation in a blender interrupts mating (rather than acting at some other stage) is supported by experiments in which Hayes (1957) used Hfr parents sensitive to phage T6 and killed the donor population by treatment with the phage at various times after mating had been initiated. The same kinetics of transfer are obtained whether conjugation is prevented by phage killing of one parent or by physical separation of mating pairs by agitation. The rationale of the interrupted mating experiments, and the evidence that agitation interferes only with genetic transfer, has been reviewed extensively by Jacob and Wollman (1961).

Another observation supporting the idea that the donor chromosome is transferred in a linear manner is the effect of temperature. The efficiency of zygote formation falls as the temperature is varied from its optimum value of 37°C. Wollman, Jacob and Hayes (1956) noted that a reduction in temperature slows the rate of chromosome transfer and reduces the number of prototrophs. At 37°C, the $thr^+$ and $leu^+$ markers enter recipient cells at 8 minutes after transfer has begun, and the $lac^+$ marker enters at 17 minutes. At 32°C, however, the $thr^+$ and $leu^+$ markers enter at 18 minutes and $lac^+$ enters at 38 minutes; transfer thus takes about twice as long, the proportionality between increases in time for transfer of each segment suggesting that the donor chromosome enters the $F^-$ cell at a uniform rate which depends upon the energy available.

The linear transfer from the origin of only a part of the donor chromosome imposes restrictions upon which selected and unselected markers can be utilized. In order to map unselected markers by kinetic analysis, it is necessary to choose selected markers close to the origin of transfer; selection for more distant markers would mean that the recombinants will show similar frequencies for all intervening markers. The unselected markers followed in the HfrH × $F^-$ cross represented in Figures 2.5 and 2.6 all are part of the class transferred at high frequency (generally considered to be above 1% representation in the prototrophs). Other markers, located farther from the origin, are transferred only at lower frequencies; by using conditions that minimize interruptions to mating, increases can be obtained in the frequencies of transfer of distal markers, but the same relative gradient always exists. Wollman and Jacob (1958) showed that the markers *trp, his, str* and *met,* all transferred by HfrH at low frequencies, enter recipients only after 33, 59, 90 and 115 minutes; under these conditions, transfer of the entire chromosome should therefore take about two hours, which is why it is so rarely observed. The kinetic analysis of time of entry and ultimate recombination frequencies is thus usually utilized only for the high frequency markers; only these loci, residing not too far away from the origin of transfer, can be mapped in these experiments, since under normal conditions the more distant markers are transferred rarely, if at all, and thus display recombination frequencies too low to analyze.

### *Efficiency of Genetic Transfer and Recombination*

Donor genetic markers usually are not expressed until the donor DNA carrying them has been integrated by recombination into the recipient genome; and then the delay before phenotypic expression is observed will depend upon whatever segregation and phenotypic lags may be involved. Observations of *zygotic induction,* however, depend upon the time when a donor marker enters the recipient. The E.coli K12 strain and most of its derivatives are lysogenic for phage lambda, that is, the phage genome is carried in an inert form, the prophage, as part of the host chromosome. When an Hfr strain that has lost its λ prophage and is therefore nonlysogenic is crossed with a lysogenic $F^-$ recipient, the prophage segregates like any other genetic marker, according to the map position that it occupies near *gal.* Similar results are obtained when both parents are lysogenic but carry different mutant prophages that may be distinguished. But when the Hfr parent is lysogenic and the $F^-$ parent does not carry a prophage, transfer of the phage genome from donor to recipient is followed by immediate induction of its development in the $F^-$ cell, resulting ultimately in lysis and release of phage particles. This is zygotic induction; the frequency of induction is measured as the number of infectious centers (plaques) per 100 initial Hfr cells. (Its cause is that the nonlysogenic recipients lack the repressor protein synthesized by prophage which maintains the inert state, so that the incoming prophage is expressed immediately upon entry into the nonlysogenic cell.)

The existence of zygotic induction illustrates the unidirectional transfer of markers in conjugation, since it is displayed only when it is the Hfr parent that carries the prophage and the $F^-$ parent that lacks it. The frequency of zygotic induction is usually less than complete, however, because only some of the recipients acquire the prophage; recombinants can be formed by other recipients, which gain some donor genes but in which transfer fails to proceed as far as the prophage. The surviving recombinants thus demonstrate that only a partial transfer of the donor chromosome takes place in these conjugation events.

Entry of a donor marker into a recipient is not sufficient to ensure its expression since it must first be integrated into the resident genome by genetic recombination; the efficiency of integration is thus critical in establishing the number of recombinants. Zygotic induction does not depend upon recombination, however, but takes place in every non-lysogenic recipient that gains a prophage. Comparison of crosses in which zygotic induction takes place with those in which it does not therefore allows estimates to be made of the relative roles played by the series of events preceding zygote formation and the successive events which create genetic recombinants; by using the frequency of zygotic induction as a standard, it is possible to measure how many cells are involved in zygote formation and to estimate the efficiency of integration.

In one series of experiments, Wollman, Jacob and Hayes (1956) compared the frequency with which $\lambda$ is transferred in Hfr $\lambda^+ \times F^- \lambda^-$ crosses with the frequency of recombination for the early donor markers $thr^+$ and $leu^+$. The frequency of occurrence of $thr^+$ $leu^+$ prototrophs achieved in the cross Hfr $thr^+leu^+\lambda^+ \times F^-$ $thr^-leu^-\lambda^+$ is about 20% of the number of parental Hfr cells. Taken by itself, this value simply sets a minimum value for the frequency of zygote formation, since we do not know what proportion of zygotes yields recombinants. But when the $F^-$ parent in this cross is not lysogenic for lambda, that is, the mating is Hfr $thr^+leu^+\lambda^+ \times$ $F^-$ $thr^-leu^-\lambda^-$, so that zygotic induction takes place whenever the $\lambda$ prophage enters the recipient, about 50% of the number of possible zygotes is destroyed by zygotic induction. The number of $thr^+leu^+$ prototrophs is reduced only by half to 10%. This means that the 50% of the cells in which zygotic induction took place must have included half of the cells that gained the $thr^+leu^+$ markers from the donor; zygote formation must therefore have been sponsored by all the Hfr cells in the population. Conjugation promoted by Hfr donors is therefore a very common occurrence, contrasted with the rare events sponsored by $F^+$ cells.

Although the $thr^+$ and $leu^+$ markers are transferred to about 100% of the zygotes, only some 20% yield the $thr^+$ $leu^+$ prototrophs; the efficiency of integration for these markers is therefore only about 20%, so that in the majority of cells zygote formation is not succeeded by genetic recombination. The $gal^+$ marker, located just before prophage $\lambda$, enters about 60% of zygotes, but is found among the recombinants derived from only about 5% of total zygotes; its integration efficiency is therefore about 8%.

Since only half of the zygotes are destroyed by induction, the other half must never gain the $\lambda$ locus from the donor, presumably because spontaneous chromosome breakage place after transfer has been initiated but before the lambda prophage is reached. This model for spontaneous breakage during transfer is supported by the results of a series of crosses performed by Jacob and Wollman (1956a) using the prophages $\lambda$, 21, and

421. The position of each prophage on the chromosome can be mapped by conventional analysis of crosses between Hfr $phg^-$ and $F^-$ $phg^+$, using the prophage as an unselected marker that differs between the parents. In the reverse crosses between Hfr $phg^+$ and $F^-$ $phg^-$, the frequency of zygotic induction for each prophage depends on its map position; the farther the prophage is located from the origin of transfer, the lower the frequency with which it displays zygotic induction. Since these frequencies depend directly upon genetic transfer and are not influenced by possible variations in integration efficiencies, they imply that the representation of any donor marker in the progeny reflects its map position; the likelihood that a spontaneous break occurs before its transfer depends simply on its distance from the origin.

### *Surface Properties of Sexual Mating Types*

The ability of Hfr or $F^+$ cells to act as donors depends upon specific surface properties that are acquired by the bacterial cell as a result of the presence of the F factor. Whether the factor is free or integrated does not appear to influence these properties. One early observation of the difference between Hfr and $F^+$ bacteria on the one hand and $F^-$ strains on the other was that of Loeb (1960) that certain bacteriophages can lyse only cells of E.coli with the ability to act as donors. Many phages possess this male-specific selectivity which means that they can infect only donor strains; these include the RNA phages f2, MS2, R17, M12, which form an immunologically related group, the RNA phage $Q\beta$, and the filamentous phages f1, fd and M13 that contain single stranded DNA genomes (for review see Curtiss, 1969).

The characteristic surface properties of donor bacteria reside in the F pili, or sex fimbriae. Pili in general are considered to comprise any morphologically distinct filaments extruded from bacteria, other than flagella (for review see Brinton, 1965, 1971); the most common of E.coli is the type 1 pilus, present in many copies per cell. Donor cells possess an additional type of pilus, the F pilus or sex pilus; Crawford and Gesteland (1964) showed that R17 adsorbs to pili present on donor but not on recipient bacteria and Brinton, Gemski and Carnahan (1964) identified the donor-specific pili by their ability to adsorb the phage M12. When the ♂-specific RNA phages bind to E.coli cells they do so by adsorption at random sites along the length of the sex pilus; when present in sufficient concentration they generate the picture seen in Figure 2.7. Male specific DNA phages have a comparable reaction, but bind instead to the tip of the pilus, as first demonstrated for f1 by Caro and Schnoss (1966).

All Hfr and $F^+$ strains possess F pili as well as type 1 pili. Brinton (1965)

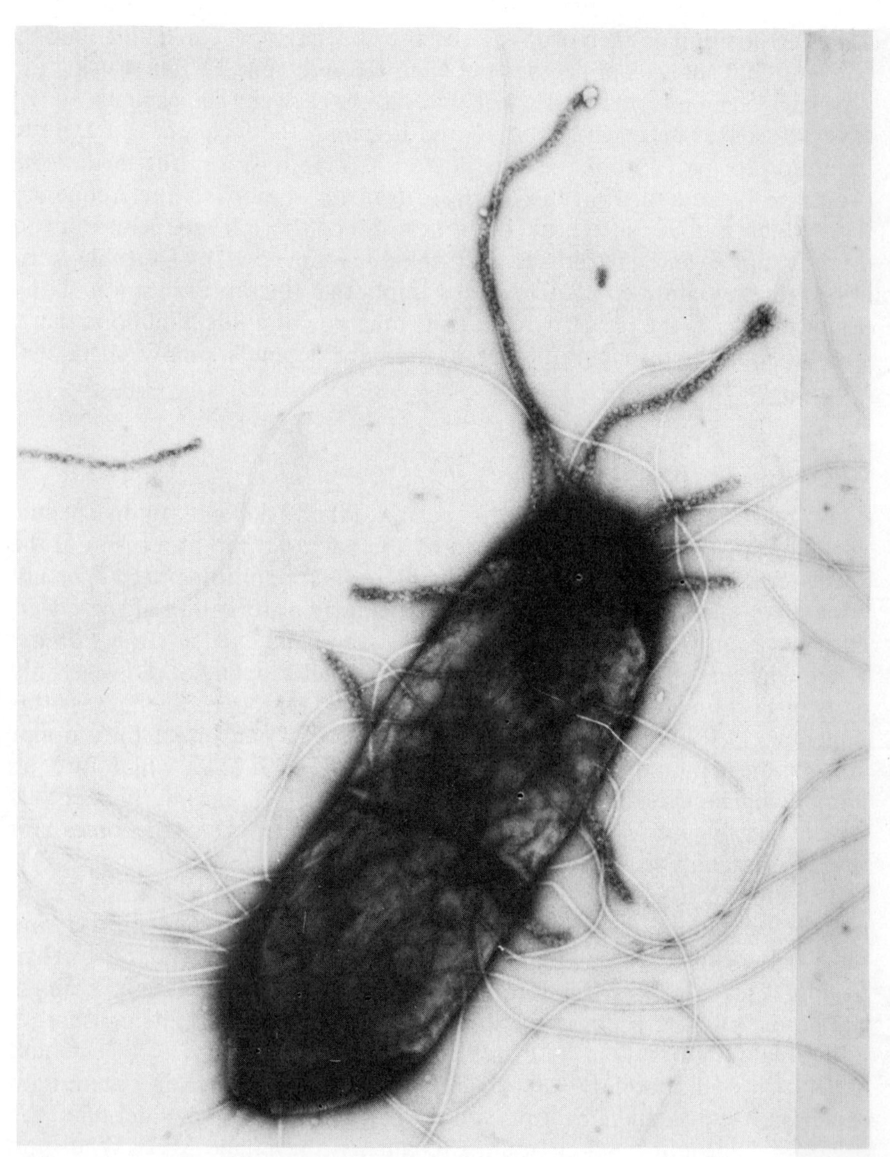

**Figure 2.7:** Sex pili of E.coli visualized by binding to male-specific RNA phage. The phage particles are about 250 Å in diameter and bind along the length of the pilus. Photograph kindly provided by A. Lawn, data of Datta, Lawn and Meynell (1966).

noted that, in contrast with the larger numbers of type 1 pili, there are only a few F pili per donor cell, often only one but never more than three. The length of an F pilus may vary somewhat from cell to cell but has an average of between 1 $\mu$m and 2 $\mu$m. Curtiss et al. (1969) reported that the number of F pili per cell is influenced by the conditions of bacterial growth and may be from one to five depending upon the environment. Since rapidly growing cells may contain a greater number of copies of the sex factor and bacterial chromosome (see Chapter 3), this is consistent with the idea that a single pilus may be synthesized by each F factor present in the cell, whether free or integrated.

Although both Hfr and $F^+$ bacteria possess F pili, they may be distinguished by the effect of treatment with acridine orange or acriflavine. Hirota (1960) observed that when a strain carrying the $F^+$ factor is diluted to a low concentration in nutrient broth containing acridine orange and is incubated overnight, it is converted to the $F^-$ state. These cells are not phenocopies, for the recipient status is stably inherited by their progeny; they appear to have lost the sex factor. Acridine orange has no effect on Hfr strains; it can "cure" only $F^+$ strains of the sex factor. This is consistent with the idea that the sex factor of $F^+$ bacteria is in an autonomous condition and thus can be cured by acridine, whereas the sex factor of Hfr bacteria is part of the bacterial chromosome and thus inviolate. Acridine is effective only with growing bacteria, which is consistent with the idea that its action may be to prevent replication of the autonomous sex factor; its effect probably lies in inhibiting the initiation of cycles of replication (see Chapter 3). Use of acridine orange has now become a standard test to see whether the F factor in a given cell strain is autonomous (susceptible to curing) or integrated (cannot be cured).

In reviewing the isolation and characterization of the F pilus, Brinton (1971) noted that F pili are rather insoluble and tend to form large aggregates; isolation therefore required the development of gentle methods that dissolve these aggregates without changing the structure of the pili themselves. Isolation of the pili was assisted also by the discovery that they are released into the medium in large numbers and by the use of bacterial mutants that produce an increased number of pili. With these special variations, the F pili were purified by the same methods used to purify viruses. The pilus represents an assembly of a single type of protein subunit, called F pilin, which is about 11,800 daltons in size, and has an isoelectric point of pH 4.15. Its polypeptide chain contains two molecules of phosphate and one of glucose, apparently the only modifications. Wendt, Ippen and Valentine (1966) had previously reported that the F pilus constitutes a number of repeating structural units, each of which can bind a ♂-specific RNA phage particle.

The F pili grow perpendicular to the cell surface by assembly of preformed F pilin subunits, a model supported by the observation of Brinton (1971) that use of chloramphenicol to stop protein synthesis in donor cells does not prevent continued assembly of F pili. When the pilus reaches a characteristic length of about 1 $\mu$m, a process which takes some 4–5 minutes, it tends to separate from the cell and to appear in the growth medium. Synthesis of another pilus then commences. Some methods of visualizing the sex pili led to observations of terminal "knobs"; Brinton observed these only very rarely but noted that similar structures can be produced in vitro from subunits of F pilin by appropriate treatment of the pili. The F pilus is essentially a hollow rod of external diameter 85 Å and with an axial hole of 20–25 Å diameter. Brinton proposed that it may consist of two parallel protein rods each comprising an assembly of F pilin subunits. Beard and Connolly (1975) found that the outer membrane of the cell has about $7 \times 10^4$ copies of a protein of the same molecular size as F pilin and this may therefore constitute the preformed pool.

The F pilus is not the only structure of its type, for Meynell, Meynell, and Datta (1968) observed that certain F-like R factors produce pili that are very similar but not identical; these pili are recognized by the same ♂-specific phages that are adsorbed by the pili produced by the sex factor. When an F factor and an F-like R factor are present and expressed in the same cell, a mixed pilus assembled from both types of subunit may be produced. Variations in the sex pilus may therefore take place that do not affect its assembly or function. On the other hand, certain plasmids of the I-type produce I pili, which appear to serve the same role for these plasmids as that served by the F pili for the sex factor; and their construction is different for no mixing occurs between I pili and F pili subunits (see Chapter 3). Although the F pilus is assembled from a single protein subunit, at least 10 genes on the sex factor are necessary for its production; some may be concerned with modifications of the protein subunit and perhaps others are necessary for assembly (see Chapter 3).

The presence of sex pili is essential for conjugation. Brinton (1965) reported that when the F pili are removed mechanically from donor cells by blending at high speed, the ability of the cells to form competent mating pairs is lost; and it is then recovered in parallel with the reappearance of microscopically visible F pili. Curtiss et al. (1969) noted that the number of mating pairs, and hence the number of recombinants, in a mating population shows a good correlation with the number of F pili. Visualization of cultures containing mating pairs shows that donor and recipient cells may be connected by an F pilus; an example, in which the pilus is revealed by adsorbed RNA phage particles, is shown in Figure 2.8.

The role of the pilus in mating, however, was a subject of debate for

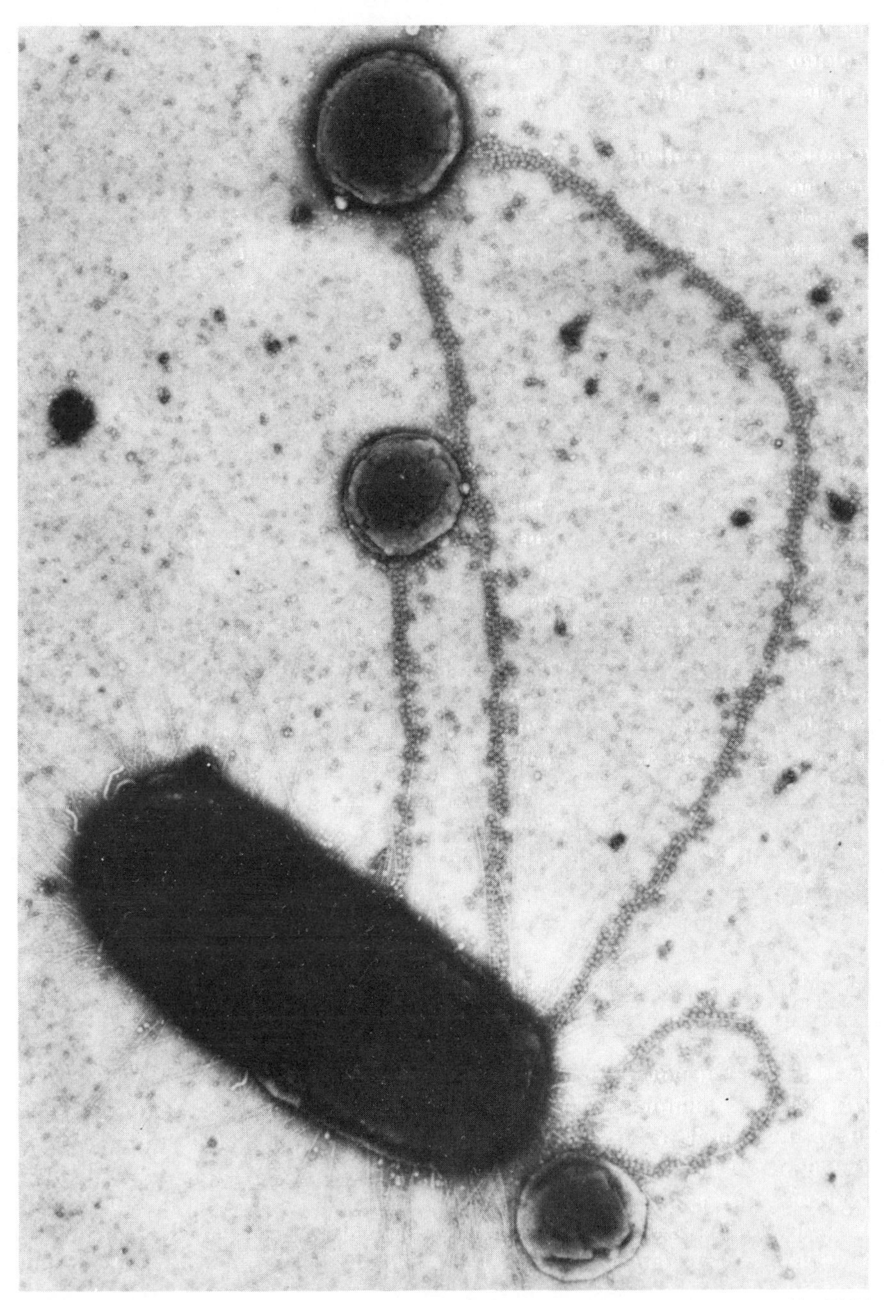

**Figure 2.8:** E.coli cells connected by sex pili visualized by adsorption of ♂-specific RNA phages. Photograph kindly provided by R. Curtiss.

many years. The correlation between presence of the pilus and donor activity clearly suggested that it is the sex pilus which makes first contact between donor and recipient cell. The early studies of Anderson, Wollman and Jacob (1957), however, suggested that donor transfer takes place by means of structures termed conjugation bridges, visualized in mating pairs where two bacteria were joined by a fairly rigid structure in a region of fusion of their membranes, of diameter some 0.1–0.3 $\mu$m. But such structures have since been seen also in recipient populations and Brinton (1965) proposed that the sex pilus itself provides the means by which the donor genetic material is transferred; that is, the donor DNA travels through the axial hole of the pilus from donor to recipient, in a manner analogous to the injection of phage DNA into a cell through the tail of the phage particle. This model for pilus function is consistent also with the adsorption of ♂-specific phages along its length, since their nucleic acid genomes must then reach the male cell by traveling down the axial hole (that is, in the reverse direction from that of the donor DNA transferring to an $F^-$ recipient). Ou and Anderson (1970) proposed that there might be two types of structure that can connect donor and recipient cells; by isolating mating pairs with a micromanipulator, they compared the production of recombinants from pairs loosely connected only by a sex pilus and from pairs more intimately linked by close association. The tightly connected pairs were about twice as fertile as the loosely connected pairs. Brinton (1971) noted that wall to wall pairs are much less common in mating populations than pilus-connected pairs; the number of donors estimated by genetic means is about the number of pilus-connected pairs, which implies that they must be responsible for most conjugation. During mating, the sex pilus remains attached to the donor cell for much longer than the usual 5 minutes, for up to 60–100 minutes in uninterrupted Hfr matings. Perhaps the presence of DNA in the pilus stabilizes it. Another indication that the F pilus is sufficient for mating is provided by observations of changes in the cell surface of donor cells. By examining the antigenic responses of the different mating types of E.coli, Ørskov and Ørskov (1960) demonstrated that donor bacteria possess a surface determinant absent from recipients; Hfr and $F^+$ bacteria may therefore be characterized as phenotypically $f^+$ whereas $F^-$ bacteria are phenotypically $f^-$. Ishibashi (1967) confirmed that the $f^+$ antigen corresponds to the F pilus. Brinton (1971) noted that antibodies against the F pilus prevent donor activity, but antibodies against the cell wall do not do so, which implies that the F pilus may be the only structure necessary for donor activity.

The surface of the recipient cell also is important in mating. The presence of the F factor in a cell prevents it from behaving as a recipient because DNA from the donor cell is unable to enter; this phenomenon, surface exclusion, is mediated by a surface component specified by the sex factor (see Chapter 3). Surface proteins specified by the bacterial genome also are necessary for uptake of donor DNA; Skurray, Hancock and Reeves (1974) isolated *con*$^-$ mutants that are defective as recipients in conjugation because they are unable to form mating pairs. These cells have much reduced amounts of two of the proteins of the outer membrane; since *con*$^-$ cells show normal growth, these proteins cannot be essential for viability.

Mating usually has been thought to take place by the formation of *mating pairs,* which represent the conjunction of a single donor and recipient cell. However, Achtman (1975) reported that mating mixtures contain aggregates of cells that are heterogeneous in size, ranging from 2–20 cells, and which appear to be engaged in transferring DNA from donors to recipients. The existence of these *mating aggregates* seems to demonstrate that conjugation may take place in cellular structures more complex than a single mating pair.

## Interactions between Sex Factor and Bacterial Chromosome

### *Genetic Transfer by Different Hfr Donors*

Three criteria can be used to follow transfer of genetic markers from Hfr donors to F$^-$ recipients. In interrupted matings, the order in which the donor markers appear in the recombinants reflects their closeness to the origin of transfer. The frequency with which any donor marker is found in the progeny also depends on how close is its map position to the origin. And linkage relationships may be established between selected markers and unselected markers which lie distal to them. All these methods of analysis suggest the same order and relationship between donor markers and support the construction of a single linkage map for the HfrH chromosome, which is transferred in linear manner from donor to recipient.

By isolating Hfr donors from different strains of E.coli K12, Jacob and Wollman (1957) were able to show that the sequential transfer of a linear molecule from donor to recipient is characteristic of all Hfr types. In each strain there is a gradient of frequencies of marker transfer which correlates with the time of first appearance of each marker in the progeny. But different markers are transferred at high frequency by each strain. Figure 2.9 shows the orders in which markers are transferred by several Hfr strains.

Comparing these sequences demonstrates that all markers of E.coli are

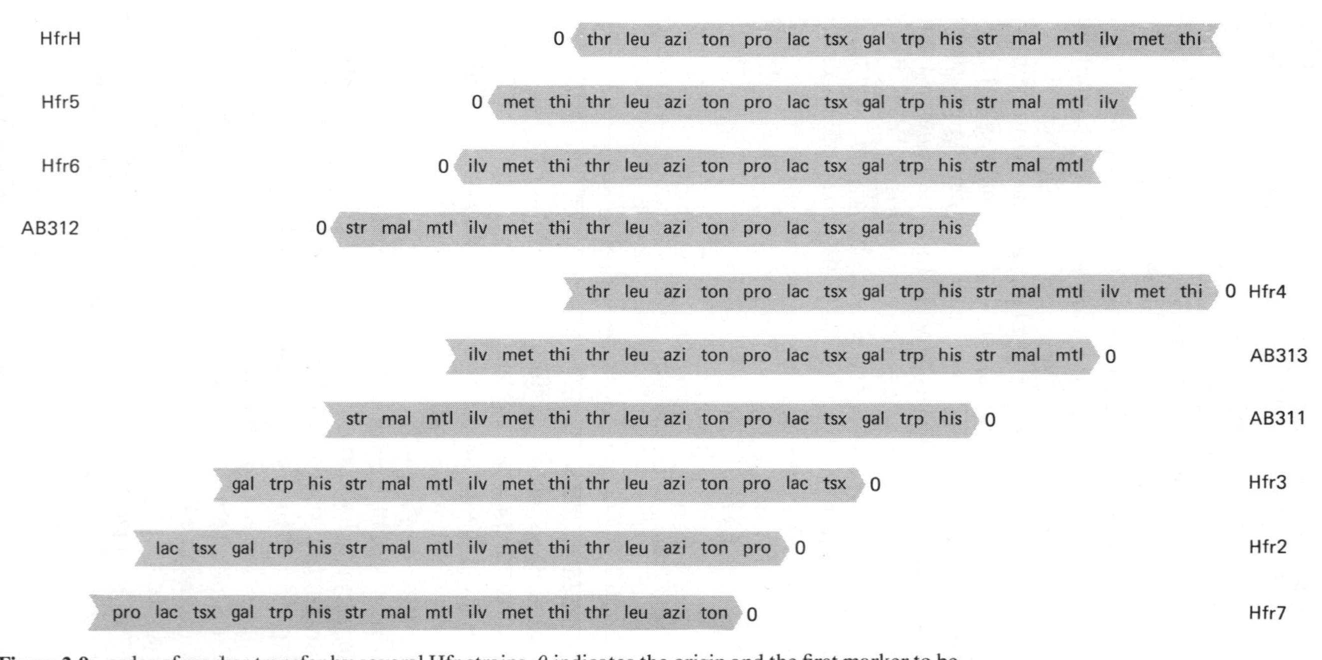


**Figure 2.9:** order of marker transfer by several Hfr strains. *0* indicates the origin and the first marker to be transferred is that at the head of the arrow. The markers listed for each donor represent the complete chromosome. The four strains represented at the top of the figure all transfer markers from left to right as drawn here. The six strains at the bottom of the figure all transfer markers from right to left as drawn here. The same relative relationships between markers are displayed by all strains. HfrH was isolated by Hayes (1953b), Hfr2-7 by Jacob and Wollman (1958, 1961); AN311-313 by Taylor and Adelberg (1961). These data were summarized by Jacob and Wollman (1961).

linked in a single group (for the sequence shown for each Hfr donor represents its entire chromosome); but it is not possible to arrange the sequences of all Hfr donor chromosomes as a simple linear map. Two features are apparent from the data of Figure 2.9. Genetic transfer may take place in either direction; whereas the first four strains shown in the figure all transfer *thr* before *leu,* for example, the last six strains all show the reverse order, with transfer of *leu* preceding transfer of *thr*. With the exception of loci close to the origin, the same relationship is shown between markers in all Hfr strains; comparison of HfrH and Hfr5, for example, shows the same order of loci from *thr* to *ilv,* the sole difference being that *met* and *thi* are the last two markers transferred by HfrH but are the first two transferred by Hfr5. The strains analyzed in Figure 2.9 are arranged in an order which illustrates the relationship between them.

The sequences in which markers are transferred by these and other Hfr strains since discovered all can be reconciled by construction of the circular genetic map shown in Figure 2.10 (see Jacob and Wollman, 1961; Hayes, 1968). This explains why markers at opposite extremities of the map of one Hfr strain are closely linked in other strains. The origins and directions of transfer of the strains represented in Figure 2.9 all are marked on the circular map of Figure 2.10; these strains all show similar behavior, except that AB311, AB312, and AB313 transfer distal markers at somewhat higher frequencies than the other Hfr strains. The reason for this is not known.

In mapping experiments with several strains, Taylor and Thoman (1964) were able to increase the sensitivity of detection of early recombinants by plating directly after the interrupted mating, that is by omitting the dilution step. This more rapid detection of the early recombinants reduced the period of time required for transfer of the complete chromosome to a duration of 90 minutes. The numbers shown in Figure 2.10 represent the time in minutes when each part of the circular map is transferred by HfrH under these conditions of transfer, that is with the time for complete transfer estimated at 90 minutes. More recently, however, in order to provide a uniform basis which is consistent with later experiments, Bachmann, Low and Taylor (1976) have revised the estimate of the time taken for total transfer to 100 minutes. The map shown in Figure 2.11 gives the time of transfer of all known markers of E.coli K12 according to this scale. The number of markers now identified has become so great that the circular form in which the map formerly was presented (see Taylor and Trotter, 1972) has had to be replaced by a linear representation. Details of the construction of this map are given by Bachmann et al. A feature of the map is that in some regions there are many markers,

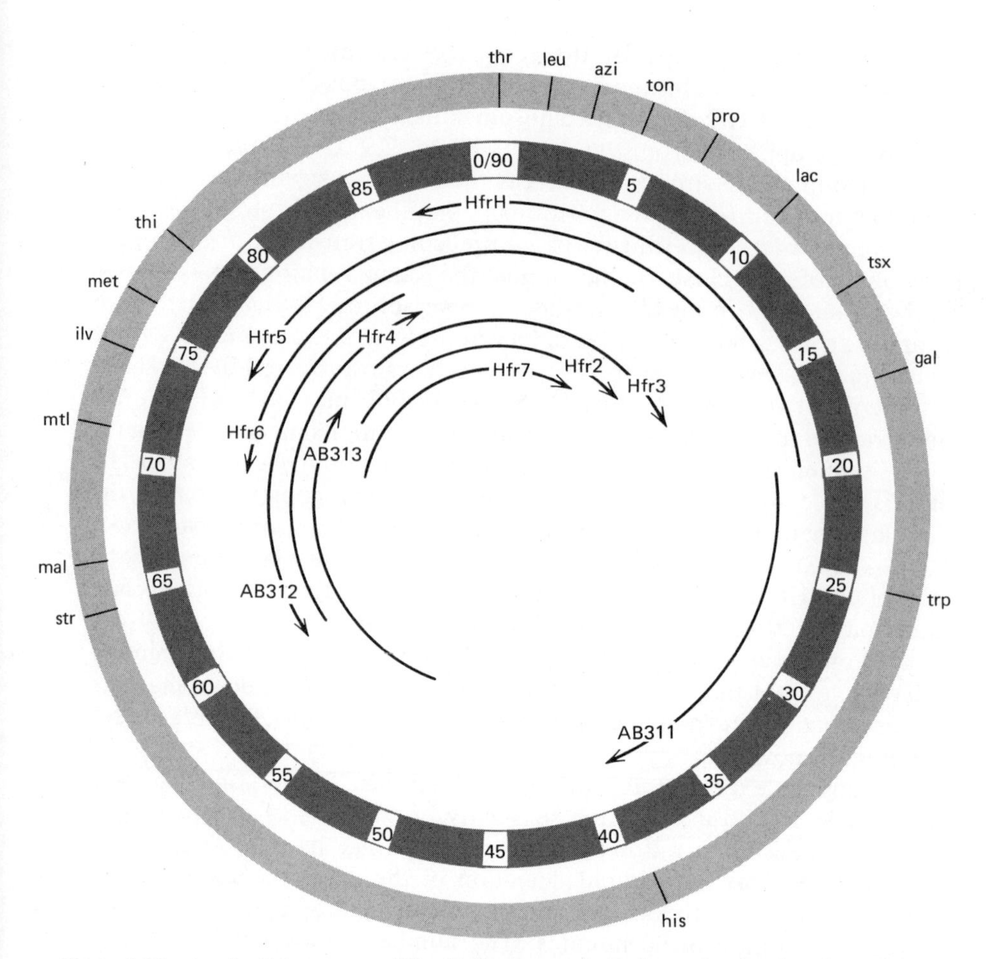


**Figure 2.10:** circular linkage map of E.coli. The outer circle shows the map locations of the markers followed in the Hfr x $F^-$ crosses reported in Figure 2.9. The inner circle shows the time of transfer, relative to HfrH and with *thr* identifying zero time; under the conditions then used as standard, transfer of the complete chromosome took 90 minutes at 37°C. Markers may thus be identified by their position on the HfrH transfer map in minutes. The origins of the Hfr strains of Figure 2.9 are indicated by the arrow heads; the tails of HfrH and Hfr2-7 indicate very roughly the region of the chromosome that is transferred at high frequency. AB311-313 transfer all markers at frequencies of this order of magnitude.

while others possess few identified genes; whether this has any significance for genetic organization in E.coli is not clear.

There is some variation in transfer rates using different donors (Broda and Collins, (1974); but these may be compensated by converting measured distances by comparison with the standard distance between the same markers on the HfrH map of Figure 2.11. As is implied by the circular map of Figure 2.10, although the time at which any marker is transferred depends upon the donor type, the relationship between any two markers remains similar in all crosses. (This is to exclude the case, of course, where the two markers are separated by the origin.) And, of course, the existence of these different Hfr strains means that an appropriate donor can be selected to transfer at high frequency the markers of whatever part of the chromosome is to be examined in a mapping experiment.

Several criteria can be used to estimate genetic linkage in E.coli. When recombinants are selected for a proximal marker and then examined for the presence of a donor marker more distant from the origin, the location of the unselected donor marker may be determined by its time of transfer or frequency of representation in recombinants, provided that an appropriate Hfr strain is used for transfer. This technique is restricted to markers that are transferred at reasonably high frequency, that is which lie in the part of the map proximal to the origin. Such mapping locates a marker within a few (say less than 2–3) minutes, but is not sufficiently precise to determine the relationship between loci lying closer together than this. For finer mapping it is therefore necessary to use other techniques.

Conjugation may be used to determine the relationship between two close markers by selecting recombinants for the marker that is more distal from the origin of transfer. All these recombinants must have received the preceding sequence of the chromosome, carrying the other (more proximal) marker. These recombinants then are examined by replica plating for the presence of the *recipient* marker at the more proximal locus; this requires separation of the two donor markers, that is, it depends on the probability that a single crossover event occurs between them, and so fits the traditional criterion for estimating linkage. The percentage of recombinants gives the distance between the two loci in map units (one unit equals 1% recombination). The same recombination frequency is observed for all parts of the chromosome (Verhoef and De Haan, 1966). This technique is limited to loci that lie within about 10 minutes of each other; the probability of recombination between loci separated by distances greater than this ensures that they are inherited independently, that is with linkage of 50% (for review see Susman, 1970).

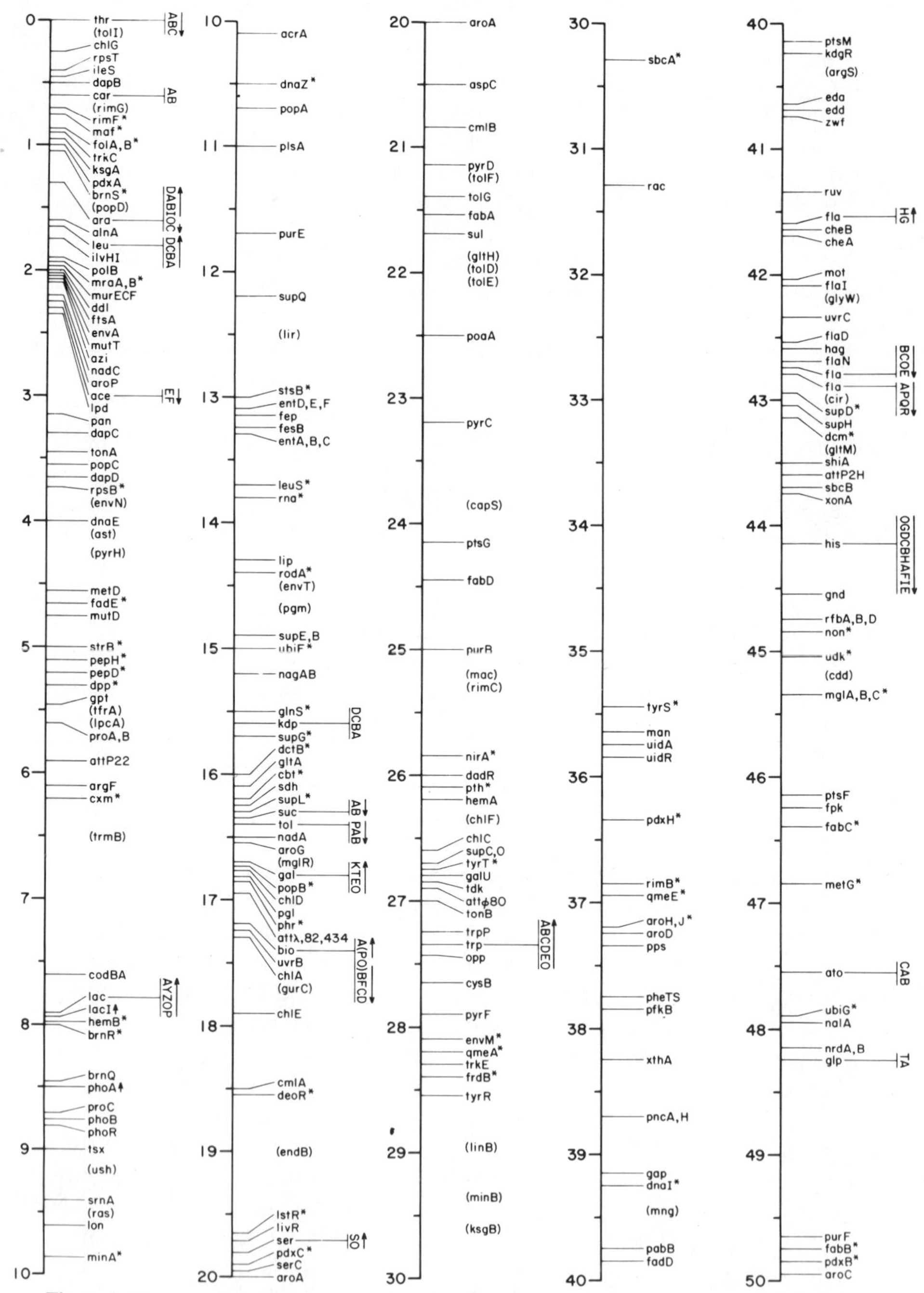


**Figure 2.11:** map of the chromosome of E.coli K12. Although shown here in linear form for reasons of space, the map is circular with the loci *thrABC* located at the point 0/100. Data of Bachmann, Low and Taylor (1976).

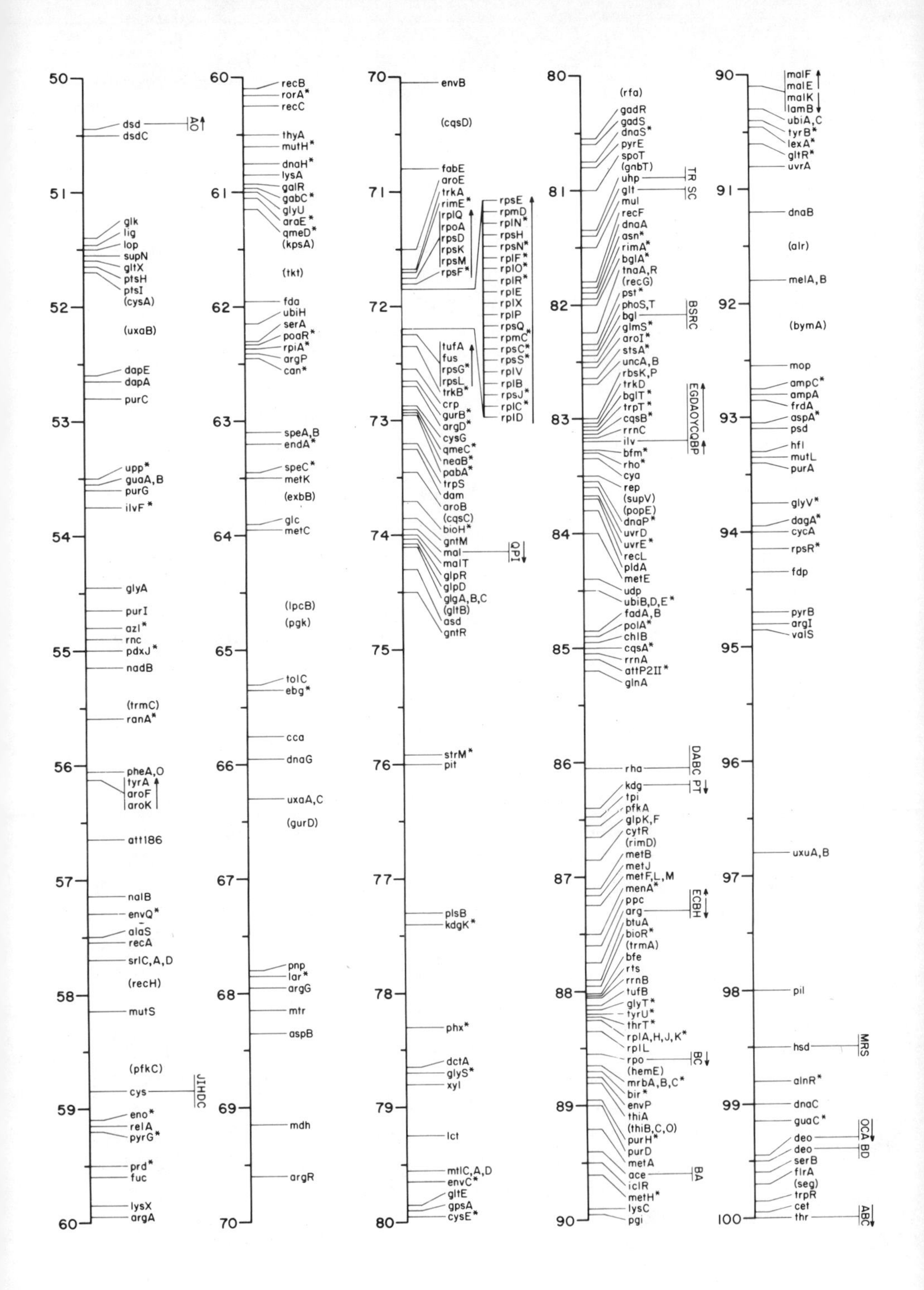
50
dsd
dsdC
AO
51
glk
lig
lop
supN
gltX
ptsH
ptsI
(cysA)
52
(uxaB)
dapE
dapA
purC
53
upp*
guaA,B
purG
ilvF*
54
glyA
purI
azl*
rnc
55
pdxJ*
nadB
(trmC)
ranA*
56
pheA,O
tyrA
aroF
aroK
att186
57
nalB
envQ*
alaS
recA
srlC,A,D
(recH)
58
mutS
(pfkC)
cys
JIHDC
59
eno*
relA
pyrG*
prd*
fuc
lysX
argA
60
60
recB
rorA*
recC
thyA
mutH*
dnaH*
lysA
61
galR
gabC*
glyU
araE*
qmeD*
(kpsA)
(tkt)
62
fda
ubiH
serA
poaR*
rpiA*
argP
can*
63
speA,B
endA*
speC*
metK
(exbB)
64
glc
metC
(lpcB)
(pgk)
65
tolC
ebg*
cca
66
dnaG
uxaA,C
(gurD)
67
pnp
lar*
68
argG
mtr
aspB
69
mdh
argR
70
70
envB
(cqsD)
fabE
71
aroE
trkA
rimE*
rplQ
rpoA
rpsD
rpsK
rpsM
rpsF*
rpsE
rpmD
rplN*
rpsH
rpsN*
rplF*
rplO*
rplR*
rplE
rplX
rplP
rpsQ*
rpmC*
rpsC*
rpsS*
rplV
rplB
rpsJ*
rplC*
rplD
72
tufA
fus
rpsG*
rpsL
trkB*
crp
73
gurB*
argD*
cysG
qmeC*
neaB*
pabA*
trpS
dam
aroB
(cqsC)
bioH*
74
gntM
mal
QPI
malT
glpR
glpD
glgA,B,C
(gltB)
asd
gntR
75
76
strM*
pit
77
plsB
kdgK*
78
phx*
dctA
glyS*
xyl
79
lct
mtlC,A,D
envC*
gltE
gpsA
cysE*
80
80
(rfa)
gadR
gadS
dnaS*
pyrE
spoT
(gnbT)
uhp
TR
81
glt
SC
mul
recF
dnaA
asn*
rimA*
bglA*
tnaA,R
(recG)
pst*
82
phoS,T
bgl
BSRC
glmS*
aroI*
stsA*
uncA,B
rbsK,P
trkD
bglT*
trpT*
83
cqsB*
rrnC
ilv
EGDAOYCQBP
bfm*
rho*
cya
rep
(supV)
(popE)
dnaP*
84
uvrD
uvrE*
recL
pldA
metE
udp
ubiB,D,E*
fadA,B
polA*
chlB
85
cqsA*
rrnA
attP2II*
glnA
86
rha
DABC
kdg
PT
tpi
pfkA
glpK,F
cytR
(rimD)
metB
87
metJ
metF,L,M
menA*
ppc
arg
ECBH
btuA
bioR*
(trmA)
bfe
rts
rrnB
88
tufB
glyT*
tyrU*
thrT*
rplA,H,J,K*
rplL
rpo
BC
(hemE)
mrbA,B,C*
bir*
89
envP
thiA
(thiB,C,O)
purH*
purD
metA
ace
BA
iclR
metH*
lysC
90
pgi
90
malF
malE
malK
lamB
ubiA,C
tyrB*
lexA*
gltR*
uvrA
91
dnaB
(alr)
melA,B
92
(bymA)
mop
ampC*
ampA
frdA
93
aspA*
psd
hfl
mutL
purA
glyV*
dagA*
94
cycA
rpsR*
fdp
pyrB
argI
95
valS
96
uxuA,B
97
98
pil
hsd
MRS
alnR*
99
dnaC
guaC*
deo
OCA
deo
BD
serB
flrA
(seg)
trpR
cet
100
thr
ABC

The distance between two loci on the E.coli chromosome may be expressed in minutes of transfer, units of recombination, or length of DNA (see Wood, 1968). The position of a marker usually is described by its time of transfer on the HfrH map shown in Figure 2.11. Since the entire chromosome of $2.8 \times 10^9$ daltons or $4.6 \times 10^6$ base pairs can be taken as equivalent to 100 minutes of transfer, the rate of transfer is about $30 \times 10^6$ daltons or 50,000 base pairs per minute. Put another way, it takes 2 seconds to transfer a gene of $10^6$ daltons. In early experiments it was usually reckoned that one minute is equivalent to 20 recombination units, so that one recombination unit would be about 2500 base pairs and the whole chromosome would comprise about 1800 map units; more recent experiments, however, suggest that the recombination frequency per unit distance may be somewhat smaller. Of course, any such relationship holds only as an average for large distances and does not apply to short regions, where the frequency of recombination may be distorted by local effects which alter the relationship between map unit and length of DNA.

Two classes of structure for the bacterial chromosome might explain these results. Each Hfr strain might possess a linear chromosome with the order of markers shown in Figures 2.9 and 2.10, presumably derived at some point from a circular structure. Or the chromosome might be a circular duplex of DNA, corresponding directly to the map of Figure 2.10. That the chromosome is in fact circular is suggested by both genetic analysis and direct visualization.

By showing that proximal and distal markers are linked on the genetic map, Taylor and Adelberg (1961) were able to suggest that it must be circular. These experiments used as donor an Hfr strain transferring *ton*$^r$ and *pro*$^-$ as proximal markers and *xyl*$^+$ and *met*$^+$ as distal markers (this is the orientation and sequence of transfer shown for HfrH in Figure 2.10). Recombinants were selected for the presence of one of the distal markers and then examined for the presence of the two unselected proximal markers. The *met* locus proved to be linked to *ton* and less closely linked to *pro;* the *xyl* locus is linked less closely to the proximal markers but also showed a closer relationship with *ton* than with *pro*. This is consistent with an order of markers in the progeny: *xyl-met-ton-pro*. The recipients in these experiments were phenocopies of the same Hfr strain used to provide the donor, that is in which *xyl-met* are distal and *ton-pro* are proximal; that the distal and proximal markers are linked in this donor strain implies that the chromosome must be circular and that it must be broken into a linear structure for transfer by cleavage between the distal and proximal markers.

The conclusion that the chromosome takes a circular structure is supported also by the autoradiographic data of Cairns (1963), in which a replicating E.coli chromosome was visualized directly as a tangled circle. The consequences that this has for the replication of the bacterial chromosome are reviewed in Volume 1 of Gene Expression.

The noninfectious nature of the Hfr genotype, contrasted with the transferable nature of the $F^+$ genotype, suggests that the F factor postulated to determine these states may be in a different condition in Hfr and $F^+$ bacteria. The inheritance of the Hfr genotype from parent to progeny is consistent with the idea that the F factor behaves as part of the bacterial chromosome in these cells; whereas in $F^+$ cells it behaves as a free agent, able to multiply independently of the bacterial chromosome and able also to transfer between cells. The different order in which markers are transferred in each Hfr strain can be explained by the concept that the F factor is integrated at different positions in their chromosomes; and the orientation of integration determines the direction of chromosome transfer. The recipients in Hfr $\times$ $F^-$ crosses rarely gain the Hfr type; however, those progeny that are Hfr inevitably possess the markers transferred at lowest frequencies in the cross. By selecting for the most distal markers, it is possible therefore to isolate increased numbers of Hfr progeny. What this suggests is that transfer of the donor chromosome starts at the site where the F factor is integrated; the chromosome is broken to generate a linear structure for genetic transfer and the orientation of the F factor determines that one side of the break provides the origin while the other contains the last sequences to be transferred, including the F factor itself (or some part of it essential for conferring the Hfr genotype). This model is illustrated in Figure 2.12. The research which led to the formulation of this model has been reviewed in some detail by Jacob and Wollman (1958a, 1961).

Implicit in this model is the concept that chromosome transfer is mediated by the sex factor through its ability to recombine with the bacterial chromosome. Transfer of genetic markers in $F^+$ $\times$ $F^-$ crosses was at first attributed to the formation of Hfr strains in the $F^+$ population by spontaneous insertion of the F factor into the bacterial chromosome. If a variety of Hfr types arises in the $F^+$ population, each will transfer a characteristic set of markers at high frequency. Given the low frequency of occurrence of Hfr cells and the transfer by each Hfr type of different markers at high frequency, the population as a whole should transfer all markers at rather low frequency. However, when Broda (1967) and Curtiss and Stallions (1969) examined the frequencies of marker transfer by $F^+$ populations, they found that they did not fit with those that would be predicted from the characteristic transfer patterns of the Hfr strains identified in these experiments. Less than 20% of transfer from $F^+$ populations appeared to be due to the formation of stable Hfr strains. One explanation, of course, would be that the sex factor also can interact transiently with the bacterial chromosome to promote transfer without giving rise to a stably integrated strain. Such an interaction might perhaps be analogous to that responsible for the chromosome transfer mediated by certain F-like R factors, which can transfer chromosome markers at low levels but which do not appear to integrate into the chromosome (see Chapter 3). And finally we should note, of course, that the transfer of host markers is quite distinct from the transfer of the sex factor itself, which takes place infectiously at frequencies very much greater than those at which chromosome markers are transferred by $F^+$ donors.

### *Insertion of F into the Chromosome*

The behavior of $F^+$ and Hfr strains of E.coli is consistent with the idea that the F factor resides in $F^+$ cells in an autonomous state, independent of the bacterial chromosome, whereas in Hfr cells the factor is integrated into the chromosome, at a site that is characteristic for each Hfr strain. The concept that a single event is responsible for the formation of an Hfr strain is supported by the results of a fluctuation test performed by Jacob and Wollman (1956b) to follow the appearance of Hfr cells in an $F^+$ population; the conversion takes place as a mutationlike event, that is, it arises with a random probability with regard to time in any $F^+$ cell and then is inherited in a stable manner by the (Hfr) progeny (see Jacob and Wollman, 1961). Most $F^+$ populations are able to give rise to a variety of Hfr lines, although some $F^+$ strains have been obtained that do not seem able to do so (see Curtiss and Renshaw, 1969). The probability that an Hfr cell will be generated from an $F^+$ population is about $10^{-5}$ per cell per generation; Hfr cells vary in stability and revert to the $F^+$ state with characteristic probabilities that are generally about the same order of magnitude as their probability of formation (see Adelberg and Pittard, 1965).

That there are only a limited number of sites on the bacterial chromosome where the sex factor may be integrated is suggested by the results obtained by Broda (1967). In these experiments, Hfr cells were isolated from $F^+$ populations by a technique involving a dilution step to ensure that each Hfr isolated is derived from a single $F^+$ colony. This prevents repeated isolations of the same strain. Selection for strains that transfer the markers *thr leu* at early times produced six independent isolates, all with origins in the same region, between *pro* and *lac*. Strains transferring *trp* early fell into three groups. Hfr strains with similar, presumably identical, origins gave similar frequencies of reversion to $F^+$; those with different origins gave characteristically different frequencies. These results suggest that Hfr strains arise by nonrandom integration of F into the bacterial chromosome, at specific sites.

The insertion of F is analogous to the integration of certain (lysogenic) bacteriophages into the host chromosome, which creates a prophage that can be carried as a chromosomal locus; the insertion and excision of phage lambda has been analyzed in some detail and is discussed in Chapter 5. A model to account for this change in state was proposed by Campbell (1962) and can be applied to the sex factor as well as to lysogenic phages. To maintain the integrity of the sex factor (or phage) when it recombines with the much longer bacterial chromosome, the factor as well as the chromosome was ascribed a circular structure. The single recombination event depicted in Figure 2.12, presumed to take place between sites of homology on factor and chromosome, can then generate a single structure in which the factor is inserted as a linear sequence into the chromosome. As the figure implies, a reversal of these same events can then release a circular factor from the integrated state.

Evidence to support this model for the integration of the sex factor is now considerable, and includes physical characterization of its structure as well as genetic analysis of chromosome transfer (for review see Scaife, 1967; Hayes, 1968). Questions that it raises include the nature of the homology between factor and chromosome; and we may extend this to ask whether and how integration differs at each site of insertion. Another important question is what gene functions are needed to support insertion and excision and whether they are coded by the factor and/or by the host cell. We may ask also what functions are implicated when the factor sponsors transfer of the host chromosome to an $F^-$ recipient; in the figure, the origin is located within the factor, which implies that upon chromosome transfer some of the factor is transferred first, with other factor sequences following last—these distal sequences must be essential if a recipient is to become an Hfr type. A final question posed by this model for factor insertion is what mechanisms are responsible for suppressing autonomous replication in the integrated factor.

The term *episome* was introduced by Jacob, Schaeffer and Wollman (1960) to describe genetic elements, such as the sex factor, that are able to exist in alternative states, autonomous of the bacterial chromosome or integrated into it. The term *plasmid,* on the other hand, was originally used to describe elements that exist in the cell only as extrachromosomal genetic units, which do not have the ability to become integrated. These terms are no longer always used with these precise meanings, however, and sometimes have been taken to be synonymous. This trend has in part

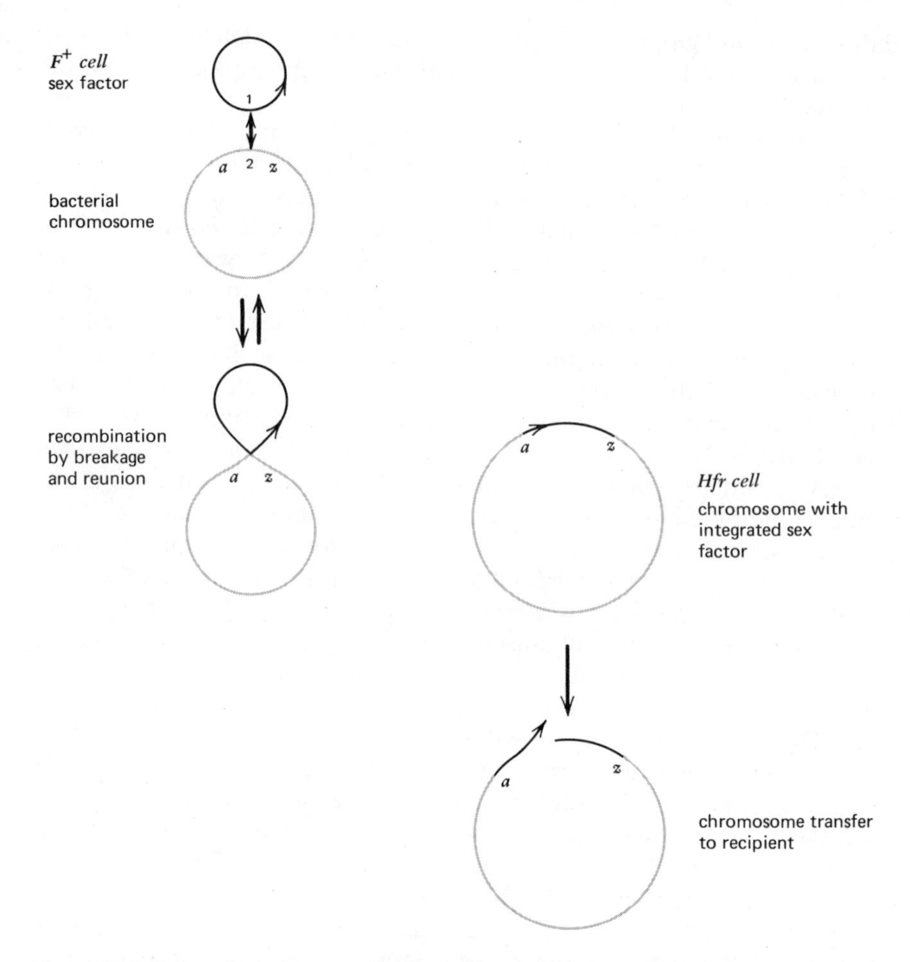


**Figure 2.12:** integration of the sex factor in the bacterial chromosome. The circular sex factor is represented by the dark circle, with the arrow marking the origin for transfer; the bacterial chromosome is the lighter circle. A single recombination event between the homologous regions *1* (in the sex factor) and *2* (in the chromosome), indicated by the double headed arrow, generates the twisted circle shown at the left of center; this is shown in an untwisted form on the right. The sex factor is thus integrated as a linear region between the bacterial markers *a* and *z*. This can be reversed and a free circular sex factor released by the reverse single recombination. To transfer the chromosome from the integrated state, the sex factor breaks at the origin as shown in the lower part of the figure; *a* is a proximal and *z* is a distal marker.

been encouraged by the isolation of mutations which render episomes unable to integrate into the bacterial chromosome, so that strictly speaking the mutant should be described as a plasmid rather than episome.

### *Excision of F and Formation of F′ Factors*

The distinction between Hfr and $F^+$ bacteria is clear. Although both possess an F factor, produce F pili, and can act as donors for host markers, the differences in the state of the sex factor mean that the Hfr transfers the host chromosome at high frequency with a specific orientation whereas the $F^+$ transfers the sex factor efficiently but transfers chromosomal markers less frequently. Another type of donor was first isolated by Adelberg and Burns (1960), who were working with the P4x Hfr strain, which transfers *pro, leu, thr* as early markers and *lac* as a late marker (the sequence shown for Hfr2 in Figures 2.9 and 2.10). The progeny of P4x Hfr × $F^-$ matings are $F^-$, except when they are selected for the distal donor marker *lac*$^+$—the *lac*$^+$ recombinants are usually Hfr type because the sex factor appears to be transferred as a distal marker and thus is closely linked to the *lac* locus. A subclone of this strain, P4x-1, however, had the unusual property that recombinants gaining early chromosome markers acquired also the capacity to behave as genetic donors. When used as donors, these recombinants transfer the host chromosome with the same orientation as the parent Hfr strain, although at a somewhat lower frequency. The ability of recombinants gaining early markers also to act as donors is perpetuated through a series of crosses with $F^-$ recipients. While chromosome transfer by these cells resembles that of the ancestral Hfr strain, the sex factor appears to be independently and rapidly transferred in the manner characteristic of $F^+$ donors.

This strain thus appears to carry a modified sex factor, termed F′, with characteristics intermediate between Hfr and $F^+$ types; chromosome transfer has a specific orientation but sex factor transfer is infectious. A model to account for these features is shown in Figure 2.13; this supposes that an illegitimate genetic exchange took place when the F′ factor of the P4x-1 strain was excised from the chromosome. Instead of reversing integration as shown in Figure 2.12 by recombination between the original sites, a genetic exchange might take place between a site located within the sex factor and a site to one side of it on the chromosome. This *type I exchange* generates an F′ factor which carries certain chromosomal genes and lacks some of the material of the F factor; and it leaves a bacterial chromosome which has gained some of the material of the sex factor but has lost certain genes. In the exchange shown in Figure 2.13, the F′ factor carries the distal chromosomal gene *z*.

As a free sex factor, the F′ episome is transferred infectiously from F′ populations to $F^-$ populations, just like the F factor of $F^+$ cells. But because the F′ factor carries bacterial genes, upon transfer to an $F^-$ recipient it is able to pair preferentially with the corresponding region of the bacterial chromosome, as illustrated in Figure 2.14. When recombination sees insertion of the factor, it mediates chromosome transfer from this locus; the frequency of such transfer is, of course, greater than that shown by $F^+$ populations but is lower than that of Hfr strains in which the factor is integrated at this site. Adelberg and Pittard (1965) quoted the probability of integration in one F′ strain to give an Hfr as $10^{-1}$ per division cycle, in contrast with a rate of $10^{-5}$ per cell per generation for an $F^+$ strain. This demonstrates the importance of homology between factor and chromosome for insertion. The intermediate donor property is inherited whenever an $F^-$ population is infected with F′ cells; recipients that receive the F′ factor become (infectious) donors of the F′ and in chromosome transfer behave like the ancestral strain. This explains the dual properties of the cells derived from the P4x-1 strain.

Transfer of an F′ factor sees the appearance in the recipients of the functions coded by the bacterial genes that it carries. Thus if the F′ factor carries a *lac*$^+$ operon, for example like the factor studied by Adelberg and Burns which resulted from excision of an F factor inserted between *pro* and *lac*, the $F^-$ *lac*$^-$ recipient cells that are mixed with the F′ population should become *lac*$^+$. This marker, formerly a distal locus of the ancestral Hfr strain, is thus transferred at high frequency by the F′ cells (see Scaife, 1967). The F′ factor carrying the *lac*$^+$ genes may be described as F *lac*$^+$. Two sets of genetic markers may be transferred, then, when an F′ population is mixed with $F^-$ cells: the markers carried by the F′ factor are transferred infectiously as part of it; and the markers proximal to its preferred site of integration are transferred whenever the factor is inserted and mobilizes the host chromosome for transfer.

A bacterium gaining an F′ factor becomes diploid for the bacterial genes carried by the episome. That transfer of the host chromosome by the F′ factor results from its insertion by recombination between these homologous bacterial sequences has been confirmed by several genetic experiments. The consequences of the insertion of the factor can be followed when it carries an allele different from that of the chromosome. Figure 2.14 shows that insertion of the F′ factor introduces into the bacterial chromosome the second set of bacterial genes that it carries; the resulting chromosome has one set of these genes immediately adjacent to each side of the factor, that is, one set is proximal and one is distal. When the chromosome is transferred, one set of the genes formerly carried by the factor should act as an early marker and one as a very late

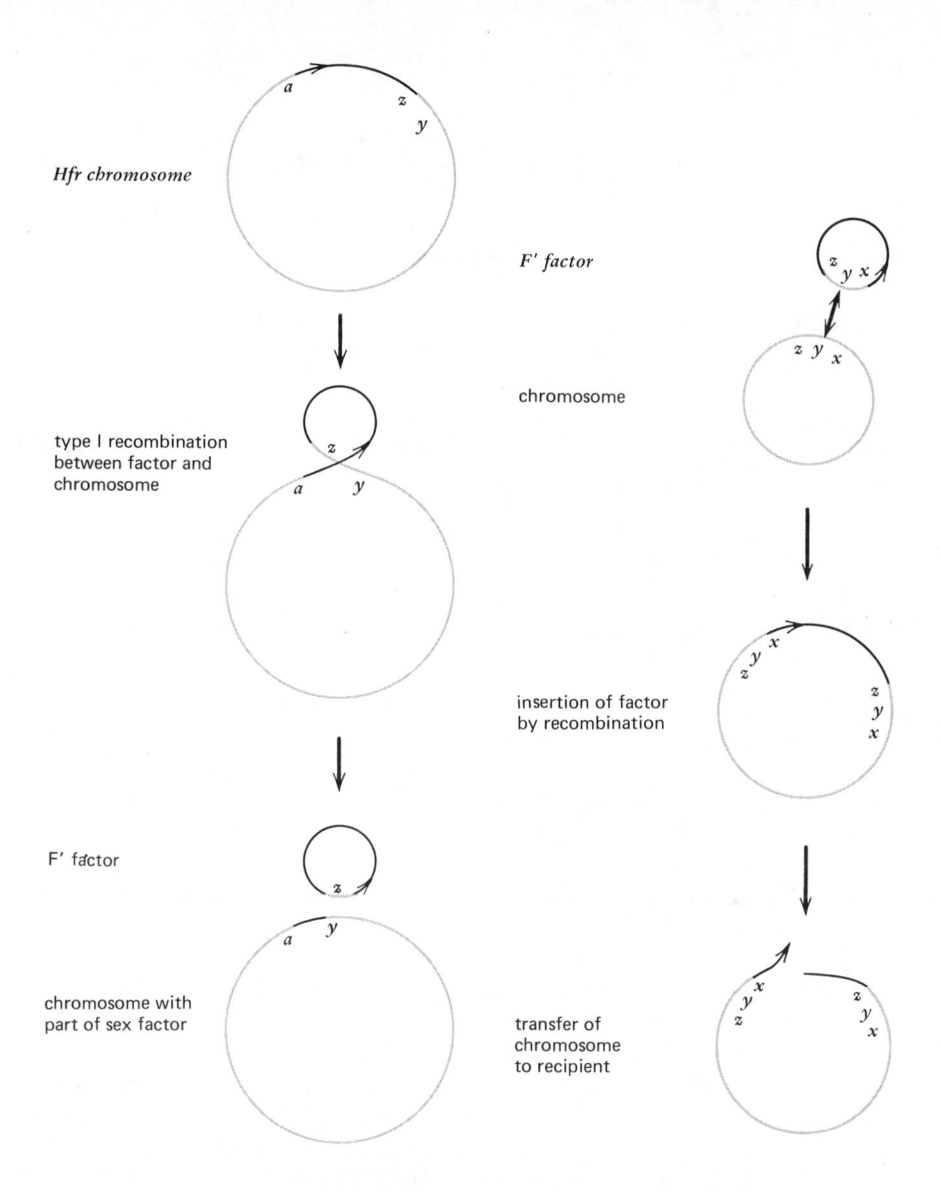


**Figure 2.13:** type I excision of an integrated F factor to generate an F′ factor. Recombination takes place between a site on the sex factor and a site on the bacterial chromosome between *z* and *y*. The factor that is released carries the distal bacterial *z* marker; the chromosome that remains possesses some sequences of the sex factor, but has suffered a deletion for gene *z*.

**Figure 2.14:** insertion of an F′ factor into a new host chromosome. The F′ factor carries the bacterial genes *x,y,z* and pairs with this region of the bacterial chromosome. A single recombination event inserts the factor and the resulting chromosome possess both sets of *x,y,z*, one transferred as a proximal and one as a distal marker.

marker. Of course, only a few of the cells in an F′ population transfer the host chromosome to a recipient population; most donors (infectiously) transfer only the F′ factor. Thus when an F *lac*$^+$ population is crossed with F$^-$ *lac*$^-$ recipients, it is not possible to test for the transfer of *lac*$^+$ as a chromosomal early marker, for it is transferred rapidly as part of the F′ factor. By constructing an F *lac*$^-$ factor, however, Scaife and Gross (1963) were able to demonstrate the transfer as a proximal marker of the host *lac*$^+$. As Figure 2.15 shows, which of the alleles *lac*$^+$ and *lac*$^-$ is transferred first and which last depends upon whether the site of recombination is between the mutant site and the origin of transfer or on the other side of the mutation; in the former case, a *lac*$^+$ gene is transferred proximally (although in the ancestral *Hfr* strain it was a distal marker). And by examining recipients which gain chromosomal genes but do not gain the F′ factor, Scaife and Gross (1963) and Cuzin and Jacob (1963) were able to show that either *lac*$^-$ or *lac*$^+$ may be transferred as the proximal marker.

The type I genetic exchange that generates the class of F′ factor shown in Figure 2.13 changes the genetic constitution of the host cell, whose chromosome gains some sequences of the F factor but loses the bacterial genes that are inserted into the episome. This has two consequences for the parent cell. One is that the episome carries the only copies of the genes it has acquired from the chromosome; thus if the episome is an F *lac*$^+$ factor, it should be impossible to obtain viable cells by curing with acridine during growth on medium when the *lac*$^+$ function is essential. The loss of the F *lac*$^+$ episome in these circumstances leads to death of the cells (see below).

A second consequence is that the sequences the host chromosome has gained from the F factor should be homologous with any future F factor introduced into the cell and should therefore provide a site for preferential integration (the homology at this site is much greater than at other potential sites of integration). This was tested by Adelberg and Burns (1960) in an experiment in which they cured P4x-1 of its F′ factor by treatment with acridine orange and then used the cured cells as recipients in a cross with F$^+$ cells. The cells became donors which transferred *pro, thr, leu* at high frequency in the order characteristic of the ancestral Hfr strain from which P4x-1 had been derived. Figure 2.16 illustrates this process and shows how the new F factor preferentially integrates at the site of the original insertion—the new Hfr is similar to the ancestral strain, but (of course) has a duplication of some of the F sequences. A similar observation of the memory of integration sites was made by Richter (1961). Adelberg and Burns therefore suggested that episome sequences left in the bacterial chromosome by type I excision of an F′ can be regarded as

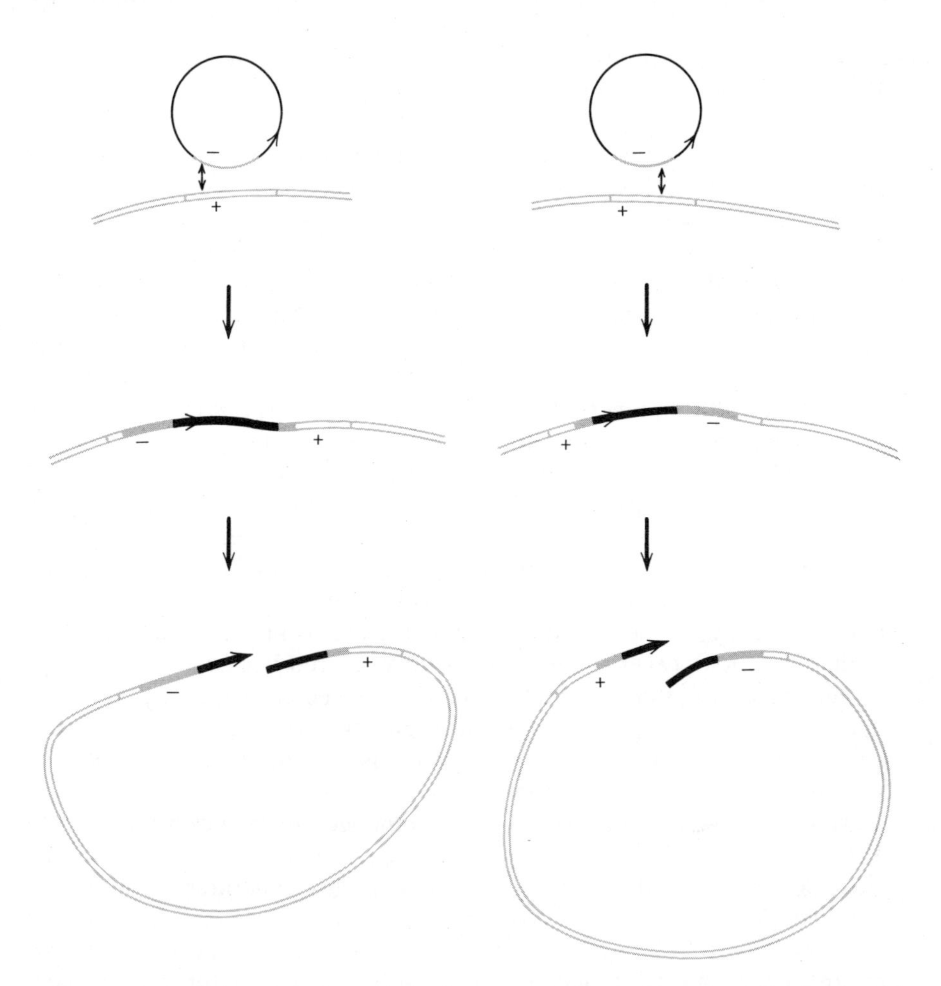


**Figure 2.15:** insertion of F′ factor into chromosome of a heterogenote. The factor and chromosome carry different alleles of some bacterial gene; the site of mutation is indicated by a (−) in the factor and by a (+) in the chromosome. If recombination takes place on the side of the mutation opposite to the origin (left), the (−) allele becomes a proximal marker and the (+) allele becomes a distal chromosome marker. If recombination takes place between the mutant site and the origin of the sex factor (right), the (+) allele becomes a proximal chromosome marker and the (−) allele is a distal marker. In this case, insertion of a factor carrying a negative allele is responsible for transfer from the resulting *Hfr* cell of the chromosomal plus allele.

constituting a sex affinity site (*sfa*), which is inherited as a genetic locus. This view is borne out by the results of an experiment in which the $sfa^+$ strain was cured with acridine and used as recipient for an Hfr donor; selection for the donor markers in the region of the *sfa* led to the loss of ability to preferentially integrate the sex factor at this site, since the Hfr chromosome lacked it.

Two classes of event can generate an F′ factor. The type I exchange between a sequence within the integrated factor and a sequence within the chromosome shown in Figure 2.13 produces a factor containing distal chromosome markers and a chromosome containing some sequences of the episome. Such factors can be isolated readily by their ability to transfer at high frequencies the markers that they carry, which previously were transferred only at very low frequencies as distal markers. A type I exchange involving a site in a proximal chromosome region would generate a factor carrying proximal chromosome markers, again leaving a chromosome containing some sequences of the episome. Such factors are more difficult to isolate directly because the markers that they carry and therefore transfer at high frequency previously were transferred also at high frequency by the Hfr strain. By using recipients carrying the $recA^-$ mutation which prevents integration of transferred DNA (see later), Low (1968) observed that F′ factors can be selected which carry early markers; a transferred chromosome segment cannot generate recombinants in these recipients, but an F′ factor is an autonomous unit able to multiply and express the genes that it carries.

Another class of event, the type II exchange illustrated in Figure 2.17 also is possible; this involves an exchange between two chromosomal sites, located on either side of the factor. This generates an F′ factor which carries genes derived from both sides of the site of insertion, that is, it carries both proximal and distal markers; and the chromosome retains no episome sequences but suffers a deletion for the markers gained by the epsiome. Such episomes may be produced by selection for factors carrying either proximal or distal markers. An F′ factor generated by a type II exchange was isolated by Broda, Beckwith and Scaife (1964). They used a donor in which $lac^+$ is transferred as a distal marker; after mating the donor cells with a recipient population for 60 minutes, a time too short to allow the distal $lac^+$ gene to be transferred as a chromosome marker, they isolated $lac^+$ recombinants. These resulted from the formation and spreading through the population of an F $lac^+$ factor, termed F13. A large proportion of these recombinants also carried the proximal markers $ade^+$ and $tsx^r$; and when the F′ factor of the progeny was transferred to fresh $F^-$ recipients, it transferred both $ade^+$ and $tsx^r$ as well as $lac^+$.

If this F′ factor was generated by the type II excision illustrated in

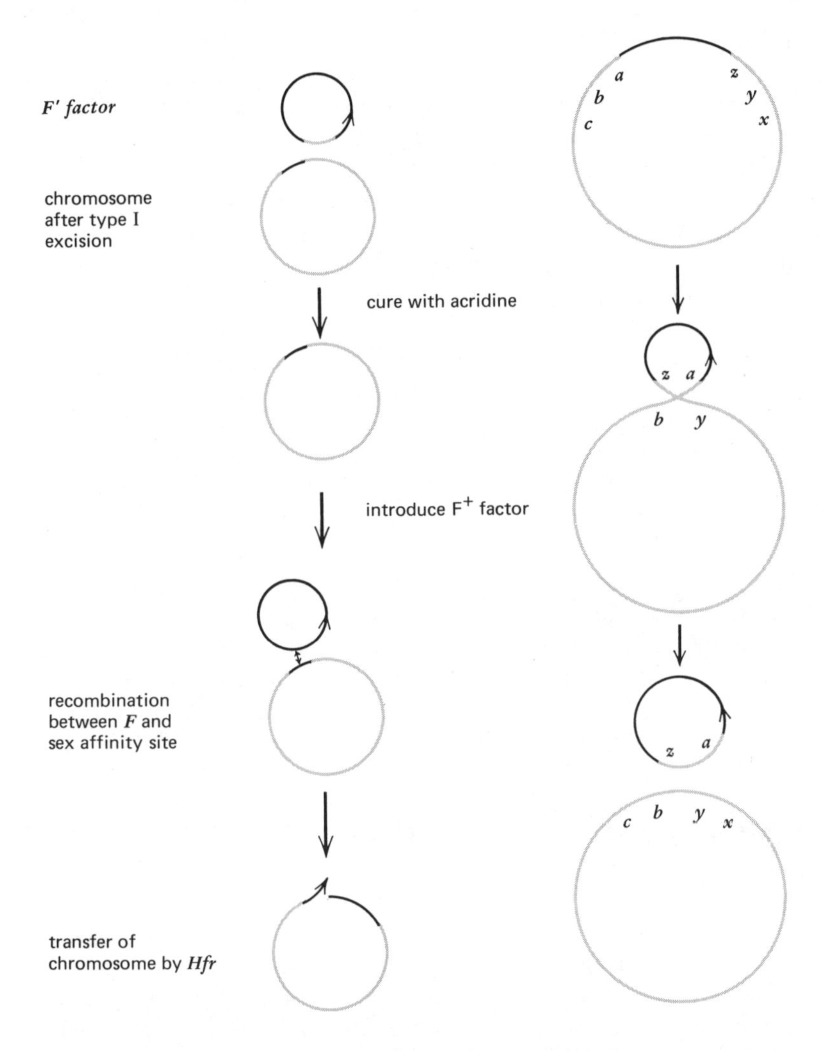


**Figure 2.16:** properties of bacterial strain that has suffered type I excision of sex factor. When the F′ factor is removed by curing with acridine, a new F factor can be introduced. This pairs preferentially with the sex affinity locus created by the type I excision of the F′. Recombination results in the insertion of the new F factor at the same site where the F′ had previously been located.

**Figure 2.17:** type II excision of F factor from the bacterial chromosome. A recombination event takes place between two bacterial sites, one on each side of the integrated factor; in the figure one site is between *z* and *y* and the other site is between *a* and *b*. This generates an F′ factor carrying the distal marker *z* and the proximal marker *a*; the bacterial chromosome has suffered a deletion for both genes. No part of the sex factor remains in the bacterial chromosome.

Figure 2.17, the bacterial chromosome should have lost the *ade, tsx,* and *lac* loci during excision, and with them the ability to integrate an F factor at this site. Scaife and Pekhov (1964) demonstrated that this F′ factor does not seem to be able to integrate readily at its ancestral site, although when transferred into other $F^-$ recipients (which possess this chromosome region), it sponsors chromosome transfer with its ancestral orientation. Its failure to do so in the parental strain must thus be due to deletion of the sequences usually recognized at this site by the sex factor. Cuzin and Jacob (1964) also isolated a strain in which deletion prevented insertion of F′ at its ancestral site.

In both type I and type II excisions material is lost from the bacterial chromosome. Scaife and Pekhov (1964) confirmed that this makes it impossible to derive $F^-$ cells under appropriate selective conditions, a conclusion reached also by Pittard and Ramakrishnan (1964). In further studies, Scaife (1966) demonstrated that when cells are grown on medium in which a $lac^+$ genetic function is required, attempts at curing F13 with acridine orange do not produce any $F^-$ cells; the only $lac^+$ survivors represent Hfr cells in which the F′ factor has been integrated into the bacterial chromosome. Since this F′ factor was generated by a type II excision, although the sites of integration varied they never corresponded to the original site. In these Hfr strains, the integration of F13 at these various sites always resulted in transfer of $ade^+$ as an early marker; so transfer following integration starts with the bacterial material carried by the original F′ factor.

Because a type II F′ factor possesses both proximal and distal markers, on entering a new recipient possessing both these regions it should be able to undergo a double recombination. This explains why cells carrying type II F factors often revert to an Hfr state identical to that of the ancestral strain; two recombination events, one in the proximal and one in the distal marker region, can insert the F factor at its original site, freeing fragments that in toto correspond to the duplicate set of markers, and which presumably are lost at cell division. This double recombination thus generates a haploid Hfr strain (see Hayes, 1968).

When a reciprocal exchange occurs within a circular structure, two consequences are possible, depending on the type of reunion that follows the breakage. As Figure 2.18 demonstrates, a homodirectional reunion causes the formation of two circular structures; this corresponds to the type II excision shown in Figure 2.17. A heterodirectional reunion, however, causes an inversion, in which the sequence containing the sex factor is replaced in the bacterial chromosome in the opposite orientation to its previous one. Both have the result that a donor is produced in which previously late markers are now transferred early, in the first case as part

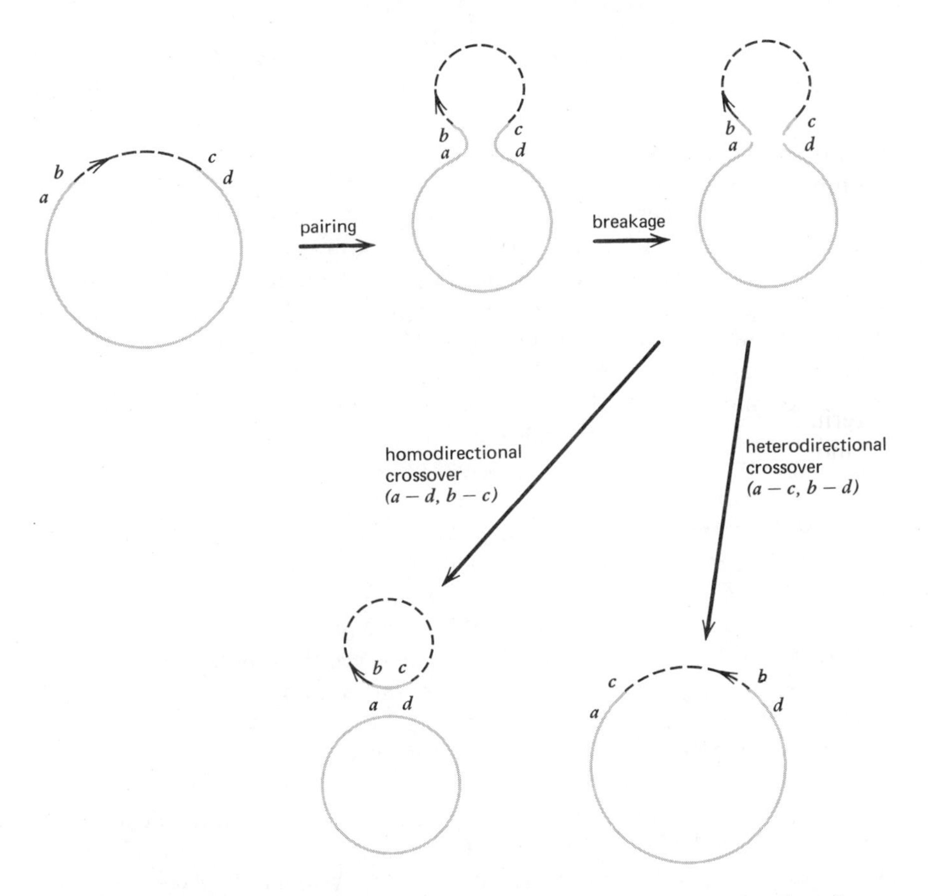


**Figure 2.18:** recombination between homologous sites on a circle. Two types of reunion can take place after pairing between *ab* and *dc* has been succeeded by breakage to give the structure shown at the upper right. A homodirectional crossover involves reunion between *a* and *d* and between *b* and *c*; when the two sites are located on either side of an integrated sex factor, as shown in the figure, this results in a type II excision of F′. (The orientation of the origin in the released F′ differs in this figure from that shown in Figure 2.17 and other preceding figures because there the event is depicted by means of formation of a twisted circle whereas here the free ends are shown.) A heterodirectional crossover involves reunion between *a* and *c* and between *b* and *d*; this results in an inversion of the sequence containing the F factor. Both the formation of F′ and the inversion of the integrated sex factor result in early transfer of markers that previously were late in transfer; the F′ factor carries the distal marker *c* and the inverted sex factor transfers the chromosome in a reversed order, so that *d* and succeeding markers are proximal instead of distal.

of the F′ factor, in the second during chromosome transfer by the inverted sex factor. In a study of derivatives in which late markers are transferred early, Berg and Curtiss (1967) identified transposition strains of this type and confirmed that they result from an inversion of the sex factor which reverses the order in which the chromosome is transferred. Transpositions are about equally common with the formation of type II F′ factors. The sites of exchange implicated in these transpositions appear to be random. They must therefore occur between sequences that fortuitously resemble each other; and following pairing and breakage, there is an equal chance of either homodirectional or heterodirectional reunion.

Many F′ factors have now been isolated. A factor carrying any particular marker may most easily be selected by using a donor strain in which it is transferred distally; after allowing only a short time for mating, any recombinants gaining this marker must have done so via generation of an F′ factor. Whatever form of selection for the factor is used, either a type I excision or type II excision may generate an episome carrying the selected markers. Type I excision is more frequent than type II excision (for review see Low, 1972).

The primary strain in which an F′ factor is generated remains haploid for the genes that are transferred from the chromosome to the episome. When a secondary strain is generated by introducing the F′ factor into a recipient that has a chromosomal copy of the genes that it carries, the bacterium becomes a partial diploid for this region. If the factor and chromosome carry different alleles, the bacterium may be described as a *heterogenote*. Its genetic constitution might, for example, be characterized as $lac^-$/F $lac^+$, indicating the presence of a $lac^-$ allele on the chromosome and a $lac^+$ allele on the episome. As might be expected from the homology between the sequences that define the diploid state, recombination to generate chromosomal $lac^+$ and episomal $lac^-$ is fairly frequent; heterogenotes therefore are not too stable. Heterogenotes also may suffer conversion to the homogenote in which both chromosome and episome carry the same allele. In principle, these changes might be prevented by introducing a $recA^-$ allele into the chromosome, since the $recA^+$ function is necessary for these recombination events (see below); but this mutation has deleterious effects upon the cell and so reduces the potential of this method for increasing stability.

### *Circular Structure of the Sex Factor*

A circular structure was first proposed for the sex factor on genetic grounds. Like the lysogenic phages, the factor is a small genetic element able to integrate into the much larger bacterial chromosome; this insertion can be accommodated by a single recombination event if the factor is circular. The model proposed by Campbell (1962) to explain the integration of lysogenic phages, and applied in Figure 2.12 to the sex factor, is able also to explain the derivation of F′ factors by type I excision (Figure 2.13) or by type II excision (Figure 2.17). The prediction that this model makes about the sex factor is that it comprises a circular duplex of DNA, with a region (or regions) that can be recognized by a small number of sites on the bacterial chromosome, at which recombination can take place. We know also that the sex factor contains an origin for transfer, at which the donor chromosome of an Hfr strain presumably breaks so that it may be transferred in a linear manner to a recipient. Because recipients acquire the Hfr mating type only when the entire chromosome is transferred, the orientation of the integrated sex factor must be such that at least some (essential) regions of it are transferred at the tail of the donor chromosome.

Early estimates for the size of the sex factor were not very accurate, but suggested that it comprises a length of DNA corresponding to 0.5–3.5% of the length of the bacterial chromosome (Silver, 1963; Driskell-Zamenhof and Adelberg, 1963). A problem in isolating the sex factor at this time was that in E.coli host cells its DNA could not be distinguished from that of the bacterial genome; but Falkow and Citarella (1965) overcame this difficulty by transferring the factor into Serratia marcescens or Proteus mirabilis, when it forms a band of DNA different in density from that of its host. By isolating two F′ factors carrying different sets of bacterial genes, they were able to measure their homology, which corresponded to 1.9% of the bacterial chromosome ($53 \times 10^6$ daltons) and presumably represents the sequences of the factor itself.

One difficulty in relying upon experiments such as these is that the factor might suffer changes upon introduction to a new bacterial host. A technique for isolating F *lac* DNA in E.coli was developed by Freifelder and Freifelder (1968a), however, and essentially relies upon providing conditions in which neither male nor female cells can incorporate exogenous thymine into chromosomal DNA; they used donor cell mutants unable to utilize exogenous thymine, and the females were inhibited in DNA synthesis by ultraviolet irradiation. When such cells are mated to allow transfer of the F *lac* factor (or of F itself) the sex factor replicates immediately after transfer and thus incorporates thymine. By using donor cells carrying F *lac* but lacking the chromosome *lac* region, it is possible to reduce the efficiency of chromosome transfer, so that only the sex factor passes from donor to recipient. Using this technique, Freifelder and Freifelder (1968b) reported that F *lac* DNA is about $55\ (\pm 8) \times 10^6$ daltons and that F DNA is $35\ (\pm 7) \times 10^6$ daltons.

As identified by this labeling protocol, the sex factor can be found in the form of covalently closed circles of DNA. Freifelder (1968a) used two techniques that identify circular DNA by its unusual sedimentation properties under appropriate conditions. If denatured circles are centrifuged in alkali, they sediment at a rate 3–4 times faster than a single strand at the same molecular weight; and a single strand break causes a sudden transition to the more slowly sedimenting form. If duplex molecules are centrifuged to equilibrium in CsCl density gradients containing ethidium bromide, they band at a position which is less dense than that occupied in CsCl alone; the intercalation of ethidium bromide into the duplex makes the decrease in density much less for covalently closed circles than for linear strands. Again the introduction of a single strand break causes an abrupt transition. Both techniques suggest that F *lac* DNA contains 40–60% circles, with the linear molecules probably produced by breakage of circles during excision (see Freifelder, Folkmanis and Kirschner, 1971). Several factors, including F, F *lac,* F *gal,* were shown to exist as circles; by utilizing the properties characteristic of the circular state, Freifelder (1968b) estimated the size of F to be $45 \times 10^6$ daltons and that of F *lac* as $74 \times 10^6$ daltons.

Visualization in the electron microscope allows the length of a nucleic acid molecule to be estimated more accurately than is possible in centrifugation studies, where conformation as well as size may influence the sedimentation rate. Of course, this technique is most readily applied when the nucleic acid can be isolated in bulk, for example from a virus particle. Heteroduplex mapping makes it possible to gain more information about a molecule and to delineate some of its sequences; this technique demands two related molecules which differ in some sequences and utilizes these differences to provide structural features which can be recognized and mapped. This allows the consistent identification of such structural features in a preparation to be correlated with genetic differences. The principle that underlies heteroduplex mapping is that when two preparations of DNA are denatured and then renatured together, two classes of molecule are produced: homoduplexes, in which a parental DNA strand has paired with its original complementary strand; and heteroduplexes, in which a single strand of one parental molecule has paired with the complementary single strand of the other parent. The homoduplex molecules should consist solely of duplex nucleic acid, formed by an exact reconstitution of the original molecule; the heteroduplex molecules comprise duplex regions that identify sequences present in both parental molecules, and single strand regions that represent sequences present in one but not in the other parent.

After renaturation of two denatured preparations, the molecules are spread for electron microscopy. They are first mixed with a basic protein, usually cytochrome c, which binds to the nucleic acid and gives a structure that can conveniently be visualized (for review see Davis, Simon and Davidson, 1971). The nucleic acid-protein preparation may be spread in either of two ways. In aqueous spreading medium, duplex nucleic acid sequences appear as elongated threads but single strands suffer random interactions and thus become condensed into bushes. This identifies the positions of the single stranded regions relative to the duplex regions, whose contour lengths can be measured. In spreading medium containing appropriate concentrations of formamide and salt, the random collapse of the single strands is prevented and so both duplex and single strand regions form elongated threads. The single strand regions appear thinner in diameter than the duplex regions, and so both single strand and duplex contour lengths can be measured.

Heteroduplex mapping is a particularly appropriate technique to use for defining sequences of the sex factor because so many F′ variants are available to provide suitable preparations of parental molecules that are in part identical and in part different. In the experiments of Sharp et al. (1972), covalently closed circles of F and F′ molecules were nicked to introduce single breaks in each strand, the duplexes were denatured in NaOH and single strands recovered, and then the F preparation was annealed with the F′ preparation. Heteroduplexes form whenever an F strand renatures with a complementary F′ strand and were visualized by spreading in a medium containing 50 $\mu$g/ml cytochrome c and 45% formamide.

Two types of heteroduplex structure can be predicted, depending on whether the F′ factor arose by a type I or type II excision. Figure 2.19 shows that in both types of heteroduplex the F′ strand carries some bacterial sequences that are missing from the F factor strand and thus form a single stranded loop unable to find a complement with which to pair. After a type II excision, the F′ factor contains all of the sequences of F; and so the F strand of the heteroduplex is perfectly paired and only the F′ strand bears a single strand extrusion. This thus appears as an *insertion loop* (also sometimes known as a *deletion loop,* since the same result is produced by an insertion into one strand or a deletion in the other). In a type I excision the same single strand sequence corresponding to the insertion of bacterial DNA is seen in the F′ strand. But part of the F factor is left in the bacterial chromosome after a type I excision, so that the F′ strand has suffered a deletion where F sequences have been replaced by the bacterial sequences gained from the chromosome; this means that the corresponding sequences in the F strand find no partner in the F′ strand. A single strand loop of these sequences therefore forms on the F strand

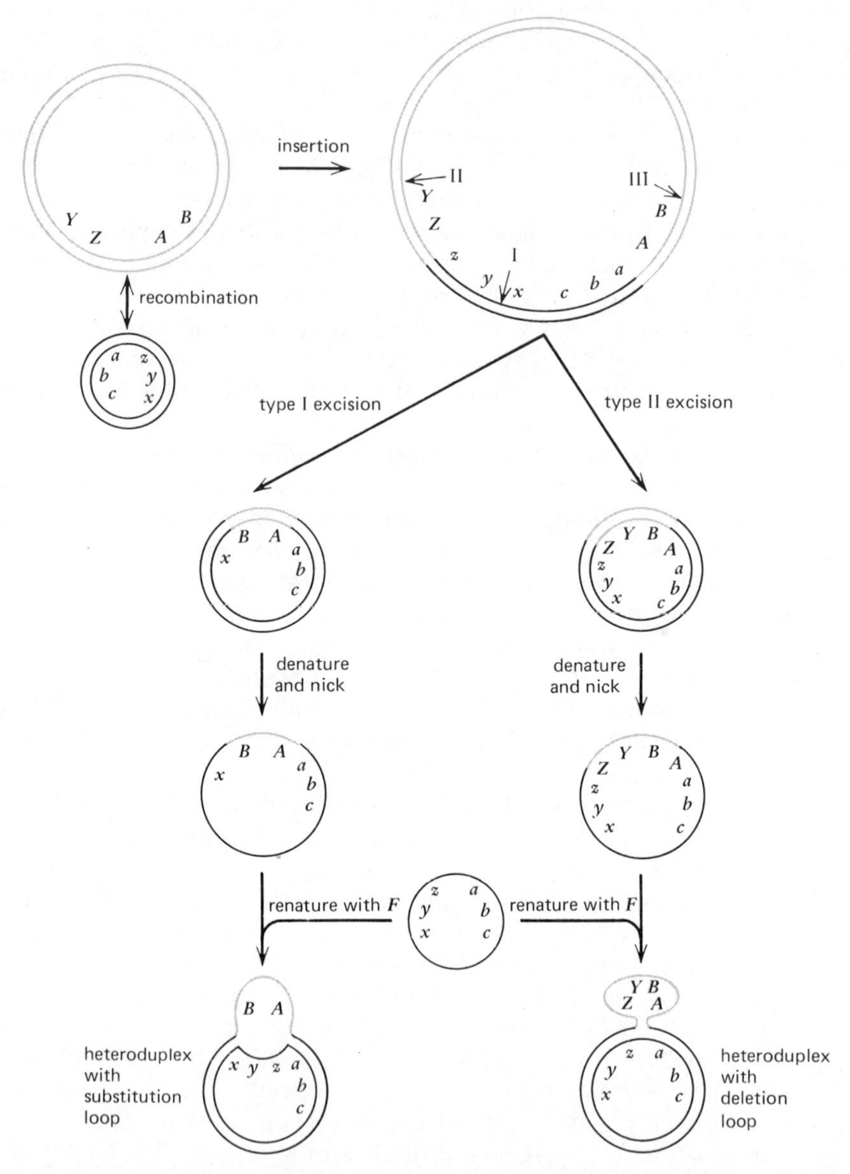


**Figure 2.19:** heteroduplex mapping of F′ factors derived by type I and type II excisions. The F factor is inserted by a recombination event involving a bacterial site between *A* and *Z* and a site on the factor between *a* and *z*. A type I excision takes place by recombination between sites I (between *y* and *x*) and III (beyond *B*); this generates an F′ factor containing *AB* and lacking *yz*. When this factor is denatured, nicked, and annealed with single strand nicked F DNA, a substitution loop is formed in which *yz* of the F strand is single stranded and *AB* of

opposite the single strand loop of bacterial sequences on the F′ strand. A type I excision is therefore marked by a *substitution loop* in which single strands extrude at one site from both partners of the circular duplex.

The structure of the heteroduplex formed by any F′ factor under study with the F factor gives only restricted information since their differences are likely to be confined to the short loops depicted in Figure 2.19. From such a structure it is possible to tell only whether a type I or type II excision has occurred and how much bacterial DNA has been gained or factor DNA lost in the process. To provide coordinates to map the sequences of an F′ factor, it is necessary to form heteroduplexes with various reference F′ factors that possess certain defined structural features; which reference factors are most useful for mapping any particular unknown F′ structure depends upon which sequences of F may be missing from the F′ and at which site on it the bacterial DNA has been inserted.

The heteroduplex formed between the F factor and the F450 factor has been taken to define an arbitrary coordinate system for describing the maps of sex factors. Figure 2.20 illustrates the structure of this heteroduplex. The total contour length of the F strand corresponds to 62.6 ($\pm$ 1.6) $\times$ $10^6$ daltons, which equals 94,500 bases and is sometimes described as 94.5 kb (kilobases). The F450 strand pairs perfectly with the F strand except for a length of 2800 bases that has been deleted at the site of a bacterial insertion of total length 80,000 bases. The coordinate system defines a starting point at one of the junctions between bacterial DNA and sex factor DNA on the F450 factor. Sequences of the F factor are numbered in thousands of base pairs, measured clockwise from the site of insertion as indicated in the figure: they bear the suffix F to distinguish them from the bacterial sequences, which are given the suffix B. Since the sex factor is circular, the starting point thus bears the coordinates 0/94.5 F: the region of the sex factor that has been deleted in F450 is described as 0–2.8 F; and the bacterial sequences that have been substituted are described as 0–80.0 B. The sequences of all sex factors may be described in terms of the F coordinate system defined by the F/F450 heteroduplex so that any given sequence of the factor always has the same coordinates. However, the bacterial sequences present on each F′ factor usually are given a coordinate system in which O B defines one junction with the sex factor and numbering is then continued in thousands of base pairs towards the

---

the F′ strand is single stranded. A type II excision takes place between sites II (beyond *Y*) and III (beyond *B*); this generates an F′ factor containing all the sequences of F and in addition the bacterial markers *ZYAB*. When this factor is denatured, nicked, and annealed with single stranded nicked F DNA, a deletion loop is formed in which ZYAB of the F′ strand is single stranded opposite the region between *z* and *a* of the F strand.

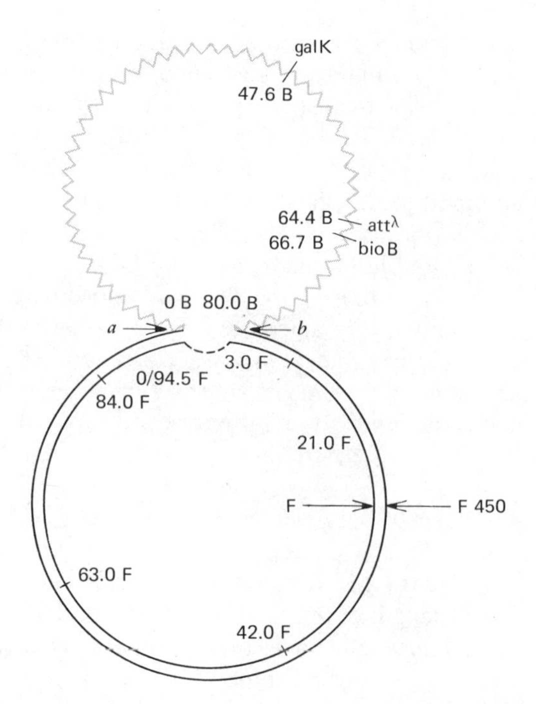


**Figure 2.20:** heteroduplex structure of F/F450. This defines the coordinate system used to describe the map of the sex factor, in which sites are described by distance in thousands of base pairs from the point 0/94.5 F. The bacterial DNA of the F′ factor is described in the same way, but with a suffix B. Data of Sharp et al. (1972).

other junction; thus when the same bacterial sequence is present on two different F′ factors of independent origins, it may have different B coordinates.

In order to assign map locations to sequences on the sex factor, it is necessary to have reference points against which their positions can be measured. (Although the distance of a sequence from, for example, the insertion in F450 can be determined by forming a heteroduplex with this factor, this does not define a position unambiguously because it might lie on either side, clockwise or counterclockwise, of the insertion.) The structure of one of the reference factors, F8-33, that has been especially useful for heteroduplex mapping is illustrated in Figure 2.21. The F8-33 factor can be used to map virtually any F′ factor because it has three distinctive structural features, comprising the small insertions *c* and *d* at 91.0 F and 35.2 F which are seen as deletion loops, and a bacterial insertion of 22,000 base pairs replacing the sex factor sequences from 8.5 to 16.3 F which is seen as a substitution loop. These features can be used

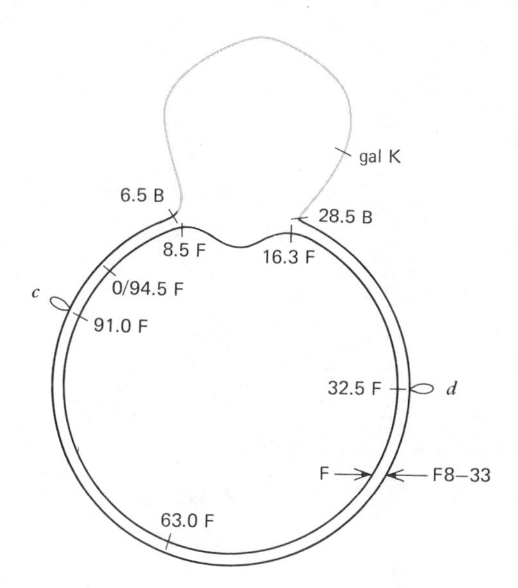


**Figure 2.21:** the F/F8-33 heteroduplex. Two small deletion loops, *c* and *d*, are formed at 91.0 F and 32.5 F. A bacterial insertion has replaced the sex factor material from 8.5 F–16.3 F; this forms two single stranded substitution loops emerging from the duplex. The bacterial DNA is numbered from 6.5 B–28.5 B because the F8-33 factor is derived from F8, in which the bacterial DNA/sex factor junction is given the coordinate zero. Data of Ohtsubo et al. (1974a).

to orientate insertions and deletions in other factors (except, of course, those located in the region 8.5–16.3 F), since their positions relative to the characteristic single strand loops of F8-33 unambiguously define their map locations.

The mapping of factor F(Δ0–15) is illustrated in Figure 2.22. The structure of the heteroduplex with F8-33 shows that the sole change in F(Δ0–15) is the deletion for which it has been named; it lacks the material from 0–15.0 F. The heteroduplex that would be formed between F and F(Δ0-15) also is shown in the figure, but this could not be used for mapping because it would reveal only that a deletion of 15,000 bases had occurred, and would not allow the position of the deletion to be inferred. Since this factor is viable, apparently suffering no defect in comparison with F, the deleted region must be nonessential. This factor was originally known as F(Δ0-14.5) because the deletion was thought to end at 14.5 F; more recent measurements have suggested an endpoint of 15.0 F. Contour length measurements are, of course, subject to some variation with spreading conditions and on occasion assignments of map positions by this tech-

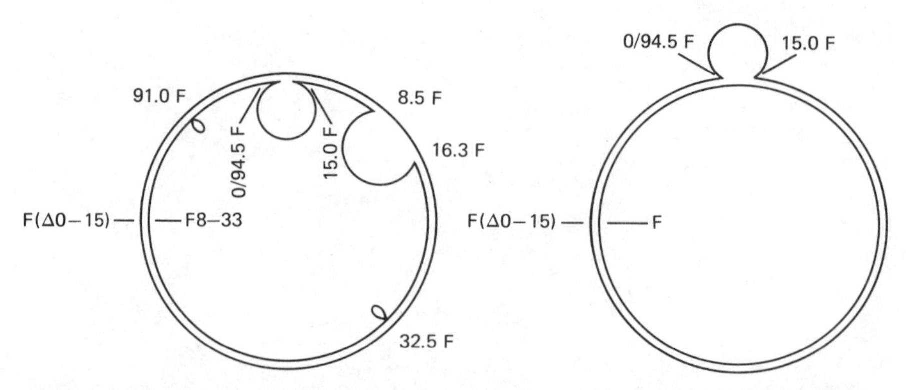


**Figure 2.22:** structure of sex factor F(Δ0-15). The heteroduplex of F(Δ0-15) with F8-33 shows one feature additional to those seen in the F/F8-33 heteroduplex; a deletion loop corresponds to the region from 0–15.0 F. If a heteroduplex were constructed between F(Δ0-15) and F, the two factors would differ only in the deletion of 0–15.0 F in the smaller factor. Data of Sharp et al. (1972).

nique may therefore be revised; the values given here are those obtained in the most recent measurements and may therefore sometimes show small discrepancies from earlier published values.

Mapping of the F *lac* factor commonly used in genetic experiments (F42) is shown in Figure 2.23. Its heteroduplex with F8-33 shows that it has suffered a deletion from 0–2.8 F and has an insertion at this site of 51,700 bases, making a total contour length of 143,200 bases. An electron micrograph of the F *lac*/F8-33 heteroduplex is shown in Figure 2.24. It is striking that F *lac* and F450, which were derived from different Hfr strains in which the F factor was integrated at separate sites, have the same deletion (within experimental error) of 2800 bases from 0–2.8 F, where varying lengths of bacterial DNA have been inserted. Two other F′ factors, F100 and F152 (also known as $F_1$*gal* and $F_2$*gal*), since have been shown to possess this feature. This suggests that there is a hotspot on F at 0/94.5 F which is used for inserting the factor into the bacterial chromosome, so that a sequence on the factor at this site is likely to recombine with any one of several bacterial sequences. Both F *lac* and F450 correspond to typical type I excisions and the common length of their deletion of F material suggests that the site at 2.8 F may be a hotspot for recombination leading to this class of excision event. Insertion and excision can take place at other sites, as seen in the F8-33 factor where a region of 7800 bases of F from 8.5–16.3 F is deleted at the site of insertion; this suggests that insertion took place at 8.5 F and excision involved 16.3 F. The factor F210 also appears to utilize sites in this region, inserting at 8.5 F and

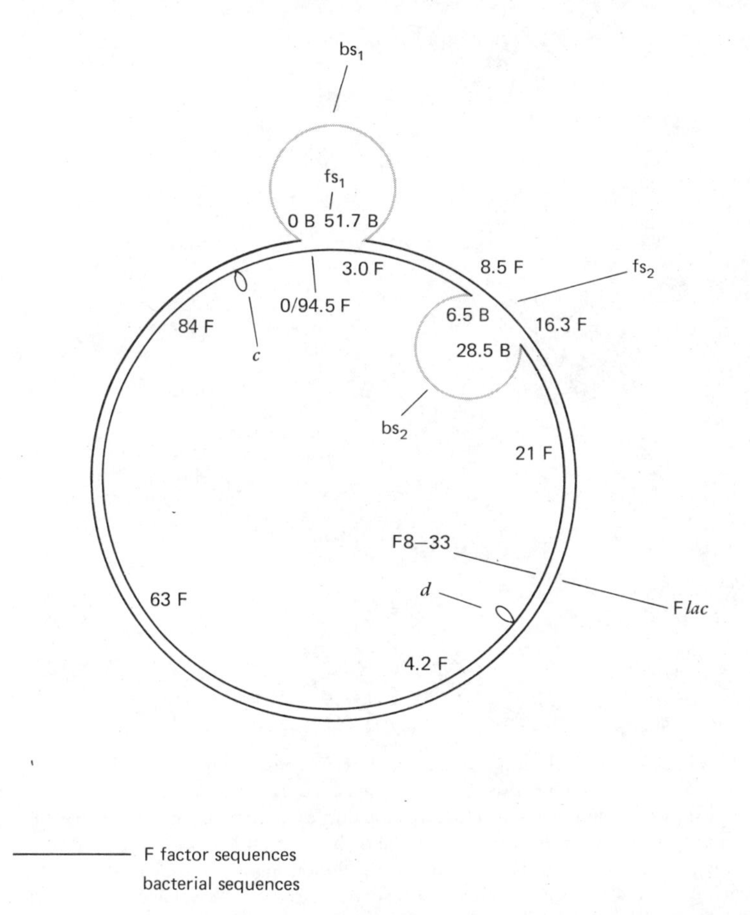


**Figure 2.23:** structure of the F8-33 / F *lac* heteroduplex. The strand contributed by F8-33 is identified by the two deletion loops *c* and *d* and by the bacterial insertion seen as the single stranded loop *bs*2 which replaces the material from 8.5–16.3 F (causing this sequence on the F *lac* strand to form the single strand loop *fs*2). In the F *lac* strand the region from 0–2.8 F has been replaced with a bacterial sequence of 51,700 bases; the two substitution loops are described as *fs*1 and *bs*1. Data of Sharp et al. (1972).

excising at 12.0 F. The material deleted in all these F′ factors occurs within or extends just beyond the nonessential region of 0–15.0 F defined by the F(Δ0-15) factor and the structure of the F8-33 factor suggests that the region from 0–16.3 F can be classified in this manner. This is necessary to explain why these F′ factors are viable as autonomous episomes in spite of the loss of some of their genetic material. It therefore seems likely that integration and excision events commonly take place in this region.

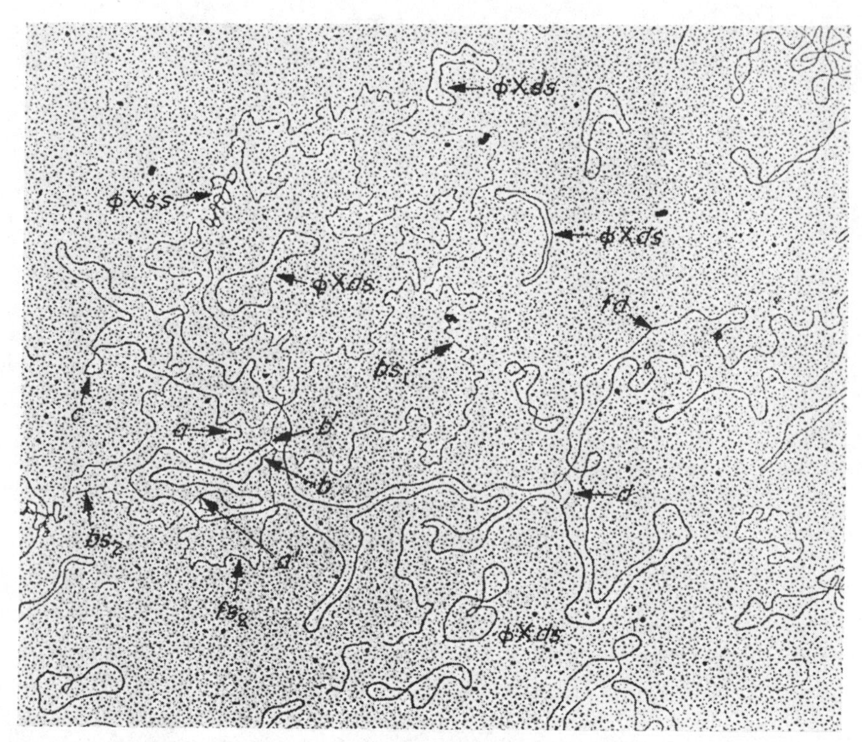


**Figure 2.24:** electron micrograph of the F8-33/F *lac* heteroduplex. For identification of the sequences compare with the illustration of Figure 2.23: *c* and *d* mark deletion loops and *bs* 1, *bs*2, *fs*2 are substitution loops; *fd* indicates the duplex sequences of the annealed F factors; *a* and *a'* mark the junctions between duplex and single stranded DNA at 0/94.5 F–0 B and 2.8 F–51.7 B, respectively; *b* and *b'* indicate the junctions 8.5 F–6.5 B and 16.3 F–28.5 B. Controls of $\phi$X 174 DNA are seen as both single stranded ($\phi$X ss) and double stranded ($\phi$X ds) molecules. Data of Sharp et al. (1972) kindly provided by Norman Davidson.

The corollary of this concept, of course, is that if an integration and type I excision were to result in the deletion of genes essential for sex factor function, the F' factor would not be viable and would be lost with cell growth. The structure of the F factor suggested by these mapping experiments is summarized in Figure 2.25.

Type II factors also have been identified by electron microscopic mapping of heteroduplexes. One factor that has been studied in detail is F14, an unusually large F' factor that carries the bacterial genes located between 74.7 and 78.7 minutes on the standard E.coli map. F14 is relatively stable in the strain in which it was first isolated, which is haploid for the genes carried on the factor, but is extremely unstable when transferred into other hosts to generate partial diploids. Ohtsubo et al. (1974a) showed

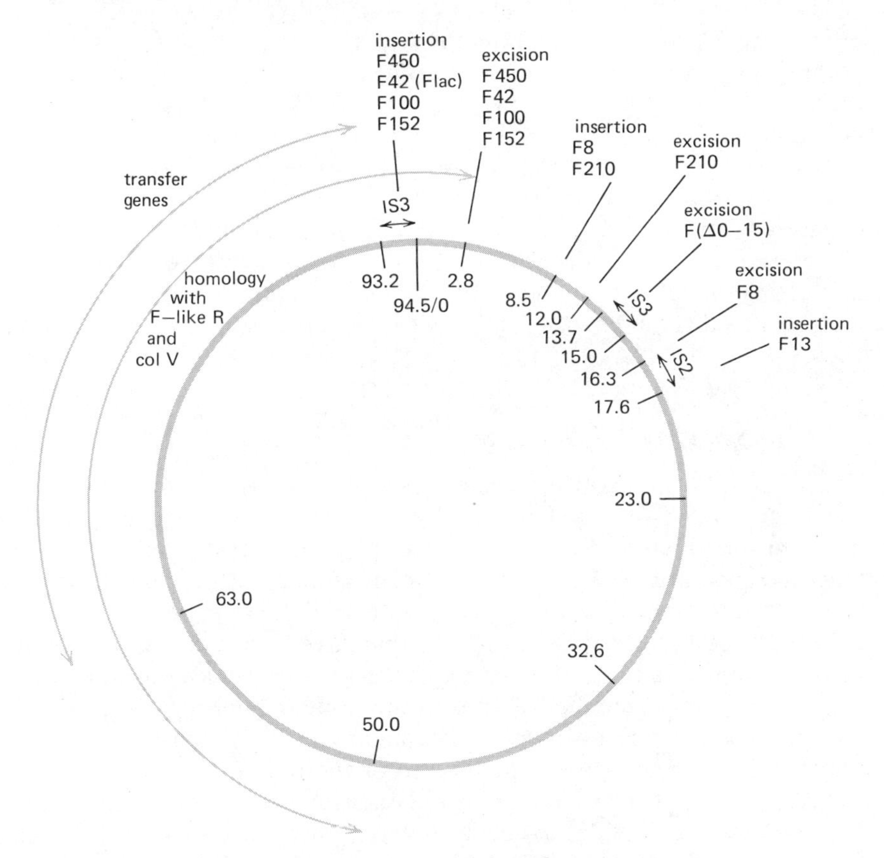


**Figure 2.25:** map of the F factor.

that all the sequences of the F factor itself are present in F14, demonstrating that it must have arisen by a type II excision; in addition to the entire sequence of F, the factor carries 210,000 base pairs of bacterial DNA. At both the junctions of sex factor and bacterial DNA the sequence 2.8–8.5 F is found; the source of the duplication is not known, but its presence means that reciprocal recombination between the repeated copies should be able to occur. This may explain the tendency of F14 to segregate into the F factor and a sequence representing the bacterial insertion. Details of the structure of this factor have been reported by Lee et al. (1974), Ohtsubo et al. (1974b) and Deonier et al. (1974); in general terms they confirm the model for type II excision which supposes that recombination takes place between two bacterial sequences on either side of the sex factor.

The use of F′ factors to identify the regions of the sex factor has made it possible to perform denaturation mapping of the sex factor itself. Various techniques can be used for denaturation mapping, but the general principle is that DNA is spread for electron microscopy under conditions that encourage partial denaturation, so that some sequences are single stranded while most remain duplex. The sequences that are preferentially denatured comprise regions rich in A-T base pairs. By using this technique, Hsu (1974) obtained a denaturation map of the sex factor; it is interesting that almost all the A-T rich regions giving rise to single strand loops are located in the dispensable region where integration and excision occurs.

### *Insertion Sequences in Sex Factors*

The sex factor indulges in several types of sequence rearrangement. A free factor may be inserted into the bacterial chromosome to form an Hfr donor strain: such events seem to take place at a restricted number of chromosomal sites and to involve a small number of sites on the factor itself. An integrated factor may give rise to an F′ factor by either type I or type II excision. In the (viable) factors that have been characterized, type I excision appears to involve recombination between a chromosomal site and one of a small number of sites in the nonessential region of the sex factor. Type II excision involves recombination between two bacterial sequences. Questions about the nature of the recombination events responsible for sex factor integration and excision can be considered in two classes: what nucleotide sequences are involved; and what enzyme systems are responsible?

Sites that may be involved in plasmid recombination have been identified in the form of *insertion sequences*. Three types of insertion sequence so far have been characterized and it is entirely possible that others remain to be discovered. These sequences are found both in the host chromosome and in plasmid and phage DNA. Sex (F) factors and drug (R) factors both carry insertion sequences, and in both types of plasmid it is likely that these sequences are involved in recombination events. We shall discuss here the origin of the insertion sequences and the implications of their presence in sex factors; their presence in R factors and the mobilities of drug determinants will be discussed in Chapter 3.

Insertion sequences were first identified as strong polar mutations in the E.coli *galT* gene which differ from other known polar mutations. Polarity usually is caused by substitution or frameshift mutations that create a nonsense codon. Translation is terminated prematurely at the mutant site and, in addition to inactivating the protein product of the mutated gene, this causes cessation (in strong polar mutants) or reduction (in weak polar mutants) in the expression of subsequent genes in the same operon. But the strong polar mutants in *galT* have a different origin. By comparing the densities of λ*gal* phages carrying wild type and mutant galactose operons, Jordan, Saedler and Starlinger (1968) demonstrated that the phages with these mutant operons have an increased density. Upon reversion to wild type, the density is shifted back to the usual value. No such changes are seen with polar point mutations. This suggests that the unusual polar mutations are caused by insertion of DNA into the *galT* gene.

Several further such mutations have since been identified, including those mapping in the control (operator/promotor) region of the galactose operon (Saedler et al., 1972), in the lactose operon (Malamy, 1970) and in phage lambda (Brachet, Eisen and Rambach, 1970). That the mutations in the galactose control region represent insertions was confirmed by Hirsch, Saedler and Starlinger (1972), who examined the heteroduplexes formed between transducing phages carrying the mutant *gal* operons and a control λ*gal* phage which has a deletion in another part of the galactose operon (this provides a deletion loop that can be used to map the positions of the insertions). All four mutations at the control region could be visualized as single stranded insertion loops, apparently all located at the same site (within the limits of the electron microscopic technique). The contour lengths of the insertion loops were consistent with estimates for the amount of added DNA based upon the increase in density of phages carrying each mutant operon.

The insertions in the galactose operon and in phage lambda fall into two size classes. The best estimates for their lengths are derived from measurements of contour lengths, either of the single stranded insertion loops visualized as such (this is less accurate) or of duplex regions formed by annealing of the inserted sequences (the most precise method). The two insertions in *galT* (originally described as *N116* and *N102*), three of the insertions in the *gal* control region (known as *OP128*, *OP141*, and *OP306*), and an insertion in the *x cII* region of lambda (*r14*; see Chapter 4) all are close to 800 nucleotides in length. A fourth insertion in the galactose control region (*OP308*) and another insertion in lambda (*r32*) are somewhat longer; original estimates were from 1400–1500 base pairs but more recent experiments have suggested a value of 1300 nucleotides.

Do the various insertions comprising each size class represent the same or different sequences? Hirsch, Starlinger and Brachet (1972) examined the insertions in the galactose operon and phage lambda by electron microscopy of heteroduplexes. Figure 2.26 illustrates the structures that may be formed when denatured DNA from two phages carrying insertion sequences is annealed. If the two DNA preparations carry different inser-

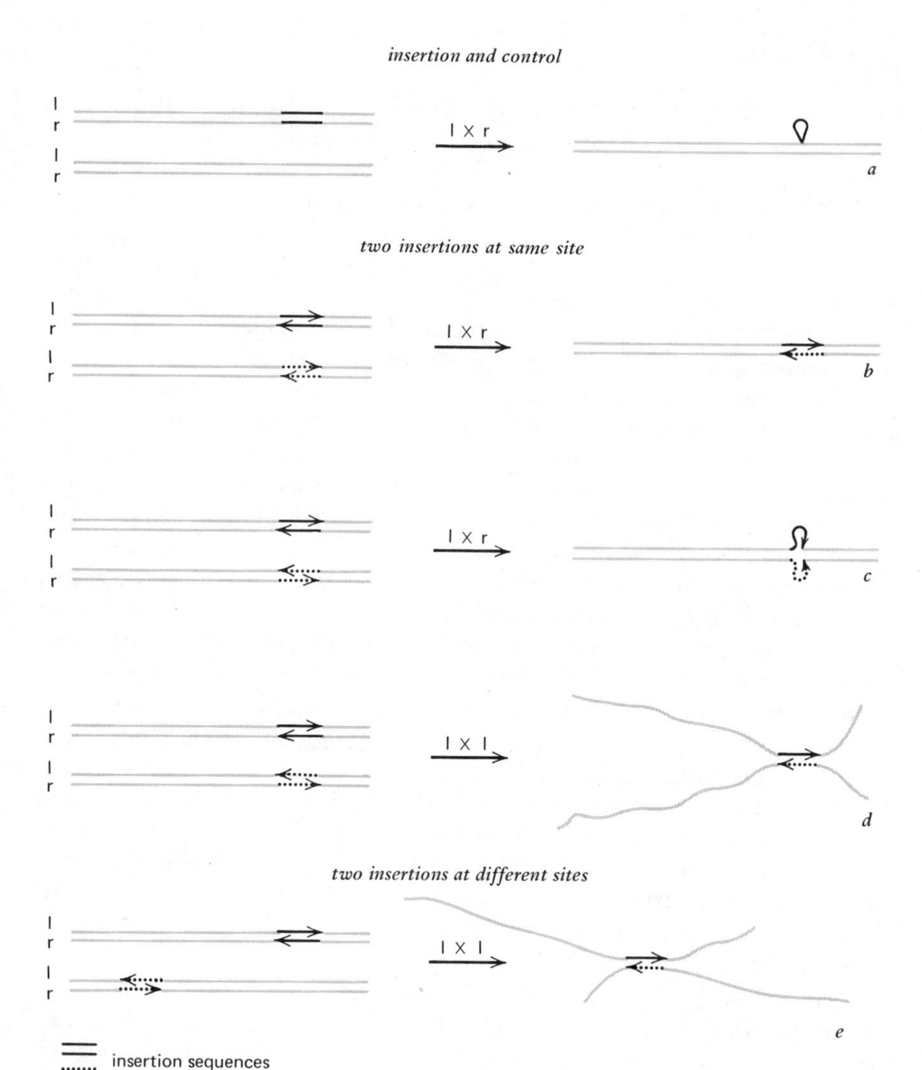


**Figure 2.26:** structures formed in electron microscope heteroduplex mapping of insertion sequences. The insertion sequences are carried either in the lambda genome or in the bacterial sequences of a transducing phage so that separated *l* strands and *r* strands may be obtained. The arrow heads indicate the orientation of the insertion. (*a*) When one strand of a phage carrying an insertion is annealed with the complementary strand of a phage lacking the insertion, a small single stranded loop is seen at the site of insertion. (*b*) When the complementary strands of two phages carrying the same insertion at the same site are annealed, a perfect duplex is formed if both insertions lie in the same orientation. (*c*) If insertions at the same site lie in opposite orientations, two single stranded loops are seen

tion sequences, there is of course no hybridization between them. If they carry the same insertion sequence, the reaction depends upon whether the two copies of the sequence are organized in the same or opposite orientations. If they lie in opposite orientations, they can anneal to form a duplex region when the *l* strands of both phages or the *r* strands of both phages are hybridized; since the two *l* strands of the λ or λ*gal* phages have no homology, this is easily visualized as the only duplex region formed. If the two insertions are identical and lie in the same orientation, the results may be more complicated. There is then no annealing when two *l* strands or two *r* strands are hybridized, but a duplex is formed when an *l* strand and an *r* strand are annealed. When two mutants represent the same insertion at the same site, a simple duplex is formed. But when the same insertion is found at two different sites, then cross annealing may take place and since there is also a general hybridization reaction between the complementary *l* and *r* sequences, this may generate tangled structures that are difficult or impossible to interpret in detail (these are not shown in the figure).

These experiments showed that each size of insertion represents a single sequence; the 800 base pair insertion is described as IS1 and the 1300 nucleotide insertion as IS2. IS1 is inserted in the same direction in the galactose mutants *OP128, OP141* and *N116*; it is inserted in the opposite direction in the galactose mutants *OP306* and *N102* and in λ*r14*. Since the four *gal OP* mutations all appear to be located at the same site, this means that IS1 may be inserted at a given site in either direction. When inserted at *gal OP*, IS1 has the same polar effect in either orientation; presumably it therefore carries either a symmetrical signal responsible for creating polarity or has at least two signals in opposite orientations. IS2 also appears to insert at the same site in *gal OP*.

Insertion sequences mapping at four sites in the *lacZ* gene were placed on λ*lac* phages and examined by heteroduplex mapping by Malamy, Fiandt and Szybalski (1972) and Fiandt, Szybalski and Malamy (1972). By annealing DNA from the λ*lac* phage with DNA of a λ*gal* phage carrying IS1, they showed that short insertions of about 800 bases in *lacZ* all represent IS1. A longer insertion, then measured at about 1200 base pairs and now thought to be about 1300 nucleotides long, represents a new insertion sequence, IS3.

The DNA comprising the insertion sequences has been isolated in two

---

when complementary strands are annealed; this is the same result observed whenever insertions are not of the same sequence. (*d*) Insertions at the same site in opposite orientation provide the only annealing sequences when two *l* strands (or two *r* strands) are hybridized. (*e*) When the same insertion is found at two sites in reverse orientation, again it represents the only duplex formed in annealing of two *l* or two *r* strands.

series of experiments. Schmidt, Besemer and Starlinger (1976) took advantage of phages carrying IS1 or IS2 in opposite polarities to isolate the insertion sequences as the sole regions able to form duplex DNA when the denatured phage genomes are annealed. Gel electrophoresis and electron microscopy of the isolated IS DNA showed that IS1 is 820–850 base pairs long and IS2 has a length of 1300–1350 base pairs. Ohtsubo and Ohtsubo (1976) developed a method for the isolation of inverted repeat sequences, essentially relying upon their ability to renature very much more rapidly than other sequences. Applied to the insertion sequences, this identified lengths for the duplex stems of the isolated IS DNAs of 710–730 base pairs for IS1, 1220–1270 base pairs for IS2, and 1290–1340 base pairs for IS3. The availability of isolated DNA molecules corresponding to the insertion sequences may provide a useful probe for locating them readily in genetic elements and also may make possible the determination of their sequences by direct means.

The isolation of insertion sequences by virtue of their polar effects in the *gal* operon implies that they must carry signals that influence gene expression. In at least one case polarity has been shown to result from the premature termination of transcription and it is possible that this applies to all instances of this phenomenon (see Adhya, Gottesman and De Crombrugghe, 1974). This premature termination appears to take place at sites subject to the action of the rho termination factor; rho-dependent termination is suppressed by certain bacterial mutations and a study of the effect of IS sequences in such mutant hosts may lead to definition of the types of control signal that they carry. For example, Reyes, Gottesman and Adhya (1976) have shown that certain bacterial mutations able to suppress polarity are effective upon IS1 but not IS2 or IS3; and the polarity of an insertion probably of the IS2 type is suppressed by the *N* function of phage lambda, which is an antiterminator (see Chapter 4). Further experiments of this type may resolve the relationship between the control signals carried on the various insertion sequences and their connection with the signals utilized in the bacterial cell.

What is the source of these insertion sequences? In a search for small plasmids carrying just these sequences, Saedler and Heiss (1973) were unable to find any covalently closed circles of appropriate size in E.coli. But when denatured bacterial DNA was annealed with the DNA of λ*gal* phages carrying either IS1 or IS2, the extent of hybridization was greater than that shown with control phages lacking the IS sequences. The increase corresponds to the presence of about 6–9 copies of IS1 and 4–6 copies of IS2 in the chromosomes of the E.coli K12 strains that were used. This demonstrates that these sequences are present in bacterial DNA, although we do not yet know how much variation there may be in the number of copies present in different strains and in the sites that are occupied. But it seems likely that IS sequences may be inserted into DNA at any one of a large number of locations and, indeed, may be mobile in the sense that it is possible to identify sites that have gained or lost these sequences. The nature of this mobility is quite unknown. The annealing seen between copies of an IS found at different sites implies that the same linear sequence always is maintained; this demonstrates that there is no circular permutation of the insertion sequence. Thus if insertion and excision involves a circular form of the IS, a specific site on it must be involved in the reaction.

The idea that IS elements may be active in recombination is supported by the identification of these sequences in F and R plasmids at sites that are likely to be implicated in recombination events. Both IS2 and IS3 occur on the sex factor; IS1 is not present. The locations of the insertion sequences are shown in Figure 2.25 and their coincidence with the sites at which insertion and excision appears to have taken place in several F′ factors immediately suggests that they are implicated in genetic exchanges between plasmid and bacterial DNA.

A role for IS2 in recombination between plasmid and chromosome was suggested by an analysis of factor F13. This factor possesses all the sequences of the sex factor itself and thus appears to have been generated by a type II excision; it has a sequence of 290 kb of chromosomal DNA inserted at the point 17.6 F. Hu, Ohtsubo and Davidson (1975) demonstrated that F13 carries two copies of IS2: one occurs on the plasmid DNA at 16.3–17.6 F and had previously been known as the sequence $\epsilon\zeta$; the second, lying in the same orientation, occurs at the other end of the bacterial insertion, that is at 289.9–290.2 B. Thus a copy of IS2 lies at each junction of plasmid and bacterial DNA.

Earlier electron microscopic analysis of F factors showed that a sequence of 1300 bases is repeated at two positions of the sex factor, 93.2–94.5 F and 13.7–15.0 F. This was originally termed $\alpha\beta$. By hybridizing the DNA of F8-33 with the denatured strands of a lambda phage carrying IS3, Hu et al. (1975b) demonstrated that $\alpha\beta$ corresponds to IS3. The two copies of IS3 are located at sites implicated in insertion and excision; several factors appear to have been inserted at the IS3 located at 93.2–94.5 F, and the deletion factor F(Δ0-15) appears to have been deleted for the sequence lying between this IS3 and the copy at 13.7–15.0 F.

The common occurrence of IS3 in bacterial DNA is emphasized by the analysis of factor F13 performed by Hu, Ohtsubo and Davidson. In addition to the two IS3 elements on the plasmid sequences, F13 carries a further three copies in its bacterial DNA, one in the same orientation as the two elements of the plasmid and two in the reverse orientation. The

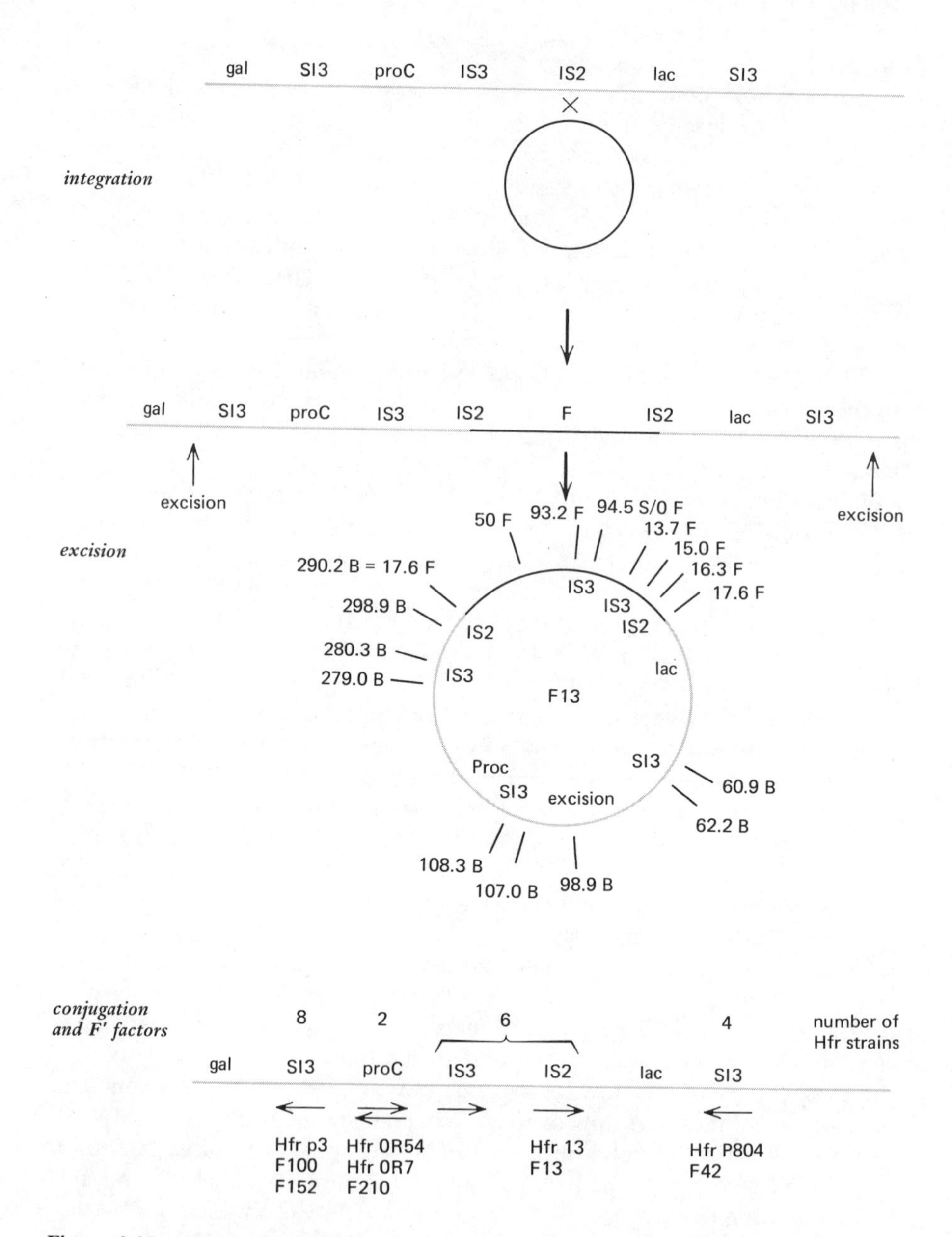


**Figure 2.27:** reconstruction of the chromosome region represented in F13. The factor carries 290,200 base pairs of bacterial DNA, with the IS sequences and genetic markers indicated; from the structure of the factor it is possible to deduce which sites must have been involved in insertion and excision, as illustrated. Note that the bacterial map has been drawn in the reverse direction from the usual convention (in order to retain the same orientation of

structure of F13 is illustrated in Figure 2.27. Several other F′ factors carry some of the same bacterial sequences that are present in F13. The factor F42-1 (essentially the same as F42) carries the bacterial sequences 60.9 B −8.2 B of F13, replacing the sequences 0–2.8 F. This suggests that the parental Hfr was generated by an insertion that took the form of a genetic exchange between the IS3 at 93.2–94.5 F of the plasmid and the IS3 at position 60.9–62.2 B on the bacterial chromosome as measured on the coordinates of the F13 bacterial sequences; F42 must then have been generated by a type I excision involving the sites 2.8 F and 8.2 B. Similar mapping of the bacterial sequences present on the factors F100 and F152 suggests that their parent Hfr was formed by genetic exchange between the IS3 at 93.2–94.5 F of the sex factor and the IS3 present at the bacterial position corresponding to the coordinates 107.0–108.3 B of F13. The excisions that generated F100 and F152 must have involved exchange between the site 2.8 F and bacterial sites beyond the marker *gal* (which is carried on both factors).

Another line of evidence consistent with the idea that plasmid insertion takes place at IS sequences is provided by analysis of the Hfr strains that have their origins of transfer within the bacterial region that is carried on F13. The positions of the origins and the directions of transfer of twenty strains are summarized in Figure 2.27. The origins are clustered at 4–5 sites and in all except two of the strains the position of the origin (within the limits defined by genetic mapping) and the direction of transfer can be accommodated by the idea that a sex factor has been integrated in the appropriate orientation by an exchange occurring at an IS2 or IS3 sequence of the bacterial DNA. The analysis by heteroduplex mapping of F′ factors derived from some of these Hfr strains also is summarized and in each case supports the assignment of the IS sequence on bacterial DNA as the insertion site and also demonstrates that an appropriate IS sequence on the plasmid was involved in the exchange (see above and Figure 2.25).

---

the sex factor as that shown in Figure 2.25). When IS sequences are present in the reverse orientation from those of the factor as represented in Figure 2.25, they are shown as SI2 or SI3. From the properties of other F′ factors carrying parts of the same bacterial region carried by F13, it is possible to determine the sites at which integration of F must have taken place in the parent Hfr strains. These sites are summarized in the lowest part of the figure, which shows the direction of transfer of the parent Hfr and indicates the F′ plasmids upon which this analysis is based. In each case many other Hfr strains also appear to start transfer from the same point, and the number utilizing each site is indicated. Several Hfr strains appear to represent insertion at either the IS2 site utilized in Hfr 13 or at the IS3 site immediately to its left; in these cases it is not known which site is involved. Only two strains (Hfr OR54 and OR7) involve insertion at sites other than IS2 or IS3. This map is not to scale. Data of Hu, Ohtsubo and Davidson (1975) and Hu et al. (1975b).

In summary, three sets of data suggest that IS2 and IS3 sequences on bacterial DNA represent hotspots where the sex factor is inserted by means of a recombination that takes place with a matching IS sequence on the plasmid. With F′ factors derived by type II excision, both the bacterial-plasmid junctions generated by insertion can be directly visualized; in factor F13 each has an IS2 sequence. (In factor F14 each junction has a copy of the sequence 2.8–8.5 F; the origin of the duplication and its relationship to the insertions that take place at IS sequences is not clear.) With F′ factors derived by type I excision, one of the junctions represents the original insertion and the other is generated by excision; in the factors F42, F100, F152, the junction representing insertion occurs at the location of a plasmid IS3 sequence and (by comparison with the bacterial sequences of F13) the site involved on bacterial DNA may be inferred to be an IS element. With Hfr bacteria the origin and direction of transfer seem nearly always to coincide with the position and orientation of an IS sequence. This suggests that the recombination event involves exchange between plasmid and bacterial IS sequences aligned in the same direction.

What systems may be responsible for insertion and excision of F factors is not known; nor is there evidence on whether all or only part of the IS sequence is required for the recombination. Genetic exchange between the two IS sequences at the termini of an integrated sex factor presumably can release the F factor from the Hfr chromosome; how variations in the frequencies of insertion and excision at the hotspots may be caused remains to be determined. In an F′ factor derived by type II excision, recombination between the two IS sequences at each bacterial-plasmid DNA junction should release the plasmid from its bacterial sequences; a low rate of such reversion appears to occur with F13 (implying exchange between IS2 sequences) and a somewhat higher rate is seen in F14 (due to reaction between the two copies of the 2.8–8.5 F sequence). Excision of F′ factors is not well defined but does not appear to be correlated with the presence of IS sequences; it is therefore possible that type I and type II excisions rely upon systems different from the insertion/excision at hotspots.

Two qualifications to the model of insertion via recombination between IS sequences are worth emphasizing. The apparent mobility of the insertion sequences in bacterial DNA (see also Chapter 3) suggests that in different strains of E.coli the hotspots may be located at different positions; indeed, this may explain why two of the Hfr strains shown in Figure 2.27 do not represent insertion at an IS sequence present in the bacterial strain from which F13 was derived–an IS sequence might have been present in the strain in which they were generated. And it is not proven that all the copies of IS2 or of IS3 are identical. Hybridization analysis demonstrates that the sequences annealing together are close enough to complementarity to form a duplex under the conditions used; but it remains possible that different copies of an IS sequence might possess sequences that although closely related enough to anneal nonetheless are not identical.

## Chromosome Transfer and Recombination

### *DNA Synthesis During Conjugation*

Transfer of the bacterial chromosome from donor to recipient is under the control of an integrated sex factor and commences at a site, the transfer origin, located within the factor. As Figure 2.12 shows, a break must be made in the circular chromosome at this origin so that the chromosome can be transferred as a linear structure. Only the origin and whatever part of the sex factor lies between it and the chromosomal material is transferred early; most of the factor is transferred at the tail of the chromosome. Hfr cells remain viable after mating, of course, and so must have retained one complete copy of their chromosome while transferring another copy (or part of one) to the recipient cell. Accounting for the source of the chromosome that is transferred by the donor cell thus has been a critical question in resolving the mechanism of transfer.

Two general classes of model can be constructed for chromosome transfer, as first formulated by Jacob, Brenner and Cuzin (1963). One is to suppose that the chromosome to be transferred may have been synthesized previously in the donor, so that one of the two chromosomes in the cell is transferred to the recipient and one remains. This may leave unresolved the problem of what signal initiates the replication that produces a chromosome copy for transfer, since the speed with which initiation of transfer follows upon mixing of donor and recipient populations presumably would demand that this event precede the start of mating.

An alternative is to suppose that transfer may require concomitant replication in the donor. In its original form this model postulated that one of the replica chromosomes might be transferred into the recipient as it is synthesized; more recently another idea has been suggested, that only one of the strands of the DNA chromosome is freed by replication for passage to the recipient. These models both imply that replication for transfer is controlled by the sex factor and not by the chromosomal replication system. Contact between donor and recipient presumably activates the F replication system and initiates DNA synthesis at the origin in the F factor (the origin for chromosomal replication remaining quiescent). Such a system could be related topologically to the architecture of the cell so that the replication drives donor genetic material (whether double or single stranded) through the connecting bridge into the recipient cell.

Several approaches have been used to try to determine the mechanism of chromosome transfer. We can consider the questions that are asked by each experimental design in four classes. By preferentially inhibiting DNA synthesis in one of the partner cells, it is possible to ask whether DNA synthesis is necessary in either donor or recipient during conjugation. A clear conclusion seems to be possible that DNA synthesis is not needed in the recipient; and replication appears to be necessary in the donor cell during conjugation. By using donor cells whose DNA is labeled with heavy or radioactive isotopes, it is possible to ask whether the transferred DNA consists of preexisting donor duplex molecules, duplex molecules consisting of one preexisting strand and one strand newly synthesized during the mating, or a single old or newly synthesized strand. Because only the transfer of duplex DNA was considered when early experiments were performed, some of these are able only to exclude the proposal that a preexisting duplex is transferred and cannot distinguish between transfer of a hybrid duplex (one old and one newly synthesized strand) and transfer of a single preexisting strand. By showing that a preexisting duplex is not transferred, such experiments imply that replication must be implicated in conjugation, although they do not reveal its role. The state of donor DNA during passage cannot be determined directly; but by isolating donor molecules after their transfer to recipient cells, it is possible to ask whether they represent both strands or only one strand of DNA. The evidence is clear here that the donor DNA in the recipient represents the genetic information of only one donor strand and chemically is in the single strand state. A final question is whether the single stranded donor DNA is integrated as such into the recipient chromosome or whether it is first converted to a duplex state by synthesis in the recipient of its complement; since replication in the recipient does not appear to be necessary for conjugation, it is probable that the single donor strand replaces the homologous recipient strand in the chromosome.

Many experiments have implicated DNA synthesis in conjugational transfer, but the conclusions drawn from them have not been consistent, due principally to some of the assumptions that were made at the time in interpreting the results. In the early experiments of Jacob, Brenner and Cuzin (1963), Hfr cells containing DNA labeled with heavy and radioactive isotopes were mated in light nonradioactive medium with $F^-$ bacteria possessing light nonradioactive DNA. After mating, the radioactivity present in DNA in the recipient cells identifies the donor material that has been transferred. When DNA was recovered from the recipients and centrifuged on a density gradient, the radioactive label was found at a hybrid position corresponding to one heavy and one light strand instead of at the original heavy position representing two heavy strands of donor DNA. What this experiment demonstrates directly is that the two strands of the DNA constituting the donor chromosome before conjugation have been separated by the time that donor strands were recovered from the recipients. Because only a short time (20 minutes) was allowed for mating, thought to be too brief to allow integration and replication of a duplex DNA molecule after transfer to a recipient, the conclusion inferred from these experiments was that replication of DNA must take place in the donor cells as a prerequisite for transfer. Of course, this leaves open the question of whether replication of any chromosome segment is simultaneous with its transfer, and provides a driving force for chromosome passage, or whether it may precede transfer by a brief period. It was assumed in interpreting these experiments that donor DNA is transferred in duplex form to the recipient; these results were therefore taken to imply that one of the daughter chromosomes produced by replication in the donor cell passes to the recipient. However, these results are consistent also with an alternative model supported by more recent research: replication takes place in the donor cell but only a single preexisting donor strand is transferred to the recipient, where it is converted into duplex form (either by synthesis of a complement or by insertion into the recipient chromosome, possibilities not resolved in this experiment).

The conclusion that a newly replicated (duplex) daughter chromosome is transferred from donor to recipient was inferred also from the results of Gross and Caro (1966), although some of their experiments also are consistent with the transfer of a single strand. In these experiments the DNA of Hfr cells was labeled with $^3$H-thymidine before or during conjugation; and DNA synthesis was prevented in the $F^-$ recipient by using an adenine requiring strain starved for adenine, so that the only incorporation of the label should take place in the donor cells. By using the rod shaped cells of E.coli K12 as the donor strain, with the ovoid shaped cells of E.coli C as the recipient strain, the two parental types could be distinguished morphologically, making it possible to use autoradiography to estimate the gain of radioactive label by the recipients. Because E.coli DNA replicates semiconservatively, transfer of a fully labeled culture to nonradioactive medium for just one generation produces chromosomes with one labeled and one unlabeled strand (see Volume 1 of Gene Expression); when such donors are mated with recipients in unlabeled medium, they transfer radioactivity to only half the number of cells that become labeled in crosses with donors both of whose DNA strands are labeled. But the amount of radioactivity gained by each cell that becomes labeled is the same in both experiments. This demonstrates that only one of the two previously existing parental donor strands is transferred to the recipient. This observation is again consistent either with the original interpretation that one of two daughter chromosomes is transferred or with the concept that only one of the donor chromosome strands passes to the recipient.

Results that do not appear consistent with both these models, however, were obtained in two other experiments. When labeled donor cells were mated with unlabeled recipients in the presence of $^3$H-thymidine, twice as many grains were found over the recipient cells as when the mating was performed in unlabeled medium. The obvious interpretation of this result is that replication takes place during mating and that one of the two daughter chromosomes (consisting of one old and one newly synthesized strand) is transferred to the recipient. Another experiment taken to support this conclusion was provided by mating unlabeled donor and recipient cells in nonradioactive medium; the addition of $^3$H-thymidine 20 minutes after the start of mating caused an immediate appearance of label in the recipients. To accommodate these results in a model postulating transfer of only one donor strand, it would be necessary to suppose that either a preexisting single donor strand or a newly synthesized single donor strand may thus provide the material transferred to the recipient. Of course, it is difficult to be definite in interpreting these experiments because the recipient cells were starved for adenine, a treatment that may affect other processes as well as DNA synthesis, possibly including their ability to accept donor DNA.

What these experiments clearly demonstrate is that the two preexisting strands of the donor chromosome have been separated when donor material is examined in the recipient under conditions intended to exclude replication after transfer; this constitutes good inferential evidence that DNA replication is required in the donor cell for conjugation. Experiments in which direct attempts have been made to show that inhibition of DNA synthesis in donor cells prevents chromosome transfer have in general yielded results consistent with this conclusion. By using mating pairs derived from strains that are sensitive or resistant to the inhibitor of DNA synthesis, nalidixic acid, for example, Fisher and Fisher (1968) showed that chromosome transfer is prevented by inhibition of DNA synthesis in the donor strain but is not influenced by the ability of recipients to synthesize DNA. A general problem in the use of drugs that inhibit DNA synthesis, however, is that it is not always clear that any inhibition applies solely to DNA synthesis (other cellular processes also may be influenced by the protocols that are used). Although consistent with the conclusion that DNA synthesis is needed in and only in the donor, such experiments therefore do not provide a satisfactory proof of this model.

Experiments in which DNA synthesis is halted by incubating temperature sensitive mutants at high temperature should in principle be more satisfactory. But it is necessary first to identify suitable bacterial mutants. An experiment reported by Bonhoeffer et al. (1967) which used a temperature sensitive replication mutant appeared to suggest an alternative role for replication: DNA synthesis seemed to be necessary in the recipient rather than in the donor parent. However, it has since become apparent that this mutation causes inhibition of replication of the chromosome under its normal control system, but does not prevent the synthesis of DNA under control of the sex factor replication system; recipients with this mutation appear to be defective in the formation of recombinants not because they are unable to replicate the donor DNA, but because the mutation inhibits integration of the transferred material (see Cuzin, Buttin and Jacob, 1967; Curtiss, 1969; Stallions and Curtiss, 1971). In order to find a bacterial mutant that can be used to demonstrate the need for DNA synthesis in the donor, it is therefore necessary to identify a gene whose product is essential for the replication directed by the F factor as well as for bacterial replication. Now that the enzymes involved in the various stages of DNA synthesis are better characterized, with defined mutations available at many loci (see Volume 1 of Gene Expression), it should be possible to obtain appropriate mutants with which to perform a definitive experiment to confirm that replication is necessary in the conjugating donor cell. To test the role of DNA synthesis in the recipient cell is more difficult, because this may be less demanding in its requirements for bacterial functions (see below).

The predictions made by three models for conjugational transfer are illustrated in Figure 2.28. The first model, suggesting that one of two previously existing duplex chromosomes is transferred from donor to recipient, clearly is refuted by data showing that only one of two preexisting strands enters the recipient. These data do not resolve the other two models, however, both of which require DNA synthesis in the donor: the second postulates transfer of a newly synthesized duplex chromosome (consisting of one newly synthesized and one preexisting strand); and the third sees transfer of only a single preexisting strand.

Transfer of a newly synthesized duplex chromosome would imply that DNA replication takes place by the same mechanism that characterizes the semiconservative replication usually occurring in each division cycle. This takes place, of course, by discontinuous synthesis of DNA; short

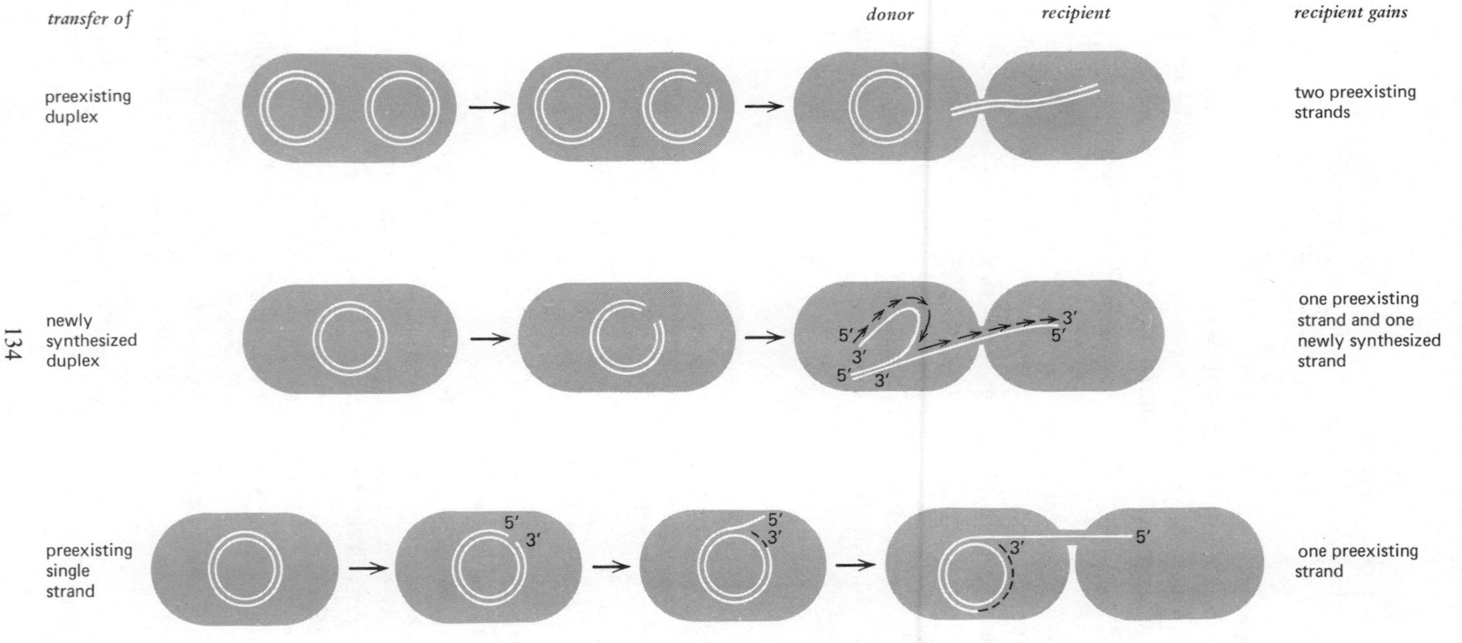


**Figure 2.28:** models for DNA synthesis in conjugation.

lengths of DNA, on average about $10^6$ daltons in size, are synthesized from 5′ to 3′ on each of the preexisting strands as the chromosome unwinds. Synthesis of each discontinuous fragment starts with a primer consisting of RNA which is removed before the fragments are linked into a covalent strand of newly synthesized material. Replication thus requires whatever enzyme synthesizes the RNA primer, the replicase that extends this primer from its 3′ end, any enzyme involved in filling gaps left between adjacent fragments, and the polynucleotide ligase that undertakes the final step of sealing the single strand breaks. Since all known DNA polymerases are capable only of extending DNA chains from 5′ to 3′, utilizing a primer with a free 3′-OH end, it is clear why such a mechanism must be invoked for the DNA strand that grows overall in the direction 3′ to 5′; short fragments must be synthesized from 5′ to 3′ as the duplex chromosome unwinds and must subsequently be linked together. It is not obvious why discontinuous synthesis should apply to the other strand, which grows overall in the same direction, 5′ to 3′, in which synthesis occurs; it is not certain beyond doubt that this strand is synthesized discontinuously, but the weight of evidence favors the model that both strands grow discontinuously rather than the alternative that one strand grows discontinuously and one continuously. Semiconservative replication is discussed in detail in Volume 1 of Gene Expression.

The last model shown in the figure is the rolling circle constructed by Gilbert and Dressler (1968); this takes advantage of the circular structure of bacterial DNA to argue that the introduction of a nick in one strand should allow the free 3′-OH end that is generated to be used as a primer for synthesis of a new strand. As the new strand is synthesized, it displaces the original parental strand on the other side of the nick, thus freeing a sequence of parental DNA with a free 5′ terminus. This model is described as the rolling circle because it can be visualized by imagining that the chromosome rolls round counterclockwise as the new strand is synthesized to displace the preexisting outer strand. Because the new strand is growing in the direction 5′ to 3′, it could be synthesized continuously by simple extension of the primer 3′-OH terminus; the original form of the model postulated that it is this continuous growth which is responsible for displacing the parental strand, but of course it is possible to consider alternatives in which this strand grows by discontinuous growth. If the new strand grows continuously round the circle, the only enzyme needed for its synthesis would be a DNA polymerase capable of 5′ to 3′ strand extension. If a complement were to be synthesized for the strand that is displaced from the circle, discontinuous synthesis becomes necessary since the overall direction of its growth would be 3′ to 5′; the full panoply of enzymes involved in semiconservative replication would be needed for this synthesis. These features of the rolling circle model are illustrated in Figure 2.29. Applied to conjugation, the rolling circle model could accommodate transfer of either a duplex or single strand of donor DNA; if a single strand is transferred, the topography of the growing point may be different from that of a semiconservatively replicating chromosome and this might permit continuous growth of the strand around the circle even if this in fact is not possible in semiconservative replication.

One prediction of this model is that it should be possible to produce a tail of single stranded DNA containing more than one genome's worth of information; for when the circle has rolled round once completely, the continuation of synthesis for a second revolution identical to the first can simply release another length of chromosome, and so on ad infinitum so that in each cycle the strand that is being synthesized displaces the strand that was synthesized in the preceding cycle. This model is supported by a genetic observation implying that more than one round of transfer can take place if mating is not interrupted. By selecting recipients for a proximal marker, Fulton (1965) found that two types of recombinant genotype occurred with respect to unselected distal markers; one appeared to have resulted from entry of the proximal marker before the distal marker (the usual situation), but the other appeared to result from entry of the proximal marker after the distal marker. This of course is precisely the situation predicted to occur if a rolling circle continues through a second cycle.

The interpretation originally drawn from the results of Jacob, Brenner and Cuzin (1963) and Gross and Caro (1966) supposed that replication is initiated in the donor at the start of mating, presumably at the origin in the sex factor. This was thought to generate two replica chromosomes, each of which contains one preexisting and one newly synthesized strand; the transfer of one of these chromosomes to the recipient as it is synthesized, as shown in the second model in Figure 2.28, would be consistent with the experimental data (for review see Hayes, 1968). The rolling circle model illustrated in the third part of the figure also is consistent with these data on transfer of preexisting strands; this model supposes that only a single preexisting strand of DNA is transferred from donor to recipient. According to this model, the donor DNA synthesis that is necessary for conjugation takes place to maintain the duplex structure of the chromosome that remains in the donor cell; that is, one parental strand is displaced and transferred to the recipient and DNA synthesis is necessary to replace it in the donor chromosome. This model predicts that newly synthesized DNA should not be transferred to the recipient (at least not unless conjugation proceeds for long enough to allow a second genome to be transferred); if this model is correct, therefore, data implying the transfer of newly

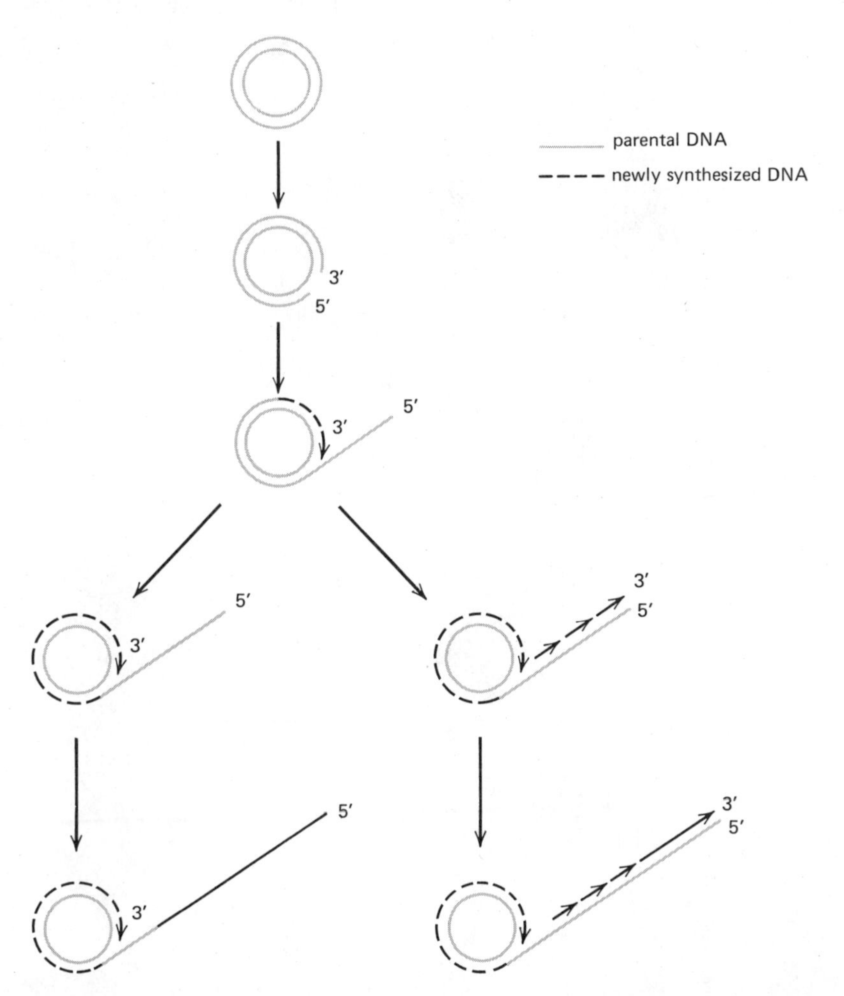


**Figure 2.29:** the rolling circle model for replication. A nick is introduced in the outer strand of the circular duplex. Synthesis of new DNA starts at the 3′ end, which is shown to grow continuously (imagine that the inner circle is revolving in a counterclockwise direction). The left part of the figure shows displacement of the parental strand with a free 5′ end, at first less than a genome in length and then containing more than one genome of information as a second revolution commences. The right part of the figure shows the need for discontinuous synthesis if a complement is synthesized for the displaced strand; at first short discontinuous segments are produced and subsequently these are joined covalently together. After Gilbert and Dressler (1968).

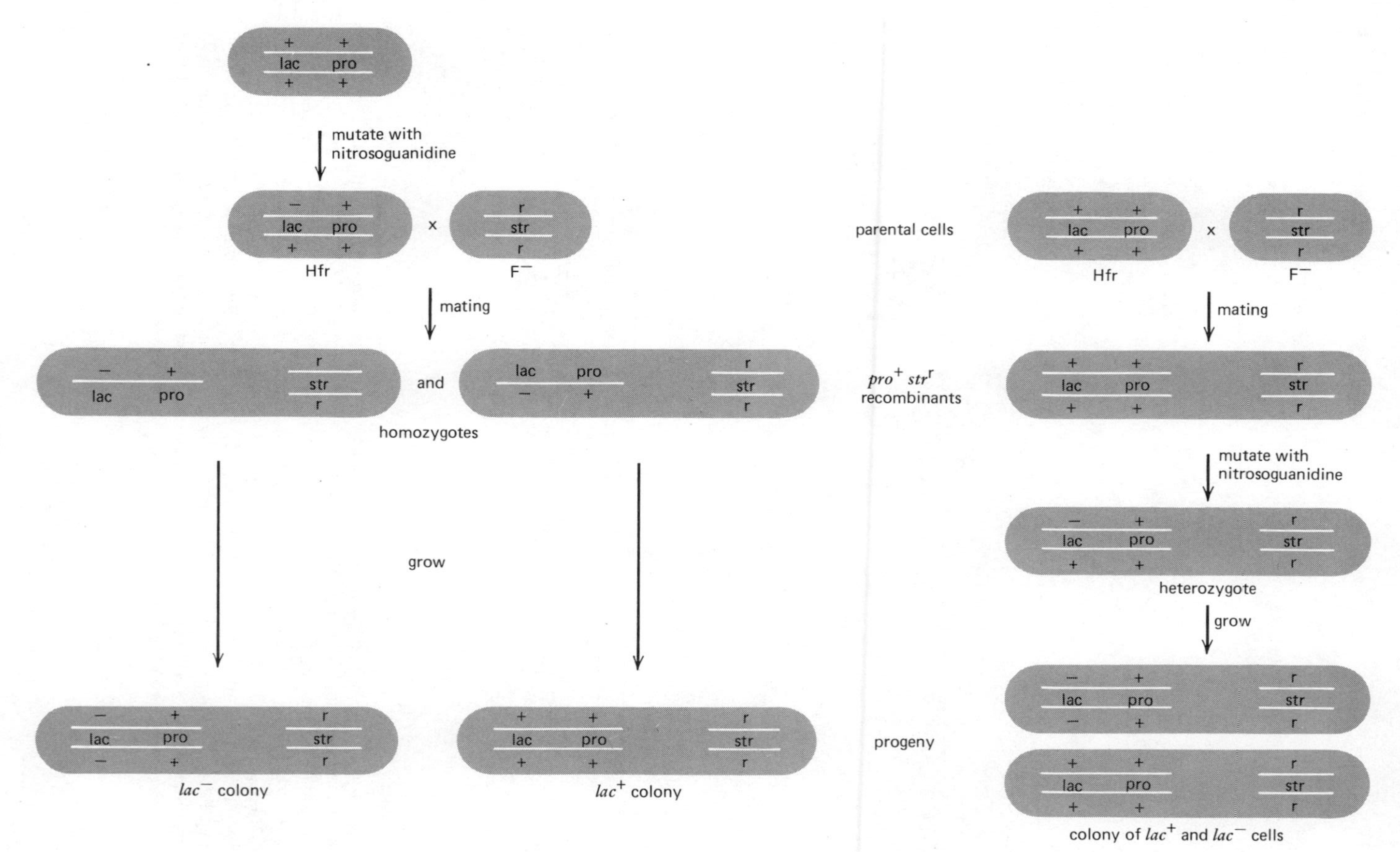
mutate with
nitrosoguanidine
Hfr
F⁻
mating
and
homozygotes
grow
lac⁻ colony
lac⁺ colony
parental cells
pro⁺ strʳ
recombinants
progeny
heterozygote
colony of lac⁺ and lac⁻ cells
lac
pro
str

synthesized material must be in error. Another model consistent with all these results, of course, is to suppose that a duplex daughter chromosome (generated by a rolling circle or movement of a conventional semiconserva- tive growing point) is indeed transferred from the donor as it is synthe- sized, but that one strand is degraded immediately upon entry into the recipient; however, this is a wasteful and therefore unpleasing model.

### *Transfer of a Single Preexisting Donor Strand*

Genetic analysis shows that the information of only one strand of the donor chromosome is transferred to the recipient. Vielmetter, Bonhoeffer and Schutte (1968) used a system in which Hfr cells carried the alleles *lac*$^+$ and *pro*$^+$ and the recipients had deletions of these loci but carried a *str*$^r$ allele. Recombinant clones were then selected for the *pro*$^+$ *str*$^r$ genotype and then scrutinized for the allele present at the *lac* locus (that is the gene introduced on the donor DNA), as illustrated in Figure 2.30. When mu- tants of the type *lacI*$^-$ were induced by treating the recipients with the mutagen nitrosoguanidine immediately after the *lac* genes had been intro- duced, all the recombinants that contained *lacI*$^-$ segregants also con- tained *lacI*$^+$ segregants; this suggests that nitrosoguanidine mutates only one strand of DNA so that the progeny of a treated cell may contain *lacI*$^-$ cells representing the mutated strand and *lacI*$^+$ cells representing the nonmutated strand. But when *lacI*$^-$ mutants were induced in the donor cells immediately before transfer, all the recombinants that contained *lacI*$^-$ mutants were homozygous for this allele; this suggests that the information of only one donor strand is transferred to the recipient during conjugation.

Biochemical analysis of the strands of the sex factor or chromosome transferred during mating suggests that only one strand physically is transferred to the recipient. The DNA of phage λ is a convenient probe to use for chromosomal sequences because its strands can be distinguished by their reaction with poly(U,G); because the *l* (left) and *r* (right) strands bind different amounts of poly(U,G) they band at different densities in CsCl gradients after reaction with the polynucleotide. When a radioac- tively labeled Hfr donor carrying a prophage λ is mated with an unlabeled

---

**Figure 2.30:** transfer of donor gentic information. Hfr *lac*$^+$ *pro*$^+$ cells were mated with F$^-$ *str*$^r$ cells and the *pro*$^+$ *str*$^r$ recombinants were selected. Right: mutation of the recombinants with nitrosoguanidine induces a change in only one DNA strand and thus generates a heterozygote; this gives colonies which contain both *lac*$^+$ and *lac*$^-$ cells. Left: mutation of the donors with nitrosoguanidine before mating changes only one strand of the duplex, so that recombinants are homozygous for either one of the two genotypes; this gives colonies which contain only *lac*$^+$ or only *lac*$^-$ progeny.

$F^-$ recipient that does not carry the phage, zygotic induction causes reproduction and release of phage particles when the prophage enters the recipient. The released phage particles can then be isolated, DNA extracted, and the strands separated on CsCl after reaction with poly(U,G); the position of the radioactive label reveals which preexisting donor strand(s) has been transferred to the recipient. Figure 2.31 illustrates an experiment using this protocol that was performed by Rupp and Ihler (1968) and which shows that a single, specific strand is transferred by either Hfr or F′ donors. With an Hfr donor transferring DNA in the clockwise direction, so that the order of transfer of markers around the prophage is *gal*-λ-*bio,* almost all of the radioactivity in the λ DNA recovered from the recipient is in the *l* strand. With an F′ donor that carries the λ prophage, and transfers the bacterial markers in the order *bio*-λ-*gal,* it is the *r* strand that is labeled in the phage progeny recovered from the recipients. Because the orientation of the λ strands is known, it is possible to conclude from these results that the single preexisting donor strand that is transferred to the recipient always enters with a 5′ end. The same conclusion was suggested by the results reported by Ohki and Tomizawa (1968); when a labeled F′ factor carrying λ was transferred to recipients and the transferred λ DNA assayed by its ability to hybridize with the purified strands of λ DNA, only one labeled strand was recovered from the recipients, corresponding to transfer starting with a 5′ end. These results suggest that transfer is initiated when a nick is made in one of the strands of DNA comprising the sex factor, and the 5′ end that is generated leads the attached strand of DNA into the recipient. The orientation with which the F factor is inserted into the chromosome thus determines which strand of the bacterial chromosome is transferred to the recipient.

Only one of the strands of prophage lambda therefore is transferred to recipients by Hfr or F′ donors. This is consistent with the rolling circle model but does not directly prove that the donor DNA is transferred in the form of single strands. By using minicells of E.coli as recipients, however, it is possible to retrieve donor DNA after mating in a state which presumably is the same as the condition in which it was transferred. Minicells are formed by a mutant of E.coli which undertakes an unusual type of cell division; one product is a normal sized cell but in addition anucleate spheres are formed, each about one-tenth of the volume of a normal cell. The yield of these minicells is about one to three times the number of normal size cells; and because of their small size they can be separated by differential centrifugation on sucrose gradients. They contain RNA and protein in ratios similar to those of normal cells, but possess little or no DNA. All donor mating types can transfer DNA into minicells, where it can immediately be identified since there is no resident

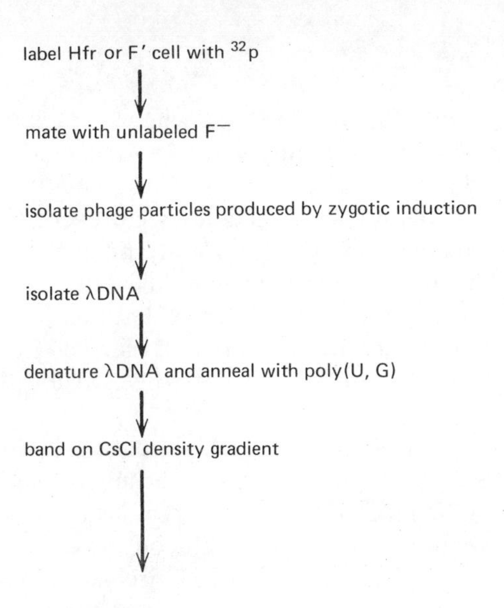


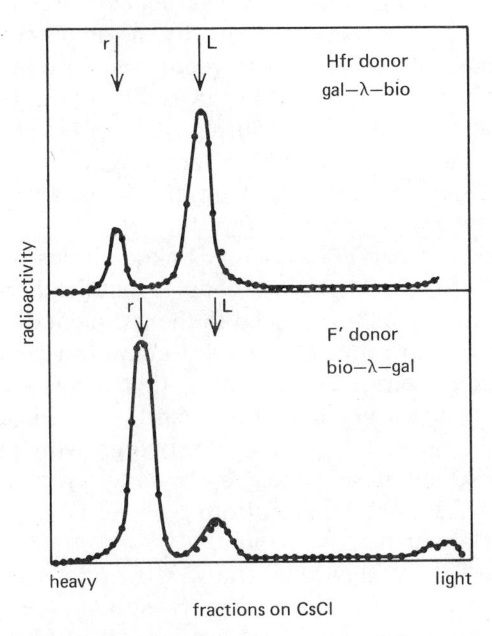


**Figure 2.31:** assay of preexisting donor strands transferred to recipient. When the Hfr donor transfer markers in the order *gal*-λ-*bio*, only the preexisting *l* strand of λ is found in the recipients; when the F′ donor transfers markers in the reverse order, *bio*-λ-*gal*, only the preexisting *r* strand of λ is found in the recipients. Data of Rupp and Ihler (1968).

genome. Cohen et al. (1968) found that when minicells are isolated immediately after a 90 minute mating with Hfr donors they contain donor DNA entirely in the single stranded state; after mating with F′ donors, both single stranded and duplex molecules are present. This suggests that donor DNA is transferred only as single strands; the bacterial chromosome transferred by an Hfr cell remains in this state, but an F′ strand may sometimes be converted to duplex form. The ability of the F′ single strands to reform duplex molecules probably derives from the transfer of the entire genetic element, contrasted with the transfer of only part of the Hfr chromosome.

Only one strand of the sex factor itself should be transferred in matings between $F^+$ and $F^-$ strains, a prediction that has been tested by Vapnek and Rupp (1970). When the DNA of an $F^+$ strain was radioactively labeled before mating, about 20% of the label could be recovered from the recipients in the form of covalently closed circles sedimenting rapidly on alkaline sucrose gradients. As can be seen from the upper part of Figure 2.32, the strands of the sex factor can be separated by their reaction with poly(U,G), after the preparation has been nicked to release linear molecules. The center panel shows that when only the donor cells were labeled before mating, most of the label was recovered in one of the strands obtained from the recipients. As the lower panel shows, when instead the recipients were labeled before and during mating, virtually all the label was found in the other strand of the sex factor. This implies that a single strand of the sex factor is transferred from the donor and that its complement is synthesized as soon as it enters the recipient, after which both strands are closed to generate covalently linked circles of duplex DNA.

One implication of the rolling circle model is that the displaced preexisting strand of donor DNA is transferred to the recipient as fast as the new strand is synthesized to replace it. Under conditions when it takes 90 minutes to transfer the entire chromosome, this corresponds to a rate of synthesis of 50,000 bases per minute. When the E.coli chromosome is replicated during its normal division cycle, two growing points complete one replication in 40 minutes; this corresponds to a rate of movement of just under 60,000 bases each minute. Both types of DNA synthesis therefore take place at about the same rate.

That replication for conjugational transfer is under the control of the sex factor replication system, and does not respond to the system controlling replication of the bacterial chromosome, was first suggested by Jacob, Brenner and Cuzin (1963), who found that acridines exert a specific inhibition on conjugational transfer. If acridine orange is added to mating mixtures at zero time, it prevents transfer; if added at later times its effect is much less pronounced. This is consistent with the idea that

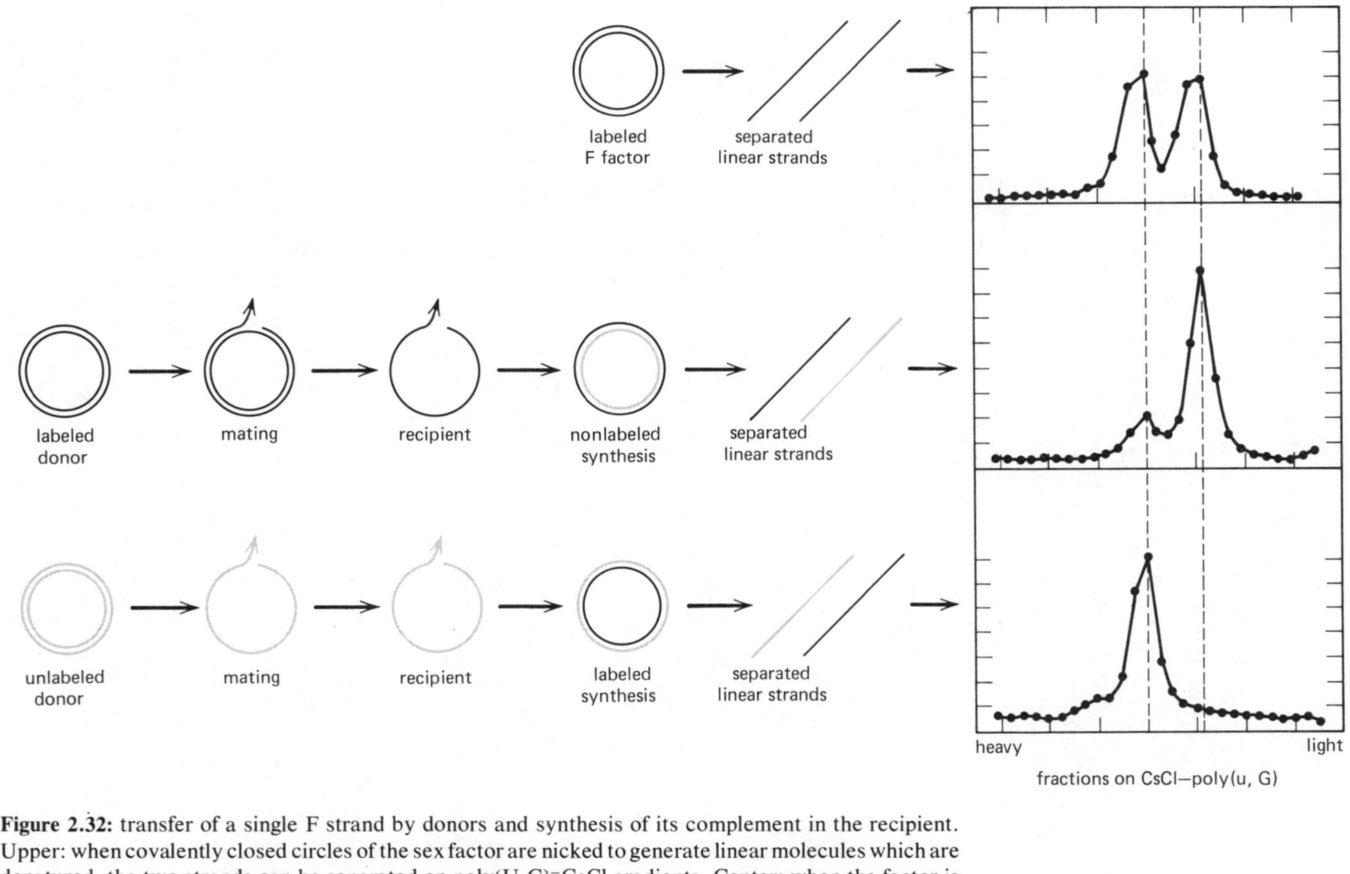


**Figure 2.32:** transfer of a single F strand by donors and synthesis of its complement in the recipient. Upper: when covalently closed circles of the sex factor are nicked to generate linear molecules which are denatured, the two strands can be separated on poly(U,G)$^-$CsCl gradients. Center: when the factor is labeled in the donor before transfer, only one labeled strand is recovered from the recipients. Lower: when the factor is labeled in the recipient during and immediately after mating, only the other strand is recovered in labeled form. Data of Vapnek and Rupp (1970).

acridines act to prevent initiation of replication by the F factor system, but do not inhibit the continuation of a synthetic cycle once it has begun (see Chapter 3).

The single strands transferred by Hfr donors and by $F^+$ or $F'$ donors may suffer different fates upon entering recipients. The DNA strand transferred by the Hfr donor represents part of the bacterial chromosome, possessing only the origin of transfer of the F factor together with any part of the factor lying between the origin and the bacterial material. This strand probably lacks the ability to support autonomous replication and so must be integrated into the recipient chromosome by recombination or otherwise is lost when the cell divides (see below). The DNA transferred by $F^+$ or $F'$ donors, on the other hand, represents a complete genetic element, and therefore is able to synthesize its complement and reform the original sex factor. It is reasonable to suppose that the same origin is used for transfer of a bacterial chromosome promoted by an integrated sex factor and for transfer of autonomous $F^+$ or $F'$ factors; since single strands are transferred in both instances, it is likely that they are generated by the same mechanism, that is the rolling circle. This model thus supposes that initiation of transfer is the same in all donor mating types, with a nick introduced at the origin in the sex factor to allow DNA synthesis from the free 3′ end with transfer into the recipient of the free 5′ end; replication follows the same course for all attached sequences, so that bacterial DNA in an $F'$ factor, or the whole bacterial chromosome in an Hfr cell, is treated as part of the factor. Because the $F^+$ and $F'$ factors are limited in length, however, all their material may readily be transferred to a recipient where the complete genetic unit is reconstituted; the large size of the bacterial chromosome ensures that in most matings only a small part of it enters the recipient, where it must rely upon the option of recombination if it is to contribute to the genotype.

### *Integration of Donor DNA*

The single strand of donor DNA that enters a recipient must recombine with the resident chromosome if it is to be maintained in the cell and used for the expression of genetic information. Two models can in principle be constructed for its recovery: synthesis of the complementary strand might convert it to a DNA duplex which then recombines with the recipient duplex genome; or the single donor strand might directly replace homologous material in the recipient chromosome. Since the DNA transferred by an Hfr donor may represent any part of the bacterial chromosome, and includes only a leading sequence from the F factor, it must usually lack the bacterial origin for initiating replication and presumably is thus unable to support replication proper. Although this does not exclude the possibility that some other system for DNA synthesis might synthesize the complementary strand (an undertaking less complex than semiconservative replication of duplex DNA), the single stranded state of such DNA after transfer to minicells, contrasted with the ability of $F^+$ and F′ factors to support some replication, is consistent with the concept that transferred chromosomal DNA remains unreplicated.

Since experiments in which DNA synthesis is inhibited in either donor or recipient usually have been designed to follow the transfer of DNA, they demonstrate only that replication is necessary in the donor and that it is not needed in the recipient cell for the acceptance of DNA; this casts no implication on whether DNA synthesis is involved in the stages subsequent to transfer, that is, those concerned with recombination. To test whether transferred single strands enter the donor state in the recipient because they displace material in the recipient chromosome or because they are replicated requires mating to be performed under conditions in which donor DNA and recipient DNA both are identified by labels which are distinct from each other and also from any DNA synthesized after mating. Such experiments in general suggest that recombination takes place by insertion of the donor single strands into the recipient chromosome.

By labeling donor DNA with $^{3}H$-thymidine and recipient DNA with the heavy isotope $^{15}N$, Oppenheim and Riley (1966) were able to use mating in light, unlabeled medium to follow the fate of the donor DNA. When DNA was extracted from recipient cells after mating and banded on CsCl density gradients, the radioactive label was found at increased density; this marks the association of preexisting donor DNA with preexisting recipient DNA. Denaturation to yield single strands, however, released the $^{3}H$ label from the density label, showing that covalent linkage had not been established between donor and recipient material. Oppenheim and Riley (1967) confirmed that covalent linkage is not established even after a couple of hours is allowed for mating in liquid culture when growth conditions remain those not supporting bacterial multiplication; but when the culture is diluted to stimulate division, donor material becomes covalently linked to recipient material. This implies that the final stage of recombination, linking the two parental DNA molecules, occurs only under conditions allowing cell growth, a somewhat surprising observation.

In other experiments using both radioactive and density labels, Siddiqui and Fox (1973) obtained data consistent with the idea that a single donor strand is integrated into the recipient, and they too found a lack of covalent linkage between donor and recipient DNA for periods of up to three hours after mating. By using a recipient that is temperature sensitive in replication, they found that donor material apparently associates with the recipient chromosome only when replication is possible. The role of any such replication cannot be determined from these experiments.

Little is known of the processes by which a single donor strand may be inserted into the duplex recipient genome. The first step must certainly be pairing between a short region of the single strand and a complementary sequence in the chromosome; this requires the recipient chromosome to unwind at the region of pairing. Virtually nothing is known about this process; it is not clear how the homologous regions of recipient and donor material may recognize each other (this is a general problem in recombination, for both duplex DNA molecules of procaryotes and chromosomes of eucaryotes), what length is required to establish the initial region of pairing, or what enzyme functions are implicated. The order and nature of subsequent events is not established but two models which illustrate some of the possible pathways for integration are shown in Figure 2.33. These models represent extremes: intermediates with features of both can readily be constructed. However, they illustrate most of the events that may be implicated in recombination (although with either model the order in which they occur may be different from that illustrated in the figure).

Once an initial pairing has taken place, its extension along the chromosome could allow the donor strand to replace one resident strand. This model is shown on the left of the figure. Pairing takes place at a single site where the recipient strand to be replaced is nicked; exonucleolytic degradation of one recipient strand then makes available the complementary recipient strand for pairing with the single donor strand. Observations that markers located more than 10 minutes apart are not linked in conjugation suggest that a single insertion event of this nature usually extends for less than $30 \times 10^6$ daltons ($5 \times 10^5$ bases). Since the implication of interrupted mating experiments is that the donor DNA strand is transferred as a single covalently linked molecule, the limitation on length of the integrated sequence must be due to events in the recipient, perhaps breakage of the donor strand during the recombination process. Enzymes that assist pairing, nicking activities, and nucleases may therefore be implicated in the early stages of recombination.

If the length of donor strand that is inserted is less than the length of the recipient strand that has been removed, whatever gap exists must be filled by DNA synthesis; this demands a repair type function, in which the 3′-OH terminus on one side of the gap is used as a primer for strand extension by a DNA polymerase. Of course, if the length of the inserted donor strand corresponds precisely to the length of displaced recipient strand, this action is not necessary. At this final stage the DNA molecule

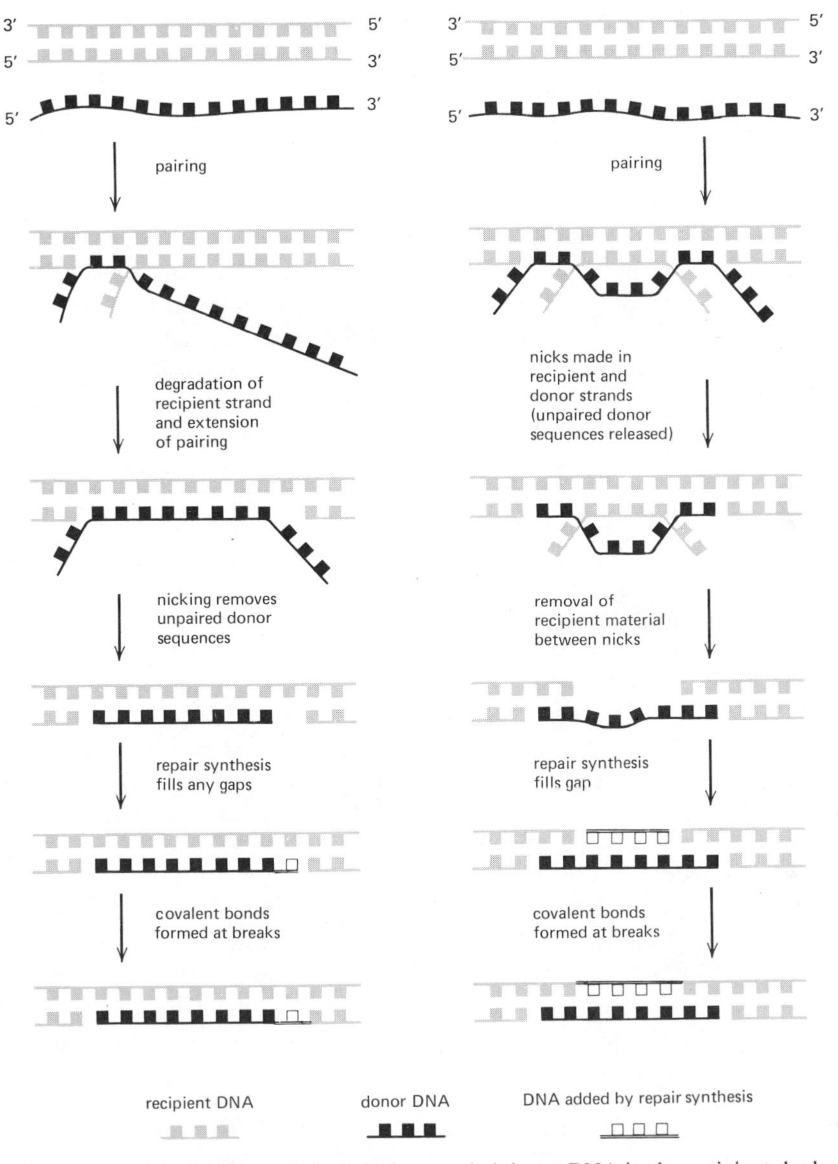


**Figure 2.33:** models for integration of single stranded donor DNA in the recipient duplex. Left: after initial pairing at one site, a recipient strand is degraded and replaced by the donor strand. Right: after initial pairing at two sites, both recipient strands are degraded and replaced by the donor strand.

represents a recombinant in which the recipient and donor strands have not yet been covalently linked since there are single strand breaks at each end of the inserted DNA. This should display 3′-OH and 5′-P termini and so can be linked by the action of the enzyme polynucleotide ligase.

An alternative model is to suppose that initial pairing takes place at two sites on the DNA. In a model proposed by Curtiss (1969) and shown on the right of the figure, the initial reaction is succeeded by removal, between the sites of pairing, of both recipient strands and not just that replaced by the donor strand. This is illustrated by the introduction of nicks with subsequent removal of intervening material but might take place by any of several mechanisms; these include degradation by nuclease action of the two strands separately or by a single enzyme. The net effect of this step is to leave the chromosome with a single stranded region provided by the donor strand. Repair synthesis must then fill this gap; this might depend upon a single DNA polymerase which continuously extends the free 3′-OH end by adding nucleotides one at a time with a specificity dictated by the complementary strand (see Volume 1 of Gene Expression). The final step in producing covalent recombinants is again the sealing of all single strand breaks by polynucleotide ligase.

These two models do not represent exclusive pathways, of course, but are useful in visualizing possible reactions that may be implicated in recombination. And as illustrated, they make different predictions subject to experimental test. The first model, postulating that just one strand of the recipient DNA is replaced by the donor DNA, is closely analogous to the models for the formation of hybrid DNA by which recombination between duplex molecules of DNA probably occurs. These postulate that both DNA duplexes are nicked in one strand at corresponding sites. A single strand unwinds from each nick and base pairs with the complementary sequence that is revealed in the other parental duplex. Nicking and crosswise pairing of the other two strands then generates a recombinant molecule in which duplex lengths derived from different parental molecules are connected by a length of hybrid DNA consisting of one strand from each parent. The formation of hybrid DNA is discussed in detail in Volume 1 of Gene Expression.

If the two parental DNA sequences differ in the region in which the hybrid DNA is formed the two strands may each carry different genetic information. If a point mutation has taken place in one parental molecule, for example, leading to the substitution of one base pair by another, the hybrid DNA will be mispaired at this site with some other combination replacing the usual G-C or A-T pair. Mispaired bases are subject to correction by repair systems that appear to recognize distortions in the usual structure of the double helix and which may excise one strand of the

DNA and replace it with a sequence whose synthesis is dictated by the complementary strand and which therefore forms a perfect duplex. These repair systems are described in Volume 1 of Gene Expression.

Models in which hybrid DNA is formed in recombination (either between duplex DNA molecules or by the single strand insertion illustrated on the left of Figure 2.33) thus predict that when two parental DNA molecules differ at some site, the hybrid DNA for this region will be heterozygous, carrying the information characteristic of both parents. If correction is mediated by an excision-repair system, then the DNA may become homozygous (for either parental marker, depending on which strand is excised and replaced). If repair does not take place before the DNA is replicated, the two strands will separate, each will synthesize the appropriate complementary strand, and the product will be two daughter DNA molecules each of which is homozygous for the information of one parent. Models which postulate that recombination following conjugation takes place by displacement of a recipient strand by the homologous donor strand therefore carry the genetic implication that the recipient cell should be heterozygous for the donor markers that it gains; this can in principle be tested experimentally by determining whether mixed clones in fact are derived from a mating. Of course, the frequency with which a heterozygous duplex remains in this condition may depend upon the efficiency of the repair systems of the cell. A biochemical prediction of this model is that the donor strand should become associated with preexisting recipient DNA, forming a duplex region which consists of one strand of the DNA preexisting in each parent with little or no newly synthesized material. These are general predictions, of course, which are not affected by details of the particular model invoked to explain the formation of the hybrid DNA; although Figure 2.33 illustrates a single initial pairing site for the formation of hybrid DNA, it is possible to construct a model in which pairing starts at two separated sites and then spreads between them (that is, with the initial sequence of events depicted on the right, but with pairing between donor and recipient single strands occurring between the sites, instead of the removal of the donor duplex depicted).

Models which propose that the single donor strand is converted into a duplex before recombination make the genetic prediction that the recombinant chromosome should be homozygous in sequence, with both strands carrying donor information at the site of an insertion; the corresponding biochemical prediction is that the preexisting single donor strand should have as its partner a strand newly synthesized in the recipient after mating. Precisely the same predictions are made by the model shown on the right of Figure 2.33, which differs only in its timing of events; this model suggests that pairing takes place at two sites between donor and recipient DNA; when the recipient strand is removed between the sites of pairing, the remaining single donor strand is converted into a duplex by repairlike synthesis of complementary DNA. This model thus essentially proposes that pairing and insertion precedes repair replication rather than the reverse. The model is therefore rendered less likely by the same arguments that mitigate against models proposing replication of the donor single strand upon entry into the cell.

But the model is useful in illustrating a situation that may arise when the donor DNA carries sequences that have been deleted from the recipient genome. In this situation, generation of a recombinant molecule containing the additional donor sequences clearly demands that pairing takes place independently at sites on either side of the deletion; if recombination takes place by the formation of hybrid DNA, the extension of pairing across the deletion will leave a loop of donor DNA representing the additional donor sequences extruded at the site of deletion. To generate a covalently linked duplex molecule containing these sequences therefore requires the introduction of a nick in the nondisplaced recipient strand; this produces the same structure as that shown in the right side of Figure 2.33 after the removal of recipient material—these recipient sequences will simply never have been present in the duplex of the resident genome. Extensive gap filling must therefore take place to synthesize the complement to the additional donor sequences. Some variation in the mechanisms of recombination can therefore be visualized, depending on the relationship between donor and recipient sequences.

Zygotic induction provides a phenomenon which is at first sight difficult to reconcile with the idea that single donor strands are not converted to the duplex state prior to recombination. For when the strands of a lysogenic bacteriophage are transferred into a recipient lacking the phage, entry into the lytic cycle, culminating in the production of mature phage particles, appears to take place rapidly in all cells. This implies that the single strand of the transferred DNA prophage must have been converted into a duplex which then provided the template for the first of many rounds of semiconservative replication. However, the phage represents an autonomous genetic unit and (when its expression is not repressed) is able to support independent replication. Of course, this is similar to the fate of the sex factor itself when an $F^+$ or $F'$ episome is transferred. This suggests that these elements not only are capable of independent semiconservative replication of their duplex genomes but also retain the ability to support DNA synthesis when they are in the single stranded state. Conversion of the single strands to the duplex form presumably demands only a repair like synthesis proceeding by continual extension from a primer; how a suitable primer sequence is provided is not known and we can only speculate on whether it might involve an initiation event at the same origin that is responsible for semiconservative replication of the duplex.

### *Enzyme Activities Involved in Recombination*

Several types of enzyme activity must be necessary for recombination. The initial pairing reaction may require an activity capable of "melting" DNA—this might be provided by a protein analogous to the gene-32 protein of phage T4, which can preferentially maintain A-T rich regions of DNA in a denatured state (see Chapter 6). A "nickase" activity able to make single strand breaks in the recipient duplex must be required at an early stage. Nuclease activities must be essential, although at present it is not possible to specify the precise roles that they take; however, their actions must be concerned with degrading the sequences that are replaced in the recipient genome. Of course, a single enzyme may have more than one of these activities. The model on the left of Figure 2.33, for example, illustrates a sequence of events in which a nick is made and the adjacent sequence of the recipient strand is then degraded; this could be achieved by the action of a nickase followed by an exonuclease but an alternative is to invoke a single enzyme with both endonuclease and exonuclease activities. Finally repair synthesis and ligase activities are necessary to achieve covalent linkage between the donor and recipient material.

Isolation of mutants defective in recombination has identified several loci whose products presumably comprise recombination enzymes; but it is not yet possible to correlate any enzyme activities with specific stages in recombination. The *rec*$^-$ mutants of E.coli were first isolated by Clark and Margulies (1965) as variants of an F$^-$ strain that were no longer able to produce haploid recombinants after conjugation with an Hfr strain; that is, these mutants cannot integrate the donor single strand into the recipient genome (see Clowes and Moody, 1966). Cells with any one of these mutations are defective not only in recombination but also display increased sensitivity to both ultraviolet and X-irradiation. This suggests that the enzymes of this system may undertake repair of DNA damaged by irradiation as well as participating in a pathway for recombination.

Since repair involves the removal of DNA sequences that are mispaired or represent structural distortions, such as the covalently linked thymine dimers generated by ultraviolet irradiation, it is not surprising that the mechanisms involved in repair and recombination may overlap. Repair of damaged DNA in E.coli is mediated by two principal systems: the *uvr* system undertakes excision-repair, that is to say that one strand of damaged DNA is removed from the chromosome and replaced by synthesis of a sequence matching the complementary strand; the *rec* system appears to act only after replication has generated two daughter chromosomes and apparently retrieves information from one in order to rectify damage in the other. The *uvr* system appears to be involved only in repair since *uvr*$^-$ mutants are defective in the correction of damage induced by ultraviolet irradiation but suffer no defect in recombination. The *rec* system appears to participate in both repair and recombination, since the introduction of a *rec*$^-$ mutation into a *uvr*$^-$ cell prevents the residual repair that takes place in *uvr*$^-$ cells that possess a functional *rec*$^+$ system. The activities of the *uvr* and *rec* systems, and of other loci in which mutation inhibits repair, are discussed in Volume 1 of Gene Expression.

The first group of *rec*$^-$ mutants that were identified map at three loci, *recA, recB, recC*. Mutants in *recA* are described as "reckless," for they are very sensitive to ultraviolet irradiation, which causes them to degrade a much larger proportion of their genome than is degraded in wild type cells. Mutants in *recB* or *recC* are "cautious" and their breakdown of DNA after irradiation is comparable to that of the wild type parental strain. The product of the *recB* and *recC* genes is an ATP-dependent DNAase, exonuclease V. Cells which are either *recB*$^-$ or *recC*$^-$ lack this enzyme activity; the enzyme consists of two polypeptide subunits, one specified by each gene (Barbour et al., 1970).

Exonuclease V displays several activities in vitro. It acts as an exonuclease on either single stranded or double stranded DNA with no preference apparent for either 5′ or 3′ terminus. It can also act as an endonuclease to make scissions in single stranded DNA. An additional reaction is its ability to act as an ATPase. Its mode of action as an exonuclease is processive; after binding to one DNA molecule it continues to degrade it, releasing short 3-4 base oligonucleotides (Goldmark and Linn, 1972; Karu et al., 1973; Mackay and Linn, 1974; for review see Clark, 1973; Radding, 1973). What nucleolytic action is exercised by the enzyme in vivo is not known. The product of the *recA* gene has not been identified but the reckless nature of *recA*$^-$ mutants suggests that the *recA*$^+$ protein usually may act to limit the activity of the *recBC* nuclease (exonuclease V). This idea is supported by the response to irradiation of *recA*$^-$ *recB*$^-$ or *recA*$^-$ *recC*$^-$ double mutants, which exhibit low levels of breakdown (Barbour and Clark, 1970).

Mutations in *recB* or *recC* may be suppressed by mutations at either of two further loci, *sbcA* and *sbcB*. Kushner et al. (1971) showed that *sbcA* mutants possess elevated levels of an ATP-independent exonuclease which may act as an analogue of the *recBC* product (exonuclease V) to restore the function of the *recBC* pathway. The nature of this mutation is not known. Mutants in *sbcB* lack the enzyme exonuclease I; thus the absence of exonuclease I prevents recombination in cells that lack the *recBC* nuclease (exonuclease V). Exonuclease I degrades single strands of DNA from 3′ to 5′, releasing mononucleotides, and causing the disappearance of free 3′-OH termini. Kushner et al. (1972) proposed that this enzyme action must prevent recombination when exonuclease V is absent from the cell. A complicating feature in interpreting the properties of these mutants is the existence of *xonA* mutations; these also lack exonuclease I and share with *sbcB* mutations the ability to suppress the deficiency in repair caused by the *recB*$^-$ or *recC*$^-$ mutation, but they do not suppress the deficiency in recombination. Presumably the *sbcB* and *xonA* mutations differ either quantitatively or qualitatively in their effect upon exonuclease I activity.

A model to account for the relationship between the *rec* genes and the *sbc* suppressor genes has been proposed by Clark (1973) and postulates that there are two pathways for recombination in E.coli. The *recA* gene product is implicated in both because *recA*$^-$ cells lack virtually all ability to form recombinants when used as recipients for conjugation. One of the pathways relies upon the *recBC* nuclease; since *recB*$^-$ or *recC*$^-$ cells retain a residual recombination activity of the order of only 1% of wild type, this pathway must be responsible for most of the recombination that takes place. The residual activity may be undertaken by a second, minor pathway, called the *recF* pathway after one of the (hypothetical) loci involved (see Figure 2.34).

Support for the idea that there is a second pathway has been provided by the experiments of Horii and Clark (1973) to identify mutations in it. The principle of these experiments was to isolate Rec$^-$ mutants from Rec$^+$ cells of the genotype *recB*$^-$ *recC*$^-$ *sbcB*$^-$; because these cells are mutant in both *recB* and *recC*, reversion at these loci is not likely. Some of the Rec$^-$ cells recovered after a mutagenic treatment are *recA*$^-$ and some are *sbcB*$^+$ revertants; the remaining Rec$^-$ cells should possess mutations in loci of the *recF* pathway. Any *sbcB*$^+$ revertants can be identified simply by screening for exonuclease I in cell extracts. Mutants of the *recA*$^-$ class can be distinguished by using the Rec$^-$ cells as recipients in crosses with two Hfr donors, isogenic except that one is *recA*$^-$ and one is *recA*$^+$; if the recipients are Rec$^-$ because of a mutation in *recA*, only the *recA*$^+$ donors can restore activity whereas if they possess a mutation in some other *rec* gene, both donors will be equally effective.

Eleven new *rec*$^-$ mutants were identified by this protocol, falling into four groups so far identified, *recF, recK, recJ, recL*. By virtue of the method used to isolate these mutants, we know that (to take *recF* as an example) *recBC*$^-$ *sbcB*$^-$ *recF*$^-$ cells are Rec$^-$ in phenotype. Cells of the genotype *recBC*$^-$ *sbcB*$^+$ *recF*$^-$ also are Rec$^-$, which is consistent with the idea that the loci *recBC* and *recF* identify two different recombination pathways and that the *sbcB* locus is involved in neither. A plausible role for the exonuclease I enzyme specified by *sbcB*$^+$ would be to connect these pathways; in the absence of the enzyme (*sbcB*$^-$ genotype) recombination would proceed through the *recF* pathway, but the presence of the enzyme (*sbcB*$^+$ genotype) would convert the DNA molecule into an intermediate that proceeds through the *recBC* pathway (see Figure 2.34). Thus in *sbcB*$^+$ cells, exonuclease V is essential as recombination must proceed through the *recBC* pathway; but if exonuclease V is absent (that is *recB*$^-$ or *recC*$^-$ alleles are present), the deficiency can be overcome by the *sbcB*$^-$ mutation which by inactivating exonuclease I stops converting DNA for the *recBC* pathway and allows recombination instead to continue by the *recF* pathway. Cells bearing only the *recF*$^-$ mutation (that is

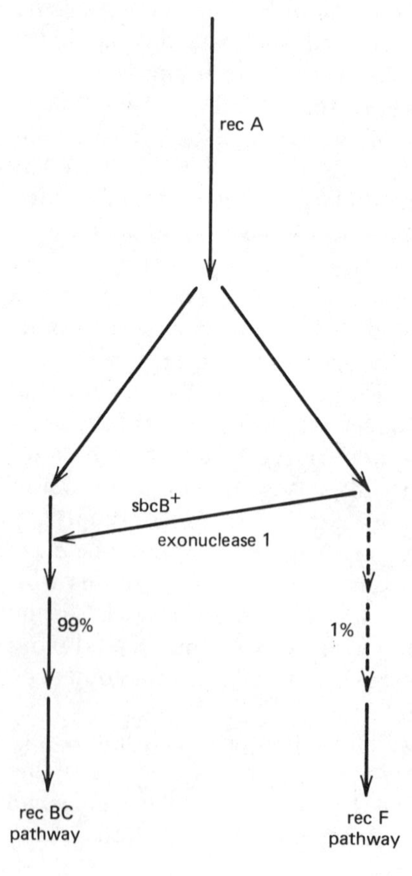


**Figure 2.34:** recombination pathways. The *recA* gene is needed for all recombination. The *sbcB*$^+$ gene product, exonuclease I, shunts recombination intermediates from the *recF* to the *recBC* pathway and when all genes are wild type some 99% of recombination is undertaken through this pathway, and only 1% through the *recF* pathway. Mutation to *sbcB*$^-$ thus prevents recombination intermediates from being shunted into the *recBC* pathway; instead recombination must proceed through the *recF* pathway. Thus a *recBC*$^-$ mutation, which prevents completion of recombination through this pathway, is suppressed by the *sbcB*$^-$ mutation which allows utilization of the *recF* pathway.

which are *recBC*$^+$ *sbcB*$^+$) are Rec$^+$ in phenotype; this implies that the *recF* pathway is indeed independent of the *recBC* pathway. However, the *recF*$^-$ cells are more sensitive to ultraviolet irradiation than the *recF*$^+$ parents, so that the product of this gene may play some unique role in repair. The *recL*$^-$ mutants give results similar to those obtained with *recF*$^-$ mutants, but are not stable so that any conclusion about their role can be only tentative. The *recK* and *recJ* mutants have not been characterized.

One difficulty in establishing the roles played in vivo by the products of the *rec* genes is that problems in cell viability are encountered, especially with certain double mutants. When recombination is measured simply by determining the number of recombinant colonies relative to the original parent populations, *recA*$^-$ cells show a very low level, less than 0.01% of wild type, and *recB*$^-$ or *recC*$^-$ cells show about 0.3–2.0% of wild type recombination levels. But the detection of recombinants in these experiments depends upon cell survival and multiplication. An alternative protocol was therefore established by Birge and Low (1974), who made use of two *lacZ*$^-$ mutations, each of which reduces $\beta$-galactosidase levels to a very low activity. The two mutations do not complement, but there is considerable recombination after conjugation and the resulting enzyme activity can readily be measured. When a *lacZ*$^+$ gene is transferred into a *rec*$^+$ recipient as a control, enzyme activity can be detected after 30 minutes. When one *lacZ*$^-$ gene is transferred into a *rec*$^+$ recipient possessing the other *lacZ*$^-$ allele, active enzyme is present after 75 minutes. Recombination to produce a DNA molecule that can be transcribed and translated thus takes of the order of 45 minutes. With recipients that are *recB*$^-$ or *recC*$^-$, the enzyme level reaches near wild type within about the same period even though the number of viable recombinants is only 0.3–3.0% of the *rec*$^+$ value. This implies that there is no defect in the early stages of recombination in *recBC*$^-$ cells, which are able to produce recombinant DNA that can be transcribed and translated; but the absence of the enzyme presumably makes some later step in recombination defective, with the result that cells are inviable. One speculation on the sort of event that might be involved is that the DNA may have structural deficiencies that do not interfere with its expression but which cause irrevocable damage to the genome in the next replication cycle. With *recA*$^-$ recipients, no active enzyme is produced, confirming the idea that this locus is involved in some step necessary for all recombination.

It is impossible to construct viable cells with certain combinants of mutant loci implicated in repair and recombination. DNA polymerase I has several catalytic activities in vitro that might be utilized in repair pathways, including the ability to undertake repair synthesis and to degrade DNA. Gross, Grunstein and Witkin (1971) found that they could not construct viable *polA*$^-$ *recA*$^-$ cells; this implies that the functions of these genes may be related in such a way that one of them must be present in wild type form for survival of E.coli cells. The double mutant *polA*$^-$ *recB*$^-$ also is inviable according to the results obtained by Monk and Kinross (1972), but Strike and Emmerson (1972) were able to isolate viable cells of the genotype *polA*$^-$ *recB*$^-$ *sbcA*; this implies that to be viable a cell must possess one of the enzymes DNA polymerase I, exonuclease V, and the ATP-independent exonuclease apparently specified by the *sbcA* mutant. Some mutants in the polynucleotide ligase are conditional lethal as such, but by using the less damaging *lig-4* mutation, Gottesman et al. (1973) were able to show that *polA*$^-$ *lig-4* double mutants are inviable. These mutant strains were constructed with the intention of elucidating the role in DNA metabolism of the enzyme DNA polymerase I (see Volume 1 of Gene Expression); they demonstrate that its absence in conjunction with defects in any one of a number of other loci (*recA, recB, lig*) is lethal to the cell, but of course this does not demonstrate what molecular activities are implicated. It is clear that there are many loci whose products are concerned with repair and recombination activities; and there may be significant overlaps between the various pathways so that one can to some extent substitute for another in the case of mutation. When more than one pathway is incapacitated, the cell may become inviable through its failure to be able to undertake certain catalytic activities; but at present it is not possible to define these pathways and to resolve the activities which they comprise.

The *rec* system(s) appear to constitute the general capacity of the E.coli cell for recombination. In additon to the integration of single stranded DNA transferred by conjugation, the *recA* locus is implicated in the retrieval of information from one daughter chromosome to rectify damage in another (see Volume 1 of Gene Expression). To what extent this recombination-repair overlaps with the functions implicated in single strand integration is not known; at present it is known only that the *recA* locus is involved in both pathways. Another function in which the *recA* function participates appears to be the recombination event that inserts an F factor into the bacterial chromosome. Clowes and Moody (1966) first reported that F$^+$ or F*lac* cells with a *rec*$^-$ mutation transfer their chromosome at only some 1% of the level supported by *rec*$^+$ cells. Wilkins (1969) noted that F*lac* strains transfer the chromosome at 10–50% of wild type levels if they carry *recB*$^-$ or *recC*$^-$ mutations and at less than 0.01% if they carry a *recA*$^-$ mutation. This implies that the *recA*$^+$ activity is essential and that the *recBC*$^+$ enzyme also may be implicated in the insertion of the sex factor into the chromosome; the corollary of this implication, of course, is to lend support to the concept that integration is necessary for chromosome transfer. Willetts (1975) noted a fourfold reduction in the frequency of chromosome mobilization by *recB*$^-$ mutants and a further tenfold reduction in *recB*$^-$ *sbc*$^-$ *recF*$^-$ mutants; this implies that chromosome mobilization uses the same systems that are responsible for integration of donor DNA in the recipient genome after conjugation. The failure of *recA*$^-$ cells to support chromosome transfer by F*lac* was confirmed by DeVries and Maas (1971), who also noted that very occasionally the F*lac* factor can integrate into the chromosome in an Hfr *recA*$^-$ strain; this insertion takes place at less than 0.1% of the frequency with which it occurs in *recA*$^+$ cells. Integration takes place in a region homologous with the bacterial genes carried on the episome; it is not known whether the *recA*$^-$ mutation is slightly leaky or whether some other very inefficient system is responsible for this integration. The implication of the *recA* locus in single strand integration, recombination-repair, and sex factor insertion suggests two possible roles for its gene product: this might be an enzyme catalyzing a step common to all these processes; or it might be a control function that regulates other functions, including activities necessary for all three pathways.

### *Location of Crossovers*

Viewed in formal genetic terms, recombination between a complete chromosome and an incomplete fragment requires an even number of exchanges if the integrity of the chromosome is to be maintained. (This contrasts with reciprocal recombination between two complete chromosomes, where any number of breakage and reunion events may take place without altering the total genetic content of each chromosome.) In terms of the molecular models of Figure 2.33, the requirement for an even number of exchanges is provided by the breakages and donor-recipient reunions that take place at each end of the length of the insertion.

For any donor marker to be inserted into the recipient chromosome therefore requires a genetic exchange on either side of it. This explains the reduction in the frequency of appearance of very proximal markers in the recombinants, a phenomenon noted by Pittard and Adelberg (1964). In general, a crossover appears to have a more or less constant probability of occurring in any particular chromosome region, so that the efficiency with which most markers are integrated is reasonably constant. But when a marker is located very close to the origin of transfer, its insertion demands an exchange in the limited region between this site and the end of the molecule; such an exchange may be expected to occur less frequently and so a smaller number of recipients gains donor markers located very close to the origin. Low (1966) thus found that markers very close to the origin appear in recombinants less frequently than expected from extrapolation of the transfer gradient. In following the appearance in recombinants of the arginine genes transferred by F′14 donors (F′14 is an unusually large episome carrying many bacterial genes), Glansdorff (1966) obtained results that at first sight appeared to be caused by a transposition in the sequence of *arg* genes; but the relevant genes in fact had reduced integration frequencies relative to other markers due to their proximal position—this caused a change in their linkage with adjacent markers.

An observation related to the reduction in integration of very proximal markers is that a crossing over event may be obligatory very close to the origin. The probability of recombination in different chromosome regions was measured by Pittard and Walker (1967) in experiments in which the progeny of Hfr × $F^-$ matings were selected for one marker and then examined for unselected markers proximal to it (that is lying between the selected marker and the origin). Two Hfr donors were used, both transferring markers with the same orientation but from origins some distance apart. The first has the order of transfer *pro-ilv-tna-xyl-mal,* with *ilv* transferred at 25 minutes; the second has the order of transfer *tna-xyl-mal,* and since *ilv* and *tna* are less than 1 minute apart, *tna* is transferred almost immediately as a proximal marker. After mating for 90 minutes, recipients were selected for the donor *mal* marker and then examined for the preceding markers. From these frequencies it is possible to calculate the probability that crossovers will take place between *ilv* and *tna,* between *tna* and *xyl,* or between *xyl* and *mal.* With the first strain, in which the earliest of these markers, *ilv,* is transferred after 25 minutes, the probability of recombination is constant for all three regions at about 2% per minute. (This is much lower than values obtained in earlier experiments, often quoted to be as high as 20%.) But with the second strain, in which *tna* is a highly proximal marker, the frequencies were: *ilv-tna* 72.5%; *tna-xyl* 9.0%; *xyl-mal* 2.8%. The recombination frequency is therefore very much increased between *ilv* and *tna,* that is effectively between the origin and *tna.* This suggests that a crossover in the highly proximal region may be an integral feature of conjugation. This means that markers close to but more than 1 minute from the origin do not show the reduction in recombination frequency that might be expected from the limited distance between their positions and the end of the donor DNA, because the obligatory crossover ensures their integration.

The occurrence of obligatory pairing very close to the donor origin provides one possible explanation of the reduction in frequency of integration of the markers located within 1 minute of the end of the DNA; the obligatory events taking place in this region might inhibit the incorporation of this DNA into the recipient genome. Curtiss (1969) has suggested that such a reaction might be necessary to help "pull" the donor DNA into the recipient. According to this model, $F^+$ and $F'$ factors can transfer the DNA of the episome without any aid from the recipient because only a small length is involved; similarly the initial length of the chromosome transferred by an Hfr donor demands only the "pushing" force of the donor, but transfer of greater chromosome lengths requires assistance from the recipient. Early experiments suggested that energy production is necessary for conjugation only in the donor, but more recent results have shown that both partners play an active role in transfer (see Curtiss et al., 1968). The role of the recipient could thus be some sort of pulling function mediated through an obligatory pairing reaction. And certainly it is possible to construct models for recombination that would produce the observed results.

DNA can be transferred by conjugation from E.coli to strains of Salmonella. Using S.typhosa, Johnson, Falkow and Baron (1964) noted that a large increase in the frequency of recombinant formation is obtained with recipients that are hybrids containing some E.coli DNA as well as the Salmonella genetic material. But only E.coli sequences homologous with the chromosome region that is transferred proximally are effective. As Curtiss et al. (1968) observed, a reduction in frequency of recombinant formation in matings between E.coli cells may be obtained if the recipient carries a large deletion in the region homologous with the proximal donor DNA. These results are consistent with the idea that some form of interaction between proximal donor material and the homologous recipient sequences is implicated in conjugation.

That the nature of the bacterial genome itself is not pertinent in chromosome mobilization and transfer is implied by the ability of the sex factor to interact with the Salmonella chromosome in a manner similar to its interactions with the E.coli chromosome. Many different Salmonella strains now carry the sex factor and $F^+$, Hfr, and $F'$ strains of S.typhimurium and S.abony all now are available (for review see Sanderson et al., 1972).

CHAPTER 3

# Plasmids

This chapter is concerned with the reproduction and expression of three classes of plasmid: F factors, R factors, and colicinogenic factors. We have already considered the ability of the F factor to promote sexual conjugation and the consequences that ensue from its insertion into the bacterial chromosome. The replication of the sex factor in its autonomous state is most readily followed by the use of F′ factors carrying bacterial markers; and these factors have been used for genetic analysis of the system responsible for transfer of F between cells. The R factors, some of which are closely related to the F factor, were first identified through the drug resistance markers that they carry: the interactions between R factors and between some R factors and the F factor are central to the analysis of plasmid reproduction and expression. The colicinogenic factors are named for their ability to produce colicins, extracellular molecules that have an antibiotic function, and these Col factors fall into several categories, some of which are interestingly related to the F factor and the R factors. Finally we shall consider the chimeric plasmids that can be constructed by inserting a sequence of foreign DNA from any source into plasmid DNA in vitro.

## Replication of Autonomous Sex Factors

### *Mutants in the Sex Factor Replicon*

The bacterial chromosome and the sex factor constitute independent systems of replication. Jacob, Brenner and Cuzin (1963) proposed the term *replicon* to describe an autonomous genetic element that replicates as a single unit. A replicon may be characterized by its possession of an *origin* at which replication commences and a control system that acts on the initiation of replication. In its simplest form the control system might comprise a regulator gene whose product acts at a *replicator* locus (which may or may not be coincident with the origin). The original formulation of the model postulated a positive control system in which the regulator protein is an initiator whose action is necessary in order to switch on the cycle of replication. Models of negative control have also been proposed, in which the regulator is a repressor protein: replication then is initiated unless the repressor is present to prevent it. It is worth emphasizing that what characterizes a replicon is its independent system for controlling the initiation of replication; once initiated, replication of plasmids appears to be undertaken by host enzymes. In addition to the control of replication, another feature of the replicon must be some mechanism to ensure the equitable distribution of replicas to the daughter cells at division (see Jacob, Ryter and Cuzin, 1966; Novick, 1969). Attachment to the cell membrane is often invoked to provide this mechanism; and one variation of the positive control model is to suppose that initiation depends upon the provision of membrane sites to which the duplicating chromosomes can attach.

In the autonomous state the sex factor represents an independent replicon. When it is integrated into the bacterial chromosome it becomes unable to replicate independently and is replicated only as part of the host genome. If the sex factor replicon is needed to initiate replication only when the factor is autonomous, it should be possible to isolate mutants that are no longer able to replicate autonomously but which still can be replicated as part of the bacterial chromosome. In order to maintain the episome, it is necessary to use a conditional lethal mutation (that is, one which prevents replication under certain, nonpermissive conditions, but which allows the episome to be perpetuated under other, permissive conditions). Jacob, Brenner and Cuzin (1963) therefore isolated temperature sensitive mutants in which episome replication functions are defective at high but not at low temperature: the episome can be maintained in cells grown at low temperature, but its mutant properties are revealed in cells incubated at high temperature. Using the protocol illustrated in Figure 3.1, bacteria carrying an F*lac* episome were screened to find mutants in which the *lac*$^+$ function of the episome is present at 30°C but is absent at 42°C. (Lac$^-$ bacteria form white colonies on the indicator agar, EMB-lactose, whereas Lac$^+$ colonies are red. So by growing *lac*$^-$/F*lac*$^+$ bacteria on this medium, it is possible to isolate mutants simply by their loss of color.)

The clones that were normal at 30°C but mutant at 42°C fell into two classes. The first formed homogeneous white colonies at 42°C and homogeneous red colonies at 30°C; when mutants that have first been isolated at 42°C are transferred to 30°C, the white colonies turn red. These bacteria carry a thermosensitive mutation in the *lac* operon of the F*lac* factor. The second class of mutants produces red colonies at 30°C and these turn white at 42°C; at both temperatures there are segregants with

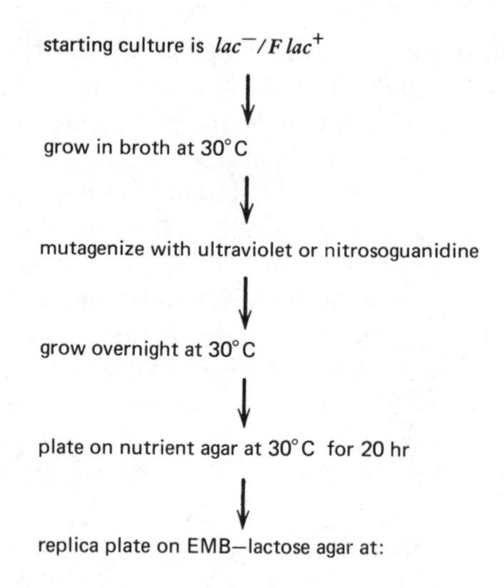


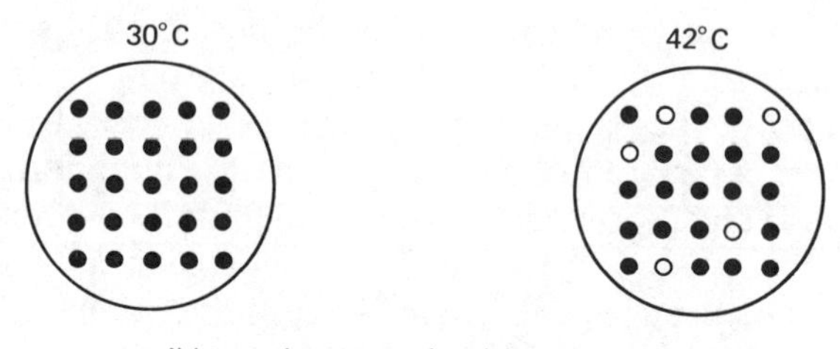


**Figure 3.1:** protocol for isolating mutants that are temperature sensitive for replication of F*lac*.

the reverse Lac phenotype. When clones that were isolated as white colonies at 42°C are incubated at 30°C, they remain white. These bacteria must represent mutants in which the episome multiplies normally at 30°C (giving red Lac$^+$ colonies), but ceases to reproduce at 42°C, causing dilution out of the F*lac* factor as the bacteria divide, with irretrievable loss of the *lac*$^+$ genes (and thus generating white Lac$^-$ colonies that remain Lac$^-$ on transfer to the lower temperature). This interpretation was confirmed by showing that the Lac$^+$ colonies grown at 30°C were able to act as genetic donors, whereas the colonies grown at 42°C and then reisolated at 30°C could not do so.

Do mutations that prevent replication of the sex factor at high temperature affect the episome itself or the bacterial chromosome? This can be tested by transferring F*lac* from the mutant cells grown at 30°C, in which the factor is maintained, into fresh $F^-$ recipients. If the mutation lies in the episome, it will be displayed in the new host cells. Conversely, the presence of mutations on the bacterial chromosome can be examined by isolating $F^-$ segregants that have lost the F*lac* factor at high temperature; when a new episome is introduced it should be lost only if the mutation is bacterial. Of ten thermosensitive mutants, six proved to be located in the episome and four in the bacterial chromosome. This confirms that bacterial as well as episomal functions are implicated in the replication of the factor.

Thermoresistant $lac^+$ bacteria are found in the mutant cultures and these fall into two classes. Some carry an $Flac^-$ factor and must therefore have resulted from recombination between the original $Flac^+$ factor and $lac^-$ bacterial chromosome to reverse the locations of the two *lac* alleles. Some are Hfr bacteria in which the thermosensitive $Flac^+$ factor has been integrated into the bacterial chromosome. These Hfr lines often are unstable and at low temperature may tend to segregate the original thermosensitive F*lac* factor. The Hfr cells transfer the bacterial chromosome with an origin close to the *lac* locus; one set of *lac* genes is transferred as a proximal marker and the other as a distal marker (this is the situation illustrated in Figure 2.14). Some of these Hfr clones display reduced efficiencies of transfer. Their properties thus hint that the functions of the F factor that are concerned with its autonomous replication may overlap with those involved in conjugational transfer.

The F factor replicon includes not only the DNA of the episome itself but also any other sequences carried by it, as is illustrated by the replication under F control of autonomous F′ factors, some of which are very much larger than the $F^+$ factor. In certain circumstances, even the entire bacterial chromosome may be replicated under the control of the sex factor. To investigate the relationship between the episome replication system and that of the bacterial chromosome, Nishimura et al. (1971) used a temperature sensitive *dnaA* mutant that is defective in the initiation of chromosome replication at high temperature. Infection of these cells with the F factor does not influence the temperature sensitivity of the strain; thus there is no complementation of replication functions between the autonomous F factor and the bacterial chromosome. But the infected cells display an increased frequency of reversion to temperature resistance; and all except one of the revertants isolated in these experiments were of the Hfr type, that is, the infecting F factor had been integrated into the bacterial chromosome. That the Hfr temperature resistant strains retain the *dnaA* mutation was shown by transduction of this locus into other strains; the only revertant in which the *dnaA* allele had back mutated to wild type was the single line that had remained $F^+$. In later experiments with a *dnaA* strain carrying an integrated factor R100-1, Bird, Chandler and Caro (1976) showed that at high temperature replication is initiated at an origin within the factor instead of at the usual bacterial origin.

This *integrative suppression* of the temperature sensitive mutation can be explained by a two step process: first the F factor is integrated into the bacterial chromosome to generate an Hfr; and then the integrated F replication system becomes nonrepressed and substitutes for the defective system of the bacterial chromosome. Consistent with this model is the observation that when Hfr derivatives of *dnaA* strains are prepared at low temperatures, they display a greater rate of reversion to temperature resistance than that of the $F^+$ strains (because they need undertake only the second step). The requirement for integration before the F factor can mediate reversion to temperature resistance was confirmed directly by the use of $F^+$ *dnaA* strains carrying the *recA*$^-$ mutation; their rate of reversion is no greater than that of the control $F^-$ cells, because the F factor is unable to integrate into the chromosome without a *recA*$^+$ function. And a demonstration that it is the replication system of the F factor which is responsible for bacterial replication in the Hfr *dnaA* temperature resistant strains was provided by the use of acridine orange: at low temperatures, when the *dnaA* mutation is not expressed and replication is under bacterial control, acridine has no effect; but at high temperature, when the F system must be responsible for replication, acridine prevents replication of the bacterial chromosome and inhibits growth of the strain.

Integrative suppression can overcome the deficiencies only of mutants in initiation. Temperature sensitive mutations in genes affecting DNA synthesis per se cannot be relieved by this means. (Suppression of a bacterial replication mutation by integrative suppression therefore may be taken as evidence that it affects initiation; see Table 8.5.) This is consistent with the idea that the F replicon possesses its own control system, but that once replication has been initiated it may rely largely upon bacterial functions for the synthesis of DNA. During integrative suppression the F factor takes over the control of replication only at the high temperature at which the bacterial system is inactive, whereas at low temperature the bacterial control continues to be dominant; this poses the question of the nature of the event usually responsible for repressing the integrated F replicon and the reason for its nonrepression when the bacterial system is inactivated.

In these experiments, cells in which integrative suppression had occurred were detected by following the growth of colonies at the nonpermissive (42°C) temperature. This leaves open the possibility that in fact all cells with an integrated sex factor might be able to synthesize DNA at the nonpermissive temperature, but that some might be unable to form colonies and would therefore escape detection. Tresguerres, Nandadasa and Pritchard (1975) therefore followed DNA synthesis directly by labeling *dnaA* mutant Hfr cells with $^3$H-thymidine; their results suggest that the chromosomal initiation defect is suppressed in all Hfr strains, that is, inactivation of the host replication system by the *dnaA* mutation is in itself sufficient to release the sex factor from the inhibition that usually applies to its replication in the integrated state. This implies that the functioning of the bacterial replication control system may be necessary to repress the integrated (although not the autonomous) sex factor; in the absence of the bacterial function(s) the repression is relieved. This suggests that components of the two control systems may interact, although the nature of this interaction, and in particular the interesting question of how a distinction is drawn between the autonomous and integrated states, remains to be established.

### *Initiation of Sex Factor Replication*

Autonomous sex factors appear to share with the bacterial chromosome the characteristic that initiation of replication cycles is related to the cell growth rate and thus controls the number of copies of the replicon in the cell. And another common feature is that the two copies of the replicon produced by a cycle of replication are evenly segregated when the cell divides, so that both daughter cells receive a copy.

The replication system of the bacterial chromosome is responsible for a bidirectional replication that starts in all strains (irrespective of mating type) at an origin located at about 70 minutes on the standard map. Growing points progress in both directions from this origin until they meet at a terminus located at about 30 minutes. Cooper and Helmstetter (1968) reported that replication of the complete chromosome requires a fixed period of about 40 minutes; and a constant time of about 20 minutes separates the end of a replication cycle from the ensuing cell division. These parameters are constant only for cells growing at doubling times of 60 minutes or less; for in slower growth, conditions become limiting and these periods occupy times proportional to the length of division cycle.

The initiation of cycles of replication is controlled by the growth rate of the cell, so that cells dividing more frequently initiate cycles at shorter intervals of time. Because a cycle of replication always starts 60 minutes before the division to which its completion is linked, it is only at doubling times of 60 minutes or more that initiation of a replication cycle takes place in the same division cycle in which it is completed. At doubling times between 60 and 30 minutes a replication cycle is initiated in the division cycle preceding that in which it is completed; at doubling times between 30 and 20 minutes a replication cycle is initiated two generations before its completion. The chromosomes of these cells must therefore have suffered one or even two initiations before completing a current cycle. They may therefore possess multifork structures, with two or four origins but only one terminus.

The number of chromosome origins shows a constant ratio to the cell mass at the time of initiation and Donachie (1968, 1969) therefore suggested that a new cycle of replication is initiated when a critical cell mass per origin is achieved. Two classes of model can explain this type of result. Helmstetter et al. (1968) and Pierucci (1969) suggested that an initiator protein might be synthesized at a rate depending upon both the number of origins and the growth rate of the cell; initiation would result when a critical amount of protein has accumulated in the cell. Pritchard, Barth and Collins (1969) suggested that an inhibitor protein synthesized at a specific time in the cell cycle might be diluted out with growth; initiation would result when its concentration falls below a critical level. The link with the number of origins can be provided in both models by postulating that the gene coding for the regulator protein is located close to the origin.

These mechanisms ensure that the number of copies of the bacterial replicon is always appropriate for the division rate. Thus in the period between two cell divisions, the chromosome number exactly doubles. The chromosomes of the parent cell are evenly distributed to the daughter cells at division, so some mechanism must exist to establish their distribution. The bacterial chromosome appears to be linked to the cell membrane, possibly at several sites but very probably at least at the origin, and it has often been suggested that this linkage might provide the distribution mechanism. Some suggestive evidence is provided by one mutant which is defective in the initiation of DNA synthesis and appears to have altered membrane proteins, implying that some relationship exists between the membrane and the replication cycle. This concept is interesting in light of suggestions that the sex factor may possess membrane attachment sites. The bacterial replication cycle and its control is discussed in Volume 1 of Gene Expression.

The early experiments of Stouthamer, De Haan and Bulten (1963) suggested that only a single episome enters a recipient cell as the result of transfer from the donor and that it then replicates to reach an equilibrium level (presumably one which is appropriate for the cell growth rate). There is then a constant probability in each generation of segregating $F^-$ cells and this implies that the rate at which the episome replicates maintains only a small number of copies per cell; this varied between one and two, depending on whether the factor had replicated yet during the current cell cycle. De Haan and Stouthamer (1963) reported also that segregation of $F^-$ cells immediately after recipients each have received a single factor is sufficiently rare to suggest that there is usually an even segregation of the copies present into the daughter cells.

That the events responsible for initiation of episome replication may be similar in nature to those controlling replication of the bacterial chromosome was suggested by the results of Cooper (1972) and Davis and Helmstetter (1973). Replication of an F*lac* factor can be followed by the increase in synthesis of $\beta$-galactosidase that takes place when the *lacZ* gene is replicated; if the bacterial chromosome is *lac*$^-$, enzyme synthesis doubles immediately following replication. The F*lac* factor is sufficiently small for the time of enzyme doubling to be taken as an indication of the time of replication of the factor; if the sex factor is replicated at the same speed as the bacterial chromosome, it should take less than 4 minutes for the F*lac* factor to be reproduced. Replication of the bacterial chromosome can be followed by the incorporation of labeled thymidine, which doubles in rate at the time of initiation of a new cycle of replication.

At several different steady state growth rates, the same relationship always was observed: the sex factor replicated at a time in the division cycle just before the time when a new cycle was initiated by the bacterial chromosome. When cells are shifted from a slow growth rate into a medium promoting more rapid growth, chromosome replication is soon initiated; again this displayed a constant relationship with replication of the sex factor. This relationship applied to the growth rates at which chromosome initiation is controlled by cell mass (that is, doubling times less than 60 minutes). These results suggest that the same dependence upon mass controls the initiation of both bacterial chromosome and sex factor, so that the number of copies of the sex factor in the cell is the same as the number of chromosome origins.

This model predicts that the number of copies of the episome should fluctuate with the stage of the cell cycle. Thus newly born, slowly growing cells have a single episome which replicates during the cycle to give an older cell that has two copies; as the growth rate increases, replication of the episome takes place at earlier and earlier times in the division cycle, so that new born cells have two copies and older cells have four copies. The estimates of 1–2 episomes per cell made by Jacob and Monod (1961) and Revel (1965) therefore agree with the number predicted for more slowly growing cells. Frame and Bishop (1971) found a greater number in more rapidly growing cells.

In contrast with the idea that sex factor replication is controlled in the same manner as the bacterial chromosome, however, Pritchard, Chandler and Collins (1975) found that the time when sex factor replication is initiated is not constant relative to the time of initiation of host chromosome replication. They pointed out that the measurement of β-galactosidase levels when cells are grown at different growth rates may be influenced by growth conditions as well as by replication per se and they therefore used a technique in which the rate of replication is controlled by the level of thymine provided in the medium of thymineless strains of E.coli. In these experiments no change in *lac* expression could be correlated with changes in the rate of replication; these results imply that replication of the sex factor is not coupled to cell mass. Of course, there is a possible difficulty in interpreting these experiments also, which is that replication control might not be unaffected by the partial thymine deprivation that is involved.

From these conflicting conclusions, it is clear that the number of copies of the sex factor present per cell is rather low, probably 1–2 in less rapidly growing cells. However, the nature of the control system is not yet clear. But since the number of copies of the sex factor in a bacterium is low, some mechanism must exist to ensure even segregation to progeny cells. Hohn and Korn (1969) noted that during curing with acridine orange, the sex factor segregates with the chromosome with which it was originally associated in the ancestral cell. A similar observation was made by Cuzin and Jacob (1967). One model to explain this result would be to postulate that some topological organization links the episome with the bacterial chromosome. The chromosome of E.coli has a folded structure with many loops, whose structure each appears to be independently maintained (see Volume 1 of Gene Expression). Kline and Miller (1975) reported that at least 50% of the DNA of the F factor can be recovered from $F^+$ cells in the form of covalently closed circles associated with the folded chromosome. If this association is a specific one, it might explain cosegregation with the host chromosome and the even distribution to daughter cells.

### *Incompatibility Between Sex Factors*

The existence of F′ factors has made it possible to follow the life cycles of autonomous sex factors, since the expression of bacterial markers carried by the factor may provide a way to determine its presence. The F*lac* factor has proved particularly useful in view of the genetic analysis that has been developed on the lactose operon, whose activity can therefore readily be measured. Jacob and Monod (1961) noted that a fully induced F*lac* diploid (that is with one copy of the lactose chromosome on the episome and one on the chromosome) produces 2.5–3 times more $\beta$-galactosidase than an $F^-$ strain. Revel (1965) noted that when an F*lac* factor is introduced into an $F^-$ *lac*$^-$ recipient, about 1.5–2 times as much $\beta$-galactosidase is produced as in an $F^-$ *lac*$^+$ strain. This level of synthesis suggests that only one or two copies of the episome are present for every copy of the bacterial chromosome (see above).

An autonomous sex factor cannot coexist in the cell with an integrated sex factor. Maas and Maas (1962) and Maas (1963) noticed that they were unable to transfer F*lac* into an Hfr strain; and Scaife and Gross (1962) found that when $F^+$, $F^-$ or Hfr cells are mixed with an F*lac* population, different types of cell are found among the *lac*$^+$ progeny from each of the three crosses. Virtually all the *lac*$^+$ cells derived from F*lac*$^+$ × $F^-$*lac*$^-$ or F*lac*$^+$ × $F^+$ *lac*$^-$ crosses are susceptible to acridine; that is, they become $F^-$ upon curing and so the *lac*$^+$ gene must be carried by the infecting F′ episome. But the *lac*$^+$ cells derived from F*lac*$^+$ × Hfr *lac*$^-$ crosses are not susceptible; the *lac*$^+$ genes are therefore no longer carried on an autonomous episome but must have been integrated into the chromosome by recombination. This conclusion was confirmed by showing that the *lac*$^+$ genes could be transferred in conjugation by the cells only as a chromosome marker. This implies that it must be impossible for the F*lac* factor to multiply in an Hfr cell; the only way the *lac*$^+$ genes can be passed on to the progeny is by insertion into the genome.

A detailed study of the ability of the F*lac* factor to enter cells of different mating types has been carried out by Dubnau and Maas (1968). After a 30 minute mating period, the frequencies of *lac*$^+$ progeny per recipient were

| | |
|---|---|
| F*lac*$^+$ × $F^-$*lac*$^-$ | 60–80% |
| F*lac*$^+$ × $F^+$*lac*$^-$ | 30–50% |
| F*lac*$^+$ × Hfr *lac*$^-$ | 15–30% |

Some 90–100% of the colonies obtained with $F^+$ or Hfr recipients appeared highly variegated, often appearing as small *lac*$^+$ papillae on *lac*$^-$ colonies; by contrast, only 1–10% of the *lac*$^+$ offspring from $F^-$ recipients were variegated. This phenomenon, also noted previously by Scaife and Gross, suggests that the F*lac* factor may be transferred efficiently to all recipients but is not inherited by many of the descendents of the $F^+$ and Hfr recipients. Hfr cells effectively inhibit the multiplication of any F factor that is introduced by recombination. With $F^+$ cells, there is an approximately even chance that either the previously resident $F^+$ factor or the newly introduced F*lac* factor will replicate; only one can remain.

That the genes carried by the F*lac* factor can be inherited by descendants of Hfr recipients only if they become integrated into the chromosome has been confirmed by determining their state after transfer. The *lac*$^+$ progeny of F*lac*$^+$ × F$^-$ *lac*$^-$ crosses themselves act as typical F*lac* donors in further crosses, that is, they possess an autonomous F′ factor; whereas the *lac*$^+$ progeny of F*lac*$^+$ × Hfr *lac*$^-$ crosses are Hfr cells in which *lac*$^+$ is a chromosome marker. That recombination is needed to recover *lac*$^+$ strains from the Hfr recipients was confirmed by experiments using *recA*$^-$ Hfr and F$^-$ strains. (The *recA*$^-$ mutation completely prevents any recombination between transferred donor material and the recipient chromosome.) With *recA*$^-$ F$^-$ recipients the same results are obtained as with *recA*$^+$ F$^-$ recipients: acquisition of the *lac*$^+$ genotype results from the presence of an autonomous sex factor. But with *recA*$^-$ Hfr recipients, no *lac*$^+$ colonies are produced; this implies that unless recombination is permitted to insert the *lac*$^+$ genes into the chromosome, they cannot be passed on to the progeny, that is, the F*lac* factor is unable to replicate in the Hfr cell.

What is the fate of an F*lac* episome after it enters a recipient cell? By following the protocol illustrated in Figure 3.2, Dubnau and Maas demonstrated that it is able to replicate in F$^-$ recipients but that in Hfr recipients the F*lac* factor becomes diluted out as the cells divide because it fails to replicate. The figure shows that in the F*lac*$^+$ × F$^-$*lac*$^-$ control cross, the proportion of *lac*$^+$ cells increased from 60% immediately after mating to a stable plateau of 90%. The F*lac* factor therefore replicates in these cells so that each progeny cell gains a copy at division. But with the F*lac*$^+$ × Hfr *lac*$^-$ cross, the proportion of *lac*$^+$ colonies declined with cell division, the decline following the theoretical curve expected of a gene which is neither replicated nor destroyed so that the proportion of *lac*$^+$ cells is halved in every generation. The decline presumably levels off after three generations because of the presence of a growing number of cells in which the *lac*$^+$ gene has been integrated by recombination with the chromosome and so is inherited by their progeny.

Two classes of mechanism can be invoked to explain why only one copy of the sex factor can be maintained in a bacterial cell. One is to suppose that this is a consequence of the operation of the sex factor replication control system. Since the replication control system of the factor is repressed in the integrated state but allows replication when the factor is autonomous, this raises the question of whether the same mechanism is responsible for the inability of an autonomous factor to coexist with an integrated factor and for the incompatibility between two autonomous factors. If the mechanism responsible for preventing independent replication of the integrated sex factor also prevents the replication of any new factor that enters the cell, this would be equivalent to the superinfec-

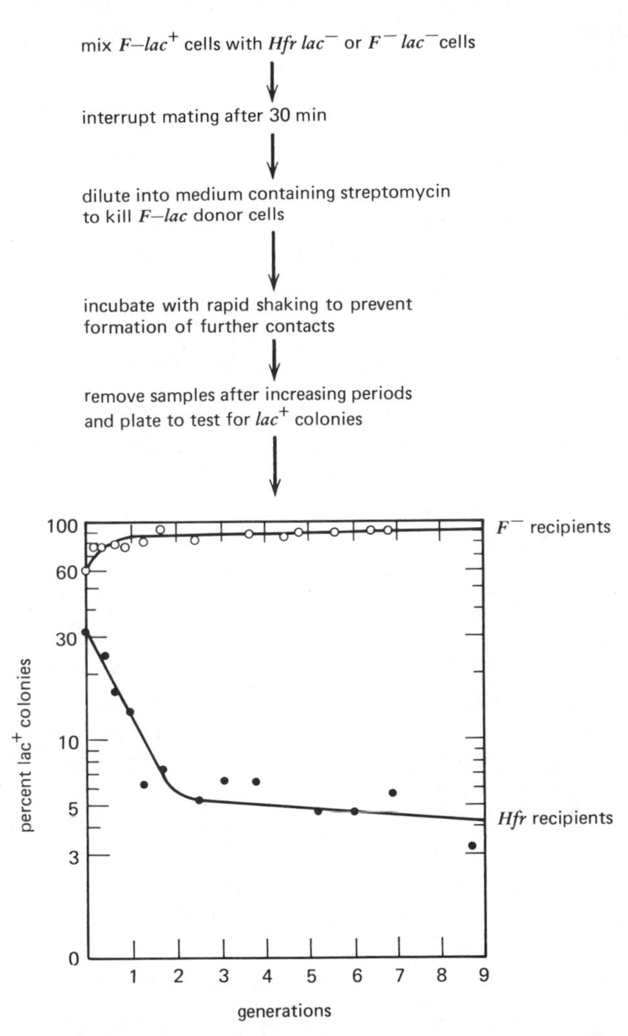


**Figure 3.2:** fate of F*lac* in Hfr and F⁻ recipients. The F*lac* factor cannot replicate in Hfr cells and is diluted out with growth, the proportion of Lac⁺ cells therefore halving every generation. A plateau in the decline is reached due to the presence of cells in which the *lac*⁺ genes have been integrated into the bacterial chromosome and therefore are inherited by all progeny. In the control experiment with F⁻ recipients, the F*lac* factor rapidly infects most of the cells of the population and then is stably inherited. Data of Dubnau and Maas (1968).

tion immunity conferred by lysogenic prophage against an infecting phage (see Chapter 4). If the incompatibility between two autonomous factors is attributed to the system controlling the initiation of episome replication, what is responsible may be the limitation on the number of copies of the factor able to exist in the cell; the introduction of another factor then can be taken to represent the premature addition of another copy, which may postpone any new initiation events until segregation at cell division has reduced the number of episomes to the correct value. According to this model, the relationship between the incompatibility exercised by an integrated factor with that resulting from the presence of an autonomous factor may depend upon the effect of integration on the episome replication system.

An alternative type of model is to suppose that there is some topological explanation: a single attachment site, with which the factor must be associated if it is to be replicated, might exist at the cell membrane; and one factor, whether present in the free or integrated condition, might occupy this site and thus prevent a second factor from achieving the association necessary for replication. Although there is evidence that autonomous sex factors segregate normally at cell division and may be associated with the host chromosome, one argument against the idea that such an association might be responsible for incompatibility is provided by the observation that it is possible to create double male strains that possess two integrated sex factors in a single bacterial chromosome. Such strains may be obtained by crossing one Hfr with a second Hfr maintained in the phenocopy condition; they transfer markers equally well from either origin and display the usual immunity to extrachromosomal sex factors (Clark, 1963; Falkingham and Clark, 1974).

That the system responsible for incompatibility is controlled by the F factor itself is suggested by the success of Maas and Goldschmidt (1969) in isolating mutants of an Hfr strain that allow superinfection with an F*lac* factor. The principle of these experiments was to mix a population of Hfr $lac^-$ cells, in the phenocopy condition, with a large excess of F$lac^+$ cells and then to isolate the $lac^+$ progeny. The usual outcome of such a cross is that any stable $lac^+$ cells have the episome genes integrated into the bacterial chromosome; in order to prevent this outcome the Hfr cells carried the $recA^-$ mutation. A $lac^+$ strain isolated from this cross was able to transfer chromosome markers only in the usual order characteristic of the parent Hfr (implying that the F*lac* factor had not become integrated to give a double male strain); and any $lac^+$ offspring derived by crossing this strain with $lac^-$ recipients always were of the F*lac* type (and so were not the result of chromosome transfer of an integrated $lac^+$ marker).

The F*lac* factor transferred from this strain retained all its usual characteristics, and so did not appear to have suffered mutation; but the Hfr properties had changed. When the Hfr/F*lac* strain was cured of the autonomous sex factor by growth in acridine orange, it lost its ability to undertake chromosome transfer; this was regained when an F*lac* factor was reintroduced and this factor again could be stably maintained in the cells, which then were able to act as Hfr as well as F*lac* donors. The integrated sex factor of this Hfr strain thus appears to have suffered a mutation which prevents it from acting as a chromosome donor, and which also allows the coexistence with an autonomous F′ factor; the F′ factor then is able to supply whatever function is defective in the integrated factor, thus supporting transfer by either sex factor.

A locus responsible for the incompatibility between an integrated factor and an autonomous factor has been identified by the use of deletions of the integrated episome that allow survival of an autonomous factor in the cell. Willetts (1974) reported that the locus *inc* maps in the region that probably specifies replication functions and which appears to be transferred early in conjugation. It is likey that it will prove to code for a protein function that is part of the replication control system of the factor (see later). Pfister, DeVries and Maas (1976) have identified an $inc^-$ mutation on an integrated factor which relieves incompatibility with autonomous factors, but which does not affect incompatibility between two autonomous factors (when the mutant sex factor is released from the bacterial chromosome). Although this does not necessarily imply that different gene products are involved in the two types of incompatibility, it does demonstrate that there may be differences between them, possibly residing in quantitative rather than qualitative effects.

Host functions are involved in the incompatibility between autonomous factors. Using a $recA^-$ host to prevent the formation of recombinants, Blas, Thompson and Broda (1974) isolated bacterial strains carrying the markers of two different F′ factors. These appeared to be bacterial mutants able to contain more than one F′ factor in the cell, although only certain pairs of tested factors were able to coexist. The basis of this effect is not yet known.

## Constitution of R Factor Replicons

### *Isolation of Drug Resistance Factors*

Bacterial strains that had gained resistance to more than one drug in a single step were first identified when Shigellae isolated from human cases of dysentery were examined in Japan during the late fifties. Following use of only one of four drugs commonly used in treatment, streptomycin, sulfonamide, chloramphenicol and tetracycline, Shigellae resistant to more than the single drug often were obtained, apparently as the result of a single event. Strains resistant to all four drugs appeared more common than strains resistant only to some. Many other strains carrying multiple drug resistance have since been isolated, in other countries as well as in Japan; resistance to kanamycin and neomycin and to the penicillins may be added to some or all of the first four drug resistances.

In papers published in Japanese, Ochiai et al. (1959) and Akiba et al. (1960) suggested that strains of E.coli resistant to several drugs might be implicated in generating drug resistant Shigellae in the human intestinal tract; that is, multiple drug resistance might be developed in E.coli cells and then passed to Shigellae. Treatment with any one of the drugs to which these cells are resistant effectively creates a situation in which bacteria carrying the (multiple) resistance possess a selective advantage. This idea was supported by experiments in which multiple drug resistance was transferred from resistant E.coli to Shigellae and from resistant Shigellae to E.coli; only the drug resistance markers appeared to be transferred and conjugation was implicated in the mechanism of transfer because cell free filtrates always were inactive. Subsequent experiments showed that multiple drug resistance can be transferred to many other types of bacteria by mixed culture, including virtually every member of the family Enterobacteriaceae (Mitsuhashi et al., 1960a,b; Harada et al., 1960; for review see Watanabe, 1963a).

In the first report of this system in English, Watanabe and Fukasawa (1961a) summarized a series of papers that they had originally published in Japanese and in which they showed that resistance is transferred only by cell to cell contact in mixed cultures. A cell that acquires resistance by such conjugation becomes a donor able to pass the markers on to fresh recipients in the same way. Transfer of drug resistance by conjugation takes about 15 minutes and the transferred resistance is expressed phenotypically very rapidly. (This contrasts with the much slower phenotypic expression of resistance conferred by chromosomal mutations.) The nature of resistance acquired in this manner clearly is not specific for any bacterial strain, although the level at which it is expressed depends upon the host cell. The mechanisms of resistance have been summarized by Davies and Rownd (1969): resistance to tetracycline ($tet^r$) is acquired by virtue of inability to take up the drug, possibly because of the presence of a mechanism induced by tetracycline itself; chloramphenicol resistance ($cam^r$) depends upon an enzyme, chloramphenicol acetyltransferase, which is synthesized constitutively and performs an O-acetylation with acetyl-CoA as cofactor; resistance to ampicillin ($amp^r$)

is conferred by a penicillinase activity, more properly known as $\beta$-lactamase, which is synthesized constitutively; resistance to the aminoglycosides streptomycin (*str*$^r$), kanamycin (*kan*$^r$) and neomycin (conferred by the *kan*$^r$ gene), depends upon a variety of enzymes that are able to modify these antibiotics.

Resistance is infectious since a small number of resistant cells rapidly converts a large number of cells to the resistant phenotype (that is, at a rate faster than that of cell multiplication). Resistance may be lost spontaneously from a culture, generating segregants sensitive to the drugs. Watanabe and Fukasawa (1961b) reported that multiple drug resistance can be eliminated by treatment with acridine dyes, although elimination is less efficient than with sex factors. This suggests that the genetic element responsible for transfer of drug resistance markers may be a plasmid. The term *R factor* describes the infectious unit carrying multiple drug resistance; this may be considered to consist of two parts, the resistance transfer factor, abbreviated to *RTF*, which is responsible for infectious transfer, and the drug resistance markers, generally described as constituting the *r* component. (More recent analysis of this drug factor has shown that the r component carries all the drug resistance markers except that for tetracycline, which appears to be carried by the RTF; see later). The origin of the drug factor is not known but it is generally presumed that the resistance markers were acquired in a single step from the host chromosome of some unknown bacterium (for review see Watanabe, 1963a).

A strain resistant to four drugs (*sul*$^r$, *str*$^r$, *cam*$^r$, *tet*$^r$) usually transfers all the resistance markers, apparently as a single unit. But spontaneous segregation may occur, most often to generate the factors RTF *sul*$^r$ *str*$^r$ *cam*$^r$ and RTF *tet*$^r$. The frequency of spontaneous segregation varies with the host cell and is highest in Salmonella and lowest in E.coli. In Salmonella, it may be high enough to permit clonal analysis of the segregants; Watanabe (1963a) reported that when 500 cells containing RTF *sul*$^r$ *str*$^r$ *cam*$^r$ *tet*$^r$ were incubated at 37°C for 24 hours, 75% of the clones remained resistant to all four drugs and 25% lacked resistance to some or all of them; the most common factor remaining was RTF *tet*$^r$. The sequences present in the r component may therefore change; presumably all the drug resistance markers might on occasion be lost from an R factor, leaving an RTF lacking drug, which would not be detected.

Recombination can take place when E.coli cells are superinfected with two R factors, each of which carries different drug determinants. Thus upon infection with both RTF *tet*$^r$ and RTF *sul*$^r$ *str*$^r$ *cam*$^r$, Mitsuhashi et al. (1962) were able to recover a four-drug resistance factor. Similarly, superinfection with both RTF *tet*$^r$ and RTF *cam*$^r$ allows the factor RTF *tet*$^r$ *cam*$^r$ to be recovered. The frequency with which these recombinants are generated is low, but their occurrence again demonstrates that the association between RTF and r is subject to change, it being possible for the RTF either to gain or to lose drug resistance markers.

In addition to transfer by conjugation in mixed culture, drug resistance can be transferred by transduction of Salmonella typhimurium LT2 with phage P22 or of E.coli with phage Plkc. Although the drug resistance markers usually all are transferred jointly by conjugation, they may be separated in transductional transfer, presumably because the transducing phage picks up only part of the R factor. Watanabe and Fukasawa (1961c) reported that the resistant transductants of S.typhimurium cannot transfer their resistance to sensitive recipients by conjugation (in contrast with cells that gain the resistance markers by conjugation); transductants of E.coli, however, may be able to act as donors in conjugation. The reason for this difference may be related to the amount of DNA that can be carried by the transducing phage; P22 carries a shorter length of DNA in transduction and may therefore pick up only part of the R factor, whereas phage Plkc carries a greater length and may be more prone to pick up the entire R factor. Resistant transductants inherit the resistance in a stable manner through many generations and so the markers must either be able to exist in a stable extrachromosomal state independent of the RTF component or must have been integrated into the bacterial chromosome (for review see Watanabe, 1963a; see also Watanabe et al., 1968).

A relationship between one of the early Japanese factors, R222, and the F factor was first implied by the observation of Watanabe and Fukasawa (1961a, 1962) that the presence of an F factor reduces the ability of a recipient to accept R222. And when an Hfr or $F^+$ cell receives an R222 factor, its fertility is suppressed: as judged by either donor activity or the production of F pili, the F factor ceases to function. Hfr cells that carry R222 show activities in recombination with $F^-$ cells that are reduced to 1% of their previous donor activity (although when transfer does take place its characteristics are unaltered). That the suppression of F activity is due to the presence of the R factor is demonstrated by the recovery of ability to transfer the F factor (in $F^+$ cells) or to transfer the bacterial chromosome (in Hfr cells) in drug-sensitive revertants that have lost the R factor. The presence in the same cell of an F factor and R factor does not inhibit transfer of the R factor, however, which takes place at the same frequency as that supported by cells that carry only the R factor. Transfer of the R factor is thus independent of transfer mediated by the F factor; and the suppression of activity represents a unidirectional effect in which the R factor represses the F factor.

Another indication of a relationship between the F factor and R222 was provided by analysis of the properties of drug-resistant transductants of Salmonella. Harada et al. (1963, 1964) found that infection with the F*lac* factor allowed transductants that had not been able to transfer their drug resistance by conjugation to do so. The frequency of transfer of drug-resistant markers was about $10^{-5}$, compared with a frequency of transmission of F*lac* of about $10^{-2}$. Recipient cells that received both F*lac* and drug-resistant markers as a result of this conjugation were then able to act as donors for a further round of conjugation; recipients that gained only the drug resistance markers were unable to transfer the markers, that is showed the same inability as the original transductants in the absence of the F*lac* factor. The cells gaining only the drug resistance markers were able to pass them on to their progeny, however, so that these markers appear to be maintained in a stable state in which they are unable to transfer to other cells, but can do so upon introduction of an F factor. The explanation originally proposed for these results was that a complex might be formed between the drug resistance markers and the F*lac* factor; however, Hirota, Fujii and Nishimura (1966) observed complementation between an R factor mutant unable to support transfer and either a free or integrated sex factor, and suggested that this might also explain the earlier results (see also Watanabe, 1963a). The implication of the conclusion that some drug-resistant transductants can be complemented by the F*lac* factor is that they were deficient in conjugation because of the absence (or inactivity) of genes that also are found in the F factor.

### *Circular Structure of R Factors*

The isolation of R factor DNA was first achieved by use of a method developed to isolate the sex factor (see Chapter 2). When transferred to a host such as Proteus mirabilis in which the buoyant density of the plasmid DNA is distinct from that of the chromosome, R factor DNA can be isolated in the form of satellite bands. Some of the R factors that exist as single genetic units in E.coli appear to suffer rearrangements of their components, the RTF and r determinants, when Proteus cells are grown under certain conditions; transfer into Proteus is therefore useful also for investigating the relationship between the RTF and r components of these factors. The buoyant density of R factor DNA is not different from that of E.coli DNA, however, and so isolation of the R factor in the form of the single unit in which it exists in E.coli requires the use of other methods.

Early experiments demonstrated that bacteria carrying R factors possess circular molecules of DNA that are absent from strains lacking the factors; it is now clear that all factors take the form of circular DNA (see Møller et al., 1976). By utilizing the characteristic centrifugation properties of circular molecules, which under appropriate conditions display enhanced buoyant density and increased sedimentation velocity, it is possible to isolate R factor DNA from any host. One technique that is commonly used to isolate circular molecules is density gradient centrifugation in the presence of ethidium bromide. Ethidium bromide intercalates in the DNA duplex to decrease its density; however, the decrease is much less with covalently closed circles of DNA, so that circular molecules can be isolated in ethidium bromide-CsCl gradients by their increased buoyant density relative to other DNA molecules. Two approaches can be used to determine the size of the circular molecule. Covalently closed circles that are supercoiled sediment more rapidly than open circles, which have single strand scission(s) preventing the occurrence of supercoiling; linear molecules sediment more slowly yet. The sedimentation coefficients of these forms, especially those of the closed circle and open circle, can be used to calculate the molecular weight of the DNA (for review see Clowes, 1972). An alternative is to rely upon direct visualization of the DNA by measuring the contour length of circles after spreading for electron microscopy; this measurement yields what appears to be the most accurate determination of the length of the molecule.

Data on the sizes of some classes of R factors are given in Table 3.1. By using ethidium bromide, Cohen and Miller (1969) isolated closed circular molecules of factor R1 from E.coli. In addition to the covalently closed supercoiled molecules obtained by this procedure, they obtained more slowly sedimenting open circular molecules, in which single strand scission(s) had been introduced. From the sedimentation velocities of the two forms, 85S and 51S, it is possible to calculate a probable size for the factor of $63 \times 10^6$ daltons. Its contour length on a grid was 33.1 $\mu$m, which would correspond to a size of $65 \times 10^6$ daltons. The coincidence of these two values lends confidence to this estimate for the size of R1 DNA. In addition to these molecules, there was a smaller number of circles, about 3–6% of the total number, with a contour length of only about 5.5 $\mu$m.

A similar technique was used by Nisioka, Mitani and Clowes (1970) to isolate the factor R100 and various derivatives of it from E.coli. (The same factor has been variously described as 222, R100 and NR1; we shall use the term R100 here, although the original terminology is given in the table.) Figure 3.3 shows density gradients on CsCl run in the presence of ethidium bromide; an $R^-$ strain displays only the peak corresponding to host DNA, but an $R^+$ strain possesses in addition a second, denser peak that corresponds to the covalently closed circles of the R factor. The contour length of the R100 DNA of 33.6 $\mu$m suggests a molecular weight of $66 \times 10^6$ daltons.

Another technique that can be used to identify R factor molecules relies

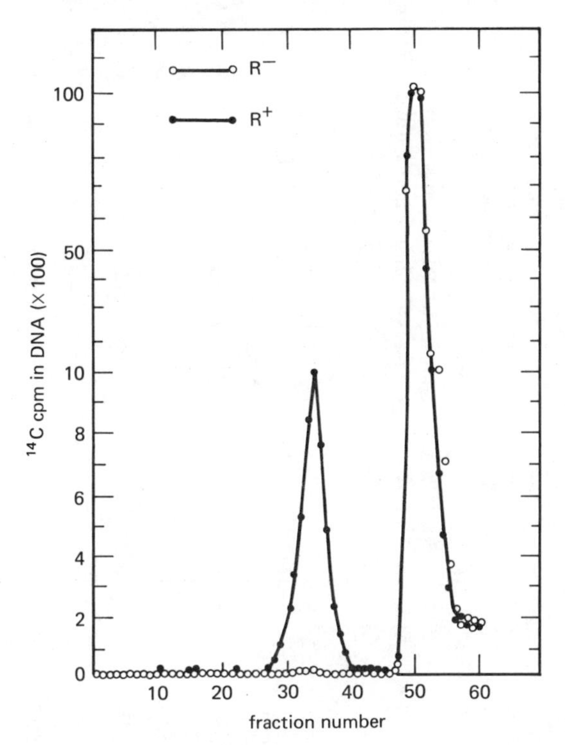


**Figure 3.3:** separation of R factor DNA in E.coli by centrifugation through ethidium bromide–CsC1 density gradients. In the control $R^-$ cells there is only a single peak, representing E.coli chromosomal DNA, density usually about 1.555 g/cm$^3$. In the $R^+$ cells that carry factor 222/R3N (see Table 3.1) there is an additional peak, usually of density about 1.588 g/cm$^3$. The gradients shown above were run by Nisioka, Mitani and Clowes (1970); very similar data were obtained by Cohen and Miller (1969).

upon isolating DNA from minicells (which contain no chromosomal DNA). Inselburg (1971) reported that the DNA of R100 can be found in minicells produced from a strain carrying this plasmid: covalently closed circles could be identified by density gradient centrifugation; and electron microscopy showed the presence of closed circular, open circular and linear molecules.

R factor DNA can be specifically labeled by the same technique developed by Freifelder and Freifelder for isolating sex factor DNA (see Chapter 2). By using this system, Falkow et al. (1971) demonstrated that the label enters two molecular forms of R1 that can be resolved by centrifugation through a neutral sucrose gradient: Figure 3.4 shows that about one-third is found in a peak sedimenting at 75S and about two-thirds

**Table 3.1:** circular lengths of R factors

| Factor | buoyant density ($g/cm^3$) | | contour length ($\mu m$) | sedimentation | | size in daltons | copies per chromo-some | reference |
|---|---|---|---|---|---|---|---|---|
| | E.coli | P.mira-bilis | | closed | open | | | |
| 222 | 1.710 | | 33.6 | | | $66 \times 10^6$ | 1–2 | Nisioka, Mitani and Clowes (1970) |
| | — | | 31.0 | 77S | 52S | $62 \times 10^6$ | — | Inselburg (1971) |
| 222/R3 | 1.710 | | 33.5 | | | $65 \times 10^6$ | 2–3 | Nisioka, Mitani and Clowes (1969) |
| 222/R3 | | 1.708 | 28.5 | | | $54 \times 10^6$ | — | Nisioka, Mitani and Clowes (1970) |
| | | 1.711 | 35.8 | | | $68 \times 10^6$ | — | |
| | | 1.717 | 6.4 | | | $12 \times 10^6$ | — | |
| 222/R3N | 1.710 | | 30.3 | | | $59 \times 10^6$ | 1–2 | Nisioka, Mitani and Clowes (1970) |
| 222/R1 | 1.709 | | 25.6 | | | $50 \times 10^6$ | 1–2 | Nisioka, Mitani and Clowes (1970) |
| R100-1 | — | | 28.2 | | | $55 \times 10^6$ | — | Sheehy et al. (1973) |
| R100-1 | | — | 19.9 | | | $39 \times 10^6$ | — | Sheehy et al. (1973) |
| | | — | 26.3 | | | $51 \times 10^6$ | — | |
| | | — | 5.7 | | | $11 \times 10^6$ | — | |
| R1 | 1.710 | | | 85S | 51S | $63 \times 10^6$ | | Cohen and Miller (1969) |
| | 1.710 | | 33.1 | | | $65 \times 10^6$ | | |
| | | | 5.5 | | | $11 \times 10^6$ | | |
| R1 | 1.711 | | | 75S | 50S | $65 \times 10^6$ | 1–2 | Silver and Falkow (1970a) |
| R1 | | 1.709 | 28.0 | | | $55 \times 10^6$ | | Cohen and Miller (1970a) |
| | | 1.711 | 33.0 | | | $65 \times 10^6$ | | |
| | | 1.717 | 5.5 | | | $10 \times 10^6$ | | |
| R1-RTF | 1.709 | | 28.0 | | | $55 \times 10^6$ | | Cohen and Miller (1970b) |
| R1-RTF | 1.709 | | | 64S | 44S | $52 \times 10^6$ | | Silver and Falkow (1970b) |

**Table 3.1:** (continued)

| Factor | buoyant density (g/cm³) | | contour length | sedimentation | | size in | copies per chromo- | reference |
|---|---|---|---|---|---|---|---|---|
| | E.coli | P.mirabilis | ($\mu$m) | closed | open | daltons | some | |
| R15 | 1.708 | | 22.3 | | | $44 \times 10^6$ | 1–2 | Nisioka, Mitani and Clowes (1970) |
| R15 | | 1.709 | 18.3 | | | $35 \times 10^6$ | | Nisioka, Mitani and Clowes (1970) |
| R28K | | 1.710 | 21.4 | | | $42 \times 10^6$ | 2–3 | Kontmichalou, Mitani and |
| R6K | | 1.704 | 12.8 | | | $24 \times 10^6$ | 13–28 | Clowes (1970) |

When the buoyant density of a factor was determined, the value is shown in the appropriate column; when the factor was isolated in minicells and no value was determined, a dash in this column indicates the appropriate source.

The R factor 222 carries resistance to four drugs, streptomycin, sulfonamide, chloramphenicol, tetracycline; 222/R3 is an early variant that is resistant only to the first three drugs, having lost its resistance to tetracycline. The similar sizes of 222 and 222/R3 suggest that a small change may be responsible for the loss of tetracycline resistance. Factors 222/R3N and 222/R1 represent independent isolates; 222/R3N carries the same three-drug resistance as 222/R3 while 222/R1 carries resistance only to tetracycline. Their reduced sizes suggest that they may have arisen by deletions of 222. The factor R100-1 is a derepressed variant of factor 222 (the original factor has been described as 222, R100 and NR1); the reduction in size of the two larger components suggests that it may have arisen by a deletion located largely or entirely in the RTF part. This factor was isolated from minicells of E.coli (first column) or minicells of S.typhimurium (shown in the P.mirabilis column).

The factor R1 carries resistance to ampicillin, kanamycin, streptomycin, sulfonamide, chloramphenicol; the R1 RTF was isolated by examining cells that do not possess drug resistance for covalently closed circles but which either have been mated with resistant cells or carry the sensitivity to phage characteristic of R1 (see text).

The factor R15 is a *fi*$^-$ factor that is different in origin and behavior from the two *fi*$^+$ factors (222 and R1). It does not dissociate into separate bands in Proteus. Factors R28K and R6K also are *fi*$^-$

in a peak sedimenting at about 50S. These peaks would correspond to covalently closed and nicked open circles of duplex DNA of about 64 × $10^6$ daltons. The circularity of the R factor DNA was confirmed by sedimentation on an alkaline sucrose gradient; when DNA is denatured to single strands, circles sediment 3–4 times faster than single strands with a molecular weight of half that of the circle. Two fractions were apparent after denaturation, again with the more rapidly sedimenting peak representing about one-third of the total label. The introduction of single strand breaks by X-irradiation caused the predicted discontinuous shift from the more rapidly sedimenting peak to the more slowly sedimenting form.

These sets of data all therefore suggest that the factors R100 and R1 both have a size of about 65 × $10^6$ daltons (about $10^5$ base pairs) and can be isolated from E.coli in the form of covalently closed and nicked open circles of duplex DNA.

When R factors are transferred into Proteus mirabilis, they may form more than one band on CsC1 density gradients. The host chromosome has a G+C content of almost 40%, with a corresponding buoyant density in

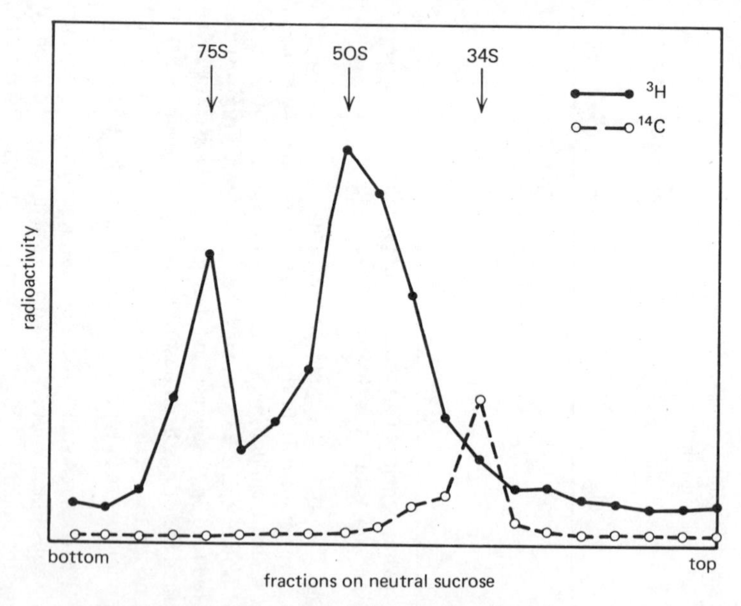


**Figure 3.4:** sedimentation of R factor DNA in neutral sucrose. R1 DNA was labeled with $^3$H-thymine by the Freifelder protocol and extracted after 60 minutes. The 75S peak represents covalently closed circles and the 50S peak represents open circles bearing a nick(s) in one strand. The $^{14}$C label resides in a control of λ DNA, which sediments at 34S in linear duplex form. Data of Falkow et al. (1971).

CsCl of about 1.698 g/cm³. Early experiments resolved two satellite bands of R factor DNA, obtained with several factors, including R100 and R1. Falkow et al. (1966) reported that bands at 1.710 g/cm³ and 1.716 g/cm³, representing G+C contents of 50% and 56%, were present in amounts of about 7% of the main band (chromosomal) DNA. Mutant factors that had lost some of their drug resistance markers displayed changes in the proportions and/or densities of these bands, as well as a reduction in the total amount of R factor DNA; these mutants therefore probably represented deletions. Similar bands in Proteus were observed by Rownd, Nakaya and Nakamura (1966), occupying density positions of 1.712 g/cm³ (52% G+C) and 1.718 g/cm³ (58% G+C), and representing about 8% of the total DNA. The proportions of these bands again varied with the factor and with the conditions of growth of the host cells. Curing with acridine orange caused loss of the bands, consistent with the idea that they represent plasmid DNA.

The nature of the drug factor DNA present in these bands has been a matter of some controversy. We shall discuss later the effect upon the bands of changing the growth conditions of Proteus and the nature of the rearrangements of the RTF and r determinants that can take place in these conditions. Here, however, we shall be concerned with those experiments that identified the RTF and r components in these bands and thus with the division of the R factor into the sequences of its constituent units. In this context, it does not matter whether these isolated components represent the major types of molecule present in each band, although we should note that later reports imply that other types of molecule may be found in these bands.

The three density bands shown in Figure 3.5 were isolated for factor 222/R3 (a derivative of 222 = R100 that has lost tetracycline resistance) by Nisioka, Mitani and Clowes (1969). The figure also shows a similar density gradient run with DNA extracted from E.coli cells carrying this plasmid. As summarized in Table 3.1, the E.coli cells possess a single band of 1.710 g/cm³, whereas the cells of P.mirabilis possess bands of 1.708 g/cm³, 1.711 g/cm³ (this most closely corresponds to the single band in E.coli) and 1.717 g/cm³. A similar analysis with the factors R1 and R6 was performed by Cohen and Miller (1970a) and, as Table 3.1 reports, virtually identical results were obtained. By isolating the DNA of each peak, it is possible to examine the molecules by electron microscopy to determine what sizes of circular molecules are present. Each Proteus band predominantly contained only a single circular size class: the heaviest band, 1.717 g/cm³, contained only molecules of about 6 $\mu$m in length; the lightest band, 1.709 g/cm³, contained molecules of about 28 $\mu$m in length and also some of size 33 $\mu$m; and the intermediate band, 1.711

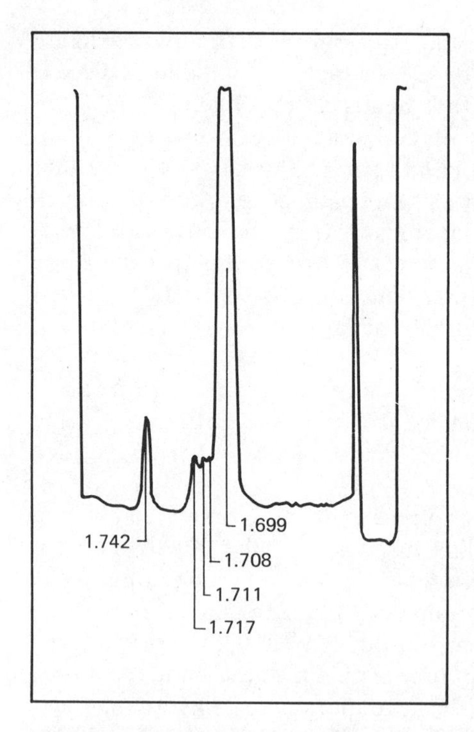


**Figure 3.5:** buoyant density of R factor DNA. The DNA of P.mirabilis carrying factor 222/R3 was centrifuged through a CsC1 density gradient. The peak at 1.699 g/cm³ represents the Proteus chromosomal DNA; the peak at 1.742 g/cm³ is a control of the highly G-C rich phage SP01 of B.subtilis. The R factor DNA forms three peaks, 1.708 g/cm³, 1.711 g/cm³, 1.717 g/cm³. In other experiments, only two peaks of R factor DNA, of 1.710 g/cm³ and 1.717 g/cm³ have been resolved, in which case the 1.710 g/cm³ peak presumably contains the molecules present in both the less dense R factor peaks shown above. Data of Nisioka, Mitani and Clowes (1969).

g/cm³, principally contained the longest molecules, of about 33 $\mu$m, but also had some of 28 $\mu$m. The single buoyant density band in E.coli contained only molecules of the longest class.

The combined length of the 6 $\mu$m and 28 $\mu$m molecules found in the heaviest and lightest density classes in Proteus corresponds well with the length of the longest molecules in Proteus and with that of the single class in E.coli. This suggests that the R100 and R1 factors exist in E.coli as circular molecules of length 33 $\mu$m and buoyant density 1.710 g/cm³. In Proteus, however, only some molecules were found in this condition; many appeared to have dissociated into component elements of 6 $\mu$m and 1.717 g/cm³ and of 28 $\mu$m and 1.709 g/cm³. This relationship is consistent with both the lengths and buoyant densities of the molecules in all three size classes, since the joining together of these molecules should generate a species with the combined length of both and an intermediate buoyant density. (Since the smallest species is sometimes observed in E.coli, dissociation into these components must presumably take place occasionally in this host also.)

When $R^+$ cells are mated with $R^-$ recipients, the progeny usually are selected for one of the drug resistance markers carried by the plasmid.

Since ability to transfer between cells lies with the RTF component, this procedure selects for transfer of the entire R factor (that is RTF + r), for the RTF does not carry the selected drug resistance determinants and the r component cannot transfer between cells. (We shall see later that the RTF component in fact carries the tetracycline marker, but that all other markers appear to be carried on the r determinant; thus this statement remains true for selection for all drug resistance markers except that for tetracycline.) Cohen and Miller (1970b) argued that if $R^+$ Proteus cells are used as donors and the recipients are not selected for drug resistance, then some should gain the RTF alone. These cells can be characterized by their drug sensitivity, in spite of which they possess circles of DNA. By using several successive matings between resistant Proteus and sensitive E.coli, to increase the probability of transfer, they were able to identify drug-sensitive cells carrying circular molecules of buoyant density 1.709 g/cm$^3$ and length 28 $\mu$m. This suggests that the 28 $\mu$m molecule represents the RTF component and the 6 $\mu$m molecule represents the r component.

Another approach for isolating the RTF was developed by Silver and Falkow (1970a,b), who made use of derepressed R factors that are able to transfer between cells at much increased frequencies (see later). By using the labeling technique originally developed by Freifelder for the sex factor, they were able to identify closed circles of factor R1 in E.coli, with a density of 1.711 g/cm$^3$ and a size of $65 \times 10^6$ daltons. They then isolated from this strain of E.coli a variant that was sensitive to all of the drugs against which R1 confers resistance, but which remained sensitive to the phage MS2 (this RNA phage is able to lyse cells carrying either the F factor or certain R factors, including R1). The idea that this variant had lost the drug resistance determinant r but had retained the RTF component was supported by the observation that it possessed a DNA fraction of buoyant density 1.709 g/cm$^3$ and size $55 \times 10^6$ daltons.

These experiments therefore suggest that the R1 factor can exist as a circular molecule of about $65 \times 10^6$ daltons (roughly 100,000 base pairs). In Proteus it may be obtained in the form of an RTF component of about $55 \times 10^6$ daltons (82,000 base pairs), which alone remains infectious, and an r component of about $12 \times 10^6$ daltons (18,000 base pairs), which although not infectious can be maintained in the cells. Dissociation of the R factor into RTF and r components does not take place often in E.coli, but may occur upon occasion. In terms of gene numbers, the RTF could carry about 85 genes of average length 1000 base pairs and the r component could include about 15 genes of this size.

Analysis of variants of R100 suggests that deletions may on occasion be responsible for the loss of drug resistance markers. The factor R100, then described as the isolate 222, was one of the early Japanese factors carrying resistance to four drugs, streptomycin, sulfonamide, chloramphenicol and tetracycline. The variant 222/R3 was an early isolate that had lost tetracycline resistance. As Table 3.1 shows, Nisioka, Mitani and Clowes (1970) found that the length of this factor is the same as that of parent strain (within experimental error), suggesting that a small change is responsible for the loss. We shall see in the next section that tetracycline resistance may be lost by insertion of a short length of DNA close to the *tet*$^r$ marker. Two other factors, 222/R3N and 222/R1, were isolated by Nisioka et al. for changes in the ability to confer drug resistance; 222/R3N has lost tetracycline resistance and 222/R1 has lost all the drug resistance markers except that for tetracycline. Both proved to be smaller than the parent factor, apparently representing substantial deletions. The importance of the isolation of such variants is that they allow heteroduplex mapping to be applied to define the physical locations of the drug resistance markers.

One factor that has been used in many studies is the derivative R100-1, which is derepressed and as a result transfers more frequently between cells (see later). This variant has been studied by Sheehy et al. (1973), who isolated it from minicells. In E.coli minicells the factor forms a single class of circular molecule of some 28 $\mu$m in length; in minicells of Salmonella typhimurium it dissociates to generate two additional sizes of circle, one of almost 6 $\mu$m and one of 20 $\mu$m. Comparison with the analysis of the parent factor (R100) in E.coli and in P.mirabilis suggests that R100-1 has arisen by a deletion that is largely or entirely located in the RTF component, reducing its length from 28 $\mu$m to 20 $\mu$m.

Another R factor system has been studied by Anderson (1968); S.typhimurium of phage type 29 may carry resistance to the four drugs ampicillin, sulfonamides, streptomycin and tetracycline. Unlike R100 and similar (*fi*$^+$) factors, these drug resistance markers do not form a single linkage group; they are of the *fi*$^-$ type and are unrelated to the sex factor. Resistance to sulfonamides and streptomycin forms a single group as these markers are never separated; resistance to either ampicillin or tetracycline is independently acquired. All the markers are transmitted infectiously. Infectious *amp*$^r$ or *sul*$^r$ *str*$^r$ is transferred to recipients of E.coli K12 with a frequency of about $10^{-2}$ in overnight crosses, and is then transferred from these cells with the same frequency in further crosses; transfer of these two groups is independent as the number of recipients receiving both *amp*$^r$ and *sul*$^r$ *str*$^r$ is only about 10% of the number receiving either group alone. Infectious *tet*$^r$, however, is transferred from its host to E.coli K12 recipients at the more infrequent level of $<10^{-6}$ under the same conditions, but the cells that receive the *tet*$^r$ marker become very good donors, showing a frequency of $>5 \times 10^{-1}$ in further crosses (Anderson and Lewis, 1965).

Crosses involving either the *amp*$^r$ or *sul*$^r$ *str*$^r$ determinants can produce recipients that possess the drug resistance markers but are no longer infectious, or which have the infectious capacity but are no longer drug resistant (this is in addition to the recipients that gain both infectious capacity and drug resistance). Anderson proposed that the S.typhimurium phage 29 host cells contain as independent elements an infectious transfer unit termed Δ (this, of course, is directly analogous to RTF), and three independent elements carrying drug resistance markers, *amp*$^r$, *sul*$^r$ *str*$^r$, and *tet*$^r$. The factor Δ promotes infectious transfer between cells, either as an individual element or in association with one of the drug resistance elements; the low and stable frequency of transfer of *amp*$^r$ and *sul*$^r$ *str*$^r$ suggests that Δ associates only transiently with these elements for transfer, dissociating after entering the recipient cell; reassociation must therefore take place before another transfer can occur. Cells that gain *amp*$^r$ or *sul*$^r$ *str*$^r$ but fail to gain Δ can be recognized by their possession of noninfectious drug resistance; cells that gain Δ but do not receive the drug resistance elements remain drug sensitive, but the presence of Δ can be revealed by crossing them with recipients that carry noninfectious *amp*$^r$ or *sul*$^r$ *str*$^r$, for this converts these recipients into infectious carriers of their drug resistance marker. Anderson (1968) reported that in crosses where no selection is imposed for drug resistance, more than half of the recipients may receive only Δ; however, this frequency depends on the recipient, an effect that is not understood. The low initial frequency of transfer of *tet*$^r$ and the much higher retransfer frequency suggests that the association of Δ and *tet*$^r$ may take a different form; association may occur less frequently, but then is stable so that the Δ-*tet*$^r$ element does not dissociate and retransfers efficiently. Cells that gain *tet*$^r$ therefore always also gain Δ.

A transductional analysis carried out by Anderson and Natkin (1972) supports this model. Phage Plkc was able to transduce Δ-*tet*$^r$ as such and was able also to transduce a stable Δ-*amp*$^r$ factor as a single unit (this factor was formed by ultraviolet irradiation of the phage 29 type Salmonella); but when E.coli cells contained Δ and *amp*$^r$ and *sul*$^r$ *str*$^r$ as independent entities, each was transduced separately.

In the host bacterial cell, each of the four host elements Δ, *amp*$^r$, *sul*$^r$ *str*$^r$ and *tet*$^r$ must represent an independent replicon that is maintained in the cell line by replication and then segregates at division. All four elements must be compatible. An incompatibility effect suggesting that only one copy of the *sul*$^r$ *str*$^r$ element can be maintained in the cell has been reported by Anderson et al. (1968); this made use of an unusual element in which *amp*$^r$ appeared to have replaced *str*$^r$ on the *sul*$^r$ *str*$^r$ element. The two elements *sul*$^r$ *str*$^r$ and *sul*$^r$ *amp*$^r$ could not coexist in the same cell line, suggesting that usually only one copy of the *sul*$^r$ *str*$^r$ element may be present. The presence of Δ did not affect this relationship.

This model for the Δ transfer element and the three drug resistance elements has been directly confirmed by examination of the covalently closed circles present in appropriate cell lines of E.coli. Milliken and Clowes (1973) showed that cells carrying Δ alone possess only a single circular species, of a contour length of 29.1 μm, which corresponds to a size of about $60 \times 10^6$ daltons. Cells carrying only the *sul*$^r$ *str*$^r$ element have circles of contour length 2.9 μm, presumably representing less than ten genes. Cell lines that carry both Δ and *sul*$^r$ *str*$^r$ display both types of circle but do not have any circles large enough to represent covalent linkage between them. Estimates of the amount of plasmid DNA suggest that there is only one copy of the Δ element per bacterial chromosome, consistent with its proposed role, but contradict the apparent restriction on the number of *sul*$^r$ *str*$^r$ elements by showing that there appear to be 14 per bacterial chromosome. In cells that carry Δ and *amp*$^r$, circles of 29.5 μm and 2.7 μm were present; there appeared to be no circles representing covalent linkage of the two elements, although the results do not completely exclude the possibility that one or two longer circles might represent such a form. In cells to which Δ-*tet*$^r$ had been transferred, there were circles only of 32.3 μm, suggesting that the element is of about the same size as the other two drug resistance elements, but that for transfer it becomes covalently linked to Δ. Although all of similar size, the three drug resistance elements therefore vary in the nature of their association with Δ as predicted by the genetic analysis; the failure to observe any covalently linked Δ-*amp*$^r$ or Δ-*sul*$^r$ *str*$^r$ molecules implies that Δ can mobilize these elements for transfer without achieving covalent linkage, perhaps relying upon a hydrogen bonded association (for review see Helinski, 1973).

### *Insertion Sequences in F-like R Factors*

The sequences of several R factors have been analyzed by heteroduplex mapping. These factors all are related to the sex factor and have a substantial region that is largely homologous with a part of the F sequence that includes the transfer genes. These *F-like* R factors all have closely related sequences. All three classes of insertion sequence are found in the R factors; and in particular they appear to be implicated in the junction between the RTF section and the r determinant. Insertion sequences also are found in other regions of these drug factors, sometimes associated with a change in drug resistance.

Three R factors have been studied in detail. The factor R6 carries resistance to streptomycin, sulfonamides, chloramphenicol, kanamycin and tetracycline; the variant R6-5 has lost the ability to confer resistance to tetracycline. The factor R1 carries resistance markers for streptomycin, sulfonamides, chloramphenicol, kanamycin and ampicillin. The factor R100 (also known as 222 and NR1) is one of the most widely investigated; this is the classical four-drug factor conferring resistance to streptomycin, sulfonamides, chloramphenicol and tetracycline. When Sharp, Cohen and Davidson (1973) formed heteroduplexes between R6 and F or between R1 and F, they found that more than 90% of the sequences lying between 49.6 F and 3.0 F (that is from 49.6 F through 94.5/0 F to 3.0 F) are homologous with sequences present in both R factors. The same homology is seen with the factor R100 and also with the factor ColV-K94. The sex factor can therefore be considered to comprise two classes of region. Part of its genome is shared with other plasmids and this includes the region in which the genes responsible for transfer are located (see Figures 2.25 and 3.10); by virtue of this homology with the sex factor, these plasmids may be described as the F-like class. The remaining sequences of the sex factor are unique to it.

The structure of R6 is depicted in Figure 3.6. The coordinates of the R factor are given in kilobases. The starting point for their assignment is the coordinate 91.1, which is common to the sex factor and the F-like R factors; each R factor is individually given coordinates counting back from 91.1 until the locus zero is reached. This means that the same sequence on two related R factors may have a different coordinate on each (see Figure 3.7); when describing a position on an R factor, it is therefore necessary to specify the factor.

By forming a heteroduplex between the isolated RTF determinant and a complete R factor, it is possible to distinguish the sequences of the two components: the RTF anneals to form a duplex structure but the r determinant shows as a large deletion loop. The RTF occupies the sequences outside 3.0 and 29.0 on R6; the length of the RTF is therefore 71,000 base pairs and the length of the r determinant is 26,000 base pairs. Similar lengths are found in the other factors. The RTF was originally isolated by the deletion of 22,500 base pairs from R1, so this must be the length occupied by the r determinant in this factor; since the total length of R1 is 86,300 base pairs, the length occupied by its RTF sequences must be 64,000 base pairs. The lengths of the RTF and r determinants measured in these experiments represent the most accurate estimates yet obtained;

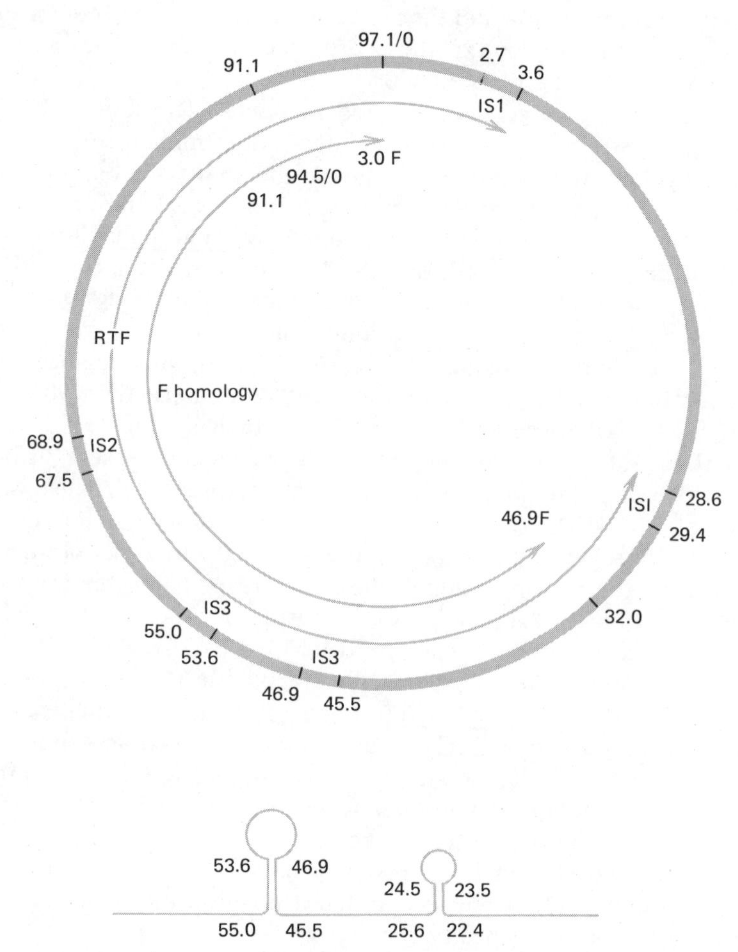


**Figure 3.6:** structure of factor R6. The factor comprises a circular duplex of 97,100 base pairs; its coordinates are marked in kilobases with reference to the point 91.1, which is identical in F and the F-like R factors. The RTF corresponds to the sequences outside 3.0–29.0; both junctions between RTF and r determinant therefore possess IS1 elements. The part of the RTF between 32.0 and 97.1 is largely homologous with the sex factor sequences from 46.9–3.0 F, although there are many substitutions within this region which distinguish the sequences of the drug and sex factors (see Figure 3.7). An IS2 element lies at 67.5–68.9 within the RTF. Two IS3 elements organized in opposite orientation lie at 55.0–53.6 and at 46.9–45.5. In denatured R6 DNA these may form the inverted loop structure drawn in the lower part of the figure; a second inverted loop is formed by annealing between sequences at 25.6–25.4 and at 23.5–22.4. In factor R6-5 a third IS3 element is inserted within the large loop, to occupy the R6-5 coordinates 51.7–50.3 (this means that any sequence between 51.7 and 0 on R6 DNA differs by 1.4 from the coordinates it is ascribed on R6-5 DNA). Data of Hu et al. (1975).

they are slightly different from, but reasonably close to, the earlier estimates of contour length.

As is evident from these measurements, some variation occurs in the lengths of the RTF and r determinants. The heteroduplexes formed between the R factors confirm that although their sequences are largely homologous, there may be substitution and deletion loops distinguishing the genomes. Figure 3.7 shows an illustration of the heteroduplex formed between R6-5 and R1 which demonstrates that relief from the general homology is seen at several sites. The same is true of the heteroduplex between R6-5 and F, which shows extensive homology between the corresponding regions of the two factors, but with insertions and deletions that render these regions nonidentical and change their overall length.

Although the R100 factor was isolated in Japan and R6 was isolated in Germany, their derivatives R100-1 and R6-5 proved to be almost completely homologous when examined by heteroduplex mapping; R6-5 possesses three insertions in addition to the sequences of R100-1. Presumably the Japanese factor migrated to Europe and gained some additional sequences before its isolation in the form of the R6 factor. This close relationship but nonidentity between the drug resistance factors emphasizes the point that the occurrence of evolutionary changes in R factors may be frequent.

Analysis of two factors has demonstrated that the marker for tetracycline resistance resides on the RTF. A *tet*$^{s}$ derivative of R100 was found to represent a deletion of the parent factor; this identifies the location of the marker for tetracycline resistance. The R1 factor, which is not resistant to tetracycline, also lacks this sequence, which takes the form of an *inverted repeat*. The factor R6 and its derivative R6-5 differ only in the insertion of a small additional sequence in the loop of this inverted repeat; this insertion must therefore be responsible for the loss of tetracycline resistance by R6-5. This locates the marker for tetracycline resistance within the loop at 53.6–46.9 on the R6 factor; this lies within the region of the RTF (see Figure 3.6). Heteroduplex mapping of R factors that carry deletions for other drug resistance markers has shown that these all reside on the r determinant.

The presence of insertion sequences in the R factors was examined by Hu et al. (1975a). By annealing R6 with an excess of the phage λ*r14*, which carries an IS1 element (see page 121), they were able to visualize about 1% of the R6 strands as R6-λ*r14* heteroduplexes. The short duplex region of 800 base pairs identified the IS1 element on the R factor; this occurs at two locations, 28.6-29.4 and 2.7-3.6. These two IS1 sequences map (within experimental error) at the junctions of the RTF and r deter-

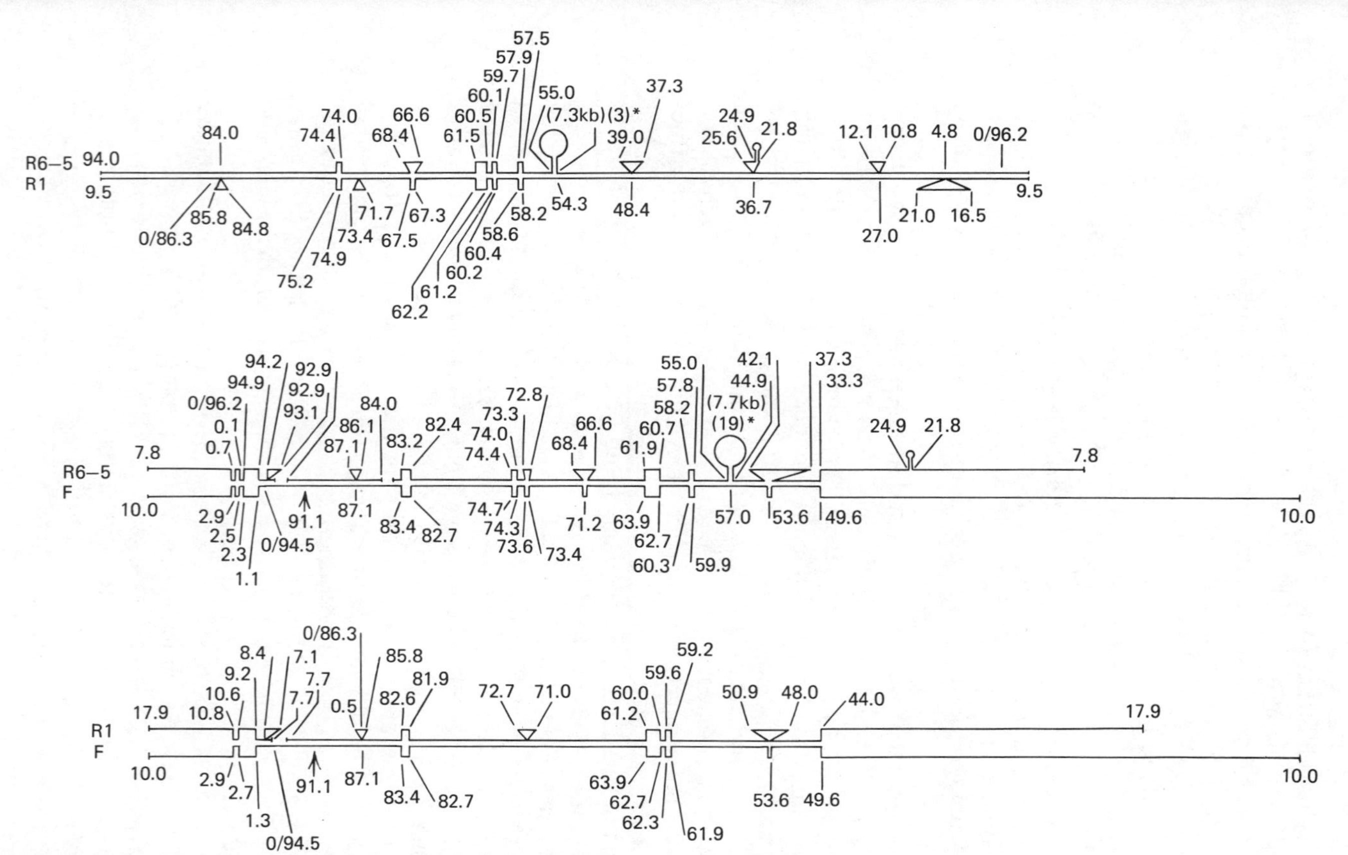


**Figure 3.7:** structures of the heteroduplexes formed between R6-5 and R1, R6-5 and F, and R1 and F. Although in reality circular, these are represented as linear structures. Substitutions are indicated by rectangles and deletions by triangles. (The coordinates given in this figure are not immutable and are subject to revision as more accurate measurements are made; they may therefore be small discrepancies between the values given here and those quoted in more recent work.) Data of Sharp, Cohen and Davidson (1973).

minants, as illustrated in Figure 3.6. Since the positions of the RTF-r junctions are identical in the factors R6, R1, R100 (and in their derivatives), this implies that IS1 elements may constitute these junctions in all these plasmids. This suggests that recombination between IS1 elements may provide the mechanism by which an R factor can dissociate into its RTF and r determinants (and also by which other forms of the factor may arise: see below). The parallel with the role for insertion sequences in the sex factor implied by their presence at junctions with chromosomal DNA is obvious (see page 125).

A single IS2 element is found on R6, at the position 67.5-68.9. Whether this has any special significance for the physiology of the plasmid is not known.

Following denaturation and renaturation, the DNA of R6 may generate the two looped structures illustrated in Figure 3.6 by intramolecular annealing. Each consists of a single stranded loop joined by a duplex region to the remaining single stranded regions of the DNA. The formation of this structure is due to the presence of an inverted repeat; that is the same sequence is found but in opposite orientation at (for example) the positions 25.6-24.5 and 23.5-22.4. No wider significance has been attached to the sequences that comprise the inverted repeat of this smaller loop. But the sequences that anneal to form the duplex stem of the larger loop have been shown by Ptashne and Cohen (1975) to represent IS3. This means that R6 must carry two IS3 elements, at 53.6-55.0 and at 46.9-45.5 in opposite orientations. Hu et al. (1975a) showed that a sequence identical to that of the stem of the large loop comprises the insertion that is present in the variant R6-5; this therefore represents a third IS3 element, inserted at the R6 coordinate 51.7. This is the insertion that identifies the position of the marker for tetracycline resistance; the loss of the drug resistance that is caused by this IS3 element is analogous to the polar effect that IS elements exert in bacterial operons (see page 124).

### *Translocation of Drug Resistance Markers*

Mobility has been associated with several drug resistance markers. An insertion conferring resistance to tetracycline was apparently gained by phage P22 upon growth in host cells of S.typhimurium carrying a *tet*$^r$ R factor. Kleckner et al. (1975) demonstrated that the *tet*$^r$ marker carried by the phage takes the form of an 8.3 kb insertion which represents a sequence flanked by an inverted duplication, that is, the structure formed upon denaturation and renaturation is the duplex stem and single stranded loop illustrated in Figure 3.6 for the region associated with tetracycline resistance in R6. Independently isolated *tet*$^r$ variants of P22 carried the same insertion located at different sites in the phage genome. By using the *tet*$^r$ phage variants to infect cells under conditions when the phage cannot maintain itself in the host, it is possible to isolate rare bacterial strains that have gained the *tet*$^r$ marker. Some of these strains are auxotrophs, that is, they have lost the ability to express some nutritional bacterial function; and revertants to prototrophy that regain the bacterial function always lose their resistance to tetracycline. This suggests that the *tet*$^r$ marker can be inserted at several sites in the bacterial genome, to cause loss of whatever function is specified by the gene in which it inserts; and excision can take place to restore exactly the original sequence of the bacterial gene. Another situation in which the *tet*$^r$ marker was transferred to the bacterial chromosome was provided by Foster, Howe and Richmond (1975), who attempted to force two incompatible derivatives of R100-1 to coexist in the same cell; the drug factors carried the markers *cam*$^s$ *tet*$^r$ and *cam*$^r$ *tet*$^s$ and cells were selected for Cam$^r$ Tet$^r$. The stable derivatives that were isolated all carried the marker *tet*$^r$ on the bacterial chromosome.

The unit carrying the *tet*$^r$ marker, the sequence flanked by the two inverted IS3 elements, therefore appears to comprise a mobile determinant, apparently displaying the capacity for insertion and excision that is held by the insertion sequences themselves. This capacity has been described as *translocation* and the unit that is transferred has correspondingly been termed a *translocon*. The translocon carrying the marker for tetracycline resistance has been given the abbreviation TnD.

Another translocon, TnA, carries resistance to ampicillin. This drug resistance is associated with the enzyme TEM-$\beta$-lactamase; the same enzyme is specified by many different, not necessarily related, plasmids that carry the marker *amp*$^r$. Hedges and Jacob (1974) demonstrated that plasmids gaining this marker show an increase in molecular weight of about $3 \times 10^6$ daltons (4.5 kb); and Heffron et al. (1975) showed that a sequence of this length is common to a variety of plasmids that specify the TEM-$\beta$-lactamase. The TnA element has been visualized by electron microscopic mapping following its transfer to the plasmid RSF1010; Heffron, Rubens and Falkow (1975) generated plasmid recombinants from cells carrying R64-1 (which includes TnA) and RSF1010 (which confers resistance to streptomycin and sulfonamides) and examined the resulting variants of RSF1010 that carry ampicillin resistance. RSF1010 is a small plasmid of $5.5 \times 10^6$ daltons; its *amp*$^r$ derivatives have a size of $8.7 \times 10^6$ daltons. Heteroduplex mapping showed that the insertion of 4.5 kb that confers resistance to ampicillin can be located at any one of at least twelve sites within a $2 \times 10^6$ dalton sequence of RSF1010 DNA; insertions probably are not isolated within the remaining part of the molecule because they would inactivate essential genes. Electron microscopic analysis shows that the insertion is bounded by inverted repeat sequences; these appear to be rather short, less than 100 base pairs in length. Bennett and Richmond (1976) demonstrated that translocation of an element carrying *amp*$^r$, that may be identical to TnA, is independent of *recA* function.

Another drug resistance determinant that has been identified in the form of an insertion is the marker *cam*$^r$, which codes for the enzyme chloramphenicol acetyltransferase. Gottesman and Rosner (1975) isolated two independent variants of lambda carrying this marker; each has a single insertion loop of about 2.6 kb when examined by heteroduplex mapping, the insertion lying at different sites in the two variants. No inverted repeats were seen in these experiments, although they might be present if rather short. Translocation of this element was independent of host *rec* and phage *red* functions.

Kanamycin resistance also may be transferred as part of a longer sequence. Berg et al. (1975) isolated a phage transducing the marker *kan*$^r$ after lysis of cells carrying an R factor that specifies this drug resistance. Two such phages specified the enzyme neomycin-kanamycin phosphotransferase and possessed insertions of 5.2 kb, which were bounded by inverted repeats of 1.45 kb; the two insertions were located at different sites on phage lambda.

The largest translocated element yet identified is TnC, which carries the drug resistance markers for trimethoprim and streptomycin and has a size of about 15 kb. Barth et al. (1976) showed that this sequence can be translocated between plasmids, when it may be inserted at any one of a number of sites on a given plasmid, and also may be acquired by the bacterial chromosome at an apparently specific site. Its translocation is independent of *recA* function. It is not yet known whether this sequence is bounded by inverted repeats.

A recombinational event that appears to be related to the movement of translocons has been observed by Kopecko and Cohen (1975). Using *recA*$^-$ hosts carrying the small plasmid pSC101, which has the marker *tet*$^r$, and the large plasmid pSC50, which has the markers *amp*$^r$, *cam*$^r$, *str*$^r$, they isolated clones possessing recombinant plasmids that carried the two markers *tet*$^r$ and *amp*$^r$. These recombinants were generated by the insertion of the entire sequence of pSC101 into a site on pSC50 at which a large part of the pSC50 sequences were deleted; that is, pSC101 was substituted for a large sequence of pSC50. The site on pSC50 at which this event occurs is unique; any one of at least four sites on pSC101 can be used. At the site of the substitution, the recombinant plasmid possesses an inverted repeat of about 130 base pairs. It seems likely that the *recA*-independent process responsible for the insertion may be related to that responsible for translocation of drug resistance markers.

The events responsible for the mobility of the translocon elements carrying drug resistance markers have not yet been defined. The mobility of these elements shares with the three insertion sequences, IS1, IS2, IS3, an independence of known bacterial or phage recombination functions. While the functions responsible for translocation remain to be identified, therefore, it is tempting to speculate that the mobile elements may code for enzymes that undertake insertion and excision. In the case of the three insertion sequences, the lack of homology between their DNA would imply that in this case each must specify a different enzyme. How the mobility of translocons that include IS elements may be related to the mobility of the insertion sequences themselves is not known; nor is it possible yet to say whether there is a common mechanism or a variety of mechanisms for translocation of different elements.

Since the insertion sequences and translocons maintain their integrity when transferred from one site to another, the ends of each mobile element must be distinct; it is possible that it is these terminal sequences that represent the critical feature in mobility. The identification of inverted repeats at the extremities of several translocated sequences hints that this feature may have some functional significance, although its nature is not yet clear. Perhaps it may be related to the stability of the unit; recombination between the inverted repeats by a conventional genetic exchange will not release the element from the genome containing it, but simply reverses its orientation. Translocation presumably does not involve a free form of an IS or Tn element, but may be accomplished by direct transfer from one genome to another. The evidence that is available suggests that translocation involves the insertion of the mobile element at sites that are not random although their selectivity may be low. Excision may restore exactly the original sequence of the genome, or may in some instances cause deletions. A similar capacity to insert and excise at many locations in a foreign genome, without causing any alterations of base sequence, is displayed by the bacteriophage mu; it is not known if there is any similarity in the events involved. Whatever the nature of the process by which translocation is accomplished, however, it seems likely that the existence of translocons carrying drug resistance markers may make an important contribution to the reassortment of these markers that is seen among the R factors.

### *Reorganization of R Factor Components*

Sex and drug factors can be isolated as satellite bands of buoyant density distinct from that of the main band representing chromosomal DNA when they are perpetuated in Proteus mirabilis. However, the conditions in which the Proteus cells are grown influence the densities and proportions of the bands that are present. Earlier experiments equated each density band with a single type of DNA molecule; but more recently it has been shown that more complex molecules may be generated by a rearrangement of the components of the R factor. Although of undoubted importance in establishing the nature of the components of the R factors, the earlier experiments therefore do not resolve the issue of the state in which R factors are maintained in Proteus. It is possible that a density band may contain more than one type of structure derived from an R factor. To define the status of the R factor components in Proteus therefore requires a determination of which is the predominant class of molecules in each band, a point that may be difficult to establish, especially since larger molecules are likely to suffer preferential breakage relative to smaller molecules during extraction.

Early experiments identified two buoyant density bands representing R factor DNA in Proteus, in the ranges 1.710–1.712 g/cm$^3$ and 1.716–1.718 g/cm$^3$. Further analysis resolved the lighter band into two bands, although these usually were not well distinguished from each other. Electron microscopy of the molecules comprising each band identified species of 28 $\mu$m in the lightest (1.708–1.710 g/cm$^3$) band, molecules of 33 $\mu$m in the intermediate (1.710–1.712 g/cm$^3$) band, and molecules of 6 $\mu$m in the heaviest (1.717–1.718 g/cm$^3$) band. These experiments, summarized in Table 3.1, therefore equated the small 6 $\mu$m molecule with the r determinant, the 28 $\mu$m species with the RTF component, and the 33 $\mu$m circle with the complete R factor. These measurements are reasonably consistent with the subsequent heteroduplex maps that show the r determinants of R1 and R6 to be 22,500 and 26,000 base pairs, respectively, the RTF elements to be 64,000 and 71,000 base pairs, and the composite factors therefore to possess genomes of 86,000 and 97,000 base pairs.

If the F-like R factors are able to dissociate into their RTF and r components, each element must constitute an independent replicon in the dissociated state. The composite R factor then must possess two replication control systems, one of which is inactivated so that the complete molecule replicates only under a single control. Host-plasmid interactions must influence replication control since different amounts of these R factors are maintained in the cell during growth in E.coli and in Proteus.

When R factor DNA is isolated in the form of covalently closed circles, usually some of the molecules are supercoiled although some may not be twisted; we do not know whether the plasmid DNA is supercoiled within the cell or whether this occurs during isolation (for review see Helinski, 1973). Not all of the plasmid DNA is recovered in the form of covalently closed circles and estimates for the proportion in this state usually range from 25–60%. When the molecular size of an R factor is known, the number of copies present relative to the host chromosome can be calculated by comparing the total radioactive label in the plasmid DNA with that in the bacterial DNA. Isolation in E.coli depends upon the circular state and so any such estimate in this host represents a minimum since it does not take account of any plasmid DNA molecules that may be broken. Estimates of the amount of DNA of the factor R100 and several of its variants made by Nisioka, Mitani and Clowes (1970) show clearly that, in spite of these qualifications, the number of copies of the plasmid is rather low, apparently one or two per host chromosome. This is consistent with a model in which replication of the plasmid DNA is initiated at a frequency related to that of the bacterial chromosome, limiting the number of copies of plasmid DNA in the cell. This is known as *stringent* replication control.

A somewhat different situation exists in Proteus mirabilis, however, where the number of copies of the plasmid DNA may be very much greater. In the early experiments of Falkow et al. (1966) and Rownd et al. (1966), the satellite bands represented about 8% of the total DNA, which was then thought to correspond to some ten or so copies of the plasmid. Rownd (1969) performed a density transfer analysis of R100 replication in Proteus, analogous to the Meselson-Stahl experiment to demonstrate semiconservative replication (see Volume I of Gene Expression), and could not identify cycles of replication; more than one model for replication control is consistent with these data, but the general conclusion that they suggest is that replication is not limited in Proteus to a single synchronous initiation event in each cycle. Another aspect of R factor existence in Proteus is that replication may continue for several hours in the stationary phase, after chromosome replication has ceased; this generates a very large number of copies of the plasmid genome, with the result that stationary phase cells contain a much greater proportion of plasmid DNA than cells in exponential growth. (In E.coli, plasmid replication halts together with that of the chromosome.) The control of R factor replication in Proteus may be said to be *relaxed*.

The identification of more than one density band of R factor DNA in Proteus raised the possibility that the RTF and r components might be under different types of replication control. Since the r determinant is much smaller than the RTF component, the apparent rough equality in the amounts of DNA present in the high and lower density bands suggested that there might be more copies of the r element than of the RTF or composite R factor. Clowes (1972) proposed that in Proteus the r element is under relaxed control whereas the RTF element is under stringent control; the complete R factor is under the control of its RTF component.

Conditions of growth may change the proportions of the density bands in Proteus. Rownd and Mickel (1971) noted that culture of Proteus cells in drug-free medium for extended periods results in the accumulation of R100 DNA in a single band of density 1.712 g/cm$^3$, representing about 8% of the total DNA. When the cells are transferred to medium containing chloramphenicol, over the next 40 generations there is a shift through an intermediate density peak to one of 1.718 g/cm$^3$, which finally includes all the R factor DNA. This phenomenon has been termed *transition*; it is reversed by *back-transition* when the cells are transferred from drug-containing to drug-free medium.

Two types of explanation have been proposed to account for the transition. The first class of model supposes that the observed density bands represent single components of the R factor; the 1.712 g/cm$^3$ peak represents the RTF element and also any composite R factor genomes that are supposed to be under RTF control; the 1.718 g/cm$^3$ peak represents the r determinant. In this case, differences in the control of the RTF and r replicons might be responsible for their different responses to the addition of chloramphenicol. Punch and Kopecko (1972) proposed a model relying upon the postulate that transition is a consequence of the inhibition of protein synthesis that is brought about by chloramphenicol. This model suggested that both the RTF and r replicons are negatively regulated by a repressor protein that prevents initiation of cycles of replication; but replication of the RTF has an additional control, a requirement for an activator protein that initiates cycles of replication. Inhibition of protein synthesis then preferentially prevents replication of the RTF. Obviously many variants of this model are possible: its principal feature is the proposal that transition represents an increase in the number of r determinants while the number of RTF and R factor units remains constant.

An alternative model is to suppose that transition represents a reorganization of the components of the R factor. Rownd and Mickel (1971) proposed that chloramphenicol causes transition by establishing a selection for Proteus cells with increased levels of resistance to the drug. In this original model they suggested that in drug-free medium the 1.712 g/cm$^3$ peak represents the RTF replicating under relaxed control to produce multiple copies, whereas the r determinant replicates under stringent control to generate only one copy per chromosome (giving a peak at 1.718 g/cm$^3$ that is too small to detect). It now seems, however, that the 1.712 g/cm$^3$ peak in fact represents undissociated complete R factors, but the same basis for the model remains if it is supposed that the complete R factor replicates under RTF control whereas the dissociated r determinant would replicate in a much more restricted manner. In this case, in drug-containing medium cells would have an advantage if they could gain more copies of the r determinant; this might be accomplished if more than one copy of the r determinant is able to associate with each RTF. The 1.718 g/cm$^3$ peak that accumulates in the presence of chloramphenicol therefore was taken to represent large molecules consisting of an RTF associated with many r determinants, replicating under the relaxed control of the RTF. The number of r determinants that would have to associate with a single RTF to produce the observed density shift is very large, about 20.

A detailed study of the R factor molecules present in Proteus cells grown under various conditions has been made by Perlman and Rownd (1975). They found that when P.mirabilis was grown in drug-free medium, a single band of 1.712 g/cm$^3$ was observed in both stationary and exponential phase, although it was present in a greater proportion after entry to stationary phase. DNA extracted from this density band sedimented on sucrose gradients as a single peak of 48.5 S. Electron microscopy showed that all the molecules were circular, with a contour length of 37 $\mu$m, corresponding to a weight of 65 $\times$ $10^6$ daltons. This result contrasts with the data summarized in Table 3.1, in which R factors were dissociated into more than one buoyant density band following transfer into Proteus; the perpetuation of R factor DNA as a single band, representing the complete (RTF + r) unit, demonstrates therefore that dissociation into the RTF and r components does not follow inevitably when the factor is maintained in Proteus. Hashimoto and Rownd (1975) reported that dissociation of R factor DNA to occupy more than one density band occurs in only a minority of the Proteus cells grown on drug-free medium. This conclusion implies that the individual elements identified in earlier experiments may have represented a minority of factors that dissociated into their RTF and r components rather than reflecting a general dissociation of the R factor.

Following the addition of chloramphenicol, Perlman and Rownd observed that the single 1.712 g/cm$^3$ band disappeared, to be replaced by a band of 1.718 g/cm$^3$; in some experiments a band of intermediate density, 1.715–1.716 g/cm$^3$ could be seen, being most evident when cells had proceeded to stationary phase. When plasmid DNA extracted from the 1.718 g/cm$^3$ band was sedimented through a sucrose gradient, it behaved as linear molecules of a large size, about 150 $\times$ $10^6$ daltons (225,000 base pairs); electron microscopy confirmed that most of this material comprises large linear molecules, mostly 75–100 $\mu$m long. Their buoyant density suggests that these molecules represent plasmid DNA rather than chromosomal contaminants; and a reconstruction experiment in which labeled host DNA extracted from $R^-$ cells was mixed with unlabeled DNA extracted from $R^+$ cells suggested that contamination should be low. The linear structures probably are generated by breakage during preparation of large circular molecules; although their presence demonstrates that plasmid DNA longer than a single R factor can be generated under the conditions of transition, this does not reveal their sequence components.

Although the linear molecules comprised most of the high density fraction, some circular molecules also were seen. In the 1.718 g/cm$^3$ band, these included circles of 8 $\mu$m, corresponding to the r determinant, and circular multiples of this length; these might correspond to poly-r multimers of up to five r determinants. In the intermediate density bands, circular molecules were seen with lengths appropriate for structures representing an RTF associated with more than one r determinant: a 1.713 g/cm$^3$ density fraction possessed molecules of 46 $\mu$m (which might be RTF-2r); a 1.715 g/cm$^3$ density fraction contained molecules of 54 $\mu$m (corresponding to the length of RTF-3r) and of 62 $\mu$m (RTF-4r).

To confirm the nature of these sequences, Perlman et al. (1975) obtained denaturation maps of the R factor and its components. At a pH of 10.7, the R factor displays about 20 small sites of denaturation; by examining the RTF component itself (obtained by selecting smaller plasmids that retain *tet*$^r$ but have lost the other drug resistance determinants), it was possible to show that all these sites of denaturation are located in the RTF, so that the G-C rich r determinant remains undenatured in these conditions. By comparing the denaturation maps of the R100 factor and a derivative that had lost the *tet*$^r$ marker, it was possible to align the denaturation map with the heteroduplex map obtained previously by Sharp, Cohen and Davidson (1973). By increasing the pH to 11.1, a denaturation map was obtained for the r determinant, displaying some 4–5 sites of denaturation; some of the DNA of the RTF component then becomes single stranded. The difference in the G-C proportion in the RTF and r determinants thus means that it is difficult to obtain conditions that allow both the RTF and the r sequences to be visualized in the same experiment.

The 8 $\mu$m circular molecules of the 1.718 g/cm$^3$ peak of cells that had undergone transition all displayed the characteristic denaturation pattern at pH 11.1 that defines the r determinant. The circles with lengths that are multiples of 8 $\mu$m displayed a repetition of this pattern. This confirms that poly-r multimers are produced during transition. Circles present in the fractions of intermediate density were examined at pH 10.7 and proved to comprise one RTF genome with an undenatured length of DNA corresponding to a multiple of the length of the r determinant. This confirms the production of RTF-poly-r factors. Some small linear molecules were mapped, and proved to comprise poly-r sequences.

The principal conclusion suggested by these experiments is that transition is accompanied by a change in the structure of the R factor: this represents the association together of many r determinants, arranged head to tail, which may or may not be associated with an RTF element. The predominance of linear molecules that must represent breakage products of the dense fraction DNA makes it difficult to obtain a quantitative estimate of whether the principal products of transition are RTF-poly-r or poly-r factors. However, the disappearance of the peak at 1.712 g/cm$^3$ during transition implies that all the RTF elements must become associated with several r determinants. This association can be reversed during back-transition, when the number of r determinants associated in the multimers declines. However, the final product of back-transition is the composite R factor (RTF + r), in which the RTF component retains one r determinant; this implies that the nature of the association between the RTF and r components may be different from the association prevailing between linked r determinants.

The nature of the process by which several r determinants become associated is not at all clear. If transition and back-transition occur by the same reversible mechanism, a recombination event must be responsible for the addition or loss of each r determinant in the complex structure. But how does the number of r determinants increase relative to the number of RTF elements? This presumably involves a preferential replication. In this case, transition must involve a dissociation of the original R factor into its components in order to generate a source of the r determinants; an increase in the number of these determinants then might be followed by their association into complex multimers. Back-transition then might take place by dissociation of r determinants from the multimers, with the number of free r determinants declining because they fail to be replicated.

To resolve the events that occur in transition and back-transition, in particular it is important to know whether the principal product of transition is the RTF-poly-r or the poly-r structure. If transition represents selection for cells carrying an increased number of r determinants, why should there be any advantage in generating a poly-r structure rather than simply increasing the number of monomeric r determinants? On the other hand, selection for the RTF-poly-r structure might result if the RTF replicon, and any r determinants under its control, were able to replicate more freely than the isolated r replicon. This essentially is the original model for transition; but it raises the problem of how the number of r determinants increases relative to the number of RTF elements. To suggest a preferential replication of the r determinant is not consistent with the idea that r determinants are placed under RTF control to increase their total number. At present it seems therefore that there are internal inconsistencies in the models to explain transition (and see page 221).

## Transfer of R and F Factors

### *Transfer and Replication of R Factor DNA*

By analogy with the sex factor, it seems likely that R factor DNA is transferred from donor to recipient in the form of a linear single strand; the complementary strand must then be synthesized in the recipient and the two ends joined to generate a circle. By centrifuging denatured R factor DNA on gradients of poly(U,G)-CsCl, Vapnek, Lipman and Rupp (1971) showed that the two strands can be separated in a manner analogous to the separation of F factor strands (see Chapter 2). When donor cells are labeled before mating, only one labeled strand is found in the recipients, a result identical to that obtained when transfer of the sex factor was followed by the same procedure. Both F-like and I-like R factors (see later) show the same type of transfer. When covalently closed circles were isolated shortly after mating by use of density gradients containing ethidium bromide, about 25% of the R factor DNA was found in this form; this value is similar to the proportion of F factor DNA that can be extracted as circles under similar conditions and therefore suggests that complement synthesis and circle formation may be equally efficient with R factors after transfer.

By using the Freifelder system to label transferred R1 DNA, Falkow et al. (1971) were able to follow the transfer and subsequent replication of the factor. When the R factor DNA was extracted after mating and sedimented through sucrose gradients, both covalently closed and nicked open circles were obtained (see Figure 3.4). The presence of these two and only these two peaks raises two questions. Are R factors present in their host cells in the forms of both covalently closed and nicked open circles, or are the open circles derived from the closed circles by nicking during extraction? And from the sedimentation coefficients of these two peaks, 75S and 50S, it is possible to calculate that linear molecules of duplex R1 DNA should sediment at a rate of about 43S; does the absence of such a peak mean that the R1 DNA, although presumably linear during transfer, is subsequently maintained only in circular form?

The results of using a pulse chase procedure to follow the transfer and subsequent replication of R1 DNA are shown in Figure 3.8. Falkow et al. allowed R1 donors to mate with $R^-$ recipients in medium containing $^3$H-thymine; after 15 minutes an excess of unlabeled thymine was added to terminate the labeling and then the cells were incubated for a further period. Samples of DNA were extracted immediately upon the termination of labeling and after a further 10 minutes and 15 minutes of incubation; the figure shows the sedimentation of the labeled DNA through a neutral sucrose gradient. After 15 minutes of mating there is a prominent

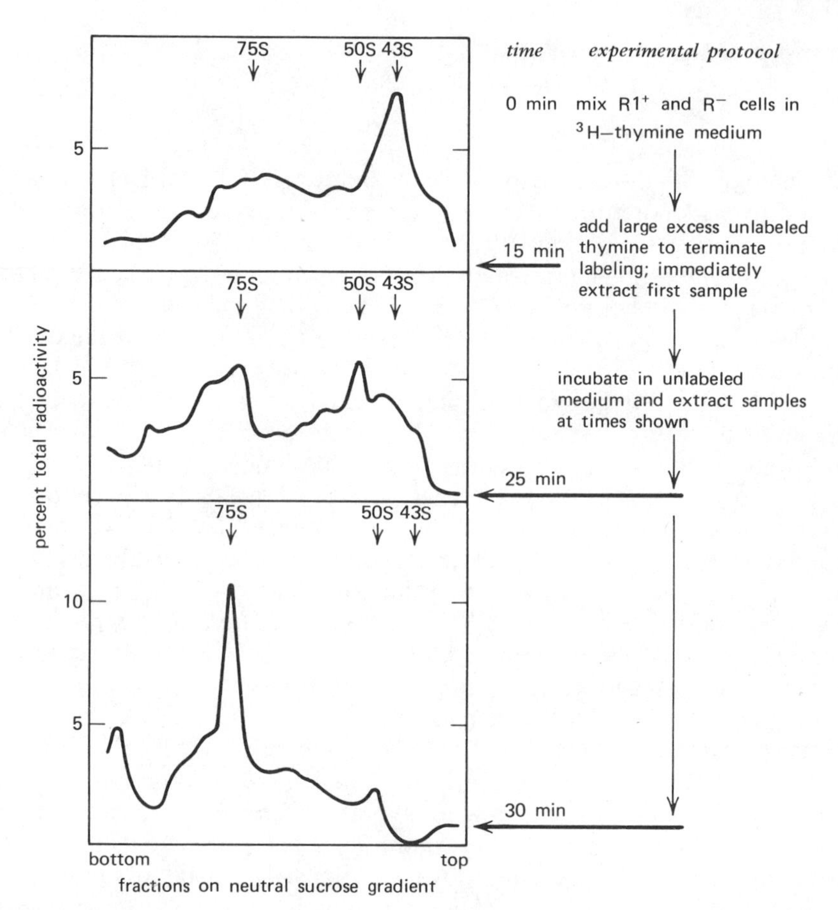


**Figure 3.8:** sedimentation analysis of R factor DNA after mating. A mating population of cells transferring R1 DNA was labeled with $^3$H-thymine for 15 minutes of mating; this labels the complement to the transferred single strand, that is, the strand synthesized in the recipient. Labeling was terminated after 15 minutes by the addition of excess unlabeled thymine. Samples of DNA were extracted and sedimented at 15 minutes and subsequent times. The 43S peak represents linear molecules, the 50S peak corresponds to open circles, and the 75S peak represents covalently closed circles. The R factor DNA is converted during the experiment from linear molecules through an open circular intermediate to closed circles. Data of Falkow et al. (1971).

peak at 43S which represents linear molecules of duplex R1 DNA; these must have been generated by synthesis of the complement to the single strand that was tranferred (it is the complementary strand synthesized after entry into the recipient that carries the radioactive label in this experiment). After 25 minutes (that is, after a chase of 10 minutes) the 43S peak is less prominent and peaks are clear at 50S and 75S, representing open and closed circles. Five minutes later, the 43S peak has entirely disappeared, the 50S peak is much reduced in size, and most of the R factor DNA is found in the 75S peak of covalently closed circles. Linear molecules thus represent a transient form of the factor found only immediately after transfer; and open circles represent an intermediate in the formation of the covalently closed circles that appear to comprise the state of the mature R DNA.

Factors such as R1 which are under stringent control and present at only 1–2 copies per chromosome in E.coli must segregate evenly at cell division if both daughter cells are to gain the plasmid. One obvious mechanism which might be implicated in such segregation is membrane attachment; there is good evidence that the bacterial chromosome itself is attached to the cell membrane (see Volume 1 of Gene Expression) and there is some evidence that F factor segregation may follow chromosome segregation (see earlier). Shull et al. (1971) reported that when R DNA is transferred to minicells most of it is found in the membrane fraction that is isolated by use of sarkosyl detergent in the presence of $Mg^{2+}$ ions; this technique generates a band on sucrose gradients that contains only membrane-sarkosyl crystals, in which DNA should be present only if it is specifically attached to the membrane. This should exclude association during extraction (see Volume 1 of Gene Expression).

When a nonionic detergent, Brij 58, was used to lyse cells that had first been treated with lysozyme, Falkow et al. (1971) found that most of the cellular DNA was present in the membrane fraction isolated on a sucrose gradient. But when the cells carried the R1 plasmid, most of the plasmid DNA behaved as free covalently closed circles; this suggests that in the mature condition, R factors represent free circles of DNA, but of course does not preclude the possibility (consistent with the minicell results) that attachment to the membrane takes place for replication, after transfer or during the cell cycle. When the experiment in which Brij detergent was used to break cells was repeated at various times after mating took place in the presence of $^3$H-thymine, DNA could be found in the membrane fraction. At 5 minutes after mating, 20% of the R1 DNA was present in the membrane fraction; this proportion rose to 60% at 10 minutes after mating and then declined in parallel with the increase in free open (50S) and closed (75S) circles. The kinetics of this conversion are shown in Figure 3.9; they are consistent with the idea that transferred single strands are attached to the membrane in the recipient, converted to linear 43S duplex molecules on the membrane, and then suffer covalent linkage of the ends of one strand to form an open circle; the open circle is then released from the membrane into the cell, after which the nick in one strand is subsequently sealed by a ligase activity. Of course, it is tempting to speculate that membrane attachment might also be implicated in the regular replica-

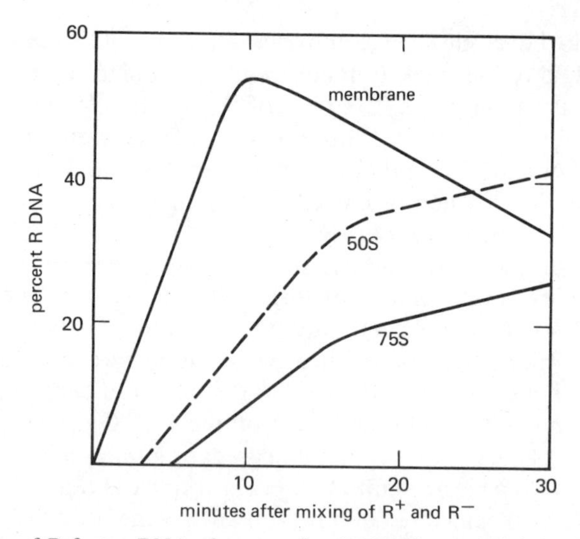


**Figure 3.9:** state of R factor DNA after transfer. R1 DNA was labeled with $^3$H-thymine during mating and extracted after various times with Brij detergent. Sedimentation on a neutral sucrose gradient resolved the amount present in a membrane fraction, as open (50S) circles or closed (75S) circles. The kinetics of accumulation of each form are shown above. The DNA initially associates with the membrane and then is released as free open circles that are later converted to free covalently closed circles. Data of Falkow et al. (1971).

tion cycle within the cell and might be linked to the segregation of R factor DNA into daughter cells.

Where is the nick(s) located in the open circles of R factor DNA? If open circles represent an intermediate of normal replication as well as of the conversion of the transferred linear strand to covalently closed circles, it is reasonable to suppose that the open circles isolated from cells carrying the factor may possess nick(s) at a specific site(s). Morris, Hershberger and Rownd (1973) reported that the open circles of R12 DNA (a derivative of R222) appear to be attached to a protein-containing structure in Proteus via one strand that carries a nick(s). They found that on alkaline gradients of CsCl the R12 DNA formed three fractions, one representing each single (linear) strand and one representing covalently closed (duplex) circles (these are more resistant to denaturation than open circles or linear molecules). When the pH was increased from 10.0 to 12.0, however, the fraction of covalently closed (duplex) circles declined from 85% to zero, and a marked asymmetry developed in the separated single strands, with a much smaller amount present of the less dense of the two bands. That this was not caused by nicking during preparation was implied by the failure of exogenously added R12 closed circles to suffer nicking. The missing single strand can be recovered when the fraction of cellular debris that is found at the meniscus of the centrifuge tube is treated with pronase; both chromosomal and R12 DNA are released by this treatment, and the R12 DNA shows the reverse strand asymmetry to that displayed by the free DNA. As might be expected from this result, treatment of cells with pronase during initial extraction of the R12 DNA abolished the asymmetry so that both strands were found in equal proportions on alkaline gradients. By denaturing isolated open circles and separating the circular and linear single strands on alkaline sucrose gradients, they showed that the linear strand corresponds to the less dense strand that is found in the protein-containing structure under conditions of alkaline preparation. Morris et al. therefore suggested that the covalently closed circles of R12 DNA are attached to a protein-containing structure in Proteus; under less alkaline conditions the DNA is released from the protein, but under more alkaline conditions the DNA is released as open circles in which the strand attached to protein is nicked. The nature of any protein attachment in E.coli must be different, since increases in the pH at which R12 DNA was extracted did not have this effect with this host. These results are analogous to those obtained with colicinogenic factors, which can be isolated in the form of relaxation complexes, in which a protein is attached to the covalently closed duplex of DNA and treatment with pronase may cause a nick to be made in one strand (see later). Such structures may represent intermediates that are common to the replication cycle of many plasmids.

The same techniques that are used to follow transfer and replication of R factors can be applied to transcription. By isolating RNA from cells carrying the factor R538-1*drd*, Vapnek and Spingler (1974) showed that only the strand that is heavy in poly(U,G)-CsCl appears to be transcribed: about 60% of this strand is represented in RNA sequences in the cell. This strand is of course that transferred from donor to recipient in conjugation, suggesting in this sense a parallel between the R factor and negative strand phages (in which the infecting strand is the one that is transcribed).

### *Inhibition of F Fertility by R Factors*

R factors can be divided into two groups by the effect that they have upon the expression of the F factor (whether autonomous in the $F^+$ or $F'$ state, or integrated in the Hfr state). Most of the R factors isolated in the early years in Japan suppress the expression of the F factor. For example, Watanabe, Fukasawa and Takano (1962) found that 18 out of 22 tested R factors conferred upon $F^+$ cells resistance to phage f1 or f2, in contrast to their previous susceptibility; this indicates that the R factor has prevented the F factor from synthesizing the F pili. Such factors also may be detected by their effect in preventing the sex factor from transferring either itself infectiously (when in the $F^+$ or $F'$ state) or transferring chromosome markers (when in the Hfr state). This type of R factor is termed $fi^+$, indicating that it is able to exercise *fertility inhibition*. The second type of R factor, represented by 4 of the 22 samples in this experiment, does not suppress sensitivity to ♂-specific phages (that is synthesis of F pili) or donor activity in sex factor or chromosomal transfer; these are called $fi^-$, indicating that they do not exercise fertility inhibition. Some differences in the drug resistance markers carried by $fi^+$ and $fi^-$ factors were noted by Watanabe et al. (1964), but the isolation of further factors since has shown that resistance to any drug may be carried by either $fi^+$ or $fi^-$ types of R factor.

The suppression of F factor activity that is caused by $fi^+$ R factors appears to result from the action of a single gene in the R factor. The R100 factor (also known as R222 or NR1) is able to inhibit several male properties in $F^+$ or Hfr cells of E.coli K12. Hirota et al. (1964) showed that fertility itself, as measured by donor activity, is reduced; agglutinability in anti-$f^+$ sera is reduced; and sensitivity to phage f1 and to the RNA phages is inhibited. Egawa and Hirota (1962) obtained a mutant of this factor, the variant R100-1, which has lost the ability to suppress expression of the sex factor; R100-1 was isolated by selecting an Hfr cell line that had high fertility in spite of the presence of the R factor (which could be detected by retention of drug resistance).

The R100-1 factor lacks all of the inhibitory effects upon the expression of the sex factor that are displayed in male E.coli strains by the parent R100 factor (Hirota et al., 1964). This suggests that all these phenotypic changes result from the action of the same R factor gene(s). The inhibitory effect of the R factor could be explained by synthesis of a repressor protein which acts to prevent expression of the appropriate F factor functions; and by analogy with the repressor mutants of the lactose operon, the R100-1 mutation has been described as $i^-$ (although, of course, these experiments do not prove that the mutation inactivates a repressor protein). The *i* gene product must be active upon either autonomous or integrated sex factors since $F^+$ and Hfr donor activities both are suppressed. That the *i* gene might be located close to tetracycline resistance on the R factor was suggested by the loss of the $tet^r$ marker together with the loss of ability to inhibit male expression; as Table 3.1 shows, it has since become apparent that the mutation in R100-1 is a deletion of sequences from the parent R100 factor.

We should emphasize that the effect of the $i^-$ mutation in causing loss of the fertility inhibition previously exercised by the *fi*$^+$ R100 factor is not equivalent to converting the factor to one of the *fi*$^-$ class; the naturally isolated *fi*$^-$ R factors differ in several ways from the *fi*$^+$ type, including compatibility group and restriction activities (see later). The difference between the *fi*$^+$ and *fi*$^-$ types of R factor thus is substantial and they may show little or no homology of nucleic acid sequences.

### *Chromosome Transfer by R Factors*

R factors appear able to promote transfer of chromosome markers only at very low frequencies, usually of the order of 1% of the frequency supported by an F$^+$ strain, that is $10^{-7}$–$10^{-8}$ recombinants are generated for every R$^+$ donor cell. However, when transfer takes place it displays characteristics common to the transfer of chromosomal markers by the sex factor. This suggests the concept that transfer frequencies may be low with the R factors because only a small minority, of the order of 1%, of the cells in the population can support transfer; but when they do so the mechanism may be similar to that sponsored by the F factor.

Several R factors tested by Sugino and Hirota (1962) gave frequencies for selected recombinants of less than $10^{-8}$. Transfer promoted by the R factor shares with transfer mediated by F$^+$ donor cells the property that no region of the chromosome appears to be transferred at preferential frequencies (although it is difficult to be sure of such conclusions because the number of recombinants is so very small). However, when two R factors were introduced into the ♀3 strain which has a sex affinity locus (*sfa*), much higher recombination frequencies were achieved. This suggested that there might be similarities of sequence between these R factors and the F factor, such that the F sequences remaining in the ♀3 strain to constitute the *sfa* locus have some homology with the R factors. This idea has since been supported, of course, by the demonstration that half of the F factor is homologous with part of certain *fi*$^+$ R factors. In the R-mating that takes place when R$^+$ donors are crossed with R$^-$ recipients, alleles of the recipient parent are predominant in the recombinants, which suggests that a unidirectional transfer takes place analogous to that supported by F factors.

Mutants of the R factor can be isolated which are unable to transfer either the factor itself or chromosome markers from R$^+$ to R$^-$ cells. Hirota, Fujii and Nishimura (1966) found that when they selected mutants that were no longer infective in transfer of R, they were also inactive in transfer of chromosome markers by R-mating. And when they selected mutants that were no longer able to transfer chromosome markers by R-mating, all except two were no longer infectious and so also had lost their ability to transfer the R factor. Reverse mutation of either type of mutant leads to reacquisition by the factor of both characters. This suggests that chromosome transfer is mediated by the same system responsible for transfer of the factor itself.

(The two exceptional mutants were no longer able to transfer chromosome markers but were still infectious and therefore able to transfer R itself. When they were tested by removing R from the cell by treatment with acridine and then introducing a new $R^+$ or $F^+$ factor, in either case there was a reduction in mating capacity. These must therefore comprise chromosomal mutations which inhibit transfer of chromosome markers by either the sex factor or R factor. That the R factor of the mutant cells was not mutated was confirmed by transferring it to fresh $R^-$ cells, where it sponsored infectious R transfer and conjugational chromosome transfer at the frequencies characteristic of R factors.)

Some combinations of independently isolated nontransmissible factors become transmissible when harbored in the same host, implying that more than one gene is involved and that complementation can take place. The mutant factors exist in the autonomous state since they are still susceptible to acridine curing; and this implies that the autonomous replication system can be divorced from conjugal fertility and infectivity. When mutant F-like R factors are introduced into $F^-$ cells, no change takes place in their properties; but upon entry into Hfr or $F^+$ cells, infectivity is restored (with three exceptions). Thus irrespective of its state, the F factor can complement the mutant functions of the nontransmissible R factors. When a nontransmissible mutant R factor is transferred from an $F^+$ cell, it continues to be transmissible if the recipient also gains the $F^+$ factor, but loses this ability if it alone enters the recipient. This confirms that complementation by a diffusible product is occurring.

### *Derepression of R Factor Fertility*

The low frequency of transfer sponsored by R factors makes it difficult to investigate its mechanism and its relationship with the transfer promoted by the sex factor. Two systems have been used to overcome this problem. One is an *hft* system, an abbreviation for high frequency transfer, in which $R^+$ factors promote high levels of transfer, comparable to those displayed by $F^+$ factors. Another is the use of derepressed mutants of the R factor which support conjugation at increased frequencies, again close to those typical of F factors.

The high frequency transfer system was first developed by Watanabe (1963b) by analogy with a similar system previously developed for transfer of Col I in S.typhimurium (see later). Cells that have just received R factors, that is, the recipient population immediately after conjugation, can act as highly competent donor cells for retransfer of the R factor; this ability is lost when the culture is allowed to multiply for a few generations. Transfer of the R factor itself, and also of chromosome markers, thus is inefficient in established cultures but takes place at much higher frequencies from new $R^+$ cultures. The same observation has been made with both *fi*$^+$ and *fi*$^-$ factors, which implies that their mechanisms responsible for limiting transfer to the low frequency typical of established cultures may be similar.

Cultures of $R^+$ cells do not display the $f^+$ antigen, that is, they do not possess F pili. The attempts of Datta, Lawn and Meynell (1966) to visualize pili in $R^+$ cells showed that F-piliated cells are absent or very rare when established $R^+$ cultures are scrutinized by electron microscopy. However, some bacteria sensitive to ♂-specific phages may be present in $R^+$ cultures and the proportion is much increased in newly established hft cultures. F-like pili can be seen on these $R^+$ cells under hft conditions; the number of F-like pili per cell in this state may be greater than the number present in $F^+$ or Hfr cultures. The proportion of cells with pili is roughly equal to the proportion able to conjugate.

When 30 R factors chosen at random were tested for ability to confer sensitivity to ♂-specific phages, that is to specify F-like pili on $F^-$) bacteria, Meynell and Datta (1966) found that only *fi*$^+$ factors can do so. (These are the F-like *fi*$^+$ factors: other *fi*$^+$ factors may specify other types of pilus, as do the *fi*$^-$ factors; see below.) The much increased ability of these *fi*$^+$ R factors to confer sensitivity to ♂-specific phages under high frequency transfer conditions demonstrates that they carry the genetic information necessary to specify the F-like pilus, but that in established cultures this information is not expressed.

This suggests a model for the control of R factor expression in which an established *fi*$^+$ factor synthesizes a repressor protein that inhibits the production of pili and the ability to act as donor in either self-transfer or chromosome transfer. Conjugation can be supported only by those few cells in which R factor expression is able to take place. The phenomenon of high frequency transfer suggests that when an R factor enters a new host cell it is able to express the information necessary for pilus formation and for conjugation; it must take several generations before repressor synthesis reaches levels sufficient to inhibit expression. The suppression of F factor expression by *fi*$^+$ R factors then can be explained by supposing that the R factor repressor acts upon the sex factor as well as upon the R factor DNA. The occurrence of high frequency transfer in newly infected cultures of *fi*$^-$ factors suggests that these synthesize a repressor in the same way, although its specificity of action is of course different.

The production of pili by cells carrying R factors has been followed by

Nishimura et al. (1967), who used a strain of E.coli which lacks the usual surface appendages; this makes it simpler to visualize any pili specified by the plasmid. Pili synthesized under the direction of sex factors can be characterized, of course, by their adsorption of ♂-specific phages; for example, the RNA phages such as MS2 or f2 adsorb along the length of the pilus whereas the DNA phage M13 adsorbs to the tip of the pilus. No pili were produced in $F^-$ cells carrying the wild type factor R100. But pili were produced by cells carrying the mutant factor R100-1. This is consistent with the idea that R100-1 carries a mutation of the type $i^-$ which inactivates the repressor protein that prevents expression of R or F factors. The pili specified by R100-1 are very similar morphologically to those specified by the F factor, although they may be distinguished by a slightly lower affinity for the ♂-specific RNA phages.

Two types of pili specified by plasmids have been well characterized on E.coli host cells. Lawn et al. (1967) found that all cultures of $F^+$, $F'$, or Hfr cells possess the F pili. Under conditions when F-like $fi^+$ R factors are expressed, that is, in newly infected high frequency transfer cultures of wild type factors, or in cultures infected with the R100-1 mutant, pili similar to F pili are produced. Another factor producing such pili is Col V. These pili have been described as F-like, since they display serological cross reaction with the F pili specified by the sex factor, but appear to be distinguished by slight differences. Beard, Howe and Richmond (1972) have since found that the pili from cells carrying F or these $fi^+$ R factors band at different densities in CsCl; cells carrying both types of plasmid form mixed pili, presumably derived from the subunits of both the F pilus and the F-like pilus, and these band at an intermediate density in CsCl.

A second, distinct type of pilus, the I-like pilus, named for the factor Col Ib which causes its synthesis, is morphologically similar in construction to the F-type pilus, but does not adsorb ♂-specific phages; this type of pilus can be characterized by its ability to adsorb two other phages, If1 and If2, described by Meynell and Lawn (1968). The I-like pili are produced by high frequency transfer cultures of Col Ib, Col Ia, and by certain $fi^-$ R factors; although not all identical, all the I-like pili show serological cross reaction (analogous to the reaction with F and F-like pili). In cells carrying both F-like and I-like plasmids, both types of pilus are formed separately: their subunits do not mix. In their ability to specify pili, all the plasmids studied in these experiments were related either to F or to Col Ib, although other types since have been identified. Since the function of all types of pilus is to enable cells carrying them to act as genetic donors, Lawn et al. (1967) suggested that they should be described by the common term, sex pilus. The general conclusion suggested by these and subsequent experiments is that each conjugative plasmid specifies a pilus: pili can be classified into several groups, and the members of each group are related but not necessarily identical. This implies that there may be polymorphism in the genes responsible for specifying each type of pilus.

Mutants of the F-like *fi*$^+$ factors that are defective in their repression of the transfer system have been isolated in two ways. The R100-1 $i^-$ mutant was selected by loss of its ability to repress the F factor (see above). Another technique for isolating such mutants was developed by Meynell and Datta (1967), who selected directly for *fi*$^+$ R factor variants able to express their own conjugation functions at high frequency. This cannot be done either by screening simply for increased transfer of chromosome markers (because the frequencies of transfer are too low) or by detecting increased transfer of R itself (because the frequencies are too great). They therefore used the ♀3 (*sfa*$^+$) strain as host, enabling both R and F factors to promote chromosome transfer with increased frequency, in particular showing an increase in the frequency of transfer of the *met*$^+$ locus near the *sfa* site. The experimental protocol was to infect ♀3 cells with a wild type R factor, mutagenize with nitrosoguanidine or ultraviolet irradiation, and isolate clones that transfer *met*$^+$ to recipients. One possible flaw in this procedure is that the ♀3 cells might possess a defective F factor, so that the mutant R factors isolated for their ability to transfer *met*$^+$ might in fact comprise recombinants with this defective factor. Another procedure also therefore was introduced to isolate mutants: the use of microcolonies reduced the frequency of transfer of R to a more manageable level, so that it was possible to isolate mutants transferring the drug resistance markers at increased frequencies. The properties of the variants isolated by both procedures suggested that in fact both represent mutants of the R factor. All the mutants increased both the frequency of transfer of the R factor and the frequency of transfer of chromosome markers by about 300-fold; all displayed increased susceptibility to plaque formation by ♂-specific phages, implying that F-like pili were produced. These mutants are described as *drd* (for derepressed).

Both the R100-1 ($i^-$) and various R*drd* mutants were used by Cooke and Meynell (1969) to examine the transfer of chromosome markers. In all cases recombinants were generated at frequencies similar to those produced by F$^+$ cells. This suggests an analogy between R-mediated and F-mediated transfer of chromosome markers, with the lower frequency of transfer by the R factor being due to the repression of its activity in most of the cells that carry it. R100-1 displayed no preferred site of attachment; derepressed derivatives of seven R factors, some *fi*$^+$ and some *fi*$^-$, showed the same property. In these and other experiments, it was not possible to obtain Hfr lines in which R factors became stably integrated into the host chromosome, a difference from the properties of the sex factor. However, by using host cells of the *dnaA*$^{ts}$ type in which bacterial replication can occur at high temperature only if a resident plasmid displays integrative suppression, Nishimura, Nishimura and Caro (1973) were able to isolate cells in which F-like R factors were integrated into the bacterial chromosome. Many subsequent experiments that attempted to use this highly selective technique to isolate integrated I-like R factors did not succeed, suggesting the conclusion that these factors are unable to insert into the chromosome. However, Datta and Barth (1976) succeeded in isolating a strain in which an I-like plasmid has accomplished integrative suppression and it now seems likely that most conjugative plasmids can do so. But of course, it remains true that these plasmids do not display the ready formation of Hfr strains and F′ plasmids that is characteristic of the sex factor; this remains a difference between the sex factor and other plasmids.

Unlike other R factors, the factor R1 resembles F′ factors in exhibiting a preferred site of attachment. The wild type factor R1 yields *try*$^{+}$ recombinants at a frequency of about $10^{-5}$, much higher than the frequency of $10^{-7}$ to $10^{-8}$ at which other markers are found in recombinants. The low frequency of transfer prevents examination of this phenomenon with the factor R1 itself, but Pearce and Meynell (1968) were able to analyze chromosome transfer by this factor by making use of an R1*drd* derivative. By the usual criteria used to examine transfer from Hfr strains—frequencies of recombinant formation for different selected markers, times of entry of selected markers, segregation for unselected markers—they showed that the R1*drd* factor transfers chromosome markers with the same sequence and orientation characteristic of an Hfr strain with an origin close to *try*. Since this feature is characteristic of the factor and not of the cell, it is the reverse of the situation represented by the ♀3 strain. The R1 factor must possess some homology with a site on the bacterial chromosome close to the *try* locus, presumably because at some time it has picked up bacterial DNA. However, this homology leads only to the preferred transfer characteristic of an F′ factor and not to the establishment of a stable integrated factor.

### *Superinfection Immunity of R and F Factors*

Two sex factors are unable to infect the same cell, a phenomenon that has been termed *superinfection immunity*. Similar behavior is displayed by R factors, although these fall into different classes: the members of a class display superinfection immunity but members of different groups can infect the same cell. Superinfection immunity comprises two independent mechanisms: *surface exclusion* and *plasmid incompatibility*. Surface exclusion refers to the inability of a plasmid to enter cells already carrying another plasmid of the same group; this phenomenon appears to be mediated at the surface of the cell. Plasmid incompatibility refers to the inability of two plasmids of the same compatibility group to coexist stably in the cell; this phenomenon appears to be mediated via the replication system of the plasmid, which apparently functions in such a way that only one replicon of any given compatibility group can be maintained in the E.coli host cell. The two effects are independent since plasmid incompatibility may or may not be accompanied by surface exclusion, a relationship that has been reviewed by Novick (1969).

Surface exclusion is exerted at the level of transfer of plasmid DNA from donor to recipient cell. The independence of plasmid incompatibility and surface exclusion has been demonstrated by observations that surface exclusion can occur in situations when incompatibility is not displayed. Thus conjugal transfer of a *fi*$^+$ R factor into a *fi*$^+$ recipient may be inhibited under conditions where superinfection can be achieved when transduction is used to transfer the second factor so that it enters by the phage uptake system. This suggests that it is conjugal pair formation that is inhibited by surface exclusion (see Watanabe, Sakaizumi and Furuse, 1968; Novick, 1969). And surface exclusion prevents the transfer of DNA from an Hfr donor to an F$^+$ cell, but can be alleviated by phenocopy conditions that do not relieve plasmid incompatibility.

Experiments to demonstrate directly that surface exclusion represents inhibition of transfer of DNA can be performed by providing conditions under which there is no plasmid incompatibility. Sheehy, Orr and Curtiss (1972) achieved this situation by following the transfer of DNA to minicells derived from different parental strains. Since the minicells contain no DNA, they cannot display plasmid incompatibility and any inhibition of conjugation must be due to surface exclusion alone, presumably as a result of the presence of surface components inherited from the parental cell. Minicells derived from F′ parents have a much reduced capacity to accept DNA from an Hfr donor; surface exclusion is more efficient when the parental cells are in exponential growth than when they have reached stationary phase. Starvation of growing cells for amino acids considerably reduces surface exclusion; thus by growth in appropriate conditions it is possible to make phenocopies that are Sex$^-$ in spite of their *sex*$^+$ (surface exclusion positive) genotype. Of course, this is precisely the protocol that has been developed in order to cross two strains both carrying the F factor. An F-like R factor in the recipient also is able to inhibit the transfer of DNA from an Hfr donor. A resident F-like R factor also inhibits transfer of a similar factor from donor cells. These results show that inhibition of transfer is responsible for reducing the ability of a donor cell to contribute genetic material to a recipient that carries a related plasmid. The most likely stages at which this interaction might occur are the formation of initial contacts between donor and recipient or the conversion of these contacts into effective pair formation. However, it is possible also that it may be the mobilization for transfer of the donor material that is prevented.

Surface exclusion is associated with pilus type in the sense that only plasmids of the same pilus type show surface exclusion, although two plasmids of the same type need not necessarily display exclusion. An obvious question therefore is whether the pilus is involved in preventing the transfer of DNA. Early models tended towards the view that the presence of the pilus on the recipient cell might prevent mating. However, the lack of any need for a pilus on the recipient is shown by the inability of F′ cells to transfer DNA to minicells derived from an $F^+$ parent that lack the factor and its pili. This implies that the minicells extruded by $F^+$ bacteria must possess some surface component, independent of the sex pilus, that prevents successful conjugation with F donor cells. Although the sex pilus of the recipient is not needed for surface exclusion, an indication that the pilus of the donor is important is provided by the observation of Willetts and Maule (1974) that surface exclusion may fail to be exerted against donor cells that carry two plasmids responsible for the formation of "mixed" pili. Another component of the donor cell that may be involved has been identified by the observation of Beard and Bishop (1975) that disruption of the mucopeptide layer on the donor cell (although not on the recipient cell) abolishes surface exclusion. The *traS* gene of the sex factor in the recipient is needed for the expression of surface exclusion (see below); these results therefore suggest that the mucopeptide layer and perhaps the pilus of the donor in some way interact with the *traS* product of the recipient. The nature of this interaction is not clear, but one possible implication of the involvement of the donor mucopeptide layer is that surface exclusion acts to prevent exit of DNA from the donor rather than inhibiting entry into the recipient.

Two incompatible plasmids can survive only transiently in the same cell and segregate at cell division. The first classification of R factors according to incompatibility corresponded to their division into the $fi^+$ and $fi^-$ classes. Watanabe et al. (1964) demonstrated that two $fi^+$ factors were unable to survive in the same cell; similarly two $fi^-$ factors could not coexist in the same strain. However, there was no impediment to the joint presence in the same cell of one $fi^+$ and one $fi^-$ factor (for review see Watanabe, 1967). In fact, because the presence of a $fi^+$ factor usually is detected by its inhibition of F-fertility, it is possible that $fi^-$ factors also might be present in many of the strains isolated as carriers of *fi*⁺ factors; Romero and Meynell (1969) examined 65 such strains for susceptibility to phage If1, which specifically attacks cells with the I-pili specified by I-like R factors, and found that 16 carried an I-like *fi*⁻ factor in addition to their *fi*⁺ factor. Similarly there is no impediment to the stable coexistence of an F factor with an R factor, although of course a *fi*⁺ R factor will suppress the fertility of the F factor.

R factors and other plasmids may be characterized by their ability or inability to suppress F factor fertility and by the type of pilus that they specify. The most common types of plasmids initially identified were either *fi*⁺ in fertility inhibition and F-like in pilus formation or were *fi*⁻ in fertility inhibition and I-like in pilus formation. However, these characteristics are not necessarily associated since some R factors that are *fi*⁺ and I-like have been identified by Grindley and Anderson (1971). And within the large groups of plasmids characterized either by common fertility-inhibition type or by common pilus formation, there are many compatibility groups; however, it so happened that the *fi*⁺ and *fi*⁻ R factors that were examined in early analyses fell into only one compatibility group for each type.

Four compatibility groups were identified among naturally occurring F-like plasmids by Hedges and Datta (1972). Group FI includes the F factor, Col V2, Col V3, and the unusual R factor R386 isolated by Dennison (1972); this R factor is the only one identified in the same compatibility group as the sex factor and it may have arisen from recombination between a progenitor R factor and the sex factor. Group FII includes R1, R100, and many other *fi*⁺ R factors; this essentially is the classical group of the first identified *fi*⁺ R factors. Group FIII comprises the colicinogenic factors ColB-K98 and ColB-K166. Group FIV is provided by the factor R124, an unusual *fi*⁺ factor since it is able to restrict phages, an ability in general displayed only by the *fi*⁻ R factors (see later). Another *fi*⁺ factor, R62, belongs to one of the I-like compatibility groups and probably evolved by illegitimate recombination between R factors of different groups.

An initial division of *fi*⁻ R factors into I-like and other pilus types was made by Hedges and Datta (1971). Out of 26 tested *fi*⁻ R factors, 20 were I-like; the remaining 6 were not sensitive to I-specific phages and so must produce other type(s) of pilus. To distinguish the types of pilus produced by these factors requires the isolation of further phages with appropriate specificities. However, these non-I-pilus plasmids were divided into the two new compatibility groups N and P by determining whether transfer could occur between pairs of cells carrying different factors and by testing pairs for the ability to survive in the same cell line. By examining the factors that determine I-pili, Hedges and Datta (1973) defined several incompatibility groups; I$\alpha$ (which contains some of the classical $fi^-$ I-like factors), I$\beta$, I$\omega$, and I$\gamma$. Thus plasmids determining I-pili fall into several compatibility groups just as do plasmids determining F-pili. Other compatibility groups have been identified by Hedges and Datta (1971) and Chabbert et al. (1972); see also Gasson and Willetts (1975).

Plasmids in the same compatibility group presumably must be related by their possession of certain genes in common. A measure of the relationship between two factors can be obtained by hybridization analysis to determine the proportion of DNA sequences that is common to both plasmids. In general the plasmids belonging to a single compatibility group show sequence homologies, whereas plasmids in different compatibility groups show less (if any) relationship with each other (see Guerry and Falkow, 1971; Grindley, Humphries and Anderson, 1973; Ingram, 1973). Recombination usually appears to take place only between plasmids that are related; in early experiments, for example, Watanabe et al. (1964) found that their $fi^+$ factors could recombine when two were present in the same cell, but these plasmids did not recombine with $fi^-$ factors. Unusual recombination events may generate "atypical" factors, such as R62; Guerry, Falkow and Datta (1974) showed that although this factor is $fi^+$, only 2–3% of its sequences are related to the sex factor, whereas 57% are related to Col I and an I-type R factor, and 25% are related to an N-type plasmid. Since there is little homology of sequence between I-type and N-type plasmids, illegitimate recombination probably was involved in the formation of R62. Although apparently of the $fi^+$ type, this factor inhibits F factor transfer by a mechanism different from that of any other of the $fi^+$ F-like R factors (Willetts and Paranchych, 1974; Gasson and Willetts, 1975). Unusual factors generated in this way may therefore possess features usually found only in plasmids of different compatibility groups.

### *Incompatibility and Copy Number*

Two classes of model have been proposed to explain why two incompatible plasmids cannot coexist in the same cell and these are directly analogous to the models constructed to explain superinfection immunity of F factors (see earlier). The first is a positive control model based upon the original formulation of the replicon model by Jacob, Brenner and Cuzin (1963). There may be specific sites on the cell membrane to which the plasmid must be attached before it can replicate; if there is only one site (relative to the number of copies of the bacterial chromosome) for each compatibility group, only one member of a group can survive in the cell. The second is a negative control model in which the plasmid synthesizes a repressor protein whose action is to limit replication to a single initiation event (relative to the number of host chromosome initiations) in each cell cycle; a model of this type for controlling host replication has been proposed by Pritchard, Barth and Collins (1969). Applied to plasmid replication, this model allows only one copy to survive in the cell because the introduction of another copy by transfer in effect postpones the next initiation event until after the two copies have segregated at division to restore the usual copy number.

Both these models were originally formulated to account for the relationship between chromosome replication and the cell division cycle, that is, to explain how the number of chromosome origins remains constant in proportion to the cell mass (see Volume 1 of Gene Expression). They may therefore be applied directly to plasmids that replicate under stringent control, that is, which appear to be present in only one copy per chromosome origin. In both models, incompatibility is in effect a consequence of the mechanism of control of replication. Modification of the models is necessary to explain the control of replication of plasmids under relaxed control, which may display much greater copy numbers, for example, about 20 for Col E1 in E.coli. The model of positive control by attachment to a membrane site requires that there should be the appropriate number of membrane sites, which duplicates once in each division cycle to maintain a constant number of copies in the cell. Negative control by a repressor can be accommodated by supposing that the affinity between the repressor and the plasmid DNA is lower for high copy number plasmids. Repression therefore is achieved only after a greater concentration has built up in the cell, establishing an equilibrium at which there is an appropriate number of copies of the plasmid. The introduction of further features then may be necessary to explain incompatibility for high copy number plasmids, since neither model in its simplest form excludes a newly entered plasmid from joining the pool of those previously in the cell.

Changes in the copy number of a plasmid might be introduced on the first model by altering the number of membrane sites and on the second model by influencing the affinity of repressor for plasmid DNA. In general, at present it seems easier to explain data on replication and incompatibility in terms of the negative control model, although alternative models certainly remain possible. (With the sex factor, repressor models can more readily explain the incompatibility of Hfr chromosomes with autonomous plasmids; see page 168). Definition of the control of replication and incompatibility will require the identification of the regulator proteins involved and the sites at which they act; at present such resolution is approached only with the sex factor (see below).

Several lines of evidence indicate that plasmid incompatibility results from the inability of more than one plasmid genome to replicate in the cell. The experiments of Dubnau and Maas (1968), for example, discussed earlier and illustrated in Figure 3.2, show that additional plasmids appear to be diluted out with growth, implying that they are unable to reproduce. Such results, of course, do not reveal the stage of the replication cycle that is inhibited. Different experiments have suggested different stages. On the one hand, LeBlanc and Falkow (1973) compared the fate of R1 DNA introduced into compatible recipients (which lacked any plasmid) and into incompatible recipients (which carried the RTF); cells were labeled during mating and the R1 DNA examined after various periods of chase with unlabeled medium. In compatible recipients, most of the DNA was first found in a form sedimenting at 44S, but after a 20 minute pulse was largely converted from this linear state into open circles of 50S and closed circles of 75S. But when the recipients were incompatible, replication appeared to halt at the intermediate stage and took much longer; 75S closed circles were obtained only after 90–150 minutes of incubation. The rate of replication of R1 DNA appeared in general to be some ten times slower in the incompatible than in the compatible recipient. On the other hand, Saitoh and Hiraga (1975) found that sex factor DNA is converted into covalently closed circles just as well in male phenocopy recipients as in females; this suggests that incompatibility takes place only at a subsequent step, such as at the initiation of cycles of (vegetative) replication. If conversion of the entering plasmid to its mature form is inhibited, the existing plasmid might have an advantage for survival in the cell; if initiation of vegetative cycles is inhibited, each plasmid should have the same probability of remaining in the cell. Equal opportunity of survival often is found; for incompatibility most commonly is reflected in an even probability that either of two incompatible factors will be lost from a transient heterozygote. Thus when an $F^+$ and an F*lac* factor are present in the same cell, each factor is carried by about 50% of the immediate progeny. With some pairs of plasmids, however, one may preferentially dislodge the other; no explanation is known for this phenomenon, but in these instances simple dilution due to inhibition of vegetative initiation cannot provide an explanation.

Copy number mutants of R1 have been isolated by Nordstrom, Ingram and Lundback (1972) in cells that display increased resistance to all the drugs whose markers are carried by the factor. That this is the result of the presence of an increased number of copies of plasmid DNA per cell was shown by measuring the proportion of DNA found in covalently closed circles; this increased from the wild type equivalent of about one copy of plasmid DNA per bacterial chromosome to a level of 2–4 copies per chromosome. Uhlin and Nordstrom (1975) noted that some copy mutants decrease incompatibility while others increase it; this might be explained if the first class has a mutation reducing the affinity of a repressor for its operator (which should increase the total number of copies of the plasmid able to exist in the cell), while the second class represents a mutation reducing the affinity of the operator for the repressor (which increases the number of copies of the mutant plasmid but also increases the concentration of repressor, which therefore is more active upon other, wild type copies of the plasmid introduced into the cell). Of course, while this speculation supports a repressor model, the characterization of putative regulator protein and operator mutants by determining their dominance with wild type alleles is necessary before a model can be constructed. Copy number mutants of R100 have been isolated by Morris et al. (1974).

Although incompatibility is a plasmid function, it may rely upon the utilization of certain host functions, those concerned with replication. Thus Palchoudhury and Iyer (1971) found that both surface exclusion and plasmid incompatibility were reduced between two F′ factors when E.coli cells temperature sensitive in DNA replication were incubated at high temperature. And Davis and Vapnek (1976) have isolated a bacterial mutant that has an increased resistance to antibiotics because of a chromosomal mutation that increases the copy number of R538-1*drd* from 5 to 13 copies per host chromosome origin.

An obvious question about R factor replication is whether the RTF or r component is responsible for controlling initiation of the complete factor. Since transfer from donor to recipient is a property of the RTF, it seems clear that this replicon must function in the transfer mode during conjugation. Analogy with the sex factor suggests that this requires the introduction of a nick at the origin for transfer, followed by rolling circle replication to provide the strand that is transferred to the recipient (see Figure 2.29). In F-like R factors, the RTF component shares considerable homology with the sex factor and may therefore utilize the same transfer origin. For vegetative replication, however, it is not clear how the two component replicons interact, although it has generally been supposed that one is active and the other is suppressed. There is no a priori reason to suggest which component fills which role.

One experimental approach is to determine by visualization of replicating molecules where the origin is located. Perlman and Rownd (1976) have obtained replicating DNA of R100 from Proteus, using conditions in which the concentration of DNA precursors is limited to decrease the rate of chain elongation; this appears to increase the proportion of replicating R factor DNA. When the plasmid DNA is spread for electron microscopy under partially denaturing conditions it is possible to distinguish the RTF and r components (see earlier). The replicating molecules take the form of $\theta$ structures in which replication can be seen to have started at a single origin and may be either unidirectional or bidirectional. About half of the molecules visualized had commenced replication at an origin in the RTF component; the other half had utilized an origin in the r determinant. No molecules using both origins were seen. From these results it is therefore not at all clear how the replication of the R factor is controlled in Proteus; in particular, it is puzzling that either origin may be used on different molecules but not on the same molecule.

### *Transfer Genes of the Sex Factor*

Mutants of the sex factor that are defective in the ability to transfer plasmid DNA in conjugation can be isolated by selecting F*lac* variants that no longer are able to transfer their $lac^+$ operon to recipient cultures. Achtman, Willetts and Clark (1971) selected several mutants by this approach; these mutations were given the description $tra^-$. Most of these mutations are leaky (that is some expression of wild type function occurs, although at a much reduced level) and these can be perpetuated by transfer (at low levels) by selection for the $lac^+$ genes; the nonleaky mutations can be perpetuated by transiently introducing an F*his* factor into the F*lac* cell, so that the wild type functions of the F*his* can complement the mutant function of the F*lac*. Some of the $tra^-$ mutations are ambers (nonsense mutations causing premature termination of protein synthesis) which can be suppressed by using an appropriate $su^+$ strain to allow protein synthesis to continue past the mutant site, so that these mutants can be perpetuated in the $su^+$ hosts.

Complementation analysis to separate these mutations into genes cannot be performed directly by forming stable heterozygotes because of the incompatibility between F factors. However, it is possible to form transient heterozygotes, that is, to introduce two mutant F factors into the one cell and examine the complementation between them immediately, before they segregate. Both conjugation and transduction have been used to generate such heterozygotes. In their conjugational analysis, Achtman, Willetts and Clark (1972) used stationary phase conditions so that recipient cells carrying a $tra^-$ F*lac* factor entered the $Sex^-$ phenocopy state in which surface exclusion is inhibited. In order to perform this experiment, the donor cells must be able to transfer another $tra^-$ mutant into these recipients (in spite of the defect in the transfer system of the plasmid); so in these crosses amber *tra*$^-$ mutants were used to provide the donors, since they can transfer efficiently from *su*$^+$ host cells. By using donor cells that were sensitive to phage T6 but recipients that were resistant, it was possible to use the phage to kill the donor cells after mating. The recipient cells were then grown for one generation and tested for the ability to transfer *lac*$^+$ alleles into an F$^-$ recipient culture; they can do so only if the mutations in the two *tra*$^-$ plasmids are located in different genes. And it is then necessary, of course, to check that the transferred F*lac* episome is of the *tra*$^-$ type and is not a *tra*$^+$ recombinant generated by exchange between the two F factors.

In a second series of experiments, Willetts and Achtman (1972) used transduction with phage P1 to generate transient heterozygotes; strains carrying *tra*$^-$ F*lac* derivatives were transduced with P1 into strains carrying other *tra*$^-$ mutants. The transductants were then examined for their ability to transfer F*lac* to a Lac$^-$ P1$^r$ strain. This technique has the advantage that any *tra*$^-$ plasmid can be used as the donor, so that it is possible to perform the reciprocal crosses with nonsuppressible mutants (whereas with the conjugation technique only the suppressible amber mutants could be used as donors).

The results of the conjugational and transductional analysis identified a total of eleven cistrons (for review see Willetts, 1972a). These fall into four phenotypic groups. The largest group comprises mutants in eight cistrons, *traA, traB, traC, traE, traF, traG, traH, traK,* which are resistant to all three of the ♂-specific phages that have been tested against them, f2, Q$\beta$, and f1. These mutants are unable to support the formation of mating pairs and appear to have lost the ability to produce F pili. A ninth member of this group, *traL*, has since been identified (Helmuth and Achtman, 1975). Although the F pilus is constructed from a single protein subunit (known as pilin), this may carry modifications and presumably these genes include those specifying enzymes concerned with modification or assembly functions as well as the structural gene coding for the pilus protein subunit.

The second phenotypic class of mutants represents a single complementation group, *traD*. These mutants are resistant to phage f2 (although the phage remains able to adsorb to the F pilus) but they remain sensitive to phages Q$\beta$ and f1. The nature of the product specified by this gene is not known.

The third class of mutants remains sensitive to all three phages and so must be deficient in some function necessary for transfer other than those concerned with pilus formation. This includes mutants in two loci, *traG* and *traI*. Of the mutants in *traG*, five fall into the first phenotypic group and lack pili, whereas five fall into the third group and continue to manufacture the pili. Thus *traG* might specify a bifunctional protein, one needed both for conjugation and for pilus formation; or there may really be two cistrons in this complementation group.

The fourth group is provided by mutants in *traJ*. All the mutants except those in *traJ* retained the ability to show surface exclusion. Since the *traJ* mutants lack pili, this confirms that the pilus is not needed on the recipient for surface exclusion. As *traJ*$^-$ mutants both lack pili and fail to show surface exclusion, they are pleiotropic, a phenomenon which raises the possibility that *traJ* may be a regulator gene.

Transfer defective mutants also were isolated and mapped by Ohtsubo (1970). These comprised F*gal* factors that carried deletions extending from the material of the sex factor into the bacterial genes. These mutants displayed low or zero fertility when transfer was tested in either the F′ or Hfr state; most had lost their sensitivity to ♂-specific phages and these mutants lacked pili. The functions lost in these deletion plasmids could be provided by an F-like R factor, for the derepressed factor R100-1 could restore the conjugal fertility lost by the sex factor; the conjugation sponsored by cells carrying an F deletion plasmid and R100-1 takes place with the characteristics typical of the F factor, showing that it represents complementation and not transfer by the R factor itself. Ohtsubo, Nishimura and Hirota (1970) tested the F deletion strains for complementation with various R factor mutants by using the phage Plkc to transduce the mutant R factors into cells carrying the F deletion plasmids. This identified six complementation groups, described as *ABCDEF*, in the sex factor for which corresponding functions could be identified in the R factor; two further F factor complementation groups, *G* and *H*, could not be equated with functions in the R factor. Ippen-Ihler, Achtman and Willetts (1972) showed that the Ohtsubo loci *ABCDEF* correspond to the *tra* genes, *traI, traD, traG, traF, traC, traE*, respectively.

The complementation of F*lac tra*$^-$ mutants with R100-1 or R19-1, both derepressed variants, was tested by Willetts (1971) and Alfaro and Willetts (1972); only *traI* and *traJ* were not complemented. Thus *traA, traB, traC, traD, traE, traF, traG, traH, traK*, are not plasmid specific but are carried by both F and F-like R factors: these genes can be taken to define the F-type transfer system. The genes *traI* and *traJ* are plasmid specific in the sex factor, and presumably whatever functions correspond to them in the R factor also must be plasmid specific. Willetts (1970) and Alfaro and Willetts (1972) found that no complementation takes place between F-like and I-like plasmids, which must therefore carry entirely different transfer systems.

By inserting the F*lac* factor into the bacterial chromosome in the *gal* region, close to the attachment site of phage lambda, it is possible to obtain deletions; for when the phage is induced, it picks up material extending into the region of the integrated sex factor, leaving a variable amount of material. Ippen-Ihler, Achtman and Willetts (1972) used such deletions to map the *tra* genes; they obtained the same order as that found by Ohtsubo (1970). Further genetic mapping data have been reported by Helmuth and Achtman (1975). The positions of the *tra* genes on the circular map of the F factor are shown in Figure 3.10. Physical mapping of the sex factor has become possible with the development of the technique of restriction fragment analysis; this consists of cleaving the F factor

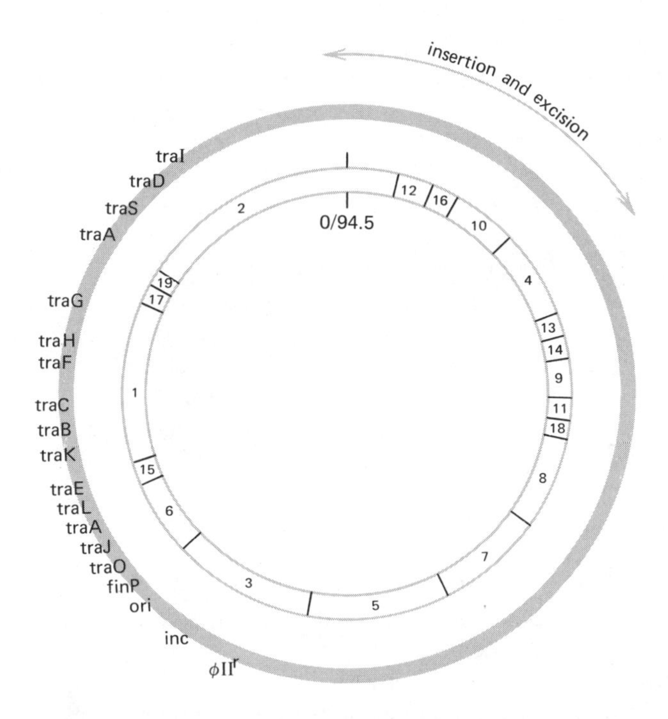


**Figure 3.10:** the sex factor (map not strictly to scale). The outer map gives the positions of known genes. The locus $\phi II^r$ governs resistance to the ♀-specific phage $\phi$II (which is related to T7); mutation at this locus allows the phage to propagate on E.coli cells carrying the F factor, a capacity it otherwise lacks. Other loci are discussed in the text. Details of the structure of the insertion region are given in Figure 2.25. Transfer starts at the origin *ori* and proceeds counterclockwise. Insertion in the bacterial chromosome using the region indicated (0–15.0 F) thus generates an Hfr strain in which *inc* is an early marker and the *tra* gene block is transferred late. The inner map represents the positions and approximate sizes of the 19 fragments generated by cleavage with Eco RI nuclease. Data of Ippen-Ihler, Achtman and Willetts (1972) and Skurray, Nagaishi and Clark (1976).

DNA with restriction enzymes whose specificity ensures that only a small number of cuts is made, at particular sites (see Volume 1 of Gene Expression and below). The enzyme Eco RI releases 19 fragments from F DNA, and these can be separated on the basis of their size by electrophoresis on agarose gels. Figure 3.10 shows the Eco RI restriction map of the sex factor. Individual fragments may be examined for their possession of particular gene functions by attachment to another plasmid, which then is examined for its ability to complement F factor mutants. Skurray, Nagaishi and Clark (1976) have reported experiments using this approach.

Polar effects show that transcription of the *tra* genes proceeds from *traA* to *traI*. (Polar mutations have two effects: the product of the gene in which the mutation lies is inactivated; and the amounts synthesized of the products of subsequent genes in the same transcriptional unit are much reduced. Polarity may be caused by nonsense mutations or by insertions and may be taken to identify genes that are transcribed into a single polycistronic messenger. The basis of polarity is discussed in Volume 1 of Gene Expression; see also page 403.) Earlier experiments showed that mutants in *traK* may exert a polar effect upon the expression of the genes *traB—traI*; and the more recent data of Helmuth and Achtman (1975), relying upon an examination of the polarity caused when phage mu is inserted into a *tra* gene, demonstrate that polar mutations in each of the *tra* genes reduce expression of the subsequent genes. Another observation suggesting that all the *tra* genes comprise a single operon is that none of them is expressed in *traJ*$^-$ mutants, consistent with the idea that the *traJ* product is a regulator protein that controls the expression of the transcriptional unit. As a single operon, the genes *traA-traI* comprise a large transcriptional unit of about 30,000 base pairs, almost a third of the sex factor. At a rate of 50 bases per second, this would take 10 minutes to transcribe. If each gene is ascribed an average size of 1000 base pairs, then less than half of the total length of the transcriptional unit is identified by the known *tra* genes; and, indeed, some spaces lacking gene functions can be seen on the genetic map. Whether further genes are located in this operon remains to be established.

The loci responsible for surface exclusion and plasmid incompatibility have been mapped by Willetts (1974), who examined the exercise of these functions by sex factor strains possessing deletions of different regions. Surface exclusion is lost when a deletion removes the region between *traG* and *traD*; so the locus *traS* can be mapped between these two cistrons. It is now known to lie between *traA* and *traD*. Although *traS* is not necessary for transfer, it comprises part of the *tra* operon; perhaps this location offers some selective advantage in reducing matings between cells that both carry the sex factor. The *traS*⁻ strains continue to synthesize the F pilus, another confirmation that the component(s) required for surface exclusion in the recipient do not include the pilus. It is interesting that the genes concerned with pilus formation are grouped at the proximal end of the operon while three genes not concerned with pilus formation, *traS, traD, traI,* lie together at the distal end.

Four different classes of surface exclusion system have been identified by Willetts and Maule (1974), in all cases expressed coordinately with the plasmid transfer system. These systems are carried by the factors:

| | |
|---|---|
| Sfx I | F factor |
| Sfx II | Col V2, R538 |
| Sfx III | Col VB*trp,* R1-19 |
| Sfx IV | R100, R136 |

This implies that the F factor and these Col and R factors, although they appear to share features of their transfer systems, must differ in their *traS* alleles. There is some interaction between the systems since Sfx III and Sfx IV are expressed in the presence of each other and in the presence of Sfx I or Sfx II, but Sfx I and Sfx II can be expressed only in the presence of each other and not in the presence of Sfx III or Sfx IV.

A locus responsible for incompatibility was identified by examining the ability of deletion strains to retain a transferred F*lac* factor. A site *inc* mapped between $øII^r$ and the transfer origin *ori* (see below). This means that the *inc* gene must be transferred very early by strains that integrate at the typical sites shown in Figure 2.25 (see also Figure 3.10 and below). Presumably *inc* is one of the genes of the sex factor that is concerned with replication. This supposition is supported by the observation of Timmis, Cabello and Cohen (1975) that a fragment of the sex factor, generated by the association of some of its Eco RI fragments and representing about 8000 base pairs in total, carries all the functions needed for replication and also displays incompatibility with F. The cluster of replication genes probably lies close to and counterclockwise from *inc*.

The location of the origin for transfer has been investigated by Willetts (1972b), using a transient heterozygote technique. When F*his* is transferred into a *tra*⁻ F*lac* mutant strain, the F*lac* is transferred as the result of complementation; that is, the *tra*⁺ genes of the F*his* element have been able to substitute for the defective functions of the mutant. But when an Hfr donor is crossed with a *tra*⁻ F*lac* recipient, no increase in the transfer of F*lac* is found at early times. This implies that the *tra* genes of the integrated sex factor in this Hfr strain cannot be transferred as early markers. Since the orientation of the sex factor in this Hfr donor is

*pro*——————$\phi II^r$ ——*traJ*————*traI*————*lac*———*bio*
early marker late markers

this means that the origin must lie to the left of *traJ*.

A more detailed mapping of the origin was performed by introducing a $tra^+$ F*lac* factor into a series of Hfr strains that are $tra^-$ because they carry deletions that extend into the integrated sex factor for various distances. By using strains that are also $recA^-$, the insertion of the F*lac* factor was prevented, so that any transfer of chromosome markers that takes place must be due to complementation between the integrated and autonomous sex factors. The ability to transfer the early marker $pro^+$ immediately after entry of the F*lac* factor provided a test for complementation; this is therefore a transient heterozygote technique. All Hfr strains carrying deletions that enter the sex factor from the right and terminate within the *tra* genes can transfer $pro^+$ efficiently. But when the deletion extends past all known transfer genes, transfer of $pro^+$ is abolished. This indicates formally that a *cis*-dominant site necessary for transfer lies to the left of *traJ* (the site is *cis*-dominant, that is, it must be located on the integrated factor DNA, since it cannot be complemented); although there is no proof that this is the origin (*ori*), this is the most likely function for this site to have. It lies between *traJ* and $\phi II^r$.

The integration and excision sites that are shown on the circular F factor maps of Figures 2.25 and 3.10 would be consistent with the integration of the sex factor so that the *tra* genes are transferred only very distally. However, in a test for the transfer of *tra* genes as early markers, Broda, Meacock and Achtman (1972) found that other sites also may on occasion be used for integration. Hfr donor cells sensitive to phage T6 were mated with phenocopies of $tra^-$ F*lac* cells, after which the donor cells were killed with phage T6; the recipients then were mated with a fresh culture of $F^-$ cells and should be able to transfer the defective F*lac* factor only if $tra^+$ genes were transferred at early times. Most Hfr strains behaved like that investigated by Willetts and did not transfer $tra^+$ genes; but in a few cases all the *tra* genes were transferred very quickly. The *tra* genes that were examined included *traJ, traF, traH* and *traI*; and an Hfr strain transferring one always transferred all the others. Thus the *tra* genes are transferred as a block, usually as late markers but occasionally at early times. This indicates that insertion does not take place within the *tra* genes, but that some Hfr strains result from an event which places these genes as early rather than later markers.

By using two Hfr mutant strains that differ in their extent of deletion only in that one covers *ori* whereas the other does not, Reeves and Willetts (1974) tested the ability of other plasmids to initiate transfer from *ori*. When a plasmid able to recognize *ori* is introduced into the Hfr cell retaining the origin, the ability to sponsor chromosome transfer is restored, but of course the control strain which has lost *ori* is not complemented by the same plasmid. The plasmids Col V2 and Col VB could act upon the Hfr sex origin; the plasmids R100, R1-19, and R538-l*drd* could not do so. This classification corresponds with the specificities of the *traI* products (*traI* is the only *tra* gene apart from the pleiotropic *traJ* that is plasmid specific). Reeves and Willetts therefore suggested that the role of the *traI* gene may be concerned with the initiation of DNA transfer; perhaps it codes for an endonuclease that recognizes the origin and makes a nick to commence conjugal transfer.

The origin identified in these experiments represents a *cis* dominant site necessary for transfer. This leaves as an open question the site utilized as the origin for vegetative replication, which remains to be identified. That this site is different from the transfer origin is suggested by the isolation by Guyer and Clark (1976) of *tra*$\Delta$ F factor strains which possess lengthy deletions removing *tra* genes and also a site that is cis dominant for transfer, probably the transfer origin *ori*. But the *tra*$\Delta$ sex factors are able to replicate vegetatively, which implies that they retain the origin for vegetative replication.

### *Interaction of F and R Factor Transfer Control Systems*

Derepressed R mutants fall into two categories, resolved by the experiments of Meynell and Cooke (1969) and Cooke, Meynell and Lawn (1970), in which they examined the ability of various R*drd* factors to suppress chromosome transfer by the integrated sex factor of the E.coli strain HfrC. A wild type ($R^+$) factor, of course, virtually abolishes the donor ability of the HfrC cell. The ability of the R*drd* mutants to do so must be judged by their effect upon transfer of a highly proximal marker, *pro*$^+$ in these experiments, since the derepression allows the R factors themselves to sponsor chromosome marker transfer at increased frequency.

The presence in the HfrC cell of some derepressed variants of R1 or of R100-1 fails to repress the Hfr donor ability; that is, the frequency of transfer of *pro*$^+$ is much higher than that of other markers. The production of F pili is not inhibited by these mutants. These were initially characterized as of the $i^-$ type, that is to say that they behave as though they have lost the ability to synthesize a repressor protein that acts upon both F and R to prevent expression of transfer functions.

The second class of derepressed R1 mutants continues to abolish HfrC donor properties; for transfer of *pro*$^+$ is no greater than that of any other marker and is due only to the activity of the Rl*drd* factor. HfrC cells carrying these R1*drd* mutants lack F pili and instead possess the closely related R pili. The F transfer system is thus repressed in these cells but the R system of the derepressed factor continues to function. These mutants were initially characterized as $o^c$; that is, they behave as though resulting from an operator mutation in the R factor which prevents the R*drd* DNA from responding to the repressor; but since the repressor protein is still made, it can act upon the F factor.

However, characterization of mutants as repressor-negative or operator-constitutive requires a dominance test. The $i^-$ type of mutation should be recessive, since the introduction of another (wild type $i^+$) factor leads to the production of repressor protein able to act upon the $i^-$ R*drd* factor (and upon the sex factor also, of course). The $o^c$ type of mutation should be cis-dominant, that is should maintain the factor carrying it in the derepressed state irrespective of what other factors are present in the cell, but should not affect these other factors since the effect of the mutation is to prevent the $o^c$ R*drd* factor from being recognized by the repressor. The original definition and experimental distinction between these types of mutation is discussed in Volume 1 of Gene Expression.

One indication that this classification of the two types of R*drd* is not correct is provided by the results of Silver and Cohen (1972). When temperature-sensitive derepressed R factors were isolated by their derepressed phenotype at 42°C but repressed phenotype at 37°C, they proved to fall into both the "$i^-$" and "$o^c$" classes. The existence of a temperature-sensitive mutation shows that the locus in which it is located codes for protein; for temperature sensitivity results from the inactivation of the protein at increased temperatures. Operator mutations thus are not temperature sensitive since they change the sequence of bases that provides a recognition site in DNA. This observation thus argues that both types of derepressed mutation occur in genes coding for repressor proteins: the "$i^-$" type is not plasmid specific since both R and F become derepressed, but the "$o^c$" type is plasmid specific since it affects only the R factor and not the F factor (see Figure 3.12).

Since it is possible to isolate mutants of the R factor that no longer repress the F factor, it should be possible to isolate mutants of the F factor that are no longer subject to repression. Frydman et al. (1970) therefore mated a donor strain carrying both an F factor and a *fi*$^+$ R factor with F$^-$ recipient cells or with recipient cells carrying a *fi*$^+$ R factor. Any F mutants no longer repressed by R should be transferred from these donors at rates much greater than those of wild type F factors; and the selection is increased further when a *fi*$^+$ R factor is present in the recipient, since only nonrepressible F mutants should be able to be expressed in these cells. By using an F*lac* donor strain, a mutation of this nature was isolated in an autonomous factor; and by isolating an HfrC donor strain able to transfer *pro*$^+$ at high frequency, a similar type of mutant was obtained in the integrated sex factor. Although these mutants were then thought to be of the $o^c$ type, no classification can be achieved without a dominance test.

Irrepressible mutants of F*lac* that continue to be transferred at high frequency from cells also carrying a *fi*$^+$ R factor were also isolated by Finnegan and Willetts (1971). In these experiments an F*lac* culture was mutagenized and mated with an R100$^+$ culture; cells carrying both factors were selected by their Lac$^+$ Tet$^r$ phenotype. These cells were then tested for their ability to transfer *lac*$^+$ and for sensitivity to the ♂-specific phage MS2. Cells in which both factors are wild type show poor transfer of F*lac* and are resistant to MS2; that is the expression of the F factor is suppressed. Cells that show a high frequency of *lac*$^+$ transfer and are sensitive to MS2 carry mutants of the sex factor which are no longer repressed by the R factor. These mutants showed wild type levels of transfer, normal pilus formation, and normal surface exclusion in the presence of the R factor; so all these properties were coordinately released from repression by the R factor.

To test whether these mutations are located in a regulator gene coding for protein or in an operator, transient heterozygotes were constructed. The procotol used is depicted in Figure 3.11 and requires two successive matings. In the first mating, the mutant F*lac* factor is introduced into a strain that carries both F*his* and R100, after which the parent donor cells are killed by their susceptibility to phage T6. The intermediate strain produced by this mating is then tested immediately for its ability to transfer F*lac*, F*his* or R100 to fresh (F$^-$) recipients; in this mating, the recipients are resistant to streptomycin so that by plating on appropriate media containing the drug it is possible simply to count the number of colonies carrying F*lac* (these are Lac$^+$), carrying F*his* (the His$^+$), or gaining R100 (determined by the Tet$^r$ phenotype). Some of the results obtained in these experiments are given in Table 3.2.

Because the intermediate strain carries two F′ factors, it is not stable; this is why it is only a transient heterozygote and so must be tested for retransfer immediately after gaining the F*lac* factor. The first set of figures in the table shows the retransfer of wild type F*lac* factor from various recipients. With an F$^-$ recipient, only the F*lac* factor is retransferred, of course. The mutual incompatibility between F*lac* and F*his* usually means that in a strain transiently carrying both factors, each is transferred as shown here with about the same frequency. Repression of transfer of both F*lac* and F*his* is established in the F*his* R*100* recipients; both are transferred at only about 1% of the level displayed by the F*his* recipients. Transfer of the R factor is poor in all matings.

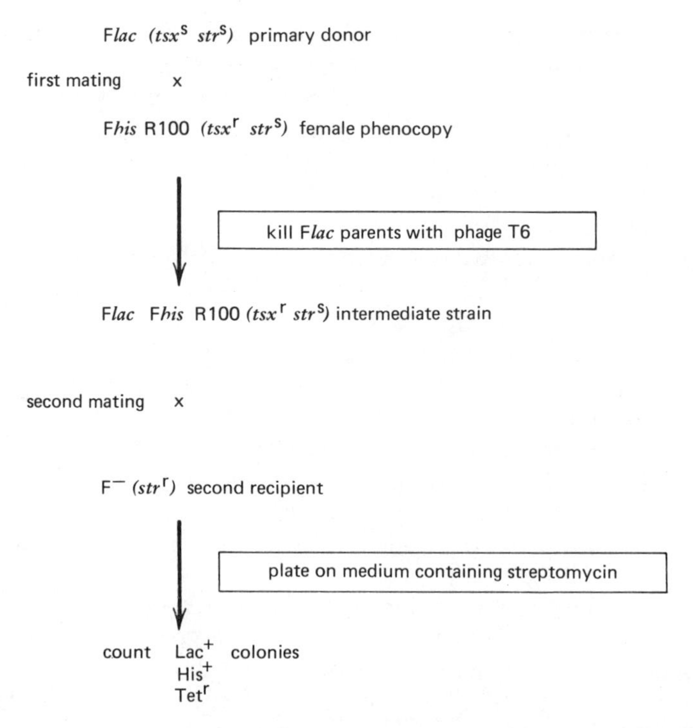


**Figure 3.11:** transient heterozygote technique for testing dominance. In the first mating an F*lac* factor is introduced into an F*his* R100 recipient which is in the female phenocopy condition; the donor cells are sensitive to phage T6 and are killed with it before the next mating. Immediately after the intermediate strain has been generated, its ability to transfer all three of the plasmids is tested by mating with an $F^-$ recipient strain; the recipients are resistant to streptomycin so that recombinants can be counted by plating on appropriate media containing the drug. Results of this analysis are shown in Table 3.2.

The irrepressible F*lac* mutants fall into two classes, originally described as *traP* and *traO*. The *traP* gene has since been renamed *finP*. Both were isolated, of course, by their ability to transfer at high frequency from cells stably carrying both the mutant F*lac* factor and a $fi^+$ R factor (not shown in the table). The nature of the mutations should be revealed by the ability of each mutant to retransfer in the presence of both F*his* and R100. The repression of wild type F*lac* in the F*his* R100 intermediate recipients shows that these cells possess all the wild type regulator proteins necessary to prevent transfer of the sex factor. If the mutant F*lac* is defective in its ability to synthesize a repressor protein, therefore, this function should be provided by the F*his* factor and the mutant should become subject to repression by the R factor; transfer should not be

**Table 3.2:** plasmid transfer from transient heterozygotes

| primary donor | intermediate recipient | progeny per 100 donor cells | | |
|---|---|---|---|---|
| | | $Lac^+$ | $His^+$ | $Tet^+$ |
| *Flac* | $F^-$ | 20 | — | — |
| *Flac* | *Fhis* | 100 | 140 | — |
| *Flac* | *Fhis* R100 | 1 | 3 | 0.6 |
| *Flac finP* | $F^-$ | 19 | — | — |
| *Flac finP* | *Fhis* | 120 | 110 | — |
| *Flac finP* | *Fhis* R100 | 1 | 1 | 0.2 |
| *Flac traO* | $F^-$ | 25 | — | — |
| *Flac traO* | *Fhis* | 190 | 180 | — |
| *Flac traO* | *Fhis* R100 | 14 | 20 | 0.7 |

Transfer was performed by the protocol shown in Figure 3.11; that is, a plasmid was introduced from the primary donor strain into the intermediate recipient and then tested for transfer into fresh recipients. The values shown on the right are the results of this retransfer; the numbers of $Lac^+$ and $His^+$ progeny are expressed relative to the number of transient heterozygotes and the number of $Tet^r$ progeny relative to the total intermediate cell population. Data of Finnegan and Willetts (1971).

possible. This condition is fulfilled by the *finP* mutants, which are able to transfer from $F^-$ recipients (see Table 3.2), from F*his* recipients (see Table 3.2), from R100 recipients (not shown); but which cannot retransfer from F*his* R100 recipients (see Table 3.2). The *finP* class of mutation thus is recessive to the presence of the (wild type) F*his* factor. The most probable explanation is that inhibition of F transfer by a $fi^+$ R factor requires two protein functions: one is coded by the R factor and the second is coded by the F factor. The *finP* F*lac* mutants thus must be unable to produce the F-coded component, but of course this can be provided by the F*his* factor.

An alternative model for the *finP* type of mutation would be to suppose that the F*his* factor is inhibited from transfer by the R factor and physically blocks some site to which the F*lac* mutant would have to bind in order to retransfer; such blocking would require F*his* DNA and a protein product of the R factor. The models can be distinguished by genetic analysis since the first predicts that it should be possible to isolate mutants of the operator-constitutive type; these should represent inactivation of the ability of some site on the F factor DNA to recognize the transfer inhibitor produced by the conjunction of the F and R repressor protein functions. Operator mutations should be *cis*-dominant; this dominance would be difficult to explain on the blocked site model.

The *traO* class of mutants fulfills this condition; the F*lac traO* mutant factors can retransfer from the F*his* R100 intermediate recipient at the levels shown in Table 3.2. The *traO* mutants thus are unresponsive to the protein repressor functions coded by F*his* and R100. A significant feature of the *traO* mutation is that it is dominant, for the F*his* factor also is released from repression and is allowed to transfer. These experiments therefore do not provide the formal proof of cis-dominance, that is the demonstration that *traO* controls only the expression of the DNA of the F*lac* factor, which would be necessary to confirm operator status. This implies that if *traO* is an operator it controls the expression of gene(s) whose products can sponsor transfer of the F*his* factor as well as of the F*lac* factor; thus constitutive function of the *traO* F*lac* factor provides the protein function(s) necessary and sufficient for transfer of any F factor present in the cell. The *traO* mutations map very close to or possibly within *traJ* (in the counterclockwise direction as shown in Figure 3.10); *finP* lies between $\phi II^r$ and *traK* (Willetts, Maule and McIntire, 1976).

Since the transfer systems of the $fi^+$ F-like R factors and the F factor itself are closely related, it is reasonable to suppose that they lie under a common type of control. Thus we may suppose that two protein components are necessary for inhibition of fertility of either F or R factors. The ability of the $fi^+$ R factors to suppress F fertility suggests that they code for a product which is necessary for suppression of F and for self-repression of R; this protein is therefore nonplasmid specific and was originally termed I by Finnegan and Willetts (1971). This is the product of the gene characterizing the $fi^+$ type of plasmid, variously described as $i^+$, $fi^+$, and $fin^+$. Gasson and Willetts (1975) introduced the terminology *finO*, which we shall follow here.

The model for the interaction of F and $fi^+$ R factors illustrated in Figure 3.12 proposes that the nonplasmid specific FinO product (I protein) interacts with the plasmid specific $P_F$ protein coded by the *finP* gene of the F factor; this generates a transfer inhibitor that prevents the expression of F. By analogy, the R factor may possess a gene of the *finP* type, but whose protein product, $P_R$, is specific for the R factor; the interaction of FinO and $P_R$ might be responsible for the self-repression of R.

When an F*lac* factor is transferred to an R100 strain, it can be retransferred at high frequency immediately after its reception, that is, in spite of the presence of the established R factor. This suggests that production of the $P_F$ component, or its interaction with the FinO component, must be a slow process. Of course, this is directly comparable to the high frequency transfer of R factors that can take place immediately after they have

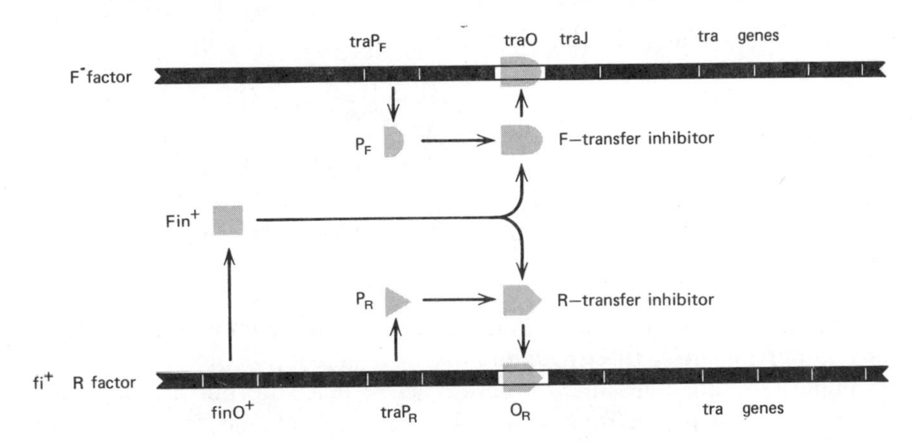


**Figure 3.12:** model for control of F and R factor transfer systems. The R factor specifies a product $\text{Fin0}^+$ (or I) which can interact with two plasmid specific proteins: together with $P_F$, coded by the *finP* gene of the F factor, it forms the F-transfer inhibitor, which acts at *tra0* to prevent expression of the adjacent gene *traJ*; together with $P_R$, presumably coded by a counterpart in the R factor to the sex factor *finP* gene, it forms the R transfer inhibitor, which acts at some operator site on the R factor. Mutation in the *fin0*$^+$ gene should relieve inhibition of both F and R transfer, since the transfer inhibitor for neither plasmid can be made; this mutation should be recessive since the introduction of a wild type R factor provides a $\text{Fin0}^+$ product able to interact with $P_F$ and $P_R$. The "i⁻" mutants of R factors fit these predictions. Mutation in the *finP*$_R$ gene should relieve inhibition of R transfer, since the R-transfer inhibitor cannot be assembled, but should not affect inhibition of F transfer; these should therefore comprise plasmid specific recessive mutations, since introduction of an R factor with a functional *finP*$_R$ gene restores ability to make the R-transfer inhibitor. The "$o^c$" mutants of R factors fit these predictions. Derepressed R mutants also should be produced by mutation in the site $O_R$ so that it no longer recognizes the R-transfer inhibitor; their dominance properties cannot be predicted without knowing what functions may be controlled by $O_R$. The effects of mutations in the F factor are described in Figures 3.13 and 3.14. Note that *tra0* is close to *traJ* and that the other *tra* genes map together in a block at the right; the positions of other loci are arbitrary and their illustration here does not imply that they occupy these positions in the F and R factors.

entered a cell; to accommodate the similarity between the F and R systems, it is necessary only to postulate that it is the synthesis of $P_R$, or its interaction with FinO, that is slow (that is instead of supposing that there is a single repressor product whose synthesis is slow). Since high frequency transfer systems can be established for several types of plasmid, implying a general similarity in control systems, the existence of two proteins necessary for transfer inhibition may be a common phenomenon.

If the FinO protein is nonspecific, then it should be able to inhibit the expression of a variety of plasmids. Finnegan and Willetts (1972) therefore tested several *fi*$^+$ plasmids for their ability to repress three mutant (formerly *fi*$^+$) plasmids that had lost their fertility inhibition, R1-19, R100-1, R136*i*$^-$, and three naturally occurring *fi*$^-$ plasmids of the F-like type, F, ColV-K94, ColVB*trp*. Each of the *fi*$^+$ plasmids ColB-K98, R1, R100, R136 inhibited expression of all of the mutant and naturally *fi*$^-$ plasmids with which it could stably coexist; all the stable strains were resistant to ♂-specific phages. The *finO*$^+$ products of these four plasmids thus appear to be functionally identical and able to act upon several other plasmids.

The specificities of the P products can be investigated by measuring the frequency of retransfer from intermediate cells that contain the P product of another plasmid. Intermediate cells carrying R1, R100, or R136, for example, should contain $P_R$ products as well as the nonspecific *finO*$^+$ product. Since the retransfer of the three mutant and three naturally occurring *fi*$^-$ plasmids was in general not inhibited in these cells, the P product must be plasmid specific. But R136 inhibited the retransfer of R100-1; and R100 inhibited the retransfer of R136*i*$^-$. Factors R100 and R136 therefore must possess the same P product. Finnegan and Willetts identified four types of P product in these experiments. Plasmids F and ColV-K94 appear to have the same P product; they are incompatible, specify identical F-pili, and have the same transfer system. Plasmid ColVB*trp* has the same transfer system as the sex factor but possesses a different P product. Factor R1 possesses a different P product; factors R100 and R136 share another P product. Although there is some correlation between incompatibility, transfer systems, and transfer inhibition, there is no fixed relationship between them.

The combined action of the $FinP_F{}^+$ and FinO$^+$ products causes the same multiple changes in the phenotype of cells carrying the sex factor as does the *traJ*$^-$ mutation: that is transfer is inhibited, pilus formation ceases, surface exclusion is prevented. A model for the action of the $FinP_F{}^+$/FinO$^+$ transfer inhibitor is therefore to suppose that it acts to block the expression of *traJ*; formally this action might take place at any stage from transcription or translation to protein function. However, the existence of the *traO* mutations suggests that the transfer inhibitor acts upon an operator element, an idea which implies that control of transcription is the most likely mechanism; this would mean that *traO* is an operator controlling the transcription of *traJ*. Figure 3.13 illustrates the consequences of mutation in the *finP* gene.

The effect on the *tra* genes of the inhibition mediated by FinO$^+$/$FinP_F{}^+$ was investigated by Finnegan and Willetts (1973) in a series of experiments in which they constructed mutants of the type F*lac traO*$^-$ *traX*$^-$, where *traX* is any one of the transfer genes (except *traI* or *traC*). The dominance of the *traO* mutation means that the transfer-inhibitor cannot

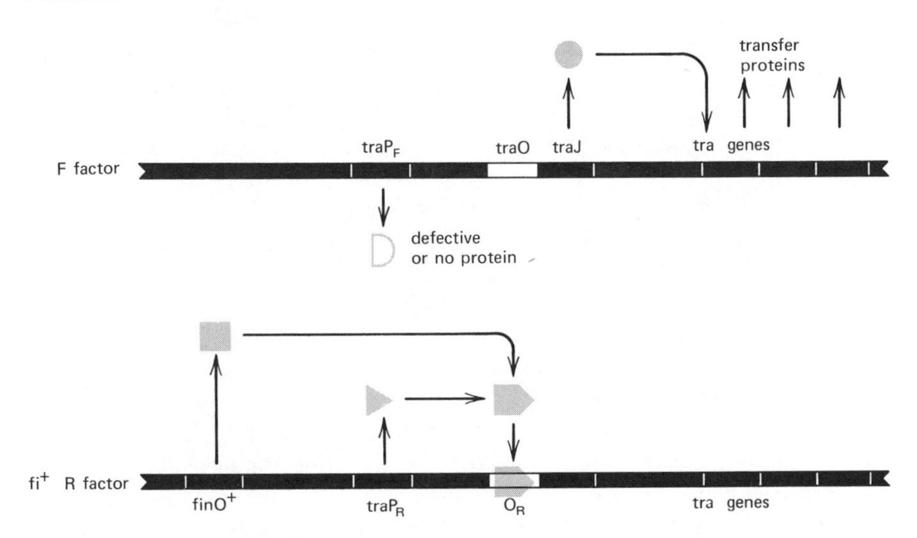


**Figure 3.13:** model for the effect of *finP* mutations in the sex factor. If the *finP* gene produces no protein or one which is defective, the F-transfer inhibitor is not formed. This mutation is plasmid specific since it has no effect on formation of the R-transfer inhibitor; so the R factor remains repressed. Since *tra0* is not bound by the F-transfer inhibitor, the adjacent gene *traJ* is expressed. The product of *traJ* acts as a regulator protein able to switch on synthesis of the transfer proteins; its mode of action is not defined. The *finP* type of mutation is recessive, since introduction of a wild type sex factor provides a functional gene able to produce $P_F$ and thus contribute to the F-transfer inhibitor.

inhibit transfer of the F*lac* episome. If this F*lac* factor fails to be transferred in a transient heterozygote, this must therefore be a consequence of the inactivity of its *traX*$^-$ gene. The transfer efficiency of the mutant F*lac* was therefore measured in transient heterozygotes carrying the three plasmids:

F*lac traO*$^-$ *traX*$^-$
F*his traO*$^+$ *traX*$^+$
R100 *finO*$^+$

Transfer of the F*lac* factor will be inhibited only if the *finO*$^+$ gene of the R100 factor directly inhibits the expression of the *traX*$^+$ gene of the F*his* factor; in this case neither of the F factors can provide the TraX$^+$ protein (for the gene is mutant in F*lac* and repressed in F*his*). However, if the *finO*$^+$ gene does not repress *traX*$^+$, then the F*his* plasmid will provide this function and the F*lac* plasmid will be transferred. If the *traX*$^+$ gene is only indirectly inhibited by *finO*$^+$, then the F*lac* plasmid can supply whatever function is directly inhibited by *fin*$^+$, thus allowing expression of the F*his* *traX*$^+$. The only gene in which mutation prevented transfer of F*lac* in these experiments was *traJ*. This demonstrates that the FinO/$FinP_F$ transfer inhibitor acts directly to prevent the expression of *traJ*, and this in turn prevents expression of the transfer genes. Although these results formally do not imply which stage of expression of *traJ* is inhibited by the FinO/$FinP_F$ regulator, the most probable interpretation is that the transfer inhibitor acts at *traO* to prevent the transcription of *traJ*.

These experiments provide formal proof that *traO* exercises cis dominant control of *traJ*. When the *traX*$^-$ mutation represents some gene other than *traJ*, the *traJ*$^+$ gene of the F*his* plasmid is repressed by the FinO/$FinP_F$ transfer inhibitor, so that expression of the transfer genes requires the product of the *traJ*$^+$ gene of the F*lac* plasmid. This can be synthesized because the *traO*$^-$ mutation is dominant, preventing inhibition of the F*lac* *traJ*$^+$ gene by the transfer inhibitor. That the *traO*$^-$ mutation is cis dominant (and not trans dominant) is shown by the situation in which the *traX*$^-$ mutation lies in *traJ*. In this case no transfer occurs because the *traJ*$^+$ gene of F*his* is inhibited by the transfer inhibitor; that is to say that the *traO*$^-$ mutation of the F*lac* factor cannot turn on its expression. This implies that *traO* acts only upon the adjacent *traJ* gene of its own DNA. The effect of *traO* must be confined to the *traJ* gene and does not extend to the other genes, for if they were expressed constitutively under its control, transfer would be independent of the *traJ*$^+$ allele of the F*his* plasmid and so would not be prevented by the transfer inhibitor. The consequences of mutation in *traO* are illustrated in Figure 3.14.

Of course, the conclusion that *traO* directly controls only *traJ* raises the question of how *traJ* controls the other *tra* genes whose expression is necessary for transfer; if *traJ* codes for a regulator protein whose action is necessary for expression of the *tra* genes (so that it is a positive control molecule, as contrasted with the negative control exercised by a repressor), then mutants in its presumed site of action should be cis dominant for transfer deficiency; no such mutants have yet been found. Another implication of these results is that *traJ* must be plasmid specific; for in spite of the similarity between the F transfer system and the F-like R systems, this function clearly could not have been provided by R100 in the triple plasmid intermediate strain.

The *finO*/*finP* system represents only one way in which a *fi*$^+$ R factor may inhibit the fertility of the sex factor. Gasson and Willetts (1975) distinguished several different mechanisms of fertility inhibition by examining the abilities of *fi*$^+$ plasmids of several different compatibility groups to repress the transfer of F factors of different genotypes. The *fi*$^+$ R factors tested in groups FII, FIII, FIV and N show *finO*/*finP* type inhibition: that is release from the inhibition is gained when the sex factor carries a *traO*$^-$ or *finP*$^-$ mutation. The *fi*$^+$ R factors of compatibility

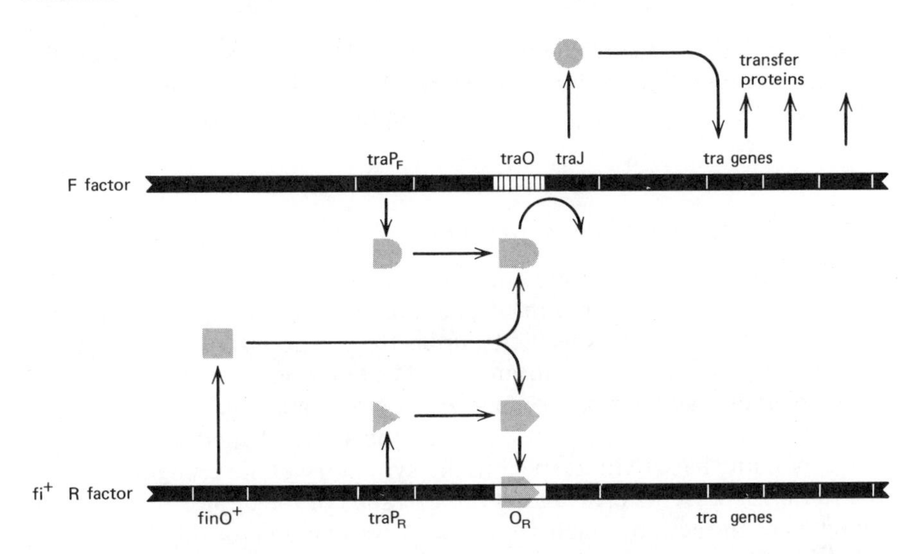


**Figure 3.14:** model for the effect of *tra0* mutations in the sex factor. A mutation in *tra0* which prevents recognition of the F-transfer inhibitor allows the adjacent gene *traJ* to be expressed and the *traJ* product then turns on the transfer genes. Since *traJ* is plasmid specific, the R factor remains repressed. The *tra0* mutation is *cis*-dominant, that is it acts upon only the *traJ* gene adjacent to it. Since, however, *tra0* mutants are dominant in causing transfer of all sex factors present in the cell, the result of expression of *traJ* is to switch on the transfer genes and cause transfer of any sex factor.

groups FI, Iω and X, however, are able to inhibit the transfer of *traO*$^-$ and *finP*$^-$ sex factors; no mutation that releases F from this inhibition has yet been identified, and so the mechanism(s) of this inhibition remain unknown. Within this class there is a difference in that an Iω plasmid prevents surface exclusion but X and FI plasmids do not; the implication is that the Iω *fi*$^+$ mechanism switches off the entire transfer operon of the sex factor, but the X and FI mechanisms leave at least the *traS* gene able to function. An FI plasmid is distinct in being the only factor that inhibits transfer but allows the sex factor to produce sex pili; all the other *fi*$^+$ plasmids prevent the production of pili. Another mechanism is identified by the R62 plasmid of the Iα compatibility group: this inhibits the transfer of *traO*$^-$ and *finP*$^-$ sex factors, but a mutation in the F factor designated *traQ*$^-$ allows the sex factor to transfer (Willetts and Paranchych, 1974). Since the *finO*/*finP* system acts to prevent expression of *traJ*, it should be unable to inhibit the transfer of *traJ*$^-$ mutants; all the plasmids displaying other systems of fertility inhibition, however, were able to inhibit transfer of *traJ*$^-$ sex factors (to allow transfer to occur at all, these plasmids also carry a further mutation conferring partial expression of the transfer operon in the absence of the *traJ* product). This implies that the fertility inhibition brought about by these plasmids represents a direct action on some stage of *tra* gene expression subsequent to the involvement of the *traJ* regulator.

### *Plasmid Modification and Restriction Systems*

The existence of modification and restriction systems was revealed when it was discovered that growth of phage lambda in either E.coli K or B strains confers a strain specificity upon the phage DNA. Each of these bacterial strains possesses an enzyme system which recognizes certain short, palindromic base pair sequences in DNA and acts either to modify or to restrict them. Modification takes the form of methylation of the corresponding base (with respect to the symmetry of the recognition site); restriction represents cleavage of both strands of the duplex in the recognition site, an action which is followed by degradation of the cleaved DNA by other, nonspecific nucleases. Any DNA in the cell, whether that of the chromosome or of a plasmid or phage, that possesses an appropriate recognition sequence is either modified or restricted at it.

The operation of this system essentially distinguishes the DNA of the host cell from that of a foreign source. The chromosome of a B-type or K-type cell is methylated on both strands at each of the appropriate recognition sites; the enzyme complex responsible for modification and restriction cannot bind to the modified sites so that they are immune from restriction. Replication produces daughter chromosomes each of which is methylated on the old strand but not on the newly synthesized strand. The enzyme complex recognizes these sites and methylates the unmodified strand. This sequence of events applies to phage grown in the cell as well as to the bacterial chromosome; thus lambda grown on cells of E.coli B carries the B-modification and is immune from restriction when it infects further B-type cells. But any DNA that is not modified on either strand is recognized by the enzyme complex and restricted, that is, cleaved, at the site. Thus λ DNA produced by growth on any strain other than E.coli B does not carry the appropriate modification and is degraded upon entering a B-type cell. The phage is thus said to be restricted in the cells upon which it can grow. Because this system is coded by the host cell and was first observed with phage DNA, it was described as the host modification and restriction system. This phenomenon is discussed in Volume 1 of Gene Expression.

The E.coli B and K modification and restriction systems are very similar in their construction and are coded by allelic genes. Each consists of three polypeptide subunits. The component coded by the *hss* gene is responsible for recognizing the site of action on DNA that has one or two unmodified strands. The component coded by the *hsm* gene is responsible for methylation; and the component coded by the *hsr* gene is responsible for cleaving DNA. Restriction appears to be the prerogative only of the complete tripartite complex, but modification may be undertaken either by this complex or by the complex of the *hss* and *hsm* gene products. The B and K systems each have different recognition sites.

Two further systems have been identified in E.coli 15; the host chromosome carries genes that specify the A system and a plasmid carries genes that specify the 15 system. The two systems are independent and additive; they have not been characterized as fully as the B and K systems. A further E.coli system is carried by phage P1; this phage is closely related to plasmid 15 and these two systems are allelic in the same way as those of type B and type K. Several other modification and restriction activities have been identified in various strains of Hemophilus. All modification and restriction enzymes share the characteristic that the recognition site is palindromic, that is, that it has twofold rotational symmetry, so that the same sequence is given by reading either strand in one (for example the 5′ to 3′) direction.

Two types of modification and restriction system are carried by R factors. An early observation of the presence of such a system in *fi*$^-$ R factors was made by Watanabe et al. (1964); they found that the efficiency of plating of phages T1, T7, and λ was reduced when the phage DNAs were introduced for the first time into cells carrying certain *fi*$^-$ R factors. The reduction is due to restriction of most of the phage DNA molecules. Those phage molecules that survived, however, displayed a much increased efficiency of plating when used to infect further cells carrying the *fi*$^-$ factor; these phage DNAs must have been modified instead of suffering restriction. These *fi*$^-$ factors therefore must carry a system able to both modify and restrict DNA possessing the appropriate recognition site.

The patterns of restriction and modification led Watanabe et al. (1966) to conclude that there are (at least) two types of modification and restriction system carried by these R factors. Most of the naturally occurring *fi*$^+$ R factors do not specify modification and restriction systems. But two R factor systems were resolved by Bannister and Glover (1970): RI was carried by the *fi*$^+$ factor R124; and RII was carried by three *fi*$^-$ factors. A characteristic of the B-type and K-type systems is that when restrictionless (r$^-$) mutants are isolated, about half lack modification (r$^-$m$^-$) and about half retain it (r$^-$m$^+$); the r$^-$m$^-$ mutants are *hss*$^-$ and cannot recognize the DNA sites, whereas the r$^-$m$^+$ mutants are *hsr*$^-$ and are unable to cleave DNA. An apparent equality of r$^-$m$^-$ and r$^-$m$^+$ phenotypes was observed among restrictionless mutants of these two R factor systems.

However, Yoshimori et al. (1972) reported that all the $r^-$ mutants that they were able to isolate of either RI or RII retained the ability to modify DNA. The previous apparent equality may perhaps have been due to deletions generated in the R factor that caused the loss of separate modification and restriction activities. The absence of single mutants of RI and RII that are $r^-m^-$ implies that there is no recognition polypeptide common to modification and restriction in either of these systems.

The sequences recognized by several of the restriction enzymes have been determined, including the recognition sites of the RI and RII nucleases; for discussion see Volume 1 of Gene Expression. In the nomenclature that has been agreed for restriction enzymes, these two nucleases are known as Eco RI and Eco RII. The Eco RI and Eco RII enzymes share the characteristic that they make staggered breaks in DNA; that is, the two strands are cleaved some bases apart. This generates single stranded sequences at the ends of the two duplex molecules produced by the cleavage; since these are complementary and are produced when the enzyme reacts upon any DNA, restriction cleavage of two different DNA molecules with either of these enzymes can be followed by cross hybridization between the free single strand ends to link together covalently quite different DNA species. This technique can be used, for example, to insert a sequence of eucaryotic DNA into a bacterial plasmid. (This is not the mode of action of all restriction endonucleases; an enzyme of H.influenzae, for example, breaks both strands at the same site.) The recognition site of the Eco RI enzyme was determined by Hedgpeth et al. (1972) and constitutes an eight base pair sequence in which the final bases are A-T pairs in either orientation; this is

```
                            ↓
3′  A/T – C – T – T – A – A – G – T/A  5′
5′  T/A – G – A – A – T – T – C – A/T  3′
              ↑
```

The recognition site of the Eco RII enzyme was determined by Bigger et al. (1973) and Boyer et al. (1973) and is the five base pair sequence

```
                             ↓
3′     G – G – T – C – C      5′
           *       *
5′    ↑C – C – A – G – G      3′
```

The arrows indicate sites of cleavage and the asterisks in the RII site indicate the bases which are methylated by the modification activity.

## Colicinogenic Factors

### *Production and Adsorption of Colicins*

The term "colicin" was originally introduced to describe antibiotic substances produced by certain strains of E.coli, although it has since become apparent that the factors responsible for colicin production may be carried also by other bacteria. Colicins are defined as antibiotics that are highly specific, may be produced by certain strains of intestinal bacteria, and act to kill other strains of bacteria in a manner which does not rely upon lysis. Much of our early knowledge of colicins was provided by the work of Fredericq; for a review see Fredericq (1957). Colicins are proteins, although some have been isolated in association with other types of structure, such as lipopolysaccharides. The factors responsible for their synthesis are plasmids; a colicinogenic strain is therefore one that carries a colicin factor and may on occasion thus produce colicin.

One technique developed to demonstrate colicinogeny was to stab colicinogenic strains on agar plates, incubate, sterilize, and then seed an indicator (sensitive) strain over the surface. Regions where growth of the indicator strain is inhibited then surround the sites where colicin-producing colonies had grown. Colicins are released from colicinogenic bacteria in concentrations that vary according to the type of colicin and the bacterial host strain. Sensitivity to colicins is determined by a series of receptors, each of which allows its host bacterium to adsorb only certain colicin(s). Thus bacteria become resistant to a colicin when they lose the receptor to which it adsorbs. Sensitivity to some colicins is related to sensitivity to certain bacteriophages; for example, mutants selected for resistance to colicin K or for resistance to phage T6 always are resistant to both the colicin and the phage; the same receptor must be involved. Colicins were therefore originally classified by their specificity for sensitive bacteria; all those colicins to which resistance is conferred by loss of a certain receptor fall into the same group. By isolating mutants resistant to one colicin and then testing the susceptibility of the strain to other colicins, Fredericq identified some twenty groups (for review see Fredericq, 1957; Reeves, 1965; Nomura, 1967a; Hardy 1975).

Bacteria may cease to be susceptible to killing by a colicin in any one of three ways. Mutants that have lost the cell wall receptor are *resistant* since the colicin cannot be adsorbed. Mutants that continue to be able to adsorb the colicin, but which are not killed by it, are said to be *tolerant*; these are defective in some stage of killing subsequent to adsorption. Acquisition of the plasmid factor responsible for synthesis of a colicin confers *immunity* to it; colicinogenic cells adsorb colicin as readily as noncolicinogenic cells, but the killing action is inhibited. (Note that immunity to colicin killing should not be confused with the superinfection immunity that governs the ability of plasmids to survive in a cell line.)

Immunity has been used to distinguish between different colicins that use the same receptor; for example, two colicins were originally classified as the I type by their common adsorption, but the presence of each factor confers immunity only to its own colicin, thus defining colicins Ia and Ib. Similarly colicins E2 and E3 share a receptor but differ in immunity. The terms Ia and E2, for example, therefore describe the colicin protein itself, that is, the killing agent. The nucleic acid plasmids which carry the information to specify these colicins are described as the colicin factors Col Ia and Col E2. It addition to describing adsorption group and immunity, it may be useful to indicate the origin of a colicin factor; thus Col E2-P9 indicates the Col E2 factor derived from Shigella sonnei strain P9 (for review see Nomura, 1967a). The naming of the E group colicins is anomalous because some bacterial mutants were obtained that were resistant to all three E colicins; however, Hill and Holland (1967) identified mutants resistant only to E2 and E3 and thus showed that E1 must utilize a different receptor (although it might share a component with the E2/E3 receptor).

In a colicinogenic bacterial population, only a small proportion of the cells produces colicin. Ozeki, Stocker and De Margerie (1959) developed two techniques to follow the release of colicin E2 by individual bacteria. In one series of experiments, single colicinogenic bacteria produced small clear spots in the growth of a sensitive strain; since these spots contained of the order of 6000 bacteria, at least this number of colicin molecules must be released in a single event. Another set of experiments assayed the bacteriacidal action of products released into microdrops of broth by single colicinogenic bacteria; these experiments showed that the cells releasing colicin are inviable. Only a small (random) fraction of the colicinogenic bacteria manufacture the killing colicin; and these cells thus are killed in the process.

Several colicins have been prepared in purified form. In each case the biologically active agent is a protein. Colicins which are related in their use of the same receptor generally appear to take a similar physicochemical form (although their modes of killing may be quite different). Konisky (1972) found that both the Ia and Ib colicins are proteins of about 80,000 daltons; they have the same amino acid composition, isoelectric point, some antigenic cross-reactivity, and similarities in tryptic fingerprints. Colicins E2 and E3 are proteins with somewhat similar characteristics, both about 60,000 daltons (Herschman and Helinski, 1967); the E1 protein is of a similar size but otherwise has characteristics rather different from E2 and E3 (Schwartz and Helinski, 1971). This is consistent with the observations that show E2 and E3 to share a common receptor but E1 to have a different receptor. Herschman and Helinski (1967) reported that colicins K, V and A all are associated with lipocarbohydrate and may form part of the surface antigens of the cells carrying them.

Receptors in the cell wall have been identified for several colicins. Weltzien and Jesaitis (1971) and Sabet and Schnaitman (1971) showed that the cell wall fractions of sensitive cells bind to colicins E3 or K; the receptor binding the colicin is destroyed by agents that attack proteins. The receptor may be a complex of protein and lipopolysaccharide, in keeping with its location in the cell wall; it is entirely absent from resistant strains. Using $^{125}$I labeled colicin Ia, Konisky and Cowell (1972) showed that only sensitive strains can bind the protein; resistant strains have lost the capacity to do so and also are unable to bind colicin Ib. Another confirmation that both Ia and Ib bind to the same receptor is provided by the observation that Ib competes with $^{125}$I-Ia for binding. Colicinogenic and noncolicinogenic cells adsorb the same amount of colicin, a demonstration that immunity is mediated by events at a stage subsequent to uptake. At doses at which from 0–50 colicin molecules are bound per cell, the killing dependence on colicin concentration follows a Poisson distribution; this implies that each single colicin molecule has a fixed probability of killing a sensitive cell. Single hit kinetics have been obtained for killing by several colicins, so this probably a general mode of action (see Nomura, 1967a). Konisky and Liu (1974) characterized the envelope component able to bind the I colicins; it is about 307,000 daltons, is sensitive to trypsin, insensitive to nuclease or lipase activities. The maximum number of colicin Ia molecules that can be bound by a single cell is high, about 5000, which implies that there are many receptors present in the cell wall.

An extensive study of resistance to colicins has been reported by Davies and Reeves (1975a,b), who suggested that the colicins can be classified into two general groups. Group I (also described as A) consists of colicins A, E1, E2, E3, K, L, N, S4, and X; group II (or B) comprises colicins B, D, G, H, I, M, S1, V, and Q. Bacteria selected for resistance to a colicin in group I (A) may or may not be resistant or tolerant to other colicins in this general group, but are never resistant to any colicin classified in general group II (B). The reverse is true when mutants are isolated for resistance to one of the colicins in general group II (B). The mutants resistant to group I (A) colicins fall into 21 phenotypic classes according to their cross resistance patterns (including those of both the resistant and tolerant types); the mutants resistant to group B colicins fall into 9 such classes. The suggestion that the colicins may fall into two general evolutionary groups also has been made by Hardy, Meynell and Dowman (1973) on the basis of relationships in their abilities to bind to cells. Hardy (1975) has noted that where the plasmid DNA has been isolated, the group I (A) plasmids are small ($10^6$–$10^7$ daltons) whereas the group II (B) plasmids are somewhat larger (about $10^8$ daltons); it remains to be seen if this will prove to be a general rule.

These and other recent studies make it clear that the distinction between resistant and tolerant mutants is less meaningful than had originally been supposed; for both types of mutant may arise in a similar manner. Mutants resistant to a single colicin may indeed represent a specific loss of its receptor protein. However, this cannot be true of mutants resistant to several colicins which bind to different receptors; in this case the mutation must block the presence of all these receptors in the membrane. This latter class of resistant mutant is similar to several previously isolated classes of tolerant mutant. Among the tolerant mutants characterized by Nagel de Zwaig and Luria (1967) and by Nomura and Witten (1967) were some classes tolerant only of one colicin, while others were tolerant of different colicins. The second class included the mutations *tolII* and *tolIII* which cause changes in the sensitivity of the cell membrane to disruption by agents such as deoxycholate or EDTA. Many other tolerant and resistant mutants have since been shown to possess alterations in membrane structure. These include both mutants insusceptible to a single colicin and those refractory to several colicins (Whitney, 1971; Bernstein, Rolfe and Onodera, 1972; Eriksson-Grennberg and Nordstrom, 1973; Davies and Reeves, 1975a,b).

This suggests that some of the resistant or tolerant mutations result in a change or loss of some component involved in assembly of the cell membrane, with the result that certain colicin receptors either are no longer able to take their place at the surface of the cell (in resistant mutants) or if present are inhibited in their ability to transmit colicin or effector molecules from the receptor site (in tolerant mutants). The specificity of the effect of the membrane change can vary very greatly. This interpretation implies that resistance and tolerance should be regarded as related mechanisms that may simply reflect the stage at which action of the colicin is prevented.

Each colicin has a characteristic killing action. Colicin K inhibits DNA, RNA, and protein synthesis; inhibition may be prevented by the addition of trypsin after colicin binding to the cell, which argues that the colicin may act indirectly at the cell wall receptor site to set in train the series of events characteristic of its killing action; this colicin probably does not enter the cell (see Nomura, 1963). Colicins E1 and K both inhibit active transport; Fields and Luria (1969a, b) found that they lower ATP levels in sensitive cells and suggested that their primary effects may be to inhibit oxidative phosphorylation, that is, energy production. Colicins Ia and Ib also have the multiple effects of inhibiting ATP production and protein and nucleic acid synthesis (see Nomura, 1967a).

It was at first thought that all colicins act indirectly, by binding to a receptor at the cell wall and initiating a characteristic series of events that leads to cell death (see Nomura, 1967b). No colicin was therefore thought to enter the cell. However, more recent experiments show that the action of colicin E3 is direct. Killing by E3 involves inhibition of protein synthesis (for review of early work see Nomura, 1967a). Senior and Holland (1971) and Bowman et al. (1971) found that when sensitive cells are incubated with E3, their ribosomes are inactivated because about 50 nucleotides are cleaved from the 3′ end of the 16S rRNA. Boon (1971, 1972) and Bowman, Sidikaro and Nomura (1971) then found that the same cleavage of rRNA takes place when ribosomes are incubated with colicin E3 in vitro. Cleavage takes place only with ribosomes as the substrate and not with free rRNA; colicin E3 therefore behaves as a highly specific endonucleolytic ribonuclease. Its specificity is indeed remarkable; it seems likely that colicin E3 recognizes a specific base sequence of the 16S rRNA in the structure of the ribosome.

Crude preparations of colicin E3 do not cleave the rRNA; the colicin protein must be purified before it becomes active as an endonuclease. This suggests the presence of some component inhibiting the colicin action. This component can be obtained only from cells carrying the Col E3 plasmid and may provide the immunity conferred by the presence of the factor in the cell. Jakes, Zinder and Boon (1974) and Sidikaro and Nomura (1974) purified this substance as a protein of about 9500 daltons. Titration experiments in which the amount of E3 immunity protein was varied but the amount of colicin E3 remained constant suggested that E3 immunity protein and colicin E3 interact in vitro in a stoichiometric fashion; use of an immunological assay also detected a complex of the two proteins. This explains the occurence of immunity breakdown, a phenomenon noted by Levisohn, Konisky and Nomura (1967): immune cells may be killed by very high levels of colicin. Presumably when sufficient colicin is present to bind all molecules of the immunity protein, the cells become susceptible to colicin killing. Although the action of colicin E3 on ribosomes is prevented in vitro by addition of E3 immunity protein, colicin E3 which has been incubated with immunity protein remains active when incubated with sensitive cells; presumably the colicin but not the immunity protein can be taken up by the cells. Immunity protein thus does not cause any irreversible inactivation of colicin E3. Sidikaro and Nomura (1975) confirmed that the structural gene for the E3 immunity protein lies on Col E3 DNA by demonstrating that it is synthesized in vitro under the direction of this template.

Although colicins E2 and E3 are related proteins that use the same receptor, their modes of action are quite different. Colicin E2 killing involves degradation of bacterial DNA. Ringrose (1970) reported that the degradation takes place in three stages: an endonuclease nicks single strands of the DNA duplex; the DNA then breaks into fragments due to double strand breakage; and finally an exonucleolytic attack degrades these fragments. Saxe (1975a) reported that a somewhat similar series of events occurs to lambda DNA when cells carrying closed circles of the phage DNA are treated with colicin E2, although in these experiments it was not possible to distinguish single strand breaks as a predecessor for the double strand breaks.

Is the breakage of host DNA initiated by an enzyme activated by the binding of colicin E2 to its receptor, or does the colicin itself enter the cell and possess the endonuclease activity? Almendinger and Hager (1972) reported that endonuclease I may be the enzyme which initiates the degradation of host DNA. Osmotic shock causes enzymes of the periplasmic space to be released from E.coli cells; amongst these is endonuclease I and the extent to which it is released can be controlled by the conditions of the osmotic shock. The resistance of (sensitive) cells to E2 killing increases in proportion to the amount of endonuclease I that is released; the binding of colicin E2 to shocked and unshocked cells remains unaltered. Almendinger and Hager therefore suggested that the binding of E2 to receptor sites at the cell surface promotes transport of endonuclease I from the periplasmic space to the interior of the cell. This would imply that the action of the colicin is indirect, in contrast with direct action of colicin E3, which binds to the same receptor. On the other hand, Saxe (1975b) has demonstrated that purified colicin E2 introduces single strand scissions into supercoils of $\lambda$ DNA in vitro. However, it remains possible that this activity might be due to a contaminating enzyme and so it remains to be seen whether colicin E2 is indeed an endonuclease in vitro and might be able to act as such in vivo.

## *Transmission of Col Factors*

Colicin factors fall into two groups with respect to transmissibility. Some factors, the best studied of which are Col V2 and Col Ib behave as plasmids able to transfer infectiously between cells; these may also transfer chromosome markers (see Ozeki and Howarth, 1961; Clowes, 1961). These factors are of the same order of size as the sex factors and $fi^+$ R factors and possess either the same or analogous transfer systems. The small Col E factors, however, are not transmissible and pass between cells only when conjugation is sponsored by some other factor. This distinction was first drawn by Ozeki, Stocker and Smith (1962), who identified determinants for five colicins able to transfer in mixed cultures, in contrast with the inability of Col E or Col K plasmids to do so (for review see Hardy, 1975).

Col I factors are self repressed, like the R factors. Stocker, Smith and Ozeki (1963) found that when a Col I broth culture of S.typhimurium LT2 is incubated with a noncolicinogenic acceptor strain, $< 0.1\%$ of the recipient bacteria become colicinogenic during a 1 hour incubation; but after 18 hours about 50% of the recipient cells carry Col I. Only a small proportion of the recipients acquire col I in the brief transfer period because a very small proportion, about 0.02% of the colicinogenic bacteria, are competent donors. Longer periods of incubation allow infectious spread of colicinogeny because newly colicinogenic cells are good donors. The high level of competence in donors that have recently acquired Col I persists for 2–7 generations; in these HFC (high frequency colicinogeny transferring) cultures, 30–100% of the cells are competent donors. Monk and Clowes (1964a,b) reported that Col I conjugation is independent of that promoted by the sex factor; and the efficiency of the self-repression system varies with the bacterial host, being less stringent in E.coli (about 1% competent donors in an established culture) than in S.typhimurium (only 0.02%).

Strains of S.typhimurium LT2 carrying Col E2 alone are not able to transfer it. However, Smith, Ozeki and Stocker (1963) found that HFC cultures of Col I can transmit Col E1 and Col E2. In crosses between an Hfr Col E1 strain and a noncolicinogenic $F^-$ factor, Clowes (1963) found no relationship between transfer of Col E1 and any chromosome position; the factor therefore appears to be extrachromosomal and must take advantage of the formation of mating pairs to transfer between bacteria. In experiments of this sort, all of the colicin factors Col E1, Col E2, Col I, Col V, appeared to be extrachromosomal since their transfer showed no correlation with any chromosome marker (Nagel de Zwaig, Anton and Puig, 1962; Nagel de Zwaig and Puig, 1964).

A relationship between Col V factors and the sex factor was implied by the observations of Kahn and Helinski (1964, 1965) that Col V cells are susceptible to ♂-specific RNA phages and so must produce F-type pili. Further, Col V and F are mutually exclusive; and Col V promotes transfer of chromosome markers. Differences reported between some Col V strains were explained by the report of McFarren and Clowes (1967) that there is more than one Col V factor. The factor Col V1 is nontransmissible; by contrast, factors Col V2 and Col V3 are transmissible, can transfer chromosome markers, and allow cells to adsorb ♂-specific phages. In superinfection experiments, Col V2 usually eliminates F, whichever was the resident factor, and this unidirectional relationship is reversed with Col V3, which usually is eliminated by F. Both Col V2 and Col V3 resemble F in being fully expressed and not repressed in a culture. Transfer of chromosome markers from *rec*$^-$ Col V2 or Col V3 cells is much reduced, as it is with cells carrying F; but chromosome marker transfer by Col I must use a different system since the introduction of a *rec*$^-$ gene has no effect. In a study of chromosome mobilization by several plasmids, Moody and Hayes (1972) found that Col Ib*drd* is indifferent to the state of the *recA* locus, whereas mutation in *recA* abolishes transfer by F, Col V2, Col V3, and other plasmids. These experiments therefore indicate an important difference between the Col Ib transfer system and that of the other plasmids. The Col V2 and Col V3 plasmids display many of the characteristics of the sex factor; Meynell, Meynell and Datta (1968) therefore suggested that they probably arose by recombination of F with the colicin determinants. A difference between F and the Col V2 and Col V3 factors lies in their ability to integrate into the chromosome; although Col V2 and Col V3 do not transfer all chromosome markers at equal frequencies, an indication that they may have preferred sites of integration, stable strains with a factor integrated in the manner of an Hfr are not obtained. Kahn (1968) was able to isolate $Hfr_V$ strains from a Col V factor which showed a highly preferential integration between *xyl* (transferred as a proximal marker) and *mal* (distally transferred); but these strains differed from sex factor Hfr strains in continuing to transfer the Col V factor rather frequently.

### *Replication of Col Factors*

Colicin factors are plasmids which take the familiar form of circles of duplex DNA. The physical state of the Col E factors has been investigated in the greatest detail, in both E.coli and Proteus mirabilis. In E. coli they can be isolated as covalently closed circles on CsC1-EtBr gradients; Bazaral and Helinski (1968a) reported that Col E1 DNA is about $4.2 \times 10^6$ daltons and Col E2 and Col E3 DNA both are about $5 \times 10^6$ daltons in size. The closed circles of Col E1 DNA sediment at 23S on sucrose gradients. In Proteus, Col E1 DNA has been isolated as a satellite band of buoyant density distinct from the chromosome; both closed and open circles are present. Bazaral and Helinski (1968b) found that in addition to 23S circles of $4.2 \times 10^6$ daltons, which correspond to those seen in E.coli, 31S dimers of $8.5 \times 10^6$ daltons and 37S trimers of $12.7 \times 10^6$ daltons are present; treatment with chloramphenicol or starvation with amino acids causes a decline in the number of monomers and an increase in the number of dimers, trimers, and also higher order forms. Figure 3.15 compares the states of Col E1 DNA in stationary phase E.coli and Proteus cells. Goebel and Helinski (1968) observed that the accumulation of longer molecules appears to depend upon replication and so presumably results from the production of daughter molecules longer than the parent, perhaps by a rolling circle type of replication, rather than by a recombinational joining of monomers. These observations thus emphasize again that the control of plasmid replication may depend upon the bacterial host.

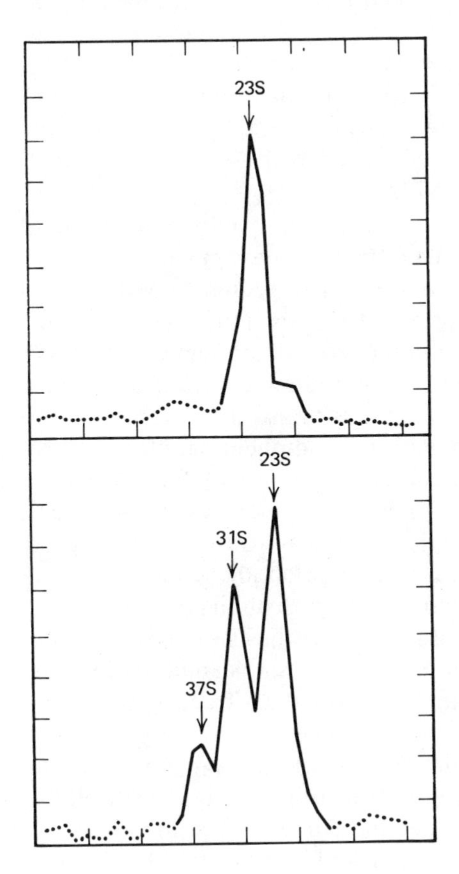


**Figure 3.15:** Col E1 DNA in E.coli and P.mirabilis stationary cultures. In E.coli there is only a single peak of 23S, representing the monomeric form ($4.2 \times 10^6$ daltons); in Proteus there are also further peaks of 31S and 37S representing dimers and trimers. Data of Goebel and Helinski (1968).

By using minicells of E.coli, Inselburg and Fuke (1971) and Oka and Inselburg (1975) were able to isolate replicating molecules of Col E1 DNA. Incorporation of a pulse label identifies a fraction that sediments more rapidly than covalently closed circles on a sucrose gradient; this corresponds to a fraction which has a buoyant density on EtBr-CsCl gradients between the densities of closed and open circles. Heating at 90°C releases pulse labeled 5-13S single strands from the replicating intermediates; these may be Okazaki fragments (about 10 such pieces would be sufficient to account for the entire length of Col E1 DNA) or they may represent initiation fragments. Some of the released single strand fragments have an increased density, which is lost upon treatment with ribonuclease or alkali, an indication that replication may start with an RNA primer (see Volume 1 of Gene Expression). Using a system in which Col E1 DNA is replicated in vitro, Sakakibara and Tomizawa (1974b) obtained similar results, and were able to calculate that 6S fragments that they identified as intermediates in replication possess about 20 ribonucleotides in their total length of some 400 bases. An interesting feature of Col E1 replication that distinguishes it from the host chromosome (and from the larger plasmids) is that it displays an absolute requirement for the enzyme DNA polymerase I; this is shared with certain other small plasmids (Kingsbury and Helinski, 1973).

RSF1030 is a nonconjugative R factor, carrying ampicillin resistance, of some $5.5 \times 10^6$ daltons; its replication control system appears to be very similar to that of Col E1, although its homology of sequence with the Col factor is less than 1%. Crosa, Luttropp and Falkow (1975) reported that RSF1030 continues to initiate and complete rounds of semiconservative replication in E.coli cells treated with chloramphenicol, a property also displayed by Col E1 (see below). This replication is inhibited by rifampicin. During replication in the presence of chloramphenicol, closed circular molecules containing RNA accumulate. Treatment with alkali or ribonuclease generates open circular molecules which can be separated into a circular single strand and a full length linear strand. This suggests that each of the closed circles contains a ribonucleotide sequence on one of its strands, apparently on either with equal probability. Thus a single RNA primer sequence may be needed for the synthesis of each strand of the plasmid DNA. This may be a common feature in plasmid replication.

Electron microscopy of the replicating intermediate fraction of Col E1 DNA has identified molecules of a biforked form, such as the example shown in Figure 3.16, which would correspond to the conventional movement of a replicating fork(s) around the circular genome. Inselburg and Fuke (1971) also found some dimers and trimers, in both closed and open circular form. That in E.coli, as in Proteus, replication rather than

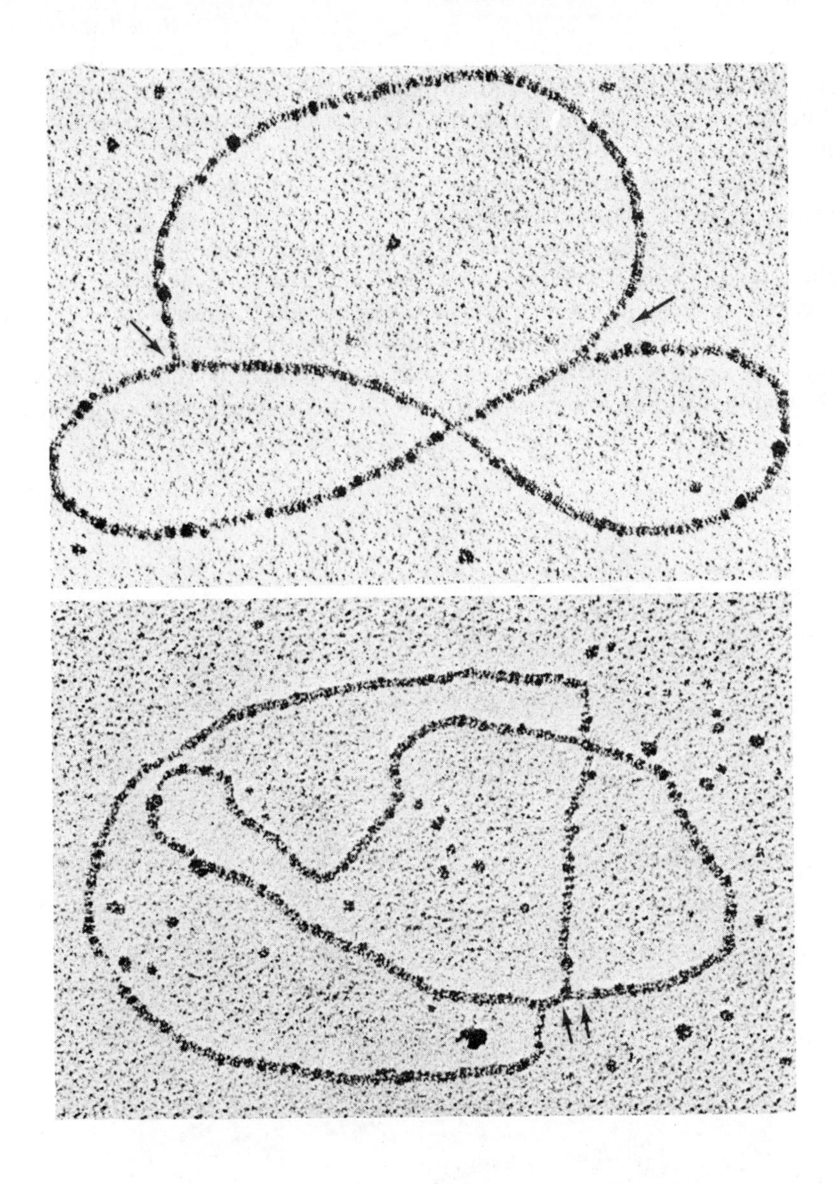

**Figure 3.16:** circular replicating Col El DNA. The upper molecule is about half replicated; the lower molecule is almost completely replicated. Both molecules are twisted but can be seen to correspond to $\theta$ structures. The branch points are marked with arrows. Data of Tomizawa, Sakakibara and Kakefuda (1974).

recombination is involved in the formation of these molecules, is implied by the observation of Inselburg (1973a) that the introduction of a *recA* mutation does not prevent the formation of catenated DNA.

An origin for replication has been identified by use of the restriction nuclease Eco RI. This enzyme cleaves Col E1 DNA at a single site, converting the circular molecule to a linear form of the same length. When replicating DNA is treated with the enzyme, the replicated and unreplicated sequences can be distinguished by the two Y shaped forks which separate them; the positions of these forks can then be determined relative to the ends of the linear molecule. An example is shown in Figure 3.17. Tomizawa, Sakakibara and Kakefuda (1974) used a system in which

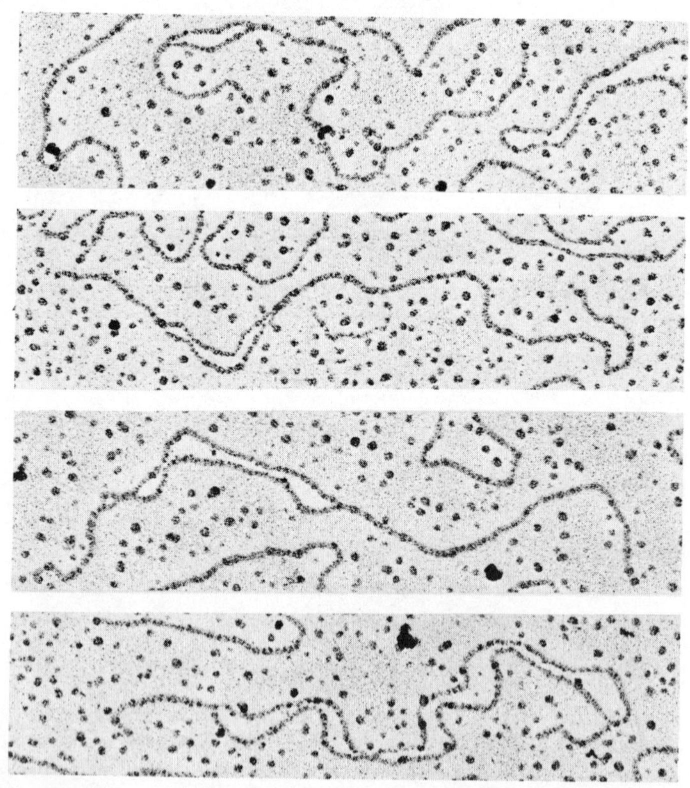

**Figure 3.17:** replicating molecules of Col El DNA treated with Eco RI nuclease. The length of replicated DNA increases in the four molecules shown from top to bottom. The left branch of the replicated region is always the same distance from the end of the molecule; this identifies the origin. The right branch moves progressively towards the end; this identifies the replicating fork. Data of Tomizawa, Sakakibara and Kakefuda (1974).

closed circular molecules of Col E1 DNA initiate and complete rounds of semiconservative replication in vitro; when examined just after replication has begun, linear molecules have a small loop of replicated DNA about 20% of the distance away from one end; this identifies the origin. Extensive replication takes place only in one direction, through the 80% of DNA on one side of the loop, since the left end of the loop remains fairly fixed in position while the right end (as usually aligned) moves. This suggests that replication is unidirectional, with a single replicating fork moving away from the origin. With replicating molecules extracted from minicells, Inselburg (1974a) also found that the position of one end of the loop was fairly constant while the other moved.

If replication is unidirectional, the terminus should be adjacent to the origin. If the left end of the loop moves at all, an idea that is rendered unlikely but is not excluded by the results, the terminus should be a corresponding distance away from the origin, that is, should be closer to the left end (as usually aligned) of the molecule. Using the in vitro system, Sakakibara and Tomizawa (1974a) substituted NMN for the essential cofactor NAD; this inhibits the polynucleotide ligase of E.coli so that the open circles formed by replication cannot be closed. On subsequent incubation with appropriate components, the gap can be closed to generate a closed circle. Cleaving the open circular molecules with Eco RI nuclease showed that the gap lies 20% distant from the left end of the molecule; this is very close to the initiation site, measured in these experiments as 17% distant from the left end. Thus termination takes place at or very close to the origin; replication is therefore most probably unidirectional.

The covalently closed circle of the Col E1 plasmid can be isolated from E.coli in the form of a DNA-protein complex. The purified closed and open circles of Col E1 DNA sediment at 23S and 17S respectively on sucrose gradients. Clewell and Helinski (1969) found that by using conditions of gentle lysis, they could obtain DNA-protein complexes sedimenting slightly faster, at 24S and 18S. Figure 3.18 compares the complexed and free forms of Col E1 DNA. The 24S form tends to decay to the 18S form; and agents that act on proteins, such as trypsin, pronase, SDS, convert the 24S complex to the 18S or 17S condition, although such treatment has no effect on free 23S supercoiled Col E1 DNA. The shift in buoyant density of the 24S complex from that of the free DNA suggests that about 240,000 daltons of protein are associated with the DNA (Clewell and Helinski, 1970c).

The 24S DNA-protein complex contains covalently closed circles of DNA and the 18S DNA-protein complex contains open circles. Because the transition from 24S to 18S that is induced by attack on the protein

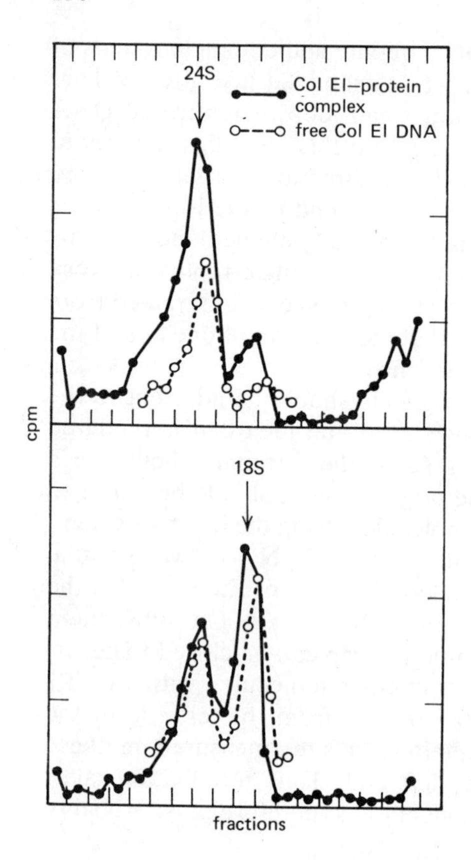


**Figure 3.18:** relaxation complexes of Col El. Under conditions of gentle lysis Col El can be isolated in the form of 24S complex shown in the upper panel; this sediments slightly faster than the 23S supercoiled circular DNA. Upon standing (or after treatment with protease activities) the 24S complex is converted to the 18S complex shown in the lower panel; this sediments slightly faster than the 17S free open circles of DNA. Data of Clewell and Helinski (1969).

component thus can be seen as a relaxation of the supercoiled structure of the DNA, these have been described as *relaxation complexes*. Clewell and Helinski (1970a) reported that the Col E2 and Col E3 plasmids exist as relaxation complexes very similar to the Col E1 complex; Blair et al. (1971) observed that there are differences in the characteristic extents of relaxation induced by pronase, SDS, or alkali (Col E1 DNA is converted into open circles to a greater extent than is Col E2 or Col E3 DNA). Clewell and Helinski (1970b) isolated a relaxation complex of the much larger plasmid DNA of Col Ib-P9; the protein-DNA complex sedimented at 76S compared with the 75S of free supercoiled Col Ib DNA. Kline and Helinski (1971) isolated a relaxation complex of the F factor that sedimented at 84S, compared with the free DNA at 80S. Relaxation of these large molecules takes place in a manner similar to that observed with the small Col E complexes. The existence of relaxation complexes therefore may be a general phenomenon.

When the strands of DNA of the 18S relaxation complexes are denatured and sedimented on an alkaline sucrose gradient, a circular strand and a linear strand are recovered. Clewell and Helinski (1970c) found that the two fractions of Col E1 DNA preferentially hybridize together, which suggests that one specific strand is nicked in the relaxation from closed to open circles. Blair et al. (1971) separated the linear and circular strands of relaxed Col E1 DNA and Col E2 DNA by equilibrium centrifugation in poly(U,G)-CsC1; Figure 3.19 shows that the linear fraction is derived

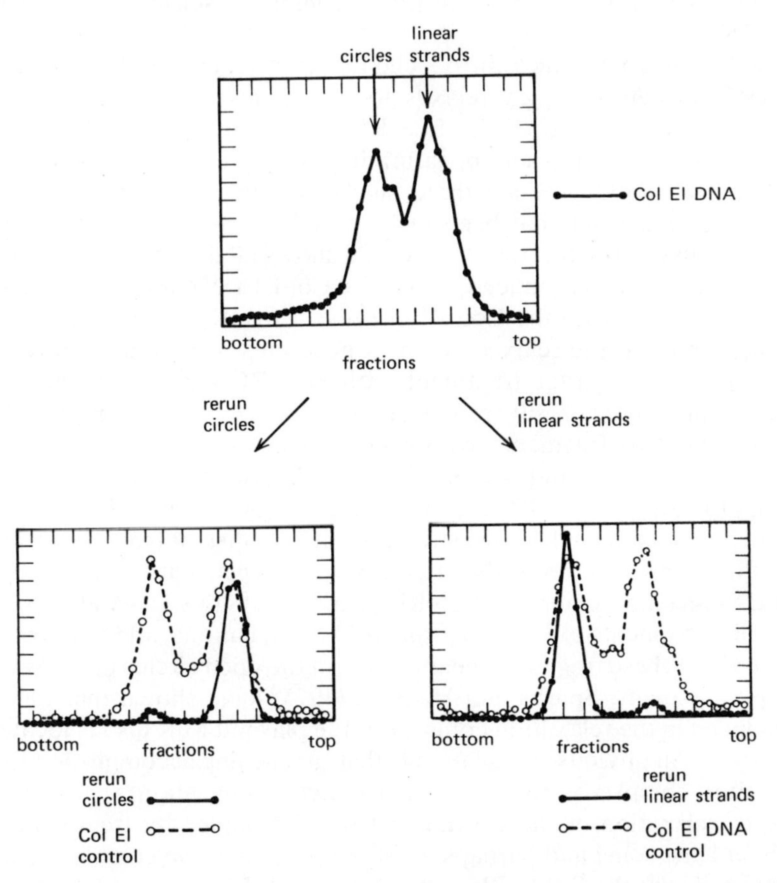


**Figure 3.19:** specificity of the nick in relaxed Col El DNA. The upper panel shows the separation of the strands of nicked Col El DNA into a circular (more rapidly sedimenting) and a linear fraction on an alkaline sucrose density gradient. The lower left panel shows the circular molecules rerun on poly(U,G)-CsC1 in the presence of a control of both strands of Col El DNA. The circular molecules almost entirely represent the lighter strand. The lower right panel shows the linear molecules rerun on poly(U,G)-CsC1 in the presence of a control of both strands of Col El DNA. The linear molecules almost entirely represent the heavier strand. Data of Blair et al. (1971).

predominantly from the heavy strand and the circular fraction largely represents the light strand.

Two classes of model might explain the relaxation of DNA in these complexes: there might be a preexisting nick in the "closed" circles of DNA, held together by a protein whose inactivation thus opens the circle; or the DNA may be covalently closed and the complex may include a latent endonuclease that is activated by agents inducing relaxation. The latter model seems more likely, especially in view of the observation that heat treatment renders the F relaxation complex insensitive to relaxation by SDS.

The location of the nick introduced in one strand when the supercoiled E1 DNA-protein complex relaxes to the open circular form has been determined by the use of the Eco RI restriction enzyme. If the break takes place at a unique site, denaturation of the linear product generated by Eco RI should produce three single stranded molecules: an intact linear strand that has not been nicked; and two fragments of the other strand produced by the nick. Figure 3.20 shows alkaline sucrose gradients of the single strands produced by relaxed Col E1 DNA and by relaxed Col E1 DNA treated with the Eco RI enzyme. Lovett, Guiney and Helinski (1974) found that the relaxed complex generates one circular strand and one linear strand; after treatment with Eco RI nuclease, there is one linear strand (presumably produced by cleavage of the circular single strand) and two fragments whose combined size is that of the linear strand. The linear strand is about $2.1 \times 10^6$ daltons; the two fragments are $1.7 \times 10^6$ and $0.4 \times 10^6$ daltons, which suggests that the relaxation complex induces a nick in one strand about 20% distant from the cleavage site, that is, presumably at the origin. A similar specificity is shown by the Col E2 relaxation complex; Eco RI nuclease cuts this DNA at two sites, releasing fragments of $3.85 \times 10^6$ and $0.35 \times 10^6$ daltons, and one strand of the larger of these fragments bears a nick in the open circles of the relaxed complex. Lovett, Sparks and Helinski (1975) have shown that the nick introduced in the relaxation complex of the plasmid R6K also is located at the origin. An obvious speculation is that the nicking accomplished by the relaxation complex is the signal that initiates replication.

After relaxation of the Col E1 or Col E2 complex by treatment with SDS, at least some and perhaps all of the protein of the complex remains associated with the DNA. Blair and Helinski (1975) demonstrated that the Col E2 relaxed complex bands in CsC1 at a lower buoyant density than the DNA of the plasmid; the reduction corresponds to the association with each genome of about 180,000 daltons of protein. When the relaxed complex is centrifuged on alkaline CsC1, only up to 50% of the DNA displays a reduced buoyant density; when this DNA is recovered and

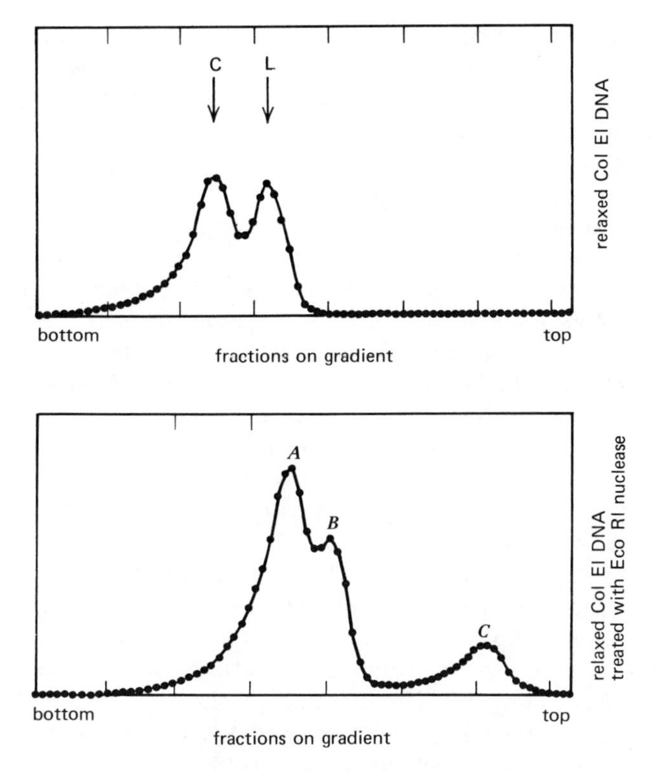


**Figure 3.20:** site of the nick introduced by relaxation of Col El DNA. The upper panel shows the two strands, circular (C) and linear (L), separated on alkaline sucrose gradients after open circles have been denatured. The lower panel shows the three fragments obtained after the relaxed Col El DNA has been treated with Eco R1 nuclease; A represents a linear strand of $2.1 \times 10^6$ daltons (the product of a single break in the circular strand C of the upper panel); B and C represent the products of the linear strand L of the upper panel, and are fragments of $1.7 \times 10^6$ and $0.4 \times 10^6$ daltons. Data of Lovett, Guiney and Helinski (1974).

recentrifuged through CsC1 in the presence of poly(U,G), it proves to comprise the H strand. This is the strand that is nicked. Lovett and Helinski (1975) demonstrated that the protein remaining associated with the nicked strand of Col E1 DNA is a single species of molecular weight 60,000 daltons; the other proteins of the relaxation complex, which include two major components of 17,000 and 11,000 daltons as well as several minor components, are lost at this stage of isolation of the denatured DNA. Since the buoyant density shift of the denatured strand corresponds to association with about 60,000 daltons of protein, this suggests that a single molecule of the protein is associated with each nicked strand of DNA.

The association of the protein with DNA through the steps of purification implies that it is tightly bound to the nucleic acid. Guiney and Helinski (1975) showed that the protein is bound at the 5′ end of the nicked strand, for although the 3′ end of the complex of protein with single strand DNA remains accessible to exonucleases, the 5′ end is no longer susceptible to exonucleolytic degradation. When the complex is digested with trypsin, most of the protein is degraded, but a small peptide remains associated with the 5′ end of the DNA; since the degradation of the protein would be expected to release any noncovalently bound polypeptide, this implies that the protein may be bound to the single strand DNA by a covalent linkage at or close to the 5′ end. Pronase is able to remove all the protein from DNA, as seen by the susceptibility of the pronase-treated complex to 5′ exonuclease. Since it is unlikely that pronase would be able to rupture an amino acid-nucleotide bond, this implies that the exonuclease may release the terminal nucleotide still attached to its amino acid; the isolation and characterization of this species would provide proof that the protein is covalently linked to DNA. During replication, Col E1 DNA remains in the form of a supercoiled, that is, closed, circle and this implies that the role of the proteins of the relaxation complex may be to undertake the nicking and resealing of the circular duplex during replication, perhaps providing a swivel that allows the DNA to unwind. The functions of the proteins of the complex, and in particular of the 60,000 dalton protein tightly associated with DNA, remain to be established, but this model would predict that they include not only endonuclease but also ligase activities.

The control of replication of Col E plasmids in E.coli is different from that of the F-like plasmids; there may be several copies of a Col E plasmid for each bacterial chromosome. Clewell and Helinski (1972) found that the number depends upon the growth conditions; for the presence of glucose reduces the amount of Col E1 DNA, possibly an effect mediated by catabolite repression. There may be more than 20 copies of Col E1 DNA in a growing cell and the number does not change under conditions in which the amount of chromosomal DNA per cell varies. The addition of chloramphenicol or starvation for amino acids prevents the bacterial initiation of replication and thus brings DNA synthesis to a halt; however, Clewell (1972) found that under such conditions Col E1 DNA continues to replicate and may reach a level of 3000 copies per cell. In a study of the replication of Col E1 DNA during this accumulation, Bazaral and Helinski (1970) suggested that molecules are drawn at random from a pool for semiconservative replication, a mechanism analogous to that displayed by some phages. These observations emphasize that the system that regulates Col E1 replication is different from the type of system controlling replication of the bacterial chromosome or of F-like (or other single copy) plasmids; instead of ensuring that there is a single initiation event per cell cycle, it establishes an equilibrium level of the plasmid in the cell (see page 219 and below). Another indication of the difference in Col E1 replication is the observation that cells cannot be cured of this plasmid by acridine orange (Kahn and Helinski, 1964).

Some plasmid incompatibility is displayed between the three Col E factors, although the magnitude of the effect is not as great as that displayed by, for example, the F-like plasmids (see Inselburg, 1974b). In view of the close relationship between colicins E2 and E3 that is implied by their adsorption to the same receptor, homology between the Col E2 and Col E3 factors might be expected. By forming heteroduplexes between denatured Col E2 and Col E3 DNA, Inselburg (1973b) showed that in 40% formamide, 90% of their lengths can anneal; a single substitution loop represents the remaining 10% of their lengths. When the concentration of the denaturing agent, formamide, is increased to 70%, however, making the conditions for hybridization more stringent, this loop enlarges to represent 20% of each genome. Inselburg therefore suggested that a nonhomologous region of 10% of each genome lies adjacent to a partially homologous region representing another 10% which can anneal in low but not in high formamide. The two factors must differ in at least part of the length of the gene coding for the colicin protein and in the genes coding for their respective immunity proteins. Since these colicin proteins are about 60,000 daltons in size and the E3 immunity protein is about 9500 daltons, a total length of about 2000 nucleotide pairs must code for these two functions. The genome represents about 8000 base pairs so that it seems likely that the entire difference between these factors may lie in these two functions. About five or six other functions could be included in the genome although none has been identified yet. That not all the genes of Col E1 are essential for survival of the plasmid is indicated by the isolation by Hershfield et al. (1976) of a mini-Col E1 plasmid which consists of about 3350 base pairs, 90% derived from Col E1, and which is able to replicate in E.coli. This plasmid shows the usual property of continuing replication in the presence of chloramphenicol. This suggests that a sequence of not more than half of Col E1 DNA is sufficient to specify all the necessary replication functions.

## Chimeric Plasmids

### *Insertion of DNA in Plasmids*

The existence of a variety of restriction enzymes that cleave DNA at specific nucleotide sequences has made possible two important and related advances in analyzing both procaryotic and eucaryotic genomes: the enzymes can be used to obtain a restriction map of a defined DNA molecule; and a restriction fragment representing a specific part of a genome then may be inserted into a suitable plasmid and perpetuated as part of it.

A very large number of restriction enzymes now has been characterized and each enzyme cleaves DNA at a specific short (usually 6–8 base pairs) sequence. When a genome such as that of a virus or phage is treated with a given restriction enzyme, it may be cleaved into a number of fragments, which can be separated on acrylamide gels according to their sizes. Each restriction enzyme generates a specific pattern of restriction fragments with a particular DNA molecule. By recovering the fragments generated by one enzyme and testing their susceptibility to cleavage by another enzyme it is possible to relate the fragment patterns produced by the two enzymes and thus to order the fragments into a map. This map can be related to the genetic map, for example, by making use of deletion variants that lack genetically defined segments of the genome and which therefore give rise to shorter or absent restriction fragments representing the region that is deleted. By using appropriate enzymes, sometimes in combination, it is possible to isolate any desired part of the genome. One example of the usefulness of this approach is the restriction enzyme mapping of the operator regions of phage lambda, which is discussed in Chapter 4.

Some restriction enzymes make staggered cuts that generate short complementary single stranded sequences at the ends of the separated duplex fragments (see page 242). The Eco RI enzyme is a notable example and leaves the single stranded ends:

N N N N T T A A 5′      3′ N N N N
and
N N N N 3′      5′ A A T T N N N N

As complementary sequences, the protruding ends can renature by base pairing. When two different molecules are cleaved with this enzyme, crosswise pairing of the complementary single strands generates a new DNA molecule, which can be given covalent integrity by using ligase to seal the single strand breaks that remain at the point of annealing.

Certain small plasmids are cleaved at only one site by the enzyme Eco RI, thus converting the circular genome into a linear molecule with protruding single strand ends. A fragment from another DNA molecule that has been cleaved with Eco RI nuclease then may be annealed with these ends to form a circle that in effect possesses an insertion of foreign DNA. The construction of such *hybrid* or *chimeric plasmids* is illustrated in Figure 3.21. If the point of insertion does not lie in an essential site of

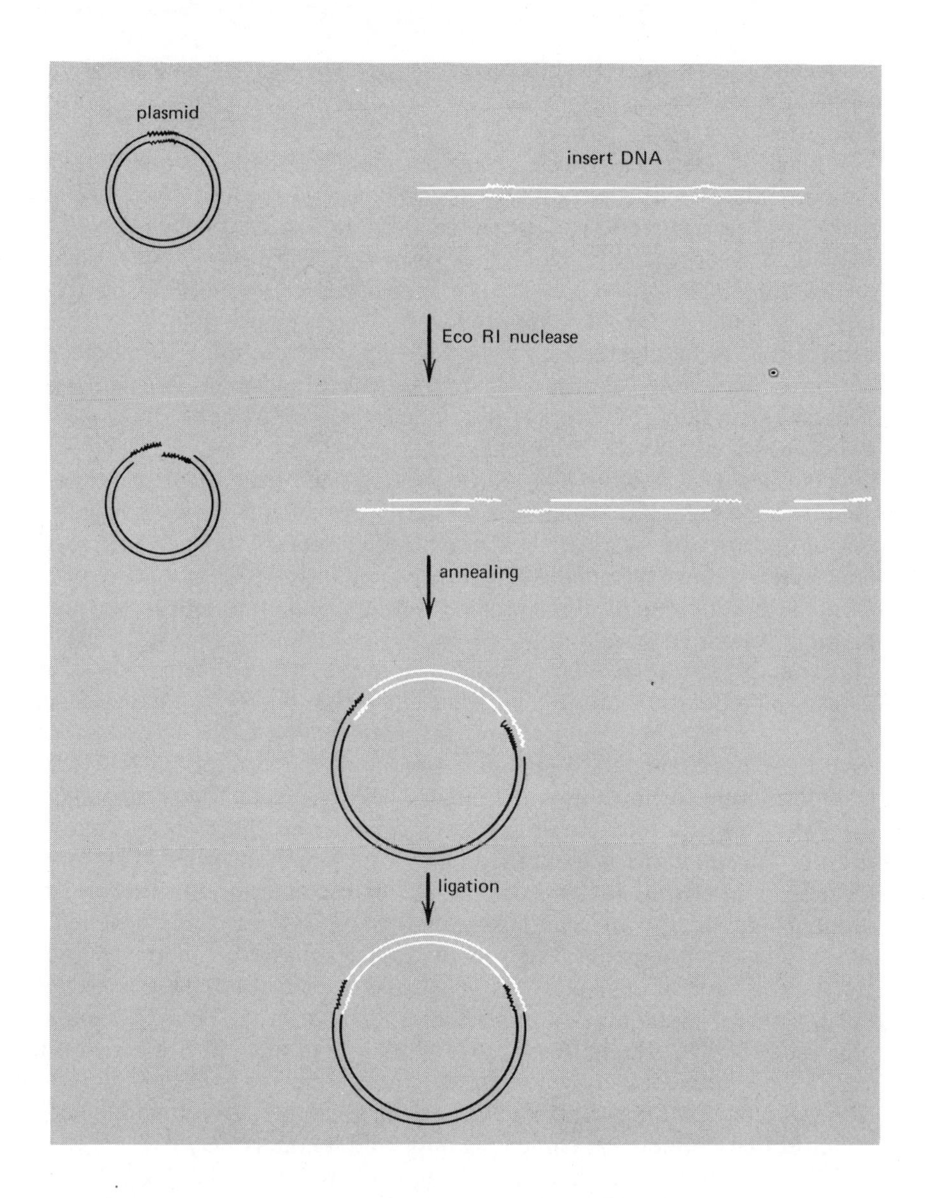


**Figure 3.21:** formation of chimeric plasmids by the Eco RI cleavage technique. Eco RI recognition sites are marked by jagged lines. Cleavage of the plasmid at a single site generates a linear duplex with protruding complementary four base ends. A fragment of foreign DNA cleaved by Eco RI then can anneal with these ends to reform a circular duplex that has single strand interruptions at the sites of joining. The interruptions can be sealed with ligase to generate a chimeric plasmid.

the plasmid, the chimeric DNA may be replicated in a bacterial host; a clone of bacteria carrying this plasmid then may be obtained and the chimeric DNA extracted and analyzed.

When a single isolated restriction fragment has been used to form the chimeric plasmid, it is necessary only to characterize it to determine that there has been only one insertion (since it would of course be possible for two or more copies of the fragment to be inserted by junction between their protruding single ends). This can be done by determining the size of the chimeric plasmid DNA, compared with the original plasmid DNA, by sedimentation or electron microscopy. And of course, it is possible to separate the inserted sequence from the original plasmid DNA simply by cleavage with Eco RI nuclease, after which the sizes and proportions of the components can be determined on gels.

It is possible to obtain defined restriction fragments only from certain sources, such as the relatively small genomes of bacteriophages or eucaryotic viruses or distinct segments of larger genomes (such as the genes coding for ribosomal RNA or for the histones) that can be isolated as such. The cloning of other sequences of a cellular genome, procaryotic or eucaryotic, in general requires what has been described as a "shotgun" experiment. The total DNA of the genome is cleaved with Eco RI nuclease, mixed with a suitable plasmid DNA that has been cleaved at one site with the enzyme, and used to transform E.coli cells; it is not necessary to seal the single strand interruptions in the chimeric DNA before transformation, because this is done by cellular ligase following entry of the DNA into the cell. Cells are then cloned and the chimeric plasmid present in each clone is examined. The total length of inserted DNA can be determined from the size of the circular chimeric plasmid; and the number and sizes of individual inserted fragments can be determined by examining the fragments released by Eco RI digestion of the chimeric DNA. This type of approach has been used to place both procaryotic and eucaryotic DNA sequences on plasmids (Cohen et al., 1973; Chang and Cohen, 1974; Hershfield et al., 1974; Morrow et al., 1974; Kedes et al., 1975). Any restriction enzyme that generates cohesive ends can be used in this way; the enzymes SalI and BamI, for example, have been used for constructing chimeric plasmids (Hamer and Thomas, 1976).

A disadvantage in using the Eco RI enzyme to cleave the DNA that is to be inserted into the plasmid is that the same recognition site always is cleaved; thus if an Eco RI cleavage site exists within a gene that is to be cloned, it becomes impossible to obtain a chimeric plasmid carrying the intact gene. An alternative approach is the poly(dA)-poly(dT) annealing method which is illustrated in Figure 3.22. The plasmid that provides the vehicle for cloning is cleaved with an enzyme that recognizes only a single

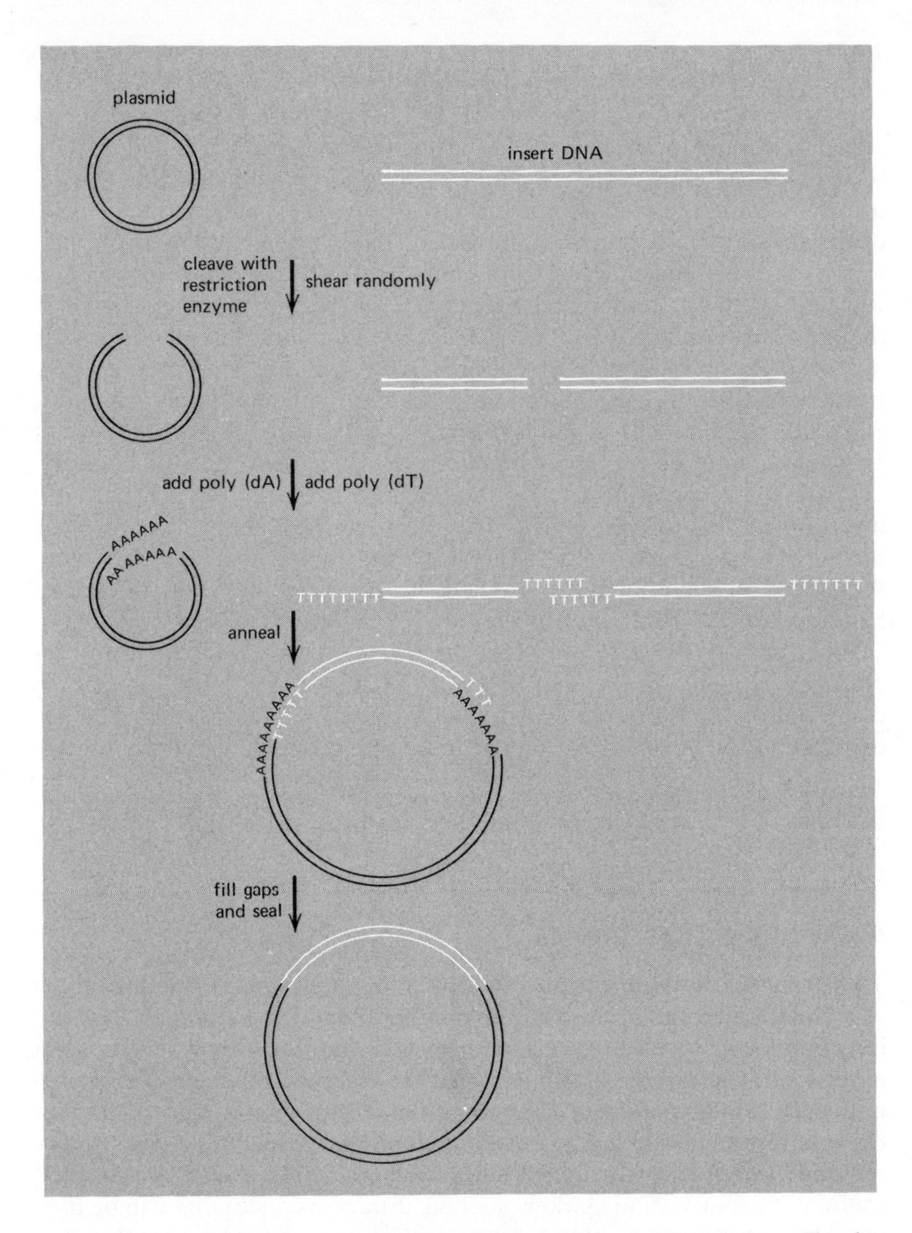


**Figure 3.22:** formation of chimeric plasmids by the poly(dA):poly(dT) technique. The plasmid is cleaved at a single site and the DNA to be inserted may be sheared randomly. Terminal deoxynucleoside transferase is used to add a poly(dA) segment to the 3′ ends of the plasmid DNA and to add a poly(dT) segment to the 3′ ends of the foreign DNA. Upon annealing a single plasmid genome can interact with a single insert fragment. Any gaps left because the poly(dA) and poly(dT) sequences are not of the same length must be filled and the remaining single strand interruptions sealed by ligase.

site; often Eco RI is suitable. The DNA that is to be inserted in the plasmid is randomly sheared to a specified average size. Poly(dA) is added to the 3′ ends of the plasmid by using the enzyme deoxynucleoside transferase and the precursor dATP. And poly(dT) is added to the 3′ ends of the sheared DNA preparation by the same procedure. The two preparations are then annealed to form chimeric DNA molecules in which an inserted sequence terminated by poly(dT) is bound to the two ends of the plasmid terminated by poly(dA). This technique has the advantage that a single plasmid genome can gain only one insert of foreign DNA, contrasted with the possibility that more than one may be joined end to end when the Eco RI method is used. A disadvantage is that the inserted fragment cannot be retrieved simply by cleavage with Eco RI nuclease, because the recognition site on the plasmid has been abolished by the insertion of the foreign DNA. This protocol has been used to insert procaryotic sequences into eucaryotic viruses and to place eucaryotic sequences on bacterial plasmids (Jackson, Symons and Berg, 1972; Lobban and Kaiser, 1973; Wensink et al., 1974; Maniatis et al., 1976).

When it is possible to preselect the fragment that is inserted into the plasmid, cloning of the chimeric DNA can be followed directly by confirmation of the structure of the inserted sequence. In a shotgun experiment, however, isolation of a chimeric plasmid carrying a specific sequence of the genome requires the use of a selective technique to detect the desired inserted sequence, for example, hybridization with a specific messenger RNA to identify chimeric plasmids carrying the complementary DNA (Grunstein and Hogness, 1975).

### *Plasmid Vehicles for Cloning*

Two principal requirements establish whether a plasmid is suitable for use as a cloning vehicle: it must be possible to convert the circular DNA to a linear molecule by cleavage at only one site; and this site must lie in some nonessential part of the plasmid so that the insertion of foreign DNA does not cause loss of viability. When these conditions are fulfilled, a chimeric plasmid reproduces in the host bacteria in the same manner as the original plasmid. Two further features are desirable in the plasmid. It is extremely useful if it carries some marker (such as drug resistance) that can be used for selection. And it is helpful if it can replicate under relaxed rather than stringent control, because this enables larger quantities of the chimeric DNA to be recovered.

Because of their size, most of the larger plasmids are cleaved more than once by Eco RI, preventing the insertion of foreign DNA by the Eco RI cleavage and annealing technique. An attempt to derive a smaller viable plasmid from the large drug resistance factor R6-5 was reported by Cohen and Chang (1973). The DNA of R6-5 is about 97,000 bases and consists of covalently closed circles that sediment at 75S, with a lesser proportion of open circles sedimenting at 53S. It carries the same resistance to several drug markers as the parent factor R6, with the exception that resistance to tetracycline has been lost because of the insertion of IS3 at the *tet*$^r$ marker (see earlier). When R6-5 DNA is used to transform $CaCl_2$-treated E.coli cells (this transformation system is discussed in Chapter 1), the bacteria acquire the full panoply of drug resistance markers carried by the plasmid; some also gain tetracycline resistance, at a frequency greater than the usual reversion rate, presumably because of loss of the IS3 located at this site. Shearing R6-5 DNA in a stirrer progressively reduces the size of the DNA. When the sheared DNA is used to transform E.coli, the cells appear to acquire resistance only to tetracycline. The plasmid DNA isolated from these cells represents closed circles that sediment at 27S; in open form the circles sediment at 21S. This corresponds to a size of $5.8 \times 10^6$ daltons, confirmed by electron microscopy that showed circular molecules of $3 \mu m$, that is, a length of about 9000 base pairs. This plasmid was originally named Tc6-5 and has since been renamed pSC101; it cannot be transferred by self-conjugation but can be mobilized by other plasmids with this capacity.

Figure 3.23 shows the electrophoresis on agarose gels of the fragments generated by cleaving R6-5 and pSC101 DNA with Eco RI nuclease. The DNA of R6-5 is cleaved into 12 fragments, three of which have molecular

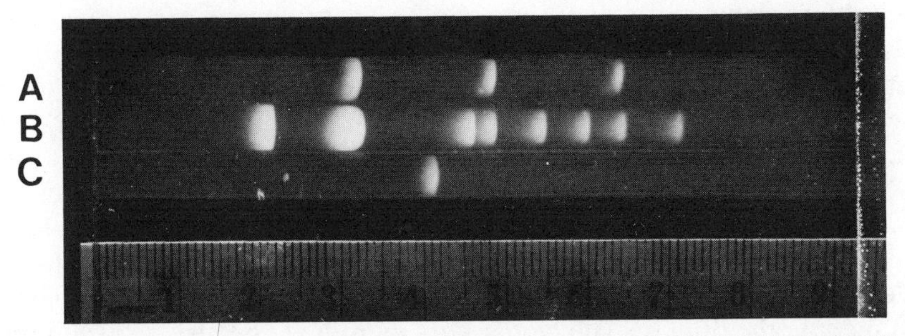


**Figure 3.23:** agarose gel electrophoresis of Eco RI digests of pSC101, R6-5 and pSC102. The DNA of each plasmid (A = pSC102, B = R6-5, C = pSC101) was cleaved with Eco RI, electrophoresed upon the agarose gel, and stained with ethidium bromide to allow visualization by fluorescence under long wavelength ultraviolet. The molecular weights of the fragments are as follows: pSC101 = $5.8 \times 10^6$ daltons; R6-5 fragment I= 17, II-III = 9.6, 9.1, IV = 5.2, V = 4.9, VI = 4.3, VII = 3.8, VIII = 3.4, IX = $2.9 \times 10^6$ daltons; pSC102 = fragments III, V and VIII of R6-5. Data of Cohen et al. (1973).

weights of less than $10^6$ daltons and are not seen on the gel. The heaviest fragment, at the top of the gel, has a weight of $17 \times 10^6$ daltons; and the next band actually contains two fragments that are not fully separated. The sum of the weights of the nine fragments seen on the gel and the three smaller fragments is $61.5 \times 10^6$ daltons, which corresponds to the size of the complete plasmid DNA. The DNA of pSC101 is cleaved at only a single site, generating a linear molecule with a mobility corresponding to a size of $5.8 \times 10^6$ daltons. This molecule does not correspond to any of the fragments released by cleavage of R6-5. This would appear to suggest that the random shearing and recircularization that generated pSC101 occurred in such a way as to generate a new fragment spanning only one of the Eco RI cleavage sites. But in lieu of a demonstration that pSC101 in fact possesses a segment of the sequences of R6-5, another possibility must be that pSC101 is not in fact derived from R6-5 but was fortuitously isolated in the experiment intended to obtain viable sheared fragments of R6-5. The relationship between pSC101 and R6-5 might be defined either by heteroduplex mapping or by analysis with other restriction enzymes; until such experiments are reported, the origin of pSC101 must remain an open question.

A demonstration by restriction mapping of the relationship between two plasmids is shown in Figure 3.23 for pSC102 and R6-5. The plasmid pSC102 was derived in an experiment in which R6-5 was cleaved with Eco RI nuclease and the resulting mixture of fragments used to transform E.coli. When examined for the drug resistance markers carried by the parental R6-5 plasmid, one clone resulting from the transformation was resistant to kanamycin, neomycin and sulfonamide, but not to tetracycline, chloramphenicol and streptomycin. These cells carried a plasmid whose closed circles sedimented at 39.5S, with DNA displaying a contour length of 8.7 $\mu$m upon electron microscopy. This corresponds to a molecular weight of $17 \times 10^6$; and the three fragments that are generated by Eco RI cleavage of the DNA have a combined weight of $17.4 \times 10^6$ daltons. The figure shows that the mobilities of these fragments correspond exactly to the mobilities of fragments III, V and VIII of R6-5. This implies that these fragments were able to anneal within the cell to form circles which were ligated; the information present on these fragments must include the drug resistance markers identified and also an origin for replication (which might be that of either the RTF or r determinant).

As vehicle for cloning, pSC101 has the advantage that selection is possible for the tetracycline resistance marker; it has the disadvantage that it replicates under stringent control, with only a small number of copies per host chromosome in the cell. Another plasmid that can be used for cloning is Col E1, which offers the advantage of replication under relaxed control, with about 20 copies present for each bacterial chromosome; and the addition of chloramphenicol halts host chromosome replication but allows Col E1 DNA replication to continue until some 1000–3000 copies per cell have been produced. Hershfield et al. (1974) demonstrated that Col E1 DNA gives a single band on agarose gels following cleavage with Eco RI nuclease. This has a mobility corresponding to 4.2 $\times 10^6$ daltons; since this is the same as the size of the circular molecule, the implication is that only one cut is made by the enzyme. Foreign DNA can be inserted at this site and then is perpetuated as part of the plasmid. The chimeric plasmid Col E1-*trp*, in which the foreign DNA was derived from the bacterial tryptophan operon, cannot produce colicin E1 although it retains immunity against it. This implies that the site recognized by Eco RI nuclease may lie in the gene responsible for specifying the colicin. The retention of immunity, which can be used as a selective technique, implies that immunity to and production of colicin must be determined by different genes. Col E1 DNA has also been used as a cloning vehicle by Clarke and Carbon (1975), using the poly(dAT) technique to link plasmid and foreign DNA.

Bacteriophages as well as plasmids may be used as cloning vehicles. Phage lambda provides a useful genome to use in cloning experiments because a large number of copies of the phage is generated in a lytic cycle. Lambda DNA possesses five sites that are cleaved by Eco RI nuclease, but variants of the phage lacking some or all of these sites can be constructed by isolating multiple mutants that are resistant to restriction (Rambach and Tiollais, 1974; Thomas, Cameron and Davis, 1974; Murray and Murray, 1974). Two of the RI fragments (generated by cleavages at three of the recognition sites) contain sequences that are not essential for successful phage infection. By following the protocol illustrated in Figure 3.24 it is possible to prepare a phage that lacks these fragments and has only a single Eco RI cleavage site spanning the deleted region: first a mutant phage lacking the other two RI sites is obtained, and then this is cleaved and reconstituted without the two nonessential fragments. This generates the λ*gt-O* phage. Foreign DNA may then be inserted at the single remaining RI site without inactivating any essential phage function. In fact, the insertion is necessary for phage viability. The reason for this dependence is that the lambda genome must be greater than a minimum length in order to complete the assembly of the phage particle; the λ*gt-O* DNA is below this limit and therefore can complete infection only when a length of DNA of the appropriate size is inserted in it. Actually, because it is not possible to propagate the λ*gt-O* phage, cloning with lambda is more readily accomplished by using the λ*gt-B* or λ*gt-C* phage, which carry the B and C fragments respectively in addition to the terminal fragments.

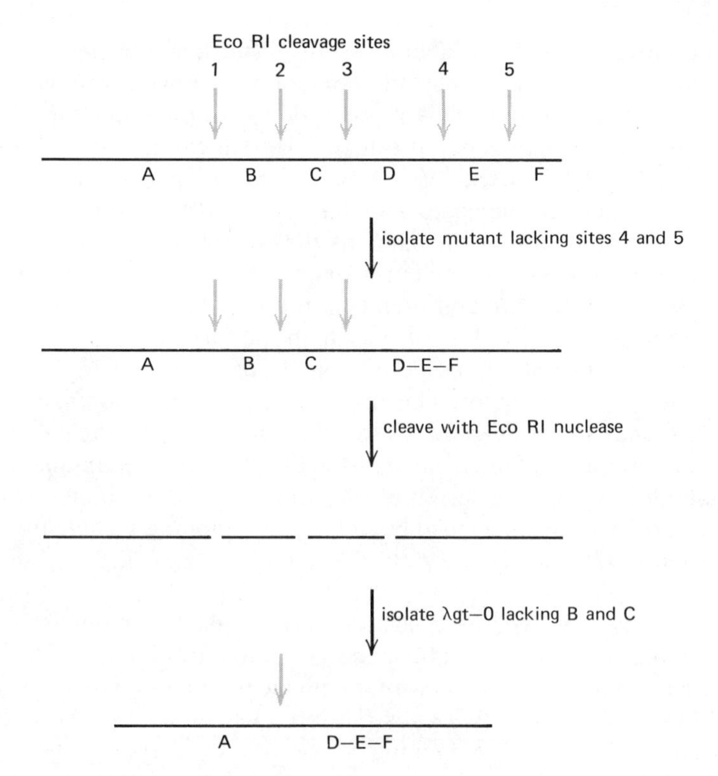


**Figure 3.24:** Eco RI cleavage of λ DNA and the preparation of λgt-0. Cleavage sites are indicated by the shaded arrows; the fragments released are indicated by letters A-F.

Upon cleavage with Eco RI, these release the terminal fragments A and D-E-F, which can anneal with a foreign DNA to form a viable phage. Cameron et al. (1975) have used this technique to clone λ DNA carrying the E.coli ligase gene; and Kramer, Cameron and Davis (1976) have cloned yeast genes inserted in λ DNA.

*Replication of Composite Plasmids*

In addition to rearranging the sequences of existing plasmids and inserting foreign DNA into them, it is possible to use the Eco RI cleavage technique to construct new plasmids that contain components from naturally occurring but unrelated plasmids that usually do not recombine. Timmis, Cabello and Cohen (1974) have constructed the interesting composite plasmid pSC101-Col E1 by transforming E.coli cells with DNA of the two plasmids that was cleaved with Eco RI, mixed, and ligated: the composite plasmids can be isolated by screening for colonies with resistance to tetracycline, immunity to colicin E1, and inability to produce the colicin. A typical clone possessed a plasmid of $10 \times 10^6$ daltons, that is, about 16,000 base pairs, which appears to represent one genome of Col E1 DNA linked to one genome of pSC101 DNA; this is plasmid pSC134. Figure 3.25 shows that when pSC134 is cleaved with Eco RI, it generates equimolar proportions of two DNAs with the mobilities of Col E1 and pSC101. In a control colony that had gained resistance to tetracycline and immunity to colicin E1, but had retained ability to produce the colicin, the two genomes were present as independent and not linked entities.

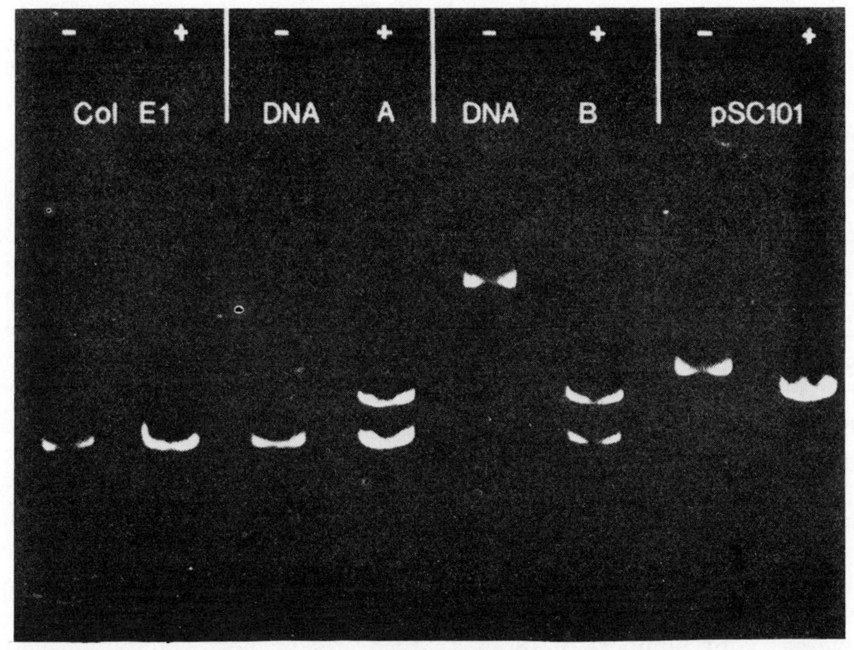


**Figure 3.25:** characterization of pSC134 by Eco RI cleavage. In each case a minus indicates that DNA was extracted without Eco RI treatment and a plus indicates that DNA was cleaved with the enzyme. The Col El and pSC101 columns show the characteristic mobilities on agarose gel electrophoresis of the parent plasmids. The DNA A preparation represents a clone that gained *tet*ʳ and was immune against and also produced colicin E1; this has both plasmids. The DNA B preparation represents a clone that gained *tet*ʳ and immunity against colicin E1 but could not produce the colicin; this has a much larger DNA (the slowly electrophoresing band in absence of Eco R1) which is cleaved by the enzyme to generate bands with the mobilities of Eco RI-cleaved Col El and pSC101 DNA. Data of Timmis, Cabello and Cohen (1974).

What mode of replication is followed by pSC134? It is possible to investigate whether both its replicons are capable of activity by providing conditions in which only one of them can function. In the presence of chloramphenicol, Col E1 DNA can replicate while pSC101 synthesis is inhibited. In the absence of DNA polymerase I, that is in *polA*$^-$ cells, pSC101 DNA can replicate since it is independent of this enzyme, but Col E1 DNA is unable to do so since it requires polymerase I activity. When cells are treated with chloramphenicol, pSC134 replicates in the same manner as Col E1: in spite of the cessation of bacterial replication, its synthesis continues. This means that the pSC101 replicon, although unable itself to function in these conditions, can be replicated as a covalent part of the molecule containing the Col E1 replicon. In *polA*$^{ts}$ cells incubated at high temperature, pSC134 is replicated in the same way as at the low permissive temperature; thus although an independent Col E1 replicon is lost from the cells under these conditions, as part of pSC134 it can be replicated by virtue of its covalent attachment to the pSC101 replicon. In appropriate conditions, either of the component replicons of pSC134 therefore is able to support replication of the entire chimeric DNA.

How does pSC134 replicate in cells that allow both its component replicons to function? Col E1 DNA replicates unidirectionally from an origin located about 20% of the distance from the Eco RI cleavage site; pSC101 DNA displays the same type of unidirectional replication, from an origin about 50% of the distance from the Eco RI cleavage site. When Cabello, Timmis and Cohen (1976) examined replicating molecules of pSC134 by electron microscopy, they found that (with only one exception) all possessed only a single replicating eye; cleavage with Eco RI demonstrated that this lies in the Col E1 part of the composite plasmid, with the origin located the usual 20% from the cleavage site. This implies that the pSC101 origin is suppressed in the joint plasmid, with replication initiating only at the Col E1 origin.

Under the conditions used in these experiments, in the absence of any restrictions on replication the Col E1 and pSC101 plasmids displayed average copy numbers of 18 and 5, respectively, per bacterial chromosome. The same numbers were seen when both plasmids coexisted independently in the same host cell. Under nonrestrictive conditions pSC134 is present at 16 copies per bacterial chromosome, consistent with the idea that it replicates under the control of the Col E1 replication apparatus. In *polA*$^-$ cells, however, the content of pSC134 is only 6 copies per bacterial chromosome, consistent with the use of the pSC101 replicon control. These results support the idea that it is some feature of the replication apparatus that is responsible for establishing the characteristic copy number of each plasmid.

Does pSC134 retain the incompatability of its parental plasmids? By using appropriate restriction fragments, Cabello et al. prepared composite plasmids in which Col E1 and pSC101 carried the new drug resistance markers *amp*$^r$, *kan*$^r$, or *str*$^r$. The presence of these markers allows each plasmid to be examined for compatibility with pSC134. Both Col E1 and pSC101 are incompatible with pSC134. Thus the composite plasmid expressed the incompatibility of both parental plasmids.

One model that explains these results is to suppose that plasmid replication is under negative control in which the copy number is determined by the relative affinity of the repressor for the DNA. This is essentially a modification of the model of Pritchard, Barth and Collins (1969) for control of host chromosome replication, which postulates that the concentration of a repressor synthesized by the plasmid is diluted by cell growth in such a way that the number of chromosome origins is proportional to the cell mass (see Volume 1 of Gene Expression). If the association constant governing the reaction between repressor and plasmid DNA is lower, then a greater number of copies of the plasmid will be present in the cell before repression is established. This model predicts that when a pSC134 plasmid is first introduced into a new host cell it can replicate from either (or both) origins. The association constant for the pSC101 DNA-repressor interaction is such that an equilibrium is established at about 6 copies per host chromosome origin. But the repressor for the Col E1 replicon has a lower association constant and thus replication continues to take place from this origin until an equilibrium level of about 20 copies per host chromosome origin is established. The pSC101 component of pSC134 continues to be replicated as part of the chimeric plasmid, although its own repressor now is present in the cell at a concentration that prevents utilization of its origin; and, of course, the repressor also prevents the replication of any independent pSC101 plasmid introduced into the cell, so that pSC134 is incompatible with pSC101 as well as with Col E1. Of course, other models also can be constructed to explain these data, but this is probably the simplest.

CHAPTER 4

# Phage Lambda: Infective Pathways

## Organization of λ DNA

### *Life Cycle of Lambda*

The small bacteriophage lambda comprises a duplex DNA of about $31 \times 10^6$ daltons—about 46,500 base pairs. As a *temperate phage,* two options are open to lambda upon infection of its host bacterium, E.coli. The λ DNA may behave in a *lytic* or *intemperate* manner and enter the *lytic cycle;* that is to say, phage genes are expressed and phage DNA is replicated to reproduce many more phage particles, which eventually are released by lysis of the cell. As an alternative to lytic development, lambda may instead exist in harmony with its host cell in the noninfectious manner known as *lysogeny*. Lysogenic bacteria carry a phage DNA in the form known as *prophage,* when the phage genome is integrated as a part of the bacterial chromosome. By virtue of their possession of prophage, lysogenic bacteria are *immune* to superinfection by further phage particles of the same type. As part of the bacterial chromosome, prophage is inherited in exactly the same manner as bacterial genes; but the prophage may be *induced,* for example, by ultraviolet irradiation, to enter the lytic cycle, in which case the phage DNA is released from the bacterial chromosome. The early work leading to this view of the dual life cycle of lambda was reviewed by Jacob and Wollman (1953). The choice of whether a phage enters the lytic cycle or forms a lysogen upon infecting a host cell depends upon the conditions of infection and the genotypes of phage and bacterium.

Our discussion of phage lambda is divided into two chapters. In this chapter we shall be concerned with the exercise of the choice between lysogeny and lytic development and thus with how each pathway is controlled; this is to say that the topic of this chapter lies with the control of transcription of lambda. In the next chapter we shall discuss the accomplishment of development, encompassing the recombination between lambda and the bacterial chromosome as well as the replication and recombination of vegetatively multiplying λ DNA and its assembly into mature particles.

## *Consequences of Phage Infection*

Infection of a bacterial culture with an excess of a phage to which it is sensitive results in lysis of the bacteria. The ratio of phage particles to bacteria is usually described as the *multiplicity of infection.* With lytic phages, which are able only to enter the lytic cycle and cannot establish lysogeny, addition of a phage suspension (at a high multiplicity of infection, that is $>1$) to a culture of sensitive bacteria growing in liquid medium results in a clearing of the culture after all the bacteria have been infected and lysed. After prolonged incubation, the usual turbidity may be regained; this is due to the presence in the population of bacteria resistant to the phage, which were not susceptible to infection and thus can reproduce to repopulate the culture. With temperate phages, however, the clearing of the culture is only transient and is rapidly overtaken by the reappearance of turbidity, in this case due to the presence of cells in which lysogeny was established; because of their immunity to subsequent lytic infection, the lysogenic bacteria reproduce and thus restore the turbidity.

The activity of a phage preparation is usually assayed by examining its ability to form *plaques* in a bacterial culture grown on agar. Phage particles are mixed with an excess of bacteria in suspension (that is, at a very low multiplicity of infection), so that each phage particle infects a single bacterium. When the mixture is poured onto an agar plate, it hardens so that the bacteria are immobilized in fixed positions. Lytically infected bacteria burst at the end of the lytic cycle to release a large number of progeny phage particles; these particles infect neighboring bacteria, which suffer the same ultimate fate of lysis. The spread of phage generations through neighboring cells finally is halted by a decline in bacterial metabolism, the result being the production of clearings in the confluent bacterial culture; these clearings are known as plaques. With lytic phages, the plaques usually are clear, since all the bacteria within them have been lysed. With temperate phages, the plaques are turbid, since they contain some bacteria which are lysogenic. With both lytic and temperate phages, changes in plaque morphology have provided an important approach to studying the physiology of the phage. Since each plaque is descended from a single phage particle, the number of plaques estimates the activity of the original phage population; this is described as the *plaque titre,* calculated as the number of *plaque forming units* (pfu). The *efficiency of plating* (eop) can then be calculated as the ratio of the plaque titre to the number of particles present in the original preparation.

### *Genetic Map of Vegetative Lambda*

Lambda includes at least 35 genes, 20 concerned with producing the head and tail proteins of the mature phage particle, 9 with replication, recombination and lysis, and 6 which are regulators. Four of the regulators are concerned with establishing or maintaining lysogeny; three are implicated in controlling the proper sequence of lytic development (one regulator is used in both lysogeny and the lytic cycle).

An early indication that the lambda genome comprises a single molecule of DNA was provided by the demonstration that all the known markers of the phage formed a single linear linkage group (Jacob and Wollman, 1954; Kaiser, 1955).The first mutants to be isolated concerned the type of plaque formed by the phage; various mutants influencing plaque size or morphology were useful in genetic mapping and the *c* (clear plaque) mutants identify three genes concerned with the establishment and maintenance of lysogeny. Another marker was provided by a host range mutant. When bacteria are selected for resistance to phage infection, they usually are unable to adsorb the phage due to changes in the cell surface; phage *h* (host range) mutants can then be isolated by their ability to infect these cells. This type of mutation, first isolated by Appleyard et al. (1956), is useful because of the ease with which it can be assayed. The early mapping of phage lambda has been reviewed by Campbell (1971).

Mutations in genes whose products are essential for lytic development often are lethal. One technique to utilize such mutants was developed by Jacob, Fuerst and Wollman (1957), who took advantage of the dual life cycle of lambda. Although essential for lytic development, these genes are not expressed in the lysogenic state, when the phage genome is reproduced in an inert state as part of the bacterial chromosome. It is therefore possible to mutate bacteria carrying a prophage, induce lytic development, and isolate the *defective lysogens* by their inability to lyse the host cell. The nature of the function that has been mutated then can be investigated by determining the ability of each defective lysogen to carry out specific functions, such as replication of phage DNA or production of components of the phage particle.

Systematic mapping of all the genes of lambda started with the isolation of a series of conditionally lethal mutants. *Conditional lethal mutants* are identified by their ability to grow only under certain, *permissive* conditions; in other, *nonpermissive* conditions the mutation is lethal to the phage. A mutant phage of this type can therefore be perpetuated by growth under permissive conditions; and the presence of the mutation and its map position can be determined by virtue of the failure to grow under nonpermissive conditions. The first conditional lethal mutants of lambda were isolated by Campbell (1961) as phage variants able to grow on one but not on another strain of E.coli. Since mutants of phage T4 previously had been isolated with precisely the same ability to grow on the first strain and inability to grow on the second strain, it was clear that some function of the first (permissive) strain was able to *suppress* this type of mutation in either phage.

These conditional lethals all since have been shown to carry amber nonsense mutations which can be suppressed in host cells carrying amber or ochre suppressors (see Volume 1 of Gene Expression). Thus a phage carrying one of these mutations is unable to grow in a nonpermissive $su^-$ host (one unable to provide a tRNA capable of reading the protein chain terminating amber codon), but can grow in a permissive $su^+$ host (which provides a suppressor tRNA that recognizes the amber codon and thus allows completion of synthesis of the protein coded by the mutant gene). Two phages each carrying one of these mutations can thus be crossed by performing a mixed infection on the permissive host (E.coli C600: a variant of strain K12) and then plating the progeny on both the permissive strain and on nonpermissive cells (strain W3350: a variant of E.coli B). The map distance between the two mutations then may be calculated as:

$$\frac{2 \times \text{W plaques} \times 100}{\text{C plaques}}$$

where *W plaques* represents the number of wild type recombinants and *C plaques* corresponds to the number of total progeny; the factor of 2 compensates for the selection of only one class of recombinants. The application of this mapping technique to phage T4 is illustrated in Figure 6.3.

These mutants were originally described as *hd* (for host dependent defective) and were later renamed in the form *susA-R,* with *A-R* identifying the complementation groups into which the mutants could be divided. Thus describing a mutant as *susX* is equivalent to saying that it is $X^-$ when grown upon a nonpermissive ($su^-$) host, but behaves as $X^+$ upon a permissive ($su^+$) host. The genetic functions of lambda are now most often described simply as genes *A-R*. Gene *A* identifies the left end of the genetic map (as conventionally written) and gene *R* identifies the right end. In addition to the 18 genes identified in these experiments, further conditional lethals of this type were isolated by Parkinson (1968).

Another type of conditional lethal is the temperature sensitive mutation. These are usually identified by phages which can grow normally at a permissive temperature of 30°C but which are unable to grow at a nonpermissive temperature of 42°C. The systematic analysis of such mutants carried out by Mount et al. (1968) brought the total number of genes identified by conditional lethal mutations to 24, comprising *S*, *T*, *U*, *V*, *W*, *Z*, as well as *A-R*. All genes in which conditional lethal mutations have been found thus are named by a capital letter; we can consider these to be the *essential genes*, necessary for propagation of the phage by lytic development. The *nonessential genes* are identified by the impairment of their functions through mutation and are described according to their roles, which may be concerned with lysogeny, integration and excision, and vegetative recombination.

The three clear plaque genes, *cI*, *cII*, *cIII*, represent regulator functions: *cI* codes for a repressor protein which maintains lysogeny; and *cII* and *cIII* specify proteins that are needed to establish lysogeny. Mutations in these genes are not lethal, since they interfere with lysogeny but do not prevent lytic development. One useful mutation is a temperature sensitive variant, $cI_{857}$, in which lysogeny can be maintained at 30°C but not at 42°C. The gene *cro* provides another regulator protein, which is involved in controlling repressor synthesis and in switching off certain genes during lytic development. The genes *int* and *xis* are concerned with integration and excision of prophage; and *red*$\alpha$ (also known as *exo*) and *red*$\beta$ are concerned with vegetative recombination. Gene *gam* is involved in both recombination and replication. A locus that is less well characterized is *rex; rex* mutants of lambda prophage lose their ability to prevent growth of *rII* mutants of T4 and it has been suggested that this mutation identifies a gene separate from *cI* rather than another function of the repressor protein. And in addition to the essential and nonessential structural genes, there are of course several control loci on λ DNA, including operators, promotors, and terminators.

Figure 4.1 displays the order and functions of the genes of the vegetative lambda genome. Accurate determination of map distances by genetic recombination is difficult because of variations in recombination frequency that depend upon the conditions in which phage crosses are performed. The distances shown in the figure were calculated by Campbell (1971) from genetic data obtained in several laboratories. The scale of the map is in recombination units; 1% recombination equals one unit. Although the variations that are seen in recombination frequencies at the molecular scale mean that genetic map distance cannot be equated with physical distance, the relative arrangement of markers by recombination is consistent with the order and relative distances obtained by physical mapping of lambda DNA. Recombination mapping of lambda is discussed in more detail in Chapter 5.

Does this map include all the genes of lambda or may some functions

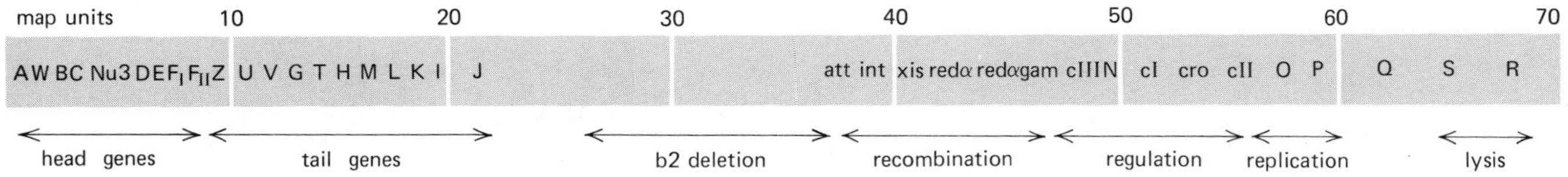

**Figure 4.1:** genetic map of the structural genes of vegetative phage lambda. The positions of loci are those determined by genetic mapping, as summarized by Campbell (1971), but are only approximate as drawn here. A recombination unit represents a frequency of exchange of 1% between two sites (as corrected for positive interference: see Chapter 5). The total length of DNA is 31 x $10^6$ daltons = 46,500 base pairs, but because of variations in recombination frequency it is not possible to convert map units into physical distance. However, this map of the genetic relationship between markers is consistent with physical studies of λ DNA. Genes fall into groups with functions illustrated (and *Q* is a regulator).

remain to be discovered? It seems unlikely that any essential genes have not been identified. There is little room for the insertion of further functions in the region of the head and tail genes. No essential functions can lie between *J* and *int* because the λ*b2* strain possesses a deletion of this entire region, yet is able to complete cycles of lytic development (λ*b2* is discussed in more detail later). Most of the region between *P* and *Q* and most of that between *R* and the end of the genome can be deleted without impairing viability (see Herskowitz, 1973). Of course, it remains possible that some nonessential genes remain to be identified; indeed, by comparing the proteins synthesized by λ and λ*b2*, Hendrix (1971) has concluded that several genes may lie in the region covered by the *b2* deletion. No significance can be ascribed to any such functions, however, until mutants are identified with altered phenotypes. This leaves the total number of lambda genes as 26 essential functions and 9 nonessential functions.

The lambda map comprises clusters of genes coding for related functions; during phage development, genes whose products are needed together may be expressed coordinately through a small number of transcriptional controls. The concept of early and late functions in lambda development was introduced by Jacob, Fuerst and Wollman (1957) on the basis of their characterization of a series of defective λ prophages. After induction, *early mutants* fail to replicate their DNA; whereas *late mutants* complete replication but are defective in later stages of lytic development. That this division of functions corresponds to the temporal sequence of gene expression was suggested by the experiments of Dove (1966) and Weigle (1966), in which bacteria were infected and the DNA was extracted after various intervals and assayed for infectivity. DNA synthesis precedes the formation of infectious DNA and mature phages. If chloramphenicol is added at 10 minutes after infection, by 70 minutes DNA synthesis has reached its usual plateau but no infectious DNA or mature phage is made. Thus the proteins synthesized during the first 10 minutes after infection can support replication at the normal level; this suggests that the early genes are fully expressed during this period. Synthesis of new proteins is required after 10 minutes if phage production is to be completed; this suggests that the late genes must be expressed after this time.

Early mutants were mapped into five loci by Eisen et al. (1966). The three loci *N*, *O*, *P*, represent structural genes; mutants of the classes *x* and *y* identify control elements. In experiments to follow the synthesis of λ DNA and mRNA, Joyner et al. (1966) and Dove (1966) showed that mutants in *N, O,* or *P* are defective in both functions; this suggests that replication to produce more copies of the genome may be necessary for transcription of late genes. Subsequent experiments have shown that the products of genes *O* and *P* are implicated directly in replication. These genes are adjacent on the map of lambda and are coordinately expressed. Gene *N* is a regulator which codes for a protein needed to allow the phage to express further functions. $N^-$ mutants are unable to express any of the functions of the lytic pathway. A further group of early genes, discovered only subsequently, comprises the cluster concerned with recombination functions, *int*, *xis*, *redα*, *redβ*, *gam*. Mutants in these genes are not defective in lytic infection since *int* and *xis* are concerned with exchanges with the chromosome and the *red* genes with vegetative recombination. Expression of both the replication and of the recombination groups of genes is controlled by gene *N*.

The experiments of Joyner et al. and of Dove showed that mutants in the late genes *A-M*, *Q*, *R* all are able to synthesize normal amounts of DNA. A deletion strain that lacks genes *A-J* appears to replicate and transcribe the remaining part of the genome to the usual extent; this provides a striking demonstration that these genes specify proteins concerned only with stages of development subsequent to replication. Mutants in gene *Q* show a reduction in the level of mRNA synthesis and display a pleiotropic defect in all late functions; since gene *Q* is not implicated in replication, this suggests that *Q* may be a regulator whose product is needed in order to transcribe the late genes.

Functions can now be ascribed to all the late genes. At the end of the period of phage development, the infected host cell is lysed and the assembled mature phage particles are released. Jacob and Fuerst (1958) first obtained mutants that are normal in development except for their inability to lyse the host cells. Campbell and Del Campillo-Campbell (1963) and Harris et al. (1967) observed that temperature sensitive mutants in gene *R* produce an altered endolysin; this gene product has sometimes been described (incorrectly) as lysozyme, but has the endopeptidase activity of cleaving between amino acids rather than the muramidase activity of lysozyme (Taylor, 1971). Mount et al. (1968) and Reader and Siminovitch (1971a,b) identified a further gene involved in lysis, *S*, which lies adjacent to *R*; $S^-$ mutants are unable to lyse cells, implying that *S* provides positive control of lysis, but the molecular action of the gene product is not yet known. Campbell and Rolfe (1975) have suggested a role for gene *rex* in lambda infection, of inhibiting lysis, to bring the action of gene *R* product under negative as well as positive control.

A component of the tail of the phage particle can be recognized by its antigenic reaction. Dove (1966) observed that all mutants except those in genes *J*, *N*, *O*, *P*, *Q* are able to produce high levels of phage tail antigen. Since *N* and *Q* are regulator genes and *O* and *P* are involved in replication, this suggests that *J* may be the gene (or one of the genes) coding for the tail antigen. Mount et al. (1968) showed that mutants in *J* both fail to display the serum reaction typical of tail antigen and also do not produce a functional host range protein; this suggests that both the serological and host range activities of the tail reside in the protein coded by *J*. Thus *J* can be equated with the gene previously identified by *h* mutants: its protein must be part of the tail of the phage.

The genes which code for the structural proteins of the phage head can be distinguished from those coding for tail components by complementation for assembly in vitro, a technique first developed for phage T4 (see Chapter 6).The principle of this approach is that structural mutants can be divided into head donors and tail donors: head donors are defective in one of the tail genes and can produce functional phage heads; tail donors are defective in one of the head genes but can produce functional phage tails. Because assembly of the phage particle is possible in vitro, mixing an extract from cells infected with a head donor with an extract of cells infected with a tail donor leads to the production of active phage particles; whereas of course no complementation occurs when two extracts of different head donors or of different tail donors are mixed.

An analysis of lambda defective mutants was performed by Weigle (1966). The procedure is to induce lysogens with ultraviolet irradiation and then to lyse the cells by addition of chloroform after 90 minutes; the lysates of two different mutants are then mixed together and tested to see whether active phage particles can be produced. In such pairs of mixed lysates, any combination of mutants *A—E* shows no increase in the phage titer of the extract; a similar result is observed for pairs of mutants *G—M*. But any mutant of the *A—E* group can produce infectious phages (that is, an increase of at least 100-fold in the titer) when its lysate is mixed with the extract from any mutant of the group *G—M*. Sedimentation of the extracts showed that *A—E* contain completed tails, whereas *G—M* contain assembled heads. Another criterion for distinguishing head from tail donors is to examine the host range of the assembled particles in subsequent infection; this is conferred by the tail donor. Parkinson (1968) identified four further structural mutants and showed by assembly complementation analysis in vitro that mutants in *W*, *C*, *F* are tail donors and mutants in *Z*, *U*, *V* are head donors. Subsequently Boklage, Wong and Bode (1973) showed that the *F* mutants in fact fall into two genes, $F_{\mathrm{I}}$ and $F_{\mathrm{II}}$; and a further head gene, *nu3*, was identified by Ray and Murialdo (1975). Thus there is a group of nine adjacent genes *A W B C nu3 D E* $F_{\mathrm{I}}$ $F_{\mathrm{II}}$ whose products are concerned with assembly of the head, lying next to a cluster of eleven adjacent genes, *Z U V G T H M L K I J*, whose products are implicated in tail assembly. The functions of these genes in phage maturation are discussed in Chapter 5.

We can therefore now divide the lambda genes into four groups. A group of genes concerned with regulation of lysogeny and lytic development, comprising *cIII*, *N*, *cI*, *cro*, *cII*, lies in the center of the genetic map. Genes *O* and *P* to the right form an early group concerned with replication; and the regulator gene *Q* is expressed as part of this group. Genes *int*, *xis*, *redα*, *redβ*, *gam* to the left comprise a group concerned with recombination functions. The late genes *S* and *R* are concerned with lysis, *A—F* with the structural components of the head, and *Z—J* with tail components; these form a single cluster of late functions when the ends of the linear form of the phage (as depicted in Figure 4.1) join together to generate a circle.

### *Structure of Linear λ DNA*

When isolated from mature phage particles, λ DNA comprises a linear duplex of $30.8 \pm 1.0 \times 10^6$ daltons; its biophysical properties have been reviewed by Davidson and Szybalski (1971). Lambda DNA can be fractionated by physical methods in two ways: it can be broken into "left" and "right" halves of the linear duplex; and the two single strands of the duplex DNA can be separated.

When a break is introduced into the lambda linear duplex, the left and right halves of the molecule can be separated by their different buoyant densities. Skalka (1966) found that the left half is richer in G-C pairs (which comprise about 55% of the nucleotides) and thus has a greater density than the right half (which has only about 45% G-C pairs). The technique used to fractionate λ DNA into its two halves is to stir a solution fast enough to break all the molecules once in a fairly short time. This generates a distribution of molecules of average length half that of λ DNA, but ranging from 0.35–0.65 times the length of lambda (see also Davidson and Szybalski, 1971; Egan and Hogness, 1972).

Another observation that has proved very productive, especially in studying transcription, is that of Hradecna and Szybalski (1967) and Doerfler and Hogness (1968a,b); the single strands of λ DNA can be separated by centrifugation on poly(I,G)-CsCl or on alkali by their different densities. Originally named W or H and C or L, they are now known as *l* and *r* (since the *l* strand is transcribed towards the left and the *r* strand is transcribed towards the right; see later). After a break has been introduced in the lambda linear duplex, the left and right halves of each single strand can be separated.

Lambda DNA can be isolated in both linear and circular forms. Hershey, Burgi and Ingraham (1963) first observed that when λ DNA is extracted from infected cells it falls into several fractions when sedimented on a sucrose gradient. Linear monomers sediment at 32S;

more rapidly sedimenting forms include circular monomers and also some dimers and trimers. Two types of circular monomer can be distinguished: open circles, which are not covalently sealed on one strand; and the closed circles first isolated by Young and Sinsheimer (1964), which are covalently sealed on both strands and thus may form supercoils.

Changes in the state of λ DNA can be followed after infection. Bode and Kaiser (1965b) and Ogawa and Tomizawa (1967) observed that immediately after infection a $^{32}P$ label in λ DNA sediments on a sucrose gradient at the rate characteristic of the DNA extracted from mature phage particles. The label in this peak declines when the infected cells are incubated for some time before the λ DNA is extracted; the label first appears in a peak sedimenting 1.13 times faster and then transfers into a peak sedimenting 1.55 times faster. Figure 4.2 shows the state of $^{32}P$ labeled infecting λ DNA at 0, 10, and 45 minutes after infection; a sample of $^{3}H$ labeled mature λ DNA was included as a control. The species found in mature phage particles, also the form in which virtually all the DNA is found immediately after infection, can be denatured into two linear strands. This suggests that it is a linear duplex of DNA. The intermediate form can be denatured into one linear strand and one circular strand; this suggests that it represents circular duplex molecules open on one strand. Denaturation of the final form yields only circular single strands; this suggests that, consistent with its rapid rate of sedimentation, it comprises twisted closed circular molecules. This model is confirmed by the effect of nicking the closed circles, which converts them into the open circles of the intermediate form.

What is responsible for the interconversion of lambda between linear and circular forms? Hershey et al. (1963) proposed that each monomeric linear duplex possesses two *cohesive ends*, which may either react together to form a circle, or which may interact with other molecules of λ DNA to form polymers. When Hershey and Burgi (1965) fragmented λ DNA into left and right halves and allowed the two fractions to react, they were able to form linear molecules of the normal length. These full length molecules sedimented at the same buoyant density as that of native lambda; this suggests that the left half is able to join only to the right half and vice versa. (Molecules consisting of two left halves would be much denser and molecules consisting of two right halves would be much lighter than native λ DNA.) An obvious model to explain this observation is to suppose that the two cohesive ends of the molecule represent complementary single stranded nucleotide sequences. In this case circle formation would proceed through the stages illustrated in Figure 4.3. Formation of hydrogen bonds between the two cohesive ends would generate a circular molecule with a break on each strand; closing of the two breaks

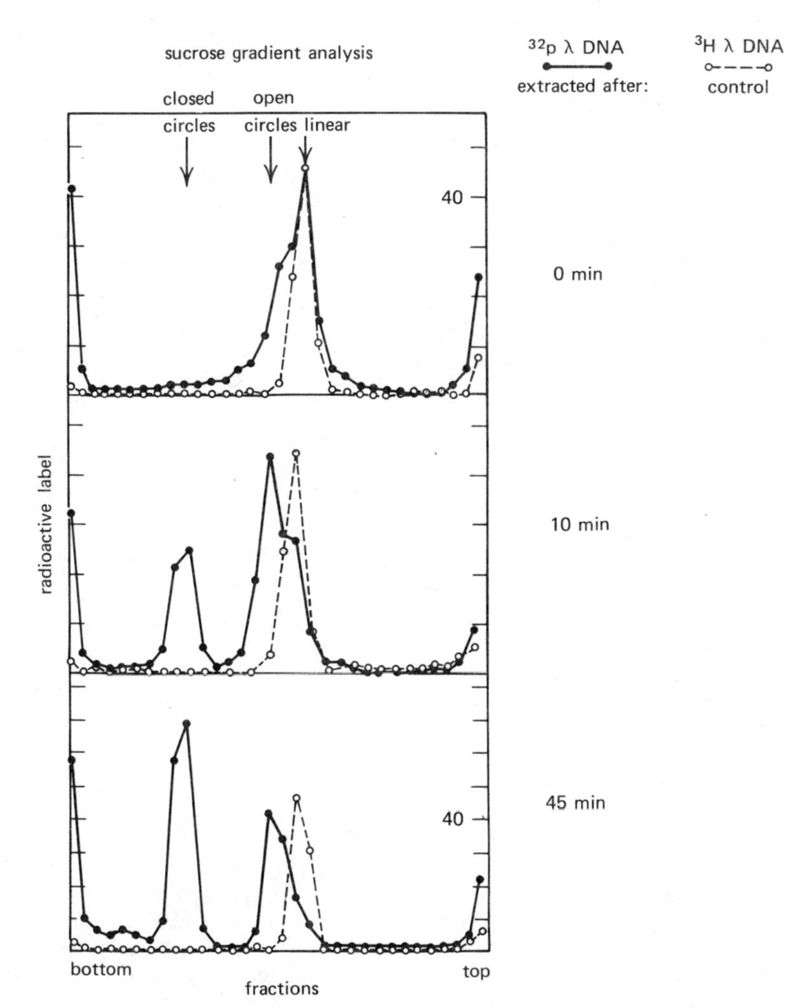


**Figure 4.2:** sedimentation of λ DNA during infection. Sensitive cells were infected with $^{32}$P labeled lambda and the DNA extracted after 0, 10, or 45 minutes. This DNA was sedimented on a neutral sucrose gradient together with a control of $^{3}$H labeled DNA extracted from mature phage particles. Immediately after infection most of the DNA remains in the position characteristic of mature DNA; then it is converted through an intermediate peak of open circles into a final peak of closed circles. Data of Ogawa and Tomizawa (1967).

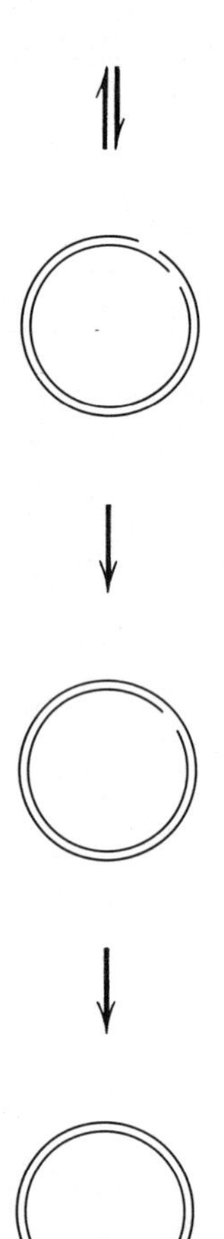

**Figure 4.3:** linear and circular forms of λ DNA. The linear form has two protruding single strand ends which can anneal with each other to form a circle with a break on each strand. A covalently closed circle can be formed when both breaks are sealed.

(by enzyme action to form covalent bonds) might take place simultaneously to generate a closed circular molecule, or might proceed one at a time thus creating an intermediate with one closed and one open strand (as illustrated in the figure).

The importance of the cohesive ends is shown by the results of assaying the biological activity of λ DNA: only molecules possessing a cohesive end are infective. An assay for infective λ DNA which has been very useful in establishing the biological activity of different preparations was developed by Kaiser (1962). The principle of this technique, as illustrated

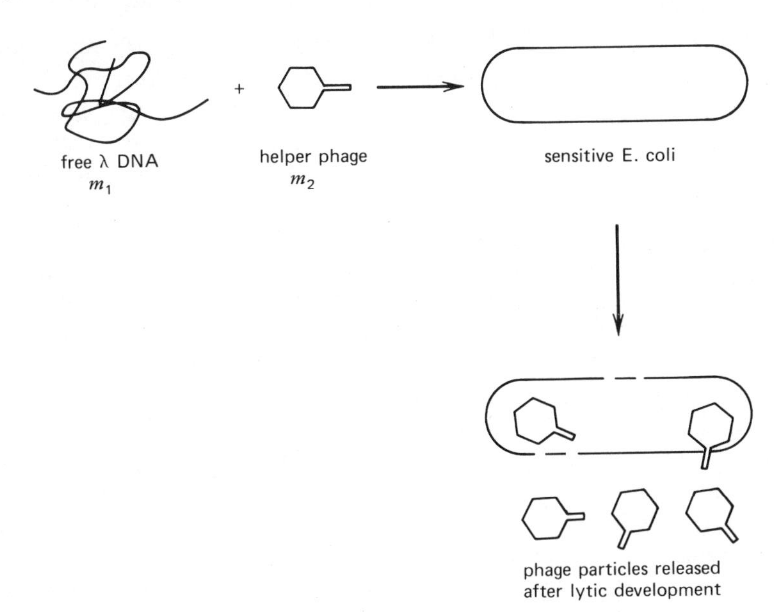


**Figure 4.4:** infectivity assay for λ DNA. Free λ DNA carrying a marker ($m_1$) is mixed with helper phage carrying a different marker ($m_2$) and used to infect sensitive E.coli cells. The phage particles released after lytic development are examined for the marker of the free DNA.

in Figure 4.4, is to infect sensitive cells simultaneously with the preparation of λ DNA to be tested and with a "helper" phage. In the presence of the helper phage, the λ DNA is taken up by the cells. Using helper phage and added DNA that carry different genetic markers, it is possible to assess the infectivity of the DNA preparation by following the appearance in the progeny phage of markers carried only by the isolated DNA. One type of marker that is useful is a plaque variant, since then its activity can be assessed simply by plating on an appropriate indicator strain where the plaque morphology is altered if the marker is present in the progeny. Calculations show that this assay detects about 0.5% of the added DNA molecules (for review see Kaiser, 1966). The basis of this assay is thus the incorporation of sequences of the added DNA into the progeny phage.

The infectivity of different forms of λ DNA can be determined by using them in this assay. Kaiser and Inman (1965) reported that when the linear duplex molecules extracted from phage particles are stored in 2M ammonium acetate at 5°C, closed molecules are generated. These molecules display a much reduced infectivity as judged by their ability to contribute markers to a helper phage. Heating causes the (open) circular molecules to return to the linear form and restores their infectivity. When linear molecules were broken into left halves and right halves, both were able to contribute genetic markers to the progeny of the helper phage. This suggests that a single cohesive end is sufficient to establish infectivity in this assay. (How the markers of the added DNA are incorporated into the phage progeny is not entirely clear.)

In a further experiment, Kaiser and Inman fragmented λ DNA into pieces each about one-sixth of the total length. This preparation is able to contribute only markers located at the ends of the vegetative genetic map. Thus genes *A* and *B* can be contributed by the extreme left fragment and genes *Q* and *R* can be contributed by the extreme right fragment; the other fragments lack cohesive ends, however, and so are unable to contribute their markers to the helper phage. The high activity of genes *A*, *B*, *Q*, and *R* in the six-fragment pieces thus suggests that the physical ends of the linear duplex correspond to the markers identified at each end of the linear genetic map.

This approach to genetic mapping was extended by Egan and Hogness (1972), who isolated the right halves of lambda and then fragmented them further by restirring at higher speeds. The fragments with cohesive ends were isolated by virtue of their ability to bind to left halves that had been labeled with BUdR and thus had a much increased buoyant density. Heating then allowed the right fragments to be recovered; and they were then fractionated into fragments of increasing length by zone sedimentation. The positions of genetic markers were then determined by using these fragments in an infectivity assay and seeing at which (increasing) length each marker was contributed to the helper phage. The results gave positions on λ DNA for those markers tested which agreed with the genetic map locations shown in Figure 4.1. This analysis of point mutations, together with the heteroduplex mapping of deletions (see later), suggests that the genetic map reflects fairly accurately the physical organization of λ DNA.

The idea that the cohesive ends are complementary single strands was supported by experiments in which Strack and Kaiser (1965) examined the effects of DNA polymerase I and exonuclease III upon λ DNA. DNA polymerase I can add nucleotides stepwise to a free 3′-OH end, extending a primer to form a duplex; exonuclease III removes 5′ mononucleotides stepwise from the strands of a duplex that terminate in 3′-OH ends. The infectivity of λ DNA is abolished when DNA polymerase I is allowed to add nucleotides; and infectivity is restored when exonuclease III is allowed to remove the added nucleotides. As illustrated in Figure 4.5, these results suggest that the protruding single strands at each end of the molecule terminate in 5′ ends; thus infectivity is lost when DNA polymerase I fills in the ends and is regained when exonuclease III

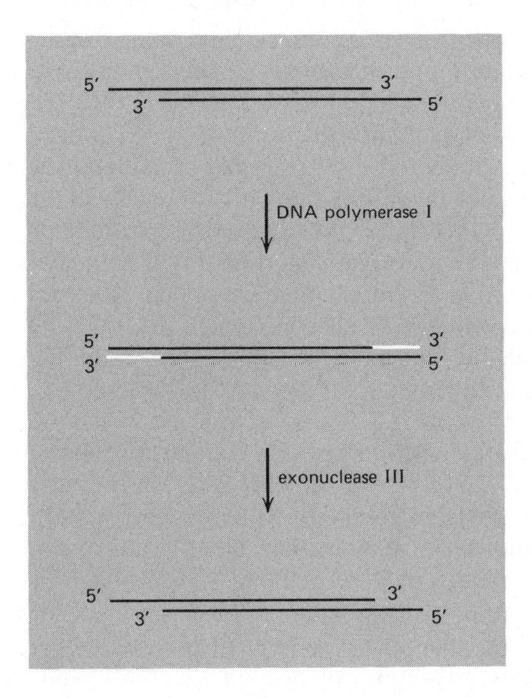


**Figure 4.5:** removal and restoration of cohesive ends. DNA polymerase I extends the 3′ ends to fill in the duplex up to each 5′ end (bases added are shown in grey). This causes loss of infectivity. The added bases can be removed by exonuclease III to restore the protruding cohesive single stranded ends.

restores them. In these experiments, about 20 nucleotides appeared to be added and removed at each end. Treatment of native λ DNA with exonuclease III also abolishes infectivity, although it may be restored by treatment with DNA polymerase; this suggests that the activity of the cohesive ends is lost when the single strand length is too great, again being restored when the appropriate length is regenerated. Wu and Kaiser (1968) showed that the same inactivation and reactivation by polymerase and exonuclease is displayed by the isolated halves of λ DNA; this suggests that both cohesive ends possess the same structure.

Are the two cohesive ends the same length? The open circles of λ DNA can be covalently sealed by the action of polynucleotide ligase alone. Since ligase is able only to form covalent bonds between adjacent nucleotides, this observation implies that there is only a single strand break and no nucleotide gap in the open circles (see Gellert, 1967; Gefter et al., 1967; Kaiser and Wu, 1968). To achieve this result, the two cohesive ends must be exactly the same length.

That the two cohesive ends represent complementary sequences was established by quantitative studies of the nucleotides incorporated by the action of DNA polymerase I. Wu and Kaiser (1968) reported that the reaction reaches a plateau after incorporation of: 13 dGMP; 13 dCMP; 7 dAMP; 7 dTMP. Although the total number of nucleotides incorporated in this reaction has since been shown to be 12 at each end rather than 20, the equality between dGMP and dCMP and between dAMP and dTMP makes the point that the two cohesive ends probably are complementary to one another. (An alternative explanation would be that each cohesive end contains sequences that are self-complementary.)

If the two cohesive ends have the same length and are complementary, the first nucleotide added to one 3′ end must be the same as the 5′ terminal nucleotide present at the other end of the molecule. By using polynucleotide kinase to attach $^{32}$P-phosphoryl residues to the 5′ termini, and then hydrolyzing the DNA to 5′ mononucleotides, Wu and Kaiser (1967) were able to show that the label enters (only) equal amounts of AMP and GMP. The use of half molecules showed that the left end terminates in 5′ G and the right end terminates in 5′ A. Wu and Kaiser (1968) then showed that when each nucleotide is tested alone (that is, in the absence of the other three nucleotides) for its ability to be incorporated into λ DNA, a single dAMP or 4 dGMP residues can be added; dCMP and dTMP are not added. This suggests that one dAMP can be added to the 3′ left end and four dGMP residues to the 3′ right end, consistent with the 5′ terminal bases at the respective opposite ends.

This approach has been extended into an analysis of the complete sequence of bases added at the right end. Wu and Kaiser found that when dCTP is added as well as dGTP, a total of 3 dCMP residues and 7 dGMP residues could be incorporated. Nearest neighbor analysis showed that all the dCMP residues were adjacent to GMP residues, suggesting that the initial run of dGMP bases is succeeded by an interspersion of dCMP and dGMP. Using these techniques, Wu (1970) and Wu and Taylor (1971) obtained the complete sequence shown in Figure 4.6. The sequence added in vitro at the right 3′ end must (of course) be complementary to the protruding 5′ terminal single strand at the right end and must be identical to the protruding 5′ terminal single strand at the left end; this is the basis on which the cohesive sequences shown in the figure were calculated. The total length of each cohesive end is 12 bases (for review see Yarmolinsky, 1971).

### *Formation of Prophage*

When sensitive bacteria are infected with lambda, those cells in which the phage enters the lytic cycle ultimately are lysed to yield the next genera-

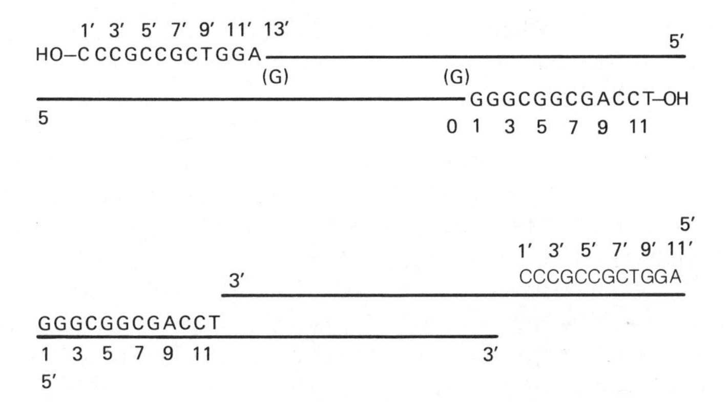


**Figure 4.6:** cohesive ends of λ DNA. The upper molecule shows the sequence of nucleotides incorporated into the cohesive ends by DNA polymerase I. The lower molecule shows the corresponding sequences of the protruding cohesive ends. Data of Wu and Taylor (1971).

tion of phage particles. Cells in which lysogeny is established, however, gain a phage genome which remains part of their genotype. The presence of this phage genome results in immunity to superinfection by further lambda particles and also may be revealed by the induction of phage upon appropriate treatment of the lysogenic cells. The retention of the ability of lysogenic bacteria to yield phage through many generations (in effect indefinitely) was noted by Lwoff and Gutmann (1950), when they suggested the term *prophage* to describe the inert state in which the phage is carried. Jacob and Wollman (1953) observed that the time course of induction following ultraviolet irradiation is the same as that seen when sensitive bacteria are infected with phage, which suggests that events succeeding induction are the same as those of the vegetative cycle after infection. Thus lysogeny is reversible and a phage released from its restraints enters the lytic cycle.

Lysogeny behaves as a genetic factor of the bacterium: a prophage can be mapped at a specific location upon the bacterial chromosome. Early experiments showed that lysogeny for lambda is closely linked to the galactose genes (Lederberg and Lederberg, 1953; Wollman, 1953), a conclusion which has since been supported by the detailed mapping of the region in which lambda is carried. In crosses between two lysogenic strains, each of which carried different mutations in the λ prophage, Appleyard (1954) was able to show that these behaved as factors linked to *gal*; and by transductional mapping Jacob (1955) was able to map the distance from lysogenic λ prophage to the neighboring genes. The position of the prophage on the bacterial chromosome is a characteristic of each lysogenic phage; using the technique of zygotic induction, in which prophage is induced when a lysogenic donor cell transfers its DNA by conjugation to a nonlysogenic recipient (see Chapter 2), Jacob and Wollman (1956a) were able to map the positions of lambda and other prophages.

Several types of model for the association of prophage with the bacterial chromosome were proposed in early years, ranging from synapsis to physical integration. The difficulties in understanding how a small phage genome might be inserted in the much larger bacterial chromosome stood in the way of integration models until Campbell (1962) proposed the model shown in Figure 4.7. If λ DNA takes a circular form (a proposal originally taken from the example of the circular bacterial chromosome and since substantiated by physical analysis of λ DNA), a single reciprocal recombination event can insert the phage genome so that it becomes a linear sequence in the host chromosome. Excision of prophage, yielding a circular phage molecule, might take place simply by reversal of the same series of events, that is, by another reciprocal recombination involving the same sites.

The original form of the model postulated that the sites at which recombination occurs on the phage and bacterial genomes might represent identical nucleotide sequences, recombination occurring as a consequence of their homology. However, it appears rather that the two sites take different (but, of course, specific) sequences. Recombination between them is catalyzed by two phage functions: integration requires the product of the gene *int*; excision requires the products of genes *int* and *xis*. The loci which are involved are described as *att* sites; thus the site on the bacterial chromosome at which lambda integrates is denoted as $att^{\lambda}$ (the sites for other phages are described in the same way, such as $att^{80}$, $att^{434}$, $att^{21}$); and the site on λ DNA at which insertion occurs is written simply as *att* on the lambda map. The *att* sites on the phage DNA and bacterial DNA, respectively, are shown in Figure 4.7 as $P.P'$ and $B.B'$. If the recombination event takes place between $P$ and $P'$ on the phage and between $B$ and $B'$ on the bacterial chromosome, the prophage is separated from the bacterial sequences by the two recombined sites $B.P'$ and $P.B'$.

All of the genetic predictions made by this model have been confirmed. A consequence for the bacterial chromosome of the insertion of the phage genome is that the distance between the bacterial markers flanking the attachment site should be increased in lysogenic strains. This was confirmed by Signer (1966) in a P1 transductional analysis of the linkage between the markers on either side of the insertion site of phage ø80. A consequence for the phage genome is that the order of markers will be different for prophage than for the linear form of the vegetative phage (assuming that the site used for insertion does not coincide with the site

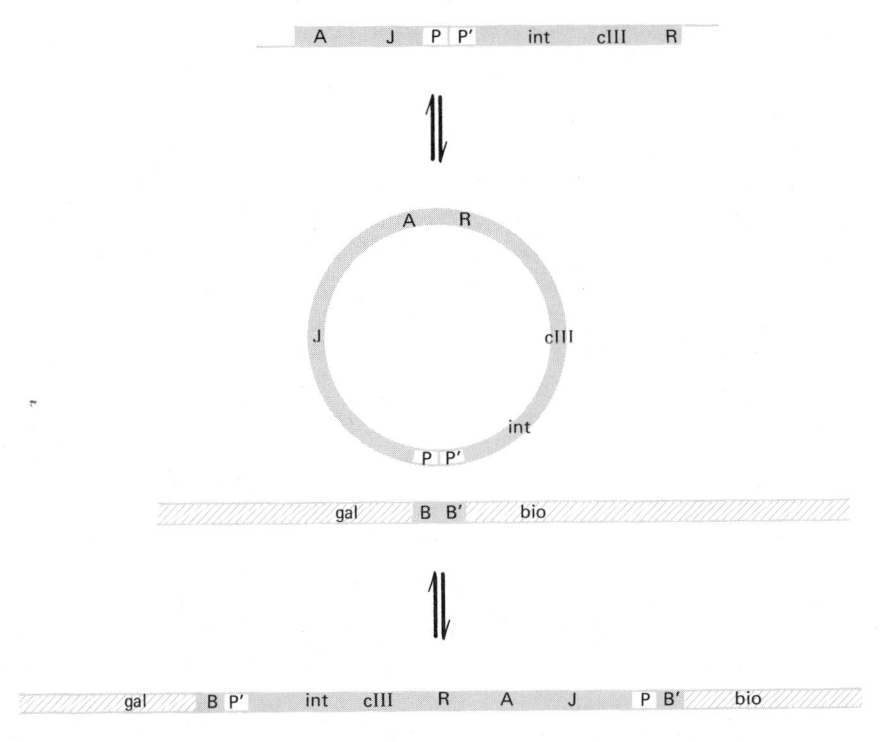


**Figure 4.7:** model for integration of phage lambda into the bacterial chromosome. The phage must be able to take a circular form, with an attachment site *P.P'* that interacts with the site *B.B'* on the bacterial chromosome. A single reciprocal crossover event occurs at the point between *P* and *P'* and between *B* and *B'*. This inserts the phage into the bacterial chromosome as a linear sequence, separated from the bacterial markers by the rearranged attachment sites *B.P'* and *P.B'*. The attachment site on the phage is depicted between *int* and *J*, which therefore become the terminal markers of the prophage. Comparison of the linear sequence of the vegetative phage (upper) and the prophage (lower) shows that integration results in permutation of sequence. The $att^{\lambda}$ site on the E.coli chromosome lies between the markers *gal* and *bio*.

where the linear ends cohere into the circle). As can be seen from Figure 4.7, if the attachment site *P.P'* lies between *int* and *J* these two markers become the terminal genes in the prophage, contrasted with the linear sequence from *A* to *R* of the linear vegetative phage.

Several experiments have confirmed the prediction that the order of prophage markers should be a permutation of the order of linear vegetative markers. In crosses between vegetative phages (that is by using mixed infections and examining the progeny for recombinants), Calef and Licciacardello (1960) mapped the order of genes: *J cIII R*. But in crosses

| gal | chlD ■ | int | redα | N | cI | cII | O | P | Q | R | A | B | C | D | E | F | G | H | M | L | K | I | J ■ | bio | chlA |
|---|---|---|---|---|---|---|---|---|---|---|---|---|---|---|---|---|---|---|---|---|---|---|---|---|---|
| + | + | − | − | − | − | − | − | − | − | − | − | − | − | − | − | − | − | − | − | − | − | − | − | − | − |
| + | + | ? | ? | − | − | − | − | − | − | − | − | − | − | − | − | − | − | − | − | − | − | − | − | − | − |
| + | + | + | + | + | + | + | + | − | − | − | − | − | − | − | − | − | − | − | − | − | − | − | − | − | − |
| + | + | + | + | + | + | + | + | + | − | − | − | − | − | − | − | − | − | − | − | − | − | − | − | − | − |
| + | + | + | + | + | + | + | + | + | + | + | − | − | − | − | − | − | − | − | − | − | − | − | − | − | − |
| + | + | + | + | + | + | + | + | + | + | + | + | − | − | − | − | − | − | − | − | − | − | − | − | − | − |
| + | + | + | + | + | + | + | + | + | + | + | + | + | − | − | − | − | − | − | − | − | − | − | − | − | − |
| + | + | + | + | + | + | + | + | + | + | + | + | + | + | + | − | − | − | − | − | − | − | − | − | − | − |
| + | + | + | + | + | + | + | + | + | + | + | + | + | + | + | + | + | + | + | + | + | + | + | − | − | − |
| + | + | + | + | + | + | + | + | + | + | + | + | + | + | + | + | + | + | + | + | + | + | + | + | − | − |
| − | − | ? | ? | + | + | + | + | + | + | + | + | + | + | + | + | + | + | + | + | + | + | + | + | + | + |
| − | − | − | − | − | + | + | + | + | + | + | + | + | + | + | + | + | + | + | + | + | + | + | + | + | + |
| − | − | − | − | − | − | − | − | − | − | − | − | − | − | − | − | − | + | + | + | + | + | + | + | + | + |

between lysogens (treating these markers as though they were bacterial factors), they found the permutated order: *cIII R J*. This suggests that the attachment site *P.P'* lies between *J* and *cIII*.

The isolation of deletions that extend from bacterial sequences into the prophage has provided a powerful technique for examining the genetic structure of the prophage. The principle of this approach is to isolate a series of mutants in each of which the deletion extends for a different distance into the prophage. These mutants constitute *defective lysogens* because the deletion of essential phage functions makes them unable to complete a lytic cycle upon induction. However, which functions remain intact in the defective lysogen can be tested by inducing the cells and superinfecting with a phage that carries a mutation in just one of its genes. If the induced prophage can provide the function that is missing in the superinfecting phage, it will be possible to complete lytic development; if this function has been deleted then the induced prophage will not be able to complement the superinfecting mutant phage.

The attachment site of phage ø80 maps close to the tryptophan locus in E.coli; using a phage that represents a hybrid between λ and ø80 but which integrates at this site, Franklin, Dove and Yanofsky (1965) isolated a series of strains in which deletions had extended from *trp* into the prophage of the lysogen. A similar analysis for lambda was performed by Adhya et al. (1968), who took advantage of the location on either side of $att^{\lambda}$ of genes whose deletion confers resistance to chlorate (which is toxic to E.coli under anaerobiosis): *chlA* is closely linked to *bio* on the right of $att^{\lambda}$ and *chlD* is linked to *gal* on the left of $att^{\lambda}$. Defective lysogens may therefore be obtained in the form of chlorate resistant strains where the deletion enters the prophage sequence from either side.

The series of deletion strains isolated by virtue of their chlorate resistance is shown in Figure 4.8. When the genes that have been deleted in each strain are identified by the inability of the prophage to contribute their products to a mutant infecting phage, it is clear that the genes absent in each case form a coherent group of adjacent loci. Thus by representing the prophage as the permutation of the linear vegetative sequence that is shown in the figure, it is possible to depict the upper ten strains as

---

**Figure 4.8:** deletion mapping of the gene order of prophage lambda. A (+) indicates that a marker in the prophage could be contributed to a superinfecting phage mutant in this gene; a (−) indicates that the prophage marker could not be contributed, that is, that it has been deleted. A (?) indicates that the marker could not be tested. The sequence remaining in each strain is shaded; the sequence deleted is shown in outline. The end points of two deletions are not certain (strains 2 and 11). Most deletions end between known markers in adjacent genes; however, the seventh and ninth end between known markers within a gene (*C* and *J*, respectively) so that these genes are marked as (/). Data of Adhya et al. (1968).

deletions that enter from the left and terminate at increasing distances along the prophage, while the lower three strains are deletions that enter from the right and again extend for varying distances into the prophage. The sequence that can be deduced for the prophage from both the *trp* and *chl* deletion strains thus represents the predicted permutation, with the attachment site on the lambda phage located between *int* and *J*. The attachment sites are represented in the figure by the black squares; two strains (the tenth and eleventh) are able to contribute all tested markers to a superinfecting phage, but nonetheless appear as defective lysogens. These may therefore represent deletions of the attachment sites which make the prophages unable to be released from the bacterial chromosome.

Direct evidence for the insertion of the phage genome in a permutated order has been provided by the heteroduplex mapping experiments of Sharp, Hsu and Davidson (1972). The episome F450 (see Figure 2.20) carries the sequence of bacterial markers *gal-att*$^{\lambda}$*-bio* and it is possible to obtain the plasmid F450($\lambda^+$) which carries a lambda genome inserted at *att*$^{\lambda}$. The sequences of F450($\lambda^+$) were mapped by forming the di-heteroduplex F450($\lambda^+$): F: λ*b2b5* in which the sequences of F are identified by a duplex circle of 94,500 base pairs, the bacterial sequences remain single stranded, and the lambda sequences form a smaller duplex circle. The location of the junction between prophage and bacterial DNA can be identified by its relationship to the position of the *b2* and *b5* deletions used as markers. The positions of the marker deletions showed that the sequences of the prophage represent a permutation of those of the linear lambda genome and identified the position of the attachment site as 14.2% of the lambda genome to the left of the *b5* marker (which covers the region of the *imm*$^{21}$ sequences), that is, about 57% from the left end of the vegetative lambda map. This agrees well with the location of *att* inferred from genetic experiments.

### *Transduction by Lambda*

Certain temperate phages can act as vectors able to transfer host markers between bacterial cells, a phenomenon observed by Zinder and Lederberg (1952) in Salmonella with phage P22 (then known as PLT22). They suggested the term *transduction* to describe genetic exchanges involving the transfer of only one part of the genome; and this term has since come to refer only to phage-mediated transfer of host markers. Transducing phages can be divided into two classes, according to the range of host markers which they are able to transfer.

In *generalized transduction,* observed by Zinder and Lederberg for P22 and characteristic of most transducing phages, any host gene may be transferred; a lysate produced either by infection of sensitive bacteria or by induction of lysogenic cells contains phage particles which have a low transducing activity for each host gene, for example of the order of 1 transduction per $10^5$ plaque forming units. Because any single phage particle can carry only a comparatively short length of bacterial DNA, generalized transduction provides a useful technique for fine structure mapping; two host markers can be transduced together only if they are close on the genetic map. For such markers, the frequency of joint transduction provides a measure of the map distance between them. In this aspect, the use of transduction for fine structure mapping is analogous to mapping by transformation, although it does not suffer from the problem of spurious linkage, since only a small proportion of a phage lysate is transducing contrasted with the potential activity of every DNA molecule in a transforming preparation (for review see Hayes, 1968).

In *specialized transduction* the phage is able only to transduce a small number of bacterial genes, those lying close to its attachment site. An important characteristic of specialized transduction is that the transducing phage cannot be prepared by infecting sensitive cells but only by inducing lysogenic cells. In terms of the Campbell model for integration, this immediately suggests that the transducing phage is generated by excision of the prophage through an "illegitimate" genetic recombination in which some bacterial sequences become part of the phage genome. This model for the origin of specialized transducing phages has indeed been confirmed in detail by testing its predictions on the construction of the transducing phage DNA (for review see Franklin, 1971a).

Transduction by phage lambda, first characterized by Morse, Lederberg and Lederberg (1956a,b), represents the best defined specialized system. In these experiments, lambda carrying the *gal* genes was prepared by inducing $gal^+$ $\lambda$ lysogens, isolating the phage particles from the lysate, mixing the particles with $gal^-$ bacteria, and isolating $gal^+$ bacterial transductants on indicator medium (on EMB-galactose agar, $gal^-$ colonies appear white whereas $gal^+$ colonies show a purple stain). This protocol yields about 1 transducing particle in $10^6$ and is known as low frequency transduction (lft).

The $gal^+$ bacterial clones isolated as a result of low frequency transduction fall into two classes. A minority is stable for the $gal^+$ genotype and appears to have been produced by replacement of the $gal^-$ gene in the bacterial chromosome with the $gal^+$ gene introduced by the transforming phage. Most of the $gal^+$ transductants, however, continue to segregate $gal^-$ progeny at a high rate, about $2 \times 10^{-3}$ per cell per generation. These $gal^-$ segregants are indistinguishable from the parental $gal^-$ bacteria; they have no transducing activity and display low frequency transduction when used as recipients. This implies that the *gal*$^-$ gene of the bacterial chromosome remained in these cells upon transduction and that the newly gained *gal*$^+$ gene of the phage can be lost. In other words, the major class of *gal*$^+$ transductants appears to be heterozygous for *gal*; such partial diploids may be described as *heterogenotes* of the type *gal*$^-$/λ-*gal*$^+$, where the *endogenote* is the *gal*$^-$ of the chromosome and the *exogenote* is the *gal*$^+$ introduced by the transducing phage.

Lysates prepared by induction of *gal*$^-$/λ-*gal*$^+$ heterogenotes (the heterogenotes usually take the form of lysogens) have characteristic properties: they yield lower titers of phage; but the number of transducing particles is much greater. This is high frequency transduction (hft), with transducing phages representing 0.1–0.5 of the particles in the lysate. The nature of high frequency transduction was revealed in experiments in which Arber et al. (1957) and Campbell (1957) examined the properties of the phage particles in the hft lysate. When sensitive bacteria are infected with particles from the hft lysate at low multiplicities of infection (instead of the high multiplicities used in earlier experiments), each infected bacterium receives only a single phage. The transductants become immune to superinfection but cannot be induced to liberate phage. This suggests that they are *defective lysogens*: they have gained a phage which is able to lysogenize but which lacks some essential function and therefore is unable to proceed to the completion of lytic development. This suggests that a transducing phage is generated by a process in which it gains bacterial markers but loses some of its own genome. The lambda phage variants that transduce the *gal* genes can thus be described as λ*dg* or λ*dgal,* where the *d* indicates that the phage is defective as the result of losing some of its genes and the *gal* indicates that it carries this bacterial marker.

The Campbell model for the integration and excision of phage genomes can accommodate the generation of defective transducing phages in the form shown in Figure 4.9. Relying upon the basic idea that a single reciprocal recombination event can insert and excise circular genomes, Campbell (1962) proposed that recombination between a site within the prophage and a site within the adjacent bacterial sequences should produce a phage carrying bacterial sequences in place of some of its own sequences. Recombination with a bacterial site to the left of the inserted prophage generates a λ*gal* phage whereas recombination with a bacterial site on the right leads to the production of a λ*bio* phage. This model predicts that a λ*gal* phage should lack sequences from the right end of the prophage; a λ*bio* phage should lack sequences from the left end of the prophage. Defective transducing phages thus originate by the same sequence of events responsible for type I excision of sex factors (see Figure

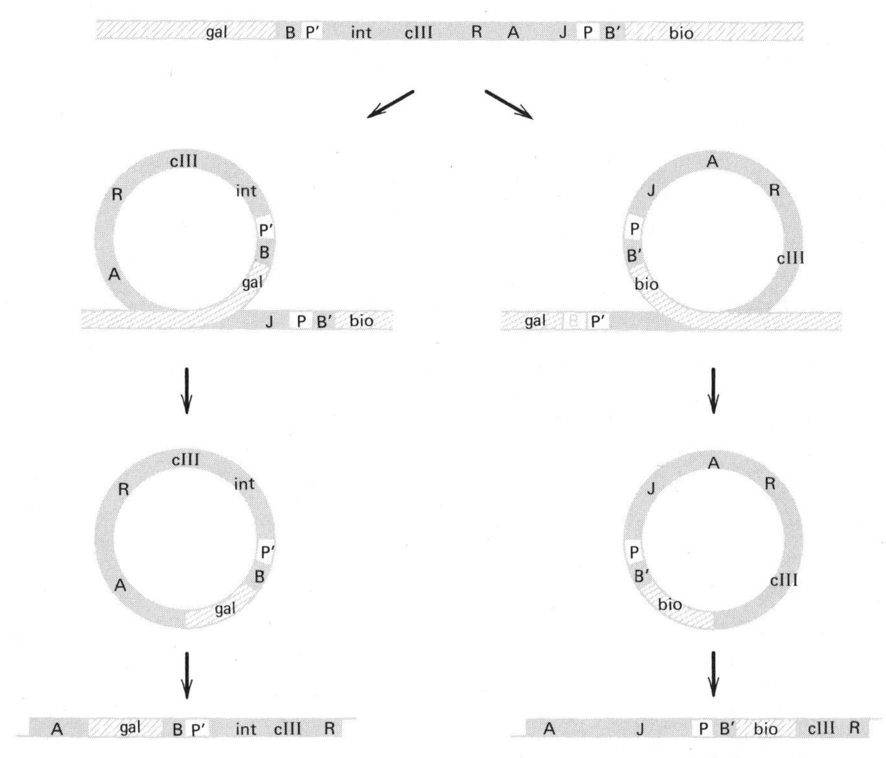


**Figure 4.9:** model for excision of defective transducing phage in lysogenic cells. A reciprocal recombination event takes place between a site within the prophage and a site in the bacterial chromosome on either the left or on the right. Excision to the left yields a phage carrying the bacterial *gal* marker and leaving behind in the bacterial chromosome some of the right-end prophage sequences, that is genes A-J. Excision to the right yields a phage carrying the bacterial *bio* marker and leaving left-end prophage sequences in the bacterial chromosome. This is the same model shown for type I excision of sex factors in Figure 2.13.

2.13). Although we have discussed λ*gal* transducing phages because these were characterized in early experiments, λ*bio* phages display corresponding properties.

The regions lost from λ*dgal* transducing phages can be determined by examining the ability of the defective phages to complement mutants that are defective in a single gene. Campbell (1959) used a series of amber mutants for this purpose, conducting mapping experiments in two ways. In one series of experiments, single lysogens carrying only a λ*dgal* prophage were induced and infected with a λ*sus* phage. The lysate was then examined for wild type recombinant phages in which the *sus* allele

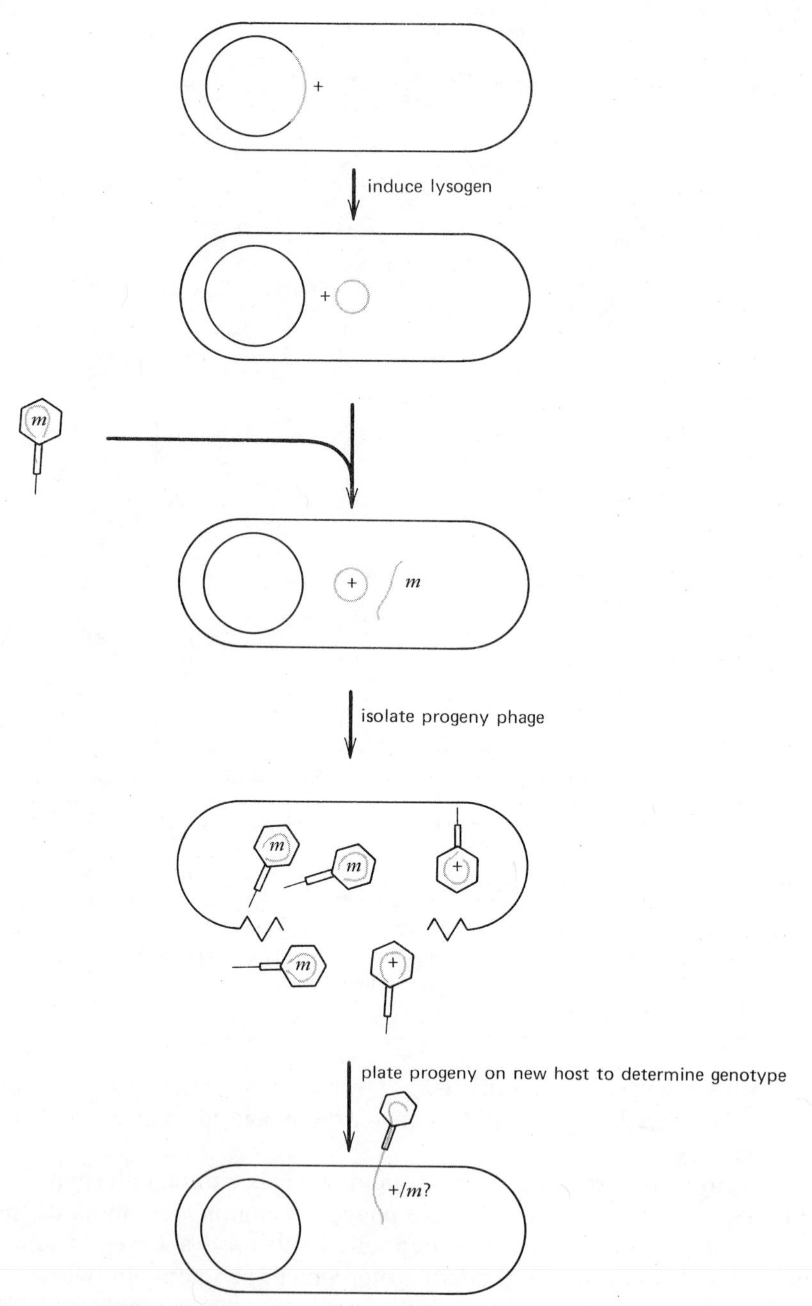


**Figure 4.10:** marker rescue technique. A mutant phage is used to superinfect an induced lysogen; the progeny phage are isolated and then plated on a new host to determine whether they carry the genotype of the original infecting phage or have gained (rescued) a marker

has been replaced by a wild type allele from the λ*dgal*. This is therefore a *marker rescue* technique, as illustrated in Figure 4.10, since what is measured is the ability of the λ*sus* phage to "rescue" a wild type allele from the λ*dgal* genome. Superinfecting each induced λ*dgal* lysogen with a series of λ*sus* mutants reveals which mutant sites are matched by wild type alleles in the λ*dgal* phage (these can be rescued to give wild type recombinants) and which mutant sites occur at loci that have been deleted in the λ*dgal* genome (these cannot be rescued). This identifies the sequences that have been deleted in each λ*dgal* lysogen that is tested. It is necessary to induce the lysogenic cells before superinfection since otherwise the superinfecting phage is prevented from entering the lytic cycle by the immunity of the lysogen. The need for induction was avoided in a second series of experiments, otherwise in principle the same as the first, which used a series of defective λ*dgal* lysogens which carried the immunity system of another phage instead of that of lambda, and could thus simply be infected with λ*sus* mutants and examined for production of wild type recombinants. The λ*dgal* phages isolated from different lft events possess deletions which extend for varying distances from gene *J* towards gene *A* as the prophage map is drawn in Figure 4.8 (for review see Campbell, 1964). This suggests that any one of many sites, perhaps any sequence, between *J* and *A* is able to participate in the recombination event which excises the transducing phage. The identification of missing sequences in the λ*dgal* phages demonstrates directly the cause of their defective genotype. (We should note in parenthesis that this type of technique examines only one class of the excised transducing phages: those that are able to form heterogenotes. This does not exclude the possibility that other structures are formed by illegitimate excision but are not detected in these experiments.)

The characteristics of the low frequency and high frequency transduction systems are summarized in Figure 4.11. The generation of a transducing phage is a rare event, so that in the original isolation represented by the development of the lft system, only one phage particle in $10^5$ in the lysate is of the form λ*dgal*. When *gal*$^-$ bacteria are infected with the particles of the lft lysate, however, all the cells that are selected by their Gal$^+$ phenotype must have gained a λ*dgal*$^+$ phage genome. Because the λ*dgal*$^+$ particles form only a small part of the lft lysate, all the bacteria

---

from the induced prophage of the lysogen. If the prophage carries a wild type gene (+) as illustrated, this may substitute for the mutant *m* in some of the progeny: phages in which this has occurred may be assayed by plating the progeny in suitable conditions when only those genomes gaining the (+) allele are able to survive. If the prophage does not carry a wild type allele for the gene that is mutant in the infecting phage, no rescue can take place.

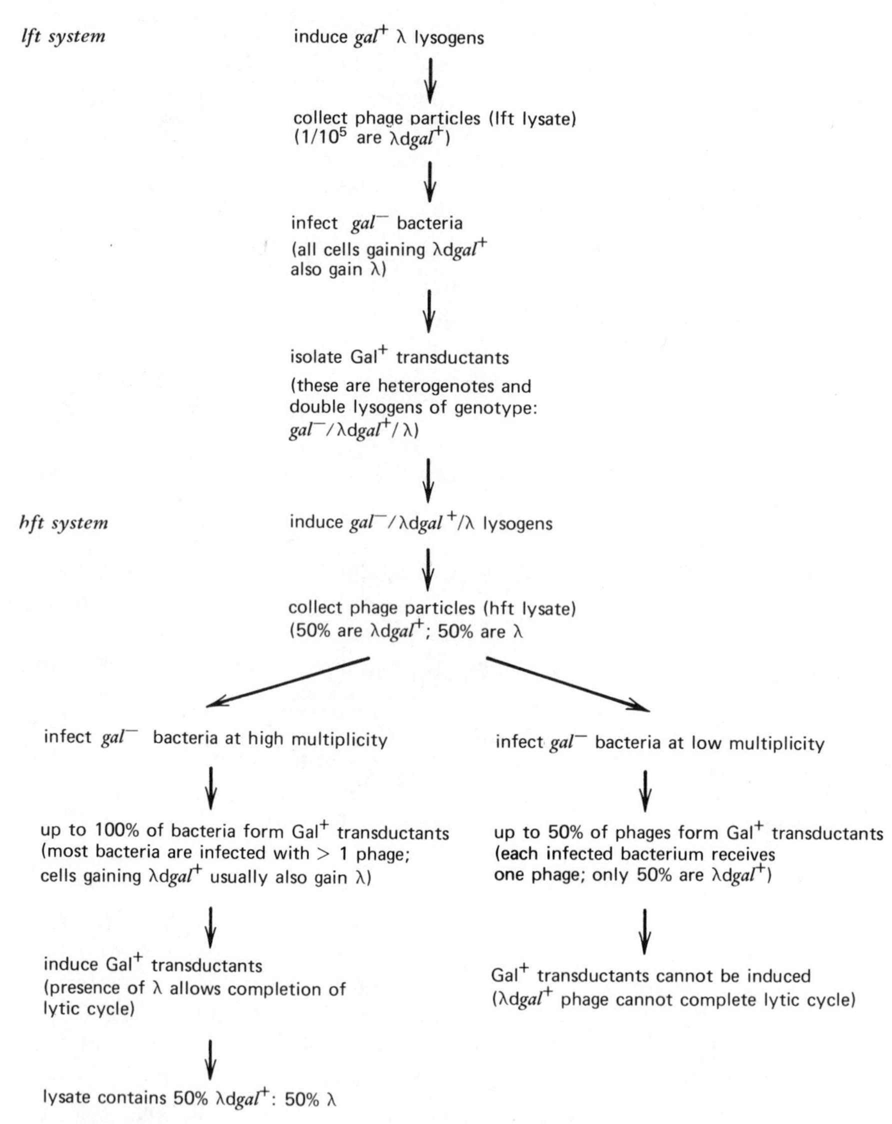


**Figure 4.11:** characteristics of low frequency and high frequency transduction.

that gain a λ*dgal* genome also gain a λ genome. Both phage genomes enter the lysogenic state, so that the $gal^-/gal^+$ heterogenotes are double lysogens carrying both λ$dgal^+$ and λ.

Upon induction, these heterogenotes therefore generate an hft lysate; for although the λ$dgal^+$ genome is defective, the presence of the λ genome allows both phages to complete the lytic cycle and contribute particles to the lysate. The hft lysate therefore contains equal numbers of λ$dgal^+$ and λ phages (the proportion of 0.5 λ$dgal^+$ genomes thus contrasts with the proportion of $10^{-5}$ in the lft lysate). When the hft lysate is used to infect $gal^-$ bacteria at high multiplicities, so that most bacteria are subject to infection with more than one phage particle, as many as all of the bacteria may form $Gal^+$ transductants and those that do so usually gain a λ genome as well as the transducing λ$dgal^+$ genome. The induction of these transductants thus again generates an hft lysate.

After infection with such hft lysates, there are usually a few heterogenotes which are immune to superinfection but cannot be induced: these must have gained a λ$dgal^+$ genome but have failed to gain a λ genome. It was the presence of these heterogenotes which prompted experiments to characterize hft lysates by low multiplicity infections of fresh recipients. Under these conditions, when each bacterium is infected with not more than one phage particle, up to 50% of the phages are able to give rise to $Gal^+$ transductants; this is consistent with the idea that half of the particles in the hft lysate are λ$dgal^+$ and half are λ, derived from induction of double lysogens. The $Gal^+$ transductants resulting from low multiplicity infection are heterogenotes which are immune to superinfection, revealing the presence of a phage genome, but which cannot be induced to yield an active lysate. These bacteria therefore carry a single, lysogenic λ$dgal^+$ genome.

This model implies that the hft lysate can be perpetuated through high multiplicity infections because the wild type λ genome that is present acts as a *helper phage*; in low multiplicity infections the λ$dgal^+$ phage enters cells alone, so there is no helper function. This idea was confirmed by an experiment in which Campbell subjected $gal^-$ cells to mixed infection with a low multiplicity of a $gal^+$ hft lysate and a high multiplicity of nontransducing λ. In spite of the low multiplicity of the λ$dgal^+$ phages, it was then possible to prepare an hft lysate by inducing the $Gal^+$ transductants. The lysate contained the nontransducing λ as well as the λ$dgal^+$, suggesting that the transductants were double lysogens carrying both types of phage.

What is the basis for the helper effect? It was at first thought that the nondefective λ genome provides functions whose genes have been deleted from the defective λ*dgal* phage. As Figure 4.9 implies, the functions lost from $\lambda dgal$ are those located at the right end of the prophage map, that is the late genes. This explains why the rare lysogens formed by $\lambda dgal$ alone cannot be induced to yield active lysates, but upon induction of a double lysogen carrying $\lambda^+$ as well as $\lambda dgal$ it is possible for both phages to complete the lytic cycle and give rise to mature phages. However, $\lambda dgal$ phages initially form lysogens only very inefficiently, and this deficiency does not appear to be due to the deletion of necessary functions since $\lambda dgal$ phages usually carry the genes necessary for recombination (which are located in the left part of the prophage map). The very large increase in integration efficiency when $\lambda^+$ is added to a $\lambda dgal$ preparation appears instead to be the result of a structural effect: the $\lambda^+$ phage is able to integrate normally at $att^\lambda$ and in so doing provides a site at which the $\lambda dgal$ phage then can integrate. That is, whereas $\lambda dgal$ integrates inefficiently into nonlysogenic cells carrying $att^\lambda$, it can integrate efficiently into a bacterial chromosome carrying a prophage where $att^\lambda$ has been replaced by the two sites $B.P'$ and $P.B'$ at the phage-bacterial junctions. Subsequent experiments have shown that the double lysogen carries the two prophages in tandem, with the $\lambda dgal$ phage inserted at the left of the $\lambda$ phage, that is at the site $B.P'$. The nature of this integration is discussed in Chapter 5 in the perspective of the recombination events responsible for insertion and excision of lambda. In summary, we can therefore see that the $\lambda$ helper phage plays two roles in perpetuating hft systems: its integration provides a site at which $\lambda dgal$ can insert; and upon subsequent induction it provides the gene functions necessary for the lytic cycle that have been deleted from $\lambda dgal$.

The demonstration that there is genetic variation in the phage sites implicated in the illegitimate excision of transducing phages is matched by the observation that the overall length of the transducing phage genome is variable. Particles of phage lambda band at a single position on density gradients which corresponds to their relative content of DNA and protein. However, Weigle, Meselson and Paigen (1959) showed that an hft lysate gives two bands: one band represents plaque forming ($\lambda$) particles and the other includes the transducing phages. This provides a direct confirmation of the nature of the phages that constitute an hft lysate. Figure 4.12 shows that when different hft lysates are compared, the position of the helper phage particles remains constant but the position of the transducing phage particles varies; the heavier particles possess an increased amount of DNA (of up to 8% more than $\lambda^+$) and the lighter particles have a decreased DNA content (by up to 14% less than $\lambda^+$). Thus the region of the phage genome that is deleted may be replaced by either a longer or shorter region of bacterial sequences.

**Figure 4.12:** densities of lambda transducing phages. Different hft lysates were centrifuged to equilibrium. In each case there is one band found at the density characteristic of wild type lambda and a second band at a different density, which may be either lighter or heavier than the wild type band. The band on the far right is a density marker of lambda grown in $^{15}N$. Data of Weigle et al. (1959).

## Repression of Prophage

### *Synthesis of Repressor*

Lysogeny represents a remarkable condition in which a phage can coexist with its host as part of the bacterial chromosome. Lysogenic bacteria can be identified by two characteristic features: *immunity* and *induction*. In the normal course of events, the prophage may be carried indefinitely in its inert state, when it is inherited like any other bacterial marker. But the presence of prophage is sufficient to confer immunity against superinfection, so that lysogenic cells are not susceptible to lytic infection by further phage particles of the same type. And although in general stable, the prophage can be induced by appropriate treatment to enter the lytic cycle, which results in destruction of the cell and release of progeny phage particles. Any model for lysogeny must therefore explain the stability of the inert state of the prophage, its ability to prevent a superinfecting phage from entering the lytic cycle, and the release from the lysogenic state upon induction.

The inert state of the integrated prophage and the immunity to superinfection of lysogenic cells both result from the operation of a single control system. The sole action of the prophage genome is to synthesize a repressor protein which prevents expression of all phage genes except its own. Sufficient repressor is synthesized by a prophage to act upon any superinfecting phage as well as upon its own genome; this prevents the expression of the functions that the superinfecting phage needs for lytic development or for integration and it is therefore diluted out of the strain with growth. Induction of a lysogen occurs when the repressor present in the cell is inactivated, allowing expression of functions necessary for excision and lytic development. Induction occasionally occurs spontaneously, but may be provoked by ultraviolet irradiation or addition of mitomycin C.

Considered in terms of the repressor protein, we can ask three questions about the lysogenic condition: what controls the synthesis of repressor; how does the repressor prevent expression of all other prophage genes; and how may the repressor be inactivated to allow release from the lysogenic condition? The lysogenic state is attained in two stages: establishment and maintenance. *Establishment* of lysogeny requires two events: the phage genome must be inserted in the bacterial chromosome; and synthesis of repressor must commence in order to prevent expression of lytic functions. Establishment of repressor synthesis requires the participation of three control genes. *Maintenance* of lysogeny requires only the continuation of repressor synthesis; this is achieved solely by the transcription and translation of a single gene. The function of the repressor protein is to bind to two operators; these operators control transcription of adjacent genes whose products are needed for the expression of the genes responsible for lytic development. By binding to these two operators, the repressor thus prevents transcription of all lambda genes except its own. Little is known about how repressor protein is inactivated during induction and the inducing effects of ultraviolet irradiation or addition of mitomycin C are not well characterized. Induction has largely been studied with a temperature sensitive variant of lambda in which the repressor is inactivated at increased temperatures. Following heat induction by the inactivation of preexisting repressor, synthesis of further copies of the repressor is prevented by the product of one of the first lambda genes to be expressed; this is the *turnoff* function. Thus the level of repressor in the cell is set by three control systems: establishment, maintenance, turnoff.

### *Functions of the* c *Genes*

The lambda genes concerned with establishing lysogeny were identified by the isolation of mutants unable to lysogenize. The turbidity of plaques of wild type lambda is due to the development of colonies of lysogenic bacteria within the region of lysed bacteria. However, clear plaques occur at a frequency of about $10^{-4}$; these represent spontaneous mutants, known as *c* (for clear plaque), which are unable to lysogenize and thus form plaques of lysed cells uncontaminated by lysogenic cells. Treatment with mutagens increases the proportion of clear plaques.

Three different classes of clear plaques were distinguished by Kaiser (1957):

*cI* mutants show complete lysis and thus produce entirely clear plaques (except for any cells derived from bacteria that are resistant to infection by lambda). These mutants can form lysogens only at frequencies $<10^{-5}$;

*cII* mutants develop plaques which have at their center a cluster of colonies of bacteria lysogenic for λ*cII* mutants. These mutants can form lysogens at frequencies of $10^{-4}$–$10^{-5}$;

*cIII* mutants develop plaques that are almost completely covered by bacteria lysogenic for this mutant type. They can form lysogens at frequencies of $10^{-2}$–$10^{-1}$;

$\lambda^+$ (for comparison) varies in the efficiency of lysogen formation with the experimental condition, but may show a frequency $\sim 4 \times 10^{-1}$.

Mixed infection of a host cell with $\lambda^+$ and any $\lambda c$ mutant yields cells lysogenic for both phages: so the wild type alleles are dominant. Mutants of each class map in a group; mutants in the same class do not complement but pairwise combinations of mutants of different classes show complementation. Each class thus appears to represent mutations in a single gene. When bacteria are infected with pairs of mutants, it is possible to isolate lysogenic cells carrying *cII* or *cIII* mutant prophages; but lysogens carrying a single *cI* mutant prophage cannot be obtained. This suggests that all three *c* genes are needed to establish lysogeny but that *cII* and *cIII* are not needed for its maintenance.

The idea that only *cI* is needed to maintain lysogeny is supported by the genetic analysis of immunity performed by Kaiser and Jacob (1957), using a series of temperate lambdoid phages, all active on E.coli K12, each displaying a different immunity. Bacteria lysogenic for any one of these phages are thus sensitive to all the others. Since genetic recombination occurs between pairs of these phages upon mixed infection, it is possible to map the location of the sites responsible for immunity by determining the genotypes of the recombinants. Crosses of $\lambda$ with three other phages, *434*, *21*, and *82*, all yield similar results when recombinants with the immune specificity of lambda are selected: the recombinants must possess a region in the center of the lambda genetic map, therefore known as the *immunity region*. To retain the immune specificity of lambda, it is necessary only for the recombinants to possess the $\lambda cI$ gene and a short length of genetic material on either side of it. This shows that the *cI* gene, together with these short adjacent sequences, is sufficient to maintain $\lambda$ immunity; this result therefore implies that the product of the *cI* gene acts within the short lambda sequence retained in these recombinants, that is, close to the *cI* gene itself.

When the reverse demand is made, for the immunity of one of the other phages, it is possible to derive recombinants derived entirely from lambda except for the substitution of a short length of DNA conferring the selected immunity. Two of these recombinants, known as $\lambda imm^{434}$ and $\lambda imm^{21}$, have proved extremely useful in analyzing the control of lambda lysogeny and lytic development. The short length of the lambda genome that is substituted by the other phage in each of these two hybrids is shown in Figure 4.13; $\lambda imm^{434}$ has the phage *434* sequence corresponding to *cI* and *cro* and a short length of adjacent material; and $\lambda imm^{21}$ has a slightly longer substitution, replacing the lambda material from gene *N* to *cII* with the DNA of phage *21*. Because of the substitution, $\lambda$ and $\lambda imm^{434}$ are heteroimmune phages; it is therefore possible to infect a $\lambda$ lysogen with $\lambda imm^{434}$, a useful technique in genetic analysis. (The ability to divorce immunity from other genetic markers makes it useful to charac-

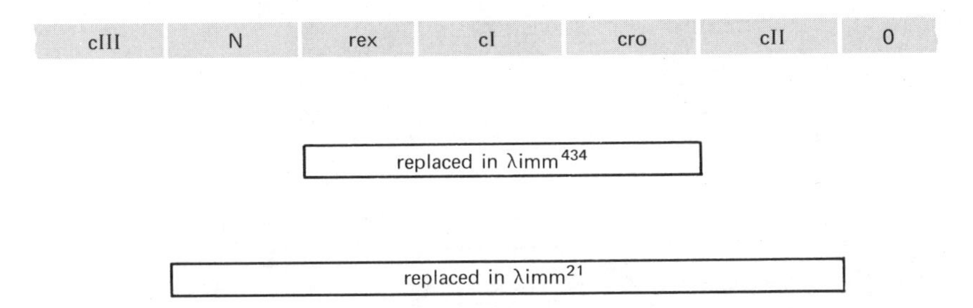


**Figure 4.13:** the immunity region. The genes *rex-cI-cro* of λ are replaced in the hybrid λ*imm*$^{434}$; this is known as the immunity region. The longer region *N-rex-cI-cro-cII* is replaced in λ*imm*$^{21}$. In this figure *rex* is assumed to be a separate gene but in most subsequent figures it will omitted since it is not clearly characterized.

terize pairs of phages as *homoimmune* or *heteroimmune* depending on whether they have the same or different immunities: λ and *434* are heteroimmune but λ*imm*$^{434}$ and *434* are homoimmune.)

By using heteroduplex mapping to compare the sequences of λ, *434* and *21*, Simon, Davis and Davidson (1971) have shown that the left arms of λ and *434*, extending from gene *A* to gene *J*, are identical; the left arm of *21* is identical with that of λ between genes *Z* and *J* and the two sequences are partially homologous from genes *A-F*. There are many substitutions of sequences between the three phages in the remaining parts of the molecule, although there are short homologous regions (where recombination could take place to generate hybrids such as λ*imm*$^{434}$ and λ*imm*$^{21}$). The relationship between phages λ, *434* and *21* revealed by this heteroduplex mapping is illustrated in Figure 4.14.

The idea that the *cI* region alone is responsible for the specificity of immunity immediately suggests a model in which *cI* codes for a repressor that is responsible for maintaining the prophage in an inert state and which also prevents the expression of a homoimmune superinfecting phage (see Jacob and Campbell, 1959; Jacob and Monod, 1961). The site(s) at which the *cI* repressor acts also must be located within the limited region that is nonhomologous between λ and λ*imm*$^{434}$. (We should note that the gene *cro*, also replaced by *434* genetic material in the λ*imm*$^{434}$ hybrid, had not been identified when these experiments were performed; at this time *cI* was the only known gene covered by the substitution.) The idea that *cI* codes for a protein product was suggested by the isolation of temperature sensitive variants by Sussman and Jacob (1962) and by the isolation of nonsense mutants subject to suppression by bacterial genes by Jacob, Sussman and Monod (1962). Lieb (1966) also studied temperature sensitive variants of lambda in which the prophage is

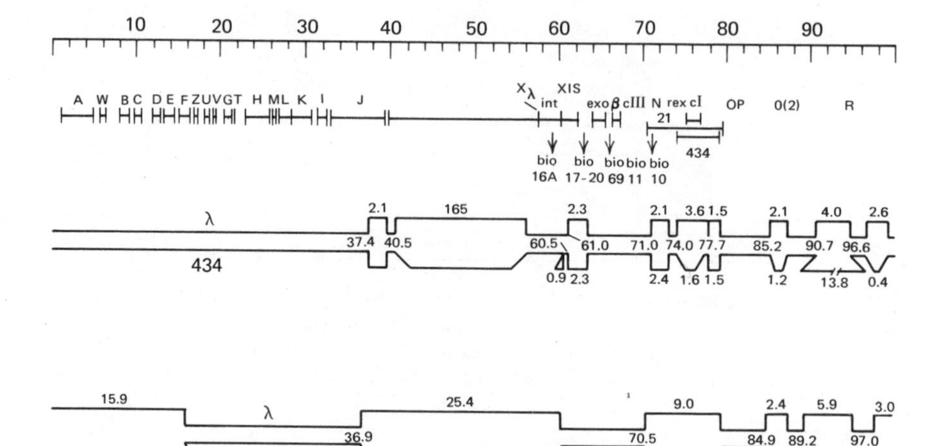


**Figure 4.14:** heteroduplex mapping of homology between λ and *434* and between λ and *21* DNA. Regions that form a duplex are indicated by parallel lines and regions of nonhomology appear as substitution loops. Distances are marked as a percentage of the length of λ DNA. The λ and *434* genomes are homologous from *A-J* and display short homologous sequences in the remaining regions. The λ and *21* genomes are partially homologous from *A-E*, since a duplex can be formed here under less stringent conditions than those used to construct the above map, are homologous from *F-J,* and then show short homologous sequences in the right arm. The substitutions that occur in $\lambda imm^{434}$ and $\lambda imm^{21}$ are marked on the λ map and appear to represent recombination occurring at sites that are homologous between the genomes. Data of Simon, Davis and Davidson (1971).

induced by a rise in temperature; all mapped in *cI*, confirming that *cI* codes for protein and implying also that it may be the only such function concerned in the maintenance of lysogeny.

Are the *cII* and *cIII* genes expressed by the prophage? Bode and Kaiser (1965a) tested the expression of prophage functions by the technique of *heteroimmune infection*. A bacterium lysogenic for λ is infected with a $\lambda imm^{434}$ phage. The reason for using the heteroimmune hybrid phage is that a homoimmune λ phage would be subject to repression by the prophage; although otherwise identical with λ, the substitution in its immunity region allows the hybrid $\lambda imm^{434}$ to infect and lysogenize the host, generating a double lysogen for λ and $\lambda imm^{434}$. And of course, since $\lambda imm^{434}$ represents λ genes except in the immunity region, it can be used for complementation tests with λ. (Although it is possible also for the $\lambda imm^{434}$ superinfecting phage to enter the lytic cycle, we are concerned here only with those cells in which it establishes lysogeny; other experiments have utilized the ability of a heteroimmune phage to enter the lytic cycle and have ignored lysogenic products.)

In these experiments the prophage λ was wild type, that is carried $cI^+$, $cII^+$, $cIII^+$ alleles. The heteroimmune superinfecting phage carried mutants in either *cII* or *cIII*, that is possessed the genotype $\lambda imm^{434}$ $cII^-$ or $\lambda imm^{434}$ $cIII^-$. The hybrid phage therefore is unable to lysogenize by relying upon its own genetic functions; but it will be able to lysogenize if *cII* or *cIII* proteins are available in the cell because they are synthesized by the prophage. In order to detect the formation of lysogens, the $\lambda imm^{434}$ mutants also carried a $gal^+$ operon while the bacterium was $gal^-$. Thus selection for $Gal^+$ bacteria following the heteroimmune infection allows any new lysogens to be isolated. Both the $\lambda imm^{434}$ $cII^-$ and $\lambda imm^{434}$ $cIII^-$ phages were unable to form lysogens; thus the *cII* and *cIII* gene products are needed for the establishment of lysogeny but (since they are not expressed by the prophage) are not needed to maintain lysogeny.

## *Transcription of Early Genes*

The repressor protein maintains the prophage genome in its inert state by preventing the expression of all genes except *cI*. Analysis of the RNA present in cells carrying lambda shows that this repression is exerted at the level of transcription. In early experiments, Isaacs, Echols and Sly (1965) measured the amount of RNA able to hybridize with λ DNA after cells were infected with $\lambda cI^-$ mutants. With sensitive cells, 8% of the extracted RNA could hybridize with λ DNA; the same result was obtained when the infected cells carried a $\lambda imm^{434}$ prophage. But only 0.2% of the RNA hybridized when the recipient cells were lysogenic for $\lambda^+$, $\lambda cII^-$ or $\lambda cIII^-$. This demonstrates that the repressor acts within the immunity region to prevent transcription of RNA from either prophage or a superinfecting homoimmune phage genome; and it confirms that *cII* and *cIII* do not contribute to the repression of a superinfecting phage.

Two types of information identify the sites where the repressor acts. Mutants which are not subject to repression occur in the operators to which repressor binds; and mutants which are unable to express certain sets of genes occur in the promotors where RNA polymerase binds. Operator and promotor mutations both are identified as regulatory by their cis dominance. And the sequences that are transcribed upon induction of lytic development can be defined by measuring the ability of the RNA to hybridize with preparations of λ DNA; by implication, these results show where transcription must be initiated and where repressor must act. By using physical methods, it is possible to separate the left and right arms and the *l* and *r* strands. By using deletion strains of the phage, often in the form of λ*gal* or λ*bio* variants where the bacterial sequences have replaced a well defined part of the phage genome, it is possible to assign more precisely the sequences that are transcribed.

Immediately upon induction of a lysogen or after infection of a sensitive cell, the amount of λ-specific mRNA rises rapidly. Lytic development can be divided into two general phases: the early period represents the time before replication of phage DNA; the late period comprises the time from replication to maturation. Different genes are expressed during these two periods, for the λ messengers present at increasing times after the start of lytic development hybridize with different regions of λ DNA. Lambda genes can be divided into three classes on the basis of such experiments: *immediate early genes* and *delayed early genes* together comprise those transcribed during the early period; *late genes* are those expressed during the late period. The immediate early genes are transcribed by the host RNA polymerase upon induction or infection: these are *N* and *cro*. The product of gene *N* is needed for transcription of the delayed early genes, *cIII, red, xis, int* on the left side of the immunity region, and *cII, O, P, Q* on the right side of the immunity region. The product of gene *Q* is needed for transcription of the late genes. An important consequence of this cascade regulation, in which an immediate early product is needed for delayed early expression, and a delayed early product is needed for late expression, is that the entire phage may be maintained in an inert state by repression of the immediate early functions.

The organization of the early units of transcription that are controlled by repressor is summarized in Figure 4.15. Using separated *l* strands and *r* strands of λ DNA in hybridization reactions shows that early messengers are transcribed from both *l* and *r* but late messengers are transcribed only from the *r* strand. The orientation of transcription from the two strands was deduced by Taylor, Hradecna and Szybalski (1967) by examining the effects of *x* mutations. Mutants in the *x* region, which maps to the left of *cII,* show cis dominant lack of expression of genes *cII, O, P.* The *x* class of mutation is associated with a block in transcription of RNA complementary to the *r* strand. If the *x* mutation is taken to identify a promotor, then transcription of the *r* strand must proceed from left to right; correspondingly, transcription of the *l* strand must proceed in the opposite direction, from right to left. Mutations of the *sex* class show an analogous cis dominant lack of expression of gene *N;* Nijkamp, Bøvre and Szybalski (1970) showed that this is correlated with a block in transcription of the *l* strand in this region. The effects of *x* and *sex* mutations thus are consistent with the idea that they identify two promotors, $p_R$ and $p_L$, respectively: $p_R$ initiates transcription of early mRNA to the right through

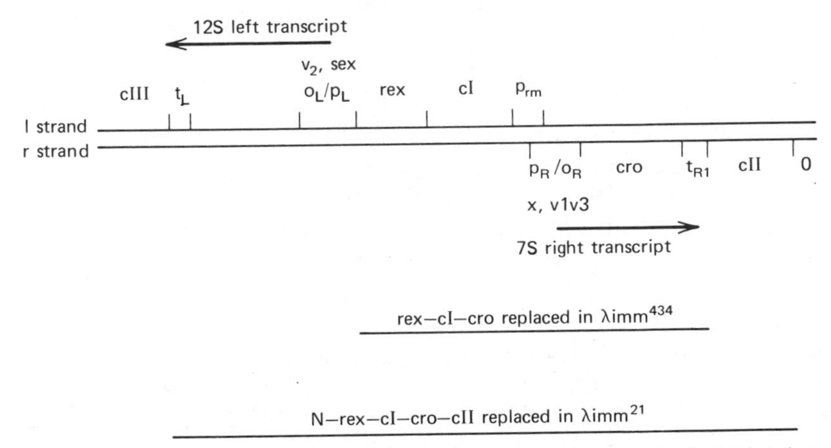


**Figure 4.15:** units of immediate early transcription. The left transcript is initiated at the control site $o_L/p_L$, which is identified by the virulent operator-constitutive mutant *v2* and the promotor mutant *sex1*. The left strand of gene *N* is transcribed to give a 12S messenger. Transcription terminates at a site $t_L$. The right transcript is initiated at the control site $p_R/o_R$, which is identified by the operator mutant *v1v3* and promotor mutants of the *x* type. The right strand of gene *cro* is transcribed into a 7S messenger. Transcription terminates at a site $t_{R1}$. (The delayed early genes lie to the left of *N* and to the right of *cro*; transcription of these genes requires the product of *N*, which allows transcription to continue to the left beyond $t_L$ and to the right beyond $t_{R1}$. This continuation of transcription generates early messengers which include sequences transcribed from both immediate early and delayed early genes.)

*cro, cII, O, P;* $p_L$ initiates transcription of early mRNA to the left through *N*. A third transcriptional unit is provided by the *cI* gene itself; Taylor et al. (1967) showed that the λ mRNA of lysogenic cells hybridizes with the *l* strand so that *cI* must be transcribed from right to left. Another promotor, $p_{rm}$ (for repressor maintenance) must therefore lie to the right of *cI*.

An important advance in following the transcription of lambda was made possible by the development by Roberts (1969b, 1970) of a system for transcribing λ DNA in vitro into products that could be equated with specific genes. Synthesis of RNA on most DNA templates in vitro can be persuaded to start at specific sites but then proceeds for a distance which depends upon the conditions of incubation; random termination thus sees the production of a very dispersed distribution of RNA molecules of various sizes. But termination at specific sites on λ DNA can be accomplished in the presence of a protein factor, *rho,* which Roberts isolated from (uninfected) E.coli cells. In the absence of rho, the RNA transcribed from λ DNA in vitro remains associated with its template; in the presence of rho, the RNA is released and is much smaller in size. Identification of the RNA products was aided by using as template the DNA of λ*b2,* which has a deletion of the region from *J* to *int* and therefore displays a lower background of nonspecific transcription.

When rho is present during transcription in vitro, the principal RNA products are the two species shown in Figure 4.16, one sedimenting at 12S and the other at 7S. Hybridizing these two RNAs with the separated strands of the DNA of λ, λ*imm*$^{434}$, and λ*imm*$^{21}$ identified the regions from which they were transcribed. Because there is no homology between the different immunity regions, the effect of the substitution in λ*imm*$^{434}$ is to delete the region of genes *cI* and *cro* for hybridization purposes; in λ*imm*$^{21}$ the entire region from *N* to *cII* is unavailable for hybridization. The figure demonstrates that 12S RNA can hybridize with the *l* strands of both λ and λ*imm*$^{434}$ but not with that of λ*imm*$^{21}$; this implies that it is synthesized to the left from the region of the *N* gene. The 7S RNA hybridizes with the *r*

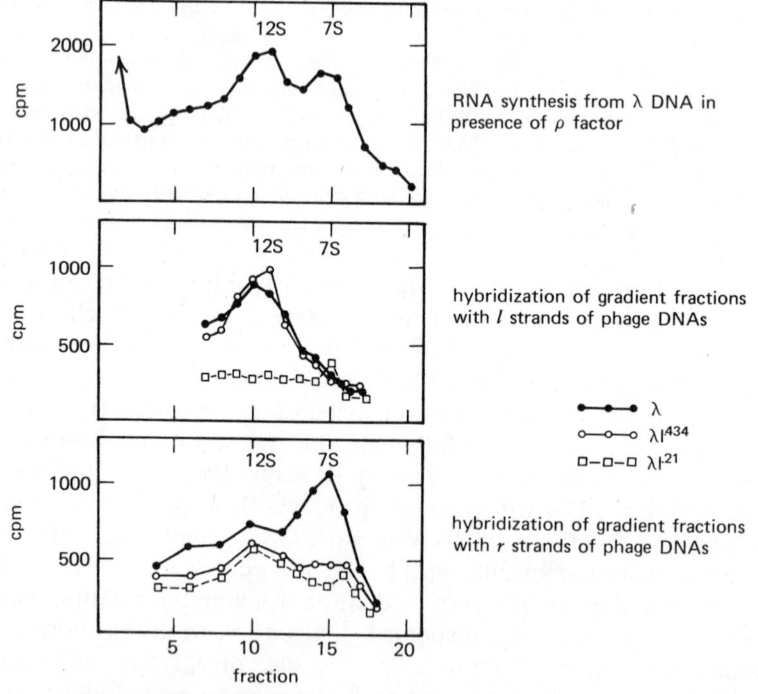


**Figure 4.16:** transcription of immediate early λ mRNA in vitro. The upper panel shows that the two principal species sediment at 12S and 7S on a sucrose gradient. The center panel shows that 12S RNA hybridizes with the *l* strands of λ and λ*imm*$^{434}$ but not with λ*imm*$^{21}$; 7S RNA does not hybridize with *l* strands. The lower panel shows that 7S RNA hybridizes only with the *r* strand of λ and not with λ*imm*$^{434}$ or λ*imm*$^{21}$; 12S RNA does not hybridize with *r* strands. Data of Roberts (1969b).

strand of λ DNA, but not with that of either λ*imm*$^{434}$ or λ*imm*$^{21}$; this implies that it must be synthesized to the right from an area lying entirely within the immunity region.

That this transcription in vitro reflects events that occur in the infected cell is suggested by the effects of the *sex1* and *x13* mutations. As Figure 4.17 shows, in the in vitro system λ*sex* DNA can be transcribed into 7S RNA but fails to yield 12S RNA. This supports the idea that the *sex1* mutation lies in the $p_L$ promotor and prevents the initiation of transcription from it. Thus $p_L$ may be taken as the starting point for leftward transcription of 12S RNA in vitro and in vivo. Figure 4.18 shows that the *x13* mutation has a comparable effect on rightward transcription; for λ*x13* DNA displays normal synthesis of 12S RNA but has a much reduced peak of 7S RNA. Analysis of these peaks shows that 12S RNA is derived from its usual source but that there is no RNA corresponding to the 7S species derived by rightward transcription. Thus the *x13* mutation may be taken to lie in the promotor $p_R$, at which rightward transcription commences.

The $p_L$ and $p_R$ promotors identified by the *sex* and *x* mutations provide the only sites at which transcription of early mRNA can commence. The units of transcription depicted in Figure 4.15 represent the immediate early genes, transcribed by host RNA polymerase upon infection of a sensitive cell or following destruction of repressor protein during induc-

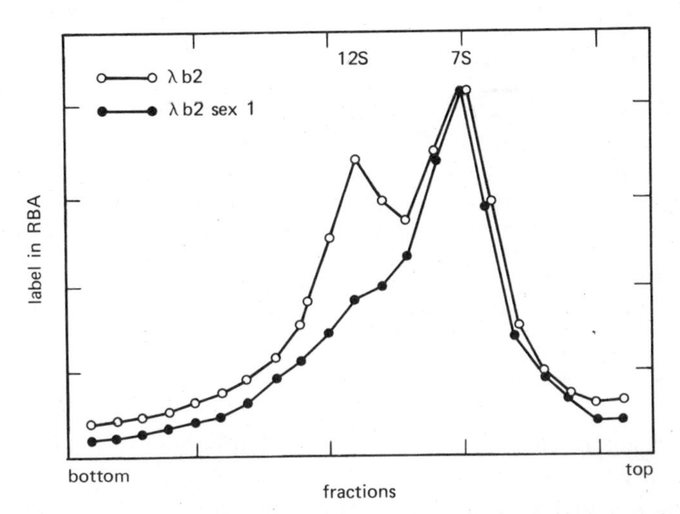


**Figure 4.17:** transcription in vitro of λ*sex* DNA. A template of λ DNA synthesizes two predominant peaks of RNA, sedimenting at 12S and 7S on a sucrose gradient. A template of λ*sex* DNA fails to transcribe the 12S RNA. In order to reduce the background synthesis of RNA, both the λ$^+$ DNA and the λ*sex* DNA were derived from phages carrying the *b2* deletion which removes most of the genome between *J* and *int*. Data of Roberts (1969b).

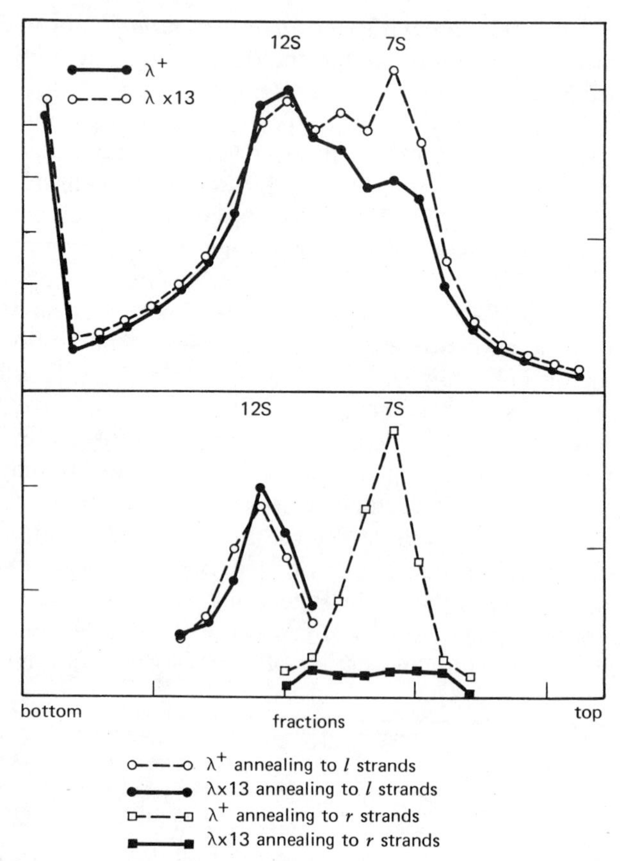


**Figure 4.18:** transcription in vitro of λ*x* DNA. The upper panel shows a sucrose gradient analysis of the RNA transcribed from a $\lambda^+$ DNA control and from the DNA carrying the *x13* mutation; λ*x13* DNA synthesizes much less RNA sedimenting in the 7S part of the gradient. The lower panel shows a hybridization analysis of the 12S and 7S fractions: 12S RNA from either template is able to anneal to *l* strands; but while 7S RNA from the λ template can anneal to *r* strands, the λ*x13* template does not make any such RNA. Data of Roberts (1969b).

tion of a lysogen. Genetic analysis shows that no genes other than *N* and *cro* are expressed upon infection unless an active protein product is made by translation of the *N* message. This suggests that the rho factor can terminate transcription both in vitro and in vivo at sites $t_L$ at the end of the *N* gene and $t_{R1}$ at the end of the *cro* gene; subsequent work has shown that the action of *N* protein is to counteract the rho-dependent termination to allow transcription to continue past the termination sites into the delayed early genes to the left of *N* and to the right of *cro* (see later). The sites $p_L$ and $p_R$ thus are *early promotors* which serve first to initiate the transcription of the 12S and 7S immediate early messengers and then (when termination is prevented) to give rise to messengers which carry delayed early as well as immediate early sequences. Repression of the two early promotors thus prevents expression of all phage functions except *cI*, which is transcribed independently.

### *Repression of Early Transcription*

The repressor protein coded by the *cI* gene was isolated by Ptashne (1967a,b) in experiments which made use of cells carrying a λ*ind*$^-$ prophage; the *ind*$^-$ mutation in *cI* prevents the prophage from being induced by ultraviolet irradiation. It is therefore possible to use ultraviolet irradiation to inhibit protein synthesis of the host cell; and when such a host is superinfected with many wild type phage particles per bacterium, immunity is maintained and all the phage genomes are restricted to synthesizing the *cI* product. In these conditions, synthesis of λ repressor can be increased to the point where it constitutes an appreciable part of total cell protein synthesis. The repressor was then isolated by using $^3$H-leucine to label proteins synthesized in cells superinfected with λ*cI*$^+$ and using $^{14}$C-leucine to label cells superinfected with λ*cI*$^-$ mutants. After mixing extracts from the two types of cell, a component labeled with $^3$H but not with $^{14}$C was identified as a fraction on DEAE-cellulose columns. The protein isolated by this protocol is missing from all cells infected with *cI* nonsense mutants and is produced in modified form by temperature sensitive and missense mutants. Mutations which map at the extreme ends of the *cI* gene modify the same polypeptide chain. The phage *434* repressor was subsequently isolated by Pirrotta and Ptashne (1969) by a similar experimental protocol. The polypeptide specified by the *cI* gene of λ consists of a chain of about 27,000 daltons; the phage *434* repressor monomer is slightly smaller, about 26,000 daltons. Each repressor can bind specifically to the DNA of its phage. The ability of the λ repressor to bind to λ DNA contrasts with its inability to bind to the DNA of λ*imm*$^{434}$, which confirms directly that its site of action lies within the short region that is nonhomologous between the phages. Similarly, the *434* repressor can bind to λ*imm*$^{434}$ DNA but not to λ DNA.

The sites at which the repressor acts were identified by Ptashne and Hopkins (1968) in experiments which made use of *virulent* mutants of lambda. The multiple mutant λ*vir* was originally isolated by Jacob and Wollman (1954) as a variant able to superinfect and grow upon λ lysogens. Its failure to be repressed by the prophage genome of the lysogen suggests that its mutations may prevent it from being recognized by the repressor protein. Ptashne and Hopkins showed that λ*vir* carries three mutations: *v2* is located to the left of *cI;* and *v1* and *v3* are located close together to the right of *cI*. By recombination they were able to isolate phages of the genotypes λ*v2*, λ*v3*, λ*v1v3* (but not λ*v1* alone, which was isolated only later).

The ability of these mutant phages to express functions that normally are repressed in lysogens was determined by mixed infection: λ lysogens were infected with either λ*v2* or λ*v1v3* together with one of the heteroimmune phages $\lambda imm^{434}\ N^-$ or $\lambda imm^{434}\ O^-$. Progeny particles derived from the $\lambda imm^{434}$ infecting phage (detected by virtue of their *434* type immunity) can be produced only if the λ*v* mutant phage can provide the function that is defective in the $\lambda imm^{434}$ infecting phage. The results showed that λ*v2* is able to provide *N* product but not *O* product; while λ*v3* is able to provide *O* product but not *N* product. (λ*v1v3* replicates extensively under these conditions and thus cannot be tested for differential release of functions from repression because the effect of replication is to increase the number of templates available for transcription.) Introduction of the *v2* mutation thus allows λ to express gene *N* constitutively (that is, in spite of the presence of repressor); the *v3* mutation (and, by implication, the double *v1v3* mutation) results in constitutive expression of gene *O*. That virulence results from constitutive transcription was confirmed by the demonstration of Sakakibara and Tomizawa (1971) that λ*v2* lysogens synthesize mRNA from the *l* strand and λ*v1v3* lysogens transcribe the *r* strand. Two types of mutation might have this result: creation of a new promotor free from the influence of repressor; or inactivation of the operator at which repressor binds.

Repressor complexed with λ DNA sediments more rapidly than free repressor, so that centrifugation through a density gradient can be used to follow the binding of repressor to the DNA. This was the technique used to show the specificity of repressor in binding to λ DNA but not to $\lambda imm^{434}$ DNA. When Ptashne and Hopkins mixed the DNA of λ*vir* with the repressor protein, they found that much less binding took place than with $\lambda^+$ DNA; and λ*v2* and λ*v1v3* each bound about half as much repressor as λ DNA. These data are shown in Figure 4.19; they suggest that the single mutation *v2* and the double mutation *v1v3* each separately decreases the affinity of λ DNA for the repressor. The virulent phages therefore behave as operator constitutive mutants; *v2* can be taken to identify the left operator, $o_L$, which controls transcription from $p_L$; and *v1v3* must lie in the right operator, $o_R$, which controls transcription from $p_R$. As shown in Figure 4.15, the left control site can thus be represented as $p_L/o_L$ and the right control site as $p_R/o_R$, where this makes no implica-

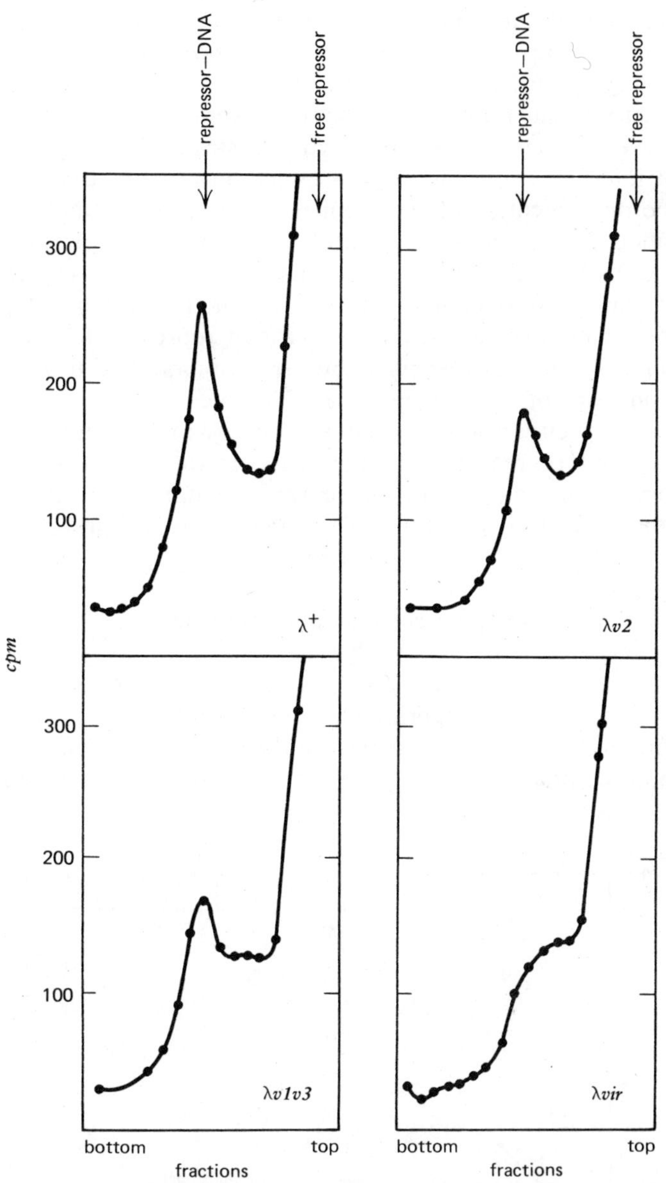


**Figure 4.19:** Repressor binding to DNA of wild type lambda and operator constitutive mutants. Phage DNA and radioactively labeled repressor were mixed and sedimented through a 7.5–30% glycerol gradient; repressor bound to DNA sediments more rapidly than free repressor. With either λ*v1v3* or λ*v2* the amount of repressor sedimenting in the form bound to DNA is reduced by about half; with λ*vir* it is reduced very greatly. Data of Ptashne and Hopkins (1968).

tion about the relationship between the two sites, that is whether they are adjacent or overlap. Both $o_L$ and $o_R$ must be inactivated to create a virulent phage.

Another technique for purifying the repressor protein was developed by Chadwick et al. (1970). Starting with a strain carrying a defective prophage that can excise and overproduce a large amount of repressor due to the multiplication of the number of copies of λ DNA in the cell, they purified the repressor by its ability to bind to nitrocellulose filters carrying λ DNA. (This technique was first developed for isolating the *lac* repressor and is discussed in Volume 1 of Gene Expression.) Figure 4.20 shows the sigmoid binding curve of purified repressor to λ DNA. The same shape of curve is displayed by repressor binding to λ DNA that contains only one of the two operators. This suggests that tight binding of the repressor to either operator requires two or more subunits to bind to the DNA. The λ repressor forms aggregates in solution, including dimers and tetramers; and dilution of the repressor before it is added to DNA much reduces its ability to bind. This suggests that dilution pushes the equilibrium

$$\text{monomer} \rightleftharpoons \text{dimer (oligomer)}$$

to favor monomers, which are unable to bind to the operators. The initial rate of binding in these experiments was second order, which suggests that a dimer may be the active form. On the other hand, Ordal and Kaiser (1973) reported that a twofold dilution decreases the concentration of

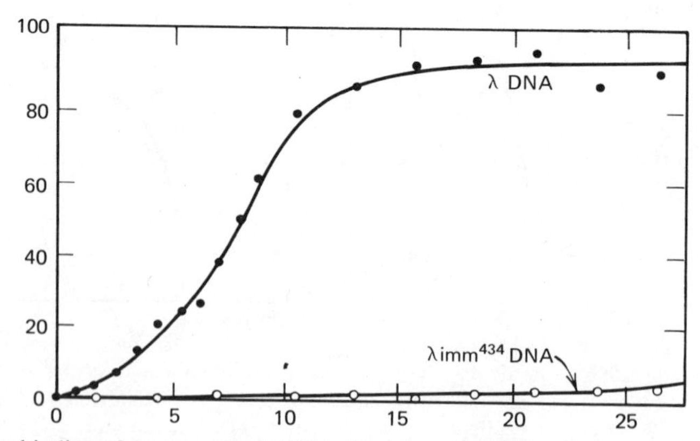


**Figure 4.20:** binding of repressor to λ DNA. The sigmoid binding curve suggests that more than one subunit of repressor is needed for binding. The lambda repressor does not bind to DNA of λ*imm*$^{434}$. Data of Chadwick et al. (1970).

repressor by 2.4 times, an indication that dimers may not be the only form of aggregate involved. The forward rate constant for the repressor-operator interaction is $3 \times 10^{10}$ mol-sec$^{-1}$, which is too fast to be accomplished by recognition resulting from random diffusion (see Ptashne, 1971); the same conclusion has been reached for the repressor-operator interaction in the lactose operon (see Volume 1 of Gene Expression). It is difficult to interpret this observation in terms of molecular mechanisms. The half life of the lambda repressor-operator complex depends on both salt concentration (like the interaction in the lactose operon) and temperature (unlike the lactose system); in 0.05 M KCl at 20°C it is 7 minutes.

Purified repressor protein can inhibit transcription of λ DNA in vitro. Using the rho-dependent transcription developed by Roberts (see Figure 4.16), Steinberg and Ptashne (1971) demonstrated that repressor inhibits the synthesis of 12S RNA and 7S RNA. Figure 4.21 shows a sucrose gradient analysis and hybridization analysis of the products of transcribing λ DNA or λ*imm*$^{434}$ DNA in either the absence or presence of repressor. In the absence of repressor, λ DNA transcribes the RNA species sedimenting at 12S and 7S on a sucrose gradient; and these hybridize with the isolated *l* and *r* strands, respectively. When repressor is added, much less RNA is present in these regions of the sucrose gradient; and hybridization analysis confirms that there has been almost complete repression of transcription of these species. As a control, the figure also shows that lambda repressor has no effect upon synthesis of the 12S and 7S RNA species by the DNA of λ*imm*$^{434}$, another demonstration that the operators at which it acts must lie within thc immunity region.

The DNA of virulent λ strains is less effectively repressed in vitro. With wild type λ DNA, a greater concentration of repressor is needed to cause complete repression of 7S RNA synthesis than is necessary to repress entirely the synthesis of 12S RNA. This suggests that $o_R$ may have a slightly lower affinity for repressor than that of $o_L$. The virulent mutations greatly reduce the affinities of the operators for repressor. At repressor concentrations adequate to repress synthesis of RNA from $\lambda^+$ templates, λ*v2* DNA continues to transcribe 12S RNA, although synthesis of 7S RNA is repressed. To achieve complete repression of 12S transcription from λ*v2* (that is repression to the same low background level displayed by $\lambda^+$ DNA) requires about a tenfold increase in the concentration of repressor. The mutations *v1* and *v3* exert a similar effect upon transcription of 7S RNA: λ*v3* DNA requires more repressor than $\lambda^+$ DNA to inhibit 7S RNA synthesis; and λ*v1v3* DNA continues to transcribe normal amounts of 7S RNA in the presence of a repressor concentration sufficient to inhibit a $\lambda^+$ DNA template completely, with only a large increase in repressor concentration causing cessation of 7S RNA production. The *v3* mutation has

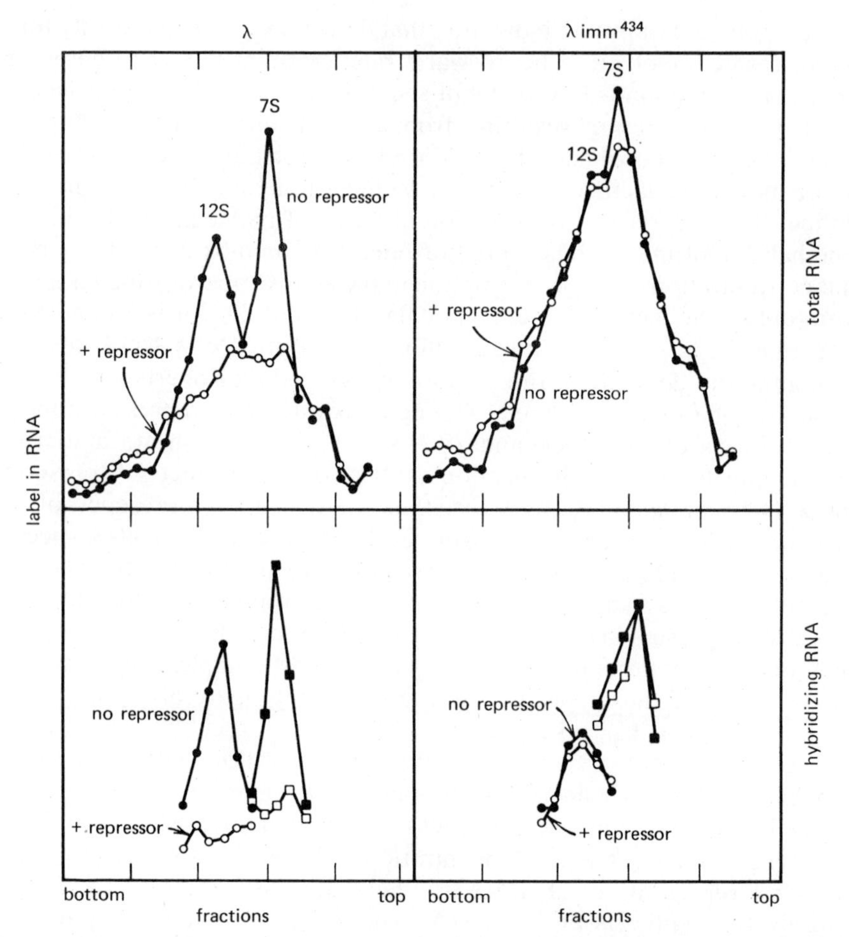


**Figure 4.21:** repression of early lambda transcription. The left panels analyze the RNA transcribed from λDNA. The upper panel shows a sucrose gradient analysis of the RNA transcribed in the absence and presence of repressor; the addition of repressor inhibits production of 12S RNA and 7S RNA. The lower panel shows the ability of RNA in the 12S part of the gradient to hybridize with isolated *l* strands and of RNA in the 7S part of the gradient to hybridize with isolated *r* strands; the addition of repressor almost completely abolishes transcription of 12S *N* mRNA and 7S *cro* mRNA. The right panels show the results obtained when λ*imm*[434] DNA is used as template; lambda repressor has no effect as judged by either the sucrose gradient or hybridization analysis. Data of Steinberg and Ptashne (1971).

only a partial effect on the binding activity of $o_R$ and the double mutation *v1v3* is needed to relieve repression at $o_R$ as effectively as is achieved by the single mutation *v2* at $o_L$. Operators carrying the virulent mutations *v2* (in $o_L$) or *v1v3* (in $o_R$) therefore have severely reduced abilities to bind repressor; this must mean that in a lysogenic cell they are not repressed by the concentration of repressor protein that can be produced by the prophage (for review see Ptashne, 1971).

Virulence requires the introduction of mutations sufficient to prevent repressor from binding to both the left and right operators. Ordal and Kaiser (1971) noted that virulent phages arise at frequencies of $4 \times 10^{-5}$ from λ*v1v3*, at $10^{-6}$ from λ*v2v3*, but at only $10^{-11}$ from λ*v2*. This corresponds to the introduction of a single mutation into λ*v1v3*; since the double mutation *v1v3* inactivates $o_R$, this suggests that a single mutation is sufficient to inactivate the left operator, a conclusion consistent with the properties of λ*v2* itself. But two mutations appear to be necessary to inactivate the right operator. Thus λ*v2v3*, which has an inactive left operator and a partially active right operator, requires only a single, second mutation in $o_R$; whereas λ*v2*, which has only an inactive left operator, requires two mutations in $o_R$ and thus yields virulent phages at a much lower rate.

Although λ*v2v3* carries mutations in both operators, it is not virulent; for it does not form plaques on a λ lysogen and forms turbid plaques on a sensitive strain. Ordal and Kaiser therefore isolated further virulent mutants by plating λ*v2v3* on lysogenic cells and isolating particles from the plaques formed; these plaques represent phages that have gained another mutation in $o_R$. The mutants were divided into two classes on the basis of their ability to plate on cells carrying λ*dv* (this is a variant of lambda containing only some of the genome and which behaves as a self-perpetuating autonomous plasmid present in several copies for cell; see page 373). *Supervirulent* phages were defined by their ability to form plaques in the presence of λ*dv*. *Restricted virulent* phages cannot plate on λ*dv* although they are able to plate on lysogens; the classical virulent phage λ*vir* (λ*v1v2v3*) falls into this category. Actually, it turns out that the basis for this characterization rests with the response to Cro protein rather than repressor protein, so that this terminology provides what is an operational definition rather than one reflecting affinity for repressor protein (see page 374).

The supervirulent and restricted virulent phages differ in the new mutation introduced into the right operator of the λ*v2v3* strain; these types of mutation have been denoted *vs* and *vr*, respectively. One of the supervirulent mutations, *vs326*, has been studied in some detail. The *v3* mutation reduces the affinity of $o_R$ for repressor by 4-fold. The *vs326* mutation by itself reduces the affinity of $o_R$ for repressor by 20-fold. But when the two mutations are carried in cis arrangement, that is as λ*v3vs326* DNA, repressor binding to $o_R$ is reduced by 4000-fold. Ordal (1973) showed that the degree of constitutive expression in vivo of functions transcribed from $p_R$ correlates well with the ability of the mutant DNA genomes to bind repressor in vitro. This demonstrates the reliance that can be placed upon the reaction between operator and repressor in vivo.

The *v3* mutation in $o_R$ maps to the left of two mutations, *x3* and *x7*, in $p_R$. Yet two further mutations, *vs326* and *vr*, map to the right of *x3* and *x7*. And another mutant in $p_R$, *x13*, maps further to the right. We can thus construct an order for mutant sites in $o_R$ and $p_R$:

*v3* – *x3*, *x7* – *vs326*, *vr* – *x13*.

This shows an interpenetration of sites known to affect the operator and promotor and therefore suggests that the sequences recognized by repressor and by RNA polymerase may overlap at least in part.

Two models can be envisaged for the ability of repressor to prevent transcription. The repressor may simply provide a physical block to the progress of RNA polymerase; in this case, promotor and operator should be adjacent in the order *p-o-structural genes*. Or repressor may compete with RNA polymerase for binding to DNA; this could be accomplished if there were some overlap in the sequences recognized by the two proteins. Indirect evidence suggests that there may be competition in the left control site. In principle it is possible to ask whether λ DNA carrying repressor protein can bind RNA polymerase and then suffer transcription; but this experiment is difficult to carry out because RNA polymerase binds to other sites (many of them incorrect) as well as to $p_R$ and $p_L$. The experiment can be performed in reverse, however, to see whether binding of RNA polymerase to the promotor inhibits the binding of repressor to the operator. Chadwick et al. (1970) utilized the *sex1* mutation to compare binding to a normal left control site and to one with a mutant promotor. Either λ or λ*sex1* DNA was incubated with RNA polymerase; labeled repressor was then added and the complex sedimented through a glycerol gradient to measure the amount of repressor bound to DNA. In the absence of RNA polymerase, both λ and λ*sex1* DNA bind the same amount of repressor; but in the presence of RNA polymerase, λ*sex1* DNA binds more repressor than λ DNA. This suggests that reduction in the ability of $p_L$ to bind RNA polymerase increases the ability of $o_L$ to bind repressor. This is consistent with the idea that polymerase and repressor compete for binding to sequences that overlap at the left control site. The same model for the control sites is therefore suggested by biochemical analysis of the left site and genetic analysis of the right site.

### *Organization of the Operators*

Determination of the lambda operator sequences has been made possible by two techniques: restriction enzymes can cleave λ DNA into specific fragments; and DNA containing operator sequences binds specifically to the repressor. Restriction enzymes cleave duplex DNA across both strands at sites comprising a specific nucleotide sequence (usually 6–8 base pairs long). The occurrence of recognition sites in DNA is a matter of chance and different restriction enzymes each therefore cleave a given DNA substrate at a different series of sites. Fragments derived by restriction cleavage of lambda can be mapped by their ability to hybridize with DNA extracted from λ strands carrying defined deletions or substitutions. In this way it is possible to construct a map of the phage genome (or of any other DNA of small size) in terms of the fragments produced by each restriction enzyme. The presence of operator sequences in fragments derived from the immunity region can be determined by the nitrocellulose filter assay; for any DNA fragment containing a repressor binding site is retained on the filter. By measuring the abilities of these fragments to bind repressor, it is therefore possible to determine the locations of the operators on the restriction fragment map. The sequences recognized by repressor can be isolated by binding the operators with the protein and using nucleases to digest all but the length of DNA in the binding site that is protected by repressor. The availability of isolated operators has made it possible to determine their nucleotide sequences; and determination of the corresponding control sequences in phage DNAs carrying mutations in the operators and promotors makes it possible to deduce which sequences serve for repression and for initiating transcription.

Fragments of DNA containing the operators can be isolated by sonicating λ DNA to pieces of 1000 base pairs and retaining the repressor-binding molecules on nitrocellulose. Using this technique, Pirrotta (1973) found that the two operators, which are some 2400 base pairs apart on the genome, are located on different fragments. By utilizing λ DNA bearing a mutation or deletion in either one of the operators, it is therefore possible to isolate selectively the DNA containing the other operator. No repressor binding fragments are found in the DNA of λ*imm*$^{434}$. Recognition of the operator takes place only with duplex DNA; denaturation of the fragments before repressor binding prevents recovery of the operators.

By rebinding the isolated operator-containing DNA fragments with varying amounts of repressor and then digesting all but the protected sequences with DNAase, Maniatis and Ptashne (1973a) demonstrated that the two operators possess complex structures. The length of DNA protected by repressor against nuclease degradation increases with the ratio of repressor to DNA: the smallest protected fragment can be isolated at rather low ratios (0.05 repressor dimers per operator) and the largest fragment is first apparent at ratios of about 10 and becomes predominant by a ratio of 30. As the concentration of repressor dimers is increased from 0.05 to 30, six protected fragments can be isolated, with lengths according to acrylamide gel electrophoresis of 35, 45, 60, 75, 90 and 105 base pairs. Similar results are obtained with preparations derived separately from either one of the two lambda operators, an indication that both possess the same type of structure. (These results contrast with the isolation of fragments of the lactose operator, which comprises a single sequence protected by the repressor through a range of concentrations; for review see Lewin, 1974b.)

By examining the pyrimidine tracts of these fragments, Maniatis and Ptashne showed that they form a hierarchy; the sequence of the smallest fragment is present in all the fragments and further lengths of DNA appear to be sequentially added to it as the sequence protected is increased by addition of more repressor protein. Although both operators show the same type of hierarchical organization, their nucleotide sequences must differ since they generate nonidentical pyrimidine tracts. A greater concentration of repressor protein is required to protect $o_R$ compared with $o_L$, so the right operator must have an affinity for repressor protein lower than that displayed by the left operator, a conclusion consistent with the earlier indirect experiments of Steinberg and Ptashne (1971).

The model which Maniatis and Ptashne suggested to account for these results postulated that at each operator a dimer of repressor must first bind to a primary sequence of some 30 base pairs, denoted S1 at the left operator and S1′ at the right operator. As more repressor protein becomes available, monomers bind one at a time to further, secondary segments of DNA each about 15 nucleotides long, which lie adjacent to the primary sequences; these are denoted S2-S6 at the left site and S2′-S6′ at the right site. Among the fragments protected from nuclease digestion by repressor, those of 35, 60, 90 base pairs are predominant. Maniatis et al. (1975) therefore suggested the revision of this model depicted in Figure 4.22 in which repressor dimers bind to lengths of 30 base pairs. The first dimer protects sites designated as $o_L1$ and $o_R1$, which correspond to S1 and S1′, respectively; the second dimer at each operator protects sites denoted $o_L2$ and $o_R2$, which correspond to the sites previously described as S2-S3 and S2′-S3′; the final dimer binds to sites $o_L3$ and $o_R3$, equivalent

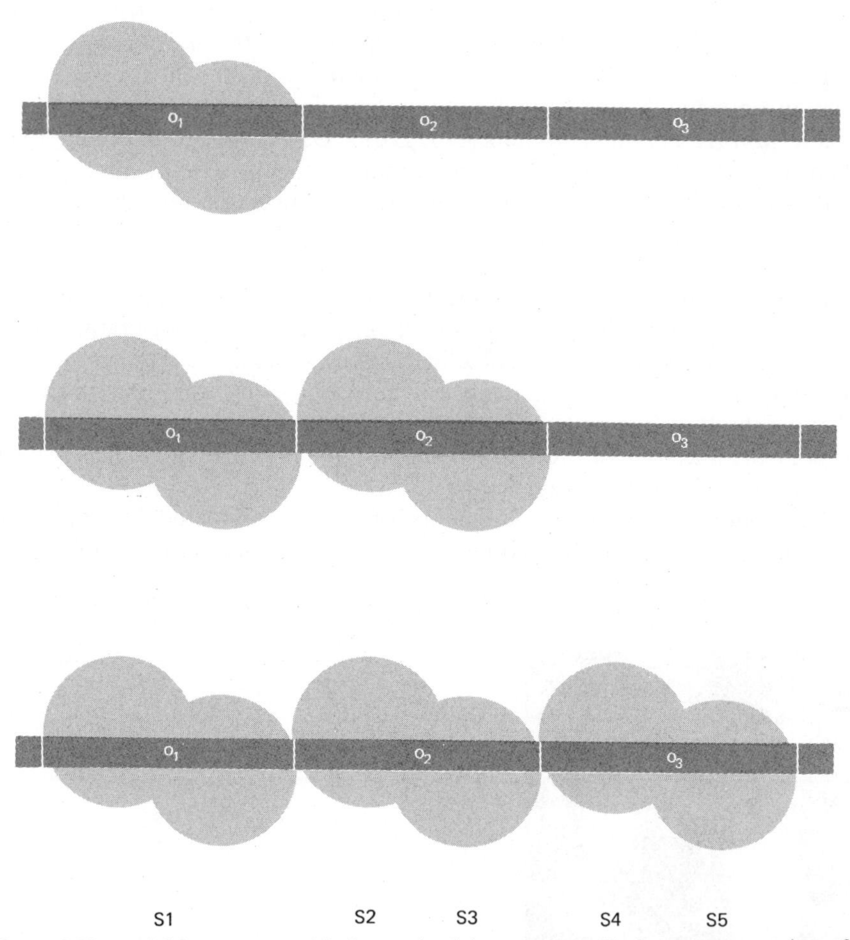


**Figure 4.22:** model for repressor binding to lambda operators. Each operator consists of three sites, each of 30 base pairs; *o*1 corresponds to S1 or S1′ (see text), *o*2 corresponds to S2-S3 or S2′-S3′ (as indicated), *o*3 corresponds to S4-S5 or S4′-S5′. Initially a repressor dimer binds to *o*1, then a second dimer may bind to *o*2, finally a third dimer may bind to *o*3. Occasionally monomers instead of dimers may bind at *o*2 and *o*3, in which case they fill the next half site (one of S2-S5) to become available.

to S4-S5 and S4′-S5′. The isolation of some fragments of 45 and 75 base pairs and, at very high repressor concentration, of 105 base pairs, suggests that occasionally monomers may bind instead of dimers at *o*2 and *o*3.

The fragments produced by cleaving $\lambda$ DNA with restriction endonucleases can be separated on polyacrylamide gels according to size. The enzyme *Hpa II* derived from the bacterium Hemophilus parainfluenzae cleaves more than 50 sites and Allet (1973) mapped these fragments by comparing the restriction patterns of λ DNA with those of phages carrying deletions or substitutions. Three fragments represent the immunity region between *N* and *cro*. The *Hin II* endonuclease derived from the bacterium Hemophilus influenzae cleaves λ DNA into about 35 fragments and several cuts are located in the immunity region. The positions of the cuts introduced by these two enzymes have allowed the repressor binding sites to be characterized.

The presence of operator sequences in these fragments can be tested by mixing a restriction enzyme digest of λ DNA with repressor protein, passing through a nitrocellulose filter, retrieving the DNA bound to the filter, and subjecting it to gel electrophoresis. Figure 4.23 shows the results obtained by Maniatis and Ptashne (1973b) by using this protocol. Two of the Hpa fragments, of 2000 base pairs and 350 base pairs, can bind repressor; and four of the Hin fragments, of 1125, 550, 375 and 320 base pairs, are able to bind repressor.

The operator-containing segments of λ DNA isolated by repressor binding also are cut by the Hin enzyme. Each of the fragments of 35, 60,

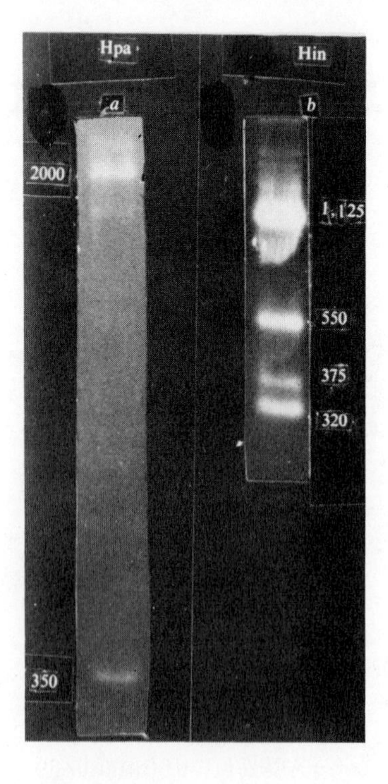


**Figure 4.23:** restriction fragments containing repressor binding sites. $^{32}P$ labeled λ DNA was digested with either the Hpa or Hin nuclease and mixed with repressor; the fragments then able to bind to nitrocellulose filters were recovered and electrophoresed through a 4% acrylamide gel. The figure shows autoradiographs of the gels; the numbers indicate the sizes of the fragments judged from the distance that they have moved. Data of Maniatis and Ptashne (1973b).

75, 90 and 105 base pairs that can be obtained by protecting the left operator against degradation by DNAase suffers a single cut. One of the pieces that is produced from each fragment is about 30 base pairs long and appears to represent the primary binding site $o_L1$. The second piece corresponds to the remainder of the fragment and comprises whatever parts of the secondary sites $o_L2$-$o_L3$ were present. All the secondary repressor binding sites must therefore lie on one side only of the primary site. Similar results were obtained with fragments derived from the right operator.

Since the Hin enzyme splits the primary ($o_L1$ and $o_R1$) from the secondary ($o_L2$-3 and $o_R2$-3) sites in each operator, the four repressor binding fragments produced by its cleavage of λ DNA (Hin 320, 1125, 375, 550) must include $o_L1$, $o_L2$-3, $o_R1$, and $o_R2$-3. To order the primary and secondary sites with respect to the lambda map, Maniatis et al. (1973) subjected these four fragments to nuclease digestion in the presence of excess repressor. The Hin 320 and Hin 550 fragments, which are known to include sequences of the structural genes *N* and *cro*, respectively, each generated a duplex of 30 base pairs; these correspond to $o_L1$ and $o_R1$. The Hin 1125 and Hin 375 fragments each gave rise to much longer protected sequences, comprising $o_L2$-3 and $o_R2$-3. The primary binding sites $o_L1$ and $o_R1$ must therefore be adjacent to the structural genes *N* and *cro*, respectively, with the secondary binding sites $o_L2$-3 and $o_R2$-3 located on the side opposite to the structural genes. This suggests the map shown in Figure 4.24. The binding affinities of the sites for repressor vary in the order:

Hin 320 ($o_L1$) > Hin 550 ($o_R1$) > Hin 1125 ($o_L2$-3) > Hin 375 ($o_R2$-3)

This agrees with demonstrations that the left operator has a greater affinity for repressor than the right and shows that the affinity of each primary binding site is greater than the adjacent secondary binding sites. For half maximal binding, $o_L2$-3 require 3–5-fold more repressor than $o_L1$; and $o_R2$-3 require 10–15-fold more repressor than $o_R1$.

From the map shown in Figure 4.24, it is clear that the immunity region is much larger than can be accounted for by the coding sequence of the *cI* gene. The distance between $o_L$ and $o_R$ is about 2400 base pairs, compared with the 750 base pairs that would be needed to code for the repressor subunit of 27,000 daltons. Certainly this leaves room for another gene; and, of course, there is some evidence that the *rex* function may be a separate gene adjacent to *cI*. And it is clear also that the *cro* gene is rather small, about 280 base pairs, with a coding potential of about 10,000 daltons of protein. Consistent with this, a protein that Folkmanis et al. (1976) have identified as the *cro* product has a subunit weight of about 9000 daltons (see page 370).

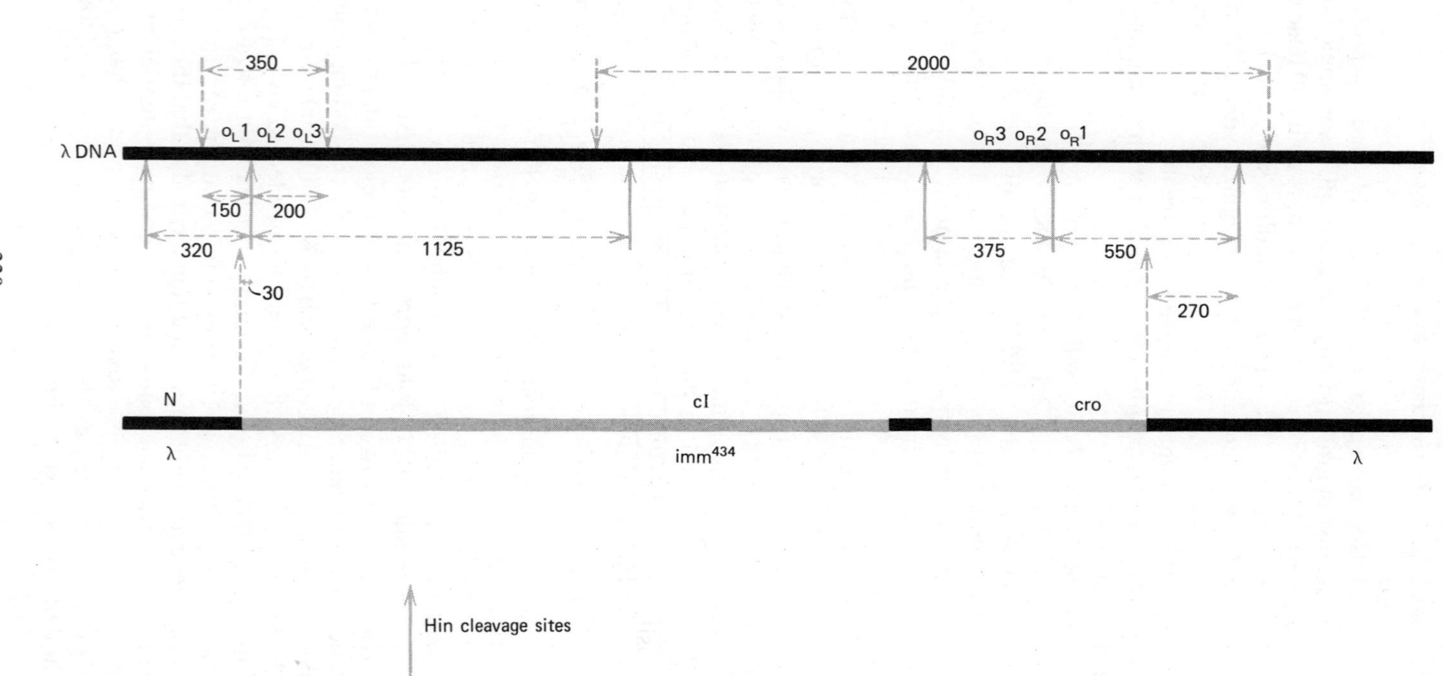

Hpa cleavage sites
350
2000
$o_L1$ $o_L2$ $o_L3$
$o_R3$ $o_R2$ $o_R1$
λ DNA
150
200
320
1125
375
550
~30
270
N
cI
cro
λ
$imm^{434}$
λ
Hin cleavage sites

## *Location of the Early Promotors*

The conclusion that the left promotor overlaps the operator follows from observations of the ability of RNA polymerase to bind to restriction fragments containing operator sequences and from the location on these fragments of *sex* mutations. Sequencing of the control site region has confirmed this idea.

The *sex1* mutation, the first to identify the left promotor, lies within the operator sequences bound by the repressor. Maniatis et al. (1973) and Maurer, Maniatis and Ptashne (1974) showed that the Hpa 350 fragment containing the left operator is able to bind RNA polymerase in vitro when restriction fragments are mixed with the enzyme, passed through nitrocellulose filters, and then recovered and assayed. But the Hpa 350 fragment derived from λ*sex1* DNA displays a 3–4-fold reduction in affinity for the enzyme. The Hpa 350 fragment derived from a wild type revertant of the *sex1* mutation binds polymerase as efficiently as does wild type λ DNA. This suggests that the *sex1* mutation lies in the binding site for polymerase.

Hpa 350 fragments are cleaved by the Hin enzyme into fragments of 150 and 200 base pairs; the Hpa/Hin 150 fragment contains $o_L1$ and the Hpa/Hin 200 fragment contains $o_L2$-3 (see Figure 4.24). But the Hpa 350 fragments derived from λ*sex1* DNA are no longer cleaved by the Hin enzyme. And only one change is seen when Hin digests of complete λ DNA and λ*sex1* DNA are compared: Figure 4.25 shows that the fragments of 325 and 1125 base pairs are replaced by a single fragment of 1450 base pairs. The *sex1* mutation must therefore lie within the nucleotide sequence recognized by the Hin endonuclease and has the effect of abolishing enzyme recognition. Allet and Solem (1974) reported that the mutation *sex3* has the same effect as *sex1*; and the mutation *x3* in $p_R$ prevents Hin cleavage in the right operator between $o_R1$ and $o_R2$. On the other hand, the mutation *x13* does not prevent Hin enzyme from cleaving

---

**Figure 4.24:** sites of action of Hpa and Hin nucleases in the lambda immunity region. The four cuts made by the Hpa enzyme (shown above the λ DNA) release a fragment of 350 base pairs containing $o_L$ and a fragment of 2000 base pairs containing $o_R$. The six cuts made by the Hin enzyme (shown below the λ DNA) generate four fragments which can bind repressor, of 320 base pairs containing $o_L1$, of 1125 base pairs containing $o_L2$-3, of 375 base pairs containing $o_R2$-3, and of 550 base pairs containing $o_R1$. The joint action of Hpa and Hin further cuts the Hpa 350 fragment to generate Hpa/Hin fragments of 150 base pairs containing $o_L1$ and of 200 base pairs containing $o_L2$-3. The *sex1* and *sex3* mutations prevent the Hin enzyme from cutting between $o_L1$ and $o_L2$-3; the *x3* mutation stops Hin from cutting between $o_R1$ and $o_R2$-3. Below the lambda genome is shown the structure of λ*imm*$^{434}$, in which shaded sequences mark the extent of *434* substitution in the black lambda sequences; the positions of the λ genes are indicated on this map. Data of Maniatis and Ptashne (1973b) and Maniatis et al. (1973).

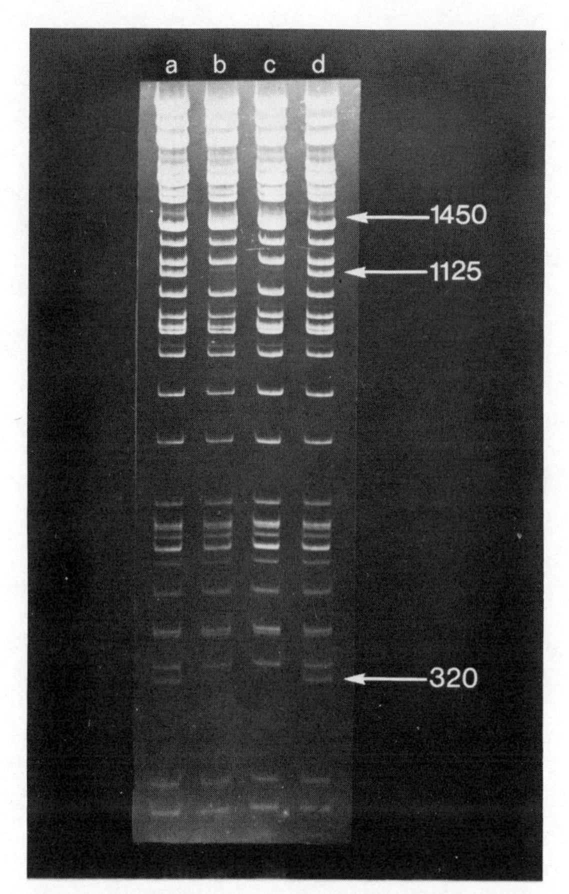


**Figure 4.25:** Hin digestion patterns of DNA from λ, λ*sex1*, and two λ*sex*$^+$ revertants of λ*sex1*. λ DNA (gel *a*) has two fragments of 1125 and 320 base pairs which are replaced by a single fragment of 1450 base pairs in the λ*sex1* digest (gel *b*); one of the revertants (gel *c*) retains the 1450 base pair fragment and the other (gel *d*) has restored the recognition site for Hin. Data of Maurer, Maniatis and Ptashne (1974).

the right control site, confirming that the sequence affecting polymerase binding and initiation is somewhat longer than the enzyme recognition site. Since the Hin site occupies the same position in each operator and in each may be altered by promotor mutations, the relationship of the promotor to the operator presumably must be the same in both the left and right control sites.

That Hin recognition sites lie within the polymerase binding sites is suggested directly by the observation that addition of low amounts of

RNA polymerase protects λ DNA against the cuts usually introduced by the Hin enzyme in the left and right operator sequences; no other Hin cleavage sites are protected. Figure 4.26 shows the results of an experiment in which Maurer et al. (1974) increased the concentration of RNA polymerase until the fragments of 1125 and 320 base pairs were replaced by a fragment of 1450 base pairs (precisely the same change caused by introduction of the *sex1* mutation shown in Figure 4.25); and the 550 and 375 base pair fragments were replaced by a single fragment of 920 base pairs. (At higher concentrations of polymerase, two further, unidentified sites also are protected.) The experiments shown in Figures 4.23, 4.25 and 4.26 demonstrate that a single nucleotide sequence is part of the sites recognized by the Hin enzyme, RNA polymerase, and repressor protein. The presence of Hin recognition sequences in promotors may be a common phenomenon, since Allet et al. (1974) reported that in λ the promotors $p_L$, $p_R$, $p'_R$ (the last is a late promotor) all are protected by polymerase against Hin cleavage, as are promotors for E.coli RNA polymerase found in the DNA templates of SV40 and adenovirus.

Where is transcription initiated? By using the DNA of λ*b2* as a template for RNA polymerase in vitro—the *b2* deletion removes a nonessential part of the λ genome and reduces background transcription—Blattner and Dahlberg (1972) developed a system in which RNA chains start and grow

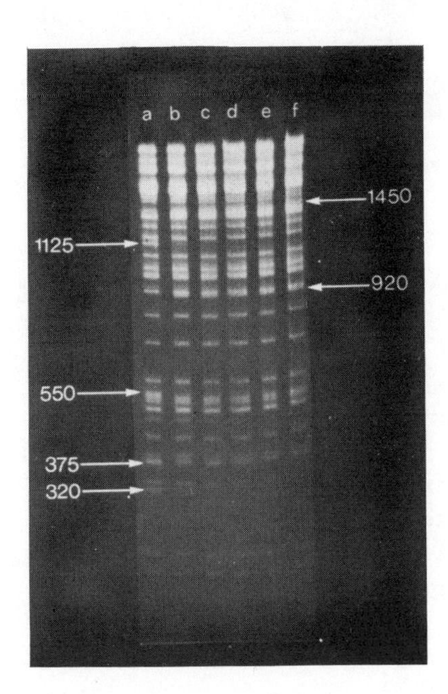


**Figure 4.26:** protection of Hin cleavage sites in $o_L$ and $o_R$ by RNA polymerase. $^{32}P$ labeled λ DNA was digested with Hin enzyme in the presence of concentrations of RNA polymerase from 0–2.5 μg/μg λ DNA. Increasing concentrations of polymerase (from gels *a* to *f*) are able to protect two Hin sites against cleavage, so that the fragments of 1125 and 320 base pairs are replaced by one of 1450 base pairs and the fragments of 550 and 375 base pairs are replaced by one of 920 base pairs. Data of Maurer, Maniatis and Ptashne (1974).

to about 175 nucleotides in 0.4 minutes at 25°C. Four different RNA species are produced by this reaction. By testing the ability of these molecules to hybridize with the DNA of various deletion strains, each RNA has been located on the λ genome. More than 85% of the RNA originates from the promotors $p_L$ and $p_R$; the remaining RNA synthesis takes place from two other promotors, located to the right of the immunity region. The messenger transcribed from $p_L$ has been isolated and partially sequenced.

The leftward message (or more precisely, the sequence of the first 175 nucleotides of 12S *N* mRNA synthesized in this system) is transcribed from a startpoint in λ DNA very close to the boundary between λ and *434* DNA in the hybrid phage $\lambda imm^{434}$. As Figure 4.27 shows, the two strains of λ*bioN2* and $\lambda imm^{434}$ share a short overlapping region of λ homology lying just to the left of $p_L$. The length of the homologous region can be estimated by measuring the length of double stranded DNA formed when the genomes are annealed with each other and visualized by heteroduplex mapping; it is about 215 base pairs. The 175 nucleotides of the $p_L$ promoted mRNA hybridize equally efficiently with the isolated *l* strands of both λ*bioN2* and $\lambda imm^{434}$. This demonstrates that the entire sequence of 175 bases must lie within the 215 nucleotide distance of overlap, that is just to the left of the λ/*434* boundary in $\lambda imm^{434}$. Confirmation that the $p_L$ promoted RNA is initiated on the λ side of the boundary is provided by the observation that the $p_L$ RNA transcribed from $\lambda imm^{434}$ DNA has the same sequence as that synthesized under direction of λ DNA itself; and the 5′ end of the $p_L$ RNA remains protected from degradation with ribonuclease when it is hybridized with the DNA of $\lambda imm^{434}$.

Using the λ*bio30* deletion strain, which has a shorter overlap with $\lambda imm^{434}$ of only 120 ± 20 nucleotides, reduces the extent of hybridization of the $p_L$ promoted RNA to only 60% of its previous value. In this strain, the biotin deletion extends further towards the right and therefore cuts into the RNA molecule from its 3′ end. After hybridizing the $p_L$ RNA with the *l* strands of λ*bio30*, treatment with ribonuclease degrades all unpaired segments of RNA. The hybridized part of the RNA remains resistant to this treatment and can be eluted from its DNA partner. Characterizing the fingerprints of the protected RNA showed that they represent the 120 nucleotides of the 5′ terminal end of the molecule. As this is very close to the length of the overlap of the λ*bio30* DNA with $\lambda imm^{434}$, the startpoint for transcription must be 0 ± 20 bases from the λ/*434* boundary in $\lambda imm^{434}$.

Since the *sex* mutations identify the location of $p_L$ within the immunity region of lambda, that is to the right of the λ/*434* boundary, the conclusion that transcription starts to the left of the boundary implies that the

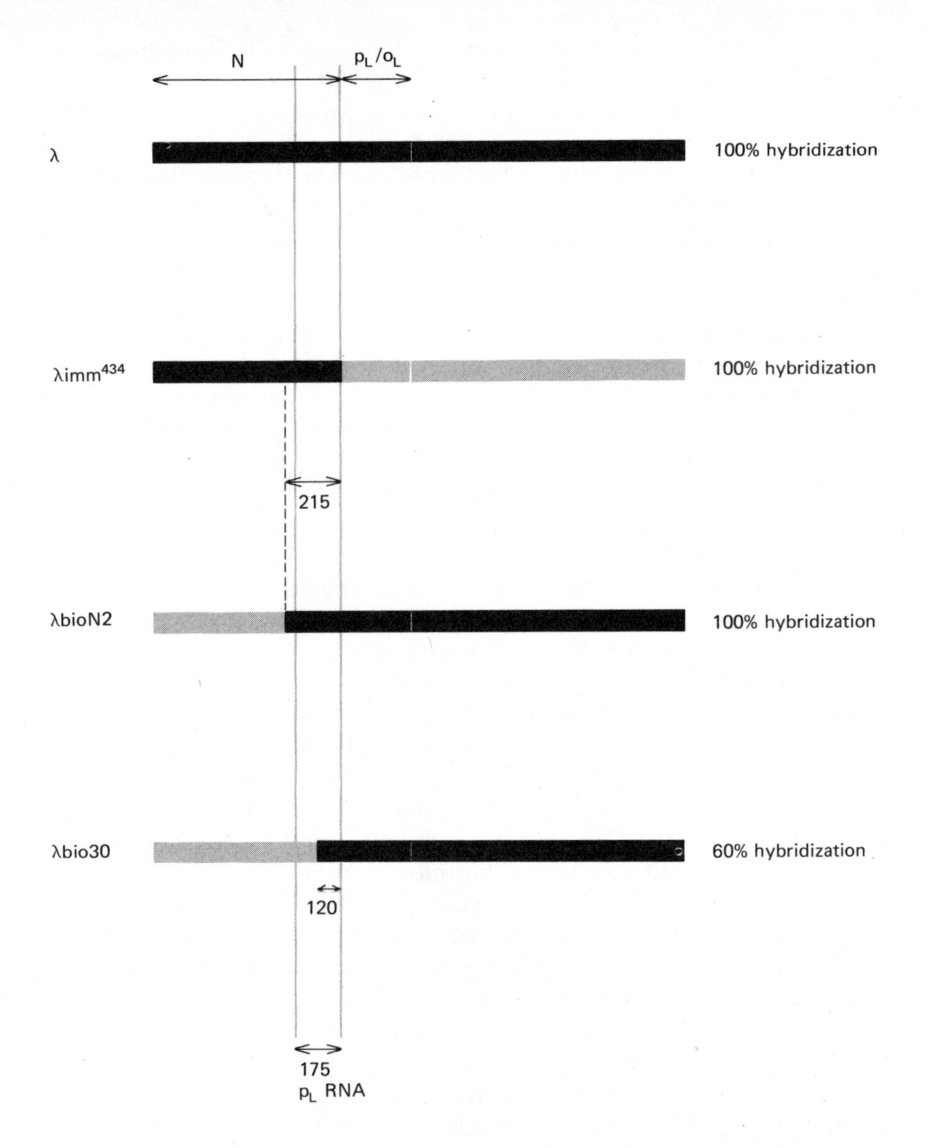


**Figure 4.27:** location of the $p_L$ promoted *N* mRNA. Black indicates regions present in λ DNA: shaded regions are deleted or substituted in variant strains. The 175 nucleotide sequence of $p_L$ RNA sythesized in vitro hybridizes equally well with λ DNA (defined as 100%), λ*imm*$^{434}$ DNA, and λ*bioN2* DNA. The only region common to all three phages is the 215 nucleotide sequence indicated immediately to the left of the λ/*434* boundary in λ*imm*$^{434}$. With λ*bio30* the deletion of λ material extends further through N, leaving only 120 nucleotides to the left of the site of the λ/*434* boundary; the reduction in hybridization to 60% between $p_L$ RNA and λ*bio30* suggests that this 120 nucleotide sequence represents the 5′ end of the N message. Transcription must therefore start within 20 nucleotides to the left of the λ/*imm*$^{434}$ boundary (since the heteroduplex mapping affords a resolution of these distances only within 20 nucleotides). Data of Blattner and Dahlberg (1972).

sequences at which polymerase binds and at which it initiates transcription may not be identical. We may therefore ask what distance separates the binding and initiation sites. The first attempt to measure this distance was that of Blattner et al. (1972); the principle of their approach was to measure the frequency of genetic recombination between the *sex1* point mutation and the ends of two deletions, λ*bioN2* (see Figure 4.27) and λ*bio3h*, which end at different points. The ratio of the frequencies of genetic recombination between *sex1* and the two deletions gives the ratio of the distances between *sex1* and the ends of the two deletions; this can be described as $a/b$. Heteroduplex mapping of the two strains reveals the distance between the ends of their deletions; this can be described as $a-b$. From these two values it is possible to deduce $a$ and $b$, that is the distance of *sex1* from the deletion ends; since the start of the $p_L$ message can be equated with the λ/*434* boundary in λ$imm^{434}$, which is a known distance from the end of the deletion in λ*bioN2* (see Figure 4.27), it is possible to calculate the distance between the start of the message and the *sex1* mutation. This appeared to be almost 200 bases. However, subsequent experiments showed that the deletion in λ*bioh3* had been assigned an incorrect end point; and by using other deletion strains, Blattner et al. (1974) recalculated the distance as 24–62 base pairs.

The ability of restriction fragments to hybridize with the DNA of λ$imm^{434}$ can be used to measure the distance of the left operator sites from the λ/*434* boundary. Maniatis, Ptashne and Maurer (1973) measured the homology between the Hpa 350 fragment and λ$imm^{434}$ DNA by hybridizing *r* strands of Hpa 350 with *l* strands of λ$imm^{434}$ DNA. The DNA sequences which remained single stranded because of their inability to hybridize were degraded with a nuclease; the remaining duplex molecule electrophoresed on polyacrylamide gels as a band of apparent length 120 ± 20 base pairs. The homology between the Hpa 350 fragment (derived from λ DNA) and the λ$imm^{434}$ DNA should lie in material immediately to the left of the λ/*434* boundary; the boundary may therefore be located as 120 nucleotides to the right of the left end of the Hpa 350 fragment, as shown in Figure 4.24. The Hpa/Hin 150 fragment derived from Hpa 350 consists of two parts: 30 bases which represent $o_L1$; and 120 bases which represent material to the left of $o_L$. This 120 base sequence must therefore be the part of Hpa 350 lying immediately to the left of the λ/*434* boundary. Since transcription of the *N* message starts at or within 20 nucleotides of the λ/*434* boundary, and we can now see that the left operator occupies a position immediately to the right of the boundary, the distance from the start of transcription to the *sex1* mutation (that is the Hin cutting site) must be the 30 nucleotides that comprise $o_L1$.

*Nucleotide Sequences of the Operators*

The nucleotide sequences of both the left and right control regions have been determined; and from them it is possible to deduce the sites that are recognized by repressor and by polymerase. The sequence of $o_L1$ was determined by Maniatis et al. (1974) in an experiment in which a single stranded polynucleotide chain including this region was synthesized in vitro. The substrate for in vitro synthesis was prepared by denaturing Hin 1125 fragments and annealing them with purified *l* strands of λ DNA. This produces a molecule which is double stranded in the region corresponding to Hin 1125 (see Figure 4.24) but which has long single (*l*) strand ends protruding from the duplex. DNA polymerase I of E.coli can be used to extend the Hin 1125 *r* strand from its 3′ terminus into $o_L1$ and *N*, using the protruding λ *l* strand as template. The newly synthesized chain can be released from the Hin 1125 primer fragment by cleavage with Hin enzyme followed by denaturation. The chain extension reaction was performed under conditions in which three deoxynucleoside triphosphates and one ribonucleotide triphosphate are provided for DNA polymerase. One of the deoxynucleotide triphosphates carries a $^{32}P$ label in the $\alpha$ position; and the ribonucleotide triphosphate is either rGTP or rCTP, so that the product can be cleaved into oligonucleotides by ribonucleases T1 or A, respectively. Then its sequence can be determined by conventional means. A sequence of 59 bases was identified. The first three bases, G A C, correspond to half of a site recognized by the Hin enzyme, which is the palindromic sequence:

5′ G T C G A C 3′
3′ C A G C T G 5′

The last 26 bases correspond to the initial sequence of the *N* gene messenger determined in the experiments of Blattner and Dahlberg (1972). The first 33 bases therefore include the high affinity repressor binding site of $o_L1$.

A sequence of 74 bases of $o_R$ was determined by Maniatis, Jeffrey and Kleid (1975) in experiments which took advantage of another restriction enzyme, Hph, isolated from Hemophilus parahemolyticus. By exposing Hin 375 and Hin 550 fragments to a series of restriction enzymes, it is possible to identify any enzymes which introduce cuts in these regions: the Hph restriction endonuclease proves to introduce cuts 45 bases to the left and 29 bases to the right of the Hin cleavage site in $o_R$. Combined use of the two enzymes (in either order) thus generates Hin/Hph 45 and Hin/Hph 29 fragments. These fragments were labeled by the application of three techniques. A general label can be obtained simply by using λ DNA that has been labeled with $^{32}P$ before it is cleaved with the restriction enzymes. The 5′ termini of the molecules can be specifically labeled by enzymic phosphorylation with labeled ATP in vitro; the order of enzyme cleavage relative to the time of labeling can be used to control which 5′ ends gain the label. And the nick translation activity of DNA polymerase I (which relies upon its ability to bind to single strand breaks in DNA) can be used to introduce labeled nucleotides into the fragments. The strands of the fragments were then separated and sequenced by pyrimidine tract and nearest neighbor analysis. The 29 base fragment was analyzed directly by this means and the 45 base fragment after it had been split into several pieces with an endonuclease. The total sequence thus represents 74 bases of the right operator, from $o_R1$ (essentially the 29 base pair fragment) through part of $o_R3$.

Another approach was used by Pirrotta (1975) to sequence the right operator. This protocol starts with the 110 base pair fragments obtained by using pancreatic DNAase to digest λ DNA carrying only the right operator, in the presence of excess repressor. These fragments can then be transcribed into an RNA product, using labeled nucleoside triphosphate precursors, which can be sequenced. At low concentrations of nucleoside triphosphates, RNA polymerase cannot initiate transcription, but can elongate a chain from a primer. By providing a suitable primer, transcription can therefore be compelled to start only at specific site(s). Transcription primed by the triplet ApApA started almost entirely at only one site; and the polymerase then terminated at (not entirely) random sites, generating products that increase in size and are useful for sequencing. From these products, a sequence of 77 nucleotides could be determined. This sequence was orientated with respect to the operator in two ways: smaller operator fragments, obtained by redigestion of the full length operator in the presence of limiting amounts of repressor, gave rise to smaller transcription products representing part of the 77 base sequence; and treatment of the full operator length with Hin enzyme also gives shorter fragments whose transcripts represent defined parts of the operator. The sequence determined in these experiments comprises the 77 bases immediately adjacent to the start of the messenger transcribed from the *cro* gene.

The sequences for $o_R$ obtained in both series of experiments are completely in agreement; and the $o_R$ sequence shows many similarities with the sequence of $o_L$. The total sequences now reported in detail represent 77 bases of $o_R$ and 61 bases of $o_L$; a sequence of 28 bases beyond the 33 previously determined in $o_L$ was obtained by application of the same methods used by Maniatis et al. (1975) to sequence $o_R$, although assignment of the last 14 bases remains tentative. Further sequences have been summarized by Ptashne et al. (1976). In Figures 4.28 and 4.29, the two operator sequences are aligned to demonstrate their similarities, so that the $o_L$-*N* sequence is reversed from its usual orientation on the lambda map.

Extensive symmetries are seen in these sequences and it is not immediately obvious which nucleotides constitute the sites recognized by the repressor. Figure 4.28 illustrates the presence of three axes of symmetry in the first sequence that was obtained, that of $o_L1$. Symmetries are displayed about axes through the G-C base pair at position 17, between the base pairs of positions 18-19, and about the base pair at position 20. In all cases the symmetry is partial rather than complete, although it is extended if considered in terms of pyrimidine bases versus purine bases at the positions not included in the symmetry shown (and this may be important if at certain base positions the repressor recognizes whether a pyrimidine or purine is present rather than identifying one base out of the four). As the figure demonstrates, the symmetry means that the same sequence is obtained (at the symmetrical positions) by reading either the top strand or the bottom strand in the same direction (for example, 5′ to 3′). In formal terms, there is said to be twofold rotational symmetry about the relevant axis; but symmetrical sequences of this nature, for example those that constitute the sites recognized by restriction enzymes, often simply are said to be palindromic.

The sequences of $o_L$ and $o_R$ are summarized in Figure 4.29. Two criteria can be used to deduce the sequences recognized by repressor: virulent mutations, which reduce the affinity of the operators for repressor, must be located in recognition sites; and there should be similarities between the binding sites in the two operators and, indeed, between the multiple binding sites in each operator. The positions of six virulent mutations and the base substitutions involved were summarized by Maniatis et al. (1975) and are shown in the figure. By comparing the sequences in $o_R1$-3 and $o_L1$-2, they proposed the model shown in the figure for repressor binding. Each recognition site comprises a sequence of 17 base pairs, displaying twofold rotational symmetry about an axis through the ninth (central) base pair. The sites of all virulent mutations fall into these proposed recognition regions. The symmetry in each binding site is only partial and it corresponds to the first axis of symmetry drawn in Figure 4.28; the two other symmetries are not associated with all the recognition sites and their function (if any) is not known.

The recognition sites are separated by varying numbers of bases, 7 between $o_R1$ and $o_R2$, 6 between $o_R2$ and $o_R3$, 7 between $o_L1$ and $o_L2$, 3 between $o_L2$ and $o_L3$ (although this last is more tentative). Repressor

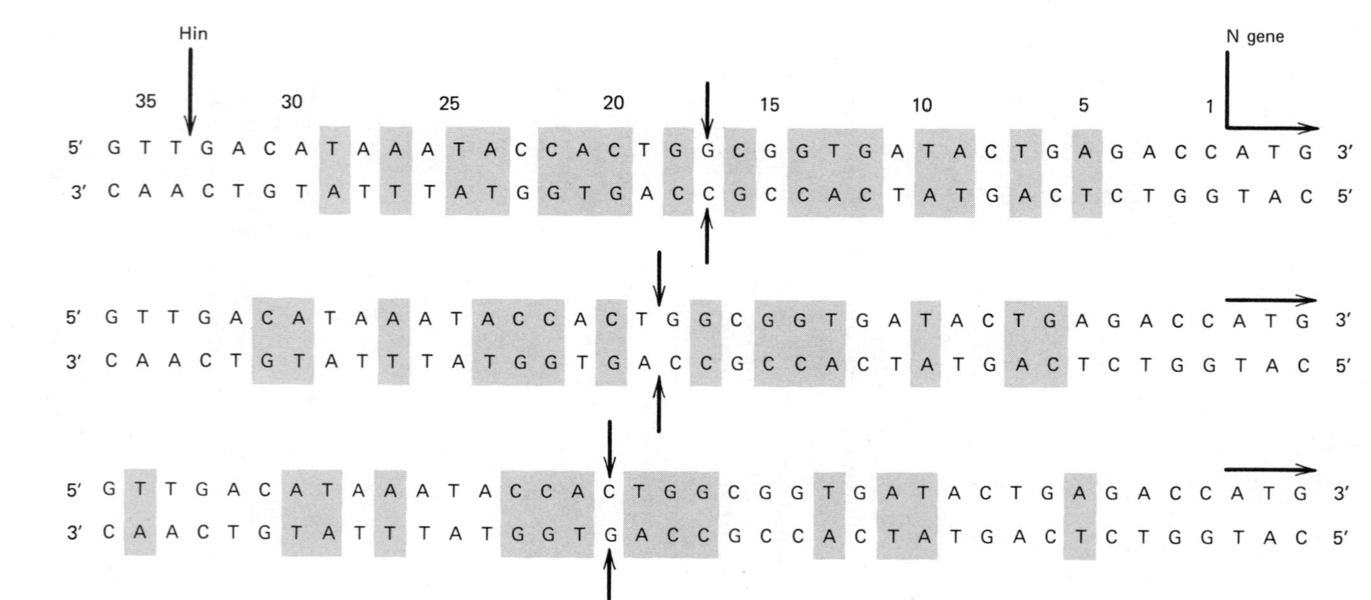


**Figure 4.28:** symmetries in $o_L$ 1. Three axes of partial symmetry can be identified in the region between the Hin cleavage site and the N gene, through position 17, between positions 18 and 19, and through position 20. The axes are identified by the arrows; the base pairs that are symmetrical about them are shaded. Data of Maniatis et al. (1974).

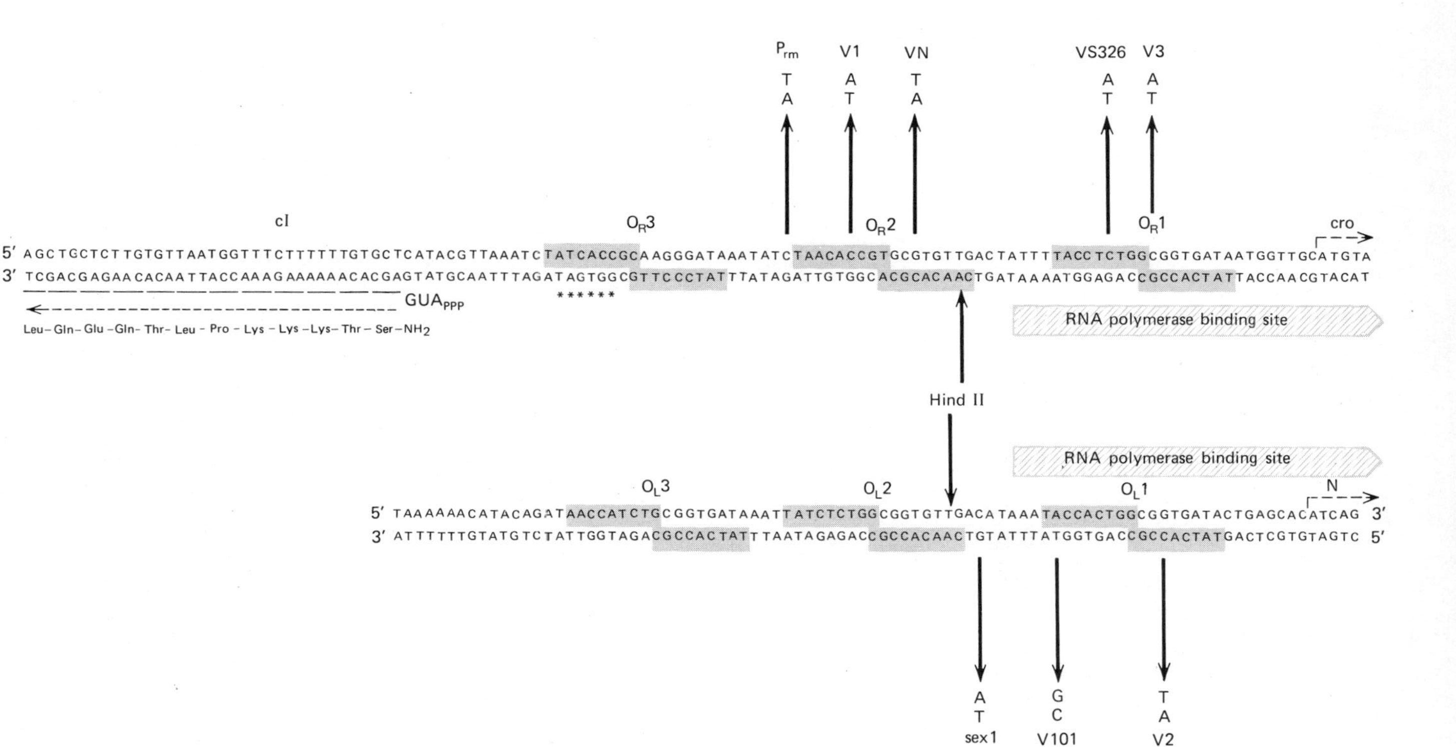


**Figure 4.29:** sequences of $o_R$ and $o_L$. The sequences of $o_R$ and $o_L$ are aligned to show their similarities. (The direction of $o_L$ is reversed from its usual presentation on the $\lambda$ map). The positions and base substitutions of some control mutations are shown. The pairs of overlapping shaded sequences indicate proposed recognition sites for repressor. Each is 17 base pairs long with an axis of (partial) twofold rotational symmetry through the central base pair; this is the same as the first axis depicted in Figure 4.28. Data of Maniatis et al. (1975) and further results summarized by Ptashne et al. (1976).

bound to DNA appears to protect a sequence longer than the recognition site itself and Maniatis et al. proposed that the 35, 60, and 90 base pair protected fragments, respectively, contain 1, 2, and 3 recognition sites. This model proposes that dimers of repressor bind to each site, perhaps in the symmetrical manner depicted symbolically in Figure 4.22. The less frequent addition of monomers to half sites could explain the protection of fragments of 45, 75 and 105 base pairs. When more than one dimer of repressor is bound, the entire length of DNA between the structural gene (*N* or *cro*) and the last repressor subunit is protected from attack.

Twofold rotational symmetry is present in the operator of the *lac* operon as well as in the lambda operators: the *lac* operator consists of a 35 base pair sequence with extensive symmetry about an axis through the central base pair (see Dickson et al., 1975). This single symmetrical sequence represents the entire *lac* operator; its relationship with the promotor is similar to, but not identical with, that seen in lambda: the promotor is adjacent, overlapping the operator for a distance of 8 base pairs, and transcription starts within the operator. Repression of the lactose operon and of lambda may therefore be exercised by a similar mechanism.

The palindromic nature of the nucleotide sequences recognized by both the lactose and the lambda repressor hints that symmetry may provide a general way for proteins to recognize specific DNA sites (although it cannot provide the only mechanism since the RNA polymerase recognition sequence is not symmetrical; see below). The general consequences of using symmetrical sequences for recognition sites have been discussed by Sobell (1973). One possible mechanism suggested by the symmetry is to suppose that hairpins are formed; these might provide distinct structural features contrasted with the uniformity of the linear duplex. A hairpin would be formed if the sequence on one strand of DNA were to pair with its (adjacent) complement on the same strand instead of with the complement on its partner strand. The lack of perfect symmetry means that hairpins formed by such rearrangements in the lambda operator sites would not be very stable. But this model is excluded for lambda by the observation of Maniatis and Ptashne (1973a) that supercoiled and linear molecules of $\lambda$ DNA differ only slightly in their affinities for repressor; this implies that no unwinding takes place when repressor binds to $\lambda$ DNA, making it very probable that DNA is recognized in its duplex state.

An alternative idea is that symmetry in recognition sites in DNA may match a symmetry in the binding protein, in the case of lambda so that one subunit interacts with each half site of DNA, the model depicted in Figure 4.22 and implicit in the representation of Figure 4.29. Some indication of the restraints that exist on variation in the repressor subunit recognition sequences is gained by comparing the sequences of the wild type half sites and also those of the half sites in virulent mutants. Figure 4.30 shows an alignment of half site sequences, that is, with all the single strand shaded sequences of Figure 4.29 written in the 5′ to 3′ direction. The last (ninth) base pair in each sequence thus represents the axis of symmetry. The only difference between the two high affinity binding site sequences ($o_L1$ and $o_R1$) is the alternation of A with T at position 5 of the nontranscribed strand. Presumably it is this change that is responsible for the greater affinity for repressor of $o_L1$ compared with $o_R1$. A greater variety of changes are found at the remaining, lower affinity, binding sites, although it is interesting to note that the tentative sequence for the first half site of $o_l3$ is identical with that of $o_l1$.

The figure summarizes the number of times each base is found at each position. No variation is seen at positions 2, 4 and 6; and at position 1 a purine is always present. Greater variation is seen at positions 3, 5, 7 and 8; and at position 9 (the axis of symmetry) all four bases are found, although usually a G-C pair occupies this site. Four of the virulent mutations, *v2, vs326, vN* and *v1,* change the C present at invariant position 6 to an A; all these mutations greatly reduce the ability of repressor to bind to their respective sites. The mutation *v3* introduces a T at position 8; this mutation again reduces repressor binding, but it is interesting to note that a T does occur at this position in the transcribed strand of $o_R3$. The mutation *v101* introduces a G at the invariant position 2, but appears to have a much weaker effect in reducing repressor binding. Of course, it is not possible to predict the consequences of individual changes in each binding site since their effects may depend upon the neighboring sequences. It is interesting, however, that all the virulent mutations change the base composition (that is, cause transitions between G-C and A-T pairs). Perhaps the base composition of the operators is important; the sequences between the recognition sites are extremely A-T rich, as are the adjacent sequences at the ends of each recognition site, whereas the internal base pairs in the recognition sites are G-C rich.

### *Sequences Recognized by RNA Polymerase*

The nucleotide sequence protected by RNA polymerase can be identified by experiments following the same general protocol as that used to isolate the repressor binding sites: DNA bound to polymerase is digested with nuclease and the protected fragments are recovered. Walz and Pirrotta (1975) isolated fragments of $p_R$ by this approach, starting with repressor protected fragments obtained from λ DNA carrying only the right operator. After binding with limiting amounts of RNA polymerase above

| Wild – Type Sites | | | | Mutant Sites |
|---|---|---|---|---|
| $o_L1$ | t | T A T C A C C G C | $V_2$ | T A T C A **A** C G C |
| | nt | T A C C A C T G G | —$V_{101}$→ | T **G** C C A C T G G |
| $o_R1$ | t | T A T C A C C G C | —$V_3$→ | T A T C A C C **T** C |
| | nt | T A C C T C T G G | —$V_{s326}$→ | T A C C T **A** T G G |
| $o_L2$ | t | C A A C A C C G C | | |
| | nt | T A T C T C T G G | | |
| $o_R2$ | t | C A A C A C G C A | —$V_N$→ | C A A C A **A** G C A |
| | nt | T A A C A C C G T | —$V_1$→ | T A A C A **A** C G T |
| $o_R3$ | t | T A T C C C T T G | | |
| | nt | T A T C A C C G C | | |
| $o_L3$ | t | T A T C A C C G C | | |

| total | | | | | | | | |
|---|---|---|---|---|---|---|---|---|
| $T_9$ | $A_{11}$ | $T_6$ | $C_{11}$ | $A_8$ | $C_{11}$ | $C_6$ | $G_9$ | $C_5$ |
| $C_2$ | | $A_3$ | | $T_2$ | | $T_4$ | $C_1$ | $G_4$ |
| | | $C_2$ | | $C_1$ | | $G_1$ | $T_1$ | $A_1$ |
| | | | | | | | | $T_1$ |

**Figure 4.30:** comparison of operator half sites. Each nine base sequence shaded in Figure 4.29 is written here and aligned in the 5′ to 3′ direction; *t* represents the transcribed strand and *nt* represents the nontranscribed strand. The number of times each base appears at each position is given below the sequences. The positions of substitutions found in six virulent mutants are shown in the shaded circles. Data of Maniatis et al. (1975).

20°C (which allows tight complexes to be formed), the fragments protected against nuclease digestion were precipitated with ethanol and analyzed by gel electrophoresis. They take the form of 40–45 base pair lengths of duplex DNA and are unable to rebind to polymerase; they are not susceptible to Hin enzyme.

When the fragments, still bound to polymerase, are treated with DNAase and nucleoside triphosphates are added, the polymerase starts transcription and DNA is digested behind it as it progresses. This means that only a single product is made from a unique starting site. The longest RNA molecule produced in this reaction is 18 nucleotides long and corresponds to the usual sequence of RNA initiated from $p_R$. Thus the polymerase-protected fragments comprise roughly 20 base pairs to the left of the startpoint for transcription and the first 20 base pairs that are transcribed into mRNA. The entire length of the region protected by polymerase in $p_R$ was determined by Walz and Pirrotta by using dinucleotide and trinucleotide primers to transcribe the fragments into RNA products that could be sequenced. Of the sequence of 40 base pairs that was determined, 25 lie in the primary repressor binding site ($o_R1$) and 15 project to the right into the *cro* gene. Sequences of similar length in the same position are protected at the promotor of phage fd and at the A3 promotor of phage T7 (Schaller, Gray and Herrmann, 1975; Pribnow, 1975). Sequences in corresponding positions are known in the left control site of lambda, the lactose operon, and SV40, although it has not been confirmed directly that they are protected by polymerase. The relationship of the RNA polymerase binding sites to the λ operators is shown in Figure 4.29.

A summary of the sequences immediately surrounding the site where transcription is initiated is given in Figure 4.31. There is no striking homology among these seven sequences. No relationship is evident between the starting sequences of the messengers, although if there were one it would be difficult to interpret since it would not be obvious whether it was involved in transcription or translation. Pribnow (1975) has proposed that a series of seven bases

5′ T A T Pu A T G 3′
3′ A T A Py T A C 5′

lying at positions 6-12 or 7-13 to the left of the startpoint may be involved in aligning the polymerase at the initiation site; this sequence, or one related to it by only one or two substitutions, is present in many of the known promotor sequences. There is no consistent twofold rotational symmetry in the sequences shown in this figure; perhaps this is not

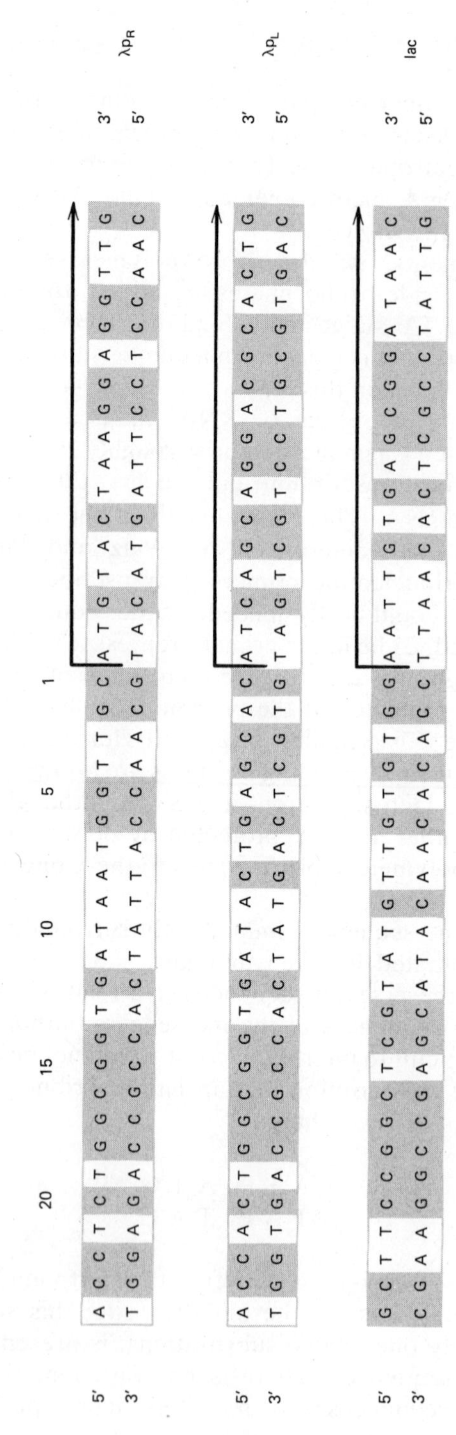

20
15
10
5
1
5' ACCTCTGGCGGTGATAATGGTTGCATGTACTAAGGAGGTTG 3'
3' TGGAGACCGCCACTATTACCAACGTACATGATTCCTCCAAC 5'
λpR
5' ACCACTGGCGGTGATACTGAGCACATCAGCAGGACGCACTG 3'
3' TGGTGACCGCCACTATGACTCGTGTAGTCGTCCTGCGTGAC 5'
λpL
5' GCTTCCGGCTCGTATGTTGTGTGGAATTGTGAGCGGATAAC 3'
3' CGAAGGCCGAGCATACAACACACCTTAACACTCGCCTATTG 5'
lac

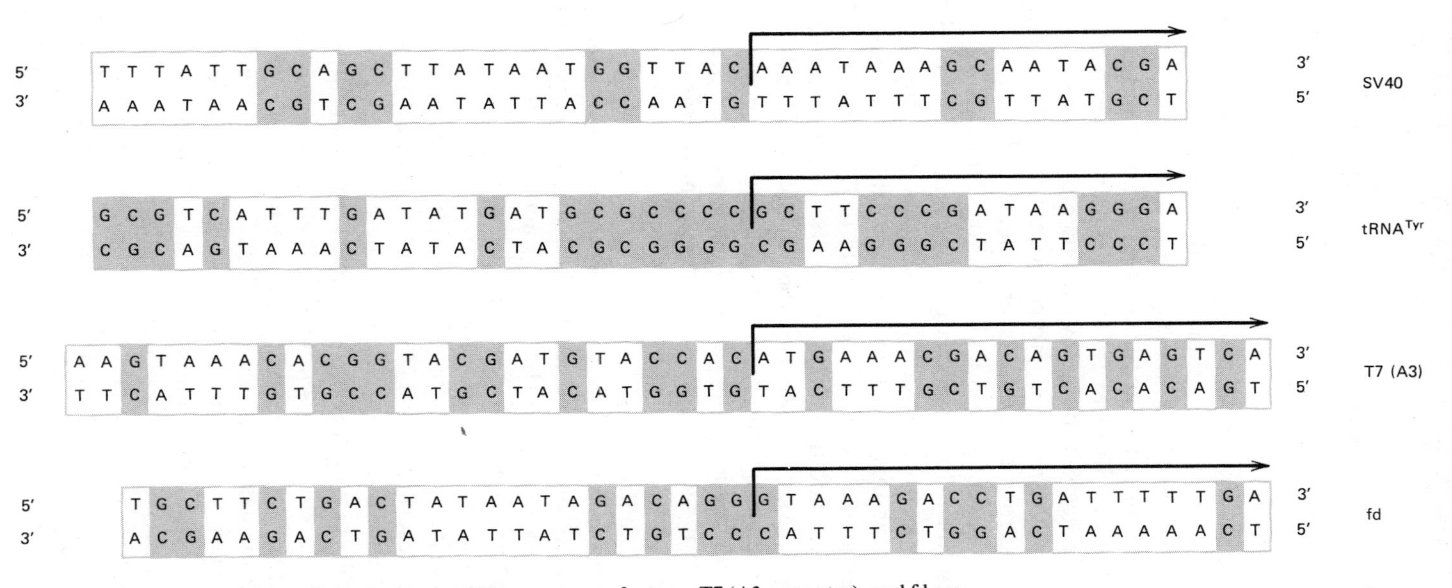


**Figure 4.31:** sequences at transcription startpoints. The sequences for $\lambda\, p_R$, T7 (A3 promotor), and fd are those protected by RNA polymerase in tight binding complexes. The other sequences are those at corresponding positions in $\lambda\, p_L$, *lac*, SV40 and the gene for $tRNA^{Tyr}$ in E.coli. The arrow indicates the startpoint and direction of transcription. Bases are numbered to the left of the startpoint. To facilitate comparisons between the sequences G-C base pairs are shaded and A-T pairs are unshaded. Data of Walz and Pirrotta (1975a), Maniatis et al. (1975), Dickson et al. (1975), Dhar et al. (1974), Sekiya and Khorana (1974), Pribnow (1975) and Schaller, Gray and Herrmann (1975).

surprising since RNA polymerase must bind with the particular orientation which leaves it able to engage in transcription of only one strand. This supports the idea that symmetry is not necessary for interaction with sequences in DNA per se, but perhaps reflects the symmetrical nature of the regulator proteins. It may well prove to be necessary to obtain mutants that are defective in initiating transcription before it is possible to define the bases involved in setting the startpoint.

A significant feature of the sequence protected by polymerase in $p_R$ is that it does not extend to the Hin cleavage site. Yet polymerase binding to λ DNA protects this site against cutting by the Hin enzyme and cleavage also is abolished at this site by the *x3* promotor mutation. This discrepancy suggests that the promotor consists of at least two types of site, involved, respectively, in recognition and initiation. Hin cleavage in $o_R$ and $o_L$ is prevented only by much larger concentrations of RNA polymerase than are used to recover the fragments protected against nuclease by the enzyme; in this condition it is reasonable to suppose that polymerase molecules may cover a greater length of DNA, with binding taking place to sequences additional to those recovered in the nuclease protection experiments. And the effect of the *x3* mutation implies that interaction at the Hin cleavage site may in fact be an essential step in polymerase action; the identical effect of *sex1* and *sex3* at $o_L$, and the same location of the Hin site relative to the startpoint of transcription, support the conclusion that the action of polymerase involves the same events at both left and control sites.

A two step model for the interaction of polymerase with the promotor is illustrated in Figure 4.32. The enzyme first binds to a *recognition site* that includes the sequence at which the Hin enzyme acts: this binding may take the form of a "loose" complex. The enzyme then moves along the DNA to an *initiation site,* where it may form a "tight" complex. Movement between the two sites occurs without transcription—for this reason it has sometimes been described as polymerase *drift*—and the synthesis of RNA starts at a point within the initiation site. The characteristics of the binding reaction of RNA polymerase to DNA are consistent with the idea that two stages involving different tightness of binding precede initiation itself (see Volume 1 of Gene Expression). We may suppose that when low concentrations of polymerase are provided under appropriate conditions, all the enzyme molecules are able to move from the recognition to the initiation site, so that only the initiation sequence is protected against degradation by nucleases. When the polymerase concentration is much higher, it may be possible or necessary for molecules to remain at the recognition site, conferring protection against Hin cleavage.

One implication of this model is that we are left with no information

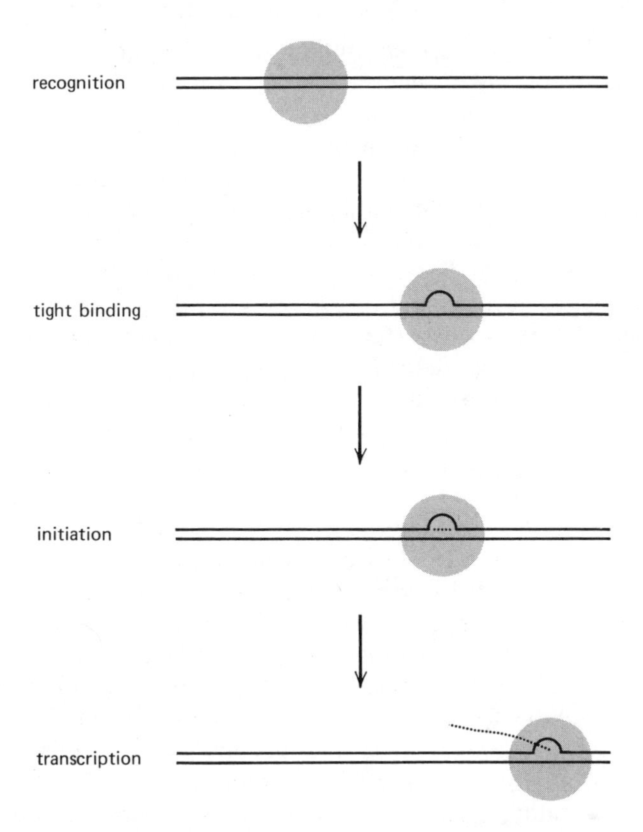


**Figure 4.32:** model for interaction of RNA polymerase with the promotor. First the enzyme binds loosely to a recognition site; then it moves (without transcription) to the initiation site, where tight binding is established, probably involving melting a few base pairs. In the presence of nucleoside triphosphates, transcription can be initiated; and RNA synthesis probably occurs at a localized region of melting which moves along the DNA.

about the sequence that is initially recognized by RNA polymerase. Presumably a specific sequence comprises the recognition site; and a second sequence then serves to align the polymerase at the initiation site after its movement subsequent to initial binding. The only formal restriction that we can infer about these sites is that their combined length should be at least 12 base pairs; for this is the minimum length of DNA in a bacterial genome which ensures that the sequence will not occur by chance many times elsewhere (see Volume 1 of Gene Expression). Since two sites are involved, their combined length is probably somewhat greater than this.

How may the recognition and initiation sites be identified? The conditions that are necessary for nuclease protection experiments may make it impossible to recover any sequence other than that including the tight binding initiation site. (Although the lack of homology between the sequences so far obtained at this position means that the sequence involved in aligning the polymerase for initiation in any case cannot be identified.) Promotor mutants should exist in both sites; it is reasonable to suppose that mutation either to prevent initial recognition or to prevent subsequent alignment at the initiation site should prevent transcription from taking place. To distinguish the nature of promotor mutations, it may therefore be necessary to determine whether they lie within the sequence protected against nuclease degradation or reside some distance to one side. A recognition site is therefore defined operationally by promotor mutations that do not lie in the sequence protected by polymerase at the startpoint. Perhaps homologies will be found between the sequences at these positions in different templates. The separation of recognition and initiation sites occurs in the lactose operon as well as in lambda; for RNA polymerase binding to the lactose operon requires mediation of the CAP protein (activated by cyclic AMP) at a site immediately adjacent to the region at which polymerase forms an initiation complex (see Lewin, 1974a; Dickson et al., 1975). Definition of the relationship between the sites may require the development in each system of means for preventing the polymerase from moving from the recognition to the initiation site as well as the exhaustive isolation of promotor mutants.

How does binding of repressor prevent polymerase from initiating transcription? The intermingling at the right operator of mutations which appear to inhibit recognition of DNA by polymerase and by repressor suggests that the recognition site must overlap with the operator sequences. This is confirmed by the location of the Hin cleavage site, which must lie within the polymerase recognition site, between operator sites recognized by the repressor. The model shown in Figure 4.29 locates the Hin cleavage site within the sequence recognized by repressor in $o_R2$ and $o_L2$; this means that polymerase recognition must certainly be mutually exclusive with repressor binding to the secondary operator sites. And it seems very probable that even if the recognition site does not formally overlap with the primary ($o1$) sites, stereochemical constraints should make polymerase recognition and repressor binding to $o1$ incompatible. Since the primary repressor binding sites overlap with the initiation region protected by polymerase, it is clear that even repressor binding only to the primary sites should make polymerase drift and initiation impossible, even if recognition could occur. This suggests the model illustrated in Figure 4.33, in which binding of repressor prevents binding of RNA polymerase.

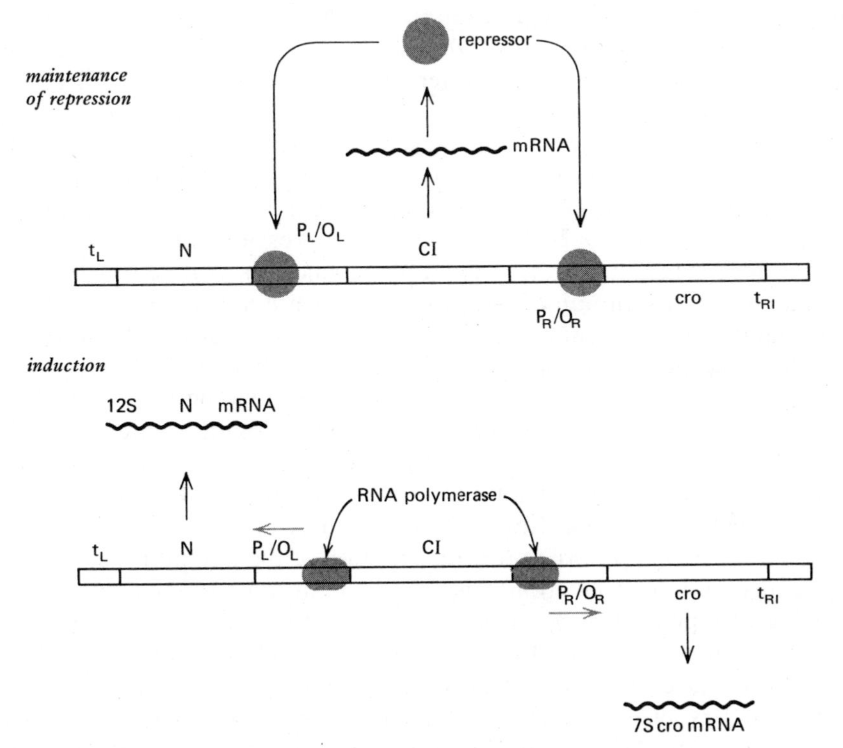


**Figure 4.33:** repression and induction of prophage. The prophage genome is repressed when the repressor protein binds to the left and right operators; in the repressed state only the *cI* gene is transcribed and translated. Binding of repressor at $o_R$ may be necessary for transcription of *cI*. Induction occurs when repressor fails to bind to the operators; RNA polymerase is then able to bind to the recognition sites of the promotors and to move to the initiation sites from which 12S and 7S RNA messengers are transcribed. Transcription is terminated at sites at the ends of the *N* and *cro* genes.

From the highly constitutive effect of the *v2* mutation in $o_L1$, it is clear that reducing the ability of repressor to bind to the left primary site is sufficient to allow polymerase to engage in recognition and initiation. This observation therefore yields no information about the role of secondary binding sites on the left. On the other hand, mutations in $o_R1$ are not sufficient to allow constitutive expression of the adjacent *cro* gene; it is necessary for there also to be a mutation in $o_R2$. This argues that both the primary and at least one secondary binding site are necessary to establish complete repression of rightward transcription. If the six virulent mutations that have so far been mapped prove to be typical, then it is not clear what functions are exercised by the remaining sites in which no mutations have been found (see Figure 4.29). That $o_R3$ is not needed for repression at $o_R$ is indicated also by the observation of Smith et al. (1976) that a deletion removing $o_R3$ and material to its left does not impinge on repression of the rightward operon. At present it is therefore clear that binding to $o_L1$ is necessary to repress leftward transcription and that binding to both $o_R1$ and $o_R2$ is necessary for complete repression of rightward transcription; what roles $o_L2$-3 and $o_R3$ play in repression is not established.

One final point about these control sequences is that repression and initiation of transcription of the early operons are not the only functions residing in the DNA. Initiation of leftward and rightward transcription at $p_L$ and $p_R$, respectively, may be inhibited by the binding of Cro protein in the control regions (see page 379); and the terminating ability of RNA polymerase initiating at these promotors is altered by the action of the N protein (see page 402). In addition to the presence of these sites affecting early transcription, the promotor for the *cI* gene, $p_{rm}$, lies within the region of $o_R$, which implies that RNA polymerase may bind in another orientation, one that is appropriate for transcription of the left strand of *cI*; and the transcription of *cI* from this promotor may be stimulated by the binding of repressor at $o_R1$ while inhibited by its binding at $o_R3$ (which suggests a function at least in this process for $o_R3$; see page 367), while Cro protein also may inhibit transcription from $p_{rm}$ (see page 370).

In summary, then, it is clear that multiple repressor binding takes place at both the left and right control sites and that this inhibits binding of RNA polymerase for transcription from $p_L$ and $p_R$. However, the relationships between the sites with which polymerase interacts, the repressor binding sites, and the role of multiple (rather than single) repressor binding sites, remain to be defined. Also it remains to be seen how polymerase binds at $p_{rm}$ for leftward transcription of *cI* and how repressor binding at the sites of $o_R$ affects this process. And in addition to the binding of polymerase and repressor protein, both control sites include the sites of action of Cro protein and N protein, whose interactions with RNA polymerase also have yet to be elucidated.

## Control of Repressor Synthesis

### *Lysogeny and Lytic Development*

The choice of whether an infecting lambda phage enters the lytic pathway or establishes lysogeny is determined by a subtle interplay of regulator proteins. Both the establishment of lysogeny and entry into the lytic pathway require functions coded by the delayed early genes; it is therefore only after expression of these functions that the two pathways diverge and it becomes apparent whether the subsequent events will be those of lysogeny or lytic development.

Establishment of lysogeny requires two events: insertion into the bacterial chromosome; and synthesis of repressor. Insertion requires the product of the left delayed early gene *int* (the other left delayed early genes are *xis*, concerned with excision, and the *red* genes concerned with vegetative recombination); and establishment of repressor synthesis requires products of both right and left delayed early genes, *cII* and *cIII*. Transcription of the *cI* gene then leads to the synthesis of repressor, which binds to the left and right operators and prevents further expression of all early genes.

Entry into the lytic cycle requires the products of the right delayed early genes *O, P* and *Q*. Both *O* and *P* are essential for replication of the phage DNA; *Q* codes for a regulator protein which promotes transcription of the late genes from the promotor $p'_R$ (which is not under the control of the early operators). To enter the lytic cycle, it is necessary also to prevent repressor synthesis: this inhibition is achieved by the product of the *cro* gene. The *cro* product turns off transcription of the delayed early genes; since the *cII* and *cIII* proteins are unstable and decay rapidly, this has the effect of preventing the initiation of transcription of repressor. Since *cro* represents the sole immediate early right gene, its expression is inhibited by repressor. Thus the *cro* product and the repressor are mutually antagonistic: synthesis of *cro* protein prevents (indirectly) the initiation of transcription of *cI* and thus prevents establishment of lysogeny; synthesis of repressor protein prevents transcription of early genes and thus switches off synthesis of *cro* protein. The antagonism between these two regulators therefore is central to setting the balance between lysogeny and lytic development. Figure 4.34 illustrates the divergence between the two pathways and shows the steps which follow from the decision between them.

The timing of the action of these regulators is important in determining the pathway followed by an infecting phage. Upon infection, only the immediate early genes, *N* and *cro*, are expressed. Synthesis of *N* protein then allows expression of the delayed early genes, necessary for both lysogeny and lytic development. Utilization of the functions coded by the *int, cIII, cII* genes leads to expression of *cI* and thus to lysogeny; utilization of the products of genes *cro, O, P, Q* leads to expression of the late genes and thus to lytic development. Which series of events predominates may depend upon the relative times and efficiencies of action of the repressor protein and *cro* protein. Both proteins turn off the transcription

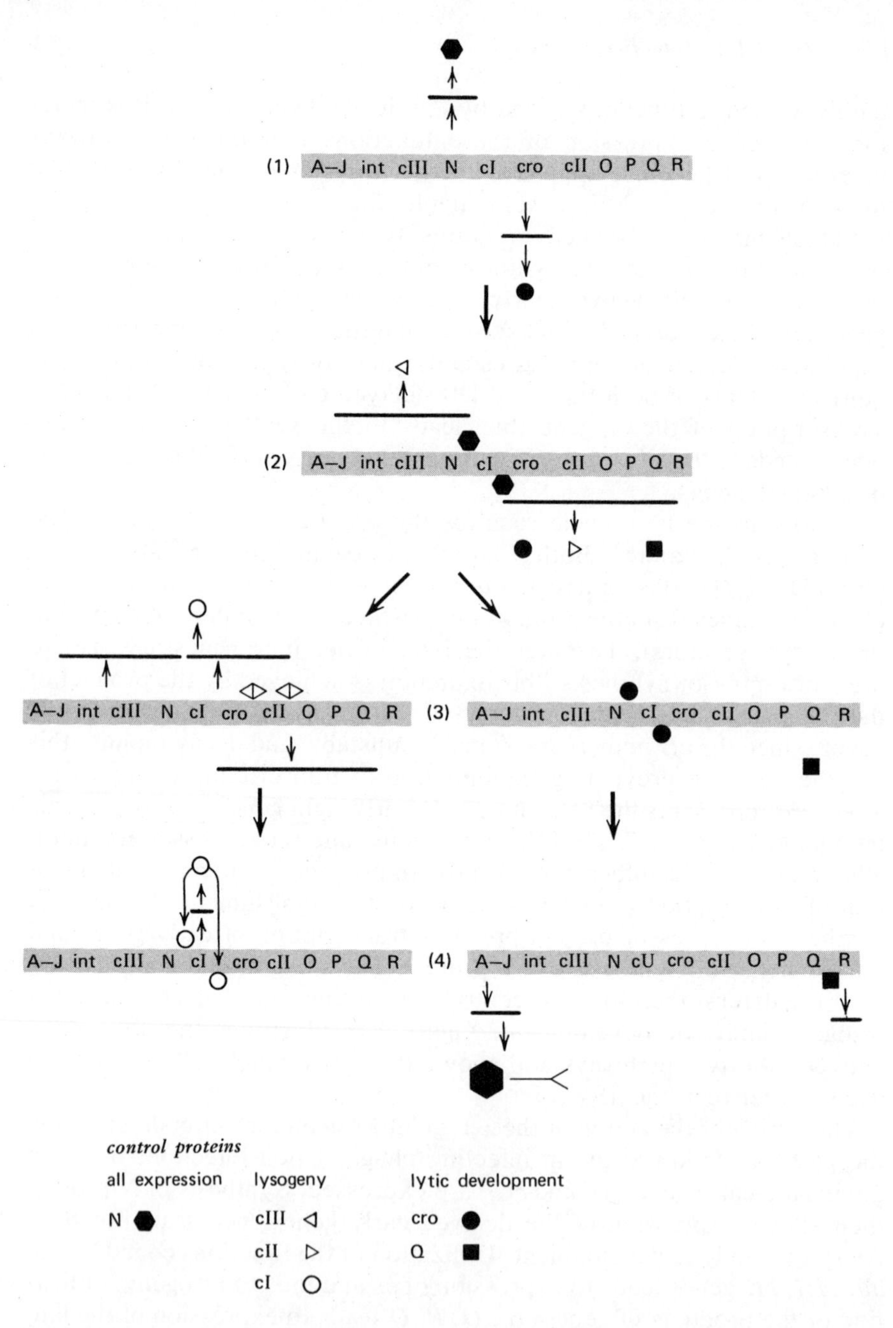


**Figure 4.34:** control of the lysogenic and lytic pathways following infection. (*1*) Immediate early expression: genes *N* and *cro* are transcribed and translated. (*2*) Delayed early expression: the *N* protein acts at the left and right control sites to allow RNA polymerase to continue transcription past *N* and *cro*.This produces the *cII*, *cIII* and *int* functions that are

of early genes, although repressor does so completely whereas *cro* leaves some residual expression (which may be necessary for lytic development). Although *cro* is an immediate early gene, the indirect nature of its inhibition of repressor synthesis ensures that the circuits for mutual repression between *cro* and *cI* do not come into play until the time of expression of the delayed early genes. If the *cro* protein achieves its effect before sufficient repressor protein has been synthesized, the phage may enter lytic development; on the other hand, if enough repressor is available, lysogeny may be the consequence.

### *Establishment of Lysogeny*

Although the maintenance of lysogeny requires only the presence of repressor protein, establishment of lysogeny when λ DNA infects a host cell requires in addition the functions *cII, cIII, cy*. Mutants in *cII* and *cIII* show the complementation behavior expected of genes which code for proteins and the $cII^-$ and $cIII^-$ alleles are recessive to the wild type; this suggests that the $cII^+$ and $cIII^+$ genes specify proteins whose activities are needed for the initiation of transcription of *cI*. The activities of these proteins are no longer necessary once lysogeny has been established, for the rare lysogens formed by $cII^-$ and $cIII^-$ phages are stable (see earlier).

This model is supported by direct measurements of the amount of repressor synthesized following infection of E.coli cells by λ DNA. Echols and Green (1971) assayed repressor by its ability to bind labeled λ DNA to a nitrocellulose filter; Reichardt and Kaiser (1971) used a radioimmune assay which measures the ability of extracts of repressor to inhibit the binding of purified $^{125}$I-repressor to an antirepressor serum. Both series of experiments showed that there is little repressor synthesized during the first 5 minutes after infection: then there is a rapid rate of synthesis for the next 10 minutes; and this in turn is succeeded by the low rate of synthesis characteristic of cells in which lysogeny has been estab-

---

needed for lysogeny; and it produces the *O, P* and *Q* functions that are needed for lytic development. The immediate early genes, *N* and *cro*, continue to be transcribed and translated. (*3*) Establishment of lysogeny versus commitment to lytic development. Left: the *cII* and *cIII* proteins are needed to allow transcription of *cI* by extension of the RNA starting between *cII* and *O*; they may act close to this site and also at *cy* between *cII* and *cro*. This yields message which is translated into repressor protein. Right: the *cro* protein turns off transcription of the delayed early genes, including *cII* and *cIII*. (Some residual expression remains.) The *cII* and *cIII* proteins decay rapidly and so are unable to initiate transcription of *cI*. Protein Q remains present. (*4*) Maintenance of lysogeny versus late development. Left: the establishment of repressor synthesis is succeeded by maintenance, when only *cI* is transcribed and the repressor protein binds to the early operators to prevent transcription of early messengers. Right: the *Q* protein allows late transcription to continue past a site between *Q* and *R* and the heads and tails of the phage particle are produced.

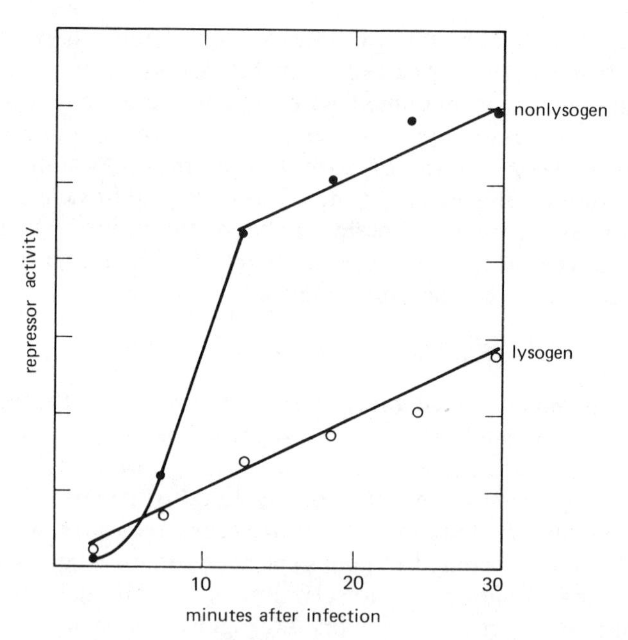


**Figure 4.35:** rate of synthesis of repressor protein after infection of lysogenic and nonlysogenic cells. Data of Echols and Green (1971).

lished. Figure 4.35 shows the biphasic rates of synthesis of repressor; when a nonlysogen is infected, synthesis starts at a rapid rate which then declines to the same rate seen in the lysogenic cells that provide the control. Figure 4.36 plots the accumulation of repressor relative to total cell protein (which is itself increasing with growth). Mutants in *cII, cIII, cy* all display much reduced rates of accumulation following infection of a cell population; this reflects their reduced ability to establish lysogeny. That these functions are needed *only* to establish lysogeny is shown by measurements of the synthesis of repressor by cell lines that are lysogenic for $\lambda^+$, $\lambda cII^-$, $\lambda cIII^-$ or $\lambda cy$; all synthesize repressor at the same low rate that is seen in infected cells after lysogeny has been established.

The two rates of repressor synthesis define the two systems concerned with control of lysogeny. The *establishment* system requires the functions *cI, cII, cIII, cy* and achieves a high rate of repressor synthesis shortly after infection of a host cell. This system ceases to operate once lysogeny has been established. The *maintenance* system requires only the function of *cI* and synthesizes repressor at a low rate after lysogeny has been established. This system therefore is adequate only to maintain and cannot establish lysogeny.

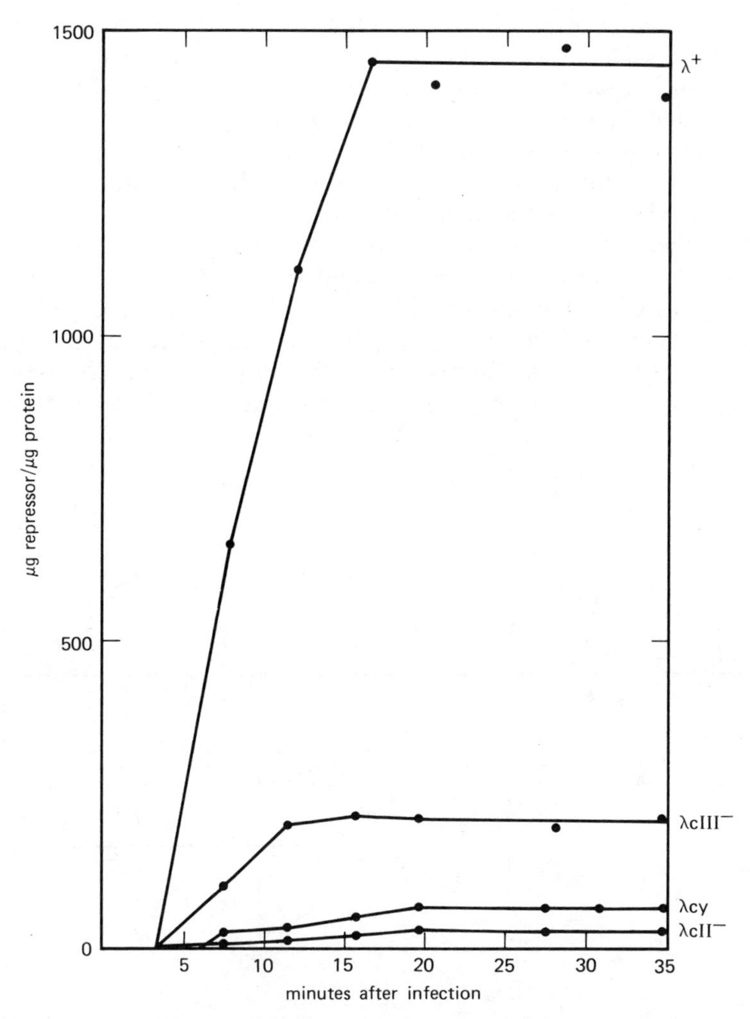


**Figure 4.36:** accumulation of repressor after infection. All strains show biphasic accumulation: there is a lag of about 5 minutes; rapid accumulation of repressor for about 10 minutes; and then the repressor level stays constant. Ability to synthesize repressor is greatly reduced in *cIII*⁻, *cy*, *cII*⁻ mutants; cell survival (that is the proportion of lysogenic cells) is reduced to 20% of wild type level in *cIII*⁻ mutants and to 5% of wild type in *cII*⁻ mutants. Data of Reichardt and Kaiser (1971).

Mutants in *cy* do not show the complementation behavior typical of genes which code for proteins, but are cis dominant, trans recessive as expected of a control element in DNA which is recognized by protein. The *cy* region maps between *cro* and *cII*, that is just to the right outside the immunity region (see Figure 4.37). A genetic study of *cy* mutants performed by Wulff (1976) and McDermit et al. (1976) identified seven sites at which the cis dominant genotype can occur; according to the length of the fine structure recombination map, the *cy* region is rather short, of the order of 12 base pairs. In addition to the *cy* mutations, two further types of mutant event can occur in this part of the genome. A mutation called *cin* allows lysogeny to be established in spite of the introduction of *cy, cII* or *cIII* mutations; the *cin* mutation is cis dominant, that is, permits lysogeny only if a functional *cI* gene is carried on the same

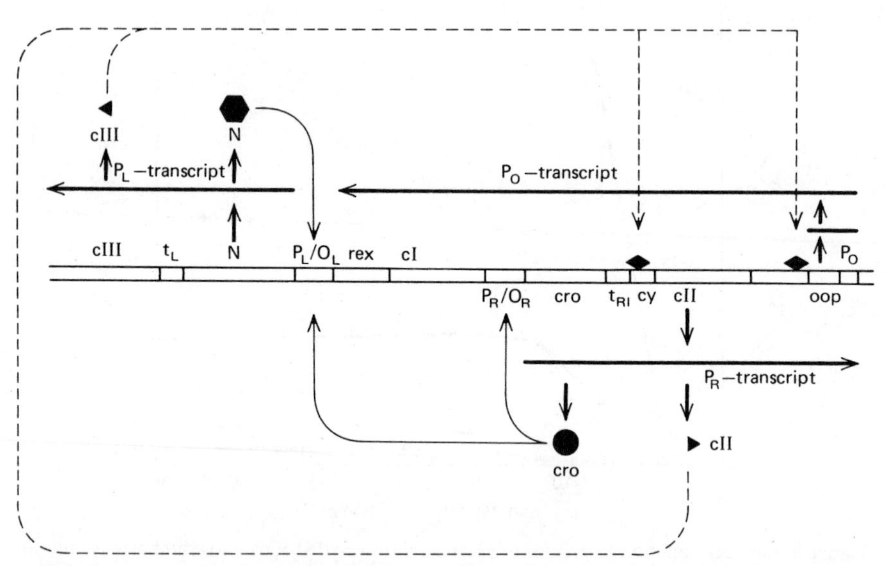


**Figure 4.37:** control circuits during the establishment of lysogeny. The *oop* RNA is synthesized on the *l* strand from the promotor $p_0$ located to the right of *cII*; this requires the phage *O* and *P* functions as well as host replication functions (details not shown). Extension of *oop* RNA into the *cI* gene requires the *cII* and *cIII* functions; these may act as antiterminators at two sites: the normal terminus of *oop* RNA and the *cy* region. The transcript of *cI* is probably terminated at the left end of the *rex* gene (the *t* site has not been identified yet). Since the *cII/cIII* and *O/P* functions are required for synthesis of the $p_0$ transcript, establishment of lysogeny can take place only when *N* function allows transcription to continue past $t_L$ (as illustrated) and past $t_R$ (action of the protein not shown). Note that the *cro* gene also is expressed during this period; and Cro protein acts at both the left and right control sites.

lambda genome. The *cin* mutation maps just to the right of the $\lambda/imm^{434}$ boundary, very close but to the left of the *cy* mutations. A further type of mutation, called *cnc*, maps just to the right of *cin* and appears to have as its sole effect a cis dominant masking of the Cin mutant phenotype.

The formal implication of this genetic analysis is that there is a site in the region of *cy, cin* and *cnc* at which it is possible for mutation to influence expression of the *cI* gene on the same molecule of λ DNA. One obvious model suggested by this conclusion is that this site may be a promotor where transcription of *cI* is initiated for the establishment of lysogeny; this has been described as $p_{re}$ (*re* stands for repressor establishment). According to this model, the *cII* and *cIII* gene products would be needed for the initiation of transcription at $p_{re}$ by host RNA polymerase. However, another source for the transcription of *cI* during establishment has been identified by Honigman et al. (1976). A small RNA called *oop* is initiated in the region of the origin for replication (see Hayes and Szybalski, 1973); it is 81 nucleotides long, its promotor has been termed $p_o$, and it may function as a primer for DNA replication. Measurements of the ability of the RNA containing the immunity region transcript to hybridize to appropriate deletion strains of λ DNA identified its startpoint at the same site as that of *oop* RNA (81% from the left end of the genome). This suggests that the *cI* gene is transcribed by extension of the *oop* RNA, a model which may be confirmed by showing that the 5′ nucleotide sequence of the establishment transcript is identical with the *oop* RNA sequence.

This model suggests that the role of the *cII* and *cIII* products in establishing lysogeny may be to act as antiterminators to allow polymerases initiating at $p_o$ to proceed through to the immunity region. The *oop* RNA is terminated at or very close to the $\lambda/imm^{21}$ boundary. This identifies one site where antitermination must take place. The *cy* mutations may identify another: they might prevent the *cII* and *cIII* functions from exercising an antitermination function at a termination site in this region. A puzzling implication of this model, however, is that it would be reasonable to expect similar types of mutation to occur in both the termination sites; their absence implies some difference in the nature of the sites.

Consistent with its postulated role in replication, *oop* RNA cannot be synthesized by $O^-$ or $P^-$ phages or in infection of $dnaB^-$ bacteria. Support for the idea that *oop* RNA represents the leader sequence of the establishment transcript thus is provided by the observation that the same mutations prevent transcription of the immunity region during the establishment period. The implication that phages defective in the *O* and *P* replication functions should be unable to lysogenize is supported by the direct observations of such deficiencies made by Ray and Skalka (1976).

Mutations affecting the synthesis of *oop* RNA have not been obtained. However, Honigman et al. (1975) have isolated *sar* mutations which map in the region of the origin; these are unable to replicate or to support transcription when grown on *cro*-constitutive lysogens (an effect which has yet to be explained). An interesting effect of the mutation is to suppress *cy* mutations in cis array; this depends on the presence of active *cII* and *cIII* functions. Although the reason for the suppression is not clear, this result supports further the idea that transcription of the *cI* gene involves events in the region of the origin.

To summarize the functions involved in the establishment of lysogeny, two sets of delayed early gene products are needed: the replication functions *O/P* are required for synthesis of *oop* RNA; and the regulator functions *cII/cIII* must intercede to allow extension of *oop* RNA towards the *cI* gene. Although the interactions of these functions are not completely defined, a model for the establishment of lysogeny that shows the control circuits in operation during this period is illustrated in Figure 4.37.

In lysogenic cells, only about 2% of the lambda genome is transcribed and this corresponds entirely to sequences within the immunity region. This value is about the size expected of the *cI* gene and thus is somewhat smaller than the total length of the immunity region, which is about 5% of the complete genome. Repressor can be synthesized at the normal level from prophages which have suffered deletion of the region from *P* to within *cro* (and thus have lost $p_0$); this implies that the promotor for maintaining repression, $p_{rm}$ (*rm* stands for repressor maintenance) lies within the immunity region to the right of *cI*. Transcription from this promotor does not require any of the products needed for transcription via $p_0$; however, it does require the presence of active repressor (see below). It is not known what is responsible for terminating transcription at the left of *cI*.

The location of the two promotors shown in Figures 4.34 and 4.37 implies that the messenger transcribed from $p_0$ should contain a transcript of the *l* strand of both *cro* and *cI* whereas the messenger starting at $p_{rm}$ should contain a transcript only of the *l* strand of *cI*. (Since the *rex* gene is so poorly characterized, it is omitted from this and other discussion here.) Since *cro* is transcribed from the *r* strand as an immediate early gene, this means that part of the $p_0$-promoted message should be a transcript of the *cro* anticoding strand; presumably this is meaningless and is not translated. The location of $p_0$ thus implies that *cro* is transcribed in both directions: rightward during early development; and leftward during establishment of lysogeny. We do not know how the topological problems that this may pose are solved and what role these conflicting transcriptions might play in controlling lambda development.

The prediction that *cro* anticoding strand RNA should be transcribed during and only during establishment of lysogeny has been tested by Spiegelman et al. (1972), who made use of a prophage deletion which retains a sequence of λ DNA extending only over the *cIII-cro* region. This prophage is able to transcribe RNA from the *r* strand of *cro*. When labeled RNA from this source is hybridized with unlabeled RNA extracted from cells newly infected with λ, some of the labeled *r* strand *cro* RNA forms duplex structures—it must have annealed with a transcript of the *l* strand of *cro*. Using this assay shows that the synthesis of *l* strand *cro* RNA follows kinetics related to the accumulation of repressor protein; it first increases and then declines. We may therefore assume that during the establishment of lysogeny the *l* strand *cro* RNA is carried on the same RNA species which carries the transcript of *cI*. That the *cII* and *cIII* proteins are part of a control system that acts to allow the extension of $p_0$-transcription is supported by the observation that mutants at these loci produce much decreased amounts of *l* strand *cro* RNA. That a different promotor, located between the *cro* gene and *cI*, must be used during the maintenance of lysogeny is implied by the failure of lysogenic bacteria to synthesize *l* strand cro RNA.

In these experiments the synthesis of repressor protein appeared to be proportional to the extent of synthesis of message from $p_0$ or $p_{rm}$, suggesting that the establishment and maintenance messengers are translated to yield repressor with about the same efficiency. This would ascribe the difference in the rates of repressor synthesis during establishment and maintenance to differing efficiencies of transcription from the two promotors. On the other hand, Walz, Pirrotta and Ineichen (1976) and Ptashne et al. (1976) have pointed out that the $p_{rm}$ transcript starts with the codon for initiating protein synthesis (see Figure 4.29); it must therefore lack the sequence preceding the initiation codon that has been characterized as a strong ribosome binding site in many procaryotic messengers. Such a sequence occurs just to the right of the startpoint and therefore should be present in the $p_0$-initiated message. They therefore suggested that the principal reason for the difference in repressor synthesis during establishment and maintenance may be due to differences in the efficiencies with which these messengers are translated.

One obvious idea for the role of the cII and cIII proteins is to suppose that both must act to allow extension of the $p_0$ started transcription. However, a hint that these proteins have different functions is provided by the different frequencies with which λ$cII^-$ and λ$cIII^-$ mutants can lysogenize: $cII^-$ mutants are able to do so much less frequently than $cIII^-$ mutants, consistent with their greater reduction in repressor synthesis (see Figure 4.36). Reichardt (1975a) noted that $cII^+$ function is absolutely necessary for synthesis of repressor, whereas under appropriate (rich broth) growth conditions it is possible for λ*cIII*⁻ to produce about 50% of the amount of repressor made by $\lambda^+$. The lower dependence on the *cIII* product is paralleled by the specificities of the two functions: *cII* is specific for λ and is unable to promote lysogeny by $\lambda imm^{21}$ whereas *cIII* is not specific and is able to promote lysogeny by the heteroimmune phage. This suggests that the *cII* product may act directly at a site within the region of nonhomology between λ and $\lambda imm^{21}$ whereas the action of *cIII* may be indirect, mediated by a host cell system.

Two host proteins have been implicated in influencing the frequency of lysogeny. The first was identified by Hong, Smith and Ames (1971), who reported that mutants in S.typhimurium of the $cya^-$ type (lacking adenylate cyclase) or $cap^-$ type (lacking the cyclic AMP receptor protein) display much reduced frequencies of lysogen formation by phage P22; Grodzicker, Arditti and Eisen (1972) reported the same observation for λ and its E.coli host. When activated by cyclic AMP, the CAP (cyclic AMP receptor) protein directs RNA polymerase to initiate transcription at certain promotors; this constitutes the system for catabolite repression. Since the cyclic AMP concentration varies inversely with the level of glucose, increase in the glucose level depletes the cell of cyclic AMP and inhibits transcription of those operons, such as *lac* and *gal,* that are susceptible to catabolite repression (see Volume 1 of Gene Expression). The formal implication of these results is that transcription from some promotor(s) activated by the cAMP/CAP complex encourages the establishment of lysogeny; this might either be a promotor(s) in the bacterial chrmosome or might be a phage promotor. Isolation of bacterial strains carrying mutations that reduce the frequency of lysogeny (that is, by selecting cells that give plaques when infected by λ) largely yields $cya^-$ and $cap^-$ mutants, a suggestive indication that the cAMP/CAP complex acts directly on the phage genome (otherwise further mutants should be found in the bacterial genes which are activated by cAMP/CAP). That it is the *y* locus which is the subject of attention from the bacterial system is suggested by the properties of lambda mutants able to continue to lysogenize $cya^-$ $cap^-$ strains with high efficiency; these possess mutations in the *y* region. When *cy* mutants were thought to identify $p_{re}$, this was taken to suggest that cAMP/CAP acts at the phage promotor. A clear interpretation now cannot be made until the nature of the *y* region is defined.

The second host function which influences establishment of lysogeny is identified by the $hfl^-$ mutations characterized by Belfort and Wulff (1973); mutants in this gene form lysogens with high frequency when infected with λ. Since the mutants are recessive, they appear to represent loss of the *hfl*$^+$ function which therefore usually must antagonize the establishment of lysogeny. All the mutants which have been isolated in this class fall into a single complementation group. Catabolite repression is normal in *hfl*$^-$ cells.

The cAMP/CAP complex and the *hfl*$^+$ product thus have opposite effects. That the role of the phage *cIII* product is concerned with an interaction at the level of these two host functions is suggested by the ability of $\lambda$*cIII*$^-$ mutants to establish lysogeny in the mutant host cells. In *hfl*$^-$ cells, $\lambda^+$ and $\lambda$*cIII*$^-$ phages form lysogens at virtually the same frequency. This suggests that the need for the *cIII* product is abolished by the *hfl*$^-$ mutation; in other words, the role of *cIII* protein may be to antagonize the *hfl*$^+$ product. On the other hand, in *cya*$^-$ or *cap*$^-$ host cells, $\lambda$*cIII*$^-$ shows a very great drop in the frequency of lysogen formation compared with that of $\lambda^+$. This suggests that the *cIII* product becomes essential whenever the cAMP/CAP complex cannot be formed.

Little can be said about how these three functions influence the establishment of lysogeny. The phage *cIII* and the host cAMP/CAP act to promote lysogeny; the host *hfl*$^+$ acts to inhibit lysogeny. Whether all three functions lie in one pathway or in different pathways is not yet clear; and the only function for which any molecular action is known is the cAMP/CAP complex. What these results do clearly imply, however, is that bacterial functions influence the decision between lysogeny and lytic development, perhaps in relationship to the conditions of cell growth.

In addition to their role in lysogeny, *cII* and *cIII* appear to influence the course of lytic development. McMacken et al. (1970) noted that $\lambda$*cII*$^-$ or $\lambda$*cIII*$^-$ mutants commence synthesis of late products (represented by tail protein and lysozyme synthesis) some 5 minutes earlier than $\lambda^+$ phages. Court, Green and Echols (1975) demonstrated that the ability of the *cII/cIII* functions to delay late gene expression is mediated via their action on establishment, for *cy* mutants display a cis dominant advance in the synthesis of lysozyme. The activation of the *cII/cIII* genes has two consequences: the anticoding strand of the *cro-cII* region is transcribed and this may perhaps inhibit rightward early transcription, needed for synthesis of a protein; and repressor protein is synthesized by translation of the message initiated at $p_o$, and this might tend to cause a delay in the time at which late genes are expressed when the phage is proceeding into the lytic cycle.

### *Control of Repressor Maintenance*

Expression of the *cI* gene by both the establishment and maintenance systems is under a delicate control. To turn on transcription of *cI* initially requires the mediation of the systems represented by the phage genes *cII* and *cIII* and/or related host functions and is accomplished by extension of the *oop* RNA initiated at the promotor $p_o$. Since $p_{rm}$ remains inactive in these conditions, we may assume that some other event is required to allow transcription from the maintenance promotor. Transcription via the establishment system must be turned off if the phage is to enter the lytic cycle. This is accomplished by the indirect action of *cro,* which prevents transcription of *cII* and *cIII;* since their protein products are unstable, their rapid decay is paralleled by inability to synthesize repressor via the establishment system. Expression of *cI* during the period of establishment is thus subject to both positive and negative control: the positive *cII/cIII* control system(s) ensure directly that *cI* can be transcribed only when these products are present; and the negative *cro* function indirectly prevents transcription.

Transcription from $p_{rm}$ requires the presence of active repressor protein; this constitutes a positive control system, in effect a positive feedback loop in which the presence of repressor is necessary for its own continued synthesis. Transcription from $p_{rm}$ also may be subject to negative control by *cro*: the Cro protein may directly inhibit transcription of *cI* from this promotor. Thus the *cro* gene antagonizes the expression of *cI* by either the Maintenance or establishment systems, directly in the first instance, indirectly in the second.

The parameters governing expression of *cI* by the establishment and maintenance systems demonstrate how repressor production is switched from one system to the next. Transcription of *cI* from the $p_o$ promotor results in the production of repressor protein. The repressor then binds to the operators and switches off transcription of *cII* and *cIII,* thus preventing its own synthesis via the establishment system. However, the presence of repressor now allows transcription to commence from $p_{rm}$, creating the positive feedback loop which represents the sole expression of the prophage genome in established lysogens: all phage functions are repressed, except *cI* (and *rex*), and it is the synthesis of repressor protein alone that is sufficient both to maintain repression and to ensure its own continued production. The switch from $p_o$ to $p_{rm}$ is accompanied by a decline of 7–8-fold in the rate of repressor synthesis. Destruction of repressor then leads to prophage induction, since the inactivation not only removes the repressor present in the cell, but also prevents continued synthesis from the $p_{rm}$ promotor.

Much of the analysis of the control of repressor synthesis has made use of the mutant $\lambda cI_{857}$, which carries a temperature sensitive mutation: the repressor protein is active at 30°C but is inactivated by an increase in temperature to 42°C. Prophage may therefore be induced by raising the temperature of cells lysogenic for $\lambda cI_{857}$: in order to avoid lysis of the host cell, the prophage usually carries other mutations which block lytic development. Eisen et al. (1970) observed that after cells lysogenic for $\lambda cI_{857}$ (carrying $N^-$ and $O^-$ mutations to prevent lytic development) have been incubated at 40°C, they do not recover immunity when they are returned to 30°C. Since the repressor protein itself can renature to recover its activity after the temperature shift, this implies that the increase in temperature has in some way prevented the synthesis of repressor upon the return to low temperature. Two events are involved in loss of the ability to synthesize repressor: first the inactivation of the thermolabile repressor at high temperature causes transcription of *cI* from $p_{rm}$ to cease; and then the absence of repressor allows synthesis of a phage function (Cro) which prevents the resumption of transcription even when repressor is reactivated by a reduction in temperature.

The *cro* gene was identified by the observation that the introduction of an *x* mutation allows $\lambda cI_{857}$ lysogens to regain immunity when they are shifted from 40°C back to 30°C. The reason for this effect was identified in an experiment in which Eisen et al. constructed partial diploids carrying a $\lambda cI_{857}\ N^-$ prophage (by itself unable to regain immunity) and a $\lambda cI_{857}\ N^-\ x$ prophage (by itself able to regain immunity). After growth at 41°C, the double lysogens do not recover immunity when they are returned to 30°C. Thus failure to recover immunity is dominant over the ability to do so. This implies that the $\lambda cI_{857}\ N^-$ prophage synthesizes a diffusible product that prevents resumption of repressor synthesis from both prophages. The ability of single $\lambda cI_{857}\ x$ lysogens to regain immunity must therefore be due to their failure to make this product; since the *x* mutation inactivates $p_R$, the gene responsible must be part of the operon under the control of this promotor. This gene was named *cro*, the term that we shall follow here; the same locus has been identified in other experiments (by virtue of other properties) in which it was named *tof*, *fed* or *Ai*.

The *cro* function is specific for the immunity of the phage and this was demonstrated by experiments in which Eisen et al. constructed two double lysogens carrying prophages of different immunities. These strains possessed a $\lambda cI_{857}$ prophage and a $\lambda imm^{434}$ prophage carrying a temperature sensitive mutation in its *cI* gene; in each strain one of the prophages carried an *x* mutation and the other possessed a functional $p_R$ promotor. In both strains only the immunity of the mutant carrying the *x*-inactivated $p_R$ promotor was recovered at 30°C after growth at 41°C. Thus the Cro protein synthesized by the phage carrying the active $p_R$ promotor in each case can turn off the synthesis of repressor by its own phage but cannot act on the other phage. This means that Cro is specific for the immunity of the phage that carries it; $\lambda$ and $\lambda imm^{434}$ must carry different *cro* genes.

This places the *cro* gene within the immunity region. Another observation that supports this location is that the *c17* mutation, which lies to the left of *cII* and causes constitutive expression of *cII-O-P* (apparently by creating a new promotor) does not act on *cro;* this implies that *cro* lies between *cI* and *c17*.

Mutants of the $cro^-$ type were isolated by their ability to recover immunity when $\lambda cI_{857}$ lysogens are shifted from 41°C to 30°C; these mutants stop synthesizing repressor at high temperature but recover the ability to do so upon a return to low temperature. At 30°C, the $cro^-$ mutants almost always form lysogens rather than entering the lytic cycle; this suggests that the absence of the Cro protein allows repressor synthesis always to be established, so that the functions necessary for vegetative growth are repressed. As expected from these results, $cro^-$ is recessive to $cro^+$. The absence of Cro protein thus directs the phage into the lysogenic pathway illustrated on the left of Figure 4.34.

The initial failure to synthesize repressor when $\lambda cI_{857}$ is shifted from low to high temperature is due solely to the cessation of transcription of *cI* caused by the absence of active repressor protein. This idea is supported by several analyses of the transcription of *cI* mRNA during heat induction, including those of Spiegelman et al. (1970), Kourilsky et al. (1970) and Kumar et al. (1970). One technique for assaying *cI* mRNA is to use a two step hybridization protocol: first lambda RNA is hybridized to the DNA of $\lambda dv$, which corresponds only to the region from $o_L$ to $Q$ and thus preselects sequences including the immunity region; and then this RNA is eluted and hybridized with the isolated *l* and *r* strands of λ and $\lambda imm^{434}$. The RNA hybridizing to λ but not to $\lambda imm^{434}$ *l* strand DNA in the second step is the transcript of *cI*.

Before induction of $\lambda cI_{857}$, more than 95% of the λ mRNA is derived from the *l* strand of the immunity region, in agreement with earlier results demonstrating that this is the sole region transcribed in lambda lysogens. At 1 minute after a shift to 42°C, the synthesis of *l* strand immunity message falls by 50%; and by 5 minutes it displays a level less than 5% of the lysogenic value. There is a concomitant large increase in transcription from both the *l* and *r* strands outside the immunity region. In $\lambda cI_{857}$ lysogens that carry both an $N^-$ mutation (to block expression of all genes outside the immunity region) and an *x* or $cro^-$ mutation (to prevent synthesis of Cro protein), destruction of the thermolabile repressor has the same effect in eliminating the synthesis of *l* strand immunity message. This means that the decline in repressor synthesis upon a shift to high temperature does not depend upon the synthesis of any new phage protein; this is consistent with the idea that repressor activates its own expression from $p_{rm}$ so that destruction of repressor inhibits transcription of *cI*.

Measurements of repressor activity in $\lambda cI_{857}$ lysogens also show a decline with growth at 42°C. But such experiments do not reveal whether activity alone is reduced or whether synthesis of further molecules of repressor is inhibited. However, the amount of repressor protein present in lysogenic cells has been measured directly by the antigenic assay developed by Reichardt and Kaiser (1971). Although $\lambda cI^+$ strains possess the same amount of repressor at either 30°C or 42°C, lysogens of $\lambda cI_{857}$ *x* or $\lambda cI_{857}$ $cro^-$ contain little repressor protein after growth at 42°C. Since neither of these $\lambda cI_{857}$ prophages can produce Cro protein, the defect in repressor production must lie with the destruction of active repressor protein.

This series of experiments also suggested that the site at which repressor must act to promote its own synthesis may overlap the right operator. When $\lambda cI_{857}$ lysogens were superinfected with $\lambda v1v3$ $cII^-$ phages, the superinfecting phage was unable to produce repressor although the lysogenic cells already contained the active repressor (at 30°C) needed for transcription from $p_{rm}$. (Any repressor made by the superinfecting phage could be distinguished from that of the prophage by its lack of thermolability; it would have to be made by the maintenance system because the $cII^-$ mutation prevents use of the establishment system.) The problem in interpreting this experiment is to decide whether the inactivity of expression of $\lambda v1v3$ from $p_{rm}$ is due directly to the *v1v3* mutations or indirectly to the resulting constitutive synthesis of Cro protein. These experiments suggest that the effect may be direct, because it can be seen at early periods, at times before Cro inhibits the synthesis of repressor (see also Reichardt, 1975b). However, proof that the *v1v3* mutations directly inhibit the repressor maintenance system requires the demonstration that a $\lambda v1v3$ $cro^-$ phage cannot maintain the synthesis of repressor. The conclusion that $\lambda v1v3$ mutants cannot utilize the maintenance system would imply that $p_{rm}$ may overlap with $o_R$, so that repressor binding in the right operator might be necessary both to prevent transcription of the right operon and to promote transcription of *cI* from $p_{rm}$.

The location of $p_{rm}$ has been defined by the isolation of a mutant that appears to inactivate the maintenance system. Yen and Gussin (1973) isolated mutants which are sensitive to infection by phage T4 *rII* (this is selection for the $Rex^-$ phenotype) and sensitive to superinfection by λ (this is selection for the $CI^-$ phenotype). One of these mutants could be characterized as $p_{rm}^-$ by its cis dominant lack of expression of both *rex* and *cI* functions in lysogenic maintenance. Since this mutant can be complemented by *cy* mutants unable to express the establishment system, it must possess a $cI^+$ gene and can establish lysogeny; however, its defect in $p_{rm}$ means that it cannot maintain the lysogenic state unless another phage is present to provide a continuous supply of repressor.

This mutation in $p_{rm}$ maps to the left of *v1v3* but to the right of *cI*. Smith et al. (1976) mapped the site of mutation more precisely by showing that it lies between the ends of two deletions that remove the *cI* gene and end at different points within the right control region. The alteration of sequence caused by the mutation was identified by isolating the part of $o_R$ DNA remaining in a deletion strain by its protection upon repressor binding; this fragment then was hybridized with the mRNA transcript that initiates at $p_o$ and the RNA sequence that hybridizes was determined. The $p_{rm}$ mutation causes a change in the sequence of the last four base pairs of $o_R2$ (reading from right to left in Figure 4.29). On the other hand, Meyer, Kleid and Ptashne (1975) reported that according to direct sequence studies the mutation induces the change shown in this figure, a substitution of the G-C pair immediately to the left of $o_R2$ by an A-T pair. Both results therefore make it clear that $p_{rm}$ overlaps with $o_R$ but differ in whether the site of this mutation lies within or just outside a repressor binding region. Whichever its location, Yen and Gussin showed in their original isolation that the mutation has little or no effect on the ability of repressor to inhibit transcription of the rightward genes in vivo. Of course, the location of a single mutation does not define the limits of the promotor; and we do not know whether it influences initial recognition or stable binding of the polymerase. As shown in Figure 4.29, transcription from $p_{rm}$ is initiated with the AUG codon that starts the *cI* gene, and this implies that a stable binding site for the polymerase must cover this region (see Walz, Pirrotta and Einechen, 1976; Ptashne et al., 1976).

The location of $p_{rm}$ within $o_R$ is consistent with the concept that repressor may control its own synthesis. We have seen already from in vivo experiments that the presence of repressor is necessary for transcription of *cI* message from $p_{rm}$. Another effect has been observed in vitro by Meyer, Kleid and Ptashne. A 790 base pair fragment containing the right control region and the beginning of both genes *cI* and *cro* can be generated by cleavage with the Hae III restriction nuclease. When transcribed by RNA polymerase the products are two RNA chains of about 110 bases, which hybridize with the λ *l* strand and thus appear to represent transcripts of *cI*, and a single chain of about 300 bases, which hybridizes with the *r* strand and corresponds to gene *cro*. Synthesis of the *cI* transcripts, which probably differ in length because they are terminated a few bases apart, does not take place on a template derived from the $p_{rm}$ mutant phage. This confirms that the effect of the mutation is to prevent initiation of *cI* transcription. Addition of small amounts of repressor to the in vitro system stimulates the synthesis of *cI* RNA about threefold; but the addition of greater concentrations represses this transcription. The specificity of this effect is indicated by the observation that, although these concentrations of repressor block transcription of *cro* mRNA from a wild type template, they do not prevent its production when the template has operator-constitutive mutations. When these mutations took the form of the *v1v3* changes in $o_R1$ and $o_R2$, repressor presumably was unable to bind efficiently to these sites, although remaining able to bind to $o_R3$. This suggests that the repression of *cI* transcription may result from repressor binding at $o_R3$ (for review see Ptashne et al., 1976).

These results suggest an autoregulating model for repressor synthesis. The need for active repressor if *cI* is to be transcribed from $p_{rm}$ may indicate that repressor binding at $o_R1$ is necessary to allow RNA polymerase to initiate transcription; at the same time, this repressor binding prevents transcription from $p_R$. The nature of the interaction stimulating transcription from $p_{rm}$ while preventing access of polymerase to $p_R$ is not known. Then when the concentration of repressor increases, binding at $o_R3$ prevents transcription from $p_{rm}$, thus reducing the synthesis of repressor and lowering the concentration in the cell. In this way, the repressor concentration will be prevented from becoming excessive by the turnoff of transcription that occurs at high levels. This model leaves the role of $o_R2$ ambiguous; and, of course, it relies on the assumption that the repression seen in vitro reflects events occurring in vivo.

What is the nature of the inhibition of repressor synthesis caused by the *cro* gene product? We have seen that it is the presence of an active *cro* gene that is responsible for inhibiting the resumption of repressor synthesis in heat induced $\lambda cI_{857}$ lysogens. This effect might be exerted in either of two ways. If resumption occurs by reactivation of $p_{rm}$ as repressor renatures with the shift to lower temperature, Cro must act to prevent transcription from $p_{rm}$. Whereas if resumption were to occur by reactivation of the establishment system, the effect of Cro upon *cII* and *cIII* synthesis (see below) might be responsible. That resumption does not involve the establishment system is implied by the construction of the experiments that we have discussed earlier, since the prophages carried $N^-$ mutations (to prevent entry to the lytic cycle) and so should not have been able to express the delayed early genes, including *cII* and *cIII*. This implies that resumption of repressor synthesis must take place from $p_{rm}$, so that the inhibitory effect of the Cro protein must be exercised at this site.

The properties of a deletion strain isolated by Castellazzi, Brachet and Eisen (1972) support this conclusion. The deletion $\Delta H1$ has its left end between $p_R/o_R$ (as identified by *x13* and *v1v3*) and a marker located in *cro;* its right end lies beyond gene $P$. The region from *cro* to $P$ is therefore deleted. Although $\lambda cI_{857}\delta^{H}1$ thus lacks the sites controlling repressor establishment it is able to recover immunity when it is returned to low temperature after a prolonged incubation at high temperature. This demonstrates that resumption of repressor synthesis does not take place through $p_0$ and presumably therefore involves $p_{rm}$. The $\Delta H1$ deletion removes *cro,* of course, but does not remove the site at which *cro* acts; for if the double lysogen $\lambda cI_{857}\Delta H1$ / $\lambda cI_{857}$ $N^-$ $O^-$ is incubated at high temperature, immunity is not recovered upon a return to low temperature. The product of the $cro^+$ gene of the second prophage must therefore be able to prevent synthesis of repressor from both prophages. This suggests that the site at which *cro* acts in $\lambda\Delta H1$ must lie between $cI$ and *cro,* that is, its action at this site must control transcription from $p_{rm}$. We shall see below that the sites at which *cro* acts when it turns off transcription of the early genes probably can be equated with $o_L$ and $o_R$; thus its binding in the right control site may have two effects, preventing transcription to the left from $p_{rm}$ and inhibiting transcription to the right from $p_R$. Which nucleotide sequences comprise the two binding sites is not known, but one possible model is to suppose that *cro* might compete with the repressor, in which case the success of repressor in binding to $o_R$ would activate $p_{rm}$ and inactivate $p_R$ whereas the success of *cro* would inactivate both $p_{rm}$ and $p_R$.

One of the most interesting features of gene control in lambda is that repressor protein and Cro protein bind to sites within the same control regions, but display antagonistic effects on determining the pathway that the phage follows. Although repressor protein was isolated and its actions in vitro characterized at an early stage, Cro protein has been isolated only more recently. Folkmanis et al. (1976) purified from infected cells a 9000 dalton protein that in dimeric form binds specifically to the immunity region of λ DNA. (Repressor binding was eliminated in these experiments by extracting proteins from cells infected with a $\lambda cI^-$ $N^-$ phage which should be unable to synthesize any repressor.) Cro protein is similar but not identical to repressor in its binding to λ DNA: the binding of both proteins is reduced by the constitutive mutations *v1, v2, v3;* but the *vs*326 mutation in $o_R$ eliminates repressor binding while leaving unimpaired Cro binding. This confirms suggestions that the two proteins recognize the same general regions of λ DNA, but that although their binding sites may overlap they are probably not identical. Clearly it will be revealing to determine which sequences are recognized by Cro and how they are related to the binding sites identified for repressor protein.

### *Turnoff of Repressor Establishment*

The antagonism between *cro* and repressor usually is expressed during the period after infection when the phage must decide whether to establish repressor synthesis (and enter lysogeny) or proceed to express late functions (and enter the lytic cycle). Two types of event have been followed as the consequence of *cro* action: expression of the early genes is much reduced; and repressor synthesis is prevented. Both probably result from a single action of the *cro* protein: binding to the left and right operators. This binding reduces expression of all early genes to a low level, of the order of 10% of the former level. The action of *cro* in turning off the expression of early genes can be seen directly during lytic development by a decline in their transcription which takes place at about the same time that late genes begin to be expressed; this can be seen as a control to shut off the expression of these genes when further synthesis of their products is no longer necessary for lytic development. The effect of *cro* upon repressor synthesis in these circumstances appears to result from its inhibition of early gene expression; the *cII* and *cIII* proteins are no longer synthesized and their capacity to support transcription initiated at $p_0$ declines as they decay rapidly. It is probably this indirect effect of *cro* in preventing transcription via $p_0$ which is responsible initially for directing the phage into lytic development rather than lysogeny, but it is difficult to know to what extent inhibition of transcription from $p_{rm}$ also may be important.

An early observation which revealed the effect of *cro* upon early gene expression was that *x* mutants in $p_R$ overproduce the enzyme exonuclease during lytic development; this enzyme is coded by the left early gene *redα*, previously described as *exo*, and can be assayed fairly easily. It is a delayed early gene whose expression depends upon the activity of the product of gene *N*, necessary for expression of all delayed early genes (see below). Synthesis of exonuclease usually rises when the delayed early genes are expressed and then declines later in development. Pero (1970) proposed that the failure of exonuclease synthesis to decline in $p_R$ mutants reflects the existence of a gene transcribed from this promotor which is responsible for turning off the expression of exonuclease later in development. This gene was named *tof*.

In support of this idea, Pero observed that the overproduction of exonuclease is recessive: double lysogens carrying both an *x* mutant and a wild type prophage are able to turn off the synthesis of exonuclease. A $\lambda imm^{434}$ prophage cannot be substituted for the $\lambda^+$ prophage in the double lysogens, which demonstrates that *tof* must be located in the immunity region. Although the $\lambda imm^{434}$ genome cannot turn off exonuclease synthesis by λ, it can regulate synthesis of enzyme from phage *434*; thus both λ and *434* possess a *tof* gene, which in each case is specific for the phage and must therefore act in the region of nonhomology, that is, in the immunity region. The *tof* function maps in the same position as the *cro* function and their functional identity has been confirmed by showing that $cro^-$ mutants accumulate large amounts of exonuclease. A genetic analysis performed by Takeda et al. (1975b) showed that *tof* and *cro* mutants map in the same region, immediately to the right of the *x13* mutation in $p_R$. Other mutants in this gene have been described as *fed* (Court and Campbell, 1972) or as *Ai* (Kumar et al., 1970), depending upon the criteria used for their isolation. We shall continue to use the description *cro*.

The site where *cro* acts to turn off exonuclease synthesis has been determined by examining phages with deletions in the left arm for their ability to regulate expression of the *exo* (*redα*) gene in the presence of a $cro^+$ gene. Pero (1971) found that phages carrying deletions which extend from the right of the immunity region into *cI* continue to turn off exonuclease. Since the site of action of *cro* must lie within the immunity region, this means that it must be located between *cI* and *N*. That the *cro* recognition site may lie in the left operator is suggested by the observation of Sly et al. (1971) that the *v2* mutation reduces the ability of *cro* to turn off transcription of *l* strand mRNA. Thus *cro* may share with repressor the ability to bind in the left control site to prevent polymerases from initiating transcription at $p_L$.

The inhibition of early transcription by *cro* has been directly followed by Kumar, Calef and Szybalski (1970). Figure 4.38 shows the synthesis of RNA able to hybridize with the *l* strand of the region *int-cIII,* that is, the left delayed early genes. In a $cro^+$ phage, a peak is reached at 5 minutes, after which the rate of synthesis declines to a rate that is about 25% of the peak level. In a $cro^-$ phage, however, synthesis continues at the peak level at least up to 25 minutes after induction. Since transcription of the left early genes is initiated at $p_L$, this is consistent with the idea that *cro* acts at the left control site to inhibit transcription. Its effect is less stringent than that of repressor, however, since it leaves a residual level of expression in contrast with the complete inhibition exerted by repressor.

In these experiments, *cro* did not appear to have any effect upon synthesis of early RNA from the *r* strand. Yet *cro* inhibits the synthesis of both *cIII* and *cII,* left and right early genes, respectively (see below).

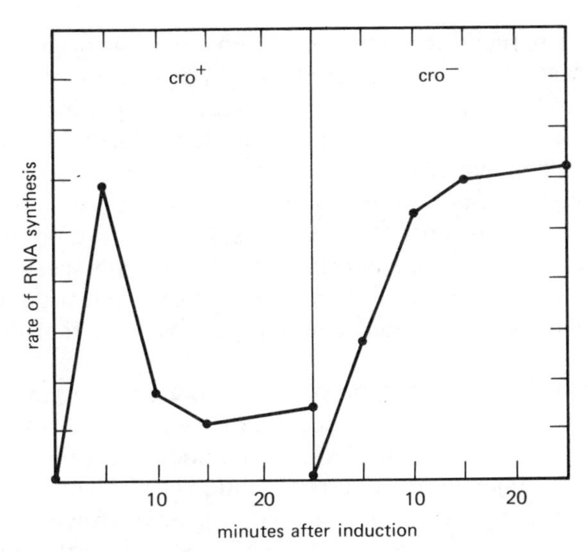


**Figure 4.38:** rate of transcription of left early genes after induction of $cro^+$ and $cro^-$ phages. The figure shows the proportion of radioactivity incorporated in a 2 minute pulse that was able to hybridize with the *l* strand of the region from *int* to *cIII*. Data of Kumar, Calef and Szybalski (1970).

When Takeda, Matsubara and Ogata (1975a) characterized a series of $cro^-$ mutants for their ability to influence early transcription, however, they found two classes: although all the $cro^-$ mutants overproduced early message, one group transcribed the right early message at wild type level, while the other group greatly overproduced the transcript initiated at $p_R$. This suggests that Cro protein acts as both the left and right control sites, but presumably in such a way that mutation can impair its affinity for the left site while leaving the protein able to bind to the right site.

As a protein that represses rightward early transcription, Cro inhibits its own synthesis. Whether this constitutes an autogenous control loop important in the early stages of infection depends upon several factors that are not known, for example the stability of Cro protein. By the later stages of infection, of course, synthesis of Cro protein has ceased either because early transcription has given way to late transcription (upon commitment to the lytic cycle) or because repressor synthesis has been established (for achievement of lysogeny). However, a situation in which *cro* expression does appear to provide a feedback loop is represented by the existence of λ*dv* plasmids. Plasmids of the λ*dv* type are derivatives of lambda which represent only a small part of the genome and may be present in many copies (of the order of 50) in each carrier cell; they behave as self-replicating autonomous elements. The original λ*dv* plasmid isolated by Matsubara and Kaiser (1968) carried the immunity region and some surrounding material. In more recent experiments, Matsubara (1972, 1974a,b, 1976), Matsubara and Takeda (1976) and Berg (1974) have shown that when plasmids of the λ*dv* type are isolated they always contain a "core" region comprising the sequence $p_Ro_R$-*cro-O-P*. Other sequences may be present in individual isolates but are not necessary for self-perpetuation of the plasmid. Because of their possession of the core region, the λ*dv* plasmids all contain the origin for replication, the genes (*O* and *P*) needed for replication, and a regulator (*cro*) that controls their expression.

Phage lambda cannot grow on cells carrying λ*dv*. Originally it seemed that this might be due to the presence in the cell of an excess of repressor protein caused by expression of a large number of copies of the *cI* gene. This was the rationale for the characterization of virulent mutants by their ability to grow on cells carrying λ*dv* (see page 323). However, the demonstration that λ*dv* plasmids may lack the *cI* gene implies that this is not the basis of the interference effect; and Matsubara and Berg both have proposed that it resides rather with the expression of the *cro* gene. As a plasmid, the perpetuation of λ*dv* in the cell may be controlled by the synthesis of *cro,* which represses the operon containing its own gene and also the genes needed for replication of the plasmid; this constitutes an autogenous circuit. The level of the plasmid in the cell, and the ability of other λ phages to grow upon the carrier cell, thus is influenced by mutations affecting the efficiency with which Cro protein can repress rightward transcription. Thus virulent mutants, which have a decreased ability to bind repressor protein, may be able to grow on λ*dv* carrier cells because the mutation also reduces the affinity for Cro protein. The implication is that the DNA sequences recognized by Cro and repressor protein must overlap.

The effect of *cro* upon the establishment of repressor synthesis is shown in Figure 4.39. After infection with a $cro^+$ phage, repressor accumulates rapidly until about 25 minutes after infection, when its synthesis is inhibited. A $cro^-$ phage, however, continues to synthesize repressor for a much longer period (although of course the accumulation of repressor will itself eventually cause the rate of synthesis to drop by switching expression from $p_o$ to $p_{rm}$). A $cro^-$ $cII^-$ phage cannot produce repressor, of course, because it is unable to activate the establishment system. We should note one important point about the timing of the inhibition of repressor synthesis: it is not possible to compare directly the times at which events take place in infections performed under different conditions. The conditions of growth may greatly influence the time

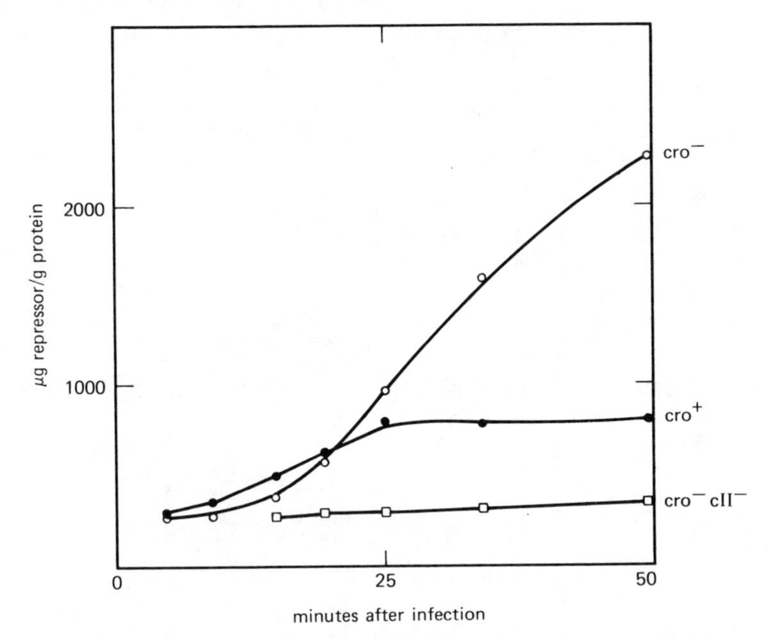


**Figure 4.39:** synthesis of repressor protein after infection. With both *cro*⁺ and *cro*⁻ phages repressor synthesis commences at a rapid rate after a short lag. Repressor synthesis ceases after about 25 minutes in *cro*⁺ phages but continues in *cro*⁻ phages. Mutants in *cII* are unable to commence synthesis of repressor. This infection was carried out at 30°C; an increase in temperature shortens the period of development and at 37°C repressor synthesis is inhibited after about 15 minutes. Data of Reichardt and Kaiser (1971).

occupied by the stages of lambda development. One variable, for example, is temperature; the inhibition of repressor synthesis that occurs at 25–30 minutes in Figure 4.39, which represents an experiment in which cells were incubated at 30°C, occurs some 10 minutes earlier, at 15–20 minutes after infection, when the cells are incubated at 37°C. It is therefore possible only to compare the *relative* timing of events unless infections are performed under identical conditions.

One general problem in understanding the action of *cro* during development is apparent, however, whatever conditions are used. As an immediate early gene, *cro* is one of the first two functions expressed upon infection; yet it does not inhibit expression of the early genes and the synthesis of repressor until after some delay. Although the results of Figures 4.38 and 4.39 cannot be directly compared, it is clear from these and other experiments that there is brief lag (of the order of 5–10 minutes) in inhibition of left early mRNA synthesis, whereas a somewhat longer period intervenes before repressor synthesis is inhibited.

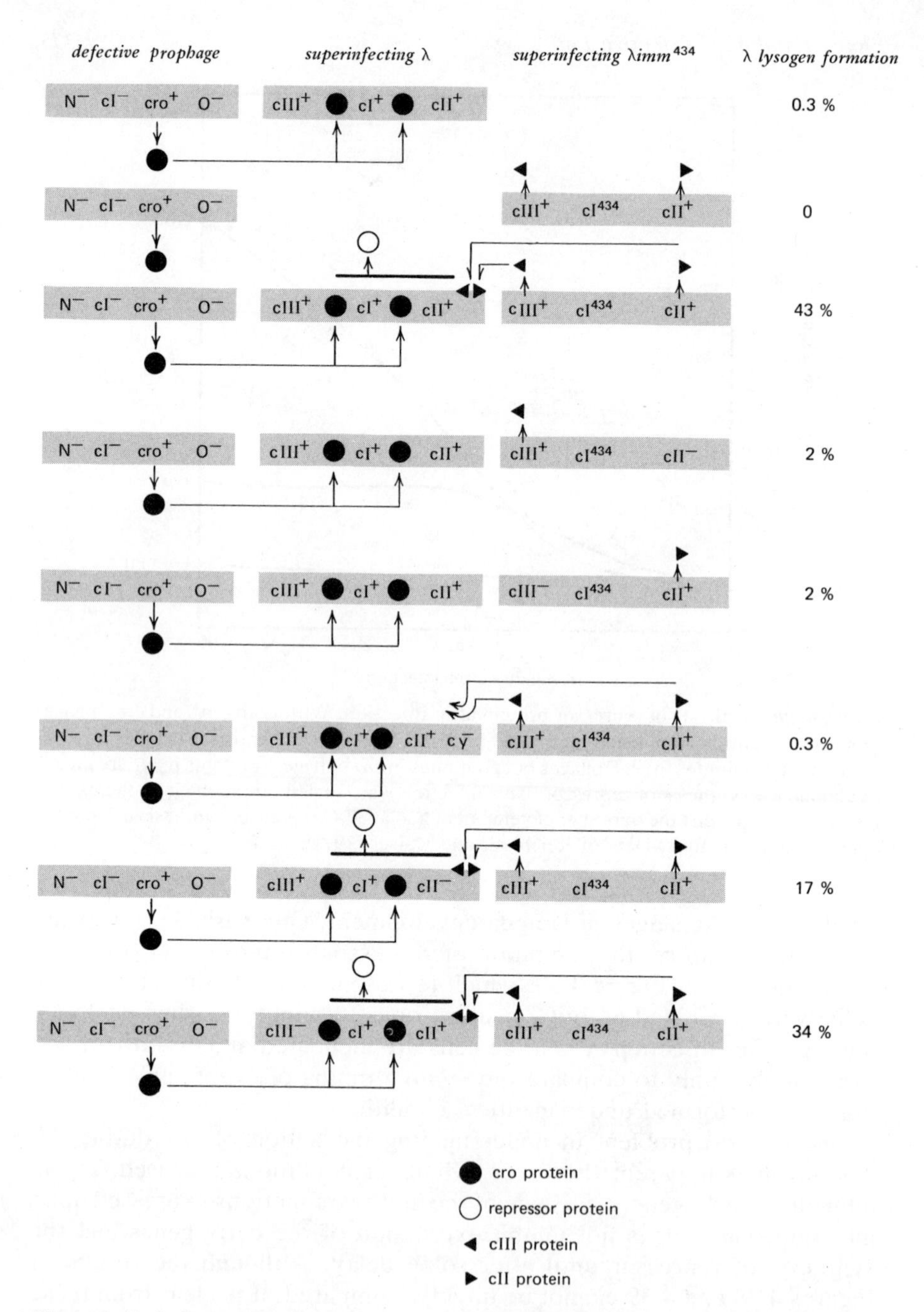


**Figure 4.40:** indirect effect of *cro* upon establishment of repressor synthesis. Defective lysogens carrying the phage genome $N^- \ cI^- \ cro^+ \ O^-$, which can express only the phage *cro* function, were superinfected with the λ and λ$imm^{434}$ phages whose genotypes are shown. The left side of the figure depicts the molecular interactions which occur: Cro protein is

No explanation is yet apparent for the delayed action of *cro* in inhibiting early transcription. That Cro protein acts more slowly than N protein is clear from the data of Figure 4.38. The RNA sequences assayed in these experiments are those of the left delayed early genes, which can be synthesized only in the presence of *N* protein. The figure therefore demonstrates that N must act to promote synthesis of this message before Cro acts to inhibit its transcription. Of course, we can see that this sequence of events must be necessary for phage development, since otherwise *cro* would immediately inhibit expression of all early genes and thus would prevent an infecting phage from entering either lysogeny or the lytic cycle. Yet *N* and *cro* are expressed simultaneously upon infection. Their sequence of action presumably is dictated by the nature of their interactions with the left and right early control sites (where N as well as Cro acts; see below). For example, one speculation would be to suppose that several molecules of RNA polymerase can bind at $p_L$ and at $p_R$ immediately upon infection and that N protein can modify them to allow continuation of transcription into the delayed early genes; Cro protein might then have to compete with further molecules of polymerase for access to the promotor, its successful binding preventing new polymerases from commencing transcription while leaving those previously bound able to do so. However, resolution of the paradox must wait until the N and Cro proteins have been isolated and their molecular actions defined.

The long delay before cessation of repressor synthesis following infection suggests that the effect of Cro is indirect; the timing is consistent with the idea that Cro directly inhibits early transcription, causing synthesis of cII and cIII proteins to cease. Transcription from $p_0$ can then continue into *cI* only while the previously made cII/cIII proteins are available; to fit this model, these proteins should have rather short half lives. When a phage enters the lytic pathway, the inhibition of $p_0$-initiated transcription by decay of cII/cIII appears to occur in time to prevent activation of the maintenance system, which thus probably functions only after a phage has become committed to lysogeny.

That the action of *cro* following infection is to inhibit indirectly the *cII/cIII* repressor establishment system was confirmed by the experiments of Echols et al. (1973) which are summarized in Figure 4.40. The design of these experiments was to test the ability of a superinfecting

---

always present to bind to the superinfecting lambda phage and prevent both leftward and rightward early transcription; but when the lambda phage has a wild type site at the *cy* locus and the $\lambda imm^{434}$ phage is able to provide both *cII* and *cIII* proteins, repressor can be synthesized by the lysogenic establishment system. The frequency of lysogen formation by the *lambda* superinfecting phage is shown on the right. Data of Echols et al. (1973).

lambda phage to lysogenize in a defective lysogen which expresses *cro* constitutively. The frequency of lysogen formation can be taken as a direct measure of the ability of the superinfecting lambda phage to synthesize repressor protein.

Cells carrying the defective phage genome $N^- \ cI^- \ cro^+ \ O^-$ synthesize Cro protein but are unable to express other phage functions. As the first line of the figure shows, virtually no lysogens result when these cells are superinfected with a $\lambda c^+$ phage; the Cro protein that is present in the cell is able to prevent any synthesis of repressor protein by the superinfecting λ phage. When $\lambda imm^{434}$ is used instead as the superinfecting phage, it is not subject to repression by the *cro* of the defective lysogen, but of course it is not possible to form any lysogens of the λ immune specificity, as reported in the second line. But mixed superinfection of the defective lysogen with $\lambda c^+$ and $\lambda imm^{434}$ results in a high frequency of lysogen formation. This implies that expression of the $imm^{434}$ phage provides functions that allow its infecting partner λ phage to synthesize repressor. This confirms that *cro* does not directly inhibit repressor synthesis from λ in this situation. In formal terms, this experiment therefore demonstrates that the product(s) of the genes which are repressed in the lambda phage by *cro* can be supplied by the $\lambda imm^{434}$ phage. The obvious interpretation of this conclusion is to suppose that the $\lambda imm^{434}$ genome provides cII/cIII proteins able to activate the establishment system of the λ phage; the λ phage cannot itself supply these functions because their expression is inhibited by the λ-specific Cro present in the cell. This model is illustrated in the third line of the figure.

If *cII* and/or *cIII* are the genes repressed by *cro*, then $\lambda imm^{434} \ cII^-$ and/or $\lambda imm^{434} \ cIII^-$ should be unable to sponsor lysogen formation by a partner superinfecting lambda phage. The next two lines of the figure show that very few lysogens are produced by mixed infection with $\lambda c^+$ and either $\lambda imm^{434} \ cII^-$ or $\lambda imm^{434} \ cIII^-$. This demonstrates that both the *cII* and *cIII* genes are repressed by *cro*, implying that *cro* acts upon both leftward and rightward early transcription. Confirmation that expression of *cI* in this situation depends upon these functions is provided by the next experiment, which shows that a $\lambda cy^-$ phage cannot be complemented by $\lambda imm^{434} \ c^+$. The final two experiments provide controls which show that $\lambda cII^-$ and $\lambda cIII^-$ phages can form lysogens in the presence of $\lambda imm^{434} \ c^+$; this confirms that it is the *cII* and *cIII* genes of the $\lambda imm^{434}$ superinfecting phage which directly allow the lambda superinfecting phage to lysogenize.

Since both the *cII* and *cIII* genes are transcribed as part of the early messengers transcribed from $p_L$ and $p_R$, it is reasonable to suppose that

Cro can bind at both the left and right control sites to repress their promotors. In this, its action is similar to the repressor protein itself. In addition to inhibiting left and right early transcription, we know that Cro can inhibit expression of $p_{rm}$; the simplest model for its action is to suppose that its binding at the right control site has two effects, inhibiting transcription from both $p_{rm}$ and $p_R$. In this, its action differs from that of repressor, which stimulates $p_{rm}$ while inhibiting $p_R$. One implication of these experiments, implicit in the illustration of Figure 4.40, is that binding of Cro at the right control site does not prevent transcription which has started at $p_o$ from proceeding through *cI*. Indeed, if Cro protein bound at the right control site were able to prevent polymerase from reading through this region, *cro* would display direct rather than indirect inhibition of repressor establishment. We cannot yet interpret the failure of Cro to inhibit polymerase readthrough in terms of its molecular action, except to say that this argues against models where repression is achieved by creating a physical block to polymerase progress but may be consistent with models supposing that repression involves competition between the regulator protein and polymerase for an initial binding site.

An essential corollary of the idea that *cro* inhibits repressor establishment indirectly by repressing early transcription is that one or both of the *cII* and *cIII* protein products must be unstable; otherwise even a transient expression of these genes might produce sufficient activity to maintain transcription of *cI* from $p_o$. The results of an experiment of Reichardt (1975a) are consistent with this model. Cells were infected with $\lambda imm^{434}$ and then after varying intervals superinfected with $\lambda cII^-$ or $\lambda cIII^-$ phages (all these phages carried $O^-$ mutants to prevent them from replicating and destroying the host cells); the amount of lambda repressor that was synthesized was measured by the antigenic assay at 25 minutes after infection with the λ phage. The synthesis of repressor reflects the ability of the $\lambda imm^{434}$ phage to provide the *cII* or *cIII* protein that the λ phage cannot produce for itself. Figure 4.41 shows that the production of λ repressor declines rapidly as the delay between addition of $\lambda imm^{434}$ and addition of λ is increased; a control experiment with $\lambda c^+$, which is able to provide its own *cII* and *cIII* functions, shows a much smaller decline. This experiment shows directly that the ability of the $\lambda imm^{434}$ phage to provide active *cII*/*cIII* proteins declines with time after infection; that this is due to turnoff of these genes by the *434 cro* is shown by the absence of any decline when the $\lambda imm^{434}$ phage carries a $cro^-$ allele. These results therefore demonstrate that Cro causes a rapid decline in *cII*/*cIII* protein activities; this is most simply explained by supposing that inhibition of early transcription is followed by rapid decay of these proteins.

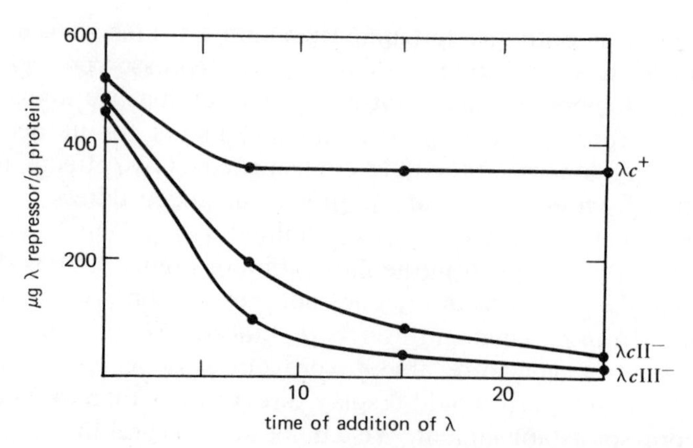


**Figure 4.41:** decline in *cII* and *cIII* expression after infection. At zero time cells were infected with λ*imm*$^{434}$; after intervals varying from 0-25 minutes, the cells were superinfected with λ$c^+$, λ$cII^-$ or λ$cIII^-$. The content of repressor was assayed at 25 minutes after the addition of the lambda phage and indicates the ability of λ*imm*$^{434}$ to provide *cII* and *cIII* functions at the time when λ$cII^-$ or λ$cIII^-$ was added (the λ$c^+$ infection is a control). Data of Reichardt (1975a).

## Regulation of Lytic Development

### *Sequence of Gene Expression*

Lytic development of lambda represents an ordered process of gene expression which depends upon the sequential actions of two positive regulator functions. In formal terms we may describe this as a cascade regulation: for first of all gene *N* must function to allow expression of the delayed early genes; and then one of the delayed early genes, *Q*, must act to allow expression of the late genes. Positive regulation describes a relationship where gene(s) are expressed only if the regulator protein is active; thus mutation of gene *N* or gene *Q* results in inability to express the delayed early or late functions, respectively. The cascade relationship between the two regulators ensures that late functions are expressed only if delayed early functions previously have been expressed. Of course, in order for this lytic pathway to be followed, the functions necessary for lysogeny must be repressed by the negative regulator *cro*. Negative regulation describes a relationship where gene(s) are expressed unless the regulator (repressor) protein is active to prevent their transcription; thus mutation in *cro* creates inability to repress these functions and directs the phage towards lysogeny.

To summarize the functions of the lambda genes that we have described earlier (see page 280), the distinction between early and late genes was first drawn by Jacob et al. (1957): mutants that are defective in early genes cannot replicate DNA; late mutants are defective in stages subsequent to replication. Eisen et al. (1966) identified genes *N, O,* and *P* as the conditional lethal mutants that are able to prevent lytic development. Joyner et al. (1966) and Dove (1966) reported that mutants in *A-M, Q* and *R* replicate normally. The in vitro complementation experiments of Weigle (1966) and Parkinson (1968) demonstrated that the group *A-F* codes for structural components of the head and the group *Z-J* specifies tail components; and Harris et al. (1967) and Mount et al. (1968) showed that *S* and *R* are concerned with lysing the host cell at the completion of development.

Analysis of the control of the lytic cycle of lambda has therefore been concerned largely with defining the actions of genes *N, O, P* and *Q*. Gene *N* can be identified as positive regulator by the failure of $N^-$ mutants to express other functions. The products of *O* and *P* appear to be directly implicated in the replication of λ DNA; since $O^-$ and $P^-$ mutants are unable to express late functions, DNA replication must be necessary for turning on the expression of the head, tail and lysis genes. Since $Q^-$ mutants exhibit normal replication but suffer a pleiotropic defect in all late functions, *Q* appears to be a positive regulator which turns on expression of late genes.

Several early lines of evidence suggested that *N* is a regulator gene which exerts positive control over lambda development. One indication lies in the pleiotropic results of $N^-$ mutation: Protass and Korn (1966) and Radding and Shreffler (1966) reported that $N^-$ mutants are unable to synthesize exonuclease (coded by the left gene *redα*, previously known as *exo*) and endolysin (coded by the right gene *R*). Subsequent experiments have shown that $N^-$ mutants are unable to express any lambda functions other than those of the immunity region (for review see Echols, 1971). That *N* provides a regulator protein was suggested by the results of Radding and Echols (1968), who found that weak suppression of amber mutations in the *N* gene allows synthesis of normal amounts of exonuclease. Since this weak suppression allows synthesis of only a small amount of functional *N* protein, this argues that *N* may provide a regulator protein that need be present in only small quantities to turn on the expression of exonuclease (and presumably the other genes which it controls).

Genetic analysis shows that expression of the *N* gene is directly controlled by the lambda repressor. The technique of heteroimmune infection was used by Thomas (1966, 1970) to determine which functions in a lysogenic prophage can be turned on by a superinfecting phage and which functions remain repressed. The principle of this use of the technique is to use a superinfecting phage that is defective in some gene and therefore unable to complete its lytic cycle; it can do so only if the missing function is supplied by the lysogenic prophage. The appearance of plaques thus indicates that the function missing from the superinfecting phage can be expressed by the lysogen; an alternative use of the technique is to measure the ability of the prophage to provide functions needed to allow the superinfecting phage to lysogenize (see page 310). The superinfecting phage must be heteroimmune in order to avoid repression by the prophage. To exclude the spurious complementation that may result from recombination between the prophage and superinfecting phage, it is necessary to use conditions that inactivate both the host and phage recombination systems; this can be achieved by using *rec*⁻ cells and *red*⁻ phages. Under these conditions, all the late genes of the prophage can complement deficiencies in the heteroimmune superinfecting phage; the presence of repressor protein therefore does not directly inhibit late gene expression (see Figure 4.50). From this we may infer that late genes are turned on at a site that lies outside the immunity region, that is, at a new promotor (now known as $p'_R$). But a $\lambda imm^{434}$ prophage cannot enable a superinfecting $\lambda N^-$ phage to complete its lytic cycle; the expression of gene *N* is therefore directly controlled by repressor protein. Similarly $\lambda red^-$ and $\lambda int^-$ superinfecting phages cannot be complemented by the $\lambda imm^{434}$ prophage; a current interpretation of this result is that transcription of these genes must start at $p_L$, the promotor in the immunity region controlled by repressor. Mutants in gene *Q* are leaky and therefore are difficult to analyze by this approach (see page 404); and the results obtained with superinfecting $\lambda O^-$ and $\lambda P^-$ phages were not unambiguous but appeared to display lack of complementation.

One assay for *N* activity is whether exonuclease enzyme can be synthesized; exonuclease is coded by a delayed early gene, so that although it is usually synthesized at early times, in $N^-$ mutants there is no production of the enzyme. This assay has the advantage that exonuclease activity can readily be measured in extracts in vitro. Figure 4.42 illustrates the protocol of experiments in which Luzzati (1970) examined the action of the *N* gene by superinfecting $\lambda N^-$ lysogens with a $\lambda imm^{434}\ red\alpha^-$ phage (this was in fact a $\lambda imm^{434}\ bio$ phage in which the biotin genes had replaced the *red* genes). Exonuclease can be synthesized in these cells only if the *N* gene of the $\lambda imm^{434}\ red\alpha^-$ phage is able to substitute for the defective function of the $N^-$ gene of the λ prophage. When $\lambda N^-$ lysogens are simply infected with the $\lambda imm^{434}\ red\alpha^-$ phage, there is no synthesis of exonuclease. However, if the lysogens are induced before they are superinfected (the prophage cannot proceed into the lytic cycle because of the deficiency in *N*), exonuclease synthesis occurs. This demonstrates

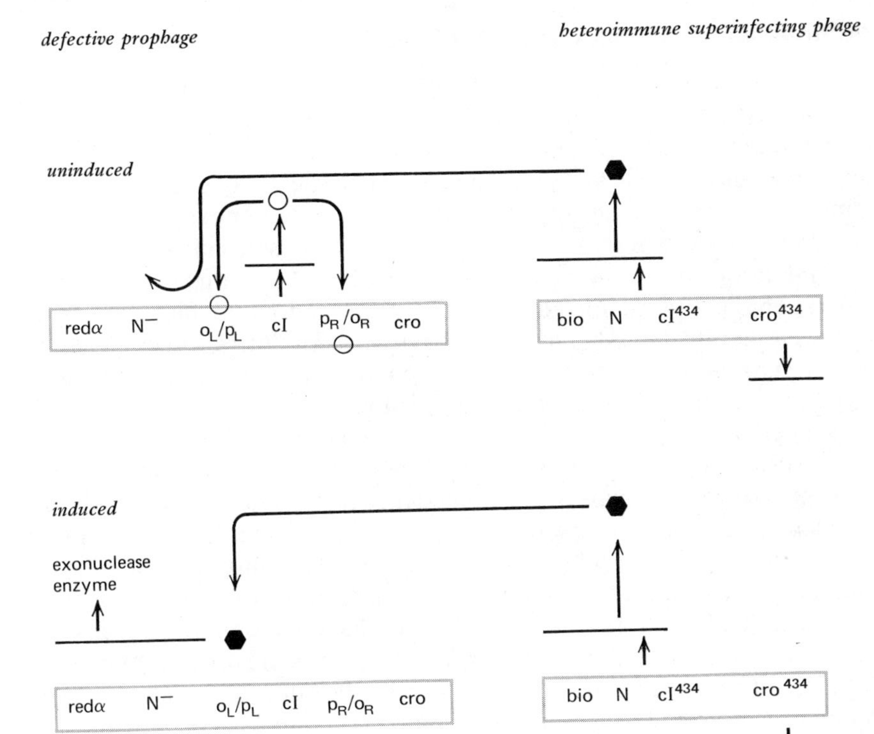


**Figure 4.42:** action of *N* on left delayed early genes. A $\lambda imm^{434}$ superinfecting phage is not able to support transcription of the left delayed early genes when the prophage is repressed. When the prophage is induced, in the presence of the *N* provided by the $\lambda imm^{434}$ phage it is possible to synthesize exonuclease enzyme; this means that transcription of the left delayed early genes must start at $p_L$. Note that this does not imply any particular site of action for the *N* protein.

that *N* function is necessary for exonuclease expression but that the *N* product can act only after repression has been lifted; in view of other data (see below) we can interpret this result as the figure shows to mean that transcription of the left delayed early genes must start at the $p_L$ promotor controlled by the repressor. In other words, this means that messages carrying delayed early sequences must start with the immediate early sequences. This leaves open the question of where the *N* protein acts, except to exclude models supposing that it might (for example) provide a new polymerase activity able to commence transcription at the delayed early genes.

*Positive Control of Delayed Early Genes*

Immediately upon infection of a sensitive cell, the amount of lambda specific mRNA rises rapidly. Early experiments showed that lambda development is accompanied by switches in transcription. Taylor, Hradecna and Szybalski (1967) showed that the early messages found immediately succeeding infection hybridize preferentially with the *l* strand of λ DNA, whereas the late messages synthesized at 30–40 minutes after infection hybridize predominantly with the *r* strand DNA. These experiments led to the further analyses of transcription which have demonstrated that the lambda messengers synthesized during infection can be divided into three classes, which correspond to the classification of genes suggested by analysis of conditional lethal mutants. The immediate early messengers are transcribed from genes *N* and *cro* upon infection; they can be transcribed by $N^-$ mutants or in the presence of chloramphenicol. The early messengers contain both immediate early and delayed early sequences; extension of transcription into the delayed early genes requires the product of gene *N* and thus may be inhibited by introduction of an $N^-$ mutation or addition of inhibitors of protein synthesis. Late messengers are transcribed from genes *S-J* when the phage genome is in a circular state; this takes place after replication and requires the product of gene *Q*. The general course of transcription during lambda development is outlined in Figure 4.43, which summarizes the data of Szybalski et al. (1970). The *l* strand transcript represents the left early genes and rises to a peak at about 5 minutes after the start of development, then declining (due to action of *cro*). The *r* strand transcript from $p_R$-*P* represents the right early genes; only an increase is seen in these experiments, with no decline. The transcript of the *r* strand from *Q-J* reflects expression of the late genes; this increases rapidly after 10 minutes of development.

By using the isolated *l* and *r* strands of the DNA of appropriate deletion and substitution strains as hybridization probes to assay the sequences of lambda RNA species, it has been possible to follow in more detail the transcription of the three classes of genes. Two strains that are especially useful in following the transition from immediate early to delayed early expression are $\lambda imm^{434}$ and $\lambda imm^{21}$ (see Figure 4.13). Hybridization with the *l* strand of $\lambda imm^{21}$ identifies transcripts derived from the region to the left of gene *N*, that is the left delayed early genes; hybridization with the *l* strand of λ assays all leftward genes, so that the difference between λ and $\lambda imm^{21}$ assays transcripts of the left immediate early gene *N*. Similarly, hybridization with the *r* strand of $\lambda imm^{434}$ identifies transcripts derived from the region to the right of *cro*, corresponding to the rightward delayed early genes; hybridization with λ *r* strands again assays all rightward

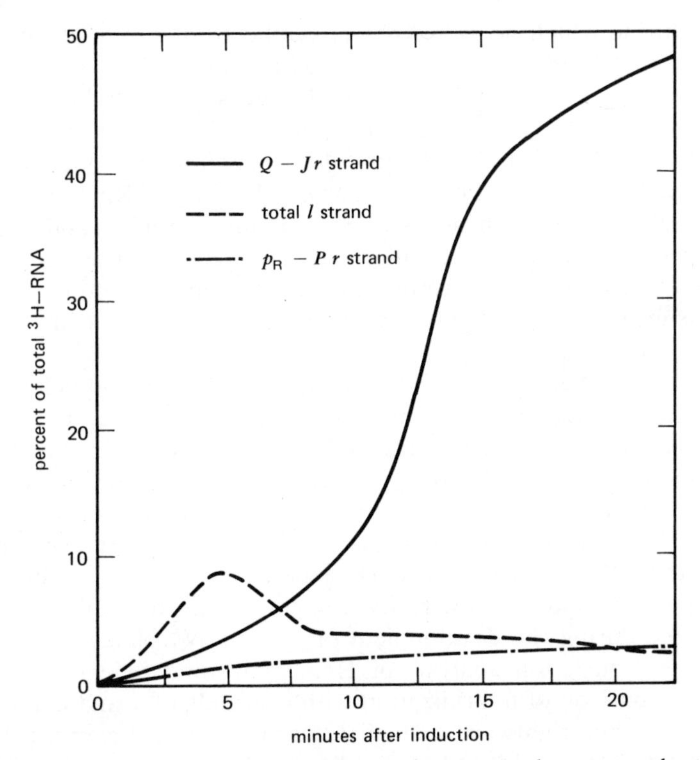


**Figure 4.43:** transcription during lambda development. A $\lambda cI_{857}$ lysogen was heat induced and the cells labeled at various times after induction with 1 minute pulses of $^3$H-uridine. The RNA was then extracted and hybridized with the DNA of appropriate phages to identify the messengers corresponding to the *Q-J r* strand (late genes), total *l* strand (from which only the left early genes are transcribed), $p_R$-*P r* strand (right early genes). Data of Szybalski et al. (1970).

genes, so that the difference between λ and $\lambda imm^{434}$ *r* strand hybridization provides a measure of the transcription of the right immediate early gene *cro*. Of course, at later times the *r* strand transcripts also include late messages, but these can largely be eliminated from the hybridization assay by using DNA from the deletion λ Δ *A-J*. To follow precisely the switch from immediate early to delayed early transcription, it is therefore necessary to compare hybridization with the *l* strands of λ and $\lambda imm^{21}$ and with the *r* strands of λ and $\lambda imm^{434}$.

Transcription at very early times has been followed by Kourilsky et al. (1970, 1971) in a series of experiments in which $\lambda cI_{857}$ lysogens were induced by raising the temperature to 42°C and labeled RNA was extracted at intervals over the next 3 minutes. For the first 10 seconds there is a lag during which no new RNA is synthesized. From 10–30 seconds the RNA that is made can hybridize with λ and λ*imm*$^{434}$ but not with λ*imm*$^{21}$. After 30 seconds the lambda messengers hybridize almost equally well with the DNA of all three phages. These experiments suggest the general conclusion that the first 10 seconds after induction are occupied by the destruction of repressor and binding of RNA polymerase, after which transcription starts at the immunity region and then progresses away from it. Hybridization with isolated *l* and *r* strands showed that the extension of transcription increases at the same time with each strand, suggesting that regulatory events occur simultaneously in both directions of transcription.

Division of the early genes into the immediate early and delayed early classes can be accomplished by two criteria: we have seen that the immediate early genes are expressed at once upon infection whereas expression of the delayed early genes can be prevented either by addition of chloramphenicol or by introduction of an $N^-$ mutation. This suggests that *N* codes for a protein which must be synthesized to enable lambda to express the delayed early genes. That the N protein acts at the level of transcription is shown directly by comparing the messengers that are synthesized during the development of $N^+$ and $N^-$ phages. Figure 4.44 shows a hybridization analysis of Heinemann and Spiegelman (1970); from a comparison of hybridization with *r* strands of λ and λ*imm*$^{434}$ it is apparent that the rightward immediate early genes are expressed to the same extent by $N^+$ and $N^-$ phages, whereas the right delayed early genes are expressed only by phages carrying an $N^+$ allele. An analysis of hybridization with λ *l* strands, which assays transcription from all the leftward early genes, shows that synthesis of this RNA also is highly dependent upon $N^+$ function, implying that the transition from leftward immediate early to delayed early transcription requires N protein. By comparing hybridization of lambda RNA with isolated *l* and *r* strands of λ, λ*imm*$^{434}$ and λ*imm*$^{21}$, Kumar et al. (1970) have shown directly that both leftward and rightward delayed early genes can be transcribed only in the presence of $N^+$ function; they reported also that chloramphenicol has precisely the same effect as $N^-$ mutation. The analysis by hybridization assays of transitions in transcription during lambda development has been reviewed by Szybalski et al. (1969, 1970).

The phase of immediate early gene expression is very brief because transcription and translation of the *N* gene leads very rapidly to transcription of the delayed early genes: the *N* function behaves as a positive regulator whose protein product switches on transcription of the delayed early genes. Two types of model might explain its action: reinitiation or antitermination.

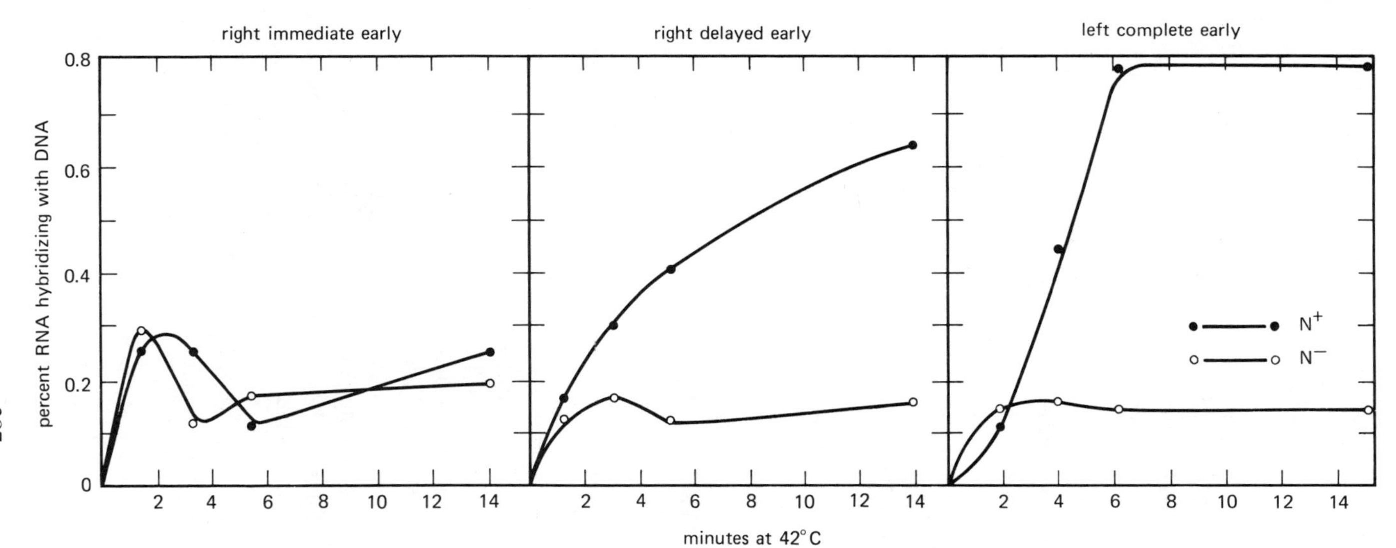


**Figure 4.44:** dependence of early transcription on $N$ function. Each experiment compares the RNA synthesized by an $N^+$ phage with that transcribed from an $N^-$ mutant. Lysogens carrying a prophage with the $cI_{857}$ allele were induced at 42°C and pulse labeled with $^3$H-uridine after various intervals. The labeled RNA was hybridized with the isolated $l$ or $r$ strands of λ and $\lambda imm^{434}$ (see Figure 4.13). The left panel shows the difference between the extents of hybridization with the $r$ strands of λ and $\lambda imm^{434}$; this identifies the transcript of the immediate early right gene *cro*. The center panel shows hybridization with the $r$ strand of $\lambda imm^{434}$; this identifies transcripts of the delayed early genes from *cII* through part of $O$ (the prophage carried a deletion removing material from within and beyond $O$). The right panel shows the extent of hybridization with $l$ strands; this corresponds to transcripts of the complete left early region. Data of Heinemann and Spiegelman (1970).

Reinitiation models suppose that transcription of the delayed early genes starts at new promotors that cannot be recognized by host RNA polymerase alone but that can be utilized in the presence of N protein. These promotors would thus differ from the early promotors, $p_L$ and $p_R$, which are recognized by the host polymerase of the uninfected cell. This model would require the presence of a leftward delayed early promotor located immediately before *cIII*, the first left delayed early gene, with a corresponding promotor for the rightward delayed early genes lying before *cII*.

Positive regulation of this type is directly analogous to the negative regulation exerted by repressor proteins: whereas the repressor prevents RNA polymerase from binding to its promotor, the positive regulator is necessary if polymerase is to do so (for review see Volume 1 of Gene Expression). Two forms can immediately be envisaged for such a positive regulator protein: it may provide an ancillary activity necessary to enable RNA polymerase to bind to a particular class of promotors (an example of this type of mechanism is provided by the cAMP/CAP activator characterized most fully in the lactose operon of E.coli); or it may modify the RNA polymerase of the host (an example is provided by the changes that are introduced in polymerase in sporulating bacteria). A third possibility is that the positive regulator gene might code for a new RNA polymerase, one that recognizes the appropriate class of promotors; this type of mechanism appears to be used by phages T3 and T7 (see Chapter 7).

Bacterial RNA polymerase can be separated into two components: the core enzyme, of subunit structure $\alpha_2\beta\beta'$, which possesses the activity to catalyze RNA synthesis; and the sigma ($\sigma$) factor which participates only in initiation and appears to function by stabilizing the binding of RNA polymerase to the promotor (see Volume 1 of Gene Expression). RNA polymerase might therefore be directed to recognize new promotors instead of or in addition to its previous promotors by the substitution of a phage coded factor for the host coded factor of the enzyme. The product of the *N* gene might therefore be a new sigma factor able to alter the specificity of RNA polymerase. We have dwelt upon this model because it has seemed likely that it might be utilized in phage infection; at one time it was proposed for transcription of late lambda genes (see below) and in phage T4 development (see Chapter 6).

Antitermination models are suggested by the contiguity of immediate early and delayed early genes. The immediate early transcription of genes *N* and *cro* alone implies that transcription must be terminated at the ends of both genes; the termination sites are described as $t_L$ and $t_{R1}$. Since the delayed early genes in both the leftward (*cIII-int*) and rightward (*cII-Q*) directions are immediately adjacent to genes *N* and *cro*, respectively, the failure of polymerase to terminate at $t_L$ and $t_{R1}$ would allow enzyme molecules that had initiated transcription at $p_L$ and $p_R$ to readthrough into the delayed early genes. Thus *N* could act as a positive regulator by preventing transcription from terminating at $t_L$ and $t_{R1}$. We can imagine two models for such antitermination: the antiterminator might act at the *t* sites themselves; or it might act at the promotors to modify RNA polymerase so that it fails to terminate when it later reaches the *t* sites.

The general features of models for reinitiation and termination are illustrated in Figure 4.45. Since reinitiation models postulate that the delayed early promotors are different from the immediate early promotors, they predict that delayed early and immediate early sequences should be found on different messenger molecules. Antitermination models propose that transcription starts only at a single class of early promotors, so that readthrough past the termination sites generates messengers that carry both immediate early and delayed early sequences. In lambda this would imply that all early transcription must start at the promotors $p_L$ and $p_R$; since these promotors are controlled by their adjacent operators, antitermination models predict that both delayed early and immediate early transcription is subject to control by repressor, whereas reinitiation models imply that only the immediate early genes are directly subject to this repression. This suggests that there may be an advantage for the phage in utilizing antitermination rather than reinitiation. The decision on lysogenic versus lytic development is made only after expression of the delayed early genes; thus if the phage enters the lysogenic pathway it must switch off the delayed early genes. Antitermination models imply that this may be achieved directly by synthesis of the repressor protein, whereas reinitiation models would require the assistance of another mechanism, for example, instability of the N protein. (In fact, there is some evidence to suggest that N protein has a short half life, although this would be relevant to this aspect of control only if *N* functions in reinitiation; see Reichardt, 1975b.)

### *Readthrough Past Transcription Termination Sites*

The results of experiments to test the predictions made by reinitiation and antitermination models suggest that *N* codes for a protein that suppresses rho-dependent termination. We have seen already that delayed early genes in a repressed prophage cannot be trans induced by a heteroimmune superinfecting phage; this means that the *N* protein is unable to exert its function when repression is maintained at the early control sites. This is therefore consistent with an antitermination model in which $p_L$ and $p_R$ serve as the promotors for both immediate early and delayed early

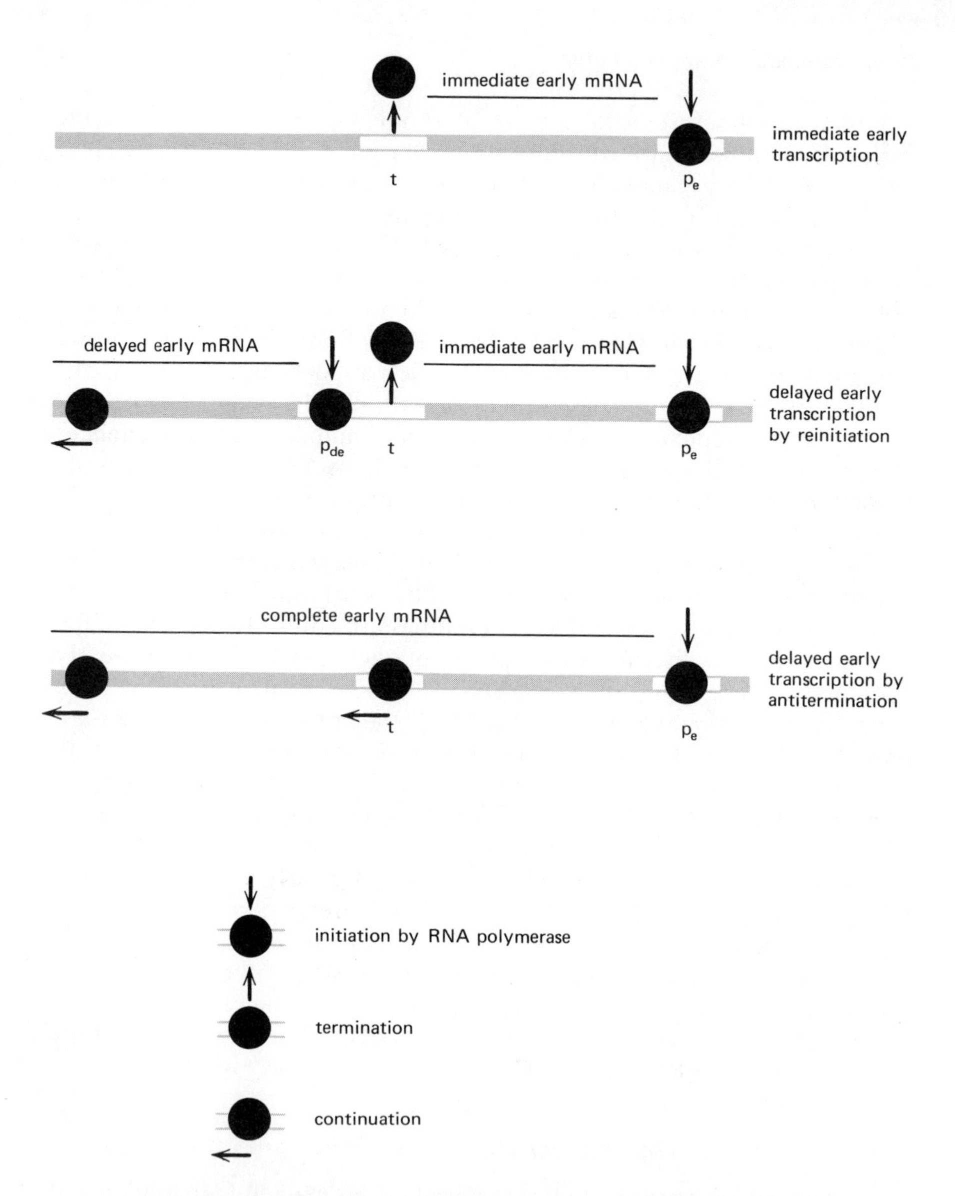


**Figure 4.45:** models for transcription of delayed early genes. The immediate early genes are transcribed by a host polymerase which initiates transcription at the promotor $p_e$ and terminates at the site $t$. Reinitiation models propose that a modified polymerase is able to bind at the promotor $p_{de}$ to initiate transcription of the delayed early genes while immediate early genes may continue to be transcribed as before; thus delayed early sequences are carried on messengers different from those representing the immediate early sequences. Antitermination models suppose that polymerase continues to initiate at $p_e$ but is enabled to continue transcription past the termination site; thus delayed early sequences are carried on messages which must also carry the immediate early sequences.

transcription. This conclusion is supported by the observation that *sex* and *x* mutations display cis dominant reduction of expression of the leftward and rightward delayed early genes, respectively. Hybridization analyses of the messengers synthesized at early times after cells are infected with $\lambda^+$ or $\lambda N^-$ phages suggest that *N* function allows immediate early messengers to be extended to carry the delayed early sequences. The mechanism of termination at the ends of the immediate early genes has been studied in vitro, when addition of rho factor generates the 12S *N* message and 7S *cro* message. The mechanism of antitermination has been inferred from the abilities of polymerases initiating at different promotors to terminate at the same rho-dependent site.

In a series of experiments performed to characterize the early messengers, Kourilsky et al. (1968, 1971) and Portier et al. (1972) isolated the RNA molecules synthesized shortly after heat induction of $\lambda cI_{857}$ in either the presence or absence of chloramphenicol. Attempts to determine the sizes of these RNAs by acrylamide gel electrophoresis encountered difficulties due to cleavage during preparation, but it is clear that the messages transcribed in the absence of chloramphenicol are somewhat longer than those produced in the presence of the inhibitor. Hybridization with isolated *l* and *r* strands shows that in each direction there appears to be a short molecule transcribed within or close to the immunity region in the presence of chloramphenicol, while the longer molecule that can be transcribed when the drug is absent contains sequences extending further away from the immunity region. Competitive hybridization between the two leftward transcripts showed that the longer molecule contains the sequences present in the shorter message. These results are therefore consistent with an antitermination model.

The events involved in termination have been followed in the in vitro transcription system developed by Roberts (1969b, 1970). When λ DNA is transcribed in vitro by E.coli RNA polymerase, products corresponding to specific genes cannot be isolated (see page 313). But when rho factor is added, the principal products are two RNA species sedimenting at 12S and 7S. Figure 4.16 shows that the 12S message corresponds to a transcript of the left strand of *N* while the 7S message is a transcript of the right strand of *cro*. The only effect which rho factor displays in this system is to cause transcription to terminate at the ends of these genes, at the sites $t_L$ to the left of *N* and $t_{R1}$ to the right of *cro*. That the transcription terminated at $t_L$ and $t_{R1}$ has started at $p_L$ and $p_R$, respectively, is shown by Figures 4.17 and 4.18, which demonstrate that mutations of the *sex* and *x* types prevent transcription of the 12S and 7S messages.

Evidence for the existence of a second site for the termination of rightward transcription is provided by the expression of the mutant $\lambda c17$.

The *c17* mutation lies between *cro* and *cII*, close to the *y* region. In either the prophage state or immediately upon infection, λ*c17* mutants express genes *O* and *P* constitutively. The *c17* mutation thus overcomes the usual dependence on gene *N* for replication, as seen in the ability of λ*c17* $N^-$ mutants to replicate extensively (Sly and Rabideau, 1969). By using the DNA of λ*c17* as a template for transcription in vitro, Roberts (1969a) showed that the *r* strand of the *O-P* region is represented in the RNA products to a very much greater extent than is seen when wild type λ DNA is used. This suggests that the *c17* mutation may create a new promotor; RNA polymerase molecules cannot be prevented by repressor from binding at this site and therefore transcribe the *cII-O-P* genes constitutively. Roberts (1970) demonstrated that two extra peaks are generated when λ*c17* DNA is transcribed in the rho-dependent in vitro system. Figure 4.46 shows that the RNA synthesized from λ*c17* DNA falls into classes sedimenting at 7S, 12S, 16S and 22S; only the 12S and 7S messengers are seen when λ DNA is transcribed. Hybridization of the isolated RNA fractions with separated *l* strands and *r* strands of λ DNA confirms that the 12S RNA is a leftward transcript, while the 16S and 22S RNAs are rightward transcripts like the 7S RNA (not shown in the figure).

Why does the *c17* mutation generate two new peaks of RNA? When the λ*c17* DNA is transcribed in this in vitro system in the absence of rho factor, the RNA products display the dispersed size distribution characteristic of RNA transcribed from λ DNA under these conditions, but with the addition of a peak at 22S. This suggests that the 16S peak is generated by the action of rho factor, but that the reaction is not completely efficient so that some RNA polymerase molecules continue past the rho dependent termination site until they reach another site where a termination event that does not depend upon rho generates the 22S RNA. Apparently specific termination of transcription occurs on some DNA templates when the reaction is performed in high salt concentration (see Volume 1 of Gene Expression) and it is possible that a reaction of this type may be responsible for termination of the 22S RNA. The constitutive expression of genes *O* and *P* but not *Q* in λ*c17* mutants suggests that the rho-dependent termination site lies between genes *P* and *Q*; this has been described as $t_{R2}$ by Szybalski et al. (1970). This is consistent with the size of the 16S RNA.

Since *Q* is expressed as a delayed early gene under the positive regulation of *N*, a polymerase molecule that initiates transcription at $p_R$ must be enabled by *N* function to proceed past $t_{R1}$ through *cII, O, P* and then past $t_{R2}$ into gene *Q*. No termination site for this transcription beyond *Q* has been identified in vivo; nor has any site been identified for termination of leftward transcription beyond $t_L$. If any such sites exist, they must of

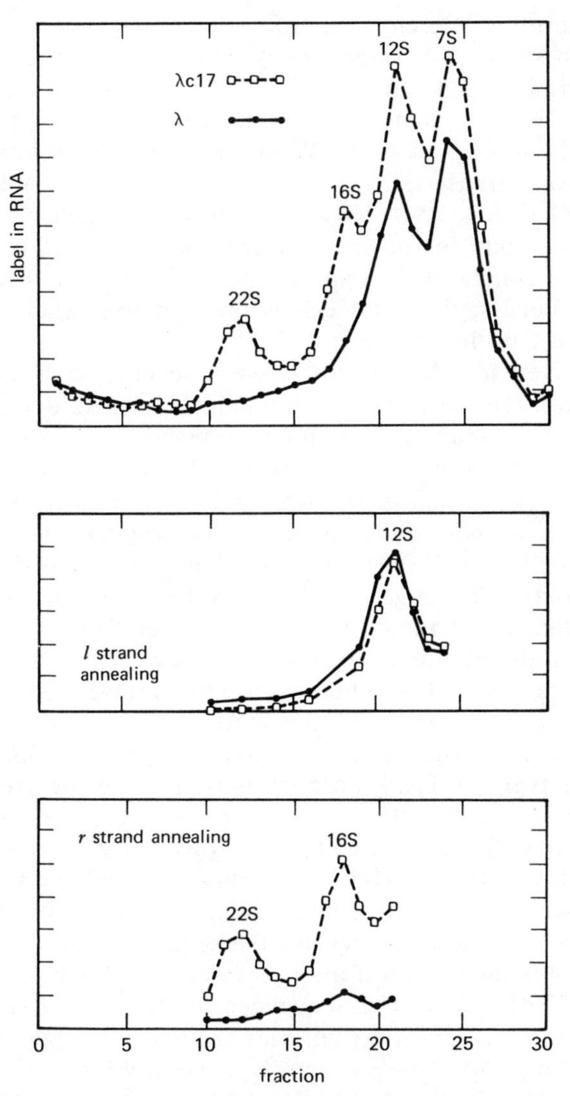


**Figure 4.46:** effect of rho factor on the transcription of λ*c17* DNA in vitro. The DNA of λ is transcribed into 7S and 12S RNA; DNA of λ*c17* is transcribed into 7S, 12S, 16S and 22S RNAs. The 12S RNA hybridizes only with the *l* strand of λ DNA; the 16S and 22S (and the 7S: not shown) RNAs hybridize with the *r* strand. Data of Roberts (1970).

course comprise a different class from the rho dependent *t* sites where *N* function can suppress termination. We do not know the location of the site that terminates the rightward 22S RNA in vitro and there is no evidence on whether it might be utilized in vivo or may represent an artifact of the in vitro system. When the phage has entered its circular state, rightward transcription past *Q* may therefore be able to continue through the late genes *S-J*; leftward transcription, generating a message that does not code for proteins, might also continue through the late genes, but of course in the opposite direction, from *J* towards *S*. If no termination occurs, the polymerases proceeding in each direction may therefore meet in the late genes.

Cells infected by $\lambda N^+$ phages possess several species of RNA that hybridize with the *l* strand and, although these were detected in earlier experiments to characterize the in vivo transcripts, their relationship has been defined only more recently. Lozeron, Dahlberg and Szybalski (1976) observed that about half of the RNA initiated at $p_L$ consists of a large species of about 36S, which appears to correspond to the region from gene *N* to *att*; this is the species denoted as $l_3$ by Kourilsky et al. (1968, 1971). The only other large molecules are found in fractions of 23S and 12-14S and may be degradation products derived from $p_L$ RNA by loss of material towards the 3′ end. In cells infected with $N^-$ phages these large transcripts are replaced with the 12S product transcribed in vitro, representing the region from $p_L$ to $t_L$; this is the $l_2$ species. In both $N^+$ and $N^-$ infection, there is a large peak of RNA sedimenting at 4-5S. This appears to comprise two species, whose proportions may be changed by the conditions under which the cells are grown. Determination of the sequences of these RNAs showed that the $l_1^\circ$ RNA represents up to the first 71 nucleotides of the $p_L$ RNA transcribed in vitro; its 3′ terminal end varies, probably because some nucleotides may be lost by exonucleolytic degradation. The $l_1$ RNA represents the sequence from the 89th nucleotide to the 190th nucleotide of the $p_L$ transcript and it appears to be more stable than the $l_1^\circ$ RNA. The sequences of the $l_1^\circ$ and $l_1$ RNAs are not found within the $l_3$ transcript; but they represent the initial sequence of the $l_2$ (12S) transcript. The positions of these RNA species are shown in Figure 4.47. The relationship between them suggests that the delayed early $p_L$ RNA synthesized in $N^+$ infection is cleaved into three species, $l_1^\circ$, $l_1$ and $l_3$; the gap between the end of $l_1^\circ$ and the start of $l_1$ implies that a small fragment must be lost here during cleavage. The same cleavage events occur in $N^-$ infection, when the $l_1^\circ$ and $l_1$ RNAs can be detected as well as the $l_2$ (12S) RNA that represents the uncleaved transcript from $p_L$-$t_L$. Cleavage is caused by the enzyme RNAase III that processes T7 mRNA and rRNA and which is discussed in more detail in Chapter 7. The role of this processing in the subsequent stages of gene expression is not yet clear.

The transcriptional units of lambda are summarized in Figure 4.47, which shows the phage in the circular state. The two transcripts needed for lysogeny are derived from the left strand and start at $p_o$ and $p_{rm}$. The immediate early leftward transcript is the 12S *N* message; when termination is suppressed at $t_L$ this messenger may be extended further, apparently indefinitely. The immediate early rightward transcript is the 7S *cro* message terminating at $t_{R1}$; as the figure shows, the *cro* region may therefore be transcribed in both direction. In wild type λ DNA, the $t_{R2}$ site presumably is not utilized for termination since $t_{R1}$ is sufficient to prevent polymerases from proceeding in the rightward direction. In the presence

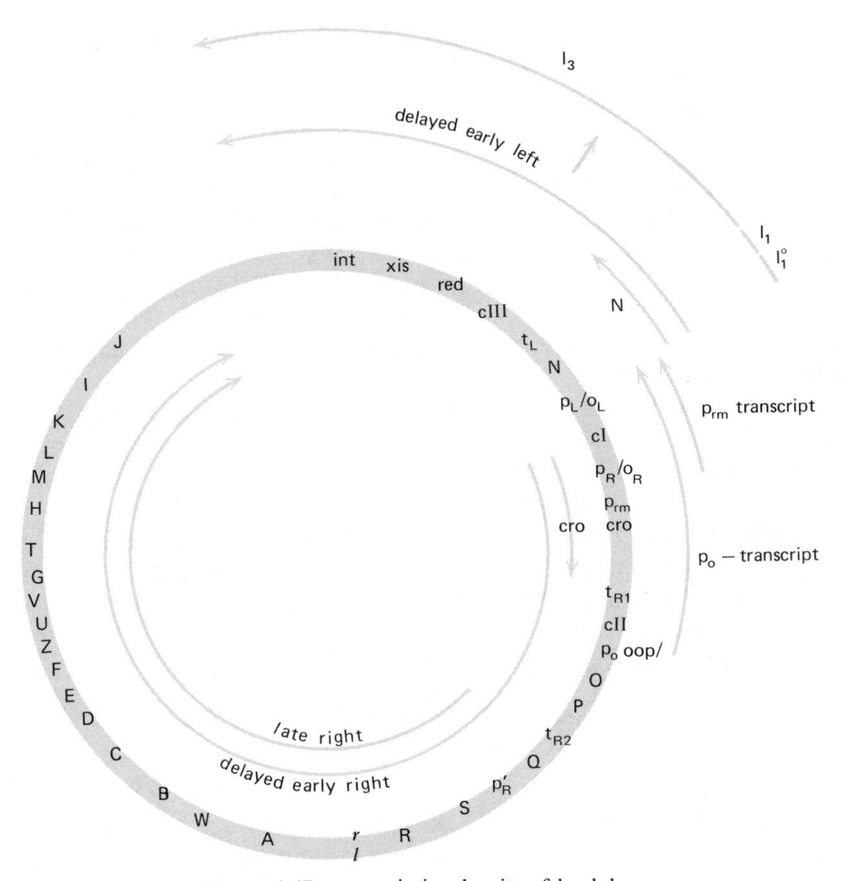


**Figure 4.47:** transcriptional units of lambda.

of *N* protein, transcription proceeds past both $t_R$ sites, presumably again indefinitely. Although the late genes are transcribed in the proper direction by the early polymerases that proceed rightward past *Q*, the extent of transcription from $p_R$ is not great enough to substitute for the transcription that initiates at $p'_R$, so the continuation of early transcription supports only a residual level of late gene expression. Presumably this difference reflects the need for only small amounts of the regulator (*cro* and *cII*), recombination (*int, xis, red*) and replication enzyme (*O* and *P*) proteins, compared with the much larger amounts required of the structural components of the phage particle.

One approach to defining the action of a regulator gene is to isolate mutants that are independent of it. Mutants that are able to form plaques in spite of their lack of an *N* gene have been isolated in two series of experiments. Starting with λ*c17* $N^-$ mutants, Butler and Echols (1970) and Hopkins (1970) isolated *byp* (*N*-bypass) mutants; the *byp* mutation appears to lie between *P* and *Q*. Thus λ*c17 byp* mutants probably can proceed through lytic development because the combination of the two mutations allows expression of *O, P* and *Q*: the *c17* mutation allows constitutive expression of *O* and *P*; and the *byp* mutation allows expression of *Q* and thus of the late genes. In earlier experiments, Court and Sato (1969) isolated a *nin* (*N* independent) mutation, which takes the form of a deletion of much of the material between *P* and *Q*. Mark (1973) has confirmed that *byp* and *nin* mutations increase transcription of *Q* and the subsequent genes. These mutations probably render $t_{R2}$ inactive, allowing all transcription through *P* to continue through *Q*, so that *N* function is not needed. The difference between *byp* and *nin* is that the *byp* mutation confers independence of *N* only in conjuction with *c17* whereas *nin* appears to be effective by itself, leaving unanswered the question of how replication is achieved in the absence of *N* in λ*nin* phages.

Another approach to investigating the action of *N* is to isolate bacterial host cells in which *N* fails to function: the nature of the antagonism might be useful in establishing the role of the regulator. Georgopoulos (1971) isolated *groN* bacterial mutants in which growth of lambda is arrested, apparently because *N* product fails to function. However, the nature of the *groN* mutation has not yet been established.

### *Site of Action of the Antitermination Function*

Direct confirmation that it is the rho factor that terminates transcription of the immediate early genes during infection can in principle be provided by following infection of mutant host cells that are defective in rho. Attempts to isolate rho mutants have been based on the rationale that polarity may be caused by the termination of transcription and that mutants suppressing polarity therefore may specify altered rho factors. (The phenomenon of polarity and the basis for supposing that it is caused by premature termination of transcription is discussed in Volume 1 of Gene Expression; and see page 403 below.) When rho factor was purified from the original polarity suppressor strain, *suA*, Richardson, Grimley and Lowery (1975) found that it was unable to terminate transcription at some sites usually recognized by the factor. By isolating mutants unable to terminate transcription at a polar site in the *gal* operon, Das, Court and Adhya (1976) obtained a conditional lethal strain in which rho is defective; the mutation maps in the same region as *suA*. While $\lambda N^-$ cannot grow on the wild type parent host strain, it is able to grow on the rho mutant. A similar conclusion was reported by Korn and Yanofsky (1976), who isolated *psu* mutants able to suppress polarity and stated that some map in the *suA* region, possess altered rho factors, and allow the growth of $\lambda N^-$. Other bacterial mutants in this region able to support the growth of $\lambda N^-$ have been isolated by Brunel and Davison (1975) and Inoko and Imai (1976), although their rho factors have not been examined. While a more complete genetic characterization of the relationship between these mutants will be necessary to define the gene(s) coding for rho, these results support the conclusion that this function resides in the *suA* region of the E.coli genome and may be mutated so as to allow suppression of polarity to occur by virtue of failure to terminate at some sites; this may allow growth of $N^-$ mutants of lambda that lack the antitermination function. These host mutations may affect differentially the ability to terminate at different classes of site, at least as represented by the IS sequences that cause polarity, and comparisons between their rho factors may therefore yield further information on the nature of the activity of the termination factor (see Reyes, Gottesman and Adhya, 1976). Of course, rho may not be the only function that affects polarity and termination; and further such functions may be identified by obtaining mutants of bacterial strains diploid for the rho factor that either suppress polarity or allow growth of $\lambda N^-$ phages. Indeed, that mutations other than those in rho can affect antitermination is shown by the report of Lecocq and Dambly (1976) that the need for *N* and *cro* functions can be overcome by mutations in the bacterial RNA polymerase.

Two models for the action of *N* function in antagonizing rho-dependent termination are illustrated in Figure 4.48. We may assume that rho acts directly at the *t* site to cause RNA polymerase to terminate transcription. One model for antitermination is to suppose that N protein also acts at the *t* site, in some way preventing rho from interacting with RNA polymerase. This model predicts that the synthesis of N protein should

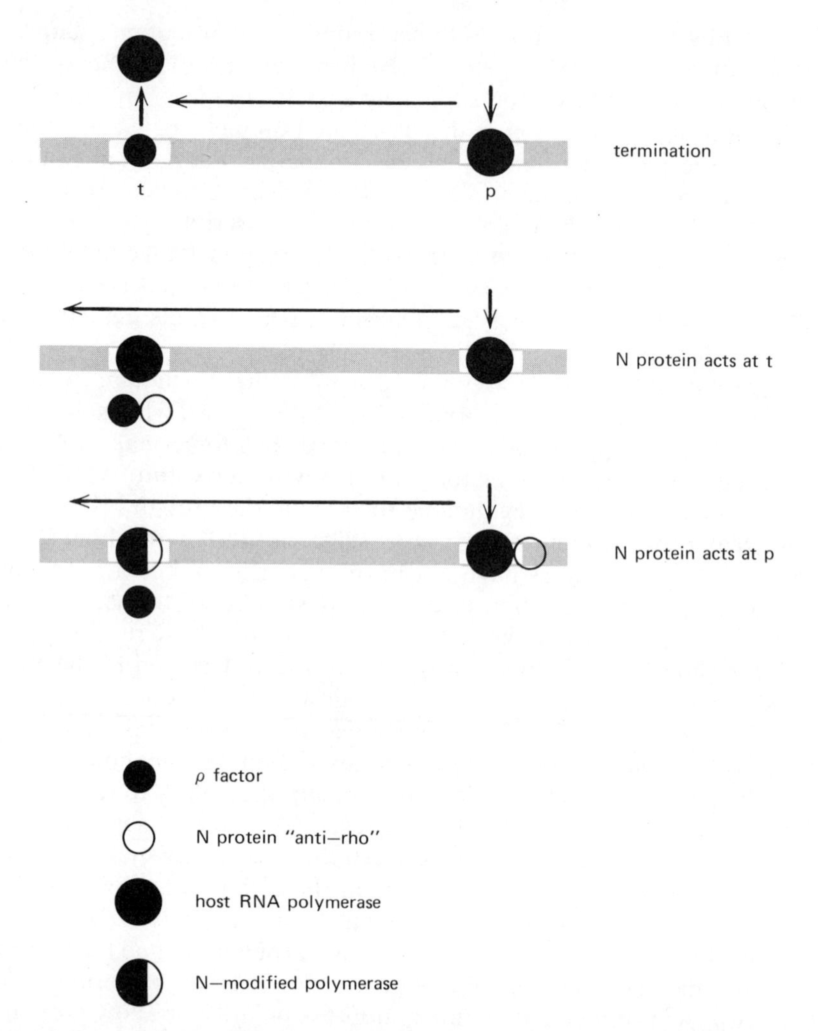


**Figure 4.48:** models for antitermination. Transcription is terminated when the rho factor acts at *t*. The *N* protein might act at *t* to counteract the action of rho, thus allowing RNA polymerase to read through the termination site. Or the *N* protein might act at *p* to modify RNA polymerase so that it is able to continue transcription past rho-dependent termination sites.

suppress termination at all rho-dependent sites in the infected cell. An alternative model is to suppose that *N* product acts at the sites $p_L$ and $p_R$ where transcription is initiated, in some way modifying RNA polymerase so that when it subsequently reaches the *t* site it fails to interact with rho and continues transcription. In formal terms this model predicts that only polymerases initiating transcription at promotors that are targets for *N* function should be immune from rho-dependent termination. With λ DNA this means that polymerases initiating at $p_L$ and $p_R$ should fail to terminate transcription at rho dependent *t* sites, but polymerases initiating transcription at any other promotors should continue to suffer rho-dependent termination.

The site of action of *N* protein has been investigated by using phages such as λ*trp* and λ*gal* in which bacterial genes have replaced some of the phage genes in the left arm. The bacterial genes may be expressed in either of the two ways illustrated in Figure 4.49. When the bacterial genes have retained their operator/promotor control site, they may be independently transcribed, subject to the usual control by induction or repres-

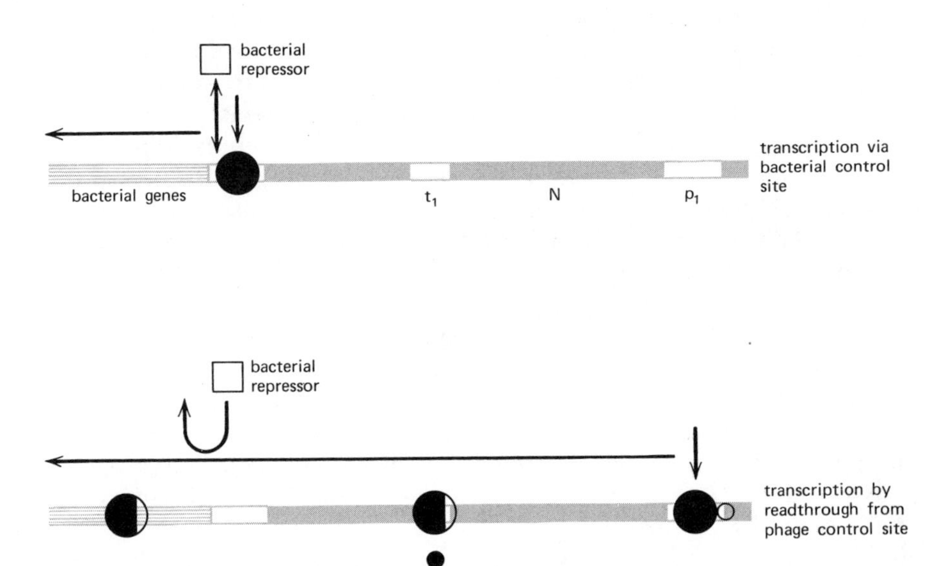


**Figure 4.49:** models for expression of bacterial genes in the lambda left arm. RNA polymerase may recognize the bacterial promotor in which case expression is subject to the usual regulation, for example, may be prevented by the repressor of the bacterial operon as shown. RNA polymerase that has initiated transcription at $p_L$ may read through into the bacterial genes so long as *N* function is present to prevent termination at $t_L$; this read through transcription cannot be influenced by the factors that regulate the bacterial operon, for example, it is not inhibited by bacterial repressor.

sion. They may also be transcribed by RNA polymerase molecules that have initiated transcription at $p_L$ and which read through the intervening phage genetic material into the bacterial genes. The two classes of transcription usually appear to be additive. The extent of transcription from the bacterial promotor can be assayed relative to its absence under conditions that repress expression of the bacterial operon. Since the readthrough transcription initiates in the phage immunity region, it is subject to inhibition by lambda repressor and can occur only when *N* function is present to allow transcription to proceed beyond $t_L$. When the bacterial genes have been inserted into the phage without their operator/promotor control site, they can be expressed only by readthrough transcription, a situation which can be recognized by the complete failure of their expression to respond to bacterial regulation. In this case, the phage genes represent functions under delayed early control. An advantage of such systems is that the bacterial enzymes often can be assayed very easily.

When a bacterial operon is carried on a phage genome, its genes may therefore be expressed in conditions in which initiation does not take place at the promotor of the operon. This expression is known as *escape synthesis* and may in fact result from either of two mechanisms: polymerase may readthrough into the bacterial genes from a phage promotor; or replication of the transducing phage may increase the number of copies of the bacterial structural genes, so that their repressor protein is titrated out by the increase in the number of operators (clearly this happens only when the gene for the repressor is not also carried on the phage). Adhya, Gottesman and De Crombrugghe (1974) have made use of escape synthesis of the enzyme of the galactose operon of a λ*gal* phage in order to define the site of action of the *N* protein. The λ*gal* phage carried $O^-$ or $P^-$ mutations to prevent escape synthesis by replication, so that any production of enzyme activity not due to initiation at the *gal* promotor can be attributed to readthrough from a phage promotor. To amplify the readthrough escape synthesis, the phage genome was used in the prophage state, carrying mutations that block excision. The prophage possessed the $cI_{857}$ allele, so that escape synthesis by readthrough could be initiated by thermal induction.

The *gal* operon lies to the left of the prophage genome and is transcribed in the same direction as the leftward early genes of lambda; thus it can be expressed by readthrough transcription that initiates at $p_L$ upon thermal induction, and which depends upon *N* function. The galactose operon can be induced by addition of D-galactose or D-fucose (which prevent the *gal* repressor from binding to its operator) and its expression requires an active cAMP/CAP complex. When fucose is added to cells maintained at 32°C, the specific activity of galactokinase, the enzyme specified by the *galK* gene, increases to 3.2 from its basal level of 1.1. Since the phage is repressed at this temperature, this represents transcription only from the *gal* promotor. Table 4.1 shows that when the cells are incubated at 41°C in the absence of fucose, the level of enzyme activity rises to 7.3; since the *gal* promotor is repressed in these conditions, this expression represent solely readthrough transcription. Upon incubation at 41°C in the presence of fucose, the level rises to 10.4. This represents expression due to initiations at both the *gal* and $p_L$ promotors. This suggests that expression by induction at the *gal* operon and by readthrough from the phage promotor is additive; that is, the amount of enzyme synthesized depends simply on the number of polymerases that transcribe the structural gene.

That the escape synthesis is due to readthrough from $p_L$ is suggested by its absence when the prophage has a *sex* mutation. The distance between $p_L$ and *galK* is considerable; it comprises at least the sequence

$p_L$-*N-cIII-red-xis-int-phr-blu-chlD-galE-galT-galK*

which represents about six phage genes and at least six bacterial genes. Galactokinase activity appears about 12 minutes after the prophage is induced, consistent with the idea that escape synthesis requires the polymerase to start at $p_L$ and traverse the distance to the *gal* operon. The

**Table 4.1:** readthrough synthesis of galactokinase

| bacterial promotor activated (by addition of 1mM fucose) | phage promotor activated (by induction at 41°C) | specific activity of galactokinase | |
|---|---|---|---|
| | | *gal*$^+$ operon | *galE* polar mutant |
| − | − | 1.1 | 0.06 |
| + | − | 3.2 | 0.07 |
| − | + | 7.3 | 3.9 |
| + | + | 10.4 | 4.1 |

A minus in the first column indicates that the bacterial promotor was repressed and a plus indicates that it was induced by addition of fucose; a minus in the second column indicates that the phage promotor was repressed by holding the cells at 32°C and a plus indicates that it was activated by thermal induction. In all cases the specific activity of galactokinase was measured 30 minutes after start of the experimental protocol. Data of Adhya, Gottesman and De Crombrugghe (1974).

phage must have an $N^+$ function and galactokinase synthesis is shut off after about 40 minutes in *cro*$^+$ phages. Very large transcripts that may correspond to the $p_L$-*galK* message also have been directly recovered.

To investigate the ability of *N* protein to antagonize rho factor, Adhya et al. made use of a lysogen in which the *galE* carries a polar mutation at which *rho* terminates transcription in vitro. The mutation appears to take the form of the insertion IS2 which includes a *t* site recognized by rho factor; its effect is therefore to prevent synthesis of galactokinase from *galK*. When fucose is added to the cells of this lysogen at 32°C, galactokinase activity fails to appear. Polymerases that initiate transcription at the *gal* promotor must therefore be unable to proceed past the insertion. When escape synthesis is induced by heating the cells to 41°C in the absence of fucose, galactokinase is synthesized, its specific activity reaching a level of 3.9 units. Polymerases that initiate transcription at $p_L$ must therefore be able to proceed past the insertion in *galE* because *N* protein has been produced and is able to exert an antitermination effect.

If the *N* protein acts at the site of insertion, as implied in the first model of Figure 4.48, then it will be able to prevent all polymerases from terminating there, whether they have initiated at $p_L$ or the *gal* promotor; but if the N protein acts at $p_L$, as illustrated in the second model of the figure, it will be able to act only on polymerases engaged in readthrough transcription. When the polar lysogens were induced at 41°C in the presence of fucose, so that transcription of the *gal* genes takes place from both the galactose promotor and $p_L$, the level of galactokinase activity remained virtually the same as that seen at this temperature in the absence of fucose (see Table 4.1). This suggests that escape synthesis can proceed past the insertion but that induction of *galK* from the bacterial promotor continues to be prevented by the insertion. In other words, whether a polymerase terminates at the insertion depends upon whether it has initiated transcription at the galactose promotor or at $p_L$. The ability of $p_L$-initiated polymerase alone to continue past the rho-dependent termination site in *galE* suggests that N acts at $p_L$ to modify the enzyme so that it fails to react to rho at *t* sites. This therefore adds another lambda regulator protein to those which act in the immunity region.

Polarity describes the situation in an operon when a mutation in one gene prevents or reduces the expression of subsequent genes (see Volume 1 of Gene Expression). Polarity often results from the introduction of a nonsense mutation in an early gene of an operon, in which case the mutated gene is unable to produce a functional protein because translation ceases at the nonsense codon, and the subsequent genes (although not themselves mutated) produce much reduced amounts of (wild type) proteins. Although the direct effect of the nonsense mutation is to cause premature termination of translation, its polarity is the consequence of a defect in transcription: the amount of messenger RNA representing genes subsequent to the mutated cistron is much reduced. The existence of mutations in one gene which are polar for the expression of another can thus be taken as evidence that the two genes are part of a single transcriptional unit. There has for some years been a debate about whether the disappearance of distal regions of the message is due to degradation because the ribosomes have dissociated at the nonsense mutation, or whether transcription and translation are related in such a way that transcription ceases at or soon after protein synthesis terminates at the nonsense codon.

Polar nonsense mutations in the *galE* gene were tested by Adhya et al. for their effect upon the expression of *galK* by either induction or escape synthesis. These mutants behave in precisely the same way as the polar insertion: they prevent production of galactokinase by induction at the *gal* promotor, but do not prevent readthrough by polymerases that start at $p_L$. Rho factor can terminate transcription in vitro at a site at the end of *galE*; Adhya et al. therefore suggested that the cessation of translation at the polar mutation in some way allows rho factor to act at this site, so that polymerases starting at the *gal* promotor terminate transcription although polymerases starting at $p_L$ are allowed to continue. Franklin (1974) obtained similar results when she compared the expression of the tryptophan genes in a λ*trp* phage under conditions of induction versus escape synthesis. The polarity of nonsense mutations is suppressed by polymerases reading through from $p_L$, whereas when polymerases initiate at the *trp* promotor the enzymes coded by the distal genes fail to be synthesized. This implies that nonsense mutations are polar because they cause rho-dependent termination of transcription (although the reason for this effect is not clear) and that the function of N protein is to antagonize rho by a mechanism that enables messages starting at $p_L$ to be distinguished from those starting at $p_{trp}$.

A similar conclusion for the control of rightward early transcription is suggested by the demonstration that gene *R* product cannot be synthesized when RNA polymerase initiates at the mutant promotors *c17* (located to the left of *cII*) or *ri*$^c$ (located within gene *O*), even in the presence of *N* gene function (Friedman and Ponce-Campos, 1975; Friedman et al., 1976). By contrast, polymerases initiating at $p_R$ can express all the late genes. This suggests that termination at $t_{R2}$ prevents progression into genes *Q, R, S* unless the polymerase has initiated at a site where N protein can act to prevent the subsequent action of rho.

*Control of Late Transcription*

The late phase of lambda development comprises the period following replication. At about 15 minutes after infection, a large rise is seen in the transcription of RNA from the *r* strand. By hybridizing this late RNA with *r* strands derived from phages with appropriate deletions or substitutions of lambda regions, Nijkamp, Bϕvre and Szybalski (1970) demonstrated that it is derived from the late genes, *S, R, A-J*. The course of its synthesis is shown in Figure 4.43. Following transcription of the late genes, the proteins of the head and tail of the phage particle are produced in large amounts, DNA is packaged into the heads, and enzymes are produced to lyse the host cell and release the mature phage progeny (see Chapter 5).

The first suggestions that *Q* may be a regulator gene were provided by the observations of Joyner et al. (1966) and Dove (1966) that $Q^-$ mutants synthesize a reduced amount of mRNA and are deficient in expression of all late functions. The same defects are shown by $O^-$ and $P^-$ mutants, which fail to replicate DNA, but $Q^-$ mutants exhibit normal replication. This suggests that *Q* may code for a positive regulator protein which is needed in order to transcribe the late genes. The dependence of late gene expression upon replication raises the question of whether there is some formal link in the sense that only replicating DNA may be able to act as the template for late transcription, or whether the relationship is quantitative. Hybridization analyses have shown that $O^-$ and $P^-$ mutants transcribe the same classes of mRNA at the same relative times as wild type lambda genomes, although the amount of late message is much reduced (Oda et al., 1969; Nijkamp et al., 1970). This suggests that regulation of lytic development remains unaltered in $O^-$ and $P^-$ mutants but that the production of progeny is hampered by a reduction in the synthesis of late gene products which is caused by the failure of replication to increase the number of copies of the template available for transcription (see Echols, 1971).

The late genes of a prophage can be trans induced by a heteroimmune superinfecting phage, in contrast with the inability of early prophage functions to be induced. Thomas (1970) therefore inferred that early functions are under the direct control of repressor protein, but that late functions are not themselves subject to the action of repressor and are maintained in an inert state in the prophage because the early genes are not expressed (see page 382). Formally this means that some function in the heteroimmune phage is able to activate the late genes of the prophage at a site where repressor is unable to act (see Figure 4.50). The positive regulator function is identified by the observation that only $Q^+$ superinfecting phages can induce the prophage late genes. Its site of action was

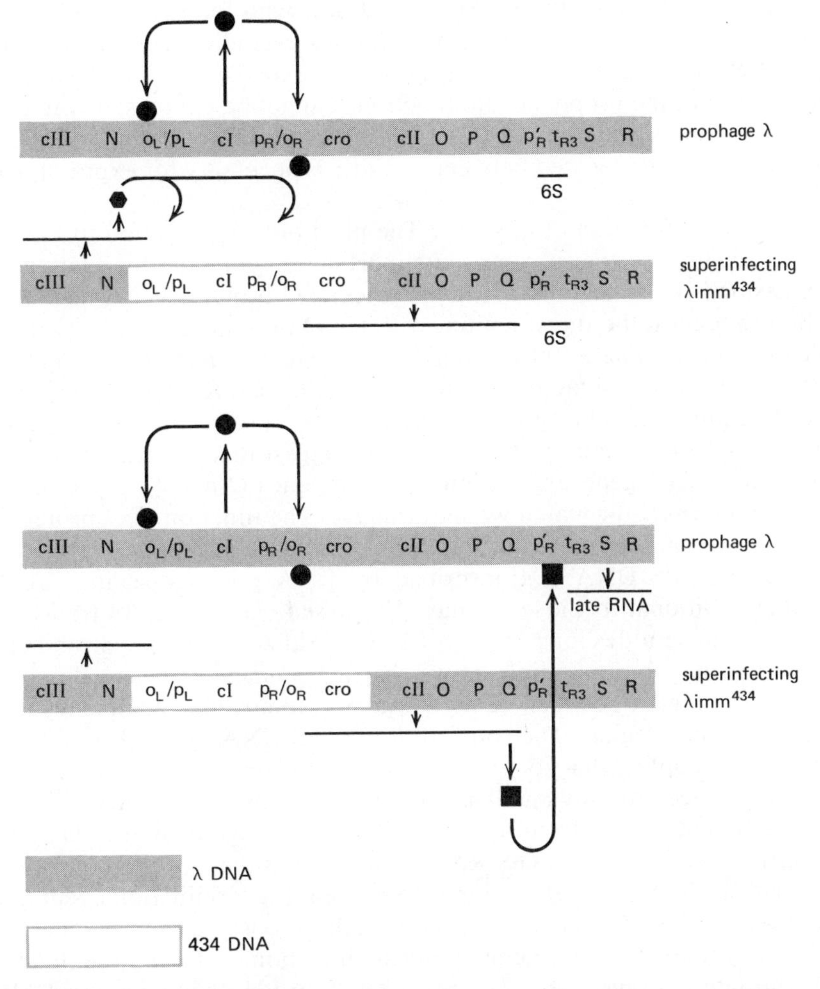


**Figure 4.50:** trans induction of λ prophage functions by superinfecting λ*imm*$^{434}$. Because the superinfecting phage is heteroimmune from the prophage, it is not repressed and can enter the lytic cycle. If it carries a mutation in an essential gene, it will be unable to complete development unless the (wild type) allele present in the prophage can be induced. The upper part of the figure shows that early lambda genes cannot be trans induced by N protein of λ*imm*$^{434}$ because their transcription must start at the $p_L$ and $p_R$ promotors which remain repressed by the lambda *cI* protein. The lower part of the figure shows that the late λ genes can be induced by the *Q* protein of the λ*imm*$^{434}$ phage because *Q* acts to continue transcription at a site not directly under the control of lambda repressor.

identified by Herskowitz and Signer (1970a), who followed trans induction of a series of lysogens carrying deletions extending from the left into genes *OPQSR*. Deletions covering *Q* but not extending into *S* allow trans induction; but the prophage late functions cannot be expressed when the deletion extends into gene *S*. This experiment therefore demonstrates that a cis dominant site located between *Q* and *S* is necessary for expression of all late genes.

What is the function of this site? The most obvious function to ascribe to it is to suppose that it represents a promotor, $p'_R$, at which late gene expression is initiated only in the presence of Q protein. This was indeed at first thought to be the situation, with the Q protein ascribed either the role of acting as a new RNA polymerase able to recognize this promotor, or that of providing a factor able to redirect the host RNA polymerase so that it acquires the ability to recognize the late promotor. But the more recent results obtained by Roberts (1975) suggest that there are two sites essential for late gene expression between genes *Q* and *S*: a promotor, $p'_R$, and a terminator which we may call $t_{R3}$. The function of *Q* protein is to act as an antiterminator at the second of these sites.

When lambda DNA is transcribed by RNA polymerase in vitro, a product additional to those we have discussed earlier is a 6S transcript about 200 nucleotides long. The 6S RNA hybridizes with the *r* strand of λ DNA; and deletion mapping shows that it becomes unable to do so as soon as the region between genes *Q* and *S* is removed. This locates it in the late control region. The transcription of 6S RNA by the E.coli RNA polymerase implies that its promotor is recognized by the host enzyme without the need for any ancillary function coded by the phage. A similar species is synthesized from the DNA of $\phi$80, a phage related to lambda. Roberts suggested that the 6S RNA is initiated at the late promotor, $p'_R$, and that the function of Q protein is to prevent the termination event that follows only 200 bases later (see Figure 4.50). According to this model, 6S RNA is synthesized continuously during infection, but can be extended into the late *r* strand message only after Q protein becomes available. Termination of 6S RNA in vitro does not depend upon rho factor, though this does not prove, of course, that $t_{R3}$ is not recognized by rho factor in vivo. Whether or not rho is involved in termination, $t_{R3}$ appears to define a new class of termination site since it is (presumably) ineffective in the presence of Q protein but is not suppressed by N protein, whereas $t_{R1}$ and $t_{R2}$ represent a class of *t* sites suppressed by N but not by Q protein. Whether Q protein acts at $p'_R$ to modify the activity of RNA polymerase or recognizes the $t_{R3}$ site directly is of course not known. However, this model implies that the regulator proteins of lambda have only two types of action: the negative control functions of *cI* and *cro* represent repressor proteins; and the positive control functions of *N*, *Q*, and probably *cII*/*cIII*, are accomplished by antitermination proteins.

The results of heteroimmune infection show that Q protein can act in trans (illustrated in Figure 4.50). However, Echols, Court and Green (1976) demonstrated that it is more active in cis, since cells infected with phages of the genotypes $Q^+ R^+$ and $Q^- R^-$ produce more endolysin (the product of gene *R*) than cells infected with the phages $Q^+ R^-$ and $Q^- R^+$. This implies that Q protein is able to activate late transcription more efficiently from the genome that codes for it. Proximity may explain this effect: Q protein newly synthesized on an mRNA may be closer in the cell to the genome from which the mRNA is transcribed and may therefore have a greater probability of finding its recognition site on this DNA. A puzzling feature of the interactions between regulator proteins and DNA is how recognition of nucleotide sequences occurs so rapidly, and it is interesting to speculate that the need to overcome the problems posed when regulator proteins must diffuse through the cell may be reflected in the propensity for regulator genes to be located close to the sites upon which their proteins act. (A discussion of the recognition of the lactose operator by its repressor is given in Volume 1 of Gene Expression and, more recently, by Lin and Riggs, 1975.) On the other hand, the extreme example of cis-acting proteins is provided by the *A* function of phage $\phi$X174 which is discussed in Chapter 8; the A protein is completely cis acting, although the concentration in the cell is great enough to imply that diffusion to find a template should not be a problem.

The biological implication of the action of *Q* within the late region is that since repressor is unable to suppress the transcription initiated at $p'_R$, the synthesis of Q protein commits the phage to the lytic pathway. As the early genes continue to be transcribed unless turned off by Cro, the Q protein presumably does not interfere with the recognition of $p_L$ and $p_R$ by the host polymerase.

When the phage genome is in the circular state, the late genes form a single contiguous block. The identification of $p'_R$ as the only site that is essential for the action of *Q* protein implies that the late genes are transcribed into a single message from *S* to *J* as depicted in Figure 4.47. We do not know whether this large message remains intact to be translated in this polycistronic state or whether it is cleaved into smaller messengers for translation. Some experiments to isolate late messages have identified several different size classes, but of course it is difficult to exclude the possibility that breaks may have been introduced during extraction of the RNA. On the other hand, Gariglio and Green (1973) have identified a 60S RNA that is synthesized late in infection and appears to correspond to the entire *A-J* region. The products of the late genes are not synthesized in equimolar proportions, which may imply some control of translation, most probably of the initiation of polypeptide chains coded by the late message. Several polarity groups have been identified among the late genes, which confirms that the genes in each group are translated from the same message, but does not show whether different groups are part of the same mRNA (see Chapter 5).

Mutants in $Q$ are very leaky and are able to produce about 10% of the number of progeny of phages that emerge from $\lambda^+$ infection. This suggests that there is a residual expression of the late genes of about 10% of the level promoted by $Q$ protein. The nature of this residual expression has been investigated by Dambly and Couturier (1971), using assays of endolysin, the product of the $R$ gene. Table 4.2 shows that the introduction of a $Q^-$ mutation results in a large decrease in the production of endolysin; the introduction of an $N^-$ mutation abolishes all synthesis of endolysin, including the residual expression.

The effect of the $N^-$ mutation might be direct, because $N$ function is required for the residual expression of late genes, or might be indirect, because replication is necessary and is prevented in $N^-$ phages. When cells are infected with both $\lambda N^-$ and $\lambda imm^{21}\ Q^-\ R^-$ phages, the production of endolysin measures the ability of the $\lambda N^-$ phage to express its late genes when it is able to replicate, since the $O$ and $P$ replication functions are provided by the $\lambda imm^{21}\ Q^-\ R^-$ phage. The $\lambda imm^{21}$ phage cannot provide $N$ function because its substitution includes this gene. (That the $\lambda imm^{21}\ Q^-\ R^-$ phage is itself unable to synthesize any endolysin is confirmed by the control experiments shown in the last two lines of the table.) No endolysin is produced in the $\lambda N^-$-$\lambda imm^{21}\ Q^-\ R^-$ mixed infection, which implies that the $N^-$ mutation directly prevents residual expression of the late genes.

Residual expression in $Q^-$ phages might be due to either of two mecha-

**Table 4.2:** residual expression of late genes

| infecting phages | endolysin synthesized |
|---|---|
| $\lambda^+$ | 650 |
| $\lambda\ Q^-\ R^+$ | 42 |
| $\lambda\ N^-\ R^+$ | 2 |
| $\lambda\ N^-\ R^+ + \lambda imm^{21}\ Q^+\ R^-$ | 560 |
| $\lambda\ N^-\ R^+ + \lambda imm^{21}\ Q^-\ R^-$ | 2 |
| $\lambda imm^{21}\ Q^-\ R^+$ | 82 |
| $\lambda imm^{21}\ Q^-\ R^-$ | 2 |
| $\lambda imm^{21}\ Q^+\ R^-$ | 2 |

Data of Dambly and Couturier (1971).

nisms: leakage in the synthesis of *Q* product might allow enough active protein to be made to support a low level of late gene expression; or if no active *Q* protein is made, there must be a pathway that functions entirely independently of *Q*. Although the $\lambda imm^{21}$ $Q^-$ $R^-$ phage is unable to act in trans to induce synthesis of endolysin from the $\lambda N^-$ phage, the table shows that a $\lambda imm^{21}$ $Q^-$ $R^+$ phage by itself displays the usual residual expression. This implies that the leakiness of $Q^-$ mutants is not due to the formation of any active *Q* protein, but results from the operation of a *Q*-independent pathway which acts only in cis and depends upon gene *N*. The obvious interpretation for this observation is to suppose that the early transcription initiated at $p_R$ is permitted by *N* function to continue through the late genes at a low level. One possible speculation about the responsible mechanism is to suppose that *N* supports a rather inefficient antitermination at $t_{R3}$.

Mutants of lambda that can complete lytic development in spite of the absence of the λ *Q* gene have been given the general name *qin* (for *Q*-independent). The best characterized were isolated by Sato and Campbell (1970) and shown by Fiandt et al. (1971) to carry related substitutions of foreign DNA for the region *Q-S-R*; another mutant known to be able to express late genes independently of *Q* function is the variant λ*p4*, which turns out to have a substitution involving the same sequence found in the *qin* mutants. Herskowitz and Signer (1974) showed that λ*p4* appears to contain a function analogous to *Q*, because a λ*p4* superinfecting phage can trans induce the late genes of a $\lambda imm^{434}p4$ prophage. Since λ*p4* can lyse cells, it is reasonable to suppose that it also carries functions able to substitute for the lambda genes *S* and *R*. Many further mutants of the *qin* type that carry a substitution of DNA related in sequence to that found in λ*p4* have been isolated from infection of several bacterial host strains with $\lambda Q^-$ variants by Henderson and Weil (1975b) and Strathern and Herskowitz (1975). The most likely explanation for the nature of the substitution is that the chromosomes of these bacteria carry genes, called *qsr'*, that are analogous to the corresponding functions in lambda and which can replace them; the source of the *qsr'* genes may reside in a cryptic prophage, and they are absent from a bacterial strain lacking other genes attributed to this prophage.

### *Balance Between Lysogeny and Lytic Development*

The balance of forces which direct an infecting phage towards lysogeny or lytic development cannot be resolved to predict which pathway will be followed by any individual infecting phage. The two antagonistic regulators are the repressor coded by the *cI* gene and the Cro protein. The paradox in their actions is that both bind to the left and right early control sites with the effect of inhibiting the initiation of transcription from the early promotors. Yet the action of repressor directs the phage into lysogeny whereas the action of Cro channels it towards the lytic cycle.

Although the two regulators differ in that repressor activates its own expression from $p_{rm}$ whereas Cro antagonizes transcription from this promotor, this difference explains only how the two regulators maintain the synthesis or repression of *cI* once it has been established: it does not explain why one regulator prevails in creating this situation. Repressor and Cro have a common effect in inhibiting expression of early functions, including the *cII* and *cIII* genes whose products are necessary to establish repression. Thus no matter whether the phage enters the lysogenic pathway or lytic development, transcription of *cI* from $p_0$ will be prevented as the production of cII/cIII proteins is prevented when either Cro or repressor acts at the early control sites. Which pathway is followed therefore depends upon whether the cessation of transcription of *cI* from $p_0$ is succeeded by activation of $p_{rm}$ by repressor, leading to the repression of all phage genes except *cI*, or whether $p_{rm}$ fails to become active and *Q* protein commits the phage to lytic development by initiating transcription of late functions from $p'_R$.

The central interaction among these control systems may therefore be the need for repressor protein to activate its own synthesis from $p_{rm}$. Since Cro protein represses its own synthesis, it is not clear that it could sustain repression of transcription from $p_{rm}$ (this would depend upon the concentration of Cro protein in the cell and its stability); but failure of repressor protein to activate transcription at $p_{rm}$ provides an adequate control to allow a phage entering the lytic cycle to be free of inhibition by the repressor. The critical decision in establishing lysogeny versus entering the lytic pathway may therefore be quantitative. If enough repressor protein has been synthesized to bind the operators by the time Cro turns off expression of *cII* and *cIII* and their activity ceases, lysogeny may be established by the activation of $p_{rm}$; the repressor will prevent expression of all phage functions, including *cro,* and so will establish the situation in which lysogeny is perpetuated. If insufficient repressor has been synthesized, $p_{rm}$ will not be activated, with the result that repressor synthesis will cease with the inhibition of transcription via $p_0$; *Q* protein may activate the late promotor $p'_R$ and commit the phage to lytic development. In this context it may be important that Cro inhibits early transcription less effectively than the repressor; the residual expression of early genes allowed by Cro may explain why *cro* does not inhibit lytic development. Since repressor and Cro protein bind to the same early control sites and thus influence the same promotors, one speculation is that the antagonism between them might extend to competition for DNA, a concept consistent with the idea that the relative concentrations achieved by the two proteins may determine the fate of the phage. The balance of decision may therefore be predictable only statistically on a population basis, since fluctuations in transcription and translation in individual cells may influence the pathway followed by the phage.

CHAPTER 5

# Phage Lambda : Development

In this chapter we shall be concerned with the processes by which reproduction of phage lambda is accomplished, that is, with the functions of all lambda genes except those implicated in the control of transcription. In the last chapter we have seen that the decision on which pathway an infecting phage will follow is made during expression of the delayed early genes, whose products are necessary for both lysogeny and lytic development. Here we are concerned with the execution of the decision, which requires a delayed early function to achieve lysogeny and both delayed early and late functions to complete the lytic cycle.

To establish the lysogenic state in which lambda may be perpetuated as part of the bacterial chromosome, a second action is necessary in addition to the initiation of repressor synthesis: the phage genome must be inserted into the host chromosome by recombination. The corollary is that when a lytic cycle starts with the induction of prophage, the phage genome must be excised from the bacterial chromosome. These actions are accomplished by the leftward delayed early genes *int* and *xis*.

For a phage to complete the lytic cycle, DNA must be replicated and then assembled into mature phage particles. Replication requires the rightward delayed early genes *O* and *P* as well as some host functions. During the lytic cycle, recombination takes place between the phage genomes present in the cell and this is the function of the leftward delayed early *red* genes of the phage and of the *rec* genes of the host. For successful replication it is necessary for the host *recBC* nuclease to be inhibited later in the cycle; this appears to be the function of the leftward delayed early gene *gam*, which therefore provides a link between the replication and recombination systems. And although the recombination functions are described as nonessential, in certain conditions, when replication is prevented, they become essential in order to provide an alternative route for maturation of λ DNA.

The final steps in phage development concern the assembly of the lambda particle, which requires construction from their protein components of the head and tail and maturation of the replicated DNA. These functions are accomplished by the products of the late genes, some of which are structural components of the head and tail and others of which are concerned with aiding the assembly process. The maturation of linear phage DNA from its replicating form is an integral part of this process.

## Site Specific Recombination

### *Models for Recombination*

Genetic recombination formally describes the rearrangement of markers that produces new genotypes following a cross between two parental genomes differing at two or more loci; it therefore encompasses mating in phages and bacteria as well as in eucaryotes. At the molecular level we can define recombination as the production of new DNA molecules that carry genetic information derived from both parents. A feature common to all recombination events is that they involve physical rearrangement of the material of the parental DNA molecules, rather than the copying of information present in parental DNA. We can distinguish at least three types of systems for recombination: classical genetic recombination involves exchange between homologous sequences of duplex DNA; bacterial transformation is achieved by the substitution of a single strand of DNA in a duplex molecule; and the recombination event involved in integration and excision of lambda takes place between specific but nonhomologous duplex sequences. All recombination events encounter certain common problems in linking together the two parental molecules of DNA and so may utilize somewhat analogous interactions.

Recombination between chromosomes in eucaryotic cells takes place by what has been described as *breakage and reunion*; a single event generates reciprocal recombinants, each of which is a linear chromosome containing a duplex of DNA in part derived from one parent and in part from the other. At the level of the chromosome the recombination event appears to take the form of the introduction of a break in each partner at the same site followed by a crosswise reunion. At the level of DNA the reaction takes place by the formation of a hybrid duplex. If breaks are made in two corresponding single strands of the duplex molecules, each strand can unpair and then pair with its complement in the other chromosome. Figure 5.1 shows that this strand exchange generates two reciprocal lengths of hybrid DNA, each comprising one strand from each partner. The fate of this intermediate structure then depends upon which two strands are involved in the second breakage. If the same two strands are broken again, the original parental duplex sequences are restored, except that a length of hybrid DNA is present. But breakage and crosswise reunion of the other strands generates recombinant chromo-

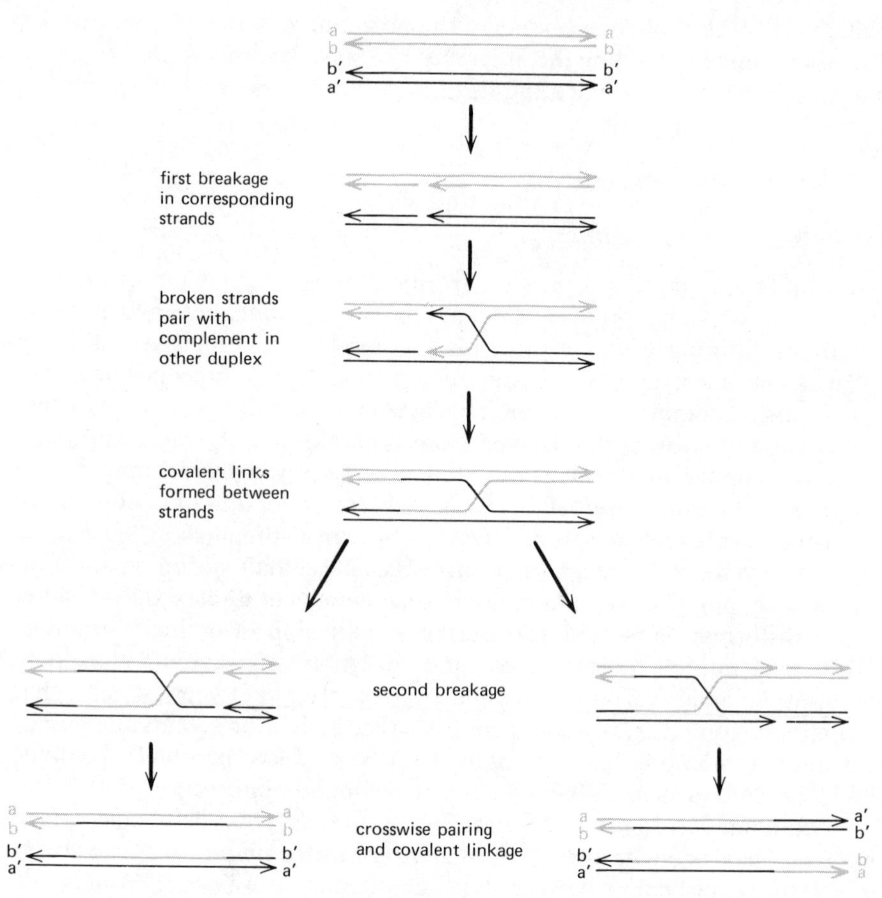


**Figure 5.1:** model for reciprocal recombination between duplex DNA molecules by formation of hybrid DNA. The first step is the introduction of nicks in strands of the same polarity (polarity is indicated by the arrows) in the two duplex molecules. These strands then unpair and pair with the complementary sequences in the opposite duplex. The gap between the original strand and the newly paired strand on each duplex is then sealed. A second breakage is required to release the two DNA duplexes from their crosswise linkage. If the second breakage occurs in the same two strands implicated in the first exchange, as illustrated on the left, there is no recombination; but although one chromosome remains constructed of strands *a,b,* and the other chromosome of *a',b'*, each carries a region of hybrid DNA. If the second breakage takes place in the other two strands, as illustrated on the right, reciprocal recombinants are produced; each represents a linkage of strands *a,b* to strands *a',b'* and possesses a sequence of hybrid DNA in the region of the crossover event. This model was proposed by Holliday (1964).

somes, each of which represents lengths of duplex DNA from the two parents, linked by a length of hybrid DNA. (We therefore see that the region of hybrid DNA has the same type of structure as that produced by the integration of donor DNA following bacterial transformation.)

The consequence of this type of event is that distant markers are indeed recombined by breakage and reunion, as proposed in the classical models, yielding reciprocal recombinants. Markers located close together, however, may suffer recombination as the result of correction events that occur on hybrid DNA. When the two parental DNA molecules differ in sequence in the region in which the hybrid DNA has been formed, the two strands of the hybrid will be mismatched at the mutant site(s). If no correction occurs, the parental sequences will segregate when the hybrid DNA is replicated, each parental strand serving as template for synthesis of a complement. If correction occurs, the hybrid duplex may be converted to the sequence of one of the parents, so that the two parental sequences are not represented equally in the progeny of the cross. When two mutant sites lie close together, if hybrid DNA covering only one of them is corrected, the correction event creates a recombinant between them. If the hybrid DNA is part of a structure in which recombination has been achieved (the second breakage involved the two strands not involved in the first breakage), then recombination between the close markers is correlated with recombination between distant markers flanking the two close markers. If the hybrid DNA is part of a structure in which the pathway has not led to recombination (the second breakage involved the same two strands as the first breakage), then the close markers suffer recombination but distant markers remain in their parental arrangement. Studies with fungi show that most recombination between close markers can be attributed to this second class of event. This description constitutes only a brief summary of the events involved in eucaryotic recombination, of course, and the process of general recombination through the formation of hybrid DNA is discussed extensively in Volume 1 of Gene Expression; here it is necessary simply to say that different types of event may be responsible for recombination at the molecular and gross levels. We may summarize two pertinent features which apply to all recombination events: the sequences of DNA which are involved in the recombination must be brought together; and the specificity of the interaction between them is assured by relying on base pairing between complementary single strands of DNA.

### *Integration and Excision*

Recombination in phage lambda may be sponsored by three different systems. *Vegetative recombination* occurs during the lytic cycle, in the population of replicating molecules of λ DNA, and represents the conventional interaction between homologous duplex sequences. Either of two systems may undertake vegetative recombination: the Rec system of the host; or the Red system of lambda. We shall discuss later the types of recombination event that are implicated in vegetative recombination and the interactions that take place between the Rec system and the Red system. *Site specific recombination* is the process responsible for the integration and excision of lambda; this takes the form of a reciprocal breakage and reunion between a specific site on lambda and a site on the bacterial chromosome.

We have seen already that lambda may exist in either circular or linear form and that the relationship between vegetative phage and lysogenic prophage may be accounted for by the Campbell model: a single recombination event inserts the circular phage genome as a linear prophage within the bacterial chromosome. Excision of the prophage takes place by reversing this exchange. This model is recapitulated in Figure 5.2.

When this model was first formulated, it seemed likely that the sites on the phage genome and bacterial chromosome would represent homologous sequences; but there is now considerable evidence to show that they are nonhomologous. For the purposes of bacterial genetics, the site on the bacterial chromosome at which lambda integrates is described as $att^{\lambda}$ and the site on the λ DNA simply as *att* or $att^{\phi}$. For the purposes of discussing the molecular events that are involved, we may describe these sites as *attB* on the bacterial chromosome and *attP* on the phage.

As the figure demonstrates, if we suppose that breakage and reunion takes place at a specific point in each attachment site, we can consider each site to consist of two elements, or half sites. The structure of *attB* can therefore be represented as $B.B'$, where $B$ and $B'$ are the half sites and (.) identifies the point of crossover. In the same way, *attP* can be written as $P.P'$. Insertion of the phage into the bacterial chromosome then generates two hybrid attachment sites; at the left terminus of the prophage *attL* has the structure $B.P'$; and at the right prophage terminus *attR* has the structure $P.B'$. In formal terms we can describe the rearrangement that takes place with integration and excision by the equation

$$\underset{P.P'}{attP} + \underset{B.B'}{attB} \underset{\text{excision}}{\overset{\text{integration}}{\rightleftarrows}} \underset{B.P'}{attL} + \underset{P.B'}{attR}$$

One of the consequences of this model is illustrated in Figure 5.2: the transducing phages λ*gal* and λ*bio* carry attachment sites that are different from that of wild type λ. The illegitimate recombination that is

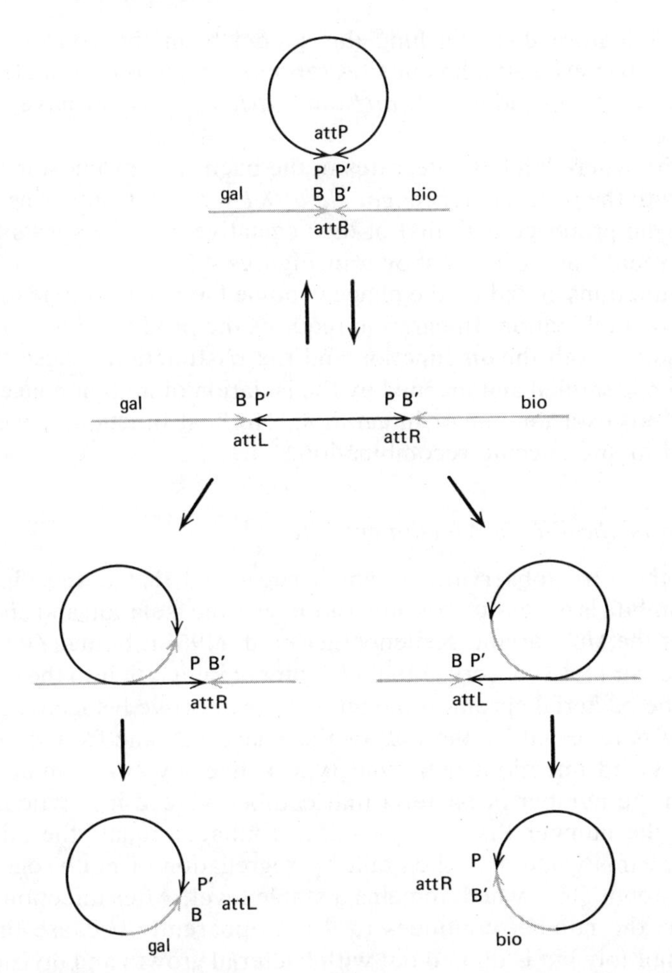


**Figure 5.2:** integration and excision of lambda. The circular phage genome carries *attP*, comprising the elements *P.P′*; the bacterial chromosome carries *attB*, comprising the elements *B.B′*. A single breakage and reunion inserts lambda as a linear prophage, with a left terminus *attL* comprising the elements *B.P′*, and a right terminus *attR* comprising the elements *P.B′*. Formation of transducing phages by illegitimate recombination generates species λ*gal* carrying *attL* or λ*bio* carrying *attR*.

responsible for excision of the transducing phage involves an exchange between sites on the phage genome and bacterial chromosome which substitutes some bacterial DNA for some phage DNA. When λ*gal* is formed, the phage carries *attL* (*B.P′*) and the site *attR* remains in the bacterial chromosome; when λ*bio* is formed, the phage carries *attR* (*P.B′*) and the site *attL* remains in the bacterial chromosome. This is one important feature distinguishing this model from the concept that the phage and bacterial attachment sites represent regions of homology; for if *attP* and *attB* were identical, *attL* and *attR* also would have the same sequence.

The site where lambda integrates in the bacterial chromosome lies in a region with the gene sequence: *gal blu att*λ *bio uvrB*. Comparing the gene order in the prophage with that of the vegetative phage suggests that *attP* lies between *J* and *cIII* as shown in Figures 4.7 and 4.8.

Two functions coded by the phage genome have been implicated in site specific recombination. Integration requires the product of gene *int*; excision requires both the *int* function and the *xis* function. These functions have been identified and mapped by the isolation of mutant phages unable to carry out excision and/or integration. No host functions appear to be involved in site specific recombination.

### *Location of the Phage Attachment Site*

One of the early observations which suggested that a specific site on phage lambda is necessary for integration was the isolation and characterization of the λ*b2* variant. Kellenberger et al. (1961a, b) and Zichichi and Kellenberger (1963) observed that λ*b2* cannot integrate into the $att^{\lambda}$ (*attB*) site on the bacterial chromosome but suffers abortive lysogeny. Abortive lysogeny is revealed by the loss of the phage genome from cells which have survived infection; with wild type phage, by 50–60 minutes after infection the number of bacteria that can be induced into lytic infection (that is, the number that has gained a prophage) equals the number of survivors; this value may then fall, by segregation of nonlysogenic bacteria, to about 50%, which remains a stable level. After infection by λ*b2*, however, the number continues to drop, apparently because the phage cannot multiply but is diluted out with bacterial growth and division. This therefore represents an abortive lysogenization in which the phage enters the cell, is repressed and thus establishes immunity to superinfection, but fails to integrate and therefore is not replicated.

The defect in λ*b2* is a deletion which removes about 17% of the DNA of the phage particle, which as a result has a decreased density. By comparing the genetic map of markers in λ*b2* with that of $\lambda^{+}$, Jordan (1964) showed that in λ*b2* the markers on either side of the deletion are closer together; the deletion is located between *J* and *cIII*. The only deficiency which the λ*b2* variant suffers is its inability to integrate; thus no essential genes can lie within the region covered by the *b2* deletion.

Two possible models for the role of the *b2* region in integration were distinguished in the experiments of Campbell (1965). One model is to suppose that the *b2* region includes gene(s) specifying protein(s) that are necessary for integration. An alternative is to suppose that the *b2* region includes the site which is necessary for insertion of the phage, that is *attP*. Lying between *J* and *cIII*, the *b2* deletion covers precisely the region in which *attP* is expected to lie according to the permutation of the phage map seen in the prophage.

Upon mixed infection with $\lambda b2$ and $\lambda^+$, the $\lambda b2$ phage can establish lysogeny. But the lysogens always carry both the $\lambda b2$ phage and the helper $\lambda^+$ phage: that is they are double lysogens. The failure of $\lambda b2$ to establish single lysogens in mixed infection is open to two interpretations. If the *b2* deletion removes gene(s) coding for integration proteins, it must be necessary for these proteins to be made continually after integration; otherwise the presence of these functions prior to integration would be sufficient, so that a $\lambda^+$ phage might help a $\lambda b2$ genome to establish lysogeny without integration of the wild type phage being necessary. If the *b2* deletion removes or alters *attP*, the presence of the $\lambda^+$ phage must be necessary for structural reasons; that is, the integration of $\lambda^+$ in some way creates a site at which $\lambda b2$ can insert in spite of its deficiency in *attP*. (We shall return later to the issue of how double lysogens are formed.)

The two models were distinguished experimentally by using a strain of E.coli that is diploid for the phage attachment site. When this strain forms a double lysogen, there are two possible modes of integration. The two phages may occupy different sites, that is, lie in trans array; or they may both integrate at one site, either adjacent to each other or with one inserted within the other. Models postulating that the *b2* region codes for enzymes able to act in trans predict that a *b2* phage should be able in mixed infection to insert equally well in trans or in cis; according to the model which postulates a cis dominant structural role for the *b2* region, it should be possible for the $\lambda b2$ phage to integrate only in cis, that is, at the site occupied by the $\lambda^+$ phage. Campbell found that the two phages always insert at the same site; formally this means that the site coverd by the *b2* deletion represents a cis dominant locus and we may interpret this observation to imply that *attP* is located there.

### *Relationship Between Phage and Bacterial Attachment Sites*

What is the relationship between the phage and bacterial attachment sites? Are they identical, representing short homologous sequences within which recombination occurs; or are they different, as depicted in Figure 5.2, in which case there must be four types of attachment site, *attP, attB, attL, attR*? As the figure demonstrates, if the phage and bacterial attachment sites are different, the transducing phages λ*gal* and λ*bio* should carry the sites *attL* and *attR*, respectively, instead of the site characteristic of wild type phage, *attP*. The observation that λ*gal* does not integrate efficiently in nonlysogenic cells, even under conditions when the necessary integration function is provided, therefore bears out the suggestion that its attachment site is different from that of $\lambda^+$. This result implies that the interaction between attachment sites depends upon their constitution: the usual integration reaction, $P.P' \times B.B'$, proceeds efficiently, but the integration of λ*gal*, requiring the reaction $B.P' \times B.B'$ is much less efficient.

This observation was expanded by Gingery and Echols (1968) in a study of integration at the prophage attachment sites. By using lysogens carrying prophages in which either the right or the left attachment sites had been deleted, they were able to follow the ability of λ to insert at the *attL* or *attR* site remaining. A $\lambda^+$ phage can indeed integrate at *attL* or *attR*, although it does so less efficiently than at *attB*; this therefore suggests that the reactions $P.P' \times B.P'$ and $P.P' \times P.B'$ can be sponsored less effectively by the integration system than the usual interaction of $P.P' \times B.B'$. This again suggests that each attachment site is different and points to the concept that the efficiency of the recombination reaction is characteristic for each pair of attachment sites (for review see Signer, 1968; Gottesman and Weisberg, 1971).

In the experiments in which λ*b2* was first characterized, this deletion variant was unable to integrate in nonlysogenic cells; the demonstration that this trait is cis dominant implied that it represents a structural effect, probably a deletion of the attachment site. A modification of this interpretation was made necessary by the observation of Fischer-Fantuzzi (1967) that λ*b2* can integrate in a host strain carrying a "cryptic" prophage; the pertinent feature of this strain is that it carries an *attR* site. Gingery and Echols confirmed this observation and demonstrated that λ*b2* is indeed able to integrate in any strain carrying *attR*, although it is unable to integrate at *attL* or *attB*. Consistent with the idea that the attachment sites are different, this suggests that whatever features of *attP* are found in λ*b2* can support a site specific recombination with *attR* although not with *attL* or *attB*. The ability of λ*b2* to engage in site specific recombination demonstrates that it must possess the crossover point; heteroduplex mapping studies (see below) have shown that the *b2* deletion cuts into the attachment site from the left, that is *attP* lies at the right extreme of the deletion, and this implies that λ*b2* must possess the element to the right of the crossover point, that is $P'$. The deletion must therefore have removed the element $P$ and brought a new sequence to lie on the left of the crossover point. We may therefore represent the attachment site in λ*b2* as Δ.P′.

When nonlysogenic cells are infected with λ*dgal*, transductants are formed only rarely; but the efficiency of transduction is greatly increased (more than a hundredfold) if a $\lambda^+$ phage is added. As we have already mentioned in discussing transduction in Chapter 4, the original interpretation of this effect was to suppose that the λ helper phage provides functions that have been deleted from the defective transducing phage (and it is true, of course, that provision of such functions is necessary to allow defective transducing phages to complete the lytic cycle upon induction). Integration of the transducing phage also requires the mediation of a helper phage and this event is distinct from the assistance which a helper may give a defective phage in completing a lytic cycle. This type of integration event has provided a useful system for investigating the nature of the attachment sites.

The helper phage necessary for integration can be provided in either of two ways: as a coinfecting phage added to nonlysogenic cells together with the defective transducing phage; or as a heteroimmune prophage. Is the role of the helper phage structural or does it assist integration by providing gene products that cannot be synthesized by the defective transducing phage genome? Usually a λ*dgal* phage carries the replication and recombination functions and is deleted for material lying at the right end of the prophage map, that is, the late genes needed for completion of the lytic cycle. Guerrini (1969) distinguished between the alternative roles for the helper phage by following the integration of λ*dgal* into a cell carrying the heteroimmune prophage λ$imm^{434}$ *dgal*, where the prophage carries exactly the same deletion as that carried by the transducing phage. Lysogen formation then displays an efficiency close to 100%, compared with the 1% lysogen formation shown by λ*dgal* in sensitive cells. This suggests that the presence of a resident prophage is in itself sufficient to allow the defective transducing phage to integrate; this implies that the role of the helper is entirely structural.

The products of transduction by λ*dgal* in the presence of a coinfecting helper phage nearly always are double lysogens, with the transducing λ*dgal* genome and the helper λ phage genome integrated at the same site. The absence of any single lysogens carrying only the transducing phage again suggests that the role of the (coinfecting) helper is structural. We can distinguish two possible arrangements for the prophage genomes in the double lysogen, depending upon whether they are inserted in tandem so that the two prophages lie adjacent to each other, or whether one phage is inserted within the sequence of the other. Figure 5.3 shows that these two arrangements can be distinguished by the different predictions that

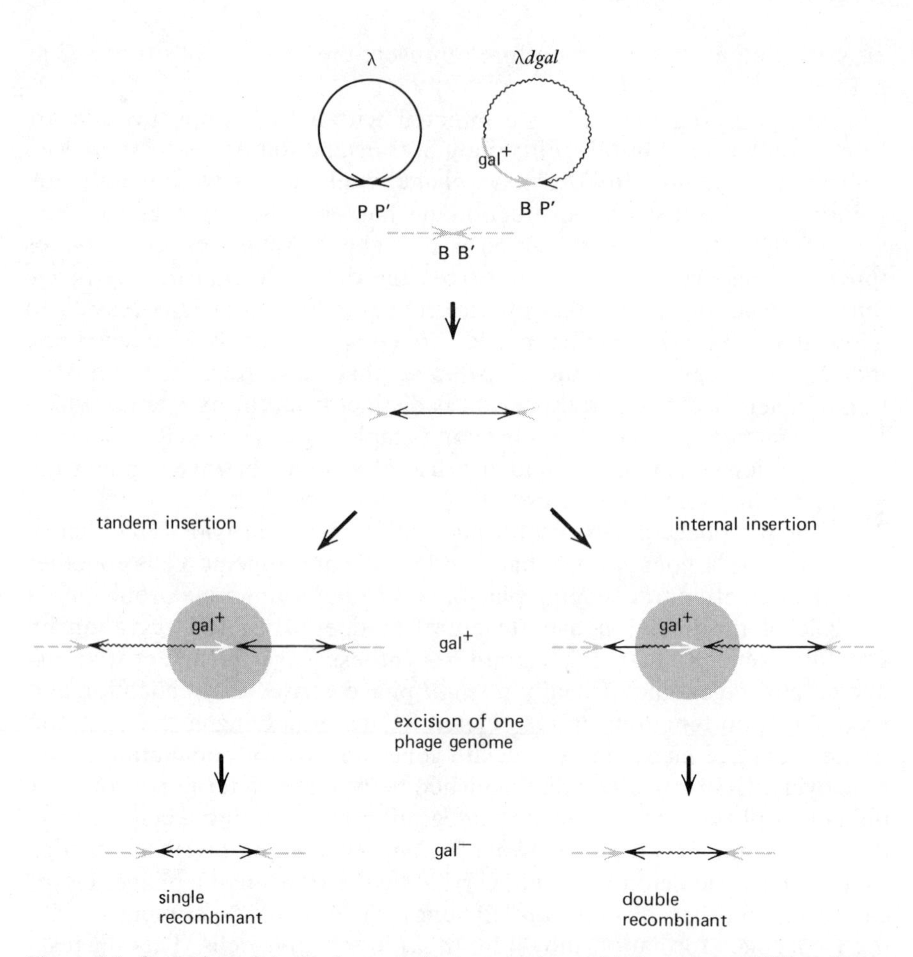


**Figure 5.3:** formation of double lysogens by transducing and helper phages. When phages $\lambda^+$ (carrying $P.P'$) and $\lambda dgal^+$ (carrying $B.P'$) are added to sensitive Gal⁻ cells, selection for the Gal⁺ phenotype largely yields double lysogens. The figure shows two models in which the wild type phage integrates first, forming a prophage with termini $B.P'$ and $P.B'$, and this enables the transducing phage then to integrate. The defective phage might integrate in tandem with the wild type phage, to the left as illustrated or to the right (not shown), or might insert within the wild type sequence. Selection for the Gal⁻ phenotype then isolates cells that have lost the *gal*⁺ gene carried by the transducing phage. This may be accomplished either by loss of the original transducing phage genome, leaving a single lysogen for $\lambda^+$, or by a single recombination event involving homologous sites in the two phage genomes; the genome that is lost in this manner is indicated by the shaded circle. In a tandem insertion the genome that remains is a single recombinant; in an internal insertion the remaining genome represents a double recombinant.

they make about the products that are generated when a single recombination event between homologous regions of the two prophages leads to excision of one complete phage genome.

These predictions were tested in a series of experiments in which Guerrini isolated double lysogens by selecting for Gal$^+$ transductants following mixed infection with $\lambda dgal$ and $\lambda^+$; and then selection for Gal$^-$ revertants allowed isolation of cells that had lost the transducing marker. Two types of lysogen are found amoung the Gal$^-$ cells. Those with the parental markers of the $\lambda^+$ phage can be generated simply by loss of the original transducing phage, irrespective of the mode of double insertion. The second class is derived by a single event involving recombination between homologous sites in the two genomes; this generates a lysogen carrying the marker array characteristic of a single recombinant if insertion was tandem, but carrying a double recombinant arrangement of markers if insertion was internal. Guerrini observed that the Gal$^-$ segregants almost always display parental marker or single recombinant arrangements. This implies that the transducing and helper phage genomes must insert end to end in the bacterial chromosome. (It also excludes models in which the cohesive ends of the linear forms of the two phages might join, to allow a circular dimer to be formed which might be inserted as a unit, since dimer formation and insertion would produce the same apparent internal arrangement as that depicted on the right of the figure.)

The helper effect therefore requires either the presence of a resident heteroimmune phage or simultaneous infection with a lambda phage that is itself able to lysogenize. In both situations the role of the helper appears to be fulfilled by its integration into the bacterial chromosome, which apparently provides a site at which the defective transducing phage can integrate (in contrast with its inability to integrate at *attB*). A segregation analysis of the double lysogens resulting from these events showed that the defective transducing phage always lies to the left of the helper genome in the tandem array. This is the situation illustrated on the left of Figure 5.3; it may be distinguished from the situation in which the defective phage lies to the right of the helper genome by virtue of the different genetic structures of the prophage remaining after loss of one genome by a single recombination event (not illustrated in the figure). Since the helper phage presumably inserts first, this implies that the defective transducing phage integrates only into the region of the left attachment site of the prophage.

The Campbell model illustrated in Figure 5.2 predicts that $\lambda dgal$ should carry *attL* (*B.P'*). The inability of $\lambda dgal$ to integrate in sensitive hosts demonstrates, as we have already seen, that the integration reaction occurs only rather inefficiently between $B.P'$ and $B.B'$. The ability of λ*dgal* to integrate at the left attachment site of a helper prophage suggests that the reaction between $B.P'$ and $B.P'$ can take place at an efficiency similar to that of the classical insertion reaction, that is $P.P' \times B.B'$. This observation supports the conclusion that the efficiency of site specific recombination depends upon the constitution of the attachment sites that are involved; and as we have seen, this must mean that the phage and bacterial sites are not identical, so that we can indeed view them as comprising half sites which can reassort to give the (different) attachment sites of the prophage. One consequence of this model, then, is that excision cannot take place simply by a reversal of the interactions of recombination: for integration involves the reaction between *attP* and *attB*, which may be represented as $P.P' \times B.B'$; excision involves the reaction between *attL* and *attR*, which may be described as $B.P' \times P.B'$. Just as the various integration reactions that we have described may take place with frequencies depending upon the sites that are involved, so may the characteristics of excision reactions depend upon the relationship between the sites that must recombine.

Another system in which recombination between various attachment sites can be followed is provided by the occurrence of site specific recombination during vegetative growth. For these experiments it is necessary to use phages carrying *red*$^-$ mutations, which inactivate the vegetative recombination system, and host cells which carry *rec*$^-$ mutations to inactivate the E.coli recombination system. Echols, Gingery and Moore (1968) and Weil and Signer (1968) found that under *red*$^-$ *rec*$^-$ conditions no recombination could be detected following mixed infection by pairs of parent phages with mutations in *A* and *L*, or *L* and *J*, or *N* and *P*. In each of these pairs, both the mutant genes lie on the same side of *attP* with respect to the linear map. The lack of any recombination in each of these crosses confirms that the site specific recombination system cannot act in these regions, in which the Red and Rec systems are responsible for all vegetative recombination. But when the two parents in the mixed infection carry mutations in *J* and *N*, some recombinant progeny are found in spite of the inactivity of the Red and Rec systems. This residual recombination is accomplished by the site specific system, which therefore must be able to promote genetic exchange between the *attP* sites of two phages in a lytic cross. The site of the specific exchange can be identified by determining where the residual recombination event occurs in these phages; and since λ*b2* also displays this residual recombination, the crossover site cannot lie within the *b2* deletion. Signer and Weil (1968b) showed that the crossover site in λ*b2* lies to the right of the deletion, which supports the idea that only the left part of *attP* is deleted in the *b2* strain.

By following recombination between phages carrying markers spanning the *att* site under conditions when only the site specific system is functional, it is possible to measure the frequency of recombination between each pair of attachment sites. Signer and Weil (1968b) and Signer, Weil and Kimball (1969) used $J^-$ and $R^-$ parents and selected for wild type progeny; a lambda phage provided a $P.P'$ site, a λ*bio* phage carried a $P.B'$ site, λ*b2* has the site that can be described as $\Delta.P'$, and λ*b2 bio* accordingly possesses a site that can be characterized as $\Delta.B'$. Each pairwise cross displayed a characteristic frequency of recombination, with a fiftyfold variation in the range of frequencies. Data obtained from a series of experiments performed by Echols (1970) are given in full later in Table 5.1, and from these we may note the frequencies of recombination found between several crosses involving the sites *attP, attB, attL* and *attR*:

| | |
|---|---|
| $P.P' \times P.P'$ | 1.1% |
| $P.P' \times B.B'$ | 2.9% |
| $P.P' \times B.P'$ | 10.0% |
| $P.P' \times P.B'$ | 1.0% |
| $B.P' \times P.B'$ | 10.0% |
| $B.P' \times B.B'$ | 0.07% |

From these (and other) experiments, it is clear that there is no simple pattern to be discerned in the frequencies that are characteristic of these crosses. For example, a change in one half site, from $P$ to $B$, decreases the frequency of recombination from 2.9% for $P.P' \times B.B'$ to 0.07% for $B.P' \times B.B'$; yet the same change in half site increases the frequency of recombination from 1.1% for $P.P' \times P.P'$ to 10.0% for $P.P' \times B.P'$. (Since these are vegetative crosses, the same result is obtained no matter which parent carries which *att* site, so that $P.P' \times B.P'$ describes the same cross as $B.P' \times P.P'$.) From these results we can draw two important conclusions: since a change in one half site can have a drastic effect, both elements of the attachment site are important in determining the frequency of recombination; and the arrangement of the elements as well as their constitution must be important, since the same four elements are present in the crosses $P.P' \times B.B'$ and $B.P' \times P.B'$ but their recombination frequencies are 1.1% and 10.0%. An even more striking example of the effects of a rearrangement of the partnership between half sites is given by the observations of Signer, Weil and Kimball (1969) that in crosses involving λ, λ*b2* and λ*b2 bio*, $\Delta.P' \times P.B'$ occurs with a frequency of 15% but $P.P' \times \Delta.B'$ occurs only at 0.25%. A further point established by the results of Signer et al. is that the frequency is not set simply by the least reactive element, since the cross $\Delta.P' \times P.P'$ occurs at 9% frequency whereas the crosses $P.P' \times P.P'$ and $\Delta.P' \times \Delta.P'$ occur at 2.0% and 1.0%, respectively. And, of course, it is implicit in these series of experiments that the attachment sites are different rather than identical.

To support these conclusions, it is necessary to confirm that the variations in recombination frequency depend entirely upon structural changes in the attachment sites and are not due to any changes or deletions in genes coding for the proteins involved in site specific recombination (these might have been introduced when the transducing phages used to provide some of the attachment sites were generated). Signer, Weil and Kimball resolved this point by performing triparental crosses with the phage system of λ, λ*bio*, λ*b2* and λ*b2 bio*. With two phage crosses $A \times B$ and $B \times C$ that have different recombination frequencies, a structural model predicts that the triparental cross $A \times B \times C$ should continue to yield the same frequencies of recombination between $A \times B$ or $B \times C$, whereas if protein factors are involved all crosses might be expected to display one frequency no matter which *att* sites are involved. The observation that each pair of attachment sites displays its characteristic frequency of recombination in a triparental cross therefore argues that the frequency is established only by the sites that are implicated.

The Campbell model predicts that site specific recombination should be reciprocal. Signer and Weil (1968b) therefore examined the products of crosses in which recombination was due only to the activity of the site specific system. When single bursts were examined, the reciprocal recombinants were found effectively in equal proportions in the progeny, a contrast with the result that is observed when vegetative recombination is sponsored by the general recombination system (see later).

### *Functions of int and xis Genes*

The independence of the system that promotes integration was first demonstrated by the experiments of Signer and Beckwith (1966) to compare the integration of phages φ80 and φ80*dlac*. The φ80*dlac* phage is able to insert at the lactose operon in the presence of a $rec^+$ system; this insertion thus relies upon the homology existing between the *lac* genes of the phage and those of the bacterium. On the other hand, φ80 is able to integrate at its usual attachment site in $rec^-$ strains; and in mixed infection a helper φ80 can promote the insertion of φ80*dlac* at the $att^{80}$ site. These experiments demonstrated that some system other than the host recombination enzymes must be responsible for phage integration. The same conclusion could be drawn from the characteristics of the excision seen in superinfection curing. When a φ80*dlac* phage has been inserted at the $att^{80}$ site, it can be "cured" by superinfection with $\lambda h_{80}$, a phage with λ immunity but the insertion specificity of $\phi 80$. By contrast, a $\phi 80$*dlac* inserted at the *lac* site cannot be cured by superinfection with a heteroimmune phage of the appropriate attachment specificity. We shall return to the phenomenon of superinfection curing later; but we may note now that these results demonstrate that the excision reaction is restricted to phages of the same attachment specificity, again implying the existence of an independent system responsible for site specific events.

The phage functions implicated in integration and excision have been identified by mutations which inhibit these activities. Mutants in the *int* gene were isolated in two series of experiments. Gingery and Echols (1967) isolated two classes of point mutations of lambda unable to integrate; all were complemented by $\lambda b2$, an indication that they do not represent loss of the attachment site. One set of the mutants, then described as *intB*, have since proved to be concerned with repression rather than integration per se. The other class, originally described as *intA* and now known as $int^-$, can be complemented in trans during mixed infection by $\lambda int^-$ and $\lambda int^+$ phages; that is, single lysogens that are $\lambda int^-$ can be obtained. (This contrasts with the inability of $\lambda b2$ phages to form single lysogens in helped infection, where the cis dominance of the integration negative character indicates the presence of a structural defect.) The isolation of temperature sensitive mutants in *int* provided another indication that the gene codes for a protein function. Gottesman and Yarmolinsky (1968a) isolated *int* mutants by selecting the abortive lysogens produced by phage strains unable to establish true lysogeny. These mutants fell into the same single cistron as those isolated by Gingery and Echols. The failure to isolate other mutants in both series of experiments hints that *int* may be the only gene needed for phage integration. Both $\lambda$ and phage *434* can help $\lambda int^-$ mutants to integrate, whereas $\phi 21$ and $\phi 80$ cannot do so; since $\lambda$ and *434* have the same attachment specificity whereas $\phi 21$ and $\phi 80$ insert elsewhere, this suggests that the Int function is site specific. Int function cannot be provided by a repressed prophage, which implies that its expression lies under immunity control; this conclusion is supported by the subsequent identification of *int* as a delayed early gene lying between *b2* and *cIII*. Lysogens which carry only $\lambda int^-$ prophages are defective in excision, giving much reduced yields of phage; this suggests that the Int function is required for both integration and excision.

Neither the Rec system of the host nor the Red system of the phage plays any role in site specific recombination. The effect of the Rec system can be tested simply by comparing the ability of lambda to integrate and excise in $rec^+$ and $rec^-$ hosts; no effect is seen. Phage mutants in vegetative recombination were isolated by Echols and Gingery (1968) and Signer and Weil (1968a) by their inability to perform marker rescue; Signer et al. (1968) identified two genes, *redα* and *redβ*. The first codes for exonuclease protein and is identical with the locus previously described as *exo*; the second codes for the β protein, which interacts with exonuclease at least in its action in vitro. The activities in vegetative recombination of the enzymes coded by these loci are discussed later. Their possible role in site specific recombination was investigated by Echols, Gingery and Moore (1968) and Weil and Signer (1968) in the experiments we have already mentioned that compared the abilities of *red*$^+$ and *red*$^-$ phages to sponsor recombination in mixed infections between various pairs of mutant phages. When the two parental markers both lie on one side of the attachment site, a *red*$^-$ mutation completely abolishes recombination; the formation of recombinant progeny occurs only when the markers lie on opposite sides of the attachment site, and is then catalyzed by the site specific system. Echols (1970) demonstrated that this residual recombination requires the activity of the *int* gene, confirming that it represents expression of the same system responsible for integration and excision. These results therefore demonstrate that general recombination is catalyzed only by the Red system whereas site specific recombination is supported only by the Int system.

The conclusion that insertion and excision represent different interactions is supported by two observations: the arrangement of half sites is critical in site specific recombination and is different for insertion ($P.P' \times B.B'$) and excision ($B.P' \times P.B'$); and while the product of the *int* gene is needed for both insertion and excision, the product of another gene, *xis*, also is necessary for excision (but not for insertion). The *xis* gene was identified by two sets of experiments. Kaiser and Masuda (1970) used superinfection curing as an assay system; as we have noted previously, when a lysogenic bacterium is infected with a heteroimmune phage with the same attachment site specificity as the prophage, many of the bacteria surviving superinfection have lost their original prophage. This is *superinfection curing*, a phenomenon which represents a way for lysogenic cells to lose their prophages without suffering the consequences of a lytic cycle. The abilities of three phages to support superinfecting curing were

| *superinfecting phage* | *curing of prophage at λ att site* | *curing of prophage at 21 att site* |
|---|---|---|
| λ$^+$ | 81% | 0.05% |
| λ*bio16A* | 37% | 0.05% |
| λ*bio7-20* | 0.05% | 0.05% |

The lack of curing of a prophage at the $\phi 21$ attachment site provides a control. A wild type lambda phage can cure a prophage at the λ attachment site very efficiently, λ*bio16A* also is efficient and carries a deletion that extends into *int*, so that the presence of an *int* gene in the superinfecting phage does not appear to be necessary. This suggests that enough Int protein may be synthesized by the prophage to support this reaction. (Although *int* is a delayed early gene subject to repression, it may suffer some constitutive expression at a very low level (see later); although this is not adequate to support (for example) integration of a superinfecting phage, it does appear to be sufficient for the superinfection curing reaction.) The deletion phage λ*bio7-20* carries a longer deletion which ends between *int* and *red*α; its inactivity in curing a prophage at $att^{\lambda}$ suggests that a gene lying in this region is necessary for excision and that its product cannot be supplied by a repressed prophage. This experiment therefore locates the *xis* gene between *int* and *red*α.

Another approach was followed by Guaneros and Echols (1970), who isolated mutants able to integrate but unable to excise; these can be identified by their persistence as prophage after a brief thermal derepression (a treatment which produces efficient excision of wild type lambda). All the mutants fall into one complementation group, which complements with $int^-$ mutants, and maps between *int* and *red*α.

The involvement of the *int* and *xis* genes in site specific recombination was studied by Echols (1970) in a series of experiments using vegetative crosses between phages carrying markers on either side of the attachment site. Phage λ itself provides a $P.P'$ site, λ*gal* provides a $B.P'$ site, λ*bio* provides a $P.B'$ site, and λ*gal bio* provides a $B.B'$ site. [Phages of the λ*gal bio* class were first isolated by Guerrini (1969), by integrating λ*dgal* to form a prophage and then helping the prophage to excise by superinfection with λ; the occurrence of the same type of illegitimate recombination event that creates λ*bio* by abnormal excision of a λ prophage then creates the double transducing phage, which carries the right attachment site of the former λ*dgal* prophage, $B.B'$, the *gal* genes of the prophage, and the *bio* genes gained from the adjacent bacterial chromosome.]

Table 5.1 gives the results of a series of crosses performed under various genetic conditions (always in $rec^-$ hosts). The $P.P' \times P.P'$ cross, which represents a mating between two λ mutants, shows 8.2% recombination when the Red system is functional; this is reduced to the residual level of 1.1% when both phages carry $red^-$ mutations. This latter value therefore represents the extent of recombination supported by the site specific system. Int function clearly is necessary for this reaction since $int^-$ phages suffer a reduction in frequency of recombination to 0.04%;

**Table 5.1:** site specific recombination in vegetative crosses

| parental phages | *att* sites | recombination genes | | | recombination percent |
|---|---|---|---|---|---|
| | | *red* | *int* | *xis* | |
| $\lambda J^- \times \lambda P^-$ | $P.P' \times P.P'$ | + | + | + | 8.2 |
| | | − | + | + | 1.1 |
| | | − | − | + | 0.04 |
| | | − | + | − | 0.6 |
| $\lambda P^- \times \lambda A^-$*gal bio* | $P.P' \times B.B'$ | + | + | + | 6.7 |
| | | − | + | + | 2.9 |
| | | − | − | + | 0.01 |
| | | − | + | − | 1.9 |
| $\lambda J^- \times \lambda P^-$*gal* | $P.P' \times B.P'$ | + | + | + | 17.0 |
| | | − | + | + | 10.0 |
| | | − | − | + | 0.05 |
| | | − | + | − | 4.4 |
| $\lambda P^- \times \lambda A^-$*bio* | $P.P' \times P.B'$ | + | + | + | 6.3 |
| | | − | + | + | 1.0 |
| | | − | − | + | 0.07 |
| | | − | + | − | 0.4 |
| $\lambda A^-$*gal* $\times \lambda P^-$*bio* | $B.P' \times P.B'$ | + | + | + | 9.3 |
| | | − | + | + | 10.0 |
| | | − | − | + | 0.05 |
| | | − | + | − | 0.03 |
| $\lambda P^-$*gal* $\times \lambda A^-$*gal bio* | $B.P' \times B.B'$ | + | + | + | 5.3 |
| | | − | + | + | 0.07 |
| | | − | − | + | 0.03 |
| | | − | + | − | 0.05 |

The recombination frequencies are calculated from the percent of $\lambda^+$ phages produced by each cross. Data of Echols (1970).

Xis function does not appear to be necessary since $xis^-$ phages achieve a recombination frequency of 0.6%, more than half of that displayed by $int^+$ $xis^+$ phages.

The interaction between the sites involved in the wild type insertion event, $P.P' \times B.B'$, clearly depends upon Int but does not require Xis. Similarly the crosses representing insertion of lambda into *attL* ($P.P' \times B.P'$) or into *attR* ($P.P' \times P.B'$) both require Int but not Xis. The variation in the frequencies of recombination in these Int-promoted reactions supports the idea that recognition of the elements present in each attachment site is important in setting the efficiency of reaction. The reaction *B.P'* × *B.B'* represents the insertion into $att^{\lambda}$ of λ*dgal* (that is *attL* × *attB*); this is very weak even under $int^{+}$ $xis^{+}$ conditions, which confirms that a structural defect and not the lack of site specific recombination enzymes is responsible for the poor integration of λ*dgal*. The excision reaction is represented by the cross λ*gal* × λ*bio*, *B.P'* × *P.B'*, and this is the only one of the crosses performed in these experiments which requires Xis as well as Int. All site specific recombination reactions therefore require Int. And Int alone appears to be sufficient for the four crosses in which one of the sites is *attP* (× *attP*, × *attB*, × *attL*, × *attR*). From these experiments it appears that the presence of Xis is necessary only when *both* sites have a "recombinant" arrangement of elements, that is, in the cross between *attL* and *attR*, *B.P'* × *P.B'*.

### *Expression of Int and Xis Functions*

Both the *int* and *xis* genes are delayed early functions which are turned on by leftward transcription shortly after the start of development and then subsequently are turned off. The two genes are therefore transcribed and translated coordinately either following infection by vegetative phage or as the result of induction of prophage. Yet in spite of the synthesis of both Int and Xis proteins in each situation, infection leads to the integration event promoted by Int but induction is followed by the excision event promoted by joint action of Int and Xis. How are these opposing recombination events achieved while the *int* and *xis* genes are coordinately expressed?

The activity of the site specific recombination system can be measured in three ways: by the ability of a phage to lysogenize; by the ability of a prophage to excise; and by recombination between phages in mixed infection (when both the Red and Rec systems are inactive). The frequency of recombination between any pair of *att* sites varies with the assay that is used, but in general there is a reasonable correlation between the activity displayed in each system. However, Weisberg and Gottesman (1971) noted that two reactions show unusual variation in the frequency of recombination. The integration reaction of *attP* × *attB* is relatively inefficient when measured by vegetative recombination compared with lysogenization or excision; for this effect there is at present no explanation. And the excision reaction of *attL* × *attR* is much less efficient when measured by lysogen formation than when assessed by excision or vegetative recombination; they suggested that this latter effect might be due to a reduction in the level of Xis activity in the insertion assay system. This effect might explain the preferential directions of the insertion and excision reactions following infection and induction, respectively.

The relatively poor recombination between $B.P'$ and $P.B'$ in the lysogen formation assay could be explained if Xis, which is necessary for this reaction, is less stable than Int protein. This would lead to a series of events in which a phage integrating by the Int/Xis dependent reaction $B.P' \times P.B'$ forms a prophage with attachment sites $P.P'$ and $B.B'$; if Xis protein decays more rapidly than Int, the disappearance of Xis will be followed by irreversible excision of the prophage by the Int dependent reaction $P.P' \times B.B'$. What this model implies is that under the conditions of lysogen formation the reaction $P.P' \times B.B'$ is the favored direction because it requires only Int protein, whereas the counter reaction $B.P' \times P.B'$ requires both Int and Xis functions and therefore is prevented by the decay of Xis. Of course, this is precisely a situation which will drive the usual integration reaction $P.P' \times B.B'$ in the forwards direction. The implication of this model is that efficiency of site specific recombination in the lysogen formation assay will depend upon what effect is exerted on the reverse reaction (excision of prophage) by the decay of Xis. For example, this explains why λ*gal* always inserts at the left attachment site in a lysogen. Integration at the left site involves the reaction $B.P' \times B.P'$ which is independent of Xis and generates the double lysogen $B.P'$ λ*gal* $B.P'$ λ $P.B'$ which is stable. Integration at the right involves the reaction $B.P' \times P.B'$ which we have just discussed; the product is the double lysogen $B.P'$ λ $P.P'$ λ*gal* $B.B'$ in which the λ*gal* prophage is subject to excision by Int promoted recombination as soon as Xis function has decayed.

A useful system for studying site specific recombination has been developed through the isolation of phage variants that carry two attachment sites. Shulman and Gottesman (1971) reported that induction of a lysogen of λ*b2* yields a dense phage which transduces the markers *bio* and *uvrB* (which lie to the right of $att^{\lambda}$ on the bacterial chromosome). Production of this phage requires either the Red or Rec system. The transducing phage displays a low level of reversion to yield a phage apparently the same as λ*b2*; this reversion depends on Int. The series of events which appears to be responsible for the generation of this type of phage is shown in Figure 5.4. First the entire phage genome is excised by a reaction equivalent to the class II excision of sex factors (see Chapter 2). Since no bacterial markers derived from the left of the phage attachment site could be identified in the phage, the two bacterial sites involved in the excision

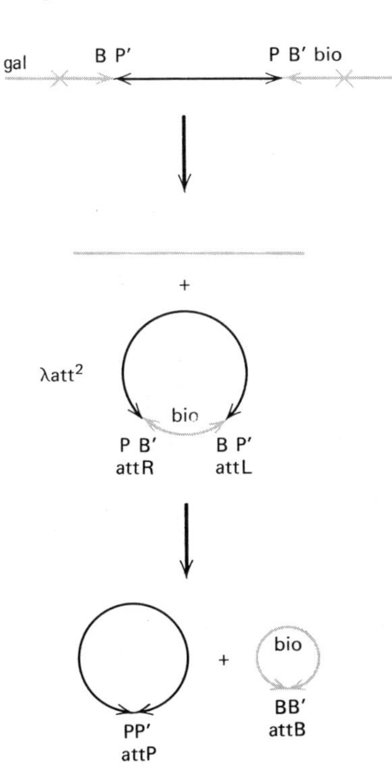


**Figure 5.4:** formation and segregation of $\lambda att^2$. Recombination between the two bacterial sequences (X) on either side of the prophage excises a circular phage that carries both *attL* and *attR*. (This is directly analogous to the type II excision of F factors illustrated in Figure 2.17.) A site specific recombination between the two *att* sites on $\lambda att^2$ generates a λ phage carrying *attP* and a circle of bacterial DNA (with the marker *bio*) carrying *attB*.

must lie close to the left side of *attL* and beyond *bio*. As the figure demonstrates, the phage produced by this type of excision carries two *att* sites: one is *attL* and the other is *attR*. This class of phage is therefore described as $\lambda att^2$. (Since the *b2* deletion removes the *P* element, these sites would be B.P′ and Δ.B′ when $\lambda att^2$ is derived from a λ*b2* prophage.) Reversion to yield the original phage lambda might then take place by an intramolecular site specific recombination between the two *att* sites on $\lambda att^2$; this should yield a phage identical to the original progenitor of the prophage and a bacterial circle carrying *attB*. The structure of the $\lambda att^2$ phage has been confirmed by heteroduplex mapping, which showed also that three independently isolated $\lambda att^2$ phages all were generated by an excision event involving the same two sites on bacterial DNA (Fiandt et al., 1976).

In lysogens of λ where excision is blocked by $int^-$ or $xis^-$ mutations, no $\lambda att^2$ phages are derived, probably because the length of DNA would be too great to package in a lambda head. To overcome this difficulty, Shulman and Gottesman used a prophage with a small genome; this carried $imm^{21}$ instead of $imm^{\lambda}$ and possessed two deletions in the *b2* region which do not extend into the attachment site. When this lysogen is induced and the phage products isolated, dense phages carrying bacterial DNA in addition to the phage sequence can be isolated. When these phages infect new hosts, two major density bands are found in the DNA (provided that the Int/Xis system is functional); one represents $\lambda att^2$ and one represents $\lambda$ that has lost the bacterial genes. It is possible also to detect a peak that may correspond to the excised bacterial genes. The revertant phage has an *attP* attachment site as judged by its ability to insert into fresh hosts. Confirmation that the reversion reaction is *intra*-molecular was provided by the use of mixed infections involving two $\lambda att^2$ phages carrying different markers; no recombination between them was found.

The activities of the Int and Xis proteins can be measured by following the conversion of $\lambda att^2$ to $\lambda$. This involves the reaction $B.P' \times P.B'$ which requires both Int and Xis functions, so that when the $\lambda att^2$ phage carries a mutant in either *int* or *xis*, the reaction can be made to depend upon the level of Int or Xis protein supplied by a complementing phage. These experiments confirmed that a repressed prophage is unable to contribute either function. By inducing a $\lambda cI_{857}$ lysogen and then superinfecting the induced cells at various intervals after derepression with $\lambda att^2$ $imm^{21}$ $int^-$ or $\lambda att^2$ $imm^{21}$ $xis^-$, the activities of Int and Xis proteins could be followed with time: the extent of reversion from $\lambda att^2$ to $\lambda$ depends upon the amount of the appropriate protein present at the time of superinfection. The results of Figure 5.5 demonstrate that Xis protein decays much more rapidly than Int protein. The cause of this more rapid decay is entirely unknown.

We have seen already that this differential stability implies that infection of sensitive bacteria will lead to integration. The insertion reaction requires recombination between $P.P'$ and $B.B'$, for which only Int function is necessary; to reverse the insertion requires recombination between the prophage sites $B.P'$ and $P.B'$, for which both Int and Xis are necessary. Decay of Xis protein therefore prevents reversal of integration. Following induction, excision occurs at the beginning of the prophage cycle when both the necessary Int and Xis functions are present. By the time that Xis has decayed, many replicating phage genomes have been produced, so that the cell is committed to the lytic cycle and the continued presence of Int protein cannot reverse the reaction.

The expression of *int* and *xis* may not be completely coordinate, for Shimada and Campbell (1974a,b) demonstrated that there is a very low level constitutive expression of *int*, but not of *xis*, in repressed prophages. In an abnormal lysogen in which $\lambda$ is inserted within the *trpC* gene, they detected a low level of constitutive synthesis from *trpB*, the next gene to

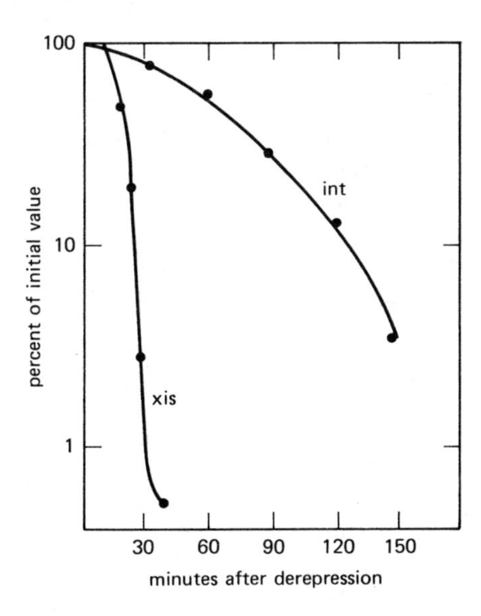


**Figure 5.5:** stability of Int and Xis functions after transient derepression. Cells carrying $\lambda cI_{857}$ were heat pulsed for 6 minutes and infected after varying intervals with $\lambda att^2$ $imm^{21} int^-$ or $\lambda att^2$ $imm^{21}$ $xis^-$. The proportion of $\lambda att^2$ phage which was converted to λ was then measured. Data of Weisberg and Gottesman (1971).

the left. The promotor responsible for this expression, which probably results from transcription reading into *trpB* from the phage, appears to lie between *att* and *red*. Mutants that enhance this effect, described as $int^c$, can be isolated, and these lie close to the promotor, $p_I$, and act only in cis arrangement. Whether this low level constitutive expression plays any function in the lambda life cycle is not yet clear.

Integration and excision represent asymmetrical recombination systems in the sense that the first requires Int whereas the second requires Int and Xis. Guaneros and Echols (1973) showed that the two reactions can be distinguished by their thermolability: integrative recombination is thermolabile whereas excisive recombination is not. This accounts for earlier observations that the establishment of lysogeny is thermosensitive. If the Int protein is sufficient for integrative recombination, this implies that the protein is thermolabile when it must act alone, but that in excision the function of the Xis protein in some way stabilizes Int against thermal inactivation. An alternative explanation would be that an additional protein is necessary for integration and that it is thermolabile; this would restore symmetry to the site specific recombination system in the sense that Int and an additional protein would be involved in both integration and excision. The identification only of $int^-$ mutants in experiments to isolate strains defective in integration, however, argues against the existence of an unidentified integration function.

*Structure of the Attachment Sites*

The distinction between integrative and excisive recombination, and the varying frequencies with which site specific recombination occurs between phages carrying attachment sites derived from different sources, argue that the phage and bacterial attachment sites, and therefore also the prophage attachment sites, all represent different sequences. This conclusion implies that each of the four half sites is distinct and that their arrangement is critical in recognition of these sequences by the Int and Xis functions. But this does not exclude the possibility that homology may exist between the attachment sites at the point of genetic exchange. A critical question about their structure is therefore whether the attachment sites are completely lacking in homology or whether they include homologous sequences. Two possible models which distinguish the consequences of these types of structure are illustrated in Figure 5.6.

If there is no homology between the phage and bacterial sites, recombination must involve scission of the duplex between $P$ and $P'$ in *attP* and between $B$ and $B'$ in *attB*; this must be followed by crosswise reunion of the duplex ends. This model therefore requires site specific recombination to proceed via an enzyme mechanism able to recognize duplex sequences of DNA, cut across the duplex, and rejoin the opposing ends. Since the half sites may be present in several different arrangements, this model requires that the Int and/or Xis proteins are able to exercise this action at more than a single pair of sequences. No enzyme is known which is able to undertake reactions of this nature. (Although restriction enzymes recognize specific sequences their action is in each case confined to a single short sequence in one duplex and the enzymes do not rejoin the cut ends.)

An alternative model is to suppose that each attachment site contains a "core," a short region that is the same in all sites. If the core is represented as $O$, then the structures of the four attachment sites can be written as:

| | |
|---|---|
| *attP* | $P\ O\ P'$ |
| *attB* | $B\ O\ B'$ |
| *attL* | $B\ O\ P'$ |
| *attR* | $P\ O\ B'$ |

The core sequence thus replaces the (.) which marked the crossover point. This form of structure implies that recombination could take place by a conventional exchange within the core sequence; that is, an unpairing of strands across the $O$ region could be followed by crosswise pairing to generate hybrid DNA. This might take place either by a recombination event allowed to occur simply at any point along the core, following the

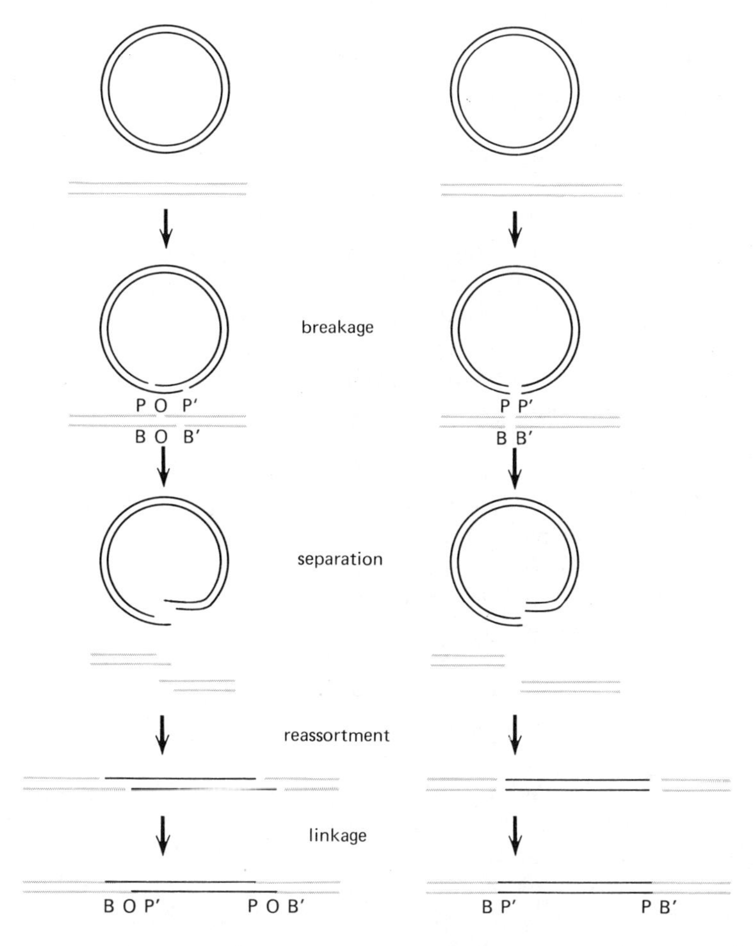


**Figure 5.6:** two models for site specific recombination. The common core model (left) supposes that the same short sequence, *O*, lies between *P* and *P'* and between *B* and *B'*. Recombination takes place within this sequence. The model illustrated shows the introduction of staggered single strand breaks at each end of *O*, followed by separation to yield single strand ends which can anneal crosswise to form hybrid DNA. This is a conventional model for recombination through the formation of hybrid DNA with the variation that the breaks generating the complementary single strands are made at specific sites. An alternative model (right) supposes that there is no common sequence in *attP* and *attB*, so that recombination must take place at a single nucleotide bond separating *P* from *P'* and between *B* and *B'*. This implies that cleavage across the duplex is followed by crosswise joining of the double stranded ends.

general model of Figure 5.1; or this recombination might take a more specific form, perhaps relying upon the introduction of staggered breaks in each strand at the extremities of the core, as illustrated in Figure 5.6. In this case the rejoining action would involve pairing between complementary single strand ends, a reaction directly analogous to the formation of circles of λ DNA from the linear form by pairing and joining of the cohesive ends. The specific enzymes of site specific recombination then would be required to include endonuclease activities able to make the staggered breaks; and the demands for variety in sequence recognition would be eased since the common core sequence would always lie on one side of the break. We should emphasize that this model in no way alters the remarkable feature of site specific recombination, that it involves recognition of nonhomologous sequences, since it remains true that the sequences of the half sites, $P$, $P'$, $B$, $B'$, control the reaction; its advantage is that it provides a mechanism which allows site specific recombination to share with all other recombination events a reliance on complementary base pairing to accomplish crosswise reunion.

Two types of approach have been used to investigate the possible existence of homology between the attachment sites: physical techniques can detect homology by the ability of single strands to anneal to form duplex structures; and genetic techniques can be used to detect the existence of heterozygosity following site specific recombination.

Electron microscopic mapping of the heteroduplexes formed between λ and transducing variants locates the position of the phage attachment site. As Figure 5.2 shows, each transducing phage consists of λ DNA in which part of the phage genome has been replaced by bacterial sequences. When a single strand from a transducing phage is annealed with its complement from a λ phage, a duplex structure therefore forms along the length of λ DNA that is common to both phages, but each strand extrudes a single stranded sequence at the site of substitution. One end of the substitution loop corresponds to the point where the illegitimate crossover occurred between phage and bacterial DNA and the other end corresponds to the prophage-bacterial junction, that is to the point of crossover in the attachment site. If this crossover always occurs at a fixed position, this junction should be found at the same location in all transducing strains; but if there is a region of homology within which recombination can occur at any position, the position of the junction may vary in different transducing phages. Davis and Parkinson (1971) demonstrated that the substitutions in λ*gal* and λ*bio* strains all have a common end point located at 57.4% of the distance along the phage from the left end. Since the substitutions of bacterial DNA lie in opposite orientation in these two strains, the length of any region of homology where crossover occurs in the attachment site should be visible as a difference in the position of the end points of the substitutions in the two types of strain (compare the structures of λ*gal* and λ*bio* illustrated in Figure 5.2). But both classes of substitution appear to end at the same position; these measurements have an error of ± 100 base pairs, which sets this as a maximum value for the length of any region of homology in the attachment sites.

When deletion mutants are derived from phages carrying an $int^+$ gene, the deletions often end at the position 57.4%. When the deletion phages are derived from $int^-$ mutants, however, the dependence on this position is abolished. This again argues that the crossover point in *attP* can be located at 57.4%; and it suggests that deletions in $int^+$ strains may arise by an interaction which depends upon the site specific recombination system, although in $int^-$ strains some other mechanism must be responsible.

By constructing a phage carrying the site *attB*, Davis and Parkinson were able to examine the homology between *attB* and *attP* in heteroduplexes formed by annealing this phage DNA with λ DNA. None could be detected. This therefore sets an upper limit of about 20 base pairs to the length of any region of homology between the phage and bacterial attachment sites, since two single stranded complementary sequences longer than this should have been able to anneal in these conditions.

Different genetic predictions are made by models which postulate that there is a common core region in all attachment sites and those which suppose that there is no homology between any pair of attachment sites. These can be distinguished by the fates of mutations in the attachment sites. A mutation occurring in one of the half elements of any attachment site will have the same pattern of inheritance according to either model. For example, a mutation occurring in the *P* element of *attP* can be transferred by recombination into the *P* element of *attR*; but it can never be followed into *attB* (*B.B'*) or *attL* (*B.P'*). But a mutation occurring in a common core can be transferred into any one of the four attachment sites. A genetic test of the two models is therefore to see whether *att* site mutations can be found which are able to enter any attachment site.

It is the formation of hybrid DNA between common cores which allows a core mutation to pass between attachment sites. Figure 5.6 illustrates the processes by which recombination may occur according to models postulating the presence or absence of a common core. Whereas the common core relies upon the formation of hybrid DNA, a lack of homology compels reliance upon a rejoining of duplex ends. The elements *P*, *P'*, *B*, *B'*, all remain intact in either model; but when recombination involves a common core, the cores of both recombinant sites comprise hybrid DNA, one strand derived from each parent DNA. Figure 5.7 traces the events which follow from recombination between attachment sites that

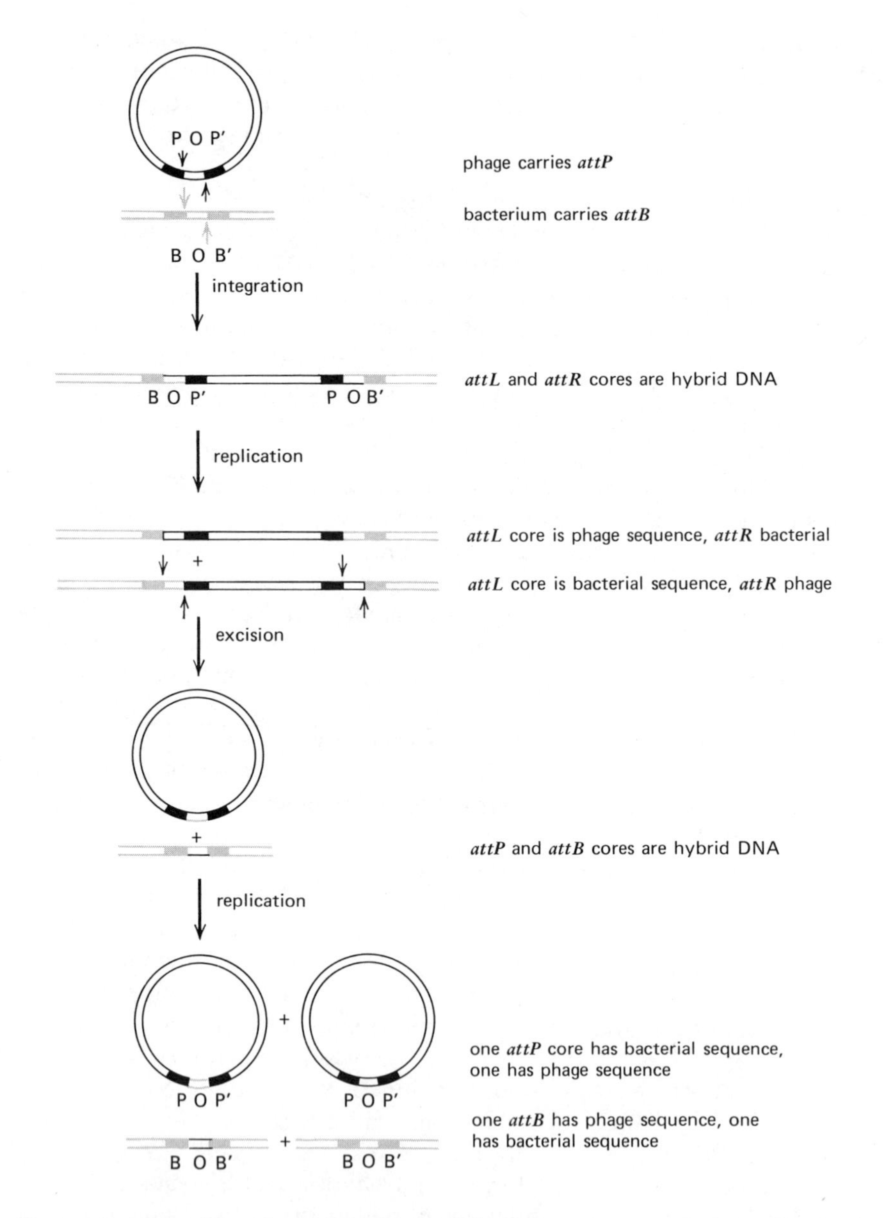


**Figure 5.7:** implications of common core model for attachment sites. The same sequence, $O$, is present in all attachment sites, so *attP* can be written as $P\ O\ P'$ and *attB* can be represented as $B\ O\ B'$. Recombination might be achieved by the model of Figure 5.6, in which staggered breaks are introduced at the ends of the core region, as indicated by the short vertical arrows; crosswise reunion of complementary strands then gives core regions consisting of hybrid DNA in which one strand is derived from the phage and one strand from

possess a common core. The recombination between *attP* and *attB* which inserts the phage into bacterial DNA generates cores in *attL* and *attR* both of which are hybrid DNA. If a mutation was present in the core of *attP*, then both the prophage cores will be heterozygous for it, with the strand derived from the phage carrying the mutant sequence and the strand from the bacterium carrying the wild type sequence. A single round of replication removes the heterozygosity as each strand acts as template for synthesis of a strand complementary to it; this produces one genome where *attL* has the phage sequence and *attR* has the bacterial core sequence, with the reverse sequence derivation applying for the other genome. Thus the product of replication is one genome with the mutation in *attL* and one genome with the mutation in *attR*.

The figure demonstrates that excision of the prophage then generates hybrid cores in the excised phage and bacterial chromosome. In both these products, each core consists of one strand carrying the sequence of the original phage core while the other carries the sequence of the original bacterial DNA. Replication again segregates any heterozygosity, producing two phage genomes in which one has the sequence of the original phage while the other has the sequence of the original bacterial core. Similarly one of the *attB* sites has the core characteristic of the original bacterial site whereas the other carries the core of the original phage. Thus a mutation in the core of the original *attP* site can be transferred into *attL* and *attR* of the prophage by insertion and then into *attB* by excision, provided that replication is allowed to take place to segregate the heterozygosity between successive insertions and excisions. Of course, this model assumes that the mispairing which must exist in heterozygous DNA is not corrected by repair systems. The model can be tested by examining *att* mutations: the detection of any mutations that can be transferred between all four attachment sites would imply that recombination must take place in a region of homology; whereas the failure to isolate

---

the bacterial chromosome. The half elements $P$, $P'$, $B$, $B'$ retain their original duplex structure. Thus if a mutation is present in the core of *attP*, both *attL* and *attR* will be heterozygous for it; the heterozygosity will be removed by replication to give genomes each of which has one core derived from the original *attP* and one with the sequence of *attB*. (Information derived from the original phage genome is represented by a heavy line and that derived from the original bacterial chromosome by a light line.) Since in each genome *attL* and *attR* bear cores of different origin, excision again generates *att* sites with hybrid DNA, as shown in the figure for the lower replicate prophage genome. Replication segregates the heterozygosity to yield one phage with the *attP* characteristic of the original phage and one phage with an *attP* whose core is derived from the original bacterial sequence; a similar segregation is seen in the bacterial chromosome, where one *attB* has the sequence of the original bacterial chromosome and the other has a core with the sequence of the original phage core.

any such mutations would render the existence of a core region less likely.

By using the $\lambda att^2$ system, it is possible to select for mutants with defective attachment sites by isolating $\lambda att^2$ phage lines that retain only the higher density peak (representing phage carrying bacterial DNA) when they are perpetuated for several generations of growth under conditions that permit site specific recombination. If inability to revert is due to mutations present in one or both of the attachment sites of $\lambda att^2$, it should be cis dominant. Shulman and Gottesman (1971) tested putative *att* mutants by subjecting cells to mixed infection with the original $\lambda att^2$ and a $\lambda att^{2-}$ mutant unable to revert to the λ form (see Figure 5.4). In 10 of 14 isolates the bacterial genes were excised only from the $\lambda att^2$ parental phage and not from the mutant. These mutants display a residual level of reversion from $\lambda att^{2-}$ to λ of about 1% and it is therefore possible to examine the products to determine the pattern of inheritance of the mutation.

An attachment site mutation in $\lambda att^2$ might lie in either *attL* or *attR*. Shulman and Gottesman (1974) used two techniques to locate some of these mutations. The first was to make vegetative crosses between $\lambda att^2$ carrying the mutation and λ*bio*, which carries *attR* only. If the ($\lambda att^2$) recombinants carrying the *attR* of λ*bio* are normal, the mutation in $\lambda att^2$ must have been in the *attR* site that has been replaced; if these recombinants are mutant, the mutation must be in *attL*. This test demonstrated that strain $\lambda att^2 24$ bears a mutation in *attR* and phage $\lambda att^2 6$ bears a mutation in *attL*. A second test is to use complementation: a cell is infected with two different $\lambda att^{2-}$ phages and if one is $attL^-$ and one is $attR^-$ it should be possible to obtain efficient site specific intermolecular recombination between them. But if both lie in the same site this is not possible. This test confirmed that the mutations *att*24 and *att*6 are carried in the different types of *att* site.

To determine the segregation patterns resulting from the inefficient site specific recombination that takes place in these mutants, Shulman and Gottesman scored the phage and bacterial excision products for the presence of *att* mutations. The assumption on which this assay rested was that the presence of the mutation in *attP* or *attB* would depress the integration efficiency in an event involving it. The phage produced by intramolecular site specific recombination in $\lambda att^2$ can be isolated directly by its reduced density and then scored for its ability to integrate into new recipients carrying a wild type *attB*. The excised bacterial segment cannot be isolated directly in this way. Instead cells were selected for insertion of the $\lambda att^{2-}$ phage by general recombination (this involves selection for the bacterial genes carried by the phage in appropriate conditions), and then the selected lysogens were cured of their prophage by site specific recombination; this leaves the bacterial product, which carries the *attB* site derived from $\lambda att^{2-}$ reversion, as part of the bacterial chromosome. These cells can then be tested for their ability to support integration by $\lambda^+$. The results are shown in Table 5.2. It is clear that both *att*6 and *att*24 can yield $att^-$ phage and bacterial sites; this implies that the mutations must lie in a sequence common to *attP* and *attB*.

Can both mutations also be isolated in *attL* and *attR*? This was tested by reconstruction experiments to generate $\lambda att^2$ phages from the mutant $attP^-$ and $attB^-$ sites. Thus phage was inserted and then excised by general recombination to yield a $\lambda att^2$ strain. The phage $\lambda att^2$R24 was derived from the crosses *attB*24 × $attP^+$ or *attP*24 × $attB^+$; it has the same properties as the original $\lambda att^2$24 mutant. The phage $\lambda att^2$L24 was derived by inserting $\lambda attP^+$ into a cell carrying *attB*24 and then excising the $att^2$ variant. And a phage with two mutant attachment sites, $\lambda att^2$LR24-24, was derived by crossing *attB*24 × *attP*24. The *att*6 mutation behaves in the same manner. Either of these mutations can therefore be crossed into any one of the four *att* sites.

If hybrid DNA forms over the core region in site specific recombination, replication of a genome carrying a hybrid core should give one wild type and one mutant genome; whereas the homozygous DNA that would be the intermediate if there were no common core would give only one type, either mutant or wild type. This prediction was tested by isolating the phage excised from $\lambda att^2$R24 grown in a *dnaB* host where λ replication is prevented. The progeny were then tested on new recipients for their ability to integrate and displayed the genotypes

29 $att^+$
95 $att^-$
9 $att^+$ and $att^-$

Thus 7% of the progeny are heterozygous.

This result demonstrates that heterozygotes do occur, implying that there must be a region of homology in the attachment sites which gives rise to hybrid DNA structures as a result of site specific recombination. But the frequency of heterozygosity is very much lower than the expected 100%. This implies that either an efficient correction mechanism is functioning or that detection of heterozygotes is hindered for some unknown reason.

Another situation where the predictions of the common core model are not fulfilled is the excision from either $\lambda att^2$L6 or $\lambda att^2$R24, which should yield equal numbers of mutant and wild type phage genomes; but as Table

5.2 shows, the proportions observed are highly asymmetric. Again this might be due to the involvement of correction mechanisms that remove heterozygosity.

These conclusions rely upon events occurring at a single site; for *att*6 and *att*24 appear to represent the same mutation, the first in *attL* and the second in *attR*. Shulman, Mizuuchi and Gottesman (1976) found that another five independently isolated point mutations lie in the same site in *O*, which may therefore be a hot spot for mutation. No point mutations in the unique elements *B*, *B'*, *P*, *P'* could be isolated; all the mutations affecting these elements represented deletions removing one of flanking sequences of the *att* site. One possible implication of this surprising observation is that mutations in *B*, *B'*, *P*, *P'* might not depress excision in $\lambda att^2$; and this would demand a revision in the view of the roles of these elements that we have presented. Although it is clear from these results that a region of homology is implicated in site specific recombination, its role is not yet defined, and the nature of the core and of the unique elements remains to be determined. The best prospect of elucidating the relationship between the components of the *att* sites at present appears to lie with attempts to determine their nucleotide sequences directly.

Although site specific recombination usually involves only sites composed of the elements *B*, *B'*, *P*, *P'*, the ability of $\lambda b2$ to participate in integration with *attR* and to recombine in vegetative crosses demonstrates that sites of the form $\Delta .P'$ can be recognized. A demonstration that site specific recombination can utilize sequences other than those of the phage, bacterial and prophage attachment sites was provided by the study which Shimada, Weisberg and Gottesman (1972, 1973) conducted of inte-

**Table 5.2:** segregation of mutations in attachment sites

| | | | genotype of excision products | | | |
|---|---|---|---|---|---|---|
| | attachment sites | | phage | | bacterial | |
| phage | *attL* | *attR* | $attP^+$ | $attP^-$ | $attB^+$ | $attB^-$ |
| $\lambda att^{2+}$ | + | + | 38 | 0 | 20 | 0 |
| $\lambda att^2 6$ | − | + | 27 | 1 | 0 | 20 |
| $\lambda att^2 24$ | + | − | 18 | 16 | 19 | 1 |

The genotypes of the excision products of $\lambda att^2$ phages were tested by measuring the ability of the phage to integrate into new ($attB^+$) recipients and assaying the ability of the bacterial site to support integration of a new ($attP^+$) phage. (+) indicates a wild type site and (−) indicates a mutant site. Data of Shulman and Gottesman (1971).

gration in bacterial host strains deleted for *attB*. Rare lysogens can be obtained from these cells when lambda inserts elsewhere. A comparatively small number of sites (about ten) is used (relatively) frequently for this type of insertion. The inserted prophages show the same permutation of markers as usual; and excision, although inefficient, requires the Int and Xis functions. This recombination therefore appears to occur at *attP* and involves the site specific system. In some cells the site of insertion lies within a gene, for insertion creates an auxotroph and excision restores prototrophy; this can be used to select for insertion and excision at such sites. These events reveal the specificity of site specific recombination, since the insertion and excision reactions produce no change in nucleotide sequence. A practical consequence of such insertion is that it makes possible the isolation from these lysogens of new types of transducing phage, where lambda carries bacterial genes that lie adjacent to the new insertion site.

These results suggest that in a bacterial host possessing *attB*, the affinity of the phage for its attachment site is very much greater than for any other sequence on the chromosome, so that *attB* is the only site at which insertion occurs. Deletion of this sequence provides the opportunity for recognition of *secondary sites* which have a much lower affinity for the phage. Shimada, Weisberg and Gottesman (1975) investigated the nature of these secondary sites by obtaining transducing phages from the secondary lysogens; these carry attachment sites in which the elements $B$ and $B'$ have been replaced by the corresponding sequences in the secondary sites. If we describe the structure of the secondary sites by the general form $\Delta O \Delta'$, then the transducing phages carry either $\Delta O\ P'$ or $P\ O\Delta'$. If the secondary sites represent a single sequence which is repeated at different places in the bacterial chromosome, all the transducing phages of either class should carry the same variant of the attachment site; if the secondary sites represent different sequences, each transducing phage will have characteristically different properties.

Variations in their properties were observed among each of the two classes of transducing phage and this therefore suggests that the elements of the secondary sites, $\Delta$ and $\Delta'$, each may be represented by a different sequence at each secondary site. The core of the secondary sites appears to be the same as the core of the usual attachment sites; for any difference between the cores of two recombining attachment sites should behave as a mutant of the type of *att*6 or *att*24 which enters all the successive types of site. The failure of insertion and excision at a secondary site to produce any change in *attP* thus demonstrates that the same core is involved in the reactions $P\ O\ P' \times \Delta O \Delta'$ and $\Delta O\ P' \times P\ O\ \Delta'$. Sequences which can function as secondary sites may be present on the phage DNA as well as in the bacterial chromosome. The transducing phages produced from some secondary site lysogens resemble in structure and function certain λ deletion mutants; this suggests that these deletions may arise by a site specific recombination of the type $P\ O\ P' \times \Delta O \Delta$, where both sites represent sequences of the phage DNA.

### *Site Specific Recombination In Vitro*

The general nature of the interactions involved in site specific recombination is clear from the genetic analysis of attachment sites and the *int* and *xis* genes: reciprocal recombination takes place between two very short homologous duplex sequences, recognition of the sequences on either side of this core is crucial in controlling the reaction, and either the Int protein alone or the Int plus Xis protein may be necessary according to the types and arrangement of the elements that are present. To define the sequence of events and the roles of the Int and Xis proteins requires the characterization of intermediates of recombination. One approach is to develop systems for accomplishing site specific recombination in vitro.

An assay for recombinant DNA has been developed in the form of a two step transfection protocol. The principle of the procedure formulated by Nash (1975a) is to form recombinant DNA in the first step; then the phage DNA is extracted and converted to mature phages in competent cells unable to carry out any recombination. This offers the advantage that the first step may be conducted in any appropriate system and imposes no requirement for completion of a lytic cycle; all recombinant products may therefore be examined, including those that would be inviable. In the second step, the product DNA of the first step is added to spheroplasts together with a helper DNA.

Two series of experiments have shown that the recombinant step is independent of concomitant transcription and translation. Nash used a system in which intramolecular recombination is followed in a phage which carries both *attP* and *attB*; this variant is $int^-$ and the Int function is provided by a heteroimmune prophage. Gottesman and Gottesman (1975a) induced lysogens to synthesize early gene products and then added a $\lambda att^2$ phage whose recombination depends upon the proteins synthesized by the lysogen. In both cases it is therefore possible to allow the cells to synthesize Int and Xis proteins and then to halt any further transcription and/or translation by the addition of suitable inhibitors; the superinfecting phage then is added, time is allowed for the formation of recombinants, and the putative recombinant DNA is extracted for assay in the spheroplast. Inhibition of transcription and translation does not interfere with site specific recombination, in contrast with the suggestion of Davies et al. (1972) and Inokuchi et al. (1973) that concomitant transcription plays a role in the recombination reaction.

In a further utilization of this system, Nash (1975b) and Gottesman and Gottesman (1975b) used the transfection assay to investigate the nature of the DNA products of systems for in vitro recombination. In the λ*attP-attB* system used by Nash, the formation of recombinants requires Int protein and is inhibited by Xis protein (this is not due to reversal of the recombination but appears to be caused by direct interference). In the λ$att^2$ system used by Gottesman and Gottesman, the formation of the λ recombinant can be assayed by the use of EDTA. Because of the size difference between the λ$att^2$ and λ phages, the λ$att^2$ phage is very sensitive to disruption by EDTA whereas the λ phage is resistant; the progeny phages produced by the spheroplasts therefore can be readily assayed for the presence of recombinants by replating in the presence of EDTA, since only the λ recombinant progeny can survive this treatment. In this system, which involves a recombination between *attL* and *attR*, the products of both the *int* and *xis* genes must be present in the in vitro extract. Another system previously developed by Syvanen (1974) relied upon the inability of linear DNA to be packaged into mature phages: recombination of a linear molecule to insert markers into a circular molecule thus can be assayed by appearance of the markers in phage particles. However, since these experiments did not demonstrate dependence on Int and Xis proteins, it is not certain that only site specific recombination occurred in vitro.

The potential of the in vitro systems is that it is possible to purify crude extracts for their ability to support recombination in vitro; this should allow the Int and Xis proteins to be obtained in purified form, and for any other necessary components of the reaction to be identified if they exist. The sequence of events may be determined from the structures of intermediates produced during the in vitro recombination reaction. If site specific recombination occurs by the introduction of staggered breaks illustrated in Figure 5.6, it should be possible to isolate intermediates with cohesive ends; and in this case it might be possible to determine the sequence of the region of homology by procedures similar to those used to sequence the cohesive ends of the vegetative phage.

## General Recombination

### *Vegetative Recombination in Lambda*

The construction of maps by classical genetic analysis relied upon the concept that the frequency of recombination between two markers depends only upon their distance apart on the chromosome. Thus the equation

$$\text{percent recombination} = \frac{\text{total number of (reciprocal) recombinants}}{\text{total number of progeny}} \times 100$$

defines the distance apart of two loci. Map distance is usually expressed simply in terms of percent recombination (1 recombination unit equals 1% recombination). The recombination frequency that can be measured between two markers has a maximum of 50%, which is achieved when they are independently inherited, either because they lie on different chromosomes or because they lie so far apart on one chromosome that genetic exchanges between them are very frequent.

Recombination events do not take place independently in adjacent regions. When three markers are followed, the frequency of double recombinants should be the product of the frequencies of occurrence of the single recombinants if events in the two adjacent regions are independent. In most eucaryotes, however, the frequency of double recombinants is much reduced compared with expectation. This is due to what has been described as *positive interference*: the reduction in probability that two exchanges will occur in the same vicinity. The occurrence of one recombination event therefore appears to inhibit the occurrence of another in the same region.

The reverse effect is observed in lambda and other phages (and also in some fungi): by analogy with the nomenclature of eucaryotic genetics, this has been called *negative interference*. In spite of this awkward description, the phenomenon observed is that more than one genetic exchange is likely to occur in the same region. That is, the occurrence of one recombination event increases the probability that another will take place close by. Two forms of negative interference are seen in phage crosses (see also Chapter 6). Between distant markers, negative interference reflects the probability that more than one exchange event will occur; this is sometimes known as *generalized negative interference*. Between close markers, recombination in lambda seems to take the form of a cluster of exchanges in the region where the recombination event occurs (for review see Signer, 1971). The negative interference caused by this clustering is known as *highly localized negative interference*. A crossover between two distant markers in lambda thus generally proves to be a cluster of exchanges when markers close to the site of exchange are scored.

Construction of a genetic map for lambda requires compensation for negative interference, since its result is to make the recombination frequency measured between two markers much smaller than the distance calculated by summing the recombination frequencies seen with intervening markers. Amati and Meselson (1965) therefore introduced a factor termec the *interference index*: this is the ratio by which the number of double recombinants exceeds that expected for independently occurring exchanges. Thus

$$i = \frac{R_{1,2}}{R_1 \times R_2}$$

where $R_{1,2}$ is the observed frequency of double recombinants and $R_1$ and $R_2$ are the single recombination frequencies observed in adjacent regions.

Figure 5.8 plots the value of $i$ against map distance (given by the sum of the two single recombination frequencies $R_1$ and $R_2$). Even for distant markers, $i$ takes a value due to generalized negative interference that is great enough to be important in calculating distances, about 3. For close markers, when highly localized negative interference occurs, $i$ may take much greater values, about 50.

To place individual crosses on a uniform basis and thus allow such data to be used to extend the map of lambda to represent the whole genome, Amati and Meselson introduced the function

$$R = R_1 + R_2 - 2i\, R_1\, R_2$$

where R gives the map distance between two loci and where $R_1$ and $R_2$ represent the frequencies of recombination in two adjacent regions that together make up the distance being measured. The use of this expression requires the assumption that the relationship between $i$ and distance shown in Figure 5.8 is constant over the lambda genome and may therefore be applied to all crosses. A further assumption, implicit in the use of recombination frequencies to represent map distance, is of course that exchanges take place at random in the sense that a cluster of exchanges is equally likely to occur at all sites. In other words, use of the interference index compensates for the tendency of one exchange to occur near another, but would not compensate for a preferential localization of exchanges at recombination hot spots.

The consequence of using this expression to derive map distances is to expand the lambda map. Amati and Meselson calculated the length of the whole genome as 70 units; this compares with the recombination frequency measured between terminal markers of 15% by Jacob and Wollman (1954) and Kaiser (1955). Another example is recombination

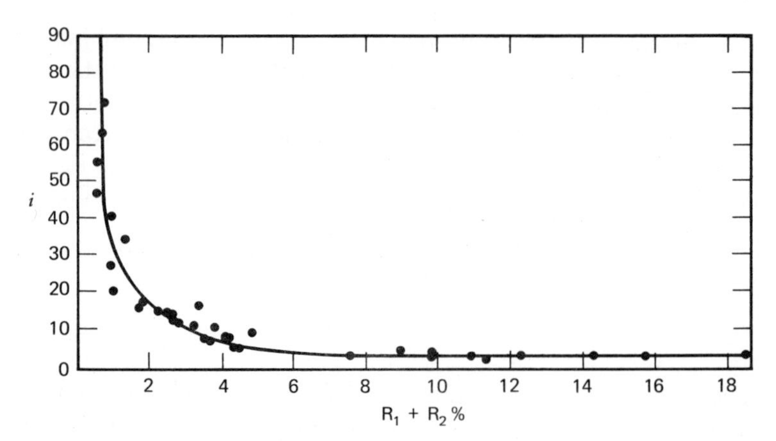


**Figure 5.8:** dependence of the interference index *i* on map distance ($R_1 + R_2$%). Data of Amati and Meselson (1965).

between *B* and *J*, which displays a frequency of about 4%; but correction for negative interference increases this value to about 19 (see Campbell, 1971). Of course, absolute recombination frequencies vary with the conditions of each cross and cannot be directly compared. One approach is to normalize data to the map constructed by Amati and Meselson by comparing the recombination obtained between two markers with the distance on the map; the values obtained in other crosses may then be adjusted proportionately. The data assembled on this basis by Campbell (1971) form the map shown in Figure 4.1, which provides a reasonable fit with data on the physical relationship between markers that have been obtained from electron microscopic mapping.

One of the important features of phage crosses is that they represent a situation best approached in terms of population genetics. As was first observed by Visconti and Delbruck (1953): "one may attempt to explain the genetic findings . . . by the idea that mating occurs repeatedly during every intrabacterial cycle of phage growth." From this concept has been derived the treatment of phage genetics as a population system in which recombination takes place among the cellular population of (replicating) genomes. We shall take up this view in the next chapter, since it was first developed for, and is most important in, T-even phage genetics, where many rounds of mating may occur in one phage cycle. The general principle, however, may be summarized in the view that vegetative phage genomes recombine pairwise and at random with respect to partners during the period while the population of phage genomes is being expanded by replication; genomes are withdrawn from this pool at random for incorporation into particles.

In lambda mating occurs during most of the latent period: recombinants are found among the first mature phages and Wollman and Jacob (1954) demonstrated that the proportion of recombinants increases with time during the second part of the latent period. Kaiser (1955) observed that if mating occurs at random, when a two factor cross is made with equal input of both parents:

$$r = \tfrac{1}{2} \cdot m\, p$$

where r is the observed frequency of recombination, m is the average number of rounds of mating, and p is the probability of recombination between two markers. This expression thus simply says that the probability that any given mature phage particle will be a recombinant (that is r) is the product of the probability that mating occurred between it or its ancestors and a phage of opposite parental type (that is ½m) and the probability that mating resulted in recombination (that is p). (This expression is only approximate because it does not take account of mating between recombinants. But this is small in lambda and for the purposes of this calculation may be ignored.)

This equation supposes that two reciprocal recombinants are formed in each recombination event. If only one recombinant is formed, the probability that mating generates a recombinant that is observed then becomes ½p so that the equation is

$$r = \tfrac{1}{4}\, m\, p$$

Which form of the equation is used therefore depends upon whether recombination is reciprocal or nonreciprocal.

In principle m can be determined from the case where p is known. Kaiser used the triple cross $m_6\, c\, + \times + + mi$ ($m_6$ was an early mutation in *W*, *c* lies in *cI*, and *mi* lies at the right end of the vegetative map). When progeny were selected for recombination between *c* and *mi*, the allele at the unselected locus was freely inherited since 50% were $m_6$ and 50% were + type. This means that the value of p for recombination between $m_6$ and *c* is ½. Since the observed frequency of recombination between $m_6$ and *c* was 0.13, this means that m must take the value of 0.52 or 1.04 depending upon which form of the equation is used. This means that the number of rounds of mating is 0.5–1.0 according to whether recombination is reciprocal or nonreciprocal.

A similar result was calculated by Amati and Meselson (1965) using a different approach. They found that the total number of recombinant clusters in each progeny lineage of lambda is about 0.3. There is an average of about 2.2 genetic exchanges per cluster. (Clusters may include up to 3 or even 4 exchanges.) This suggests that about 0.7 recombination events, that is, rounds of mating, take place in the average progeny lineage.

Formal genetic analysis therefore shows that each lambda genome on average suffers less than one recombination with another phage genome during the lytic cycle, but this single event may result in several closely related exchanges. This establishes the general nature of the system which must be characterized in terms of the molecular interactions that occur between recombining genomes and the enzymes that are responsible.

### *Formation of Heterozygotes*

The first question to be asked about the mechansim of recombination in lambda is whether it involves physical exchange of genetic material or copying of information from the two parental genotypes. The first attempt to resolve this issue was made in the experiments of Kellenberger, Zichichi and Weigle (1961a) and Meselson and Weigle (1961). In both sets of experiments the parental phages were characterized by physical properties and recombinants were then identified either by physical or by genetic analysis.

The buoyant density at which lambda bands in CsCl is established by its relative composition of DNA and protein. Wild type lambda particles band at 1.508 g/cm$^3$; particles of the deletion strains λ*b2* and λ*b5* display densities of 1.491 and 1.501 g/cm$^3$ which reflect their loss of DNA; the double mutant carrying both deletions bands at 1.483 g/cm$^3$. When cells are subjected to mixed infection with λ*b2* and λ*b5*, it is therefore possible to identify each parental class (that is the two single deletion strains) and each recombinant class (wild type and double mutant) among the progeny particles by virtue of their densities. Using this system, Kellenberger et al. found that when the DNA of one parent was labeled with $^{32}$P, the label appeared among the progeny phages. This demonstrates the incorporation of parental genetic material into recombinants.

Another method to identify parental phage, which has been widely used, is density labeling with the heavy isotopes $^{13}$C and $^{15}$N. The DNA of the labeled phage then bands at a heavy position due to its possession of two heavy strands; this can be described as HH. By using such phage to infect bacteria grown on normal (light) medium, Meselson and Weigle were able to show that replication is semiconservative. The first round of replication generates genomes of hybrid density, with DNA that can be described as HL since one (parental) strand is heavy and one (newly synthesized) strand is light. Light genomes of type LL can be generated only by subsequent rounds of replication.

Following a mixed infection when the heavy parental phage population includes two different genotypes, the progeny genomes can be scored for genetic recombinants. Discrete amounts of parental DNA appeared in the recombinants, including some that were >50% heavy, that is, whose strands had not been separated by replication. In further experiments Meselson (1964) found recombinants comprising entirely heavy (HH) DNA. These must have been generated by breakage and reunion of the original infecting parental genomes. Confirmation that recombination can take place without replication has been provided by experiments which show that recombinants may be formed during infection of cells in which phage DNA synthesis is blocked. When McMilin and Russo (1972) used density labeled $O^-$ and $P^-$ parental phages to infect cells that were temperature sensitive for replication (growing at the high nonpermissive temperature), they found that some unduplicated DNA matures to give progeny particles; among the heavy (HH) progeny were recombinants. Although none of these experiments excludes the possibility that a certain amount of DNA synthesis is associated with recombination, the ability of parental phage DNA to recombine without replication demonstrates that recombination involves physical exchange of genetic material and not copying of information.

In these experiments all the recombination systems of the phage (Int and Red) and of the host (Rec) were presumably functional. To analyze the events resulting from operation of each system, Kellenberger and Weisberg (1971) reexamined recombination between λ*b2* and λ*b5* in conditions when only one system was active, since the others carried mutations. All three gave the same result: recombination takes place by breakage and reunion between DNA genomes of the two genotypes.

Is vegetative recombination reciprocal? The consequences of reciprocal and nonreciprocal recombination depend upon the types of structure that are involved. Figure 5.9 demonstrates that reciprocal recombination between circles generates a dimeric circle—essentially one circle has been inserted in the other. Nonreciprocal recombination between circles generates a linear dimer (essentially the two ends that would have been joined in reciprocal recombination are left unlinked). When linear molecules are involved, reciprocal recombination generates two linear recombinants whereas nonreciprocal recombination generates one linear recombinant and two fragments. Between a linear molecule and a circular molecule, reciprocal recombination inserts the circle into the linear molecule, whereas nonreciprocal recombination in effect breaks the circle and attaches a fragment of the linear molecule to one end.

Lambda DNA is present in several forms, including circles and oligomers, in infected cells and so it is not possible to say which form is involved in recombination or to distinguish the molecular forms of the product. The

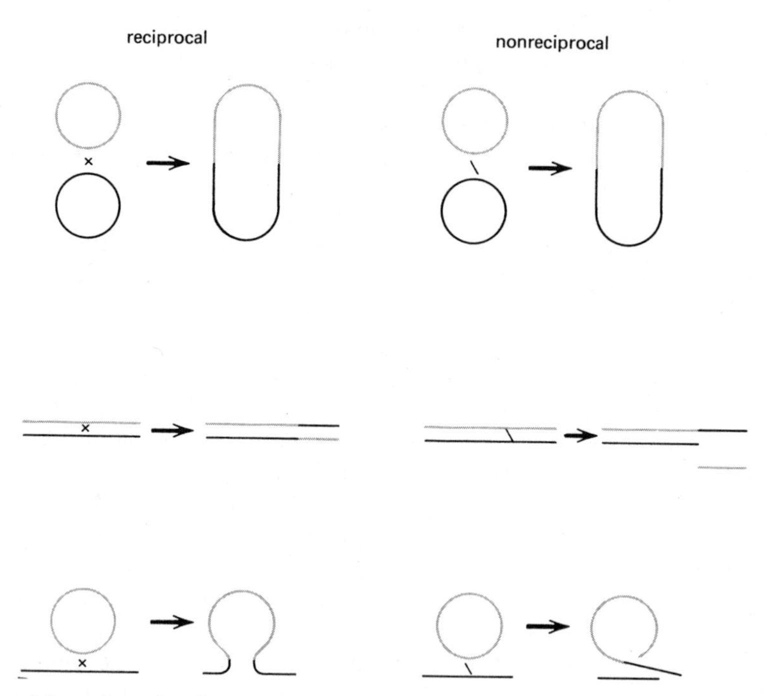


**Figure 5.9:** reciprocal and nonreciprocal recombination involving circular and linear DNA.

situation is complicated also because DNA matures from oligomeric forms into the monomeric linear molecule of the mature particle (see later): thus an examination of recombinant progeny is concerned with sequences which have matured from the oligomeric intermediate and not with the immediate products of recombination. Attempts to investigate reciprocity have therefore rested upon the assumption that if recombination is reciprocal the progeny of an infected cell should include both recombinant types at equal frequency.

In order to examine the reciprocity of recombination, it is necessary to analyze the products of a single exchange event, as for example is possible in the tetrad analysis of some fungi (see Volume 1 of Gene Expression). The demonstration that reciprocal recombinants occur with equal frequency from a statistical analysis of the progeny of many crosses implies that either recombination is reciprocal or that opposing nonreciprocal events occur with equal frequency. Thus although reciprocal recombinant genotypes are generally found in equal frequencies in the overall output of a phage cross—indeed, this is a necessary condition for constructing the genetic map—this does not make any implication about the nature of the individual recombination event. Although even individual cells represent a population of phages, in lambda the number of rounds of mating is small and this makes it reasonable to expect that a correlation between reciprocal recombinants should be observed in individual cells. Such analysis is performed by means of a *single burst* experiment, when the population of infected cells is diluted to the point where each sample contains only one infected cell, whose progeny then may be analyzed. (The design of single burst experiments is discussed in more detail in Chapter 6).

In early experiments when all the phage and host recombination systems were functional, Jacob and Wollman (1955) and Kaiser (1955) observed no such correlation. A striking demonstration which provides its own control was the observation of Weil (1969) that in single bursts, recombination sponsored by the Int system is strictly reciprocal (as demanded by the Campbell model; see earlier) whereas recombination promoted by the Red system is nonreciprocal. This comparison demonstrates that nonreciprocity in Red recombination cannot be attributed to some complicating factor of the phage-host system, for example concerned with maturation; rather must it reflect the nature of the recombination event itself. The Rec system of the host, however, may generate reciprocal recombinants as observed by Herman (1968) for recombination between the E.coli chromosome and a resident sex factor; and Meselson (1967) observed reciprocal recombination between two lambda prophages, one carried on the bacterial chromosome and one on an F′ factor (presumably due to the Rec system since the Red and Int genes should have been repressed). These experiments suggest the tentative conclusion that Rec recombination is reciprocal but Red recombination is nonreciprocal. We should stress that this applies to the recombination between distant markers that results from breakage and reunion at a DNA site between them; recombination between very close sites may be achieved by different mechanisms through the action of systems that act on hybrid DNA.

What is the structure of a recombinant genome in the region where genetic exchange has occurred? If recombination takes place by the formation of hybrid DNA, as illustrated in Figure 5.1, the recombinant DNA should be heterozygous whenever the hybrid DNA extends over a site which differs in the parental genotypes. Such mismatched DNA may provide a substrate for cellular correction systems, which may act to restore homozygosity by removing the bases of one strand and replacing them with bases complementary to those of the other strand. If the region of hybrid DNA is long enough to extend over several markers, recombination between them may occur when a correction system acts upon some but not upon other of the mismatched sites. For example, if the two strands of a recombinant DNA are represented by the marker sequences:

A B C D E F G H I J K L M N O P Q R s t u v w x y z
A B C D E F G H I J k l m n o p q r s t u v w x y z

where capital letters represent one parent and lower case letters represent the other parental genotype, the markers *K*/*k* to *R*/*r* form a length of hybrid DNA. Recombination has been achieved on both strands of the DNA between *distant* markers, that is those lying outside the region of hybrid DNA, by virtue of the breakage and reunion event; thus the alleles *A-J* of one parent have been linked to alleles *s-z* of the other parent. But there has been no recombination between those markers that lie within the region of hybrid DNA. Recombination between these *close* markers, however, results whenever only some of them are acted upon by correction enzymes. For example, if the sequence *n o p* of the lower strand is corrected to *N O P* the region of hybrid DNA has the sequence

K L M N O P Q R
k l m N O P q r

This correction event has two consequences. The ratio of the alleles *N O P*/*n o p* is no longer 1 : 1; and on the lower strand a recombination has been achieved between *N O P* and the other markers. Genetic analysis of fungal systems has suggested that recombination between very close markers usually is the consequence of correction events, in contrast with the recombination by classical breakage and reunion that applies for more distant markers. (A relationship between recombination frequency and the distance apart of markers is displayed in corrective recombination because the probability that two markers will be simultaneously corrected, in which case there is no recombination between them, increases as the distance decreases.) Recombination by correction and by breakage and reunion is discussed in Volume 1 of Gene Expression.

One important point follows from the conclusion that recombination between close markers is due to corrective events. Hybrid DNA may be formed without recombination, as illustrated on the left of Figure 5.1, for example, generating a structure of the form

A B C D E F G H I J K L M N O P Q R S T U V W X Y Z
A B C D E F G H I J k l m n o p q r S T U V W X Y Z

This has exactly the same heterozygous sequence as that of the recombinant DNA drawn above, and may therefore generate recombinants in the region *K/k* to *R/r* by precisely the same corrective events; however, there is no recombination between the distant markers outside the region of hybrid DNA. To determine which sort of structure has given rise to recombination between very close markers, it is therefore necessary to examine outside flanking markers on each side to see whether they have undergone recombination or not.

The proportion of heterozygotes which is formed in lambda crosses is low, usually of the order of 0.1% of the progeny. Two effects may be responsible. When recombination takes place early in lambda development, many of the heterozygotes may be removed when their strands segregate for replication. And of course correction systems may act on the mismatched DNA to restore homozygosity. Kellenberger, Zichichi and Epstein (1962) were able to increase the proportion of heterozygous progeny by ultraviolet irradiation of the host bacterium; this treatment inhibits replication and its effect may thus depend at least in part upon preventing removal of heterozygotes. When phages that were heterozygous for a $c/c^+$ marker were examined for flanking markers, 70% were recombinants. Heterozygosity is therefore associated with recombination, in general support of the idea that recombination occurs by formation of hybrid DNA. (The remaining 30% of heterozygotes may include double recombinants and/or genomes in which hybrid DNA was formed without recombination). An estimate for the length of the heterozygous region was derived by comparing heterozygosity for distant markers (heterozygosis at the two sites is independent) with that of closer markers, when the increase in the frequency of double heterozygotes shows how often the hybrid DNA extends over both sites. The length of hybrid DNA is very variable and in these experiments displayed an average length of about 10% of the lambda genome, of the order of 5000 base pairs.

The association of genetic heterozygosity with recombination lends credence to models for pairwise exchange of single strands such as that illustrated in Figure 5.1. But although such models for recombination have seemed likely for some time, little is known about the molecular nature of the series of events which may be responsible for this breakage and reunion. More recently, however, structures which appear to represent the joining of two DNA molecules at homologous sites have been observed by Valenzuela and Inman (1975). These *anomalous junctions* were identified as connections between two lambda genomes, probably able to occur at random positions but always located at sites that are homologous in the two connected genomes. The structure of these junctions is illustrated in Figure 5.10; they take the form of a box of single

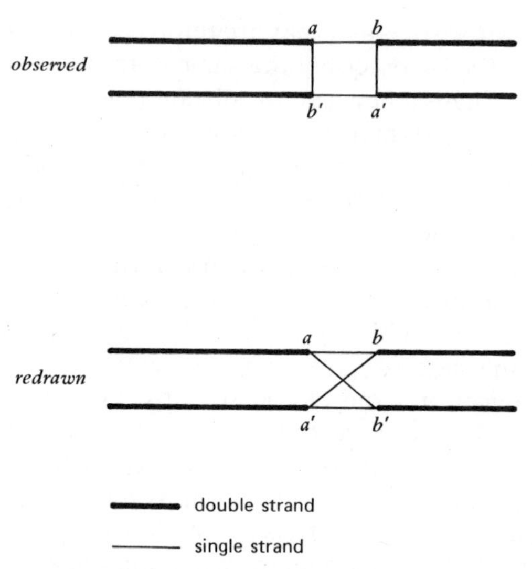


**Figure 5.10:** structure of anomalous junctions. The upper structure illustrates the junction observed as a connection between two λ DNA molecules at homologous sites; a box of single strands appears to connect the two duplex molecules, which are orientated in reverse polarity (*a* and *a'* represent homologous sites; *b* and *b'* are homologous). The lower illustration depicts the result of redrawing the structure so that both duplex molecules are in the same orientation; this generates a junction in which single strands appear to cross over between the duplexes in the configuration displayed by the model for recombination shown in Figure 5.1. Data of Valenzuela and Inman (1975).

strands connecting two duplex sequences orientated in opposite polarity. If the structure is redrawn in order to make the polarity of both genomes the same, the junction takes the form of a crossover of single strands connecting the duplexes; this of course is precisely the sort of structure which might correspond to an intermediate of recombination. It is not clear whether it is significant that all (14) observed anomalous junctions involved duplexes orientated in reverse polarity.

Anomalous junctions have been observed in two types of λ DNA. In replicating circles they might be the result either of recombination between the two daughter duplexes or could be produced by migration of a replicating fork (by winding the growing point backwards to reproduce the original duplex between parental strands and generate a duplex between the two newly synthesized strands). As connections involving nonreplicating circles, however, they seem likely to be the product of recombination. Although no proof of the origin of the anomalous junctions is possible, it seems probable therefore that they represent intermediates of recombination. Similar structures have been observed in the fused dimers of phage $\phi$X174 that appear to be recombination intermediates (see Figures 8.12 and 8.13).

A detailed examination of the structure of recombinant λ DNA has been performed by White and Fox (1974, 1975a,b). These experiments were designed to determine the structure of the unduplicated recombinants produced by three factor crosses involving closely linked markers. By using amber mutations in *O* and *P*, which map in the order *Oam29—Pam3—Pam80*, replication is prevented and so any recombinants must contain only parental DNA. At high multiplicities of infection up to 5% of the input phages are able to mature. When the two parental types of phage are of different density, one carrying a heavy label in its DNA and one with a genome of normal (light) density, the amount of DNA contributed by each parent to a recombinant can be assessed directly by its density on a CsCl gradient. The absence of replication is necessary to ensure that the density is not reduced by the synthesis of light DNA after infection; in addition to the mutations in the phage replication functions, replication was prevented by carrying out crosses at high temperature in a host which is temperature sensitive for phage replication. In a cross between heavy *Pam3* and light *Oam29 Pam80* phages, the double recombinants ($+^{am29}$ $+^{am3}$ $+^{am80}$) sedimented in a broad density band, corresponding from those with 15% of the heavy parental DNA and 85% of the light parental DNA to those with the reverse combination. Since the two outside markers lie only 3% of the lambda genome apart, this suggests that this recombination is associated with a breakage and reunion event of the classical kind, that is, in which each end of the recombinant λ DNA is derived from a different parent and a region of linkage between them represents hybrid DNA. (If these recombinants resulted from the formation of hybrid DNA without recombination between outside markers, these results would imply that the minimum length of a single strand which could be inserted to form the hybrid sequence as shown in Figure 5.1 is 30% of the genome. Breakage and reunion between duplex DNA is a more probable explanation. Thus the density analysis can be taken as a biochemical counterpart to the usual genetic control which examines recombination between flanking markers.)

The three markers in *O* and *P* are closely linked and the most likely model is to suppose that their recombination therefore involves correction events occurring in a sequence of DNA which covers either two or all three of the sites. If this hybrid DNA has been generated by a genetic exchange between duplexes, it must lie between duplex regions derived from each parent. Four types of intermediate can be imagined to result from the exchange event; these are depicted in Figure 5.11. The DNA to

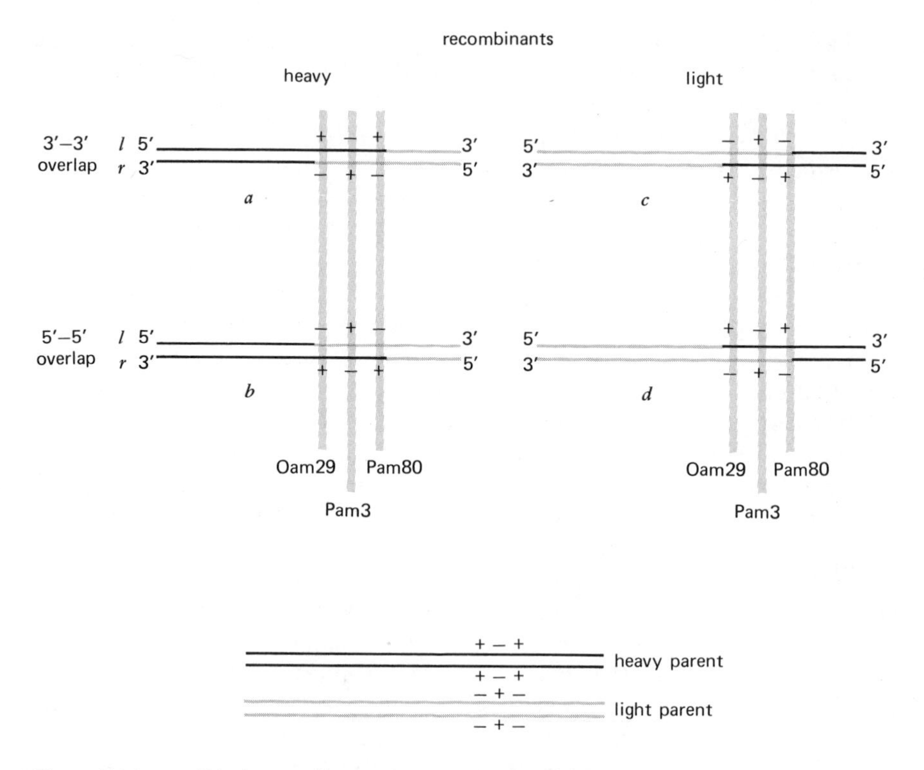


**Figure 5.11:** possible intermediates of recombination between heavy density $+^{am29}$ *Pam3* $+^{am80}$ lambda and light density *Oam29* $+^{am3}$ *Pam80* phage. After White and Fox (1974).

the left of the exchange site may be derived from the heavy parent (structures *a* and *b*) or from the light parent (structures *c* and *d*); since the site of exchange (the *O-P* region) lies near the right end of lambda, recombinants of heavy density should be derived from the first type and recombinants of light density from the second type. The overlap may be organized in either of two orientations, which are distinguished by the positions of the junctions between the parental sequences on each strand. The overlap shown in structures *a* and *c* may be described as 3′-3′, for if the intermediate shown were disassembled into the two parental parts, each would comprise a duplex with a single strand protrusion ending in a 3′ terminus. The overlap shown in structures *b* and *d* may be described as 5′-5′, since the single stranded ends would terminate with 5′ groups.

The genetic constitution of the hybrid DNA depends upon which parental genome provides which strand. Thus the mispairing that is present at the heterozygous sites can take two forms among the four intermediate structures. Suppose for example that the wild type sequence at the *0am29* site is a G-C pair, orientated with guanine on the *l* strand and cytosine on the *r* strand; suppose the mutant sequence has an A-T pair, with adenine on the *l* strand and thymine on the *r* strand. Then the *l-r* +/− combination of structures *a* and *d* represents a G-T mismatched pair; whereas the *l-r* reverse −/+ combination of structures *b* and *c* represents an A-C mismatch. Since the activity of the correction systems and their orientation of action (which strand is corrected) may depend upon the particular mismatches that are present, these two types of intermediate may therefore give rise to different classes of recombinant.

If both types of overlap, 3′-3′ and 5′-5′, occur as intermediates in recombination, the heavy intermediates (*a* and *b*) will include the same two types as the light intermediates (*d* and *c*). No difference should therefore be seen in their patterns of heterozygosity. But if only one type of overlap is utilized as an intermediate in recombination, the heavy and light density classes will have different genetic constitutions: for a 3′-3′ overlap the heavy class will have structure *a* whereas the light class will have structure *c*; for a 5′-5′ overlap, heavy recombinants must be derived from *b* and light recombinants from *d*. The idea that only one type of overlap is permitted may explain the differences in heterozygosity which White and Fox observed in the two density classes. When double recombinants of the class $+^{am29}$ $+^{am3}$ $+^{am80}$ were isolated in the heavy fraction, 75% were homozygous +++/+++ and 22% were heterozygous +++/ + *am3* + (with other heterozygous classes represented to a lesser extent). In the light fraction 43% were homozygous +++/+++ and 41% were heterozygous +++/*am29* + +. This difference cannot be explained in molecular terms without knowing details of the actions of the correction systems, but its general implication is that the intermediates upon which these systems act may be different in the heavy and light density classes.

Reconstruction experiments suggested that this result reflects the occurrence of only 3′-3′ intermediates in recombination. White and Fox prepared two types of heterozygote by annealing the separated *l* and *r* strands of the two parental phage types. Cells were then transfected with these hybrid λ DNAs and the wild type double recombinants examined for heterozygosity. The annealing product

| | | | | | |
|---|---|---|---|---|---|
| *l* | + | | Pam3 | | + |
| *r* | | *Oam29* | | + | Pam80 |

has the same genetic constitution as intermediates *a* and *d* of Figure 5.11. It gives rise to the same progeny classes that are found in the heavy recombinants. This correlation suggests that heavy recombinants may be derived from a structure with 3′-3′ overlap (*a* of Figure 5.11). On the other hand, the annealing product

| *l* | Oam29 | + | | Pam80 |
|---|---|---|---|---|
| *r* | + | | Pam3 | + |

has the same gentic constitution as intermediates *b* and *c* of Figure 5.11. It gives rise to the same progeny classes that are found in the light density recombinants. The light recombinants also therefore appear to be derived from a structure with 3′-3′ overlap (*c* of Figure 5.11).

Several general conclusions are suggested by these results. The broad density distribution suggests that recombinants may contain an appreciable length of hybrid DNA; for example, recombinants which have 50% of their DNA from each parent, presumably corresponding to an exchange in the center of the λ DNA, may be heterozygous in the *O-P* region which lies within 20% of the right end. This conclusion is supported by observations that the recombination in *O-P* is often associated with heterozygosity at the flanking markers *cI* and *R*. And in other experiments involving recombination between density labeled $O^-$ phages, Russo (1973) observed that some 16% of the progeny that are heterozygous at *cI* also are heterozygous at *R*; this implies that the length of the heteroduplex may not infrequently extend from *cI* to *R*, a distance of about 25% of the lambda genome (11,000 base pairs).

The correlation between genetic constitution and density suggests that recombination occurs through intermediates which possess only the 3′-3′ type of overlap. The generation of recombinants from intermediates in which heterozygous DNA extends over the mutant sites implies that correction systems are responsible for this recombination between closely linked markers; and since recombination demands that correction takes place at only some of the closely linked sites, the length of the sequence which is excised and repaired must be comparatively short. This conclusion is consistent with the detailed pattern of heterozygosity that is observed. The high negative interference seen in lambda recombination then might be explained by supposing that the limiting event in recombination is the formation of hybrid DNA; the occurrence of several correction events, each occupying a short region of the lengthy heteroduplex, would produce the cluster of tightly linked exchanges associated with each event.

### *Phage Recombination (Red) Functions*

Under normal conditions the general recombination functions of lambda are nonessential: that is, phages carrying mutations in them are able to complete the lytic cycle and generate progeny. To isolate mutants defective in vegetative recombination, Echols and Gingery (1968) and Signer and Weil (1968a) used a marker rescue technique. In principle mutants were isolated by their inability to perform marker rescue. These *red*$^-$ mutants display a tenfold reduction in the frequency of rescue in Rec$^+$ hosts; there is a further twofold drop to an overall twentyfold reduction in Rec$^-$ hosts. This implies that the phage Red system and the host Rec system are able independently to undertake vegetative recombination during the lytic cycle, although most of the recombination appears to be due to the action of the phage system.

Two proteins have been identified with the general recombination system of the phage. An exonuclease activity was first characterized by the ability of extracts from infected cells to degrade DNA in vitro. Radding and Shreffler (1966) showed that antiserum against the purified activity has two components: the $\alpha$ component was directed against the exonuclease enzyme; and the $\beta$ component reacted with an associated protein, described as $\beta$ protein. Radding et al. (1967) mapped the gene responsible for exonuclease synthesis, then called *exo;* and in a further study Manly et al. (1969) demonstrated that different genes are responsible for the production of the two proteins.

The *red*$^-$ mutations fall into three complementation groups, which may be identified with these two proteins. Signer et al. (1968) and Shulman et al. (1970) reported that *redA* and *redC* failed to synthesize the lambda exonuclease enzyme whereas *redB* mutants do not produce $\beta$ protein. Complementation between *redA* and *redC* appears to be intracistronic, probably reflecting the assembly of active enzyme from identical subunits. The two *red* genes can therefore be described as *red*$\alpha$, coding for exonuclease and previously known as *exo*, and *red*$\beta$, coding for $\beta$ protein. Some *red*$\alpha$ mutants produce an inactive protein antigenically related to exonuclease; and some *red*$\beta$ mutants produce a protein immunologically related to $\beta$ protein (see Radding, 1970). Further confirmation of the assignment of *red*$\alpha$ to exonuclease was provided by the observation that four temperature sensitive *red*$\alpha$ mutants produce an exonuclease that is heat labile in vitro.

The isolation of only two classes of recombination-defective phage mutants suggests that the two *red* genes alone are sufficient to code for the phage general recombination system. But of course, this does not exclude the possibility that host functions also are involved in this pathway; however, if any further such components exist, at present they remain unidentified. The Rec pathway provides the only host functions implicated in phage recombination, but appears to be independent rather than part of the Red pathway. What then is the role of exonuclease and $\beta$ protein in genetic recombination?

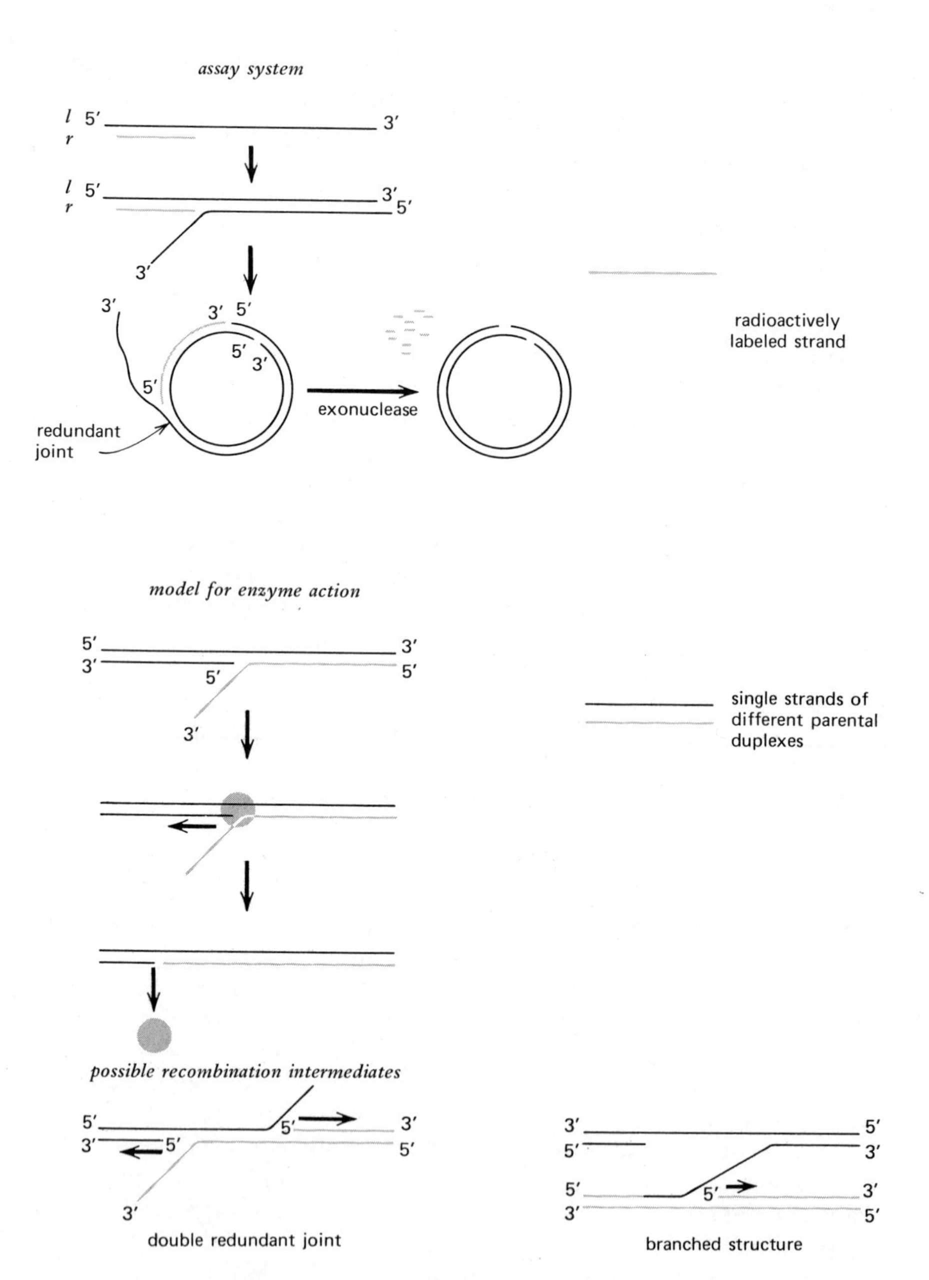


**Figure 5.12:** action of lambda exonuclease. In the assay system, a fragment of the *r* strand was annealed to the left end of the *l* strand to give the first structure shown; this was annealed with a complete *r* strand; the cohesive ends were allowed to unite. This generates a hydrogen bonded circle with a single redundant joint. Exonuclease degrades the redundant *r*

Exonuclease is composed of identical subunits of 24,000 daltons; $\beta$ protein has a subunit of 28,000 daltons. The two proteins interact together and a complex of both attacks DNA more effectively than exonuclease enzyme alone; no independent activity has been discovered for the $\beta$ protein and it is therefore possible that its role is in some way to control the activity of the exonuclease component. Carter and Radding (1971) reported that with conventional DNA substrates the action of exonuclease is to degrade each strand of a linear duplex from a free 5′-phosphate end, releasing 5′ mononucleotides; this action is processive since the enzyme continues along a single molecule until its action is complete. The enzyme is about 100 times less active with substrates of single stranded DNA than duplex DNA; but Sriprakash et al. (1975) found that its activity with single strands depends upon the length, being very much greater as the chain length is reduced. In this context Radding (1973) noted that the enzyme is able to digest the single stranded cohesive ends of λ DNA (which are 12 bases long).

Exonuclease does not act at nicks in circular DNA. But Cassuto and Radding (1971) demonstrated that it is able to act at *redundant single strands* and from this observation proposed a role for the enzyme in recombination. The substrate which they prepared to test the action of enzyme in vitro is depicted in Figure 5.12; a λ *l* strand was annealed to a radioactively labeled 3′ terminal fragment of the *r* strand after which a complete *r* strand was added and the structure allowed to form a circle by reaction between the cohesive ends. The hydrogen bonded circle includes a redundant single strand, which essentially can be considered the result of displacement of one end of the complete *r* strand by the *r* fragment previously annealed to the *l* strand. The enzyme is able to release the radioactive label from the circle (in the form of mononucleotides), but does not degrade the unlabeled material; this suggests that it halts at the single strand gap left when the redundant sequence has been removed. This forms the basis of the model for its action in recombination shown in the figure in which it binds to a redundant single joint, degrades one strand of the duplex from the free 5′ end, and ceases this exonucleolytic activity as soon as it has proceeded far enough to allow the redundant single strand to be incorporated into the duplex. This leaves a nick or gap to be

strand sequence on the circle to allow assimilation of the displaced *r* strand. The model for enzyme action suggests that the enzyme binds to a redundant joint with a free 5′ end and then degrades one strand of the duplex from this free end until a single strand gap is created by assimilation of the formerly free single strand. Two possible intermediates in recombination are duplex structures with a double redundant joint and the branched structure shown; exonuclease might act as shown by the arrows, in the first case regenerating a duplex structure and in the second allowing migration of the branch point along the lower genome.

filled by polymerase and/or ligase. Since the action allows the redundant single strand to become part of the duplex, it can be described as *single strand assimilation*. When the substrate possesses two redundant single strands, the demand of the enzyme for a free 5′ end results in the creation of 3′-3′ overlap in the heteroduplex sequence.

From these in vitro experiments it is not possible to define the specific role of the enzyme in lambda recombination, but this type of enzyme activity has an obvious part to play in recombination and some examples of possible intermediates upon which exonuclease might act are illustrated in Figure 5.12 (for review see Radding, 1973). These experiments suggest the general conclusion that the role of exonuclease is to convert the unorthodox structures that represent intermediates of recombination into the usual duplex structure. Unless some further unreported activities lie with the exonuclease enzyme itself or with the $\beta$ protein, however, this conclusion implies that further enzyme functions (presumably provided by the host) must be necessary to complete lambda recombination.

### *Role of the Gam Function*

In addition to the *red* genes involved in recombination and the genes *O* and *P* required for replication, a further phage function has been implicated in the reproduction of lambda DNA. This was defined by a mutation originally described as $\gamma$ and now known as *gam*. The role of the product of the $gam^+$ gene is to inhibit the activity of the RecBC nuclease; and this is necessary if lambda replication is to proceed in the proper mode during the later part of the lytic cycle (see later). The Gam function is therefore involved both with the host recombination system and with phage replication.

The *red* genes are not essential for growth of lambda since either type of $red^-$ mutant grows well in either $rec^+$ or $rec^-$ hosts. However, the existence of the *gam* gene was first revealed by the inability of certain mutants to grow on $recA^-$ host cells. Zissler, Signer and Schaeffer (1971a) described these as $fec^-$, named for their feckless phenotype. Each $fec^-$ mutant carried two point mutations; one inactivated the *red* system, with either $red\alpha^-$ or $red\beta^-$ equally effective; and the other identified the new gene by its possession of the $gam^-$ mutation. The *gam* gene occupies the region between $red\beta$ and *cIII;* and $gam^-$ mutants are recessive to $gam^+$, suggesting that *gam* codes for a diffusible protein. This location suggests that *gam* is expressed as a delayed early gene.

The role of the *gam* gene is suggested by the nature of an interaction observed between lambda and the unrelated phage P2. Wild type $\lambda^+$ does not form plaques on lysogens carrying a P2 prophage; it is therefore said to be *spi*$^+$ (sensitive to P2 interference). Mutants of lambda can be isolated for their ability to grow on P2 lysogens; these are described as *spi*$^-$ and they carry mutations which inactivate the *red* system and the *gam* function. [The feckless phage carrying *red*$^-$ *gam*$^-$ point mutations tend to form plaques on P2 lysogens that are smaller than usual, that is they display some alleviation of the interference by P2 but do not appear fully Spi$^-$. When Zissler, Signer and Schaeffer (1971b) isolated *spi*$^-$ mutants able to form normal (large) plaques on P2 lysogens, they found that these comprised deletions extending from *att* to *gam*. They therefore suggested that in addition to removing *red* and *gam* functions, the deletion covered a further gene, δ or *del* (apparently close to *xis*), whose mutation also is necessary to achieve the *spi*$^-$ genotype. Henderson and Weil (1975a) also have shown that deletions extending from *att* to *gam* confer the *spi*$^-$ genotype. It seems likely, however, that the difference between these deletions and the *red*$^-$ *gam*$^-$ double point mutations is that the former completely remove *red* and *gam* functions whereas the latter may be leaky enough to allow sufficient residual expression to prevent full acquisition of the *spi*$^-$ genotype. This interpretation is supported by the results of attempts of Barta and Zissler (1974) and Barta et al. (1974) to isolate point mutations in *del*. These followed the rationale of isolating *spi*$^-$ mutants from *red*$^-$ *gam*$^-$ double point mutants; a further point mutation conferring the Spi$^-$ phenotype was identified in the region to the left of the *b2* deletion. This mutation was described as *del*$^-$, but it is of course different from the *del* gene originally inferred to lie near *xis*. Although these results show that further mutations in lambda may be able to influence its growth, the failure to identify by point mutation a *del* gene between *att* and *cIII* supports the interpretation that only *red* and *gam* functions are involved in the Spi phenotype, but that complete inactivation (that is, deletion) is necessary to achieve the *spi*$^-$ condition, whereas the *fec*$^-$ genotype can be produced by point mutations (which may be slightly leaky).]

The inability of lambda to plate on P2 lysogens may be overcome in either of two ways: acquisition of the *spi*$^-$ genotype by lambda which makes it insusceptible to interference from P2; or introduction of an *old*$^-$ mutation in P2, which makes it unable to interfere with lambda. Phage P2 strains that carry the *old*$^+$ gene kill host cells that are *recBC*$^-$ but can coexist with hosts that possess the *recBC* nuclease (Lindahl et al., 1970; Sironi et al., 1971). This suggests that the basis of interference between P2 and λ may lie with this interaction with the host *rec* system. Perhaps lambda does not multiply in P2 lysogens because some λ function renders the cells phenotypically RecBC$^-$ and the *old*$^+$ function of P2 then kills the cells. Mutation of the λ function (achieved in *spi*$^-$ phages) or of the P2 *old* function then allows lambda to grow on P2 lysogens.

A direct interaction between lambda and the host *rec* system was noted by Unger and Clark (1972), who found that λ*spi*$^-$ multiplies poorly in *recA*$^-$ hosts but normally in *recA*$^-$ *recB*$^-$ hosts. This effect is not displayed by λ*spi*$^+$, which grows normally on both *recA*$^-$ and *recA*$^-$ *recB*$^-$ hosts. The *recA*$^+$ product is known to exercise what appears to be a restraining influence on the activity of the *recBC* nuclease; and this suggests that λ*spi*$^-$ may grow poorly in *recA*$^-$ cells because it is attacked by the *recBC* enzyme, an event which of course does not occur in the double mutant *recA*$^-$ *recB*$^-$. This implies that λ*spi*$^+$ possesses some gene whose product inhibits the *recBC* enzyme activity. The ability of either class of *red*$^-$ mutation, *redα*$^-$ or *redβ*$^-$, to contribute to the *spi*$^-$ genotype, together with their known involvement in phage general recombination, suggests that the exonuclease and β protein do not interact directly with the RecBC enzyme.

Both genetic and biochemical experiments show that Gam protein inhibits the RecBC enzyme. When Unger, Echols and Clark (1972) measured recombination between phages with mutations in *red* or *gam* that were crossed in *rec*$^+$ or *recB*$^-$ hosts, they observed the frequencies:

| | *rec*$^+$ | *recB*$^-$ |
|---|---|---|
| *red*$^+$ *gam*$^+$ | 3.2% | 2.4% |
| *red*$^-$ *gam*$^+$ | 0.4% | 0.04% |
| *red*$^-$ *gam*$^-$ | 2.2% | 0.06% |

The *red*$^+$ phage can recombine in both hosts, although the frequency is reduced by the *recB*$^-$ muation, presumably because this inhibits the RecBC recombination pathway. The introduction of a *red*$^-$ mutation greatly reduces the extent of recombination in either host. But the addition of a *gam*$^-$ mutation restores recombination ability in *rec*$^+$ cells; this suggests that the Gam function inhibits the host *recBC* pathway, and that the mutation allows the pathway to function in recombination. No restoration of recombination is seen in the *recB*$^-$ host, of course, because the pathway is inactivated by the lack of RecBC enzyme.

Cells infected with λ lack exonuclease V (Rec BC nuclease) activity when assayed in vitro; Unger and Clark (1972) found that λ*gam*$^-$ phages do not have this effect. Sakaki et al. (1973) isolated the component of λ infected cells that is able to inhibit the RecBC nuclease in vitro; this is a protein which is effective in preventing all of the known degradative activities of the nuclease. Karu et al. (1975) showed that the Gam protein consists of two identical subunits of 16,500 daltons. The *redα*, *redβ*, *gam* and *cIII* genes map adjacent on the lambda genome; comparison of the sizes of their protein products with the length of λ DNA known to occupy this region from heteroduplex mapping of deletions suggests that they account for all the genetic material from *redα* to *cIII*, that is, there can be no further functions to be identified in this region.

The roles of mutations in the *red* genes and in *gam* in conferring the *fec*$^-$ genotype presumably lies in the interaction of recombination systems. When the *recA* gene is mutated, the *recBC* enzyme is able to attack λ DNA if *gam* has been mutated. Presumably a functional *red* system either prevents this attack or is able to rescue λ DNA that has been attacked; a *red*$^-$ mutation therefore must be introduced before the *fec*$^-$ genotype is achieved. The role of mutation in *gam* in contributing to the *spi*$^-$ genotype is clear; λ*gam*$^-$ fails to convert the P2 lysogen to the RecBC$^-$ condition in which it is susceptible to the action of the *old*$^+$ gene. The role of mutation in *red* in achieving this state is not clear.

### *Recombination Without Replication*

During a normal lytic cycle recombination and replication occur concurrently and are succeeded by incorporation of λ DNA into mature particles, with the consequence that the immediate products of recombination cannot be followed. A certain amount of DNA synthesis appears usually to be associated with a recombination event; but recombination can take place in the absence of replication. When replication is prevented, only some of the parental genomes mature, and recombination becomes essential for their incorporation into progeny phage. Under these conditions it is possible to investigate events such as the site of crossover, the length of heteroduplex DNA, the roles of different recombination systems.

The extent to which DNA synthesis is possible in infected cells can be controlled by incubating hosts that are temperature sensitive for replication at temperatures between the permissive (low) and nonpermissive (high). Using this technique, Stahl et al. (1972b) found that when two density labeled parents are crossed under conditions allowing some DNA synthesis, the recombinants are shifted away from the heavy density of the parents. The amount of DNA synthesis can be assessed from the extent of the density shift. The position of the recombination event can be determined from the genetic constitution of the recombinants. Under Red$^+$ Rec$^+$ conditions, the largest amount of new DNA synthesis is associated with recombinants in the central region (*J-cI*), a smaller amount with recombination in the left arm, and the least new synthesis takes place with recombination in the right arm. Synthesis of new DNA is therefore associated with recombination, but may vary in extent with the particular recombination event that is involved. McMilin and Russo (1972) calculated that the average amount of new DNA synthesis associated with a recombination event corresponds to about 5% of the lambda genome. When Stahl et al. repeated their experiments at increased temperatures which completely prevented replication, they observed a reduction in the proportion of recombinants in the central region. This suggests that the DNA synthesis is necessary for recombination and that those events usually associated with a greater extent of synthesis are more susceptible to inhibition when DNA synthesis is prevented. This makes the implication that the nature of the recombination event may be influenced by the absence of replication.

The need for recombination when replication is prevented was suggested by experiments in which Stahl et al. (1972a) crossed two density labeled parents under conditions allowing restricted replication. Figure 5.13 shows that the progeny fall into three density classes, corresponding to genomes with the constitution HH, HL and LL, respectively. The HH class represents parental genomes that have found their way into the progeny without the benefit of any replication. In formal terms the HL class comprises genomes consisting of one parental strand and one newly synthesized strand; these probably result from a single round of replication, although since successive rounds continue to generate HL as well as LL duplexes it is not possible to exclude the presence in this class of some genomes that may have undergone more than a single round of replication. Phages in the LL class contain genomes consisting of two newly synthesized strands and must therefore have passed through at least two rounds of replication.

The number of progeny phage in the LL peak is relatively unaffected by recombination activities. But the number of progeny in the HH and (to a lesser extent) the HL peaks is influenced by the activity of the recombination systems. The proportion of progeny in each of the three peaks remains the same in conditions when all recombination systems are functional and after inactivation of the Rec system alone. But inactivation of the Red system sees a decline in the sizes of the HH and HL peaks; the addition of a *rec* mutation to give *red*$^-$ *rec*$^-$ conditions then increases the decline; and when all three recombination systems are inactive there is a very large fall in the number of HH progeny.

The preferential effect which eliminating recombination has upon the HH peak suggests that its maturation may follow a different pathway from that of the LL peak. The usual course by which λ DNA enters mature phage lies through replication of the parental genome to generate progeny genomes which are utilized in the assembly of mature particles; this

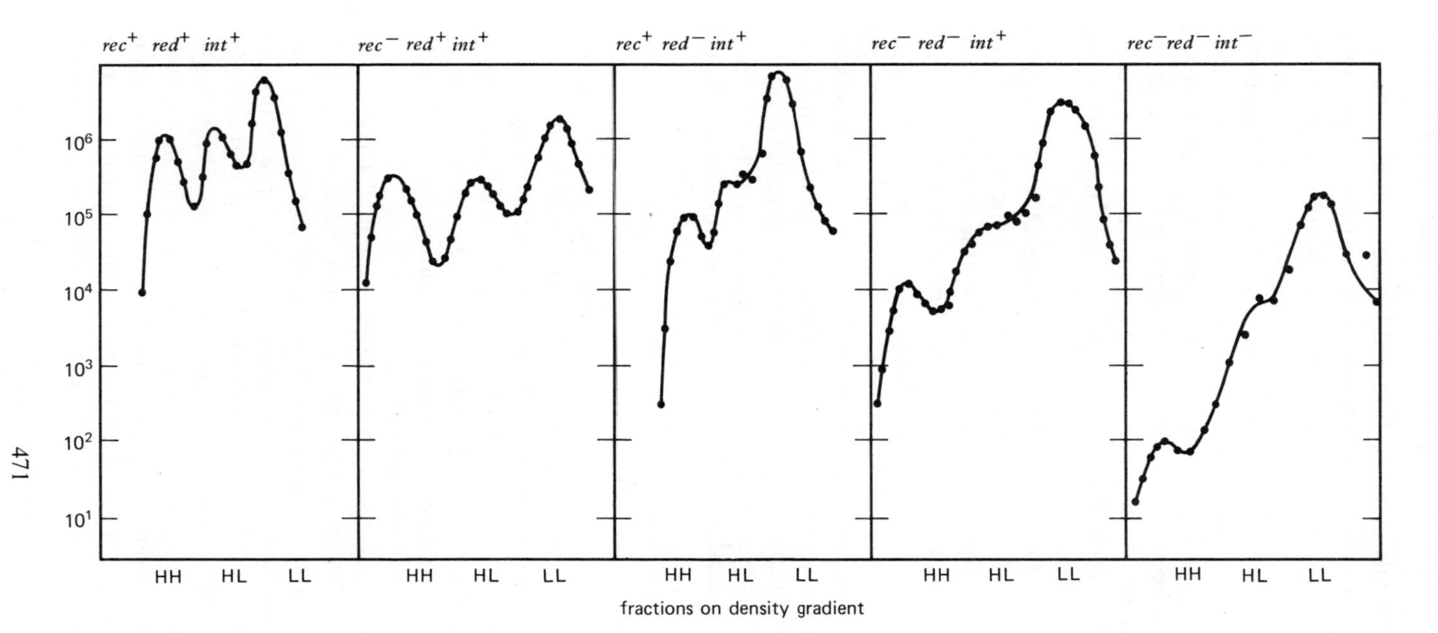


**Figure 5.13:** effect of recombination systems on formation of progeny from parental DNA. Bacteria growing in light medium were infected with parental phages that had been grown in heavy medium; the progeny were separated into fractions on a CsCl density gradient. The HH peak corresponds to unduplicated parental DNA, HL represents genomes that have probably undergone only a single round of replication, and LL is derived from genomes that have undergone two or more rounds of replication. Data of Stahl et al. (1972a).

involves at least two rounds of replication and is represented by the LL fraction. An alternative which is seen in the absence of replication is for a recombination event to occur; this in some way allows unreplicated parental DNA to become a precursor for incorporation into the phage particle. This is represented by the HH fraction. Thus a lambda genome can find its way into progeny in either of two ways: it may be replicated through at least two cycles; or it may be recombined. For parental genomes that have not been replicated. recombination therefore represents the only route by which they may enter progeny; and this explains why inactivation of the recombination systems reduces the number of progeny in the HH fraction.

Support for the concept that recombination is essential for the maturation of unreplicated λ DNA is lent by measurements of the proportion of recombinants in each of the density fractions. In the experiments of Figure 5.13 the parental phages comprises a mixture of *susJ* and *cI⁻susR* genotypes, so that the *sus⁺* recombinants could be assayed by plating on an *su⁻* host strain and their *cI* allele assessed by the morphology of the plaques. The recombinants that are detected by this means are, of course, only a proportion of those that may provide the HH peak, since the essential recombination event might take place between parental genomes of the same genotype, may lie outside the *J-R* region, or may generate classes that are not measured (*susJ susR* or double recombinants). But a comparison between the progeny of the wild type and *rec⁻ red⁻* suggests the the HH peak may consist entirely of recombinant genomes. In both cases the proportion of *sus⁺* recombinants in the LL peak is less than 1%. But in the HH peak it is 1% under wild type conditions and 25% in the *rec⁻ red⁻* condition when all recombination is due to the activity of the Int system, and must therefore take place at *att*. That the crossover was located between *J* and *cI* was confirmed by showing that all these recombinants were *cI⁺*. Since these represent only some of the possible recombinant progeny, this high frequency suggests that all phage in the HH peak have undergone a recombination event.

What maturation pathway is followed by the HL progeny? These might recombine either before or after replication; if they have recombined before replication they should include no heterozygotes, whereas recombination following replication should generate the same proportion of heterozygotes as seen in the HH peak. The same heterozygote frequency was measured in both peaks, which suggests that both density classes are generated by the same type of event; that is, their maturation depends upon recombination and this may take place in the absence of replication (generating the HH peak) or after a single round of replication (generating the HL peak). In other words, much of the HL peak may consist of genomes that have replicated only once; this cannot be sufficient to allow maturation, which must therefore depend upon recombination. The proportion of $sus^+$ recombinants in the HL peak in the $rec^-$ $red^-$ cross takes the value of 4%, intermediate between that of the HH and LL peaks, and supports the idea that many of these genomes have followed this course of events; others, however, may of course represent parental strands that have replicated through more than one partner and have therefore matured through the replication pathway. This interpretation suggests that the pathway for maturation via replication demands more than one round of replication.

In the absence of replication, a cross between a heavy density labeled parent and a light parent generates recombinant progeny whose density depends upon the site of crossover. That is, the density of any particular recombinant phage can be taken to indicate the position at which recombination occurred; for example, a recombinant whose density is shifted 20% of the distance from the HH towards the LL peak must have undergone a recombination event at a site 20% from one end of the genome. Figure 5.14 gives the results of a cross between heavy *tsA* phage and light $cI^-$ *tsR* phage performed by Stahl et al. (1974) in conditions when $O^-$ and $P^-$ mutations prevent phage replication and the host cells carry a temperature sensitive *dnaB* mutation and are incubated at high temperature. The $ts^+$ recombinants can therefore be generated by a crossover event occurring anywhere along the lambda genome and their density should be lighter the closer the recombination event lies to the right end of lambda; whether the crossover is to the left or right of *cI* can be determined by examining which parental allele is present in the recombinants. If exchanges occur with equal frequency at all sites between *A* and *R* the density distribution would be flat, with all values from heavy to light equally represented in the $ts^+$ recombinants. Any hot spots where recombination takes place preferentially should be evident as peaks at corresponding densities.

The distribution of total progeny in the heavy *tsA* × light $cI^-$ *tsR* cross falls into two fractions: a heavy fraction carrying the $cI^+$ allele; and a light fraction displaying the $cI^-$ allelle. These correspond to the parental phage classes, with the shift towards the center of the distribution in each case due to the introduction of material from the opposite parental type. In the absence of DNA replication, all these progeny should have undergone a recombination event; the correlation between density and the distribution of the *cI* alleles supports the idea that density reflects the site of crossover. The $ts^+$ recombinants can be divided into three classes according to which *cI* allele is present. The $ts^+$ $cI^+$ recombinants are biased towards the heavy side of the distribution, as expected for crossovers occurring

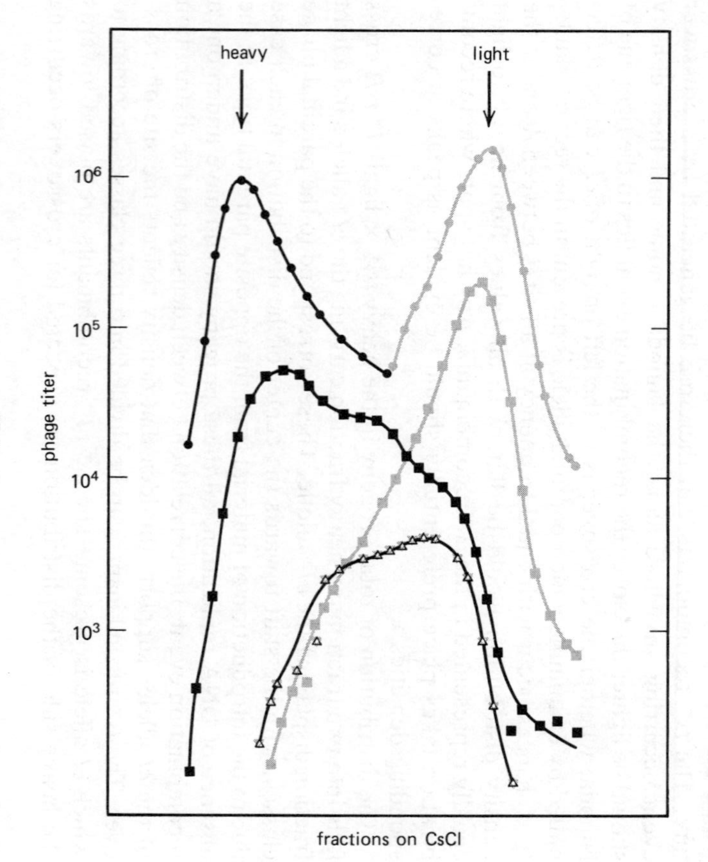
heavy
light
10⁶
10⁵
10⁴
10³
phage titer
fractions on CsCl

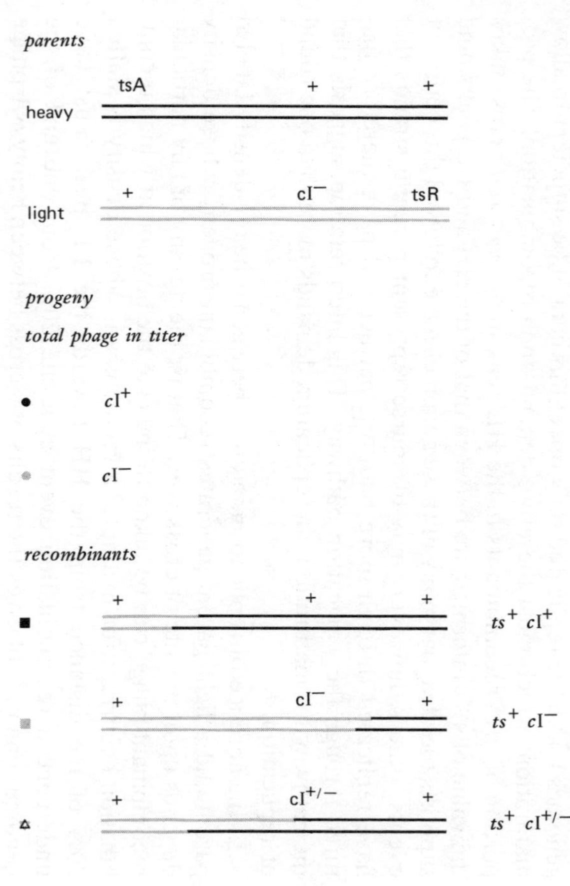
parents
tsA
+
+
heavy
+
cI⁻
tsR
light
progeny
total phage in titer
cI⁺
cI⁻
recombinants
+
+
+
ts⁺ cI⁺
+
cI⁻
+
ts⁺ cI⁻
+
cI⁺/⁻
+
ts⁺ cI⁺/⁻

between *A* and *cI*; and the $ts^+$ $cI^-$ recombinants are biased towards the lighter density fractions, corresponding to crossover locations between *cI* and *R*. The predominance in these recombinant classes of phages at densities close to the parental heavy and light values, compared with the smaller number showing appreciable density shifts towards central values, suggests that recombination in this system tends to take place preferentially at the ends of the lambda genome. The terminal emphasis of crossovers is consistent with the observation that lack of replication tends to have a greater effect on recombination events in the central part of the lambda map. The $ts^+$ $cI^+/cI^-$ heterozygotes display a very broad distribution of densities; since all of these must be heteroduplex over the *cI* gene, this supports the idea that the length of hybrid DNA may vary widely and often is appreciable. (We should note that these experiments and those of White and Fox discussed earlier have been interpreted as due solely to biparental recombination; however, Stahl and Stahl (1976) have shown that some of the molecules that are heterozygous at both *cI* and at *R* represent the results of events involving three parents. This does not alter the basic conclusions, but makes construction of detailed models less rigorous.)

Using this system it is possible to investigate the contributions that the different phage and host systems make to vegetative recombination. The introduction of an $int^-$ mutation does not influence the density distribution, confirming that the Int system usually plays a small or no role in vegetative recombination. Inactivation of the *red* system yields a much flatter density distribution of $ts^+$ recombinants, with preferential crossing over at the left end of lambda abolished and the right terminal preferential crossing over much reduced. A $recA^-$ mutation severely reduces all crossing over and leaves only recombinants that are the result of crossing over at the right terminus, presumably due to the action of the Red system. This therefore implies that in the absence of replication it is the Red system that is responsible for the preferential crossing over at the termini. This is consistent with the previous conclusion of Stahl et al. (1972b) that DNA synthesis occurs to the greatest extent with recombination spon-

---

**Figure 5.14:** site of recombination between parental phages in absence of replication. One parental phage carried a temperature sensitive mutation in gene *A*; the other was temperature sensitive in gene *R* and also carried a $cI^-$ mutation. The *tsA* phage carried a heavy density label; the $cI^-$ *tsR* phage was of normal density. Both phages carried the mutation *Pam 80* which blocks replication and the host cells were temperature sensitive in *dnaB* and were incubated at the high (nonpermissive) temperature. Progeny were assayed both for total phage (of both $cI^+$ and $cI^-$ genotype) and for $ts^+$ recombinants carrying the $A^+$ and $R^+$ alleles. The right side of the figure illustrates the structures of the parental and recombinant phages and the left side shows the analysis on density gradients of the progeny phage. Data of Stahl et al. (1974).

sored in the central part of the lambda map by the Red system in *recA*$^-$ cells; DNA synthesis is also associated, but to a smaller extent, with Red sponsored recombination in the terminal regions and with all Rec sponsored events. The preferential crossing over at termini that the Red system undertakes in the absence of replication therefore suggests that the replication associated with these recombination events is necessary, so that their frequency is reduced in its absence.

The genetic map of lambda constructed from standard crosses displays the same relationship between markers as the physical map determined from electron microscopy of heteroduplexes. This correspondence suggests that under normal conditions, that is, when replication and recombination proceed simultaneously, recombination must take place uniformly along the lambda genome. Under standard conditions a *red*$^-$ mutation reduces recombination considerably whereas a *recA*$^-$ mutation has much less effect. This implies that most of the recombinants are produced by the Red system, which must therefore be expected to act evenly along λ DNA. Two models would be consistent with this uniform activity: recombination may occur with equal probability at each nucleotide position; or there may be hot spots at which recombination occurs preferentially, but which are equally distributed along the chromosome. The differences in the extent to which DNA synthesis is associated with Red sponsored recombination in different parts of λ DNA, and the preferential crossing over towards the termini in the absence of replication, hint that there may be some variation in the mechanism of action of the Red system at different sites. The molecular actions of the Red proteins and any associated host functions remain to be defined.

Since most recombinants are produced by the Red system, only a small contribution is made to recombination frequencies by the Rec system, whose activity can be studied only when the phage system is inactivated. To achieve conditions in which only the Rec system is active, McMilin, Stahl and Stahl (1974) performed crosses between phages that had been deleted for the lambda recombination genes. Using the system in which a density labeled parent carried a mutation in *A* and a light parent carried a mutation in *R*, in the absence of replication they were able to examine the density distribution of the progeny: this essentially comprises a hot spot survey, in which any hot spots should be visible as peaks above the otherwise flat density distribution. Such preferential sites for action of the Rec system might pass unnoticed when the Red system also is functional (and would not distort the linearity of the lambda map) since they would comprise only a small proportion of the total recombination.

With the deletion λ*b1453,* which lacks the material from *att* to *gam,* a density peak was found which corresponds to approximately a fifteenfold increase in recombination at a site some 10% of the distance from the right end of lambda. With the substitution strain λ*bio1*, in which bacterial genes are substituted for the same region of the phage, a density peak was found corresponding to an increase in recombination at a site probably located within the bacterial substitution. Are these preferential locations for crossing over associated with the absence of DNA synthesis or do they represent hot spots for Rec action irrespective of replication? Using appropriate genetic markers it is possible to assess whether recombination has increased in the region of the map containing the hot spot and in both cases the same increase was seen in the presence of replication.

These hot spots appear to be the result of mutation. When Henderson and Weil (1975a) isolated $Spi^-$ deletions (which must lack the *red* and *gam* genes) they obtained a series of phage strains able only to form minute plaques; but these phages were able to gain the ability to form large plaques due to the acquisition of mutations. One such mutation occurs close to the right end of lambda; and since λ*b1453* is a $Spi^-$ deletion of this type, Lam et al. (1974) investigated the possibility that it might carry this mutation. By constructing strains that were isogenic apart from the mutation, they were able to show that the presence of the mutation and hot spot coincide. The mutation is described as $chi^-$ and wild type lambda as $chi^+$. Thus in a cross between two $chi^+$ phages no Rec hot spot is seen; in a cross between two $chi^-$ phages preferential recombination is seen at the hot spot. Dominance can be tested by crossing a $chi^-$ with a $chi^+$ phage; the $chi^-$ mutation is dominant since the hot spot is seen. In all these experiments, recombination is due entirely to the Rec system since the phage strains lack the lambda recombination genes. These phages are therefore $Spi^-$; when $chi^+$ they form small plaques and when $chi^-$ they form large plaques. The reason for the effect on plaque size may be that with these strains the recombination undertaken by the Rec system is necessary for maturation; the increase in recombination frequency caused by the $chi^-$ mutation therefore increases the number of progeny.

Do $chi^-$ mutations occur only at one or at several sites? Henderson and Weil (1975a) isolated $chi^-$ strains from several $Spi^-$ deletions and determined whether the mutations were able to recombine by crossing them. Recombinants between a $J^-$ phage carrying one $chi^-$ mutation and an $R^-$ phage carrying another, independently derived $chi^-$ mutation were scored for plaque size: large plaques represent the parental $chi^-$ type; and small plaques can be produced only by $chi^+$ recombinants, formed when the two $chi^-$ mutations are different. This test distinguished three $chi^-$ mutations, all located in the *J-R* region. The frequency of recombination at the hot spot when two $chi^-$ mutants are crossed depends upon their orientation, which hints that there may be some asymmetry in the recombination. Although the Spi$^-$ deletions form large plaques only after acquisition of *chi*$^-$ mutations, the Spi$^-$ phages that are produced by substitution of *bio* genes for the *att-gam* region are able to form large plaques; Henderson and Weil showed that this is because the bacterial insertion carries a *chi* site. This suggests that *chi* sites may occur in the E.coli chromosome as hot spots for the action of the Rec system.

The conclusion that *chi*$^-$ mutations represents base changes that create a sequence with a high affinity for the Rec system is supported by the demonstration of Stahl, Craseman and Stahl (1975) that they may occur at four locations on the lambda chromosome as hot spots for recombination. Because it is difficult to map mutations which influence the rate of recombination, they were located by deletion mapping: *chiA* lies between *L* and *att*, *chiB* between *att* and *P*, *chiC* is apparently located within *cII*, and *chiD* lies between *Q* and *S* and is the site towards the right end that was first identified. The nature of these sequences and the effects which they have upon recombination are likely to prove useful in defining the action of the Rec system.

## Replication of Phage DNA

### *Early Replication of Circular Monomers*

The DNA of the mature phage particle is a linear molecule of some 46,500 base pairs, with a molecular weight of $31 \times 10^6$ daltons and a length of about 16$\mu$m when visualized by electron microscopy. It is this linear form which provides the infecting genome at the beginning of a lytic cycle and which represents the mature genomes packaged into phage particles at the end of the cycle. During the course of infection, however, the state of $\lambda$ DNA changes as it passes through the replicative intermediates. The replication cycle falls into two parts: immediately following infection $\lambda$ DNA replicates as a circular molecule; but later in development it replicates in the form of a rolling circle which is the precursor for generating the mature linear form. Other intermediate molecules may be generated by the recombination systems.

The first step following infection is the conversion of the linear infecting DNA first into an open circle and then into a closed circle (Bode and Kaiser, 1965b; Ogawa and Tomizawa, 1967). This conversion depends upon the cohesive ends of the linear DNA described by Hershey and Burgi (1965) and takes place by the events described in Chapter 4: first the complementary protruding single strands that comprise the cohesive ends anneal to generate an open circle, whose integrity is maintained by the base pairing between the cohesive ends; and then ligase action converts the open circle into a closed circle in which both strands are covalently continuous.

Cohesive ends are essential for infectivity since λ DNA is active in the Kaiser helper assay (see Figure 4.4) only if it possesses at least one cohesive end (Kaiser and Inman, 1965). As cohesive ends are characteristic only of the mature linear duplex, the infectivity in this assay of DNA extracted from cells at various times after infection can be equated with the number of mature genomes. Figure 5.15 shows the results of Young and Sinsheimer (1967) in following the state of λ DNA during the lytic cycle. Immediately following infection, replication of λ DNA commences and then continues until many phage equivalents have been synthesized; this generates the continuously rising curve shown for DNA synthesis. However, the immediate result of infection is to cause a decline in the activity of λ DNA in the helper assay; only after a long eclipse period does infective DNA reappear and then increase in amount. This implies that the cohesive ends of λ DNA are lost immediately following infection, as

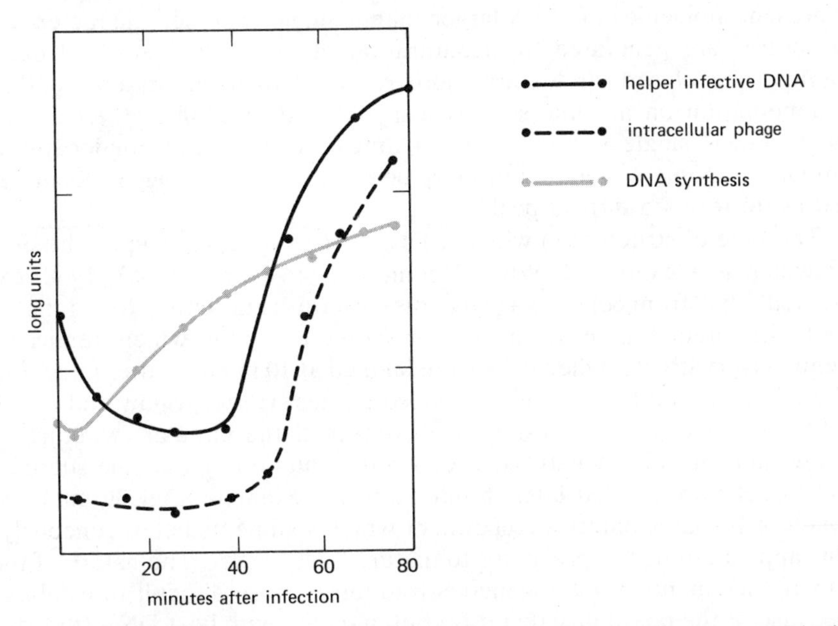


**Figure 5.15:** production of λ DNA following infection. The host cells were treated with mitomycin C to prevent bacterial replication. Synthesis of phage DNA starts immediately, there is an eclipse period before the appearance of DNA that is infective in the helper assay, and a longer lag before the production of intracellular phage particles. Data of Young and Sinsheimer (1967).

mature DNA is converted to replicating intermediates; the absence of cohesive ends in the replicating structures was confirmed directly by Tomizawa and Ogawa (1968), who isolated early replicating DNA and showed that it does not react when mixed with labeled linear DNA possessing cohesive ends. The linear duplex form is generated from replicating DNA only in the later part of infection; and the appearance of phage particles follows the production of infective DNA after a short lag. During infection λ DNA therefore follows a course in which it exists as replicating intermediates until the time of maturation of progeny particles when it reappears in linear duplex form.

Two periods of replication can be defined by the forms taken by λ DNA. Sedimentation through sucrose gradients distinguishes several components of λ DNA. Under neutral conditions the linear duplex sediments the most slowly, open circles sediment at a rate 1.13 times that of the linear duplex, and closed circles sediment at a rate 1.5 times that of the linear duplex; these three monomeric forms are characterized by sharp peaks. In addition a heterogeneous peak is found, sedimenting in the range from 1.1–2.0 times that of the linear duplex, and this appears to represent molecules of DNA larger than a single genome. Characteristic structures are generated by denaturation of linear duplexes and open circles, and closed circles are more resistant to denaturation, so that sedimentation on alkaline sucrose can be used to confirm these structures: linear single strands, circular single strands, and nondenatured circular monomers sediment in sharp peaks; and the heterogeneous material again forms a diffuse peak.

The type of structure in which λ DNA is found depends upon the time at which it is examined. When Young and Sinsheimer (1967, 1968) extracted DNA from cells at various times after infection, they found that at first the predominant form is a monomeric circle which replicates semiconservatively; when cells were labeled at 10 minutes after infection, the label entered both open and closed circles. Carter, Shaw and Smith (1969) obtained similar results and confirmed the circular structure of these molecules by denaturation and sedimentation on alkaline sucrose. When cells are labeled later in infection, for example after 30 minutes, much of the label enters a component which sediments heterogeneously; this appears to be the precursor to mature linear DNA. The nature of the rapidly sedimenting heterogeneous fraction is more difficult to establish because of the possibility that it is contaminated with host DNA (this can be excluded with the early circular forms because they have the size expected for lambda monomers). By inhibiting host replication it is possible to reduce the likelihood of contamination and Carter et al. (1969) and

Kiger and Sinsheimer (1969) showed that the heterogeneous fraction contains molecules longer than a single lambda genome. We shall return in the next section to the structure of late replicating DNA and here shall be concerned only with the early replicating circular form. (The use of *early* and *late* to describe the periods of lambda replication is an operational definition based upon the structures found by labeling at different times and is of course unrelated to the early and late periods defined in expression of phage genes.)

The entry of an early label into circular DNA implies that this is the form in which replication commences. Circular DNA may be replicated either unidirectionally (with a single growing point moving round the circle from an origin to a terminus adjacent to it) or bidirectionally (with two growing points moving away from a common origin to meet at a terminus 180° round the circle). Both types of structure appear similar under the electron microscope: the circle contains two Y-shaped branch points and the two branches between them (which represent newly synthesized replicas) are of equal length. This structure is illustrated in Figure 5.16. When replication is unidirectional one branch point must be stationary and represents the origin while the other represents the moving replication fork; in bidirectional replication both branch points represent replication forks and the origin should lie equidistant between them. Interpretation of such electron micrographs therefore requires further information. For E.coli DNA this was obtained by measuring the frequencies with which different genes are present in growing cells, since those close to the origin should be replicated first and thus are present in more copies; this approach demonstrated that the bacterial replication is bidirectional (see Volume 1 of Gene Expression). For λ DNA direct visualization is possible, as indeed it is also for some plasmids (see Chapter 3).

Lambda DNA engaged in early replication can be isolated by a buoyant density technique. Tomizawa and Ogawa (1968) used light phages to infect E.coli cells growing in heavy medium; the parental phage genomes thus can be described as LL DNA, genomes replicated once are HL, and after the second round of replication genomes can be isolated that are HH. During the first round of replication DNA that is partially replicated should occupy a density position between LL and HL; during the second round the replicating DNA should be found between HL and HH. Of 42 replicating molecules isolated in this way on CsCl density gradients, 30 were circular when visualized by electron microscopy; and 25 possessed two branch points. These have been described as $\theta$ structures (for review see Kaiser, 1971). The length of DNA comprising the circle was 17 $\mu$m,

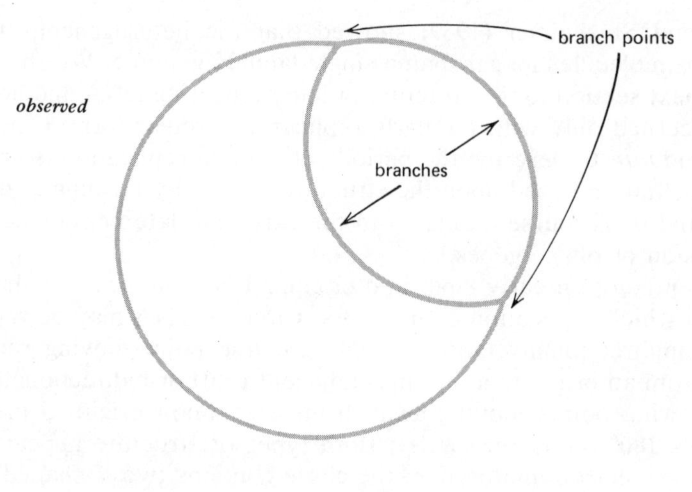


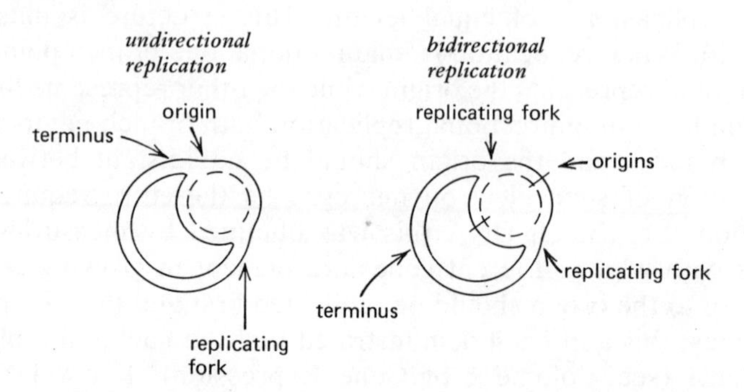


**Figure 5.16:** the $\theta$ form of replicating DNA. The circular molecule observed by electron microscopy has two branch points and the branches between them are of equal length. The length of the DNA outside the branch points plus that of either branch equals the length of linear $\lambda$ DNA. Two interpretations are possible. In both the branches represent replicated daughter duplexes and the DNA outside the branch points is yet to be replicated. If replication is unidirectional one branch point represents the origin and one the single replicating fork; if replication is bidirectional both branch points represent replicating forks and the origin lies equidistant between them.

corresponding to one lambda genome; and the distance between the two branch points is always the same on each branch, suggesting that these represent the newly replicated daughter strands.

To determine whether the $\theta$ structure represents a unidirectionally or bidirectionally replicating circle, it is necessary to have a reference point on the lambda circle against which the positions of the branches can be judged. This is provided by the denaturation map of lambda which can be constructed by partially denaturing linear λ DNA, when sites that are rich in A-T base pairs separate into single strands first. Because the right half of lambda is the richer in A-T pairs, this denaturation map takes the form of a characteristic series of single strand loops in the right half, while the left half remains largely in duplex form. Schnos and Inman (1970) constructed a denaturation map by exposing linear λ DNA to high pH and this is shown in Figure 5.17.

When they isolated a buoyant density fraction expected to contain early replicating DNA, they found that 40% of the molecules were linear (probably corresponding to contaminating bacterial DNA) and 60% were circular. Some 60% of the circles were branched monomers and about 1% were complex, consisting of molecules longer than a single lambda genome. The circular monomers had a contour length of 17.5 $\mu$m, characteristic of λ DNA under these conditions, and displayed the same denaturation pattern generated by linear DNA. The positions of the branch points then can be determined by reference to the denaturation map. The most common pattern is for the two branch points to be equidistant from a point located at 14.4± 1.3 $\mu$m (81.7 ± 2.9%) from the left end of lambda; this

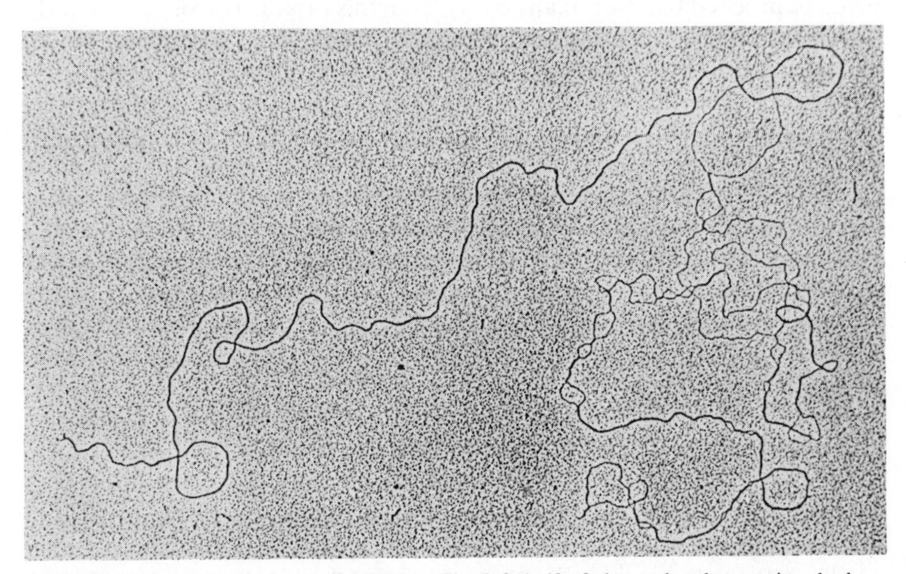

**Figure 5.17:** denaturation map of λ DNA. The left half of the molecule remains duplex, except for the terminus. Several single stranded regions are formed in the right part of the molecule; these are seen as single stranded loops in the electron micrograph and are represented by the black squares on the linear map drawn below. Data of Inman and Schnos (1974), kindly provided by Dr. R. B. Inman.

suggests that both branch points represent replicating forks moving to the right and left (clockwise and counterclockwise) from an origin at the 81.7% position. Termination then occurs when the two growing points collide, that is, rather than at a unique genetic site (Valenzuela, Freifelder and Inman, 1976). Occasionally the left branch point of a replicating circle does not appear to have moved but is located in the region of the origin, while the right branch appears to have moved; the reverse also has been seen. These results suggest that replication may on occasion take place unidirectionally, in either direction.

A very similar location for the early origin was identified by Stevens, Adhya and Szybalski (1971) from a different approach. They examined the ability of prophages with various deletions to be replicated under conditions when the necessary phage and host functions were provided; in this situation waves of replication appear to start within the integrated prophage and extend into the adjacent bacterial chromosome. Replication was measured by following the amount of λ DNA able to hybridize with labeled mRNA representing different parts of the phage genome. Deletion of the region between 79.1–81.5% from the left end prevents this replication. In formal terms this region includes a cis acting site necessary for replication: this is presumably the origin. In this system replication appears to proceed bidirectionally away from this origin. The origin for early replication, *ori*, can therefore be located at a site about 81% from the left end of the genome.

The replication fork is the point at which the nonreplicated parental duplex joins the two daughter chromosomes; each of the parental strands extends covalently from parental to daughter DNA, although its partner changes at the replicating fork from the other parental strand to a newly synthesized strand (see Figure 5.16). Inman and Schnos (1971) reported that single strand connections often appear to join the parental to the daughter duplexes at the branch points. The length of the single strand connection varies from 0.04–0.40 μm and appears to represent a length of unwound DNA on average about 500 bases long which is awaiting synthesis of its complement.

### *Rolling Circle Replication*

The first suggestion that the precursor to the linear DNA of mature lambda particles is a complex molecule came from the work of Smith and Skalka (1966). Pulse labeled DNA extracted from infected bacteria was separated into three classes by sedimentation: a linear monomer; a supercoil (closed circle); and a form sedimenting up to twice as fast as the linear molecule. This last form contained most of the label, which could be chased from it into the mature linear duplex. The supercoils do not appear to provide precursors for mature DNA. Smith and Skalka suggested that the rapidly sedimenting material (which corresponds to the heterogeneous peak we have already mentioned) represents *concatemers*—molecules that consist of more than one lambda genome covalently joined end to end.

What type of structure is the concatemer and how might it be generated? The two simplest concatemeric structures are linear multimers and multimeric circles: a linear concatemer consists simply of at least two phage genomes joined end to end, and in a circular concatemer the two free ends also have been covalently joined. Figure 5.18 depicts possible ways to generate such structures. Annealing between the cohesive ends of two linear molecules generates a linear dimer, which can form a circle by intramolecular reaction between the two free cohesive ends. Reciprocal recombination between two monomeric circles directly generates a dimeric circle; a similar event involving a monomeric circle and linear duplex generates a linear dimer. Further additional recombination events are needed to extend these dimers into higher oligomers. Concatemers also may be generated by replication of a rolling circle. As illustrated in detail in Figure 2.29 with reference to conjugation, the circle itself should correspond to a unit genome but the tail may be extended indefinitely to generate a concatemeric structure.

Since sedimentation rates do not unambiguously define structure, it is necessary first of all to show directly that this fraction contains DNA longer than the unit length of lambda. One criterion is that upon denaturation it releases single strands of greater than unit length; another is to characterize the molecule in duplex form by electron microscopy. In attempting to visualize the structures of the concatemers, it is necessary to distinguish the λ DNA from any contaminating bacterial DNA and also to minimize changes caused during isolation (such as breakage of circular structures). Density labeling can be used to separate phage and host DNA and the proportion of the concatemers may be increased by use of phage mutants. But the most satisfactory approach is positively to identify DNA as corresponding to the phage genome: when the DNA is in circular form it can identified as lambda since its length should be a multiple of the unit genome; and linear multimers can be characterized by denaturation mapping.

One difficulty in analyzing the large heterogeneous molecules is that as intermediates of maturation they may be expected to turn over. But the observation first made by Dove (1966) that λ DNA does not mature to the linear form in certain head mutants provides a way to increase the proportion of this component. These head mutants produce normal amounts of λ

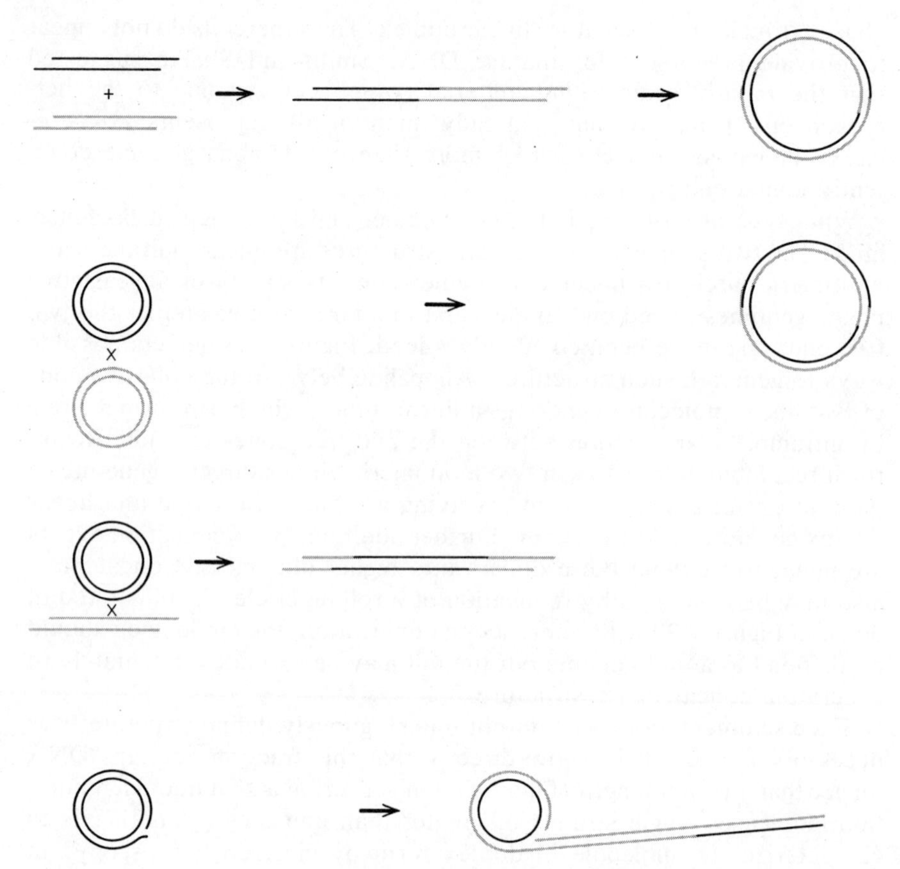


**Figure 5.18:** models for generation of concatemers. End to end joining of the cohesive ends of two linear monomers generates a linear dimer; further monomers can be added in the same way to generate longer concatemers. A circular molecule can be generated by joining of the two free ends of the linear form. Reciprocal recombination between monomeric circles generates a circular dimer; and between a circle and linear molecule generates a linear product (which of course could circularize by end to end joining). Production of longer concatemers by recombination requires the occurrence of additional genetic exchange events. Replication of a rolling circle generates a structure in which the circular duplex is monomeric in length but the tail may be of any length depending upon how long replication has proceeded.

DNA but it is not infective in the helper assay. This implies that they cannot accomplish a step of maturation which produces cohesive ends and they may therefore be expected to accumulate the precursor to mature DNA. In a series of experiments utilizing this approach, Weissbach, Bartl and Salzman (1968) reported that late in lambda infection there are prominent peaks of rapidly sedimenting DNA and of mature linear DNA; but with a $\lambda A^-$ mutant no mature DNA is formed and the rapidly sedimenting material alone accumulates. Examination of this fraction by electron microscopy demonstrated its heterogeneous nature: its constitution was

| | |
|---|---|
| linear molecules < unit length | 6% |
| linear monomers | 11% |
| circular monomers | 17% |
| linear molecules > unit length | 60% |
| circular molecules > unit length | 7% |

The important conclusion suggested by this analysis is that two thirds of the molecules in this fraction are longer than the monomeric length; this leaves open the question of whether they are generated by replication or recombination.

The accumulation of large DNA in maturation defective mutants implies that this is the precursor to mature linear DNA. This concept is lent further support by the results obtained by McClure and Gold (1973) with a temperature sensitive head mutant. Rapidly sedimenting molecules accumulated at high temperature but disappeared when the temperature was shifted down to allow production of linear monomers. When McClure, MacHattie and Gold (1973) examined the structure of the rapidly sedimenting material that accumulates at high temperature they were able to identify both circular and linear molecules longer than the unit length of lambda.

The concatemers are produced by replication. Skalka (1971) showed that the long linear molecules still are formed when *recA*$^-$ cells are infected by $\lambda A^-$ *red*$^-$ *int*$^-$ phages, a procedure which ensures the absence of recombination functions. When Skalka, Poonian and Bartl (1972) examined the rapidly sedimenting fraction formed under these conditions, they found 60% linear monomeric molecules, 20% linear molecules shorter than unit length, and 20% linear molecules of up to 4 times the unit length. Denaturation mapping showed that the ends of both the unit length and longer linear molecules are random; thus the monomers represent permutations of the usual lambda sequence. The absence of cohesive ends is expected since the $A^-$ mutation inactivates the Ter nuclease responsible for generating them; the origin of the randomly permutated monomers is not clear. When $\lambda A^+$ is used instead of $\lambda A^-$, similar results are obtained but about 30% of the monomers have the sequence characteristic of mature DNA. The important conclusion suggested by these experiments is that concatemers can be generated under conditions when both the host and phage recombination systems are inactive, and the phage system for generating cohesive ends is absent, so that the long molecules must be the product of replication.

A similar analysis of the late replicating DNA was reported by Wake, Kaiser and Inman (1972). Lysogens were grown in heavy (BUdR) medium, mitomycin C added to inhibit host replication, and $^3$H-thymidine added to replace the heavy BUdR just before induction. The DNA isolated later in infection with LL strands should therefore represent phage genomes that have passed through at least two cycles of replication. With head mutants unable to mature DNA, about 60% of the label sedimented rapidly on neutral sucrose; and on resedimentation in alkali about 60% of the DNA sedimented more rapidly than linear single strands. Electron microscopy showed that almost all the molecules were linear, extending for up to 4 unit lengths; the remainder were monomeric circles. Denaturation mapping showed that the long molecules represent repetitions of the lambda genome; and their ends do not coincide with the cohesive termini.

The most obvious way to reconcile the linear structure of the concatemers with their origin as replicating intermediates is to suppose that they are derived by breakage of rolling circles; in this case the absence of such structures under the electron microscope must be taken to imply that they are especially susceptible to breakage during preparation. The idea that lambda replicates as a rolling circle during the second period of replication is supported by the visualization of such structures. Takahashi (1974) infected cells grown in heavy medium with light phage and isolated replicating molecules by their intermediate buoyant density. At 15 minutes after infection some 85% of the replicating molecules take the form of $\theta$ structures. At 30 minutes almost all the replicating molecules appeared to take the structure of the rolling circle. When these molecules were examined under conditions of partial denaturation, most of the tails of the rolling circles possessed a free end corresponding to the early origin. Since these structures were observed in cells when recombination was prevented, they must be products of replication. In similar experiments when recombination was permitted, Takahashi (1975) identified dimeric, trimeric and longer circular molecules; denaturation mapping confirmed that these represent lambda molecules joined end to end. This is consistent with the idea that concatemers can be generated by recombination as an alternative to the replicative system.

Electron microscopic evidence for rolling circles also has been obtained by Bastia, Sueoka and Cox (1975). At up to 20 minutes after infection the predominant replicating intermediate appears to be the $\theta$ structure. After 20 minutes rolling circles can be visualized, with tails extending from <1 to 5.5 unit lengths. (The observation of tails of greater than unit length is essential to demonstrate that the structure is a rolling circle and is not generated by breakage of a $\theta$ molecule.) The denaturation map of the rolling circles shows a polarity in the direction of replication, with most circles rolling in the counterclockwise direction; but some molecules do appear to replicate in the reverse direction.

The origin used to initiate replication of a rolling circle should be represented by the sequence at the free end of the tail. In these experiments about 25% of the tails terminated at the position of the early origin. Bearing in mind the possibility of breakage of the tail during isolation, it seems likely that the origin for early replication also may be used for late replication. The same conclusion was suggested by Takahashi (1976), who observed also that 40% of the rolling circles appear to replicate in one direction and 60% in the other direction. The switch from early to late replication thus may be accomplished by a change in the nature of the events that take place at the origin when a replication cycle is initiated.

### *Transition from Early to Late Replication*

Immediately upon infection the linear molecule of $\lambda$ DNA is converted to a closed circular form. This is then replicated, most commonly bidirectionally but on occasion unidirectionally, from the early origin *ori* to generate circular progeny. This mode of replication is then largely succeeded by a form which generates concatemers: replication of a rolling circle produces long tails which consist of many lambda genomes. The production of concatemers is essential for maturation of $\lambda$ DNA, for only a concatemeric structure can serve as precursor to the linear monomers that are packaged into mature phage particles. The usual precursor to mature $\lambda$ DNA is the concatemeric tail of the rolling circle; but in the absence of replication, concatemeric molecules, largely circular, are generated by recombination and may serve as precursors. The need for a concatemeric precursor thus explains why recombination becomes essential for maturation when replication is prevented. We shall consider later the process by which mature linear genomes are derived from concatemeric precursors; here we shall be concerned with the transition from early to late replication.

The occurence of recombination in the pool of replicating DNA, and the need for concatemers as precursors to mature DNA, reflect the relationship between replication and recombination in the reproduction of λ DNA. A connection between these systems is displayed in the conditions that are necessary if the transition from early to late replication is to be accomplished. In a study of the effect of recombination functions upon the transition, Enquist and Skalka (1973) demonstrated that the presence of the Gam function is essential. Synthesis of DNA is reduced to about half of the wild type by *red*$^-$ or *gam*$^-$ mutations; no reduction is seen in the period of early replication so this effect appears to represent a reduction of late replication. Using phages carrying *A*$^-$ mutations to cause accumulation of late intermediates, Enquist and Skalka confirmed the previous result that in a *recA*$^-$ host linear monomers and concatemers are found, with no circular structures. This pattern is taken to represent late replication, on the assumption that under these conditions rolling circles represent the late intermediates in the infected cell but have been broken during extraction.

Infection with λ*A*$^-$ *red*$^-$ showed that inactivation of the Red system in itself has no effect. But after infection with λ*A*$^-$ *red*$^-$ *gam*$^-$ the late replicating DNA consists largely of circular monomers, half closed and half open; this is in effect the early replication pattern. The absence of Gam function therefore appears to prevent the transition from early to late replication. In a λ*A*$^-$ *gam*$^-$ infection the results appear intermediate, since both linear molecules and circles are seen; this suggests that the presence of the Red system is able to help overcome the inability to generate concatemers by replication that is caused by the *gam*$^-$ mutation. This, of course, is consistent with the idea that recombination may provide an alternative pathway for generating concatemers; its practical implication is that it may be necessary to inactivate the recombination systems in order to follow the role of the Gam function.

Quite different results are obtained in *recA*$^-$ *recB*$^-$ hosts which lack the RecBC nuclease: λ*A*$^-$ and λ*A*$^-$ *gam*$^-$ phages generate the same late replicating intermediates, that is, linear monomers and concatemers. This demonstrates that in the absence of the RecBC nuclease it is no longer necessary to provide Gam function in order to make the transition from early to late replication. These results imply that the RecBC nuclease prevents the transition from early to late replication, presumably by degrading some structure that is part of the late pathway. Early replication is not inhibited by the nuclease, consistent with the conclusion of Pilarski and Egan (1973) that circular molecules are not susceptible to this enzyme. The substrate for the enzyme might be the rolling circle itself (for example the tail may be susceptible to attack) or some precursor to it. The transition from early to late replication therefore requires Gam function to inhibit the RecBC nuclease so that rolling circles may be generated. The efficiency of the enzyme in inhibiting late replication is seen in the complete absence of any late replicating intermediates under *recA*$^-$ *red*$^-$ *gam*$^-$ conditions. The presence of either a Red$^+$ or Rec$^+$ system allows the production of some concatemers in spite of the restriction to early replication caused by a *gam*$^-$ mutation, consistent with the results discussed earlier on maturation in the absence of replication.

The stage of replication at which the Gam function and (by implication) the RecBC nuclease are involved has been defined more precisely by the experiments performed by Greenstein and Skalka (1975) with a $\lambda$*gam*$^{ts}$ mutant. Cells incubated at the permissive temperature of 32°C were labeled and then raised to the nonpermissive temperature of 41°C late during development; in effect this protocol activates the RecBC nuclease when the temperature is raised. If the enzyme acts upon an intermediate which precedes the rolling circle, its activation after late replication has begun should not cause degradation of the DNA; but if its substrate is the late replicating DNA itself, activation of the nuclease at any time should result in degradation of DNA. Lambda DNA was degraded more extensively when the label in $\lambda$*gam*$^{ts}$ was chased at 41°C compared with 32°C (there was no degradation at 32°C in $\lambda$*gam*$^+$ infection which implies that the *gam*$^{ts}$ product is partially defective even at low temperature); this suggests that some product of late replication is directly attacked by the RecBC nuclease.

Analysis of the state of late replicating DNA before and after cells infected with $\lambda$*gam*$^{ts}$ were incubated at increased temperature confirmed these conclusions. Figure 5.19 shows the results of a pulse chase experiment in which cells grown at 32°C were labeled with $^3$H-thymidine late in infection and DNA was immediately extracted from one sample while another was incubated for a further 60 minutes at 41°C before DNA was analyzed. To make this experiment possible the phage carries two further mutations, $E^-$ to cause accumulation of late intermediates and $S^-$ to delay lysis so that incubation could be prolonged. The effect of this protocol is to compare the state of late replicating DNA when Gam function is present (RecBC nuclease inhibited) with its condition after a period when Gam function has been inactivated (RecBC nuclease reactivated). On both neutral and sucrose gradients much heterogenous rapidly sedimenting material is seen in the first extract, representing the concatemeric late replicating structures; but after the RecBC nuclease has been activated, this material disappears and only the circular replicating structures characteristic of the early period are seen. This demonstrates that the concatemers are attacked directly by the enzyme and confirms that the early replicating circles are insusceptible to it. It is probably the free end of the tail of the rolling circle that provides the substrate for the enzyme.

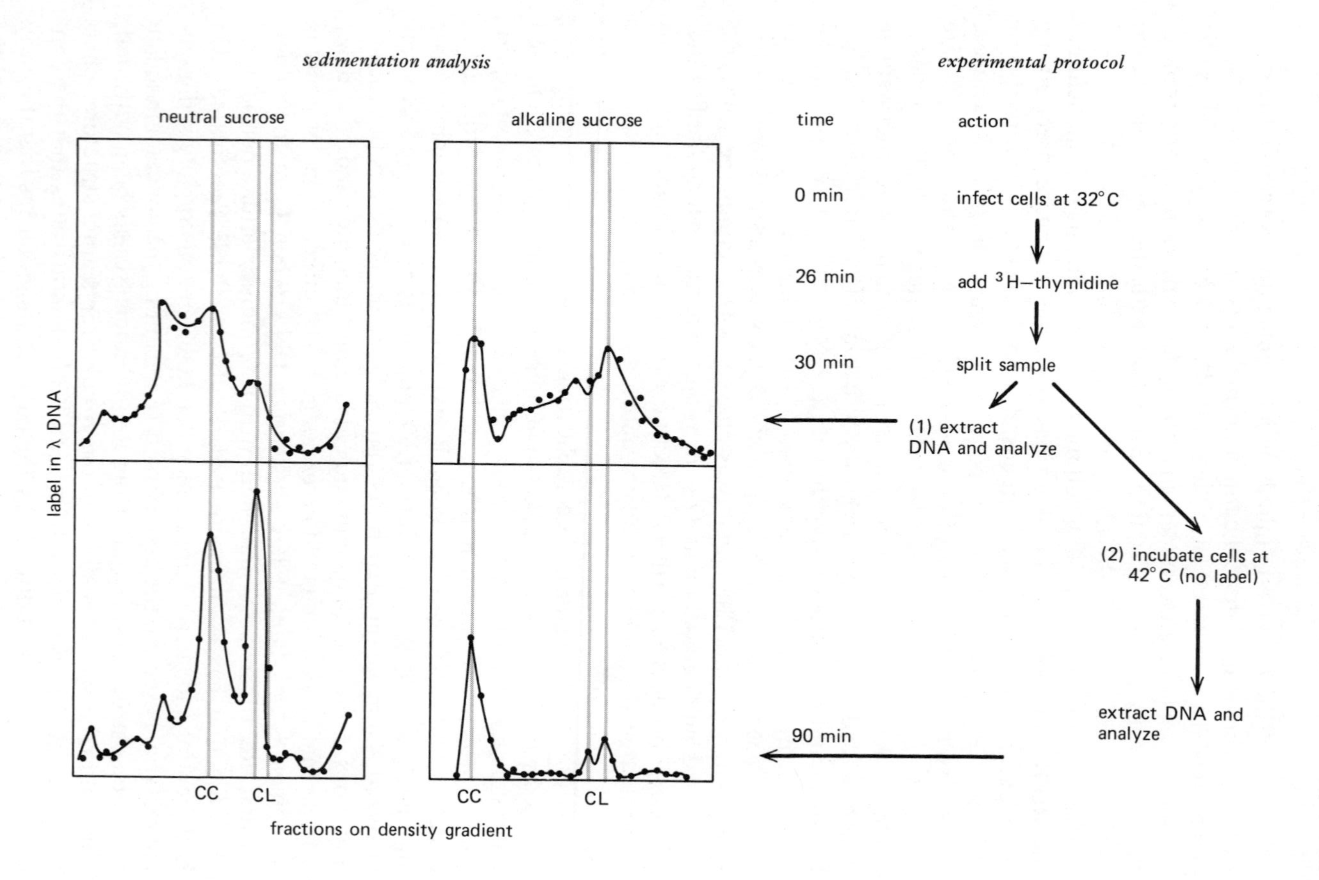
sedimentation analysis
neutral sucrose
alkaline sucrose
label in λ DNA
CC
CL
CC
CL
fractions on density gradient
experimental protocol
time
action
0 min
infect cells at 32°C
26 min
add ³H–thymidine
30 min
split sample
(1) extract
DNA and analyze
(2) incubate cells at
42°C (no label)
extract DNA and
analyze
90 min

Figure 5.20 summarizes the replication cycle of lambda. A linear monomer enters the cell and is converted to a closed circle; the circle also may be derived by excision of a prophage. During the period of early replication λ DNA replicates in circular form, usually bidirectionally, to generate circular progeny. Both open and closed circles are found; neither form has any free end susceptible to attack by the RecBC enzyme. Late replication is initiated by the conversion of circular DNA to the rolling circle; the events responsible for this transition are not known. The rolling circle is attacked via its tail by the RecBC nuclease but this degradation is prevented when the Gam protein synthesized by the phage is present to inhibit the enzyme. Replication of the rolling circle generates a concatemeric tail which provides the precursor to mature DNA. Concatemers also may be generated by recombination; the most likely form is the multimeric circle resulting from reciprocal recombination between two circles. (Any linear concatemers produced by recombination presumably would be subject to attack by the RecBC nuclease.) In the absence of replication, this recombination shunt therefore provides an alternative pathway for the synthesis of concatemers. The system responsible for DNA maturation, which is intimately linked to the processes of head assembly, can probably use any concatemeric structure as substrate for the cleavage of linear monomers terminated by cohesive ends.

### *Phage and Host Replication Functions*

Replication of λ DNA probably takes place by the same mechanism responsible for replication of bacterial DNA: short fragments are synthesized discontinuously in the 5′→3′ direction and subsequently linked covalently into the new strand (see Volume 1 of Gene Expression). Since only two phage genes are concerned with replication, phage replication must rely upon host components to provide the activities necessary for DNA synthesis. A connection between phage and host replication systems was first demonstrated by Fangman and Feiss (1969), who reported

---

**Figure 5.19:** sedimentation analysis of replicating λ DNA in presence and absence of Gam function. Cells were infected with λ*red*$^-$ *gam*$^{ts}$*E*$^-$*S*$^-$ at the low temperature at which Gam is active. A radioactive label was added during late replication and DNA immediately extracted from one sample of the cells. The remaining cells were incubated further in the absence of radioactive label at the high temperature at which Gam is inactive; the DNA extracted from these cells therefore represents a chase in Gam$^-$ conditions of the labeled DNA. Both extracts of DNA were analyzed on neutral and alkaline sucrose gradients. On the neutral gradient *L* indicates the position where a control of linear duplex DNA sedimented and *C* and *CC* mark the calculated positions where open and closed circles should sediment. On the alkaline gradient *L* indicates the position taken by a control linear single strand and *C* and *CC* mark the positions where a circular linear strand and a circular (nondenatured) duplex should sediment. Data of Greenstein and Skalka (1975).

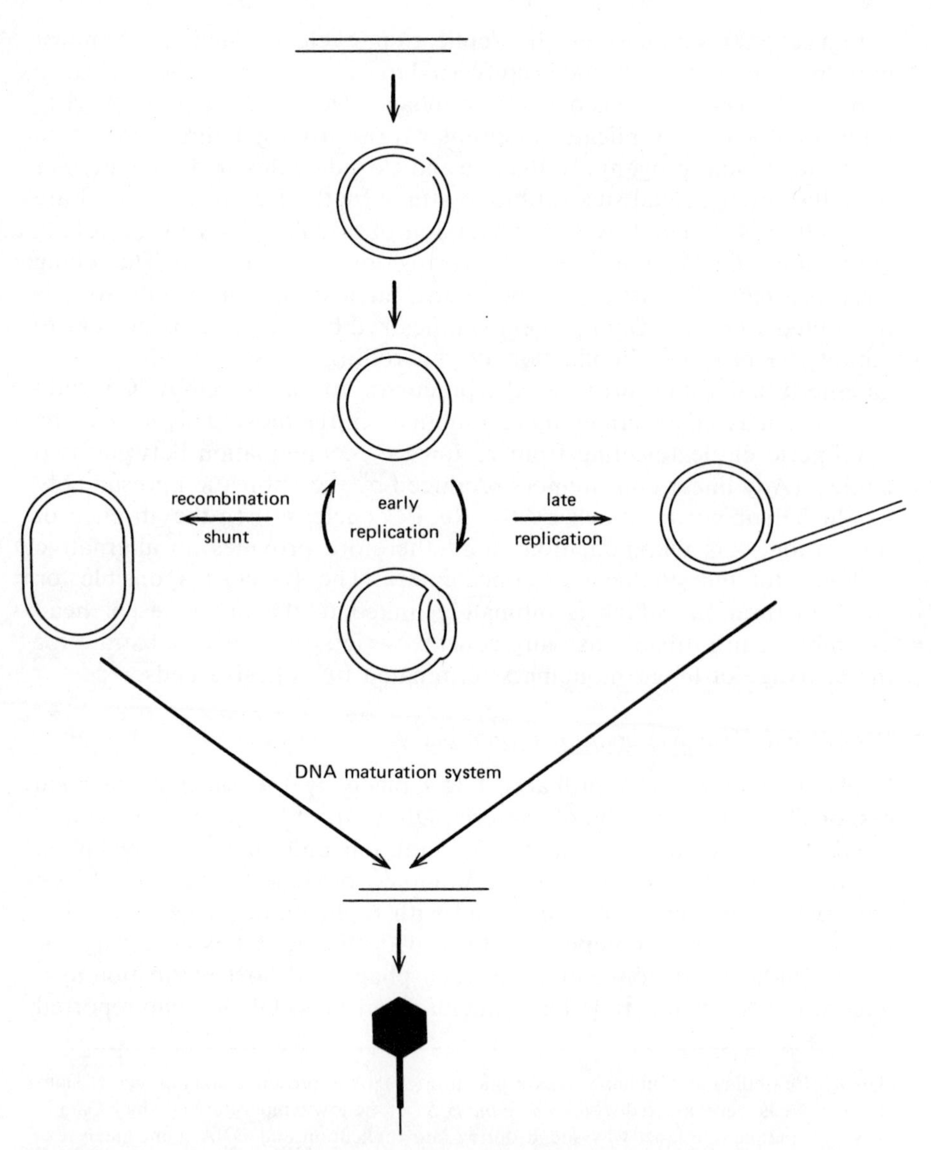


**Figure 5.20:** model for maturation of λ DNA. Upon entering the cell the linear duplex is converted to an open and then to a closed circle. In early replication the circular template is replicated from an origin located at 80% from the left end, usually bidirectionally. This generates circular molecules which do not act as precursors for mature linear duplexes. Late replication follows a switch in the form of replicating DNA to a rolling circle; it is not certain whether this always utilizes the same origin as early replication. The transition is prevented by the RecBC nuclease which degrades some structure necessary for late replication; Gam function is necessary to inhibit the nuclease and permit the switch to late replication. The

that lambda cannot replicate in cells of a temperature sensitive mutant in *dnaB* at high temperature. The inability of lambda to replicate under nonpermissive conditions in other mutants has since been reported.

The role of the phage functions *O* and *P*, which are essential for replication, is not known. However, Tomizawa (1971) was able to suggest that their activities are related. Starting with a double mutant $O^{ts}$ $P^{am}$, revertants in *P* could be isolated which are not temperature sensitive in spite of their retention of the $O^{ts}$ gene. The ability of a mutation in gene *P* to change the temperature sensitivity of the product of gene *O* implies that the O and P proteins interact.

These functions may interact also with the product of the host *dnaB* gene. Georgopoulos and Herskowitz (1971) isolated *groP*$^-$ mutants of E.coli which are unable to propagate lambda. Revertants of lambda, described as $\pi$ mutants, able to grow in the *groP*$^-$ host carry mutations in gene *P*. This suggests that the GroP function of the host usually interacts with the P function of the phage: mutation in *groP* can prevent the product of *P* from fulfilling its function in replication, but independence of *groP* can be conferred by a mutation in *P*. The *groP*$^-$ cells are temperature sensitive for DNA synthesis and it seems likely that the mutation lies in *dnaB*.

It is reasonable to suppose that the process of DNA synthesis itself is undertaken by host functions but that the phage functions are needed for control of replication. This must involve at least two activities: initiation of replication of the circular lambda genome immediately after infection; and initiation of the rolling circle replication which characterizes late replication. One possible role for the phage functions in initiating circular replication is to nick the closed duplex at the origin and McMacken, Kessler and Boyce (1975) have suggested that both the *O* and *P* gene products may be concerned in such an activity. Synthesis of the initiator *oop* RNA requires both *O* and *P* and *dnaB* functions (see page 359). Nothing is known about the events responsible for initiating the transition from early to late replication; clearly it would be informative to have mutants unable to make the transition, but it may be difficult to devise a selective technique appropriate for their isolation.

---

rolling circle generates a concatemeric tail, which serves as the usual substrate for production of mature DNA. Oligomers of $\lambda$ DNA also can be generated by recombination; these too may be utilized for cutting monomeric linear duplexes and thus recombination provides an alternative pathway for maturation when replication is prevented. Only concatemeric DNA can act as precursor for mature DNA and production of monomeric linear DNA requires the components of the phage particle head assembly system.

## Maturation of Phage Particles

### *Structural Products of the Assembly Reaction*

The particle of phage lambda comprises the structure illustrated diagrammatically in Figure 5.21, with an icosahedral head (hexagonal in outline) of diameter about 0.055 $\mu$m (550 Å) with a tail of length about 0.15 $\mu$m (1500 Å) and diameter 0.012 $\mu$m (120 Å). The head shell represents the assembly of a regular array of more than one type of protein subunit. The tail consists of about 35 stacked discs terminating in a fine fiber, about 0.025 $\mu$m (250 Å) in length with a diameter of 0.002 $\mu$m (20 Å), which appears to be responsible for host range. A connecting structure joins the head to the tail. The structure of the lambda particle and its constituents as seen in the electron microscope has been described by Eiserling and Boy de la Tour, 1965; Kemp, Howatson and Siminovitch, 1968; Kellenberger and Edgar, 1971; Harrison, Brown and Bode, 1973).

The head contains one genome of λ DNA. A very rough calculation demonstrates how compactly it must be organized in the particle. Taking the head as a spherical structure of diameter 0.055 $\mu$m corresponds to a volume of about $9 \times 10^{-5}$ $\mu m^3$; treating the DNA as a cylinder of length 17 $\mu$m and diameter 0.002 $\mu$m suggests a volume of about $5 \times 10^{-5}$ $\mu m^3$. Although

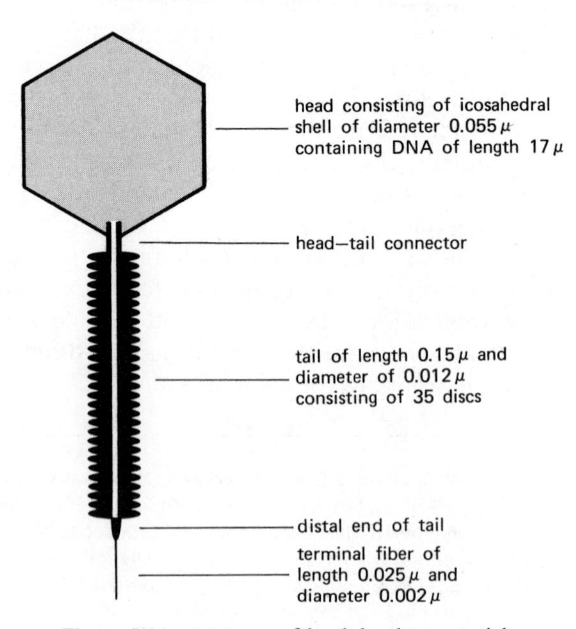


**Figure 5.21:** structure of lambda phage particle.

these figures are of course only very approximate, they indicate that the volume of the head into which the DNA must be packed is not very much greater than that occupied by the DNA. The nucleic acid must therefore be very highly condensed within the head; and we do not know how this is accomplished. Upon adsorption to a host cell, the λ DNA is ejected through the tail, which is therefore shown in the form of a tube possessing a hollow core through which the DNA may pass.

Assembly of the lambda head and tail takes place independently. The procedure for complementation of lambda assembly in vitro developed by Weigle (1966) demonstrated that cells infected with phages carrying mutations in head genes are able to make tails, whereas phages with mutations in tail genes are able to produce heads. Thus a mixture of extracts allows the assembly in vitro of active phage particles when one extract is derived from a *tail donor* (head mutant) and the other is prepared from a *head donor* (tail mutant). These experiments and those of Parkinson (1968) allowed the late morphogenetic genes to be classified into two adjacent groups: genes *A-F* are implicated in head assembly; and genes *Z-J* are concerned with tail assembly (see also Chapter 4). The map of the late genes summarized in Figure 5.22 also includes two further genes identified in subsequent experiments (see later).

In addition to mature particles, lysates of infected cells also contain other structures, including "ghost" particles that lack DNA and appear to be "empty" as judged by penetration of stain, apparently normal heads, and free tails. Lysates always contain large numbers of the structure known as *petit lambda*, abbreviated as pλ: this is similar to a small head, with a diameter of about 0.05 μm (500 Å) and has a less angular outline. It lacks nucleic acid and never is associated with a tail. The presence of pλ particles in all lysates of infected cells suggests that their formation may be an obligatory part of phage assembly. Figures 5.23 and 5.24 show electron micrographs of λ and pλ.

Lysates of certain mutants contain appreciable proportions of structures that are rarely or never observed in wild type lysates and which appear to represent aberrations of the normal assembly process. Kemp,

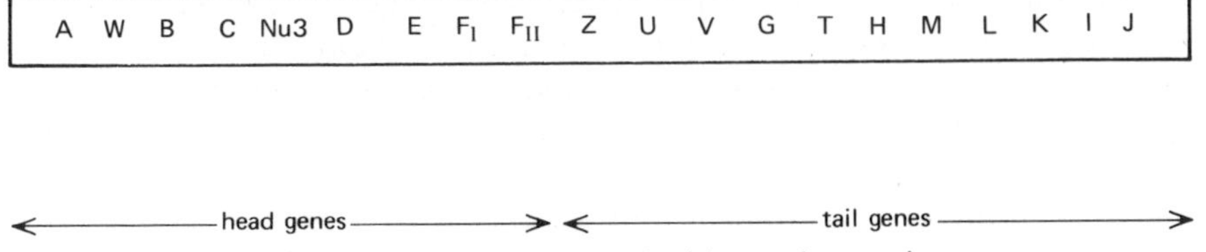


**Figure 5.22:** genes concerned with morphogenesis.

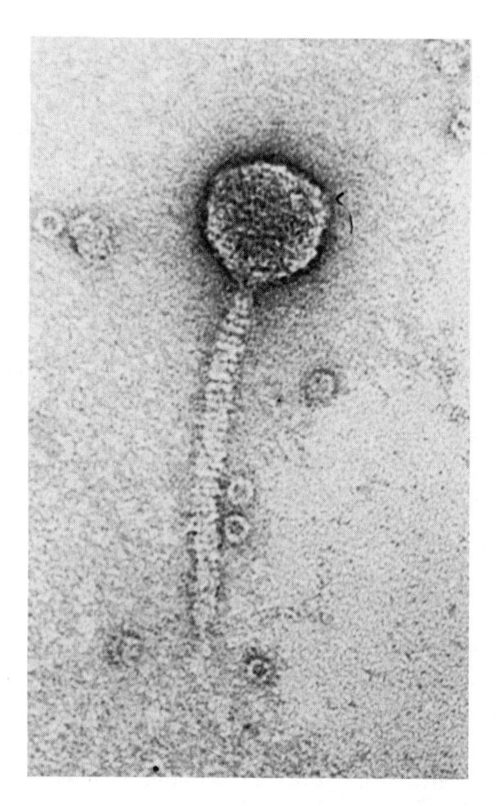

**Figure 5.23:** electron micrograph of phage lambda. Magnification × 250,000. Photograph kindly provided by A. F. Howatson.

Howatson and Siminovitch (1968) characterized these structures in an electron microscopic examination of development in lambda mutants; and Murialdo and Siminovitch (1972b) extended this study to obtain the quantitative data summarized in Table 5.3. The data given in the table represent analysis of one mutant in each class; since different mutants in the same gene may sometimes give different aberrant structures, this provides only an approximate guide to the possible morphogenetic roles of the late genes.

In wild type infection the predominant structures are completed phage particles and petit lambda; empty phages, empty heads, and free tails also are seen. Because heads containing DNA are not stable in the absence of attached tails and tend to lose their DNA, it is not possible to say whether empty heads were assembled as such or were produced by loss of DNA from full heads. The accumulation of large amounts of free heads, full or empty, can therefore be taken to indicate only that they are defective in some way which prevents association with tails to generate complete particles.

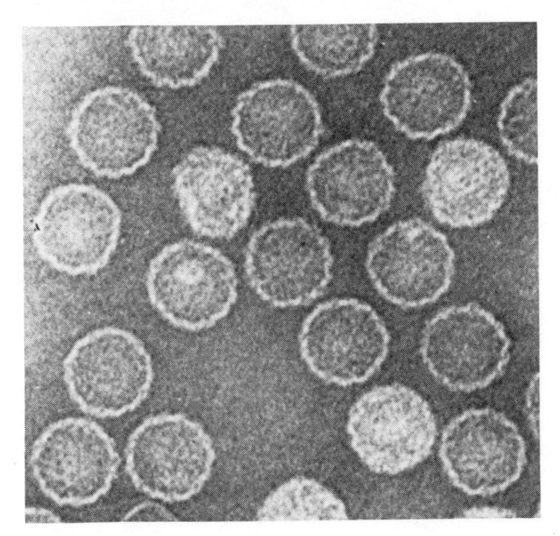

**Figure 5.24:** electron micrograph of pλ particles. Magnification x 250,000. Photograph kindly provided by A. F. Howatson.

No head structures of any type are produced by mutants in gene *E*. This argues that *E* may represent a structural gene for a major head protein in whose absence no assembly at all is possible.

*Tubular heads* are produced by mutants in *B*. Tubular heads extend as straight structures for distances much greater than the dimensions of the usual lambda particle. These appear to be equivalent to the polyheads

**Table 5.3:** assembly of head structures by head mutants

| mutant | phage particles | empty phages | full heads | empty heads | pλ | monsters | tubular | tails |
|---|---|---|---|---|---|---|---|---|
| + | 26 | 8 | — | 3.5 | 62 | — | — | 11 |
| A | — | — | — | — | 99 | — | — | 22 |
| W | — | — | 5.5 | 11 | 79 | 4 | — | 59 |
| B | — | — | — | — | 79 | 9 | 12.5 | 60 |
| C | — | — | — | — | 99 | — | — | 32 |
| Nu3 | — | — | — | 1.5 | 92 | 4.5 | — | 50 |
| D | — | 1.5 | — | 14 | 83 | — | — | 40 |
| E | — | — | — | — | — | — | — | 80 |
| $F_{II}$ | — | — | 9 | 10 | 81 | — | — | 65 |

All figures are given as the percentage of total head and head-containing structures observed (except for the E mutant where the tail number is the total counted); a dash indicates less than 0.5%. The difference between full and empty phages and heads is judged by the penetration of stain. Data of Murialdo and Siminovitch (1972b).

seen in phage T4 (see Chapter 6). The tubular heads are uniform in diameter along the tubule and fall into two classes. Figure 5.25 shows the structure of the type I tubular head, which represents an assembly of well separated ringlike structures. The tubule has a diameter of about 0.044 $\mu$m (440 Å) and the center-center spacing of the rings along the axis is about 0.011 $\mu$m (110 Å). The type I tubule sometimes terminates in an incomplete "head" analogous to a pλ structure. The figure also shows the type II tubular head in which the diameter is larger at about 0.050 $\mu$m (500 Å) and the spacing of rings is less at about 0.008 $\mu$m (80 Å). When a double mutant is constructed by combining $B^-$ with any one of $A^-$, $W^-$, $D^-$ or $F^-_{II}$, tubular heads are still produced; but they fail to appear in $B^-$ $C^-$ double mutants. This demonstrates that the products of genes *A, W, D,* $F_{II}$ are not needed for formation of the tubular head, which is accomplished by the products of genes *E* and *C* in the absence of *B*. One interpretation of these results is to suppose that *E* and *C* code for structural components of the head and that the product of *B* is in some way concerned with controlling its size and shape.

*Monsters* are structures which appear to represent imperfectly formed particles of λ and pλ. Figure 5.26 shows two examples from which it is clear that an aberrant control of shape is involved. Mutants in genes *W, B, nu3* all produce monsters. The effect of *W* mutations may be indirect, due to a polarity on the expression of *B* (see below). The lack of *B* product may thus generate either tubular heads or monsters, while the lack of *nu3* leads to assembly of monsters. Some mutants in *C* also generate monsters. The functions of genes *B, C, nu3* are therefore concerned (directly or indirectly) with shape.

Mutants in *A, D* and $F_{II}$ do not produce these aberrant forms. These functions therefore are not concerned with limiting the size of the head.

Less information is available about the roles of the tail genes. Figure 5.27 shows the long tailed type; these particles are normal apart from the length of the tail, which may be up to a hundred times the usual length. These occur only in lysates of $U^-$ mutants, so the product of gene *U* may be concerned with terminating tail synthesis at the correct length.

The important general conclusion suggested by the analysis of head structures is that the protein components of the head may assemble into more than one type of structure. Presumably the subunits of the head interact in a regular manner but the dimensions of the product are dictated by certain gene functions which are concerned with activities such (for example) as proper turning of corners. Rather than a process of self assembly in which the protein subunits ineluctably generate a single structure, lambda morphogenesis represents an assisted assembly in which certain shape-determining activities are necessary to complement the ability of the protein subunits to interact with each other.

**Figure 5.25:** tubular heads of lambda. (a) Type I tubular head terminating in a structure resembling pλ. (b) Type II tubular head. Magnification × 200,000. Data of Kemp, Howatson and Siminovitch (1968) kindly provided by A. F. Howatson.

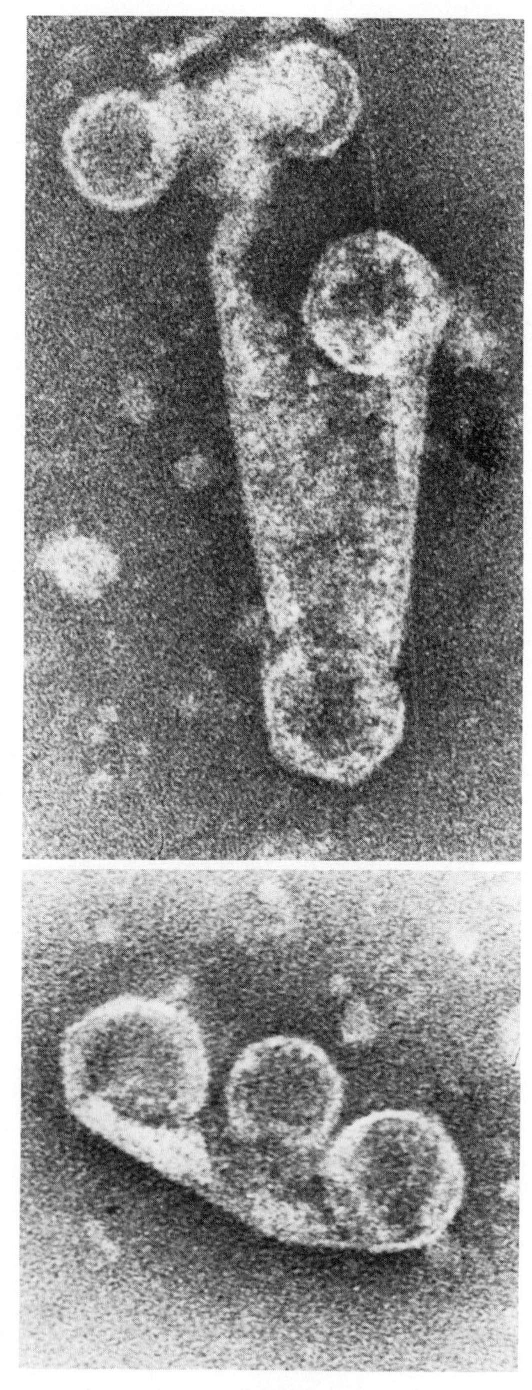

**Figure 5.26:** monsters of phage lambda. Magnification x 300,000. Photograph kindly provided by A. F. Howatson.

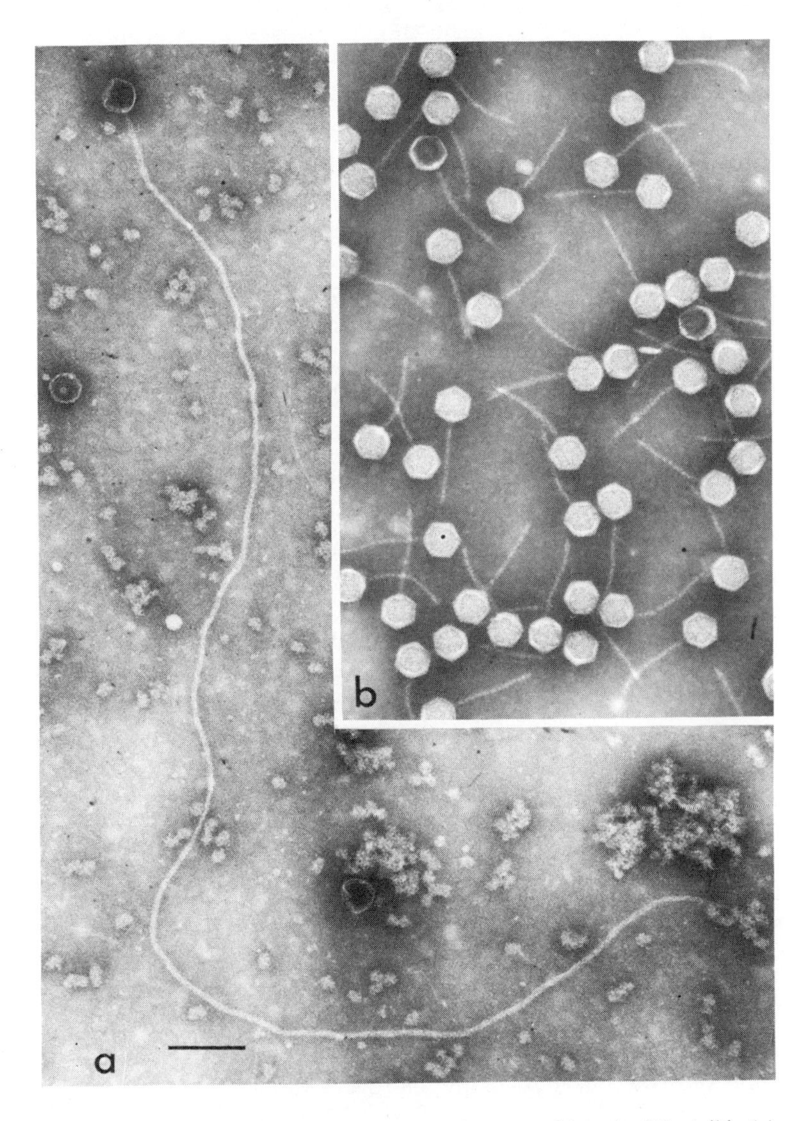


**Figure 5.27:** long tailed mutant of phage lambda. The unusual length of the tail in (*a*) can be compared with the shorter tails of the wild type particles seen in (*b*). Magnification × 75,000. Data of Kemp, Howatson and Siminovitch (1968) kindly provided by A. F. Howatson.

### *Proteins of the Late Genes*

The first step in defining the process of particle assembly is to identify the products of the twenty late genes and to divide them into those representing structural components of the head and tail and those concerned with assisting the interactions of assembly. All but four of the late gene products have now been identified; and the structural components of both head and tail have been characterized. The roles in assembly of most of the head gene products are known in outline; and the stages at which most of the tail gene products act have been identified.

The protein products of most of the late genes were identified in the experiments of Murialdo and Siminovitch (1971, 1972a) in which phage proteins were labeled with radioactive isotopes after host protein synthesis had been inhibited by ultraviolet irradiation. The protein coded by each late gene can be identified by its absence from extracts prepared from cells infected with a phage defective in this function. The results of these and subsequent experiments are summarized in Table 5.4. Only the products of *W, Z, M, I* have not yet been identified. The total length of DNA needed to code for the identified proteins is about 18,000 base pairs; the region of lambda DNA from the left end to *J* represents almost exactly 40% of the phage, that is, 18,600 base pairs. The length of DNA needed to code for the sixteen known late proteins therefore accounts for virtually all (~ 95%) of the available coding material. Even allowing for errors that may have occurred in the assignment of molecular weights, this makes two important implications: the products of genes *W, Z, M, I* all are likely to be rather small; and it is unlikely that there are any further late functions to be identified in this region.

The table includes twenty genes, two more than the eighteen assigned in the original head-tail in vitro complementation assays (see above and Chapter 4). The *nu3* gene was first identified by the loss of a 19,000 dalton protein that Murialdo and Siminovitch observed when deletions extended from *C* into *D*; this function was then called *hp*. Ray and Murialdo (1975) isolated an amber mutant not belonging to any of the previously known genes and showed that it corresponds to this function; it maps between *C* and *D* and is responsible for the substitution of the 19,000 dalton protein by a smaller (17,000 dalton) polypeptide. (This gene is now described as *nu3* because at one time there were reports of two other new genes, *nu1* and *nu2*.) Another new gene was generated by the split of *F* into two cistrons, $F_{\mathrm{I}}$ and $F_{\mathrm{II}}$. Boklage, Wong and Bode (1973) noted that mutants in *F* displayed two different phenotypes and demonstrated that the basis for this lies in the presence of two different cistrons. Work performed with mutants originally described as $F^-$ is therefore here attributed to $F_{\mathrm{I}}^-$ or $F_{\mathrm{II}}^-$ mutants depending upon which mutation was used.

**Table 5.4:** late genes and their products

| gene | molecular weight of protein (daltons) | size of gene (base pairs) |
|---|---|---|
| *A* | 79,000 | 2,150 |
| *W* | | |
| *B* | 62,000 | 1,700 |
| *C* | 61,000 | 1,650 |
| *nu3* | 19,000 | 500 |
| *D* | 11,000 | 300 |
| *E* | 38,000 | 1,050 |
| $F_{I}$ | 11,500 | 300 |
| $F_{II}$ | 11,500 | 300 |
| *Z* | | |
| *U* | 14,000 | 400 |
| *V* | 31,000 | 850 |
| *G* | 33,000 | 900 |
| *T* | 16,000 | 450 |
| *H* | 87,000 | 2,400 |
| *M* | | |
| *L* | 29,000 | 800 |
| *K* | 27,000 | 750 |
| *I* | | |
| *J* | 130,000 | 3,550 |
| total coding length | | 18,050 |

Molecular weights of proteins were determined from gel electrophoresis and assigned to genes by the absence of bands in appropriate mutant extracts. The coding length of each gene is calculated from the size of the protein. Based upon data of Becker and Gold (1975), Boklage et al. (1973), Casjens (1974), Casjens and Hendrix (1974), Hendrix and Casjens (1974a), Murialdo and Siminovitch (1972a) and Ray and Murialdo (1975).

Are the proteins of phage assembly used in the form in which they are synthesized or do some of the primary translation products represent precursors which must mature to another form? Evidence that maturation is involved in some instances was obtained by Murialdo and Siminovitch (1972a) by pulse chase experiments. Cells were labeled with radioactive amino acids at 25–29 minutes after infection and then chased with cold amino acids: phage proteins were analyzed at various intervals during the next 20 minutes. In several cases the label in a protein increased during this period, in some cases being associated with a decrease in the label present in a larger protein. It is therefore likely that several of the primary proteins are matured to further forms during assembly. The nomenclature used to describe late proteins therefore takes pX to be the immediate protein product of translation of gene *X;* and pX* represents a smaller protein cleaved from pX.

Polar relationships between some of the late genes suggest that there may be several groups of coordinate expression. A polar effect describes the situation when a nonsense mutation in one gene reduces the extent of synthesis of (wild type) protein from subsequent genes. The implication is that the genes are transcribed into a polycistronic messenger so that termination of protein synthesis at a nonsense mutation in a gene early in the sequence prevents or reduces the synthesis of products from genes subsequent to it (see Volume 1 of Gene Expression). Of course, genetic analysis shows that expression of all late functions depends upon the synthesis of Q protein to utilize the late promotor $p'_r$. And Murialdo and Siminovitch (1972a) noted that amber mutations in *Q* display much reduced production of all late proteins; their sequential order of appearance also is consistent with the idea of coordinate expression from $p'_r$. Although these results suggest that transcription of late genes proceeds from *S* to *J* as the result of a single initiation, the polar effects may reflect the existence of subsidiary clusters of common expression, for example representing the presence of weak additional promotors in the late region.

The genetic experiments of Parkinson (1968) identified three polar groups among the late genes: *W B C; V G H; H M*. Murialdo and Siminovitch (1972a) extended these observations by examining the effects of amber mutations in the late genes upon production of late proteins. They were able to identify three polar groups by this technique: *E F U; V G H; L K*. The previous experiments of Buchwald and Siminovitch (1969) showed that amber mutants in *L, I,* or *K* produce less protein from *J*. Taking account of all these experiments, the polar groups can therefore be summarized as

$$A\ W\ B\ C\ nu3\ D\ E\ F_{\mathrm{I}}\ F_{\mathrm{II}}\ Z\ U\ V\ G\ T\ H\ M\ L\ K\ I\ J$$

$$\longrightarrow \qquad \longrightarrow \qquad \longrightarrow \qquad \longrightarrow$$

The late gene products are not synthesized in equimolar amounts; there is more than a hundredfold variation in the numbers produced of the various proteins. Since variation extends within as well as between the polar groups, this observation may reflect a quantitative control that is exerted at translation in addition to any transcriptional controls. There may be a correlation between the number of copies of each structural protein utilized in construction of the head shell and the number of molecules that is synthesized; this would imply that the efficiency of utilization may be similar for all head proteins.

### *Structural Components of the Phage Particle*

Only a small number of proteins is found in the mature phage particle. Two proteins account for 95% of the weight of the head shell itself; and a single protein represents at least 90% of the weight of the stacked discs comprising the tail. Interactions between these subunits therefore must generate the regular organization that characterizes both head and tail; other proteins, present in minor amounts or absent from the mature particle itself, are implicated in the control of shape.

Separated heads and tails were isolated by Buchwald et al. (1970a, 1970b) by use of a technique in which a radioactively labeled lysate from a mutant lacking the unit to be purified was added to a lysate containing it; the disappearance of the radioactive label was the criterion for purification. Tails were purified from $E^-$ extracts and heads from $I^-$ extracts, using electron microscopy to determine the absence of pλ. Heads sedimented at 114S, corresponding to a weight of $18.0 \times 10^6$ daltons, and apparently consisted largely of a single major protein, pE. Another major protein, corresponding to about 20% of the weight of the phage, could be identified with neither the head nor tail; this is now known to be the head component pD, however, and taking account of this suggests a weight for the head of about $24 \times 10^6$ daltons. Tails sedimented at 43S, corresponding to a weight of $6.1 \times 10^6$ daltons, and consisted largely of the protein pV. The weight of the empty particle is therefore about $30 \times 10^6$ daltons; including the DNA its weight should be about $62 \times 10^6$ daltons.

Three proteins accounting for virtually all of the weight of the head shell, then described as h1, h2, h3, and now known to be pE, pD, pB*, were identified by Casjens, Hohn and Kaiser (1970); these experiments also resolved the three tail proteins described as t1, t2, t3, and known to be pV, pH* and pJ. By treating phage particles with antisera directed against pE, pD or pV, Casjens and Hendrix (1974) demonstrated that pE and pD both are distributed over the entire surface of the head (a previous idea had been that pD might be an "internal" protein) and that pV is distributed over the entire length of the tail, except for the fiber at its distal end. This confirms that it is from these proteins that the basic organization of head and tail is constructed.

In addition to the major proteins, several minor species are found in much smaller amounts on lambda heads. Casjens and Hendrix measured the relative amounts of phage proteins by gel electrophoresis and their

**Table 5.5:** protein constituents of phage particles

| gene | protein | molecular weight (daltons) | number of copies | total weight ($10^6$ daltons) | proportion |
|---|---|---|---|---|---|
| *E* | pE | 38,000 | 420 | 16.0 | 72% of head |
| *D* | pD | 11,000 | 415 | 4.6 | 20% |
| *B* | pB | 62,000 | 3 | 0.19 | 1% |
| *B* | pB* | 56,000 | 11 | 0.62 | 3% |
| *C, E* | X1 | 31,000 | 6 | 0.19 | 1% |
| *C, E* | X2 | 29,000 | 6 | 0.17 | 1% |
| $F_{II}$ | $pF_{II}$ | 11,500 | 5 | 0.58 | 3% |
| total head | | | | 22.4 | 100% |
| *V* | pV | 31,000 | 160 | 5.0 | 84% of tail |
| *H* | pH* | 79,000 | 9 | 0.5 | 9% |
| *J* | pJ | 130,000 | 7 | 0.4 | 7% |
| total tail | | | | 5.9 | 100% |

Proteins indicated by the form pX are primary translation products; proteins described as pX* represent cleavage products of the primary polypeptide; and X1 and X2 are derived by cleavage and fusion of pC and pE. Data of Murialdo and Siminovitch (1972a), Casjens (1974), Hendrix and Casjens (1974a,b) and Casjens and Hendrix (1974).

results provide the principal basis for the data summarized in Table 5.5. The major head proteins pE and pD appear to be present in about equimolar amounts. The most prominent of the remaining proteins of the head shell itself is the product of gene *B*; this is present largely in the form of the product pB* derived by cleavage of pB and this corresponds to the principal precursor-product relationship identified in the analysis of late proteins of Murialdo and Siminovitch (1972a) and Hendrix and Casjens (1975). Some uncleaved pB also is present in the head.

Electrophoresis of the head proteins reveals about 12 bands present in amounts smaller than pB/pB* and the principal two of these are X1 and X2, with weights of 31,000 and 29,000, respectively. As can be seen from Table 5.4, it is difficult to imagine that these might represent the products of new genes since there is no surplus coding capacity in the late region. Hendrix and Casjens (1974a) therefore investigated the possibility that these proteins might be derived by maturation of the primary polypeptides coded by identified late genes. Tryptic fingerprint analysis showed that X1 and X2 are very similar: X1 has two spots not present in X2; and X2 has one spot not present in X1. This suggests that X1 and X2 are derived from the same source but that X1 is a longer polypeptide. The tryptic patterns of X1 and X2 have several spots in common with pE and several in common with pC. This suggests that pE and pC might both be cleaved, the cleavage products being covalently joined in the X1 and X2 proteins.

To prove that X1 and X2 carry part of pE, Hendrix and Casjens examined the proteins synthesized by an *Eam* phage grown in a suppressor strain which inserts tyrosine instead of the wild type amino acid. This changes the position of one tryptic spot in the pE digest and the same change is seen in the digest of X2 (and by implication therefore in X1). By using $^{35}S$ label it is possible to determine the intensity of each spot in pE, pC and X2. This comparison suggests that fusion takes place on a 1:1 basis. Since pE is synthesized in amounts about 40 times greater than pC, a smaller fraction of the pE pool than the pC pool must be utilized for fusion, so that use of pC is likely to be the limiting step. X1 and X2 are not synthesized in $C^-$ infections; and their derivation from pC and pE is consistent with the observation that pC is not found on phage heads, but is present in lysates and disappears during a pulse chase in an action dependent on an active *E* gene.

The protein $pF_{II}$ is needed for the head-tail joining reaction (see below). It is probably therefore localized on the head at the point of junction with the tail. In this sense it is not part of the repetitive array of proteins that comprises the regular head shell structure.

## *Addition of Proteins to the Head Shell*

The in vitro complementation system originally developed to distinguish head and tail donors also can be used to investigate assembly of the head and tail. This extension of the assay has proved particularly useful in helping define the sequence in which the maturation functions act, making it possible to identify the precursor stages through which assembly passes. When extracts of pairs of certain head mutants are mixed, active particles are formed: the active head must have been assembled in vitro, after which it joined with a tail (whose assembly is, of course, unimpeded in both head mutants). When an $E^-$ extract is mixed with the extract derived from another mutant, the second extract provides the precursor particle and the $E^-$ extract (which lacks any type of head structure due to the total absence of the main structural protein pE) provides the functions necessary to complete maturation of the blocked precursors.

When $E^-$ extracts are mixed with $W^-$ or $F_{II}^-$ extracts, there is an increase of more than a hundredfold in the titer of active phage. Both the progeny phage particles and the incomplete precursors can be characterized. Casjens, Hohn and Kaiser (1972) demonstrated that the progeny particles generated in the mixed extracts always display the genotype $E^+ W^-$ or $E^+ F_{II}^-$; that is, their DNA is derived from the $W^-$ or $F_{II}^-$ extract. The complementing activity of the $W^-$ and $F_{II}^-$ extracts sediments at the rate characteristic of the head particle. These observations suggest that these extracts provide incomplete head shells, containing DNA, whose maturation is completed by the pW or $pF_{II}$ activity of the $E^-$ extract. Extracts of $W^-$ or $F_{II}^-$ infected cells possess infective DNA (unlike $A^-$, $B^-$, $C^-$, $D^-$, or $E^-$ extracts which accumulate the concatemeric precursor). This is consistent with the idea that the $W^-$ and $F_{II}^-$ extracts can package DNA into immature head structures and thus are defective in some later stage of assembly.

An extract from the double mutant $W^- E^-$ is just as efficient as an $E^-$ extract in complementing the $F_{II}^-$ head. This implies that the $pF_{II}^-$ head does not need pW for maturation. Thus the action of pW precedes that of $pF_{II}$. The reverse situation is found with $F_{II}^- E^-$ extracts, which cannot complement the heads assembled in $W^-$ extracts, implying that the $pW^-$ head has not yet reacted with pF. In other words, $pF_{II}$ is unable to act before pW, so that the head structures made in $W^-$ extracts are deficient in the steps catalyzed by both proteins. There is therefore an obligatory order of events in lambda maturation: pW must act to generate a head structure upon which $pF_{II}$ can act.

The $pW^-$ and $pF_{II}^-$ heads are inactive because they are unable to combine with tails. Boklage, Wong and Bode (1974) and Casjens (1974) showed that when $pF_{II}^-$ heads are incubated with $pF_{II}$ in vitro they are then able to bind to tails to generate active progeny particles. By purifying the activity of $E^-$ extracts that complements the $pF_{II}^-$ heads, Casjens was able to show that the $pF_{II}$ protein can be isolated in the same form from nonassembled extracts or from mature particles. Thus it is not modified during assembly. It is not present on the heads assembled in either $F_{II}^-$ or $W^-$ infected cells. It is present in about 5 copies per virion and may provide the head-tail connector. Extracts of $E^- S^-$ cells contain about 6000 copies of $pF_{II}$; about 700 phage particles are produced in an $S^-$ infection, so that about 3500 copies of the protein are used. Thus about 60% of the available $pF_{II}$ is utilized in assembly.

The $pF_{II}$ protein appears to be responsible for the specificity of head-tail joining. In extracts, $\lambda$ heads can be mixed with $\phi 80$ tails to generate infectious phage particles, although $\phi 80$ heads cannot join with $\lambda$ tails. When $\lambda F_{II}^-$ heads are mixed with equal amounts of $\lambda$ tails and $\phi 80$ tails in the presence of $pF_{II}$, both types of phage particle are formed (one-third have $\lambda$ tails, two-thirds have $\phi 80$ tails). This is consistent with the ability of completed λ heads to join with either type of tail. The $pF_{II}$-like activity released by disruption of φ80 heads has different properties, however; it exclusively directs the addition of φ80 tails to the φ80 head. This suggests that it is the $pF_{II}$ protein which establishes the affinity of the head for the tail structure.

At least one host function is needed for assembly of the lambda particle. E.coli cells carrying the *groE* mutation are unable to support propagation of lambda; the development of lambda in *groE* cells has been characterized by Sternberg (1973a,b) and Georgopoulos et al. (1973). No heads are made when $\lambda^+$ infects *groE* mutant cells: monsters, tubular forms, and pλ structures comprise all the headlike assemblies. Consistent with this result, no infective DNA is produced; assembly of the head is therefore halted early enough to prevent packaging of DNA. Little pB* is found in the head structures (taken as the fraction sedimenting at 60S), which implies that the *groE* function may be concerned either with the cleavage event or with the association of pB* with the head shell. Phage mutants able to form plaques on *groE* mutant cells are located in *E* (as amber or temperature sensitive mutations) or in *B*. This is consistent with the idea that the *groE* function is implicated in the interaction of pE and pB. Suppressed *Eam* phages grow better on *groE* mutant cells while suppressed *Bam* or *Cam* phages grow more poorly; one consequence of suppression is to reduce the amount of protein and this therefore suggests that a quantitative effect may be implicated. Perhaps the levels of pE, pB, pC are important in some interaction with GroE in such a way that the absence of GroE protein upsets the normal balance; this might be restored by reducing the amount of pE (enhancing growth) or might be made worse by reducing the amount of pB and/or pC (reducing growth). Hendrix and Casjens (1974a) noted that X1 and X2 are not found on the pλ particles produced by infection of *groE* mutant hosts (it is necessary to examine pλ because of the absence of mature head shells) but pC is found to replace them. This suggests that nascent head structures possess binding sites for unprocessed pC so that its cleavage and fusion with pE may take place on the head structure; the GroE protein may be involved in this process. While the role of *groE* therefore is not yet clear, it can be seen that it represents a host function that is involved in some of the maturation events involving proteins pE, pC, pB. T4 assembly also involves *groE* (see *page 648*).

Another function involved in these maturation steps is *nu3*. Ray and Murialdo (1975) showed that lysates of cells infected with an amber mutant in *nu3* contain pλ and monster structures. Since the only other mutants to produce monsters are ambers in *B* and (sometimes) in *C*, this hints that the functions of the three genes may be connected. The *nu3* mutant produces pC* instead of pC, due to the cleavage of 5600 daltons of material from the protein. The conversion of pC to pC* requires the presence of pE and GroE protein. When pNu3 is present, a pulse chase experiment shows that label in pB and pC declines but that pB*, X1 and X2 become labeled; a label in pNu3 itself declines, so this protein may be unstable. The implication of these experiments is that pNu3 is needed for the cleavage and fusion of pC and pE to generate X1 and X2; in the absence of pNu3, an aberrant cleavage takes place in which pC is converted to pC*. Cleavage of pB to pB* is not seen in $C^-$ extracts, so the presence of pC is necessary for this reaction. The disappearance of pNu3 is prevented by $B^-$, $C^-$ or $E^-$ mutations; this suggests that it may depend upon the assembly into the head structure (that is association with pE) of pB and pC. The kinetics are consistent with the idea that cleavage of pNu3 takes place after cleavage of pC and before cleavage of pB. These experiments therefore suggest the conclusion that pNu3 and GroE protein are implicated in the processing of pB to pB* and in the reaction of pC with pE to generate X1 and X2.

### *Structural Transitions in Head Assembly*

The central issue in analyzing the construction of the lambda head shell is the relationship between the mature head structure and that of pλ. Early ideas were based upon the concept that pλ might be generated by self assembly of pE without mediation of any of the other head proteins; in this case it might represent a by-product of assembly in wild type infections and would accumulate in mutants unable to provide the proteins needed to direct assembly into the mature head structure. More recent results have shown that there are several classes of pλ structure. The simplest does indeed consist solely of pE, demonstrating that this protein has the ability to self-interact to form a regular structure. Two further types of pλ are: intermediate forms, comprising pE associated with proteins products of certain other head genes; and a final form whose protein constitution is identical with that of the mature head except that it lacks pD. This analysis suggests that the pλ type of structure is not a by-product but rather represents a series of precursors to the mature head; the pλ structure starts by self assembly of pE, is converted by the addition of other proteins into a form lacking only pD, and finally is converted into the larger more angular structure of the mature head by the insertion of pD. The head shell is then prepared for tail joining by the actions of pW and $pF_{II}$.

It is perhaps simplest to start with the structures closest in the assembly pathway to the mature head and work back to the earlier precursors. The lack of pD, then known as hl, from all pλ structures was demonstrated by

Casjens, Hohn and Kaiser (1970). Hendrix and Casjens (1975) extended this analysis to show that pD is the only protein missing from the pλ produced in wild type infections (this is to exclude $pF_{II}$, which as the head-tail connector is not considered part of the regular repeating structure of the head shell). Similar results were reported by Hohn and Hohn (1974) and Hohn et al. (1975). Table 5.6 summarizes the data of Hendrix and Casjens. In these experiments the mass of each protein relative to the mass of pE was determined by autoradiography of proteins labeled with $^{35}S$ and separated on SDS polyacrylamide gels. This appears to overestimate the amount of X1, X2 and pNu3, which are probably rich in sulfur, by a factor of up to two. In the table the results are calculated as the number of copies of each protein relative to every 420 copies of pE, the number that appears to be present in the head of the mature particle. We should therefore emphasize that this calculation assumes the number of copies of pE to be the same in pλ as that measured for the head of the mature particle.

Two types of pλ can be separated by sedimentation through glycerol gradients. In cells infected with $\lambda^+$, $\lambda A^-$, $AB^-$ or $\lambda D^-$ the particle sediments at about 150S; in cells infected with $\lambda C^-$ or in *groE*$^-$ cells infected with $\lambda^+$, the pλ structure sediments at about 190S. The idea that there are at least two classes of pλ is supported by the data of Table 5.6. The pλ particle produced in a $\lambda^+$ infection has essentially the same constitution as the lambda head except for the greatly reduced amount of pD; in fact this is only about 5% of the level seen in the head and might represent contamination. The other head proteins, pB, pB*, X1 and X2 are present in pλ in amounts (relative to pE) that are the same as those of the lambda head. In both $A^-$ and $D^-$ mutants the composition of pλ is the same as that of the lambda head; pA and pD thus are not needed for assembly of the petit lambda structure. The protein constitution of the pλ lambda particles of $B^-$ mutants differs from wild type only in the absence of pB and pB*. This suggests that although these proteins usually are present on pλ, they cannot represent an intrinsic part of the construction of the pλ shell.

Different results are obtained in $C^-$ and in *groE*$^-$ infections. These pλ structures sediment more rapidly and possess an additional protein band of 19,000 daltons; although this has not yet been identified by mutation, it is probably pNu3. The $C^-$ mutants have reduced amounts of X1 and X2; the extent of reduction depends upon the mutant used, varying from the smaller degree given in the table (representing a mutation located close to the terminus of the gene and thus synthesizing most of the pC protein) to the complete absence of synthesis seen when the mutation lies near the beginning of the gene. The pλ of $C^-$ infections possesses entirely pB and no pB*; this confirms that pC is needed for the cleavage of pB. Similar

**Table 5.6:** relative protein constitutions of pλ particles produced by lambda mutants

| protein | $\lambda^+$ head | pλ of $\lambda^+$ | $A^-$ | $D^-$ | $B^-$ | $C^-$ | $groE^-$ |
|---|---|---|---|---|---|---|---|
| pE | 420 | 420 | 420 | 420 | 420 | 420 | 420 |
| pD | 432 | 28 | — | — | — | — | — |
| pB | 4 | 4 | 2 | 1 | — | 19 | 1 |
| pB* | 15 | 13 | 13 | 12 | — | — | — |
| X1 | 11 | 13 | 13 | 13 | 18 | 5 | 5 |
| X2 | 9 | 13 | 18 | 12 | 15 | 3 | 4 |
| pNu3 | — | — | — | — | — | 121 | 146 |
| pC | — | — | — | — | — | — | 10 |

pλ particles were obtained by sedimentation on glycerol gradients and their protein constitutions determined by gel electrophoresis. The mass of each protein was determined by autoradiography of an $^{35}S$ label. Results are calculated as the number of copies present of each protein relative to the number of pE units; the pE number is normalized to 420, the value that appears to be characteristic of the mature lambda head shell. These numbers are imprecise because the proportion of sulfur-containing amino acids may vary in the proteins; comparison with Table 5.5 suggests that the proportion of X1 and X2 is overestimated by a factor of 2; the same is probably true of pNu3. Data of Hendrix and Casjens (1975).

defects are seen in $groE^-$ infection, when pλ lacks pB* and has very little (probably no) pB, has reduced amounts of X1 and X2, but has a large number of copies of pNu3.

The two types of pλ structures distinguished by their protein constitutions and sedimentation rates display different structures. In negatively stained lysates, the different types of pλ display the same appearance; but by analyzing thin sections, Zachary, Simon and Litwin (1976) observed two distinct structures. The predominant pλ particle that accumulates in $A^-$, $B^-$ and $D^-$ mutant infections has the appearance typical of an empty structure and has been described as *prohead II*. The predominant pλ particles accumulated in $C^-$, $B^-C^-$ and *groE* mutant infections appear to contain a protein "core" and are described as *prohead I* structures. In the course of normal infection, prohead I particles accumulate first and then disappear as prohead II particles accumulate, in turn to be replaced by filled heads. These kinetics, and the protein constitutions of pλ given in Table 5.6, suggest that prohead I is the precursor of prohead II, which is the precursor to the mature head.

From the perspective of the structures observed by thin section electron microscopy, the conversion of prohead I to prohead II appears to involve the loss of its protein core. From the viewpoint of the protein constitutions of the pλ structures summarized in Table 5.6, the conversion involves the loss of pNu3 from prohead I and the addition and/or processing by cleavage and fusion reactions of pB and pC to generate pB*, X1 and X2. How may the alterations in protein constitution be equated with the change in structure?

The components of prohead I common to *groE*$^-$ and *C*$^-$ infections are the gene products pE and pNu3. Prohead I is able to bind unprocessed pB (seen in *C*$^-$ infection) and unprocessed pC (seen in *groE* infection). The *C*$^-$ prohead I protein constitution demonstrates that binding of pB does not require the presence of pC; but the cleavage to pB* does need the pC protein. The *groE* pattern shows that pC binding does not depend on GroE protein; but processing (fusion with pE and cleavage to X1 and X2) requires this host function. This suggests that pB and pC are able to bind to prohead I and that their processing occurs upon it.

What is the role of pNu3 in forming the prohead I structure? The view taken of its function depends on whether the structures formed in *nu3*$^-$ infection are considered to be precursors to prohead I or aberrant structures that cannot be utilized in head formation. Indeed, this emphasizes the general need for caution in drawing conclusions about the order of events in head assembly on the basis of structures seen in mutant infections, because it is always possible that an observed structure may be a by-product formed in the absence of some head component rather than a precursor blocked from proceeding further on the assembly pathway.

Structures consisting solely of pE were observed by Ray and Murialdo (1975) in infections with *nu3*$^-$ or *B*$^-$*C*$^-$ mutant phages. Taking the view that these structures are precursors on the maturation pathway, this would suggest a sequence of events in which the first step is the self-assembly of pE to generate a pλ type of structure, after which pNu3, pC and pB associate with this structure. Ray and Murialdo suggested that pNu3, pC and pB might form a complex together which is necessary for ensuring proper processing of pC and pB.

Although the structures formed in *nu3*$^-$ infection appear headlike upon negative staining, Zachary, Simon and Litwin (1976) observed that they take an amorphous structure, monsterlike in appearance, in thin sections. Infection with *nu3*$^-$ *C*$^-$ phage produces the same structures, suggesting that pNu3 acts before pC. The relationship of this type of structure to the normal maturation pathway is not clear. However, the only thin sections altogether failing to display any core-containing structures are those of the *nu3*$^-$-infected cells. This suggests that pNu3 may provide the core of prohead I and is lost or degraded upon the conversion to prohead II. This suggests a model in which proper head formation may require the condensation of pE around the pNu3 core, rather than self assembly of the major head protein. It is possible that pC provides a link between the core and the outer shell because $C^-$ mutants generate a much increased proportion of particles with acentric cores; perhaps pC is responsible for establishing the position of the core within the outer shell, a role that would be consistent with its later fusion with pE and cleavage to generate X1 and X2. A model showing the structures through which head assembly passes, and the proteins that are present at each stage, is illustrated in Figure 5.28.

A defect common to $B^-$, $C^-$ and *groE*$^-$ infections is the absence of pB* from the pλ structure. These mutants all form large numbers of monsters and this suggests that the role of pB* may be concerned with turning corners. If there were one copy of pB* at each of the twelve corners of the icosahedron that constitutes the mature head shell (this is close to the value suggested by experimental measurements), the role of this protein might be to ensure that the proper three dimensional relationship between the subunits of each face is achieved. Both X1 and X2 are found on the pλ of $B^-$ infections, which suggests that they may not be implicated in the shape defect that causes formation of monsters.

A demonstration that pλ can indeed provide a precursor to the mature head shell was achieved by the experiments of Kaiser, Syvanen and Masuda (1975) in which purified pλ particles were added to extracts of an induced $E^-$ lysogen. Extracted from cells infected with $A^-$ phage, the pλ preparation must contain predominantly prohead II. The pλ must provide the only source of pE in the mixture, which produces plaque forming particles. Is the pλ a direct precursor, expanded into the head shell by addition of further proteins, or is it just a storage form of pE, disassembled and then reassembled de novo into a head shell? When heavy (labeled with $^{15}N$ and $^{13}C$) pλ and light pλ were prepared and mixed, the progeny phage contained either heavy or light heads; less than 15% had the intermediate densities that would be expected if the pλ first dissociates. The pλ structure may therefore serve as a direct precursor to the mature head shell. We should emphasize at this juncture that only about 0.1% of the pλ particles are converted into active phage particles, so that although there is no evidence for any heterogeneity among the pλ preparation, this system is concerned with the conversion of only a small proportion of the potential precursors.

Two types of model may be imagined for the process of packaging DNA into the head shell: an empty shell may be constructed from protein subunits after which it is filled with DNA; or the shell may be constructed by assembly of the protein subunits around the condensed DNA. The

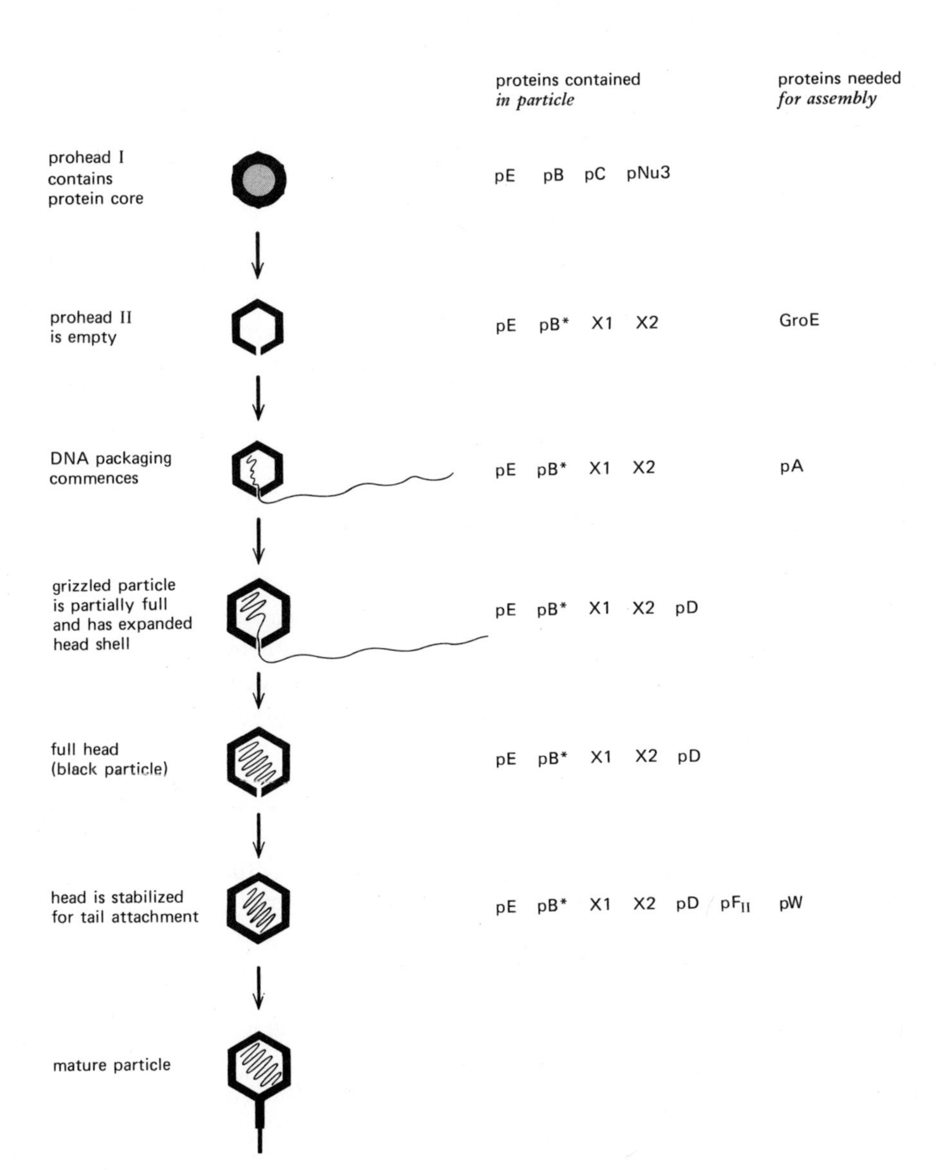


**Figure 5.28:** model for lambda head assembly.

concept that the (empty) pλ structure is the head precursor implies that DNA must be inserted either into this small shell or into an expanded mature shell derived from it. This conclusion is supported by experiments in which exogenous DNA was packaged in an in vitro extract. Hohn and Hohn (1974) isolated the activity accepting the DNA and found that it sediments at the rate characteristic of pλ; when examined under the electron microscope this fraction was rich in pλ structures. The active particles could be obtained only from infections able to generate the prohead II intermediate. But since linear monomers were packaged in these experiments, in contrast with the need for concatemeric DNA in vivo, it is not certain that packaging occurred by the physiological mechanism. On the other hand, Hohn (1975) noted that in spite of this, pA is needed for the reaction; pA provides the function that cuts concatemeric DNA (see below) and this observation therefore suggests that its role is not confined simply to cleaving DNA, as well as implying that this in vitro system may indeed be dependent on the same functions as packaging in vivo.

At what stage does DNA enter the head shell? Kaiser, Syvanen and Masuda (1975) approached this question by assaying head formation in a two stage system in which head assembly is separated from tail joining. They found that pλ particles can be converted into infectious phages first by adding an extract containing DNA, pA and pD to form head shells, and then by adding pW, $pF_{II}$ and tails to form complete particles. In the first stage, pA is needed to cleave the concatemeric precursor into mature genomes (see below) and for whatever other functions it may have in assembly, and pD is needed for expanding the pλ structure into the head shell; when the second extract is added, infectious particles are formed only if construction of the filled head shell has been successfully accomplished. Deoxyribonuclease can be added at the first stage and it is reasonable to assume that resistance to the enzyme should be acquired only when DNA has been inserted into the head. When both pA and pD were present in the first stage extract, infectious particles were formed, implying that filled heads had been formed before addition of the enzyme. When pA alone was present at this stage a lower level of infectious particle formation occurred; when pD alone was present no infectious particles were formed. This demonstrates that pD cannot act before pA, that is the head shell cannot be expanded unless DNA is packaged into it. Some packaging can take place in the absence of pD, which suggests that expansion may occur subsequent to the entry of DNA, perhaps might be triggered by it. On the other hand, $D^-$ phage infections accumulate concatemeric DNA so that no cutting of DNA can occur unless it is possible to expand the head shell. This demonstrates the close relationship between maturation of DNA and expansion of the head. Perhaps packaging of DNA commences into pλ, but may be successfully concluded, and the DNA cut into a linear monomer, only when pD is present to support expansion into the full head structure. (On the other hand, Dawson, Skalka and Simon (1975) and Dawson et al. (1976) reported that empty and filled heads have the same dimensions when thin sections of infected cells are examined; an increase in size could be seen only following lysis. This would suggest that head expansion is a consequence of changes in the environment induced at the time of lysis. But it is difficult to see how a full genome could be packaged into a structure the size of the pλ particle and Lickfield et al. (1976) reported no difference in the size of the phage head before and after lysis. Although the events involved in head expansion therefore cannot be completely defined yet, at present the genetic evidence on the roles of pA and pD appears the more persuasive and is followed in the model of Figure 5.28.)

The steps in assembly in which pA and pD are involved have been distinguished further by Lickfield et al. (1976) on the basis of studies of the particles accumulated in mutant infections. Thin section studies distinguished three types of particle according to the densities of their appearance: empty, grizzled and black. Empty particles are the usual pλ structures; grizzled particles have a center whose appearance is grizzled due to a varying concentration of material thought to be DNA; and black particles appear to contain a full complement of DNA. During a normal infection, pλ particles appear first, followed by grizzled and then by black particles. In $A^-$ mutant infection only pλ structures are seen; neither grizzled nor black particles can be produced. In $D^-$ mutant infection, grizzled but not black particles can be seen. This suggests that pA is required to initiate filling of the empty head; the grizzled particles represent heads only partially filled. Then pD is required subsequently for stabilizing the partially filled head and allowing the completion of packaging.

### *Icosahedral Structure of the Head Shell*

The head of a phage particle may be viewed as a quasi crystalline structure in the sense that it comprises a regular array of subunits. Early models for particle assembly were constructed in terms of the ability of a single type of protein subunit to self-assemble; it has since become apparent, however, that more than one type of protein subunit may be utilized and that assembly of the components comprising the head shell may be assisted by proteins additional to the structural components themselves. Head assembly raises two interesting questions: what is responsible for the proper acquisition of shape as protein subunits interact; and how is the DNA packaged into the head?

The issue of how a virus shell is constructed was discussed extensively by Caspar and Klug (1962), who observed that viruses can be considered in two categories: filamentous rods; and spherical shells. Their analysis provided a general solution for the problem of how a spherical shell with cubic symmetry is constructed. Cubic symmetry requires that all three dimensions are equivalent, that is, no direction in space can be distinguished; examples are tetrahedral, octahedral and icosahedral regular structures. The basic concept that regular organization of a head shell requires a repetition of structural interactions in which the same contacts between subunits are utilized at each point in the shell was propounded by Crick and Watson (1956). From this basis Caspar and Klug demonstrated that there is only one general way to construct a spherical shell from a large number of identical subunits, and this ineluctably leads to icosahedral symmetry. A regular icosahedron has twenty faces and twelve corners where the faces meet. Although the lambda head shell comprises more than one type of protein subunit, its construction can be regarded in the same general terms as those discussed by Caspar and Klug, albeit with some modifications.

The basic assumption of the Caspar and Klug theory is that the head shell is held together by the same types of bonds throughout but that these bonds may be deformed in slightly different ways in environments that are nonsymmetrical. The amount of nonequivalence required between units is small (within the expected flexibility of atomic bonds) and so the interactions that hold the shell together may be described as quasi-equivalent. Caspar and Klug noted that triangulating a sphere into subdivisions as equal as possible provides the basis for deriving geometrically quasi-equivalent packings. In mathematical terms the quasi-equivalent subdivisions can be enumerated by considering the triangulation of the sphere to be derived from the folding up of a planar equitriangled net into a polyhedron with icosahedral symmetry. The important conclusion of this analysis (which is too detailed to be considered here more than very briefly) is that icosahedral virus shells must have 60 T quasi-equivalent subunits, where T is the triangulation number defined by the equation

$$T = pf^2$$

where $p = h^2 + hk + k^2$, with k and h representing two integers with no common factor, and where f equals any integer. Values for T therefore fall into the series 1, 3, 4, 7, 9, 12, 13, 16, 19, 21, 25, 27, 28, etc. For the lambda head shell it became apparent at an early stage that the number of copies of pE must fall into the range of the two triangulation values T = 7 and T = 9, that is, there must be either 420 or 540 subunits in the shell (for review see Kellenberger and Edgar, 1971). The analysis given in Table 5.5

shows that there are close to 420 copies of both pE and pD in the shell, corresponding to a triangulation number of 7.

A model for the lambda head shell, which takes account of the presence of more than one type of protein subunit, has been constructed by Williams and Richards (1974). This proposes that pE is arranged in trimeric clusters and that pD is organized in clusters of hexamers and pentamers. This corresponds to a composite lattice for the head shell characterized by T = 7 levo (the levo referring to the skew orientation, a feature not found when the shell is constructed by self-assembly of a single type of subunit). Data consistent with this model on the nature of the transition from pλ to the head shell were reported by Howatson and Kemp (1975). In these experiments they studied the structures of the type I and type II polyheads and the nature of the transition between them. Electron microscopy of negatively stained preparations showed that type II tubules have the same hexagonal lattice corresponding to the T = 7 levo composite arrangement, apparently with pD grouped in clusters of hexamers (of molecular weight 66,000) and with pE grouped in trimers (of molecular weight 114,000). The type I tubule appears to correspond to pλ in consisting of pE assembled in the absence of pD, with the pE subunits probably grouped in the form of hexamers. Howatson and Kemp proposed that the transition from a type I to a type II tubule takes place when hexameric clusters of pD enter the type I lattice at the centers of the hexameric rings of pE, which then rearrange as trimers with an increase in the lattice spacing. By implication the same transition may be responsible for expansion of the pλ shell into the mature head shell. Presumably it is necessary for pB* and X1/X2 to be present for the shell to acquire the correct icosahedral structure when the rearrangement takes place.

An alternative model for the structure of the capsid has been proposed by Wurtz, Kistler and Hohn (1976) on the basis of studies of the surface organization of polyheads assembled in vitro. In these experiments, pλ particles or empty heads were dissociated in guanidium hydrochloride and dithiothreitol and then dialyzed into an ammonium carbonate buffer. Polyheads formed by precipitation can be isolated by centrifugation. The three types of polyhead obtained have different, but related, structures. The principal product derived from particles of *nu3*$^-$ infections is described as polyhead A; this also is the principal product derived from pλ particles of *groE*$^-$ or *D*$^-$ infections, where another (lesser) component is the polyhead B structure. The tubes formed by dissociation of empty heads are described as polyhead C. The polyhead A and C structures correspond to the type I and type II tubules formed in vivo. The diameter of the tubular structures was about 300 Å for the A type and 360 Å for the B and C types. The A type (as with the type I polyhead) can be taken to correspond to the structure of the prohead whereas the C type (similarly to the type II polyhead) provides an analogue for the mature capsid. Wurtz et al. suggested that the type B structure might represent an expanded polyhead, but the basis of the expansion is not clear, since these lack pD; as the B type polyhead is never more than a minor (1–5%) constituent, its role is difficult to assess and it is therefore difficult to be sure that it does indeed have an analogue in the head assembly pathway. The structures of the types A and C polyheads, however, may be expected to reflect the prohead and head structures. Wurtz et al. suggested structures in which the A polyhead consists of hexamers of pE and the C polyhead possesses trimers of pD at the centers of the pE hexameric units.

### *Cleavage of Linear Monomers from Concatemers*

Insertion of DNA into its protein coat is an orientated process in which a genome of λ DNA is cut from a concatemeric precursor; only lambda genomes in the concatemeric state can serve as precursors for the packaging reaction and the intimate link between DNA processing and head assembly is revealed by the accumulation of the concatemeric precursors in mutants blocked early in head assembly. It is, of course, this need for a concatemeric precursor which explains the importance of the switch from early to late replication and the necessity for recombination as an alternative pathway in the absence of late replication (see above).

The apparent ability of any form of concatemeric DNA to act as precursor for head assembly suggests that in formal terms the requirement of the processing reaction can be expressed as a demand for a molecule possessing two uncut cohesive end sites. The duplex sequence of the joined cohesive ends (that is, in a circular or concatemeric molecule) is described as a *cos* site; when this is cleaved the two cohesive ends, left and right, are generated. The demand for a concatemeric precursor thus can be expressed as the demand for a sequence

$$\text{———} R - cos - A \text{—————————} R - cos - A \text{———}$$

The *cos* site comprises a sequence that must be recognized by an endonuclease which makes staggered cuts on the two lambda strands. This function has been described as Ter and is coded by the *A* gene.

The activity of the Ter system was first identified by its ability to release a single lambda genome from a tandem double lysogen. The yield of phage upon induction of $int^-$ double lysogens depends upon whether the two prophage genomes lie in trans or in cis. Mousset and Thomas (1968) and Gottesman and Yarmolinsky (1968b) reported that when the two prophages lie in trans (one on the chromosome and one on an episome) the dilysogen yields phage only at low levels. When the two prophages lie in tandem array, however, there is a much increased yield of phage. This presumably corresponds to the release of a single genome by cleavage at the two *cos* sites in the tandem sequence

*att-int — R-cos-A — J-att-int — R cos-A — J-att*

In the sense that this sequence includes two *cos* sites on the same molecule of DNA, this excision is analogous to the cutting of concatemeric DNA that takes place in head assembly.

When a single genome is excised by this reaction from a tandem array of two prophages, it should carry the markers from *A-J* of the left prophage and the markers from *int-R* of the right prophage. This prediction was tested by Mousset and Thomas (1969) by examining the products of the very efficient excision of a recombinant prophage that takes place when tandem double lysogens are superinfected with a heteroimmune phage. They described this as *dilysogenic excision*: it takes place in $rec^-$ $red^-$ $int^-$ conditions to release a progeny phage genome with the genetic constitution expected if the Ter enzyme makes a cut in each of the two *cos* sites. Thus in the two cases

*gal* — $imm^{\lambda}$ $R^-$ $A^+$ — $imm^{434}$ $R^+$ $A^-$ — *bio*
*gal* — $imm^{434}$ $R^+$ $A^-$ — $imm^{\lambda}$ $R^-$ $A^+$ — *bio*

superinfection with $\lambda imm^{21}$ $red^-$ $int^-$ excised genomes that were $A^+$ $imm^{434}$ $R^+$ and $A^-$ $imm^{\lambda}$ $R^-$, respectively.

Identification of the Ter function with the product of the *A* gene was accomplished by the development of an in vitro assay for Ter activity. Wang and Kaiser (1973) prepared an extract which was able to convert covalently closed λ DNA circles into a form active in the helper assay. Since infectivity in this system requires cohesive ends, the conversion appears to represent cutting of the circles at the *cos* site. The extract is active when prepared cells infected with $\lambda A^+$ but is inactive when derived from $\lambda A^-$-infected cells. This suggests that pA represents the Ter function.

The importance of the *cos* site in maturation of λ DNA is emphasized by experiments on the ability of $\phi 80$ (which has the same cohesive ends as lambda) to mature λ DNA from the prophage state. Szpirer and Brachet (1970) showed that a λ genome can be excised and matured from a tandem lysogen by infection with $\phi 80$. Feiss and Margulies (1973) then demonstrated that $\phi 80$ is able to mature a repressed circular λ phage which has a duplication of the *cos* site, although lambda progeny are not generated when the lambda circle carries only the usual single *cos* site. This supports the idea that the requirement for concatemeric DNA in lambda development reflects a need for a precursor molecule carrying at least two *cos* sites.

Phages carrying duplications of *cos* or deletions of part of it (it is impossible to obtain deletions of the whole *cos* site, which emphasizes its importance) have proved useful in investigating the role of Ter cutting in λ DNA maturation. Isolation of duplication strains has been reported by Emmons (1974) and Feiss and Campbell (1974). When examined by heteroduplex mapping such duplications appear as extensions of the phage at either the left or right end (contrasted with the usual location of an internal duplication at a single site). This apparent mobility of the duplicated region is precisely what would be expected of a duplication including *cos*, since the presence of two *cos* sites close together means that either may be cut during maturation; which one of the two sites is cut determines whether a duplicated sequence is added to the end of the phage at either terminus. If the *cos* site is represented as comprising the two single stranded sequences L and R (that is the protruding ends at the left and right ends of the linear duplex), then the structure of a normal linear phage DNA carrying a single (cut) *cos* sequence is

The structures of the linear genomes generated by the duplication strain which carry two *cos* sequences can be described as

L —— LR ——————— R
L ——————— RL —— R

The phage carrying this duplication yields segregants at a high rate. After one lytic cycle the progeny are

19% wild type (one *cos* site: duplication lost)
80% parental (two *cos* sites: duplication retained)
1% longer (three *cos* sequences: duplication added)

Such segregation does not occur with nonterminal duplications; its occurrence can be attributed to the action of the Ter system upon a concatemeric precursor taking the structure:

λ genome —— LR — LR —— λ genome —— LR — LR —— λ genome

The number of *cos* sequences in the progeny phage depends upon which member of each pair of *cos* sites is cut. If the left member of each pair is cut, the phage has a single duplication at the left end and carries two *cos* sites; if the right member of each pair is cut the phage again carries two *cos* sites but this time the duplication lies at the right end. This accounts for the mobility of the duplication and occurs in the 80% of progeny that retain the single duplicated sequence. The wild type revertants (19% of the progeny) must be generated when the two inner *cos* sites are cut (the right member of the left pair and the left member of the right pair). The phages carrying two copies of the duplicated sequence (1% of progeny) must be generated when the two outer *cos* sites are cut.

The simplest model for the action of Ter is to suppose that the nuclease acts upon unpackaged concatemeric DNA. But this does not explain an asymmetry in the location of the duplication in the parental class of phages that perpetuated the single duplication. Some 90% or more of these carry the duplication at the left end while less than 10% carry the duplication at the right end. This suggests that there may be an asymmetry in the maturation of λ DNA from its concatemeric precursor: cutting and packaging may be compelled to proceed in a particular direction. One model is illustrated in Figure 5.29. After a cut has been made at a *cos* site, the DNA to the right of the cut is preferentially packaged into the head shell. This is to say that the left end of λ DNA enters the head first and the packaging reaction is terminated by cleavage of the right end from the next *cos* site.

This model explains the preference for left terminal duplication in the strain carrying two *cos* sites. The predominant progeny may result from the introduction of a cut at one of the members of a pair of *cos* sites, followed by packaging to the right until the first member of the next pair is encountered, when the reaction is terminated. If the *cos* site that is cut in the first pair is on the left, a left terminal duplication results; for this to be produced it is necessary to assume that cutting does not take place at the second member of this pair, perhaps because it is too close to the first site. When the *cos* site cut in the first pair is the right member, the resulting genome is wild type, with no duplication. A right terminal duplication might be produced when cutting does not take place at the first member of the second pair but proceeds to the second site.

The model illustrated in Figure 5.29 attributes different roles in maturation to the two *cos* sites. The first *cos* site is cleaved to generate a free left cohesive end which provides the starting point for packaging DNA into the head shell. The second *cos* site is cleaved to terminate the reaction while DNA is being packaged. The asymmetry between the *cos* sites was evident in the system for packaging in vitro developed by Syvanen (1975). Lambda DNA in various forms, carrying the marker $imm^{434}$, was added to an extract derived from a heat-induced lambda lysogen; any DNA

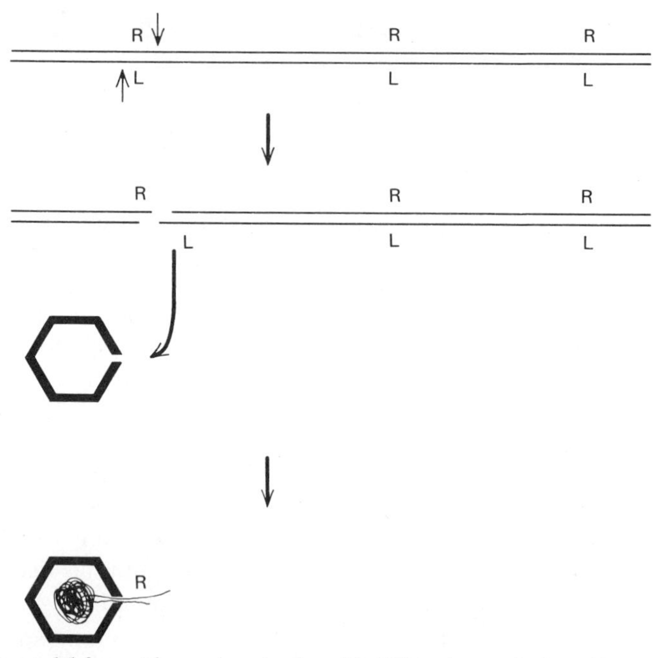


**Figure 5.29:** model for cutting and packaging of λ DNA. A *cos* site is cut to generate right (R) and left (L) cohesive ends. The left cohesive end is preferentially inserted into a head shell; insertion continues until the next *cos* site is encountered when the reaction is terminated by cleavage to generate a right cohesive end.

packaged in vitro can be distinguished from the progeny resulting from induction by the ability to propagate on a lambda lysogen conferred by its *434* immunity. Polymeric DNA formed by annealing linear monomers and sealing the gaps with ligase was utilized for the formation of phage in vitro; but linear monomers and circular monomers could not provide a substrate for this reaction. This implies that the packaging reaction in vitro has the same requirement for concatemeric precursors as that seen in the cell. When ligase was prevented from sealing the annealed cohesive ends (by addition of the inhibitor NMN, which competes with the necessary cofactor NAD), DNA could not be packaged. This implies that the *cos* sites (or in formal terms, at least one of them) must suffer cleavage of covalent bonds during maturation as an essential step in the packaging reaction.

Lambda DNA can be cleaved into fragments by the restriction endonuclease Eco R1 and Syvanen isolated the fragments carrying the left and right cohesive ends. These were then separately annealed to the

$\lambda imm^{434}$ linear monomer and tested in the in vitro packaging system. When the fragment carrying the right cohesive end alone is annealed to $\lambda imm^{434}$ to generate the structure

R1 fragment ——— *cos* —— $\lambda imm^{434}$ ——— right cohesive end

no packaging occurs. A different result is observed when the Eco R1 fragment carrying a left cohesive end is annealed to $\lambda imm^{434}$ to generate the molecule

left cohesive end —— $\lambda imm^{434}$ ——— *cos* —— R1 fragment

The $\lambda imm^{434}$ genome then can be packaged, although less efficiently than when it is in polymeric form.

These results demonstrate an asymmetry in the need for the two *cos* sites, for the relative order of the cohesive end and the intact *cos* site determines whether the DNA can be packaged. The order *cohesive end → intact cos site* allows packaging; but the order *intact cos site → cohesive* end does not. Two models might account for this asymmetry. If packaging proceeds from left to right as illustrated in Figure 5.29, packaging of the structure *left cohesive end → intact cos site* must indicate that (at least in the in vitro system) the left cohesive end can initiate packaging, which is terminated by cutting the intact *cos* site. Failure of the structure *intact cohesive site → right cohesive end* to be packaged must mean that even if the reaction starts at the intact *cos* site, it cannot be completed because the free right cohesive end cannot substitute for an intact *cos* site. This implies that cleavage of the second *cos* site which terminates packaging is an integral part of the maturation process. In formal terms it is possible to propose an alternative model in which packaging takes place from right to left, in which case the packaging of *cohesive end → intact cos site* would indicate the need for an uncut *cos* site to initiate maturation, and the failure in packaging of *intact cos site → cohesive end* would confirm that a cohesive end cannot substitute for the initial *cos* site. This model predicts that a cohesive end could substitute for a *cos* site to terminate the reaction. The two models therefore make precisely opposite predictions about the roles of the two *cos* sites in packaging. Thus these results per se demonstrate that there is an asymmetry in packaging but do not distinguish the two possible directions.

Two factors argue decisively for the first model and against the second. One is the inability of a circular monomer to be packaged, even though it possesses an intact *cos* site. This shows that it is not sufficient to initiate the reaction by cutting this site and packaging (in either direction) until the free cohesive end is reached. This implies that a second *cos* site is essential for termination of packaging, consistent with the first model. A second argument lies with the evidence derived from other experiments which point to packaging from left to right rather than the reverse. In view of these results, it is tenable to interpret events in the in vitro packaging system only in terms of the model of Figure 5.29. Although the in vitro results thus may be taken to demonstrate that termination of packaging by the introduction of cuts in an intact *cos* site is essential for maturation, they leave open the possibility that an intact *cos* site is not needed to initiate the reaction and that a free cohesive end can do so. But the accumulation of concatemers in several classes of head mutant suggests that no cutting can take place unless assembly proceeds to the point of packaging DNA into head shells. If the first cut could be made independently of packaging, free cohesive ends should be present in the accumulated DNA, but these have not been observed (see Skalka, Poonian and Bartl, 1972; Wake, Kaiser and Inman, 1972). Thus cleavage of the first *cos* site also is an intrinsic part of maturation and takes place as part of the head assembly reaction.

Evidence that the two cohesive ends of lambda are not treated symmetrically during maturation of DNA was provided by the experiments of Little and Gottesman (1971) in which they isolated phages carrying only one cohesive end. These phages were derived by induction of a lysogen in which excision could not take place by either the usual action of the Int system upon the *att* sites or by use of general recombination as an alternative. Induction therefore generates only a very small number of viable progeny phage, but the cells also release a larger number of defective particles. The defective progeny particles can be divided into two classes by their buoyant densities; the relatively lighter fraction carries the markers *gal-imm*$^{\lambda}$*-R*; and the relatively denser fraction carries the markers *A-bio* or on occasion just *A*. These genomes appear to be generated by the action of the Ter system in making a cut at the *cos* site and by a second action (of unknown nature) in making a cut in bacterial DNA either to the left or to the right of the prophage. Thus the two types of release can be represented as

```
——— gal — att — red — imm — R — cos — A — J — att — bio ———
     ↑                                ↑
    cut                            Ter cut

——— gal — att — red — imm — R — cos — A — J — att — bio ———
                                  ↑                          ↑
                               Ter cut                      cut
```

in each case generating a defective phage that carries half of the lambda genome and some bacterial sequences. Both classes of defective genome carry only one cohesive end and this type of phage has been described as λ*doc*. If the phage carries the right cohesive end it is λ*docR*; if it carries the left cohesive end it is λ*docL*.

The λ*gal* phage released by this unusual excision should carry a right cohesive end but its left end will be abnormal (depending upon the action that released it within the bacterial DNA); it therefore constitues the λ*docR* class. The λ*bio* phage correspondingly should carry only the left cohesive end; it may therefore be described as λ*docL*. This model was supported by experiments to determine directly which cohesive ends are present in the two classes of particle by measuring the ability of their DNA to anneal with BUdR labeled left and right halves.

The two types of defective particle show an important difference in their infectivity: λ*docR* particles are infectious (although they can complete a lytic cycle only if a helper phage is present to provide the functions in which they are defective); but the DNA of λ*docL* particles can be assayed only in the Kaiser helper system. There are two stages of the lambda life cycle at which the cohesive ends might be distinguished: packaging into the phage head; and injection into a new host. If the right but not the left end is needed for packaging, then λ*docL* particles might have morphological defects that are responsible for their lack of infectivity. If the right cohesive end is needed for injection, then both types of particle might be normal in construction but the λ*docL* would be functionally inert. Of course, the right cohesive end might be distinguished from the left at both stages, so that it is necessary first for packaging and then for injection.

The idea that DNA is always packaged in one direction, as illustrated in Figure 5.29, immediately suggests a difference in the origin of the λ*docL* and λ*docR* particles. Packaging of DNA into a λ*docL* particle must start with the cleavage mediated by Ter at the *cos* site. The free left cohesive end then enters the head shell and DNA continues to be packaged until finally a cleavage takes place to release a free right end. The DNA is therefore packaged in the usual direction and the difference from the normal maturation process is that no *cos* site is present to terminate the packaging reaction at the right end. The λ*docL* particles display an extremely broad range of buoyant densities, suggesting that their content of DNA is very variable; this implies that the cleavage to generate a free right end may take place at any one of many locations, perhaps at random. The origin of the λ*docR* particles may lie in packaging in the "wrong" direction, so that the right cohesive end enters the head shell first and the reaction is terminated by generation of a free left end in the

DNA beyond *gal*. The density of λ*docR* particles usually is close to that of wild type lambda, suggesting that they contain a length of DNA similar to that of the lambda genome. This might result if there were a sequence resembling that of *cos* which Ter might recognize present in the bacterial DNA to the left of *gal*. Of course, this idea leaves open a second possibility for the origin of the λ*docR* particle: Ter might act at this bacterial site, followed by packaging of DNA to the right up to the *cos* site. Sternberg and Weisberg (1975) reported that the number of λ*docL* particles exceeds the number of λ*docR* progeny by a factor of from 200–600. This supports the idea that cutting at *cos* is followed by a strongly preferential packaging of DNA to the right, that is starting with the left cohesive end, to generate the λ*docL* genome. This observation does not distinguish between the possibility that infrequent packaging in the wrong direction generates λ*docR* and the alternative that Ter infrequently recognizes (because of the difference in sequence) a site resembling *cos* in the leftward bacterial DNA, fortuitously present at about the right distance from the *cos* site of the prophage.

What is responsible for the difference in infectivity between λ*docR* and λ*docL* particles? The wide variety in the content of DNA of the λ*docL* particle suggests that it may be deficient in terminating the processing of λ DNA and thus in the construction of the particle. Sternberg and Weisberg argued that if maturation occurs by packaging DNA into the head shell until a *cos* site is reached, the absence of such a site at the correct position might mean the DNA continues to be packaged into the head to a point at which full capacity is reached and excess DNA protrudes. (From the inability to isolate transducing lambda phages containing more than a certain length of DNA we know that the lambda head is able to contain very little more DNA than is present in the lambda genome itself. The idea that the head shell has a capacity limited to a length of nucleic acid not much greater than that of λ DNA also is consistent with the calculation that the volume of the head shell is little more than the volume of λ DNA.)

If a free right end is cut to leave a protruding end only after the λ*docL* head shell is filled to capacity, tails may be unable to attach to the head with the result that functional phage particles fail to be formed. The supposition that this cleavage is a random event explains the wide variation in DNA content of λ*docL* particles. Removal of the excess DNA might restore the ability to form functional particles. When Sternberg and Weisberg treated lysates containing λ*doc* particles with DNAase, they observed a large increase in the infectivity of the λ*docL* particles. The amount of DNA in these functional particles was close to the length of the lambda genome, suggesting that removal of the excess is sufficient to allow junction with the tail. The deficiency in λ*docL* particles may therefore be attributed to their lack of tails, consistent with observations that lysates rich in λ*docL* according to genetic criteria contain large numbers of tailless heads. Of course, direct confirmation of this model requires isolation of the tailless head containing excess DNA.

In examining lysates produced by induction of an excision-defective lysogen, Sternberg and Weisberg observed that in addition to λ*doc* particles there are particles present that carry a wide variety of bacterial markers. Usually, of course, lambda is able only to engage in specialized transduction of genes adjacent to its attachment site on the bacterial chromosome, in contrast with the ability to support generalized transduction of any gene that is displayed by phages that gain bacterial DNA during infection rather than upon induction. The appearance of generalized transducing activity in the lambda lysate has the same requirement as that for λ*doc* infectivity: lysates must be treated with DNAase in conditions that allow subsequent head-tail joining. A speculative model to explain this observation is to suppose that the bacterial DNA may contain sites resembling *cos* which Ter can recognize rather inefficiently. Following Ter action at one of these sites, bacterial DNA is packaged into the phage head shell to form the same sort of particle as λ*docL*, that is, with excess DNA protruding, because there is no site in the appropriate position to terminate packaging. (The probability that two sites resembling *cos* might lie the right distance apart in the bacterial chromosome is very low.) This model thus explains why such generalized transduction by lambda is observed only under these unusual conditions.

If packaging always proceeds from left to right, the two cohesive ends of the phage genome may occupy different positions in the head shell. Two types of experiment show that the right cohesive end is distinguished from the left by its presence at the head-tail junction: under appropriate conditions it is susceptible to attack by nuclease; and treatment with certain chemical agents releases DNA whose right end is attached to the tail. These observations are consistent with the idea that the right cohesive end is the last sequence of DNA to enter the head shell.

Mature lambda particles are immune to treatment with nuclease but Bode and Gillin (1971) found that isolated heads—obtained from *J* mutants unable to make tails—can be attacked by micrococcal nuclease. When the enzyme is added to these heads and the treated preparation is then mixed with an $A^-$ extract to add tails, the yield of mature particles is normal. But the particles are defective. Their ability to inject DNA is unimpaired, but very little of the injected DNA is converted to closed circles. This suggests that the nuclease may have damaged one or both of the cohesive ends.

By isolating the DNA of these particles, Gillin and Bode (1971) demonstrated that it is infective in the Kaiser helper assay, which implies that it possesses (at least) one undamaged cohesive end. It is able to form hydrogen bonded circles in vitro, but the Tm for melting the annealed DNA is 36°C compared with the 61°C of $\lambda^+$ DNA. This suggests that the defect lies in the removal of only a few bases from one cohesive end (reducing its ability to anneal with the other cohesive end). By applying to the nuclease treated DNA the techniques developed to sequence the cohesive ends of lambda, Padmanabhan, Wu and Bode (1972) demonstrated that the defective DNA has lost the last four bases of the right cohesive end. This implies that λ DNA is packaged into the head shell in such a way that the right cohesive end is exposed at the tail attachment site.

Formaldehyde causes lysis of phage particles so that DNA pours out of the head. The DNA can often be seen to be attached to the tail and Chattoraj and Inman (1974) observed that the connection always is made to the right end of the DNA. The chemical CMC, a water soluble carbodiimide, mediates attachment of the phage particle to its DNA and Thomas (1974) observed that it is always the right end of the λ DNA that is connected to the protein shell. Treatment with formamide disrupts the phage particle and a variety of structures are seen. Among them are phages which have ejected DNA from the tail, presumably by the same mechanism that is involved in host infection, and tail-DNA complexes released from the head. Thomas showed that DNA is always ejected with the right end leaving first. Saigo and Uchida (1974) observed that it is always the right end of the DNA that remains attached to the tail in the DNA-tail complex.

These experiments imply that upon adsorption to a host cell, the right cohesive end moves down the core of the tail and is the first sequence to enter the bacterium (a conclusion which contradicts a previous study reporting that either end can inject first). Is the presence of the cohesive end at the head-tail junction important for either the structure of the particle or for proper injection? The properties of two types of phage suggest that it is needed for neither function. Particles whose heads have been treated with micrococcal nuclease to degrade part of the right cohesive end before tail joining are infective, as are the λ*docL* particles generated by nuclease treatment of the λ*docL* heads before tail joining (and these should completely lack a right cohesive end). The single stranded sequence at the right end therefore seems to exercise its function only after infection, when a circle must be formed by its reaction with the left cohesive end.

*Tail Assembly*

The lambda tail consists of two principal structures: a series of about thirty-five stacked discs of total length 0.15 μm and diameter 0.012 μm; and a fine fiber of length about 0.025 μm and diameter 0.002 μm. Eleven genes are implicated in tail assembly by the failure of mutants to form functional tails; three major proteins are found in the tail. Table 5.5 shows that there are about 160 copies of pV, 9 copies of pH*and 7 copies of pJ. Each disc probably represents a simple polymer of pV; a tetrameric structure would require 140 copies and a pentameric structure would need 175 copies, not far from the number estimated by experiment. The structural role of pH* is not known, but in view of the small number of copies present it is unlikely to be part of the structure of the discs. The fine fiber at the tip of the tail appears to make the first contact in adsorption to a host bacterium and so may be supposed to be responsible for host range; in this case it may be equated with pJ, presumably representing a polymer of this large protein.

A system for in vitro complementation between extracts from different tail mutants was developed by Kuhl and Katsura (1975). Most pairs of mutants complement to produce infectious particles and all the pairs failing to do so represent cases when both mutants are part of the same polar group, so that a polar effect may be responsible for the failure in complementation. Tail structures cannot be seen in the lysates of defective tail mutants (except those of $Z^-$ and $U^-$). To isolate tail precursors from these extracts, Katsura and Kuhl (1975) identified the fraction of a density gradient displaying the serum blocking activity, that is, the antigenic determinant represented by pJ. This means that assembled precursor structures are detected in this system only when they carry pJ; essentially, therefore, this assay follows the course of pJ as it passes through precursor structure en route to the mature tail. These experiments therefore yield no information about any structures that do not contain pJ.

Figure 5.30 illustrates the pathway for tail assembly suggested by these experiments. The serum blocking activity can be identified with a 15S fraction in $L^-$, $K^-$ or $I^-$ mutants. This fraction shows no complementa-

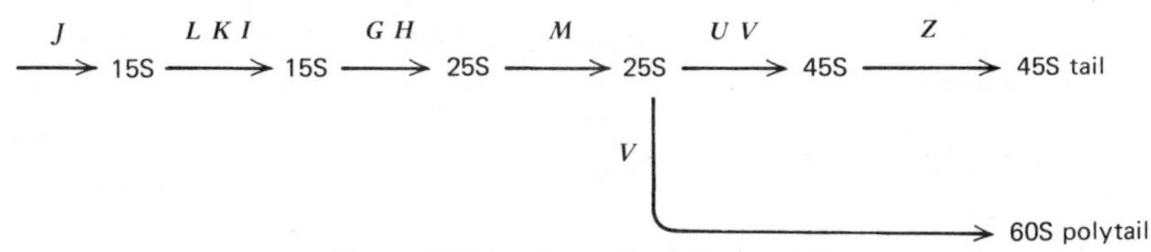


**Figure 5.30:** pathway for tail assembly.

tion activity in vitro. In $G^-$ and $H^-$ mutants the activity again sediments at 15S, but now includes complementing activities $J^+$, $I^+$, $K^+$, $L^+$, which suggests that it represents a complex of the four proteins coded by these genes. That is, it corresponds to the tail fiber associated with pI, pK, pL. In $M^-$ mutants there are peaks of serum blocking activity at 25S and 15S; both carry the $J^+$, $I^+$, $K^+$, $L^+$ complementing activities. In $V^-$ extracts the same 15S fraction is present but the 25S fraction carries $H^+$ and $M^+$ complementing activities in addition to the other four. In $U^-$ extracts the 25S peak carries only the $H^+$ complementing activity and a 60S peak carries $V^+$ and $H^+$ activities. This latter peak represents the polytail. Its presence demonstrates that the addition of pV to the previous precursor forms a taillike structure which lacks all the complementing activities except $J^+$ and $H^+$ seen on the previous precursors. This is consistent with the scheme of head assembly in which some activities associate with the head precursor and then are released at a later stage before the mature head is produced, although of course in tail assembly a much greater proportion of the activities seem to exercise this transient association. In the absence of pU, tail assembly does not terminate at the usual length, although all the additional complementing activities have been released. In $Z^-$ extracts a 45S peak is found which carries all the complementing activities: $J^+I^+$ $K^+$ $L^+$ $M^+$ $H^+$ $G^+$ $V^+$ $U^+$. This suggests that in the presence of pU the pZ protein is required to release all the transient complementing activities. In $Z^-$ extracts tails attached to heads can be seen, but the particles lack infectivity, so it should be possible to test whether such particles carry (as predicted by the pathway of Figure 5.30) all the additional tail complementing activities. It is clear from these results that pJ can be traced into the tail through a series of precursors to which new proteins are added until finally almost all the tail functions are present and must be released to generate the mature tail. It is not possible to exclude the possibility that this is not the main pathway and/or that other intermediates are important; to do so would require the same sort of experiments to be conducted but following pV activity as the means of identifying the precursors.

The tail of the phage particle plays a critical role in adsorption to the bacterial cell. The roles of the individual protein components of the lambda tail are now beginning to be elucidated by the use of in vitro systems. Lambda adsorption to E.coli depends upon the presence of a receptor which is a protein component of the outer membrane (Randall-Hazelbauer and Schwartz, 1973); it is probably coded by the gene *lamB* in which mutations may render bacteria unable to adsorb lambda (Thirion and Hofnung, 1972). Lambda particles treated with the receptor in vitro undergo a reversible interaction which does not inactivate the particle;

inactivation occurs only when chloroform or ethanol is present (see Schwartz, 1975). This suggests a two step binding reaction in which reversible contact is succeeded by irreversible association. Since DNA is not necessarily ejected when the particle is inactivated by binding to receptor, the triggering of DNA injection into the host cell must represent a separate step subsequent to irreversible binding (Roa and Scandella, 1976). By studying the temperature dependence of DNA injection, Mackay and Bode (1976a,b) showed that there may be a delay between binding and injection, again suggesting that these are separate steps. At low temperature, binding may occur without injection. Although binding requires only the presence of the lambda receptor specified by *lamB*, injection requires further host functions: Scandella and Arber (1974, 1976) found that *pel*$^-$ mutants of E.coli may prevent injection although they do not interfere with binding. Phage mutants able to infect *pel*$^-$ hosts have alterations in either gene *V* or gene *H*. This leaves the implication that pV and pH are involved in some action, presumably a conformation change, that is necessary for DNA injection. Future extensions of these experiments will no doubt lead to a detailed model for the series of events by which DNA is enabled to leave the phage head and enter the bacterium.

CHAPTER 6

# Phage T4

## Genetic System of T-even Phages

A very large number of phages infects E.coli, the best characterized bacterial host; and, of course, many other phages exist upon other bacteria. Of the phages that infect E.coli, the T phages are the most investigated of the virulent type, essentially because early in the period of phage research it was realized that progress would be impossible unless attention was focused upon a small number of systems; an account of the history of early phage research by many of the researchers involved has been given in Cairns, Stent and Watson (1966). In the other class of phages, the temperate, the lambdoid phages and especially lambda itself are by far the best characterized, and the research that we have already described in Chapters 4 and 5 naturally has relied upon many of the techniques that originally were developed to study the T phages.

The seven T phages fall into four groups. The T-even phages, T2, T4 and T6, are related in their serological properties, that is, in the components of the phage particle, and in their genetic constitutions. A general feature of T-even infection is that phage development is independent of the bacterial chromosome, which is degraded soon after infection. The T-even phages specify many of the functions needed for phage development, and this imposes a requirement for a large genome. Some of the early work on T-even phages utilized phage T2, but most of the succeeding research has been carried out with T4 and it is the life cycle of this phage with which we shall largely be concerned here.

Of the remaining T phages, T1 and T5 have not been well characterized and will not be treated here. Phages T3 and T7 are much smaller than the T-even phages; their genetic constitutions are related and their infective cycles are discussed in Chapter 7.

### *Assays of Phage Infection*

When a suspension of a virulent phage is added to a growing culture of sensitive bacteria at a multiplicity of infection greater than unity (so that there is more than one phage particle for each bacterium) the culture loses its turbidity and becomes clear as all the bacteria are infected and lysed. If the culture is then incubated for several hours further, turbidity may reappear and this represents the growth of bacterial mutants that are resistant to the infecting phage. In effect, the culture is repopulated by the progeny of the few resistant bacteria, which were present in culture before addition of the bacteriophage; that the resistant mutants arise by random mutation was of course the conclusion drawn from the results obtained by using the Luria-Delbruck (1943) fluctuation test discussed in Chapter 2.

Viable phage particles may be counted by means of a plaque assay in which a suspension of the phage is mixed with sensitive bacteria in a small volume of molten agar; a low multiplicity of infection is used to ensure that the bacteria are in great excess compared with the phage particles, so that each phage particle infects only a single bacterium. The mixture then is poured onto the surface of a nutrient agar plate and hardens to form a top layer in which the infected bacteria are immobilized. When the plate is incubated the uninfected bacteria multiply to form a confluent layer growing over the plate. But when the infected bacteria burst to release progeny phage particles, these infect adjacent bacteria which in turn are lysed to release progeny phages. The result of this chain reaction is the production of an area of clearing (limited by a decline in bacterial metabolism) which is seen as a plaque upon the bacterial lawn. The plaques produced by virulent phages usually are clear, although they may vary in size and morphology (which are useful markers subject to mutation). The number of plaques formed upon the agar plate reflects the number of *plaque forming units (pfu)* in the original phage population, known as the *plaque titer* of the suspension. The ratio of the plaque titer to the number of phage particles in the original suspension is the *efficiency of plating* (*eop*), a measure of the proportion of the phages that is viable.

The basis of T-even genetics was established by a series of experiments that began four decades ago. The kinetics of phage multiplication were first defined by Ellis and Delbruck (1939), who introduced the *one step growth experiment*. Figure 6.1 illustrates this protocol. Phage particles and bacteria are mixed, a few minutes allowed for the host cells to become infected, and then any residual free phage particles are neutralized by the addition of an antiserum directed against the phage. The mixture is then diluted to allow growth to continue and samples are assayed for their content of plaque forming units, that is, they are withdrawn from the suspension and tested by plating on agar, after various intervals of time. With phage T2 or T4 the number of plaque forming units remains low and constant for some 25 minutes after infection; this is

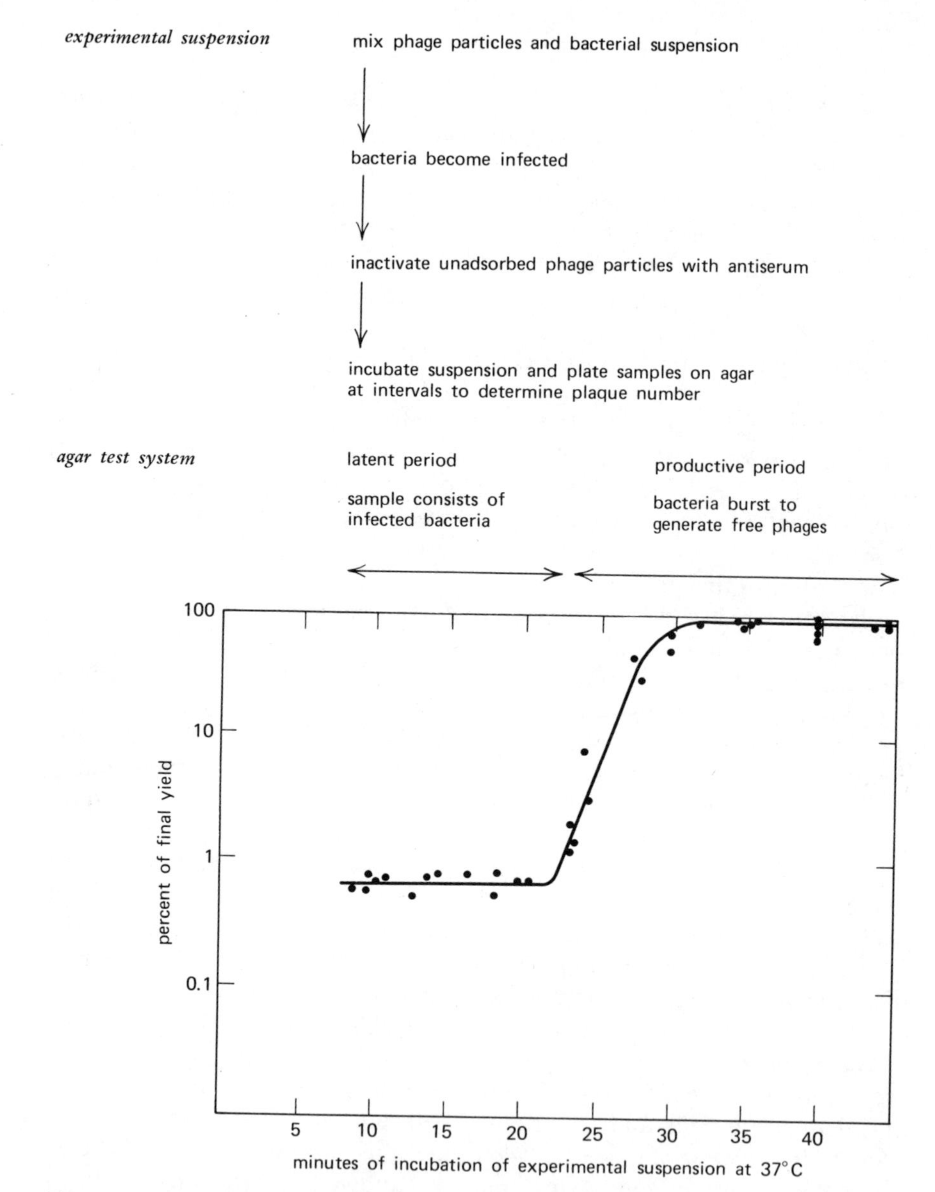


**Figure 6.1:** one step growth experiment. Infected bacteria are incubated in the experimental suspension and samples are removed and plated on agar after varying times. During the latent period the sample consists of infected bacteria, so that each plaque represents a single infected bacterium; during the productive period free phage particles have been released from completed infections, so that each plaque represents one of these phages. The phage cycle is completed in 25–30 minutes under the conditions used by Doermann (1951) to obtain the data given in the figure; the time taken to pass from the latent plaque level to the productive plateau may be due to differences in the time of adsorption and/or the length of the lytic cycle.

known as the *latent period*. Then the plaque titer rises sharply, increasing by more than a hundredfold to reach a plateau after a further 5-10 minutes.

The latent period represents the time between infection and lysis of the infected cells in the suspension culture. During this time no free phage particles are present in the culture (because any free particles have been neutralized with the antiserum). The culture therefore consists entirely of infected bacteria. Each infected bacterium may generate a plaque when a sample is assayed by plating on agar, so that during the latent period each plaque on the agar plate represents a single infected bacterium. But when samples are taken at later times, infected bacteria have lysed to release progeny phages. The samples then contain free phage particles, each of which may generate a plaque when plated upon agar. A plateau is achieved when all the infected bacteria in the origin suspension have burst to release their content of progeny phage particles.

What is the average number of phage particles liberated from each infected bacterium? Since the plaque count during the latent period reflects the number of infected bacteria in the sample, whereas the plaque titer at the productive plateau gives the number of progeny phage particles released by the bacteria, the ratio

plaque titer at plateau / plaque titer during latent period

gives this number, known as the *burst size*. In the experiment shown in Figure 6.1 it is just over 100.

What is the length of the lytic cycle? If all infected bacteria burst at the same time, the rise from the low plaque count of the latent period to the plateau of the productive period should be instantaneous. The period during which the plaque count increases therefore represents the time intervening between the first and last bursts. This range of burst times is due to two factors: variation in the time at which the initial adsorption takes place; and variation in the length of the infective cycle. The time when the plaque titer begins to rise therefore establishes a minimum value for the length of the lytic cycle and the time when the plateau is reached estimates its maximum. In the experiment shown in Figure 6.1 the duration of the infective cycle can therefore be set between 22 and 32 minutes, that is, displays an average of 27 minutes.

To investigate events taking place in a single bacterium, as contrasted with a population of infected cells, it is necessary to use conditions that allow scrutiny of the progeny of a single infection. This is achieved by means of a *single burst experiment*: bacteria are infected with phage particles, unadsorbed phages are inactivated with antiserum, and the mixture is then diluted. Samples are taken, and because of the dilution each has a low probability of containing an infected bacterium. All the samples are incubated for long enough to ensure that all the infected bacteria have burst and they are then plated to determine the plaque titer. The samples yielding no plaques identify those that failed to gain an infected bacterium; and by applying a Poisson distribution it is possible to calculate the number of samples that should have included only a single infected bacterium (and also the somewhat smaller number that should have received two infected bacteria). It is possible to correlate this distribution with the range of plaque titers and from such an analysis Delbruck (1945) showed that there is appreciable variation in the number of progeny particles released from each bacterium. The single burst experiment is important (as we have seen already in the genetic analysis of lambda discussed in Chapter 5) in allowing the recombinants released from a single cell after mixed infection to be analyzed, and its application to T4 genetics is discussed below.

### *The Infective Cycle*

The T-even particle takes the usual structure of a head based on an icosahedral design with an attached tail. Figure 6.2 illustrates its construction. The head appears to be an elongated bipyrimidal hexagonal prism about 0.100 $\mu$m (1000 Å) in length and 0.065 $\mu$m (650 Å) in width; the tail is about 0.080 $\mu$m (800 Å) long and about 0.016 $\mu$m (160 Å) in diameter. The

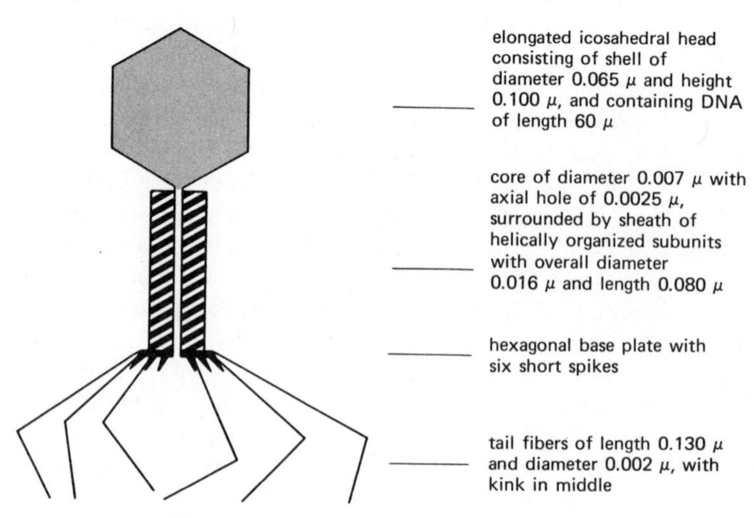


**Figure 6.2:** structure of bacteriophage T4.

tail terminates in a hexagonal base plate which has six protruding spikes and six tail fibers of length 0.130 μm (1300 Å) and diameter 0.002 μm (20 Å). The assembly from its components of the phage particle is discussed in the final section of this chapter.

The first step in infection is a random collision between phage and bacterium when the phage particle makes contact through its tail with the bacterial surface. Different phages may interact with different components of the bacterial surface. The process of phage adsorption is not well defined in detail, but it seems that the phage tail carries a lysozymelike activity which acts upon the bacterial surface to rupture some of the covalent bonds responsible for the integrity of its lipopolysaccharide components. Some of the small molecules released by this degradation then trigger a change in the structure of the tail which is responsible for injecting DNA into the cell (Weidel and Primosigh, 1958; see also Hayes, 1968).

When mature phage particles are examined, no tail fibers are seen and this has been taken to imply that they are wound around the tail. When the particles interact with the cell wall, however, tail fibers can be visualized (Kellenberger et al., 1965). There is some evidence that an interaction with zinc complexes in the cell wall may be responsible for the unwinding (Kożloff and Lute, 1957). The ionic constitution of the medium may influence the phage-bacterium interaction and some phages display requirements for specific cations, such as $Ca^{++}$ or $Mg^{++}$. Some strains of T4 require L-tryptophan for adsorption to their hosts and with these phages its presence appears to be necessary for unwinding of the tail fibers (Brenner et al., 1962). Presumably the unwinding is necessary for formation of proper contacts between phage particle and bacterium, although its role has not been accounted for in molecular terms; nor are the functions of the base plate and spikes known in detail.

The injection of DNA appears to be accomplished by a contraction of the tail. This contraction can be caused by the addition of hydrogen peroxide to isolated tails and the structure of the extended and contracted tails was visualized by Brenner et al. (1959) in a series of studies upon components purified from T2 by disaggregation of the phage particle. (It is from these results that the dimensions of the particle given in Figure 6.2 are taken.) The part of the tail that contracts is known as the sheath and the result of the contraction is to reveal a protruding core. The extended sheath has a diameter of 0.0165 μm (165 Å) with a length probably equal to that of the core of 0.080 μm (800 Å). Purified contracted sheaths can be seen to comprise hollow cylinders 0.035 μm (350 Å) in length and with a diameter of 0.025 μm (250 Å). The protruding core that is revealed by the contraction of the sheath has a diameter of some 0.007 μm (70 Å), with an axial hole of

0.0025 μm (25 Å) which is just large enough to allow passage of a duplex of DNA. If the extended sheath represents a hollow cylinder that fits closely around the core, then the volume of the sheath in its extended and contracted condition appears to remain constant. In the extended sheath some twenty-four striations can be seen, with a spacing of 0.003–0.004 μm (30–40 Å), apparently representing a helical construction. This suggests that sheath contraction may be accomplished by a compression of the subunits along the core. The result probably is to cause the core to bore through the cell wall into the bacterium, making it possible then for DNA to enter the interior of the bacterium directly from the head of the phage particle. In effect, the phage particle behaves as a microsyringe.

The DNA of the mature phage particle represents its genetic component, while the protein components represent packaging material. This conclusion was suggested by the now famous experiment of Hershey and Chase (1952), which provided one of the crucial steps in demonstrating that nucleic acid is the genetic material. In these experiments bacteria were infected with phage labeled either with $^{35}S$ (that is exclusively in protein) or with $^{32}P$ (exclusively in DNA). After mixing phages and bacteria, unadsorbed phage particles were removed from the incubation mixture by centrifugation; and then the infected bacteria were agitated in a blender to remove adsorbed phage particles from the surface. Essentially this protocol yields two fractions: empty phage coats and infected bacteria. At least 80% of the $^{35}S$ label could be recovered in the phage coat preparation (the remaining 20% probably was not removed from the bacteria by the blender step), so that very little of the phage protein enters the bacterium. But only some 30% of the $^{32}P$ label was recovered from the medium, in the form of phage particles that had not been adsorbed; the remainder of the $^{32}P$ must have entered the bacteria. When the progeny phage particles produced by the infection were examined, about 30% of the $^{32}P$ label could be recovered from them; this provides a direct demonstration that DNA of the parent phages enters the bacteria and then becomes part of the progeny phages, the pattern of inheritance expected of genetic material. A small amount of protein does appear to be injected into the bacteria together with the DNA, but it is not recovered in the progeny phages. (Formally, the limits of experimental measurement make it possible to state that less than 1% of the $^{35}S$ label appears in progeny phages; see Hershey, 1957. The injected protein(s) have low contents of S-amino acids and actually represent something less than 5% of total phage protein.) The clear implication of these experiments is that DNA is the genetic material.

The DNA of T-even phages is characterized by the presence of 5-hydroxymethylcytosine instead of cytosine and this provides one means by which the synthesis of phage DNA can be followed after infection. Since the amount of this base in the phage genome is known, it is possible by measuring the hydroxymethylcytosine content in DNA of infected cells to determine the total extent of phage DNA synthesis in terms of the number of genome equivalents that has been made (Vidaver and Kozloff, 1957). The technique of *premature lysis* allows the number of mature phage particles present at any time to be determined. When infected cells are treated with cyanide or with high concentrations of (another) phage particle, they are lysed prematurely and thus release any mature particles that have accumulated within the cell. (The lysis induced by extracellular phages provides a demonstration that they carry a lysozymelike activity.) Doermann (1951) used this technique to estimate the number of mature phages present in infected cells at various times after the start of infection. By comparing the total amount of phage DNA with the number of mature genomes, it is possible to follow the course of phage development.

Under conditions when the length of the infective cycle is 30 minutes, phage T2 DNA synthesis starts at about 6 minutes after infection and then increases until lysis (Hershey, Dixon and Chase, 1953). Infective particles begin to appear at about 12 minutes after infection, when the bacteria already contain more than fifty genome equivalents of DNA, and the number of particles increases linearly, at a rate of about five particles per minute, until the burst time (Doermann, 1951). Semiconservative replication of phage DNA should result in an exponential increase in the amount of DNA. Although DNA synthesis displays an exponential phase between 6 minutes and 12 minutes, after particles have begun to appear both DNA content and particle content display parallel linear increases for the remainder of the infective cycle. This indicates that the excess number of total compared with mature genomes remains at a constant value of about 45–50 genome equivalents. This value appears to represent a pool of replicating DNA which remains constant in size because genomes are withdrawn from it for inclusion in mature particles at the same rate that new genomes are generated by replication. Thus both the total amount of DNA and the number of mature genomes increase at a linear rate.

Since bacterial protein synthesis ceases very rapidly after T-even infection, it is possible to follow synthesis of phage proteins directly by the incorporation into polypeptide products of radioactively labeled amino acids; and again it is possible to calculate the amount of phage protein in terms of phage equivalents (Koch and Hershey, 1959). During the first few minutes after infection there is rapid protein synthesis but little or none of the product can be identified as precursors of the phage particle. The proteins synthesized during this early period represent enzymes necessary for phage DNA replication, for example such as enzymes involved in synthesis of 5-hydroxymethylcytosine, a phage DNA polymerase, a phage ligase, enzymes concerned with the degradation of the bacterial DNA. Proteins displaying the serological reactions of the mature particle appear in the infected bacteria shortly after DNA replication commences; their amount then increases until soon after the appearance of the first completed infectious particles, when a pool of precursor proteins that remains constant in size is established. The behavior of the protein precursor pool parallels that of the pool of replicating genomes, with newly synthesized protein precursors entering the pool at about the same rate that proteins are withdrawn from it to form mature particles. During this period some 60–70% of the protein synthesized in the infected cell represents precursors to the mature particle. The T-even infective cycle may therefore be divided into the same two periods as the lambda lytic cycle: during the early period preceding replication, the proteins needed for reproduction of phage DNA are synthesized; and during the late period succeeding replication, protein synthesis is directed towards assembly of the phage particle.

### *Rapid Lysis Plaque Mutants*

The classical mutations with which phage genetics started are those of plaque type and host range. Plaque type mutations may affect the size or morphology of the plaque, and the study of such variants provided one of the earliest approaches to isolating variants of T2. When all or most of the bacteria in a turbid culture are infected with T2 (or with T4), clearing does not occur for 5–6 hours, although a sample of the culture diluted into broth lyses completely after only 20 minutes. This phenomenon is called *lysis inhibition* and it appears to be due to a prevention of lysis resulting when the first particles to be liberated adsorb to infected bacteria that have not yet burst; when lysis finally does occur, the burst size is much larger than normal because the extended latent period has allowed more phage particles to be produced.

Lysis inhibition is reflected in the morphology of the plaque. When T2 (or T4) is plated on E.coli cells of strain B, the plaques that result are small and appear fuzzy due to the presence of a turbid halo; the cause of the turbidity is that only some of the infected bacteria burst before the end of the phase of active bacterial metabolism, and lysis is inhibited in the others. Wild type T2 or T4 phages that produce this type of plaque are described as $r^+$. Mutants of the $r$ (for rapid lysis) type arise in the phage population with a frequency of about $10^{-4}$; they produce large, clear plaques because the infected bacteria are not subject to lysis inhibition (Hershey, 1946).

The *r* mutants fall into three groups, designated *rI*, *rII*, *rIII*, which may be distinguished by plating on different host strains. Benzer (1957) reported that the phenotypes displayed on the host strains E.coli B, E.coli K12 (not carrying a lambda prophage) and E.coli K[λ] (a strain of K12 lysogenic for phage lambda) are

| | E.coli | | |
|---|---|---|---|
| | B | K | K(λ) |
| T4⁺ | + | + | + |
| *rI* | r | r | r |
| *rII* | r | + | − |
| *rIII* | r | + | + |

where (+) indicates the formation of wild type plaques, (r) indicates formation of the plaques typical of rapid lysis, and (−) indicates failure to form plaques. Thus the *rI* mutant type forms rapid lysis plaques on all three bacterial strains; the *rII* type is revealed as a rapid lysis mutant on E.coli B but forms a normal plaque on strain K and fails to form a plaque (or forms very small plaques similar to the type known as *minute*) on E.coli K(λ); and the *rIII* mutant forms a rapid lysis plaque only on E.coli B, appearing normal on the other two strains. For practical purposes, plating on E.coli B isolates all three types of *r* mutant and plating on the other two strains then distinguishes the three classes.

The *rII* mutant class has been the subject of an intense genetic study. Benzer (1955) mapped the first eight *rII* mutants and showed by cis/trans tests that there are two *rII* cistrons, described as *rIIA* and *rIIB*. Mixed infection is used to provide the appropriate genetic configurations. In the cis array, E.coli K(λ) cells were infected with wild type (+ +) phage together with a phage carrying both the *rII* mutations ($rII_1$ $rII_2$) whose complementation is to be tested; the plaques that are formed contain both parental types of phage, since the wild type parent is able to provide the *rII* functions missing from the double mutant. In the trans array, the two phages each carry a single mutation ($rII_1$ + ; and + $rII_2$): if the mutations lie in different genes then complementation occurs because each phage has a wild type copy of one gene, so that the same result is obtained as in the cis test; but if the mutations lie in the same gene, then both phages are defective in this function and no plaques are produced. The result of these complementation tests was to define the two groups *rIIA* and *rIIB* and analysis of further mutations isolated subsequently has confirmed that all *rII* mutants can be assigned to one of these two groups.

The difference in growth of *rII* mutants on E.coli K(λ) and E.coli B makes it possible to apply selection for recombinants and thus to isolate the rare products of recombination between very closely linked mutations; this allows an unambiguous map order to be assigned for all known *rII* mutations. Two *rII* mutants are used in mixed infection of E.coli B; both the parental phages can grow on this bacterial strain, so that the progeny consist largely of parental types, with a small number of recombinants that depends upon the distance apart of the two mutations. Samples of the progeny are then plated upon E.coli B and upon E.coli K(λ). Since all the progeny can grow on E.coli B, the number of plaques formed on this strain gives the total number of progeny; although the parental types form *r* plaques whereas wild type recombinants should form $r^+$ plaques, the proportion of the $r^+$ plaques is too small to determine directly except in cases where the two mutant sites are fairly distant. But only wild type recombinants can grow on E.coli K(λ), so that the number of plaques formed on the K(λ) cells directly identifies this recombinant class. The ratio of plaques on E.coli K(λ) to plaques formed on E.coli B therefore measures the frequency of recombination (when multiplied by a factor of two to compensate for the isolation of only one of the recombinant classes).

In a detailed mapping of the *rII* region, Benzer (1959, 1961) demonstrated that the same order of sites could be assigned by conventional mapping of point mutations and by mapping using overlapping deletions. The principle of the deletion mapping technique is that a point mutation can recombine with a deletion mutant to generate a wild type genome only if the deletion does not extend over the locus of mutation; thus by using a series of deletions, each extending for a different distance into the *rII* region, it is possible to determine the point at which the deletions extend past a site of mutation. Any given point mutation can therefore be assigned to a region lying between the ends of deletions with which it does and does not recombine; an example of deletion mapping as applied to prophage lambda is illustrated in Figure 4.8. The use of a series of 47 overlapping deletions in the *rII* region allowed each mutation to be mapped into one of the 47 regions defined by the distances between the termini of deletions of increasing length; further detailed mapping data could be obtained by crosses between point mutants.

The importance of this system was that the selection for wild type recombinants that is applied by plating on E.coli K(λ) allows even the rarest recombination events to be quantitated, so that the map distance between any pair of mutations can in principle be determined directly. Since the sensitivity of recombinant detection appeared to be limited only by the level of reversion of the mutants to wild type (recombination events occurring with frequencies greater than 0.001% should have been detectable), the failure to obtain recombination between two independently isolated mutations may be taken to mean that they lie at the same site. The 2400 mutations analyzed thus identified 304 mutant sites; the distribution of mutations among the mutant sites was not even, with some sites being mutated frequently and some infrequently. Comparison of the observed distribution with that predicted by the Poisson distribution (which gives the proportion of sites expected to be mutated each given number of times) suggested that those sites mutated more than five times represent hot spots at which mutational events occur more frequently than expected by chance. Some 60 hot spots were identified in these experiments and different mutagens generated different arrays of hot spots; the occurrence of the hot spots reflects the nucleotide constitution of the mutated site (for review see Hayes, 1968).

A critical result of this analysis was that it established unambiguously that the genetic map is linear at the molecular level, that is the results excluded the construction of genetic maps with branches. The general conclusion to be drawn from this result was that the genetic map is linear both within as well as between loci. The relationship between the *rIIA* and *rIIB* mutations defined each as a contiguous group and this is the origin of the definition of the *cistron* as the unit determined by the cis/trans test: a continuous length of genetic material all of whose sites must be present on the same molecule of nucleic acid. This definition of the gene is discussed further in Volume 1 of Gene Expression.

The concepts that mutation can take place at any base pair and that recombination may occur between adjacent base pairs also were suggested by this analysis of the *rII* system. From the number of sites that was mutated only once or twice in these experiments, it is possible to calculate that another 120 sites probably would have been identified by mutation if the experiments had been extended. This suggests that the *rII* system consists of more than 400 sites that in principle might be mutated by this approach. Since recombination can occur between any pair of adjacent sites, the unit of recombination must be the same as the unit of mutation; this concept is supported by the additivity of the genetic map, that is the frequency of recombination between adjacent sites is consistent with the overall length of the map of the *rII* region.

From this genetic analysis alone, it is not possible formally to answer the question of what is the smallest possible mutational change, especially since the only mutations that can be investigated are those causing a rapid lysis phenotype and these may represent only a proportion of the total number of sites. An obvious interpretation of these results, however, was to say that the smallest possible mutant site is probably the base pair, and recombination must therefore be able to take place between neighboring base pairs. To interpret the genetic data in terms of the physical structure of DNA requires a calibration, either by correlating recombination frequencies directly with nucleic acid distances or by determining the distance apart of mutant sites as represented by amino acids in the protein products. The variation in recombination frequencies at the molecular level makes impossible any exact and consistent correlation between map and physical distance. But a general comparison between the genetic and physical length of the *rII* region casts an interesting light on these results. By heteroduplex mapping of deletions in the *rII* region of T4, Bujard, Maizaitis and Bautz (1970) measured the length of *rIIA* as 1800 base pairs and the length of *rIIB* as 850 base pairs. This means that Benzer's genetic data concerned about 15% of the sites (base pairs) in the *rII* cistrons and from the total map length for *rII* of some eight units it can be inferred that recombination between neighboring base pairs should indeed be detectable in this system.

The *rII* system was used for the genetic analysis of frameshift mutations induced by acridines that demonstrated that the genetic code must be read in nonoverlapping triplets from a fixed starting point (Crick et al., 1961; Barnett et al., 1967; Katz and Brenner, 1975). It also provided the system in which nonsense and missense mutations initially were distinguished (Benzer and Champe, 1961, 1962) and their sequences determined (Brenner, Stretton and Kaplan, 1965). Following these genetic analyses, the obvious experiments to perform lay with analysis of the protein products of the different *rII* mutants; unfortunately it was not possible to isolate either of the proteins coded by the *rII* region and these experiments therefore were performed with other systems (see Volume 1 of Gene Expression). It is in fact only much more recently that any success has been achieved in isolating an *rII* protein product. McClain and Champe (1970) identified a peptide, released by trypsin digestion of total protein from T4 infected cells, which appeared to be derived from a protein coded by *rIIB*; this peptide was used to provide the basis for an assay to isolate the *rIIB* protein as a rather crude preparation. Infection with T4 results in several changes in the protein constitution of the cell membrane and Weintraub and Frankel (1972) and Peterson, Kievett and Ennis (1972) showed that one of the new membrane proteins is absent from cells infected with *rIIB* deletion mutants. The protein has a weight of 37,000 to 41,000 daltons, a little larger than expected from the length of the *rIIB* gene, and its presence as a protein of the membrane explains the difficulties in isolating it previously from infected cells.

### *Mutants in Host Range*

The "wild type" host range of T2 was defined (arbitrarily) by the ability of the phage to grow on cells of E.coli B. When a culture of E.coli B is plated with T2, almost all of the cells are lysed, but a small minority is not infected and subsequently grows to form colonies. These represent mutant bacteria, described as B/2, which are resistant to the phage infection (because mutations affecting the structure of the cell wall render the cell impervious to phage adsorption). The genotype of the original T2 phage population able to grow on E.coli B but unable to grow on the resistant B/2 cells is described as T2$h^+$. If these phages are plated with a culture of the resistant B/2 bacteria, there is no general occurrence of lysis, but a few plaques are produced by mutant phages that have the ability to infect the resistant bacteria. These are described as T2*h*.

To use host range as a genetic marker it is necessary to be able to distinguish the plaques produced by the T2$h^+$ and T2*h* phages. The problem is that the phage either forms or fails to form a plaque on a given host, so that a single plating does not suffice to distinguish the phage types. This led to the development of the *mixed indicator technique* devised by Delbruck (1945) and applied to phage T2 by Hershey and Rotman (1949). The T2 phage preparation is plated on a mixture of E.coli B and B/2 bacteria. The T2$h^+$ phage can form plaques only on E.coli B; thus the plaques it forms on the mixed indicator are turbid because the B bacteria are lysed but the B/2 bacteria remain uninfected and able to grow. But the T2*h* phage is able to form plaques on both the B and B/2 cell types; thus it forms clear plaques because both the bacterial types in the mixed indicator are lysed. The same technique can be used to isolate revertants from T2*h* to T2$h^+$ by plating T2*h* on the mixed indicator and isolating the rare turbid plaques that arise.

The same protocol can be used to isolate variants of the bacterial host and of phage T4; B/4 bacteria are resistant to the laboratory strain T4$h^+$ but can be infected by the mutant strain T4*h*. Thus bacteria of the B type are sensitive to both T2 and T4; B/2 bacteria are sensitive to T4 and resistant to T2; B/4 bacteria are sensitive to T2 and resistant to T4. Streisinger (1956a) demonstrated that the abilities of T2 and T4 phages to adsorb to B/2 and B/4 hosts are genetically determined by the same locus, that is, they are allelic. To avoid confusion between the genotypes of the two phages, the two genotypes of T2 are described as $h^{2+}$ and $h^2$ and those of T4 are known as $h^{4+}$ and $h^4$, instead of the less descriptive terms $h^+$ and $h$.

Using the mixed indicator technique, Streisinger and Franklin (1956) isolated a series of variants of T2. Starting with T2$h^{2+}$ they isolated the T2$h^2$ strain able to grow on B/2; and from this phage they isolated a series of strains, T2$h_i^{2+}$, T2$h_j^{2+}$ etc. again unable to grow on B/2. Obviously it is possible then to continue this cycle indefinitely. All the $h^{2+}$ variants mapped in the same region of length of about 2% recombination units; this suggests that host range may be governed by a single gene. By crossing any two of the $h^{2+}$ strains it is possible to obtain recombinants of genotype $h^2$. This demonstrates that the $h^{2+}$ strains isolated from the original T2$h^2$ strain represent mutations that may occur at different sites. But in contrast with this, all the $h^2$ strains isolated from any particular $h^{2+}$ strain are unable to recombine and must therefore represent mutation occurring at a single site. Brenner (1959) therefore suggested that $h^2$ is the *competent configuration* of the host range gene and allows the phage to adsorb to B/2; the $h^{2+}$ strains have any one of several changes in the protein which cause loss of this ability (although leaving untouched the ability to adsorb to E.coli B). This emphasizes the arbitrary nature of any assignment as wild type and mutant of the $h^{2+}$ and $h^2$ alleles.

Antiserum prepared against phage T2 contains more than one component and Lanni and Lanni (1953) first identified the tail of the phage particle as the site responsible for the interaction with neutralizing antibody (which prevents phage adsorption). Streisinger (1956a) demonstrated that the serological specificity of tail antibody is correlated with the host range; and Franklin (1961) distinguished between antiserum activities directed against the tail fibers and the sheath. These components were separated by adsorbing anti-T2 or anti-T4 sera with either the tail fibers or the sheaths purified from the phage particle. The remaining activity may then be attributed to the other component. These experiments showed that most (90%) of the killing power of anti-tail serum resides with the antifiber activity; and the host range is correlated with this component. This suggests that the *h* genes are the structural loci coding for the fiber proteins of T2 and T4 responsible for setting the host range. A complicating factor in analyzing the genetic and biochemical functions of the *h* alleles, however, is that other loci also may influence adsorption; for example, a series of mutants representing several different genes that influence the phage-bacterium interaction was isolated by Baylor et al. (1957). These activities generally remain undefined.

Phage crosses performed by mixed infection of bacteria with T2 and T4 produce some progeny in which the phenotype does not correspond to the genotype; some particles possess the T4 host range, but in a further infective cycle generate progeny with the T2 host range (Delbruck and Bailey, 1946; Novick and Szilard, 1951). Although displaying the phenotype of T4, these particles must therefore contain genetic material of the T2 genotype. A similar observation of Hershey and Rotman (1949) was that crosses of T2$h^{2+}$ × T2$h^2$ produce some particles that are phenotypically $h^{2+}$ but genetically (as judged by the host range of their progeny) are T2$h^2$. This phenomenon is described as *phenotypic mixing*.

By following the progeny of crosses between T2$h^{2+}$ and T2$h^{4+}$, Streisinger (1956b) was able to demonstrate that host range phenotypes are randomly associated with genotypes. The phenotypes of the progeny of this mixed infection were determined by plating on B/2 (plaques are formed by phages of the $h^{4+}$ phenotype) and on B/4 (plaques are formed by phages of the $h^{2+}$ phenotype). Three types of progeny were found: phage particles able to plate on B/2; phages able to plate on B/4; and particles able to plate on both B/2 and B/4. The unusual capacity of this last class to plate on bacteria resistant to T2$h^{2+}$ and on bacteria resistant to T4$h^{4+}$ must mean that these particles have the host range determinants of both $h^{2+}$ (allows plating on B/4) and $h^{4+}$ (allows plating on B/2). The existence of the $h^{2+}$–$h^{4+}$ dual phenotype thus implies that it is possible for the tail to gain subunits of each type, so that in formal terms there must be at least two sites on the tail that can undertake adsorption, and these are independently occupied by either type of determinant. This suggests that $h^{2+}$ and $h^{4+}$ proteins form a pool in the infected bacterium from which subunits are drawn at random for incorporation into complete phage particles. Similar fractions of the $h^{2+}$ and $h^{4+}$ genotypes are found in each of the three phenotypic classes; this suggests that the association of the phage genome with the tail proteins is a random event. There must therefore be a pool of T2 and T4 DNA from which genomes are withdrawn, to associate at random with protein coats and/or tails bearing either type of specificity. This general conclusion is supported by the observation of Brenner (1957) that in mixed infections using strains of either T4 or of T6 which vary in their need for tryptophan as an adsorption cofactor, the same type of phenotypic mixing is observed; particles of either phenotype may carry genomes of either genotype. The requirement for tryptophan is governed by a locus distinct from *h*, so the apparently random association between genome and proteins must apply to at least two and probably all of the tail functions. This is consistent with studies of infected cells that show the presence of pools containing many phage equivalents of nucleic acid and protein.

## *Bacteriophage Mating Theory*

The construction of a genetic map in principle requires the same approach with all organisms: mutants are crossed and the frequency of recombination between them is taken as a measure of the map distance between the mutations. (If two mutations display independent inheritance, that is, if they recombine at the maximum frequency of 50%, they must either lie on different linkage groups or must be located too far apart on the same group to display linkage; these alternatives are distinguished by extending the linkage map through testing recombination with other markers.) But the recombination frequencies determined directly by measuring the proportion of recombinants in a cross cannot simply be utilized as the map distance; it is necessary first to compensate for factors such as interference that may distort the observed recombination frequencies. The general problem in constructing genetic maps is that measured recombination frequencies are not additive; in a three point cross between the loci *a-b-c*, for example, the sum of the measured frequencies *a-b* + *b-c* may be much greater than the frequency measured directly for *a-c* recombination. The nonadditive nature of recombination frequencies has more than one cause and the purpose of introducing compensation factors is to allow construction of a map in units that are additive. The map unit therefore comprises a corrected recombination frequency. The absolute size of a map unit, that is, the number of base pairs that it represents, varies widely between genetic systems; in T-even phage crosses it is relatively high and this of course reflects the intensive recombination to which phages are subject. Distortion of measured recombination frequencies is high with the T-even phages and it is therefore necessary to introduce *mapping functions* to convert the measured frequencies into additive map units. In this section we shall be concerned with the formal analysis of bacteriophage mating, and in particular with the features of T4 crosses. The essential characteristic of T4 genetics is that it represents a population system in which several rounds of recombination occur within a single infective cycle.

One of the phenomena that seemed pertinent to early investigations of the organization of bacteriophage material was *multiplicity reactivation*. This describes the situation that is observed when ultraviolet irradiated phages are used to infect bacteria at different multiplicities of infection. When the multiplicity of infection is less than unity, so that each bacterium on average receives only a single phage genome, which is likely to have been inactivated by the ultraviolet irradiation, the burst size is very low or zero. But when the multiplicity of infection is increased, plaques are formed. Luria (1947) proposed that the phage comprises independent subunits, only some of which are inactivated in each particle; different phages are likely to suffer inactivation of different subunits. While at low multiplicities of infection the inactivation of a single subunit is sufficient to prevent completion of the infective cycle, at high multiplicities each bacterium is likely to receive more than one phage particle each of which has been inactivated in different subunits. Reassortment of subunits therefore allows active phage progeny to be formed. This was termed the recombination theory of multiplicity reactivation and, in current terms, the subunits may be equated with genes so that the formation of plaques at high multiplicities of infection represents recombination between phages carrying different mutant genes. When the theory was proposed, however, the existence of genetic recombination in bacteriophage crosses had not been discovered, and it was thought that the formation of active progeny represented reassortment of independent subunits. (More recently, Kozinski, Doermann and Kozinski (1976) have proposed that multiplicity reactivation represents recombination involving the partial genomes that are generated by replication of damaged parents, rather than a direct interaction between the parental genomes. This does not alter the basis of the interpretation that recombination and not reassortment is involved.)

The predictions of this theory were tested by Dulbecco (1952). The most important prediction in principle is that if each gene is equally susceptible to inactivation by ultraviolet irradiation, then an increase in the dose of irradiation should eventually produce a situation in which reactivation fails to occur. If each bacterium gains two phages, reactivation should cease when the dose of ultraviolet is sufficient to inactivate the same gene in each phage, making it impossible to generate wild type phages by recombination. Experimentally this can be tested by measuring the *survival curve*—the dependence of the number of phage progeny upon the dose of ultraviolet. At a given multiplicity of infection, at first no decline in survival should be seen as the dose is increased; this represents the range within which multiplicity reactivation occurs. Then a decline in survival should be seen as the dose increases; and since this represents the inactivation of phage occurring when there is no multiplicity reactivation, the rate of decline should be independent of the multiplicity of infection. This should generate a set of curves described as multiple target, because they start with a shoulder and end with a constant slope.

Dulbecco's results contradicted these predictions, for the slopes of the curves obtained at higher multiplicities of infection were smaller than the slopes at the low ($<1$) multiplicity level. This led to suggestions that recombination might not be responsible for multiplicity reactivation. It has since become apparent, however, that not all genes are equally susceptible to ultraviolet inactivation so far as multiplicity reactivation is concerned. Baricelli (1956) demonstrated that the results can be explained by supposing that the T4 genes fall into two classes, with 40% much more susceptible than the remaining 60%. Hayes (1968) noted that this can be taken to correspond to the division of genes into early and late functions. Early genes may be more susceptible because they must be expressed before replication can take place and inactivation therefore precludes the production of more copies of the genome. Late genes are utilized after replication and therefore damage, even when located at different sites in the same gene, can be repaired by recombination between the copies of the genomes produced by replication.

When it was thought that multiplicity reactivation was due to reassortment of independent subunits, models for the genetic structure of the phage generally were constructed upon the basis that each subunit must be a physically independent structure. The apparent lack of the need to postulate this type of organization during the period when it was thought that multiplicity reactivation could not be due to reassortment was one of the factors that impelled Visconti and Delbruck (1953) to construct their theory of treating phage genetics as a population system. Released from the obligation to postulate that the phage comprises many independent subunits, they endeavored instead to explain earlier genetic results on the basis that mating occurs *repeatedly* during each cycle of growth. One important result to be explained was the observation of Hershey and Chase (1951) that following triparental mixed infection (with *h, m* and *rI* markers), recombinant progeny phages are recovered that have markers derived from all three parents; this can be explained by supposing that multiple genetic exhanges take place. Another important observation explained by the theory is that there is a drift towards genetic equilibrium during the infective cycle; with distant markers the fraction of recombinants increases towards an equilibrium value of 50% when samples are examined at increasing times by inducing premature lysis. Levinthal and Visconti (1953) showed that with closely linked markers the frequency of recombinants can be increased by prolonging the time before lysis; during this period the proportion of recombinants among the mature particles increases in a linear manner. A further observation is the lack of correlation between the frequencies of occurrence of reciprocal recombinants when the products of single bursts are analyzed; instead of compelling the conclusion that recombination is not reciprocal, the Visconti-Delbruck theory allows the supposition that recombination is reciprocal but that further exchanges occurring after the reciprocal event lead to loss of the original reciprocity.

The critical basis of this theory lay in the proposal that genetic exchanges take place by recombination events in each of which a pair of genetically complete genomes is involved. The importance of this concept is obvious: it established a foundation for phage genetics which viewed the phage genome as a single entity. The postulate that repeated matings occur explains genetic observations of T4 crosses and also supports the analysis of the reproduction of phage DNA during the infective cycle. (And of course it is consistent with the explanation that multiplicity reactivation is due to recombination.) According to the population theory, parental genomes enter the bacterial cell and then multiply. Recombination takes place within the replicating pool (which as we have seen contains of the order of forty genomes); genomes are continuously withdrawn from this pool and incorporated into mature particles. This means that the progeny particles released when the bacterial cell eventually lyses represent a random selection that is only part of the total number of genomes that is synthesized. Since genomes are withdrawn after varying periods have been spent in the mating and replicating pool, different progeny therefore have suffered different numbers of rounds of mating. The theory permits analysis of this situation by supposing that matings take place pairwise and are random both with respect to choice of partner and with time. One important implication is that some matings will have no genetic consequences and (although therefore not visible in the form of recombinants) must be taken into account in analyzing the recombinants. We shall not review here the algebraic theory by which these concepts are translated into calculations of the parameters governing T4 mating, but the conclusion suggested by this analysis is that on average a phage T4 progeny genome has undergone five rounds of mating, and the mean generation time is about 2 minutes (see Visconti and Delbruck, 1953; Levinthal and Visconti, 1953). We have already seen in Chapter 5 that a similar analysis of phage lambda suggests a somewhat different conclusion; lambda progeny genomes appear on average to have suffered less than a single round of mating.

One feature of the theory is that it predicts the occurrence of negative interference, that is, the existence of a positive correlation between the likelihood that genetic exchanges will occur in two adjacent regions (see page 448). In nonmathematical terms we may say that this arises because selection for recombinants, that is, for phages that have suffered one crossover, selects for phage genomes that have been involved in more mating events than average. This form of negative interference therefore represents a statistical effect.

Another form of negative interference in T4 was observed by Chase and Doermann (1958): this is *highly localized negative interference* and is caused by the presence of multiple exchanges in very short regions of genetic material. These multiple exchanges might arise either by the occurrence of very closely linked exchanges in successive matings or by the existence of adjacent exchanges in a single mating event. In the first case one of the parents must be involved in two mating events, whereas in the second both parents participate in only a single event. Edgar and Steinberg (1958) argued that these two models predict different responses to variations in the relative frequencies of the parental phage types, since the relative multiplicities of infection should alter the occurrence of multiple exchanges produced by successive events but should not influence the frequency of exchanges generated by a single recombination event. They therefore took the absence of any effect of changes in relative concentrations of parental phages to indicate that multiple exchanges occur in single matings. Steinberg and Edgar (1961) noted that another difference should be displayed when comparing biparental crosses with triparental crosses, since only one exchange is needed to generate a biparental recombinant but two exchanges are needed to generate a triparental recombinant; the results again suggested that multiple exchanges are generated by a single event. Steinberg and Edgar (1962) observed further that the recombination resulting from the clustered genetic exchanges is not necessarily associated with recombination between outside flanking markers. These results suggest that highly localized negative interference is a consequence of the mechanism of recombination; in Chapter 5 (see page 451) we have already discussed this phenomenon as observed in phage lambda, and in both phages λ and T4 the multiple exchanges appear to represent events occurring within the length of hybrid DNA that is the immediate product of the interaction between the parental DNA molecules and which may or may not be associated with recombination between markers outside the heteroduplex region (see Figure 5.1).

Construction of a genetic map requires the use of a unit that is additive. Recombination frequencies are not additive because of the occurrence of double crossovers; thus in three point crosses of the form $a^+ b^+ c^+ \times a^- b^- c^-$, the double recombinants $a^+ b^- c^+$ and $a^- b^+ c^-$ are scored as parental types when examined for *a-c* recombination, although genetic exchanges have occurred in this interval. The occurrence of double crossovers therefore reduces the frequency of recombination observed over the *a-c* interval, with the result that the sum of the recombination frequencies measured for markers *a-b* and *b-c* is greater than the frequency directly measured for markers *a-c*. In eucaryotes this effect is to some extent compensated by positive interference, which reduces the frequency with which double crossovers occur (because the occurrence of one crossover inhibits the formation of another in the same region). This makes recombination frequencies almost additive over reasonable distances (up to about twenty-five map units). In phage crosses, recombination frequencies depart from additivity over all distances: over larger intervals, negative interference rather than positive interference is seen, so the frequent occurrence of double crossovers enhances the nonadditive effect; and over smaller intervals, highly localized negative interference greatly exacerbates this effect.

Recombination frequencies should be additive when only one genetic exchange occurs in each interval. In principle this means that a map can be constructed by measuring recombination between markers whose distance apart is small enough to exclude the occurrence of more than one genetic exchange. The negative interference seen in phage crosses makes it impossible to fulfill this condition by experiment, because interference increases as the distance between markers is reduced (as shown in Figure 5.8 for lambda markers). It therefore becomes necessary to treat observed recombination frequencies by the introduction of mathematical functions that allow for negative interference by calculating the frequencies that would be measured if it were possible to determine recombination frequences between markers close enough to be separated only by single exchanges.

The Visconti-Delbruck theory accounts for the generalized negative interference that is a statistical consequence of the need to consider phage crosses as a population system, but does not compensate for the highly localized negative interference resulting from the nature of the recombination process. In fact, two types of modification to the Visconti-Delbruck theory must be made to allow an additive map to be constructed for phage T4. The linkage group turns out to be circular (see below) and a modification of the mating theory which allows for circularity was introduced by Stahl and Steinberg (1964). Starting from the Visconti-Delbruck theory as modified for circular maps, Stahl, Edgar and Steinberg (1964) then introduced *mapping functions* to compensate for highly localized negative interference. These functions endeavor to convert all measured frequencies to map units that are additive; they are based upon various assumptions concerning the relationship between very closely spaced recombination events, such as the total distance that is involved in a multiple exchange, the spacing of individual exchanges within the cluster, and so on. It is important to emphasize that the criterion for testing the various functions that have been constructed does not lie with investigation of these assumptions but rather with their effectiveness in generating an additive map. (However, the molecular basis of recombination is now much better established and it should in principle therefore in due course become possible to take an alternative approach and construct functions based upon measured parameters that govern the relationship between individual events in a cluster of multiple exchanges.) The map unit therefore represents a theoretical value relating the distance apart of two markers and is not an experimental measurement.

The best of the mapping functions allowed an additive map to be constructed with a total number of 2500 units. This is much greater than the number that would be estimated simply by extrapolating from the frequency measured over a region about the size of a single gene. The most distant markers in the *rII* region must lie about 2500 base pairs apart and recombine with a frequency of about 8%; on this basis the 166,000 base pairs of the T4 genome would correspond to about 500 recombination units. The most distant markers in the lysozyme gene should be somewhat closer together, with a separation of about 450 base pairs, and display about 3% recombination; this would correspond to a total map length of about 1000 units. From these two examples it is clear that the total map length becomes greater when the map is constructed in terms of smaller intervals; this indeed is the effect responsible for extending the corrected map up to 2500 units, since the principle upon which it is constructed lies with visualizing recombination in terms of small regions. Precisely the same effect is seen in phage lambda when the interference index is used to correct observed recombination frequencies for the occurrence of negative interference (see page 449).

Because of the variation introduced by negative interference, observed recombination frequencies cannot be equated with physical distances along DNA. The usefulness of the map unit is that it provides an additive measure of the relative distance between markers; this is tantamount to saying that it should reflect their distance apart. This equivalence always is subject to the qualification that distortions in recombination frequency may occur at the molecular level due to the effects of particular markers; it is therefore not possible to maintain a strict relationship between recombination and nucleotide distance. But the linkage map constructed by correcting observed recombination frequencies for negative interference should in general display a relationship between markers that is similar to their separation on the nucleic acid genome. The linkage map of phage lambda seems to correspond well with the physical map constructed by measurements of heteroduplexes (see Chapter 4). The T4 linkage map fits well with the heteroduplex map, except for an apparent expansion on the genetic map of one region (covering genes *34* and *35*: see below for the construction of the T4 map), which appears to be due to the presence there of recombinational hot spot(s) (see Beckendorf and Wilson, 1972). Phage T2 appears to lack any hot spots in this region and there appear therefore to be no large scale distortions in the relationship between its genetic map and heteroduplex map (Russell, 1974). The distance apart of T-even markers measured in map units thus seems to provide a reasonable estimate of their physical relationship.

The size of the map unit can be assigned an average value for any genetic system, but varies widely between systems. In phage T4, the correspondence between a map of 2500 units and a genome of 166,000 base pairs suggests a formal equivalence of 1 unit equals 65 base pairs. In phage lambda, a map of 70 units corresponds to a genome of 46,500 base pairs, equivalent to 1 map unit for every 650 base pairs. In E.coli, measurements of recombination between unselected and selected markers during conjugation (see Chapter 2; page 94) correspond to a map of about 1800 units for the entire chromosome of $4.6 \times 10^6$ base pairs; this is equivalent to about 1 map unit for every 2500 base pairs. The map unit as such can therefore be defined only in genetic terms as a corrected recombination frequency.

Absolute recombination frequencies per unit distance cannot be determined, of course, but comparison of the frequency of recombination between markers that lie the same physical distance apart in different systems suggests that these variations in the size of the map unit reflect differences in the intensity of recombination in each system. Recombination between two markers of lambda probably lying about 2400 nucleotides apart in genes *A* and *B* occurs at a frequency of about 1.2% (when corrected for negative interference this becomes 3.5 map units: see Campbell, 1971). Recombination in T4 between two markers about 2500 base pairs apart in *rIIA* and *rIIB* occurs with an observed frequency close to 8% (correction for negative interference converts this to a value of 40 map units according to the function of Stahl et al., 1964). Recombination events in T4 therefore occur with greater frequency than those in phage lambda, 8% to 1.2% as measured for a distance of 2500 base pairs. The recombination frequency in any system is governed by the activities of whatever enzymes are involved; in a phage system this includes both phage and host functions and therefore is dependent upon the genotype of both phage and bacterium. (We have already described in some detail in Chapter 5 the reduction in recombination frequencies that results when some of the systems available for lambda recombination are inactivated.) The recombination frequency in phage crosses also is influenced by the number of rounds of mating that is permitted during the infective cycle. Several factors therefore are responsible for establishing the different relative frequencies of recombination that are observed in different phage systems.

In summary, the basic concept of the Visconti-Delbruck theory, that phage genetics represents a population system, provides the critical principle that makes it possible to analyze phage crosses. With T4 it is necessary to modify this theory to allow for the circularity of the genetic map and the phenomenon of highly localized negative interference. This allows an additive genetic map to be constructed which correlates well with the information available about the relative physical locations of markers. Nonetheless, it remains true that even now not all the assumptions of the theory can be verified; for example, the very nature of the phage system makes it impossible to determine whether recombination is reciprocal, as it is assumed to be. It is necessary therefore to remember that although the theory has in practice proved to be an effective method for analyzing phage crosses, it cannot be taken to provide a definite account of the phage life history.

### *Circular Linkage Map of T-even Phages*

The early mapping of T2 and T4 began with three types of mutant: *r, m, h*. The *r* mutants are recognized by their plaque morphology, as are *m* mutants (which form minute plaques), and the *h* mutants are recognized by their host range. The use of the *r* and *h* mutants has the advantage that all four combinations of the genes $r^+$, $r$, $h^+$, $h$, can be recognized individually by plating on a single mixed indicator plate; using this system Hershey and Rotman (1949) showed that recombination takes place between the *r* and *h* loci of T2 and demonstrated further that it conforms to the same general rules established for eucaryotes, that is, linkage can be established between loci and is a function of the position of the locus rather than of whichever particular mutations in it are used in the cross. Doermann and Hill (1953) confirmed that the same conclusions could be applied to T4. These early experiments were consistent with, but did not unambiguously support, the construction of a linear linkage map, although they made it evident that the mutants known at this time fell into only a small number (two or three) of groups, each apparently linear in organization.

By using a more sensitive test for linkage, Streisinger and Bruce (1960) demonstrated that all known markers formed a single linear linkage group in both T2 and T4. The principle of this test was to perform the cross

$$a^- \; b^- \; c^-$$

$$\times$$

$$a^+ \; b^+ \; c^+$$

and isolate the recombinants $a^- \; b^+$ and $a^+ \; b^-$, which must have been formed by a crossover in the region between the *selected* markers *a* and *b*. The recombinants then were examined to determine which allele of the *unselected* marker *c* was present. If *c* is not linked to either *a* or *b*, the $c^+$ and $c^-$ alleles should be represented with equal frequency in the two recombinant classes. But if *c* is linked to one of the markers, for example *b*, then the $c^-$ allele should predominate in the $a^+ \; b^-$ recombinants and the $c^+$ allele should be found in the majority of $a^- \; b^+$ recombinants. By this means, linkage could be demonstrated between all known markers, suggesting that the phage possesses a single chromosome.

The question of whether the T4 genetic map is linear or circular was investigated by Streisinger, Edgar and Denhardt (1964), using a variation of this sensitive linkage test. If the T4 genetic map is represented as the linear array

where $c$ is at the far end from markers $a$ and $b$, the linkage between $a$ and $c$ depends upon the form of the map. If the map is linear, $c$ must be closer to $b$ than it is to $a$. But if the map is circular, $c$ should be closer to $a$ than to $b$. These alternatives were distinguished by performing the cross $a^- b^+ c^- \times a^+ b^- c^+$; when the $a^- b^-$ recombinants were selected, 65% possessed the $c^-$ marker and 35% the $c^+$ marker. This means that $c$ is more closely linked to $a$, which suggests that the map must be circular. We should emphasize that this does not imply that the physical state of the T4 chromosome is circular and this will be discussed in the next section.

A crucial step in extending the genetic analysis of phage T4 was the introduction of conditional lethal mutations by Epstein et al. (1963) and Edgar, Denhardt and Epstein (1964). As we have already seen, conditional lethal mutations are lethal only under one set of restrictive, non-permissive conditions; but they behave as wild type under other, permissive conditions (which must, of course, be used to perpetuate the stock carrying the mutation). The two principal classes of conditional lethals comprise temperature sensitive and nonsense mutations.

Temperature sensitive mutations usually can grow and form plaques normally when propagated at 25°C but are unable to do so at 42°C (although it is possible to isolate cold sensitive mutants of bacteria which propagate normally at a higher temperature and display the mutant phenotype at a lower temperature, for example, mutants in ribosomal assembly: see Volume 1 of Gene Expression). Wild type T4 plates well at either 25°C or 42°C, although the infective cycle may be reduced in length by the higher temperature, whereas the temperature sensitive mutants display pronounced differences in plating efficiencies at these temperatures. The procedure used to isolate temperature sensitive mutants of T4 was to expose a phage preparation to a chemical mutagen, grow at low temperature to allow mutant particles to segregate, and plate with sensitive bacteria at a low multiplicity of infection at 25°C, when only small plaques develop: upon transfer to 42°C, the wild type plaques increase in size (because growth conditions are better) but the mutant plaques fail to develop further (see Edgar and Lielausis, 1964).

Nonsense mutants fall principally into two classes: amber (UAG codon) and ochre (UAA codon). (The UGA codon represents a third, less utilized class among the mutants.) All three codons are signals for the termination of translation, so that mutations creating them within a gene cause the premature termination of protein synthesis, usually resulting in a complete lack of function. But a nonsense mutation may be suppressed in an E.coli strain that possesses a mutant tRNA able to recognize the chain terminating signal; by inserting an amino acid in response to the nonsense codon, this suppressor tRNA allows protein synthesis to continue. The permissive bacterial strain carrying the suppressor tRNA is said to be $su^+$, while the wild type E.coli cell that terminates translation at the nonsense codon is described as $su^-$. Amber mutants are suppressed more effectively than ochre mutants and this made them more useful for genetic analyses of the type performed with T4 (the relative consequences for the bacterial cell of suppressing different nonsense codons are discussed in Volume 1 of Gene Expression).

Amber mutants of T4 originally were characterized by their ability to propagate on a strain of E.coli K12 now known to be $su^+$, whereas they were unable to grow on a strain of E.coli B that is $su^-$. These T4 mutants and the *sus* mutants of lambda that we have discussed in Chapter 4 belong to the same class since they display precisely the same ability to grow on the K12 strain and inability to grow on the B strain. The T4 mutants usually are described as *am* and it may be taken that an X*am* mutant is in effect $X^-$ when grown on a nonpermissive ($su^-$) host strain, but is $X^+$ when grown on a permissive ($su^+$) host. However, this situation is qualified by the observation that some amber mutants are *leaky*, that is some residual gene expression occurs on the $su^-$ strain, presumably due to a low level propagation of protein synthesis past the mutant site; to ensure complete absence of gene function it is therefore sometimes necessary to construct double mutants that possess two amber mutations in a gene. Also we should note that the amino acid inserted at the site of nonsense mutation in the permissive strain may be different from the amino acid originally present in the wild type protein. This means that suppression of nonsense mutations does not always restore complete wild type activity. Similar phenomena are observed with certain temperature sensitive mutants, where the mutant protein may have some residual activity at the nonpermissive (high) temperature and is not always fully wild type at the permissive (low) temperature.

Genetic mapping of conditional lethal mutations may be accomplished simply by comparing the ability of progeny to grow on permissive and nonpermissive hosts. The mapping of nonsense mutations in phage lambda by this approach has already been discussed in Chapter 4. Figure 6.3 illustrates the general protocol that is used in such experiments. A cross is made between the mutants by mixed infection of a permissive host; this allows both parental phage types and the recombinants to grow. The nature of the progeny then is determined by plating on two strains, one permissive and one nonpermissive. All progeny can grow on the permissive host, where the number of plaques therefore estimates the total yield. Only the wild type recombinants can grow on the nonpermissive host,

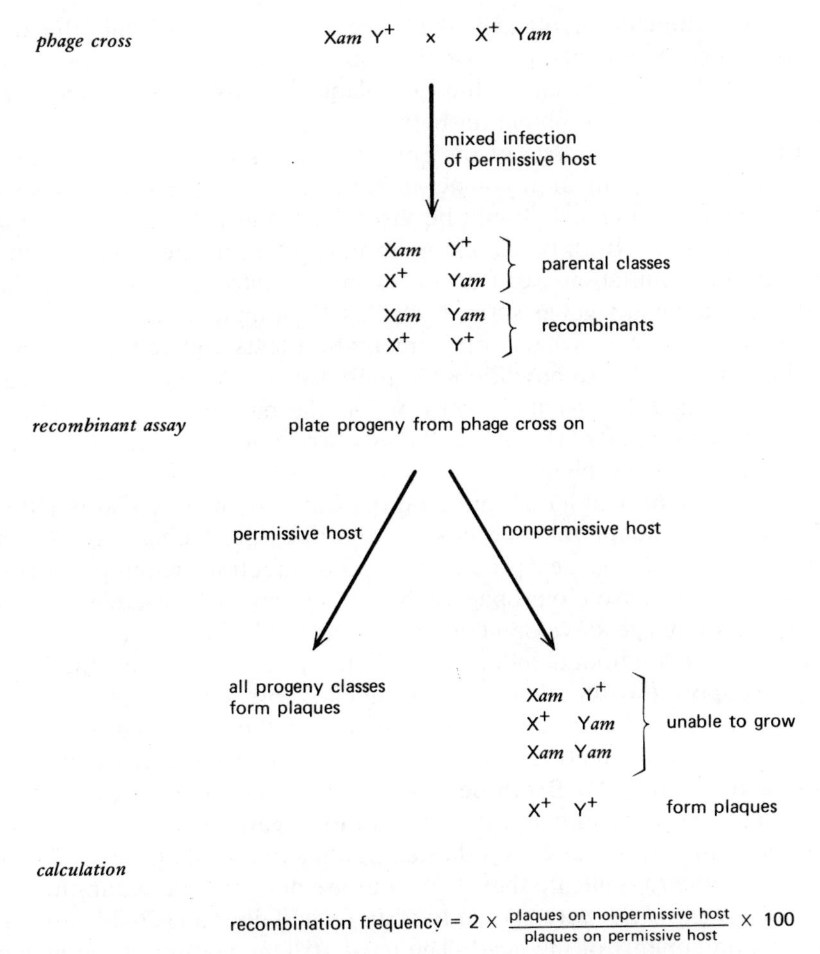


**Figure 6.3:** mapping conditional lethal phage mutations. Two mutants are crossed by mixed infection of a permissive host in which they can both grow (these are shown as ambers in the figure, but the same technique applies to temperature sensitives). The progeny contain both parental classes and both recombinant classes. To distinguish the recombinants, samples of the progeny are plated on a permissive host and on a nonpermissive host; the number of permissive plaques measures the total progeny and the number of nonpermissive plaques measures the wild type recombinants. Their ratio, adjusted by a factor of 2 to compensate for the selection of only one of the two recombinant classes, gives the recombination frequency.

where the number of plaques therefore corresponds to half the total number of recombinants (because the double mutant recombinant class is not assayed). Comparison of the two plaque counts therefore measures the frequency of recombination between the two mutations.

An important advantage of the conditional lethal system is that these mutations may occur in any gene and are not restricted to the types of character that previously could be visualized (those such as host range and plaque type). Both types of conditional lethal, temperature sensitive and nonsense mutations, can occur at only a restricted number of sites within a gene; many genes suffer both classes of mutation. Taking advantage of the ease with which complementation tests and genetic mapping can be performed with conditional lethals, Edgar, Denhardt and Epstein (1964) identified 47 essential genes of T4; the distribution of mutations among genes suggested that these should correspond to at least half of the essential loci of the phage.

When a conditional lethal mutation is located in an essential function, infection of a nonpermissive host is abortive (producing $<1\%$ of the number of viable phages produced by the infection under permissive conditions that is used to propagate the mutant) and it is possible to define the stage of phage development that is blocked. Epstein et al. (1963) characterized conditional lethals in these forty-seven genes for their ability to support the functions of host nuclear breakdown, phage DNA synthesis, synthesis of tail fiber antigen, synthesis of morphological components of the mature particle visible by electron microscopy, and lysis of infected cells. By these criteria they were able to classify the genes into two general classes: mutants in the early genes, *1, 32, 39-47,* are unable to perform DNA synthesis; mutants in the late genes, *2-31* and *33-38,* are able to replicate their DNA but are defective in maturation. Of the late genes, *34-38* were shown to code for tail fiber and *20-24, 30-31* to specify components of the head. The most striking feature of the map of these genes is the functional clustering, with related functions often occupying adjacent locations.

Since these genes were identified and mapped, further essential functions have been identified. The original series of conditional lethals was used to number the genes in clockwise array, *1-47*; the essential genes added to this map since then also have been assigned numbers, but of course these are located at various positions on the map. Figure 6.4 shows the more recent map of T4 functions according to Wood (1974), which contains 64 essential genes. Mutants defining the early genes fall into three general classes: DNA negative (sometimes known as *DO*), which are unable to synthesize any DNA; DNA arrested (often described as *DA*), which start DNA synthesis but cease after only a short time; and

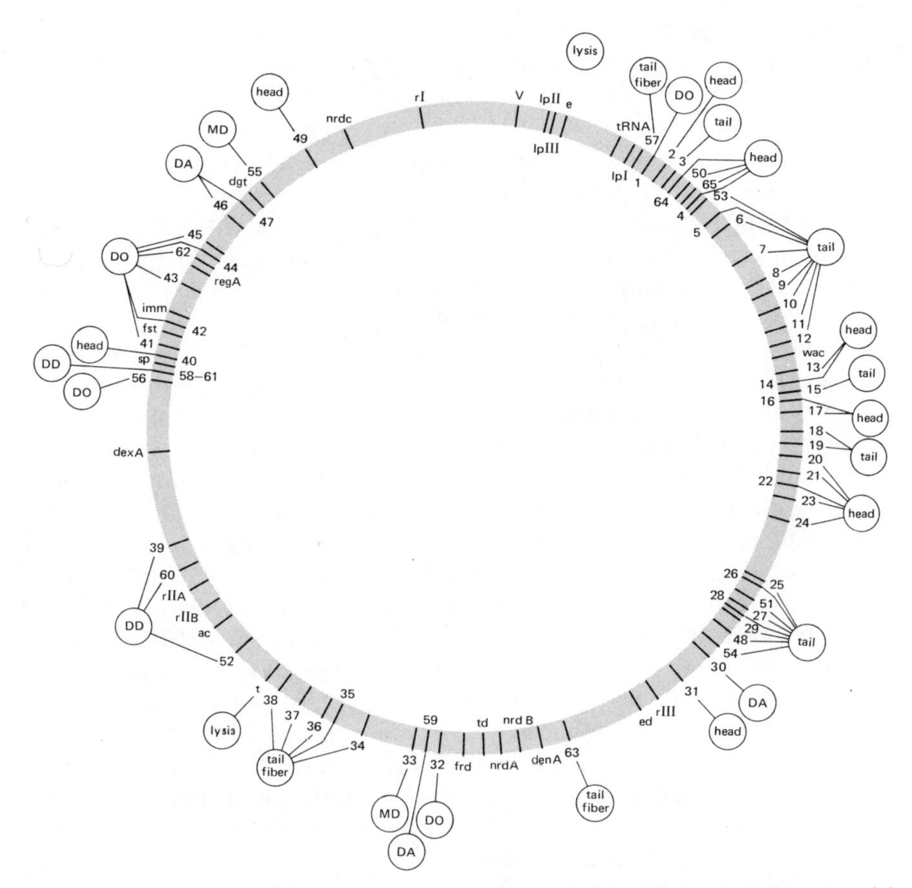


**Figure 6.4:** genetic map of phage T4. Genes that are numbered represent the essential functions identified by conditional lethal mutations; other genes are named according to the activities of their products. The early mutants that are defective in replication are shown as: DO = no DNA synthesis; DA = arrested DNA synthesis; DD = delayed DNA synthesis. The late mutants are divided into those that are: MD= maturation defective; head = unable to synthesize head structure; tail = unable to synthesize upper part of tail structure; tail fiber = unable to synthesize the fiber. Table 6.2 summarizes the loci coding for identified early phage enzyme activities.

DNA delayed (referred to as *DD*) which do synthesize phage DNA, but only after a substantial delay. Mutants in the late genes fall into the categories: maturation defective (abbreviated to *MD*), which synthesize DNA but do not express any late functions; head mutants, which fail to manufacture the phage particle head; and tail mutants, which are unable to produce the tail structure. Some of the morphological mutants can be more exactly defined by examining the structures that they produce. As is clear from the map, the essential genes by no means occupy the entire length of the map. Some nonessential functions have been identified and are shown on the map. What (presumably nonessential) functions are exercised by the remaining regions, which appear to occupy of the order of a quarter of the genetic map, is not known.

Most of the studies of T-even phages have utilized T4, but phages T2 and T6 appear to be very similar as judged by both genetic and physical studies. By developing a system of conditional lethal (amber) mutants for T2 and T6, Russell (1974) identified 52 essential genes in T2 and 45 essential genes in T6. By testing the abilities of mutants in these genes to be complemented by T4 in mixed infection, it was possible to show that all but 3 of the T2 genes and all but 3 of the T6 genes are homologous with genes of T4. This does not prove that the corresponding genes are identical in all three phages, but it demonstrates that they are at least sufficiently closely related for their products to substitute for one another. The maps of the T2 and T6 functions both display the same order and spacing of genes as T4. The three phages thus are largely similar, although they do have some differences in genetic constitution. One of the differences between the phages lies in two of the genes, *37* and *38*, concerned with tail fiber assembly; this is consistent with the early observations that the three phages display different host ranges. To argue in teleological terms, we may suppose that the occurrence of phage variants with different adsorption specificities prevents bacterial mutants from arising that are at once resistant to all three phages; the existence of three host range types may therefore reflect selective pressure on the T-even phage system to maintain diversity of adsorption.

The close genetic relationship between the three T-even phages is confirmed by the heteroduplex mapping of their genomes performed by Kim and Davidson (1974). The heteroduplexes formed by denaturing and then renaturing together any pair of the phage DNAs show extensive homology, although with some substitution loops and deletion loops; there is no evidence for any sequences of partial homology, so these differences appear to represent the substitution or addition of new sequences and not an evolutionary divergence through the accumulation of mutations. The region of greatest homology lies in the late genes, which specify the components of the mature particle, although there is a lack of homology in the two genes *37* and *38* that specify the fiber components responsible for host range; this corresponds to the difference seen in the genetic complementation tests. Many of the regions of nonhomology are larger in size than a gene; either these do not include the essential functions shown by complementation to be homologous or when they do the substituted genes must code for products able to fulfill the same function.

The extensive homology between the genomes of the three T-even phages implies that recombination between them to generate new variants should be able to take place during a mixed infection. But the number of recombinant progeny emerging from a cross between any two of these phages appears to be limited. This is a consequence of the phenomenon of *partial exclusion,* in which one infecting phage reduces the number of progeny produced by a different but related phage that infects the same bacterium (see Streisinger and Weigle, 1956). Russell and Huskey (1974) reported that when T2 is excluded by T4, the markers of T2 appear in the progeny at very different but characteristic frequencies. The recombinants that carry T2 markers are not themselves subject to exclusion by T4 when a second mixed infection is performed. Russell and Huskey therefore suggested that exclusion is due to the presence of elements in the T2 genome that are the subject of the excluding action exercised by T4. These elements lie close to the markers that are least often found in the progeny, but T2 markers lying more distant from these elements may be separated from them by recombination with T4 to generate the hybrid progeny that are not subject to exclusion. The map locations of the T2 markers that are the most efficiently excluded in these crosses suggest that there are at least three such elements in the T2 genome. The nature of the excluding action is not known and the T4 genes that mediate it have not been identified; it is tempting to speculate, of course, that the exclusion represents some action directed against DNA sequences such as nuclease attack. Partial exclusion is seen also in other T-even mixed infections: T6 appears to exclude T2; and T4 appears to exclude T6. Similar results were obtained in experiments using some newly isolated phages that appear (although they are not yet fully characterized) to belong to the T-even series.

### *Structure of the T-even Chromosome*

The construction of a single linkage map containing all known T4 genes implies that the phage chromosome must comprise a single molecule of nucleic acid. This conclusion is supported by direct measurements of the DNA extracted from mature phage particles. Early experiments to determine the number of nucleic acid molecules in the phage genome were based upon the use of sedimentation to determine the size of DNA extracted from the particle; by comparing this size with the amount of DNA known to be contained in the phage, it should be possible to deduce the number of molecules. These experiments identified DNA sedimenting at a rate corresponding to a size of not much more than $10^7$ daltons, compared with the particle content of more than $10^8$ daltons; this was taken to suggest that the genome comprised several short molecules; but shortly after, it was realized that large DNA molecules are highly sensitive to breakage caused by shearing, and it is necessary to use more gentle methods of isolation. The first demonstration that T2 DNA is a single unit was provided by the autoradiographic experiments of Rubenstein, Thomas and Hershey (1961) and Davison et al. (1961). After DNA was released from phage particles by osmotic shock, the number of units measured by visualizing the emission of $\beta$-particles from a $^{32}P$ label in the DNA corresponded to one per phage. A more sophisticated autoradiographic technique was used by Cairns (1961) to measure the length of T2 DNA, which was determined as 52 $\mu$m. A technique for allowing DNA to be spread and platinum shadowed for electron microscopy allowed Kleinschmidt et al. (1962) to measure a length of 49 $\mu$m. The more recent experiments of Kim and Davidson (1974), utilizing electron microscopy of DNA spread in the presence of cytochrome c, suggest a length of about 60 $\mu$m, corresponding to a weight of $110 \times 10^6$ daltons or 166,000 base pairs (see below); actually the weight may be greater than that calculated from the measured length by assuming that the DNA consists of the usual nucleotides, because T4 DNA is glucosylated.

Is the T4 chromosome a linear or circular duplex of DNA? The circularity of a genetic map may reflect the existence of a circular chromosome, a relationship displayed, for example, by E.coli, but does not mean that this is the only form possible for the DNA. An alternative is the existence of linear chromosomes that have circularly permutated sequences. *Circular permutation* means that each copy of the phage chromosome is a linear molecule of the same length, but the order of its genes varies as though each chromosome were generated by the introduction of a random break in a circular structure (of course, this is not necessarily the mechanism used to generate the physical structure). Thus if we suppose that a chromosome has six markers *a-f*, the circularly permutated variants might have the linear sequences

a b c d e f
b c d e f a
c d e f a b
d e f a b c
e f a b c d
f a b c d e

Although each chromosome is linear, the genetic map would be circular because any markers may occupy the terminal locations. For example, in the first chromosome *a* and *f* are terminal; but in all other chromosomes they are adjacent. The same is true of any pair of adjacent markers.

The DNA released from T4 particles appears to be linear. A model of circular permutation predicts that the ends of the linear chromosomes are random in sequence; instead of representing particular loci, they may constitute any regions of the genome. This prediction has been tested by two types of experiment. Thomas (1963) prepared $^{32}P$ labeled single strand fragments of 300 nucleotides derived from the right and left ends of the linear molecules; a similar set of molecules labeled with $^{14}C$ was derived from the entire molecule. The sequences present in these preparations were compared by measuring the relative retentions of the $^{32}P$ and $^{14}C$ labels when a mixture of the two sets of fragments was hybridized with complete single stranded molecules. With a control from phage T5 (whose chromosome and linkage group is linear), there was preferential retention of the $^{14}C$ label; this indicates that fewer sequences are present in the terminal ($^{32}P$) fragments than in the complete ($^{14}C$) fragments so that the chromosome termini must have unique sequences. But with T2 DNA the same extent of hybridization was displayed by both radioactive labels; this means that both preparations have the same set of sequences, that is, the terminal regions must contain all the sequences of the genome.

If different sequences are present at the ends of each individual linear T-even chromosome, denaturation and renaturation of the DNA should lead to the formation of circular molecules. The reactions leading to the formation of circular structures are illustrated in Figure 6.5. If the population of chromosomes in T-even phage particles is circularly permutated, the overwhelming probability is that a denatured single strand will react with a complementary single strand representing a different permutation. In this case, when a duplex is formed the break between the markers that are terminal on one strand will be bridged by the continuity of the other strand. In the first of the circular molecules depicted in the figure, the outside strand is derived from a linear molecule with termini between *a* and *f*, but the complementary sequences *a′* and *f′* are continuously joined on the inside strand; similarly the break on the inside strand lies between *c′* and *d′* but the complementary sequences *c* and *d* are covalently linked on the outside strand. In different circular molecules the breaks on each strand lie in different positions, but all the circles should have the same contour length, which represents a unit genome. When Thomas and MacHattie (1964) examined the products generated by denaturation and renaturation of T2 DNA, they observed a large number of circular molecules by electron microscopic visualization; these were not seen in a control of T5 DNA where the products remained linear. The contour length of the circles was 55 $\mu$m.

How are circularly permutated linear chromosomes produced? Since

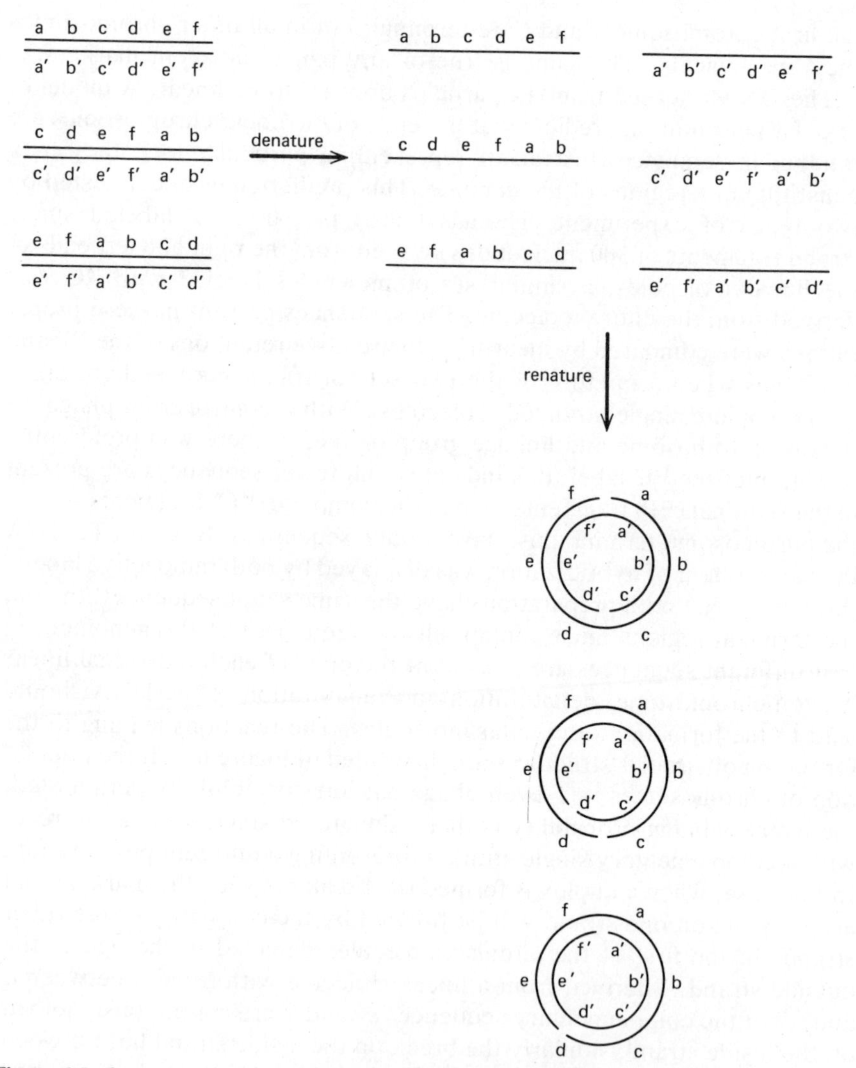


**Figure 6.5:** formation of circular molecules by denaturation and renaturation of circularly permutated linear molecules. The markers *a-f* identify one strand of the phage genome and the complementary sequences on the other strand are written as *a'-f'*.

the chromosome termini are not unique, some mechanism other than recognition of specific DNA sequences (analogous to the *cos* sites of lambda: see Chapter 5) must be involved. Also it is clear that the phage genome cannot remain in its mature state after infection, for in a mixed infection any recombination event would immediately generate products one smaller and one larger than the unit length. The nature of the process by which circularly permutated linear chromosomes are perpetuated is suggested by the observation of *terminal redundancy* in T-even phage: each phage chromosome is longer than one unit genome and possesses a duplication of the same sequence at each terminus. Since the chromosome is circularly permutated, each linear genome has a different terminally redundant sequence. Thus if the genome consists of the markers *a-f*, we may describe the sequences of a series of linear chromosomes as

a b c d e f a
b c d e f a b
c d e f a b c
d e f a b c d
e f a b c d e
f a b c d e f

In the first chromosome the sequence *a* is terminally redundant, in the second chromosome the redundant sequence is *b*, and so on. Chromosomes with these properties might be generated by cleaving linear molecules longer than a unit genome from concatemeric molecules comprising several genomes. (We have seen already in Chapter 5 that a *concatemer* consists of more than one genome covalently linked end to end in the same molecule; another term used to describe this structure is *concatenate*.) The generation of mature T-even genomes may therefore be represented as

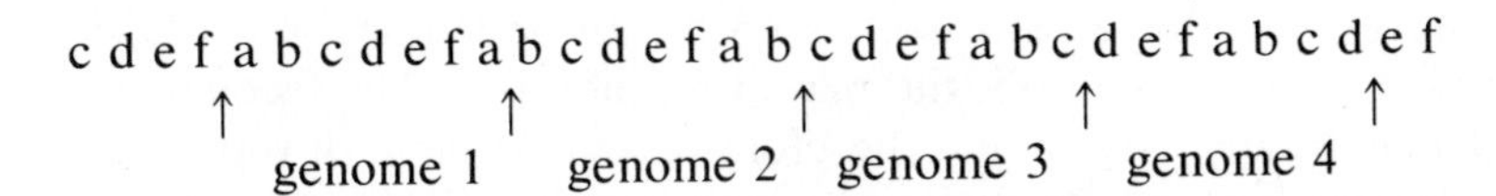


where the vertical arrows identify the sites of cleavage and their locations are controlled only by some event that depends upon the *amount* of DNA between them. The four genomes released by these cleavages are circularly permutated and terminally redundant.

The origins of the experimental investigation of terminal redundancy lie in the observation of heterozygosity in the progeny of mixed infections. Analyzing the heterozygous genomes is complicated by the discovery that there are two types of heterozygosity in the T-even phages. Heterozygosity in T2 was first observed by Hershey and Chase (1951) when they found that after a mixed infection with T2$r^+$ and T2$r$, about 2% of the progeny give rise to mottled plaques in a subsequent round of infection. Phages isolated from these mottled plaques consist of about equal proportions of $r^+$ and $r$ types. The cause of the mottling thus seems to be the presence in the same plaque of phages of two different genotypes; this implies that the original progeny phage must be heterozygous for this character, that is, carries the genetic information of both parent types. The phenomenon is not just a characteristic of the $r^+/r$ system, for it was observed at exactly the same frequency when the parental phages carried $h^{2+}$ and $h^2$ markers. The great majority of heterozygotes recovered from a two factor cross are heterozygous for only one of the markers, unless the two loci are closely linked. The length of the heterozygous region must therefore be comparatively short.

The suggestion that heterozygosity may be of two types in the T-even phage was made by Sechaud et al. (1965): internal heterozygosity is due to the formation of hybrid DNA as a consequence of the recombination process; whereas terminal heterozygosity is due to the presence of different parental alleles in terminally redundant regions. The two types of heterozygosity were distinguished by taking advantage of an observation previously made by Nomura and Benzer (1961): in crosses involving *rII* deletion mutants the frequency of heterozygosity is only about one-third of that seen in crosses involving *rII* point mutations. Sechaud et al. suggested that the reason for this reduction is that the *rII* deletion mutants are unable to form internal heterozygotes and therefore retain only the terminal heterozygosity.

Terminally redundant heterozygotes should be formed (and lost) by recombination. A phage genome is effectively diploid for the markers contained in its region of terminal redundancy, and in mixed infection the two ends of the mature genome may be derived from different parents whenever they have been separated by an odd number of genetic exchanges between the opposite parental genotypes. The frequency of terminal heterozygosity should therefore be independent of replication. But the internal heterozygosity represented in hybrid DNA should be lost by segregation of the DNA strands when replication occurs. When replication is inhibited, therefore, the frequency of internal heterozygotes should increase with time (because they continue to be generated by segregation but are no longer removed by replication), whereas the frequency of terminal heterozygotes should remain constant. With FUdR was used to inhibit replication in infected cells, the treatment caused an increase in the frequency of heterozygosity seen in mixed infections with *rII* point mutants (the heterozygosity represents both internal and terminal, presumably with only the internal component increasing in frequency); but FUdR had no effect upon the frequency of heterozygosity displayed when an *rII* deletion was used in mixed infection with wild type phage.

If measurement of a fixed amount of DNA provides the basis for cleaving mature genomes from concatemeric precursors, then the total length of DNA contained in the phage particle should not be influenced by the presence in the genome of deletions. But the presence of a deletion should increase the length of the terminally redundant region. If the normal cleavage process is

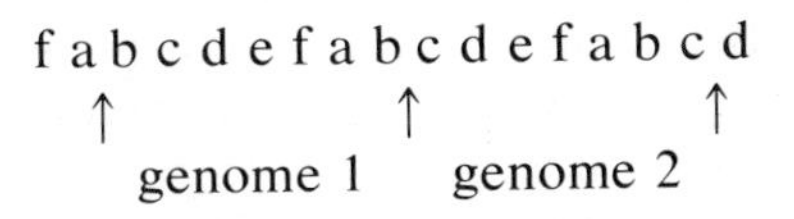


deletion of the marker *a* changes the maturation to

f b c d e f b c d e f b c d e f b
↑ ↑ ↑
genome 1 genome 2

so that the region of terminal redundancy now occupies two letters instead of one. Streisinger, Emrich and Stahl (1967) tested this prediction by investigating the effect of deletions elsewhere in the genome upon the heterozygosity of the $h^{2+}/h^{4+}$ markers. The $h^{2+}/h^{4+}$ heterozygosity behaves as though due entirely to terminal redundancy, presumably because the two alleles are sufficiently different in nucleotide sequence to prevent the formation of hybrid DNA. When deletion phages carrying these markers were crossed, the frequency of the heterozygotes increased with the length of deletion; this suggests that the length of the terminally redundant region is increased to compensate for the absence of the deleted sequences. The most likely mechanism to explain the selection of a constant amount of DNA, independent of sequence, is to suppose that packaging proceeds along the concatemer until the head of the phage particle is full with DNA; the capacity of the head must be fixed at an amount greater than the length of the unit genome.

The corollary of the idea that deletions increase the length of the region of terminal redundancy is that insertions should decrease it; and an insertion of length greater than that of the usual redundant region should make it impossible to include an entire unit genome in the phage head. Long insertions should therefore generate inviable phages. This idea is supported indirectly by experiments in which phages carrying duplications of the *rII* region were found to generate deletions more readily; the deletion presumably compensates for the duplication and thus restores the length of the unit genome (Parma, Ingraham and Snyder, 1972; Van de Vate and Symonds, 1974; Homyk and Weil, 1974). The presence in the phage of an *rII* duplication thus can be used to select for deletions; of course, only deletions that remove nonessential genes are viable. Such experiments confirm that the regions between genes *32* and *63* and between genes *39* and *56* are not essential for viability.

The properties of phages carrying deletions or duplications thus suggest that it is the amount and not the sequence of DNA that establishes the structure of the mature genome. Terminal redundancy may be inferred to be a consequence of the parameter that sets this amount at a value greater than the content of the unit genome. Direct confirmation that the same sequence is repeated at each end of the linear genome has been provided by the experiments of MacHattie et al. (1967). As Figure 6.6 shows, if the

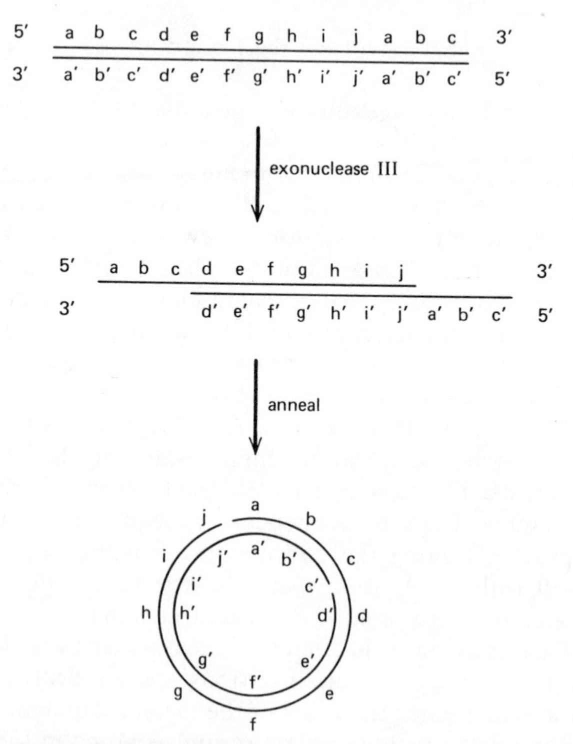


**Figure 6.6:** formation of circular molecules by generation of complementary single strand ends in regions of terminal redundancy. Exonuclease treatment degrades one strand, from 3′ to 5′, at each end of the duplex; annealing then allows circles to form.

genome is terminally redundant it should be possible to convert it into a circular structure by removing one strand of the duplex at each end; this generates protruding complementary single strands that are able to base pair. When the termini of the linear molecules were degraded in the 3′ to 5′ direction by exonuclease III and allowed to anneal, circular molecules could be seen in the electron microscope. Removal of 1–3.5% of the DNA allows circle formation (the circles are formed more rarely when less or more DNA is removed); this provides an estimate for the length of the terminal redundancy that is in the same range as that suggested by the extent of terminal heterozygosity (~1%).

Only unit circles were formed in these experiments; the absence of concatemers provides another argument to support the existence of circular permutation. If all molecules had the same terminally redundant sequence, then interactions between the ends of different molecules would occur as well as the annealing between the two ends of the same molecule; this would generate molecules containing several units linked end to end. Because the molecules are circularly permutated, however, each linear DNA has a different sequence in its terminally redundant region, so that annealing takes place only between the two ends of one molecule (see Thomas and MacHattie, 1967).

The presence of terminal redundancy means that when T-even DNA is renatured and denatured, only one copy of the redundant sequences on each strand should be able to find a partner. Thus the structure of the circular molecules shown in Figure 6.5 must be modified, because instead of a simple break on each strand of the circle there will be an unrenatured length of single strand DNA corresponding to the redundant region. An amended model for the structure of the circular molecule which takes account of terminal redundancy is shown in Figure 6.7. In the original experiments used to examine the circles formed by denaturation and renaturation, it was not possible to distinguish any single strands. MacHattie et al. (1967) therefore repeated these experiments under conditions when the single strands can be visualized in the form of condensed bushes on the circular duplex; their results confirmed the model shown in the figure.

When renatured T-even DNA is spread for electron microscopy under conditions that allow single strands to remain extended, it is possible to measure their length as well as the length of the circular duplex. The sum of these two measurements gives the total DNA content of the phage head, since the contour length of the circular duplex represents the length of the unit genome and the length of the single strands identifies the terminal redundancy. The results obtained from the heteroduplex studies of Kim and Davidson (1974) are summarized in Table 6.1. The three

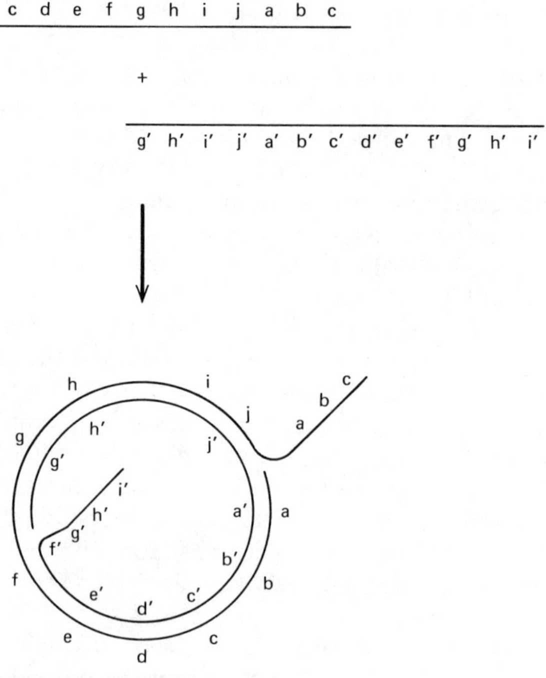


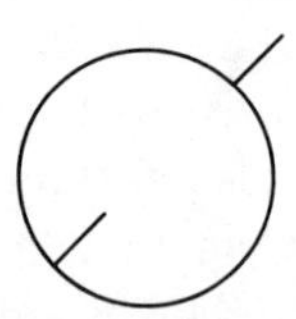

**Figure 6.7:** formation of circles with excluded single strands by denaturation and renaturation of circularly permutated terminally redundant molecules. When renaturation takes place between two complementary strands that are circular permutations of each other, and each has a (different) sequence of terminal redundancy, one of the redundant sequences of each strand is unable to participate in duplex formation and is excluded from the circle. The structure shown below depicts the form in which this molecule is visualized by electron microscopy: a circle whose thread is wider has two protruding thinner threads (which may be seen as such or may be collapsed into a bush depending upon the conditions that are used).

**Table 6.1:** genome sizes of T-even phages

| phage | unit genome | terminal redundancy | total genome | percentage terminal redundancy |
|---|---|---|---|---|
| T2 | 160 ± 2 | 8.5 ± 1.5 | 168 ± 3 | 5.3 ± 1.0 |
| T4 | 166 ± 2 | 3.0 ± 1.0 | 169 ± 3 | 1.8 ± 0.7 |
| T6 | 164 ± 2 | 3.2 ± 1.5 | 167 ± 3 | 2.0 ± 1.0 |

All genome lengths are given in kilobases. The terminal redundancy is also calculated as a percentage of the total genome. Data of Kim and Davidson (1974).

phages T2, T4, T6 vary in the length of the unit genome, with T4 being the largest at 166,000 base pairs. But the length of the terminal redundancy also varies, with T4 being the smallest at 3000 base pairs (this is sufficient to include about three genes). The result is that the total length of DNA in each of the three types of T-even phage appears to be the same (within experimental error); this lends further support to the concept that phage DNA content is determined by amount rather than by sequence and, of course, implies that the mechanism of packaging is precisely the same for each phage. An interesting feature of these results was that there appeared to be a variability in the length of the terminally redundant region, of the order of about 1% of the total genome length; since this represents up to 50% of the length of the redundant region itself, it is too great to be accounted for by experimental variation, and the most likely explanation is that it represents some small variability in the amount of DNA that is packaged into each head.

## Gene Expression in T4 Development

### *Classes of Messenger RNA*

The classification of T4 genes into early and late functions implies that there may be temporal control of T4 gene expression. Expression of the early genes is required in the period immediately following infection in order to support the replication of phage DNA; these genes code for enzymes concerned with nucleic acid metabolism. Expression of the late genes is not required for replication and their products are concerned with the subsequent stages of development, accomplishing the assembly of the mature phage particle.

Changes in the synthesis of phage proteins can be followed by pulse labeling infected cells with radioactive amino acids at different times and analyzing the appearance and disappearance of bands on polyacrylamide gels. By examining the band patterns generated by phage mutants, it is possible to equate some bands with specific gene products and thus to follow the expression of certain individual genes. In experiments using such protocols, Hosoda and Levinthal (1968) and O'Farrell and Gold (1973a) demonstrated that different proteins are synthesized during the early phase (preceding the start of replication) and the late phase (following the start of replication). During the first few minutes after infection, early proteins appear in a specific order; shortly after the start of replication, the synthesis of late proteins commences; and then synthesis of most of the early proteins ceases. This implies that at least two switches are made in T4 gene expression at the transition from the early to the late period: late protein synthesis is turned on and early protein synthesis is turned off. Early mutants of the *DO* class, unable to replicate any DNA, do not synthesize late proteins and fail to shut off the synthesis of early proteins, a concept first suggested by the analysis of certain early enzymes performed by Wiberg et al. (1962).

A counterpart to these experiments is the autoradiographic analysis of T4 RNA separated on gels by Adesnik and Levinthal (1970). The early and late patterns of RNA synthesis are different and show that during the late period new messengers are transcribed while transcription of the early messengers in general is halted. The *DO* mutants again continue to display the early pattern at late times. This implies that the principal control of T4 gene expression is exercised at the level of transcription (although it does not exclude the coexistence of translational control), and comprises both switching on and switching off expression of the phage genes (for review see Rabussay and Geiduschek, 1976).

Hybridization analysis of the RNA synthesized in the infected cell provides more detailed information about the control of transcription. Competition experiments that compare the RNA populations present at early and late times suggest that these sequences are transcribed from different genes; and detailed comparisons suggest that each of the early and late classes consists of two subclasses. The control of transcription of these RNA species can be investigated both in vivo and in vitro. Experiments in vivo have been designed to determine the effect upon transcription of mutations in particular genes or of inhibiting protein or nucleic acid synthesis: these experiments demonstrate that the first class of early genes can be expressed independently of any phage functions but that expression of the second class requires synthesis of a regulator protein coded by the phage; and transcription of late genes is coupled to replication of phage DNA and requires the activity of more than one phage regulator protein. Experiments in vitro in principle are designed to develop systems that mimic the situation in the infected cell, in which specified classes of genes can be transcribed and translated when appropriate protein activities are present: these experiments are concerned with defining the molecular interactions by which transcription is controlled, that is, the nature and role of changes in RNA polymerase and control protein activities, and in the state of the template.

The principle technique that has been used to define the different classes of messenger RNA that are found in infected cells is that of competitive hybridization, an assay that measures the extent to which the sequences of one RNA preparation are found in a second preparation. Figure 6.8 illustrates this protocol, which is based upon making a series of measurements of the ability of a constant amount of the first, labeled RNA preparation to hybridize with denatured DNA in the presence of increasing amounts of the second, unlabeled preparation. The principle of this approach is that the labeled and unlabeled RNA preparations will compete for the DNA complementary to any sequences that are present in both populations of RNA; increasing the concentration of the unlabeled (*competitor*) preparation therefore decreases the extent of hybridization by these sequences in the labeled (*competed*) preparation. To obtain competition it is necessary for all the sites on DNA complementary to the competing species to be occupied; this requires that the concentration of RNA sequences shall be in excess over the concentration of DNA sequences, and this technique therefore represents an RNA-driven hybridization reaction.

For phage and bacterial nucleic acids, hybridization is usually followed by the nitrocellulose filter assay: denatured DNA is bound to the filter, which then is incubated with the solution of RNA; RNA is bound to the filter only if it is able to hybridize with the DNA. Hybridization of the labeled RNA can therefore be followed by counting the radioactivity retained on the filter. Competition is displayed by a reduction in the amount of the radioactive label that is retained when filters are incubated with mixtures that contain the unlabeled as well as the labeled RNA preparation. Hybridization techniques are discussed further in Volume 2 of Gene Expression.

As the concentration of the unlabeled RNA preparation relative to the labeled preparation is increased, any competition between the two preparations is at first reflected in a reduction of the hybridization of the labeled species. When a sufficient excess of the unlabeled over the labeled preparation has been achieved, a plateau is seen in the level of hybridization of the labeled species, because all the sequences of the labeled preparation that also are present in the unlabeled preparation are prevented from

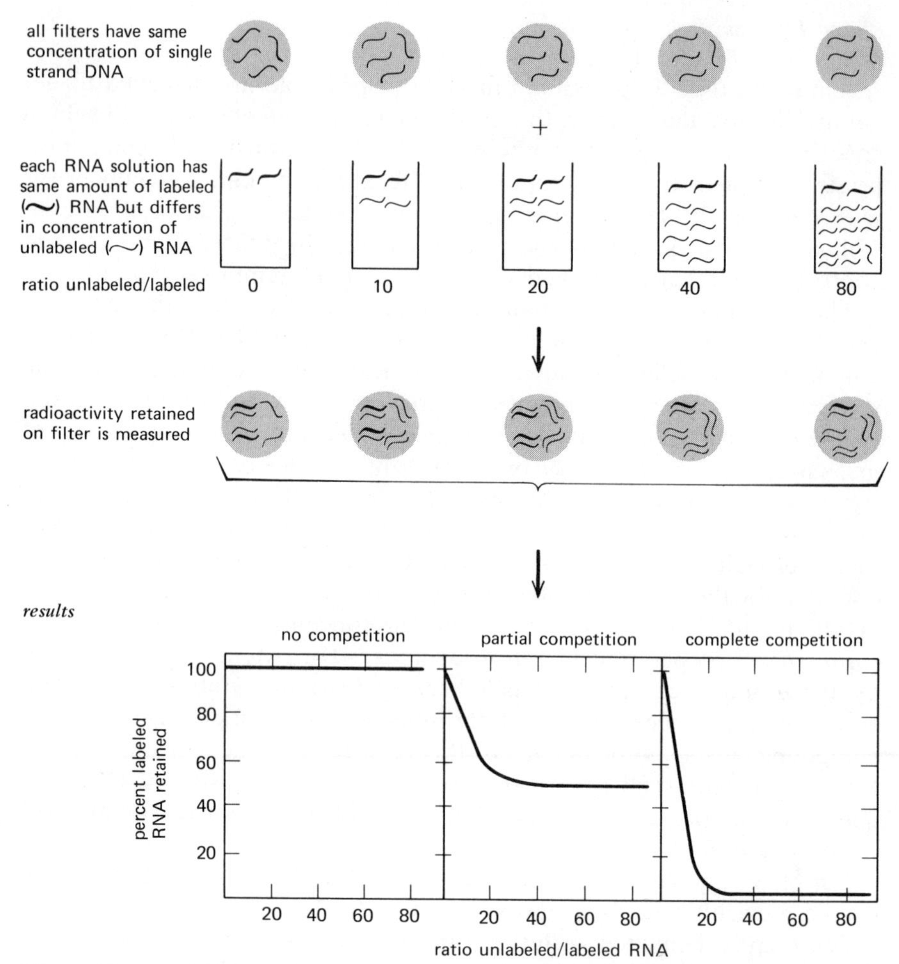


**Figure 6.8:** competition hybridization assay for RNA sequences. Filters carrying denatured DNA are incubated with a series of RNA solutions. Each solution contains a mixture of two RNA preparations, one radioactively labeled and one unlabeled; all the solutions contain the same amount of the labeled RNA but the series has increasing amounts of the unlabeled RNA. The radioactivity retained on each filter is measured and is calculated as a percentage of the label retained in the control that has only the labeled RNA preparation (the ratio unlabeled/labeled = 0). The percent retention is plotted as a function of the ratio of unlabeled to labeled RNA. Three types of result are illustrated. Left: if none of the sequences of the labeled preparation is present in the unlabeled preparation, the amount of unlabeled RNA does not influence the hybridization of the labeled RNA, which therefore remains at 100% in all incubations. Right: if all of the sequences of the labeled preparation are present in the unlabeled preparation, increasing amounts of the unlabeled preparation compete increasingly effectively for the complementary DNA and therefore completely prevent any hybridization of the labeled preparation. Center: if some of the sequences of the labeled preparation are present in the unlabeled preparation, they are prevented from hybridizing at high unlabeled/labeled ratios, reducing the proportion of radioactivity retained to a plateau that corresponds to the sequences of the labeled preparation that are not competed by the unlabeled preparation.

binding to DNA by the large excess of the unlabeled sequences that now occupies these sites. At the plateau, the proportion of the labeled RNA that continues to hybridize (that is, the percentage of its radioactive label retained on the filter) represents the sequences that are unique to the labeled preparation, that is, are not present in the unlabeled preparation. The extent of competition is given by the remaining proportion (the percentage competition equals 100 minus the percentage of label retained); since these sequences are prevented from hybridizing, they must be common to both labeled and unlabeled preparations.

The information yielded by this parameter is one sided in that it pertains only to the sequences present in the *labeled* preparation. In effect it divides the labeled sequences into two classes: those that are present in and those that are absent from the unlabeled preparation. The plateau yields no information about the *unlabeled* preparation, except, of course, to say that it contains sequences identical to a given proportion of the labeled sequences. To determine what proportion these sequences comprise of the unlabeled preparation requires an experiment in which the labels and roles of the two preparations are reversed. However, some information about the abundance of the competing sequences in the unlabeled preparation can be obtained from the excess ratio that is needed to reach the plateau: the smaller the proportion of the unlabeled sequences that is competing, the greater the excess that is required to reach the plateau. Another measure of this parameter is given by the initial slope of competition (the slower the slope, the smaller is the proportion of unlabeled sequences that is competing).

An early demonstration that different messengers are transcribed during the phases of infection was provided by the competition experiments of Hall et al. (1963, 1964). Only about half of the RNA sequences present in a preparation labeled at 19 minutes could be competed by an excess of an unlabeled preparation extracted from cells at 6.5 minutes after infection. This indicates that new RNA species are transcribed during the period from 6.5 to 19 minutes, that is, there must be temporal control of gene expression. In the reverse experiment, virtually all of the 6.5 minute RNA could be competed by an excess of 19 minute RNA; this shows that most or all of the sequences present at 6.5 minutes still are present at 19 minutes after infection.

The results of similar experiments performed by Bolle et al. (1968a) are shown in Figure 6.9. In the upper panel RNA labeled continuously from 1–20 minutes after infection was competed either with unlabeled RNA extracted at 5 minutes or with a control of unlabeled RNA extracted after 20 minutes; the 5 minute RNA competes with about 55% of the labeled RNA, compared with the 90% competition shown by the control prepara-

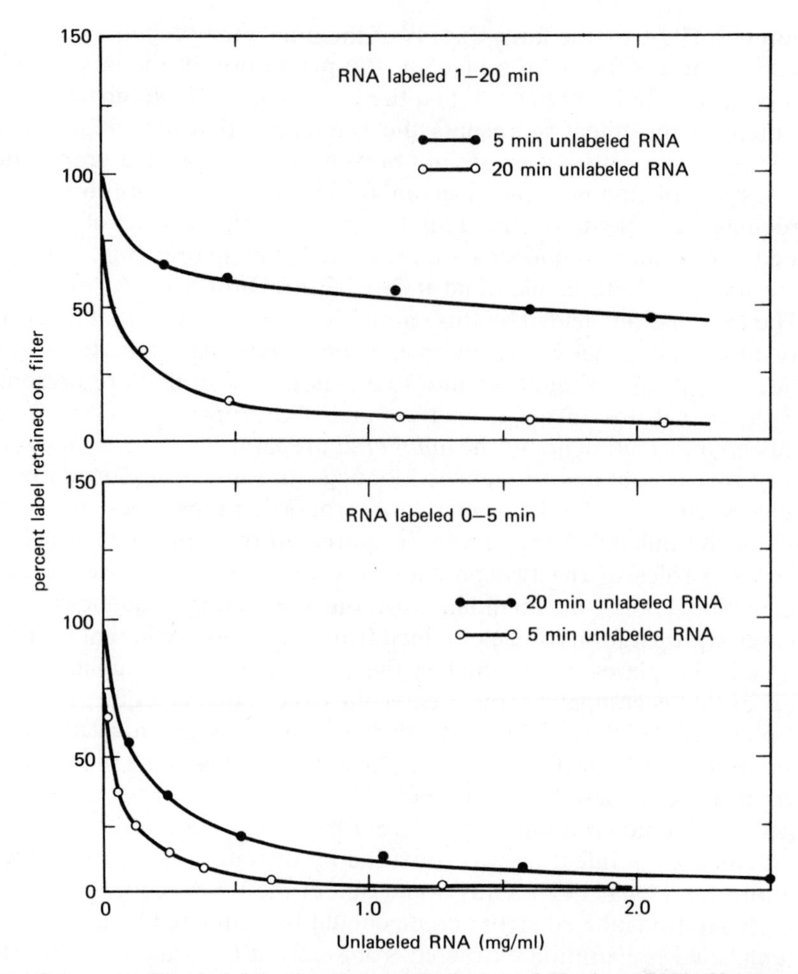


**Figure 6.9:** competitive hybridization between 5 minute and 20 minute RNA. Upper panel: 4.1 $\mu$g DNA/ml was annealed with 17.7 $\mu$g $^3$H-RNA/ml labeled from 1–20 minutes in the presence of the amounts shown of unlabeled RNA extracted at either 5 or 20 minutes. Lower panel: 5 $\mu$g DNA/ml was annealed with 1.4 $\mu$g $^3$H-RNA/ml in the presence of the amounts shown of unlabeled RNA extracted at either 5 or 20 minutes. Note that the concentrations of labeled RNA in the two experiments are different, so that the excess represented by the given concentrations of unlabeled RNA is different in the two panels. Data of Bolle et al. (1968a).

tion. Thus about 45% (perhaps slightly less when allowing for the incomplete competition of the control) of the 20 minute RNA contains sequences that are not present in the 5 minute RNA. On the other hand, the lower panel shows that a preparation labeled during the first 5 minutes of infection is competed almost equally well by a control of unlabeled 5 minute RNA and by an excess of RNA extracted after 20 minutes; the plateau of almost 100% competition in both cases indicates that all the 5 minute sequences are present in the 20 minute preparation.

These experiments were taken to define two general classes of phage messengers. The early messengers are those present at 5 minutes after infection, that is before replication commences (at about 6 minutes). The late messengers are the new sequences present in the 20 minute population, that is the part of the 20 minute RNA that cannot be competed by the 5 minute RNA. Late messenger synthesis begins at about 10 minutes. Since all of the 5 minute sequences appear to be present at 20 minutes, these experiments demonstrate that the 20 minute population consists of both early and late messengers. The same fraction of the 20 minute population is competed by unlabeled 5 minute RNA irrespective of whether the labeling is carried out continuously (the condition used for the experiments shown in the figure), by pulse labeling, or by a pulse chase protocol; this implies that the early and late sequences present at 20 minutes have similar turnover rates.

Although all the early sequences appear to be present at 20 minutes, their average concentration is lower than it is at 5 minutes. One illustration of this is provided by comparing the abilities of the unlabeled 5 minute and 20 minute RNA preparations to compete with the 0–5 minute labeled RNA. It is clear from the lower panel of Figure 6.9 that to achieve a given degree of competition requires a greater concentration of the unlabeled 20 minute RNA than of the 5 minute unlabeled RNA control; for example, 90% competition is seen at 1.4 mg/ml of the unlabeled 20 minute RNA but is reached at 0.4 mg/ml of the unlabeled 5 minute RNA. The ratio of the concentrations of the two unlabeled preparations that give the same degree of competition provides an estimate of the dilution of the early sequences in the late population; these data suggest a factor of about 3.5 (= 1.4/0.4) and other experiments have suggested values of up to about 10. Taking a factor of 5 would imply that the 5 minute sequences on average provide only about 20% of the molecules that are present at 20 minutes (formally the 20% refers to a weight average, that is, to the proportion of the total mass of RNA).

Comparing the abilities of two unlabeled preparations to compete with the same labeled RNA measures the *relative* concentration in the two preparations of the competing sequences. Thus although these experiments demonstrate that the 5 minute sequences represent only a small proportion of the sequences present at 20 minutes, they do not reveal whether this is because synthesis of 5 minute sequences has been reduced or whether it is simply a result of dilution due to the synthesis of an excess of other sequences. To determine the cause of the fall in the relative abundance of the 5 minute sequences, it is therefore necessary to know the amounts of T4 RNA present in the cell at 5 and at 20 minutes. The total level of RNA (including ribosomal and transfer species) does not appear to change during infection; and the proportion of T4-specific RNA increases only about 40% between early and late times (from 3.5% to 4.9% of total RNA). For the puposes of an approximate calculation, it is therefore reasonable to assume that the competing powers of RNA preparations extracted at different times during infection may be compared directly; this implies that the reduction in representation of 5 minute sequences in the RNA population extracted at 20 minutes is due very largely to a turnoff of early gene expression. (This calculation assumes that there is no change in messenger stability.) This is consistent with the demonstration that different mRNA species are seen on gels when the phage messengers are labeled at 5 or at 20 minutes.

A limitation of the use of competitive hybridization is that it is concerned largely with comparing the sequences that are present in greater abundance in the unlabeled preparation and does not assay any sequences that may be present in much lower concentration. The demonstration that new sequences are present in the 20 minute population therefore does not completely exclude the possibility that they are present at very low levels in the 5 minute population, but are not competed by the unlabeled 5 minute RNA because they are present at levels so low that an excess sufficient to display competition is not achieved.

To obtain a more quantitative estimate of the level of late sequences at 20 minutes compared with 5 minutes, Bolle et al. introduced the mixed competitor technique. The results of such an experiment are illustrated in Figure 6.10. Curves *a* and *c* display the same competitive hybridizations as those shown in the upper panel of Figure 6.9; a constant amount of labeled 20 minute RNA was hybridized with denatured T4 DNA in the presence of increasing amounts of either unlabeled 5 minute RNA (curve *a*) or unlabeled 20 minute RNA (curve *c*). Curve *b* displays the mixed competition experiment. The labeled 20 minute preparation was first annealed in the presence of an excess of the unlabeled 5 minute RNA sufficient to prevent the hybridization of most or all of the early sequences present in the 20 minute population. In effect this leaves available for hybridization only the noncompeted, late sequences. The properties of these late sequences then are tested by adding increasing amounts of

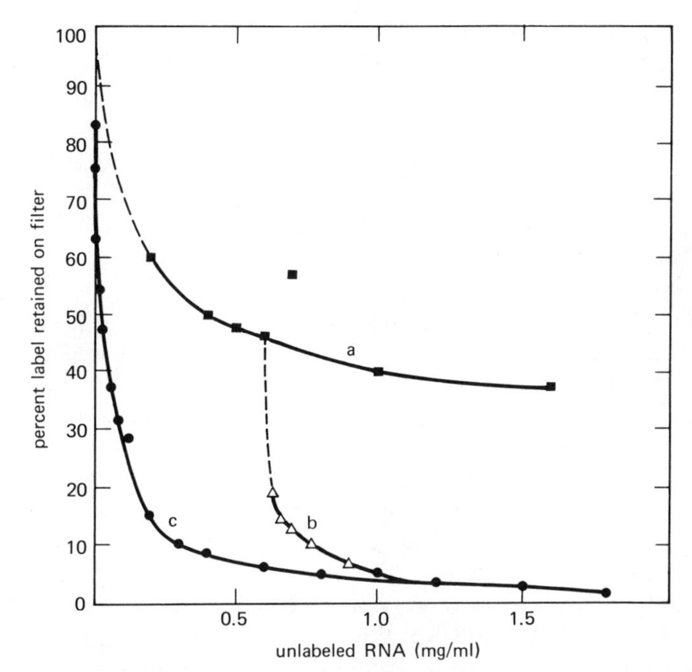


**Figure 6.10:** hybridization with mixed competitors. Curve *a* represents the annealing of labeled 1–20 minute RNA in the presence of increasing amounts of unlabeled 5 minute RNA; 2.25 $\mu$g DNA/ml was annealed with 4.5 $\mu$g $^3$H-RNA/ml in the presence of the concentrations given of unlabeled RNA. Curve *b* represents mixed competition; the same conditions as curve *a* were used except that 0.6 mg/ml of unlabeled 5 minute RNA was present and increasing amounts of unlabeled 20 minute RNA were added (the competition is plotted as a function of the total competitor present, that is unlabeled 5 minute + unlabeled 20 minute RNA). Curve *c* shows competition of the 1–20 minute labeled RNA by unlabeled 20 minute RNA. Data of Bolle et al. (1968a).

unlabeled 20 minute RNA; competition is very strong as revealed by the steep slope of curve *b*. The initial slope provides an estimate of the concentration of the late sequences in the 20 minute preparation. The concentration of these late sequences in the 5 minute preparation is given by the initial slope of the part of curve *a* after the junction with curve *b*; this can be taken to represent the effect of adding further amounts of unlabeled 5 minute RNA to the incubate of the 20 minute labeled RNA with the fixed excess of unlabeled 5 minute RNA, that is, it determines the concentration of late sequences in the 5 minute preparation. Thus the ratio of the two initial slopes estimates the relative concentration of late sequences in the unlabeled 20 minute and 5 minute populations. Depending upon how the slopes are drawn, the ratio can be calculated to lie in the range 200–4000. This means that at 5 minutes the maximum concentration of late sequences is only 0.05% of the concentration seen at 20 minutes. These sequences can therefore be defined as the *true late messengers*: they are effectively absent at 5 minutes and are synthesized by 20 minutes.

Early and late messengers appear to be transcribed largely from different strands of T4 DNA. Guha and Szybalski (1968) reported that the denatured DNA of T4 can be separated into its component strands by CsCl centrifugation in the presence of poly(U) or poly(U,G). The *l* strand is much lighter in CsCl-poly(U) than the *r* strand; in CsCl-poly(U,G) the *l* strand is denser and the *r* strand is lighter. By annealing labeled RNA with an excess of the separated phage strands, Guha et al. (1971) measured the proportion of the RNA that is complementary to each strand. At times up to 5 minutes almost all the RNA anneals with the *l* strand. Transcription of the *r* strand does not start until 11 minutes; by 20 minutes 40% of the RNA is complementary to the *r* strand, and by 30 minutes this proportion has increased to 80%. This suggests that early genes are transcribed from the *l* strand and late genes are largely transcribed from the *r* strand.

When labeled 20 minute RNA is competed by 5 minute RNA, more than half of its sequences are prevented from hybridizing (see the upper panel of Figure 6.9). Some of these sequences represent early messengers still present in the late population; these are identified by the decline in their concentration between 5 minutes and 20 minutes that is revealed by comparing the abilities of unlabeled 5 minute and 20 minute RNA preparations to compete with labeled 5 minute RNA (as seen in the lower panel of Figure 6.9). Another class of sequences present in both populations can be defined by comparing the competition for labeled 20 minute RNA of unlabeled 5 minute and 20 minute RNA preparations: considering those sequences for which both populations compete, the 5 minute preparation is less effective than the 20 minute preparation. This implies that there must be sequences common to both populations that are present at low concentration at 5 minutes but at high concentration at 20 minutes. Salser, Bolle and Epstein (1970) described these as the *quasi-late messengers*. This suggests that the 5 minute RNA population can be divided into two classes: the true early messengers decline in concentration between 5 and 20 minutes; whereas the quasi-late species increase in amount during this period. Information about these two classes is derived from comparing the abilities of excesses of 5 minute and 20 minute RNA to compete with labeled 5 minute or 20 minute RNA preparations; each of these comparisons yields information about the sequences that are present in greater abundance in the labeled preparation, that is, which can only be competed by large amounts of the unlabeled preparations.

Among the early transcripts there is a class known as the *anti-late messengers*. These are detected by their ability to form RNA-RNA duplex structures upon annealing with 20 minute RNA: they must therefore represent transcripts of the DNA strand complementary to that which is transcribed later in infection. Notani (1973) reported that by 5 minutes the anti-late sequences represent almost 2% of the total labeled RNA; and up to 30% of the RNA labeled at 20 minutes can form duplex structures when hybridized with an excess of early RNA. The late:anti-late annealing reaction seems largely or entirely to represent hybridization between late *r* strand sequences and anti-late *l* strand sequences. When 20 minute *r* strand RNA was purified by its hybridization with *r* strand DNA, almost 90% was then able to hybridize with 5 minute RNA. This implies that most of the late *r* strand messages have *l* strand counterparts in the 5 minute population. In other words, a low level of transcription of the *l* strand of many of the late genes takes place by 5 minutes. The nature of the transcriptional event leading to the synthesis of the anti-late messages is not known; no function has been identified for these species but presumably the anti-late sequences are not translated into protein.

One result of the annealing reaction between late and anti-late sequences is to interfere with measurement of the competition between early and late messengers in the hybridization assay. In effect the reaction removes some of the late sequences by complementarity instead of by competition. This presents a particular problem in analyzing quasi-late RNA. The RNA-RNA annealing reaction requires an excess of the unlabeled 5 minute RNA (only 2% of which represents anti-late sequences) and removes sequences of the 20 minute population that are abundant; this of course is precisely the assay used to assay quasi-late RNA. Proper assay of the quasi-late sequences therefore requires competition to be performed under conditions when anti-late–late RNA annealing does not occur. This in principle would be best attained by using purified *l* strand and *r* strand transcripts; that is the 5 minute and 20 minute preparations each would be separated into *l* and *r* transcripts by annealing with the denatured strands of T4 DNA, and these strand-specific preparations then would be used in competition experiments. Although competitive experiments with such preparations have not been performed, experiments that measure competitive hybridization with purified *l* strand DNA suggest that quasi-late *l* strand RNA is a real component of the 5 minute population; Guha et al. (1971) demonstrated by competition between unlabeled 5 minute RNA and labeled 20 minute RNA for *l* strand DNA that some sequences corresponding to this strand are present in the 5 minute population at much smaller levels than in the 20 minute population. Most of these sequences are under delayed early control, since much reduced competition is shown when chloramphenicol is present during infection. Quasi-late *l* strand transcripts can be assayed by this competitive hybridization because there are presumably no or very few anti-late *r* transcripts in the 5 minute population. But quasi-late *r* strand transcripts cannot be assayed in this way because of the extensive reaction between the *l* strand anti-late and the *r* strand late sequences; in the absence of experiments utilizing purified 5 minute and 20 minute *r* strand RNA, it is not clear whether a quasi-late *r* strand population exists.

### *Immediate Early and Delayed Early Transcription*

The early sequences that constitute the 5 minute RNA population appear sequentially during the first few minutes after infection. The appearance of the sequence components of the 5 minute population can be followed by testing the ability to compete with labeled 5 minute RNA that is shown by unlabeled preparations extracted after various times. The results of Salser, Bolle and Epstein (1970) that are summarized in Figure 6.11 show that by 1.25 minutes the unlabeled RNA can compete with only 27% of the 5 minute sequences; by 2.5 minutes it is able to compete with 65% of the total early sequences; and by 3.75 minutes virtually the same result is obtained as with the 5 minute control, so that all early sequences now have been transcribed. This suggests that new early sequences appear continuously during the first 3.75 minutes after infection.

The addition of chloramphenicol before infection restricts the sequences that are transcribed to those usually synthesized only by the first 1.25–2.5 minutes. When unlabeled RNA is extracted after 5 minutes of infection in the presence of chloramphenicol, it competes much less effectively with labeled 5 minute RNA from a normal infection than does the control unlabeled RNA extracted after 5 minutes of normal infection. This implies that some of the early sequences usually synthesized by 5 minutes fail to be transcribed in the presence of chloramphenicol. The CM RNA (this abbreviation describes the RNA extracted from cells treated before and during infection with chloramphenicol) is able to compete effectively against labeled RNA extracted after 1.25 minutes, reasonably well against 2.5 minute RNA, but poorly against 3.75 or 5 minute RNA. The sequences usually transcribed during the first 1.25 or 2.5 minutes therefore can continue to be synthesized in the presence of chloramphenicol, and it is the sequences expressed during the period from 2.5–3.75 minutes whose synthesis is prevented by the drug.

The use of chloramphenicol therefore divides the early sequences into two classes. The RNA that can be transcribed in the presence of chloramphenicol represents the *immediate early messengers*. The early se-

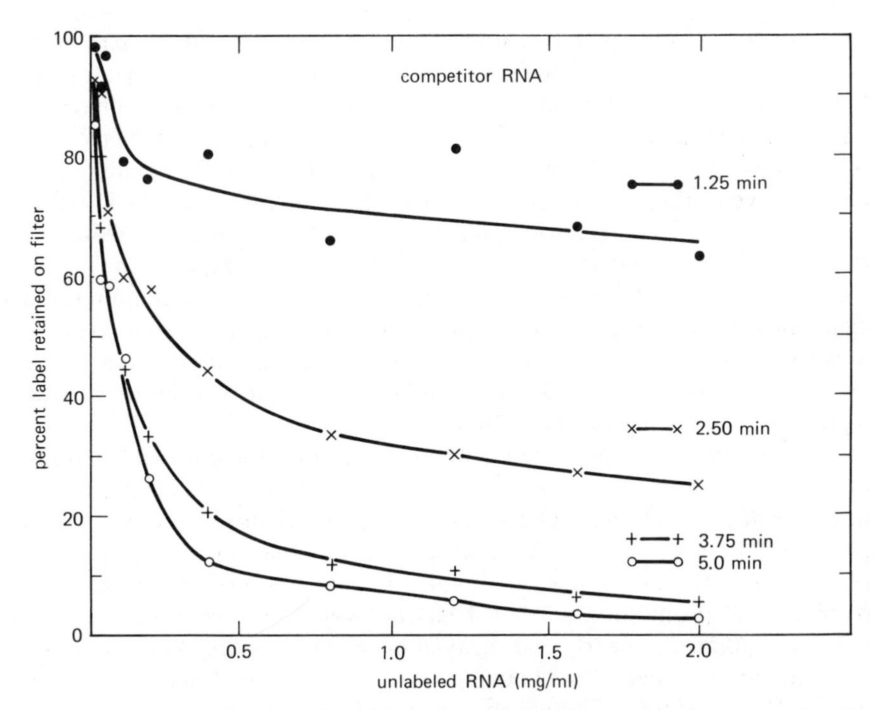


**Figure 6.11:** competitive hybridization of early RNA. 8 $\mu$g denatured DNA/ml was annealed with 1.2 $\mu$g/ml RNA labeled with $^{3}$H from 1–5 minutes; the indicated amounts of unlabeled RNA extracted at 1.25 minute intervals were present. Data of Salser, Bolle and Epstein (1970).

quences whose transcription is prevented by chloramphenicol identify the *delayed early messengers*. Immediate early messengers are transcribed upon infection; synthesis of delayed early messengers starts between 1.25 and 2.5 minutes and reaches an equilibrium level by 5 minutes. The effect of chloramphenicol implies that the immediate early messengers can be transcribed by the host RNA polymerase already present when infection commences; but transcription of the delayed early messengers must require the synthesis of some new protein(s) coded by the phage. Grasso and Buchanan (1969) supported this interpretation by demonstrating that puromycin has the same effect as chloramphenicol; this confirms that it is the inhibition of protein synthesis that prevents the transcription of delayed early genes. Chloramphenicol is able to prevent the transition from immediate early to delayed early expression only if it is added before 1 minute after infection; this means that by this time the protein needed for the transcription of delayed early genes has been synthesized. This is tantamount to saying that this protein is the product of an immediate early gene. (We should note, however, that Rabussay and Geiduschek (1976) have considered an alternative explanation for the effect of chloramphenicol: the drug might have a polar effect on transcription, that is, one in which the expression of genes distal to a promotor is preferentially inhibited relative to the more proximal genes. This interpretation would imply that there may be no need for a phage coded product in order to make the transition from immediate early to delayed early gene expression. It is to some extent supported by the observation that the presence of amino acid analogues or antibiotics that cause misreading during translation does not prevent the immediate to delayed early transition; this argues against the need for a newly synthesized protein and in favor of the possibility that any regulation is due to preformed factors).

The idea that a phage coded protein is required for the expression of delayed early genes raises the same questions about mechanism that we have already discussed for phage lambda. The mechanism of action of this protein(s) is more difficult to investigate in T4 than in lambda for two reasons: many more genes are involved, with expression taking place from several promotors, so that it has not yet been possible to identify individual immediate early and delayed early messengers whose relationship can be investigated; and no gene has been identified that is responsible for the transition. The absence of an identified regulator of delayed early gene expression contrasts with the identification of several genes whose products are needed for expression of the late genes (see below). Certainly the failure yet to identify any such locus raises questions about the nature and existence of any phage coded regulator protein.

The role of a phage coded regulator protein may be considered to take one of the two forms that are illustrated in Figure 4.45 for lambda regulation. If immediate early and delayed early genes are expressed independently, delayed early genes must have promotors that cannot be recognized by host RNA polymerase, but that are able to initiate transcription in the presence of the necessary phage coded protein. In this case the immediate early and delayed early sequences must be present on different molecules of messenger RNA. Or, if immediate early and delayed early genes are adjacent, the host polymerase usually may terminate transcription at the ends of the immediate early genes; but the phage coded protein may allow the polymerase .to read through the termination site. In this case the delayed early sequences are transcribed by extension of messengers that carry immediate early sequences; this means that the immediately early and delayed early sequences must be covalently linked on the same messenger molecules, and these molecules must be longer than those that carry only the immediate early sequences.

The characteristics of transcription of T4 DNA in vitro by E.coli RNA polymerase suggest that immediate early and delayed early genes are adjacent on the phage genome. Bautz, Bautz and Dunn (1969) observed that the RNA transcribed in the first 1 minute in vitro is 80% competed by unlabeled RNA extracted from infected cells at 1 minute and is 100% competed by unlabeled 5 minute RNA. This suggests that in vitro host polymerase initiates transcription largely at the immediate early promotors. When the in vitro reaction is allowed to continue for 30 minutes, the RNA product is 50% competed by 1 minute in vivo RNA and is 95% competed by 5 minute in vivo RNA. The transcription in vitro of delayed early genes therefore takes place at a later time than the synthesis of RNA from the immediate early genes. This lag is most easily explained by supposing that transcription in the in vitro system initiates at immediate early promotors and proceeds through the immediate early genes into adjacent delayed early genes. Similar conclusions have been reported by Milanesi, Brody and Geiduschek (1969), Black and Gold (1971) and O'Farrell and Gold (1973b).

This interpretation of the nature of the transition from immediate early to delayed early transcription in vitro is supported by the analysis of the RNA products performed by Milanesi et al. (1970) and Brody and Geiduschek (1970). When rifampicin is added to the in vitro system after 1 minute, new initiation events are prevented but RNA polymerase molecules engaged in transcription can continue to elongate the polynucleotide chains. At 1 minute the RNA products are short and correspond almost entirely to immediate early sequences. During the subsequent incubation the chains increase in length and contain a greater proportion of delayed early sequences. Because of the presence of rifampicin, the lag before the appearance of delayed early sequences cannot be explained by any delay in initiation but must be due to the time it takes for RNA polymerase to read through the immediate early sequences before it reaches the delayed early genes. After 10 minutes there was no further increase in the proportion of delayed early sequences seen in the products, although transcription in vitro continued. This implies that the immediate early and delayed early genes may be interspersed in such a way that a polymerase first transcribes one tandem (immediate early plus delayed early) unit of expression and then proceeds into the next. This emphasizes that in the in vitro systems termination events that would occur in vivo fail to take place.

Termination of transcription at the ends of the immediate early genes in vivo may be due to the action of the rho factor, which also is responsible for terminating immediate early transcription of phage lambda (see page 397). Richardson (1970) reported that the addition of rho reduced the proportion of delayed early RNA transcribed in vitro from 30% to 3%. Jayamaran (1972) followed the transcription in vitro of specific genes by annealing the RNA products with T4 DNA, degrading the unreacted single strands of DNA, and using the protected fragments of DNA in a phage transformation system to see whether they can provide functions that are defective in the accompanying phage. This assay showed that the transcripts of genes *30* and *42* have been synthesized within 10 minutes of incubation in vitro; and the transcripts of *rIIA, rIIB, 43* and *e* could be detected after 30 minutes. Addition of rho did not prevent expresssion of the presumed immediate early genes *30* and *42*, but it did prevent transcription of the delayed early genes *rIIA, rIIB, 43* and *e*. This again provides evidence that delayed early genes are transcribed in vitro by readthrough from immediate early genes and it suggests that rho-dependent termination sites exist between genes.

But the demonstration that readthrough can occur in vitro formally demonstrates only that delayed early genes lie adjacent to immediate early genes. The failure of delayed early genes to be expressed ineluctably in vivo implies that within the infected cell termination occurs at the ends of the immediate early genes. It seems probable that this is the result of the action of rho factor, but it is necessary to remember that the activity of rho in vitro depends heavily upon the salt concentration (it is active at 0.05 M KC1 but ineffectual at 0.15 M KC1); the ability to obtain rho activity in vitro at low salt concentration does not prove that appropriate conditions prevail in vivo. The demonstration that delayed early genes are located adjacent to immediate early genes leaves open the alternative mechanisms for delayed early gene expression: termination at the ends of the immediate early genes may be antagonized by a phage coded factor (which would be the same type of mechanism for delayed early expression as that seen in phage lambda); or termination may continue to occur and the phage function may sponsor initiation at delayed early promotors which cannot be recognized either in vivo or in vitro by the host polymerase.

One way to distinguish the models for delayed early transcription is to determine the sizes of the messenger RNAs that carry the delayed early sequences. The first delayed early sequences appear in vivo at 1.5 minutes; so if RNA is labeled from 1-2 minutes, it can possess only the short length of delayed early sequences synthesized in the last 0.5 minutes. If these delayed early sequences are generated by new initiation events, they should be present in short chains; but if they result from readthrough, they should be found in the longer chains of the labeled population. Brody et al. (1970) found that delayed early sequences occur in the heavier part of a density gradient; this observation favors a readthrough model.

The readthrough and new initiation models require different functions to be fulfilled by any phage coded protein(s) needed for delayed early expression: the readthrough model requires synthesis of an antiterminator protein while the new initiation model requires synthesis of a factor that allows the host polymerase to acquire the capacity to initiate at delayed early promotors (there is evidence to show that the host polymerase is needed for all T4 transcription, which excludes models relying upon the synthesis of a new enzyme). Although changes in the constitution and activity of the host enzyme have been demonstrated during T4 infection (and are discussed later) none has yet been equated unequivocally with a new initiation capacity or an altered activity in termination. The nature of the event responsible for delayed early transcription therefore remains to be established and may take either or both of the alternatives (since the models are not mutually exclusive it is possible that both apply).

A new event—which may be the synthesis of a phage coded protein—is responsible for the transition from immediate early to delayed early transcription. What is responsible for the sequential appearance of delayed early transcripts during the period from 1.5 to 3.75 minutes? Since no new events appear to be involved, the most likely explanation is that delayed early genes are organized in contiguous clusters which are read by the polymerase to generate polycistronic transcripts; the time of expression of each gene may depend simply on its distance from the promotor.

The progression of transcription through the *rII* genes can be followed in some detail because of the existence of defined deletions that extend into these genes for different distances; by using the DNA of these deletion strains in a hybridization assay, it is possible to follow the transcription of each part of *rIIA* and of *rIIB*. Using this assay, Schmidt et al. (1970) reported that *rIIA* is transcribed as a delayed early gene; its transcription begins at 1.5 minutes after infection and the first complete transcripts are present at 2.5 minutes. This corresponds to a rate of polynucleotide chain growth of about 35 bases per second at 37°C. Transcription of the first part of the adjacent gene *rIIB* can be detected also around 1.5 minutes after infection; this means that *rIIB* must have have an independent promotor. But *rIIB* also can be expressed by readthrough from *rIIA*. The RNA of infected cells was hybridized with T4 DNA containing *rIIA* sequences but lacking *rIIB* sequences; the hybridized RNA was extracted and then tested again by hybridization for the presence of *rIIB* sequences. About 60% of the RNA isolated in the first step has *rIIB* sequences, which must therefore be covalently linked to the *rIIA* sequences used to select the RNA. This suggests that *rIIB* may be transcribed by two mechanisms:

initiation at its own promotor; and readthrough from *rIIA*. It is therefore possible for a delayed early gene to be expressed by more than one mechanism.

*Coupling of Late Transcription to Replication*

True late sequences are defined as those present in the 20 minute RNA population and absent from the 5 minute population. They are therefore assayed as the proportion of sequences in any given (labeled) preparation that is competed by 20 minute RNA but is not competed by 5 minute RNA. Most of the experiments that have investigated the control of late transcription have depended upon this comparison; the fraction that is followed by this assay is often described simply as the late class, of which it is indeed the major component, although strictly speaking the assay excludes any quasi-late sequences. Salser, Bolle and Epstein (1970) reported that transcription of true late mRNA is turned on at about 11 minutes, after which synthesis continues at a constant level. Under the conditions used for these experiments, replication started at about 6 minutes; there is therefore a 5 minute lag between the start of replication and true late transcription.

Mutants of T4 defective in DNA synthesis fall into the three groups described in Figure 6.4. The *DO* mutants synthesize no or very little DNA, and this class includes mutants lacking enzymes essential for replication or for nucleic acid metabolism. The *DA* mutants initiate DNA synthesis normally but then cease replication after varying periods of time. The *DD* mutants show a delayed onset of replication but eventually accumulate DNA at the same rate as wild type T4. All three classes of mutant yield few or no viable progeny: the *DA* and *DD* mutants in particular yield much smaller amounts of progeny phage than would be predicted from the amount of DNA that is synthesized. In all three types of mutant infection the RNA labeled from 17–20 minutes is competed equally effectively by unlabeled RNA extracted at either 5 or 20 minutes from wild type infected cells. The results obtained with several of the *DO* mutants are given in Figure 6.12. This analysis implies that true late messengers are not synthesized in the mutant cells; this would explain the poor yield of progeny obtained with infections of all three types.

Is DNA synthesis necessary for the transcription of late messengers? To investigate this question, Bolle et al. (1968b) blocked DNA replication in wild type T4 infection. A critical problem in such experiments is that replication must be completely blocked, for even a small amount of escape synthesis might be sufficient to support late transcription. The host cells therefore carried a mutation in thymidylate synthetase and the

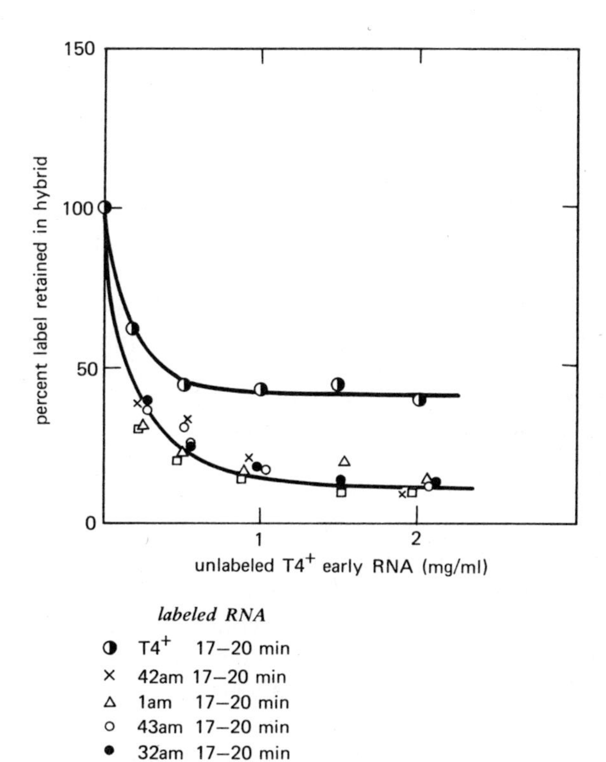


**Figure 6.12:** competitive hybridization of RNA transcribed at late times by *DO* mutant phages. RNA from cells infected with the indicated phages was labeled during the period described and competed by increasing amounts of unlabeled early RNA extracted from T4+ infected cells. Late T4+ RNA is about 60% competed by the unlabeled early RNA; late *DO* mutant RNAs are virtually completely competed by the unlabeled wild type early RNA and can be fitted to the same curve as the homologous competition between labeled T4+ and unlabeled wild type early RNA. Data of Bolle et al. (1968b).

infecting phage carried a mutation in the same function (so there should be no synthesis of this essential precursor) and also possessed a mutation preventing the degradation of host DNA (so nucleotides are not released by this reaction to become available for reutilization by the phage). The labeled RNA extracted from these infected cells at 20 minutes was equally well competed by wild type 5 minute and 20 minute RNA. The abolition of phage DNA synthesis therefore appears to confine transcription to the early sequences.

Two types of role may be imagined for DNA replication in supporting late transcription: replication may be necessary to allow the expression of some necesssary function(s); or replication may be directly coupled to late transcription, because it introduces some necessary change in the condition of the template. In the first situation, any DNA present in the cell, whether or not itself engaged in replication, should be able to express late genes if some replicating DNA is present; whereas the second model predicts that only those DNA molecules actually replicating can function as templates for late transcription. Bruner and Cape (1970) tested these predictions by superinfection rescue experiments. Cells were infected with a phage mutant in an essential late gene; the phage also carried a temperature sensitive mutation in replication, so that after a period of replication, further DNA synthesis was prevented by an increase in temperature. At this time cells were infected with a phage carrying a wild type copy of the essential late gene. Progeny can be generated only if this gene on the superinfecting phage can be expressed. No rescue was seen, implying that the nonreplicating DNA could not transcribe late genes. As a control, rescue of defective early genes could be demonstrated.

The effect upon late transcription of blocking replication can be followed by using phages that carry temperature sensitive mutations in essential replication functions. Gene *43* codes for DNA polymerase and both $pol^{ts}$ and $pol^{am}$ mutations in it have been isolated. Riva, Cascino and Geiduschek (1970a) found that when cells infected with a $pol^{ts}$ mutant are shifted up to the high temperature during the late period, the transcription of late sequences falls drastically within 2–3 minutes. Figure 6.13 shows that during a 2 minute pulse label given just before the temperature shift, 40% of the total mRNA represented late sequences; following the temperature shift this was rapidly reduced to 10%, a level which then remained stable. The assay technique underestimated the proportion of late transcription, which is probably about 50% greater than the level measured. Since the temperature shift from 30°C to 41°C increases late transcription twofold in $T4^+$ infection, with the $pol^{ts}$ mutant there is therefore a relative decrease in late transcription of about an order of magnitude. When the temperature was shifted back down to the permissive level, late transcription increased to resume its normal rate within 5–10 minutes. These results suggest that continued replication is needed to maintain normal levels of late transcription, although a lower residual level may be able to occur in the absence of replication. Similar results were obtained with other phage mutants unable to replicate DNA at high temperature. The same conclusion is suggested also by the results of Lembach, Kuninaka and Buchanan (1969), who demonstrated that late transcription is halted whenever replication is prevented by the addition of FUdR; again this implies that continued replication is necessary for late transcription.

The conclusion suggested by these results is that late transcription

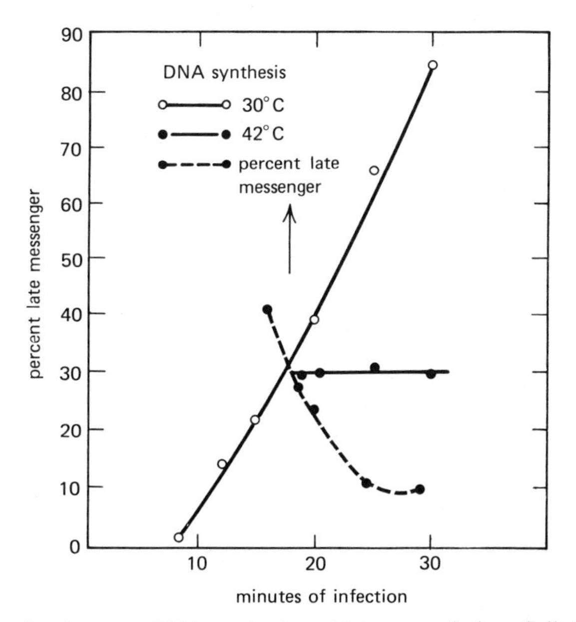


**Figure 6.13:** coupling between DNA synthesis and late transcription. Cells infected with a *43*$^{ts}$ mutant were incubated at 30°C; at the time indicated by the arrow, part of the culture was shifted up to 42°C. DNA synthesis was followed by the incorporation of labeled thymidine; it ceases immediately following the temperature shift. The late messenger is plotted as a percentage of the total message in the temperature shifted culture; just before the shift its level is 40% but it rapidly declines to a plateau of 10%. Data of Riva, Cascino and Geiduschek (1970a).

requires a special, *competent* state of the template. This is an intermediate which turns over continuously but is regenerated by replication. This model can be depicted in the form:

noncompetent DNA ⟶ mature phage particles

unknown action ↑ ↓ replication

competent DNA

What is responsible for the competence of replicating DNA? Riva, Cascino and Geiduschek (1970b) investigated this question by attempting to find conditions in which late transcription is uncoupled from replication. The principle of these experiments is to suppose that newly replicated DNA is competent and that some subsequent process abolishes its competence; if this latter function(s) can be inactivated, the DNA should remain competent in spite of the absence of continued replication.

Mutants in gene *30* of T4 lack the phage ligase and accumulate little progeny DNA; but they display almost normal late transcription. This suggests that ligase activity may be important in determining competence. This cannot be investigated simply by constructing $pol^{ts}$ $lig^{-}$ double mutants because the phage DNA is degraded in the absence of ligase. The degradation appears to be caused by an exonuclease activity, coded by genes *46* and *47*, which is involved in the breakdown of host DNA (see Wiberg, 1966; Hosoda and Matthews, 1968); this is discussed later. Cells were therefore infected with triple mutant phages of the genotype $pol^{ts}$ $lig^{am}$ $exo^{am}$ or $pol^{ts}$ $lig^{ts}$ $exo^{am}$. The $pol^{ts}$ mutation in gene *43* ensures that replication ceases immediately upon a shift in temperature to 42°C; the mutation in gene *30* means that ligase also in inactivated by the temperature shift (in the $lig^{ts}$ mutant) or is absent throughout infection (in the $lig^{am}$ mutant) so that any nicks in DNA cannot be repaired, and the $exo^{am}$ mutation in gene *46* or in gene *47* prevents nucleolytic attack from degrading the DNA. Figure 6.14 shows that upon a temperature shift up, the triple mutant ceases replication but continues late transcription at an

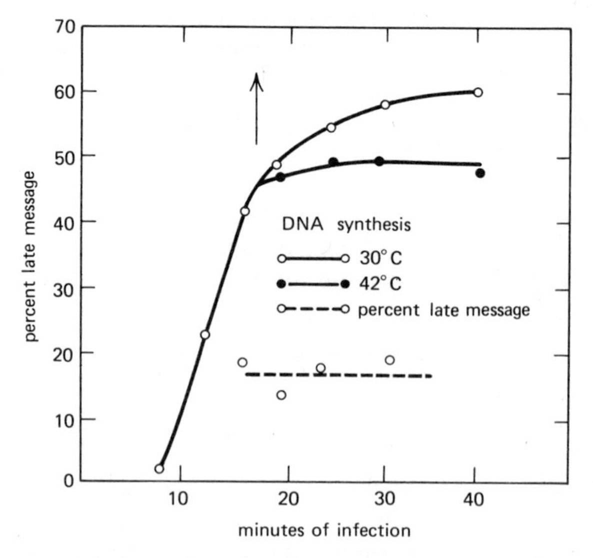


**Figure 6.14:** late transcription during infection with the triple mutant *43ts, 30ts, 46am*. Part of the culture was shifted from 30°C to 42°C at the time indicated by the arrow. DNA synthesis was followed by incorporation of labeled thymidine. Late messenger is plotted as the proportion of total message present in the temperature shifted cells; its level remains constant at about 20%. Data of Riva, Cascino and Geiduschek (1970b).

unaltered rate. (This rate is lower than that seen in wild type cells, as for example measured before the shift up in Figure 6.13, presumably because of the influence of the mutations upon the state of DNA.)

On the basis of these results, Riva et al. proposed that competence is acquired by the action of an endonuclease on DNA, either during or immediately following replication. The action of ligase is to seal the nicks created by the nuclease; and this abolishes competence. This implies that the RNA polymerase activity responsible for transcribing the late genes is able to initiate synthesis of messenger RNA only at single strand breaks; this model would explain also why inhibition of late transcription does not release polymerase to transcribe other classes of mRNA.

The importance of nicks in late transcription was demonstrated directly in the comparison made by Wu, Geiduschek and Cascino (1975) of late transcription in $pol^-$ and $pol^-lig^-$ $exo^-$ infections. At high multiplicities of infection, cells infected with these phages are able to transcribe late messages (see below). At all multiplicities, the triple mutants transcribe a greater proportion of late RNA. This is correlated with the presence of a greater number of nicks in the DNA of the triple mutant. The DNA of parent $pol^-$ $lig^{ts}$ $exo^-$ phages was labeled with $^3H$; cells were infected at 41°C, when ligase is inactive, and then later shifted down to 30°C to restore ligase activity. The DNA present in the infected cells was examined by alkaline sucrose sedimentation, when it is in the single stranded state. Since the $pol^-$ mutation ensures that there is no replication, this DNA represents the parental strands. In the absence of ligase there are extensive single strand breaks in the DNA, but restoration of ligase activity allows the molecular weight of the single stranded DNA to increase considerably. About half of the breaks introduced into the parental DNA after infection in these conditions can be repaired by ligase, that is, these must represent nicks and not gaps (for which a polymerase repair activity also would be needed). Breaks are introduced into both parental strands and the number of nicks per strand is about 10–12 at 20 minutes after infection. No nicking is observed when infection is carried out in the presence of chloramphenicol, so the endonuclease responsible must be coded by the phage genome; however, no phage function has yet been identified with this activity.

The relationship between transcription and nicking has not yet been defined. If the role of the nicks is directly to activate late transcription, it is reasonable to suppose that they must occur in the late promotor sites; this demands that the endonuclease recognizes the specific nucleotide sequences of these promotors. An alternative would be to suppose that the role of the nicking is indirect, causing some general conformational changes in the late DNA. Why does nicking usually depend upon replication? One possibility is that the enzyme reponsible for introducing the nicks is part of a complex of proteins that replicates DNA. This leaves open the question of how nicking occurs when late transcription is uncoupled from replication.

Experiments in vitro have identified phage-dependent changes that occur in RNA polymerase during infection and it is clear that these changes result in alterations of the specificity of the polymerase, that is, in the promotors that it recognizes (see below). But it is not yet clear how these changes in polymerase are related to changes in the condition of the template, a problem that is difficult to investigate because the implication of the dependence of late transcription upon the state of the template is that DNA with appropriate modifications must be used to study transcription in vitro.

The maturation defective (*MD*) mutants in genes *33* and *55* are able to synthesize DNA but are defective in late functions, failing to produce any of the structures of the phage particle. The $33^-$ and $55^-$ mutants thus behave as though unable to express any of their late genes (contrasted with head or tail mutants which show restricted defects since they fail to produce only the structures coded by the mutant gene). This deficiency is due to the absence of late transcription. Bolle et al. (1968b) found that $33^-$ mutant RNA labeled at 9.5–11.5 or at 17–20 minutes behaves like early RNA, since it is effectively competed by unlabeled 5 minute wild type T4 RNA. Similar results are observed in $55^-$ mutant infection: there is no detectable synthesis of late RNA at 20 minutes. These results suggest that the products of genes *33* and *55* are essential for late transcription.

When are the phage functions that are necessary for late transcription expressed? Pulitzer and Geiduschek (1970) found that replication and late transcription both become independent of inhibition by chloramphenicol at about 7 minutes. So any proteins needed for late transcription must have been synthesized by the time that replication starts. The gene *55* product is needed throughout the late period, for when cells infected with temperature sensitive mutants in *55* are shifted to the nonpermissive temperature at any time during the infective cycle, there is a corresponding reduction in phage yield. A shift down to the permissive temperature restores late transcription. The absence of gene *33* product also imposes a complete block on late transcription. Late transcription apparently independent of *33* and *55* function has been observed in some experiments; however, because some mutants may be leaky, allowing a certain residual level of gene expression, it is sometimes necessary to use multiple mutants in the gene in order to investigate the effect of the absence of its products. Using a double mutant in gene *33*, Notani (1973) demonstrated that the defect blocks all late transcription. In wild type T4 infection, the

20 minute population contained 12% *l* strand late RNA and 50% *r* strand late RNA (a total late proportion of 62%). Synthesis of both the *l* and *r* late transcripts was blocked under *33*⁻ or *55*⁻ conditions; and synthesis of both was reduced when a $pol^{ts}$ mutant is shifted up in temperature.

Are any other T4 functions in addition to *33* and *55* needed for late transcription? Wu and Geiduschek (1975) investigated the roles of replication genes by infecting cells with T4 *am* mutants and determining the proportion of RNA labeled at late times that is able to hybridize with *r* strand DNA. When late RNA is determined by this assay at 45–50 minutes after infection, with all *DO* mutants little late transcription can be detected under the usual conditions of infection. But at high multiplicities of infection the level of *r* strand transcription may reach an appreciable proportion, about 40% compared with the 60% seen in wild type infection. (This compared with the results obtained by Riva, Cascino and Geiduschek, 1970b: when cells are infected with the triple mutant phage $pol^{ts}$ $lig^-$ $exo^-$ at 40°C, no DNA is synthesized. But late message appears at about 15 minutes, reaches a maximum at about 30 minutes, and then declines. The effect depends upon the multiplicity of infection.) This implies that late transcription requires a sufficient amount of DNA as well as the acquisition of competence. The basis of this effect is not yet known.

In contrast with the multiplicity dependent late transcription displayed by other *DO* mutants, $45^-$ mutants synthesize little *r* strand RNA at any multiplicity value; the same result is seen with $33^-$ or $55^-$ mutants. This suggests that gene *45*, hitherto identified simply as a *DO* function, shares with genes *33* and *55* a direct involvement in late transcription. When synthesis of phage proteins was followed by autoradiographic analysis on gels, the *DO* mutants generated the same pattern in high multiplicity conditions as that displayed in wild type infections, with the exception that synthesis of some early proteins did not appear to be shut off at late times. But mutants in genes *33*, *55*, *45* and *56* all failed to synthesize late proteins even at high multiplicities of infection. This again supports the conclusion that the products of *33*, *55*, *45* all are necessary for late transcription. (The defect in protein synthesis in $56^-$ mutants appears to be a consequence of an indirect effect of the mutation which allows T4 DNA to replicate containing cytosine instead of hydroxymethylcytosine; see also Kutter et al., 1975.) In addition to the *r* strand transcription that represents the major source of late RNA, some *l* strand transcripts that are present at late times are not competed by 5 minute RNA, that is, they represent true late sequences: these correspond to some 3–5% of the total late T4 RNA. Mutations in genes *33*, *55* and *45* all prevent the appearance of this message, since all the *l* strand RNA present at late times is competed by wild type 5 minute RNA. The late transcription controlled by these three genes therefore includes that of the *l* strand as well as of the *r* strand.

If the products of these three genes are involved directly in late transcription, mutation in any one of them should prevent the replication-uncoupled transcription that takes place in the $pol^{ts}$ $lig^-$ $exo^-$ triple mutants. Wu, Geiduschek and Cascino (1975) therefore constructed the three types of quadruple mutant by adding the appropriate additional mutation to the triple mutant; in each case the further mutation prevented late transcription. The use of another quadruple mutant showed that gene *45* protein is required continually during the period of late transcription. The mutant $pol^-$ $lig^-$ $exo^-$ $45^{ts}$ displays multiplicity dependent activation of late transcription at low temperature; at high temperature there is no *r* strand transcription. If cells are infected at 30°C and then shifted to 40°C at various times during infection, late transcription declines rapidly at any time. Thus the *45* protein, like the *33* and *55* functions, plays a continuing role in maintaining late transcription. The *45* product has been identified as a component of the replication complex (see below) and therefore provides a link between the replication and transcription apparatus.

Another aspect of the control of late gene expression is the decline in early RNA and protein synthesis during the late period. We have already mentioned that Wiberg et al. (1962) first noted the disappearance of certain early enzyme activities at late times; and Hosoda and Levinthal (1968) and O'Farrell and Gold (1973a) also observed the switch off of early proteins late in infection. Figure 6.15 shows the autoradiographic analysis of proteins separated on SDS gels that was performed by Wu, Geiduschek and Cascino (1975), from which it is clear that there is a sequential switch off of early protein synthesis. The events reponsible for early gene shutoff are not yet clear, and it is likely that there is more than one cause (for review see Rabussay and Geiduschek, 1976).

In mutants that fail to replicate DNA, the shutoff of early RNA synthesis and of early protein synthesis does not occur at the usual time (Cascino et al., 1971). But Wu et al. (1975) showed that in the triple mutant $pol^-$ $lig^-exo^-$ the shutoff occurs as usual. This indicates that late transcription is necessary to turn off the expression of early genes. Thus a product of one of the genes under late control must be a regulator for early gene expression.

In the absence of replication, the shutoff of early protein synthesis may occur later in the infective cycle. This appears to be due to a second system, one that is independent of replication and is expressed at some time after the point when the product of late transcription would have turned off early gene expression. Sauerbier and Hercules (1973) noted that this second type of shutoff is eliminated if methionine is replaced by

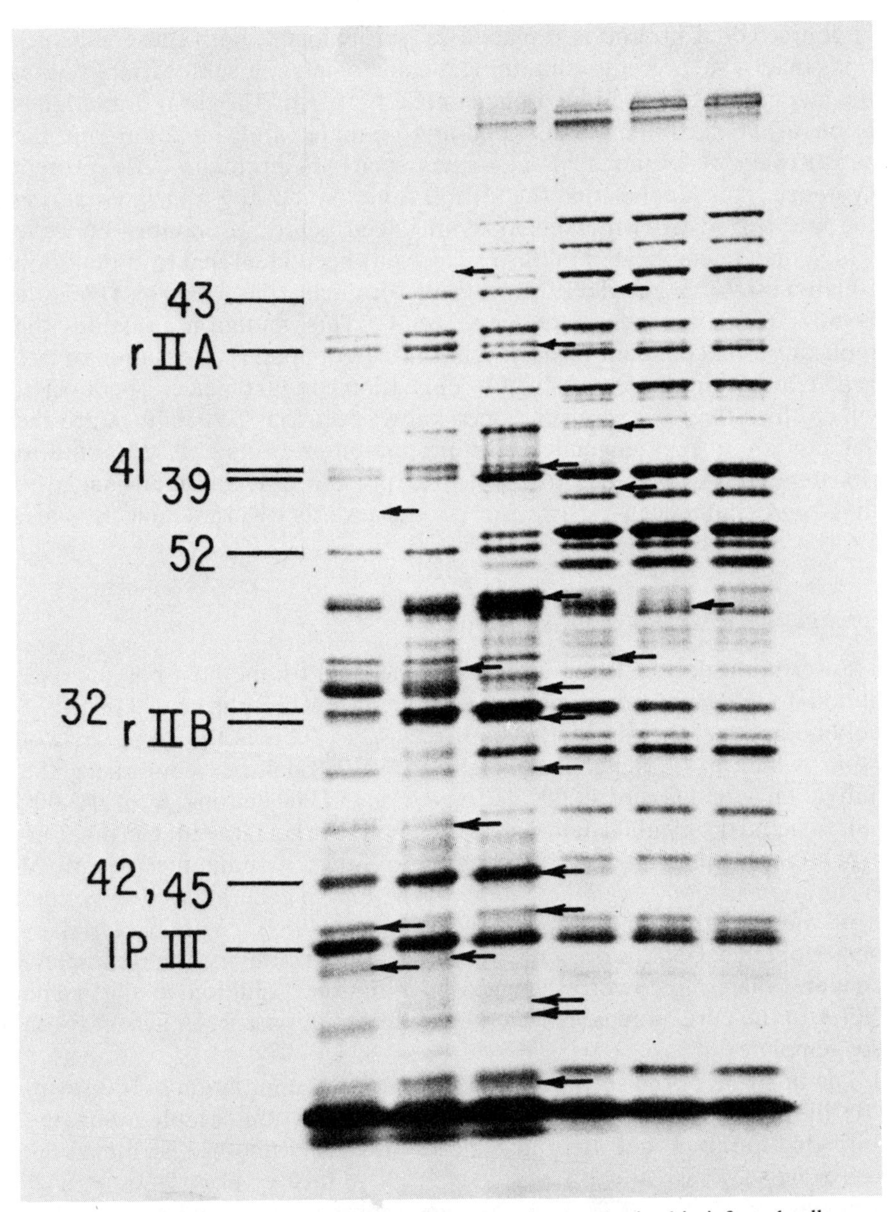


**Figure 6.15:** shutoff of early protein synthesis. Proteins synthesized in infected cells were labeled with $^{14}C$ amino acids for 5 minute periods at the various times after infection indicated. Some of the genetic functions identified with the protein bands separated on the SDS gels are indicated at the left. The changing pattern of protein synthesis is shown by the differences between the bands labeled at different times and the arrows indicate the shutting off of different early proteins at various times. Data of Wu, Geiduschek and Cascino (1975).

ethionine, or if proline is replaced by an analogue; both these substitutions inactivate proteins containing the analogues. This effect implies that the switch off is caused by a phage coded protein. The turnoff continues to occur when rifampicin is added at 10 minutes after infection; and the substitution of amino acid analogues then still prolongs early protein synthesis. This implies that the turnoff function must be translated during the late period from messenger synthesized before 10 minutes. A gene which may be implicated in this process has been identified by a mutation isolated by Wiberg et al. (1973); this was originally described as *SP62* and is now taken to locate the gene *regA*. This mutation prevents the replication-independent turnoff; it has no effect upon the cessation of late gene expression that occurs when replication can take place. The level at which this effect is exercised is not known; obvious possibilities are the degradation of messenger RNA or the inhibition of its ability to bind to ribosomes. The evidence available—at present only indirect—suggests that *regA* influences posttranscriptional events (Karam and Bowles, 1974).

### *Changes in RNA Polymerase Activity*

Transcription during T4 infection is carried out by the host polymerase, although changes in its activity are induced by the phage. E.coli RNA polymerase has the subunit structure $\alpha_2\beta\beta'\sigma$. This is the *complete enzyme*, which has a molecular weight of 495,000 daltons, comprising the individual subunits $\alpha$ of 40,000 daltons, $\beta$ of 155,000 daltons, $\beta'$ of 165,000 daltons, and $\sigma$ of 95,000 daltons (Travers and Burgess, 1969). Chromatography on phosphocellulose removes the $\sigma$ subunit, sometimes described as the *sigma factor*, from the complete enzyme to generate a *core enzyme* with the constitution $\alpha_2\beta\beta'$ (Burgess et al., 1969). With bacterial or bacteriophage DNA, core enzyme displays less activity than complete enzyme, due to a lower efficiency in initiation. Addition of the sigma factor to the core enzyme restores its ability to initiate RNA synthesis on the templates.

The function of the sigma factor is to control the initiation of transcription by the core enzyme. Sigma is released from the complete enzyme following initiation and then may be reutilized by another core enzyme aggregate (Travers and Burgess, 1969; Di Mauro et al., 1969; Gerard, Johnson and Boezi, 1972). Core enzyme therefore is the unit responsible for catalyzing RNA synthesis; sigma is involved only at initiation. Comparisons of the RNA transcribed from phage templates by core enzyme and by complete enzyme show that in the absence of sigma the transcripts represent random regions of the template, whereas in the presence of this subunit the RNA products are related to those found in vivo (Bautz, Bautz and Dunn, 1969; Bautz and Bautz, 1970; Sugiura, Okamoto and Takanami, 1970).

As might be expected from its function, core enzyme has a general affinity for DNA and the role of sigma appears to be to direct this so that the complete enzyme binds only to bona fide promotor sites; sigma does not itself bind to DNA, so that its effect upon initiation must be to influence the binding of the core complex. Addition of sigma reduces the ability of core enzyme to bind to random sites on DNA by at least two orders of magnitude (as measured by the association constant for the binding reaction); but the complete enzyme recognizes a small number of sites, presumably the promotors, with an affinity some two orders of magnitude greater than that characterizing the random binding reaction of the core (Hinckle and Chamberlin, 1972a, 1972b). Sigma factor therefore depresses random transcription but enhances specific transcription; it is responsible for the selection of sites at which the complete enzyme initiates transcription. (A more detailed discussion of the structure and function of bacterial RNA polymerase is given in Volume 1 of Gene Expression.)

The functions of the polypeptides of the core enzyme are not well defined. Mutants resistant to rifampicin (which inhibits initiation) or to streptolydigin (which inhibits elongation) occur in the $\beta$ subunit and their location identifies its gene, often described as the *rif* locus (Iwakura, Ishihama and Yura, 1973). The genes for both the $\beta$ and $\beta'$ subunits are located at this map position, about 79 minutes on the E.coli chromosome, and are organized in the array: promotor—$\beta$ gene—$\beta'$ gene (Errington et al., 1974). The $\alpha$ subunit is coded by a gene located at a different site, near 64 minutes, which may be coordinately expressed with the genes for some ribosomal proteins (Jaskunas, Burgess and Nomura, 1975). The gene for the $\sigma$ factor has not yet been identified.

An indirect experiment implying that the host enzyme continues to be involved in phage transcription was reported by Haselkorn, Vogel and Brown (1969). Using host cells from strains either sensitive to rifampicin or resistant to the drug, they tested whether the response of transcription to addition of rifampicin remains the same during phage infection as in the uninfected parental cells. Phage transcription at all times retains the response to rifampicin that is characteristic of the host cell; this demonstrates that at least the $\beta$ subunit (and probably all the other subunits) of the core remain part of the polymerase activity that transcribes phage DNA. A similar conclusion for another phage system was reported by Geiduschek and Sklar (1969). In a more direct study, Schachner and Zillig (1971) obtained tryptic fingerprints of E.coli core enzyme subunits during

T4 infection; the $\alpha$, $\beta$ and $\beta'$ subunits remained substantially unaltered, although changes in a small number of spots indicated that some modifications are made. These experiments imply that no new polymerase activity is synthesized for phage transcription, which is undertaken by a modified host enzyme.

When core polymerase is used to transcribe T4 DNA, both strands are used as templates for RNA synthesis; under these conditions initiation appears essentially to be random. But when sigma factor is present, the transcripts labeled in vitro are competed effectively by RNA extracted from infected cells at 5 minutes (Bautz, Bautz and Dunn, 1969). Transcription by complete host enzyme represents the system for in vitro gene expression that we have already discussed; the immediate early genes are transcribed initially and then the delayed early genes are transcribed after a lag which appears to represent the time taken for RNA polymerase to read through the immediate early genes. This implies that the bacterial sigma factor directs the core polymerase to initiate transcription at only the immediate early promotors of the phage DNA.

At least two changes must be made in the transcriptional specificity of the host enzyme during the course of infection. A product(s) of an immediate early gene must be responsible for switching on the transcription of the delayed early genes. This may involve acquisition of the ability to initiate at a new class of promotors (the delayed early) or abolition of the termination that occurs at the ends of the immediate early genes; or both mechanisms may be involved. An initiation model implies that some change must be made in the polymerase itself; an antitermination model could be satisfied by a change either in polymerase itself or by antagonism of the termination factor. The nature of the early change in transcriptional specificity is not yet defined. Since late genes are transcribed from the *r* strand instead of the *l* strand that is transcribed in the early period, new initiation events at late promotors must be required for late initiation. This involves a change in the state of the template, apparently depending upon the introduction of nicks; there is some evidence that polymerase modified for late transcription is unable any longer to transcribe early genes. (The classification of two changes in transcriptional specificity ignores the problem of control of quasi-late transcription.)

Since it is the host sigma factor that is responsible for recognition of the immediate early phage genes, early experiments were designed to test whether replacement of this component by a phage coded sigma factor might account for the switch on of delayed early or of late genes. No substitute sigma factor has yet been shown unequivocally to make the necessary changes in transcription specificity; but several changes in the core enzyme have been documented during the course of infection, and one consequence of these is to lower the affinity of reaction between core and sigma, very possibly with some accompanying change in transcriptional specificity. Several phage proteins form an aggregate with RNA polymerase late in infection, and although their functions are not known, this presumably alters the specificity of transcription.

Two types of change have been identified in the $\alpha$ subunit following infection. *Modification* requires phage protein synthesis and results in an increase in negative charge and a slight increase in gel mobility. Goff and Weber (1970) showed that this is due to the attachment of a phosphorus-containing nucleotide; and Goff (1974) demonstrated that the modified subunit contains one adenine nucleotide (probably 5′ AMP) per polypeptide, linked to an arginine residue. Horvitz (1974a) reported that the entire $\alpha$ subunit content of the infected cell is modified; modification occurs largely at 3–5 minutes, is complete by 7 minutes, and then remains stable. A gene responsible for this process was identified by Horvitz (1974b); a deletion of 500 base pairs in the nonessential region between genes *39* and *56* confers the *mod*$^-$ genotype. *Alteration* represents another change in the $\alpha$ subunit, also involving a phosphorylation; this occurs whenever infection is performed in the presence of an inhibitor of protein synthesis. At 2 minutes after the infection a small proportion of the $\alpha$ subunit content has been altered, but by 10 minutes the alteration has been reversed (Horvitz, 1974a). The nature and significance of alteration is unknown; it is not prevented by the *mod*$^-$ deletion and since it occurs only in abortive infection it may be unrelated to normal phage development.

The modification of the $\alpha$ subunit may be involved in reducing the ability of RNA polymerase to recognize bacterial promotors. Mailhammer et al. (1975) developed a coupled transcription-translation system for assaying gene expression in vitro; since it is the protein products that are assayed, this follows meaningful transcription rather than total RNA synthesis, which may include events due to initiation at many sites other than the true promotors. The extract for coupled gene expression was derived from a rifampicin-sensitive host strain; rifampicin was then added together with RNA polymerase extracted from a *rif*$^r$ strain. This ensures that all transcription is due to the added enzyme. Enzyme extracted from T4 infected cells at 8 minutes transcribed bacterial *lac* or *trp* genes (carried on phage DNA) much less efficiently than the enzyme extracted from uninfected cells. When the polymerase was reconstituted from its subunits, the $\alpha$ subunit of the infected cells appeared to be responsible for the reduction in effectiveness. This suggests that the $\alpha$ modification may lower the affinity of the enzyme for bacterial promotors. (This is inconsistent with an earlier suggestion of Horvitz, 1974b, that host transcription still is inhibited in the absence of $\alpha$ modification; but in these experiments a high multiplicity of phage infection was used and this might have inhibited host transcription by other means.)

Changes in the $\beta$ and $\beta'$ subunits are less well characterized than those in the $\alpha$ subunit. Travers (1970a) showed that the modification of $\beta'$ takes place late in infection and appears to represent the addition of negative charge with little or no change in molecular weight. The fingerprinting studies of Schachner and Zillig (1971) confirmed that there is a change in only one tryptic spot of $\beta'$; there appear to be several differences in the $\beta$ tryptic fingerprint. The functions of these changes are not known.

When E.coli RNA polymerase is extracted from infected cells, it appears to have much reduced host sigma activity (Bautz and Dunn, 1969; Travers, 1969). This might be due to either of two effects. The host sigma subunit may be modified or replaced by a phage coded polypeptide. Or less of the sigma factor may purify with the enzyme because of alterations that have taken place in the core subunits. There have been various reports of changes in the nature of the association between sigma and the core in vitro when enzyme is extracted from T4 infected cells, but it has been difficult to establish the extent to which this reflects physiological processes relevant to the control of transcription rather than a changed response to in vitro conditions. However, it seems likely that these changes are correlated with the activity of the enzyme in the infected cell.

In early experiments, Travers (1969, 1970b) reported the isolation of what appeared to be a new phage coded sigma factor. Addition to the core enzyme of infected cells of a ribosomal supernatant fraction stimulated the transcription of delayed early genes, measured by the competition for the RNA product of transcripts extracted from infected cells treated with chloramphenicol or after various periods of normal infection. However, these experiments were not able to demonstrate whether this stimulation was due to new initiation events. Identification of new promotors at which a putative sigmalike factor directs the core polymerase to initiate transcription is necessary in order to establish this as a mechanism of action of a new factor.

Although present in lesser amounts in RNA polymerase extracted from infected cells, sigma factor nonetheless is present in the cell. Stevens (1972) reported that when E.coli cells are labeled with $^3H$ amino acids before phage infection, the extracted polymerase contains the label in all three core subunits. In spite of the lack of sigma in this extract, when purified unlabeled sigma is added as a carrier, the labeled sigma can be extracted from the infected cells. This implies that the host sigma factor remains in the cell but is more poorly associated with the core. By determining the amount of sigma present at each stage of purification of polymerase, Stevens (1974) showed that during extraction there is a greater loss of sigma from the enzyme of infected cells. Purification involves three steps: fractionation with protamine sulfate, chromatography with DEAE-cellulose, and density gradient centrifugation. With uninfected cells the crude product of the first stage purification has 0.6 molecules of sigma per core; after the final step there are 0.4 molecules of sigma per core. When extracted from infected cells, the product of the first stage purification has 0.4 sigma subunits per core; this is reduced to 0.3 by the DEAE cellulose chromatography and to 0.15 by the density gradient centrifugation.

The change in the affinity of sigma for the core enzyme of infected cells presumably is caused by the modifications that are made in the core. This may indeed reflect a change in transcriptional specificity, but in order to assess this it is necessary to use conditions in vitro that assay comparable preparations, that is, in which the amount of sigma is the same—otherwise a reduction in activity of the infected polymerase may simply reflect the loss of sigma during preparation. Indeed, Stevens found that the specificity of transcription by the enzyme of infected cells was correlated with its content of sigma factor. Some of the fractions not associated with the core enzyme of infected cells displayed a stimulatory activity; but they inhibited the stimulatory effect displayed by the host sigma factor. This demonstrates that the activity of the polymerase may be influenced strongly by the conditions of assay and emphasizes the need to take this into account in interpreting results obtained in vitro. Another important parameter is the concentration of KCl. Stevens found that 0.1–0.25 M KCl stimulates the reaction of host enzyme with T4 DNA, but the activity of the enzyme extracted from infected cells is inhibited at KCl levels greater than 0.1M. The same result is obtained with both cores and complete enzymes. The effect of KCl may be exerted at more than one stage of transcription and is not yet clear; but the implication is that the salt dependence of the polymerase is changed after infection and this influences attempts to assay its activity in vitro.

By labeling host cells with $^{3}H$ amino acids before infection and with $^{14}C$ amino acids after infection, Stevens (1972) was able to identify new bands associated with the preexisting RNA polymerase. When the constituents of the enzyme were separated by electrophoresis on SDS gels, in addition to the usual subunits, $\alpha$, $\beta$, $\beta'$, $\sigma$, which were labeled with $^{3}H$, four new bands were labeled with $^{14}C$. All are small polypeptides, with weights of 22,000, 14,000, 12,000 and 10,000 daltons. When the $^{14}C$ label was added at varying times after infection, the 14,000 and 10,000 dalton proteins were labeled at 5–6 minutes and all of the proteins were labeled at 11–12 minutes. The 22,000 polypeptide is present at a level of 0.6 molecules per core, and its affinity for the enzyme may depend upon the presence of the sigma factor. The 14,000 protein is present at 1 copy per core and remains associated with the core upon chromatography on phosphocellulose. The 12,000 protein is present in lower amounts, 0.2 copies per core, and is lost upon phosphocellulose chromatography. The 10,000 protein is lost very readily during purification, raising a question about the specificity of its association with the polymerase.

Following infection by a *55*⁻ mutant phage, the 22,000 protein was absent from the polymerase preparation; with a *33*⁻ mutant the 12,000 protein was missing. Are genes *33* and *55* the structural loci coding for these polypeptides, or is their function to control the association of these proteins with the core? Horvitz (1973) demonstrated that *33* is the structural gene for the 12,000 dalton protein; when *33am* mutants were grown in a suppressor host that inserts tyrosine at amber codons, the number of tyrosine residues in this protein was changed. And by showing that in *55ts* infection the association of the 22,000 dalton protein with the polymerase is more labile, Ratner (1974) was able to suggest that gene *55* codes for this protein. These experiments therefore suggest that the roles of genes *33* and *55* in late transcription are mediated by the association of their products with RNA polymerase; what functions they fulfill remain to be established. The identification of the 12,000 and 22,000 dalton proteins with the products of these loci itself demonstrates that their association with polymerase reflects a phage function. The sources of the other two proteins remain unknown; to assess their significance it will be necessary to identify the genes that code for them.

Another gene whose product appears to interact with RNA polymerase is *45*. Coppo et al. (1975a,b) developed a genetic system for investigating phage-host protein interactions. Bacterial $tab^{ts}$ mutants support T4 growth only at the low permissive temperature and are blocked at high temperature; phage *com* mutants are able to grow on $tab^{ts}$ cells at the high temperature, whereas phage *k* mutants are unable to grow on $tab^{ts}$ cells even at the low temperature. This system was originally used to characterize *tabB* bacterial mutants, which interact with $com^B/k^B$ mutants in T4 gene *31*, which are concerned with head assembly. It proved impossible to isolate *tab* mutants in gene expression by starting with $T4^+$, perhaps because they would generate nonviable bacteria. But starting with a $55^{ts}$ mutant it was possible to isolate *tabD* bacteria on which the mutant phage cannot propagate even at low temperature, that is, the $55^{ts}$ mutation corresponds to the *k* type. The defect in *tabD*-$55^{ts}$ infection at 30°C is the same as that seen in E.coli$^+$–$55^{ts}$ at 40°C; this suggests that *tabD* identifies a bacterial function with which the product of gene *55* interacts. Another *tabD* mutant was isolated by using a $45^{ts}$ phage to select for the bacterial mutation at 30°C. In the *tabD*-$45^{ts}$ system, late transcription is inhibited but replication is not affected; this suggests that whatever function the gene *45* product plays in late transcription can be dissociated from its function in replication. The *tabD* mutations map in the *rif* locus. Two mutations isolated by the interaction with *55*$^{ts}$ probably map in the gene for the polymerase $\beta'$ subunit; and the mutation isolated by the interaction with *45*$^{ts}$ appears to lie in the $\beta$ subunit.

Further genetic evidence that gene *55* and *45* proteins interact with RNA polymerase is provided by the behavior of cells carrying a *rif*$^{r}$ mutation. Snyder and Montgomery (1974) reported that one RNA polymerase mutation conferring resistance to rifampicin retards T4 development, presumably because the mutation alters the $\beta$ subunit in such a way that transcription of T4 DNA is impeded. Mutations of T4 characterized as *gor* allow the phage to grow normally on the rifampicin-resistant host. Some *gor* mutations map close to the *55* gene and may lie in it; another lies within gene *45*. (Most of the *gor* mutations lie in the gene *βgt* which specifies the $\beta$-glucosyltransferase enzyme that is responsible for $\beta$-glucosylation of hydroxymethylcytosine; another enzyme is responsible for the $\alpha$-glucosylation of hydroxymethylcytosines. This implies that the presence of $\beta$-glucosylated bases inhibits the activity of the *rif*$^{r}$ polymerase. This concept is consistent with demonstrations that the substitution of cytosine for hydroxymethylcytosine causes a defect in late protein synthesis. We may speculate that $\beta$-glucosylated hydroxymethylcytosine residues in some way influence late transcription; perhaps some are located in promotor sites.)

All three of the protein products of the genes that control late transcription therefore appear to interact with RNA polymerase. The change in transcriptional specificity brought about by their interaction with the enzyme remains to be characterized in vitro. A difficulty in doing so is that presumably it is related to the change that must take place in the state of the template; no RNA synthesis in vitro has yet been performed with a template DNA competent for late transcription.

### *Protein Synthesis in Infected Cells*

Following infection with phage T4, host nucleic acid and protein synthesis is shut off, to be replaced by the transcription and translation of phage genes. Synthesis of gene products may in principle be controlled at three stages: transcription of the messenger, its degradation, and translation into protein. In the uninfected bacterial host, transcription represents certainly the principal and perhaps the only level at which gene expression is controlled. Phage T4 development also proceeds through a program of transcriptional controls. The immediate early genes of the phage are transcribed and (presumably) translated by the systems of the host cell. Product(s) of these genes then modify the transcriptional system so that further classes of phage genes can be expressed; these modifications also may be responsible for the cessation of transcription of host genes. There is some evidence that phage messengers are more stable than those of the host cell, with an average half life of 9–10 minutes compared with the host values of little more than 2 minutes (Craig, Cremer and Schlessinger, 1972); but in the absence of any evidence of discrimination between phage messengers, this does not appear to constitute a developmental control. Although phage T4 specifies certain transfer RNA species and perhaps also some ribosomal proteins, the translation of phage messengers relies largely upon the preexisting protein synthetic system of the host. Are the phage messages translated in this system as though they were endogenous species, or does their translation represent another stage at which gene expression is controlled by the introduction of modifications in the host apparatus?

The first suggestion that changes in protein synthesis may be induced by T4 phage infection was provided by the report of Hsu and Weiss (1969) that ribosomes of T4-infected cells translate the MS2 phage RNA message less efficiently (at a reduced level of 12%) than ribosomes from uninfected cells. Infection in the presence of chloramphenicol does not induce this change. The most likely stage of protein synthesis for this effect to be exerted is initiation. Ribosomes carry factors necessary for initiation that can be removed by a salt wash (usually with 1 M $NH_4Cl$). Experiments with active ribosomes reconstituted from the salt-washed preparation and the extracted initiation factors showed that it is the factors that are responsible for the decline in translation of MS2 RNA. Klem, Hsu and Weiss (1970) and Dube and Rudland (1970) confirmed this conclusion by measuring directly the formation of initiation complexes between message, ribosome, and fmet-$tRNA_f$ (the initiator tRNA); the initiation factors extracted from infected cells are much less effective in sponsoring complex formation with MS2 RNA than the factors of uninfected cells.

Three initiation factors are implicated in the formation of the initiation complex, which involves two stages: binding of the 30S ribosomal subunit to mRNA; and junction with this complex of fmet-$tRNA_f$. The role of initiation factor f1 (IF 1) is not entirely defined but appears to be to assist both stages. The function of IF 2 is to stimulate binding of fmet-$tRNA_f$ to the 30S–mRNA complex. The factor IF 3 has two roles. In its capacity as a dissociation factor, IF 3 joins a 70S ribosome to release the 30S subunit from its 50S partner. As a factor needed for 30S binding to mRNA, the presence of IF 3 is necessary for formation of the complex at the initiation site on the messenger. Following the association with this complex of fmet-$tRNA_f$ and the addition of the 50S subunit, IF 3 is released to participate in another cycle of dissociation and initiation. The role of IF 3 is therefore formally analogous to the function of sigma factor in transcription: it is needed for the formation of specific initiation complexes but does not participate in the subsequent elongation of the chain. In this capacity it must be responsible for the recognition of initiation sequences on messenger RNA, so that it is in theory possible that changes in this factor might control the specificity of protein synthesis by selecting the messengers that are to be translated. (The role in initiation of IF 3 and the other initiation factors is discussed in more detail in Volume 1 of Gene Expression.)

How specific is the change in translation capacity that is induced by phage infection? Hsu and Weiss (1969) found that ribosomes of infected and uninfected cells appeared to have equal ability to translate T4 messengers, so the reduction in their ability to recognize MS2 RNA was taken to represent a specific inhibition of their ability to recognize certain classes of messenger while leaving unimpeded the capacity to translate T4 phage mRNA. This alteration in initiation specificity appeared to lie with IF 3. Lee-Huang and Ochoa (1971) reported that the factor extracted from T4-infected cells was twice as active in translating T4 late messages as the factor from uninfected cells, was 0.6 times as active with T4 early mRNA, and was only 0.3 times as active with MS2 RNA. Goldman and Lodish (1972) confirmed the observation that ribosomes from T4-infected cells have reduced ability to translate several natural mRNAs and that this effect resides in the initiation factors; but in contrast with earlier reports, they found that this reduction applied equally to all the messengers tested, including T4 late mRNA, and therefore represents a nonspecific decrease in translational efficiency. And in experiments to characterize IF 3 from uninfected and infected cells, Schiff, Miller and Wahba (1974) observed that the same translational specificity is found in both preparations. This suggests that although in some experiments a decrease in the efficiency of initiation of protein synthesis has been observed in infected cells, there is no specific change that represents a control mechanism.

[In contrast with this conclusion, several claims have been made that IF 3 can be fractionated into different species with varying translational specificities and that the interaction of these variants with specific "interference factors" may play a role in the control of translation: changes in the IF 3 species and/or the interference factors therefore might be induced following phage infection in order to control protein synthesis. It remains necessary to explain these observations, which now appear to represent artifacts of the systems for protein synthesis in vitro.

One system that has been widely used is the translation of RNA phage message, which is discussed in more detail in Chapter 9. A specific effect induced by phage T4 was claimed by Steitz, Dube and Rudland (1970), who reported that infection induces changes in the relative frequencies of translation of the three cistrons of the phage, as well as an overall reduction in their translation. This effect appeared to reside in the initiation factors. Yoshida and Rudland (1972) fractionated IF 3 on DEAE cellulose into two major and two minor components and claimed some differences in their abilities to support initiation of the RNA phage proteins. But Lee-Huang and Ochoa (1973) reported that their preparations of IF 3 showed the same relative efficiencies with all three cistrons. The translation of the phage RNA is influenced by its conformation; for example, fragmentation results in a change in the relative frequencies of the three initiation events (Steitz, 1973). It is therefore likely that any apparent discrimination between these initiation sites resulted from artifacts of this system.

Using phosphocellulose columns, Lee-Huang and Ochoa (1971) reported that IF 3 could be fractionated into two species: IF 3$\alpha$ supported low activity with T4 late messages but high activity with MS2 RNA, whereas IF 3$\beta$ displayed the reverse affinities for these messengers. This led to the suggestion that inactivation of IF 3$\alpha$ might be utilized in phage T4 infection. But Schiff, Miller and Wahba (1974) were able to demonstrate the presence of only a single species of IF 3, the same found in both infected and uninfected cells. They suggested that the apparent isolation of more than one species might be due to damage occurring during extraction. In the absence of any demonstration that there are species of IF 3 identified as distinct proteins by (for example) tryptic fingerprints, and in lieu of any documented change in IF 3 following phage infection, we may conclude that the control of initiation of protein synthesis by this mechanism is not utilized during T4 infection. The interference factors that inhibit the action of IF 3 do not possess the specificity that was at one time attributed to them, and it is now clear that they represent an artifact due to the presence of an excess of ribosomal protein in the in vitro system (see Chapter 9). The possibility that phage infection may influence the specificity of protein synthesis through such factors therefore now also is excluded.]

Although translation in infected cells relies upon the host protein synthetic apparatus, eight new transfer RNA species are added to the existing complement during T4 infection. Comparison of the tRNA synthesized in infected and uninfected cells shows that new species are synthesized after infection with T2, T4 or T6 (Weiss et al., 1968; Tillack and Smith, 1968; Daniel and Littauer, 1970). Scherberg and Weiss (1970) identified five species by virtue of the amino acids which they accept: these are arginine, proline, glycine, isoleucine and leucine. Wilson and Kells (1972) identified another tRNA by mutations in it that suppress amber or ochre mutations in the phage; Wilson and Abelson (1972) identified the locus coding for this tRNA, which represents serine. An ochre suppressor in a glutamine tRNA was isolated by Comer, Guthrie and McClain (1974) and Seidman, Comer and McClain (1974). By using separation on acrylamide gels to analyze small RNA species labeled after infection, McClain, Guthrie and Barrell (1972) identified eight transfer RNA species and four components larger than the tRNAs. Three of the tRNAs were identified by their fingerprints as species coding for serine, leucine and glycine; the large components appear to be precursors to the tRNAs. By examining tRNA synthesis in deletion strains of T4, Wilson, Kim and Abelson (1972) demonstrated that all eight transfer RNAs map at the same locus, close to the *e* gene.

Cells infected with T4 represent a good system for examining the synthesis of transfer RNA from its precursors. Guthrie et al. (1973) identified three precursors, each of which contains the sequences of two of the T4 tRNAs. Two precursors accumulate in large amounts: one carries the sequences of $tRNA^{Ser}$ and $tRNA^{Pro}$; the other carries sequences of two transfer RNA molecules not yet identified with amino acid capacities. The third precursor is present in smaller amounts and carries the sequences for $tRNA^{Leu}$ and $tRNA^{Gln}$. The complete sequence of the precursor for $tRNA^{Ser}$ and $tRNA^{Pro}$ has been determined by Barrell et al. (1974); it possesses a total sequence of 175 bases, which is 13 bases longer than the combined length of the two transfer RNAs; and the additional bases are located at the 5′ end, between the tRNA sequences, and at the 3′ end. Guthrie (1975) has sequenced the dimeric precursor to $tRNA^{Leu}$ and $tRNA^{Gln}$; this has six additional bases, all located at the 5′ end of the molecule. Processing of these precursors to yield the mature tRNAs appears to rely entirely upon enzymes of the host. A different series of events is implicated for each precursor, and for host tRNA maturation, which also appears to involve cleavage from longer precursors, in at least some cases involving dimeric molecules (for review see Altman, 1975). The tRNA genes are characterized as nonessential functions since their deletion does not cause inviability, although it may reduce the yield of phage. At present, therefore, it is not possible to ascribe any indispensable function to the phage coded tRNA species.

Certain other changes in the protein synthetic apparatus also have been reported, but in no case is it clear that any essential function is served. Marchin et al. (1972) reported that following infection a polypeptide of 10,000 daltons is added to the host valyl-tRNA synthetase enzyme; McClain et al. (1975) identified the gene responsible, *vs*, by a mutation located between *rI* and *e*. This appears to be a nonessential function. Comer and Neidhardt (1975) showed that the modification has no discernable effect upon the enzyme activity in vitro.

Another function involved in protein synthesis was identified by the suppressor located between genes *31* and *32* that was isolated by Ribolini and Baylor (1975); this appears to have the property of suppressing all three classes of nonsense mutant. The protein coded by this locus has not yet been identified; it might influence translation either through tRNA or via the ribosome.

In summary, then, changes in the apparatus for protein synthesis are induced during T4 infection, but the necessity for them has not been demonstrated. There is no evidence to suggest that specific changes in protein synthesis are used as a control additional to the principal control that is exerted over transcription, although of course this remains possible in principle.

### Replication and Recombination

The early functions of T4 are defined by mutations that result in inability to synthesize DNA (the DNA-negative or *DO* class), a rapid cessation of DNA synthesis (DNA arrest or *DA* mutants), or a delay in the onset of replication (DNA delay or *DD* mutants). The mutants completely unable to synthesize DNA fall into two classes: enzymes concerned with the replication of phage DNA; and enzymes involved in the synthesis of nucleotide precursors. Enzymes concerned with replication include DNA polymerase (gene *43*), unwinding protein (gene *32*) and DNA ligase (gene *30*); and the products of genes *41, 44, 45* and *62* appear to comprise part of the replication apparatus. Essential enzymes in nucleotide metabolism are dCTPase (gene *56*), dCMP hydroxymethylase (gene *42*) and deoxynucleotide kinase (gene *1*). The functions of the remaining two *DO* genes, *59* and *63*, have not yet been identified. None of the functions of any of the *DD* genes, *39, 52, 58, 60* and *61*, has yet been determined. The two *DA* genes, *46* and *47*, either code for or control the activity of a nuclease that plays an essential role in both replication and recombination; many of the *DO* genes also are involved in recombination as well as replication. In addition to the essential functions identified by conditional lethal mutations, several other enzymes concerned with nucleotide metabolism are specified by the phage. And several nucleases also are synthesized during infection, but the only one to have been identified as an essential function is the enzyme of genes *46* and *47*. Table 6.2 summarizes the known enzyme activities coded by early genes.

**Table 6.2:** early gene functions of phage T4 concerned with DNA synthesis

| | gene | protein |
|---|---|---|
| *replication functions* | 30 | polynucleotide ligase |
| | 32 | unwinding protein |
| | 41 | replication complex protein |
| | 43 | DNA polymerase |
| | 44 | replication complex protein |
| | 45 | replication complex protein |
| | 62 | replication complex protein |
| *nucleotide metabolism* | 1 | deoxynucleotide kinase |
| | 42 | dCMP hydroxymethylase |
| | 46 | dCTPase |
| | $\alpha$gt | DNA $\alpha$-glucosyl transferase |
| | $\beta$gt | DNA $\beta$-glucosyl transferase |
| | cd | dCMP deaminase |
| | td | thymidylate synthetase |
| | tk | thymidine kinase |
| | nrdA-C | nucleoside diphosphate reductase |
| *nucleases* | denA | endonuclease II |
| | denB | endonuclease IV |
| | dexA | exonuclease A |
| | pse | 3′ phosphatase |
| | 46–47 | unidentified nuclease activity |
| *other* | unf | host chromosome unfolding |
| | ndd | host nuclear disruption |
| | imm | superinfection immunity |

### *Degradation of Host DNA*

Infection of a host cell with phage T4 is followed by degradation of the bacterial DNA to nucleotides that are reutilized in the synthesis of phage DNA. The DNA of phage T4 is distinguished from that of the host by the presence of hydroxymethyl-cytosine in place of cytosine; and in addition some of the hydroxymethyl groups carry an $\alpha$-glucosyl or $\beta$-glucosyl residue. The nucleases responsible for degrading bacterial DNA appear (in vitro) to act upon DNA that contains cytosine, so that it is the presence of hydroxymethyl-cytosine in the phage DNA that renders it immune from attack by the phage coded enzymes. The glucosyl residues are necessary to protect the phage DNA against restriction enzymes coded by the host bacterium.

The degradation of host DNA cannot be followed in wild type infection because of the reutilization of its component nucleotides in phage replication, but can be followed during infection with a phage mutant unable to replicate. The conversion of a preexisting label in bacterial DNA to acid soluble fragments after wild type and mutant infections is shown in Figure 6.16. During wild type phage replication the nucleotides released from breakdown of the bacterial chromosome are incorporated into phage DNA, although phage replication is not inhibited by their absence when breakdown does not occur.

Degradation of the bacterial chromosome appears to involve three types of reaction. First the compact folded structure of the chromosome is lost; this can be followed by the change that takes place in its sedimentation characteristics as it is unfolded (Tutas et al., 1974). The second reaction can be directly visualized in the bacterial cell: the centrally located nucleoid is replaced by what appear to be clumps of genetic material dispersed to the membrane (Snustad et al., 1972). And then, apparently independently of the unfolding or nuclear dispersion, an endonuclease cleaves the DNA into comparatively small fragments that are further degraded by another nuclease activity into acid soluble pieces.

A phage function involved in the unfolding reaction has been identified by Snustad et al. (1976a): the *unf* mutant is unable to unfold the host chromosome. The mutant may be interesting not only for its role in the

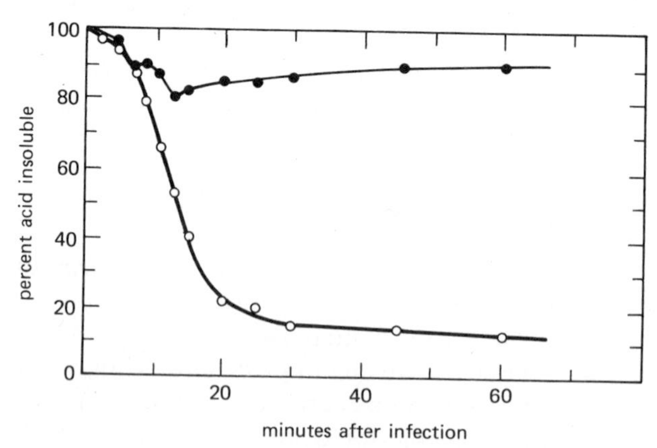


**Figure 6.16:** degradation of host DNA after T4 infection. In infection by T4$^+$ (•) no degradation is observed because any nucleotides released from host DNA are reutilized in the synthesis of phage DNA. But after infection by a *42*$^-$ mutant lacking dCMP hydroxymethylase activity (○), which does not synthesize phage DNA, the degradation of the previously exising bacterial DNA to acid soluble fragments can be followed. Data of Kutter and Wiberg (1968).

phage life cycle (which appears to be dispensable), but also because the nature of the protein activity coded by this gene may help to define the forces responsible for maintaining the folded structure. Another mutation, *ndd*, identified by Snustad et al. (1976b), is unable to accomplish the disruption of the nucleoid. Neither unfolding nor nuclear disruption is necessary for the degradation of host DNA which occurs as usual in both *unf* and *ndd* mutants. Following infection by an *unf* mutant phage, host DNA synthesis is shut off as usual; but in infection by *ndd* phage, there is a delay in the shutoff of bacterial replication from its usual time of about 5 minutes until about 10 minutes after infection. Nuclear disruption may therefore be involved in the cessation of bacterial DNA synthesis.

Mutants of T4 that fail to degrade host DNA were isolated by Warner et al. (1970) and proved to be in the gene *denA* that codes for the enzyme endonuclease II. As summarized in Table 6.3, this nuclease makes single strand breaks in DNA containing cytosine; it is inactive with a substrate of T4 DNA in which cytosine has been replaced by hydroxymethylcytosine. A small amount of degradation of host DNA is performed by *denA* mutants, but this is abolished by the mutation *denB* which removes endonuclease IV, another enzyme that is specific for DNA possessing cytosine (Souther, Bruner and Elliot, 1972). The combination of *denA* and *denB* mutations entirely prevents degradation of bacterial DNA (Kutter et al., 1975). The absence of these enzymes does not prevent the shutoff of bacterial replication which must therefore be accomplished by mechanisms independent of the degradation.

Infection by a *56⁻* mutant, which lacks the enzyme dCTPase responsible for removing cytidylic acid from the precursor pool, results in the synthesis of phage DNA that contains cytosine; this DNA is rapidly degraded so that the lack of dCTPase is lethal (Bruner et al., 1972). But this degradation does not occur if the phage also carries the *denB* mutation that removes exonuclease IV. Kutter et al. (1975) reported that whereas $56^-$ mutants manufacture small amounts of DNA that is rapidly degraded, $56^-$ *denB* double mutants are able to synthesize normal amounts of DNA. The $56^-$ *denA* double mutant, however, accumulates little more DNA that the $56^-$ mutant phage, so that endonuclease II does not appear to be involved in the degradation of the cytosine-containing phage DNA. Thus although endonucleases II and IV display somewhat similar specificities in vitro for cytosine-containing DNA (see Table 6.3), their roles are different in vivo: endonuclease II is responsible for degrading host DNA while endonuclease IV appears to act to degrade any phage DNA that contains cytosine.

Degradation of host DNA takes place in two steps. The first is cleavage of the bacterial genome into fragments of about $10^6$–$10^7$ daltons: this is the

**Table 6.3:** nucleases specified by phage T4

| enzyme | activity in vitro | reference |
|---|---|---|
| endonuclease II | introduces single strand breaks in duplex DNA that contains cytosine, cleaving on 5′ side of dCMP | Sadowski and Hurwitz (1969)<br>Sadowski et al. (1969)<br>Ray et al. (1972) |
| endonuclease IV | cleaves single stranded DNA that contains cytosine, on 5′ side of dCMP | Sadowski and Bakyta (1972)<br>Souther et al. (1972)<br>Sadowski and Vetter (1973)<br>Bernardi et al. (1976) |
| endonuclease V | introduces single strand breaks on 5′ side of thymine dimers | Minton et al. (1975)<br>Simon et al. (1975)<br>Sato and Sekiguchi (1976) |
| exonuclease A | degrades duplex DNA from 3′ end | Warner et al. (1972) |
| exonuclease B | excises thymine dimers 5′–3′ from break made by endonuclease V | Shimuzu and Sekiguchi (1976) |
| DNA polymerase | exonuclease activity degrades duplex DNA from 3′-OH terminus | Goulian et al. (1968)<br>Nossal and Hershfield (1971) |
| 3′ phosphatase | removes 3′ phosphate groups | Depew and Cozzarelli (1974) |

action undertaken by endonuclease II. This reaction was distinguished from the second stage by Kutter and Wiberg (1968), who found that $46^-$ or $47^-$ mutants accumulate these fragments instead of degrading them to acid soluble nucleotides. The implication is that the nuclease controlled by these genes is responsible for the second stage of degradation. Of course, this nuclease also has an essential role in the phage life cycle, being necessary for both replication and recombination.

Glucosylation of the hydroxymethyl-cytosine is accomplished by the enzymes DNA $\alpha$-glucosyl transferase and $\beta$-glucosyl transferase, coded by the loci *αgt* and *βgt*. Phage DNA that lacks the glucosyl residues is restricted in E.coli by a host enzyme; E.coli strains K12 and B both possess two functions, $r2,4^+$ and $r6^+$, responsible for restricting nonglucosylated DNA of phages T2, T4 and T6, respectively. These restriction functions are analogous to the host modification and restriction functions that either methylate or degrade foreign DNA (these are discussed in Volume 1 of Gene Expression); only some of the sites that fail to be glucosylated under appropriate conditions appear to be targets for the host system and the phage DNA is protected against degradation so long as one strand is glucosylated. Another restriction system specific for nonglucosylated phage DNA is specified by phage P1. The glucosylation of T-even phage DNA and the bacterial systems that act upon nonglucosylated DNA have been reviewed by Revel and Luria (1970). A plausible speculation about the evolution of these systems is to suppose that the presence of hydroxymethyl-cytosine in the phage DNA allows it to specify enzymes that degrade bacterial DNA, the host restriction enzymes then allow the bacterium to prevent infection by the phage, and the glucosylation represents a response by the phage allowing its reproduction in spite of the presence of the restriction enzymes.

### *Synthesis of Phage DNA*

Phage mutants of the conditional lethal *DO* class include those that synthesize little or no DNA although all four triphosphate precursors are present. The products of these genes may therefore comprise the replication apparatus of the phage. The need for the products of six genes has been demonstrated by means of the in vitro complementation assay developed by Barry and Alberts (1972). This system consists of gently lysed infected cells in which the endogenous DNA provides the template; when the triphosphate precursors are added, DNA is synthesized. When the cells are prepared from bacteria infected with a phage carrying a mutation in a replication gene, the need for the protein coded by this gene is demonstrated by the absence of DNA synthesis; synthetic activity can be restored by addition of an extract from cells infected with phage able to produce this protein, and this assay can therefore be used to purify the protein.

In this system, the products of genes *32, 41, 43, 44, 45, 62* all are essential. Only two of the products have been identified with enzyme activities: gene *43* codes for the T4 DNA polymerase; and gene *32* codes for the unwinding protein. However, by using this assay system Morris, Sinha and Alberts (1975) have purified the other proteins. The products of genes *41* and *44* display DNA-dependent hydrolysis of GTP and ATP, respectively, and the product of gene *62* stimulates the activity of the DNA polymerase some 3–4-fold with a single stranded template. The gene *45* protein, of course, is necessary not only in replication, but also for transcription of late mRNA; one of its properties thus is the ability to interact with RNA polymerase (see below). The six purified proteins together are able to synthesize DNA in vitro, apparently using double strand as well as single strand templates, under conditions when DNA synthesis must be initiated de novo. One view of the T4 replication system is to suppose that these proteins (and perhaps also some others) form a complex that possesses all their enzyme activities and is responsible for synthesizing DNA at the replication fork. Clearly the existence of an in vitro system offers the prospect of elucidating the activities involved in replication and their interactions with each other.

Following the early demonstration by Aposhian and Kornberg (1962) that a new DNA polymerase is synthesized in T2 infection, the enzyme coded by T4 has been isolated and characterized in some detail. In general its properties are very similar to those displayed by DNA polymerase I of E.coli, an enzyme that apparently is utilized in a repair capacity rather than replication in the bacterial host. (Bacterial DNA polymerases are discussed in Volume 1 of Gene Expression.) The T4 DNA polymerase thus is unable to initiate DNA synthesis de novo, but is able to extend a primer with a single strand as template (Goulian, Lucas and Kornberg, 1968). This action is illustrated in Figure 6.17. This implies that a separate protein function(s) must be responsible for initiation; although there is yet no experimental evidence, it would not be surprising if this proves to involve the synthesis of a short RNA primer as has been observed in the E.coli host cell (see Volume 1 of Gene Expression).

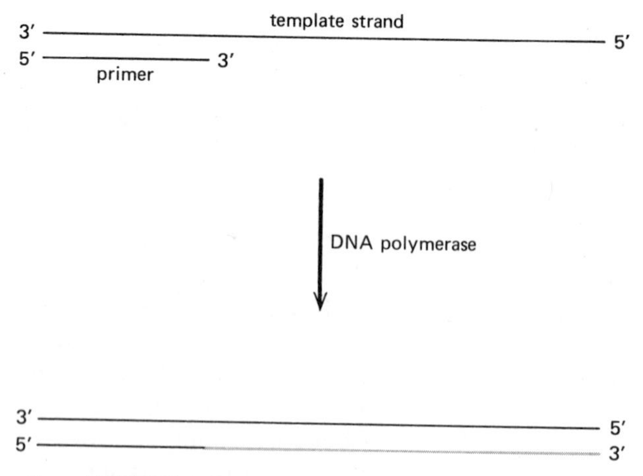


**Figure 6.17:** action of T4 DNA polymerase. Single stranded DNA can be used as template for synthesis of a new strand growing in the direction 5′ to 3′, starting from a primer which provides a free 3′-OH end.

In addition to its polymerase activity, the T4 enzyme possesses a 3′ exonuclease activity. The polymerase and exonuclease activities represent independent functions of the enzyme; Nossal and Hershfield (1971) were able to isolate an amber mutant in gene *43* which lacks polymerase but retains exonuclease activity, so that the exonuclease function must reside in the N-terminal part of the protein chain. T4 DNA polymerase lacks the 5′ exonuclease activity that is displayed by E.coli DNA polymerase I. The 3′ exonuclease activity of both enzymes may represent a proofreading function which increases the accuracy of replication by removing any mispaired bases before they are incorporated into the polynucleotide chain (Brutlag and Kornberg, 1972; Hershfield and Nossal, 1972; Huang and Lehman, 1972). Indeed, it is clear that the accuracy of replication is established by the T4 DNA polymerase, since mutations in gene *43* may have either a mutator or an antimutator effect, that is, may increase or decrease the frequency with which spontaneous mutations occur (Speyer, 1965; Hall and Lehman, 1968; Drake et al., 1969; Hershfield, 1973). The ability to isolate antimutator variants implies that it is not advantageous for the phage to reduce errors made in replication to a minimum; presumably this would impose an evolutionary cost that makes it preferable to suffer mutations at the frequency seen in "wild type" phage strains.

The gene *32* product was isolated by Alberts (1970) as one of the proteins present at early times in infected cells that is able to bind to DNA. (About twenty T4 early proteins bind to DNA and these presumably include other components of the replication apparatus.) The product of gene *32* is required for both replication and recombination, for *32*$^-$ mutants are defective in both functions. Gene dosage experiments have suggested that the 32-protein is required stoichiometrically rather than catalytically: Sinha and Snustad (1971) observed that when the number of functional copies of 32-protein was controlled by infecting cells with varying relative proportions of wild type and *32*$^-$ phages, the burst size depended upon the proportion of wild type genomes. (Interpretation of these results is complicated by the subsequent observation that 32-protein regulates its own synthesis, as does DNA polymerase, but the manner of this regulation probably means that the original interpretation remains valid: see Russel, 1973; Krisch et al., 1974.) As might be expected from this conclusion, the number of copies of 32-protein found in the infected cell is very large: Alberts (1970) estimated it to be about 10,000, which corresponds to more than 150 molecules of the protein per replication fork.

The 32-protein binds cooperatively to DNA. Alberts and Frey (1970) reported that its affinity for sites adjacent to those already occupied by another molecule of the protein is some two orders of magnitude greater than its affinity for new sites. The 32-protein also displays a complex mode of self-association, an activity that may be related to its action on DNA (Carroll, Neet and Goldthwait, 1972, 1975). The critical feature of 32-protein activity is that its affinity for denatured DNA is much greater than its affinity for duplex DNA; by virtue of its preferential affinity for denatured DNA, the protein is able to maintain single stranded DNA in this state. The effect of binding 32-protein to DNA essentially is to lower the Tm very substantially; in fact the protein is able to denature poly-dAT under physiological conditions, and therefore can generate a denaturation map by its action on A-T rich segments of DNA (Delius, Mantell and Alberts, 1972).

The 32-protein stimulates the activity of T4 DNA polymerase in vitro (Huberman, Kornberg and Alberts, 1971). Its activity has been discussed in detail in Volume 1 of Gene Expression; it is consistent with the idea that the role of the protein in T4 replication may be to make it possible to present the polymerase with a single stranded template. The in vitro ability of 32-protein is to stabilize DNA in the single stranded state, which implies that some other component of the replication apparatus may be needed to accomplish the initial unwinding of the duplex DNA. The model illustrated in Figure 6.18 thus supposes that an aggregate of 32-protein molecules maintains a region ahead of the replication fork in the single stranded state, ensuring that as DNA polymerase proceeds along the template it encounters only single strands.

The activity of T4 DNA polymerase is best visualized within the context of a replication apparatus containing other activities, including 32-protein, since the enzyme itself is unable to replicate duplex DNA, at least in vitro. Replication could be accomplished, however, by the extension from a primer of a new DNA strand, using as template the single strands maintained in the separated state by 32-protein. In this context it is interesting that Nossal (1974) observed that while T4 DNA polymerase cannot act in vitro to replicate duplex DNA possessing single strand nicks, it is able to do so in the presence of 32-protein. This suggests that the 32-protein may assist the displacement of the 5′ nicked strand. The 32-protein also influences the exonuclease activity of the enzyme: in vitro this is inhibited (Huang and Lehman, 1972). It is therefore important to remember that the activities displayed by T4 DNA polymerase within the replication apparatus may be somewhat modified from those possessed by the purified enzyme in vitro.

Synthesis of phage DNA takes place by the same mechanism of discontinuous assembly that has been well characterized for both procaryotic and eucaryotic DNA (reviewed in Volume 1 of Gene Expression). DNA is

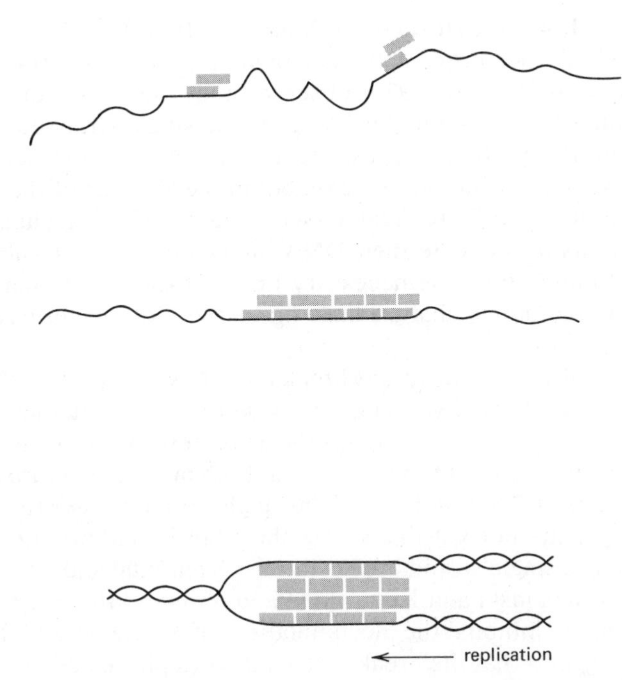


**Figure 6.18:** model for action of 32-protein. Above: monomers of the protein bind to isolated sites on DNA. Center: further protein molecules bind at sites adjacent to those already occupied. The protein has an elongated shape and its cooperative binding maintains the single stranded DNA in extended condition. Lower: model for the replication fork in which 32-protein maintains the two strands of the template DNA in separated state so that they can be used for synthesis of new chains by DNA polymerase.

synthesized in short pieces, the Okazaki fragments, in the direction 5′ to 3′ on both growing strands; these are later covalently joined together by a ligase. The activity of the T4 ligase was originally identified by the accumulation of Okazaki fragments during infection by temperature sensitive mutants in gene *30* (Okazaki et al., 1968; Newman and Hanawalt, 1968; Richardson et al., 1968). From these observations it appeared that the phage ligase must be essential for linking together the Okazaki fragments into a covalently linked chain. However, it now appears that it is possible for the host ligase to perform this reaction.

Analysis of the role of ligase is complicated by the effects of nuclease activities present in the infected cell. During infection by *lig*⁻ (gene *30*) mutants, T4 DNA is synthesized; but instead of accumulating as mature genomes, both the parental and newly synthesized DNA is degraded. This degradation is much reduced or eliminated when a mutation in gene *46* or

*47* is present. However, Hosoda and Mathews (1971) found that the *46*$^-$ or *47*$^-$ mutations do not prevent the accumulation of single strand breaks; and Prashad and Hosoda (1972) observed that the action of the *46-47*-controlled nuclease is to enlarge into gaps the single stranded nicks that are found in DNA. In the presence of the *46-47* nuclease, ligase is therefore essential to seal these nicks; but in the absence of the nuclease, DNA can be successfully replicated (see Beguin, 1973). The ability of the *30*$^-$ *46*$^-$ mutants to replicate their DNA implies that it is possible for the host ligase to undertake the necessary linking together of Okazaki fragments into a continuous chain. Phage ligase therefore is not essential for replication per se.

The source of the single strand breaks in DNA was at first a matter of some confusion: alternative suggestions were made that they were entirely the result of discontinuous synthesis of new DNA or due solely to the action of a nuclease. It now seems that both mechanisms are involved: newly synthesized DNA at first has the single strand breaks that separate Okazaki fragments not yet linked together; but in addition an endonuclease introduces single strand breaks into both parental and newly synthesized strands. Kozinski and Kozinski (1969) observed that the addition of chloramphenicol inhibits the accumulation of single strand breaks in DNA; this suggests that the breaks are made by a phage coded ''nickase'' activity, although the enzyme remains unidentified.

In addition to the proteins directly implicated in DNA replication, certain other functions, identified for their enzymatic roles in the synthesis of precursors, may be involved. These activities are responsible for the hydroxymethylation of cytosine in T4 DNA. The T4 dCTPase coded by gene *56* metabolizes the cytidylic acid present in the cell. The dCMP hydroxymethylase coded by gene *42* undertakes the first step of providing a substitute nucleotide triphosphate; this is the addition of a formaldehyde moiety to position 5 of dCMP (Flaks and Cohen, 1959). The addition of a phosphate group to yield hydroxymethyl-dCDP then is catalyzed by the kinase coded by gene *1* of the phage; the addition of a further phosphate to generate the precursor needed for DNA synthesis is accomplished by a host kinase.

An observation implying that dCMP hydroxymethylase may have another role as well as the provision of hydroxymethyl-dCMP was reported by Chiu and Greenberg (1968): a temperature sensitive mutant in gene *42*, isolated by its *DO* phenotype and therefore unable to synthesize DNA at the nonpermissive temperature, retains its hydroxymethylase activity. This suggests that the conditional lethality of the mutation is not due simply to a deficiency in the ability to supply the hydroxymethylated precursor. The role of the hydroxymethylase has been investigated by providing conditions in which replication relies upon added precursors, so that the enzyme is not needed for synthesis of hydroxymethyl-dCMP. When cells are plasmolyzed, exogenous nucleotide triophosphates can be utilized for DNA synthesis; Wovcha et al. (1973) reported that replication in this system is greatly reduced by mutations in genes *32, 43, 44* or *45*, which implies that it represents the functioning of the same system usually responsible for T4 replication. Added hydroxymethyl-dCTP is incorporated into DNA by a system derived from cells infected with $42^+$ phage; but a $42^-$ mutant remains unable to synthesize DNA in spite of this provision. This mutation has no effect on the ability to incorporate dCTP in cells plasmolyzed after infection with $42^-$ $56^-$ mutants. This suggests that the hydroxymethylase is needed not only for hydroxymethylation but also for incorporation of the hydroxymethylated precursor into DNA. In this case the enzyme might be expected to comprise part of the replication complex. The deoxynucleotide kinase product of gene *1* also appears to be necessary for functioning of the plasmolyzed system, although its enzymatic role is bypassed by the supply of exogenous precursors. Collinsworth and Mathews (1974) reported similar results, again showing a need for the *42* product in the plasmolyzed system, and observed that one mutant of gene *1* was able to support replication in these conditions while two other mutants were unable to synthesize DNA. Using a system of toluene-treated cells to achieve replication dependent upon added triphosphates, Dicou and Cozzarelli (1973) also found that the gene *1* and gene *42* activities are required even when all precursors are supplied.

Although one interpretation of these results is to argue that the products of genes *1* and *42* have some role additional to their enzymatic activity in the supply of precursors, an alternative is to suppose that the differences seen in vitro reflect conditions prevailing in the cells before they were converted for replication from added precursors; thus the number of growing points previously established in vivo might be too small in $1^-$ and $42^-$ infection to support proper replication in the in vitro system. These alternatives were distinguished by North, Stafford and Mathews (1976) by utilizing a temperature sensitive mutant of gene *42*. Cells were incubated in vivo at low temperature, to allow wild type establishment of replication, and then were plasmolyzed and incubated at high temperature to inactivate the hydroxymethylase in vitro; DNA synthesis still fails with hydroxymethyl-dCTP as substrate, implying that the deficiency is due to some direct role of the enzyme in replication. It is notable, of course, that the products of genes *32, 41, 43, 44, 45, 62* are sufficient for replication in the in vitro system of Morris, Sinha and Alberts (1975); in this system the need for the gene *1* kinase and gene *42* hydroxymethylase is bypassed by the provision of precursors. One possible interpretation of the discrepancy between the needs of the plasmolyzed and toluenized systems versus the purified proteins is that the former have retained some mechanism in which the kinase and hydroxymethylase are necessary for access of precursors to the replication complex, whereas in the latter it is possible for added precursors to be utilized directly by the replicating enzymes.

### *The Replication Cycle*

The T4 genome shares with lambda (and also with other phages) the characteristic of passing through a concatemeric state during infection: generation of concatemers is essential for the maturation of DNA into particles, which appears to proceed by packaging until the head is full of DNA, when an amount longer than the unit genome is cleaved from the concatemer. However, with T4 the concatemers appear to be generated by recombination, contrasted with lambda where they are generated by late replication of a rolling circle (see Chapter 5). Both the state in which T4 DNA is replicated and the mechanism responsible for generating the concatemers at present are not well defined.

The existence of a rapidly sedimenting form of T4 DNA was first noted by Frankel (1966a) in experiments that revealed a component sedimenting at 200S, contrasted with the 63S characteristic of mature T4 linear genomes. Figure 6.19 shows the results of Frankel (1968a, b) which demonstrate that during the period from 7 to 15 minutes after infection, that is, immediately following the start of replication, an increasing proportion of the phage DNA, both parental and newly synthesized, enters the more rapidly sedimenting component. Sedimentation of denatured DNA strands through alkaline sucrose gradients confirmed that this fraction contains strands longer than a single genome, apparently of up to some 4 genome equivalents in length. Autoradiography of the rapidly sedimenting DNA suggested an average of between 1 and 2 genomes in length. Figure 6.20 shows that a pulse label given from 10–12 minutes after infection first enters the rapidly sedimenting component and then is transferred to DNA sedimenting at the position of mature genomes. This suggests that the rapidly sedimenting material represents a precursor to the mature form, which is generated by cleavage and packaging of headfuls of DNA from the concatemers.

During replication phage T4 DNA appears to be attached to the membrane. It is always possible, of course, for an apparent attachment to be an artifact of isolation, but its demonstration by specific methods including the attachment of DNA-membrane to sarkosyl crystals (a technique discussed in Volume 1 of Gene Expression) suggests that this indeed

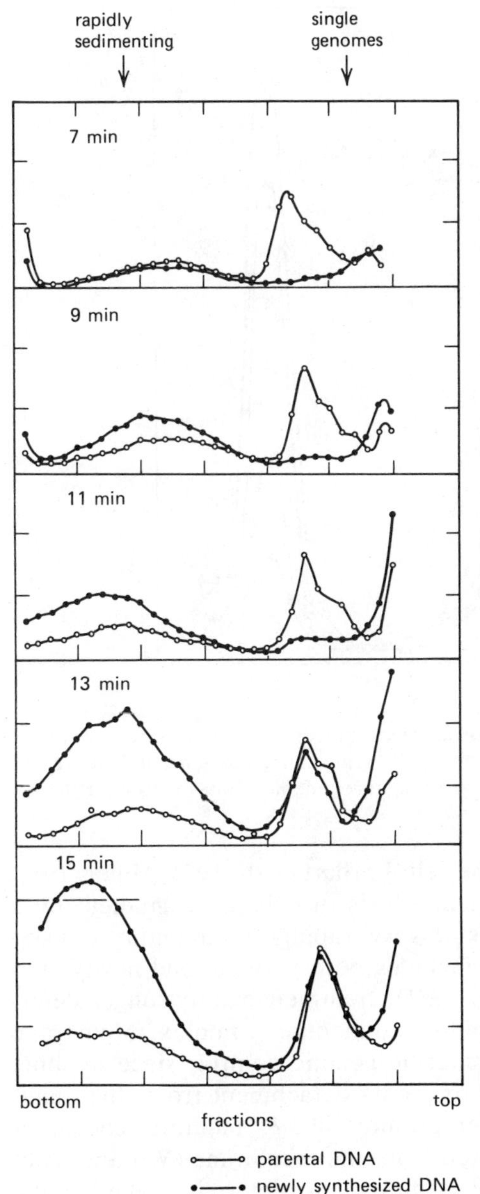


**Figure 6.19:** state of replicating T4 DNA. At varying times after infection, DNA was extracted from cells and sedimented through a sucrose gradient. Open circles show a label in parental DNA and closed circles identify a label incorporated into newly synthesized DNA. Data of Frankel (1968a).

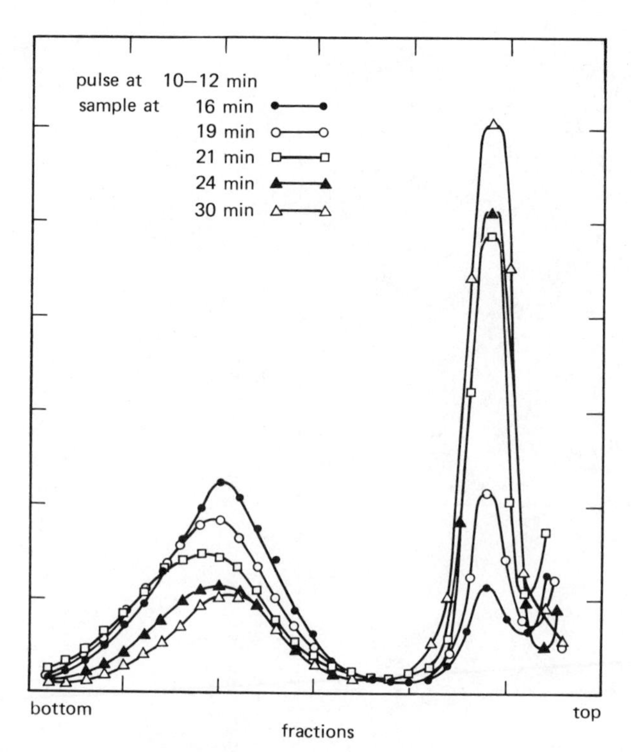


**Figure 6.20:** maturation of DNA from concatemeric precursor. A pulse label given from 10–12 minutes after infection first enters the rapidly sedimenting fraction and then over the next 20 minutes is chased into DNA of the size of mature genomes. Data of Frankel (1968a).

reflects the state of DNA within the cell (Earhart et al., 1973; Miller, 1972; Shalitin and Naot, 1971). Very gentle lysis of cells on a gradient by a nonionic detergent thus releases a very rapidly sedimenting (1000S) DNA-membrane fraction, which includes both parental and newly synthesized DNA (Shah and Berger, 1971); treatment with stronger detergents then releases the 200S rapidly sedimenting complex. Membrane attachment appears to lie under specific genetic control, since in some mutants the attachment of DNA to, or its detachment from, the membrane appears to be abnormal; for example in *59*$^{-}$ mutants replication ceases rapidly and DNA is released from the membrane (Wu and Yeh, 1974). However, in such cases it is difficult to determine whether the effect is direct or indirect and at present no particular phage coded protein has been identified as responsible for the structural association between DNA and membrane. The role of membrane attachment in replication therefore remains to be defined.

In several mutants the formation of concatemers is defective, although again it is difficult to determine whether this is a primary or secondary effect of the mutation. One of the better characterized of these observations concerns mutants in genes *46* or *47*. The nuclease controlled by these genes was first implicated in the formation of concatemers by Frankel's (1966b) observation that the 200S fraction is not generated during infection by a $46^-$ mutant. Shah and Berger (1971) later reported that shortly after replication has begun it is possible to identify 200S DNA in $46^-$ or $47^-$-infected cells, but that the DNA is rapidly released prematurely to yield the 63S form. Sedimentation of denatured strands through alkaline gradients confirmed that single strands longer than one genome can be isolated from $T4^+$-infected cells but are not generated in $46^-$ infection. The reason for the transient existence of the complex in the $46^-$ infection is not known, but the general conclusion suggested by these results is that the *46-47* nuclease is necessary for the formation of concatemers. The association of T4 DNA with the membrane also is abnormal in $46^-$ infection; and Shalitin and Naot (1971) suggested that the nuclease may be needed to recover by recombination DNA that has been released from the membrane complex.

The DNA of the mature phage particle is of course linear and appears to be replicated in this state. No satisfactory account of the replication of the linear strands has yet been given: the events responsible for initiation of replication are not well defined and there are difficulties in investigating the origins and state of replicating DNA. The first indication that T4 replication might in fact have more than one origin was given by the work of Delius, Howe and Kozinski (1971). Light bacteria were infected with phage of heavy density and partially replicating molecules isolated by their buoyant density. When examined by electron microscopy, these proved to have remained linear and to possess several replicating "eyes"; that is each molecule possesses more than one region of replicating DNA, each region equivalent in structure to the newly replicated material in the classical structure found in lambda and E.coli. Electron microscopy does not demonstrate directly whether replication within each eye is unidirectional or bidirectional, but at many of the branch points, including both in some eyes, "whiskers" could be observed; these are single strands, which must terminate in 3′-OH groups since they can be removed by exonuclease I, and their presence suggests that both branches represent replicating forks. An electron micrograph of replicating T4 DNA showing several eyes is presented in Figure 6.21. This analysis suggests that replication may be initiated simultaneously or within a brief period of time at several origins on the T4 genome.

To confirm these results by another technique, Howe et al. (1973)

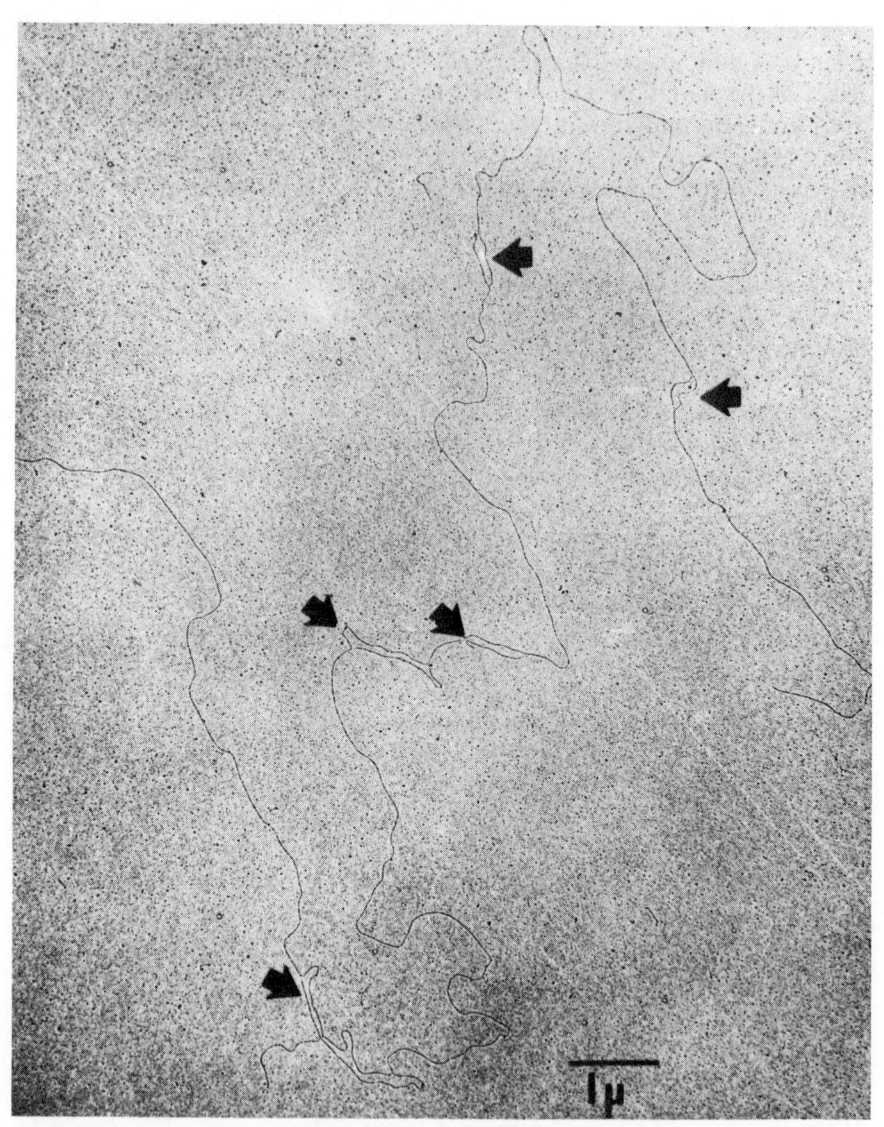


**Figure 6.21:** electron micrograph of replicating T4 DNA. The replicating "eyes" are indicated by arrows. Data of Delius, Howe and Kozinski (1971) kindly provided by A. W. Kozinski.

estimated the size of the replicating regions by shearing the partially replicated molecules and determining the buoyant density of fragments of different sizes. From the extent of the shift from light towards hybrid density, it is possible to calculate the proportion of a fragment that is newly replicated; and, of course, when fragments become shorter than the replicated length, they display fully hybrid density. The total length of newly synthesized DNA per genome can be calculated from the incorporation of a radioactive label, and proved to be greater than the length of the individual region. The ratio suggested the presence of several (3-6) replicating regions on each genome.

These results raise several practical problems in investigating the control of T4 replication. The usual criterion for defining an origin is to characterize a cis dominant site essential for replication. But when there are several origins presumably no single one of them is essential, since in the event of its deletion replication could simply proceed from the adjacent origins. An obvious question about this system concerns the specificity of initiation: how many origins exist, with what relative frequencies are they used, what is the relationship between them? The first step in defining their functions must be to identify their locations: perhaps this can be done by applying the same mapping techniques used to determine the positions of the origins for replication of plasmids (see Chapter 3).

An alternative view of T4 replication was proposed earlier by Mosig (1970) and Marsh, Breshkin and Mosig (1971). Their experiments made use of the "light particles" found in T4 phage populations: their genomes lack a sequence of length about one third of the chromosome, located at a random position. Although of course itself defective, in mixed infection with a complete helper phage the light particle can contribute markers to the progeny. Different markers are contributed by the incomplete chromosomes at different but characteristic frequencies, always less than the frequency with which the corresponding marker of the helper genome is found. The reduction was attributed to two causes: some incomplete particles might be unable to initiate replication because of the lack of an origin, in which case their markers should have a much reduced chance of contributing to progeny; and when replication occurred it appeared to be confined to only a single round. Two conclusions were drawn from this analysis. First, the markers contributed at high frequency were thought to lie near the origin, because the closer a marker lies to the origin the greater the frequency with which it should be included on a particle able to replicate. These results appeared to identify a single origin close to gene *43*, from which unidirectional replication was initiated in the clockwise sense. It is not clear how this conclusion may be reconciled with the idea that replication may utilize several origins; one possibility obviously is that there is an origin used more frequently that lies near gene *43* (for review see Miller, 1975a). A second suggestion based on these data was that replication might be confined to a single round because the incomplete genomes are not terminally redundant and might therefore be unable to perform a recombination between the termini necessary to generate a circular template. However, Kozinski and Doermann (1975) demonstrated that incomplete genomes can support repeated initiations and then replicate in the linear state. This demonstrates that circularity is not necessary for T4 replication.

All these analyses represent events in the first few rounds of replication. It is clear that during this period replication utilizes a linear template and starts from more than one origin; the utilization of circular templates, such as rolling circles, is therefore excluded in early replication (for review see Doermann, 1973). But this makes no implication about the events occurring later in the infective cycle, which might utilize a different mode of replication. Much attention has been focused on the possibility that a rolling circle might be utilized in T4 replication to generate the concatemeric precursors to mature DNA. One argument against the use of the rolling circle is that this model predicts that newly synthesized material should be linked covalently to parental DNA, but all attempts to demonstrate any such linkage have failed (for example see Kozinski, Kozinski and James, 1967). An attempt to test the possible involvement of the rolling circle has been made by Kosturko and Kozinski (1976) on the basis of genetic data. A difference between conventional replication and the rolling circle model is the prediction made for the distribution of mutant clones. When replication takes place from a fixed origin(s) which is doubled in each cycle, the clonal distribution of spontaneous mutants should correspond to exponential growth, contrasted with the linear dependence upon time expected from rolling circle replication. As early as 1951, Luria observed that the clonal distribution of spontaneous mutants corresponds to exponential growth, but since this analysis applied to the entire infective cycle, it does not exclude the possiblity of a switch from exponential to rolling circle replication at late times. Kosturko and Kozinski, however, followed the distribution of mutants induced by BUdR added at specific times after the start of infection; by following the reversion of an *rII* mutation that is increased $10^4$-fold by addition of BUdR, they were able to ensure that all the observed events were caused by the added mutagen. The distribution of mutants in clones followed the exponential pattern, suggesting that there is no change to a rolling circle later in the cycle.

The conclusion that replication does not generate concatemers implies that recombination must be responsible for their formation. Because of the circular permutation and terminal redundancy of the T4 genome, recombination between homologous sequences on two genomes can generate molecules longer than a single genome. When more than one phage genome infects a cell, the same sequence will be located at a different position on each genome; a reciprocal recombination therefore will generate one molecule longer and one molecule shorter than a single genome. When a single genome is involved, recombination between the redundant region at the left terminus of one genome with the same region at the right terminus of another replica should generate a molecule of almost two genome equivalents in length. Recombination between the terminally redundant sequences of one molecule should convert the linear genome into a circular structure. Further recombinational events involving these primary recombinants can generate longer concatemers, both circular and linear. No definitive study by electron microscopy of the T4 concatemers has yet been reported, but earlier studies identified linear molecules longer than a single genome and the studies of Bernstein and Bernstein (1973) identified circular structures.

Because many of the early genes of T4 are involved in both replication and recombination, it has not been possible to identify mutants able to replicate normally but unable to form concatemers. In the absence of such mutations, the events responsible for the formation of concatemers remain poorly defined. However, evidence that the origin of the concatemers lies with recombination has been obtained by Kozinski and Kosturko (1976). One $23^-$ mutant of T4 forms abnormal particles, containing either about two-thirds of a genome (these are the light or petite particles) or taking the form of giant heads in which several genome lengths may be packed in the concatemeric state; normal phage particles also are generated. Following a mixed infection, comparison of the genotypes of the giant and normal phages showed that most (about 68%) of the giant phages are heterozygous for a given marker; only 2% of the normal phage particles displayed heterozygosity. This suggests that the concatemers found in the giant particles contain genomes that have been linked by recombination.

### Formation of Recombinants

The formal analysis of phage genetics that we have already discussed views the content of the infected cell as a mating pool of phage genomes within which T4 replication and recombination occurs. The pool contains an equilibrium level of the order of 40 genomes: new genomes are created by replication and completed genomes are continuously withdrawn to be incorporated into mature particles. By supposing that matings occur pairwise and randomly in time, Visconti and Delbruck (1953) were able to conclude that on average a T4 progeny genome has undergone five rounds of mating. When recombinants are examined for very closely linked markers, the recombination event appears to take the form of a cluster of several exchanges; and Steinberg and Edgar (1962) were able to show that recombination generated at a cluster exchange is not always accompanied by recombination between outside flanking markers. The clustering of exchanges is reflected in highly localized negative interference and presumably reflects, as it does with phage lambda (see Chapter 5), the mechanism of recombination.

The recombination event in T4 involves the formation of hybrid DNA, in which the duplex comprises one strand from each parent, and which may therefore be heterozygous when the two parent DNA molecules differ at some site. Correction of this heterozygosity generates the cluster of tightly linked exchanges; and since hybrid DNA may be created either with or without recombination of the markers on either side of the hybrid region (see Figure 5.1), the cluster of recombination events may or may not be associated with external recombination. Heterozygosity in T2 was first observed as early as 1951, when Hershey and Chase observed that about 2% of the progeny of $r^+/r^-$ mixed infection give rise to mottled plaques. We have seen that this heterozygosity has two causes: it may reflect the presence of a length of hybrid DNA created by the recombination process; or may be due to the presence of different parental alleles in terminally redundant regions (Sechaud et al., 1965).

The generation of hybrid DNA is a critical step in the T4 life cycle, for not only is it necessary to achieve genetic recombination per se, but the recombination is needed for the formation of concatemers. Although some information is available about certain of the intermediate structures through which recombination passes, the entire process is far from defined and the roles of most of the gene products that appear to be involved remain to be determined. As we have already noted, one difficulty in elucidating their functions is that many are necessary also for replication, so that mutants fail to replicate their DNA and thus make it impossible to observe the step that may be blocked in recombination.

A central intermediate in T4 recombination has been recognized in the *joint molecule*. This structure was identified by Tomizawa and Anraku (1965) through experiments in which cells were infected with parental phages carrying different labels and DNA synthesis prevented: molecules carrying both labels can be recovered in which denaturation separates the labels. This suggests that these joint molecules represent an early stage of recombination, when the parental genomes have exchanged strands to form hybrid DNA, but the strands of the two parents have not been covalently linked. It is clear that two types of activity are required for the conversion of joint molecules to covalent recombinants: DNA polymerase must fill gaps and ligase must seal the single strand breaks that remain after the gap filling. There have been conflicting reports on whether phage coded enzymes are essential for these functions or whether host activities can be utilized. Anraku and Lehman (1969) established that *30⁻ 43⁻* double mutants are unable to perform the conversion, but either single mutant appeared able to do so. However, Miller (1975b) subsequently showed that the phage DNA polymerase is needed for the conversion, although it still appears that the host ligase may substitute for the phage enzyme.

By isolating joint molecules from cells infected with *30⁻ 43⁻* double mutants, Anraku, Anraku and Lehman (1969) were able to estimate that each such intermediate possesses about 24 gaps, of average length 300–400 bases. Prashad and Hosoda (1972) found that the creation of these gaps requires the nuclease controlled by genes *46* and *47*; its activity seems to be to enlarge single strand nicks made by the nuclease identified by Kozinski (1968). In the absence of the *46* or *47* gene function, recombination is prevented. This implies therefore that recombination requires the enlargement of nicks to generate more extensive regions of single stranded DNA.

One of the earliest products of recombination must be a branched molecule generated by the exchange of strands between the two parents. Branched molecules that may be the products of this reaction have been identified by electron microscopy by Broker and Lehman (1971). An example of a branched molecule is shown in Figure 6.22. When cells are infected with parental molecules of different buoyant densities, the hybrid density fraction that is expected to include the recombinants is enriched in branched molecules; this supports the idea that they represent intermediates in recombination. Their frequency of occurrence is influenced by mutations in genes *30, 32, 43,* but since these functions all are concerned in both recombination and replication, the roles of these products cannot be inferred from these observations.

The structures seen by Broker and Lehman suggest that *branch migration* may take place after a recombination intermediate has been generated by the formation of hybrid DNA. Branch migration was first noted as a phenomenon involving single strands in renatured circular molecules. Lee, Davis and Davidson (1970) identified molecules containing forked branches with two single strands emerging at the site of terminal repetition; this suggested the model for single strand branch migration shown in Figure 6.23. Kim, Sharp and Davidson (1972) demonstrated the existence of double strand branch migration and an example of this mechanism is shown in the figure. Broker and Lehman suggested several models for the

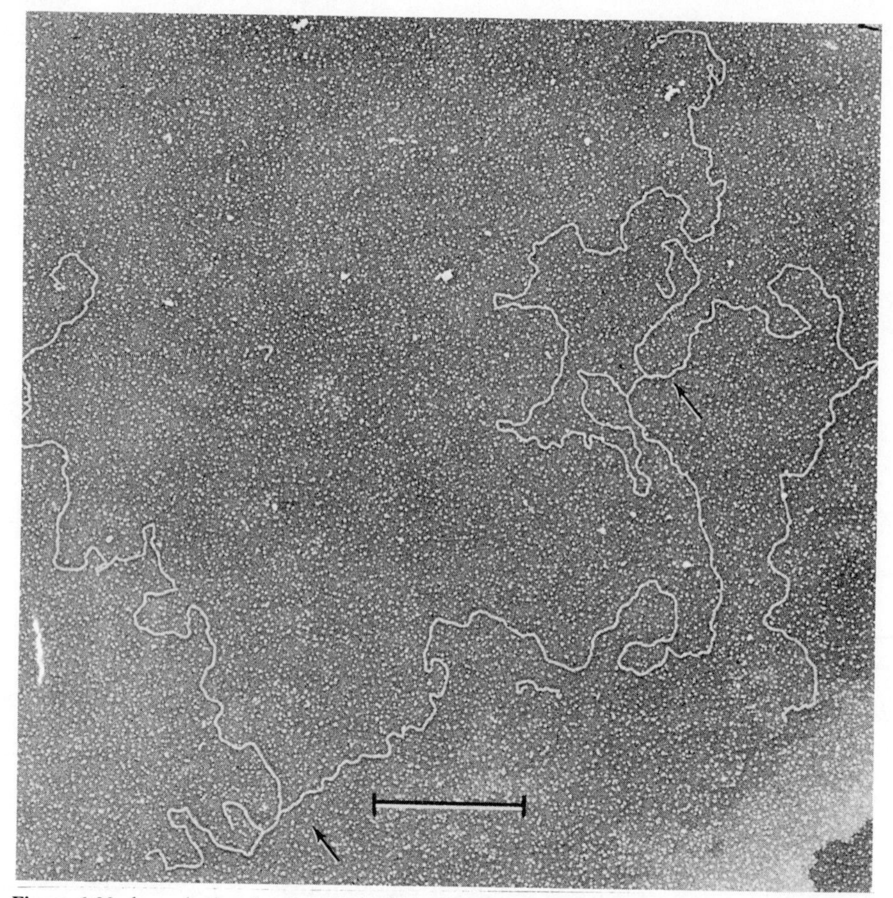

**Figure 6.22:** branched molecule of T4 DNA. The branches are indicated by arrows; the bar indicates 1 $\mu$m. Data of Broker and Lehman (1971) kindly provided by T R. Broker.

involvement of branch migration in T4 recombination. Because of the reversibility of the direction in which the branch may migrate, more definite information on the process involved has been difficult to obtain. More recently, however, recombination intermediates have been identified in which two small circular DNA molecules are fused together and these provide suitable substrates with which to study the reaction. The properties of these intermediates are discussed at the end of Chapter 8. With relevance to the involvement of branch migration in recombination, however, we may note now that in a study of branch migration in this system, Thompson, Camien and Warner (1976) have been able to calculate that the rate at which a junction may move as illustrated in Figure 6.23b is about 6000 bases per second at 37°C.

From these results it is clear that several types of enzyme activity must be involved in T4 recombination. Generating the branched molecule must involve at least an endonuclease to introduce nicks, and perhaps an exonuclease to extend them into gaps. Gene *32* protein may be involved in maintaining single stranded regions. Reduction of the branched molecules to joint structures also requires nuclease activities; and conversion of the joint molecules to yield covalent recombinants requires both polymerase and ligase activities. These processes are discussed in more detail in Volume 1 of Gene Expression. Although mutations in many of the replication functions increase or decrease the frequency of recombination, in these instances it is not clear whether the effect is a primary or secondary one and the roles of their products remain to be established (see Bernstein, 1968; Berger, Warren and Fry, 1969; Wu and Yeh, 1974; Leung et al., 1975; Shah, 1976). Recombination in T4 has been reviewed by Broker and Doermann (1975).

Several phage genes influence the susceptibility of T4 DNA to ultraviolet irradiation or damage caused by other mutagens; some of the mutants in these genes also display reduced frequencies of recombination. Table 6.4 summarizes the phage repair mutants. In general, the systems identified by these mutations appear to be analogous to host repair systems. The *v* gene codes for an ultraviolet-specific endonuclease whose general activity suggests that it plays a function in a system like the *uvr* system of the host; *v* mutants are sensitive only to ultraviolet. On the

**Table 6.4:** repair functions of phage T4

| mutation | effect | reference |
|---|---|---|
| *v* | increased ultraviolet susceptibility; codes for UV-specific endonuclease | Harm (1963); Friedberg and King (1969, 1971); Mortelmans and Friedberg (1972); Minton et al. (1975); Simon et al. (1975) |
| *x* | sensitive to UV, X-irradiation, MMS | Harm (1963); Mortelmans and Friedberg (1972) |
| *y* | similar to *x* | Boyle (1969) |
| *w* | reduced recombination | Hamlett and Berger (1975) |
| *mms* | increased susceptibility to MMS | Ebisuzaki et al. (1975) |

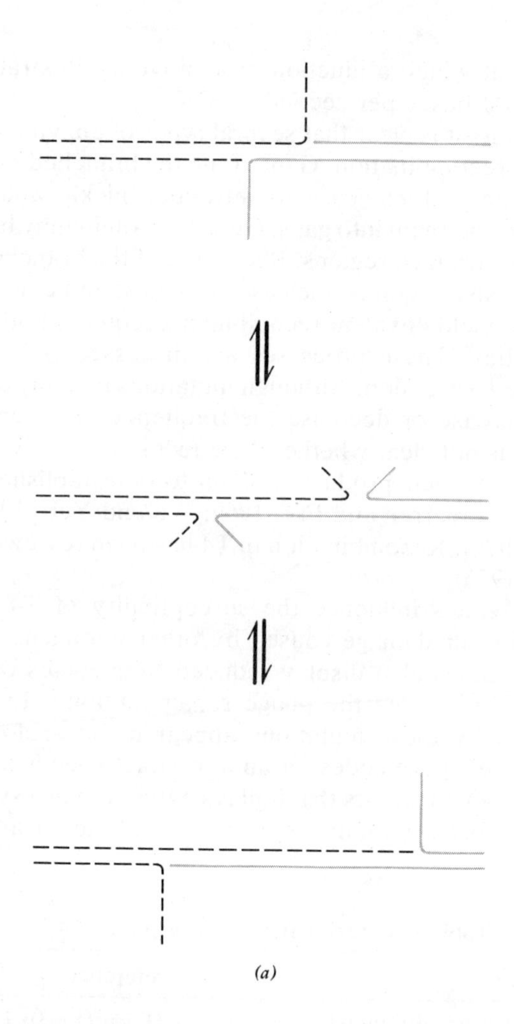

*(a)*

other hand, *x* mutants are sensitive to ultraviolet, X-irradiation and methylmethanesulfate, a response very similar to that of host *rec* mutants. These repair functions all appear dispensable in the phage life cycle and the relative roles of phage coded and host enzymes in repair of damage to T4 DNA are not defined.

## Morphogenesis of the Phage Particle

Assembly of the T4 particle takes place by the two independent pathways responsible for formation of the head and tail structures. The develop-

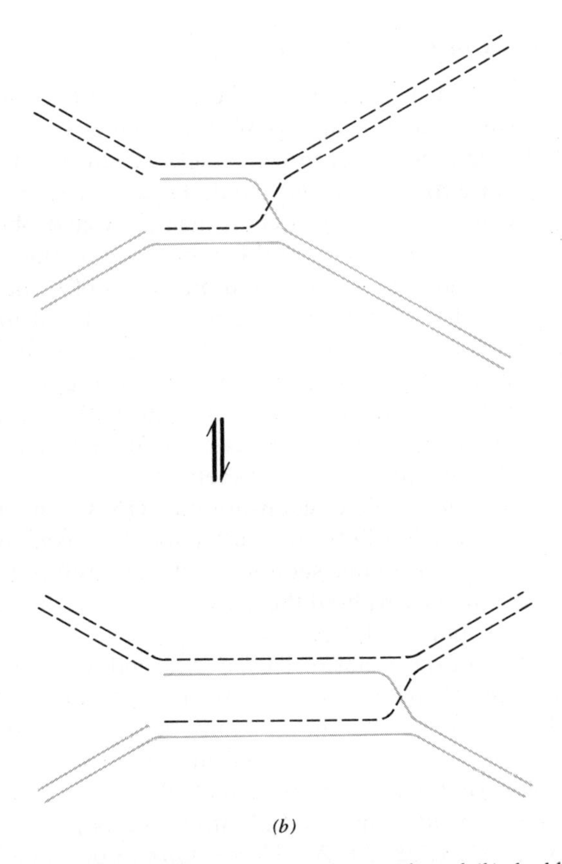



**Figure 6.23:** branch migration of (*a*) single strands and (*b*) double strands.

ment of the in vitro complementation system by Edgar and Wood (1966) showed that head mutants are able to make active tails while tail mutants can produce active heads, which are able to join in vitro to form infective particles. This approach allowed the morphogenetic genes of T4 to be divided into the two general classes involved in head and in tail formation. An analogous approach has been responsible for the successful analysis of lambda phage morphogenesis discussed in Chapter 5. Assembly of the T4 particle shares with lambda the general characteristic that it may be described as assisted self-assembly: that is, the proteins comprising the head and tail structures are capable of self-assembly, but for formation of the proper structures require the assistance of other functions that are not structural components of the particle.

### *Aberrant Structures in Head Assembly*

The structure of the T4 particle illustrated in Figure 6.2 shows that the head takes the form of an elongated bipyrimidal hexagonal prism, which can be regarded as an icosahedron that is elongated by the addition of material forming an extra equatorial band. How this shape is imposed upon the assembly of phage head protein is the principal problem of phage morphogenesis. Another important problem is posed by the condensation of DNA within the head. The head shell dimensions of some 1000 Å (0.1 $\mu$m) in length and 650 Å (0.065 $\mu$m) in width correspond to an internal volume of very roughly $2.5 \times 10^{-4}$ $\mu m^3$. The head contains T4 DNA of length about 56 $\mu$m, which must occupy a minimum volume of some $1.8 \times 10^{-4}$ $\mu m^3$. This very approximate calculation suggests the same conclusion as that inferred for phage lambda (see page 496): the genome must be packed very compactly into the head shell. The nature of the forces responsible for this remarkable condensation of DNA, which must overcome electrostatic repulsion between the charged phosphates of DNA, remains a critical question. In this section we shall therefore be concerned with what is at present known about these two problems: the acquisition of shape, and the packaging of DNA.

A large proportion of the genes of T4 is concerned with assembly of the mature particle: more than 20 genes have been implicated in head assembly and more than 30 are needed for assembly of the structures of the tail. Mutants in phage assembly were grouped into several classes by Edgar and Lielausis (1968) on the basis of the defects displayed in infected cells. Mutants defective in head formation fall into two groups. The *Y* group of mutants, in genes *20, 21, 22, 23, 24, 31, 40,* fails to produce head shells; instead these generate several aberrant variations of the head structure. This group of mutants therefore identifies the genes concerned with determining the shape of the head. The *X* group of mutants, in genes *2, 4, 50, 64, 65, 16, 17, 49,* produces heads that appear normal under the electron microscope; but the extracts are inactive when tested by in vitro complementation. These mutants must therefore be defective in stages of head maturation subsequent to the acquisition of proper shape. Also included in this group are two further mutants, in genes *13* and *14,* whose defects lie in the inability of otherwise normal heads to join with the tail. The other two groups, which we shall discuss later, are defective in tail formation (this is the *Z* group) or fail to make tail fibers (although the rest of the tail structure is normal).

The aberrant structures produced by *Y* group mutants are compared with the mature particle in Figures 6.24–6.26. The abilities of these mutants to generate the aberrant structures were characterized by

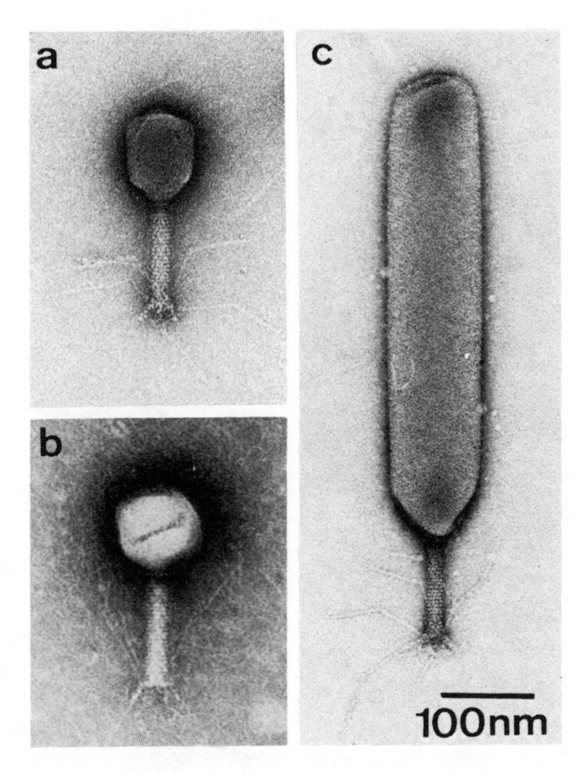


**Figure 6.24:** electron micrographs of phage T4 and head variants. (*a*) Wild type phage particle. (*b*) Phage with isometric head, produced by certain mutants in genes *22* or *23*. (*c*) Phage with giant head, produced by certain mutants in genes *23* or *24* or by treating infected cells with L canavanine. Photographs kindly provided by the groups of electron microscopy, image processing and T4 morphogenesis of the Biozentrum, Basel.

Laemmli et al. (1970b). Mutants in gene *31* produce *lumps*: these appear to comprise aggregates of head protein associated with the inner cell membrane. Mutants in several genes produce *polyheads*, which essentially comprise long tubes resembling the head shell in structure and lacking DNA (these are analogous to the tubular forms of lambda discussed in Chapter 5). These fall into two categories first identified by Favre et al. (1965). *Multilayered polyheads* consist of several concentric tubes and appear to be constructed first by the formation of a single (inner) tube and then by the aggregation on this framework of further layers of head protein. Laemmli et al. observed that only mutants in gene *22* accumulate these structures. The innermost tube of the multilayered polyhead appears to have the same dimensions as the *single layered*

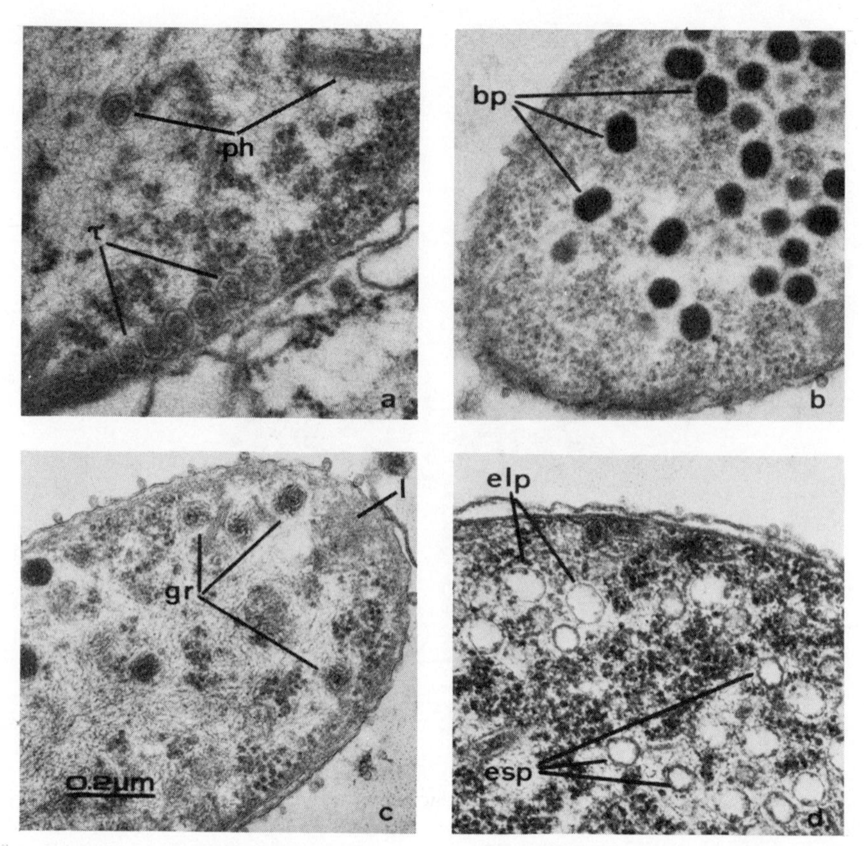


**Figure 6.25:** electron micrographs of thin sections of infected bacteria. (*a*) Tau particles [τ] and aberrant tubular polyheads [ph] accumulate in infection by *24*⁻ mutants. (*b*) Phage particles [bp] are seen in wild type infection, sectioned at various levels and orientations. (*c*) Grizzled particles [gr] can be visualized in *49*⁻ mutant infection and appear to contain only a partial complement of DNA. A lumplike accumulation of protein [1] close to the bacterial membrane also is visible. (*d*) Two kinds of head empty of DNA, [esp, empty small particles; elp, empty large particles] accumulate in infection with a *16*⁻ *17*⁻ double mutant; neither can be matured. They appear to represent abortive structures not in the usual assembly pathway. Photographs kindly provided by the groups of electron microscopy, image processing and T4 morphogenesis of the Biozentrum, Basel.

*polyheads*, which are formed by mutants in genes *20, 40, 21* and *24*, as well as by gene *22* mutants (where they are less frequent in occurrence than the multilayered form). In gene *20* and *40* mutants the single layered polyhead is the headlike product and there are usually 15–25 such structures per cell, corresponding in total to 300–500 phage equivalents of head protein; an average single layer polyhead thus is a tube of up to some 20 or

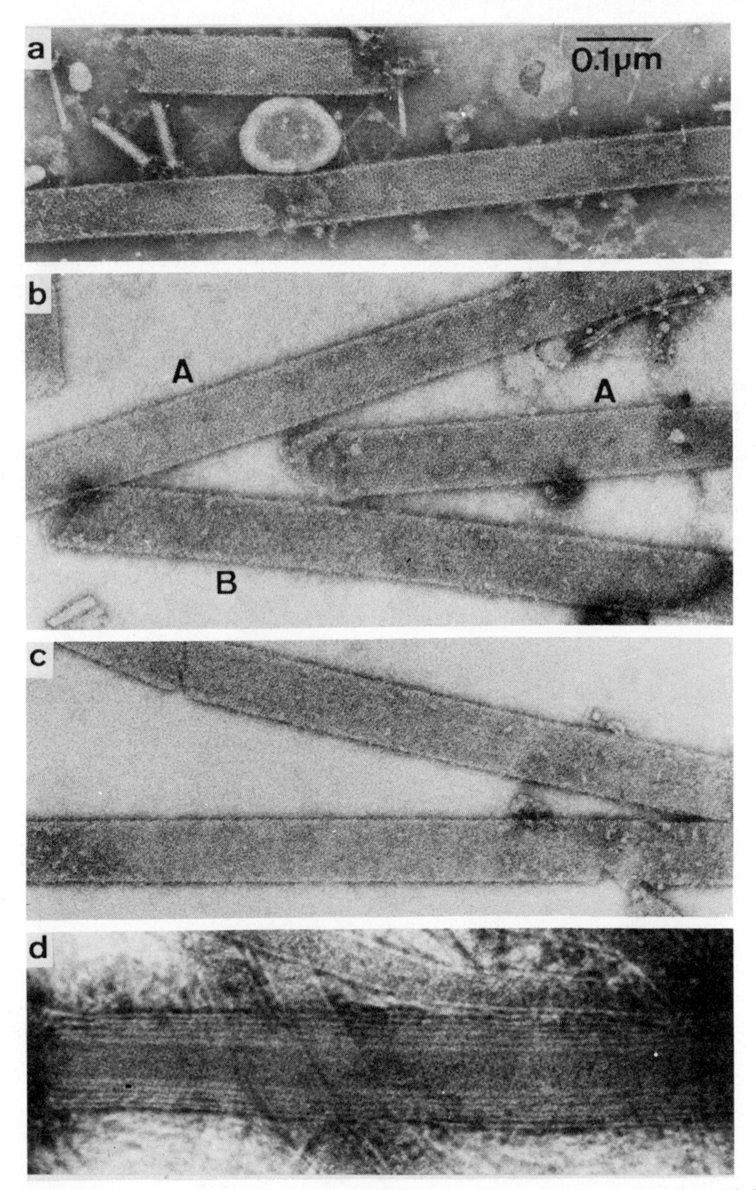


**Figure 6.26:** electron micrographs of negatively stained polyheads. (*a*) Coarse polyheads consisting of P23. (*b*) Smooth polyheads generated by the proteolytic cleavage of P23 to P23*. The A type is the immediate product of cleavage with a p6 net. The B type is generated by the addition of hoc protein at the center of the hexameric unit to give a capsomere of the (6+1) type. (*c*) The C type polyhead simulates the surface of the mature T4 capsid and is generated from the B type by incorporation of trimers of the soc protein at the lattice positions of threefold symmetry between the (6+1) units. (*d*) Multilayered polyheads produced by mutants in gene *22*. Photographs kindly provided by the groups of electron microscopy, image processing and T4 morphogenesis of the Biozentrum, Basel.

so times the length of the head. The gene *20* and *40* mutant polyheads are uncapped. Mutants in genes *21* and *24* produce some polyheads (sometimes capped) but their principal headlike product is the tau ($\tau$) particle, which was first identified by Kellenberger et al. (1968) as an (empty) headlike structure slightly smaller than the head and lacking its sharp edges and vertices. There are two types of tau particle: the isometric is approximately spherical, with a diameter of about 680 Å; the elongated is a smooth polyhedron of 520 × 715 Å.

The order in which the Y group genes act during head morphogenesis can be defined by considering the relative complexities of these aberrant structures in terms of the information needed to specify each of them. The fewer genes are needed to specify a given structure, the simpler it must be and the earlier it must lie relative to the assembly pathway; assignments can be made on the basis that a structure formed in the absence of a given gene product does not depend upon the structural information that it conveys. Based upon this concept, Figure 6.27 summarizes the order of

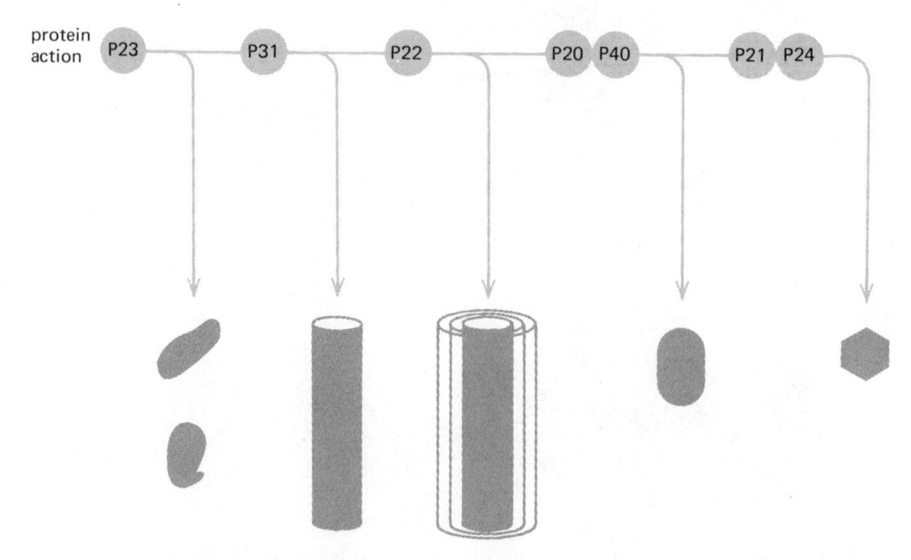


**Figure 6.27:** functions of morphogenetic genes of the *Y* group. The predominant structure generated by each type of mutant is illustrated, arranged in a hierarchy that may correspond to the order of action of the genes. Thus P23 alone is needed for the formation of lumps; P23 and P31 are needed to form multilayered polyheads; the additional action of P22 is needed to form single layered polyheads; the actions of P20 and P40 also are required to form tau particles; and P21 and P24 are needed to determine the angular structure. Note that these structures are not intermediates in assembly but represent the products of abortive assembly when the absence of an essential function diverts the normal pathway.

action of the *Y* group genes and illustrates the effects of the absence of each function.

In the absence of P23 (this designates the product of gene *23*) no head structures at all are formed since P23 is the principal protein from which the head is constructed, and in this sense lies first on the assembly pathway. In the absence of P31 only lumps are formed: P31 is therefore required for the formation of all meaningful head structures and can be assigned to the next place on the pathway. Laemmli et al. (1970a) demonstrated that a temperature sensitive mutant in gene *31* forms lumps at high temperature that may disappear upon a shift to the lower temperature. The P23 dissolved from the lumps is incorporated into phage particles, so the formation of lumps is not irreversible. Thus P31 appears to have the property of controlling the aggregation of P23 so that meaningful structures instead of formless aggregates are produced.

Since the mutants $22^-$, $20^-$, $40^-$ all produce polyheads, none of their gene products can be necessary for the formation of this type of structure. The relative order of their actions in the hierarchy of assembly is suggested by the observation that the double mutants $22^-\ 20^-$, $22^-\ 40^-$, $22^-\ 24^-$ all produce the multilayered polyheads characteristic of $22^-$ infection. This suggests that the multilayered polyhead results from the absence of P22. Since none of the other genes is needed for its production, we can say that the multilayered structure is generated by the aggregation of P23 in the presence of P31. The presence of P22 is needed for the generation of large numbers of single layered polyheads. Laemmli et al. (1970b) suggested that P22 therefore may be required for the efficient initiation of polyhead formation; in its absence it is easier to start another layer on an existing polyhead whereas in its presence it becomes more probable that formation of a new tubule will be initiated de novo. (It is of course possible to form polyheads in the absence of P22, so its action cannot be indispensable, but the frequency of initiation may be much lower than in its presence.)

The products P20 and P40 are needed for the production of tau particles, which are the predominant products in $21^-$ and $24^-$ mutant infections (some polyheads are produced but these form only a small proportion of the headlike structures). The tau particle is a closed structure and, correspondingly, the polyheads that are seen in $21^-$ and $24^-$ mutant infections may have hemispherical caps. This suggests that P20 and P40 are involved in the closing of the ends of the head structure. Comparison of the tau particle with the wild type head suggests that P21 and P24 are needed for acquisition of the angular shape of the head.

What determines the length of the head? At one time it was thought that another gene, *66*, was involved because mutants in it generated petite phage particles whose heads (and contents) of DNA are smaller than wild type since they lack the elongated (*prolate*) shape (Kellenberger, 1968; Eiserling et al., 1970). However, it has since become apparent that the mutations thought to define gene *66* in fact lie in the head protein gene *23* (Wood, 1974). This implies that the structure of the head protein may itself have a role in determining the length of the head. This conclusion is emphasized by the isolation of other mutations that affect the length of the T4 head and cause the production of both petite heads and giant heads (of increased length); these were mapped close to gene *23* by Doermann, Eiserling and Boehner (1973) and presumably lie in it. The structures produced in $24^-$ infection suggest that P24 also may be involved; and we shall see later that P22 appears to be implicated.

Certain host functions appear to be involved in phage head maturation. Coppo et al. (1973) and Takahashi et al. (1975) characterized *tabB* mutants of E.coli, in which T4 infection is abortive, apparently because cleavage of head precursor proteins (see below) does not occur under nonpermissive conditions (high temperature). Upon a temperature shift down during infection, some phage maturation is able to occur, but this requires new protein synthesis since it is prevented by the addition of chloramphenicol. Only a small proportion of the preexisting P23 is utilized after the temperature shift down, which suggests that it may have aggregated irreversibly into structures that cannot be utilized for head formation. T4 mutants characterized as $com^B$ are able to grow on *tabB* cells; and phage mutants characterized as $k^B$ are unable to grow on *tabB* cells even under permissive conditions. This system for investigating phage-host functions is in principle the same as that represented by the *tabD* host mutant that allowed the interaction of host RNA polymerase with phage functions to be investigated (see page 610). Thus the demonstration that both $com^B$ and $k^B$ phage mutants lie in gene *31* implies that the host *tabB* function must interact with the phage gene *31* function, presumably to control the aggregation of P23. The *tabB* mutations appear to reside in the same host function as the *groE* mutations that prevent assembly of the phage lambda head from the principal protein pE (see page 511). This leaves the implication that this host gene, since renamed *mop* (for morphogenesis of phage), may represent a function that is generally implicated in phage head assembly, further implying therefore that there must be features common to the assembly of at least phage lambda and T4. Other bacterial mutants that prevent T4 head assembly have been described by Simon, Snover and Doermann (1973) and Simon et al. (1975); one of these possesses an altered lipid structure in the cell membrane, which emphasizes the point that T4 head assembly may be initiated at the inner membrane.

## *Proteins of Head Structures*

Electrophoretic analysis of the proteins of T4 heads identifies a single principal protein, coded by gene *23*. Other proteins have been identified in several experiments, but their genetic origin is not always clear and their proportions relative to the major protein vary somewhat, probably because of differences in extraction procedures (see Kellenberger, 1968; Hosoda and Cone, 1970; Dickson, Barnes and Eiserling, 1970; Forrest and Cummings, 1971). Laemmli (1970) demonstrated that the product of gene *23* is present in a cleaved form, P23*, which has a molecular weight of 46,000 daltons. Its precursor, the primary product P23, has a size of 56,000 daltons. The 10,000 daltons that are cleaved to generate P23* derive from the N-terminal end of the P23 protein (Celis, Smith and Brenner, 1973). Laemmli also identified as components of the phage head shell the products P18 (69,000 daltons), P20 (63,000 daltons), P24* (43,500 daltons, cleaved from the 45,000 dalton precursor P24), and P19 (18,000 daltons).

Cleavage of head proteins to their mature forms is prevented by mutation in any one of the *Y* genes *20, 21, 22, 23, 24, 31, 40*; the implication is that the aberrant structures—lumps, polyheads, tau particles—formed by these mutants must contain P23 rather than P23*, and this has been confirmed experimentally (see below). Since the activities of all the *Y* gene products are required for efficient cleavage, it is likely that the cleavage occurs on a precursor head structure rather than with free protein molecules. This in turn hints that cleavage may be an important step in the transition from one head precursor structure to another. Onorata and Showe (1975) observed that defective lysates of the *Y* group mutants that do not cleave the protein P22 in vivo are able to do so in vitro; only $21^-$ mutant extracts are unable to do so. The proteolytic activity is associated with the *crummy* heads produced by one $23^{ts}$ mutant (these are defective in shape), although it is absent from wild type capsids. This suggests that gene *21* either specifies or controls a protease activity that is associated with head precursors but which is either inactivated or removed during maturation.

In addition to the constituents of the head shell, the T4 phage head contains internal components, which are released when the phage particle is subjected to osmotic shock. These comprise the polyamines putrescine and spermidine, acid soluble peptides, and acid insoluble proteins.

The acid soluble components were characterized by Eddleman and Champe (1966): the three oligopeptides VI, VII and II appear in infected cells during T4 development and the last two are found in mature particles. The appearance of these peptides is blocked by mutation in any of the *Y* genes. Their origin is not finally established, but one possibility is that they may be products of the cleavage of P22.

The acid insoluble proteins comprise basic species that were first shown to be injected into the cell together with the phage DNA by Hershey in 1957. Two of these *internal proteins* were identified in T4 by Stone and Cummings (1970, 1972), who showed that there are differences in the internal proteins of the various T-even phage strains. Black and Ahmad-Zadeh (1971) characterized the three species

| | molecular weight (daltons) |
|---|---|
| IPI | 13,000 |
| IPII | 9,000 |
| IPIII | 18,000 |

which together account for about 5% of the phage particle protein. Antigens prepared against these proteins do not react with the T4 phage particle, confirming that they are inaccessible in the mature structure. When released from the particle by osmotic shock, the internal proteins are bound to DNA at low salt concentrations but are detached at higher salt levels. Bachrach, Levin and Friedman (1970) reported that when particles are disrupted by heating with the detergent sarkosyl (the reagent used to prepare DNA-membrane complexes), the internal proteins remain attached to DNA isolated on hydroxyapatite. The specificity of this binding is not known.

All three internal proteins are present in the mature phage head in a cleaved form. The cleavage of IPIII was the first to be established, when Laemmli (1970) demonstrated that a protein then described as IP was present in the lower weight form of IP*. Isobe, Black and Tsugita (1976) demonstrated that IPIII has a molecular weight of 21,000 compared with the weight of 18,000 daltons of the cleaved product IPIII*. Cleavage of the protein IPII to the form IPII* represents the removal of 15 amino acids from the N-terminal end, reducing the molecular weight from 11,750 to 10,000. Since the N-terminal residue of the protein IPI* found in the mature head differs from that of its precursor IPI, cleavage of this protein also must occur. It is possible that all these cleavages involve rupture of the same peptide linkage, between glutamic acid and alanine, although this remains to be fully established and in any case there must be other requirements for recognition by the protease since these proteins also contain uncleaved glu-ala linkages. The cleavage of IPIII, like the cleavage of the head shell proteins P23 and P24, is prevented by mutation in any of the *Y* genes (and this presumably is true also for the cleavage of IPII and IPI), suggesting that internal proteins share with the outer shell proteins the characteristic that cleavage takes place only as part of a precursor structure. Another protein found within the phage head is B1*, of molecular weight 61,000 daltons derived from the 79,000 dalton precursor B1; this protein may be injected into the cell together with DNA and may be responsible for the alteration of host RNA polymerase (Paulson, Lazaroff and Laemmli, 1976). Whether it has any other role is not known.

The three IP species together with P22 appear to form a "core complex" in head assembly; when cleavage occurs the internal proteins of the core complex are converted into their mature forms and P22 is degraded into small fragments amongst which may be the internal acid soluble peptides VII and II. When late proteins are labeled during a wild type infection, P22 gains the label only transiently, since it is soon chased from the precursor pool of soluble proteins; since no large product has been detected, it appears that the protein must be degraded into small fragments (Laemmli, 1970). But there is at present no proof whether the peptides VII and II may be products of this cleavage. In the defective lysates generated by *Y* group mutants, P22 remains uncleaved. When Showe and Black (1973) purified P22 from these lysates they found the protein in two forms, one apparently representing what appears to be a complex of P22 with a product modified from it, and the second comprising a core complex of P22, IPI, IPII and IPIII, with two copies of each internal protein for every copy of P22.

Both polyheads and tau particles appear to possess a proteinaceous "core" when visualized by electron microscopy, and this appears to comprise the complex of P22 with the internal proteins (see below). Since P22 appears to play a role in the initiation of polyhead formation, this suggests the concept that it may be the protein core that is involved in initiating the formation of head structures. The role of the internal proteins in head assembly has been investigated by the isolation of mutants defective in their production; Black (1974) isolated the single mutants $ipI^-$, $ipII^-$ and $ipIII^-$ and also constructed the triple mutant strain $ip^o$ that lacks all three internal proteins. No role in head assembly has yet been found for IPI and IPII, for the mutant strains $ipI^-$, $ipII^-$ and $ipI^-$ $ipII^-$ produce heads that are normal, with aberrant forms taking the same appearance and number as in wild type infection. (Black and Abremski (1974) found one host strain in which a defective prophage restricts $ipI^-$ but not $ipI^+$ T4; probably it is the IPI injected with the infecting DNA that is needed to prevent this restriction.)

The absence of IPIII, in $ipIII^-$ or $ip^o$ infection, however, causes an increased production of polyheads. The possible interaction of IPIII with *Y* gene products can be investigated genetically by constructing double mutants of the type $Y^{am}$ $ip^o$. Black and Brown (1976) observed that with

$21^-$ *ip*° and $24^-$ *ip*° mutant infections there are more polyheads and fewer tau particles than in the single $21^-$ or $24^-$ mutant infections, which suggests that the formation of tau particles may involve an interaction of head shell functions with the core complex. The most dramatic effect, however, is seen with $22^-$ *ip*° mutants, which generate lumps similar to those seen in $31^-$ mutant infections. This supports the idea that the core complex is needed for the proper association of P23 into head structures; although either P22 or IPIII can be absent, the deficiency of both prevents the formation of any meaningful structures.

Direct investigation of the proteins present in aberrant head structures is complicated by their fragility—polyheads and tau particles disintegrate readily during isolation. Either they must be prepared very rapidly or agents that stabilize the structure must therefore be used in preparation. Laemmli and Quittner (1974) prepared polyheads from defective lysates by differential centrifugation followed by sucrose gradient sedimentation: this generated a broad peak of about 400S. Polyheads isolated after cells were lysed by freeze-thawing contain P23, P23* (in lesser amount), P22, IPIII and IPII (the presence of IPI could not be detected on the gels used to separate the proteins, but presumably its behavior follows the other internal proteins). These polyheads can be seen by electron microscopy to possess a proteinaceous core. However, when the cells were lysed with chloroform, the polyheads lacked the core and consisted only of P23. This demonstrates directly that the core of P22 and the internal proteins lies within the polyhead, just as the cleaved internal proteins reside within the mature head structure.

The presence of the core complex in the aberrant head structures implies that it may be present in precursor structures during head assembly. At the time of cleavage of head proteins, P22 must be degraded to small fragments and all three internal proteins also are shortened. Since both the aberrant polyheads and mature heads may be formed in the absence of all three internal proteins, their functions must be dispensable. The polyheads formed by (for example) $20^-$ *ip*° mutants appear identical to those of the $20^-$ mutant, apart from the lack of the proteinaceous core; the heads of *ip*° mutants are slightly more dense than wild type, as might be expected from the removal of some protein, but otherwise appear normal. The single layered polyheads formed by $20^-$ $22^-$ mutants also lack the proteinaceous core, which may imply that the binding of internal proteins in the core depends upon the presence of P22. Although there is evidence for the formation of a protein complex, P22 may be its only essential component, since each single *ip*$^-$ mutant generates heads that possess the other two internal proteins, so it must be possible to incorporate each of the three independently of the presence of the others. And since the internal proteins are dispensable they cannot be essential for DNA condensation, which is one obvious function in which they might be involved. This leaves the possibility, however, that the acid soluble internal peptides may be involved in this capacity.

### *Precursor Structures in Head Assembly*

Precursors in head assembly are difficult to isolate because of their instability, and although several intermediates have been identified the structures of these precursors have not yet been resolved in detail. Although polyheads and tau particles are aberrant structures produced by misassembly in the absence of essential components, they may be related to precursors on the head pathway and appear to undergo structural transitions related to those that occur in head assembly. Analysis of head maturation therefore has made use both of attempts to isolate and analyze structural intermediates and to determine the structures of polyheads and tau particles and their relationships.

To isolate head precursors, Laemmli and Favre (1973) introduced two techniques of preparation: spheroplasts were lysed gently in a buffer containing glutaraldehyde, which fixes the intermediates upon release from the cell; or cells were lysed in the presence of high salt concentration and the neutral polymer polyethylene glycol (which causes DNA molecules to collapse into a compact structure and thus similarly prevents the disaggregation of head precursors). When infected cells were given a pulse of $^{3}$H-leucine from 13 to 14 minutes after infection and particles were then extracted by either method, the label first entered a 300–400S peak, then passed to a 550S peak, and finally entered matured heads that sediment at 1100S. The amount of DNA in the intermediate structures could be estimated by determining their buoyant density in CsCl. The 350S particles taken from the center of the first, broad peak have no or little DNA. The 550S particles generally have about 50% of the normal DNA content; they cannot be prepared without a step of treatment with DNAase, which implies that they may be attached to T4 DNA with which they are being filled.

The broad peak of precursors sedimenting from 300 to 400S is not homogeneous, but appears to contain at least two types of particle. After a short (1 minute) radioactive pulse, the leading edge contains a fraction of about 400S which largely contains P23. This has been termed prohead I. This is converted to a particle sedimenting at about 350S, as seen in longer pulses and at the center of the broad peak. This contains mostly P23* and has been termed prohead II. The difference in the rates of sedimentation of prohead I and prohead II is about what would be expected from a reduction in weight of the particle when P23 is cleaved to P23*. Although the transition cannot be directly followed, it seems likely therefore that prohead I represents the first headlike structure and is converted to prohead II by the cleavage of P23. Since neither particle contains DNA, the cleavage step must precede packaging of the genome.

The 350S prohead II particle contains P22, which represents about 4% of its total protein, and also IPIII (presumably IPII and IPI also are present, but these could not be followed on the gels used for protein separation in these experiments). The 350S prohead II decays to the 550S particle with a half life of only 1–2 minutes. It contains less P22, representing some 1.5% of total protein, and also possesses some IPIII* in place of IPIII. Cleavage of P22 and IPIII does not appear to be complete until the 1100S mature head has been formed. This suggests that cleavage of P22 and the internal proteins takes place as the head precursor structure is being filled with DNA, a reasonable model in view of the location of these proteins within the outer shell. The 550S particle appears to represent an intermediate that is about half full with DNA; this has been termed prohead III. Its isolation as such implies that head filling may take place in two stages: first half of the DNA is inserted rather rapidly; and then the second half is inserted. In other words, packaging does not take place continuously, since in this case there should be precursor heads at all stages of fullness.

The pathway for head maturation suggested by this analysis is illustrated in Figure 6.28. It is not clear whether prohead I contains other proteins in addition to P23, but P20, P22 and IPIII all have associated with the precursor structure by the prohead II stage. By analogy with the aberrant structures formed in mutant infections, the formation of prohead I may require the presence of P31 and P40; its conversion to prohead II may require P21 and P24. Since *16*⁻ and *17*⁻ mutants are unable to fill their heads with DNA, it seems likely that P16 and P17 are needed for the formation of prohead III; and, indeed, this is confirmed by the observation that in *16*⁻ or *17*⁻ infection, the mutants accumulate 350S particles. Mutants in gene *49* accumulate half filled heads very similar to the 550S prohead III structure (see below), and this suggests that P49 is needed for the completion of head filling by which prohead III is converted into the mature head. In *49*⁻ infection the conversion of 350S particles to the 550S prohead III occurs at a reduced rate and this is consistent with a model in which the rapid half filling of prohead III is succeeded by a slower packaging of the rest of DNA in a step for which P49 is rate limiting.

Prohead I is analogous to the tau particle in representing a structure of size similar to the head but comprising P23 instead of P23*. The tau particles that accumulate in *21*⁻ or *24*⁻ mutant infections typically have

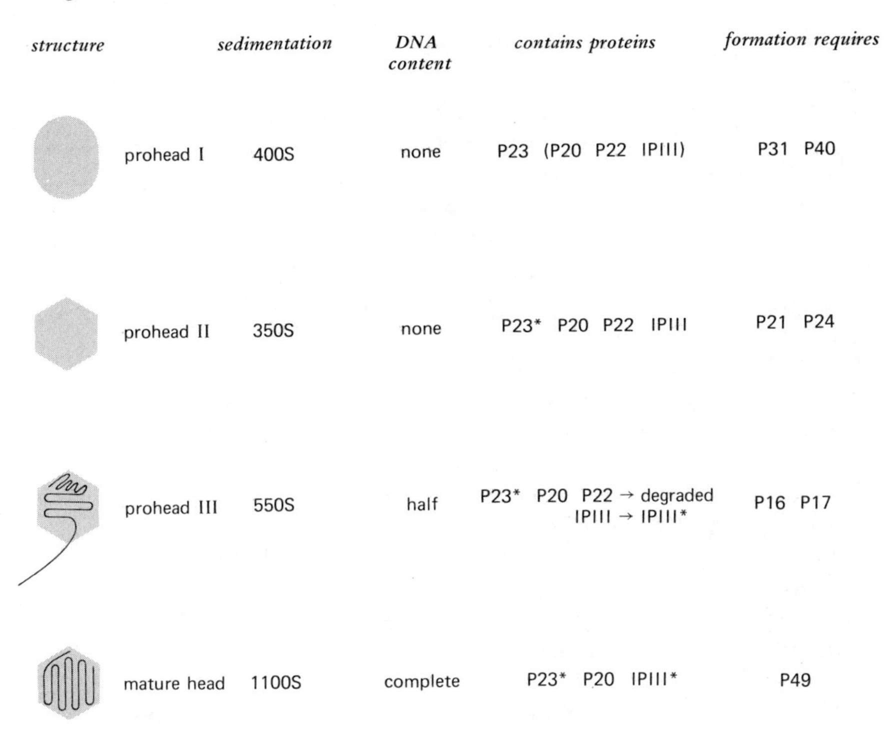


**Figure 6.28:** model for head assembly. After Laemmli and Favre (1973).

smooth outer shells of about 60–80 Å thick and in the elongated form have axes of about 520 × 715 Å, some 25% shorter than the mature head. Isometric tau particles also may be seen which appear spherical with a diameter of about 680 Å; these possibly bear the same relationship to the elongated tau particles as does the petite head of lambda to the wild type head. Most of the tau particles observed in thin sections by electron microscopy are bound to the inner cell membrane and have a proteinaceous core whose structure is not well defined (Kellenberger et al., 1968). Somewhat similar structures to the elongated tau particles were identified in wild type infection by Simon (1972), who in an electron microscopic study of particles formed during infection observed lumps, tau particles, and empty heads both attached to the membrane and free within the cell. The proportions of these structures seen during the course of infection suggested that first P23 may enter lumps, then gives rise to tau particles; these are then converted into empty heads on the membrane, and finally the heads are filled with replicating DNA within the cell. Laemmli and Favre (1973) suggested that the tau particles seen in this study represent prohead I structures, while the empty membrane-bound heads correspond to prohead III.

Are the tau particles that accumulate in *21*$^-$ or *24*$^-$ mutant infections the same structures as the prohead I particles formed during wild type infection? That they must consist of uncleaved P23 is confirmed by the observation of Laemmli (1970) that cleavage does not take place in either mutant. Laemmli and Johnson (1973) isolated the *21*$^-$ tau particles either by very rapid extraction or by using bis-methyl-suberimidate (a bifunctional cross linking agent) to stabilize them. The particles stabilized by suberimidate sediment at about 420S and lack DNA; they contain P23, P22, IPIII, an unidentified protein, and P24. After infection at high temperature with a *21*$^{ts}$ mutant, the temperature of incubation was shifted down to the permissive level. But the number of tau particles per cell remained the same and they did not appear to be converted into heads. When cells were labeled with $^{14}C$ leucine during the high temperature incubation, the label did not enter the mature heads made after a shift down and the proteins P23, P22, P24, IPIII labeled at high temperature were not cleaved. This implies that the tau particles accumulated in *21*$^-$ mutant infection are not precursors in head maturation, but instead represent a blocked form of assembly rather similar to, but not identical with, the prohead I structure.

A different result is observed with the tau particles formed in *24*$^-$ mutant infection. Bijlenga, Scraba and Kellenberger (1973) found that following their formation during infection of cells at high temperature with a *24*$^{ts}$ mutant, a shift to permissive temperature allows P23 to be cleaved to P23*. IPIII also is cleaved and P22 disappears. Of course, this does not prove that the tau particles are converted directly into heads, since it is possible that they might disaggregate into free protein that is subsequently reutilized. To exclude this possibility, Bijlenga, Brock and Kellenberger (1974) performed a density label during the high temperature incubation; light isotopes were substituted at the time of the temperature down shift, but phages were generated with heavy heads, which must have arisen by conversion of the tau particles. Comparable results using a radioactive label in protein and density label in DNA were reported by Laemmli, Teaff and D'Ambrosia (1974). The *24*$^-$ tau particles thus appear to be able to act as precursors for head assembly, and may correspond to the prohead I structure. The difference between the two sets of tau particles (which are indistinguishable by electron microscopy) may perhaps lie in the manner in which P21 functions to sponsor cleavage of the head proteins; in the *21*$^-$ infection tau particles must be formed in a way that does not allow reactivation of P21 after the reduction of temperature. This might happen if P21 must function before P24 in wild type assembly, so that in a *21⁻* mutant infection P24 acts first and thus blocks P21 activity.

What is responsible for establishing the length of the head shell? Mutants in two genes produce large numbers of petite particles, whose heads are isometric instead of elongated. The first set of mutations lies in gene *23* (these were formerly thought to identify a new gene, *66*): this implies that changes in the structure of the principal head protein can alter the elongation of the head shell. The second class of mutations was isolated by Paulson, Lazaroff and Laemmli (1976) as temperature sensitive variants of gene *22*, which at high temperature generate petite phages as well as polyheads and other abnormal structures (such, for example, as particles with two tails). The petite particles have reduced amounts of IPIII* (and could also have reduced content of IPII* and IPI*, which were not examined) and lack the internal protein B1*. This result implies that the core complex may have a role in controlling the extension of the particle. Paulson, Lazaroff and Laemmli suggested that one way in which the core complex might interact with the head shell could be through a Vernier model: if the core complex and head shell possess different repeating lengths, which must be aligned in register at the two ends of the head structure, a change in shape of either could alter the distance between the two points of alignment, thus changing the dimensions of the head. But whether or not this particular model applies to controlling the head dimensions, the implication of these results is that some type of scaffolding interaction involving both P22 and P23 must be implicated in head shape formation.

An unusual structure that may be of relevance to studies of head length determination is the *lollipop*. L-canavanine, a structural analogue of L-arginine, interferes with the growth of T4 and at high concentrations causes the production of polyheads. At lower concentrations the formation of lollipops is induced. Cummings et al. (1971, 1973) observed these as particles with very long heads, typically around ten times the usual length but up to about forty times longer; the lollipops contain DNA, have attached tails, and are infective. The principal protein of the shell is P23*, and the three IP species are present internally. Essentially the lollipop seems to be a head shell in which the relationship between the ends is distorted by excessive elongation. The formation and structure of lollipops has been reviewed by Cummings and Bolin (1976).

The involvement of P24 in determining head length was suggested by the observation of Bijlenga, Aebi and Kellenberger (1976) that infection with *24*$^{ts}$ mutants performed at temperatures intermediate between the permissive and nonpermissive generates giant phages, similar to the lollipops induced by canavanine. Giant tau particles also are produced, in addition to the tau particles and polyheads that we have already discussed. While this variety of effects makes it difficult to establish what role P24 plays in head morphogenesis, these results suggest that P24 may be involved in the formation of the end of the head, that is in the capping reaction that presumably terminates elongation. According to this view, the formation of giant phages when $24^{ts}$ mutants are used in infection at semipermissive temperatures represents a partial breakdown of the length determination mechanism, whereas the polyheads formed at nonpermissive temperatures are due to a complete breakdown.

### *Structural Transitions in Head Assembly*

The head of the mature T4 phage particle displays a somewhat smooth appearance in electron micrographs made following negative staining, from which it is not possible to determine the arrangement of subunits. However, morphological subunits—capsomers—can be seen on the surface of polyheads and the nature of their arrangement can be visualized. Since different types of polyheads have been distinguished, it is possible to define different arrangements into which P23 and P23* may be organized. The transitions that occur between these different structures may reflect the structural transitions that occur during normal head assembly.

Multilayered polyheads consist of tubular structures of different diameters, which upon lysis can be separated into the individual tubular layers that were utilized by Yanagida et al. (1970) for structural studies. The inner diameter of the cylindrical tube of the first layer of a multilayered polyhead is the same as that of the single layered polyhead, about 400 Å. The outer diameter is about 590 Å. The difference between the outer diameters of the successive layers of a multilayered polyhead is close to 190 Å, so the thickness of each layer is the same 95 Å as the thickness of the first layer. This confirms that the single layered polyhead consists of a single cylindrical shell, consisting of a single layer of protein, additional shells of the same construction simply being laid around the surface in the multilayered polyhead. (These values are converted from the measured widths of squashed polyheads, that is, which represent half the circular circumference, and are increased to take account of the shrinking that occurred during the negative staining and preparation of thin sections.)

Different types of polyhead were distinguished in gene *22* mutant lysates by Yanagida, DeRosier and Klug (1972). The most common structure seen immediately upon lysis is the "coarse" polyhead, which has a prominent surface structure. In gene *22* mutants this is "empty," that is, lacks the proteinaceous core seen in the polyheads generated by other mutants. This is the type of polyhead that was investigated in the earlier studies of Yanagida et al. (1970). Upon storage for several days at low temperature without fixation, "smooth" polyheads appear among the coarse tubes. These are about 10% wider than the coarse polyheads, in these experiments with a cylindrical diameter of about 580 Å instead of 510 Å. (Some variation in the diameters of polyheads is seen between different reports: the important point is that multilayered polyheads always can be seen to consist of successive layers of the same thickness, while there is an expansion in the diameter of the tubule of the smooth polyheads compared with their coarse predecessors.) Smooth polyheads also are seen when lysates from gene *20* mutants are examined (that is without the need for an extensive storage period).

Early studies of the surface structure of the polyhead carried out by Klug and Berger (1964), Finch, Klug and Stretton (1964) and Kellenberger and Boy de la Tour (1965) suggested the general conclusion that its construction represents a hexagonal net (although the dimensions that were then ascribed to the net have since proven to be in error and have been revised). In principle the polyhead can therefore be regarded as a plane hexagonal lattice which has been rolled up into a cylinder, as illustrated in Figure 6.29. The hexagonal unit of the net corresponds to the capsomer from which the head shell is constructed, that is an aggregate of six molecules of P23 or P23*. The different types of tubular structure that have now been distinguished all can be regarded as the result of rolling up an appropriate hexagonal lattice into the cylindrical surface, with the

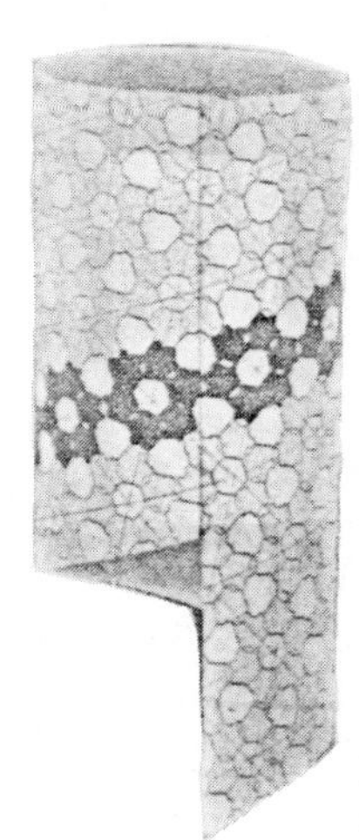

**Figure 6.29:** construction of a cylindrical tube by rolling up a hexagonal plane surface. After Yanagida et al. (1970).

variations between them accounted for by differences in the construction of the lattice, that is, in the organization of morphological units.

The electron micrograph of a negatively stained flattened tube essentially superimposes the images of both sides of the tube, making it difficult to distinguish the molecular structure. The technique of *optical filtering* can be used to reconstruct the image of each side separately and this was the basis of the studies of DeRosier and Klug (1972) on polyhead structure. Their analysis showed that the coarse and smooth polytubes have different lattice structures, the smooth corresponding to an expansion of the capsomer of the coarse structure.

Coarse polyheads may be either full, that is possessing a proteinaceous core, or empty; the only difference between them is that the filled tubes have a diameter that is smaller than the empty tubes. Otherwise the surface structure is the same. This conclusion is supported by the later studies of Steven, Aebi and Showe (1976), who found that the polyheads of *22*⁻ and *ipIII*⁻ mutants are on average some 10% smaller in diameter than those of *20*⁻, *24*⁻ or *40*⁻ mutants. However, there is also quite appreciable variation in the diameters of both the empty and full polyheads, over a range of about 100 Å. DeRosier and Klug found that the coarse tubes consist of a hexagonal net that can be described as comprising a $p6$ construction, since the capsomer is a simple hexamer; the lattice constant $a$, which can be taken to be the diameter of the capsomer, is 110 Å. The wider tubes that are derived by dissociating the outer layers of multilayered polyheads have a construction similar to the inner layer (which is identical to the single layered polyhead), but the outer layer surfaces are distorted relative to the inner surface, perhaps because they have grown on the inner tube.

The smooth empty polyheads have a lattice constant of 125 Å. This represents an expansion of the capsomer relative to the morphological unit of the coarse polyhead; the ratio of the lattice constants is about the same as the relative change in diameter of the tubules. Unlike the coarse polyheads, however, the central region of the hexamer is not penetrated by stain, suggesting that it contains additional material. The morphological unit of the smooth empty polyhead therefore is described as 6 + 1; this indicates that the nature of the central material is not known and cannot be ascribed to the same type of subunit that comprises the hexamer. Thus the morphological unit can be considered still as a hexamer, but in which the formerly empty center now has been filled. The construction of the lattice appears to depend on the conformation of the hexamer per se and does not seem to be influenced by the central filling. The surface appearance of these polyheads appears to resemble that of the capsid of ghost particles and may therefore be close to the organization of material in the mature head. Another type of smooth polyhead displayed a fine mottled appearance and appeared to possess a core of some type; its relationship to the other tubular structures is not revealed by these experiments because of the difficulty in visualizing its surface structure.

The transition from the coarse polyhead to the smooth empty tubular structure is accomplished by the cleavage and rearrangement of P23. Laemmli and Quittner (1974) observed that upon incubation at 37°C in vitro the P23 present in isolated polyheads is cleaved to P23* (except when the polyheads are isolated from *21*⁻ mutants). Laemmli, Amos and Klug (1976) took advantage of this observation to correlate the protein cleavage with structural changes. In its freshly isolated form, the coarse polyhead consists of a hexagonal lattice construction in which the hexameric capsomer is elliptical; this is described as the polyhead I structure. Upon incubation a change takes place in the organization of the capsomer so that the subunits of the hexamer lie on a circle and are more elongated towards the center of the unit. This is the polyhead II structure; the diameter of its (coarse) tubule remains unaltered by the change in arrangement within the capsomer (in these experiments being about 480 Å). Further incubation generates the polyhead III structure which corresponds to the smooth empty polyhead of wider diameter (on average 530 Å in these experiments); this has an expanded lattice unit and the hexamers also are rotated with respect to the polyhead II or polyhead I structure. Upon storage for extended periods, a polyhead IV structure was observed, apparently corresponding to the fine mottled smooth type. Examples of the filtered images of these polyheads are shown in Figure 6.30 and a representation of the hexameric organization is depicted in Figure 6.31.

Phage structures containing P23 can be distinguished from those in which the head protein has been processed to P23* by the susceptibility of the protein to extraction with SDS; P23 is readily extracted at room temperature, but P23* is scarcely extracted at all. Another criterion is that structures comprising the uncleaved protein dissociate in the presence of sodium iothalamte (Conray) whereas structures of cleaved protein are stable. According to these assays, prohead I, tau particles, and polyhead I all comprise uncleaved P23, whereas proheads II and III, mature heads, and polyheads II and III consist of P23*. Direct analysis confirmed that polyhead I is assembled from P23, whereas polyheads II and III have P23*, although the proportion of the cleaved protein may be as low as 60%; this implies that it is not necessary to cleave all the copies of the protein in the structure to achieve the structural transformation. The similarity in response to SDS or iothalamate of polyheads II and III relative to polyhead

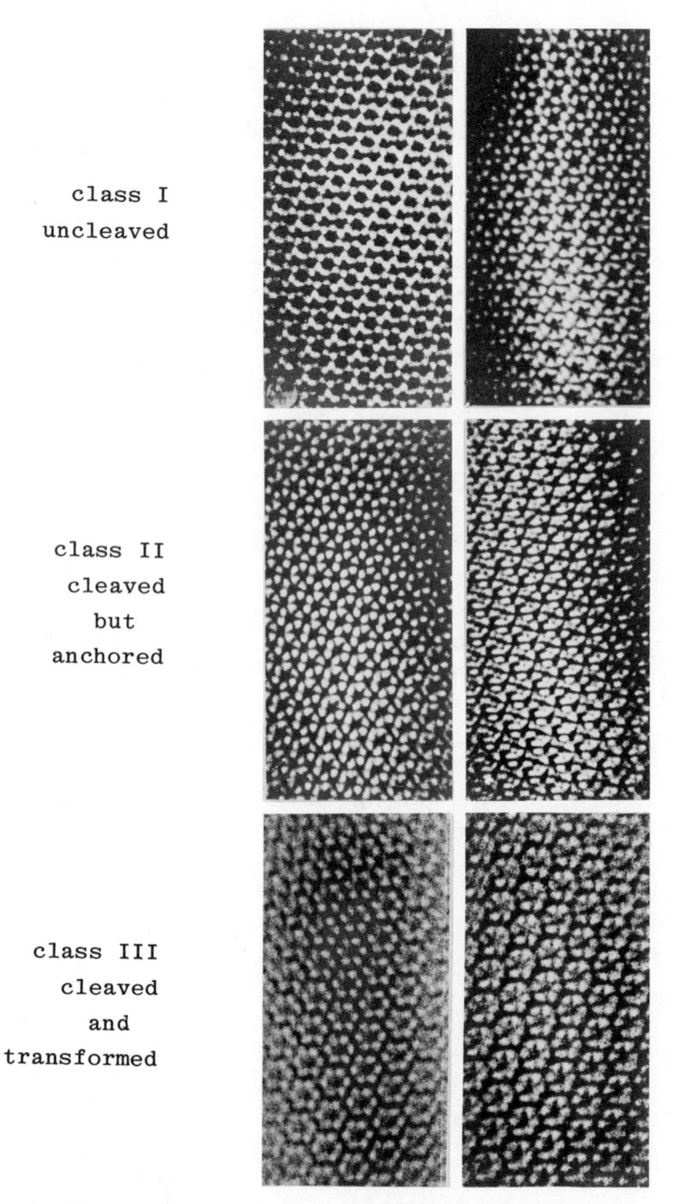


**Figure 6.30:** filtered images of T4 polyheads. The near side image is shown for type I, type II and type III polyheads stained with PTA (left) or with uranyl acetate (right). Date of Laemmli, Amos and Klug (1976).

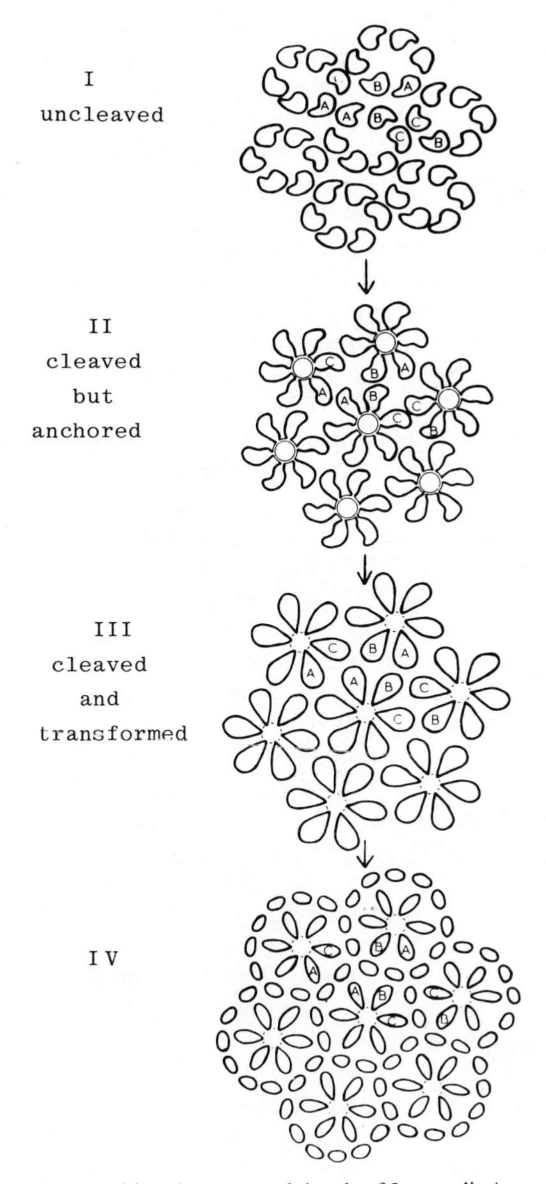


**Figure 6.31:** model for transitions between polyheads of Laemmli, Amos and Klug (1976).

I mimics the situation during normal assembly, and supports the idea that the structural transitions between the different polyhead forms may reflect transitions that occur during normal head assembly. Laemmli et al. suggested that polyhead II represents an intermediate in the structural rearrangement from the coarse to smooth organization; in this structure the molecules of P23 have been cleaved, but remain anchored in their former positions. The transition to polyhead III represents the movement of the protein subunits as the lattice is expanded.

The difference between polyhead III and polyhead IV may lie in the addition of a further protein to the outer surface. This may be the small outer capsid protein, soc, identified by Ishii and Yanagida (1975); its molecular weight is about 10,000 daltons and it appears to correspond to a protein of this size identified by Hosoda and Cone (1970) and Forrest and Cummings (1971) as a component of the phage head. It is present in all T-even phages except T2L. Another protein also identified by Ishii and Yanagida was the highly antigenic outer capsid protein, hoc, of 40,000 daltons. Neither protein appears to be essential for construction of the head shell, since T4 *soc*$^-$ *hoc*$^-$ mutants that lack both species present a normal appearance. Purified soc and hoc proteins bind to *soc*$^-$ and *hoc*$^-$ phage particles in vitro, which implies that they are able to bind to the head from the outside and need not be present at the time of assembly. The absence of the soc protein, however, is reflected in a difference in appearance of giant capsids of T2 relative to giant capsids of T4 that contain the protein, and Laemmli, Amos and Klug (1976) suggested that the addition of this protein on the surface might be responsible for the transition from the polyhead III to the polyhead IV. This idea is consistent with the observation of Ishii and Yanagida that *soc*$^-$ lollipops look very like the polyhead III structure when examined by optical filtering, but the addition of soc protein confers a much smoother surface appearance.

A similar approach of following the transitions between types of polyhead has been taken by Steven et al. (1976). Only one type of coarse polyhead was observed in these experiments; it contains P23* and is relatively unstable. Upon incubation in vitro, however, it may be converted to the more stable form of the smooth polyhead, containing P23, in an action that depends on the presence of a functional P21 protein. Three types of smooth polyhead were distinguished in these experiments; these are shown in the electron micrographs of Figure 6.26. The A type polyhead appears to be the immediate product of the coarse polyhead generated by cleavage of P23. Its conversion to the B type polyhead relies upon the addition of two proteins of about 40,000 daltons which may correspond to the *hoc* functions. The further conversion to the type C polyhead depends upon the addition of the soc protein. The A type polyhead has a p6 net; the B and C polyheads have p6+1 nets. The nature of the central material that causes the transition from p6 to p6+1 is not clearly established; but it may be the hoc protein(s).

From the results of both Laemmli et al. and Steven et al. it is clear that there is a series of well defined transitions from the coarse polyhead containing uncleaved P23 to the smooth polyhead containing cleaved P23* and the outer capsid proteins. Somewhat similar transitions were observed in each set of experiments, the principal difference lying with the polyhead structure observed to result from the first change in the coarse structure; it is likely that the polyhead II observed by Laemmli et al. is an intermediate between the coarse polyhead and the A polyhead of Steven et al., the reason for this difference in the series of events probably lying with the experimental conditions utilized in vitro. While the transitions involved in these polyhead transformations remain to be characterized in more detail in terms of the roles of individual proteins, their general nature now is clear and it is reasonable to suppose that they reflect changes occurring during head maturation.

The structure of the head of the T2 particle has been visualized directly by the use of freeze etching. Branton and Klug (1975) reported that by this method they were able to observe arrays of capsomers possessing a hexagonal substructure very similar in appearance to the optically filtered images of negatively stained polyheads. The head corresponds to the structure of an icosahedron elongated along one of its fivefold axes of symmetry, a view that has replaced the earlier concept of a hexagonal biprism (see Moody, 1965). In other words, the T-even shell can be considered as a tubular variant of an icosahedral shape generated by rolling up a hexagonal plane net. Branton and Klug observed that the head structure considered as an unelongated icosahedron (that is ignoring the additional material responsible for the elongation) is organized with a triangulation number of T = 13, that is as 780 quasiequivalent subunits (see also page 520). These can be considered to comprise 120 hexameric capsomers, and a further 12 pentameric capsomers that represent the apices of the icosahedron (there would be only 11 pentamers if one were replaced at the apex where junction with the tail occurs). The elongation corresponds to the insertion at an equatorial position on one axis of a band of a further 20 hexameric capsomers, that is, another 120 protein subunits. This model therefore predicts that the head shell should comprise 900 protein subunits. The principal protein P23* must provide the subunits of all the hexameric capsomers (a total of 840 protein molecules); it is not clear whether it or another protein provides the remaining 60 subunits that comprise the apical pentamers.

### *DNA Packaging in the Head Shell*

Two types of model can in principle be imagined for the packaging of DNA into the head shell: DNA might be condensed into a compact body that is then surrounded by a coat of protein; or an empty head shell may be assembled and then subsequently filled with DNA. As with phage lambda head morphogenesis, described in Chapter 5, assembly of the T4 head follows the latter model. The circular permutation of the T4 genome and the dependence of the extent of terminal redundancy upon genome length both argue for the sort of headful packaging model discussed earlier (see page 573) in which concatemeric DNA is packaged into the head from a random starting point until its full capacity is reached. This model also is lent support by the observation that the giant heads of lollipop particles contain continuous lengths of DNA much longer than the usual genome (Uhlenhopp, Zimm and Cummings, 1974). Although a nuclease must be involved in cleaving the headful from the longer precursor, instead of acting at specific sites (as for example happens with lambda head filling) it must in some way be activated by the completion of DNA content.

Three genes have been implicated in the packaging of DNA into the T4 head: these are *16*, *17* and *49*, and their stages of action are summarized in Figure 6.28. Similar properties are displayed by mutants in genes *16* and *17*; they produce heads of normal appearance but which lack DNA (Luftig and Ganz, 1972). This does not prove that they are required to initiate head filling with DNA, however, because heads that are not stabilized by the attachment of tails may be fragile, losing their DNA upon isolation. However, the demonstration by Laemmli and Favre (1973) that prohead II structures may accumulate in $16^-$ and $17^-$ mutant infections suggests that the absence of DNA from these head structures is due to failure to fill the head. The role of the P16 and P17 functions has not yet been defined, but appears to represent a defect in packaging per se since DNA replication is not altered.

Mutants in gene *49* accumulate half filled heads representing the prohead III structures depicted in Figure 6.28. Luftig, Wood and Okinara (1971) reported that during infection at high temperature, a $49^{ts}$ mutant accumulates particles that sediment at around 300S and possess only a small proportion of the usual DNA content, generally less than 20% of that found in wild type particles. A temperature shift down then allows the formation of active particles; this is prevented by inhibitors of replication, which argues that DNA synthesis is required for the completion of head filling. A density label given before the shift down enters mature phage particles, supporting the idea that the 300S particles are intermediates in assembly (Luftig and Lundh, 1973). Experiments of a similar design performed by Laemmli, Teaff and D'Ambrosia (1974) suggested that a somewhat greater proportion of DNA, close to 50%, is associated with the *49*⁻ head, but that some of this tends to be lost during isolation of the intermediates. These intermediates may be equated with the *grizzled particles* that can be seen in electron microscopy of thin sections from *49*⁻ mutant infected cells. Grizzled particles are visualized in Figure 6.25 and their characteristic appearance is due to their partial content of DNA. Similar particles have been seen in lambda infection (see page 519).

Replication is abnormal during *49*⁻ infection; Frankel et al. (1971) reported the accumulation of DNA sedimenting at about 1000S, very much faster than the usual concatemeric form of 200S and the mature genome of 63S (see Figure 6.19). Dewey and Frankel (1975a,b) observed that extracts from *49*⁻ mutant infected cells display a reduced activity in degrading the 1000S DNA to smaller fragments, which led to the suggestion that P49 may be involved in nucleolytic cleavage of the DNA. Suppressor mutants, falling into the two classes *fdsA* and *fdsB*, both of which are semilethal but allow *49*⁻ mutants to reproduce, do not accumulate the 1000S DNA, raising the possibility that their wild type alleles may be involved in its generation (and in creating the need for P49 in processing it). When T4 DNA is examined by the sarkosyl technique, which isolates DNA specifically associated with the membrane (see Volume 1 of Gene Expression), replicating T4 DNA is found in this fraction. Usually the DNA is released from the membrane later in infection. However, Siegel and Schachter (1973) found that this detachment does not occur in the *Y* group mutants or in *16*, *17* or *49* mutants; this leaves the implication that head filling with DNA is required for detachment from the membrane. An especially low level of detachment is shown by *49*⁻ mutants (about 1%, compared with the 5–10% displayed by the other mutant types). Kemper and Janz (1976) and Kemper and Brown (1976) showed that the very fast sedimenting (> 1000S) DNA found in *49*⁻ infection retains this property even after release in vitro from the membrane fraction: its compact structure appears therefore to be a property of the DNA itself. These experiments also showed that the DNA aggregate may reband with the sarkosyl crystals, although lacking membrane components; this casts some doubt upon the specificity of this technique when applied to very large complexes of DNA.

Some information on the structure of the concatemeric DNA found in T4 infected cells has been provided by the experiments of Curtis and Alberts (1976), who found that the gene *32* protein appears to be important in establishing the state of the DNA. When cells infected with a *32*$^{ts}$

mutant are shifted to high temperature, the T4 DNA loses its rapidly sedimenting concatemeric form and sediments instead in a range corresponding to sizes between quarter molecules and dimers. This suggests that there are single strand regions in the replicating DNA which are protected from nuclease attack by binding to 32-protein; when the 32-protein is thermolabile and is inactivated by an increase in temperature, these regions become susceptible to breakage, thus reducing the size of the intermediate. This breakdown of DNA when 32-protein is inactivated does not occur when the phage carries a mutation in gene *49*. This suggests that P49 either generates the single stranded regions to which P32 binds and/or cleaves the free single strands released when the binding protein is inactivated. That P49 is needed to generate the single strand regions is suggested by the observation that only the DNA extracted from cells infected with $49^+$ phages is susceptible to the single strand specific nuclease S1. Although the exact role of P49 is not yet clear, these results support the general conclusion that it is involved in cleavage events needed packaging a full complement of DNA into the head.

Mutants in gene *13* accumulate both filled heads and empty heads. The empty heads retain some DNA and appear to be generated by the loss of DNA from heads whose content of DNA is unstable in the absence of P13 (Hamilton and Luftig, 1972, 1976). While the role of P13 has not yet been determined in detail, this observation is consistent with the conclusion of Edgar and Lielausis (1968) that genes *13* and *14* are needed for the final stage of head-tail junction; the use of extracts suggested that P13 and P14 can act upon $13^-$ and $14^-$ heads isolated in vitro to allow them to join with tails. The inference to be drawn from the presence of empty heads in $13^-$ mutant infections then is that tail joining is needed to stabilize the presence of DNA in the head, which may otherwise be lost. This implies also that similar loss may occur from heads blocked in the earlier stages of assembly.

In summary, then, it is clear that the *Y* group gene products are needed for assembly of a head of proper shape. P16 and P17 appear to be involved in the initiation of DNA packaging, but this only proceeds to half completion unless P49 is present to support complete filling. P13 and P14 then are needed for junction with the tail.

### *Assembly of the Tail Core and Sheath*

The T4 tail includes the head-tail connector, the core to which it is attached, the surrounding sheath, and the baseplate and tail fibers. These components control the initial stages of adsorption, which are illustrated in Figure 6.32. The long tail fibers, which make the initial contact with the

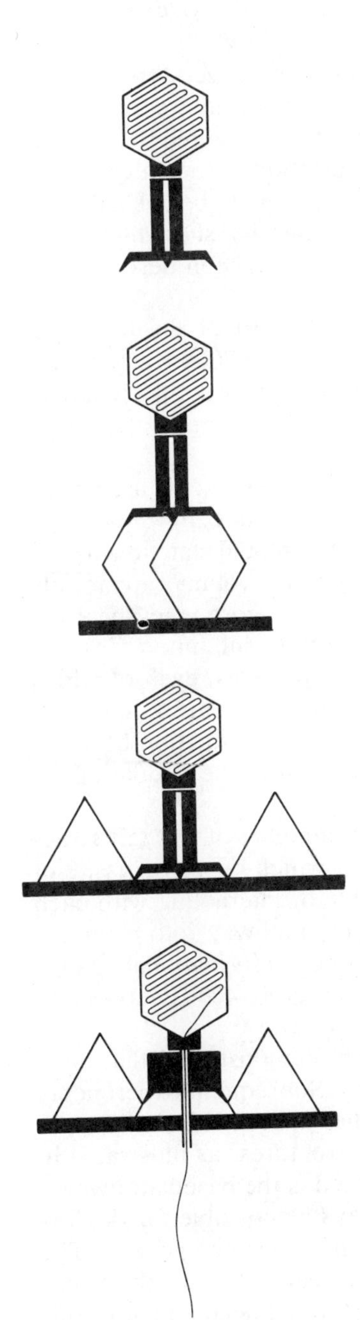

**Figure 6.32:** model for phage adsorption. In the mature phage particle fibers are wound around the tail and so are not visible; they unwind upon contact with the surface of the host cell to make the initial attachment. Then the baseplate spikes are pulled into contact with the surface, the sheath contracts so that the core penetrates the bacterium, and the baseplate is pushed away from the surface to leave the short fibers in contact with the bacterium.

host bacterium, are responsible for establishing host range. After contact has been made the spikes of the baseplate are brought into contact with the cell surface and then contraction of the sheath is triggered to drive the core into the bacterium with the consequent injection of the phage DNA. As the core enters the bacterium the baseplate appears to be pushed away from the surface and may then be held in contact by short fibers that protrude from it. The initial stages of infection have been described by Simon and Anderson (1967a).

In construction the core appears to be the simplest of the tail structures, representing a hollow cylinder of outer diameter 70 Å with an axial hole of 25 Å through which the DNA passes during injection; the length of the core is fixed at about 800 Å (see Figure 6.2). The core is assembled from a single type of protein subunit.

The sheath which surrounds the core essentially represents a contractile system, for it is able to change from its elongated state, in which it has a diameter of 160 Å and length of 800 Å, to the contracted state in which it is 250 Å in diameter but only 350 Å in length, its volume apparently remaining constant during the transition. Like the core, the sheath is constructed by assembly of a single type of protein subunit.

The tail fibers are assembled from two major proteins, each of which represents half the fiber, and also possess two minor proteins. Each of the half fibers is 690 Å long and 20 Å in diameter and the junction between them generates the characteristic kink of 156° seen in the middle of the complete fiber.

The baseplate is morphologically the most complex of the tail structures; it is constructed from about 15 proteins, which raises the issue of how so many different types of subunit capable of interacting with each other are compelled to assemble along a single pathway into a unique structure. The baseplate undergoes an interesting conformational change upon phage adsorption, from a flat hexagonal structure 400 Å across to a boatlike hexagram star 600 Å across.

The early studies of Edgar and Wood (1966) demonstrated that head and tail assembly follow independent pathways. Subsequent experiments upon in vitro complementation have shown that tail assembly is accomplished by the sequential construction of tail structures, as illustrated in Figure 6.33. The first component to be assembled is the baseplate, which represents the junction of independent pathways responsible for the formation of the outer hexagonal structure and its inner component. The core may then be assembled upon the baseplate and the sheath in turn polymerizes on the core. The tail then is joined to the head and after this union the long tail fibers are added. This order of events is obligatory: with the sole exceptions of three proteins residing on the outer surface of

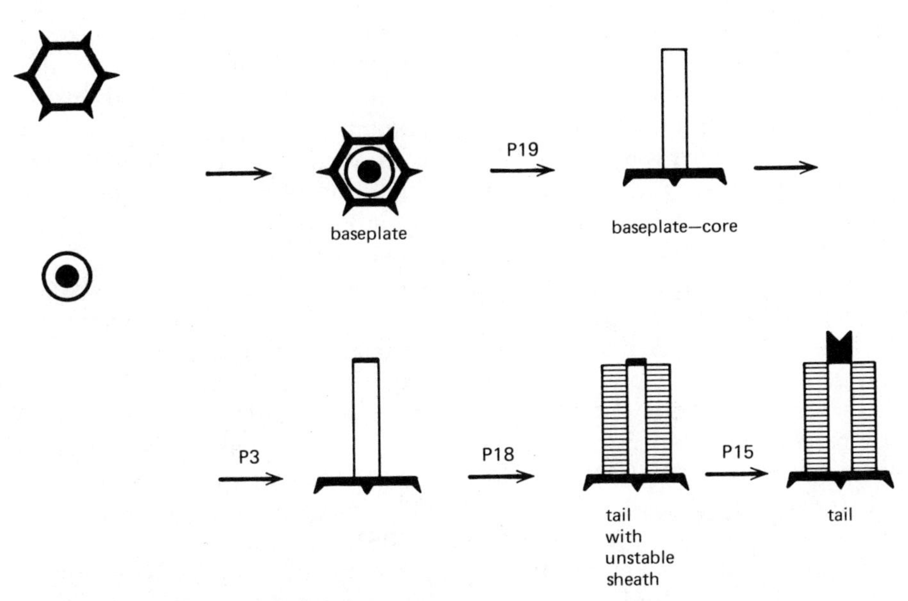


**Figure 6.33:** pathway for tail assembly. The inner and outer components of the baseplate are formed independently and then unite to generate the free baseplate. The core is formed by polymerization of P19 on the baseplate; P3 acts on the core to stabilize the sheath, which is formed by polymerization of P18; and P15 forms the head-tail connector.

the baseplate, each protein must be added to the assembling tail only at the proper point and cannot be added earlier or later. The organized nature of the interactions between assembling tail proteins is reflected in the arrangement of their genes: the tail genes fall into groups concerned with fiber formation, baseplate construction, and core and sheath assembly, the members of each group in general occupying adjacent positions on the assembly pathway.

Two principal approaches have been used to define the intermediaries of tail assembly. One is to isolate the structures that accumulate in lysates from cells infected with mutants that lack one of the tail functions (these phages usually also carry a mutation in gene *23* to prevent the accumulation of head structures that may obscure the tail intermediates). Another is to use a complementation assay. Structures isolated from an $A^-$ lysate are tested for their ability to complement an extract from $B^-$ infected cells. If the $A^-$ structure is able to do so, the product of *B* must already have acted upon it—so the $B^-$ extract can provide the remaining necessary functions. This means that the *B* product must act before the *A* product. On the other hand, if the $A^-$ structure cannot complement the $B^-$

extract, it must still need the *B* product, so that the *A* product must act before the *B* product. The reverse relationship should be seen in the reciprocal assay using a structure isolated from the $B^-$ cell and an $A^-$ extract. The structures isolated from mutants blocked in some stage of assembly, either by biochemical means or by their activity in the complementation assay, then may be analyzed for protein content. In this way the obligatory series of events in tail morphogenesis may be defined and the structural proteins present at each stage can be determined.

Mutants in gene *27* lack tails but produce active heads whereas mutants in gene *23* lack heads but produce active tails. Using these two mutants in reciprocal complementation assays, Edgar and Lielausis (1968) demonstrated that the complementing activity in the $27^-$ extracts lies in a structure sedimenting at 1200S (the free heads) and the activity of $23^-$ extracts resides in a structure of 130S (the free tails) with a minor peak at 80S. In $15^-$ or $18^-$ extracts, all the tail complementing activity sediments at 80S; this suggests that the P15 and P18 functions may be responsible for converting an 80S precursor into the 130S mature tail. Using electron microscopy to examine the structures present in $23^-$ lysates, King (1968) showed that the introduction of an $18^-$ mutation causes the accumulation of naked tails that lack a sheath, and $15^-$ or $3^-$ mutations lead to the appearance of partially sheathed tails. When extracts from cells infected with these mutants were examined by an antigenic assay for the tail, all the activity resided in a peak sedimenting at 80S. The 80S component appears to represent the core plus baseplate; the 130S tail is generated by addition of the sheath.

Gene *18* codes for the single protein that comprises the sheath. Genes *3* and *15* are needed for proper assembly of the sheath on the naked tail in order to stabilize the completed structure. When mutants that prevent construction of the core were examined for the formation of tail structures, no normal sheaths could be seen; the only sheath structures were polysheaths, which look like long contracted sheaths. They consist only of P18; P3 and P15 are not needed. This suggests that proper sheath formation occurs only in the presence of a naked tail, that is, the sheath must polymerize on the core. King and Mykalojewycz (1973) confirmed that P3 and P15 are needed for irreversible joining of sheath and core because in their absence the sheath is lost from the core during lysis, its subunits dissociating sequentially from the neck. P15 forms the *connector,* a protrusion at the end of the tail that constitutes the site for head attachment. King (1971) showed that P3 acts on the core rather than directly on the sheath. Extracts from $3^-$ infected cells are complemented by the $18^-$ tails (which lack the sheath); this implies that P3 must have acted on the core before the sheath polymerizes. When the sheath is formed on the core, then, it is necessary for it to be stabilized by the presence of P3 and P15 at the neck end.

Little is known of the mechanism by which the sheath contracts. In structural studies, Moody (1967a,b) demonstrated that the contracted sheath displays sixfold symmetry and the annuli of the extended sheath form 24 horizontal bands, on occasion with a somewhat coarse helical arrangement running up the surface of the sheath. If each annulus comprises six subunits of the 80,000 dalton P18, then each would have a weight of 480,000 daltons. A model for the contraction of the sheath was presented by Moody (1973). How individual subunits interact with each other, and how the contraction is triggered, remain to be established.

By comparing the core-baseplates found in *23*$^-$ lysates with the free baseplates that accumulate in *19*$^-$, *48*$^-$ or *54*$^-$ mutant infections, King and Mykalojewycz (1973) demonstrated that that the core is constituted solely from P19, which has a molecular weight of 21,000 daltons. The absence of cores from lysates lacking baseplates implies that polymerization is initiated on the baseplate and then continues until the correct length has been reached. Since mutants in genes *48* or *54* accumulate normal baseplates but lack cores, P48 and P54 may be concerned with the initiation of core assembly on the baseplate. How the length of the core is determined so precisely is not yet known, but this property does not appear to reside in P19 itself, since disaggregated P19 molecules can form tubes of varying length. Some other function must therefore be involved.

### *Assembly of Tail Fibers*

Mutants in any one of six genes prevent the synthesis of tail fibers. Five of these genes, *34*, *35*, *36*, *37* and *38*, are organized in a cluster, and the last, *57*, appears to have a regulatory function. Edgar and Lielausis (1965) and King and Wood (1969) showed that an anti-T4 serum contains antigenic components that react with three fiber components, called A, B and C. By determining which gene products are needed for the production of each antigen, and then identifying the position in a sucrose gradient at which each antigenic activity is found in extracts from defective mutants, the structures represented by each antigen have been identified. Their relationship on the pathway of fiber assembly is depicted in Figure 6.34. Antigen C sediments at 8S and its formation requires P37 and P38. In the presence of P36 it is converted to the antigen BC, which now carries the B reactivity in addition to the C response; this sediments at the same rate as C. In the presence of P35, the species BC is converted into BC′, which retains unaltered antigenic activities but sediments more rapidly at 9S. Antigen A requires only P34 for its formation and joins with the structure

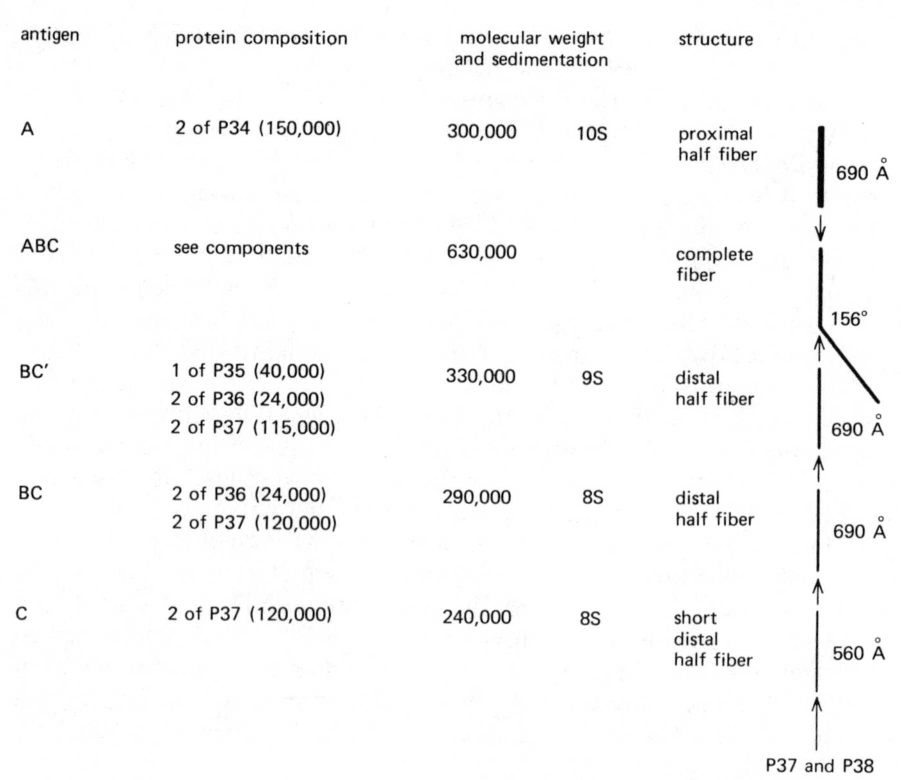

| antigen | protein composition | molecular weight and sedimentation | | structure | |
|---|---|---|---|---|---|
| A | 2 of P34 (150,000) | 300,000 | 10S | proximal half fiber | 690 Å |
| ABC | see components | 630,000 | | complete fiber | 156° |
| BC′ | 1 of P35 (40,000)<br>2 of P36 (24,000)<br>2 of P37 (115,000) | 330,000 | 9S | distal half fiber | 690 Å |
| BC | 2 of P36 (24,000)<br>2 of P37 (120,000) | 290,000 | 8S | distal half fiber | 690 Å |
| C | 2 of P37 (120,000) | 240,000 | 8S | short distal half fiber | 560 Å |
| | | | | | P37 and P38 |

**Figure 6.34:** assembly of tail fibers. The fibers consist of two half fibers, corresponding to the antigenic structures A and BC′. The A half fiber is formed by the dimerization of P34; the BC′ half fiber is formed by the dimerization of P37 to yield C (which requires the activity of P38), followed by the addition of P36 and P35.

BC′ to generate the 10S complete fiber carrying the three antigenic activities, ABC.

Electron microscopy of the two penultimate structures in fiber assembly, BC′ and A, shows that each represents a half fiber, showing the usual diameter but with a length of only some 690 Å. Ward et al. (1970) showed that A appears to have the structure of the proximal half that attaches to the baseplate, characterized by a knob at one end, while BC′ corresponds to the distal half of the fiber. This fits with the properties of mutants in the genes involved in synthesis of each half fiber. Mutants requiring cofactors for adsorption map in *34*; in the absence of the cofactor, the fibers remain bound to the sheath in their retracted position; this implies that P34 (equivalent to antigen A) may be responsible for the interaction of the fiber with the particle. On the other hand, host range mutations map in *37* (whose product is identified with antigen C); this corresponds to the location of P37 in the distal half fiber that must be responsible for making contact with the host bacterium. Also consistent with this assignment of responsibilities is the observation of Wilson, Luftig and Wood (1970) that an isolated lipopolysaccharide component of the E.coli wall, to which T4 attaches, can bind the BC′ half fiber or the BC or C half fiber structures, but not the A half fiber. This suggests that BC′ is the distal half fiber and implies further that the transition from BC to BC′ is concerned with making possible the junction with A, rather than with the host interaction.

Identification of the proteins in the precursor structures has been accomplished by performing gel electrophoresis in SDS. King and Laemmli (1971) and Ward and Dickson (1971) confirmed that antigen A is constituted solely from P34; the proximal half fiber probably comprises two molecules of P34, with a total weight of some 300,000 daltons. Antigen C contains only P37, although P38 is needed to convert the protein subunit into the antigenic structure. The C structure has a length of 560 Å and probably comprises two molecules of P37, with a total weight of about 240,000 daltons. The BC structure has the full length of 690 Å and contains P36 as well as P37; the results of Dickson (1973) suggest that there are two molecules of P36. The conversion to BC′ is accomplished by the addition of probably only a single molecule of P35 and this may be responsible for establishing the characteristic kink in the center of the completed fiber, at the junction of the two half fibers.

Polarity effects suggest that the genes that code for the fiber proteins are transcribed as two groups: *34 – 35*; and *36 – 37 – 38*. Gene *57*, which is located elsewhere, is implicated in a regulatory role by the observation that $57^-$ lysates have fiberless particles. Sometimes particles possessing only one fiber instead of the usual six are seen (Eiserling, Bolle and Epstein, 1967). This implies that one effect of P57 may be quantitative. Wood and Henninger (1969) observed that fibers appear to be attached singly and at random among the particle population; phages do not need to have all six fibers to be infectious.

### *Construction of the Baseplate*

A mature phage particle with an extended sheath possesses a baseplate that appears to be a flat hexagon with a side of 200 Å and distance from apex to apex of 400 Å. The six spikes probably protrude from the apices. In this form the baseplate has a triagonal plug at the center. Simon and Anderson (1967b) observed that when the sheath is contracted the baseplate is converted to a form in which it is boatlike instead of being flat and its shape changes to a six pointed star with a distance from apex to apex of

600 Å. A hole through the center is seen instead of the plug. This conformational change appears to be accomplished by reorganization of the protein subunits. Indeed, in tail assembly (unlike head assembly) there is no evidence for the involvement of any cleavage reactions, and both the construction of the tail and the structural transitions that occur on adsorption probably reflect rearrangements of components that do not utilize alterations in the compositions of the protein subunits.

The large number of proteins in the phage tail was established by King and Laemmli (1973) in a study that identified the genetic origin of many structural proteins. Most of these are concerned with baseplate assembly. The baseplate genes fall into two clusters: *53, 5, 6, 7, 8, 9, 10, 11* and *12* represent the major proteins; *25, 26, 51, 27, 28, 29, 48* and *54* specify the minor proteins. Some 15 proteins have been identified in isolated baseplates by Kikuchi and King (1975a,b), using extracts from cells infected with *23⁻ 19⁻* mutants that do not accumulate head structures and cannot assemble the core onto the completed baseplate. Free baseplates also accumulate in *48⁻* or *54⁻* mutant lysates, because the core cannot be assembled upon the baseplate; this suggests that P48 and P54 are concerned with the final stage of baseplate assembly that makes possible the addition of the core.

Free baseplates sediment at 70S and small peaks at this position are found in lysates produced by mutants in the second group of baseplate genes; this suggests that the hexagonal structure can be generated by the products of the first group of genes alone. Mutants in genes *53, 6, 7, 8* and *10* do not accumulate baseplatelike structures. However, these mutants generate precursor complexes that can be isolated either by sucrose gradient sedimentation or by their activity in the in vitro complementation assay. Analysis of the proteins contained in these structures suggests the scheme of sequential assembly depicted in Figure 6.35. First P10 and P11 join together and then P7, P8 and P6 are added in order to generate a 15S assembly. The 15S peak also can be seen in wild type lysates, when in addition to the first five proteins it contains P53 and P25. These may therefore be the last proteins added prior to aggregation into the hexagonal structure. The sixfold symmetry of the baseplate may be accommodated by supposing that it is assembled from six of the 15S precursor complexes. If each protein of the 15S complex were present in one copy, its mass would be about $4 \times 10^5$ daltons, compared with the estimated value of $7 \times 10^5$ daltons; this suggests that at least some proteins are present more than once.

The baseplates of the small 70S peak seen in lysates from infections with mutants in genes *5, 25, 26, 27, 51, 28* or *29* contain an outer hexagonal ring, a diffuse inner disc and (not always) a dense central plug. Star

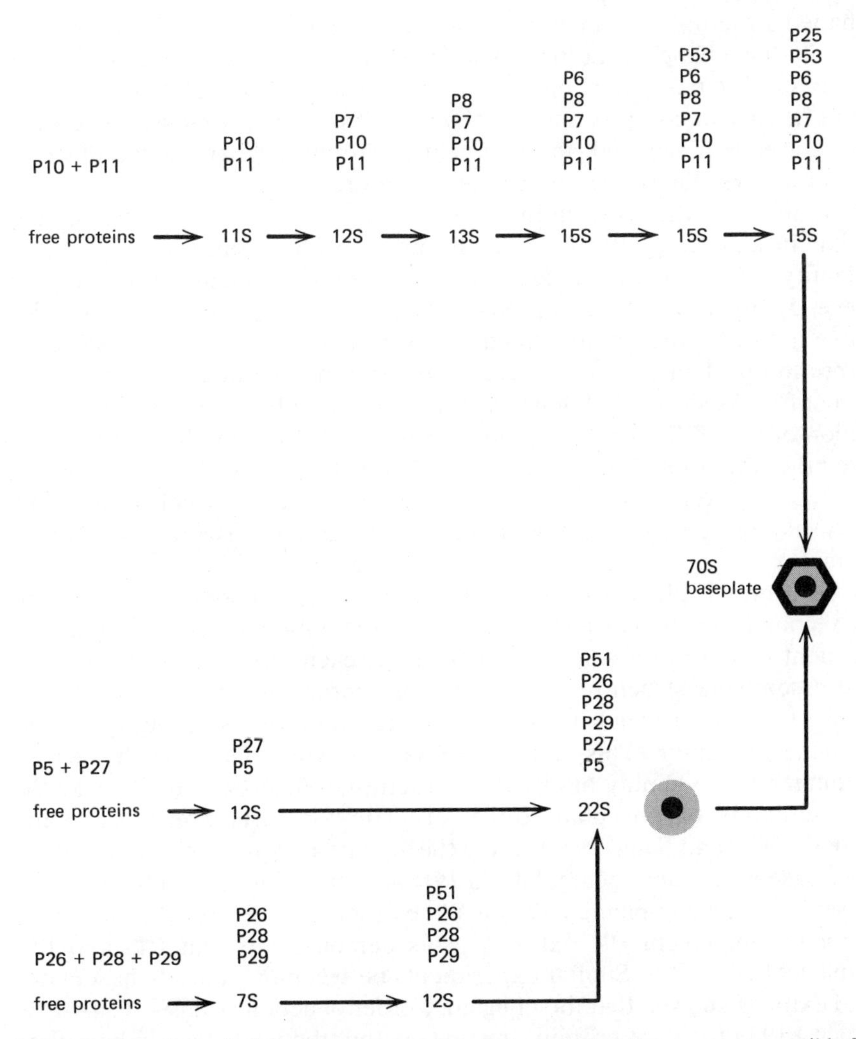


**Figure 6.35:** pathway for baseplate construction. Independent pathways are responsible for assembly of the outer hexagon and the inner core. The hexagon comprises six 15S subunits. The inner core is provided by a 22S complex. The proteins present in each precursor to the 15S complex have been identified, but the precursors in the 22S pathway have been analyzed only by activity in the complementation assay and may therefore have been acted upon by, rather than actually contain, the proteins shown.

shaped structures sedimenting at 40S also are seen; their generation may mimic the change that occurs in the baseplate upon infection. This suggests that baseplates may be formed in these mutants from the (normal) hexagonal complex and an (abnormal) inner core; because the core is not correctly assembled, these structures are unstable and break down to generate the star shaped and other structures.

By mixing extracts from infections with mutants of the second group, Kikuchi and King (1975c) used the in vitro complementation assay to identify a 22S complex able to provide the complementing activities of genes *5, 26, 27, 28, 29* and *51*. Since the proteins contained in this complex have not been directly identified, it is possible that some of these genes represent proteins that have acted upon, but that are not included in, this structure. As shown in Figure 6.37, the 22S complex may result from the union of two 12S complexes, one requiring the proteins P5 and P27 and the other requiring the functions P29, P28, P26 and P51. The 22S complex probably represents the center of the baseplate; and baseplate assembly may take place by the aggregation of six 15S complexes with a single 22S complex.

The general rule suggested by the ability of precursors to the 15S and 22S complexes to participate in complementation in vitro is that the sequential order of assembly is obligatory: each protein can be added to the assembling structure only at the appropriate point, presumably because the site to which it binds is created only by assembly of its immediate precursor. This can be seen as an evolutionary mechanism to minimize the assembly of aberrant structures. This is exemplified by the conversion of the baseplate to the baseplate-core structure. Meezan and Wood (1971) and King (1971) tested the activity in complementation of the free baseplates that accumulate in $19^-$, $48^-$ or $54^-$ mutant infection: $19^-$ baseplates can complement $48^-$ or $54^-$ extracts, but $48^-$ or $54^-$ baseplates cannot complement $19^-$ extracts. This demonstrates that P48 and P54 must act before P19. Similar experiments between $48^-$ and $54^-$ baseplates and extracts showed that the obligatory order of actions is P48—P54—P19. Since P19 is the core protein, this means that the final stage in baseplate assembly is the acquisition of ability to initiate core polymerization. That this reflects the addition of P48 and P54 (rather than their action upon other, structural proteins) was shown by the observation of Berget and Warner (1975) that they are components of mature baseplates.

An exception to the rule of strict sequential assembly, however, is provided by P9, P11 and P12. Edgar and Lielausis (1968) and King and Mykolajewycz (1973) demonstrated that these proteins appear to be able to bind to the assembling structure at any stage. This suggests that these proteins may constitute a part of the outer surface of the baseplate that remains accessible (and unchanged in conformation) during assembly. Usually P11 is added early in assembly as depicted in Figure 6.35; the addition of P11 is necessary for the binding of P12. The addition of P9 is necessary for the binding of the long tail fibers.

The function of P12 fits with the concept that these proteins can bind at any time because they join an outer surface. Particles derived from *12*⁻ mutant infection appear normal but fail to maintain their initial attachment to the cell wall. Mason and Haselkorn (1972) and Kells and Haselkorn (1974) showed that the *12*⁻ particles lack short (about 340 Å long) tail fibers; these are distinct from the long tail fibers and each particle has about six. A role for the short fibers in adsorption is postulated in Figure 6.32; and in addition to this their presence is needed for tail contraction. The addition of P12 to the *12*⁻ particles restores their activity. The sites for addition of the short tail fibers thus seem to be present during particle assembly, whereas by contrast the long tail fibers can be added to the tail only after its junction with the head. Mutants in gene *57* lack or have much reduced amounts of long tail fibers; they also lack the short tail fibers, emphasizing the pleiotropic and presumably regulatory role of P57.

In addition to the proteins coded by the genes identified by conditional lethal mutation, baseplates contain (at least) two phage coded enzymes. Kozloff et al. (1970a, b) first showed that T4 infection induces the synthesis of a folate derivative which is incorporated into the phage tail structure; and the baseplate contains also the enzyme dihydrofolate reductase. Kozloff et al. (1975a,b) identified the folate component as dihydropteroyl hexaglutamate and showed that the pteridine ring is exposed in particles lacking P11 but is not accessible in intact phage particles. An antiserum against folic acid inhibits the addition of P11 to the tail, which suggests that P11 usually binds onto the site where the folic acid is located. Baseplates isolated from defective mutants do not contain folate, which suggests that it may become irreversibly bound only when the core is added to the baseplate. The presence of the enzyme dihydrofolate reductase (the product of gene *frd*, also known as *dfr*) also can be demonstrated only after disruption of the baseplate structure and it appears to be located close to the folate and similarly protected by P11. Temperature sensitive mutants in the *frd* gene produce phages that are more heat labile; this is consistent with observations that the baseplate is the most heat sensitive part of the phage structure and implies that the enzyme is somehow involved in establishing this sensitivity. Kozloff et al. (1975c) showed that the phage enzyme thymidylate synthetase also is a part of the baseplate and again lies close to the site occupied by the folate. Since normal baseplates are assembled in the absence of these enzymes, their role is not clear. However, cofactors that inhibit or stimulate the reductase enzyme display the same effect upon adsorption to host cells and the reverse effect upon the addition of P11 to the baseplate during assembly; this implies that the reductase may be involved in some structural change helpful to injection that is inimical to assembly.

The involvement of the baseplate in tail contraction is demonstrated by the properties of temperature sensitive mutations in several baseplate genes. Yamamoto and Uchida (1973) divided these mutants into three groups. Mutants in genes *5* and *12* show no obvious morphological change but are inactivated by heating; in the temperature sensitive mutants of genes *18*, *25* and *48* the sheath contracts upon heating, but the baseplate remains at the distal end of the tail; and in mutants in genes *6*, *7*, *10*, *25* and *27* heating causes a contraction mimicking that seen in normal assembly, when the sheath contracts and the baseplate is pulled up with it. Yamamoto and Uchida (1975) then demonstrated that mutants in this third group are not able to display the heat-sensitive contraction if the long tail fibers are absent (because of the introduction of a $37^-$ mutation). Similar results with mutants in genes *5*, *6*, *7*, *8* and *10* were reported by Arscott and Goldberg (1976). Another indication that there is an interaction between the long fibers and the baseplate in contraction is provided by the observation of Dawes and Goldberg (1973) that some baseplate mutants have an extended host range; this could be achieved if their tails contract more rapidly than usual upon fiber-host contact, that is before a mistaken contact is corrected as it would be in a wild type phage.

The short tail fibers also are needed for tail contraction. Heat sensitive mutants in gene *12* lose infectivity without morphological alteration and Yamamoto and Uchida (1975) showed that they are unable to contract their tails, even upon mixing with sensitive bacteria. In a double mutant carrying heat sensitive mutations in both *6* and *12*, the lack of ability to contract the tail characteristic of the *12* mutation prevailed over the heat sensitive contraction displayed by the *6* mutant. Thus both normal and heat sensitive tail contraction require the presence of short tail fibers.

The production of fiberless particles by $9^-$ mutants suggests that P9 is needed for attachment of the long tail fibers; it is possible that it provides the sites on the baseplate to which they attach. This is consistent with the observation that P9 can be added at any stage of assembly, since presumably in this case it would lie at the outer surface of the baseplate. Particles lacking P9 provide the only exception to the rule that long fibers are needed for contraction; for $9^-$ particles are able to contract their tails. This unusual behavior is seen also in the response to inactivation of P12 in $9^-$ particles; when the P12 is temperature sensitive and is inactivated by heat, the particles retain their ability in tail contraction, in contrast with other particles whose P12 is inactive. Thus tail contraction may occur in the absence of P9 (and thus without tail fibers) and in this case does not require P12; but if P9 is present, tail fibers must be attached to the baseplate to allow contraction to occur, and the presence of active P12 is necessary. This implies that conformational changes may occur in the baseplate when P9 is added, in such a way that tail contraction is blocked unless both long and short fibers are added.

From these results it is not yet possible to construct a model for the molecular changes that occur in the baseplate upon tail contraction. However, the involvement of so many of the baseplate proteins, most of which cannot be in direct contact with either the short or long tail fibers, suggests the general conclusion that contraction represents a concerted conformational change involving many proteins; changes in the structure of any of these proteins are able to cause altered susceptibility to contraction, which may represent the transmission of the conformational change to other parts of the baseplate.

CHAPTER 7

# Phages T3 and T7

## Expression of T3 and T7 Genes

### *Organization of the Genome*

The three related phages T7, T3 and $\phi$II all generate particles with small heads and short tails. Their genomes are linear DNA molecules of about $25 \times 10^6$ daltons. Although T7 is the best characterized, all three genomes appear to be similar in organization. Phage T7 is one of the original series of T phages characterized by Delbruck (1945, 1946) and by Demerec and Fano (1945) as a single species; however, more than one variant appears to have evolved from this original isolate as demonstrated by the different buoyant densities of the particles of T7M (1.516 g-cm$^{-3}$) and T7L (1.521 g-cm$^{-3}$). The standard wild type strain of the phage now is taken as the T7L form utilized by Studier (1969).

All three of these phage genomes are terminally repetitious; Ritchie et al. (1967) found that treatment with exonuclease III, to degrade one strand at each end of the DNA molecule from the 3′-OH terminus, allows circles to form upon annealing. This suggests that the same terminal sequence is present at each end of the genome, the same conclusion suggested by a similar analysis of phage T4 DNA (see Figure 6.6). However, unlike T4 DNA, the molecules treated with exonuclease are able to form dimers, trimers and higher order circles as well as the monomers; this demonstrates that the *same* sequence constitutes the repetitious sequence present at the end of every molecule. Renaturation of denatured single strands generates only linear molecules (unlike T4 DNA which generates the circles illustrated in Figures 6.5 and 6.7). Thus there is no permutation and each T7 genome consists of the same unique sequence, with a short region repeated at each end.

These electron microscopic mapping experiments suggested a length for the repetitious region of about 0.7% of the genome, corresponding to 260 base pairs. This value was accepted until Ludwig and Summers (1975) obtained a lower estimate based upon restriction enzyme mapping. The enzyme Hind III makes only one cut in the terminal regions; this lies 40 bases from the left end and 30 bases from the right end of the phage. This leaves the implication that the region of terminal repetition must be 70 (± 20) base pairs (assuming the repeats are completely identical).

Phages T7 and $\phi$II possess the same terminally repetitious sequence while T3 differs. This relationship correlates with the abilities of the phages to recombine: in mixed infection T7 can recombine with $\phi$II but not with T3; and T3 and $\phi$II do not recombine. In fact, T7 and T3 exclude each other in attempted mixed infection. As might be expected from their ability to recombine, T7 and $\phi$II are closely related; Hyman, Brunovskis and Summers (1973, 1974) found by heteroduplex mapping that there is only about 10% homology between their genomes, the remaining sequences being identical. A similar relationship is shown between T7 and the series of related phages $\phi$I, W31 and H; similar regions are involved in the nonhomology in all of these cases, suggesting that certain genes are less restrained from diverging in evolution.

A different type of relationship is displayed between T3 and T7. Davis and Hyman (1971) found that heteroduplex mapping in low formamide reveals extensive homology, with many small substitution loops. But when more stringent conditions are imposed by increasing the formamide concentration, substantial homology is seen only at the right end of the genome. This implies that through most of the genome there is partial homology, that is, the two phages comprise related but not identical sequences. Presumably this reflects their evolution from some common ancestor.

These results offer a hint that the possibility of recombination (between T7 and $\phi$II) imposes a pressure for complete or zero homology in each region; whereas in the absence of recombination (between T7 and T3 or between T3 and $\phi$II) it is possible for base changes to accumulate by mutation. One model to explain the restriction on recombination between T7 and T3 or between T3 and $\phi$II is to suppose that a phage is viable only when both terminal regions are identical, perhaps because the ends are involved in the generation of concatemers (see later). A single recombinant between T3 and either T7 or $\phi$II would possess terminal regions derived one from each parent, which would therefore be different. The inviability of the single recombinant might exert a selective pressure against allowing recombination between these phages, although it does not seem that this can be a complete explanation for the exclusion and lack of recombination because double recombinants would be viable and should offer some escape from the selection.

Using the same approach with conditional lethals that proved successful with phage lambda (see page 277) and with phage T4 (see page 562), amber mutants have been isolated and mapped into complementation groups. By this means Hausmann and Gomez (1967), Hausmann and La Rue (1969) and Studier (1969) identified 18 essential genes in T3 and 19 in T7. These genes form a single linkage map for each phage, with high recombination, of the order of 20–40%, between distant markers. Studier numbered the T7 genes *1-19* from left to right and proposed the convention that any further genes should be given fractional descriptions that correspond to their positions. Thus the first new gene to be discovered between genes *3* and *4* would be described as *3.5*; any further gene between *3* and *3.5* would be termed *3.2*, and so on. Studies subsequent to the original characterization of conditional lethals have identified only one further essential gene, *3.5*, bringing the total in T7 to twenty functions. Genes in T3 are numbered by analogy with corresponding T7 functions; for in spite of their divergence in genome sequence the T3 and T7 maps appear to be very similar (Beier and Hausmann, 1973).

In addition to the essential genes, the phage also carries nonessential genes. These can be identified in two ways. When a phage is known to induce the synthesis of some (nonessential) protein, its gene can be identified by mutagenizing phages and then screening for mutants unable to make the protein. This approach was used by Masamune, Frenkel and Richardson (1971) to isolate mutants in the gene *1.3* that codes for the T7 polynucleotide ligase. The properties of the T7 ligase mutant emphasize the point that the lethality of a mutation may depend upon the host strain. Thus $1.3^-$ mutant phages can propagate normally on $lig^+$ E.coli, when the activity of the host ligase is sufficient to support phage reproduction; but the phage mutants cannot grow on $lig^-$ host cells, when the combination of phage and bacterial mutations results in the total absence of any ligase activity. This means that in effect the $lig^+$ and $lig^-$ host strains can be used to establish a system for the selection of conditional lethal phage mutants, where the lethality depends on the presence or absence of host ligase enzyme rather than on whether the bacteria possess the suppressor tRNA used in the selection of amber mutants. The principle of this approach was used by Studier (1973a) to select deletions of the phage ligase gene.

Another approach to the identification of nonessential functions lies with the isolation of viable deletion strains, where the deletion can be correlated with the absence of some protein synthesized during infection. Studier (1972, 1973a) isolated such variants of T7 by applying a technique first developed by Parkinson and Huskey (1971) for lambda (see page 447). This relies upon the observation that deletion phages may be more resistant to disruption by heating or by treatment with EDTA (perhaps because the smaller DNA content alters the head structure). The deletions isolated by this method fell in two regions, to the left and right of gene *1*.

The convention for describing locations on the T7 genome is to give the percent distance from the left end; 1% is about 370 base pairs. Figure 7.1 gives the locations in these terms of all known T7 genes. The viable deletions cover the region from 3–8% and from 15–24%. The genes *0.3* and *0.7* in the deleted region to the left of gene *1* and genes *1.1*, *1.3* and *1.7* in the deletions to the right of gene *1* were identified by correlating the absences of particular proteins with each deletion. These experiments therefore identify five nonessential genes. Five genes with corresponding functions are present in analogous positions in T3 (Studier and Movva, 1976).

The proteins coded by phage T7 can be identified by electrophoresis on SDS gels of the proteins synthesized in infected cells. By using mutants in each gene in turn it is possible to identify the band that is absent in each case and thus to determine the gene product. Studier and Maizel (1969) identified the products of 17 of the essential genes and Studier (1972) identified the further proteins coded by the nonessential genes. A summary of the identified proteins is given in Table 7.1. This includes the products of 23 genes, leaving unidentified the proteins of only two genes. The total weight of the proteins ascribed to known genes is some $1.1 \times 10^6$ daltons, which would require a coding length of about 30,000 base pairs. This corresponds to about 80% of the T7 genome of 37,000 base pairs. There can therefore be only a small number of unidentified genes and these might code for the proteins seen in infected cells that have not yet been attributed to genetic loci. Or, if some 20% of the genome is not used for coding purposes, all the T7 genes may have been identified (in which case the uncharacterized proteins seen in infection are not phage products).

The analysis of proteins synthesized in infected cells divides the T7 genes into three classes. The course of T7 infection reported by Studier and Maizel (1969), using a multiplicity of infection of 5–10 at 30°C, shows replication increasing rapidly after about 10 minutes with lysis beginning at about 25 minutes. (These times are much reduced, roughly by about half, if infection is performed at 37°C.) Host protein synthesis is much reduced and bacterial DNA synthesis ceases after 5–7.5 minutes. Studier (1972) demonstrated that the class I (early) proteins of genes *0.3*, *0.7*, *1*, *1.1* and *1.3* all are synthesized between 4 and 8 minutes, after which their production declines in parallel with the shutoff of host protein synthesis. The class II proteins are the products of genes *1.7*, *2*, *3*, *3.5*, *4*, *5* and *6* and are synthesized between 6 and 15 minutes, after which their production is shut off. The class III (true late) proteins are the products of genes *7-19* and are synthesized from 7 minutes until lysis.

The properties of mutants in the essential genes correspond to the same

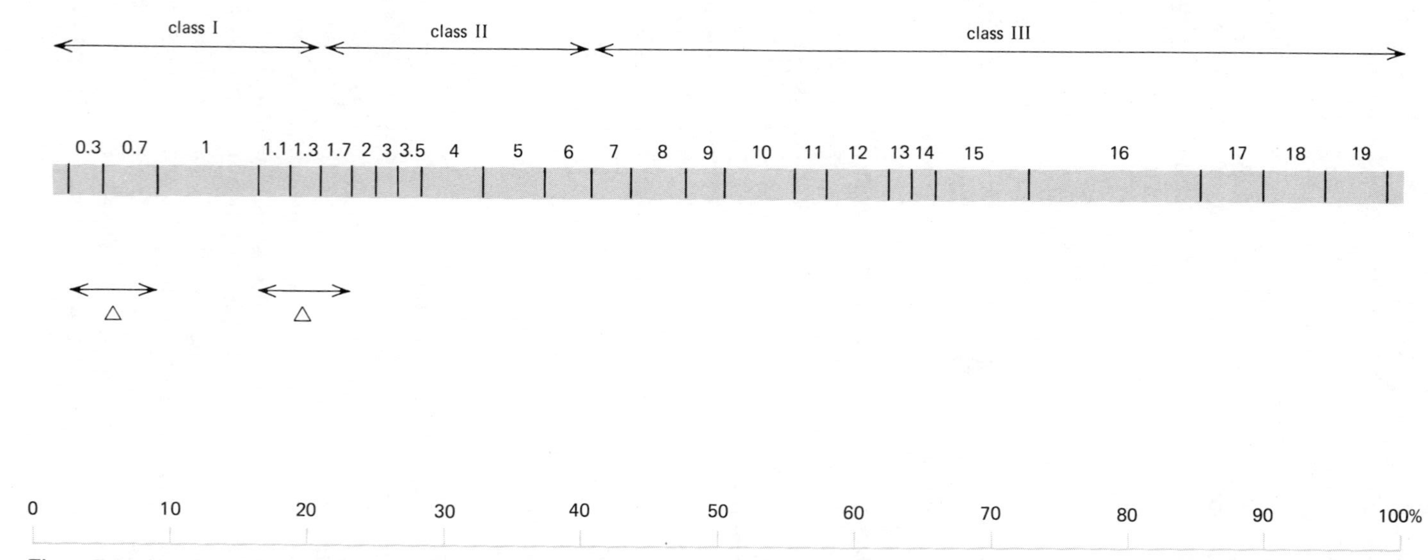


**Figure 7.1:** the T7 genome. The class I genes occupy the region from 1–20%, the class II genes lie from 20–39%, and the class III genes are located at 39–99%. The positions of the class I genes are assigned on the basis of estimates of the sizes of their messenger RNAs and from their relationship with deletions in the two regions indicated as Δ. The positions of the class II and class III genes are estimated from the sizes of their protein products (except for *2* and *18* which are not known) and are therefore more approximate.

**Table 7.1:** genes and proteins of phage T7

| gene | protein (daltons) | function |
|---|---|---|
| 0.3 | 9,000 | antagonizes host restriction |
| 0.7 | 42,000 | protein kinase |
| 1 | 110,000 | RNA polymerase |
| 1.1 | 8,000 | |
| 1.3 | 40,000 | DNA ligase |
| 1.7 | 15,000 | |
| 2 | | DNA synthesis |
| 3 | 13,500 | endonuclease |
| 3.5 | 13,000 | lysozyme |
| 4 | 67,000 | DNA synthesis |
| 5 | 84,000 | DNA polymerase |
| 6 | 31,000 | exonuclease |
| 7 | 15,000 | particle component |
| 8 | 62,000 | head protein |
| 9 | 40,000 | head assembly protein |
| 10 | 38,000 | major head protein |
| 11 | 21,000 | tail protein |
| 12 | 86,000 | tail protein |
| 13 | 14,000 | particle component |
| 14 | 18,000 | head protein |
| 15 | 83,000 | head protein |
| 16 | 150,000 | head protein |
| 17 | 76,000 | tail protein |
| 18 | | DNA maturation |
| 19 | 73,000 | DNA maturation |

three classes. Mutants in genes *1-6* display reduced DNA synthesis or fail altogether to replicate their DNA. Mutants in gene *1* are distinguished from those in genes *2-6* by their effect on protein synthesis. In all the mutants except gene *1*, the pattern of protein synthesis is the same as wild type, except for the absence of the single protein coded by the mutant gene. There do not appear to be any polar effects. But in gene *1* mutants only the class I proteins are synthesized. This suggests that the class I genes, comprising the essential gene *1* and the four nonessential genes *0.3, 0.7, 1.1* and *1.3*, represent early functions whose expression does not depend upon any phage functions, that is it is undertaken by the apparatus of the host cell. Because of their nonessential nature, no impediment to infection is seen with mutants in *0.3, 0.7, 1.1* or *1.3;* the functions of these genes are summarized in Table 7.1 and the first two are concerned with interactions with host systems and the last with providing a phage ligase.

Gene *1* must code for a protein whose activity is necessary for expression of the class II and class III genes. The class II genes provide functions necessary for DNA replication (including the lysozyme which is involved in replication and does not appear to be necessary for lysis of the host cell). Mutants in the class III genes, *7-19,* display the classical behavior of the late type: they are similar to wild type in the extent of DNA synthesis, but fail to generate infective phage particles. As is clear from Table 7.1, these genes specify proteins that are structural components of the phage particle or are concerned with the maturation of DNA.

### *Transcription of Early Genes*

The division of the T7 genome into three clusters of genes whose products fulfill related functions reflects the control of transcription during the infective cycle. As is suggested by the pattern of protein synthesis, the genes in each cluster are expressed coordinately; and this control resides at the level of transcription. The early (class I) genes are transcribed by the host RNA polymerase; while transcription of the class II and class III genes is accomplished by the product of gene *1,* a new RNA polymerase. In this section we shall be concerned with the control of early transcription, that is with the recognition of T7 DNA by the bacterial RNA polymerase and with the events responsible for processing of the RNA product and termination of its transcription. In the next section we shall consider the functions of the early gene products, including the shutoff of host and early protein synthesis; and then we shall take up the transcription of late genes by the T7 RNA polymerase, including questions on the functions of the late promotors and the nature of the distinction between the class II and class III genes.

All T7 genes are transcribed from the same strand of DNA. Summers and Szybalski (1968) demonstrated that the denatured strands of T7 DNA can be separated by centrifugation on poly(G)-CsCl; these now are known as the *l* and *r* strands. The phage-specific RNA extracted from infected cells at either early or late times hybridizes only with the *r* strand. Hyman (1971) extracted RNA from cells infected under conditions in which only the early genes are expressed (the phage carries a mutation in gene *1* or the cells are treated with chloramphenicol). Upon annealing with the *r* strand, the region from 1–19.8% of the genome enters duplex form. This agrees with the genetic definition of the early genes. As the genetic map is conventionally written, all transcription proceeds from left to right.

A single initiation site for early transcription was identified in the first studies of T7 transcription in vitro by host enzyme. Davis and Hyman (1970) visualized T7 DNA engaged in transcription by a technique in which the RNA transcript is collapsed into a bush at the location of the RNA polymerase on the template. When visualized after transcription has been allowed to proceed for varying periods of time, the size of the bush increases and its position lies farther along the template. An extrapolation of a plot of bush position versus time of transcription suggested that initiation occurs at the point 1.3%. Given the limits of this technique, these results do not prove the existence of a single promotor, but they imply that the great majority of T7 early transcripts are initiated at a site or sites close to the left end of the genome.

Where does early transcription terminate? Because of the effect upon termination of conditions such as ionic strength, it is always difficult to know whether termination observed in vitro reflects the situation prevailing in vivo or is an artifact of the system. However, there is clear evidence that the bacterial RNA polymerase terminates at about 20%. Conflicting evidence has been presented on whether the rho factor is needed for this termination event, probably due to the differing conditions under which in vitro transcription systems have been operated. Davis and Hyman (1970) observed that in the presence of rho, bushes of RNA were very rarely found past the site 18%, although in the absence of rho the polymerase could continue to the end of the template. Some inefficient rho-dependent termination also was seen at earlier sites, probably at several locations including the loci 3.4% and 7.2%. On the other hand, Millette et al. (1970) found that E.coli RNA polymerase transcribes a single RNA species from T7 DNA; this has a size of about $2.2 \times 10^6$ daltons. Millette et al. noted that it terminated only in U, but Peters and Heywood (1974) later observed termination only with the base C; termination with a single nucleotide implies that RNA polymerase ceases synthesis at a specific site. Rho factor is not needed in this system. The length of the RNA corresponds to about 6600 base pairs of DNA and is therefore the same length as the early region. Its transcription thus implies that the bacterial RNA polymerase can initiate and terminate transcription of a single RNA molecule from the early genes in the absence of any additional factors.

It is clear from several studies that the addition of rho to systems for in vitro transcription of T7 DNA by E.coli RNA polymerase results in the production of shorter transcripts due to termination within the early region (Dunn, McAllister and Bautz, 1972; Darlix, 1973; Darlix and Horaist, 1975). However, the transcription of T7 DNA by E.coli RNA polymerase in vitro, with or without rho, does not generate the pattern of RNAs seen in vivo. Dunn and Studier (1973a) confirmed that in the absence of rho a single product is transcribed, of about $2.5 \times 10^6$ daltons according to the mobility of the RNA on gel electrophoresis. The addition of rho depressed the incorporation of label into RNA and the large product was replaced by several smaller molecules. The abilities of these smaller molecules to hybridize with the DNA of deletion strains of the phage suggest that they are derived from the left end of the early region. This implies that rho is causing the premature termination of transcription; since these RNA species do not correspond with those observed in vivo, this is presumably a feature of the in vitro system rather than a reflection of events during infection, although we shall return later to the possibility that there is a low level of such premature termination in the cell.

The RNA transcribed in vivo can be separated into several fractions by gel electrophoresis. Siegel and Summers (1970) reported that more than 12 RNA molecules are made during the course of a wild type infection, but an amber mutant in gene *1* generates only a smaller number: these are the early messengers. The same RNA species accumulate when chloramphenicol is added to cells infected with wild type phage. (In this series of experiments the gene *1* message was not synthesized in the *1* mutant infection, but subsequent experiments have shown this to be atypical: in general mutation in gene *1* or addition of chloramphenicol has the same effect.) The genetic origins of the early RNA species made in vivo were identified by Summers, Brunovskis and Hyman (1973) and by Studier (1973b) and Simon and Studier (1973) by determining the ability of each RNA to hybridize with different deletion strains. The results are summarized in Figure 7.2 and suggest that each mRNA is monocistronic. By annealing the gene *1* message with the *r* strand of T7 DNA, Hyman and Summers (1972) visualized a duplex region from about 8% to 18%; this is in reasonable agreement with the estimates from deletion mapping.

From the sizes of the mRNAs estimated by their electrophoretic mobilities, it seems likely that their total length ($24 \times 10^6$ daltons) entirely accounts for the region between 1.8% and 20.2%, that is there are no gaps or are only very small gaps between adjacent messengers. The coding capacities of the gene *1* and *1.3* messengers are close to the sizes of their protein products; but the messengers for genes *0.3*, *0.7* and *1.1* are long enough to code for proteins somewhat larger than their products and may therefore possess appreciable untranslated regions. The idea that there may be untranslated regions on at least some messengers is supported also by the existence of deletions that enter genes *0.3*, *1* and *1.1* without changing the sizes of the protein products.

Deletion mapping locates the early termination signal at 20.2–20.4% (in reasonable agreement with earlier electron microscopic measurements). Simon and Studier (1973) examined the transcription in vivo of phages carrying deletions extending over the ligase (*1.3*) gene; to exclude all transcription except that undertaken by the E.coli RNA polymerase, the

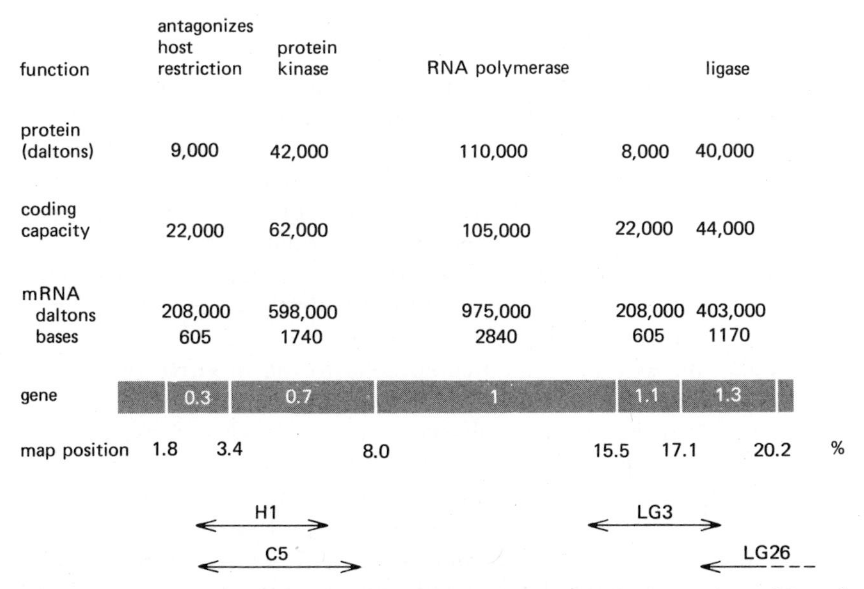


**Figure 7.2:** early genes of T7. The map positions shown for each gene are estimated from the sizes determined for the messenger RNAs and are consistent with the effects of the deletions to the left and right of gene *1*. The coding capacity of each RNA is calculated from its size and is accurate within about 10%. The sizes of the protein products are those determined by electrophoresis on SDS gels. The positions shown for the four deletions were determined by heteroduplex mapping. Data of Simon and Studier (1973).

phages also carried double amber mutations in gene *1* (single mutations may be leaky). When the deletion does not extend past the termination site, only the early genes can be transcribed by the host enzyme. When the deletion removes the termination signal, the bacterial polymerase can translate the late genes simply by continuing to read further along the template. Deletions extending past the point 20.2–20.4% confer this ability to read through. Under these conditions the class II and class III proteins appear sequentially as polymerase progresses along the template; from their times of appearance a minimum rate can be assigned for transcription by the E.coli enzyme, of about 50 nucleotides per second at 30°C (see Studier, 1972). This means it should usually take just over 2 minutes to transcribe the early genes at this temperature.

What is the relationship between the single large RNA transcribed in vitro by E.coli RNA polymerase and the several smaller RNAs that are the messengers produced early in infection? One model is to suppose that initiation in vitro reflects the situation in vivo, but that subsequent events differ. These subsequent events might be of two sorts. The RNA polymerase may transcribe a single RNA molecule representing all the early genes, as seen in vitro; and this may be a precursor which in vivo is processed by cleavages to generate the smaller monocistronic messengers. Alternatively, transcription may be terminated in vivo at sites at the end of each gene, but in such a way that the enzyme is not released from the template but is able immediately to reinitiate transcription at the beginning of the next gene (although it would not be able to bind at these initiation sites de novo). (Models which postulate that each gene possesses an independent promotor encounter the difficulty that it is necessary to suppose that only the promotor of the first gene can be recognized in vitro, although the others are utilized in vivo.)

By adding an extract of uninfected cells to the in vitro transcription system, Dunn and Studier (1973a) searched for a factor that might generate the pattern of RNA bands observed for the in vivo transcripts. The activity of the fraction that they obtained is shown in Figure 7.3. In this gel system the messengers synthesized in vivo form four bands, representing genes *1, 0.7, 1.3,* and *0.3* and *1.1* together (these last two messengers are the same size and are not separated). The in vitro transcript represents a single, much larger band; that it represents the entire early region is demonstrated by the use of deletions to the left and to the right of gene *1,* all of which reduce its size. When the isolated factor is added to the in vitro system, the single large band is replaced by the four smaller bands characteristic of the in vivo products; that these represent the same species produced in vivo is confirmed by the identical effects upon them of deletions in the genome.

This factor is an endonuclease, RNAase III, first identified in E.coli by Robertson, Webster and Zinder (1968) as an enzyme that is active upon double stranded RNA. Dunn and Studier (1973b) used an *rnc*$^-$ E.coli mutant that lacks this enzyme to test its role in T7 transcription. When the *rnc*$^-$ cells are infected with T7, instead of the usual pattern of RNAs only a single large molecule is synthesized: this is equivalent to the large transcript formed by RNA polymerase in vitro. Treatment with RNAase III in vitro converts this large molecule to the smaller RNA molecules. This suggests that the large RNA molecule represents the primary transcript of the early genes; it must be cleaved by RNAase very efficiently in vivo, before its transcription has been completed, since the precursor is never seen in normal infection. The same conclusion applies to the processing of the 16S and 23S rRNA species from their 30S precursor which also is accomplished by RNAase III (Nikolaev, Silengo and Schlessinger, 1973).

The processing model for maturation of the early messengers implies that only the RNA species derived from the far left of the precursor

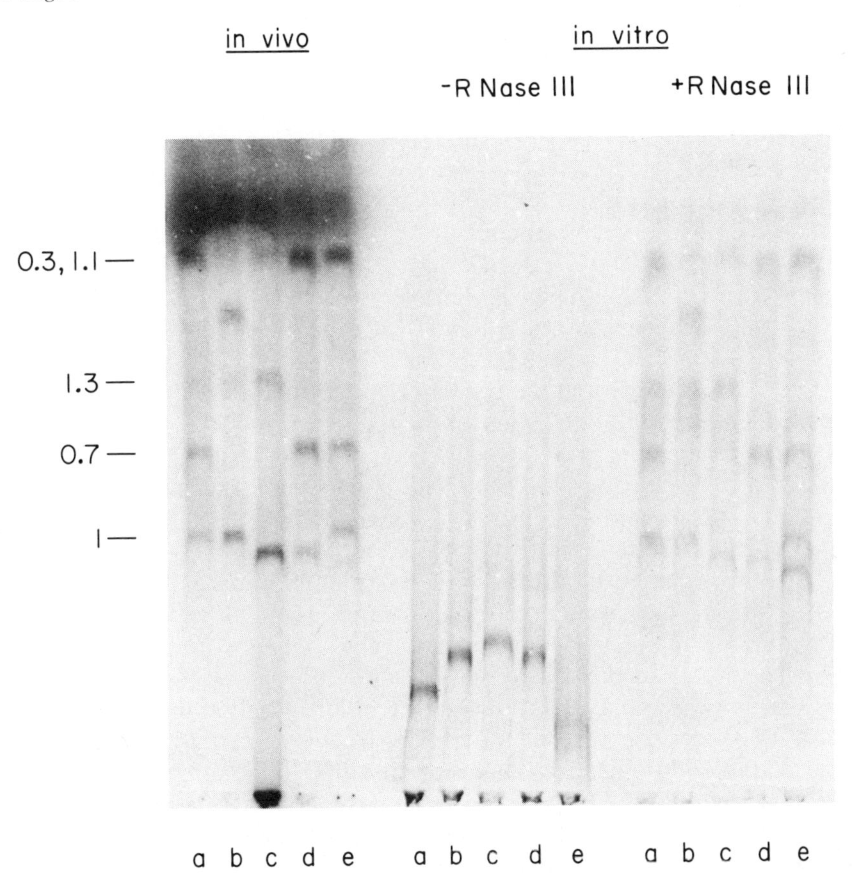


**Figure 7.3:** early transcription in vivo and in vitro. The RNAs synthesized from several strains of T7 in vivo, in vitro in the absence of RNAase III, or in vitro in the presence of RNAase III were separated by electrophoresis on SDS gels. The T7 strains are as follows: *a* wild type; *b* deletion H1, which fuses the *0.3* and *0.7* RNAs into a single species; *c* deletion C5, which eliminates the *0.7* RNA and fuses the *0.3* and *1* RNAs; *d* deletion LG3 which eliminates the *1.1* RNA and fuses the *1* and *1.3* RNAs; *e* deletion LG26 which deletes the early termination signal and fuses *1.3* RNA with the RNA to the right of the terminator. The exact positions of the deletion strains are given in Figure 7.2. Data of Dunn and Studier (1973a).

should possess the 5′ triphosphate typical of a primary transcript. Dunn and Studier (1973a) therefore labeled the RNA transcribed in vitro with $\gamma$ $^{32}$P −ATP or −GTP. The large precursor acquires label from either nucleotide, that is, it must be able to start with either pppA or pppG. In the presence of RNAase III none of the bands representing the monocistronic messengers displays the label. But in addition to these major bands there are three minor bands representing smaller RNA molecules (these are not clear in the gels of Figure 7.3 but have been assayed in other experiments). Two of these small mRNAs display a $\gamma\ ^{32}P$ label given in ATP and one obtains a $\gamma\ ^{32}P$ label from GTP. Deletion mapping shows that these RNA molecules all are derived from the region to the left of gene *0.3*. If their right (3′) termini all correspond to the start of gene *0.3*, close to the position 1.8%, then their lengths suggest that they must have been initiated at 0.5%, 1% and 1.5%, the first and last with ATP and the second with GTP. In similar experiments using a $\gamma\ ^{32}P$ label, Minkley (1974a) obtained results corresponding to locations for the three promotors of 1.0%, 1.4% and 1.8%.

These results imply that the E.coli RNA polymerase may initiate at any one of three closely related sites just to the left of gene *0.3*, then proceeding until it terminates spontaneously (that is in the absence of rho) at the point 20.2%. Cleavage of the transcript by RNAase III generates the five messenger RNAs and an "initiator" fragment whose length depends on which of the three initiation sites was used. A model which summarizes the use of these sites and the generation of mature initiator and messenger RNAs from the early precursor is shown in Figure 7.4. In addition to the three initiator fragments and five messengers, there are another five rather minor species seen on gels, 4 started with ATP and 1 with GTP. It is possible that these may be generated by other initiation and termination events that occur in vivo at rather low frequencies, perhaps reflecting some of the additional reactions seen in vitro.

Similar small RNAs and messengers have been reported by Studier and Movva (1976) for T3 infection. Together with the assignment of analogous genes in the same map order, this suggests that the organization of the early region may be the same in both phages (although the sizes and indeed specific enzyme actions of the proteins are related but not identical). Similarity of organization would imply that the use of three closely spaced promotors is followed by cleavage with RNAase III to release monocistronic messengers, a conclusion rendered the more striking by the observed divergence in genome sequences between these phages. This must leave the implication that this form of organization has some strong selective advantage for the phages.

The terminal sequences of several of the early RNA species produced in vivo have been determined by Kramer, Rosenberg and Steitz (1974) and these provide evidence in support of the cleavage model. The mRNAs for the genes *0.3*, *0.7* and *1* all possess the 5′ sequence pG-pA-pU- and the 3′ terminal sequence pC-pC-(pC)-pU-pU-pA-pU-OH. This implies that these termini may have a common origin caused by cleavage at a single recognition sequence. The gene *1.3* message possesses a

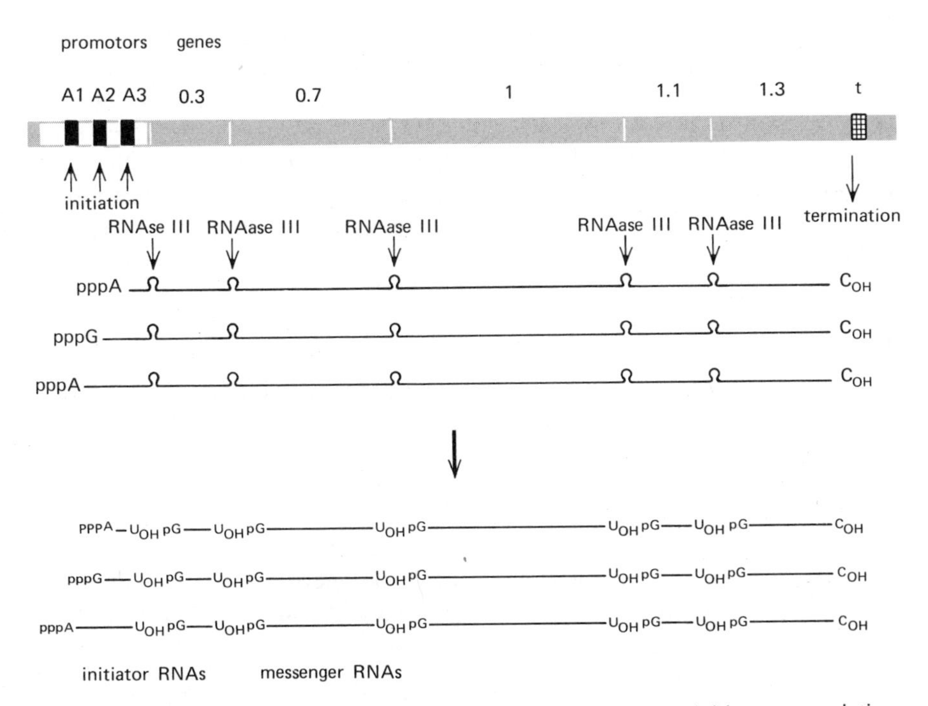


**Figure 7.4:** model for T7 early transcription. E.coli RNA polymerase initiates transcription at three closely spaced promotors, A1, A2 and A3. Elongation of the RNA chain continues to the termination point t, where the enzyme is released without involving rho factor. RNAase III cleaves the transcript at the five sites indicated by the hairpin loops to release three initiator fragments and messengers for the five structural genes. Although the complete early transcript is shown here for purposes of illustration, in vivo the cleavage sites are recognized by RNAase III so soon after their transcription that the complete precursor is not seen and only the individual products can be found. The map is not to scale; the initiator fragments are much shorter than drawn here.

different 3′ terminal sequence, which suggests that it is terminated by the release of RNA polymerase at 20.2% and not by cleavage. The gel system used in these experiments resolved only two of the three initiator fragments, which have 5′ terminal sequences of pppApUpCpG- and pppG-. The first species (prepared from infected cells) terminates in the same 3′ sequence as the messages; so does the second (synthesized in vitro). As expected if these small RNAs start at closely related but different promotors and are cleaved at the same site, their T1 RNAase fingerprints show extensive overlap. When Rosenberg, Kramer and Steitz (1974) examined the termini of the species generated by in vitro cleavage with RNAase III they found that these are the same as those seen in the messages produced in vivo; this confirms the model for their origin and shows that the in vivo messages are direct products of the cleavage reaction, that is, there does not appear to be any further processing after the cleavage. However, these results do not show whether a single scission is made in the polynucleotide chain of the precursor so that (for example) the 3′ terminus of the *0.3* message is adjacent to the 5′ terminus of the *0.7* message, or whether a fragment may be lost from a spacer region between the genes.

### *Initiation at the T7 Early Promotors*

The locations of the three early promotors have been identified by characterizing the RNA products transcribed from them in vitro. The synthesis of RNA from some templates can be stimulated by the addition of certain dinucleotides, which cause transcription to start from specific promotors. This technique has been used to prime transcription in the lambda control region (see page 338) and to identify the T7 early promotors (Minkley and Pribnow, 1973; Dausse, Sentenac and Fromageot, 1975). Minkley and Pribnow found that the T7 RNA products depend upon which dinucleotides are used, suggesting the existence of different promotors that are utilized to different extents in these conditions. The initiation sites were mapped by using deletions and fall into three classes: A close to 1%; B at 3.5%; and C at 8%. A $^{32}P$ label in the $\gamma$ phosphate group of either ATP or GTP enters the products primed at the A position; only GTP is used to initiate RNA at the B promotor and only ATP is used at the C promotor. The chains initiated at all of these promotors are propagated on the *r* strand to terminate at any one of a number of rho-independent sites. (Since the size pattern of the RNA products does not correspond with that seen in vivo, these termination sites presumably are used very infrequently or not at all in vivo.) In addition to these chains, a fourth class of products was initiated at a promotor D on the right end of the *l* strand and generated a small RNA of $8 \times 10^4$ daltons; presumably this also is an aberration of the in vitro system. As is implied by the use of either ATP or GTP to start transcription, the A class of products is not homogeneous. Minkley (1974b) demonstrated that there are in fact three promotors in this class, termed A1, A2 and A3 from left to right, A1 and A3 initiating with ATP and A2 initiating with GTP.

The identification of these promotors in the dinucleotide-primed system does not prove that they are utilized in infection. In fact, it appears that only the three A promotors are used extensively either in vivo or under stringent conditions in vitro. It is immediately apparent from their location and relative uses of ATP and GTP starters that the A1, A2 and A3 sites may correspond with the three classes of initiator fragments depicted in Figure 7.4. It is not clear whether the B and C promotors are used in vivo, but if they are this usage must be to only a very minor extent. Corresponding with the analysis of the RNA species found in vivo, Chamberlin and Ring (1972) showed that there are three sites at which E.coli RNA polymerase may initiate transcription on T7 DNA in vitro. These experiments followed the demonstration of Hinckle and Chamberlin (1972a,b) that eight molecules of RNA polymerase may be stably bound to T7 DNA at "tight" binding sites. Transcription is initiated in two steps: first a *binary complex* is formed between enzyme and DNA at a tight binding site; and then this is converted into a *ternary complex* by formation of the first phosphodiester bond. These events are discussed in Volume 1 of Gene Expression and further details have been reported by Mangel and Chamberlin (1974,a,b,c). The binary and ternary complexes are distinguished by their different susceptibilities to rifampicin: Hinckle, Mangel and Chamberlin (1972) found that rifampicin inactivates RNA polymerase bound in the form of a binary complex, but enzyme becomes resistant when it enters the ternary complex. Since inactivation by rifampicin is a second order reaction, it is possible to control the length of time that a binary complex may survive by adding to it an appropriate concentration of the drug. Chamberlin and Ring took advantage of this reaction by adding simultaneously to the binary complex an appropriate concentration of rifampicin and nucleoside triphosphates. This creates a sitation in which there are two competing reactions: inactivation and initiation. Under these conditions only stably bound polymerases can make the conversion from binary to ternary complex before they are inactivated; enzyme molecules bound elsewhere are unable to do so. In other words, this experiment measures the number of ternary complexes. Only three RNA chains can start, two with ATP and one with GTP. This poses an interesting paradox: eight polymerase molecules can be stably bound to T7 DNA but only three are able to initiate transcription rapidly enough to gain resistance to rifampicin. Perhaps more than one polymerase molecule may bind to each site, possibly by forming a "queue" at the promotor, but with only the first in a position to make the conversion from binary to ternary complex fast enough to avoid inactivation by rifampicin.

A counterpart to the approach of analyzing the ability of the polymerase to initiate transcription is to determine which sites upon the T7 DNA are occupied by the enzyme at initiation. These sites can be located by determining the abilities of restriction fragments of T7 DNA to bind RNA polymerase; and also by the protection against cleavage by Hind III at some sites that is conferred by polymerase binding. In the lambda control region, cleavage by Hind III occurs within the two promotors $p_L$ and $p_R$ and is prevented if RNA polymerase first is bound to the DNA (see page 333). This was made the basis of a technique for assaying promotors by Allet et al. (1974) and was applied to T7 DNA by Ludwig and Summers (1976). When RNA polymerase is bound to T7 DNA, cleavage is blocked at two of the Hind III sites as shown by the disappearance of four restriction fragments and their replacement by two larger fragments. This identifies the A1 promotor at 0.6% and the A3 promotor at 1.7%. When the isolated fragments of T4 DNA generated by restriction with the Hind III nuclease are tested for their ability to bind polymerase, a strong binding site is found on a fragment that contains the region around 1.5%; this identifies A2. There is therefore a difference in the sequences of the A promotors such that A1 and A3 carry a Hind III recognition sequence but A2 does not. When the concentration of polymerase used to bind the restriction fragments is increased, two further sites are identified, one in the region from 3.0–3.9% (corresponding to B) and one at 7.6–8.2% (corresponding to C). Since these sites are bound only at much higher enzyme concentrations than are necessary to bind the fragment containing A2, the B and C promotors must be much weaker. The two weak promotors lie in positions that could be used to initiate transcription of the *0.7* and *1* gene messages; although the predominant origin of these messages clearly is by cleavage from the long early precursor, the in vitro results raise the possibility that B and C might function at rather low levels in vivo. Although it is doubtful whether these promotors have any significance for the phage infective cycle, this could be tested by determining whether a phage deleted in all three A promotors could engage in any transcription in vivo.

The sequences of the regions within the A2 and A3 promotors protected by RNA polymerase binding have been determined by Pribnow (1975a,b). When the E.coli enzyme is used to transcribe T7 DNA in the presence of the dinucleotide CpA and the starter nucleoside triphosphate CTP, it initiates at the A3 promotor; and in the presence of UpC and GTP it is compelled to start transcription at the A2 promotor. The sequence of the binding site was determined by taking advantage of the readthrough past A2 and A3 that is displayed by polymerases initiating further to the left at A1. Thus the A2 or A3 DNA fragment protected by polymerase against DNAase I was isolated, denatured into single strands, and hybridized with a mixture of T7 early RNA containing molecules in which the A2 or A3 polymerase binding sequences had been transcribed. After treatment with RNAase to remove any unhybridized RNA, the RNA sequence corresponding to the polymerase binding site was recovered in the form of an RNA-DNA hybrid, from which it could be retrieved and sequenced. The position within this sequence of the startpoint for transcription was identified by comparing it with the initial sequence of the appropriate message.

The sequences of both polymerase binding sites are shown in Figure 7.5. The only pronounced homology lies in the region of the septameric sequence seven bases to the left of the initiation point, which is found in these or in closely related forms in several promotors (see Figure 4.31). It is interesting that the isolated fragments of DNA that are protected by polymerase binding are unable on reisolation to bind polymerases afresh. However, an 80 base pair Hind III restriction fragment isolated by Ludwig and Summers (1976) that contains A2 is able to bind polymerase. This leaves the implication that the protected sequence is not sufficient for recognition and that some additional sequence must be involved. In the absence of promotor mutants, it is not possible to say whether the situation may be analogous to that of lambda and of the lactose operon, where a sequence some distance upstream (about 35 base pairs) from the stable binding site appears to be involved in the recognition reaction. The three promotors are extremely close together; according to Pribnow (1975b) each pair is separated by only some 120 base pairs, a distance which is less than but close to what would be estimated from the relative lengths of the initiator RNA species (see above). Whether all three promotors are entirely independent or there is some relationship between them (for example connected with the apparent stable binding of eight polymerases at the three sites) cannot be determined from the data presently available; again it seems likely that promotor mutants will be needed to resolve this point.

## *Early Gene Functions*

The five genes that are transcribed by the host polymerase occupy the entire length of genome between the three early promotors close to 1% and the termination site at 20%. Of these functions, only gene *1* is defined as essential for the phage by the existence of conditional lethal mutations; the others are defined as nonessential by the ability of phages deleted for them to grow on the usual strains of host cell. But the nonessential genes must confer some selective advantage for phage reproduction in the wild; otherwise they would have been lost by the accumulation of deleterious mutations. This conclusion is emphasized by the relationship between T7 and T3: their genomes display extensive partial homology, presumably due to the occurrence of mutations in the two variants since their divergence from some common ancestor, but in spite of these differences in sequence there appears to be an analogous organization of the genome in each phage (so far as is at present known from incomplete evidence).

```
5′ T A A C A T G C A G T A A G A T A C A A A T C G C T A G G T A A C A C T A G C A G 3′
3′ A T T G T A C G T C A T T C T A T G T T T A G C G A T C C A T T G T G A T C G T C 5′

5′ G T A A A C A C G G T A C G A T G T A C C A C A T G A A A C G A C A G T G A G T C 3′
3′ C A T T T G T G C C A T G C T A C A T G G T G T A C T T T G C T G T C A C T C A G 5′
```

**Figure 7.5:** sequences of the A2 and A3 promotors. These sequences of 41 base pairs are protected by RNA polymerase against DNAase attack; they are A-T rich, with 25 A-T pairs in A2 and 26 A-T pairs in A3. The single shaded base pair indicates the startpoint of transcription and the horizontal arrow marks the 5′ sequence of the transcript. The shaded septamer is the sequence that occurs as a variation of T A T Pu A T G / A T A Py T A C in several promotors not far to the left of the startpoint. Data of Pribnow (1975a,b).

Surely then the maintenance of nonessential genes with analogous functions in each phage must imply some need for these functions in the phage history. Thus while these gene functions appear to be dispensable for phage growth on laboratory strains of E.coli, they may be necessary for perpetuation of the phage on other host strains.

Of the early functions, the best characterized is gene *1*, whose product is essential for the transcription of the class II and class III genes. We shall see in the next section that gene *1* codes for a new RNA polymerase activity which specifically transcribes these two classes of phage genes from a new set of promotors. Phage T3 codes for an enzyme with properties similar to those of the T7 RNA polymerase.

The product of gene *1.3* is a DNA ligase, similar in its properties to the other ligase enzymes that have been characterized in bacterial systems. Again phage T3 infection generates an enzyme similar to that of T7. The phage enzyme is not needed during infection of a host that synthesizes its own ligase, but becomes essential for infection of host cells deficient in the bacterial ligase activity.

The gene *0.3* function is concerned with antagonizing the host restriction and modification systems of E.coli K and B. Phage T7 can grow on either of these host strains in spite of their possession of restriction enzymes of different specificities (which, for example, do not allow phage lambda grown on B cells to plate on K cells). Two explanations are possible. The DNA of phage T7 may not possess the sites recognized by these enzymes; this is excluded by the observation of Eskin, Lautenberger and Linn (1973) that the purified B restriction enzyme recognizes about five cleavage sites on T7 DNA in vitro. This leaves the alternative that infection with T7 antagonizes the host restriction system. By using deletion mutants lacking gene *0.3*, Studier (1975) showed that this gene appears to code for a function of this nature. Mutants of the $0.3^-$ type are restricted; and they appear also to suffer some partial modification (as an alternative to restriction). This suggests that the *0.3* gene product may prevent both modification and restriction of T7 DNA.

The means by which the *0.3* product accomplishes this antagonism has not yet been established. Restriction and modification both are undertaken by an enzyme complex that relies upon the same (Hss) polypeptide for recognition of the sites on DNA (see Volume 1 of Gene Expression); thus the *0.3* function might act against this polypeptide. A less direct action is hinted at, however, by the observation that phage T3 induces an SAMase (S-adenosyl-methionine cleaving) coded by the gene analogous to the T7 *0.3* function. Such an activity might be related to the function of overcoming host restriction and modification since S-adenosylmethionine is an essential cofactor for both the activities of the K and B cell systems.

Indeed, Studier and Movva (1975) have shown that it is the T3 *0.3* SAMase function that is responsible for overcoming host restriction and modification, although it is possible to obtain mutants in it which have lost the SAMase activity but remain able to overcome host restriction. It should be interesting to compare the properties of the T3 and T7 proteins and this may reveal the nature of the antagonism to restriction and its relationship to the SAMase activity.

At the end of the early phase of gene expression in T7 infection, host transcription and translation are shut off together with the phage early functions. Brunovskis and Summers (1972) demonstrated that neither host nor early phage functions are shut off following infection with a mutant carrying a deletion to the left of gene *1*, apparently located in gene *0.7*. Rahmsdorf et al. (1974) observed that T7 induces a kinase activity that phosphorylates host proteins; and Zillig et al. (1975) observed that the $\beta$ and $\beta'$ subunits of the host RNA polymerase are amongst the 40 or so proteins that are phosphorylated. The time of phosphorylation coincides with the time when host transcription is shut off, raising the possibility that this may be the mechanism responsible. This is strengthened by the observation that the gene that specifies the kinase lies to the left of gene *1*; it is probably therefore gene *0.7*. Given the general nature of the kinase activity, the specificity of its involvement in the shut off effect remains to be investigated.

Genes *0.3* and *0.7* are defined as nonessential by the existence of deletion mutants that remain viable in spite of their lack of these functions. Perhaps it is therefore not surprising that Hyman, Brunovskis and Summers (1974) found that these loci are included in one of the regions susceptible to sequence variation between T7 and the related phages $\phi$I, $\phi$II, W31 and H. In fact, in the latter three of these phages RNA and protein corresponding to these genes could not be detected in infection.

Nothing is known about the function of gene *1.1;* no activities have been ascribed to its protein product and no deficiencies have been observed during infection with $1.1^-$ mutants.

The model for transcription of the early genes illustrated in Figure 7.4 predicts that their messenger RNAs should be synthesized in equimolar amounts. However, when the messengers separated on gels such as those shown in Figure 7.3 are quantitated on a molar basis, there appears always to be an excess of the messages closer to the promotor. (A difficulty in quantitating the *0.3* and *1.1* messages is caused by their failure to separate on the gels, but by using appropriate deletion strains it is possible to estimate the proportion that each contributes to the band.) Table 7.2 summarizes one set of data on the transcription and translation

**Table 7.2:** relative rates of expression of T7 early genes

| gene | molar ratios of mRNA at 5 minutes | messenger half-life measured as decline in | | calculated relative rate of mRNA synthesis | observed relative rate of protein synthesis | relative protein synthesis/ mole of message |
|---|---|---|---|---|---|---|
| | | RNA level | protein synthesis | | | |
| *0.3* | 9.6 | 4.5 min | 3.1 min | 5.0 | 15.8 | 1.8 |
| *0.7* | 1.1 | 1.1 min | 0.8 min | 2.2 | 1.3 | 1.0 |
| *1* | 1.0 | 1.6 min | 1.0 min | 1.0 | 1.0 | 1.0 |
| *1.3* | 1.7 | 4.0 min | 2.2 min | 1.0 | 4.8 | 2.2 |

All values given above are expressed relative to gene *1*. Messenger half-life was measured by adding rifampicin at 3.5 minutes after infection and following either the decline in level of each mRNA or the decline in its ability to synthesize the protein product. The relative rates of mRNA synthesis are calculated from the observed molar ratios and half-lives of each species. The relative rates of synthesis of each protein were determined by using 1 minute pulse labels of $^{14}C$-leucine and following the incorporation of label into each species. The relative rate of translation of each message is calculated by comparing the rate of protein synthesis with the molar amount of message, taking into account the synthesis of new message and decay of old message during the pulse period used to determine protein synthesis. The *1.1* message is not analyzed because it cannot be separated from *0.3* mRNA by gel electrophoresis; the values given for *0.3* mRNA therefore include *1.1* mRNA, but this should introduce only a small error because the *0.3* mRNA constitutes the great majority of the molecules. Data of Hercules, Jovanovich and Sauerbier (1976).

of the early genes. In these experiments, Hercules, Jovanovich and Sauerbier (1976) irradiated cells with ultraviolet light to suppress host transcription, infected with a T7 $1^{am}$ mutant to ensure that only early genes are expressed, and then extracted RNA or protein after labeling with radioactive precursors. This revealed a pronounced inequality in the molar ratio of *0.3* message compared with the other RNA species.

The lifetimes of the early messengers were determined by following their decay after rifampicin had been added at 3.5 minutes to block any further initiations. The rate of decay can be measured chemically (by the disappearance of mRNA from the gel) or functionally (by its ability to synthesize protein). These approaches yield similar results, but the functional half-life is always somewhat shorter than the chemical half-life, suggesting that the messenger is inactivated first by some event that does not change its size and then later is degraded. Results suggesting the same conclusion have been obtained by Yamada, Whitaker and Nakada (1974,

1975); although the stabilities that they measured were somewhat greater, the messages always were more stable chemically than functionally. We should note also that Marrs and Yanofsky (1971) reported that phage T7 RNA decays with an average half-life of 6 minutes, while the host *trp* mRNA continues to decay with its usual half-life of 80 seconds.

The stability of each T7 message is roughly proportional to its length, suggesting that there is no specificity in decay. The differences in messenger half-life in part account for the differences in molar ratios; and from these two sets of data it is possible to calculate the relative rates of synthesis of each mRNA. These suggest that messenger production declines proceeding from the promotor-proximal genes to the distal genes. This suggests that some polymerases may terminate within the early region, presumably at the ends of genes *0.3* and *0.7*, instead of continuing to the termination site beyond gene *1.3*. Taken at face value, in fact, these data suggest that only half of the polymerases that intitate at the A promotors continue into gene *0.7* and only half of those continue into gene *1*.

When the rate of synthesis of each protein is compared with the molar amount of each message, it appears that all messages are translated with roughly the same efficiency (within a factor of about 2). It is possible that the shorter messages may be translated more often. These data confirm that the control of T7 early gene expression is exercised at the level of transcription; there is no evidence for any specific control of messenger stability or translation.

Is the cleavage of the early precursor RNA by RNAase III necessary for translation of the messages or might they be translated as efficiently as part of the giant molecule? Experiments in vitro have yielded conflicting results: Hercules, Schweiger and Sauerbier (1974) found translation to be inefficient unless the messengers were cleaved, whereas Yamada and Nakada (1976) observed efficient translation in either case. An explanation for the former result may lie in the use of an $F^+$ strain of E.coli, which impedes T7 translation in vivo. The decisive experiment on this issue comes from the use by Dunn and Studier (1975) of an $rnc^-$ host lacking RNAase III. Infection proceeds normally in these cells, except for a deficiency in the production of *0.3* protein, although early protein synthesis starts a little later and continues a little longer than usual. But the early proteins *0.7*, *1*, *1.1* and *1.3* all are produced in normal amounts. The same result was obtained with translation of the large precursor in vitro, again with the exception that production of *0.3* protein was strongly stimulated by cleavage. This suggests that (apart from gene *0.3*) cleavage is not needed for successful T7 infection. As with the "nonessential" genes, the reason for the retention of the cleavage signals in the genome may be that they are necessary for phage reproduction under its natural conditions of infection, although they are dispensable in laboratory host strains. Since cleavage can altogether be suppressed in the *rnc*$^-$ host, it should be possible to obtain phage mutants which lack specific cleavage sites; and these may yield further information about their roles and about the activity of RNAase III, especially concerning the production of active *0.3* message.

### *Transcription of Class II and Class III Genes*

The role of gene *1*, the only essential function of the five early (class I) genes, is to promote transcription of the remaining genes that comprise the rightmost 80% of the T7 genome. Of these, the class II genes (*1.7-6*) specify functions needed for the replication of DNA, and the class III genes (*7-19*) code for proteins that are either structural components of the phage particle or are needed for its assembly. The difference in expression between the class II genes (active from 6 minutes only until a shutoff at 15 minutes) and the class III genes (whose transcription and translation continues until lysis) implies, however, that some function additional to the gene *1* product must be involved controlling class II and class III transcription.

Initial attempts to purify the gene *1* product suggest that it might be a new sigma factor responsible for redirecting the activity of the host core enzyme, since it purified with E.coli RNA polymerase to yield a fraction with altered transcriptional properties (Summers and Siegel, 1969). However, Chamberlin, McGrath and Waskell (1970) demonstrated that the gene *1* product is a phage-specific RNA polymerase. The enzyme is a single polypeptide chain of about 110,000 daltons which is synthesized as an early protein. Characterized in some detail by Chamberlin and Ring (1973a,b), the T7 RNA polymerase differs from the host enzyme in its insensitivity to rifampicin and streptolydigin, which explains why T7 infection becomes resistant to rifampicin at an early time. Although the conditions required for transcription in vitro by the T7 enzyme are generally similar to those established by the bacterial enzyme, the ionic conditions necessary for optimal activity are not identical. The T7 enzyme is inactive in dilute solution but is stabilized by the presence of other proteins, such as bovine serum albumin or E.coli RNA polymerase; this explains why it was at first isolated apparently in the form of an association with the host enzyme.

When used to transcribe T7 DNA in vitro, the phage RNA polymerase transcribes only the *r* strand, from the region to the right of the early genes; the lack of early transcripts among the in vitro products argues that new promotors are recognized in the class II and class III gene regions Summers and Siegel, 1970). The T7 RNA polymerase displays high specificity for its template and is only about half as active with T3 DNA as with T7 DNA (Dunn, Bautz and Bautz, 1971; Chamberlin and Ring, 1973a). This implies that the sequences of the promotors recognized by the phage enzymes are different in T7 and T3 DNA. Since the RNA polymerase enzymes coded by the two phages represent genes in similar locations and appear to be somewhat similar proteins, this system offers the interesting prospect of correlating evolutionary changes in the protein with changes in its recognition of nucleotide sequences.

The RNA species transcribed in vitro by the T7 polymerase have now been partially characterized; although less information is available about their origins than about the early messengers, it is at least clear that several promotors must be utilized. Niles, Conlon and Summers (1974) and Golomb and Chamberlin (1974a) identified six major bands by gel electrophoresis of the in vitro products. There are also some minor bands seen at somewhat lower levels. The major bands are synthesized at equimolar ratios, except for one band which has twice as much RNA and appears to represent two RNA species with overlapping mobilities. The sizes of the RNAs seen in the two series of experiments are in good agreement for the smaller classes and differ somewhat for the two larger species. The molecules classified by Golomb and Chamberlin are

| | |
|---|---|
| I | $5.5 \times 10^6$ daltons |
| II | $4.5 \times 10^6$ daltons |
| IIIa | $2.0 \times 10^6$ daltons |
| IIIb | $2.0 \times 10^6$ daltons |
| IV | $0.84 \times 10^6$ daltons |
| V | $0.44 \times 10^6$ daltons |
| VI | $0.22 \times 10^6$ daltons |

The T7 in vitro transcription system generates a rapid rate of chain growth, about 230 nucleotides per second at 37°C, compared with a rate of 50 nucleotides per second displayed under the same conditions by the bacterial enzyme. This means that even the largest RNA, whose length is more than half of that of the total length of the class II–class III region, takes only just over 1 minute to transcribe in vitro.

What is the relationship between these RNAs? Is each independently initiated at its own promotor or is only one or a small number of promotors used, with termination occurring to varying degrees after different extents of elongation? Golomb and Chamberlain (1974b) located some of the RNA species by studying the effects of reducing the length of the

DNA template by treatment with exonuclease. This results in a reduction in length of the three species I, II and IIIb and the complete disappearance of VI. This suggests that all four of these RNAs terminate very close to the right end of T7 DNA (an inaccurate estimate based on the extent of shortening locates the terminator at 98.5%); the implication then is that each RNA must have initiated at a different promotor. As illustrated in Figure 7.6, the lengths of the RNAs suggest locations for the promotors used to sponsor this overlapping transcription at 56%, 64%, 83% and 97%. In this context, it is interesting to speculate that the promotors and the terminator might coincide with some of the A-T rich regions indicated by the denaturation map to lie at 57%, 63%, 66%, 86%, 91% and 99%.

When T3 DNA is used as template for the T7 RNA polymerase, only species IIIb is synthesized in large amount; this is consistent with the demonstration by heteroduplex mapping that it is the rightmost 20% of the two phage genomes that displays close homology (these sequences may be identical). Rather small amounts appear to be synthesized of some of the other bands, which may indicate that the T7 enzyme is able to recognize the T3 promotors only rather poorly. In view of this difference, recombinants between T3 and T7 DNA (which can be obtained by forming hybrid phages under unusual conditions) can be used to test the location of the promotors. When the T7 sequences up to 40% are replaced by T3 sequences, synthesis of all the RNA products remains unimpeded. But when the T7 sequences up to 70% are replaced, only IIIb and VI continue to be synthesized. This suggests that the promotors for I, II, IIIa, IV and V RNA species all lie in the region between 40% and 70% (although this location is only approximate because the hybrid phages have not been fully characterized and the crossover points may be to the left of the 40% and 70% points, respectively). These assignments are supported by the results obtained by Niles and Condit (1975) in translating these in vitro RNA products. On the basis of the proteins synthesized in vitro, band III RNA corresponds to genes *9, 10* and *17*, presumably with IIIa coding for *9* and *10* and with IIIb representing *17*; the IV RNA also codes for the *9* and *10* proteins; and the V RNA again codes for *10* protein. This suggests that the IIIa, IV and V RNAs are the products of overlapping transcription.

These experiments therefore identify two groups of messengers, each representing transcripts of class III genes: the first group comprises the RNA species IIIa, IV and V and covers about 6000 base pairs (equivalent to 16% of the T7 genome) to the right of the 40% locus; the second group of RNA molecules I, II, IIIb and VI covers 17,000 base pairs from about 55% to 99%. If there were no overlap between the two groups, which implies that the first group represents genes *6* or *7* through *10* and the second group is transcribed between genes *11* and *19*, these transcripts could account for the expression of all of the class III genes. However, with the possible exception of gene *6*, no class II genes appear to be transcribed by the T7 RNA polymerase in vitro. Whether this preferential transcription of the class III genes in vitro reflects a difference between class II and class III transcription in vivo remains to be seen, but perhaps it is worth reiterating that the dependence of class II transcription on gene *1* shows that these genes must be transcribed by the phage enzyme. Perhaps the class II promotors have a much weaker affinity for the T7 polymerase.

In addition to the promotors in the class III region identified by the locations of the major in vitro transcripts, another promotor for the T7 RNA polymerase may lie in the early region. Gene *1.3* is unusual in that it continues to be expressed after the other early genes have been turned off; this raises the possibility that it might be transcribed not only by the host enzyme but also by the phage enzyme (which then might or might not recognize the termination signal at the end of the gene). Skare, Niles and Summers (1974) investigated this by using the DNA of deletion strains to remove from T7 RNA populations all the RNA except that corresponding to the sequences that had been deleted. The remaining RNA then can be examined for the presence of sequences complementary to the deleted region by hybridization with wild type DNA. By this means they were able to suggest that there is a site where the T7 RNA polymerase may initiate transcription in the early region, somewhere to the left of gene *1.3*, and which might be responsible for the late expression of ligase. As an early gene, ligase is in any case anomalous since the other functions concerned with replication all are class II genes; its expression in this way would therefore bring its control more in accordance with that of the functions related to it. Since no transcripts of early genes are found among the major products generated in vitro by the T7 RNA polymerase (as summarized in Figure 7.6), transcription of the ligase gene by the T7 enzyme must at best represent the activity of a promotor much weaker than those for the class III genes.

### *Translation of Phage Messengers*

Several phages are unable to propagate in cells carrying sex factors: these include T7, $\phi$II and T3, which may therefore be described as ♀-specific phages. Sex factors that prevent the replication of these phages have been described as *pif*$^+$ and permissive mutants correspondingly are termed *pif*$^-$. Morrison and Malamy (1971) observed that T7 infection of *pif*$^+$ cells is followed by the production of class I but not of class II or class

III proteins and they therefore suggested that the sex factor inhibits T7 reproduction by preventing the expression of the class II and class III genes. Linial and Malamy (1970) were unable to detect any difference in RNA synthesis in infection of $F^+$ compared with $F^-$ host cells, although the resolution of RNA species afforded on the gels used for their separation was not good enough to make definite the conclusion that the effect of the F factor lies subsequent to transcription. However, Blumberg and Malamy (1974) subsequently reported that T7 RNA extracted after infection of $F^+$ cells was as able as the RNA produced in $F^-$ cells to code for lysozyme and the gene *10* protein (measured by a radioimmune assay) upon translation in an in vitro system. This left the implication that the presence of the sex factor interferes with the translation of class II and class III messages (though not with early protein synthesis which, of course, is a prerequisite for the subsequent later translation). The question raised by these results is whether the inhibition of later T7 protein synthesis represents a specific interference with the translation of the class II and class III T7 messengers, as originally suggested by Morrison and Malamy, or whether it results from some less specific effect involving a general change in cellular metabolism.

Direct measurements of the extent of transcription during infection of $F^+$ cells have now shown that phage RNA synthesis as well as protein synthesis is reduced by the presence of the sex factor. When Britton and Haselkorn (1975a) assayed rifampicin-resistant transcription (that is, RNA synthesized by the phage polymerase), they found that the presence of the sex factor causes a decline in the concentration of phage messengers, of about fourfold for the most abundant messages. When they followed the production of RNA in a $1^{am}$ mutant, however, they observed a similar effect: at later times there is less RNA in the $F^+$ hosts compared with that found in $F^-$ host cells, although the magnitude of the effect is less than that seen upon infection with wild type phage. In a reexamination of the time course of protein and RNA synthesis in T7-infected ♂ and ♀ cells, Condit and Steitz (1975) came to the same conclusion: in a male host the labeling of RNA declines at about the time the class II and class III genes usually begin to be expressed, whereas in a female host there is no such interference. A 50% reduction in RNA synthesis occurring after the phase of early gene expression was observed also by Whitaker, Yamada and Nakada (1975), although they suggested that the effect was specific for the class II and class III genes because a much lower reduction (20%) was seen upon infection with a gene *1* mutant (but it takes longer for the effect to manifest itself in the *1* mutant infection, which probably explains why it was not observed in these experiments).

These experiments make it clear that phage transcription is inhibited

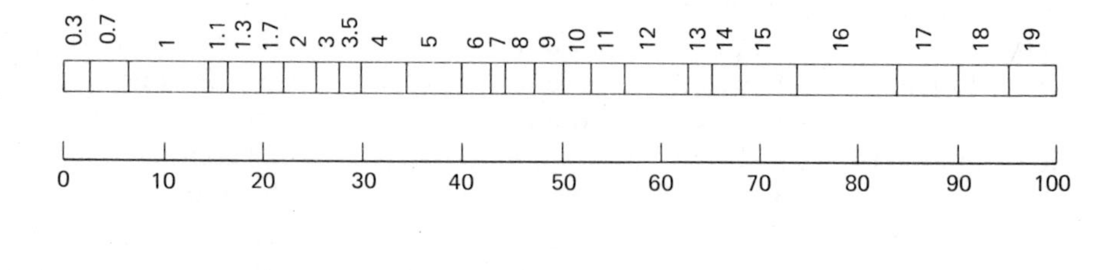


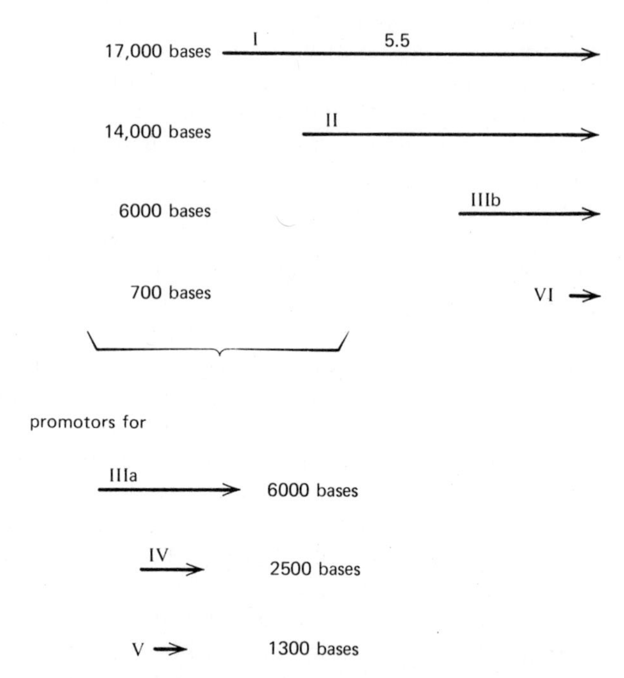


**Figure 7.6:** locations of late transcripts of T7 synthesized by the phage enzyme.

during the period of class II and class III gene expression when T7 infects $F^+$ cells. This conclusion raises two questions: how specific is the effect; and is it sufficient to explain the cessation of phage protein synthesis? The experiments with gene *1* mutants do not provide firm evidence on the specificity: although a reduction in transcription is indeed seen at later times in spite of the synthesis only of early messages, this is less than that observed in wild type infection. This might mean either that the effect is indeed at least partly specific for the late sequences or might mean the late functions are needed to induce a nonspecific effect. An answer to this question was provided by Britton and Haselkorn (1975b) in an experiment which made use of a deletion (LG37) that removes the early transcription termination signal. This allows E.coli RNA polymerase to transcribe the late genes by readthrough; but this does not relieve the inhibition of transcription caused by the sex factor. Thus transcription by E.coli polymerase is inhibited as well as that by the T7 enzyme; this implies that the inhibition is not specific. (This conclusion would be stronger if a $1^{am}$-LG37 double mutant had been used so that all transcription relied upon the E.coli enzyme whose activity could then be assessed directly; however, since no late transcription was seen in LG37 infection, the implication remains that the E.coli and T7 enzymes must be equally inhibited.) Another experiment showing from a different viewpoint that there is no specific effect exerted upon either transcription or translation was performed by Condit and Steitz (1975), who demonstrated that systems from uninfected ♀ and ♂ cells have an identical capacity to express T7 DNA in vitro; the addition of extracts from infected ♀ and ♂ cells did not change this capacity and so could not contain an inhibitor. There is therefore neither an intrinsic difference in the capacities to express T7 of $F^-$ and $F^+$ cells; and nor does any such difference develop during infection.

What then is the cause of the inhibition of T7 transcription and translation in $F^+$ cells? Condit and Steitz (1975) and Britton and Haselkorn (1975b) demonstrated that when $F^+$ cells are infected with T7 they become leaky at about 7–8 minutes (at 30°C) and release to the medium their pools of phosphorus-containing precursors for nucleic acids. In a study of the permeability of infected ♂ and ♀ cells, Condit (1975) observed that in $F^+$ cells the permeability first increases at about 8 minutes and then continues to rise; in $F^-$ cells the level remains the same as in uninfected cells until about 16 minutes, when there is a sharp rise in permeability due to cell lysis. That the permeability effect is related to the F-mediated inhibition of T7 growth is demonstrated by its failure to occur when the F factor carries a deletion that abolishes its interference with T7. These results suggest therefore that the reduction in T7 transcription and translation that occurs at the time of class II and class III gene expression is due not to any specific interference with either RNA or protein synthesis, but simply reflects a general change in host cell permeability that deprives the phage of necessary precursors. Because both RNA and protein synthesis are inhibited simultaneously, whereas there is usually about a 2 minute lag in relating their kinetics, Condit and Steitz suggested that translation does not cease simply as the result of failure to transcribe mRNA but because amino acids as well as nucleotides are lost from the cell.

Which phage and episomal genes are needed for this change in permeability? The increase in permeability continues to occur upon infection with gene *1* mutants, which implies that the necessary phage gene is an early function; on the hand, this activity cannot be equated with any one of the early genes since it continues to occur in strains carrying deletions in any of them. Evidence upon the phage contribution to this effect is therefore inconsistent. Mutants of the sex factor that do not prevent growth of T7 were first isolated by Morrison and Malamy (1971), who reported that they fell into the two groups *pifA* and *pifB*. Morrison, Blumberg and Malamy (1974) noted that these mutants have different effects upon T7 reproduction and Blumberg, Mabie and Malamy (1976) reported that only the second class cause permeability effects although both appear to interfere with T7 protein synthesis. A divorce between the permeability effect and inhibition of T7 gene expression would reopen the question of possible direct inhibition of transcription and translation; however, the *pifA* and *pifB* mutants have not yet been characterized, or shown to identify two different genes, and before their effects are investigated further it must be necessary first to determine their genetic properties. The $pif^-$ mutations lie in the region 33F—43F as does the mutant *fex* isolated by Britton and Haselkorn (1975b) which abolishes the inhibitory effect. Characterization of the product(s) of the sex factor concerned with inhibition of T7 growth should make possible a more detailed analysis of its mechanism.

## Reproduction of T7 DNA

### *Replication Apparatus of T7*

The reproduction of T7 follows the pattern familiar from lambda and T4: it commences with replication of the mature (monomeric) phage genome which then gives way to the generation of concameters. This reinforces the view that the formation of concatemeric DNA represents a genetic strategy common amongst the bacteriophages, apparently reflecting a prevalent feature of the maturation process by which DNA is packaged into the phage head. From studies of lambda and T4 it is evident that concatemers may be produced in more than one way, including rolling circle replication (lambda: see page 484) and recombination (T4: see page 634). The nature of the generation of T7 concatemers has not yet been fully established, but in lieu of the evidence sought to implicate replication of circular forms, it seems likely by inference that recombination or some other action may be involved.

The class II genes whose functions are defined to be necessary for replication by the effect of mutations in reducing or abolishing DNA synthesis code for two types of product: those concerned with the provision of precursors for phage DNA synthesis (genes *3* and *6*, which code for nucleases); and those involved in replication itself (genes *2, 3.5, 4* and *5*, which code for enzymes involved in DNA synthesis). The phage replication proteins that have been identified include a DNA polymerase (to which gene *5* contributes a subunit), the gene *4* product which may be involved in initiation events, a DNA-binding protein with properties analogous to those of the T4 32-protein (and for which no phage gene has yet been identified), ligase (coded by the class I gene *1.3*, but which can be substituted by the host enzyme), and the gene *3.5* lysozyme (which apparently is involved in replication rather than in cell lysis). In addition at least one host protein has been identified in the form of a component of the DNA polymerase. Replicating T7 DNA appears to pass through a membrane-bound form and it is possible that a complex comprising some or all of these proteins may be associated with it in this state, constituting a replication apparatus analogous to that by which phage T4 DNA is thought to be synthesized (see page 621).

Early studies on the replication of T7 demonstrated that most of the nitrogen and phosphorus of the phage DNA is derived from degradation of the bacterial chromosome (Putnam et al., 1952; Labaw, 1953). Sadowski and Kerr (1970) demonstrated that the host DNA is degraded in three stages:

First it is released from a rapidly sedimenting structure (presumably the cell membrane) at about 5–6 minutes.

Then it is cleaved endonucleolytically (at about 6–10 minutes).

Finally it is reduced to acid soluble products from 7.5–15 minutes, the period when nucleotide precursors are needed for phage DNA synthesis.

Release from the membrane does not occur in gene *1* amber mutants, suggesting that a class II or class III function is responsible. Silberstein, Inouye and Studier (1975) reported that the lysozyme product of gene *3.5* is involved; but the continued association of bacterial DNA with the membrane in lysozyme mutants does not prevent its degradation, so that the release is not a necessary step. (We shall discuss the role of lysozyme in phage replication later.) Both endonucleolytic cleavage and the subsequent exonucleolytic degradation to nucleotides are blocked by mutation in gene *3*; and the exonucleolytic degradation alone is prevented by mutation in gene *6*. Center, Studier and Richardson (1970) and Sadowski (1971) showed accordingly that the gene *3* product is an endonuclease and

Kerr and Sadowski (1972) showed that the gene *6* product is an exonuclease.

Characterization of the activity of the gene *3* endonuclease in vitro shows that it can utilize as substrate either duplex or single stranded DNA, although it displays a preference for the latter (Center and Richardson, 1970a,b). The products of this endonucleolytic breakage are fairly large pieces of DNA. These results therefore suggest a sequence of events in degradation of host DNA in which the cleavage of DNA by endonucleolytic action is necessary to generate the substrate for the gene *6* exonuclease, whose action produces the nucleotides used as precursors for phage DNA synthesis. The gene *3* endonuclease is able to degrade T7 DNA in vitro and so does not appear to possess an intrinsic specificity for the host chromosome; how it is prevented from attacking the phage DNA in vivo is not known. We shall turn later to the question of the possible involvement of these enzymes in phage reproduction per se, but suffice it for the present to note that the reason for the reduction in DNA synthesis by gene *3* or gene *6* mutants appears to lie with the resultant defect in the provision of nucleotide precursors and not with any deficiency in the replication process itself.

Replication of phage T7 DNA occurs by the usual process of discontinuous synthesis in which short Okazaki fragments grow in the direction from 5′ to 3′ by the activity of DNA polymerase and are subsequently linked together by ligase action. (Bacterial replication and the functions of DNA polymerase and ligase enzymes are discussed in Volume 1 of Gene Expression.) Using host cells of the $polA^-$ type that lack the bacterial enzyme DNA polymerase I, Grippo and Richardson (1971) were able to show that a new DNA polymerase is induced upon T7 infection and that gene *5* is necessary for its formation. Its properties are similar to those displayed by E.coli and phage T4 DNA polymerases: it can propagate deoxynucleotide chains on a single stranded template from a primer sequence but is not able to initiate their synthesis de novo. Masamune, Frenkel and Richardson (1971) showed that when T7 mutants in gene *1.3* (that lack the phage ligase) are grown on $lig^-$ host cells, 11S Okazaki fragments accumulate. Under conditions of normal growth either the phage or the host ligase appears able to undertake the joining of these fragments. These results leave open the possibility that the space between adjacent Okazaki fragments may represent gaps rather than nicks, in which case a repairlike polymerase activity may be needed in addition to the ligase. No direct investigation has been made of whether synthesis of the Okazaki fragments is initiated with RNA, as it is in the host cell, but some indirect evidence from the operation of in vitro replication systems suggests that the process may be similar (see below).

In addition to the enzymes that provide catalytic activities necessary for phosphodiester bond synthesis, cells infected with T7 possess a DNA-binding protein apparently coded by the phage. Reuben and Gefter (1973) searched for such an activity by fractionating extracts of infected and uninfected cells for protein(s) that bind to single stranded DNA. The DNA unwinding activity of E.coli, a protein of 22,000 daltons, was present in both extracts. In addition, the T7-infected cells contained a protein of 31,000 daltons. This protein is synthesized in large amounts upon infection by wild type T7; it does not appear upon infection with gene *1* amber mutants which suggests that it is coded by a class II or class III gene. Although its time of appearance is similar to that of the other enzymes concerned with DNA metabolism, which are coded by class II genes, no mutants in T7 that generate a defective DNA-binding protein have been identified and its gene therefore remains undefined. Reuben and Gefter (1974) found that the addition of this protein to the phage DNA polymerase stimulated DNA synthesis upon a T7 template by about an order of magnitude; its action appears to be stoichiometric rather than catalytic and in general its properties seem similar to those of the T4 32 protein (see page 624). However, it may be difficult to define fully its role in replication until mutants are available.

Systems for the synthesis in vitro of T7 DNA have been developed by Stratling et al. (1973) and Hinckle and Richardson (1974). Replication requires an exogenous T7 DNA template, the products of certain phage and host genes, the four deoxynucleoside triphosphates, and is stimulated also by addition of the four ribonucleoside triphosphates (which hints that RNA synthesis may be involved in the replication). Synthesis of DNA continues at a linear rate for some 20–30 minutes and the amount of DNA synthesized is about 2–3 times the amount of added DNA template; the product appears to result from semiconservative replication and is not linked covalently to the template (an indication that synthesis has been initiated de novo and does not simply represent the extension of a parental primer). The product generally takes the form of rather small lengths of DNA, which might correspond to Okazaki fragments that have failed to become covalently linked because of a deficiency in the in vitro system.

The in vitro system can be used to provide a complementation assay to purify the components necessary for phage replication. Instead of preparing the in vitro extract from cells infected with wild type T7, the host cells are infected with a mutant in one of the DNA replication functions. The in vitro system derived from these cells is inactive; but its activity may be restored by addition of a preparation made from cells infected with wild type T7. The active component of this preparation can be purified on the basis of its ability to restore function to the defective in vitro system. The phage genes essential for replication are defined by the inactivity of in vitro systems prepared from cells infected with mutants in genes *4* or *5*; in vitro extracts from cells infected with phages mutant in genes *2*, *3* or *6* remain active, which implies that the products of these genes are not involved in DNA synthesis per se (which is all that is assayed by the in vitro system), although it remains possible that they are involved at other stages of DNA reproduction, for example the formation of concatemers.

Using inactive in vitro systems prepared from cells infected with gene *4* mutants, Stratling and Knippers (1973) and Hinckle and Richardson (1975) purified the gene *4* product. This takes the form of a protein found in two peaks, corresponding to sizes of 66,000 and 57,000 daltons; presumably the second is derived by cleavage of the first. The significance of the occurrence of two forms is not known. The protein in either form stimulates synthesis of T7 DNA upon a duplex template and its addition may make possible synthesis of *l* strand DNA. Since replication is initiated not far from the left end of the genome (see below), it is the newly synthesized *l* strand that must grow largely in the overall direction 3′ to 5′ and which may therefore be the more dependent on proper initiation of Okazaki fragments. These results are therefore consistent with the idea that the gene *4* product may be involved in initiation of synthesis of the Okazaki fragments, although of course the data do not prove this conclusion and also allow other possibilities.

The host cells used to prepare these in vitro systems carried a $polA^-$ mutation in order to eliminate the high background activity that is caused by E.coli DNA polymerase I. Is it the absence of this enzyme that is responsible for the accumulation of the product as small DNA pieces, because it is needed to fill gaps between Okazaki fragments? To test this possibility Masker and Richardson (1976a,b) prepared an in vitro system containing DNA polymerase I. Its general characteristics of replication are not altered by the presence of this enzyme, except that the rate of DNA synthesis is slightly increased and the amount of product is increased by up to about twofold. But some 60% of the product is of the size of mature DNA, compared with only little more than 10% when the enzyme is absent. Is DNA polymerase I participating in the replication proper of T7 DNA or is it instead conducting an extensive repair activity? An indication that its capacity is the former is that no DNA synthesis is supported by the enzyme under conditions when the T7 system is inactive; and it does not affect the lack of covalent linkage between product and template. Although this does not prove that DNA polymerase I is necessary for T7 replication, these results at least raise the possibility that it (or some other comparable activity) may be needed to fill gaps between

Okazaki fragments. An argument against an obligatory involvement of this enzyme, however, is provided by the observation that a deficiency in DNA polymerase I does not prevent T7 replication in vivo.

The involvement of host functions in T7 replication was first demonstrated by Chamberlin (1974), who isolated the *tsnB* and *tsnC* mutants of E.coli that do not allow growth of T7. Since the mutant cells appear to grow normally, the altered functions cannot be essential for host viability. Upon infection both host mutants support normal RNA and protein synthesis but show inhibition of DNA synthesis. With *tsnB* cells DNA synthesis starts slightly later than usual and then suddenly is terminated; this suggests that a host component is needed for some intermediate stage of replication. Neither T7 nor $\phi$II can grow on the *tsnB* host but T3 is able to do so, which implies some specificity in whatever reaction is involved. With *tsnC* cells none of the three phages displays any DNA synthesis, so the host component must be an essential part of both the T7/$\phi$II and the T3 replication systems.

Using their system for in vitro synthesis of T7 DNA, Modrich and Richardson (1975a,b) purified the TsnC protein by the same approach used to purify the phage gene *4* protein: that is an extract was prepared from *tsnC* mutant cells and complemented by a fraction from wild type cells. The active component is an unusually heat stable protein of 12,000 daltons which appears to be a component of the DNA polymerase produced during infection. The active enzyme contains an equal number of subunits of the gene *5* product (a protein of 84,000 daltons) and the TsnC protein, and displays the same properties previously attributed to it when it was thought simply to be the product of phage gene *5* alone. The phage DNA polymerase activity is absent from *tsnC* infected cells, whose defect appears very similar to that displayed in gene *5* mutant infection.

What is the function in the host cell of the TsnC protein? Mark and Richardson (1976) demonstrated that it is identical with thioredoxin, a cofactor needed for the enzyme ribonucleoside reductase which converts ribonucleoside diphosphates to the deoxynucleoside form. The reduction is accompanied by the conversion of thioredoxin from the form thioredoxin $(SH)_2$ to the form thioredoxin $(S)_2$; and the latter form of the cofactor then is reconverted to the former by the enzyme thioredoxin reductase. The sequestration of host functions to serve phage purposes is not unprecedented: we shall see in Chapter 9 that the RNA phage Q$\beta$ takes charge of host proteins usually involved in protein synthesis in order to constitute the phage RNA replicase. How such proteins may serve two such quite different functions is not at present clear, but poses an interesting question in protein evolution. In the context of the dual function of thioredoxin, it is interesting that there is some evidence to suggest that enzymes involved in nucleotide metabolism may be part of the replication apparatus of phage T4 (see page 626). The viability of *tsnC* cells implies that thioredoxin is not essential for nucleotide metabolism in the bacterium; but on the other hand it is possible that the present mutations leave a residual level of the cofactor that is in fact adequate and so to resolve this point decisively it will be necessary to construct deletions of this gene.

### *States of Replicating DNA*

The replication of phage T7 has been followed by two means: replicating molecules can be isolated and examined by electron microscopy; and pulse labels of $^{3}$H-thymidine can be followed into replicating DNA and then chased into mature DNA. From such data it is clear that T7 DNA commences replication in the form of a linear molecule: two growing points move bidirectionally from an origin at about 17% from the left end. In fact, although data on the later rounds of replication are less decisive, there is no evidence to support the involvement of circular forms of DNA at any stage of T7 replication. Studies of the sedimentation properties of replicating T7 DNA suggest that concatemeric molecules are present in the pool; since the absence of circular forms means that they cannot be produced by replication of a rolling circle, it is likely that their origin lies with recombination. The concatemers appear to be attached to the cell membrane, although the significance of this attachment is not yet clear. The nature of the events responsible for the generation of the concatemers and for the maturation from them of individual genomes is not defined at the molecular level, although some of the phage functions involved in these steps have been identified. Since mutants in several of the class III genes needed for particle assembly, including *8, 9, 10, 18* and *19* are unable to generate mature genomes from the concatemers, it seems likely that the same rule applies as seen with phages lambda and T4: maturation of DNA is accomplished only as a part of head assembly (see Studier, 1972).

Molecules engaging in their first round of replication can be isolated by making use of a density technique: cells are grown in the heavy isotopes $^{15}$N and $^{2}$H and the infecting phage carries the light isotopes $^{14}$N and $^{1}$H. At the end of the first round of replication the daughter phage chromosomes are fully hybrid in density; during the round their density shows a shift from light to hybrid density. When Wolfson, Dressler and Magazin (1971) examined the partially hybrid fraction they found that the majority was Y shaped, that is in the form of linear molecules with a single replication fork. Other forms, including rods with replicating eyes, also could be seen. By using a low multiplicity of infection in order to suppress the occurrence of recombination, and thus to offer an improved likelihood that the molecules are indeed the products only of replication, Dressler, Wolfson and Magzin (1972) found that only molecules with eyes at the left end or taking the Y shaped form could be seen. Aligning the molecules with eyes to determine the point from which the eyes appear to enlarge (see also page 254) suggests that the origin is located at 17%. Figure 7.7 illustrates the course of T7 replication by which the molecules with eyes are converted into the Y shaped structures and Figure 7.8 shows examples of these forms of T7 DNA. Wolfson and Dressler (1972) noted further that single stranded regions often are seen at the forks in these structures, which supports the idea that they may represent growing points at which one of the parental strands is waiting for the synthesis of an Okazaki fragment, perhaps because the strand to be synthesized is the one that must grow in the 3′ direction. One interesting implication of the location of the origin at 17% is that any mutants with this region deleted should be unable to replicate; however, deletions across this region do not seem to cause inviability and this must leave the implication that some alternative origin is available to the phage.

Molecules larger than the mature T7 genome were first detected in infected cells by Kelley and Thomas (1969) and Schlegel and Thomas (1972), when they found that a pulse label enters a peak sedimenting more rapidly than T7 DNA but may be chased from it into DNA of the size of mature molecules. Similar observations have since been made by Center (1972) and Serwer (1974). Sedimentation of the denatured strands of this DNA in alkaline gradients has produced conflicting results: sometimes the single strands behave as expected from the length of the duplex form, but in some experiments they appear to be about the length of mature genomes, suggesting the presence of nicks between adjacent genomes. Electron microscopy of these fractions identifies only linear molecules but their exact structure remains to be established. One approach to confirming the nature of these concatemers may be to obtain denaturation maps that locate individual genomes within them.

The formation of concatemers appears to require the product of gene *2*. Center (1975) found that upon infection with gene *2* mutants DNA synthesis occurs to a level of about 60% of that of wild type infection; but instead of sedimenting more rapidly than mature DNA, a pulse label enters molecules of the mature size. This suggests that replication per se can occur without the formation of concatemers and that the gene *2* product may have the function of joining together their component genomes.

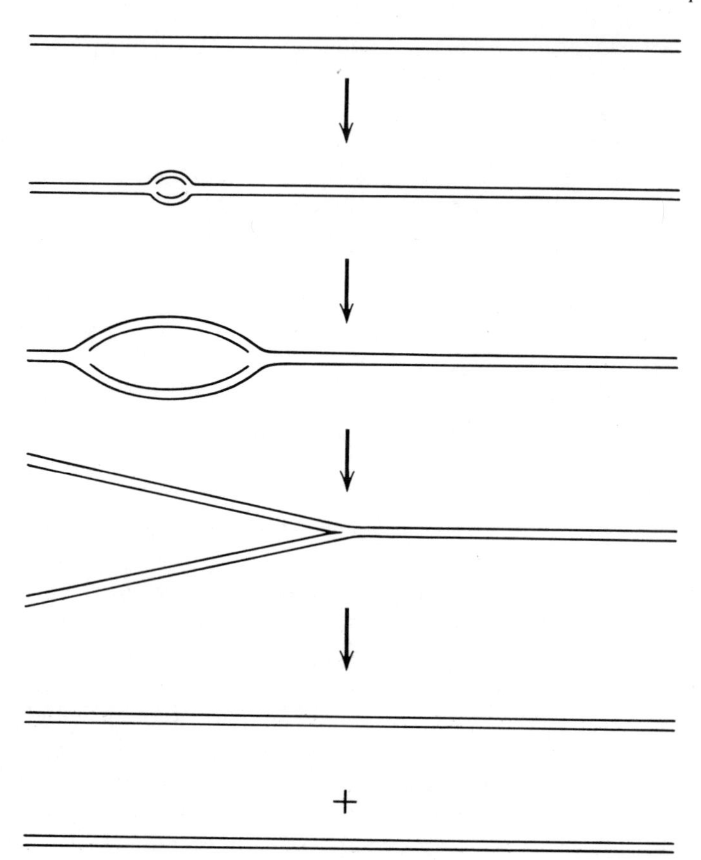


**Figure 7.7:** replication of T7 DNA. Replication is initiated at an origin about 20% from the left end of the genome and proceeds bidirectionally to generate "eyes" in which the two growing points are the same distance away from the origin. When the leftward growing point reaches the end of the linear genome, the replicating form is converted into a fork shaped structure.

Since no active particles are formed upon *2* mutant infection ($10^{-4}$ of the wild type yield), the formation of concatemers appears to be necessary for head assembly.

Maturation of concatemeric DNA into single genomes appears to require the nucleases coded by phage genes *3* and *6*, which thus provides a function for these enzymes in phage reproduction in addition to making available nucleotides by degradation of host DNA. Stratling, Krause and Knippers (1973) used very short pulse labels to identify replicating T7 DNA (if T7 replication proceeds at the same rate as bacterial replication, the complete chromosome could be duplicated in 30 seconds at 37°C so

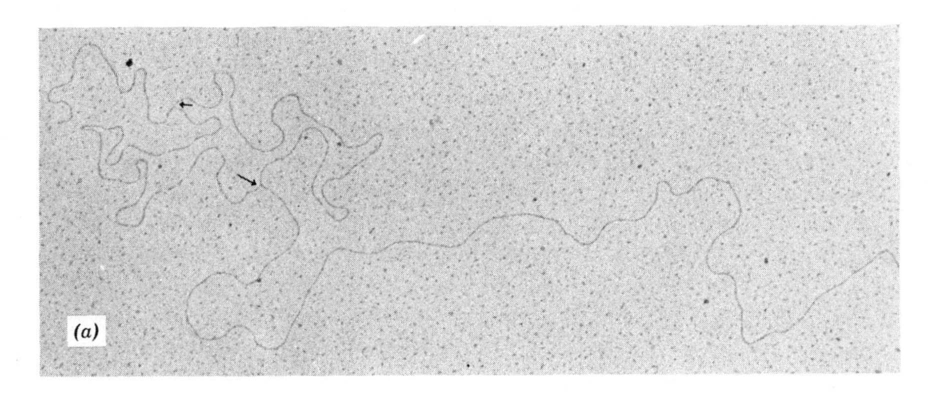


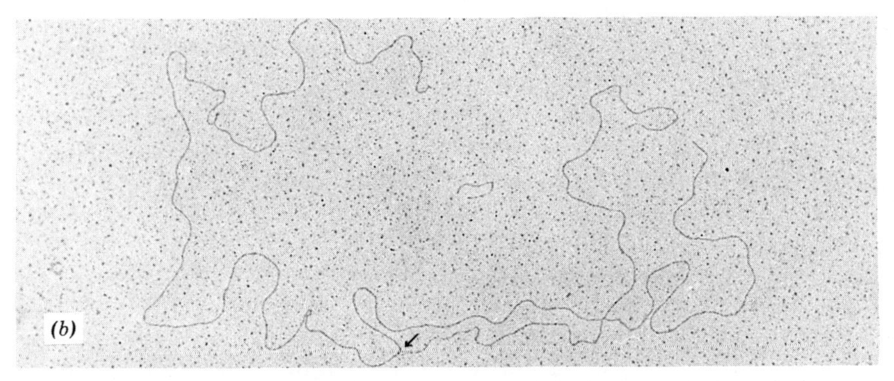


**Figure 7.8:** electron micrographs of replicating T7 DNA. The upper photograph shows an eye and the lower photograph shows a fork (the two arms of the fork have become twisted over each). The positions of the growing points are indicated by arrows. Each thread of DNA seen here represents a duplex structure. Photographs kindly provided by D. Dressler.

that a 1 or 2 minute pulse label would not specifically identify replicating molecules). The very short pulse labels enter a rapidly sedimenting fraction, probably representing concatemers of up to five or more genomes. The label then may be chased into mature T7 DNA. But in cells infected with gene *3* or gene *6* mutants, the label remains in the rapidly sedimenting DNA and does not enter molecules of mature size. This suggests that both the endonuclease and exonuclease activities are needed for maturation of DNA. While it is immediately apparent how an endonuclease may be involved, the role of an exonuclease is less clear, however, and results suggesting a different role for it have been reported by Miller et al. (1976). In these experiments stable concatemers were formed upon infection with either a $3^-$ single mutant or with a $3^-$ $6^-$ double mutant; this confirms that the endonuclease is needed for maturation of concatemeric DNA. But gene *6* single mutants failed to accumulate concatemers, instead degrading them to a size smaller than that of the mature genome. This suggests that in the absence of exonuclease, the endonuclease breaks down the concatemers; that is, if the endonuclease is active the presence of an active exonuclease becomes necessary to prevent degradation of DNA. One possible model to explain these results is to suppose that the action of the endonuclease creates a substrate upon which the exonuclease acts in genetic recombination; thus by itself the endonuclease makes breaks in DNA, but the subsequent action of the exonuclease rescues the DNA by making possible its recombination. This model implies that the endonuclease and exonuclease are needed for phage genetic recombination; since stable concatemers are formed in the $3^- 6^-$ mutants, this would mean that genetic recombination is not responsible for the formation of concatemeric DNA, which must therefore rely on some other mechanism, perhaps involving the terminally redundant sequences.

That the exonuclease is indeed needed for recombination is suggested by the observation that infection by a $6^{ts}$ mutant under semipermissive conditions results in a large decrease in the frequency with which recombinants are found amongst the (much reduced number of) progeny. The decrease in recombination does not appear to be a consequence of the reduction in phage yield, because a gene *5* mutation which reduces the yield to the same extent leaves unaltered the proportion of recombinants. While not conclusive, these experiments are consistent with the view that the exonuclease is essential for recombination in some action that is not necessary for maturation of DNA. Other evidence to support the involvement of this function and also the gene *4, 5* and *1.3* products in recombination is provided by experiments on in vitro packaging of DNA in a system developed by Kerr and Sadowski (1974). Sadowski and Vetter (1976) found that mature DNA can be packaged, but only when it first has recombined with concatemeric DNA. Presumably this reflects a need for a substrate for packaging that is longer than the single genome. Using this system as an assay for recombination, extracts from cells infected with mutants in genes *4, 5* or *6* proved to be defective; extracts from $1.3^-$ mutant infected cells were defective only if the host cells also were $lig^-$; and mutations in genes *2* and *3* had no effect. Of course, it is not certain that the in vitro system accurately reflects the events involved in recombination in vivo, but if it does the implication may be that a capacity for DNA synthesis as well as for exonuclease activity is needed for recombination.

A complex of T7 DNA with the cell membrane was isolated by Center (1973) and Pacumbaba and Center (1973), who found that it contained concatemetric DNA. Silberstein, Inouye and Studier (1975) found that during a wild type infection the phage DNA usually is later released from the membrane. But in cells infected with $3.5^-$ mutant phages, this release did not occur; nor did the release of host cell DNA from the membrane that usually is seen. What role the release of phage DNA plays, and how this is related to maturation of the concatemers, is not clear. However, the phage lysozyme mutants show a deficiency in cell lysis that is no greater than that of other class II mutants (a feature of T7 infection is that inhibition of replication blocks cell lysis); this suggests that the role of lysozyme in infection may be concerned with DNA replication rather than cell lysis, as also would be consistent with the location of its gene in the class II rather than the class III region.

Although no evidence is available on the origin of concatemeric T7 DNA, we may speculate that the repetitious sequences at the ends of linear genomes are involved. This is consistent with the idea that the redundancy is essential for phage viability and that this may be one reason why T7 and T3, which have different terminal redundancies, are unable to recombine. Watson (1972) has pointed to a problem in the replication of linear DNA that may be related to this issue. Unless synthesis of an Okazaki fragment is initiated at each 5′ terminus, the strand that must grow overall in the direction 3′ to 5′, that is by discontinuous synthesis of shorter pieces from 5′ to 3′, cannot be completed up to the end of the molecule but must fall short. The strand that grows in the other direction, from 5′ to 3′, can of course be completed simply by synthesis to the very end of the DNA. This means that a single stranded region might be left at one end of each daughter molecule; and in view of the terminal redundancy, the single strands at the left and right ends would of course be complementary. Since each replicated chromosome should have only one single strand end, left or right, annealing between complementary ends must generate concatemers. We should emphasize that this model is speculative, but it does provide a possible answer to the question of how concatemers might be generated other than by rolling circle replication or recombination.

CHAPTER 8

# Single Strand DNA Phages

## Genetic Organization

### *Structures of Icosahedral and Filamentous Particles*

The single strand DNA phages fall into two general classes: icosahedral and filamentous. The sizes of their genomes are very similar, in each case 8 or 9 genes having been identified. Their products represent the structural proteins of the virion and functions necessary for DNA replication and particle assembly. The physicochemical parameters of the particles of each type of phage are summarized in Table 8.1.

The life cycles of the two phage classes are strikingly different. The icosahedral phages follow the conventional infective cycle of adsorption, reproduction, and release of progeny particles accomplished by lysis of the host bacteria. The filamentous phages, however, do not kill their host cells: instead progeny particles are released continuously from the infected bacteria, which continue to survive. But in spite of this difference, there are many similarities between the two types of phage genome, perhaps stemming from the common problems that they must face in reproducing a single stranded DNA through the mediation of a replicating duplex form.

The two best characterized members of the icosahedral class are S13, first isolated by Burnet in 1927, and $\phi$X174, isolated by Sertic and Boulgakov in 1935; and these were shown to be serologically related by Zahler in 1958. Current investigation of these phages started with the characterization of $\phi$X174 by Sinsheimer (1959a). The phage particle sediments at about 114S and has an icosahedral structure roughly of 250 Å diameter. The particle appears to comprise twelve morphological units of diameter about 70 Å arranged on the vertices of an icosahedron (Hall et al., 1959; Tromans and Horne, 1961). No tail is evident, but spikes are present at the apices of the icosahedron and one of these may be involved in adsorption to the host. Figure 8.1 shows an electron micrograph of $\phi$X174.

Filamentous phages were discovered by their ability to adsorb to the F

**Table 8.1:** particles of single strand DNA phages

| | icosahedral ($\phi$X174) | filamentous (fd) |
|---|---|---|
| particle mass | 6.2 × $10^6$ daltons | 14.6 × $10^6$ daltons |
| DNA content | 1.7 × $10^6$ daltons (5500 bases) | 1.9 × $10^6$ daltons (6000 bases) |
| protein content | 4.5 × $10^6$ daltons | 12.7 × $10^6$ daltons |
| protein constitution | ~60 copies of pF (48,000 daltons)<br>~60 copies of pG (19,000 daltons)<br><12 copies of pH (36,000 daltons)<br>? copies of pJ (9000 daltons) | ~2400 copies of pVIII (5240 daltons)<br><4 copies of pIII (56,000 daltons) |
| particle dimensions | icosahedron of diameter 250 Å | filament of 8500 × 60 Å |

Estimates for the mass of the phage particle and the content of DNA and protein are based on physical analysis of mature phages (Sinsheimer, 1959a; Berkowitz and Day, 1976). The protein constitution is based upon the identification of virion components and the number of copies of each protein is calculated from the proportion of the protein mass it occupies; this is reasonably accurate for the fd phage but may be in error for $\phi$X174. Based on data of Burgess and Denhardt (1969), Weisbeek and Sinsheimer (1974) and Henry and Pratt (1969).

pilus present on cells carrying a sex factor (Loeb, 1960). One group of ♂-specific phages contains RNA as the genome and adsorbs along the length of the sex pilus (see page 81). These polyhedral phages, the f2 group which comprises f2, R17, MS2 and M12, are discussed in the next chapter. A second type of ♂-specific phage takes the form of a filamentous particle that adsorbs to the tip of the F pilus; Zinder et al. (1963) characterized one member of this group, f1, which is a filamentous rod about 8500 Å long and 60 Å in diameter. The closely related phage fd was isolated by Marvin and Hoffmann-Berling (1963); and M13 was characterized by Hofschneider (1963), Hofschneider and Preuss (1963) and Salivar, Tzagoloff and Pratt (1964). Figure 8.2 shows an electron micrograph of M13.

All three phages are closely related serologically but are not identical. Together they are described as members of the Ff group (F specific

**Figure 8.1:** particles of $\phi$X174 shadowed with platinum. The structure appears to be that of a regular polyhedron with a diameter of about 250Å. Data of Hall, MacLean and Tessman (1959) kindly provided by I. Tessman.

filamentous phages). Another group of filamentous phages is the If class, characterized by the ability to adsorb to the I pilus (see page 212). Although generally similar in size to the Ff phages, much less is known about the If phages; and here we shall be concerned with the three well characterized members of the Ff class. Early work on the filamentous phages has been reviewed by Marvin and Hohn (1969).

In the absence of a tail, neither the icosahedral nor the filamentous single strand phage particle adsorbs to the host bacterium by the series of events familiar from the larger phages (see pages 534 and 669). The icosahedral phage coat remains outside the infected bacterium, which

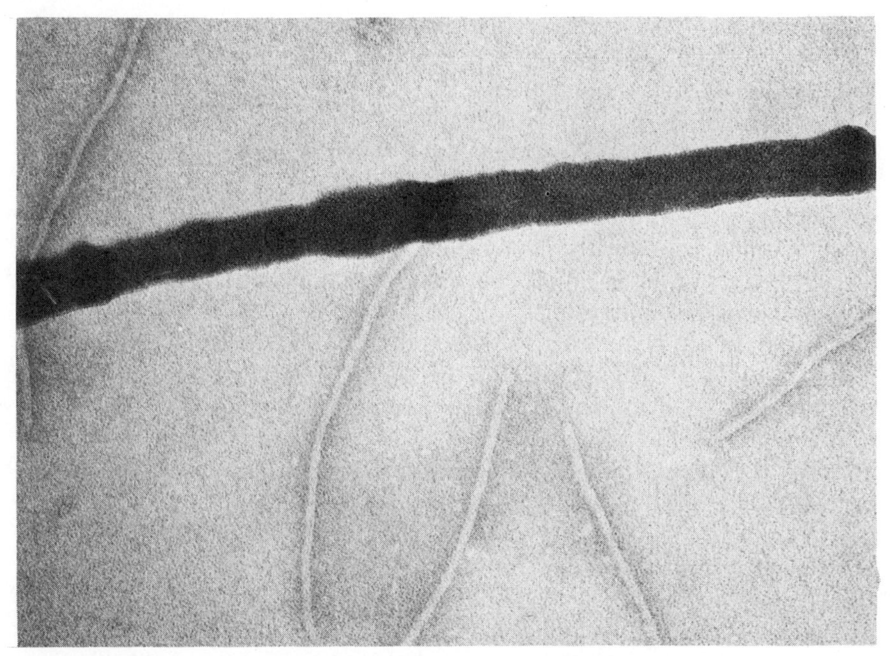

**Figure 8.2:** electron micrograph of particles of the filamentous phage M13 stained with phosphotungstic acid. The photograph is about 1 $\mu$m wide. Kindly provided by J. Griffith.

only the DNA enters. Newbold and Sinsheimer (1970a,b) showed that infection with $\phi$X174 passes through three initial stages:

*attachment* the phage particle binds reversibly to a receptor at the host cell wall;

*eclipse* binding becomes irreversible and the DNA within the particle becomes accessible to DNAase;

*penetration* phage DNA enters the host cell.

The second stage appears to be accompanied by a conformational change in the phage particle, for Brown, Mackenzie and Bayer (1971) observed that isolated cell walls can inactivate the phage and render its DNA accessible to the enzyme DNAase. Electron microscopy shows that the phage particle always is attached to the wall by only one of the icosahedral vertices; this hints that one of the twelve spikes may differ from the others in having the responsibility for adsorption. Unlike the larger phages, where the head maintains its distance from the bacterium by virtue of the tail, with $\phi$X174 adsorption involves close contact with the bacterial surface, and the particle becomes embedded in the wall to a depth of about half its diameter.

The filamentous particle is essentially a flexible rod of deoxyribonucleoprotein. X-diffraction studies show that the capsid proteins are organized into a left handed helix of 15 Å pitch with 4.5 units per turn. Each protein molecule is elongated in an axial direction and overlaps with the next, forming a tube with inner diameter 20 Å and outer diameter 60 Å. The structures of the Ff and If phage particles appear to be related but not identical in construction (Marvin et al., 1974; Marvin and Wachtel, 1975). The length of the phage particle is about half of the contour length of the single stranded DNA genome; and this comparison leaves the implication that the interior of the particle must contain a genome folded back upon itself, so that two antiparallel strands of DNA run along the length of the particle. Whether duplex regions are formed by intramolecular base pairing within the particle is not known.

Are specific regions of the genome associated with particular parts of the virion? Tate and Peterson (1974) investigated this question by sonicating f1 particles and then isolating the fragments (of about one-third to one-quarter length) that remained able to adsorb to the host bacterium. Only one end of the intact phage particle has the ability to bind to the host cell, and it is reasonable therefore to assume that the isolated fragments are derived from this end of the virion. However, their DNA displayed the same pattern of pyrimidine tracts as those characteristic of the complete genome. There appears therefore to be no sequence-specific organization of DNA in the phage coat.

Filamentous phages differ from those that we have so far discussed in their mode of adsorption: for the entire phage particle, protein coat as well as DNA, appears to enter the cell. This suggestion was first made by Trenkner, Bonhoeffer and Gierer (1967) and was investigated in more detail by Jacobson (1972). When f1 attaches to the top of the pilus, the particle cannot enter the cell via this appendage; its dimensions are too large. However, pili are no longer visible after Ff adsorption, which led to the idea that the pilus might be retracted into the cell together with the particle (see Marvin and Hohn, 1969). Jacobson observed that sex pili seem to get shorter and then to disappear as the phage enters the cell. This resorption could also be seen in uninfected cells; perhaps it is a normal feature which is utilized by the phage, with some conformational change leading to retraction being triggered when the Ff particle adsorbs to the tip of the pilus. After adsorption, the coat of the infecting particle is found at the inner cell membrane, the same site where progeny phage particles will be assembled. Whether the coat of the infecting particle is used again, is broken down to monomeric proteins which are incorporated into fresh coats, or is degraded to amino acids which are utilized in protein synthesis, is not yet clear.

Only two proteins are found in the filamentous phage coat. These are the products of genes *III* and *VIII*. The gene *VIII* protein is the major protein; and if it accounts for virtually all of the protein mass of the particle there must be about 2400 copies per coat. The gene *III* protein is present in very much smaller accounts, corresponding to 4 or less copies per particle (Henry and Pratt, 1969). The *III* protein is implicated in adsorption by the properties of M13 mutants defective in gene *III*; mutations in this gene may cause the production of particles whose ability to adsorb to the host is abnormal (Pratt, Tzagoloff and Beaudoin, 1969). Mutation in gene *III* also may cause the production of *polyphages*, very long defective particles that lack the *III* protein. This is consistent with the view that gene *III* protein is added to one end of the phage particle in a reaction which terminates its extension and which confers upon this end the ability to adsorb to the host.

### *Circularity of the Genetic Map*

The genomes of both the icosahedral and filamentous phages take the form of a single stranded circle of DNA. In his initial characterization of $\phi$X174, Sinsheimer (1959a,b) demonstrated that its DNA reacts with formaldehyde, does not show a sharp Tm with melting, lacks the A=T, G=C equality, and is inactivated by pancreatic DNAase with single hit kinetics: all of these are characteristics of single stranded DNA. Another indication of single strandedness was the observation of Tessman (1959) that mutagenesis with nitrous acid generates pure mutant clones with $\phi$X DNA, whereas with the duplex control of T4 mixed clones are generated. The circularity of the DNA was first indicated by the results of Fiers and Sinsheimer (1962a,b,c) which showed that $\phi$X DNA is not susceptible to several enzymes that require free 3′-OH or 5′-OH ends. Sedimentation analysis confirmed the circular structure and showed that a linear molecule of full length can be generated by cleavage with alkali or DNAase. Electron microscopy of $\phi$X174 DNA directly identified a circle of 1.77 $\mu$m (Freifelder, Kleinschmidt and Sinsheimer, 1964; for review see Sinsheimer, 1968). Similar experiments have shown that the genome of the filamentous viruses exists in the same state (Marvin and Schaller, 1966; Mitra et al., 1967; for review see Marvin and Hohn, 1969). With both types of phage, the isolated viral DNA remains infective when tested in a spheroplast assay only so long as it retains its circular condition.

Genetic analysis of $\phi$X174 and the related phage S13 can be performed by conventional mapping techniques, although the frequencies of recombination are rather low. Recombination in S13 was first observed by Tessman and Tessman in 1959 and this led to a series of experiments to isolate mutants and identify complementation groups. Tessman (1965) identified four complementation groups and in further experiments Tessman et al. (1967) brought the total number to seven. Only one further gene has since been identified. Baker and Tessman (1967) demonstrated that the genetic map of S13 is circular, corresponding to the physical state of the DNA. The map sums roughly to about 0.2% in recombination units.

By analysis of nonsense and temperature sensitive mutations, eight genes have been identified in $\phi$X174 (Hutchison and Sinsheimer, 1966; Dowell and Sinsheimer, 1966; Funk and Sinsheimer, 1970; for review see Sinsheimer, 1968). A further gene has been identified by its protein product; no mutations have yet been found. Following the observation by Tessman and Shleser (1963) that $\phi$X174 and S13 can recombine, complementation experiments have been performed by using mixed infections with both phages. All the mutants identified in one phage can be complemented by a function in the other phage (Jeng et al., 1970). The genetic content of the two phages is therefore closely related and the circular genetic map constructed for $\phi$X174 by Benbow et al. (1971) corresponds well with the map of S13. A uniform nomenclature has now been adopted for both phages and is given in Table 8.2 (together with the separate nomenclatures that it has superseded).

Although $\phi$X174 and S13 display similar genetic organization, the sequences of their genomes have diverged appreciably in their evolution from some common ancestor. Godson (1973) showed by heteroduplex mapping that only 5% of the sequences of the two genomes are highly homologous; the remaining regions display an average divergence of sequence corresponding to changes in more than 20% of the bases. Godson (1974b) also isolated a series of four new phages related to $\phi$X, the series G4, G6, G13 and G14; all show extensive substitutions in nucleotide sequence compared with $\phi$X174. A comparison of the products made in infected cells shows that the number of phage proteins is the same and their sizes are very similar, although not identical, in each case. Given the redundancy of the genetic code, it would be interesting to see a detailed comparison of the protein functions of $\phi$X174 and S13 to determine what proportion of the changes in sequence are reflected in changes in the gene products.

The products of eight genes of $\phi$X174 have been identified in infected cells by the usual procedure of characterizing the proteins made in large amounts after infection and determining which protein is not produced by a given amber mutant. Benbow et al. (1972) identified 13 proteins as major

**Table 8.2:** genes and proteins of $\phi$X174

| gene | former name ($\phi$X) | former name (S13) | function | genetic size | estimated protein size | sequenced protein size | bases in gene |
|---|---|---|---|---|---|---|---|
| A | VI | IV | endonuclease | 0.30% | 56,000–62,000 and 35,000 | 53,000* | 1440* |
| B | IV | II | | 0.022% | 15,000–25,000 | 18,000* | 490* |
| C | VIII | VI | | 0.015% | 7,000–34,000 | 10,000* | 280* |
| D | V | VII | particle assembly | 0.013% | 14,500–15,200 | 16,811 | 456 |
| E | I | V | lysis | 0.013% | 10,000–17,500 | 9,940 | 273 |
| J | — | — | particle component | — | 5,000– 9,000 | 4,097 | 114 |
| F | VII | I | coat protein | 0.064% | 48,000–50,000 | 46,700 | 1287 |
| G | III | IIIa | particle protein | 0.02% | 19,000–20,500 | 19,053 | 525 |
| H | II | IIIb | spike protein | 0.033% | 36,000–37,000 | 35,500 | 981 |

Data are given for $\phi$X174 but the corresponding functions in S13 are probably very similar. The original (Roman numeral) nomenclature for $\phi$X and S13 was used prior to 1970 by the Sinsheimer and Tessman groups; but both sets of phage genes now are described as *A–J*. The genetic size of each gene is that calculated from the recombination data of Benbow et al. (1971, 1972). Protein sizes (in daltons) have been estimated through gel electrophoresis by Burgess and Denhardt (1969), Burgess (1969), Linney and Hayashi (1973), Benbow et al. (1972), Godson (1971), Henry and Knippers (1974), and Farber (1976). The values for sequenced protein size and the number of bases in the gene are based upon the conclusions reported by Barrell, Air and Hutchison (1976), relying either upon the amino acid sequence of the protein or the nucleotide sequence of the gene; the values that are asterisked are maximum numbers calculated from the restriction mapping of Weisbeek et al. (1976) but Barrell et al. (1976) report a number of nucleotides for genes *A*, *B* and *C* combined of the order of 1850, corresponding to a coding capacity of 67,000.

products of infected cells, but 2–3 of these may be host functions since their production is prevented by ultraviolet irradiation of the host. The sizes of the proteins identified in these and in the similar experiments of Godson (1971) are given in Table 8.2. There is generally good agreement about the sizes of the *A*, *D*, *F*, *G* and *H* gene products, while the *B* and *C* products have been ascribed somewhat varying sizes in these and other experiments and therefore remain to be clearly characterized.

One of the $\phi$X genes has been equated with more than one protein product. Linney, Hayashi and Hayashi (1972) showed that two proteins, of 62,000 and 35,000 daltons, are missing in certain amber mutants of gene *A*. Linney and Hayashi (1973) then were able to demonstrate that mutants near the beginning of the gene cease to make the larger protein (pA′) but continue to produce the smaller protein (pA). Mutants located towards the end of the gene make neither protein. The tryptic peptides of pA are a subset of those of pA′ and this suggests that pA may result from an internal initiation more than halfway along the gene, producing a fragment corresponding to the C-terminal region. The full size protein has been shown to possess an endonuclease activity essential for replication; the smaller protein has not been investigated. The functional relationship between the two proteins therefore remains to be established.

The nature of the proteins of genes *E* and *J* has been resolved only more recently. A mutation in gene *E* causes the disappearance of a 17,500 dalton protein, tentatively identified as the product of this locus by Benbow et al. (1972). Another mutation close to this site complements poorly with $E^-$ mutations and might represent either a different gene or a mutation in *E* whose properties are different from those of the other *E* mutants. This mutant was ascribed to the new gene *J* because the known *E* mutants are defective in cell lysis but otherwise generate normal progeny; the putative $J^-$ mutant also is deficient in lysis but in addition accumulates defective particles. The mutation in *J* causes the disappearance of a 9000 dalton protein which has been reported to be a component of the virion. However, the sequencing studies on $\phi$X DNA summarized by Barrell, Air and Hutchison (1976) suggest that the second mutation in fact lies within gene *E*, which has a nucleotide sequence long enough to code for a protein of just under 10,000 daltons. Adjacent to gene *E* is a sequence of nucleotides that appears to represent a protein of just over 4000 daltons and which provides the small component of the virion previously equated with gene *J*. Although not yet identified by mutation, therefore, this latter sequence now may be taken to define the *J* locus.

Four proteins have been identified as components of the virion (Burgess, 1969; Edgell, Hutchison and Sinsheimer, 1969; Weisbeek and Sinsheimer, 1974). Benbow et al. (1972) identified these as the products of genes *F*, *G*, *H* and *J*. It has not been entirely clear which proteins are constituents of the spikes (which can be removed from the particle by 4M urea) and which comprise the phage coat. However, the *F* product accounts for most of the protein mass of the particle and represents the major coat protein. The *G* product has been ascribed to the spikes, but the quantitation of Benbow et al. suggests that it is present in the same molar amount as the *F* product (see Table 8.1); and this is supported by the observation of Siden and Hayashi (1974) that an equimolar complex can form between the two proteins. This complex appears to contain 5 molecules of *F* protein and 5 molecules of G protein and we may speculate that it is the unit from which the coat is constructed. As we have already remarked, the origin of the protein attributed to the *J* gene has been somewhat confused, and its function remains to be determined.

The H protein is involved in adsorption to the host cell, consistent with the observation that *H* mutants may display altered host range as well as defects in adsorption. In addition to its role in making contact between phage and bacterium, the H protein may be involved in assisting the DNA to penetrate the bacterial wall. Jazwinski, Lindberg and Kornberg (1975a,b) reported that the H protein can bind to the single stranded phage genome and increases its infectivity in a spheroplast assay. Since the spheroplast has no outer cell wall, this must imply some role for the H protein in the events subsequent to adsorption. The receptor for $\phi$X174 and S13 is a lipopolysaccharide component of the cell surface; and it is able to inactivate the DNA-H protein complex so that it is no longer infective. Jazwinski, Marco and Kornberg (1973, 1975) reported that H protein can be recovered from infected cells in association with the DNA. They suggested that it might be regarded as a "pilot" protein which assists the entry of the DNA into the cell; in their earlier paper, very similar results were reported for the gene *III* protein of M13, for which the same function was suggested. The reaction between H protein and its cellular receptor can be mimicked in vitro; Jazwinski and Kornberg (1975) found that an extract of spheroplasts lysed with a nonionic detergent can convert the single stranded DNA present in intact particles into the duplex form that it acquires upon infection; this reaction cannot be catalyzed by an extract of soluble enzymes that synthesizes DNA with the isolated phage genome as template (see later). This suggests that the interaction between the cellular receptor and H protein is responsible for releasing the DNA from its particle for entry into the cell.

The product of gene *A* is concerned with replication of the phage DNA; the products of genes *B* and *D* appear to be concerned with assembly of the phage particle; and the function of gene *C* is unknown at present.

The three male-specific filamentous phages, f1, fd and M13, are closely related and may for genetic purposes be regarded as variants of a single strain. While the differences between them have not been defined in detail, it seems likely that they are very minor. Early nonsense and temperature sensitive mutants of M13 were classified into six complementation groups by Pratt, Tzagoloff and Erdahl (1966); a seventh group then was identified by a further single mutation and Mazur and Zinder (1975a) showed that this represents a distinct gene. However, none of these mutations resided in the gene coding for the major coat protein. Pratt, Tzagoloff and Beaudoin (1969) therefore examined mutants generated with hydroxylamine for the production of particles with altered buoyant density (that is, in which the proportion of protein mass is altered). This led to the isolation of an amber mutant in the coat protein gene. No further genes have been identified since then. The known genes of the Ff phages are described in Table 8.3.

The products of the Ff genes have not yet been fully characterized. The proteins of genes *I*, *II*, *III*, *IV*, *V* and *VIII* have been identified as the products of in vitro expression of f1 or M13 DNA by Model and Zinder (1974) and Konings, Hulsebos and Van den Hondel (1975). The sizes estimated for these proteins in the two sets of experiments are in good agreement. Two proteins were equated with gene *IV* in the first series; the

**Table 8.3:** genes and proteins of filamentous phages (Ff)

| gene | function | protein size (daltons) | |
|---|---|---|---|
| I | ? | 35,000 | 36,000 |
| II | RF replication | 40,000 | 46,000 |
| III | minor coat protein | 68,000 | 59,000 → 56,000 |
| IV | ? | 50,000 and 26,000 | 48,000 |
| V | single strand DNA binding protein | 9,690 | 9,690 |
| VI | ? | unidentified | |
| VII | ? | unidentified | |
| VIII | major coat protein | 5,200 | 5,800 → 5,200 |
| ? | | 12,000 | |

Protein sizes were determined by characterizing the products of coupled in vitro transcription and translation of f1 (first column) by Model and Zinder (1974) or M13 (second column) by Konings, Hulsebos and Van den Hondel (1975). In the second set of experiments the in vitro products of genes III and VIII were larger than the proteins found in the mature virion, and this is the basis of the indicated precursor-product relationship.

significance of this is not yet clear. Genes *III* and *VIII* code for the minor and major coat proteins, respectively; and it may therefore be significant that in the experiments of Konings et al. the in vitro products were larger than the proteins found in the mature virions. This could indicate that a cleavage reaction occurs at some stage between synthesis of the coat proteins and assembly of the particle. The sequence of the coat protein of phage fd was determined by Asbeck et al. (1969) and corrected by Nakashima and Konigsberg (1974): it comprises 50 amino acids and only 1 amino acid change is seen in the coat protein of the phage ZJ-2, another member of the Ff group (Snell and Offord, 1972). The coats of M13 and f1 have the same overall amino acid composition and also may be identical or very similar. The product of gene *V* also has been sequenced for both phages M13 and fd and appears to be identical in each, for the only reported differences lie in a single tryptic peptide and probably represent a mistake in one sequence (Cuypers et al., 1974; Nakashima et al., 1974). The conservation of amino acid sequences between proteins coded by different members of the Ff group emphasizes the close relationship of these phages. The products of two genes *VI* and *VII*, have not yet been identified; and a 12,000 dalton protein corresponding to none of the genes (as defined by existing mutations) also has been attributed to the phage. Since only the products of genes *III* and *VIII* are found in the virion, the functions of all the remaining genes may be concerned with DNA replication and maturation.

Conventional mapping procedures can be applied only with difficulty to the filamentous phages because the yield of recombinants following mixed infection is very low—scarcely greater than the rate of reversion of point mutations. Another complicating factor is the presence of some 1–5% particles in the lysate that are diploid; roughly half of these may be heterozygous after a mixed infection. Salivar, Henry and Pratt (1967) showed that these heterozygotes are produced in every pairwise cross of M13 amber mutants; they can be distinguished from wild type recombinants because although they also form plaques on a nonpermissive host, these plaques are smaller than the wild type. There is a large variation in the proportion of heterozygotes formed in different genetic crosses. The diploid particles are twice the usual length (and cannot be dissociated into haploid particles) but contain DNA of the usual size. Each diploid particle appears therefore to contain two haploid genomes.

In view of these complications, Lyons and Zinder (1972) introduced two procedures to construct a map of f1. For intracistronic crosses (when recombination frequencies naturally are very low and difficult or impossible to measure conventionally), they took advantage of the ability of only wild type recombinants to grow when two amber mutants are crossed on a nonpermissive host. What happens is that a recombination event between the two sites of amber mutation may generate a wild type genome; and this is able to grow normally on the host cell to release progeny particles. In the absence of such a recombination event, no particles are produced. For intercistronic crosses, unselected markers (the loci determining susceptibility to the B restriction system; see Volume 1 of Gene Expression) were used so that recombinants could be distinguished from revertants; also, single bursts were analyzed so that only those cells producing recombinants were examined. Although these recombination data were not able to permit distances between sites of mutation to be determined, they allowed the mutations to be ordered into a circular array.

### *Physical Organization of the Genome*

Restriction fragment mapping allows regions of the genome to be delineated on a physical basis; comparison of this map with the genetic map identifies the size of each gene. The DNA of $\phi$X174 (in duplex form) is cleaved into a small number of fragments by each of several restriction enzymes: Hind II generates 13 fragments, Hae III produces 11 fragments, and Hpa II releases 8 fragments. The fragments in each set can be ordered by making use of the overlaps between the different sets: the fragments released by one enzyme can be treated with another and the information gained from such double digests usually allows some or all of the fragments of each set to be aligned in an unambiguous order. Sometimes it is also useful to obtain enzymes that cleave less frequently, for example only once or twice in the genome, for then it is possible to determine which of the larger number of fragments present in each restriction set is susceptible to the rare cleavage. Another technique that can be used to order fragments is to anneal the single strand derived from a single restriction fragment with an intact complementary strand of the genome. This may then be used as a primer for DNA synthesis from its 3′-OH end; by using labeled nucleotide precursors, the location of the label can be identified by cleaving the newly synthesized duplex with the restriction enzyme used to generate the primer; this identifies neighboring fragments. By means of these techniques, Lee and Sinsheimer (1974) constructed the three restriction maps of $\phi$X DNA shown in Figure 8.3.

There are in principle two ways to correlate the restriction map with the genetic map: physical correlation using deletions; or assays of marker rescue with the restriction fragments. When deletions that are genetically well characterized are available, this is the most straightforward method, for it is possible then simply to determine which restriction fragments are absent, reduced in size, or fused together. In $\phi$X, however, deletions are very rare: only one has so far been characterized (Zuccarelli, Benbow and Sinsheimer, 1972). The second approach has therefore been applied to $\phi$X restriction fragments by Weisbeek et al. (1976). In this protocol, restriction fragments derived from wild type $\phi$X DNA are denatured into single strand fragments which are then individually annealed with complete single strands derived from various strains carrying point mutations. This DNA is used to infect a spheroplast. Such an infection can produce wild type progeny when the restriction fragment carries the wild type sequences complementary to the region of the genome that is mutated on the intact strand. In principle this can therefore be considered as a marker rescue technique and the successful rescue of a mutation locates it within the restriction fragment used.

The correlation between the restriction and genetic maps determined in this way by Weisbeek et al. (1976) is shown in Figure 8.3. Of course, the positions of the ends of the genes cannot be perfectly defined because unless mutant proteins have been sequenced it is not possible to say how close the outside mutations are to each end of the gene. The size of $\phi$X174 DNA has generally been taken as 5500 bases. This estimate is in fairly good agreement with the total length obtained by summing the sizes of the restriction fragments in each of three sets: Lee and Sinsheimer reported these all total in the region of 5250–5400 bases. Consistent with these values, Barrell, Air and Hutchison (1976) stated that sequence studies show $\phi$X174 DNA to be no longer than 5400 bases.

A genome of 5400 bases has a maximum coding capacity equivalent to about 200,000 daltons of protein. However, the sum of the estimated sizes of the $\phi$X proteins is greater than this; the extent of the discrepancy depends upon which estimate is taken for each protein, but the significance of this calculation is made clearer by considering the maximum length that each gene may occupy. From the known sizes of the restriction fragments, Weisbeek et al. calculated the greatest number of bases that could lie in each gene and in most cases this is slightly smaller than would be needed to code for the observed protein. Formally this calculation suggests one of two conclusions: there may be an underlying error that causes the sizes of all the restriction fragments to be underestimated; or the sizes estimated for the proteins may be too large. Given the uncertainties about the sizes of some of the protein products, however, there are only two serious discrepancies, since in the other instances the difference between calculated maximum size and observed size appears to lie within experimental error.

The best characterized of these discrepancies concerns the region of genes *D*, *E*, *J*, where the maximum coding capacity of the appropriate restriction fragments appears to be about 25,000 daltons, which is very

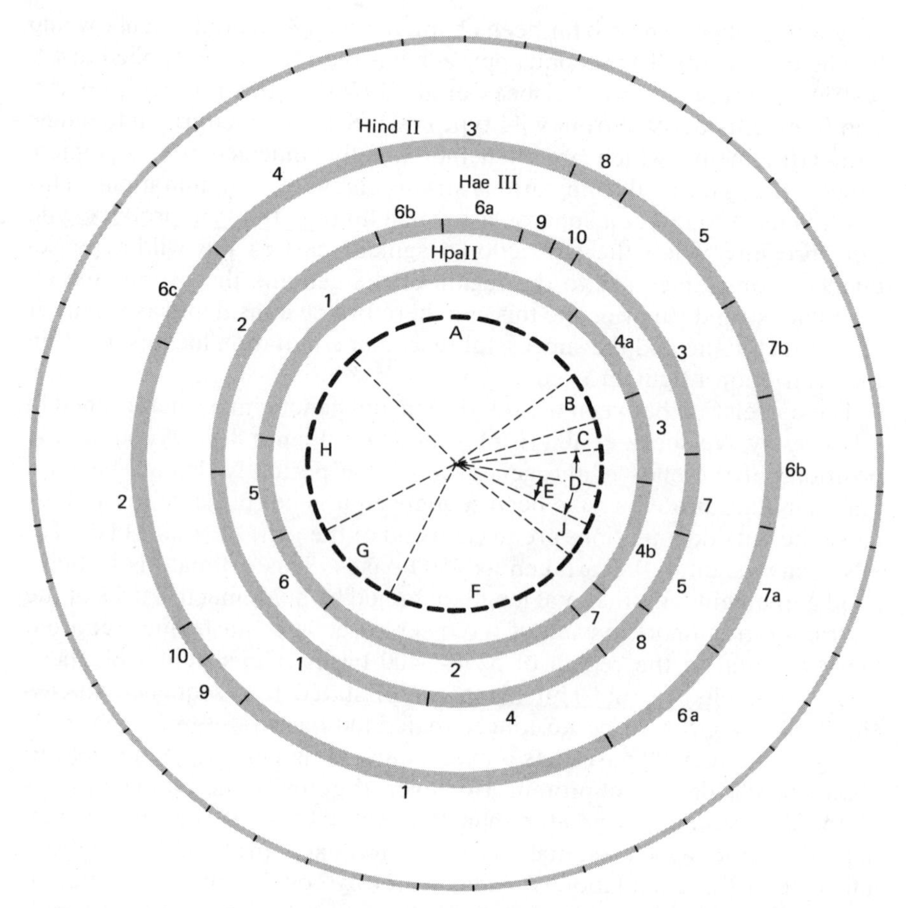


**Figure 8.3:** genetic and restriction maps of $\phi$X174. The inner map shows the position of each gene and the outer three maps show the locations of the fragments released by the enzymes Hpa II, Hae III and Hind II. The scale on the outermost circle is divided into lengths of 100 bases on the assumption that the total map length is 5500 bases. Based on data of Lee and Sinsheimer (1974) and Weisbeek et al. (1976).

close to the sum of the lowest estimates for the sizes of the *D* and *E* products. Indeed, as the result of this calculation, Weisbeek et al. were compelled to suggest that gene *J* might in fact not exist, since there appears to be no room for it and its identification rested upon a single mutation whose properties also might be interpreted as due to mutation in *E*.

In a striking demonstration of the compact genetic organization of the

$\phi$X174 genome, this problem has been resolved by the observation that the sequences of genes *D* and *E* overlap each other. By determining the nucleotide sequences of the appropriate restriction fragments, Barrell, Air and Hutchison (1976) have identified the regions of genes *D*, *E* and *J*. The sequence of the D gene was unequivocally correlated with its product by determining the sequence of the protein. The termination codon of gene *D* overlaps the initiation codon for gene *J* by one nucleotide: thus in the sequence TAATG, the first three bases provide the ochre codon terminating gene *D* while the last three bases provide the initiator for gene *J*. The following sequence can be defined as gene *J* by virtue of its equivalence with the amino acid sequence of a small protein isolated from the virion, which appears to correspond with the product previously identified with *J*. However, the single mutation that was taken to define *J* genetically lies 179 nucleotides before the *J* initiation codon, substantiating the suggestion of Weisbeek et al. (1976) that it lies in another gene.

Where does gene *E* lie? Marker rescue experiments located two mutations on restriction fragments, one the classical amber *am3* that has been taken to identify gene *E*, and the other the mutation *am6* formerly thought to identify *J* but also complementing poorly with *E*. Both lie within the sequence equated with gene *D*. By determining the sequence surrounding the mutant amber codon, it is possible to define the phase in which gene *E* is read. This proved to be displaced one base to the right (that is in the direction of reading) from that of gene *D*. By scrutinizing the sequence of gene *D* for an initiation codon (in association with a ribosome binding sequence) and a termination codon in this reading frame, it is possible to deduce that gene *E* resides *within* the latter part of gene *D*, but is read in a different phase. This suggests the remarkable conclusion that one sequence of DNA codes for two proteins.

How this situation may have evolved can be a subject only for speculation, but it is reasonable to suppose that gene *D* arose first, and that some event may later have led to the translation of part of it in a different phase to give rise to the *E* protein. Of course, the evolutionary constraints upon the sequence representing both proteins must presumably be very severe and it is difficult to know how much scope there might be for changes in sequence. This overlapping expression must mean that all mutants in gene *E* also carry alterations in the sequence coding for *D* (and vice versa within the region of overlap). Whether a given mutation affects the properties of only one or both proteins will depend upon the exact nature of the alteration (for example it may alter an amino acid in one protein but fail to do so in the other because of third base degeneracy). A possible utilization of the overlap as a control mechanism is that ribosomes translating *D* might interfere with access to the initiation codon of *E*; this predicts that mutants causing termination early in *D* might have increased expression of *E*.

A second region where a serious discrepancy exists between coding capacity and the sizes of the known proteins is that of genes *A*, *B* and *C*. The total length of this region according to the sequence studies is about 1850 nucleotides, which could code for only some 67,000 daltons of protein; this is only slightly greater than the size of the *A* protein and certainly does not appear to leave room for contiguous genes coding for the *B* and *C* proteins (even allowing for the uncertainty about their sizes). Whether the same solution of overlapping expression is adopted here remains to be seen.

The data summarized in Table 8.2 show that in most cases the sizes of the genes deduced either from nucleotide sequencing or from amino acid sequencing studies are close to those anticipated from the sizes of proteins determined by gel electrophoresis. Only genes *A*, *B* and *C* remain to be characterized. The general implication of these results is that the known genes of $\phi$X174 probably occupy virtually the entire length of its genome.

The physical and genetic maps of $\phi$X174 are colinear and proportional; the genetic size of each gene, estimated from recombination data, in all instances except gene *A* is in reasonable proportion with the size estimated for the gene. Within genes *B* - *H* the genetic map therefore provides a reasonably faithful representation of position on $\phi$X DNA. In gene *A*, however, there appears to be a hotspot for recombination which increases the frequency by about an order of magnitude. This may be related to the location of the origin of replication in this region. The hotspot is seen only in *recA*$^+$ cells; in *recA*$^-$ mutants, recombination in gene *A* shows the same frequency relative to size as that displayed by the other genes (see later).

Because of the difficulties involved in estimating genetic distances on the filamentous phage genome, recombination data provide only an order for the known mutations. The relative sizes of the genes have therefore been estimated from the sizes of their protein products and the restriction fragment map. A restriction map has been constructed with the enzymes Hap II and Hae III for M13 DNA by Van den Hondel and Schoenmakers (1975) and was correlated with the genetic map by Van den Hondel et al. (1975a,b), using both the marker rescue technique used with $\phi$X and also relying upon the abilities of isolated fragments to act as templates for synthesis of phage proteins in vitro. In similar experiments Horiuchi et al. (1975) have constructed a map for f1 and Seeburg and Schaller (1975) obtained a map for fd. The close relationship between the three Ff phages is emphasized by the observation of Seeburg and Schaller that the M13 and f1 Hap maps are identical; fd differs in the location of one cleavage site (that is, it has lost one cleavage site and has gained another elsewhere). Figure 8.4 gives the map of M13. This coincides with the previous genetic map of Lyons and Zinder (1972), except that the restriction data showed that the order of two genes, *V* and *VII*, should be reversed from that deduced from the genetic data.

The sizes of the Ff restriction fragments have not been determined on an absolute basis, with reference to an external standard, but were estimated relative to one another. This poses a problem in correlating the restriction map with the genetic map. The sizes of the Ff genes have been estimated from the sizes of their protein products summarized in Table 8.3; the sum of the mass of the identified proteins is about 208,000 daltons, equivalent to a distance of some 5700 nucleotides. However, this does not include the products of genes *VI* and *VII*, which have not yet been identified; the space left for these genes therefore depends upon the total length of the Ff DNA.

Estimates for the size of the Ff genome have varied but show that it is longer than $\phi$X, usually by about 15% (Ray, Preuss and Hofschneider, 1966; Marvin and Hohn, 1969). This suggests a length of 6300 nucleotides if $\phi$X is taken as 5500 bases. On the other hand, Berkowitz and Day (1976) by several biophysical methods measured a length for fd DNA of 5740 (+ 210) nucleotides. (If this is taken to be 15% longer than the $\phi$X genome, there then is a problem in accounting for the known coding capacity of $\phi$X since the discrepancy between protein size and maximum gene size becomes much greater.) If the estimates for the sizes of Ff proteins and this low estimate for the length of fd DNA both are correct, then only a very small part, if any, of the genome can be occupied by genes *VI* and *VII*. On the other hand, by taking an estimate for the genome of 6800 bases, and assuming that genes *VI* and *VII* are small, Vovis, Horiuchi and Zinder (1975) inferred that there must be a considerable amount of intercistronic material, which they placed between genes *II* and *IV*. Taking a genome size of 6400 bases, and assuming arbitrarily that genes *VI* and *VII* are about 250 bases each, led Van den Hondel et al. (1975a,b) to infer the presence of an unaccounted region of some 300 bases, which they located between genes *II* and *V*, with the suggestion that it might code for the 12,000 dalton protein of unidentified origin. We have gone into this detail to make it clear that the difficulties in determining the end points of the genes with repect to the restriction fragments make the Ff map at present somewhat ambiguous in certain regions. In Figure 8.4 it is assumed (arbitrarily) that the genome size is 6000 bases and that the protein sizes of Table 8.3 accurately reflect the sizes of the genes; the remaining 300 bases are divided equally between genes *VI* and *VII*. The map shown in this figure fits with the attribution of known mutations to restriction frag-

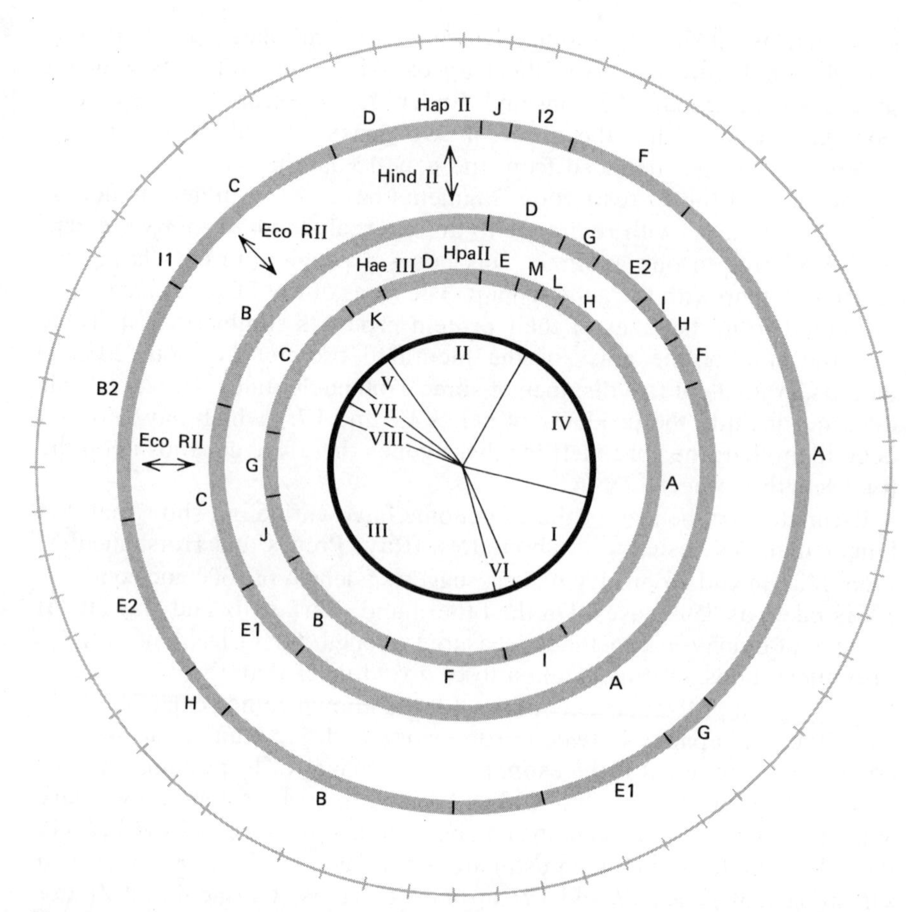


**Figure 8.4:** restriction map of M13. The inner map shows the position of each gene, calculated from the size of its protein product and the order of mutations on the map; no products have been identified for genes *VI* and *VII* which have arbitrarily been given equal sizes on the assumption that the total map length is 6000 bases (see text). The next three maps show the positions of the restriction fragments generated by the enzymes Hpa II, Hae III and Hap II; the locations of the single site cleaved by Hind II and the two sites cleaved by Eco RII also are shown. The Hpa II map of fl is identical with that of M13; the map of fd differs in that fragments F and I of M13 are found as a single fragment in fd, and fragment E of M13 is split into two fragments in fd. The Hae map of fl is very similar to the map of M13. The outermost circle divides the genome into lengths of 100 bases. Note that the correlation between the restriction and genetic maps is not precise enough to locate the ends of the genes exactly. Based on data of Van den Hondel et al. (1975a,b), Seeburg and Schaller (1975) and Horiuchi et al. (1975).

ments, but in view of the uncertainties involved in its construction we should emphasize that it may on the one hand overestimate the sizes of the genes relative to the total length (so that there is in addition some intercistronic material or another gene) or on the other side may underestimate the relative sizes of the genes (in which case the lengths of genes *VI* and *VII* would have to be even smaller).

### *Transcription of Phage Messengers*

When a virus genome takes the form of a single strand of DNA, its entry into the host cell is succeeded by synthesis of a complementary strand, which converts it into the duplex state in which replication takes place. Later during infection, single strands are regenerated by asymmetric replication of the duplex and are assembled into progeny particles. It is the duplex form of the genome that provides the template for transcription. The strand that carries the sequence of the infecting genome is known as the *viral* or *positive strand*; its partner, present only as a component of the duplex replicative form, is known as the *complementary* or *negative strand*. Either of these strands might in principle be the one copied into RNA. The virus usually is named for the strand that is the template for messenger synthesis: with a negative strand virus it is the infecting sequence that is utilized, so that the messengers carry the same sequence as the negative strand; with a positive strand virus it is the complementary strand that is the template, so that the messengers have the same sequence as the mature genome. The icosahedral and filamentous phages both are positive strand viruses; Hayashi, Hayashi and Spiegelman (1963) first showed that $\phi$X mRNA can hybridize only with the complementary strand of the replicative form duplex, so that the messenger sequences must be identical with the sequence of the mature viral single strand. The same situation prevails with the Ff phages (Marvin and Hohn, 1969).

Transcription of the single strand DNA phages does not appear to involve the temporal controls of gene expression that are seen with the larger duplex DNA phages. There is no evidence to suggest that any changes occur in the pattern of phage transcription during the infective cycle; it seems that the same messengers continue to be transcribed at all times. However, there are quantitative differences in the expression of the phage genes, since some proteins are found in much greater amounts than others in the infected cells. Formally it has not yet been shown to what extent this effect resides in preferential transcription versus translation of these genes, but it seems at least in part to reside in the transcription of more copies of the messages for the highly expressed genes.

Single strand DNA phages appear to be transcribed by the host RNA polymerase; none of the phage genes appears to influence the transcription process. Phage gene expression may therefore be defined by characterizing the promotors and terminators on the phage genome and determining the efficiencies with which they are utilized. For each type of single strand DNA genome, attempts have been made both to characterize the RNA products of transcription and to identify the sites in the genome where RNA polymerase initiates and terminates. Although this work is not yet complete, in both systems there appear to be several promotor sites but only one termination locus.

Most of our information about the transcription of the Ff phage genome is derived from studies of transcription and translation in vitro. However, the evidence available suggests that the in vitro systems may reflect with good accuracy the process of gene expression in vivo. In Ff infection much larger amounts are produced of the gene *V* and *VIII* products than of the other proteins. The products of these genes also are prominent amongst the proteins synthesized when M13 DNA is translated and transcribed in a coupled in vitro system, although the excess is not so great as found in vivo (Konings et al., 1973). This provides an indication that messengers and proteins are transcribed and translated in vitro to an extent reflecting that prevailing within the infected cell.

In the first volume of Gene Expression we have already discussed the transcription of fd DNA from the perspective of the mechanism of termination of transcription. Here we shall be concerned with the utilization of phage promotors and terminators. Takanami, Okamoto and Sugiura (1970, 1971) showed that synthesis of fd mRNA may start in vitro with either ATP or GTP. When fd DNA is transcribed by RNA polymerase alone, the ATP starter chain grows to a size of 26S, about that expected for a polynucleotide corresponding to the entire genome, while GTP starter chains are found at sizes of 17S, 13S and 10S. Takanami et al. suggested that all these messengers are terminated at a single site on fd DNA, their different sizes resulting from initiation at promotors located varying distances from the terminator. This termination occurs in the absence of rho. However, when rho was added to the in vitro transcription system, shorter transcripts of discrete sizes were produced: these appeared to result from termination sponsored by rho at five sites, with varying efficiencies as revealed by the effect upon each termination of increasing the ionic strength. As always, it is difficult to know whether rho is utilized in vivo or whether its action in vitro is an artifact of the conditions used for transcription; the RNA synthesized from fd DNA in vivo has not yet been characterized, so it is not possible to compare the in vitro products with those formed in the cell. However, recent work has concentrated upon identifying the promotors used to initiate fd transcription and the site of the single terminator recognized in the rho-independent system.

Promotors on Ff DNA have been identified in three series of experiments and their apparent locations are summarized in Figure 8.5. One approach is to identify the genes represented on each of the in vitro transcripts. Using the four RNAs transcribed from f1 DNA, Chan, Model and Zinder (1975) reported that the 26S RNA can act as template for synthesis of all the proteins translated from fd in vitro, while the three smaller RNAs each can direct the synthesis only of some of these proteins. The gene products translated from the messengers in these experiments were

| | |
|---|---|
| 26S RNA: | III, IV, II, V and VIII |
| 17S RNA: | IV, II, V and VIII |
| 13S RNA: | V and VIII |
| 10S RNA: | only VIII |

This hierarchy is consistent with the idea that there is a single terminator located at the end of gene *VIII* and that each RNA is initiated a different distance before the terminator. Since genes *III* and *VIII* are adjacent, the promotor for 26S RNA must lie at the start of gene *III* just after its junction with gene *VIII*; transcription must then proceed counterclockwise for almost a whole revolution of the genome. As shown on the innermost of thc three outer circles of Figure 8.5, the promotor for the 17S RNA must lie before the start of gene *IV*, the promotor for 13S RNA is placed within gene *II* (because this species is too large to represent only the sequences of genes *V-VIII*), and the promotor for 10S RNA must lie between gene *V* and gene *VIII*.

Another approach is to locate the starts of the RNA molecules on restriction fragments. Okamoto et al. (1975) used three enzymes to generate a series of sets of restriction fragments; using these fragments they determined which are able to initiate synthesis of RNAs and how far the RNA chains can grow before being prematurely terminated at the end of each fragment. The positions of the starting points on the genetic map of fd then could be assigned by reference to the cleavage site of Hind II. The four initiation sites identified in these experiments are shown in the central circle of the outer three rings of Figure 8.5. The only RNA whose synthesis terminated within a restriction fragment was the smallest and this identifies a terminator in the same region as that suggested for the other two series of experiments.

The locations of promotors can be defined also by determining which

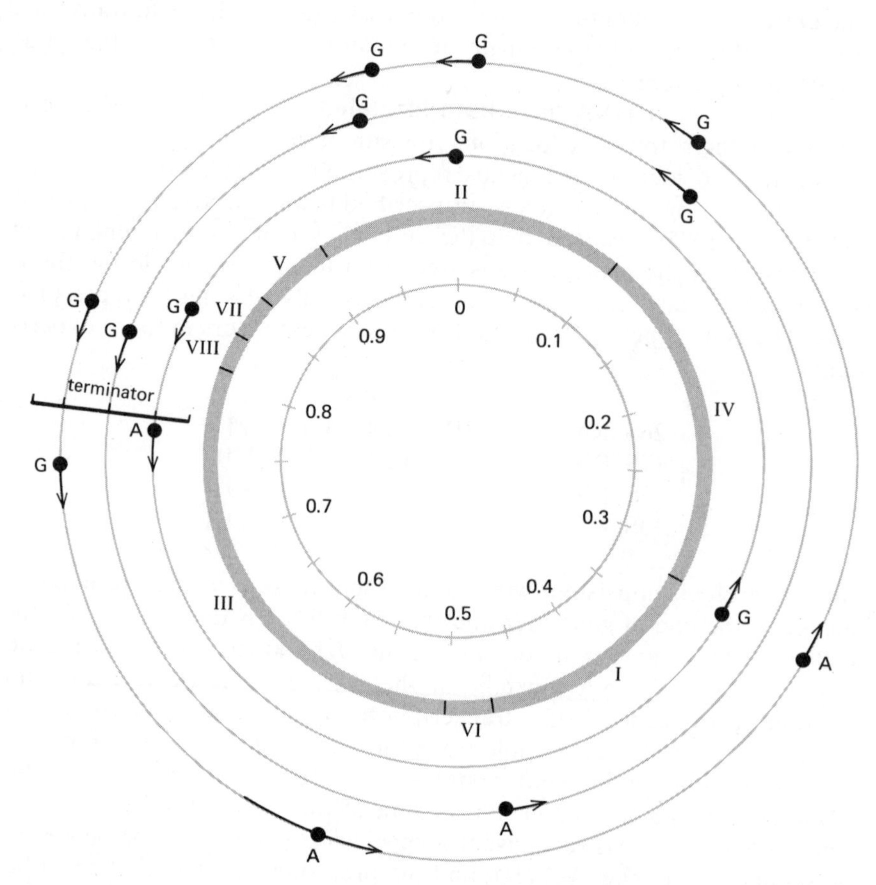


**Figure 8.5:** locations of promotor and terminator sites of Ff DNA. The two inner circles show the positions of the genes relative to fractions of the total map length (see Figure 8.4). The next three circles proceeding outward show the locations for transcription startpoints or RNA polymerase binding sites determined by Chan, Model and Zinder (1975), Okamoto et al. (1975) and Seeburg and Schaller (1975), respectively. In the first case locations are assigned on the basis of the proteins coded by each message and the size of the RNA; in the second two instances locations are determined relative to restriction fragment maps and have been transposed to the genetic map on the basis of fractional distance from the Hind II cleavage site that defines position 0.

restrictions fragments can form stable complexes with RNA polymerase, although this approach is susceptible to the problem that a promotor may be missed when more than one lies on the same restriction fragment or when a binding site is cleaved by the restriction enzyme. Seeburg and Schaller (1975) found that six of the Hpa II fragments of M13, fd or f1 DNA bind RNA polymerase, four strongly and two weakly as judged by the effect of increasing ionic strength. One of the fragments can be split into two with another restriction enzyme, both of the segments binding polymerase, so that there appears to be a total of seven binding sites. One of the seven binding sites shows a difference in affinity for polymerase in fd compared with M13 and f1 (the site at 0.97); the others all appear to be identical. The positions of these binding sites are shown on the outer circle of Figure 8.5. Termination appears to take place within the fragment Hpa-C, thus again lying in the region from 0.75–0.8.

Since these three sets of experiments were performed with different techniques, the reason for the discrepancies between the locations that they ascribe to the promotors is not clear. But all the data are consistent with (although of course they do not provide proof of) the idea that there is a single terminator for the transcription initiated at any of several promotors. If this terminator is efficient and does not allow any polymerases to read through the stop signal, transcription of gene *III* must demand the presence of a promotor very close to the terminator in the counterclockwise direction; an ATP starter locus was identified at this position in only one set of experiments. All the experiments appear to be consistent with the idea that a promotor lies a short distance before the terminator, probably to direct transcription only of gene *VIII*; another promotor may lie towards the end of gene *II*, causing transcription of a message for genes *V-VIII*. A large range of positions has been suggested for another promotor, between the terminator and gene *II* (in the counterclockwise sense). The general effect of transcription starting from several promotors and ending at a single terminator is to generate a larger number of message sequences representing the genes lying closest to the end of the transcription unit. These include genes *V* and *VIII* whose products are found in the largest amounts and this lends some confidence to the idea that this model may reflect biological reality. Of course, the excess of the gene *V* and *VIII* products compared with gene *III* (the first gene in the transcriptional unit) in vivo is much greater than the 3–4-fold effect that would be expected if all promotors were equally active; this might mean either that some promotors are stronger or that the same messenger sequence can be translated more efficiently in some contexts than others (or both).

The sequence of one of the fd binding sites for RNA polymerase has been determined by Sugimoto et al. (1975) and Schaller, Gray and Herrmann (1975). This corresponds to the promotor that lies towards the end of gene *II*. A sequence of just over 40 base pairs is protected by the polymerase against nuclease digestion and transcription starts with GTP in the middle of the protected region. Regions of twofold symmetry are present, but these are not homologous with any such organization in the other binding sites that have been sequenced. Homologies can be seen between this sequence and some of the other known polymerase binding sites; but their significance is not yet clear (see Figure 4.31).

When phage-specific messengers are isolated from cells infected with $\phi$X174, their sizes range from rather small to species of a length about that of the genome (Hayashi and Hayashi, 1970; Sedat and Sinsheimer, 1970). Confirming these earlier conclusions, Clements and Sinsheimer (1975) identified a total of some 18 $\phi$X RNA species, with molecular weights from 0.28–3.42 $\times$ $10^6$ daltons. Most of these messengers are too large to be the products of only single cistrons and must therefore be polycistronic. The total mass of these species is much larger than that of the genome and the messengers must therefore be the products of overlapping transcription. They appear to be comparatively stable, with half lives of at least 4–5 minutes; and apparently there are no changes in host RNA synthesis caused by $\phi$X infection.

In a series of studies on the in vitro transcripts of $\phi$X174 DNA, Smith and Sinsheimer (1976a,b,c) have shown that E.coli RNA polymerase appears to initiate at three sites but to terminate at only one site (using the $\phi$X template in the DNA duplex state). About 18 distinct RNA species are transcribed in vitro, and there is also heterogeneous material that does not appear to represent specific initiation and termination. The origins of the most prominent of the discrete transcripts were determined by using a method for 5′ proximal labeling of specifically initiated molecules. In this protocol, RNA polymerase is mixed with the $\phi$X DNA and nucleoside triphosphate precursors; and the reaction is allowed to proceed for 5 minutes. Then $^{32}$P labeled precursors are added for a 60 second pulse, after which the reaction is stopped by addition of EDTA and the transcription complex is isolated by gel filtration. The reaction is then restarted by addition of $Mg^{2+}$ ions and the transcription of each RNA chain proceeds to completion. The result of this procedure is to separate properly initiated transcripts from the short products of start-stop transcription and it yields RNAs which carry the radioactive label close to the 5′ end.

The properties of the most prominent discrete transcripts are summarized in Table 8.4. They range in size from 15% of the genome to up to three times the unit length. An interesting feature is that three groups of

**Table 8.4:** products of in vitro transcription of $\phi$X174 DNA

| component | molecular weight | unit lengths | initiation region | starting sequence | relative yield |
|---|---|---|---|---|---|
| A | $5 \times 10^6$ | 3 | — | — | 0.4 |
| B1 | $4.15 \times 10^6$ | 2.4 | — | — | 0.2 |
| C1 | $2.35 \times 10^6$ | 1.37 | Hin-4 – Hae-2 – Hpa-1 | pppApApAp(Ap)UpCpUp(Up)Gp | 1.0 |
| D1 | $0.63 \times 10^6$ | 0.37 | Hin-4 – Hae-2 – Hpa-1 | pppApApAp(Ap)UpCpUp(Up)Gp | 0.4 |
| B2 | $4.0 \times 10^6$ | 2.3 | — | — | 0.2 |
| C2 | $1.95 \times 10^6$ | 1.15 | Hin-8 – Hae-3 – Hpa-1 | pppApUpCpGp | 1.0 |
| D2 | $0.26 \times 10^6$ | 0.15 | Hin-8 – Hae-3 – Hpa-1 | pppApUpCpGp | 0.4 |
| B3 | $3.7 \times 10^6$ | 2.1 | — | — | 0.3 |
| C3 | $1.55 \times 10^6$ | 0.9 | Hin-6b – Hae-3 – Hpa-3 | pppGpApUp and pppCpGp | 1.7 |

These products comprise between 40–50% of the total 5′ proximally labeled material; the remaining molecules take the form of nine discrete species of weights 0.08–$3.1 \times 10^6$ daltons which are formed in much smaller amounts and a large number of heterogeneous molecules (not specifically initiated and terminated) in the weight range 2–$3 \times 10^6$ daltons. Only ATP and GTP starters are used in vivo and the second starter sequence found for C3 may therefore be an artifact of the in vitro system. Data of Smith and Sinsheimer (1976,b,c).

RNAs can be constructed in each of which the sizes of the molecules differ by one or two genome lengths. This raises the possibility that the members of each group may be initiated at the same site but that termination may be inefficient enough to allow some polymerases to pass the termination site once or twice before finishing synthesis. This implies, of course, that there is only one termination site on $\phi$X DNA.

The region in which each RNA starts can be determined by measuring the ability of the sequence carrying the 5′ proximal label to hybridize with restriction fragments of $\phi$X DNA. For these experiments the individual C and D RNAs were isolated from acrylamide-agarose slab gels; the B RNAs were recovered as a single fraction. As the table reports, the initiation region for each RNA can be localized to a single restriction fragment in each of the sets generated by the enzymes Hind II, Hae III and Hpa II. The C1 and D1 RNAs appear by this assay to have initiation sites in the same region; the C2 and D2 RNAs show a similar relationship; and the C3 RNA has another initiation site. The B RNAs bind with all these sets of fragments and so their initiation sites are not yet resolved. From the restriction map of Figure 8.3 it is apparent that the C1 and D1 RNAs must be initiated in the region of the *H–A* boundary, the C2 and D2 RNAs start somewhere near the *A–B* boundary, and the C3 RNA initiates near the *C–D* junction. These relationships are illustrated in Figure 8.6. From the sizes of the RNAs, it seems likely that there is a single terminator for transcription in the region of gene *C*. One implication of these results is that gene *C* is transcribed only by polymerases that read through the termination site.

By using a $\gamma$ $^{32}$P label, it is possible to determine the 5′ terminal sequences of the RNAs. The data available are summarized in Table 8.4. The demonstration that C1 and D1 have the same 5′ terminal sequence and that C2 and D2 also share a common starting sequence provides strong evidence in support of the idea that each pair of RNAs initiates at the same site and that they are related by additional rounds of transcription resulting from inefficient termination. Clearly it will be important to determine the initiation sites and starting sequences of the B molecules. On the assumption that groups 1 and 2 represent behavior of this sort, from the molar ratios in which the three members of each group are produced we can calculate that the polymerase has little more than a 10% chance of terminating the first time it passes the promotor, has about a 60% chance of terminating the second time, leaving a 30% chance of termination on the third occasion. The low frequency of termination on the first pass prompts the speculation that the length of the nascent RNA chain might be a relevant factor.

It is always difficult to know to what extent transcription in vitro

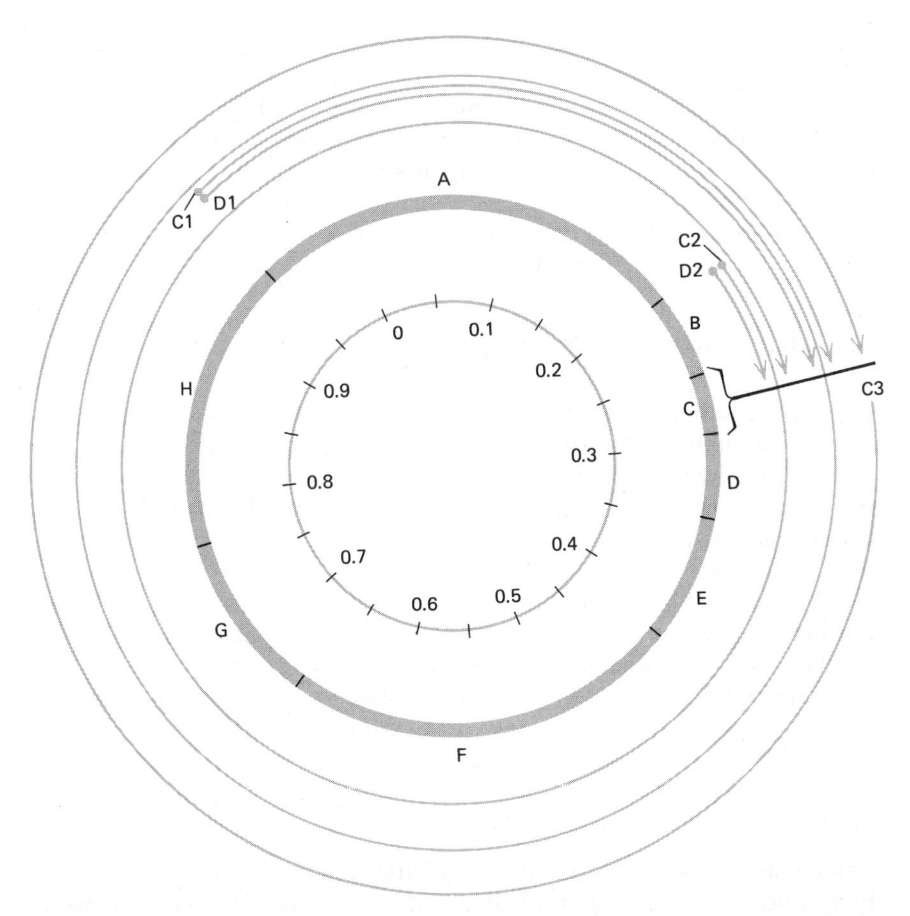


**Figure 8.6:** model for $\phi$X transcription in vitro. Three initiation sites are located in the regions of the *A-H* gene boundary, the *A-B* boundary, and the *C-D* boundary. A single termination site lies in the region of gene *C*. Termination at this site is inefficient and so some polymerases continue transcription past it for an additional cycle. Only those RNA species whose initiation points have been identified are shown on this map; but the B species (see Table 8.4) may represent initiations at the same three sites shown here but with transcription continuing for an additional round. Data of Smith and Sinsheimer (1976b,c).

represents events in vivo. Confirmation of the model shown in Figure 8.6 will therefore require the characterization of the in vivo messengers. The implication of the in vitro results is that transcription of $\phi$X174 may be accomplished by RNA polymerase alone, that is without any additional factors such as rho. This might be tested by comparing the transcription of $\phi$X174 (and of the Ff phages where the same conclusion might be inferred from the properties of the single rho-independent terminator) in wild type and rho mutant cells. Another approach to determining the exact locations of the promotors that are active in vitro is to characterize the fragments bound to RNA polymerase under stringent conditions; Chen, Hutchison and Edgell (1973) identified three promotors by this means, two of which may be the *1* and *3* promotors shown in Figure 8.6 but the third of which seems to be different.

## Reproduction of Single Strand DNA

The reproduction of single stranded phage genomes passes through three stages. The immediate problem faced by the viral strand on entering the cell is its single stranded state. This is overcome by synthesizing a complementary strand to generate a duplex circle of DNA: this is known as the *replicative form* (RF). The duplex molecule formed by synthesis of the complement to the infecting viral strand thus comprises the *parental RF*. The second phase of reproduction is accomplished by replication of the parental RF to generate *progeny RF* molecules. The third phase of synthesis sees the conversion of RF molecules into rolling circles in which the complementary strand serves as template for the continuous production of viral DNA. Unit genomes are cut from the single stranded lengths of viral DNA, acquire circular status, and enter the phage coat.

### *Conversion to the Replicative Form*

Immediately following infection, very little (if any) of the DNA of $\phi$X or Ff can be found in the mature single stranded condition. This change in state can be followed by the consequent alterations in buoyant density or sedimentation rate. With $\phi$X174 the single strand genome has a density of 1.725 g-cm$^{-3}$ and the duplex RF is distinguished by its density of 1.703 g-cm$^{-3}$ in neutral CsC1 (Denhardt and Sinsheimer, 1965b). Following alkaline denaturation of the duplex form, the infecting single strand and the complement to it synthesized after infection can be distinguished by their buoyant densities on alkaline CsC1. In neutral sucrose gradients the single strand infecting circle sediments at 24S; the duplex RF forms two peaks, at 21S and 16S (Burton and Sinsheimer, 1965; Roth and Hayashi, 1966). The 16S peak represents a duplex circle one strand of which is not covalently closed: this is known as RFII. The 21S peak is a covalently closed duplex: this is RFI. These structures can be confirmed by electron microscopy—when RFI is seen as a supercoil and RFII as an open circle—and the same forms of the RF are found following Ff infection

(Jaenisch, Hofschneider and Preuss, 1966; Brown and Dowell, 1968a,b). This suggests that in both types of infection the first step can be characterized as synthesis of a complement to the infecting strand to generate RFII, followed by closure of the newly synthesized strand to form the RFI structure.

Conversion of the infecting single strand to the replicative form relies entirely upon host functions: with both $\phi$X and Ff it is unimpeded when infection is performed in the presence of chloramphenicol to block synthesis of phage proteins (Burton and Sinsheimer, 1965; Pratt and Erdahl, 1968). Phage genes are needed for the subsequent stages of RF replication and single strand production; but no mutation in either type of phage genome prevents its conversion to the duplex state (Lindqvist and Sinsheimer, 1967b, 1968; Pratt and Erdahl, 1968).

Synthesis of the complementary strand of $\phi$X174 and M13 has been followed by a series of experiments in vitro which have led to the identification of the relevant host functions. Although the overall process by which each complementary strand is synthesized is very similar, the two phage systems show different requirements for host functions, $\phi$X demanding several gene products that are not needed for M13. The phage G4, related to $\phi$X174, demands many fewer host functions than $\phi$X.

The functions of E.coli that have been implicated in replication by mutation are summarized in Table 8.5. The *dna* series of mutations all have been isolated as temperature sensitive variants defective in replication at 42°C but uninhibited at 30°C. They fall into two classes, distinguished by their response to a temperature shift. Some mutants continue synthesis for 60–90 minutes after the temperature of incubation has been increased to the nonpermissive level; during this period the extent of DNA synthesis usually amounts to about 50% of the genome content. This appears to represent the completion of current rounds of replication without any initiation of new rounds. These mutants are therefore characterized as defective in the ability to initiate cycles of replication. The second class of mutation causes the cessation of DNA synthesis immediately upon a shift to nonpermissive temperature. These mutants appear to be defective in DNA chain elongation.

The *pol* mutations identify the three DNA polymerases of E.coli. Mutations in *polA* and *polB* do not prevent replication, implying that DNA polymerases I and II do not undertake the principal task of replicating DNA (although they may possibly be used in subsidiary roles). Mutants in *polC* were formerly identified by the *dnaE* chain elongation mutation: this locus codes for DNA polymerase III, which appears to be the enzyme responsible for synthesis of phosphodiester bonds in replication.

Bacterial replication is discontinuous, taking place by the synthesis of

**Table 8.5:** mutants in replication of E.coli

| locus | mutant phenotype | characteristics of protein | reference |
|---|---|---|---|
| *dnaA* | initiation defective;<br>altered membrane;<br>integratively suppressible | ? | Shapiro et al. (1970)<br>Shekman et al. (1972) |
| *dnaB* | elongation defective;<br>altered membrane;<br>some make shorter Okazaki fragments | ~250,000 dalton active protein 48,000 monomer;<br>~10 molecules per cell;<br>has DNA dependent ATPase activity | Siccardi et al. (1971)<br>Schekman et al. (1972)<br>Wright, Wickner and Hurwitz (1973)<br>Wickner, Wright and Hurwitz (1974)<br>Lark and Wechsler (1975) |
| *dnaC* | initiation defective;<br>integratively suppressible | 25,000 dalton protein | Wickner et al. (1973b)<br>Wechsler (1975) |
| *dnaG* | elongation defective;<br>cannot initiate Okazaki fragments | 64,000 dalton RNA polymerase | Wickner, Wright and Hurwitz (1973)<br>Bouche, Zechel and Kornberg (1975) |
| *dnaH* | initiation defective | ? | Sakai, Hasimoto and Komano (1974) |
| *dnaI* | initiation defective | ? | Beyersmann, Messer and Schlicht (1974) |

| | | | |
|---|---|---|---|
| *dnaP* | initiation defective | ? | Wada and Yura (1974) |
| *dnaS* | nonlethal;<br>accumulates fragments shorter than Okazaki segments | ? | Konrad and Lehman (1975) |
| *dnaZ* | elongation defective | 125,000 dalton protein needed for action of DNA polymerase III | Filip et al. (1974)<br>Wickner and Hurwitz (1976) |
| *polA* | nonlethal;<br>some defect in repair systems | DNA polymerase I | De Lucia and Cairns (1969) |
| *polB* | nonlethal | DNA polymerase II | Campbell et al. (1972)<br>Hirota et al. (1972) |
| *polC* | formerly *dnaE*;<br>elongation defective | DNA polymerase III<br>10–20 molecules per cell | Gefter et al. (1971) |
| *lig* | accumulates Okazaki fragments | polynucleotide ligase | Gottesman et al. (1973)<br>Konrad et al. (1973) |

short Okazaki fragments that are later covalently linked into continuous new chains; polynucleotide ligase is needed for the joining reaction. It is not yet clear what enzymes are responsible for removing the RNA primer sequence with which each fragment starts and for filling any gaps that may ensue.

A fuller discussion of the characterization of bacterial mutants affecting replication is given in Volume 1 of Gene Expression. However, the functions of the replication genes at the molecular level largely are unknown. But some of these functions are involved in the replication of single strand phages; and the use of in vitro systems for converting the viral strand to the replicative form has made it possible to identify and characterize their products.

Synthesis of the complementary strand on the circular viral template requires three processes: formation of a primer for DNA polymerase (no known DNA polymerase can initiate synthesis de novo: all rely upon the provision of a primer, as shown in Figure 6.17); elongation of a DNA chain from the primer; and closing of the gap between the terminus and origin of synthesis to form a covalently closed circular strand. The three systems that have been investigated in vitro, M13/fd, $\phi$X174 and G4, all appear to use the same host functions in the second two stages, but differ in the gene products involved in the initiation of the reaction.

Conversion of the single strand genomes of $\phi$X174 and M13 DNA into the duplex state was first accomplished by Goulian and Kornberg (1967) and Mitra et al. (1967), using DNA polymerase I to synthesize DNA and polynucleotide ligase to seal the break in the newly synthesized strand. This system included a boiled extract of E.coli, which subsequently proved to be providing a primer able to support initiation by DNA polymerase I. However, later work showed that these phages can grow on *polA*$^-$ *polB*$^-$ double mutant strains, so that neither DNA polymerase I nor DNA polymerase II is likely to be the enzyme responsible for their replication (Campbell, Soll and Richardson, 1972; Hirota, Gefter and Mindich, 1972).

Both $\phi$X174 and M13 DNA can be converted from single strand genome to duplex replicative form in vitro by extracts from E.coli; to prevent DNA polymerase I from undertaking this reaction, the extracts are prepared from cells carrying a *polA*$^-$ mutation. Which bacterial gene products are needed for synthesis of the complementary strand can be determined in two ways. The crude extract system may be purified to resolve its individual components by dividing it into fractions whose addition reconstitutes the active system. While this identifies components by their biochemical roles, it has the disadvantage that no information is gained about their involvement in bacterial and phage replication in vivo. Another approach is to prepare the extract system from cells carrying a *dna* mutation; if the extract is inactive, this gene product must be necessary for the reaction. In this case the protein can be purified by purifying from wild type cells the component that is able to restore activity to the mutant extract. While this allows the products of replication genes to be identified, it is restricted in that no information is gained about the roles of any proteins whose genes have not yet been identified by mutation. The two approaches are therefore complementary in permitting a complete resolution of the DNA synthetic system.

The enzyme responsible for synthesis of the complementary strand of DNA was identified as DNA polymerase III by the inability of extracts of *polC*$^-$ cells to replicate $\phi$X174, G4 or M13 DNA (Wickner et al., 1972b). This deficiency is overcome by addition of purified enzyme to the extracts. The form in which DNA polymerase III is active has not been completely resolved, but it seems clear that it is more complex than a single protein chain. In one laboratory, Wickner et al. (1973a) and Wickner and Kornberg (1973) characterized the active form of the enzyme as DNA polymerase III*, apparently a multimeric form of the DNA polymerase III protein which is converted irreversibly to DNA polymerase III by heating, and reported that its action requires the presence of a factor called copolymerase III*. Wickner and Kornberg (1974b) later reported the existence of the active enzyme in a form described as the DNA polymerase III holoenzyme, a complex containing subunits of DNA polymerase III* and copolymerase III*. In another laboratory Hurwitz and Wickner (1974) showed that DNA polymerase III itself is active when supplemented with two further proteins, elongation factors I and II. Wickner and Hurwitz (1976) subsequently fractionated elongation factor II into two components: the product of *dnaZ* and a further factor, III. Wickner and Hurwitz (1974) suggested that factor I might be equivalent to copolymerase III* and that the DNA polymerase III* activity represents an aggregate of DNA polymerase III with the proteins DnaZ and factor III. The situation is complicated also by reports of very different subunit weights for DNA polymerase III (see Wickner and Kornberg, 1974b; Livingston, Hinckle and Richardson, 1975). However, comparison of the components of the two laboratories' in vitro systems suggests that DNA polymerase III holoenzyme (that is DNA polymerase III* plus copolymerase III*) is functionally equivalent to DNA polymerase III plus DnaZ protein plus factors I and III. It is clear, therefore, that chain elongation results from the action of what can be regarded as a complex of proteins; and, indeed, it should not in principle be difficult to determine the biochemical equivalence of the components of the two preparations and to discover whether they may represent the products of any of the more recently isolated *dna* mutations that have not yet been investigated.

The extract systems can be fractionated into subsystems that divide the reaction into presynthetic and synthetic phases. Wickner and Hurwitz (1974, 1975a,b) identified a presynthetic reaction requiring the components of $\phi$X single strand DNA, DnaB and DnaC proteins, E.coli unwinding protein, three protein factors X, Y and Z, and ATP. Following the incubation together of these components, synthesis of the complementary strand occurs immediately upon addition of DnaG protein, DNA polymerase III, DnaZ protein and factors I and III. A similar resolution of the extract has been reported by Weiner, McMacken and Kornberg (1976), who found that the presynthetic step requires single stranded $\phi$X DNA, DnaB and DnaC proteins, unwinding protein, two protein factors i and n, and ATP. Gel filtration of this incubate then isolates an intermediate that can support the synthesis of complementary strand DNA when DnaG and DNA polymerase III holoenzyme are added. This suggests that the first step may represent the formation of a DNA-protein complex that renders the template ready for the synthetic reactions.

Although the presence of several components is necessary for formation of the presynthetic complex, they may not necessarily all be structural parts of the complex. Little is known about the roles of the DnaB and DnaC proteins, which are able to form an aggregate in the presence of ATP. The former has an ATPase activity that may or may not be pertinent (Wickner, Wright and Hurwitz, 1974; Wickner and Hurwitz, 1975a). It is interesting, however, that these two proteins are involved in the same stage of the reaction, since *dnaB* mutants are defective in elongation whereas *dnaC* mutants are defective in initiation. Nor is there any information yet available about the roles of the additional protein factors not yet equated with bacterial genes. We should mention that it is now clear that DnaA protein is not required in this reaction, although in some earlier experiments it appeared to be necessary (see Schekman, Weiner and Kornberg, 1974; Schekman et al., 1975).

Of the functions necessary for the presynthetic phase with $\phi$X174 DNA, only the unwinding protein is required for the reaction with M13 or G4 DNA (Schekman, Weiner and Kornberg, 1974; Schekman et al., 1975). Why the requirements of M13 and G4 are so much simpler than $\phi$X174 is not yet known; however, since G4 is related to $\phi$X174, it may be possible to use this difference to investigate the roles of the additional components needed for $\phi$X DNA synthesis. The host functions required for each stage of the complementary strand synthesis on M13, G4 or $\phi$X templates are summarized in Figure 8.7.

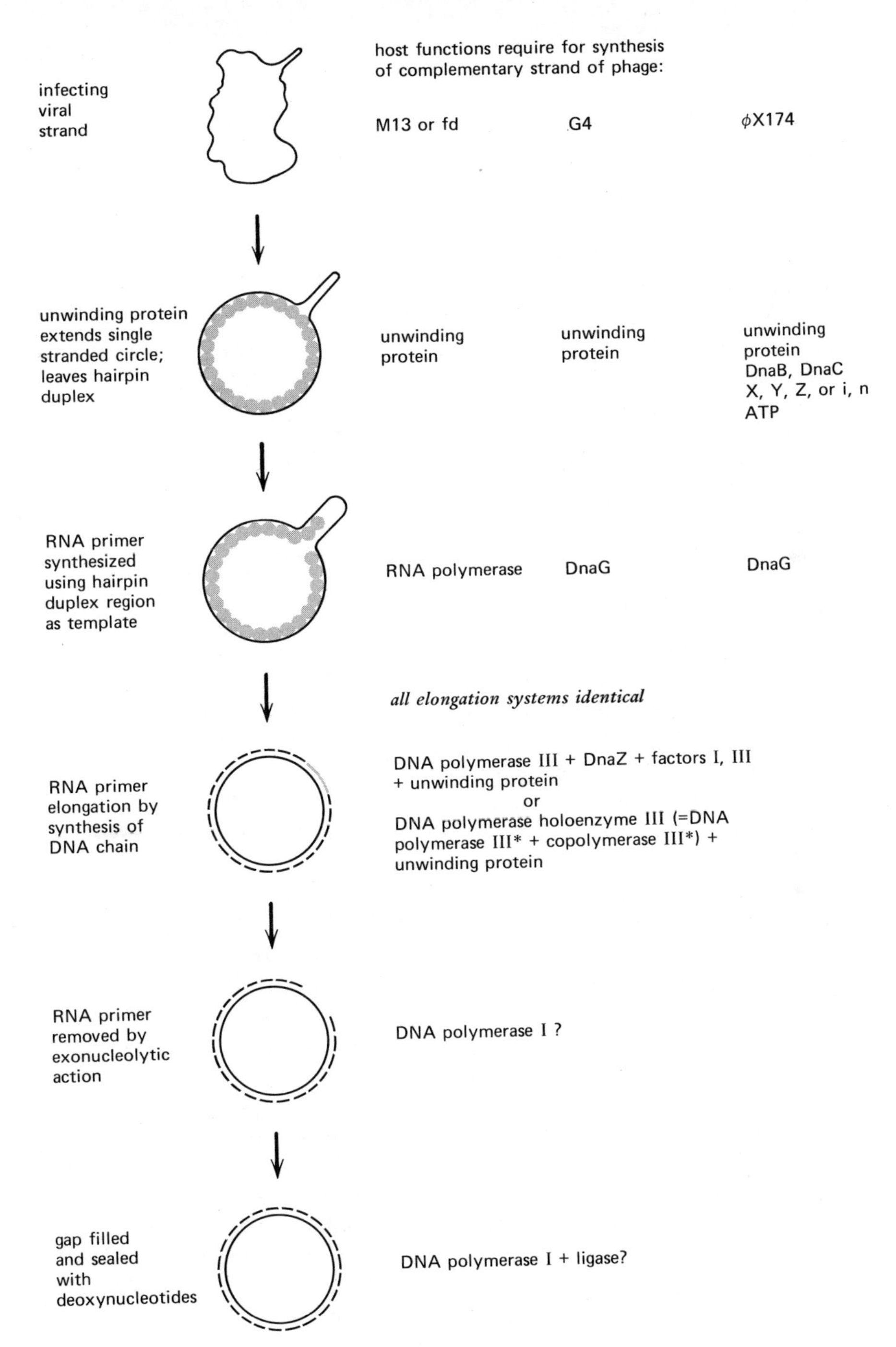


**Figure 8.7:** model for synthesis of the complementary strand of single stranded DNA phages, showing which host functions are needed in each system.

Unwinding protein is required both at the presynthetic stage and during DNA chain elongation. This protein was first isolated by Sigal et al. (1972): it has a size of some 22,000 daltons and there are about 800 copies per cell, corresponding to about 130 copies per replication fork. Its properties are analogous to those displayed by the 32-protein of T4 (see page 624 and the first volume of Gene Expression): it binds to single stranded DNA but does not bind to duplex DNA (Molineux and Gefter, 1975; Weiner, Bertsch and Kornberg, 1975). No mutant in its gene has yet been identified. We should note that it is therefore only the analogy with 32-protein that suggests a function for the E.coli unwinding protein in bacterial replication; and at present its involvement in the phage in vitro systems provides the best information about its role.

The need for unwinding protein as an essential component of the presynthetic reaction suggests that its role may be structural, to help establish the single stranded DNA in a state that makes it available as a template for complementary strand synthesis. The model for this function illustrated in Figure 8.7 supposes that synthesis of the complementary strand is initiated at a duplex region that is generated as a hairpin in the viral strand. By binding to the single stranded regions but not to the hairpin, the unwinding protein might permit DNA synthesis to be initiated only at an origin within the duplex region. The presence of hairpin structures in viral DNA was first suggested by observations that fractions corresponding to 1–2% of the $\phi$X or fd single stranded genomes appear resistant to digestion with exonuclease I (Fiers and Sinsheimer, 1962b; Schaller, Voss and Gucker, 1969). Hairpins of $\phi$X174 that Bartok, Harber and Denhardt (1975) isolated by their nuclease resistance proved to fall into two classes of 32 and 48 base pairs in length. This model predicts that a region containing such hairpin(s) should bind the protein responsible for initiating synthesis of the complementary strand (see below).

With templates of both M13 and $\phi$X174 DNA, synthesis of the complementary strand requires two types of synthetic reaction: synthesis of an RNA primer; and its extension by DNA polymerase into a DNA chain. However, different enzyme systems are involved in synthesis of the primer in the two phages. In the M13 system the involvement of the bacterial RNA polymerase that is responsible for transcription is indicated by the observation that rifampicin prevents synthesis of the replicative form in vivo when the host cell carries the $rif^s$ allele; when the host is $rif^r$ the drug has no effect (Brutlag, Schekman and Kornberg, 1971). Rifampicin also inhibits synthesis of the M13 complement in vitro. And using a system of purified enzymes, Geider and Kornberg (1974) confirmed the need for this RNA polymerase.

However, $\phi$X complementary strand synthesis is not inhibited by rifampicin either in vitro or in vivo. But the requirement for an RNA primer remains nonetheless: all four ribonucleoside triphosphates are required in the in vitro system as well as the deoxynucleotides. Schekman et al. (1972) directly identified the use of an RNA primer in the $\phi$X system by observing the transfer of a $^{32}$P label from a deoxynucleotide precursor to an rNMP residue released from the duplex molecule by alkaline digestion. This indicates that while the RNA primer of M13 is synthesized by the RNA polymerase resonsible for bacterial transcription, another RNA polymerase activity must be responsible for synthesizing the $\phi$X primer sequence. Phage G4 shares with $\phi$X174 the property that initiation is resistant to rifampicin. Synthesis of the M13 complementary strand requires the presence of E.coli RNA polymerase but does not require the *dnaG* product; the reverse demand is shown by $\phi$X and G4. Bouché, Zechel and Kornberg (1975) showed that in the presence of the unwinding protein, the DnaG protein can render G4 DNA able to synthesize its complement upon the addition of DNA polymerase III holoenzyme alone. The reaction can be divided into two steps: first G4 DNA is incubated with DnaG protein and all four ribonucleoside triphosphates; then the mixture is diluted and deoxynucleotide precursors are added together with DNA polymerase III holoenzyme. This shows that the first step provides an intermediate that is able to support synthesis by the DNA polymerase enzyme. When the RNA is labeled by use of a $^{32}$P precursor, degradation with ribonuclease T1 generates about 7 spots on a fingerprint, corresponding to a primer sequence of about 20–25 nucleotides, some half of which appear to be G residues. Thus the DnaG protein provides the priming RNA polymerase activity.

The product of initiation with DnaG and elongation with DNA polymerase III holoenzyme is an RFII structure in which the newly synthesized strand is not covalently sealed. If a single cut is made in this structure by a restriction enzyme, it generates a full length linear viral strand bound to a complementary strand which has a break at the origin. By denaturing this structure into single strands and identifying the fragment with a label incorporated into RNA, it is possible to locate the primer sequence relative to the site of restriction. Also it is of course possible to cleave the RFII with restriction enzymes that make many cuts and identify the fragment carrying the RNA label. Zechel, Bouché and Kornberg (1975) introduced a single cut into the G4 RFII structure with the enzyme Eco RI and denatured the product to release the two parts of the complementary strand. These were an almost full length fragment and a short fragment of about 5% of the length of the genome that carried the RNA label. This means that synthesis of the complementary strand is initiated at a unique site about 5% distant from the Eco RI cleavage site. A similar experiment with M13 led Tabak et al. (1974) to conclude that the RNA primer is located at a unique point close to the single site of cleavage by Hind II.

The RFII structure generated by synthesis of the strand complementary to the single stranded viral genome cannot be closed to form RFI by a ligase activity alone. This implies that a gap separates the terminus and origin of synthesis of the complementary strand. The nature of the events needed to fill the gap and close the newly synthesized strand have been investigated by Westergaard, Brutlag and Kornberg (1973) by making use of the differences in the properties of ligases and DNA polymerases of T4 and E.coli. The T4 ligase can link a DNA strand to RNA; the E.coli ligase can only link DNA strands. And T4 DNA polymerase cannot remove RNA from the duplex by $5' \rightarrow 3'$ exonucleolytic action whereas the E.coli DNA polymerase I is able to do so. The same results were obtained with both $\phi$X174 and M13. The combination of T4 DNA polymerase with T4 ligase generates an RFI structure: this is labile to alkali, showing that it contains the unexcised RNA primer. The buoyant density shift of the RFI containing RNA away from the position occupied by a wholly DNA duplex corresponds to an RNA content of about 1% of the DNA, that is of the order of 100 bases. When E.coli DNA polymerase is used to fill the gap, either T4 or E.coli ligase can seal the final break, generating a duplex RFI consisting only of DNA: the RNA must have been excised and replaced with DNA. The distance from the terminus of DNA synthesis to the origin must be fairly short, for Tabak et al. (1974) found that when the RFII structure is repaired with labeled nucleotides by DNA polymerase I, the label enters only one of the Hpa fragments and on treatment to generate pyrimidine tracts displays a low complexity. This identifies the gap as a unique and not very long site.

The steps involving the synthesis and removal of the RNA primer may be the same as those involved in initiating and connecting together the Okazaki fragments in bacterial replication. There is evidence that *dnaG* mutants are defective in the initiation of Okazaki fragments and this would support the concept that dnaG is an RNA polymerase responsible for the synthesis of a short primer sequence in bacterial replication. Why this function is needed for $\phi$X and G4 but the transcriptional polymerase is utilized by M13 and fd is not clear. The demonstration that E.coli DNA polymerase I can excise the RNA primer in vitro does not, of course, prove that this is its role in vivo, although this is certainly possible; however, the ability of bacterial cells to replicate in the absence of this enzyme, and the ability of the single strand phages to reproduce in *polA*$^-$ cells, implies that in this case there may also be some other protein able to undertake this function. Ligase is the obvious enzyme to complete the covalent integrity of the replicative form, but experiments with *lig* mutants of E.coli have not provided support for this conclusion (see Miller and Sinsheimer, 1974).

One approach to identifying the origin where synthesis of the complementary strand is initiated is to determine the location on the genome of the RNA primer. The same approach can be used to identify the terminus by filling in the gap between terminus and origin and identifying its location. We have seen that such experiments demonstrate that the origin and terminus lie in a unique and rather short part of the map. Another approach is to attempt to characterize directly the sequence of the viral strand that lies at the origin; this might be accomplished by binding the priming enzyme—RNA polymerase for M13 or fd and DnaG protein for $\phi$X or G4—to the circular single strand and then degrading sequences that are not protected by the enzyme. If the origin can be isolated in this way, it should be possible to determine directly whether its sequence is palindromic and supports the formation of a hairpin. An experiment to this end was performed by Schaller, Uhlmann and Geider (1976), who bound RNA polymerase to fd single stranded DNA complexed with the unwinding protein. The presence of the unwinding protein should ensure that only the proper initiation site is available to the polymerase. DNA not part of the initiation complex was attacked with DNAase I and Neurospora endonuclease and the fragments then degraded with exonuclease I; apparently the presence of unwinding protein does not impede the actions of these enzymes. The undigested DNA took the form of what appears to be a unique sequence of about 120 bases; addition of the four ribonucleoside triphosphates to the unwinding protein–genome–polymerase complex abolished the recovery of this sequence, implying that it may indeed represent the binding site at the origin. However, its base composition does not have the equality expected for a duplex of G=C, A=T, so that although it may contain a duplex region it cannot entirely represent a perfect hairpin. An indication that it contains duplex material is provided by its rather low rate of attack by S1 nuclease, which is highly active with single stranded DNA and not at all active with duplex DNA. The protected sequence binds to the Hpa-H fragment located in the region of the gene *II-IV* boundary (see Figure 8.4). It is interesting to note that in some of the experiments summarized in Figure 8.5 a promotor for transcription was identified in this region. Whether this is a coincidence remains to be seen.

Another interesting question is whether RNA polymerase is used to initiate replication in the same form in which it is active in transcription. In this context, Wickner and Kornberg (1974a) reported that the enzyme active in their in vitro system appeared to have been modified by the addition of a further component; however it is not clear whether this represents a necessary change or simply some artifact of the in vitro system. It will of course be interesting also to characterize the DnaG polymerase activity more fully, both for its activity in RNA synthesis and for its ability to bind a specific starting sequence at the origin of $\phi$X and G4 DNA; the possible relationship between the phage initiation signal and the initiation of Okazaki fragments also is an intriguing question.

### *Structure of Replicating Intermediates of $\phi$X DNA*

In both $\phi$X and Ff phage infection the parental replicative form is the progenitor for two types of molecule: first it is replicated to build up a pool of progeny replicative form molecules; and then positive single strands are synthesized on a rolling circle. The relationship between these two phases of synthesis is not entirely clear. One possibility suggested by earlier experiments was that the first rounds of replication to generate progeny RF molecules represent a conventional semiconservative process in which a growing point moves round the circular genome from origin to terminus with each strand of the duplex serving as template for synthesis of its complement. An alternative model suggested in more recent work is that all replication of the parental RF proceeds via the rolling circle mechanism (see Figure 8.8). Whether the rolling circle generates duplex or single stranded progeny then depends on whether a complement is synthesized for the displaced strand; this may be controlled by the presence of phage gene products. While the use of the rolling circle for single strand production can be established unequivocally, it is more difficult to distinguish the predictions made by the two models for the earlier phase of RF replication; many of the data are consistent with either model. Although phages $\phi$X and Ff place a common reliance on the rolling circles at least in the final stage of single strand production, the nature of the events that control the transition from RF replication to single strand synthesis may be different in the two types of infection.

The one step growth curve of $\phi$X174 is similar to that of other phages. The infective cycle lasts about 20 minutes at 37°C. The first 10 minutes is occupied by an eclipse period during which no intact phage particles are found in the cell. Synthesis of mature particles commences at the end of the eclipse and then continues at a constant rate until lysis. Following conversion of the infecting single strand to the replicative form, it is impossible to detect single strands at any stage. This implies that progeny single strands must be packaged into phage particles as soon as they are synthesized (Sinsheimer et al., 1962; Denhardt and Sinsheimer, 1965a,b).

Infection with $\phi X$ causes host synthesis of DNA to cease by about 12–14 minutes, but bacterial RNA and protein synthesis do not seem to be affected (Lindqvist and Sinsheimer, 1967a).

The course of DNA synthesis following Ff infection is different from that displayed by the lytic phages. After a latent period there is a rapid production of virus, but particles continue to be released until the host bacteria enter the stationary phase (Hofschneider and Preuss, 1963; Hoffmann-Berling and Maze, 1964; Marvin and Hohn, 1969). Infection with Ff does not cause any change in host replication. Early in infection a pool of RF molecules builds up and then replication turns to the production of single strands, which pass through a pool of viral strands before using incorporated into mature particles. The phage DNA pool during Ff infection contains about 200 genomes (Pratt and Erdahl, 1968; Ray, 1969; Hohn, Lechner and Marvin, 1971). Unlike $\phi X$ infection, when the phases of RF replication and single strand production appear to be distinct, in Ff infection some production of single strands can be detected even at the beginning of the RF replication phase, although it becomes predominant only later (Forsheit, Ray and Lica, 1971). Replicating Ff DNA is associated with the membrane throughout both phases (Forsheit and Ray, 1971).

The role of the original infecting strand has been a matter of some confusion. The early experiments of Kozinski (1961) implied that the infecting strand does not enter $\phi X$ progeny phage. This means that it must in some way be distinguished from the copies of it that are synthesized during infection. Knippers et al. (1969a) later found that at high multiplicities of infection parental strands can be transferred to the progeny. This suggests that a limited number of molecules, which includes all the parental infecting strands at low multiplicities of infection, is distinguished from the progeny RF; at higher multiplicities of infection some of the parental strands behave instead in the same way as the progeny RF. However, Iwaya and Denhardt (1973b) were able to detect transfer of the parental strand into progeny, even at low multiplicities of infection, but they found that to observe this transfer it is necessary to prolong the infective cycle. This may indicate that at the usual time of lysis the parental RF has not yet generated single strands, but is able to do so later.

Another distinction between the parental and progeny strands was suggested by the report of Denhardt and Sinsheimer (1965c) that when the infecting phage carries a $^{32}P$ label, the generation of progeny phage remains susceptible to inactivation by $^{32}P$ decay even at late times when the cells contain progeny RF molecules. (This experiment is performed by incubating cells to a given stage of infection and then storing them at low temperature to allow time for radioactive decay before the conditions of infective incubation are restored.) The formal implication of this result is that there is some essential late function that can be exercised only by the molecule containing the original infecting strand. This was interpreted to mean that only the parental RF could give rise to progeny single strands.

When reproduction of phage DNA proceeds by semiconservative replication to generate two daughter chromosomes from each duplex template in each generation, the clonal distribution of mutants is exponential. Phage T4 is an example (see page 634). Denhardt and Silver (1966) found that with $\phi$X the distribution is instead random, as would be expected if a master template(s) "stamps out" progeny genomes. In this case the frequency with which a mutation is represented amongst the progeny of a single burst must depend upon time in a linear rather than exponential manner.

Investigating $\phi$X replication by labeling with tritiated thymidine after infection suffers from the difficulty that host replication continues for some time. Lindqvist and Sinsheimer (1967b) therefore introduced the technique of suppressing host replication by treatment with mitomycin C before infection. They also made use of the *am3* mutation in gene *E*, which delays lysis but appears to have no other effect on the phage cycle. This therefore allows infection to continue for a longer period. Synthesis of DNA can be restricted to the phase of RF replication either by the use of appropriate phage mutants (see below) or by addition of chloramphenicol at a concentration of 30 $\mu$g/ml, a level of the drug that allows RF replication but prevents the initiation of single strand synthesis (Tessman, 1966a; Levine and Sinsheimer, 1969a). A higher level of chloramphenicol, 100 $\mu$g/ml, prevents single strand synthesis as well as RF replication. The use of the low level of chloramphenicol rests on the assumption that the RF replication which it permits takes place by the same mechanism usually responsible for RF replication.

The production of $\phi$X replicative form molecules consisting of strands synthesized after infection can be demonstrated by infecting cells with phage whose DNA carries a density label. Using BUdR labeled $\phi$X DNA, Rust and Sinsheimer (1967) demonstrated that both hybrid density and light density RF molecules are found after infection. Formally this demonstrates that the complementary strand as well as the infecting strand of the parental RF is able to act as a template for DNA synthesis. Knippers and Sinsheimer (1968) showed that a labeled parental phage strand enters a membrane fraction and remains there in the form of RFI and RFII molecules. An $^3$H-thymine pulse label given during the phase of RF replication enters the membrane fraction and then is chased out of it. Knippers, Komano and Sinsheimer (1968) showed that both the positive and negative strands of the progeny RF are labeled. Sinsheimer, Knippers and Komano (1968) used density shift experiments to show that only the membrane bound RF containing the parental strand is able to reproduce: the progeny RF are released to the cytoplasm and do not replicate. When cells growing in $^{3}$H-thymine labeled medium are infected with $^{32}$P labeled phage, the parental label enters an RF of hybrid density also carrying the $^{3}$H label; but the progeny RF molecules, identified by their possession of only the $^{3}$H label, are found at hybrid density. When the cells then are transferred to light unlabeled medium, the $^{3}$H label does not enter the hybrid density fraction but remains at the heavy position. These results are consistent with the observation of Yarus and Sinsheimer (1967) that when host cells grown under poor conditions are infected simultaneously with phages of different genotypes, usually only one of the parental genotypes is represented in the progeny. They suggested that there might be a limited number of sites in the cell (one under the conditions of poor growth used in these experiments) to which the parental RF must attach. Based on these data, Knippers et al. (1969a) suggested a model in which the parental RF replicates at the membrane, in each cycle releasing into the cytoplasm the duplicate made on the negative strand. Failure of the cytoplasmic progeny RF molecules to replicate then explains the observation that phage DNA increases in a linear instead of exponential manner (Denhardt and Sinsheimer, 1965b). Then in the subsequent phase of single strand production it is the cytoplasmic progeny RF molecules that are used for single strand synthesis.

Results suggesting a different conclusion, however, have been obtained by Iwaya and Denhardt (1973a). In these experiments cells were treated with a low level of chloramphenicol and infected with density labeled phage; then they were pulse labeled with $^{3}$H-thymidine at varying times after infection. At 5 minutes after infection the pulse label enters hybrid density; this accords with the replication of parental material. But at 15–35 minutes the label exclusively enters molecules at the light density position; this implies that the parental strand is not replicating. This experiment therefore suggests that it is the progeny RF that can replicate. The reason for the discrepancy between the two sets of experiments is not clear.

During RF replication the positive strand appears to be synthesized continuously while the negative strand is extended by discontinuous synthesis of Okazaki fragments that are later joined together. Eisenberg and Denhardt (1974) observed that when RFII molecules labeled with $^{3}$H-thymidine are subjected to a repair reaction with T4 DNA polymerase in the presence of dCT$^{32}$P, all the restriction fragments generated with H.influenzae endonuclease R contain $^{32}$P as well as $^{3}$H; this implies that they all contain gaps. McFadden and Denhardt (1975) showed that denaturation of the RF formed under *lig*$^-$ conditions releases 5-9S fragments from the negative strand, while the positive strands are of unit or longer length. Using wild type host cells, Fukuda and Sinsheimer (1976a) also characterized replicating intermediates whose positive strands are intact but with newly synthesized negative strands consisting of small pieces. By characterizing the pyrimidine oligonucleotide tracts of labeled material incorporated by repair of the gaps in the complementary strand, Eisenberg et al. (1975) showed that these sequences are less complex than those of the total genome. This suggests that the complementary strand is synthesized discontinuously and that this process generates gaps in it at particular positions that later must be filled. The newly synthesized positive strand, by contrast, appears to contain only a single gap, in the Hind-3 fragment in gene *A*. This appears to identify the origin for replication. Discontinuous synthesis of the negative strand could be accomplished by either of two models: a growing point moving round the circular RF might allow the positive strand to grow continuously while the negative strand grows discontinuously; or the (continuously synthesized) tail of a rolling circle might be a positive strand that is used as template for discontinuous synthesis of the negative strand.

The discontinuous synthesis of the negative strand during the phase of RF replication poses the question of what mechanism is used for synthesis of the complementary strand when the infecting single strand is converted to the parental RF. Two models are possible. If the first negative strand is synthesized in the same direction as the later negative strands, its synthesis also must be discontinuous. This would demand the existence of several points at which RNA primers initiate DNA synthesis, in contrast with the observed location in G4 phage DNA of a single such site. However, if the complementary strand is synthesized from a single starting point, the chain must grow in the opposite direction from that seen during duplex replication.

By the late period of single strand production, an $^3$H-thymidine label enters only growing positive strands; the synthesis of negative strands that was detectable earlier now has ceased. Komano, Knippers and Sinsheimer (1968) showed that the RFII labeled during the final phase has a closed circular negative strand and an open positive strand of unit length. Lindqvist and Sinsheimer (1968) demonstrated that single strand synthesis is initiated on RFII structures. In these experiments structures containing single strands were isolated by the use of naphthoylated benzoylated cellulose, which fractionates DNA on the basis of its single strand content. In a pulse chase experiment, a label first enters RFII and then appears in single strands (whereas at early times the label enters RFII and then is converted to RFI).

The replicative intermediates during the late phase can be identified by following the fate of a pulse label. By this means Knippers et al. (1969b) identified two fractions: the sharp 16S peak of the RFII; and a broad peak sedimenting in the range 18S-22S. Shorter pulses are found predominantly in the heterogeneous peak, suggesting that this contains the replicating intermediates. On benzoylated naphthoylated cellulose, the 18S-22S fraction behaves as predominantly duplex with some single strand material. When cells are labeled with $^{14}$C-thymidine at 2 minutes after infection and then are given an $^{3}$H label during the late period of single strand production, alkaline denaturation of the 18S-22S peak releases the $^{14}$C label in the form of circular negative strands, while the $^{3}$H label is found largely in positive strands, some of unit length but some longer, generally up to about two unit lengths. Only some of the RF molecules generated during the second phase of replication appear to be utilized for single strand production, as judged by thc fatc of a label given at the end of the eclipse period. Most of the RF molecules produced during the eclipse period are recovered in the form of RF structures. This suggests a model in which the second phase of replication generates a pool of RF molecules, usually some 20–40 in number and mostly in the closed supercoiled form; and then the final phase of replication is initiated by the introduction of a nick in the positive (viral) strand of the RFI, which generates a 3′-OH end that is used as a primer for strand elongation according to the rolling circle model. The absence of positive strands much longer than two unit lengths when these intermediates are denatured with alkali implies that as soon as a full length single viral strand has been displaced, it is cut off, formed into a circle, and packaged. The sequence of events in $\phi$X replication is illustrated in Figure 8.8.

The first suggestion that production of $\phi$X single strands might be accomplished by a rolling circle in fact was made by Gilbert and Dressler (1968) in their original proposal of the model. In support of this idea they adduced evidence that a label enters positive strands that are longer than the unit genome. Dressler (1970) showed that a pulse of $^{3}$H-thymine given during the period of single strand synthesis enters a fraction sedimenting heterogeneously, in these experiments in the range 16–30S. When intermediates of different sizes were denatured by alkali and the strands separated by buoyant density, the faster sedimenting intermediates proved to have the longer positive strands, again up to about two unit genomes in lengths (as we have noted this corresponds to the strand on the rolling circle, with a displaced tail corresponding to no more than a single genome in length). Only the positive strand could be detected by the incorporation of a radioactive label during the period of single strand synthesis, implying that the synthesis of negative strands has ceased.

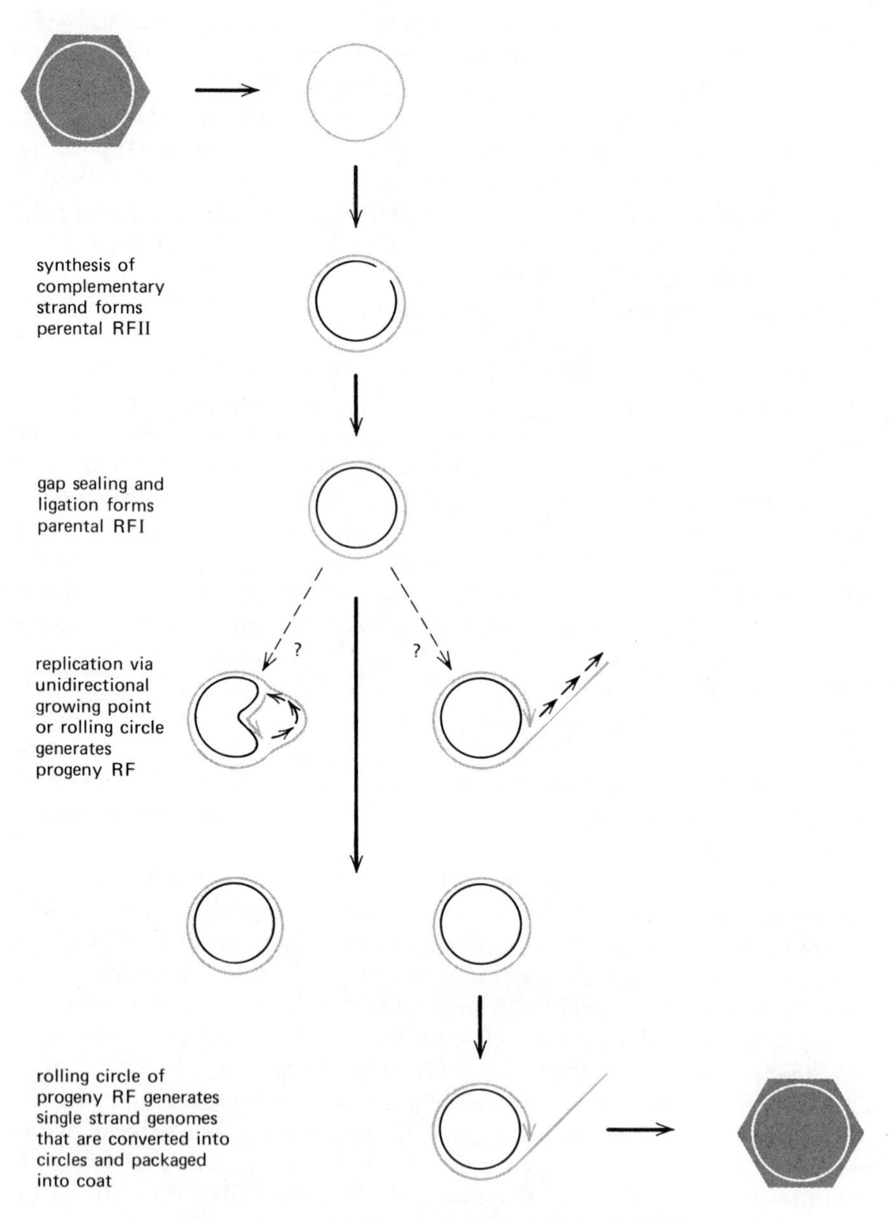


**Figure 8.8:** model for the replication cycle of $\phi$X174.

However, the negative strand could be detected by an infectivity assay and proved to be a closed circle. These rolling circles were visualized by electron microscopy and an example is shown in Figure 8.9. (We should note that at one point an alternative model for the rolling circle was proposed: Knippers, Whalley and Sinsheimer (1969) reported that it is the positive strand that remains circular and the negative strand that is synthesized by displacement from the circle; this would imply that the mature positive strands must be synthesized on the displaced negative tails. However, subsequent experiments have confirmed the proposal of the original model that it is the positive strand that is synthesized on the circular negative template; see for example Schroder and Kaerner, 1972; Baas and Jansz, 1972a).

One of the features of the rolling circle model that remains to be worked out is the reaction responsible for generating a circle from the linear single strand tail. There have been various suggestions that linear strands of $\phi$X DNA have an intrinsic ability to form circles, perhaps because of the presence of a palindromic sequence able to support formation of a hairpin (see Schekman and Ray, 1971; Iwaya et al., 1973). However, the significance of these observations is not clear and all that can be said at present is that the nature of the ring closing reaction and the functions that are responsible for accomplishing it are unknown (see Miller and Sinsheimer, 1974).

Is the rolling circle used for RF replication as well as for single strand production? A suggestion that this might be the case was made by Dressler and Wolfson (1970) on the basis that a positive strand label follows a negative strand label into RFI supercoils with a delay of some 20 seconds. This could be explained if the supercoil is derived from the tail of a rolling circle rather than from semiconservative replication around the circle. Asymmetry in labeling of the RF also was noted by Fukuda and Sinsheimer (1976a), who found that the first RFI molecules to be labeled by a pulse carry it in the negative strand; by contrast, the first RFII molecules to be pulse labeled display a radioactive positive strand. This demonstrates that during RF replication the two newly synthesized strands are not equivalent: one daughter RF contains an intact newly synthesized positive strand while the other contains discontinuous fragments of the newly made negative strand. The appearance of positive labeled strands first in the RFII can be explained by supposing that the negative strand is synthesized only after the positive strand because it grows overall in the direction opposed to synthesis of the fragments; and this may depend upon either exposure of the template (for semiconservative circular replication) or displacement of the positive tail (for rolling circle replication). The first appearance of newly synthesized negative material in RFI struc-

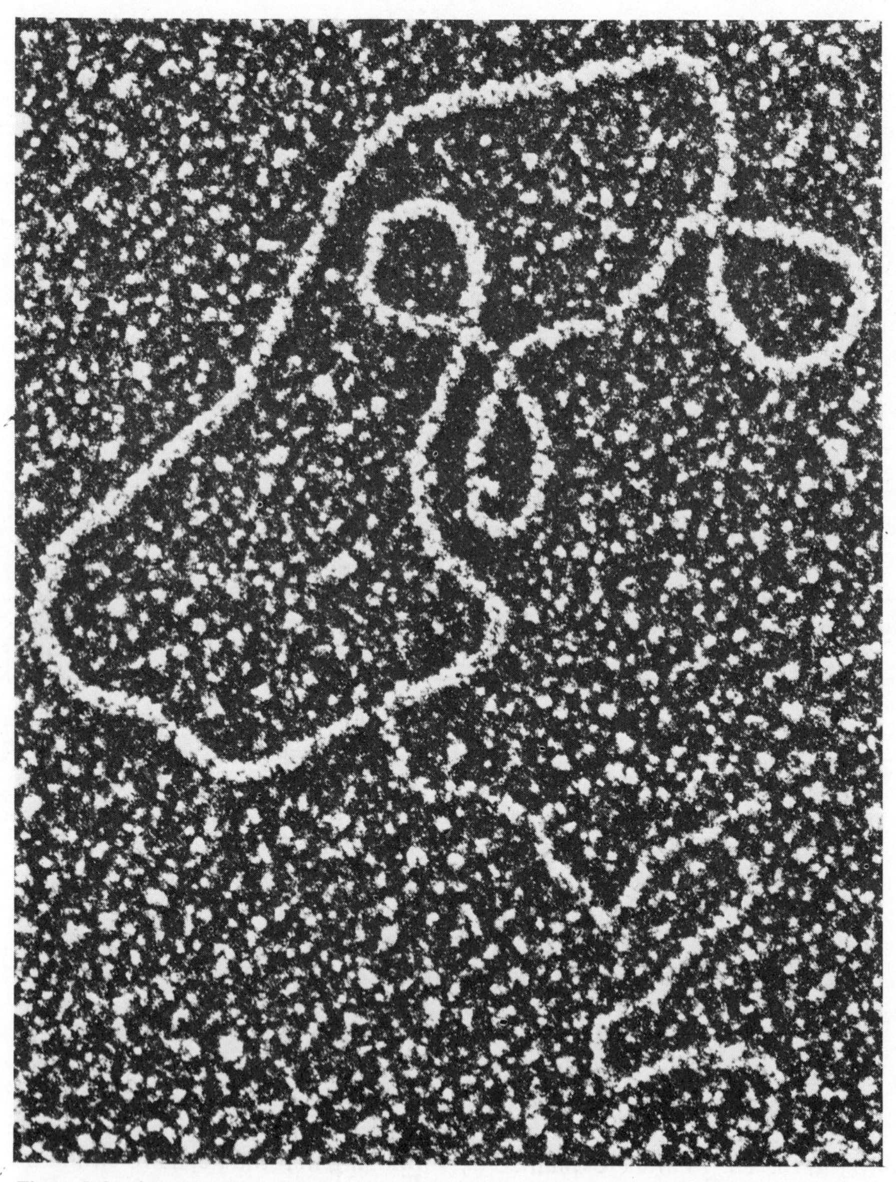

**Figure 8.9:** electron micrograph of rolling circle of $\phi$X174, kindly provided by D. Dressler.

tures could be explained if it is only the displaced tail (and its complement) of the rolling circle that gives rise to RFI, while the circle itself continues to revolve. This would mean that the progeny RFI molecules become labeled in the positive strand only one revolution after they become labeled in the negative strand. On the other hand, it is possible also to postulate that the two strands of the parental RF are not equivalent, for example because one of them is attached to the membrane and is held open; this would allow the progeny molecules to be distinguished. At present it therefore remains true that no definitive statement can be made about the nature of the replicating intermediates involved in the phase of RF replication.

Where is the origin of replication? The same origin appears to be used for both replication of the RF and production of single strands and it may be coincident also with the origin for formation of the RF from the infecting single strand. Godson (1974a) used a protocol analogous to that used to investigate the direction of protein synthesis by Dintzis (1961): after pulse labeling for different periods of time, *completed* molecules are examined. The terminus should then be the first region to display a label, and with increasing periods of labeling the radioactivity is found closer to the origin, in fact displaying a gradient from terminus to origin (the application of this technique to protein synthesis is discussed in Volume 1 of Gene Expression). For $\phi$X replication the gradient can be determined by restriction mapping; the closer a fragment lies to the terminus, the greater should be its label. Which strands are labeled can be determined by alkaline denaturation: early in the phase of RF replication, both positive and negative strands were labeled; during the phase of single strand production only the positive strands were labeled. But in both cases the label fell in a gradient which corresponds to the location of origin and terminus in gene *A*, with replication proceeding clockwise.

Another approach to defining the origin is to determine the location of the interruption in the RFII structure that represents the predecessor or product of replication. We have seen already that the newly synthesized negative strand has many gaps but that the new positive strand has a single gap in the region of gene *A* (Eisenberg et al., 1975). When Johnson and Sinsheimer (1974) isolated RFII molecules during the phase of single strand production, they found that about a third of the molecules could be covalently closed on treatment with ligase, while two-thirds required DNA polymerase I as well. Similar experiments were reported previously by Schekman et al. (1971); although the proportions that they observed for the gapped and nicked molecules were more or less reversed, they found a large increase in the proportion of gapped molecules generated by infection of *polA*$^-$ strains, an indication that DNA polymerase I may be involved in the repair reaction in vivo. Johnson and Sinsheimer found that when repair of these intermediates is performed under conditions that permit single strand displacement, the incorporation of a label into different restriction fragments can be followed as the polymerase progresses round the circle. The reaction starts in the region of gene *A* and proceeds clockwise.

An earlier technique used to locate the origin by less direct means was developed by Baas and Jansz (1972b). When heterozygous RF molecules (prepared by annealing positive and negative strands of different genotypes) are used to infect spheroplasts, competition between repair and replication determines whether the phage burst is homozygous or heterozygous. If the heterozygosity is corrected before replication, only a single phage type is seen in the burst; but if replication segregates the two strands before correction can take place, both genotypes are perpetuated to generate a heterozygous burst. A plot of the frequency of mixed bursts against map position shows a gradient; this suggests that the likelihood replication occurs before correction depends upon distance from the origin. The minimum heterozygous burst frequencies are displayed by mutants in gene *A*, and the frequency increases proceeding clockwise. By applying this technique to a series of mutations in gene *A*, Weisbeek and Van Arkel (1976) located the origin more precisely, within Hind-5, which means that it must be within about 200 nucleotides of the *A-B* boundary (see Figure 8.3). However, the gap identified in RFII molecules by Johnson and Sinsheimer (1974) and Eisenberg et al. (1975), and the terminus identified by Godson (1974a) lies close to the *A-H* boundary, in the Hae-6b fragment and close to the Hind-3–Hind-4 boundary. The reason for this discrepancy is not clear. However, one possibility is that some structural abnormality connected with the initiation of replication is located in the region of gene *A* and influences the results. In this context it is interesting that mutants in gene *A* are located in groups towards the left and the right ends of the gene, but have not been identified in the central region; whether this is related to the location of the origin is at present a matter for possible speculation.

### *Roles of Phage and Host Functions in φX Replication*

Phage genes are needed both for RF replication and for single strand production in φX infection. Lindqvist and Sinsheimer (1967b) and Fukuda and Sinsheimer (1976b) reported that mutants in genes *E* and *H* are able to replicate normally, duplicating the RF and generating single strands. Mutants in genes *B*, *C*, *D*, *F* and *G* replicate the RF but do not proceed to the stage of single strand production. Since *F* and *G* code for structural proteins of the phage particle (see Table 8.2) this implies the existence of the same relationship described for other phages: coat protein is required for synthesis of mature genomes. Mutants in gene *A* do not proceed past the stage of parental RF formation: all subsequent replication is blocked. We have seen already that the effects of the phage mutations can be mimicked by addition of appropriate concentrations of chloramphenicol: a level of 30 $\mu$g/ml prevents single strand synthesis but permits RF replication, while a level of 100 $\mu$g/ml prevents any progress from the parental RF (Tessman, 1966a; Levine and Sinsheimer, 1969a). The implication is that at 30 $\mu$g/ml the drug prevents synthesis of some or all of the products of genes *B*, *C*, *D*, *F* and *G* but does not prevent expression of gene *A*; at the higher level of 100 $\mu$g/ml synthesis of the gene *A* product also must be inhibited. The basis for this differential effect on gene *A* expression is not known.

The *A* protein is responsible for initiating replication of the RF. Francke and Ray (1971) found that in cells infected with $A^-$ mutants the replicative form accumulates in the closed supercoil of the RFI state. The *A* protein has been purified from $\phi$X-infected cells by Henry and Knippers (1974). In the form of the full size product (see Table 8.2) there appear to be some 3000 copies per cell. The protein has an endonuclease activity that converts RFI to RFII by making a single strand scission in the viral strand. The single stranded genome also can be nicked by the enzyme, raising the possibility that even in the single stranded condition there may be a duplex region, such as a hairpin, that can be recognized by the enzyme. Although the action of the endonuclease has not yet been fully characterized with regard to specificity and site of action, it seems almost certain that it acts to introduce a nick at the origin.

A striking and unusual property of the *A* gene is its cis dominance, usually a characteristic displayed only by elements in DNA that do not code for protein but are recognized by proteins. Tessman (1966a) and Levine and Sinsheimer (1969b) found that when $A^-$ mutants of S13 or $\phi$X174 are used in mixed infection together with wild type phage, very few of the progeny have the $A^-$ genotype. In other words, trans complementation does not occur: provision of a wild type *A* gene in the $S13^+$ genome does not support replication of the $A^-$ genome in the same cell. Does this mean that the *A* protein is able to act only on the genome specifying it? The large number of copies of the protein means that the basis of this effect cannot reside in some quantitative restriction, such as limits on the rate of diffusion of the protein from sites of synthesis to genomes elsewhere in the cell. If the effect is due to lack of protein activity, this means that activity is closely related to synthesis, and this must imply further that gene *A* is transcribed and translated on all replicating genomes.

An alternative is that the mutations have another effect, for example on recognition of the origin, as well as on the *A* function.

A host gene necessary for replication of the RF was identified by Denhardt, Dressler and Hathaway (1967); *rep*$^-$ mutant cells allow $\phi$X to form the parental RF but phage reproduction proceeds no further. The *rep* mutation has some comparatively minor effects on replication of the E.coli chromosome: Lane and Denhardt (1974, 1975) reported that the size and DNA content of the cell both increase, with the content of DNA per cell becoming greater; the number of replicating forks per chromosome increases and all these effects can be explained by supposing that the replicating forks move more slowly than usual, apparently at 50–60% of the usual rate.

The *rep* mutation offers an opportunity to study the early events in $\phi$X replication without the complication of the ensuing events. Francke and Ray (1971, 1972) found that when a *rep* mutant is infected with $\phi$X $A^-$ phage, only RFI is formed, whereas with $\phi$X $A^+$ phage some 1–4 RFII molecules are formed per cell. This number is independent of the multiplicity of infection; and this suggests that some cellular component may be present in limiting amounts that control the ability of parental RF to replicate. This accords with the earlier conclusions of Yarus and Sinsheimer (1967) that only some parental RF molecules give rise to progeny (and which are discussed in the previous section). In addition to the RFII molecules, which have open positive strands and closed circular negative strands, some rolling circles are formed in $A^+$ infection of *rep* cells; these may have positive strands a little longer than the genome length. Bowman and Ray (1975) found that under these conditions the positive strand of replicative form molecules is slowly degraded, with little of a parental label remaining after 47 minutes.

A more restricted degradation of the positive strand of the replicative form structure generated in *rep* cells has been observed by Baas, Jansz and Sinsheimer (1976). If DNA synthesis in *rep* cells is confined to synthesis of a complement for the infecting positive strand, a label given after infection should enter only the negative strand. But in both RFI and RFII molecules the positive strand gained about 20% of the label seen in the newly synthesized negative strand. This incorporation occurs only during infection with $A^+$ phage, suggesting that it represents synthesis of positive strand DNA at the site of the nick introduced by the gene *A* product. The origin for this incorporation can be located by determining the amount of label taken up into positive strands as a function of map position, using restriction enzymes to generate fragments of known location. This identifies an origin in the Hae-6b fragment, suggesting that in *rep* cells the usual nick is made in this region, but that abnormal events then follow.

Two types of process might be responsible for the replacement of positive strand material by newly synthesized sequences. One possibility is that parental material is degraded by nick translation, in which the preexisting positive strand is degraded from the 3′-OH terminus while being extended from the 5′ terminus. (This activity can be undertaken by DNA polymerase I in vitro.) An alternative is that replication of a rolling circle is initiated but that the tail containing the displaced parental material is degraded to restore the simple circular structure. In both cases, ligation would be necessary to generate the RFI molecules in which some of the parental strand has been replaced by newly synthesized material. Both mechanisms may in fact function. Most of the duplex structures take the form of RF molecules in which the positive strand is no longer than unit length, suggesting that the incorporation of label may be due to nick translation. But in addition to the RFI and RFII species, a fraction containing single stranded material forms a small part of the replicative structures; this may represent an abortive replication in which a rolling circle is initiated but elongation of the positive strand is somewhat limited. What controls which pathway is entered by any particular molecule is not known.

Systems for the replication of $\phi$X RFI DNA in vitro have been developed by Eisenberg, Scott and Kornberg (1976) and Sumida-Yasumoto et al. (1976). A crude fraction from uninfected cells does not support RF replication (this contrasts with the ability of such extracts to convert the mature single strand into the replicative form). However, addition to the uninfected cell extract of an extract from infected cells supports DNA synthesis with RFI DNA as template. Cells infected in the presence of high concentrations of chloramphenicol or by $A^-$ mutants do not yield active extracts; and an extract of *rep* cells is not active. The activity of this system in vitro implies that the *A* product function has been divorced from its synthesis and it will be interesting to determine how its action in vitro relates to the cis dominance seen in vivo.

Infection of mutant cells with $\phi$X has shown that some of the E.coli functions needed for host replication also are needed for $\phi$X replication, but some host genes have appeared necessary in some experiments and dispensable in others; these experiments are difficult to rely on because the temperature sensitivity of the *dna* mutations leaves open the possibility that they are leaky enough to generate enough protein activity to support phage replication, even though insufficient active protein is made to support bacterial replication. Loci that may be involved in $\phi$X replication include *dnaB*, *dnaC*, *dnaE*, *dnaG*, *dnaH* and *dnaZ*; loci whose products do not appear to be necessary are *dnaA*, *dnaI* and *dnaP* (see Taketo, 1973, 1975; Dumas and Miller, 1974, 1976; Kranias and Dumas, 1974; McFadden and Denhardt, 1974; Truitt and Walker, 1974). More definite information about the involvement and molecular functions of the various *dna* gene products will no doubt result from characterization of the in vitro system for RF replication.

What is the nature of the block to single strand synthesis imposed by mutations in the phage structural genes *F* and *G* and in genes *B*, *C* and *D*? Two types of model may be imagined. The absence of these gene products may prevent the initiation of single strand synthesis. Or single strands may be made but may be unstable without these proteins. Fukuda and Sinsheimer (1976b) tested the effects of mutations in genes *B*, *D*, *F* and *G* (the available mutants in *C* are too leaky to use in this sort of experiment); all produce about 40 RF molecules and then net synthesis ceases instead of switching to single strand production. The same results were obtained in each case. When infected cells are labeled with $^{3}H$-thymine from 2–12 minutes the radioactivity enters both positive and negative strands; during a chase the label then disappears from the positive strand. By contrast, upon infection in the presence of chloramphenicol at 35 $\mu$g/ml to prevent single strand synthesis, the label does not disappear from the positive strand. Using short pulse labels, it was possible to detect the synthesis of positive strand material at late times during the mutant infections. These results suggest that positive strands are synthesized on a rolling circle during the late period, displacing in the form of the tail a label previously incorporated into the RF molecule, but that no net synthesis is seen because the tails are degraded. The extent of synthesis by the mutants is much less than occurs in wild type infection (see also Iwaya and Denhardt, 1971). Nick translation cannot be rigorously excluded as a possible explanation, but the observation that the same results are obtained in *polA*$^{-}$ cells decreases the likelihood that this is the reason for the positive strand degradation and synthesis; it is more likely to represent a rolling circle replication rendered abortive by the absence of any one of the *B*, *D*, *F* or *G* gene products. The rate of chain elongation appears to be much slower in the mutant infections, and this also may be a consequence of the absence of these gene products.

All four of these genes appear to specify proteins whose functions are concerned with the assembly of the phage particle. Tonegawa and Hayashi (1970) observed structures of phage proteins in infected cells sedimenting at 6S, 9S and 12S. Siden and Hayashi (1974) then demonstrated that the 6S peak represents a pentamer of pG while the 9S peak appears to be a pentamer of pF. The 12S structure contains both pF and pG in the same 1:1 ratio as that of the phage particle and upon dialysis or centrifugation breaks down to give the 9S and 6S structures. This suggests that the 12S component may be the precursor structure from which the phage coat is assembled. Mutation in gene *B* prevents formation of the 12S component; instead a greater number of 9S and 6S particles are formed. This suggests that the function of pB may be to assist the conjugation of the pF and pG pentamers.

A precursor to the phage particle that contains pD has been identified by Weisbeek and Sinsheimer (1974). Mature particles sediment at 114S and ghosts at 70S; in addition, a peak at 140S can be seen late in infection by the *E* mutant used to allow prolongation of the lytic cycle. Upon storage in vitro the 140S particles generate 114S and 70S particles, showing that (at least in vitro) there is a precursor-product relationship. The 140S component contains the same proportions of the proteins pF, pG, pH, and pJ as the 114S phage particle; but in addition this fraction contains a large amount of pD. The exact proportion of pD varies rather widely, presumably because it is lost during isolation. However, there are roughly 80 copies of the 14,000 dalton pD protein present per particle. Farber (1976) showed that pD has the property of self-aggregation and does not bind to DNA. While its precise role remains to be defined, it is clear that its function is concerned with the assembly of some precursor to the mature phage particle.

## *Synthesis of Ff Single Strands*

Replication of the Ff replicative form molecule to produce single stranded genomes takes place by the same mechanism as that of $\phi$X174: the negative strand provides a closed circular template on which a positive strand is elongated in linear form to displace a single stranded tail that is cleaved into mature genomes (Ray, 1969; Tseng and Marvin, 1972a). The products of genes *III* and *VIII*, the minor and major coat proteins, respectively, are not needed for single strand production: for the production of single strands is not impeded by mutations in these genes (Pratt, Tzagoloff and Beaudoin, 1969). Nor are the products of genes *I*, *IV*, *VI* and *VII* necessary for single strand synthesis: these genes all are implicated in morphogenesis since mutations in them prevent the production of phage particles (Pratt et al., 1966). Assembly of the Ff phage particle therefore is divorced from the production of its genome, a situation different from that seen with $\phi$X174 and the larger phages that we have discussed. The products of both genes *II* and *V*, however, are essential for single strand synthesis.

Gene *II* of the Ff phages appears to be a counterpart to gene *A* of $\phi$X: it is needed for all replication of the RF form, both reproduction of duplex molecules and single strand synthesis. Gene *II* mutants thus convert infecting single strand to duplex form but perform no further replication. The *II* product is needed continuously throughout the phases of RF replication and single strand synthesis, for an increase in the temperature of incubation of during infection with $II^{ts}$ mutants brings to a halt all replication (Fidanian and Ray, 1972; Lin and Pratt, 1972; Tseng and Marvin, 1972b). If the function of the gene *II* product is to nick the positive strand, then, there must be a continuous need for such nicking to keep the template open.

Gene *V* of the Ff phages codes for a DNA binding protein with physicochemical properties analogous to those of the T4 32-protein and the E.coli unwinding protein: it binds cooperatively to single stranded DNA and is essential for the production of single strands. First characterized by Oey and Knippers (1972) and Alberts, Frey and Delius (1972), it is a protein of some 10,000 daltons present in about 120,000 copies in the infected cell. In low salt the *V* protein forms a dimer (Pretorius, Klein and Day, 1975; Cavalieri, Neet and Goldthwait, 1976). However, the *V* product differs from the 32-protein in failing to display aggregation into larger structures. When coated with the gene *V* protein, the circular single strand genome takes the appearance of a linear rod, whose structure may represent a bilayer of the protein separating parallel strands of DNA. Pratt, Laws and Griffith (1974) have characterized a DNA-*V* protein complex from M13-infected cells that has some 1300 molecules of *V* protein per genome, that is, one protein molecule for every five nucleotides.

The role of the gene *V* protein may be to bind to the single stranded positive tail as it is displaced from the rolling circle, thus preventing synthesis of a complement. According to the model illustrated in Figure 8.10, the same intermediate then may be used for both RF replication and single strand production: this is a circular negative strand which acts as template for synthesis of a linear positive strand. For RF replication a complement would be synthesized on the positive strand tail, which would then acquire circular form and be sealed by ligation; for single strand production no complement would be made, due to the presence of *V* protein. This model predicts that which pathway is followed depends upon the ratio of available *V* protein to newly synthesized DNA.

This model is supported by observations of the effects of temperature shifts upon infection with $V^{ts}$ mutants of M13. Salstrom and Pratt (1971) labeled the single stranded DNA synthesized upon infection at 30°C and then shifted the culture to 42°C; about half of the label in single strands

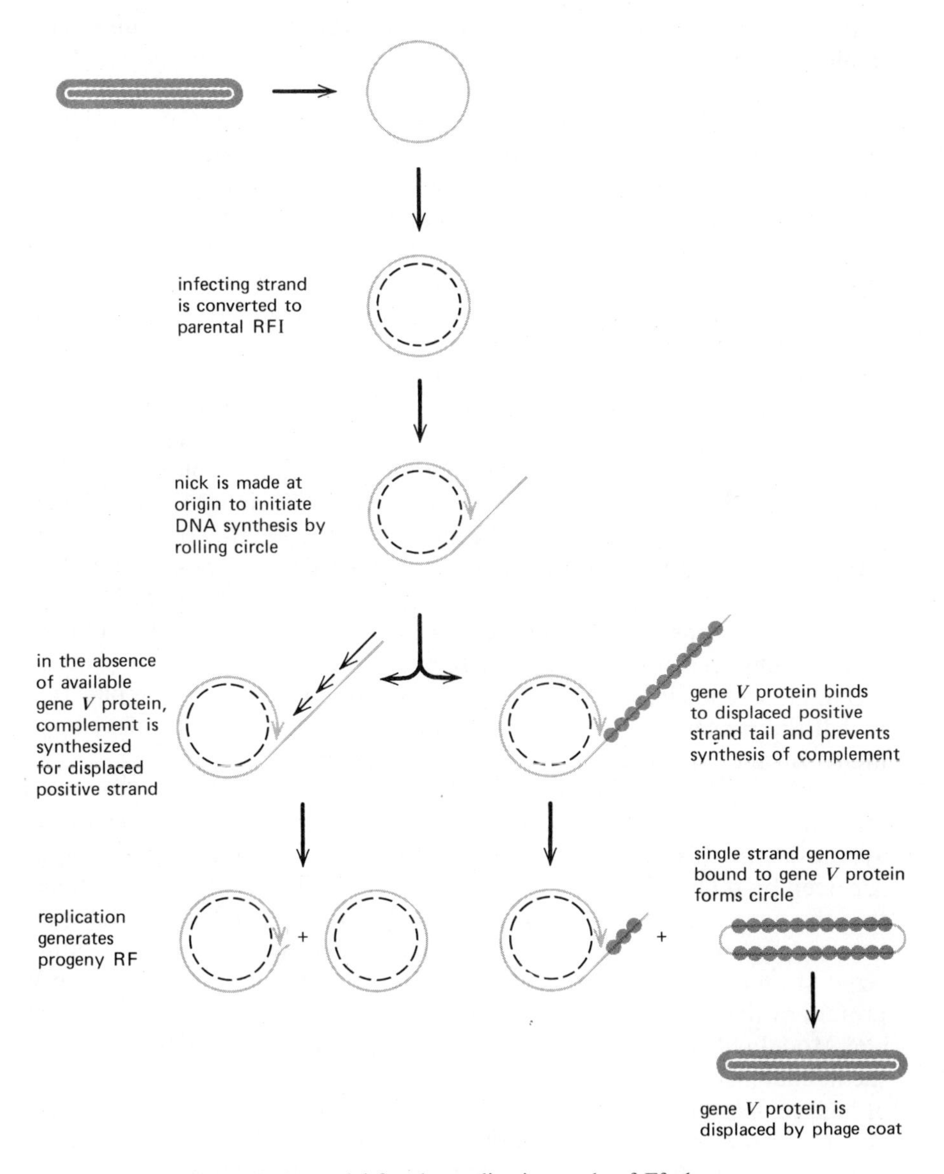


**Figure 8.10:** model for the replication cycle of Ff phages.

enters mature phages and about half is converted to RFI molecules. This implies that while some single strands are able to complete their maturation, others act as templates for the synthesis of a complement. The fate of an individual strand presumably depends upon its extent of completion and nearness to maturation. Maturation of single strands can be prevented by introducing an amber mutation in gene *IV*; it is then possible to compare directly the fate of labeled material upon a temperature shift in $V^{ts} IV^{am}$ and $V^{+} IV^{am}$ infection. Figure 8.11 demonstrates that in the former case the label in single strands declines and that in RF structures increases; whereas in the latter the single strands remain in the cellular pool.

The gene *V* protein bound to single strand genomes must be replaced with the proteins of the phage coat during morphogenesis. This action appears to be necessary to allow the *V* protein to recycle so that it becomes available for binding to further single strands. Chloramphenicol prevents replication of Ff DNA and 30 μg/ml is sufficient to inhibit either parental RF replication or single strand production (unlike the differential effect seen with ϕX). Salstrom and Pratt observed that upon the addition of chloramphenicol, single strands are maintained as such in $IV^{am}$ infection; but when the same experiment is performed with a $V^{ts} IV^{am}$ mutant and the cells are shifted to 42°C for 10 minutes to inactivate the *V* protein, the single strands instead are converted to RF molecules. The implication is that the presence of gene *V* protein prevents the synthesis of complements on the single strands to which it is bound.

A different result is seen when the synthesis of new material is followed after chloramphenicol has been added during the phase of single strand synthesis: in this case the production of single strands ceases and is replaced by replication to generate duplex DNA (Ray, 1970). This implies that the gene *V* protein present in the cell at the time of drug addition is able only to prevent synthesis of complements to the previously synthesized strands to which it is already bound; there can be no free gene *V* protein available for binding to material synthesized subsequently. Mazur and Model (1973) tested the idea that single strand synthesis depends on the availability of free gene *V* protein by infecting cells with a $II^{ts}$ mutant at high temperature and then later returning the culture to the low permissive temperature. At high temperature the parental RF is formed but does not replicate further; replication commences immediately upon a shift back to low temperature, but takes the form of single strand production. There is no replication of duplex DNA. This suggests that *V* protein has accumulated during the high temperature incubation, with the result that it prevents duplex synthesis when replication is activated by the gene *II* protein at low temperature. This interpretation is supported by the obser-

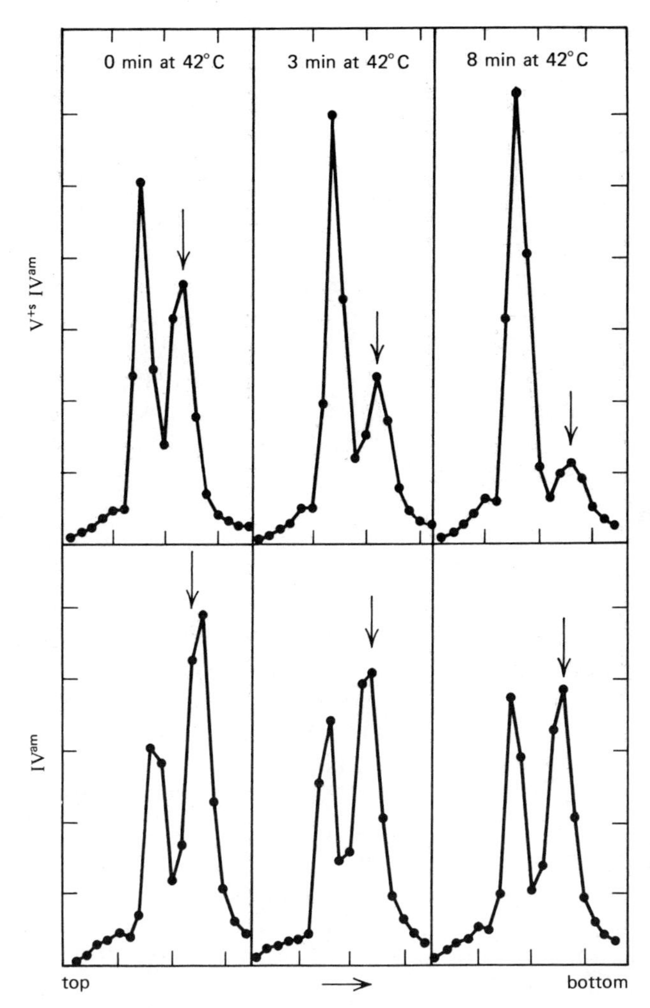


**Figure 8.11:** effect of gene *V* protein on single strand synthesis. Cells infected with M13 $V^{ts}$ $IV^{am}$ or $IV^{am}$ were labeled with $^3$H-thymidine at 30°C during the phase of single strand production and then chased with cold thymidine at 42°C; DNA was analyzed at 0, 3 and 8 minutes after the temperature shift. The arrow indicates the position of a single strand control (24S); the peak to its left is RFI (21S). The upper panel shows that inactivation of gene *V* protein by temperature shifting in $V^{ts}$ mutant infection causes the preexisting single strands to enter duplex form; the lower panel shows that the single strands remain stable when the gene *V* protein is wild type and thus is not inactivated by the temperature shift. The $IV^{am}$ mutation is present to prevent loss of single strands by maturation into particles. Data of Salstrom and Pratt (1971).

vation that when the same experiment is performed with a $II^{ts}$ $V^{am}$ double mutant, only RF replication occurs upon the temperature downshift. Mazur and Zinder (1975b) suggested that in wild type infection chloramphenicol may cause replication to switch to duplex production for two reasons: no new gene *V* protein is synthesized and all the copies previously made are unavailable because they are bound to single strands already; and this unavailable gene *V* protein cannot be reused because the phage proteins that would displace it during morphogenesis no longer are synthesized in the presence of the drug. The products of genes *I*, *IV*, *VII* and *VIII* are implicated in the latter capacity because mutations in these genes have the same effect as addition of chloramphenicol; using the technique of controlling the level of available gene *V* protein by temperature shifting cells infected with $II^{ts}$ mutants, the lack of these products results in a shift from single strand to duplex synthesis once the supply of free *V* protein is exhausted. Mutations in genes *III* and *VI* do not have this effect.

Although bacteria survive a wild type infection by Ff phage, they may be killed by an abortive infection in which progeny phage do not mature. Hohn, Von Schutz and Marvin (1971) noted that abortive killing does not occur upon infection with gene *II* mutants; this implies that it depends upon replication of the phage genome. Timmis and Marvin (1974a,b) found that the killing is in fact prevented whenever phage replication is inhibited, for example by treatment with nalidixic acid. Incubation at 42°C prevents abortive killing, which implies that the component responsible must be temperature sensitive. The component appears to be the *V* protein; the high temperature may alter its affinity for DNA. This suggests that replication without morphogenesis may influence the size of the *V* protein pool in a way that becomes lethal to the host.

Assembly of the Ff phage particle may take place at the inner cell membrane. Smilovitz (1974) detected the major coat protein at this site; and Lin and Pratt (1974) observed proteins *II, III, IV, V* and *VIII* in this fraction. This is consistent with a model in which the *V* protein–DNA complex is located at the membrane, where the *V* protein is displaced by the *VIII* and *III* proteins as the particle is assembled. Cavalieri, Goldthwait and Neet (1976) showed that the gene *VIII* protein forms dimers and possesses the ability to self-aggregate, but it remains to be seen how this is connected with its role in particle assembly.

### *Formation of Recombinants*

Both classes of single strand DNA phages display recombination: but the mechanism by which this occurs is nonreciprocal. To overcome the problem of low recombination frequencies with the Ff phage, Boon and Zinder (1969, 1970, 1971) developed a system in which two mutants in gene *II* of f1 are crossed. Both phage genomes generate the parental RF, but can proceed no further in the absence of an active gene *II* product. However, if a recombination event occurs between the two mutations to generate a wild type genome, functional protein is produced and replication and maturation then can occur. Only cells in which such a recombination event has taken place release phage particles; and these particles include genomes derived from both the products of the original recombination event. Information about the nature of the event can be gained in the usual way, by determining the genotypes represented in the progeny with respect both to the original mutations in gene *II* and for unselected mutations (usually the SB loci that identify the sites of action of the host B restriction and modification enzyme complex). Using this system, Boon and Zinder observed that the predominant type of "burst" contains the $f1^+$ recombinant phage and one of the two original parental types; usually the recombinant carries a combination of unselected markers that is different from that of the parent. Also found are "bursts" containing only $f1^+$ phages, but with the SB marker of only one parent; these may be due to reversion of the mutation in one parent or to inclusion of only phages of the $f1^+$ genotype by chance. Other types of burst are rather rare. Benbow et al. (1974) found that analysis of single bursts of $\phi$X174 reveals the same type of production of equal numbers of phages with recombinant genotype and one parental genotype.

The primary pathway for recombination in S13 or $\phi$X174 requires the *recA*$^+$ function. This pathway is stimulated by UV irradiation or thymine starvation; Benbow, Zuccarelli and Sinsheimer (1974) showed that both these treatments generate nicks or gaps and they suggested that recombination then is stimulated by "single strand aggression" in which single strands from the damaged genome initiate pairing with another genome. The existence of another, secondary system is implied by the observations of Tessman (1966b, 1968) that in *recA* mutant cells some residual recombination occurs; that this is not due to leakiness of the *recA* mutation is suggested by its failure to be stimulated by UV irradiation. When Benbow et al. (1974) compared recombination of $\phi$X174 in *recA*$^+$ and *recA* mutant cells, they found that recombination frequencies were reduced some tenfold by the absence of the *recA*$^+$ allele; this compared with a reduction of $10^3$-fold seen in host cell recombination. In all regions of the genome except one, the same relationship was observed between phage mutations in both the *recA*$^+$ and *recA* mutant host cells. The exception is the region in gene *A* which seems to undergo excessively high recombination in *recA*$^+$ cells. This suggests that there is a hot spot for the *recA* system at this site and it is tempting to speculate that this may be connected with the location of the origin for replication; for example, a gap persisting at the origin might lead to frequent initiation of recombination in this region. In *recA*$^+$ cells, mutation in phage gene *A* does not alter the frequency of phage recombination (Baker, Doniger and Tessman, 1971). This implies that the parental RF is sufficient to support normal levels of recombination: the progeny RF are not needed. In *recA* mutant cells recombination is reduced by the introduction of either a phage *A* gene mutation or a host *recB* mutation; this implies that the residual recombination pathway may utilize progeny RF molecules (an alternative that seems less likely is to suppose that the gene *A* product is directly required) and involves the RecBC exonuclease.

Structures that may be involved in $\phi$X or S13 recombination have been observed in infected cells in the form of molecules longer than a single genome. Rush and Warner (1968) observed multimeric circles corresponding to about 2–3% of the total phage DNA in S13-infected cells; these are mostly dimers and when tested in a spheroplast assay they contain a greater proportion of recombinants (0.45%) than the monomeric circles (0.03%), although this increase is much less than would be expected of an obligatory intermediate in recombination.

Two types of dimeric structure contained in this preparation were distinguished by Doniger, Warner and Tessman (1973). These comprised conventional circular dimers and catenated dimers, which represent two monomeric circles interlocked like two links of a chain. Both preparations show the same frequency of recombination when tested in a spheroplast assay; this is about three times the frequency seen in monomeric circles. We should note, however, that when a small increase of this nature is involved it is difficult to exclude the possibility that it results from the presence at low levels of a contaminating fraction that is rich in recombinants.

A suggestion that circular dimers arise by replication rather than recombination has been made by Benbow, Eisenberg and Sinsheimer (1972) on the basis of a study in which cells were infected with $\phi$X DNA of normal length together with a deletion strain. The dimers all proved to display lengths that correspond to twice the length of one parent, whereas if they represented recombinants the dimeric length should take an intermediate value corresponding to the sum of the two parent genome lengths. Also consistent with the idea that these circles do not originate in recombination is the observation that about 3% of the progeny RF are found as multimeric circles in either *recA*$^-$ conditions or when the replication of progeny RF is blocked; but in the first case most of the dimers are simple circles whereas in the second most are catenated. Benbow, Zuccarelli and Sinsheimer (1975) reported that the dimeric circles show a frequency of recombination that is enriched only about twofold compared with the monomeric dimers, while the catenated molecules show a frequency that is enriched only some fivefold. While these results suggest that circular dimers (and higher order multimers) may result from replication, with the production of catenated dimers depending on recombination, they do not prove that these are the origins of each form; certainly, however, the comparatively small increase in the proportion of recombinants in these fractions suggests that they are not obligatory intermediates in recombination, but if produced by this process rather may be products of side reactions. That catenated dimers may result from the replication of circular genomes, probably from events involved in termination, has been suggested by the isolation of such structures from cells infected with phage λ under conditions when replication is confined to the early (circular) mode (Sogo, Greenstein and Skalka, 1976).

A novel type of dimeric structure has been observed by Thompson et al. (1975) and Benbow, Zuccarelli and Sinsheimer (1975); this takes the form of a *fused dimer* or *figure 8*. This structure is compared with that of the catenated dimer in the electron micrograph shown in Figure 8.12. In the fused dimer the two circles appear to be held together by a region of base pairing and thus display a single locus of contact (where the distance of conjunction may vary but usually is fairly short), whereas in the

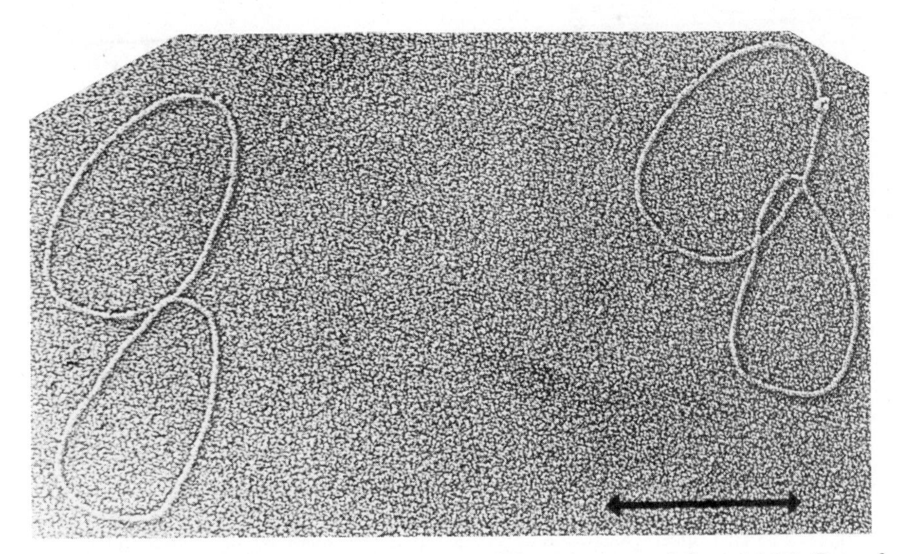

**Figure 8.12:** electron micrograph of fused dimer (left) and catenated dimer (right). Data of Thompson et al. (1975) kindly provided by I. Tessman.

catenated dimer the two circles are interlocked and therefore display two points of crossover. The idea that fused dimers represent intermediates in recombination is supported by the observation of Benbow et al. that after a mixed infection with normal length and shorter length $\phi$X genomes, 60% of the fused dimers have a contour length corresponding to the sum of the two different parental lengths. Another observation supporting this conclusion is that the proportion of fused dimers is decreased about tenfold by the introduction of a *recA* mutation.

Fused dimers have been observed also in populations of Col El genomes. Potter and Dressler (1976) took advantage of the large number of Col El DNA molecules that is generated when host cells are exposed to chloramphenicol. When closed circles were isolated from these cells by

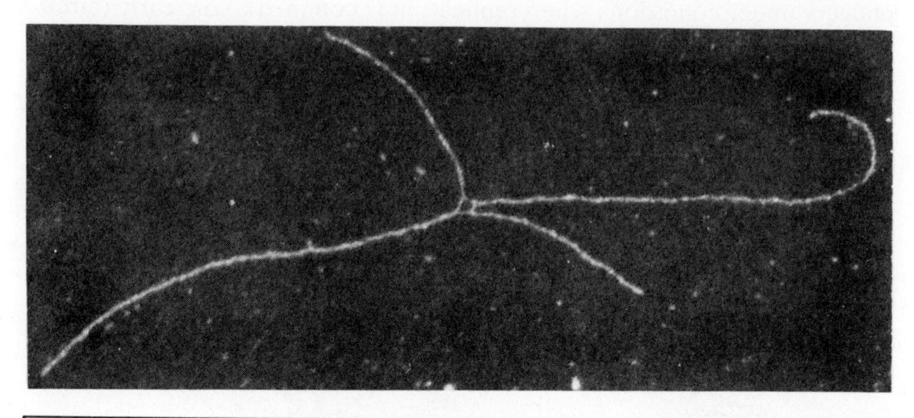

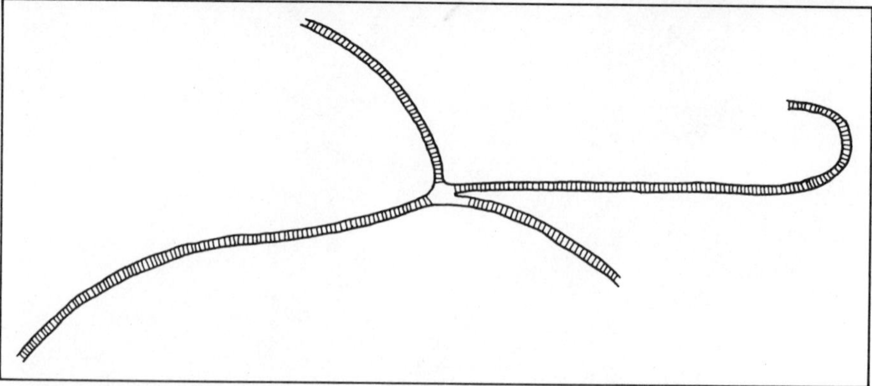

**Figure 8.13:** a chi molecule of Col El. Fused dimers were treated with Eco RI nuclease to break each partner once. The region of fusion appears to have been denatured and a box of four single strands can be seen. The lower illustration suggests an interpretation of the conjunction of the single strands. Data of Potter and Dressler (1976).

centrifugation in CsCl-EtBr gradients, 65% of the molecules represented simple monomers, 25% were of dimeric size, 8% were present as larger multimers, and the remainder were seen in linear form. Amongst the circular dimers were molecules taking the figure 8 configuration and these were distinguished from catenated dimers by treatment with the enzyme Eco RI. The Eco RI nuclease cleaves Col El DNA only once. When a fused dimer is cleaved, each genome should be cleaved once to leave a chi ($\chi$)-shaped structure in which contact between the two genomes is maintained by the base pairing in the region of fusion. When a catenated dimer is cleaved, each of the partners is converted to a linear rod and they therefore become separated.

The frequency of chi-shaped molecules in Col El DNA preparations treated with Eco RI nuclease is in the range of 0.5–3%. That the fusion between the two partners represents pairing between homologous regions is demonstrated by measuring the lengths of the arms of the chi structures: since Eco RI cleaves Col El DNA at a unique site, if the same region on each molecule is present at the point of fusion, the four arms can be divided into two pairs each of which comprises two arms of equal length. This model can be supported also by obtaining a denaturation map of the chi molecule and showing that the same region on each partner is present at the fusion. Both criteria show that the two genomes in the chi structure are held together at a region of homology.

Partial denaturation by formamide or resulting from preparative procedures allows the structure of the individual DNA strands in the fusion region to be discerned. Figure 8.13 shows a striking example in which it is clear that the two partners are held together by a crossover involving their single strands. This is clearly related to the anomalous junctions observed in lambda by Valenzuela and Inman (1975) and illustrated in Figure 5.10. Obviously it is the general class of structure to be expected from recombination by mechanisms involving the formation of hybrid DNA of the sort illustrated in Figure 5.1. Evidence that suggests these structures are the result of recombination, rather than (for example) an intermediate of replication, is provided by the observation that none could be observed in *recA* mutant cells (formally it is possible to say that their frequency of occurrence must be less than 0.01%, a reduction of at least two orders of magnitude).

We shall not now consider detailed models for the origin of these molecules; but the models considered by Benbow, Zuccarelli and Sinsheimer (1975) and Potter and Dressler (1976) make it clear that the features of $\phi$X recombination can be accounted by invoking the fused dimer as an intermediate.

CHAPTER 9

# RNA Phages

## Structure of the Phage Particle

The RNA phages fall into three classes, two of which have been well studied and the other of which appears on serological grounds to be different. The group I phages, including f2, fr, R17, MS2, M12 and $\mu$2, were first identified when Loeb and Zinder (1960) characterized phage f2 by virtue of its ability to adsorb only to bacteria carrying the sex factor; all the members of this group are very similar and sometimes it has been referred to as the f2 group, after the original isolate. Group III is identified by the phage Q$\beta$, which was first isolated by Watanabe (1964): serologically distinct from the f2 phages, the particle is about 15% larger but otherwise rather similar (Overby et al., 1966a,b). Although these two groups of E.coli bacteriophages are the best characterized, the RNA phages are not peculiar to E.coli since somewhat similar phages have been found that exist on a variety of bacterial hosts (for review see Shapiro and Bendis, 1975). Here we shall be concerned with the properties of the f2 phages and of Q$\beta$.

The f2 phage particles sediment at about 80S and the Q$\beta$ particle at about 84S. The weight of the f2 particle has been determined by a variety of physicochemical means and is close to $3.9 \times 10^6$ daltons; Q$\beta$ is about $4.25 \times 10^6$ daltons. The RNA content is about 30%, which agrees well with the directly measured values of $1.3 \times 10^6$ daltons for the f2 group and $1.5 \times 10^6$ daltons for Q$\beta$. In both cases the RNA displays extensive secondary structure which complicates determination of its molecular weight; the structure presumably plays a part in the condensation of nucleic acid in the phage particle and also appears to be important for controlling phage gene expression in infection. The phage particles of both types are rather similar, appearing polyhedral with a diameter of about 250 Å (for review see Boedtker and Gesteland, 1975).

One of the most intriguing features of the RNA phages is the small size of the genome. Just under 3600 nucleotides for the f2 phages and little more than 4000 nucleotides for Q$\beta$, this can code for only some 130,000 daltons of protein in f2 and about 150,000 in Q$\beta$. Clearly this can correspond only to a very small number of proteins. In fact it turns out that each phage genome comprises three genes, coding for a coat protein, a "maturation" protein, and a subunit of an RNA replicase able to reproduce the genome. To extend the range of protein activities available to the phage, the other subunits of the RNA replicase represent host proteins sequestered from their usual activities. The three genes are expressed to characteristic extents and the major product of infection is the coat protein, which constitutes the phage particle together with a single copy of the maturation protein and the genome. The principal interest in the RNA phages, then, is to explain how such a small genome is able to contain information sufficient for its reproduction, packaging into the protein coat, and successful infection of new hosts.

The processes involved in adsorption to the host cell are not yet well defined. Of course, it is clear that the phages bind not to the cell wall but to the sex pilus, as first demonstrated by Crawford and Gesteland (1964) for R17. We have already discussed the use of this reaction to visualize the pili in Chapter 2. Only RNA phages specific for the F pili have been found: no RNA phages able to infect cells carrying I or N pili have been reported (for review see Paranchych, 1975). This contrasts with the situation represented by the single strand DNA phages, where both Ff and If phages have been identified (see pages 212 and 726).

The f2 phages adsorb to the side of the F pilus (see Figure 2.7). We have seen that there are differences detectable in the pili specified by the sex factor and certain other F-like plasmids; this appears to be reflected in the efficiency with which the RNA phages can be adsorbed, an effect presumably residing in the molecular structure of the subunit of the pilus. Since large numbers of RNA phage particles can adsorb along the pilus, it seems possible that each pilin subunit may have the ability to bind the phage particle.

The f2 phage particle consists of 180 subunits of a coat protein of 13,700 daltons, organized with icosahedral symmetry of T = 3; the subunits probably are arranged as 20 hexamers forming the faces and 12 pentamers forming the vertices of the particle (for review see Boedtker and Gesteland, 1975). Ability to adsorb to the sex pilus appears to be conferred by the maturation protein, a subunit of 40,000 daltons present in only one copy per virion. Phage adsorption appears to follow the classical process in which the nucleic acid genome enters the cell but the protein coat remains outside, for Edgell and Ginoza (1965) showed that after adsorption the R17 coat can be released from the bacteria by shearing to remove the sex pili. The coat protein appears to be released from the pilus as an intact empty capsid (Silverman and Valentine, 1969; Paranchych, Krahn and Bradley, 1970). This contrasts, of course, with the uptake of the entire particle of the filamentous single strand DNA phage; and this difference implies that the role of the pilus in phage adsorption may not be the same in the two situations.

In contrast with the loss of coat protein upon adsorption, the maturation protein appears to enter the cell together with the RNA genome, possibly suffering a cleavage in the process (Kozak and Nathans, 1971; Krahn, O'Callaghan and Paranchych, 1972). This raises the possibility that the maturation protein may be needed not only for the phage to attach to the pilus but also for some stage of infection subsequent to uptake, a situation analogous with the postulated role of the gene *III* protein of f1 and the gene *H* protein of $\phi$X174 as a "pilot" needed for penetration of the cell wall (see page 733); also, of course, we should remember that both phages lambda and T4 inject proteins together with the genome, so this feature is not uncommon. Whether the maturation protein has any role subsequent to attachment, however, remains to be established.

When the phage particle binds to the pilus, its RNA becomes susceptible to attack by RNAase, as first shown by the observation of Valentine and Wedel (1965) that the addition of RNAase during adsorption can block infection. The conditions for binding R17 to pili have been studied by Danziger and Paranchych (1970a,b); binding is energy-dependent and appears to proceed through a series of steps in which first a reversible binding occurs and then irreversible binding takes place and the RNA is ejected. It is at the second step of irreversible binding that the RNA in the particle becomes accessible to RNAase, and we may therefore surmise that adsorption proceeds through the stages: initial (reversible) contact; irreversible binding, with a change in conformation of the particle affecting access to the nucleic acid; and injection into the cell of RNA and maturation protein, leaving behind an empty capsid of the coat protein. Only about half of an RNA phage population can attach and inject their nucleic acid into host cells; the remaining particles either interact with the bacteria and become susceptible to RNAase (about 40%) or fail altogether to bind to the cells (about 10%) (Paranchych, Krahn and Bradley, 1970; Krahn and Paranchych, 1971). The 10% entirely unable to attach represent particles lacking the maturation protein; but all the remaining particles possess the protein and thus the reason why almost half of them are defective in injection of RNA is not clear.

In summary, it is clear that adsorption involves attachment to the side of the sex pilus, followed by injection of RNA and maturation protein. A critical question remaining unanswered concerns the nature of events by which the RNA enters the cell. One possibility is that the RNA is transported along the pilus, from the site of attachment to the base in the cell membrane. Of course, this is the general type of mechanism proposed to account for the transfer of DNA from cell to cell during conjugation (see page 86); although phage RNA would have to be transported in the opposite direction from that in which DNA moves during conjugation. An alternative type of model is to suppose that pilus retraction is involved, possibly with the pilus retracting so as to bring the phage particle into contact with the surface of the cell, where injection of RNA might be triggered. This is similar to the type of process proposed to account for entry of the ♂-specific single strand DNA phages. While it would be attractive to construct a model able to propose a unified mode of action for the F pilus in all nucleic acid transport events, at present this is difficult and it seems that more than one mode of action may be involved (for review see Paranchych, 1975).

## Organization of the Phage Genome

The genetic studies that can be performed with the RNA phages are limited by the lack of recombination, which results from the absence of enzymes able to recombine RNA. Genes may be identified, however, by the isolation of conditional lethal mutants together with complementation analysis to determine the number of cistrons (for review see Horiuchi, 1975). Complementation analysis is complicated by the high rate of reversion to wild type of most mutations and by the difficulty of ensuring that both types of parental genome infect the cell (a problem caused by the apparently low number of genomes entering per cell). To overcome these difficulties, Valentine, Engelhardt and Zinder (1964) introduced the use of single bursts. Similar results are obtained with both types of RNA phage. Amber and temperature sensitive mutants of f2 fall into three complementation groups (Zinder and Cooper, 1964; Horiuchi, Lodish and Zinder, 1966; Gussin, 1966). One group is unable to replicate the phage RNA and lies in the replicase gene; one fails to form coat protein; and the other group generates a normal yield of about 10,000 particles per cell, but these are defective (they lack the maturation protein). Horiuchi and Matsuhashi (1970) identified the same three complementation groups in Qβ.

The products corresponding to these genes can be identified by screening the proteins synthesized in phage infected cells when host protein synthesis is inhibited (for example with actinomycin or rifampicin). By this means Nathans et al. (1966) and Vinuela, Algranati and Ochoa (1967) identified three phage coded proteins; and by comparing the wild type pattern with the proteins synthesized in cells infected with amber mutants, Vinuela et al. (1968) and Horiuchi, Webster and Matsuhashi (1971) equated three proteins with the three genes. In addition, a fourth protein is generated in Q$\beta$ by readthrough past the termination site of the coat protein gene (see below).

Replication of the RNA phages takes place by a process in which the infecting (plus) strand is first used to synthesize a complementary (minus) strand; then the minus strand is used as template for the synthesis of further plus strands. The dependence of this process on a phage coded function is demonstrated by the inability of mutants defining the RNA replicase gene to synthesize minus and new plus strands upon infection. With temperature sensitive mutants, a shift up halts the production of new RNA at any time, showing that the same phage function is needed for both minus and plus strand synthesis (Lodish and Zinder, 1966a). The product of the RNA replicase gene is a protein of about 63,000 daltons in the f2 phages and about 65,000 daltons with Q$\beta$.

The gene for the coat protein was first identified by the demonstration that suppression of an amber mutation alters an amino acid residue in the protein (Notani et al., 1965). The coat proteins of several of the phages have been sequenced (for review see Weber and Konigsberg, 1975). The sequences of the R17, MS2 and ZR coat proteins are identical; a single amino acid substitution occurs in f2; and fr has several (19) substitutions relative to f2. The coat protein is 129 amino acids in length and has a weight of close to 13,700 daltons. The sequence of the Q$\beta$ coat protein is more distantly related to that of the f2 group, but by assuming appropriate deletions and insertions it displays homology at several sites; 29 of its 131 residues can be aligned so that they are the same and 40% of the remaining differences can be accommodated by single base changes. The Q$\beta$ coat protein has a mass of 14,000 daltons. The group I and group III phages therefore probably are derived from some common ancestor.

The maturation protein was given this name when it was thought that its role might be to assure proper assembly of the phage particle. An alternative name is the A protein, according to a nomenclature in which the principal coat protein may be described as the B protein. Mutants with a defect in the maturation protein (A protein) gene generate defective particles that are unable to adsorb to host cells (Zinder and Cooper, 1964; Lodish, Horiuchi and Zinder, 1965); these particles may be defective not only in adsorption but also in their structure, since the RNA becomes susceptible to RNAase (for review see Horiuchi, 1975). Argetsinger and Gussin (1966) showed that the RNA of the particles is infectious when prepared by means that avoid its degradation by nuclease; and Steitz (1968a,b) showed that the only defect of the particles is their lack of a protein of 40,000 daltons present in only one copy per virion. Because the principal coat protein lacks the amino acid histidine but the A protein contains it, the presence of A protein can be measured by using histidine carrying a radioactive label.

The lack of genetic recombination means that mutations of the RNA phages cannot be ordered by making genetic crosses. The order of the genes, and of mutations within them, therefore has had to be determined by the application of techniques for determining protein and nucleic acid sequences. The genes can in principle be located by identifying the regions of the RNA molecule that contain the initiation and termination signals for the translation of each protein; comparison of their nucleotide sequences with the N-terminal and C-terminal amino acid sequences of each protein has been made for the genes of both the f2 group and of Qβ. In fact, the complete sequence of MS2 RNA now has been determined and of course this provides an unambiguous definition of the relationship between the genes. The positions of mutations within the genes can be determined by identifying the amino acid that has been mutated; this approach has been used to map amber mutations of the coat protein genes of both classes of phage. Direct sequencing of each mutant nucleic acid also is possible in principle.

The phage RNA can be split into two fragments by limited nuclease action. Intact MS2 RNA sediments at about 27S; Spahr and Gesteland (1968) identified an enzyme activity of E.coli, RNAase IV, which splits the molecule into a 15S RNA representing the 5′ terminal 40% and a 21S species representing the 3′ terminal 60%. A similar result is seen with Qβ RNA (Bassel and Spiegelman, 1967). Probably the cleavage occurs at a site that is particularly exposed in the secondary structure. The existence of the two fragments makes it possible to assign to either end of the molecule the oligonucleotides derived by more complete digestion with an enzyme such as T1 ribonuclease.

Two types of experiment made possible the original assignment of an order to the genes. Jeppesen et al. (1970) identified the initiation sites of each gene of R17 by virtue of their ability to bind ribosomes; the sequence of the ribosome binding site for each gene is known (see below) and it was therefore necessary just to determine their order. The A protein ribosome binding site lies on the 15S fragment and therefore must be in the 5′ terminal part of the molecule. While the coat protein gene also has a ribosome binding site on the 15S fragment, nucleotide sequences corresponding to stretches of the coat protein are found on the 21S fragment; this suggests that the coat protein gene spans the break between the 15S and 21S fragments. The ribosome binding site of the replicase gene lies on the 21S fragment and therefore this gene must lie towards the 3′ end of the molecule. These experiments therefore suggest the gene order

5′—A protein gene—coat protein gene—replicase gene—3′

A second approach was to obtain different size fractions of M12 RNA representing uncompleted plus strands extracted from replicative intermediates. When translated in vitro, short strands should be able to direct the synthesis only of the 5′ terminal protein, and with increasing length the strands should successively acquire the ability to synthesize the other proteins. By this means Konings et al. (1970) were able to suggest the same gene order for M12 RNA.

The sequences of various parts of the phage RNA, especially including the 5′ and 3′ terminal ends and the punctuation signals for each gene, have been determined for MS2 and also for some other phages of the f2 group as well as for Qβ (see legends to Figures 9.1 and 9.3). The entire sequence of MS2 RNA has been reported by Fiers et al. (1976), making this the best characterized of the RNA phage genomes. Figure 9.1 summarizes its structure. At both the 5′ and 3′ ends there are extensive lengths of RNA that are not translated, a conclusion that was suggested by some of the earlier sequencing studies in which punctuation signals were found to be absent from 5′ and 3′ terminal fragments (Adams, Spahr and Cory, 1972; De Wachter et al., 1971; Vandenberghe et al., 1975). These sequences amount to 129 nucleotides at the 5′ end and 171 nucleotides at the 3′ end. Between the genes there are short intercistronic regions; in fact, the sequence between the coat protein gene and the replicase gene was determined earlier when Nichols (1970) identified an oligonucleotide including the coat protein termination sequence that overlapped with the ribosome binding site for the replicase gene previously determined by Steitz (1969).

Both amber and ochre codons are used to terminate translation of the three genes; and, indeed, the coat protein is terminated with a tandem amber-ochre arrangement. This presumably has the result that there is no leakage past the coat protein termination signal. (This contrasts with the coat protein gene of Qβ RNA, which is terminated with a leaky UGA codon; see below). The assignment of single amber codons to terminate the A protein and replicase protein genes is supported by observations that infection of amber suppressor hosts increases the size of these proteins, presumably by allowing readthrough until an ochre or UGA codon is reached in the same phase (Remaut and Fiers, 1972; Atkins and Gesteland, 1973).

The coat protein and replicase genes of MS2 RNA are initiated with the codon AUG which provides the usual start signal for bacterial protein synthesis. However, the A protein gene starts with the codon GUG. At present this seems to be the only known instance where GUG initiates a

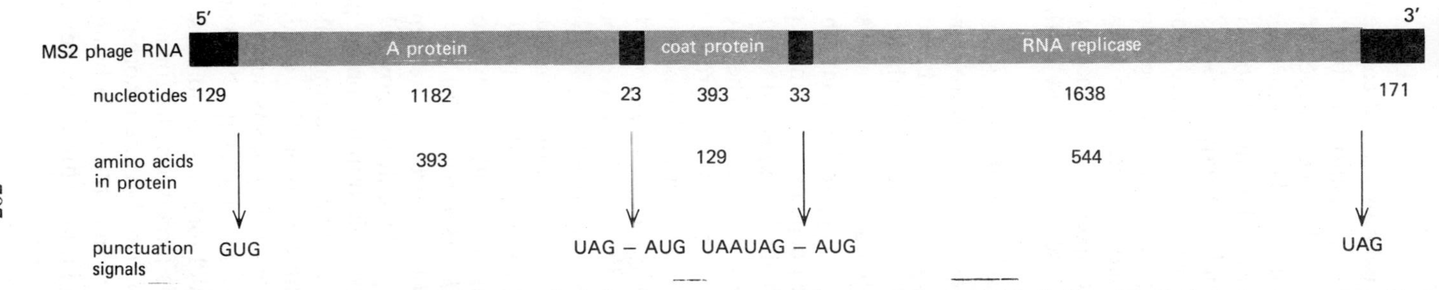


**Figure 9.1:** organization of phage MS2 RNA. Genes are shaded and nontranslated regions are indicated in black. Gene lengths include punctuation signals; the initial methionine of the nascent polypeptide chain is retained in the A protein but cleaved from the coat protein and RNA replicase gene products. The GUG codon initiating the A protein gene of MS2 is replaced by AUG in R17. Based on sequence data of Adams et al. (1969), Nichols (1970), De Wachter et al. (1971), Min Jou et al. (1972), Contreras et al. (1973), Vandenberghe, Min Jou and Fiers (1975) and Fiers et al. (1975, 1976).

natural message, although it is known that it can be used for translational restarts following premature termination in the *lacZ* gene. This codon marks one of the sites that differ between phages of the first group; in R17 the more common AUG signal is used for all three genes. The coat protein nucleotide sequence agrees with the known amino acid sequence of the protein; the A protein and replicase sequences have not yet been determined (although now they may be predicted from the nucleotide sequence). The variation in sequence between the phages comprising group I appears to be rather small, corresponding to a few nucleotide substitutions, and it may therefore be taken that the map of Figure 9.1 represents all of these genomes fairly closely (for review of the sequencing studies with MS2, see Fiers, 1975).

The total length of MS2 RNA is 3569 nucleotides; the 3213 nucleotides representing the coding sequences therefore amount to 90% of the genome. The functions of the long untranslated leader sequences at the 5′ and 3′ termini are not fully known; although these must contain binding sites for the replicase, for plus and minus strand synthesis, it does not seem likely that this accounts for their entire length. Presumably a selective advantage accrues to the phage in maintaining them, since otherwise a reduction in their length might be expected to confer an advantage in shortening the time required for replication. The intercistronic regions are utilized for control of translation; in the f2 group the coat protein may bind to a sequence including the start of the replicase gene and thus prevent its translation; and in Qβ the replicase may bind to the start of the coat protein gene to repress its expression (see below). Probably both mechanisms function in both types of phage. Also we may ask whether there may be an advantage in separating the termination codon of one gene from the initiation codon of the next in order to prevent steric interference between ribosomes translating each gene; but this contrasts with the overlapping punctuation seen in φX (see page 739).

One of the more remarkable features of the genome is its extensive secondary structure; while it is difficult to know which self complementary sequences in fact are base paired in the phage particle and/or in the infected cell, evidence that such base pairing takes place in vitro is provided by the physicochemical properties of the phage RNA. The existence of base pairing in certain regions is supported by their reduced susceptibility to ribonucleases. Min Jou et al. (1972) and Fiers et al. (1975, 1976) have constructed a model for the secondary structure of the entire phage genome which depends largely on the formation of hairpin loops but also includes some base pairing between more distant sequences that are complementary (and which may therefore bring sequences from different parts of the phage into juxtaposition) (for review see Fiers, 1975).

Their "flower" model for the coat protein gene, so called because of the appearance of the sequence when written out, is shown in Figure 9.2. As with the secondary structures proposed for the other regions of the phage, there is good evidence to support the existence of some of these base paired regions and less compelling data on others; some are advanced simply on theoretical considerations of the stability of the base paired region. However, although the model for the phage RNA may not be correct in every detail, it is clear that the formation of secondary structure is important for the phage, probably for packaging into its coat as well as for control of gene translation (see below).

The organization of the Qβ genome appears to be generally rather similar to that of MS2, although its sequence has not been determined in such detail. A map based in part upon sequencing studies and in part upon the properties of the protein products is shown in Figure 9.3. The order of genes is the same and the A protein gene starts at position 62 from the 5′ end (Staples et al., 1971). The last 32 nucleotides of Qβ RNA show some homology with MS2 and do not carry a termination signal, so that there must be an untranslated sequence of at least this length at the 3′ end (Goodman et al., 1970). Thus Qβ RNA shares with MS2 the possession of noncoding leader sequences at each end of the genome.

The proteins of Qβ, and therefore its genes, are of very similar size to MS2. The prinicipal difference between the phages is that the Qβ coat protein is not the sole product of the coat protein gene. In addition to the three principal proteins (maturation, coat, replicase), Qβ infection results in the synthesis of a fourth protein, of 36,000 daltons, and sometimes described as the IIb product. Horiuchi, Webster and Mitsuhashi (1971) observed that this protein is not synthesized in amber coat mutants, and since the genome cannot be large enough to code for it as a separate protein in addition to the other three, they suggested that it might be derived by readthrough past the termination signal of the coat protein. Moore et al. (1971) and Weiner and Weber (1971) showed that the IIb protein has the same amino terminal sequence as the coat protein; and Weiner and Weber (1973) showed that when Qβ is grown on a UGA suppressor host, the proportion of IIb relative to the coat protein is increased from 2% to 7%. This suggests that the coat protein gene terminates with a UGA codon which is leaky enough to permit occasional readthrough; the level is increased, of course, by suppression of the UGA signal. Weiner and Weber confirmed this model by sequencing a tryptic peptide from the IIb protein that contains the suppressed signal; the nucleotide sequence required to code for these amino acids must contain a single UGA as the termination codon. Since the readthrough protein is some 22,000 daltons larger than the coat protein, it must represent the

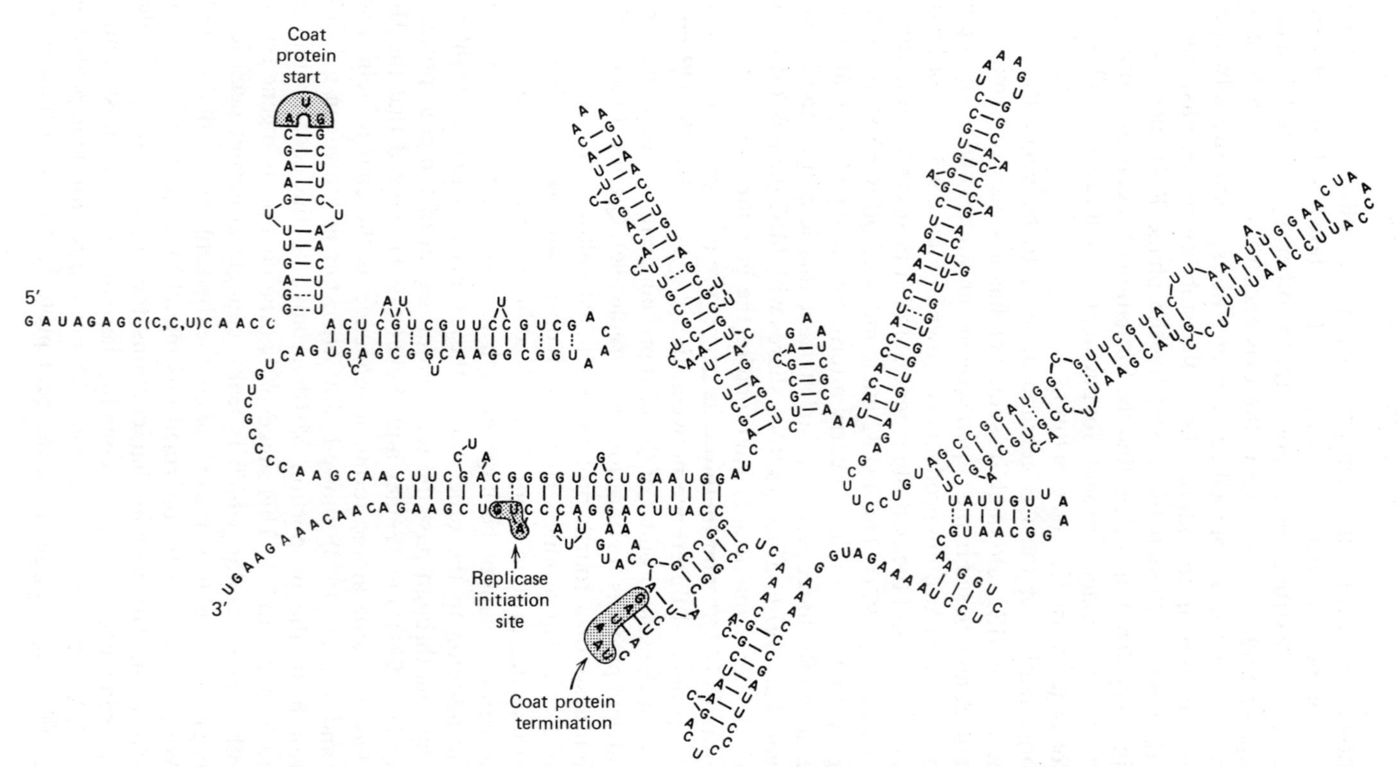


**Figure 9.2:** the flower model for the secondary structure of the coat protein gene of MS2 RNA. The coat protein initiation codon lies at the end of a hairpin structure. The replicase initiation site is base paired to a complementary sequence early in the coat protein gene. Data of Min Jou et al. (1972).

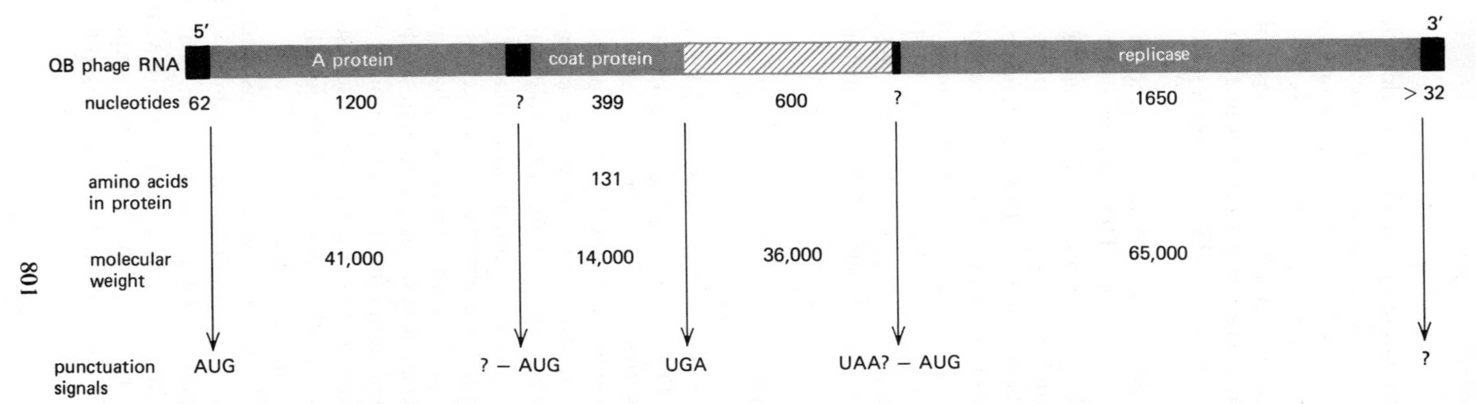


**Figure 9.3:** organization of phage Q$\beta$ RNA. Genes are shaded and nontranslated regions are in black. The sequence translated by readthrough past the coat protein termination signal is striped. The initiation signals all have been determined by sequencing the ribosome binding sites. The use of UGA to terminate the coat protein gene is inferred from the enhanced readthrough caused by UGA suppression. The termination signal for the readthrough product may be one of the ochres blocking all phases of translation in the ribosome binding site of the replicase gene. Based on data of Billeter et al. (1969), Hindley and Staples (1969), Hindley et al. (1970), Staples and Hindley (1971), Staples et al. (1971) and Weiner and Weber (1971, 1973).

translation of another 600 nucleotides. These nucleotides probably comprise the "intercistronic" region between the end of the coat protein gene proper and the start of the replicase gene; just before the initiation codon for the replicase gene there are three ochre codons blocking all phases of translation and one of these may terminate synthesis of the readthrough protein. This means that a greater proportion of the Q$\beta$ genome has a coding function: apparently about 3850 nucleotides out of the 4000 total length.

The function of the readthrough protein is not known. In principle, two types of model may be imagined. The protein itself may serve some necessary function, either physiological (for example concerned with particle assembly) or structural (it is found in the phage particle). Or the necessary function may be translation of the sequence between the end of the coat protein gene and the start of the replicase gene, that is, the movement of ribosomes per se may be important and the protein product irrelevant. However, since the phage has retained a length of 600 nucleotides that is translated during synthesis of the readthrough protein, there must presumably be some compelling reason for its presence.

### Translation of Phage RNA

The RNA phages are positive strand viruses and because their nucleic acid is RNA rather than DNA it is the phage genome itself that provides the template for protein synthesis. The parental infecting strand therefore has two functions: it must be translated into the phage coded proteins; and it must act as template for synthesis of the complementary minus strand. The minus strand has only one function: to act as template for the synthesis of further plus strands, which may be utilized for either translation or replication. As a natural template for protein synthesis that can readily be isolated in large amounts, the RNA phage genome has been used extensively as the basis of systems for protein synthesis in vitro. The significance of these experiments for the elucidation of the operations of the protein synthetic apparatus has already been discussed in Volume 1 of Gene Expression; here we shall be concerned principally with the implications for the life cycle of the phage.

All three genes of the RNA molecule can be translated in vitro. During infection the coat protein is the predominant product, with the replicase produced in smaller amounts and the maturation protein representing only a minor product. This situation is mimicked in vitro. By applying fingerprinting techniques to detect the N-terminal sequences of f2 proteins synthesized when $^{35}S$ labeled fmet-$tRNA_f$ is used to initiate synthesis, Lodish (1968c) showed that the three proteins are synthesized approximately in the ratio of 20 coat protein:5 replicase:1 maturation protein. The idea that the preferential expression of the coat protein gene depends upon the secondary structure of the phage RNA is supported by the observation of Lodish (1970b) that treatment of f2 RNA with formaldehyde—at low temperatures which cause loss of secondary structure but do not inactivate the message function—abolishes the effect: the replicase gene then is translated to about the same extent as the coat protein gene, and the maturation protein represents about 15–20% of the product instead of the usual 2–3%. Under these conditions some sequences not representing the starts of the genes are used to initiate protein synthesis, presumably because they resemble ribosome binding sites which are not usually expressed but become exposed when the secondary structure is released.

Consistent with these results, E.coli ribosomes bind almost exclusively to the initiation site of the coat protein gene of either f2 or Qβ RNA (Gupta et al., 1970; Hindley and Staples, 1969). The other ribosome binding sites can be made available by treatments which interfere with the maintenance of secondary structure; thus Steitz (1972, 1973) found that the use of R17 or Qβ fragments derived by breakage induced by $^{32}$P decay does not reveal any preference for the coat protein site: all the ribosome binding sites can be recovered from the ribosomes. That binding to the replicase initiation site depends on the structure of the rest of the molecule is suggested also by the observation of Lodish (1968b) that a fragment of f2 RNA containing the replicase gene but lacking the coat protein gene is a more efficient template for replicase synthesis. This implies that the secondary structure of the phage RNA controls access to the ribosome binding sites.

The RNA sequence bound by the ribosome can be recovered by degrading the unprotected regions of the molecule with ribonuclease: then the sequence of the protected fragment can be determined. Figure 9.4 summarizes the sequences of all three binding sites in both R17 and Qβ (for review see Steitz, 1975). Each protected sequence extends for about 20 nucleotides to the left (5′ side) of the initiation codon and for about the same distance into the first codons of the gene. A comparison of the sequences shows no striking homologies, either between the different sites of each phage or between the corresponding sites in the two phage classes. However, all of the binding sites possess a short sequence, of at least four nucleotides, in the region preceding the initiation codon, which is complementary to part of the 3′ terminal sequence of the 16S ribosomal RNA of E.coli; Shine and Dalgarno (1974) have proposed that part of the sequence GAAGGAGGU may be a feature of the ribosome binding site

```
QB A protein         U C A C U G A G U A U A A G A G G A C A U A U G C C U A A A U U A C C G G G U
                                                                   fmet  pro   lys   leu   pro   arg

R17 A protein          A U U C C U A G G A G G U U U G A C C U A U G C G A G C U U U U A G U G
                                                                   fmet  arg   ala   phe   ser

QB coat protein      A A A C U U U G G G U C A A U U U G A U C A U G G C A A A A U U A G A G A C U
                                                                   fmet  ala   lys   leu   glu   thr

R17 coat
protein A G A G C C C U C A A C C G G A G U U U G A A G C A U G G C U U C U A A C U U U
                                  ↕                                fmet  ala   ser   asn   phe
                                  G

QB replicase           A G U A A C U A A G G A U G A A A U G C A U G U C U A A G A C A G C
                                                                   fmet  ser   lys   thr   ala

R17 replicase                A A A C A U G A G G A U U A C C C A U G U C G A A G A C A A C A A A G
                                                                   fmet  ser   lys   thr   thr   lys

complementary sequence            G A A G G A G G U G A U C
E. coli 16S ribosomal RNA   3' HO–A U U C C U C C A C U A G – – –
```

which is important for its ability to be recognized by base pairing with the ribosomal RNA. All of the ribosome binding sites except that of the Qβ coat protein have this capacity; and this exceptional site has a sequence complementary to a further sequence in the ribosomal RNA. Since the sequences of the binding sites of the three genes of each phage are different, they may possess different affinities for the ribosome, so that this as well as availability in the secondary structure may be a factor in controlling the extent of translation.

The maturation protein and coat protein initiation sites of both R17 and Qβ have sequences that may be able to form duplex hairpin structures in which the AUG initiation codon is exposed on the single strand loop at the end of the duplex stem. (The possible structure of the MS2 coat protein initiation site is shown in Figure 9.2.) Whether this plays any role in recognition of the binding site is not known. In each case, the duplex stem includes the sequence complementary to the ribosomal RNA; this leaves the implication that if the hairpin exists in vivo, it must be disrupted if the ribosomal RNA is to bind to the complementary sequence. The possibility that the 3′ end of the 16S rRNA is involved in binding to mRNA is supported by observations that the ribosomal protein S1 binds specifically to it (see below); although it remains to be seen exactly how ribosome recognition occurs, at present the only messenger sequences for which there is any evidence of involvement (apart from the AUG codon itself) are those indicated in the figure for their complementarity to the ribosomal RNA.

The R17 replicase initiation site is complementary to a sequence of nucleotides early in the coat protein gene and the formation of a base paired structure between them may act as a translational control which prevents expression of the replicase gene (see below). This model implies that, at least in this instance, the formation of a duplex structure prevents ribosome binding. Whether the Qβ replicase initiation site may form a similar duplex structure is not yet known since the remaining parts of the phage RNA have not been sequenced.

The existence of translational control was first revealed by the polarity

**Figure 9.4:** ribosome binding sequences of the three genes of R17 and Qβ. The sequences of R17 and MS2 are identical, except that some strains of R17 possess a G instead of an A as indicated. All of the binding sites except the Qβ coat protein possess a sequence of at least four nucleotides that is complementary to part of the sequence of the 3′ terminal nine nucleotides of the 16S ribosomal RNA; the R17 coat protein possesses two such sequences, one of which is involved in the A-G base transition. The Qβ coat protein site possesses a pentanucleotide sequence that is complementary to the sequence of nucleotides at positions 9-13 of the rRNA. After Dalgarno and Shine (1974); data of Steitz (1969, 1972), Gupta et al. (1970), Min Jou et al. (1972), Hindley and Staples (1969), Staples and Hindley (1971), Staples et al. (1971).

of early amber mutants in the coat protein gene of f2. Lodish, Cooper and Zinder (1964) reported that the early amber mutant *sus3* makes little replicase upon infection, due to its polar effect upon the translation of the replicase gene. By contrast the later amber mutant *sus11* produces a normal amount of replicase and RNA, although not completing synthesis of the coat protein, which demonstrates that coat protein per se is not required for replicase synthesis. This last point was confirmed directly by Lodish (1968a), who demonstrated that the *sus3* polar mutation has no effect upon the ability of replicase to generate progeny plus strands. The *sus3* mutation lies in the sixth amino acid of the coat protein; *sus11* lies in amino acid 70, that is about half way along the gene. This suggests that translation of the early part of the coat protein gene may be necessary for the replicase gene to be translated; when ribosomes are released at the sixth codon the replicase gene therefore cannot be translated, but when they progress to codon 70 a sufficient part of the coat protein has been translated to allow translation of the replicase gene to be initiated.

Polarity is maintained when phage RNA is translated in vitro. When the phage RNA template carries an amber mutation, a polypeptide chain fragment replaces the full length protein product of the mutant gene; and, indeed, this was useful in establishing the assignment of genes to proteins (for review see Capecchi and Webster, 1975). The *sus3* mutation of f2 or R17 has two effects in vitro: a hexapeptide is synthesized instead of the coat protein; and virtually no replicase is synthesized (Webster et al., 1967; Engelhardt, Webster and Zinder, 1967; Capecchi, 1967; Roberts and Gussin, 1967). Faithful to the situation prevailing in vivo, the *sus11* mutant of f2 directs synthesis in vitro of a coat polypeptide of 70 amino acids and produces as much replicase protein as a wild type template. And amber mutants terminating coat protein synthesis at codons 50 or 54 of the R17 coat protein gene are not polar (Tooze and Weber, 1967). Polarity is displayed also by Qβ RNA; Ball and Kaesberg (1973) observed that an amber mutation in position 17 of the coat protein gene entirely prevents replicase synthesis, an amber mutation at position 37 allows a rather small amount ($<10\%$) of synthesis, whereas a mutation in position 86 allows some 40% translation of the replicase gene. Although apparently in contrast with the all or nothing effect seen in the f2 group RNA, the idea that there might instead be a gradient of polarity in Qβ RNA rests largely upon the properties of the mutation at position 86; although perhaps suggestive, this cannot be taken as evidence that the dependence of replicase expression on coat protein translation is significantly different in Qβ and f2.

That the polarity of the *sus3* amber mutation reflects an effect of translation upon secondary structure was confirmed by the observation of

Lodish (1970b) that *sus3* RNA is not polar when translated in the presence of formaldehyde; in these conditions the mutant and wild type RNA both translate the replicase gene to about the same extent as the coat protein gene. Together with the observation that ribosomes cannot bind directly to the initiation site of the replicase gene in intact phage RNA, but can bind to a fragment of the RNA containing this site, this suggests that with an intact phage genome the first part of the coat protein gene must be translated before synthesis of replicase can be initiated. The obvious interpretation is that translation of the start of the coat protein gene disrupts the secondary structure of the phage in such a way that when the ribosome has traversed some 40–50 codons the replicase binding site becomes available to the ribosome. A model for this effect has been suggested by Fiers et al. (1976) on the basis of possible secondary structure in the MS2 RNA sequence. Figure 9.2 shows that the sequence of codons 27–33 of the coat protein gene is complementary to a sequence which includes the AUG initiation codon and much of the ribosome binding site of the replicase gene. This raises the possibility that these sequences are base paired and that the duplex structure is disrupted when the ribosome translates this part of the coat gene, thus releasing in single strand form the replicase initiation site. This model is illustrated in Figure 9.5. Of course, there is no proof that these features of secondary structure are maintained in vivo, but the similarity of the polar effects in vitro and in vivo suggests that in general the in vitro translation system may reflect the events occurring in vivo, although the template is somewhat less active outside the cell (for review see Capecchi and Webster, 1975; Lodish, 1975).

The coat protein gene exerts a polar effect only upon the replicase gene; the maturation protein gene appears to be translated independently of events at the other two genes (Lodish and Robertson, 1969; Lodish, 1968c). However, the secondary structure of the phage may be important also in controlling the synthesis of maturation protein. Lodish (1971) reported that ribosomes of B.stearothermophilus bind only to the initiation site of the maturation protein; this implies that they do not recognize the binding site of the coat protein gene. Even though drawn from a heterologous system, this emphasizes the possibility that even when equally available, different binding sites may have different affinities for the ribosome. When the temperature of incubation is increased from 47°C to 65°C, there is a twentyfold increase in the synthesis of maturation protein, presumably because the secondary structure of the RNA has been melted, making the ribosome binding site more available. (This experiment could not be performed with E.coli ribosomes because the E.coli protein synthetic system would be inactivated at such high temper-

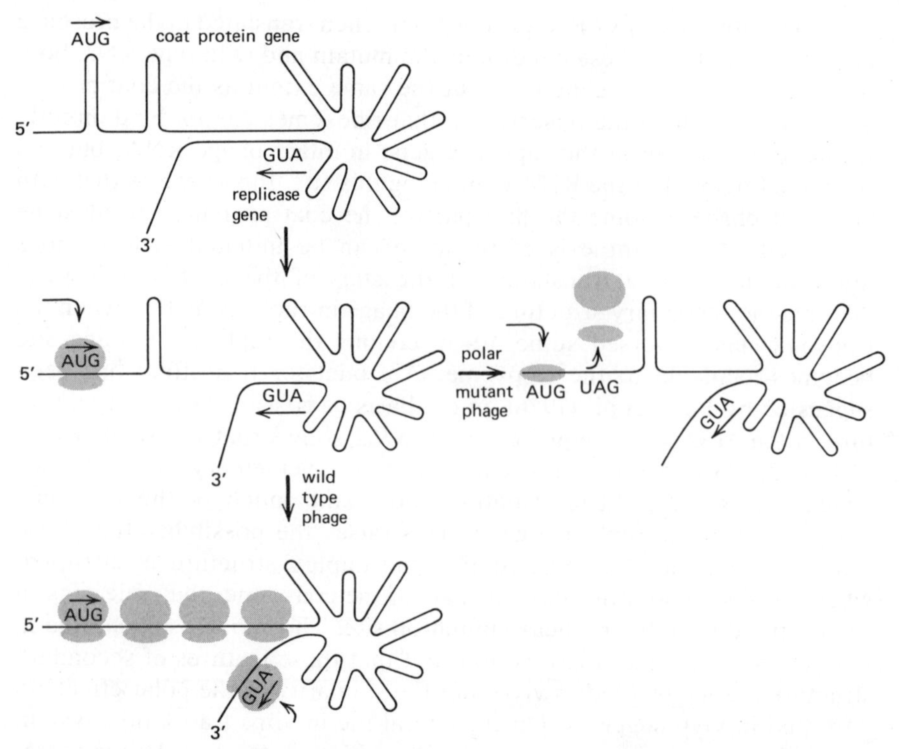


**Figure 9.5:** model for polarity in translation of f2 RNA. The intact phage RNA has a structure in which the initiation site of the replicase gene is base paired with part of the coat protein gene and therefore is unavailable for ribosome attachment. Ribosomes bind to the initiation site of the coat protein gene and commence translation only at this site. As the ribosomes proceed they disrupt the secondary structure of the phage RNA, converting it to an unpaired strand, and thus freeing the initiation site of the replicase gene. Once ribosomes have translated the first part of the coat protein gene, attachment to the initiation site of the replicase gene becomes possible. In a polar mutant, ribosomes leave the message at the mutant amber codon before they can disrupt the base paired structure between a later sequence in the coat protein gene and the start of the replicase gene.

ature; however, the thermophilic B.stearothermophilus system can continue to function.) The importance of secondary structure is strikingly illustrated by the rebinding experiments of Steitz (1973, 1975). In this protocol the binding site fragments were recovered from the ribosomes and then tested for their ability to rebind either B.stearothermophilus or E.coli ribosomes. This assays directly the ability of the ribosomes of each species to recognize the binding sites free of any effects introduced by secondary structure. As might be expected, the Bacillus ribosomes rebind only to the fragment from the A protein binding site. But E.coli ribosomes bind to the three sites roughly in the ratios 20 A protein:1 coat protein:2 replicase. This suggests that the affinities for the ribosome of the sites per se in fact lie in the reverse order from the frequencies with which they are used in vivo or in vitro. The discrepancy can be ascribed only to restriction of access by the secondary structure of the intact phage RNA.

Does the specificity of translation lie entirely in the recognition of their binding sites by the ribosomes or are other factors able to influence the choice of initiation sites? Two types of factor have been proposed for roles in discriminating between different ribosomal binding sites: the initiation factor IF3, which is necessary for binding of the 30S subunit to the message; and the "interference factor," i, which appeared to inhibit ribosome-messenger binding. The original impetus for the proposal that changes may occur in the ability of ribosomes to translate the RNA phage message was provided by studies of the observation that infection with phage T4 interferes with MS2 infection (see page 612). A series of experiments then showed that ribosomes from cells infected with T4 have a reduced ability to translate MS2 RNA in vitro and the effect appeared to reside in the initiation factors (Hsu and Weiss, 1969; Klem, Hsu and Weiss, 1970; Dube and Rudland, 1970; Lee-Huang and Ochoa, 1971). That the reduction in translational ability is specific was suggested by Steitz, Dube and Rudland (1970), who reported that when initiation factors are extracted from T4-infected cells there is a reduction in the translation in vitro of the coat protein gene relative to the other genes. Further reports on the properties of IF3 suggested that this factor might comprise more than one protein, each component having specificity for certain classes of ribosome binding sites (Lee-Huang and Ochoa, 1971, 1973; Yoshida and Rudland, 1972). In contrast with this idea, however, further studies showed that the change in ribosomes from T4-infected cells represents a generalized reduction in efficiency, applying equally to all messengers, and that preparations of IF3 from infected and uninfected cells contain only a single protein factor with constant properties (Goldman and Lodish, 1972; Schiff, Miller and Wahba, 1974).

Present evidence supports the idea that the initiation factors needed for formation of the complex between ribosome and messenger do not control the specificity of binding, in the sense that they do not discriminate between different types of ribosome binding site (see also page 614). But this does not answer the question of whether the 30S ribosome subunit or an initiation factor (presumably IF3) recognizes the binding sequence on the message. The conclusion that specificity resides solely in the ribosome was suggested by the experiments that Lodish (1970a) performed in which components of the E.coli and B.stearothermophilus systems were mixed. Both ribosomes and initiation factors are necessary for translation in vitro, but the ribosomes of one species can function with the initiation factors of the other; however, with ribosomes from E.coli the coat protein gene was recognized, and with ribosomes from B.stearothermophilus the A protein gene was translated, irrespective of the source of the initiation factors. This suggests that recognition of ribosome binding sites is a function solely of the structure of the 30S ribosomal subunit and of the nucleotide sequence of the messenger (and of course of its availability in the secondary structure).

The idea that recognition of some ribosome binding sites might be prevented in a specific manner was suggested by the isolation by Groner et al. (1972a) of an "interference factor" which preferentially inhibited translation of the MS2 coat protein and replicase genes in vitro. The nature of the i factor was identified by Inouye, Pollack and Petre (1974), who found that it is the 30S ribosomal subunit protein S1; using the purified protein they found preferential inhibition of coat protein synthesis with RNA phage templates. However, Gilbert and Kaempfer (1975b) reported that the protein inhibits translation of all R17 cistrons with equal effect. Because the factor is a ribosomal protein, assays of its activity are complicated by its presence on the 30S subunits. Van Dieijen et al. (1975) therefore determined the effect of adding S1 to in vitro protein synthetic systems whose ribosomes were depleted of the protein. No inhibition of protein synthesis was seen until an excess of the S1 protein was reached; several other ribosome proteins assayed under the same conditions also inhibited MS2 translation when present in excess. This suggests that translational repression caused by addition of S1 is an artifact of the in vitro system resulting from the presence of more copies of the protein than are needed for ribosome function. The suggestion that S1 binds to the messengers, and thus when in excess competes with the ribosomes for mRNA, is supported by observations that its inhibitory effect cannot be relieved by addition of an excess of components of the protein synthetic apparatus, but only by adding more phage template (Miller and Wahba, 1974; Gilbert and Kaempfer, 1975b). In addition to its function on the ribosome, S1 has another role: it is a component of the Q$\beta$ (and probably also the f2) RNA replicase. In this context we shall discuss its ability to bind RNA in the next section. But we should note here that S1 does appear to bind to specific nucleotide sequences and also with less specificity to sequences rich in pyrimidines. Although its role as an interference factor appears to be an artifact resulting from the presence of an excess amount of protein that binds to some initiation sites, and thus renders them unavailable to the ribosome, its preference for pyrimidines means that it might in fact bind more readily to some initiation sites and this could be the basis of its apparent (spurious) specificity of interference.

## Repression of Translation

Until the coat protein gene is translated, the replicase gene remains quiescent because its ribosome binding site has not been exposed. This effect resides in the secondary structure of the phage genome. Another interaction between these two functions is exercised by the coat protein product, which is able to bind to the start of the replicase gene and thus prevent ribosome binding. Thus translation of the coat protein gene first is necessary for replicase synthesis to be initiated; and then prevents replicase translation by the action of the coat gene product.

The first indication that coat protein might repress replicase synthesis was provided by genetic studies in which Lodish and Zinder (1966b) and Nathans et al. (1969) found that the *sus3* mutant not only loses its polarity when grown on the *su2* E.coli suppressor strain, but in fact then overproduces replicase. The same effect is displayed by the *sus11* amber mutant. Horiuchi and Matsuhashi (1970) observed a similar phenomenon with Qβ coat amber mutants that are polar in an $su^-$ host: in *su2* or *su3* suppressor hosts, replicase is overproduced. Since the overproduction is displayed in *su2* hosts, which insert glutamine, the amino acid present at the position in the wild type strains that is mutated, the basis for this effect cannot lie with an amino acid substitution, but must be quantitative. The probable explanation is that coat protein represses translation of the replicase gene; when an amber mutant is grown on a suppressor strain, sufficient translation past the mutant site is achieved to free the replicase binding site for ribosome attachment, but insufficient coat protein is synthesized to repress replicase synthesis later. Replicase is therefore overproduced. (An alternative model is to suppose that coat protein inhibits the synthesis of plus strand RNA beyond the coat protein gene, thus restricting the availability of a template for replicase synthesis; this has been proposed by Robertson, Webster and Zinder, 1968; and see Robertson, 1975).

The idea that coat protein directly represses replicase synthesis is supported by the demonstration that addition of coat protein to an in vitro synthetic system reduces the incorporation of histidine (a measure of the synthesis of maturation protein and replicase) relative to the incorporation of phenylalanine (which is present in all the phage proteins) (Sugiyama and Nakada, 1968; Eggen and Nathans, 1969). A complex that appears to be responsible for this repression was first identified by

Sugiyama, Hebert and Hartman (1967). When MS2 RNA is mixed with up to 6 coat protein molecules per genome, the coat protein binds to the nucleic acid to form a structure described as *complex I*. When larger amounts of coat protein are present, a structure called *complex II* is formed; this has the general appearance of the phage particle but is not infective.

When complex I is formed in vitro and incubated with T1 ribonuclease, a fragment of 59 nucleotides is protected against degradation by the coat protein. Bernardi and Spahr (1972) showed that the protected sequence includes the ribosome binding site of the R17 replicase gene, extending from the left (5′ side) up to the second codon of the replicase. Gralla, Steitz and Crothers (1974) showed that this fragment possesses a secondary structure of two hairpins, the first including the termination signal for the coat protein and the second containing the replicase initiation site. This suggests the model shown in Figure 9.6, in which synthesis of coat protein to reach a sufficient concentration results in the formation of complex I and thus the repression of replicase translation. This model explains why during infection replicase synthesis starts after a lag (required for translation of the early part of the coat protein gene) and then later, at 10 or 20 minutes, levels off (when the coat protein has reached a high concentration) (Lodish, Cooper and Zinder, 1964; Lodish and Zinder, 1966b). While the possibility has been raised that translation of the maturation protein is under a similar control, the evidence available is not decisive (for review see Robertson, 1975).

Although the genetic data with Qβ suggest that a similar interaction between coat protein and replicase synthesis may exist, direct evidence of the ability of coat protein to bind to a site controlling the replicase gene has not been obtained, and the only support for this model is by analogy.

An obvious topological problem is raised by the dual roles of the infecting RNA phage strand. Ribosomes translate the phage genes in the direction from 5′ to 3′. The replicase must use the genome as template for synthesis of a minus strand; the minus strand grows in the direction from 5′ to 3′ and thus moves along the template strand in the direction from 3′ end to 5′ end. What happens when a ribosome moving in one direction encounters a replicase moving in the opposite direction?

An experiment suggesting that translation and replication cannot take place simultaneously was performed by Kolakofsky and Weissmann (1971a). When the initiation complex of the 70S ribosome with fmet-$tRNA_f$ and Qβ RNA was used as a template for RNA synthesis, the replicase was unable to complete synthesis of the minus strand. This implies that the ribosome provides a block to progress of the polymerase.

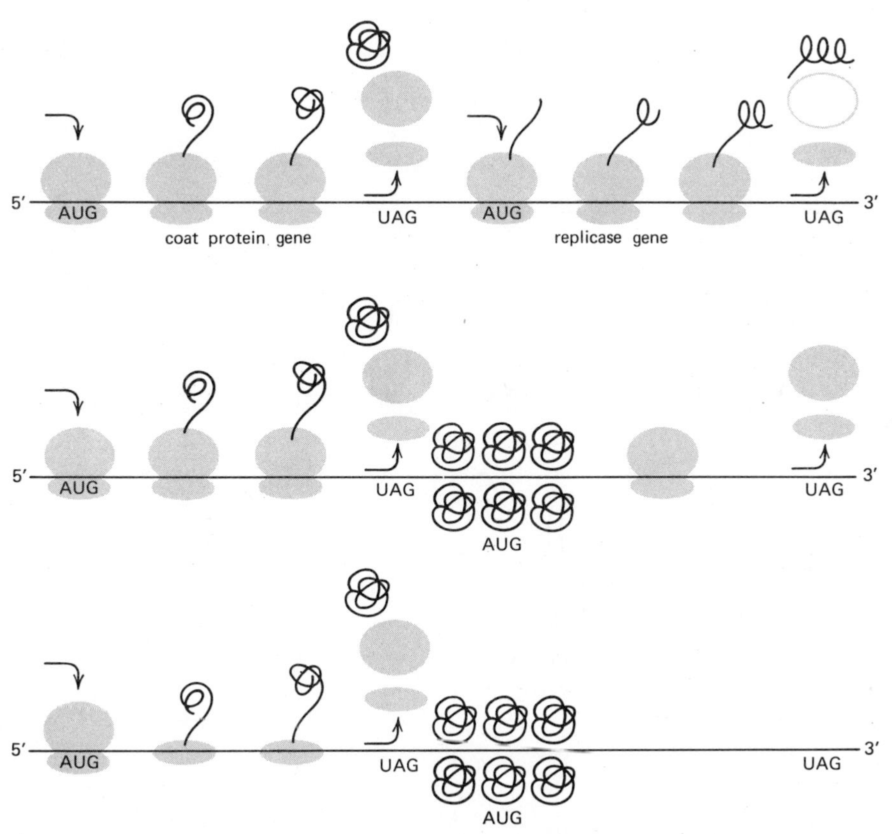


**Figure 9.6:** model for translational repression by f2 coat protein. Both the coat protein gene and replicase gene are under translation. When the concentration of coat protein becomes great enough, up to six molecules of coat protein bind to a region including the initiation site of the replicase gene; this prevents further ribosomes from attaching and ribosomes currently on the gene continue until they reach the termination codon. This clears the replicase gene of ribosomes but the coat protein gene remains under translation.

This problem appears to be overcome by an ability of the Qβ replicase to repress translation of the coat protein gene, an action which presumably leads indirectly to the inactivity also of the replicase gene. Kolakofsky and Weissmann (1971b) showed that the addition of the Qβ replicase to polysomes of Qβ RNA causes protein synthesis to cease as the polysomes are disrupted. Weber et al. (1972) identified the basis of the inhibition of translation by showing that Qβ replicase is able to bind to a site within the plus strand RNA which overlaps with the 5′ terminal half of

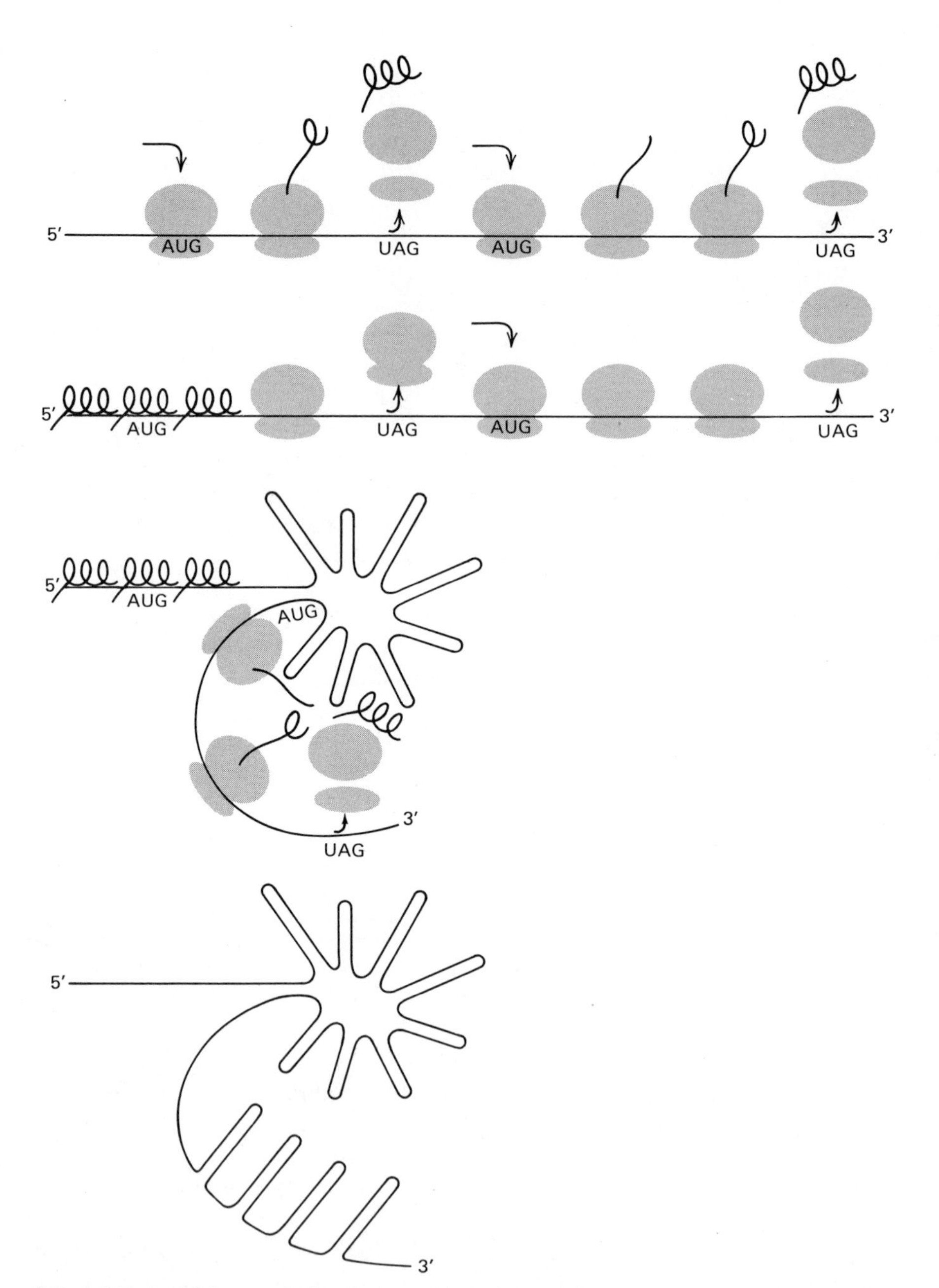


**Figure 9.7:** model for translational repression by Qβ replicase. Both the coat protein gene and the coat protein gene must be under translation before replicase molecules can be synthesized. The replicase binds to a region including the initiation site of the coat protein

the ribosome binding site of the coat protein gene. By visualizing Qβ RNA bound to the replicase, Vollenweider et al. (1976) identified two binding sites, at locations 30% and 65% from one end of the genome; Qβ RNA may be held in a loop with these two sites brought into juxtaposition by the replicase. The beginning of the coat protein gene lies about 30% from the left end of the genome and so this may represent the first site.

A model for the series of events triggered by the behavior of Qβ replicase as a translational repressor is illustrated in Figure 9.7. First the phage genome is translated into coat protein and replicase. Then the replicase binds to the coat protein initiation site, so that translation of the coat protein gene ceases when the ribosomes already on the message run off; this may lead to inhibition of translation of the replicase gene. Replicase initiating synthesis at the 3′ end of the plus strand then is free to synthesize the minus strand. (Since the repression of coat protein synthesis by replicase has been demonstrated with phage Qβ whereas the repression of replicase synthesis by coat protein has been demonstrated with phage f2, it is not proven that both control mechanisms operate with each phage. If they do, however, they may be reconciled by supposing that the replicase is an active translational repressor at rather low concentrations whereas the coat protein becomes a repressor only at somewhat higher concentrations; thus early in infection the replicase will function as repressor and as RNA synthetic enzyme to ensure replication, while later in infection the coat protein concentration will become sufficient to ensure the shutoff of the translation of replicase that is no longer needed.) Once a minus strand has been synthesized, it can act as template for synthesis of new plus strands by the replicase; these strands may be translated as they are synthesized, since replicase and ribosomes will be moving in the same direction. Indeed, Robertson (1975) has suggested that these nascent strands may provide the only template that is used in the cell for synthesis of the A protein; when f2 strands grow much beyond 500 nucleotides, they may acquire the secondary structure characteristic of the intact phage genome in which the A protein gene is translated very inefficiently if at all. When the new plus strands are complete, presumably some continue to be translated while others may be used as templates for

---

gene, preventing further ribosomes from initiating translation; ribosomes on the coat protein gene continue until reaching the termination codon and there is at this stage no impediment to continued initiation and elongation of replicase protein. When ribosomes run off the coat protein gene, the RNA acquires secondary structure and binding to the initiation region of the replicase gene prevents further initiation; ribosomes on the replicase gene run off. The phage RNA then is denuded of ribosomes and can function as template for synthesis of RNA. (The secondary structures shown in this figure are drawn for illustrative purposes only and do not reflect any particular structure known to be taken up by the phage RNA.)

synthesis of further minus strands. Obviously this model leaves unanswered problems of timing in this series of events, for example concerning the inhibition of replicase synthesis that must presumably follow the repression of coat protein synthesis; but there is evidence to suggest that many of the individual stages operate in vivo, even if it is not yet possible to say for certain how they interact to form a control circuit that governs phage translation.

### Replication of Phage RNA

Replication of the RNA phages represents a unique situation in which RNA serves as the template for synthesis of new polyribonucleotide chains. Although relying upon base pairing with the template to direct the sequence of the product, the reaction differs from all other known processes of nucleic acid synthesis in directly generating single strand rather than duplex structures. In the replication of duplex DNA, each parental strand acts as template for synthesis of the progeny strand with which it remains associated, so that both template and product are duplex except for the transient region of single strand material present at the replicating fork. Single strand DNA replication involves the synthesis of a duplex molecule from the single stranded genome; and this then replicates as duplex DNA. In rolling circle replication to generate progeny single strands, the product at the growing point remains double stranded and single stranded material is generated by displacement of the previously paired strand. Transcription of RNA from a DNA template also involves a displacement reaction: the template DNA is unwound, RNA is synthesized by base pairing with one of the template strands, and then the original DNA duplex reforms to displace the RNA transcript. Replication of the RNA tumor viruses takes place by reverse transcription to generate a DNA representing the viral genome; RNA may later be transcribed from the DNA copy. Synthesis of the DNA complement, formation and replication of duplex DNA, and transcription of RNA all involve the persistence of duplex structures. The template of the RNA phage, however, gives rise directly to a single stranded product: thus both template and product take the form of single strands of RNA and a duplex structure is formed only transiently at the growing point, a mechanism interesting for its difference from the involvement of stable duplex structures in all other nucleic acid synthesis.

In early experiments, RNA extracted from infected cells took the form of duplex molecules, resistant to single strand ribonuclease, amongst which two types of structure could be distinguished: rods representing a plus strand paired with a minus strand; and branched molecules consisting of duplexes with additional single strand protrusions (Weissmann et al., 1964a,b; Amman, Delius and Hofschneider, 1964; Francke and Hofschneider, 1966; Granboulan and Franklin, 1966). However, the presence of duplex structures appears to depend upon the removal of protein during extraction of the replicating intermediates. When Feix, Slor and Weissmann (1967) isolated the replicating structures of Qβ in the form of a 40S complex containing the template strand, progeny strand, and RNA replicase, they found that less than 10% of the product is resistant to ribonuclease specific for single strands; but upon removal of the protein by phenol extraction, more than 70% of the product becomes resistant to RNAase. Keil and Hofschneider (1973) reported that both forms of duplex structure of M12 are generated only after purification procedures that remove protein and Thach and Thach (1973) also have reported that they could not detect duplex material of R17 in lysates prior to removal of protein.

These results suggest the model shown in Figure 9.8 in which the replicase synthesizes a new strand by utilizing base pairing between the template and incoming nucleotides, but the duplex region is confined to the sequences covered by the enzyme; this implies that the enzyme breaks the hydrogen bonds between template and product as it moves along, thus maintaining both template and product in the single stranded state except for the region of the growing point. When protein is extracted, however, the complementary sequences of the plus and minus strand collapse into a duplex structure. Formation of the duplex in vivo probably is prevented by the acquisition of secondary structure by template and product as soon as the hydrogen bonds between them are broken by the enzyme.

Replication of Qβ proceeds through two stages. First the infecting plus strand is used as template for synthesis of the complementary minus strand; and then the minus strand is used as template for the synthesis of plus strands. The two reactions have different requirements: synthesis of the minus strand requires the phage replicase and a further protein factor; synthesis of the plus strand can be accomplished by the replicase alone. The production of plus strands outnumbers the production of minus strands in vivo by a factor of about 10 (Weissmann, Colthart and Libonati, 1968); in itself this is an argument against replication being semiconservative. While the only function of the minus strand is to serve as template for plus strand synthesis, the plus strand acts as template for both RNA (minus strand) synthesis and protein synthesis, and also is packaged into phage particles. One reason for the predominance of plus strand synthesis may be that plus strand products are removed from the

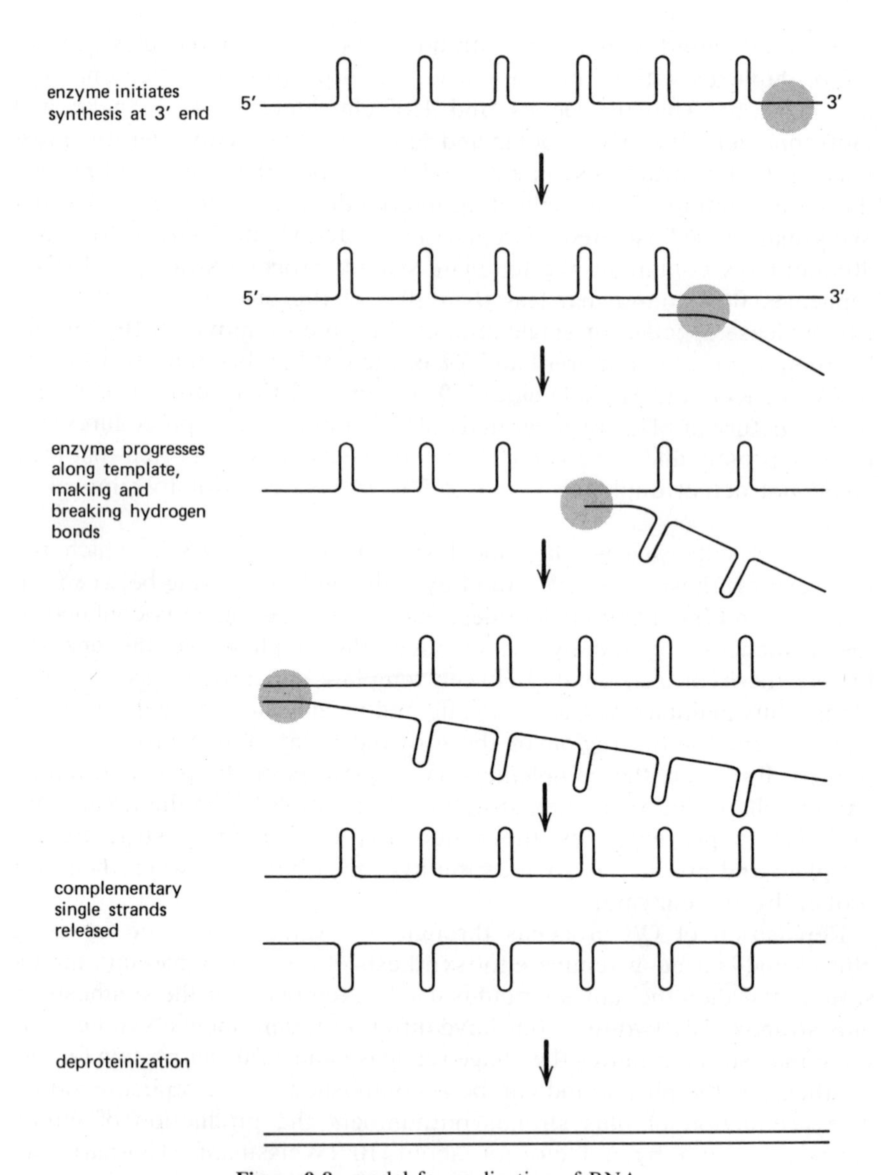


**Figure 9.8:** model for replication of RNA.

pool of templates available for RNA synthesis as they are packaged into the virion, but the minus strands continue to remain available as templates for plus strand synthesis (for review see Kamen, 1975).

The synthesis of Q$\beta$ RNA in vitro has made possible the resolution of the stages of replication and has allowed the roles of the components of the replicase to be investigated. Obviously a critical test of the fidelity of the in vitro system is given by its ability to generate progeny plus strands that are infectious; and this capacity was first demonstrated by Spiegelman et al. (1965). (Since minus strands are not infective, activity in an infectivity assay is a measure of plus strand synthesis alone.) Weissmann, Feix and Slor (1968) delineated the stages of the reaction by showing that labeled precursors first (at 2–4 minutes) enter a 40S complex which consists of parental plus strands and newly synthesized minus strands. A small proportion of the minus strands enters a 30S peak in the form of free RNA. However, most of the minus strand products are used as templates for the synthesis of plus strands; by 6 minutes of reaction this second stage is well under way, and at this time most of 30S peak of free RNA represents plus strand progeny. More than one product strand may be under synthesis on each template strand; by measuring the ratio of label in end groups to internal positions in an isolated fraction of replicative intermediates of MS2 (mostly consisting of minus strand templates and plus strand products), Fiers et al. (1969) calculated that each complex on average carries 2.5 nascent chains. In other words, each template strand carries 2 to 3 replicase molecules engaged in synthesis of the complementary strand.

Characterization of the RNA replicase has proved somewhat easier with Q$\beta$ than with the phages of the f2 group. Both replicase enzymes contain four subunits, only one of which is the polypeptide specified by the phage. Kondo, Gallerani and Weissmann (1970) and Kamen (1970) identified the phage coded component of the Q$\beta$ replicase by labeling infective cells with $^{35}S$ and mixing an extract from these cells with an extract from unlabeled uninfected cells; only one subunit of the replicase was labeled, and this must be the phage coded component. In the reverse experiment, uninfected cells were labeled and infected cells were unlabeled; the other three subunits of the replicase then were labeled, confirming their source as the host bacterium. A similar situation in which host proteins are sequestered for a phage function is seen with the DNA polymerase of phage T7 (see Chapter 7).

The three subunits that the host provides for the Q$\beta$ replicase all have been identified as components of the protein synthetic apparatus. The smallest two subunits of the replicase were identified as the elongation factors Tu and Ts by Blumenthal, Landers and Weber (1972); this relationship is supported by demonstrations of both chemical identity and functional equivalence. The largest subunit of the replicase is provided by ribosomal protein S1, also identified in some experiments as the interference factor, i (Kamen et al., 1972; Groner et al., 1972b; Miller, Niveleau and Wahba, 1974; Wahba et al., 1974; Inouye, Pollack and Petre, 1974). The constitution of the Qβ replicase therefore may be summarized as shown in Table 9.1 (for review see Kamen, 1975).

In the form consisting of four protein subunits, the Qβ replicase is able to use minus strands as templates for plus strand synthesis; but it is not able to synthesize the minus strand upon a plus strand template. Additional proteins, described as the host factors I and II, allow the enzyme to use the plus strand as template (De Fernandez, Eoyang and August, 1968; August et al., 1968; August, 1969). De Fernandez, Hayward and August (1972) purified host factor I (HFI) in the form of a hexamer consisting of subunits of about 12,000 daltons. The amount of the protein required to support synthesis of the minus strand is proportional to the amount of plus strand template; this suggests a stoichiometric reaction in which one hexamer is required for each viral genome (for review see Kuo, Eoyang and August, 1975).

The role of HFII has always been more dubious, because the need for this factor can be fulfilled by any one of several basic proteins; and in some laboratories no dependence on HFII could be demonstrated for the synthesis of minus strands (Kamen et al., 1972). Kamen, Monstein and Weissmann (1974) clarified this situation when they reported that dependence of synthesis on HFII is a feature of the preparation of template rather than of the replicase. They suggested that this factor activity becomes necessary only when the template is contaminated by an inhibitor, whose action can be overcome by the presence of basic proteins. In view of the lack of specificity for HFII, it seems that the four subunits

**Table 9.1:** structure and function of the Qβ RNA replicase

| subunit | molecular weight | source | function |
|---|---|---|---|
| I | 70,000 | ribosomal protein Sl | Qβ RNA binding |
| II | 65,000 | phage coded | chain elongation |
| III | 45,000 | elongation factor Tu | initiation |
| IV | 35,000 | elongation factor Ts | initiation |
| HFI | 72,000 (hexamer) | ribosome-associated protein | minus strand synthesis |

of the replicase and the host factor I hexamer can be regarded as constituting the Qβ replication apparatus.

The Qβ replicase in either complete or partial form (that is lacking some of the subunits) is active as a poly(G) polymerase on a poly(C) template (Hori et al., 1967). The same activity is displayed by the f2 replicase and this has been used as an assay for its purification. The replicase of the f2 phages has proved more difficult to work with than the Qβ enzyme because it is less stable and usually is strongly complexed with endogenous phage RNA. Relying upon the poly(G) polymerase activity, however, Federoff and Zinder (1971, 1972a,b, 1973) have purified an enzyme whose components all are close in molecular weight to those of the Qβ enzyme. The phage coded component is slightly smaller in f2 than in Qβ; in view of their similarity of sizes, the other subunits probably are the same host components, S1, Tu, Ts, that are sequestered for the Qβ replicase, although this has not been confirmed by demonstrating chemical identity or functional equivalence. However, in the form of the tetrameric protein, the f2 replicase cannot use either plus or minus strands as template, but requires an additional host factor (for review see Federoff, 1975). In this it differs from the ability of the Qβ replicase itself to use minus strand templates. If the bacterial components of each replicase are the same, this difference in specificity must reside in the phage component. The additional factor(s) needed for f2 replicase activity have not yet been purified and so cannot be compared with the Qβ HFI.

The capacity to dissociate the RNA replicase into its components and then to reconstitute an active enzyme is central to attempts to define the roles of the individual subunits. While this approach has been pursued successfully with the Qβ replicase, the difficulties of working with the f2 replicase have so far stood in the way of extending this analysis to the second phage system. Early work showed that centrifugation in sucrose gradients separates the Qβ replicase into two components, although these were not characterized at the time (Eikhom and Spiegelman, 1967; Spiegelman et al., 1968). Kamen et al. (1972) were able to isolate Qβ replicase in a form lacking subunit I (protein S1), apparently generated during isolation. The enzyme has nearly normal activity with the Qβ minus strand as template, but cannot use the plus strand to direct RNA synthesis. Landers, Blumenthal and Weber (1974) showed that when this enzyme is allowed to initiate RNA synthesis on poly(A,C)—a template on which it is active because of the high content of C residues—and then is centrifuged, only subunit II remains bound to the template: subunits III and IV are released in the form of the EF Tu-Ts complex. The phage coded subunit II then can elongate chains on the preincubated complexes;

this suggests that it is this subunit that is responsible for synthesis of the growing chain, a conclusion consistent with the roles in protein synthesis of the host components (none of which is involved in the synthesis of phosphodiester bonds). This is to take the view that the host functions used in the replicase are not redirected into activities involved in RNA synthesis per se, but represent functions ancillary to the essential catalytic activity of the Qβ-coded subunit.

When the dissociated I + II subunits and the III + IV complex are mixed, active enzyme does not result; however, the enzyme recovers its activity if the reconstitution is performed by mixing in 8M urea and then removing the urea. This contrasts with the simple reconstitution that is possible when subunit I is added to enzyme deficient in it. The need for EF Tu-Ts to be present with subunit II before removal of the urea by dialysis suggests that the presence of these host components during this process is necessary if the proper structure of the complete enzyme is to be formed. Presumably the enzyme must be constructed in vivo by interaction between subunit II and S1 and Tu-Ts, either under conditions using complete proteins that mimic recovery in urea, or perhaps by association with subunit II before its synthesis is complete, that is, before it has taken up its secondary structure.

Inhibitors of protein synthesis that act by blocking the functions of the elongation factors do not prevent RNA synthesis. Analogues of GTP that inhibit the ribosome-dependent GTPase activity of Tu leave RNA synthesis unimpaired. Brown and Blumenthal (1976a,b) showed that treatment with N-ethylmaleimide (which inhibits the ability of Tu to bind aminoacyl-tRNA) or with kirromycin (which prevents association of Tu-Ts and renders the GTPase activity independent of the ribosome) does not interfere with RNA synthesis. This suggests that Tu-Ts may function as a complex within the replicase, since it is only as separate subunits that the functions of Tu and Ts in protein synthesis are inhibited by these agents. This conclusion is strengthened by the observation that Tu-Ts must be in complex form to support the reconstitution of the replicase from urea; and, indeed, when the two subunits are cross linked together with the agent dimethyl suberimidate, Tu-Ts is more effective in supporting the reconstitution reaction.

Cross-linking reagents identify three types of protein interaction in the replicase. Cross-linked structures may include all four subunits, subunits I and II, or subunits III and IV (Young and Blumenthal, 1975; Blumenthal, Young and Brown, 1976). The addition of RNA changes the conformation of the replicase in such a way that cross-linking of subunits I–II to subunits III–IV no longer occurs. This effect can be produced by any RNA. Also in the presence of Qβ RNA, centrifugation separates the enzyme into RNA bound to subunits I–II and a free complex of subunits III–IV. However, all four subunits usually remain associated during the course of RNA synthesis: Tu-Ts are not released at initiation. By preparing antibodies against each of the four components of the enzyme, Carmichael, Landers and Weber (1976) showed that anti-Tu and anti-Ts remain inhibitory throughout the reaction, implying the continued presence of these components. The ability of subunit II alone to elongate RNA chains implies that, although playing an active role only during initiation, the Tu-Ts complex remains bound to subunit II in such a way that interference with it may block RNA synthesis. On the other hand, anti-S1 blocks the reaction only if added before initiation; this fits with the ability of replicase lacking subunit I to utilize minus strand templates and suggests that S1 does not interact with the other components in an important way after chain synthesis has begun.

The role of S1 as a component of the replicase may be related to its function in protein synthesis, which appears to be concerned with binding to the messenger. Dahlberg and Dahlberg (1975) reported that S1 binds to a 40 nucleotide fragment cleaved from the 3′ end of the 16S ribosomal RNA, and in particular seems to display an affinity for the sequence ACCUCC which may be involved in pairing with the complementary sequence in mRNA (see Figure 9.4). The ability of S1 to bind Qβ and R17 RNA has been characterized by Senear and Steitz (1976), using a protocol in which the RNA is mixed with the protein, digested with T1 ribonuclease, and then the fragments protected by the protein are collected on nitrocellulose filters. The sequence of Qβ RNA lying from nucleotides 38–63 from the 3′ end appears to be bound specifically; R17 RNA appears to be bound in a nonspecific manner. The sequence bound in Qβ is close enough to the 3′ end—especially when making allowance for the possibility that the secondary structure may bring the nucleotides closer to the terminus into juxtaposition with it—to suggest that binding here may be part of the initiation reaction. If the failure of the nitrocellulose assay to detect specific binding to R17 RNA is correct, then it seems possible that S1 may play different roles in Qβ and R17 RNA replication. Of course, these experiments concern binding to the plus strand; affinity of S1 for the minus strand has not been investigated.

The role of host factor I also may be concerned with binding to the RNA template. Carmichael et al. (1975) showed that this protein is associated with the ribosomes, although it has not been implicated in any role in host protein synthesis. It has an affinity for purine rich sequences, which has proved useful in its purification. (Since S1 has an affinity for pyrimidine rich sequences, HFI is attracted by purine rich sequences, and no doubt other proteins with such properties also exist, it is tempting to speculate that the presence of these proteins may generate the apparently specific interference effects that sometimes have been noted in vitro. While the role of S1 in this capacity is well characterized, the possibility that similar effects are produced by HFI and other such proteins is worth mentioning.)

The ability of HFI to bind to phage RNA has been characterized by Senear and Steitz (1976), using the nitrocellulose filter technique. Specific sequences of both R17 and Qβ RNA are protected against degradation by ribonuclease. The binding sequence of R17 RNA is a purine rich fragment derived from the replicase gene. With Qβ RNA two fragments are protected, one a sequence within the replicase gene but several hundred residues from the 3′ end, the other the same sequence near the 3′ end that is bound by S1. While the role of binding within the phage RNA is not clear, as with the binding of S1 it seems likely that the reaction with the sequence near the 3′ terminus may be significant. Whether binding at this site directly assists initiation or whether it perhaps helps to make available the 3′ terminal sequence to the enzyme remains to be seen. Since neither S1 nor HFI appears necessary for plus strand synthesis on minus strand templates, a useful control for the sequence specificity of these factors may be to see whether they can bind to the minus strand of Qβ.

A phenomenon that may cast some light on the need of Qβ replicase for specific binding sequences is the existence of small RNA molecules that can be replicated by the enzyme. Bannerjee, Rensing and August (1969) first observed that a 6S RNA is associated with a partially purified replicase preparation; it is about 110–130 nucleotides in length and by virtue of its resistance to ribonuclease appears to have an extensively base paired structure. Prives and Silvermann (1972) showed that the 6S RNA can be replicated by Qβ replicase lacking host factor and that it is active as a template only when denatured. Other small RNAs that can be replicated by the enzyme have been isolated by Kacian et al. (1972), Mills, Kramer and Spiegelman (1973) and Mills et al. (1975). One of these, MDV-1, has been sequenced and comprises a unique series of 218 nucleotides; another variant consists of an entirely different sequence. The implication seems to be that these small RNA molecules may be replicated by the enzyme although they apparently lack common sequences; thus the Qβ replicase may be able to recognize more than a single class of initiation site. And perhaps the use of minus strand RNA templates to define the sequences recognized by the enzyme in the absence of host factor may be interesting in revealing its specificity of action, which is high at least so far as intact phage RNA templates are concerned, since Qβ RNA is the only template that can be used.

# References


Abelson R.D. and Dubnau D. (1971). Fate of transforming DNA after uptake by competent B.subtilis: failure of donor DNA to replicate in a recombination deficient recipient. *Proc. Natl. Acad. Sci. USA* **68,** 1070–1074.

Achtman M. (1975). Mating aggregates in E.coli conjugation. *J. Bacteriol.* **123,** 505–515.

Achtman M., Willetts N. and Clark A.J. (1971). Beginning a genetic analysis of conjugational transfer determined by the F factor in E.coli by isolation and characterization of transfer-deficient mutants. *J. Bacteriol.* **106,** 529–538.

Achtman M., Willetts N. and Clark A.J. (1972). Conjugational complementation analysis of transfer-deficient mutants of Flac in E.coli. *J. Bacteriol.* **110,** 831–842.

Adams J.M., Spahr P.-F. and Cory S. (1972). Nucleotide sequence from the 5′ end to the first cistron of R17 phage RNA. *Biochemistry* **11,** 976–988.

Adams J.M., Jeppesen P.G.V., Sanger F. and Barrell B.G. (1969). Nucleotide sequence from the coat protein cistron of R17 bacteriophage RNA. *Nature* **223,** 1009–1015.

Adelberg E.A. and Burns S.N. (1960). Genetic variations in the sex factor of E.coli. *J. Bacteriol.* **79,** 321–330.

Adelberg E.A. and Pittard J. (1965). Chromosome transfer in bacterial conjugation. *Bacteriol. Rev.* **29,** 161–172.

Adesnik M. and Levinthal C. (1970). RNA metabolism in T4 infected E. coli. *J. Mol. Biol.* **48,** 187–208.

Adhya S., Cleary P. and Campbell A. (1968). A deletion analysis of prophage λ and adjacent genetic regions. *Proc. Natl. Acad. Sci. USA* **61,** 956–962.

Adhya S., Gottesman M. and De Crombrugghe B. (1974). Release of polarity in E.coli by gene N of phage lambda: termination and antitermination of transcription. *Proc. Natl. Acad. Sci. USA* **71,** 2534–2538.

Akiba T., Koyama T., Ishiki Y., Kimura S. and Fukushima T. (1960). Studies on the mechanism of development of multiple drug resistant Shigella strains (*in Japanese*). *Nihon Iji Shimpo* **1866,** 45–60.

Alberts B.M. (1970). Function of the gene 32 protein, a new protein essential for the genetic recombination and replication of T4 phage DNA. *Fed. Proc.* **29,** 1154–1163.

Alberts B.M. and Frey L. (1970). T4 phage gene 32: a structural protein in the replication and recombination of DNA. *Nature* **227,** 1313–1317.

Alberts B.N., Frey L. and Delius H. (1972). Isolation and characterization of gene 5 protein of filamentous bacterial viruses. *J. Mol. Biol.* **68,** 139–152.

Alexander H.E. and Leidy G. (1951). Determination of inherited traits of H.influenzae by desoxyribonucleic acid fractions isolated from type specific cells. *J. Exp. Med.* **93,** 345–359.

Alexander H.E., Leidy G. and Hahn E. (1954). Studies on the nature of H.influenzae cells susceptible to heritable changes by DNAs. *J. Exp. Med.* **99,** 505–533.

Alfaro G. and Willetts N.S. (1972). The relationship between the transfer systems of some bacterial plasmids. *Genet. Res.* **20,** 279–289.

Allet B. (1973). Fragments produced by cleavage of λ DNA with the H.parainfluenzae restriction enzyme Hpa II. *Biochemistry* **12,** 3972–3977.

Allet B. and Solem R. (1974). Separation and analysis of promotor sites in phage lambda DNA by specific endonucleases. *J. Mol. Biol.* **85,** 475–484.

Allet B., Roberts R.J., Gesteland R. and Solem R. (1974). Class of promotor sites for E.coli DNA-dependent RNA polymerase. *Nature* **249,** 217–221.

Alloway J.L. (1932). The transformation in vitro of R pneumococcus into S forms of different specific types by the use of filtered pneumococcus extracts. *J. Exp. Med.* **55,** 91–99.

Alloway J.L. (1933). Further observations on the use of pneumococcus extracts in effecting transformation of type in vitro. *J. Exp. Med.* **57,** 265–278.

Almendinger R. and Hager L.P. (1972). Role for endonuclease I in the transmission process of colicin E2. *Nature New Biology* **235,** 199–203.

Altman S. (1975). Biosynthesis of tRNA in E.coli. *Cell* **4,** 21–29.

Amati P. and Meselson M. (1965). Localised negative interference in bacteriophage λ. *Genetics* **51,** 369–379.

Ammann J., Delius H. and Hofschneider P.H. (1964). Isolation and properties of an intact phage specific form of RNA phage M12. *J. Mol. Biol.* **10,** 557–561.

Anderson E.S. (1968). The ecology of transferable drug resistance in enterobacteria. *Ann. Rev. Microbiol.* **22,** 131–180.

Anderson E.S. and Lewis M.J. (1965). Characterization of a transfer factor associated with drug resistance in S.typhimurium. *Nature* **208,** 843–849.

Anderson E.S. and Natkin E. (1972). Transduction of resistance determinants and R-factors of the Δ transfer system by phage Plkc. *Molec. Gen. Genetics* **114,** 261–265.

Anderson E.S., Keleman M.V., Jones C.M. and Pitton J.S. (1968). Study of the association of resistance to two drugs in a transferable determinant in S.typhimurium. *Genet. Res.* **11,** 119–123.

Anderson T.F., Wollman E.L. and Jacob F. (1957). Sur les processes de conjugaison et de recombinaison chez E.coli. III. Aspects morphologiques en microscopie electronique. *Ann. Inst. Pasteur* **93,** 45–455.

Anraku N. and Lehman I.R. (1969). Enzymic joining of polynucleotides. VII. Role of the T4-induced ligase in the formation of recombinant molecules. *J. Mol. Biol.* **46,** 467–479.

Anraku N., Anraku Y. and Lehman I.R. (1969). Enzymic joining of polynucleotides. VIII. Structure of hybrids of parental T4 DNA molecules. *J. Mol. Biol.* **46,** 481–492.

Aposhian H.V. and Kornberg A. (1962). Enzymatic synthesis of DNA. IX. The polymerase formed after T2 phage infection: a new enzyme. *J. Biol. Chem.* **237,** 519–525.

Appleyard R.K. (1954). Segregation of lambda lysogenicity during bacterial recombination in E.coli K12. *Genetics* **39,** 429–439.

Arber W., Kellenberger G. and Weigle J. (1957). La defectuosite du phage lambda transducteur. *Schweiz. Z. Pathol. Bakteriol.* **30,** 659–665.

Argetsinger J.E. and Gussin G. (1966). Intact RNA from defective particles of R17. *J. Mol. Biol.* **21,** 421–434.

Arscott P.G. and Goldberg E.B. (1976). Cooperative action of the T4 tail fibers and baseplate in triggering conformational change and in determining host range. *Virology* **69,** 15–22.

Asbeck V.F., Beyreuther K., Kohler H., Von Wettstein G. and Braunitzer G. (1969). Die konstitution des hullproteins des phagen fd. *Hoppe Seyler Z. Physiol. Chem.* **350,** 1047–1066.

Atkins J.F. and Gesteland R.F. (1975). The synthetase gene of the RNA phages R17, MS2 and f2 has a single UAG termination codon. *Molec. Gen. Genet.* **139,** 19–32.

August J.T. (1969). Mechanism of synthesis of bacteriophage RNA. *Nature* **222,** 121–123.

August J.T., Bannerjee A.K., Eoyang L., De Fernandez M.T., Hori K., Kuo C.H., Rensing U. and Shapiro L. (1968). Synthesis of bacteriophage Qβ RNA. *Cold Spring Harbor Symp. Quant. Biol.* **33,** 73–82.

Avery O.T., Macleod C.M. and McCarty M. (1944). Studies on the chemical nature of the substance inducing transformation of pneumococcal types. I. Induction of transformation by a DNA fraction isolated from pneumococcus type III. *J. Exp. Med.* **79,** 137–158.

Baas P.D. and Jansz H.S. (1972a). Asymmetric information transfer during φX174 DNA replication. *J. Mol. Biol.* **63,** 557–568.

Baas P.D. and Jansz H.S. (1972b). φX174 replicative form DNA replication, origin and direction. *J. Mol. Biol.* **63,** 569–576.

Baas P.D., Jansz H.S. and Sinsheimer R.L. (1976). Phage φX174 DNA synthesis in a replication-deficient host: determination of the origin of φX DNA replication. *J. Mol. Biol.* **102,** 633–656.

Bachman B.J., Low K.B. and Taylor A.L. (1976). Recalibrated linkage map of E.coli K12. *Bacteriol. Rev.* **40,** 116–167.

Bachrach U. and Benchetrit L. (1974). Studies on phage internal proteins. III. Specific binding of T4 internal proteins to T4 DNA. *Virology* **59,** 443–452.

Bachrach U., Levin R. and Friedman A. (1970). Studies on phage internal proteins: isolation of protein-DNA complexes from T2 phages and from phage infected bacteria. *Virology* **40,** 882–892.

Baker R. and Tessman I. (1967). The circular genetic map of S13. *Proc. Natl. Acad. Sci. USA* **58,** 1438–1445.

Baker R., Doniger J. and Tessman I. (1971). Roles of parental and progeny DNA in two mechanisms of phage S13 recombination. *Nature New Biol.* **230,** 23–25.

Ball L.A. and Kaesberg P. (1973). A polarity gradient in the expression of the replicase gene of RNA phage Qβ. *J. Mol. Biol.* **74,** 547–562.

Bannerjee A.K., Rensing U. and August J.T. (1969). Replication of RNA viruses. X. Replication of a natural 6S RNA by the Qβ RNA polymerase. *J. Mol. Biol.* **45,** 181–193.

Bannister D. and Glover S.W. (1970). The isolation and properties of nonrestricting mutants of two different host specificities associated with drug resistance factors. *J. Gen. Microbiol.* **61,** 63–71.

Barbour S.D. and Clark A.J. (1970). Biochemical and genetic studies of recombination proficiency in E.coli. I. Enzymatic activity associated with $recB^+$ and $recC^+$ genes. *Proc. Natl. Acad. Sci. USA* **65,** 955–961.

Barbour S.D., Nagaishi H., Templin A. and Clark A.J. (1970). Biochemical and genetic studies of recombination proficiency in E.coli. II. $rec^+$ revertants caused by indirect suppression of $rec^-$ mutations. *Proc. Natl. Acad. Sci. USA* **67,** 128–135.

Baricelli N.A. (1956). A chromosomic recombination theory for multiplicity reactivation in phages. *Acta Biotheoretica, Leiden* **11,** 107–120.

Barnett L., Brenner S., Crick F.H.C., Schulman R.G. and Watts-Tobin R.J. (1967). Phase shift and other mutants in the first part of the *rII* cistron of phage T4. *Phil. Trans. Roy. Soc. B* **252,** 487–560.

Barrell B.G., Air G.M. and Hutchison C.A. III (1976). Overlapping genes in phage φX174. *Nature,* **264,** 33–40.

Barrell B.G., Seidman J.G., Guthrie C. and McClain W.H. (1974). Transfer RNA bisynthesis: the nucleotide sequence of a precursor to serine and proline tRNAs. *Proc. Natl. Acad. Sci. USA* **71,** 413–416.

Barrington L.F. and Kozloff L.M. (1954). Action of $T2r^+$ phage on host cell membranes. *Science* **120,** 110–111.

Barry J. and Alberts B. (1972). In vitro complementation as an assay for new proteins required for phage T4 DNA replication. *Proc. Natl. Acad. Sci. USA* **69,** 2717–2721.

Barta K. and Zissler J. (1974). Role of genetic recombination in DNA replication of phage λ. II. Effect in DNA replication by gene delta. *J. Virol.* **14,** 1451–1462.

Barta K., Tavernier P. and Zissler J. (1974). Role of genetic recombination in DNA replication of phage λ. I. Genetic characterization of the delta gene. *J. Virol.* **14,** 1445–1450.

Barth P.T., Datta N., Hedges R.W. and Grinter N.J. (1976). Transposition of a DNA sequence encoding trimethoprim and streptomycin resistances from R483 to other replicons. *J. Bacteriol.* **125,** 800–810.

Bartok K., Harbers B. and Denhardt D.T. (1975). Isolation and characterization of self-complementary sequences from φX174 viral DNA. *J. Mol. Biol.* **99,** 93–106.

Bassel B.A. and Spiegelman S. (1967). Specific cleavage of Qβ RNA and identification of the fragment carrying the 3′-OH terminus. *Proc. Natl. Acad. Sci. USA* **58,** 1155–1161.

Bastia D., Sueoka N. and Cox E.C. (1975). Studies on the late replication of phage lambda: rolling circle replication of the wild type and a partially suppressed strain, Oam29 Pam80. *J. Mol. Biol.* **98,** 305–320.

Bautz E.K.F. and Bautz F.A. (1970). Initiation of RNA synthesis: the function of sigma in the binding of RNA polymerase to promotor sites. *Nature* **226,** 1219–1222.

Bautz E.K.F. and Dunn J.J. (1969). DNA-dependent RNA polymerase from phage T4 infected E.coli: an enzyme missing a factor required for transcription of T4 DNA. *Biochem. Biophys. Res. Commun.* **34,** 230–237.

Bautz E.K.F., Bautz F.A. and Dunn J.J. (1969). E.coli σ factor: a positive control element in phage T4 development. *Nature* **223,** 1022–1024.

Baylor M.B., Hurst D.D., Allen S.L. and Bertani E.T. (1957). The frequency and distribution of loci affecting host range in the coliphage T2H. *Genetics* **42,** 104–120.

Bazaral M. and Helinski D.R. (1968a). Circular DNA forms of colicinogenic factors E1, E2 and E3 from E.coli. *J. Mol. Biol.* **36,** 185–194.

Bazaral M. and Helinski D.R. (1968b). Characterization of multiple circular DNA forms of colicinogenic factor El from P.mirabilis. *Biochemistry* **7,** 3513–3520.

Bazaral M. and Helinski D. (1970). Replication of a bacterial plasmid and episome in E.coli. *Biochemistry* **9,** 399–406.

Beard J.P. and Bishop S.F. (1975). Role of the cell surface in bacterial mating: requirement for intact mucopeptide in donors for the expression of surface exclusion in $R^+$ strains of E.coli. *J. Bacteriol.* **123,** 916–920.

Beard J.P. and Connolly J.C. (1975). Detection of a protein, similar to the sex pilus subunit, in the outer membrane of E.coli cells carrying a derepressed F-like R factor. *J. Bacteriol.* **122,** 59–65.

Beard J.P., Howe T.G.B. and Richmond M.H. (1972). Purification of sex pili from E.coli carrying a derepressed F-like R factor. *J. Mol. Biol.* **66,** 311–313.

Beattie K.L. and Setlow J.K. (1971). Transformation defective strains of H.influenzae. *Nature New Biol.* **231,** 177–179.

Beckendorf S.K. and Wilson J.H. (1972). A recombination gradient in phage T4 gene 34. *Virology* **50,** 315–321.

Becker A. and Gold M. (1975). Isolation of the phage lambda A gene protein. *Proc. Natl. Acad. Sci. USA* **72,** 581–585.

Beguin C.F. (1973). Degradation of T4 DNA synthesized in the absence of phage ligase. *Virology* **52,** 488–501.

Beier H. and Hausmann R. (1973). Genetic map of phage T3. *J. Virol* **12,** 417–419.

Belfort M. and Wulff D. (1973). Genetic and biochemical investigation of the E.coli mutant hfl-1 which is lysogenized at high frequency by phage λ. *J. Bacteriol.* **115,** 299–306.

Benbow R.M., Eisenberg M. and Sinsheimer R.L. (1972). Multiple length DNA molecules of phage φX174. *Nature New Biology* **237,** 141–143.

Benbow R.M., Zuccarelli J. and Sinsheimer R.L. (1974). A role for single strand breaks in phage φX174 genetic recombination. *J. Mol. Biol.* **88,** 629–652.

Benbow R.M., Zuccarelli A.J. and Sinsheimer R.L. (1975). Recombinant DNA molecules of phage φX174. *Proc. Natl. Acad. Sci. USA* **72,** 235–239.

Benbow R.M., Hutchison C.A., Fabricant J.C. and Sinsheimer R.L. (1971). Genetic map of phage φX174. *J. Virol.* **7,** 549–558.

Benbow R.M., Mayof R.F., Picchi J.C. and Sinsheimer R.L. (1972). Direction of translation and size of phage φX174 cistrons. *J. Virol.* **10,** 99–104.

Benbow R.M., Zuccarelli A.J., Davis G.C. and Sinsheimer R.L. (1974). Genetic recombination in phage φX174. *J. Virol.* **13,** 898–907.

Bennett P.M. and Richmond M.H. (1976). Translocation of a discrete piece of DNA carrying an *amp* gene between replicons in E.coli. *J. Bacteriol.* **126,** 1–6.

Benzer S. (1955). Fine structure of a genetic region in bacteriophage. *Proc. Natl. Acad. Sci. USA* **41,** 344–354.

Benzer S. (1957). The elementary units of heredity. In The Chemical Basis of Heredity, eds. W.D. McElroy and B. Glass. *(Johns Hopkins Press, Baltimore).* pp. 70–93.

Benzer S. (1959). On the topography of the genetic fine structure. *Proc. Natl. Acad. Sci. USA* **45,** 1607–1620.

Benzer S. (1961). On the topography of the genetic fine structure. *Proc. Natl. Acad. Sci. USA* **47,** 403–415.

Benzer S. and Champe S.P. (1961). Ambivalent *rII* mutants of phage T4. *Proc. Natl. Acad. Sci. USA* **47,** 1025–1038.

Benzer S. and Champe S.P. (1962). A change from nonsense to sense in the genetic code. *Proc. Natl. Acad. Sci. USA* **48,** 1114–1121.

Berg C.M. and Curtis R. III (1967). Transposition derivatives of an Hfr strain of E. coli K12. *Genetics* **56,** 503–525.

Berg D.E. (1974). Genetic evidence for two types of gene arrangements in new λdv plasmid mutants. *J. Mol. Biol.* **86,** 59–68.

Berg D.E., Davies J., Allet B. and Rochaix J.-D. (1975). Transposition of R factor genes to phage λ. *Proc. Natl. Acad. Sci. USA* **72,** 3628–3632.

Berger H., Warren A.J. and Fry K.E. (1969). Variations in genetic recombination due to amber mutation in T4D phage. *J. Virol.* **3,** 171–175.

Berget P.B. and Warner H.R. (1975). Identification of P48 and P54 as components of phage T4 baseplates. *J, Virol.* **16,** 1669–1676.

Berkowitz S.A. and Day L.A. (1976). Mass, length, composition and structure of the filamentous bacterial virus fd. *J. Mol. Biol.* **102,** 531–548.

Bernardi A. and Spahr P.-F. (1972). Nucleotide sequence at the binding site for coat protein on RNA of phage R17. *Proc. Natl. Acad. Sci. USA* **69,** 3033–3037.

Bernardi A., Maat J., de Waard A. and Bernardi G. (1976). Preparation and specificity of endonuclease IV induced by phage T4. *Eur J. Biochem.* **66,** 175–180.

Bernstein A., Rolfe B. and Onodera K. (1972). Pleiotropic properties and genetic organization of the to1A,B locus of E.coli K12. *J. Bacteriol.* **112,** 74–83.

Bernstein H. (1968). Repair and recombination in phage T4. I. Genes affecting recombination. *Cold Spring Harbor Symp. Quant. Biol.* **33,** 325–331.

Bernstein H. and Bernstein C. (1973). Circular and branched circular concatenates as possible intermediates in phage T4 replication. *J. Mol. Biol.* **77,** 355–361.

Beyersmann D., Messer W. and Schlicht M. (1974). Mutants of E.coli B/r defective in DNA initiation: *dnaI,* a new gene for replication. *J. Bacteriol.* **118,** 783–789.

Bigger C.H., Murray K. and Murray N.E. (1973). Recognition sequence of a restriction enzyme. *Nature New Biol.* **244,** 7–10.

Bijlenga R.K.L., Aebi U. and Kellenberger E. (1976). Properties and structure of a gene *24* controlled T4 giant phage. *J Mol. Biol.* **103,** 469–498.

Bijlenga R.K.L., Broek R. and Kellenberger E. (1974). Particle containing uncleaved major head protein (P23) which is maturable into the head of phage T4. *Nature* **249,** 825–827.

Bijlenga R.K.L., Scraba D. and Kellenberger E. (1973). Studies on the morphopoiesis of the head of T-even phage. IX. Tau particles: their morphology, kinetics of appearance and possible precursor function. *Virology* **56,** 250–267.

Billeter M.A., Dahlberg J.E., Goodman H.M., Hindley J. and Weissmann C. (1969). Sequence of the first 175 nucleotides from the 5′ terminus of Qβ RNA synthesized in vitro. *Nature* **224,** 1083–1087.

Bird R.E., Chandler M. and Caro L. (1976). Suppression of an E.coli *dnaA* mutation by the integrated R factor R100.1: change of chromosome replication origin in synchronized cultures. *J. Bacteriol.* **126,** 1215–1223.

Birge E.A. and Low K.D. (1974). Detection of transcribable recombination products following conjugation in $rec^+$, $recB^-$ and $recC^-$ strains of E.coli K12. *J. Mol. Biol.* **83,** 447–458.

Black L.W. (1974). Phage T4 internal protein mutants: isolation and properties. *Virology* **60,** 166–179.

Black L.W. and Abremski K. (1974). Restriction of phage T4 internal protein I mutants by a strain of E.coli. *Virology* **60,** 180–191.

Black L.W. and Ahmad-Zadeh C. (1971). Internal proteins of phage T4D: their characterization and relation to head structure and assembly. *J. Mol. Biol.* **57,** 71–92.

Black L.W. and Brown D.T. (1976). Head morphologies in phage T4 and internal protein mutant infections. *J. Virol.* **17,** 894–905.

Black L.W. and Gold L.M. (1971). Prereplicative development of the phage T4: RNA and protein synthesis in vivo and in vitro. *J. Mol. Biol.* **60,** 365–388.

Blair D.G. and Helinski D.R. (1975). Relaxation complexes of plasmid DNA and protein. I.

Strand specific complexes of protein and DNA in the relaxed complexes of plasmids ColEl and ColE2. *J. Biol. Chem.* **250,** 8785–8789.

Blair D.G., Clewell D.B., Sherratt D.J. and Helinski D.R. (1971). Strand specific supercoiled DNA-protein relaxation complexes: comparison of the complexes of bacterial plasmids colE1 and colE2. *Proc. Natl. Acad. Sci. USA* **68,** 210–214.

Blas S.F., Thompson R. and Broda P. (1974). An E.coli mutant apparently carrying two autonomous F prime factors. *Molec. Gen. Genet.* **130,** 153–165.

Blattner F.R. and Dahlberg J.E. (1972). RNA synthesis startpoints in phage lambda: are the promotor and operator transcribed? *Nature New Biology* **237,** 227–232.

Blattner F.R., Dahlberg J.E., Boettiger J.K., Fiandt M. and Szybalski W. (1972). Distance from a promotor mutation to an RNA synthesis startpoint on phage lambda DNA. *Nature New Biology* **237,** 232–237.

Blattner F.R., Fiandt M., Hass K.K., Twose P.A. and Szybalski W. (1974). Deletions and insertions in the immunity region of phage lambda: revised measurement of the promotor-startpoint distance. *Virology* **62,** 458–471.

Blumberg D.D. and Malamy M.M. (1974). Evidence for the presence of nontranslated T7 late mRNA in infected F′(PIF$^+$) episome containing cells. *J. Virol.* **13,** 378–385.

Blumberg D.D., Mabie C.T. and Malamy M.H. (1976). T7 protein synthesis in F factor containing cells: evidence for an episomally induced impairment of translation and its relation to an alteration in membrane permeability. *J. Virol.* **17,** 94–105.

Blumenthal T., Landers T.A. and Weber K. (1972). Bacteriophage Q$\beta$ replicase contains the protein biosynthetic elongation factors Tu and Ts. *Proc. Natl. Acad. Sci. USA* **69,** 1313–1317.

Blumenthal T., Young R.A. and Brown S. (1976). Function and structure in phage Q$\beta$ RNA replicase. Association of EF Tu-Ts with the other enzyme subunits. *J. Biol. Chem.* **251,** 2740–2743.

Bodmer W.F. (1965). Recombination and integration in B.subtilis transformation: involvement of DNA synthesis. *J. Mol. Biol.* **14,** 534–547.

Bodmer W. (1966). Integration of DNAase-treated DNA in B.subtilis transformation. *J. Gen. Physiol.* **49,** suppl., 233–258.

Bodmer W.F. and Ganesan A.T. (1964). Biochemical and genetic studies of integration and recombination in B.subtilis transformation. *Genetics* **50,** 717–738.

Bode V.C. and Gillin F.D. (1971). The arrangement of DNA in λ phage heads. I. Biological consequences of micrococcal nuclease attacks on a portion of the chromosome exposed in tailless heads. *J. Mol. Biol.* **62,** 493–502.

Bode V.C. and Kaiser A.D. (1965a). Repression of the cII and cIII cistrons of phage lambda in a lysogenic bacterium. *Virology* **25,** 111–121.

Bode V.C. and Kaiser A.D. (1965b). Changes in the structure and activity of λ DNA in a superinfected immune bacterium. *J. Mol. Biol.* **14,** 399–417.

Boedtker H. and Gesteland R.F. (1975). Physical properties of RNA phages and their RNA. In RNA Phages, ed N.D. Zinder. (*Cold Spring Harbor Laboratory: New York*), pp. 1–28.

Boklage C.E., Wong E.C.T. and Bode V.C. (1973). The lambda F mutants belong to two cistrons. *Genetics* **75,** 221–230.

Boklage C.E., Wong E.C.T. and Bode V.C. (1974). Functional abnormality of lambda phage particles from complemented FII mutant lysates. *Virology* **61,** 22–28.

Bolle A., Epstein R.H., Salser W. and Geiduschek E.P. (1968a). Transcription during phage

T4 development: synthesis and relative stability of early and late RNA. *J. Mol. Biol.* **31,** 325–348.

Bolle A., Epstein R.H., Salser W. and Geiduschek E.P. (1968b). Transcription during phage T4 development: requirements for late messenger synthesis. *J. Mol. Biol.* **33,** 339–362.

Bonhoeffer F.R., Hosselbarth R. and Lehman K. (1967). Dependence of the conjugational DNA transfer on DNA synthesis. *J. Mol. Biol.* **29,** 539–541.

Boon T. (1971). Inactivation of ribosomes in vitro by colicin E3. *Proc. Natl. Acad. Sci. USA* **68,** 2421–2425.

Boon T. (1972). Inactivation of ribosomes in vitro by colicin E3 and its mechanism of action. *Proc. Natl. Acad. Sci. USA* **69,** 549–552.

Boon T. and Zinder N.D. (1969). A mechanism for genetic recombination and generating one parent and one recombinant. *Proc. Natl. Acad. Sci. USA* **64,** 573–577.

Boon T. and Zinder N.D. (1971). Genotypes produced by individual recombination events involving phage fl. *J. Mol. Biol.* **58,** 133–151.

Boon T. and Zinder N.D. (1970). Genetic recombination in phage fl: transfer of parental DNA to the recombinant. *Virology* **41,** 444–452.

Bouché J.-P., Zechel K. and Kornberg A. (1975). *dnaG* gene product, a rifampicin resistant RNA polymerase, initiates the conversion of a single stranded phage DNA to its duplex replicative form. *J. Biol. Chem.* **250,** 5995–6001.

Bowman C.M., Sidikaro J. and Nomura M. (1971). Specific inactivation of ribosomes by colicin E3 in vitro and mechanism of immunity in colicinogenic cells. *Nature New Biology* **234,** 133–137.

Bowman C.M., Dahlberg J.E., Ikemura T., Konisky J. and Nomura M. Specific inactivation of 16S ribosomal RNA induced by colicin E3 in vitro. *Proc. Natl. Acad. Sci. USA* **68,** 964–968.

Bowman K.L. and Ray D.S. (1975). Degradation of the viral strand of $\phi$X174 parental replicative form DNA in a $rep^-$ host. *J. Virol.* **16,** 838–843.

Boyer H.W., Chow L.T., Dugaiczyk A., Hedgpeth J. and Goodman H.M. (1973). DNA substrate site for the Eco RII restriction endonuclease and modification methylase. *Nature New Biol.* **244,** 40–43.

Boyle M.J. (1969). Radiation sensitive mutants of T4D. II. T4*y*: genetic characterization. *Mutat. Res.* **8,** 441–449.

Brachet J., Eisen H. and Rambach A. (1970). Mutations of phage $\lambda$ affecting the expression of replicative functions O and P. *Molec. Gen. Genetics* **108,** 266–276.

Branton D. and Klug A. (1975). Capsid geometry of phage T2: a freeze etching study. *J. Mol. Biol.* **92,** 559–566.

Brenner S. (1957). Genetic control and phenotypic mixing of the adsorption cofactor requirement in phages T2 and T4. *Virology* **3,** 560–574.

Brenner S. (1959). Physiological aspects of phage genetics. *Adv. Virus. Res.* **6,** 137–158.

Brenner S., Stretton A.O.W. and Kaplan S. (1965). Genetic code: the nonsense triplets for chain termination and their suppression. *Nature* **206,** 994–998.

Brenner S., Streisinger G., Horne R.W., Champe S.P., Barnett L., Benzer S. and Rees M.W. (1959). Structural components of bacteriophage. *J. Mol. Biol.* **1,** 281–292.

Brenner S., Champe S.P., Streisinger G. and Barnett L. (1962). On the interaction of adsorption cofactors with phages T2 and T4. *Virology* **17,** 30–39.

Bresler S.E., Kreneca R.A. and Kushev V.V. (1971). Molecular heterozygotes in B.subtilis and their correction. *Molec. Gen. Genet.* **113,** 204–213.

Brinton C.C. jr. (1965). The structure, formation, synthesis and genetic control of bacterial pili and a molecular model for DNA and RNA transport in gram negative bacteria. *Trans. NY Acad. Sci.* **27,** 1003–1054.

Brinton C.C. (1971). The properties of sex pili, the viral nature of "conjugal" genetic transfer systems, and some possible approaches to the control of bacterial drug resistance. *Crit. Rev. Microbiol.* **1,** 105–160.

Brinton C.C. jr., Gemski P. and Carnahan J. (1964). A new type of bacterial pilus genetically controlled by the fertility factor of E.coli K12 and its role in chromosome transfer. *Proc. Natl. Acad. Sci. USA* **52,** 776–783.

Britton J.R. and Haselkorn R. (1975a). Macromolecular synthesis in T7 infected F′ cells. *Virology* **67,** 264–275.

Britton J.R. and Haselkorn R. (1975b). Permeability lesions in male E.coli infected with phage T7. *Proc. Natl. Acad. Sci. USA* **72,** 2222–2226.

Broda P. (1967). The formation of Hfr strains in E. coli K12. *Genet. Res.* **9,** 35–47.

Broda P. and Collins J.F. (1974). Gross map distances and Hfr transfer times in E.coli K12. *J. Bact.,* **117,** 747–752.

Broda P., Beckwith J.R. and Scaife J. (1964). The characterization of a new type of F-prime factor in E.coli K12. *Genet. Res.* **5,** 489–494.

Broda P., Meacock P. and Achtman M. (1972). Early transfer of genes determining transfer functions by some Hfr strains in E.coli K12. *Molec. Gen. Genet.* **116,** 336–347.

Brody E.N. and Geiduschek E.P. (1970). Transcription of the phage T4 template. Detailed comparison of the in vitro and in vivo transcripts. *Biochemistry* **9,** 1300–1309.

Brody E.R., Sederoff R., Bolle A. and Epstein R. H. (1970). Early transcription in T4 infected cells. *Cold Spring Harbor Symp. Quant. Biol.* **35,** 203–212.

Broker T.R. and Doermann A.H. (1975). Molecular and genetic recombination of phage T4. *Ann. Rev. Genet.* **9,** 213–244.

Broker T.R. and Lehman I.R. (1971). Branched DNA molecules: intermediates in T4 recombination. *J. Mol. Biol.* **60,** 131–150.

Brown D.T., Mackenzie J. M. and Bayer M.E. (1971). Mode of host cell penetration by phage $\phi$X174. *J. Virol* **7,** 836–846.

Brown L.R. and Dowell C.E. (1968a). Replication of phage M13. I. Effects on host cells after synchronized infection. *J. Virol.* **2,** 1290–1295.

Brown L.R. and Dowell C.E. (1968b). Replication of phage M13. II. Intracellular DNA forms associated with M13 infection of mitomycin C treated cells. *J. Virol.* **2,** 1296–1347.

Brown S. and Blumenthal T. (1976a). Function and structure in RNA phage Q$\beta$ RNA replicase. Effect of inhibitors of EF-Tu on RNA synthesis and renaturation of active enzyme. *J. Biol. Chem.* **251,** 2749–2753.

Brown S. and Blumenthal T. (1976b). Reconstitution of Q$\beta$ RNA replicase from a covalently bonded elongation factor Tu-Ts complex. *Proc. Natl. Acad. Sci. USA* **73,** 1131–1135.

Brunel F. and Davison J. (1975). Bacterial mutants able to partly suppress the effect of *N* mutations in phage $\lambda$. *Molec. Gen. Genet.* **136,** 167–180.

Bruner R. and Cape R.E. (1970). The expression of two classes of late genes of bacteriophage T4. *J. Mol. Biol.* **53,** 69–90.

Bruner R., Souther A. and Suggs S. (1972). Stability of cytosine-containing DNA after infection by certain T4 rII-D deletion mutants. *J. Virol.* **10,** 88–98.

Brunovskis I. and Summers W.C. (1972). The process of infection with phage T7. VI. A phage gene controlling shutoff of host RNA synthesis. *Virology* **50,** 322–327.

Brutlag D. and Kornberg A. (1972). Enzymatic synthesis of DNA. XXXVI. A proofreading function for the 3′-5′ exonuclease activity in DNA polymerases. *J. Biol. Chem.* **247,** 241–248.

Brutlag D., Schekman R. and Kornberg A. (1971). A possible role for RNA polymerase in the initiation of M13 DNA synthesis. *Proc. Natl. Acad. Sci. USA* **68,** 2826–2829.

Buchwald M. and Siminovitch L. (1969). Production of serum blocking material by mutants of the left arm of the λ chromosome. *Virology* **38,** 1–7.

Buchwald M., Steed-Glaister P. and Siminovitch L. (1970a). The morphogenesis of bacteriophage lambda. I. Purification and characterization of λ heads and λ tails. *Virology* **42,** 375–389.

Buchwald M., Murialdo H. and Siminovitch L. (1970b). The morphogenesis of bacteriophage lambda. II. Identification of the principal structural proteins. *Virology* **42,** 390–400.

Bujard H., Mazaitis A.J. and Bautz E.K.F. (1970). The size of the rII region of phage T4 *Virology* **42,** 717–723.

Burgess A.B. (1969). Studies on the proteins of ϕX174. The protein composition of the ϕX coat. *Proc. Natl. Acad. Sci. USA* **64,** 613–617.

Burgess A.B. and Denhardt D.T. (1969). Studies on ϕX proteins. I. Phage specific proteins synthesized after infection of E.coli. *J. Mol. Biol.* **44,** 377–386.

Burgess R.R., Travers A.A., Dunn J.J. and Bautz E.K.F. (1969). Factor stimulating transcription by RNA polymerase. *Nature* **221,** 43–47.

Burnet F.M. (1927). The relationship between heat stable agglutinogens and sensitivity to phages in the Salmonella group. *Brit. J. Exp. Path.* **8,** 121–129.

Burton A. and Sinsheimer R.L. (1965). The process of infection with ϕX174. VII. Ultracentrifugal analysis of the replicative form. *J. Mol. Biol.* **14,** 327–347.

Butler B. and Echols H. (1970). Regulation of phage λ development by gene N: properties of a mutation that bypasses N control of late protein synthesis. *Virology* **40,** 212–222.

Buxton R.S. and Holland I.B. (1973). Genetic studies of tolerance to colicin E2 in E.coli K12. I. Relocation and dominance relationships of *cet* mutations. *Molec. Gen. Genetics* **127,** 69–88.

Cabello F., Timmis K. and Cohen S.N. (1976). Replication control in a composite plasmid constructed by the in vitro linkage of two distinct replicons. *Nature* **259,** 285–290.

Cairns J. (1961). An estimate of the length of the DNA molecule of T2 phage by autoradiography. *J. Mol. Biol.* **3,** 756–761.

Cairns J. (1963). The chromosome of E.coli. *Cold Spring Harbor Symp. Quant. Biol.* **28,** 43–46.

Cairns J., Stent G.S. and Watson J.D. (Eds.) (1966). Phage and the Origins of Molecular Biology. (*Cold Spring Harbor Laboratory: New York*).

Calef E. and Licciacardello G. (1960). Recombination experiments on prophage-host relationships. *Virology* **12,** 81–103.

Cameron J.R., Panasenko S.M., Lehman I.R. and Davis R.W. (1975). In vitro construction of phage λ carrying segments of the E.coli chromosome: selection of hybrids containing the gene for DNA ligase. *Proc. Natl. Acad. Sci. USA* **72,** 3416–3420.

Campbell A. (1957). Tranduction and segregation in E.coli K12. *Virology* **4,** 366–384.

Campbell A. (1959). Ordering of genetic sites in phage λ by the use of galactose-transducing defective phages. *Virology* **9,** 293–305.

Campbell A. (1961). Sensitive mutants of phage λ. *Virology* **14,** 22–32.

Campbell A. (1962). Episomes. *Adv. Gen.* **11,** 101–146.

Campbell A. (1963). Segregants from lysogenic heterogenotes carrying recombinant lambda prophages. *Virology* **20,** 344–356.

Campbell A. (1964). Transduction. In The Bacteria, eds., I.C. Gunsalus and R.Y. Stanier. **5,** 49–85. (*Academic Press: London and New York*).

Campbell A. (1965). The steric effect in lysogenization by phage lambda. II. Chromosomal attachment of the b2 mutant. *Virology* **27,** 340–345.

Campbell A. (1971). Genetic Structure. In The Bacteriophage Lambda, ed. A.D. Hershey. (*Cold Spring Harbor Laboratory, New York*), pp. 13–44.

Campbell A. and Del Campillo-Campbell A. (1963). Mutant of phage lambda producing a thermolabile endolysin. *J. Bacteriol.* **85,** 1202–1207.

Campbell J., Soll L. and Richardson C.C. (1972). Isolation and partial characterization of a mutant of E.coli deficient in DNA polymerase II. *Proc. Natl. Acad. Sci. USA* **69,** 2090–2094.

Campbell J.H., and Rolfe B.G. (1975). Evidence for a dual control of the initiation of host cell lysis caused by phage λ. *Molec. Gen. Genet.* **139,** 1–8.

Capecchi M.R. (1967). Polarity in vitro. *J. Mol. Biol.* **30,** 213–218.

Capecchi M.R. and Webster R.E. (1975). Phage RNA as template for in vivo protein synthesis. In RNA Phages, ed. N.D. Zinder. (*Cold Spring Harbor Laboratory: New York*), pp. 279–300.

Carmichael G.G., Landers T.A. and Weber K. (1976). Immunochemical analysis of the functions of the subunits of phage Qβ RNA replicase. *J. Biol. Chem.* **251,** 2744–2748.

Carmichael G.G., Weber K., Niveleau A. and Wahba A.J. (1975). The host factor required for RNA phage Qβ RNA replication in vitro. Intracellular location, quantitation, and purification by poly(A)-cellulose chromatography. *J. Biol. Chem.* **250,** 3607–3612.

Caro L.G. and Schnos M. (1966). The attachment of male specific phage fl to sensitive strains of E.coli. *Proc. Natl. Acad. Sci. USA* **56,** 126–132.

Carroll R.B., Neet K.E. and Goldthwait D.A. (1972). Self-association of gene 32 protein of phage T4. *Proc. Natl. Acad. Sci. USA* **69,** 2741–2744.

Carroll R.B., Neet K. and Goldthwait D.A. (1975). Studies of the self-association of phage T4 gene 32 protein by equilibrium sedimentation. *J. Mol. Biol.* **91,** 275–292.

Carter B.J., Shaw B.D. and Smith M.G. (1969). Two stages in the replication of phage lambda. *Biochim. Biophys. Acta* **195,** 494–505.

Carter D.H. and Radding C.M. (1971). The role of exonuclease and β protein of phage λ in genetic recombination. II. Substrate specificity and the mode of action of λ exonuclease. *J. Biol. Chem.* **246,** 2502–2512.

Cascino A., Riva S. and Geiduschek E.P. (1970). DNA ligation and the coupling of T4 late transcription to replication. *Cold Spring Harbor Symp. Quant. Biol.* **35,** 213–220.

Casjens S. (1974). Phage lambda $F_{II}$ gene protein: role in head assembly. *J. Mol. Biol.* **90,** 1–19.

Casjens S.R. and Hendrix R.W. (1974). Locations and amounts of the major structural proteins in phage lambda. *J. Mol. Biol.* **88,** 535–546.

Casjens S., Hohn T. and Kaiser A.D. (1970). Morphological proteins of phage lambda: identification of the major head protein as the product of gene E. *Virology* **42,** 496–507.

Casjens S., Hohn T. and Kaiser A.D. (1972). Head assembly step controlled by genes F and W in phage λ. *J. Mol. Biol.* **64,** 551–564.

Caspar D.L.D. and Klug A. (1962). Physical principles in the construction of regular viruses. *Cold Spring Harbor Symp. Quant. Biol.* **27,** 1–24.

Cassuto E. and Radding C.M. (1971). Mechanism for the action of λ exonuclease in genetic recombination. *Nature New Biol.* **229,** 13–16.

Castellazzi M., Brachet P. and Eisen N. (1972). Isolation and characterization of deletions in phage λ residing as prophage in E.coli K12. *Molec. Gen. Genet.* **117,** 211–218.

Caster J.H., Postel E.H. and Goodgal S.H. (1970). Competence mutants: isolation of transformation deficient strains of H.influenzae. *Nature* **227,** 515–517.

Cato A. jr. and Guild W.R. (1968). Transformation and DNA size. I. Activity of fragments of defined size and a fit to a random double crossover model. *J. Mol. Biol.* **37,** 157–178.

Cavalieri S.J., Goldthwait D.A. and Neet K.E. (1976). The isolation of a dimer of gene *8* protein of phage fd. *J. Mol. Biol.* **102,** 713–722.

Cavalieri S.J., Neet K.E. and Goldthwait D.A. (1976). Gene *5* protein of phage fd: a dimer which interacts cooperatively with DNA. *J. Mol. Biol.* **102,** 697–712.

Ceglowski P., Fuchs P.G. and Soltyk A. (1975). Competitive inhibition of transformation in group H Streptococcus strain Challis by heterologous DNA. *J. Bacteriol.* **124,** 1621–1629.

Celis J.E., Smith J.D. and Brenner S. (1973). Correlation between genetic and translational maps of gene *23* in phage T4. *Nature New Biol.* **241,** 130–132.

Center M.S. (1972). Replicative intermediates of phage T7 DNA. *J. Virol.* **10,** 115–123.

Center M.S. (1973). Phage T7 DNA synthesis in isolated DNA membrane complexes. *J. Virol.* **12,** 847–854.

Center M.S. (1975). Role of gene *2* in phage T7 DNA synthesis. *J. Virol.* **16,** 94–100.

Center M.S. and Richardson C.C. (1970a). An endonuclease induced after infection of E.coli with phage T7. I. Purification and properties of the enzyme. *J. Biol. Chem.* **245,** 6285–6291.

Center M.S. and Richardson C.C. (1970b). An endonuclease induced after infection of E.coli with phage T4. II. Specificity of the enzyme toward single and double stranded DNA. *J. Biol. Chem.* **245,** 6292–6299.

Center M.S., Studier F.W. and Richardson C.C. (1970). The structural gene for a T7 endonuclease essential for phage DNA synthesis. *Proc. Natl. Acad. Sci. USA* **65,** 242–248.

Chabbert Y.A., Scavizz M.R., Witchitz J.L., Gerbaud G.R. and Bouanchaud D.H. (1972). Incompatibility groups and the classification of $fi^-$ resistance factors. *J. Bacteriol.* **112,** 666–675.

Chadwick P., Pirotta V., Steinberg R., Hopkins N. and Ptashne M. (1970). The lambda and 434 phage repressors. *Cold Spring Harbor Symp. Quant. Biol.* **35,** 283–294.

Chamberlin M.M. (1974). Isolation and characterization of prototrophic mutants of E.coli unable to support the intracellular growth of T7. *J. Virol.* **14,** 509–516.

Chamberlin M. and Ring J. (1972). Studies of the binding of E.coli RNA polymerase to DNA. V. T7 RNA chain initiation by enzyme-DNA complexes. *J. Mol. Biol.* **70,** 221–237.

Chamberlin M. and Ring J. (1973a). Characterization of T7 RNA polymerase. I. General properties of the enzymatic reaction and the template specificity of the enzyme. *J. Biol. Chem.* **248,** 2235–2245.

Chamberlin M. and Ring J. (1973b). Characterization of T7 RNA polymerase. II. Inhibitors of the enzyme and their application to the study of the enzymatic reaction. *J. Biol. Chem.* **248,** 2245–2250.

Chamberlin M., McGrath J. and Waskell L. (1970). New RNA polymerase from E.coli infected with phage T7. *Nature* **228,** 227–232.

Chan T.S., Model P. and Zinder N.D. (1975). In vitro protein synthesis directed by separated transcriptions of phage f1 DNA. *J. Mol. Biol.* **99,** 369–382.

Chang A.C.Y. and Cohen S.N. (1974). Genome construction between bacterial species in vitro: replication and expression of Staphylococcus plasmid genes in E.coli. *Proc. Natl. Acad. Sci. USA* **71,** 1030–1034.

Chase M. and Doermann A.H. (1958). High negative interference over short segments of the genetic structure of phage T4. *Genetics* **43,** 332–353.

Chasin L.A. and Magasanik B. (1970). Gene expression after transformation in B.subtilis. *J. Bacteriol.* **102,** 661–665.

Chattoraj D.K. and Inman R.B. (1974). Location of DNA ends in P2, 186, P4 and λ phage heads. *J. Mol. Biol.* **87,** 11–22.

Chen C.-Y., Hutchison C.A. III and Edgell M.H. (1973). Isolation and genetic localization of three φX174 promotor regions. *Nature New Biol.* **243,** 233–236.

Chilton M.D. and Hall B. (1968). Transforming activity in single stranded DNA from B.subtilis. *J. Mol. Biol.* **34,** 439–451.

Chiu C.-S. and Greenberg G.R. (1968). Evidence for a possible direct role of dCMP hydroxymethylase in T4 phage DNA synthesis. *Cold Spring Harbor Symp. Quant. Biol.* **33,** 351–359.

Clark A.J. (1963). Genetic analysis of a double male strain of E.coli K12. *Genetics* **48,** 105–120.

Clark A.J. (1973). Recombination deficient mutants of E.coli and other bacteria. *Ann. Rev. Genetics* **7,** 67–86.

Clark A.J. and Margulies A. (1965). Isolation and characterization of recombination-deficient mutants of E.coli K12. *Proc. Natl. Acad. Sci. USA* **53,** 451–459.

Clarke L. and Carbon J. (1975). Biochemical construction and selection of hybrid plasmids containing specific segments of the E.coli genome. *Proc. Natl. Acad. Sci. USA* **72,** 4361–4365.

Clements J.B. and Sinsheimer R.L. (1975). Process of infection with phage φX174. XXXVII. RNA metabolism in φX174 infected cells. *J. Virol.* **15,** 151–160.

Clewell D.B. (1972). Nature of ColE1 plasmid replication in E.coli in the presence of chloramphenicol. *J. Bacteriol.* **110,** 667–676.

Clewell D.B. and Helinski D.R. (1969). Supercoiled circular DNA-protein complex in E.coli: purification and induced conversion to an open circular DNA form. *Proc. Natl. Acad. Sci. USA* **62,** 1159–1166.

Clewell D.B. and Helinski D.R. (1970a). Evidence for the existence of the colicinogenic factors E2 and E3 as supercoiled DNA protein relaxation complexes. *Biochem. Biophys. Res. Commun.* **40,** 608–613.

Clewell D.B. and Helinski D.R. (1970b). Existence of the colicinogenic factor-sex factor colIb-P9 as a supercoiled circular DNA-protein relaxation complex. *Biochem. Biophys. Res. Commun.* **41,** 150–156.

Clewell D.B. and Helinski D.R. (1970c). Properties of a supercoiled DNA protein relaxation complex and strand specificity of the relaxation event. *Biochemistry* **9,** 4428–4440.

Clewell D.B. and Helinski D.R. (1972). Effect of growth conditions on the formation of the relaxation complex of supercoiled colE1 DNA and protein in E.coli. *J. Bacteriol.* **110,** 1135–1146.

Clowes R.C. (1961). Colicin factors as fertility factors in bacteria: Ecoli K12. *Nature* **190,** 988–989.

Clowes R.C. (1963). Colicin factors and episomes. *Genet. Res.* **4,** 163–166.

Clowes R.C. (1972). Molecular structure of bacterial plasmids. *Bacteriol. Rev.* **36,** 361–406.

Clowes R.C. and Moody E.E.M. (1966). Chromosomal transfer from recombination-deficient strains of E.coli K12. *Genetics* **53,** 717–726.

Clowes R.C. and Rowley D. (1954). Some observations on linkage effects in genetic recombination in E.coli K12. *J. Gen. Microbiol.* **11,** 250–260.

Cohen A., Fisher W.D., Curtis R. and Adler H.I. (1968). DNA isolated from E.coli minicells mated with $F^+$ cells. *Proc. Natl. Acad. Sci. USA* **61,** 61–68.

Cohen G. (1972). Recombination in phage λ DNA half molecules. *J. Mol. Biol.* **66,** 37–48.

Cohen S.N. and Chang A.C.Y. (1973). Recircularization and autonomous replication of a sheared R-factor DNA segment in E.coli transformants. *Proc. Natl. Acad. Sci. USA* **70,** 1293–1297.

Cohen S.N. and Miller C.A. (1969). Multiple molecular species of circular R factor DNA isolated from E.coli. *Nature* **224,** 1273–1277.

Cohen S.N. and Miller C.A. (1970a). Nonchromosomal antibiotic resistance in bacteria. II. Molecular nature of R factors isolated from Proteus mirabilus and E.coli. *J. Mol. Biol.* **50,** 671–687.

Cohen S.N. and Miller C.A. (1970b). Nonchromosomal antibiotic resistance in bacteria. III. Isolation of the discrete transfer unit of the R factor R1. *Proc. Natl. Acad. Sci. USA* **67,** 510–516.

Cohen S.N., Chang A.C.Y. and Hsu L. (1972). Nonchromosomal antibiotic resistance in bacteria: genetic transformation of E.coli by R factor DNA. *Proc. Natl. Acad. Sci. USA* **69,** 2110–2114.

Cohen S.N., Chang A.C.Y., Boyer H.W. and Helling R.B. (1973). Construction of biologically functional bacterial plasmids in vitro. *Proc. Natl. Acad. Sci. USA* **70,** 3240–3244.

Collinsworth W.L. and Mathews C.K. (1974). Biochemistry of DNA-defective amber mutants of phage T4. IV. DNA synthesis in plasmolyzed cells. *J. Virol.* **13,** 908–915.

Comer M.M. and Neidhardt F.C. (1975). Effect of T4 modification of host valyl tRNA synthetase on enzyme action in vivo. *Virology* **67,** 395–403.

Comer M.M., Guthrie C. and McClain W.H. (1974). An ochre suppressor of phage T4 that is associated with a transfer RNA. *J. Mol. Biol.* **90,** 665–676.

Condit R.C. (1975). F factor mediated inhibition of phage T7 growth: increased membrane permeability and decreased ATP levels following T7 infection of male E.coli. *J. Mol. Biol.* **98,** 45–56.

Condit R.C. and Steitz J.A. (1975). F factor mediated inhibition of phage T7 growth: analysis of T7 RNA and protein synthesis in vivo and in vitro using male and female E.coli. *J. Mol. Biol.* **98,** 31–43.

Contreras R., Ysebaert M., Jou W.M. and Fiers W. (1973). Phage MS2 RNA: nucleotide sequence of the end of the A protein gene and the intercistronic region. *Nature New Biology* **241,** 99–101.

Cooke M. and Meynell E. (1969). Chromosomal transfer by derepressed R factors in $F^-$ E.coli K12. *Genet. Res.* **14,** 79–87.

Cooke M., Meynell E. and Lawn A.M. (1970). Mutant Hfr strains defective in transfer: restoration by F-like and I-like derepressed R factors. *Genet. Res.* **16,** 101–112.

Cooper S. (1972). Relationship of Flac replication and chromosome replication. *Proc. Natl. Acad. Sci. USA* **69,** 2706–2710.

Cooper S. and Helmstetter C.E. (1968). Chromosome replication and the division cycle of E.coli B/r. *J. Mol. Biol.* **31,** 519–540.

Coppo A., Manzi A., Pulitzer J.F. and Takahashi H. (1973). Abortive phage T4 head assembly in mutants of E.coli. *J. Mol. Biol.* **76,** 61–88.

Coppo A., Manzi A., Pulitzer J.F. and Takahashi H. (1975a). Host mutant (*tabD*) induced inhibition of phage T4 late transcription. I. Isolation and phenotypic characterization of the mutants. *J. Mol. Biol.* **96,** 579–600.

Coppo A., Manzi A., Pulitzer J.F. and Takahasi H. (1975b). Host mutant (*tabD*) induced inhibition of phage T4 late transcription. II. Genetic characterization of mutants. *J. Mol. Biol.* **96,** 601–624.

Cosley S.D. and Oishi M. (1973). Genetic transformation in E.coli K12. *Proc. Natl. Acad. Sci. USA* **70,** 84–87.

Court D. and Campbell A. (1972). Gene regulation in N mutants of phage λ. *J. Virol.* **9,** 938–945.

Court D. and Sato K. (1969). Studies of novel transducing variants of lambda. Dispensability of genes N and Q. *Virology* **39,** 348–352.

Court D., Green L. and Echols H. (1975). Positive and negative regulation by the cII and cIII gene products of phage λ. *Virology* **63,** 484–491.

Craig E., Cremer K. and Schlessinger D. (1972). Metabolism of T4 mRNA, host mRNA and rRNA in T4 infected E.coli B. *J. Mol. Biol.* **71,** 701–716.

Crawford E. and Gesteland R. (1964). Adsorption of phage R17. *Virology* **22,** 165–167.

Crick F.H.C. and Watson J.D. (1956). The structure of small viruses. *Nature* **177,** 473–475.

Crick F.H.C., Barnett L., Brenner S. and Watts-Tobin R.J. (1961). General nature of the genetic code for proteins. *Nature* **192,** 1227–1232.

Cummings D.J. and Bolin R.W. (1976). Head length control in T4 phage morphogenesis: effect of canavanine on assembly. *Bacteriol. Rev.* **40,** 313–359.

Cummings D.J., Chapman V.A., DeLong S.S. and Kusy A.R. (1971). Structural aberrations in T-even phage. II. Characterization of the proteins contained in aberrant heads. *Virology* **44,** 425–442.

Cummings D.J., Chapman V.A., DeLong S.S. and Couse N.L. (1973). Structural aberrations in T-even phage. III. Induction of "lollipops" and their partial characterization. *Virology* **54,** 245–261.

Curtis M.J. and Alberts B. (1976). Studies on the structure of intracellular phage T4 DNA. *J. Mol. Biol.* **102,** 793–816.

Curtiss R. (1969). Bacterial conjugation. *Ann. Rev. Microbiol.* **23,** 69–136.

Curtiss R. and Renshaw J. (1969). $F^+$ strains of E.coli K12 defective in Hfr formation. *Genetics* **63,** 7–26.

Curtiss R. III and Stallions D.R. (1969). Probability of F integration and frequency of stable Hfr donors in $F^+$ populations of E. coli K12. *Genetics* **63,** 27–38.

Curtiss R., Charamella L.J., Stallions D.R. and May J.A. (1968). Parental functions during conjugation in E.coli K12. *Bacteriol. Rev.* **32,** 320–348.

Curtiss R. III, Caro L., Allison D. and Stallions D. (1969). Early stages of conjugation in E.coli. *J. Bacteriol.* **100,** 1091–1104.

Cuypers T., Van der Ouderaa F.J. and De Jong W.W. (1974). The amino acid sequence of gene *5* protein of phage M13. *Biochem. Biophys. Res. Commun.* **59,** 557–563.

Cuzin F. and Jacob F. (1963). Integration reversible de l'episome sexuel F′ chez E.coli K12. *Comptes Rendues Acad. Sci. Paris* **257,** 795–797.

Cuzin F. and Jacob F. (1964). Deletions chromosomiques et integration d'un episome sexuel F-lac$^+$ chez E.coli K12. *Comptes Rendues Acad. Sci. Paris* **258,** 1350–1352.

Cuzin F. and Jacob F. (1967). Existence chez E.coli K12 d'une unite genetique formee de differents replicons. *Ann. Inst. Pasteur* **112,** 529–545.

Cuzin F., Buttin G. and Jacob F. (1967). On the mechanism of genetic transfer during conjugation of E.coli. *J. Cell. Physiol.* **70,** suppl. **1,** 77–88.

Dahlberg A.E. and Dahlberg J.E. (1975). Binding of ribosomal protein S1 of E.coli to the 3′ end of 16S rRNA. *Proc. Natl. Acad. Sci. USA* **72,** 2940–2944.

Dambly C. and Couturier M. (1971). A minor Q-independent pathway for the expression of the late genes in phage lambda. *Molec. Gen. Genetics* **113,** 244–250.

Daniel V., Sarid S. and Littauer U.Z. (1970). Phage induced tRNA in E.coli. *Science* **167,** 1682–1688.

Danziger R.E. and Paranchych W. (1970a). Stages in phage R17 infection. II. Ionic requirements for phage R17 attachment to F pili. *Virology* **40,** 547–553.

Danziger R.E. and Paranchych W. (1970b). Stages in R17 infection. III. Energy requirements for the F pili mediated eclipse of viral infectivity. *Virology* **40,** 554–564.

Darlix J.L. (1973). The functions of rho in T7 transcription in vitro. Europ. J. Biochem. **35,** 517–526.

Darlix J.-L. and Horaist M. (1975). Existence and possible roles of transcriptional barriers in T7 DNA early region as shown by electron microscopy. *Nature* **256,** 288–292.

Das A., Court D. and Adhya S. (1976). Isolation and characterization of conditional lethal mutants of E.coli defective in transcription termination factor rho. *Proc. Natl. Acad. Sci. USA* **73,** 1959–1963.

Datta N. and Barth P.T. (1976). Hfr formation by I pilus-determining plasmids in E.coli K12. *J. Bacteriol.* **125,** 811–817.

Datta N. and Hedges R.W. (1971). Compatibility groups among fi$^-$ R factors. *Nature* **234,** 222–223.

Datta N., Lawn A.M. and Meynell E. (1966). The relationship of F type pilation and F phage sensitivity to drug resistance transfer in R$^+$F$^-$ E.coli K12. *J. Gen. Microbiol.* **45,** 365–376.

Dausse J.-P., Sentenac A. and Fromageot P. (1975). Interaction of RNA polymerase from E.coli with DNA. Analysis of T7 DNA early promotor sites. *Eur. J. Biochem.* **57,** 569–578.

Davidoff-Abelson R. and Dubnau D. (1973a). Conditions affecting the isolation from transformed cells of B.subtilis of high molecular weight single stranded DNA of donor origin. *J. Bacteriol.* **116,** 146–153.

Davidoff-Abelson R. and Dubnau D. (1973b). Kinetic analysis of products of donor DNA in transformed cells of B.subtilis. *J. Bacteriol.* **116,** 154–162.

Davidson N. and Szybalski W. (1971). Physical and chemical characteristics of lambda DNA. In The Bacteriophage Lambda, ed. A.D. Hershey. (*Cold Spring Harbor Laboratory, New York*). pp. 45–82.

Davies J.E. and Rownd R. (1969). Transmissible multiple drug resistance in Enterobacteriaceae. *Science* **176,** 758–768.

Davies J.K. and Reeves P. (1975a). Genetics of resistance to colicins in E.coli K12: cross resistance among colicins of group B. *J. Bacteriol.* **123,** 96–101.

Davies J.K. and Reeves P. (1975b). Genetics of resistance to colicins in E.coli K12: cross resistance among colicins of group A. *J. Bacteriol.* **123,** 102–117.

Davies R.W., Dove W.F., Inokuchi H., Lehman J.F. and Roehrdanz R.L. (1972). Regulation of λ prophage excision by the transcriptional state of DNA. *Nature New Biol.* **238,** 43–45.

Davis B.D. and Helmstetter C.E. (1973). Control of F′ lac replication in E.coli B/r. *J. Bacteriol.* **114,** 294–299.

Davis R. and Parkinson J. (1971). Deletion mutants of phage lambda. III. Physical structure of attφ. *J. Mol. Biol.* **56,** 403–423.

Davis R. and Vapnek D. (1976). In vivo transcription of R plasmid DNA in E.coli strains with altered antibiotic resistance levels and/or conjugal proficiency. *J. Bacteriol.* **125,** 1148–1155.

Davis R.W. and Hyman R.W. (1970). Physical locations of the in vitro RNA initiation site and termination sites of T7M DNA. *Cold Spring Harbor Symp. Quant. Biol.* **35,** 269–282.

Davis R.W. and Hyman R.W. (1971). A study in evolution: the DNA base sequence homology between coliphages T7 and T3. *J. Mol. Biol.* **62,** 287–302.

Davis R.W., Simon M. and Davidson N. (1971). Electron microscope duplex method for mapping regions of base sequence homology in nucleic acids. *Methods Enzymol.* **210,** 413–428.

Davison P.F., Freifelder D., Hede R. and Levinthal C. (1961). The structural unity of the DNA of phage T2. *Proc. Natl. Acad. Sci. USA* **47,** 1123–1129.

Dawes J. and Goldberg E.B. (1973). Functions of baseplate components in phage T4 infection. II. Products of genes 5,6,7,8, and 10. *Virology* **55,** 391–396.

Dawson M.H. (1930). The transformation of pneumococcal types. II. The interconvertibility of type-S specific pneumococci. *J. Exp. Med.* **51,** 123–147.

Dawson M.H. and Sia R.P.H. (1931). In vitro transformation of pneumoccocal types. I. A technique for inducing transformation of pneumococcal types in vitro. *J. Exp. Med.* **54,** 681–691.

Dawson P., Skalka A. and Simon LD. (1975). Phage lambda head morphogenesis: studies on the role of DNA. *J. Mol. Biol.* **93,** 167–180.

Dawson P., Hohn B., Hohn T. and Skalka A. (1976). Functional empty capsid precursors produced by a lambda mutant defective for late λ DNA replication. *J. Virol.* **17,** 576–583.

De Fernandez F.M.T., Eoyang L. and August J.T. (1968). Factor fraction required for the synthesis of Qβ RNA. *Nature* **219,** 588–590.

De Fernandez F.M.T., Hayward W.S. and August J.T. (1972). Bacterial proteins required for replication of phage Qβ RNA. *J. Biol. Chem.* **247,** 824–831.

De Haan P.G. and Stouthamer A.H. (1963). F-prime transfer and multiplication of sexduced cells. *Genet. Res.* **4,** 30–41.

Delbruck M. (1945). The burst size distribution in the growth of bacterial viruses (bacteriophages). *J. Bacteriol.* **50,** 131–135.

Delbruck M. (1946). Bacterial viruses or bacteriophages. *Biol. Rev.* **21,** 30–41.

Delbruck M. and Bailey W.T. jr. (1946). Induced mutations in bacterial viruses. *Cold Spring Harbor Symp. Quant. Biol.* **11,** 33–37.

Delius H., Howe C. and Kozinski A.W. (1971). Structure of the replicating DNA from phage T4. *Proc. Natl. Acad. Sci. USA* **68,** 3049–3053.

Delius H., Mantell N.J. and Alberts B. (1972). Characterization by electron microscopy of the complex formed between T4 phage gene 32 protein and DNA. *J. Mol. Biol.* **67,** 341–350.

De Lucia P. and Cairns J. (1969). Isolation of an E.coli strain with a mutation affecting DNA polymerase. *Nature* **224,** 1164–1166.

Demerec M. and Fano U. (1945). Phage resistant mutants in E.coli. *Genetics* **30,** 119–136.

Demerec M., Adelberg E.A., Clark A.J. and Hartman P.E. (1966). A proposal for a uniform nomenclature in bacterial genetics. *Genetics* **54,** 61–76.

Denhardt D.T. and Silver R.B. (1966). An analysis of the clone size distribution of $\phi$X 174 mutants and recombinants. *Virology* **30,** 10–19.

Denhardt D.T. and Sinsheimer R.L. (1965a). The process of infection with phage $\phi$X174. III. Phage maturation and lysis after synchronized infection. *J. Mol. Biol.* **12,** 641–646.

Denhardt D.T. and Sinsheimer R.L. (1965b). The process of infection with phage $\phi$X174. IV. Replication of the viral DNA in a synchronized infection. *J. Mol. Biol.* **12,** 647–662.

Denhardt D.T. and Sinsheimer R.L. (1965c). The process of infection with phage $\phi$X174. V. Inactivation of the phage bacterium complex by decay of $^{32}$P incorporated in the infecting particle. *J. Mol. Biol.* **12,** 663–673.

Denhardt D.T., Dressler D.H. and Hathaway A. (1967). The abortive replication of $\phi$X174 DNA in a recombination deficient mutant of E.coli. *Proc. Natl. Acad. Sci. USA* **57,** 813–820.

Dennison S. (1972). Naturally occurring R factor, depressed for pilus synthesis, belonging to the same compatibility group as the sex factor F of E.coli. *J. Bacteriol.* **109,** 416–422.

Deonier R.C., Ohtsubo E., Lee H.J. and Davidson N. (1974). Electron microscope heteroduplex studies of sequence relations among plasmids of E.coli. VII. Mapping the rRNA genes of plasmid F14. *J. Mol. Biol.* **89,** 619–630.

Depew R.E. and Cozzarelli N.R. (1974). Genetics and physiology of phage T4 3′ phosphatase: evidence for involvement of the enzyme in T4 DNA metabolism. *J. Virol.* **13,** 888–897.

DeRosier D.J. and Klug A. (1972). Structure of the tubular variants of the head of phage T4 (polyheads). I. Arrangement of subunits in some classes of polyhead. *J. Mol. Biol.* **65,** 469–488.

DeVries J. and Maas W.K. (1971). Chromosomal integration of F′ factors in recombination deficient Hfr strains of E.coli. *J. Bacteriol.* **106,** 150–156.

De Wachter R., Vandenberghe A., Merregaert J., Contreras R. and Fiers W. (1971). The leader sequence from the 5′ terminus to the A protein initiation codon in MS2 virus RNA. *Proc. Natl. Acad. Sci. USA* **68,** 585–589.

Dewey M.J. and Frankel F.R. (1975a). Two suppressor loci for gene 49 mutations of phage T4. I. Genetic properties and DNA synthesis. *Virology* **68** 387–401.

Dewey M.J. and Frankel F.R. (1975b). Two suppressor loci for gene 49 mutations of phage T4. II. DNA and capsid maturation. *Virology* **68,** 402–417.

Dhar R., Weissman S.M., Zain B.S., Pan J. and Lewis A.M. (1974). The nucleotide sequence preceding an RNA polymerase initiation site on SV40 DNA. 2. The sequence of the early strand trascript. *Nuc. Acids Res.* **1,** 595–613.

Dickson R.C. (1973). Assembly of phage T4 tail fibres. IV. Subunit composition of tail fibres and fiber precursors. *J. Mol. Biol.* **79,** 633–648.

Dickson R.C., Barnes S.L. and Eiserling F.A. (1970). Structural proteins of phage T4. *J. Mol. Biol.* **53,** 461–474.

Dickson R.C., Abelson J., Barnes W.M. and Reznikoff W.S. (1975). Genetic regulation: the lac control region. *Science* **187,** 27–35.

Dicou L. and Cozzarelli N.R. (1973). Phage T4-directed DNA synthesis in toluene treated cells. *J. Virol.* **12,** 1293–1302.

Di Mauro E., Snyder L., Marino P., Lamberti A., Coppo A. and Tocchini-Valentini G.P. (1969). Rifampicin sensitivity of the components of DNA dependent RNA polymerase. *Nature* **222,** 533–537.

Dintzis H.M. (1961). Assembly of the peptide chains of hemoglobin. *Proc. Natl. Acad. Sci. USA* **47,** 247–260.

Doerfler W. and Hogness D.S. (1968a). The strands of DNA from lambda and related phages: isolation and characterization. *J. Mol. Biol.* **33,** 635–659.

Doerfler W. and Hogness D.S. (1968b). Gene orientation in phage lambda as determined from the genetic activities of heteroduplex DNA formed in vitro. *J. Mol. Biol.* **33,** 661–678.

Doermann A.H. (1951). The intracellular growth of bacteriophages. I. Liberation of intracellular phage T4 by premature lysis with another phage or with cyanide. *J. Gen. Physiol.* **35,** 645–656.

Doermann A.H. (1973). T4 and thc rolling circle model of replication. *Ann. Rev. Genet.* **7,** 325–342.

Doermann A.H. and Hill M.B. (1953). Genetic structure of phage T4 as described by recombination studies of factors influencing plaque morphology. *Genetics* **38,** 79–90.

Doermann A.H., Eiserling F.A. and Boehner L. (1973). Genetic control of capsid length in phage T4. I. Isolation and preliminary description of four new mutants. *J. Virol.* **12,** 374–385.

Donachie W.D. (1968). Relationship between cell size and time of initiation of DNA replication. *Nature* **219,** 1077–1079.

Donachie W.D. (1969). Control of cell division in E.coli: experiments with thymine starvation. *J. Bacteriol.* **100,** 260–268.

Doniger J., Warner R.C. and Tessman I. (1973). Role of circular dimer DNA in the primary recombination mechanism of S13. *Nature New Biol.* **242,** 9–12.

Dooley D.C., Hadden C.T. and Nester E.W. (1971). Macromolecular synthesis in B.subtilis during development of the competent state. *J. Bacteriol.* **108,** 668–679.

Dove W. (1966). Action of the lambda chromosome. I. Control of functions late in phage development. *J. Mol. Biol.* **19,** 187–201.

Dowell C.E. and Sinsheimer R.L. (1966). The process of infection with $\phi$X174. IX. Studies on the physiology of three $\phi$X174 temperature sensitive mutants. *J. Mol. Biol.* **16,** 374–386.

Drake J.W., Allen E.F., Forsberg S.A., Preparata R.M. and Greening E.O. (1969). Genetic control of mutation rates in T4. *Nature* **221,** 1128–1131.

Dressler D. (1970). The rolling circle for $\phi$X DNA replication. II. Synthesis of single stranded circles. *Proc. Natl. Acad. Sci.* **67,** 1934–1942.

Dressler D. and Wolfson J. (1970). The rolling circle for $\phi$X DNA replication. III. Synthesis of supercoiled duplex rings. *Proc. Natl. Acad. Sci. USA* **67,** 456–463.

Dressler D., Wolfson J. and Magazin M. (1972). Initiation and reinitiation of DNA synthesis during replication of phage T7. *Proc. Natl. Acad. Sci. USA* **69,** 998–1002.

Driskell-Zamenhof P.J. and Adelberg E.A. (1963). Studies on the chemical nature and size of sex factors of E.coli K12. *J. Mol. Biol.* **6,** 483–497.

Dube S.K. and Rudland P.S. (1970). Control of translation by T4 phage: altered binding of disfavoured messengers. *Nature* **226,** 820–824.

Dubnau D. and Cirigliano C. (1972a). Fate of transforming DNA after uptake by competent B.subtilis: size and distribution of the integrated donor segments. *J. Bacteriol.* **111,** 488–494.

Dubnau D. and Davidoff-Abelson R. (1971). Fate of transforming DNA following uptake by competent B.subtilis. I. Formation and properties of the donor recipient complex. *J. Mol. Biol.* **56,** 209–222.

Dubnau D., Davidoff-Abelson R., Scher B. and Cirigliano C. (1973). Fate of transforming DNA after uptake by competent B.subtilis: phenotypic characterization of radiation-sensitive recombination deficient mutants. *J. Bacteriol.* **114,** 273–286.

Dubnau E. and Maas W.K. (1968). Inhibition of replication of an F′ lac episome in Hfr cells of E.coli. *J. Bacteriol.* **95,** 531–539.

Dulbecco R. (1952). A critical test of the recombination theory of multiple reactivation. *J. Bacteriol.* **63,** 199–207.

Dumas L.B. and Miller C.A. (1974). Inhibition of phage $\phi$X174 DNA replication in *dnaB* mutants of E.coli C. *J. Virol.* **14,** 1369–1379.

Dumas L.B. and Miller C.A. (1976). Phage $\phi$X174 single stranded DNA synthesis in temperature sensitive *dnaB* and *dnaC* mutants of E.coli. *J. Virol.* **18,** 426–435.

Dunn J.J. and Studier F.W. (1973a). T7 early RNAs are generated by site specific cleavages. *Proc. Natl. Acad. Sci. USA* **70,** 1559–1563.

Dunn J.J. and Studier F.W. (1973b). T7 early RNAs and E.coli ribosomal RNAs are cut from large precursor RNAs in vivo by ribonuclease III. *Proc. Natl. Acad. Sci. USA* **70,** 3296–3300.

Dunn J.J. and Studier F.W. (1975). Effect of RNAase III cleavage on translation of phage T7 mRNA. *J. Mol. Biol.* **99,** 487–500.

Dunn J.J., Bautz F.A. and Bautz E.K.F. (1971). Different template specificities of phage T3 and T7 RNA polymerases. *Nature New Biol.* **230,** 94–96.

Dunn J.J., McAllister W.T. and Bautz E.K.F. (1972). In vitro transcription of T3 DNA by E.coli and T3 polymerases. *Virology* **48,** 112–125.

Earhart C.F., Sauri C.J., Fletcher G. and Wulff J.L. (1973). Effect of inhibition of macromolecule synthesis on the association of phage T4 DNA with membrane. *J. Virol.* **11,** 527–534.

Ebisuzaki K., Dewey C.L., and Behme M.T. (1975). Pathways of DNA repair in T4 phage. I. MMS sensitive mutant. *Virology* **64,** 330–338.

Echols H. (1970). Integrative and excisive recombination by bacteriophage lambda: evidence for an excision-specific recombination protein. *J. Mol. Biol.* **47,** 575–583.

Echols H. (1971). Regulation of lytic development. In The Bacteriophage Lambda, ed. A.D. Hershey. (*Cold Spring Harbor Laboratory, New York*), pp. 247–270.

Echols H. and Gingery R. (1968). Mutants of phage λ defective in vegetative genetic recombination. *J. Mol. Biol.* **34,** 239–249.

Echols H. and Green L. (1971). Establishment and maintenance of repression by phage lambda: the role of the cI, cII and cIII proteins. *Proc. Natl. Acad. Sci. USA* **68,** 2190- 2194.

Echols H., Court D. and Green L. (1976). On the nature of cis acting regulatory proteins and genetic organization in phage: the example of gene *Q* of phage lambda. *Genetics* **83,** 5–10.

Echols H., Gingery R. and Moore L. (1968). Integrative recombination function of phage λ: evidence for a site specific recombination enzyme. *J. Mol. Biol.* **34,** 251–260.

Echols H., Green L., Oppenheim A.B., Oppenheim A. and Honigman A. (1973). Role of the cro gene in phage lambda development. *J. Mol. Biol.* **80,** 203–216.

Eddleman H.L. and Champe S.P. (1966). Components in T4 infected cells associated with phage assembly. *Virology* **30,** 471–481.

Edgar R.S. and Lielausis I. (1964). Temperature sensitive mutants of phage T4D: their isolation and genetic characterization. *Genetics* **49,** 649–662.

Edgar R.S. and Lielausis I. (1965). Serological studies with mutants of phage T4D defective in genes determining tail fiber structure. *Genetics* **52,** 1187–1200.

Edgar R.S. and Lielausis I. (1968). Some steps in the assembly of phage T4. *J. Mol. Biol.* **32,** 263–276.

Edgar R.S. and Steinberg C.M. (1958). On the origin of high negative interference over short segments of the genetic structure of phage T4. *Virology* **6,** 115–128.

Edgar R.S. and Wood W.B. (1966). Morphologies of phage T4 in extracts of mutant infected cells. *Proc. Natl. Acad. Sci. USA* **55,** 498–505.

Edgar R.S., Denhardt G.H. and Epstein R.H. (1964). A comparative genetic study of conditional lethal mutations of phage T4D. *Genetics* **49,** 635–648.

Edgell M. and Ginoza W. (1965). The fate during infection of the coat protein of the spherical phage R17. *Virology* **27,** 23–27.

Edgell M.H., Hutchison C.A. and Sinsheimer R.L. (1969). The process of infection with φX174. XXVIII. Removal of the spike proteins from the phage capsid. *J. Mol. Biol.* **42,** 547–557.

Egan J.B. and Hogness D.S. (1972). The topography of λ DNA: Isolation of ordered fragments and the physical mapping of point mutations. *J. Mol. Biol.* **71,** 363–382.

Egawa R. and Hirota Y. (1962). Inhibition of fertility by the resistance factor (R) in E.coli K12. *Jap. J. Genetics* **37,** 66–69.

Eggen K. and Nathans D. (1969). Regulation of protein synthesis directed by phage MS2 RNA. II. In vitro suppression by phage coat protein. *J. Mol. Biol.* **39,** 293–306.

Eikhom T.S. and Spiegelman S. (1967). The dissociation of Qβ replicase and the relation of one of the components to a poly(C)-dependent poly(G) polymerase. *Proc. Natl. Acad. Sci. USA* **57,** 1833–1840.

Eisen H.A., Fuerst C.R., Siminovitch L., Thomas R., Lambert L., Pereira de Silva L. and Jacob F. (1966). Genetics and physiology of defective lysogeny in K12(λ): studies of early mutants. *Virology* **30,** 224–241.

Eisenberg S. and Denhardt D.T. (1974). Structure of nascent φX174 replicative form: evidence for discontinuous replication. *Proc. Natl. Acad. Sci. USA* **71,** 984–988.

Eisenberg S., Scott J.F. and Kornberg A. (1976). An enzyme system for replication of duplex circular DNA: the replicative form of phage $\phi$X174. *Proc. Natl. Acad. Sci. USA* **73,** 1594–1597.

Eisenberg S., Harbers B., Hours C. and Denhardt D.T. (1975). The mechanism of replication of $\phi$X174 DNA. XII. Nonrandom location of gaps in nascent $\phi$X174 RF II DNA. *J. Mol. Biol.* **99,** 107–124.

Eisen H., Brachet P., da Silva L.P. and Jacob F. (1970). Regulation of repressor expression in lambda. *Proc. Natl. Acad. Sci. USA* **66,** 855–862.

Eiserling F.A. and Boy de la Tour E. (1965). Capsomers and other structures observed on some bacteriophages. *Path. Microbiol.* **28,** 175–188.

Eiserling F.A., Bolle A. and Epstein R.H. (1967). Electron microscopic study of the structure of mutants of phage T4D defective in tail fiber genes. *Virology* **33,** 405–412.

Eiserling F.A., Geiduschek E.P., Epstein R.H. and Metter E.J. (1970). Capsid size and DNA length: the petit variant of bacteriophage T4. *J. Virol.* **6,** 865–876.

Ellis E.L. and Delbruck M. (1939). The growth of bacteriophage. *J. Gen. Physiol.* **22,** 365–384.

Emmons S.W. (1974). Phage lambda derivatives carrying two copies of the cohesive end site. *J. Mol. Biol.* **83,** 511–526.

Engelhardt D.L., Webster R.E. and Zinder N.D. (1967). Amber mutants and polarity in vitro. *J. Mol. Biol.* **29,** 45–58.

Enquist L.W. and Skalka A. (1973). Replication of phage lambda DNA depends on the function of host and viral genes. I. Interaction of *red, gam,* and *rec. J. Mol. Biol.* **75,** 185–212.

Ephrussi-Taylor H.E. (1951). Genetic aspects of transformations of pneumococci. *Cold Spring Harbor Symp. Quant. Biol.* **16,** 445–456.

Ephrussi-Taylor H. and Freed B.A. (1964). Incorporation of thymidine and amino acids into DNA insoluble cell structures in pneumococcal cultures synchronized for competence to transform. *J. Bacteriol.* **87,** 1211–1215.

Ephrussi-Taylor H. (1966). Genetic recombination in DNA-induced transformation of pneumococcus. IV. The pattern of transmission and phenotypic expression of high and low efficiency donor sites in the *amiA* locus. *Genetics* **54,** 211–222.

Ephrussi-Taylor H. and Gray T.C. (1966). Genetic studies of recombining DNA in pneumococcus transformation. *J. Gen. Physiol.* **49,** suppl., 211–231.

Ephrussi-Taylor H., Sicard A.M. and Kamen R. (1965). Genetic recombination in DNA-induced transformation of pneumococcus. I. The problem of relative efficiency of transforming factors. *Genetics* **51,** 455–475.

Epstein R.H., Bolle A., Steinberg C.M., Kellenberger E., Boy de la Tour E., Chevalley R., Edgar R.S., Slisman M., Denhardt G.H. and Lielausis A. (1963). Physiological studies of conditional lethal mutant of phage T4D. *Cold Spring Harbor Symp. Quant. Biol.* **28,** 375–392.

Eriksson-Grennberg K.G. and Nordstrom K. (1973). Genetics and physiology of a tolE mutant of E.coli K12 and phenotypic suppression of its phenotype by galactose. *J. Bacteriol.* **115,** 1219–1222.

Errington L., Glass R.E., Hayward R.S. and Scaife J.G. (1974). Structure and orientation of an RNA polymerase operon in E.coli. *Nature* **249,** 519–522.

Eskin B., Lautenberger J.A. and Linn S. (1973). Host controlled modification and restriction of phage T7 by E.coli B. *J. Virol.* **11,** 1020–1023.

Falkingham J.O. and Clark A.J. (1974). Genetic analysis of a double male strain of E.coli K12. *Genetics* **78,** 633–644.

Falkow S. and Citarella R.V. (1965). Molecular homology of F-merogenote DNA. *J. Mol. Biol.* **12,** 138–151.

Falkow S., Citarella R.V., Wohlhieter J.A. and Watanabe T. (1966). The molecular nature of R factors. *J. Mol. Biol.* **17,** 102–116.

Fangman W.L. and Feiss M. (1969). Fate of λ DNA in a bacterial host defective in DNA synthesis. *J. Mol. Biol.* **44,** 103–116.

Falkow S., Tompkins L.S., Silver R.P., Guerry P. and LeBlanc D.J. (1971). The replication of R factor DNA in E.coli following conjugation. *Ann. NY Acad. Sci.* **182,** 153–171.

Farber M.B. (1976). Purification and properties of phage φX174 gene D product. *J. Virol.* **17,** 1027–1037.

Favre R., Boy de la Tour E., Segre N., and Kellenberger E. (1965). Studies on the morphopoiesis of the head of phage T-even. I. Morphological, immunological, and genetic characterization of the polyheads. *J. Ult. Res.* **13,** 318–342.

Federoff N. (1975). Replicase of the phage f2. In RNA Phages, ed. N.D. Zinder. (*Cold Spring Harbor Laboratory: New York*). pp. 235–258.

Federoff N.V. and Zinder N.D. (1971). Structure of the poly(G) polymerase component of the phage f2 replicase. *Proc. Natl. Acad. Sci. USA* **68,** 1838–1843.

Federoff N.V. and Zinder N.D. (1972a). Properties of the phage f2 replicase. I. Optimal condition for replicase activity and analysis of the polynucleotide product synthesized in vitro. *J. Biol. Chem.* **247,** 4577–4585.

Federoff N.V. and Zinder N.D. (1972b). Properties of the f2 phage replicase. II. Comparative studies on the RNA dependent and poly(C) dependent activities of the replicase. *J. Biol. Chem.* **247,** 4586–4592.

Federoff N.V. and Zinder N.D. (1973). Factor requirement of the phage f2 replicase. *Nature New Biol.* **241,** 105–108.

Feiss M. and Campbell A. (1974). Duplication of the phage lambda end site; genetic studies *J. Mol. Biol.* **83,** 527–540.

Feiss M. and Margulies T. (1973). On maturation of the phage λ chromosome. *Molec. Gen. Genetics* **127,** 285–296.

Feix G., Slor H. and Weissmann C. (1967). Replication of viral RNA. XIII. The early product of phage RNA synthesis in vitro. *Proc. Natl. Acad. Sci.* **57,** 1401–1408.

Fiandt M., Szybalski W. and Malamy M.H. (1972). Polar mutations in lac, gal and phage λ consist of a few DNA sequences inserted with either orientation. *Molec. Gen. Genet.* **119,** 223–231.

Fiandt M., Hradecna Z., Lozeron H.A. and Szybalski W. (1971). Electron micrographic mapping of deletions, insertions, inversions and homologies in the DNAs of coliphages λ and φ80. In The Bacteriophage Lambda, ed. A.D. Hershey. (*Cold Spring Harbor Laboratory: New York*). pp. 329–354.

Fiandt M., Gottesman M.E., Shulman M.J., Szybalski E.H., Szybalski W. and Weisberg R.A. (1976). Physical mapping of phage lambda *att*[2]. *Virology* **72,** 6–12.

Fidanian H.M. and Ray D.S. (1972). Replication of phage M13. Requirement of the gene 2 protein for the accumulation of a specific RFII species. *J. Mol. Biol.* **72,** 51–63.

Fields K.L. and Luria S.E. (1969a). Effects of colicins E1 and K on transport systems. *J. Bacteriol.* **97,** 57–63.

Fields K.L. and Luria S.E. (1969b). Effects of colicins E1 and K on cellular metabolism. *J. Bacteriol.* **97,** 64–77.

Fiers W. (1975). Chemical structure and biological activity of phage MS2 RNA. In RNA Phages, ed. N.D. Zinder. (*Cold Spring Harbor Laboratory: New York*), pp. 353–397.

Fiers W. and Sinsheimer R.L. (1962a). The structure of the DNA of phage $\phi$X174. I. The action of exopolynucleotidases. *J. Mol. Biol.* **5,** 408–419.

Fiers W. and Sinsheimer R.L. (1962b). The structure of the DNA of phage $\phi$X174. II. Thermal inactivation. *J. Mol. Biol.* **5,** 420–423.

Fiers W. and Sinsheimer R.L. (1962c). The structure of the DNA of phage $\phi$X174. III. Ultracentrifugal evidence for a ring structure. *J. Mol. Biol.* **5,** 424–434.

Fiers W., Van Montagu M., De Wachter R., Haegeman G., Min Jou W., Messens E., Remaut E., Vanderberghe A. and Van Styvendaele B. (1969). Studies on the primary structure and the replication mechanism of phage RNA. *Cold Spring Harbor Symp. Quant. Biol.* **34,** 697–705.

Fiers W., Contreras R., Duerinck F., Haegeman G., Merregaert J., Min Jou W., Raeymakers A., Volckaert G., Ysebaert M., Van de Kerckhove J., Nolf F. and Van Montagu M. (1975). A protein gene of phage MS2. *Nature* **256,** 273–278.

Fiers W., Contreras R., Duerinck F., Haegeman G., Iserentant D., Merregaert J., Min Jou W., Molemans F., Raeymakers A., Van den Berghe A., Volckaert G. and Ysebaert M. (1976). Complete nucleotide sequence of phage MS2 RNA: primary and secondary structure of the replicase gene. *Nature* **260,** 500–507.

Filip C.C., Allen J.S., Gustafson R.A., Allen R.G. and Walker J.R. (1974). Bacterial cell division regulation: characterization of the *dnaH* locus of E.coli. *J. Bacteriol.* **119,** 443–449.

Finch J.T., Klug A. and Stretton A.O.W. (1964). The structure of the polyheads of T4 phage. *J. Mol. Biol.* **10,** 570–575.

Finnegan D.J. and Willetts N.S. (1971). Two classes of Flac mutants insensitive to transfer inhibition by an F-like R factor. *Molec. Gen. Genetics* **111,** 256–264.

Finnegan D.J. and Willetts N.S. (1972). The nature of the transfer inhibitor of several F-like plasmids. *Molec. Gen. Genetics* **119,** 57–66.

Finnegan D. and Willetts N. (1973). The site of action of the F transfer inhibitor. *Molec. Gen. Genetics* **127,** 307–316.

Fischer-Fantuzzi L. (1967). Integration of $\lambda$ and $\lambda$b2 genomes in nonimmune host bacteria carrying a $\lambda$ cryptic prophage. *Virology* **32,** 18–32.

Fisher K.W. and Fisher M.B. (1968). Nalidixic acid inhibition of DNA transfer in E.coli K 12. *Cold Spring Harbor Symp. Quant. Biol.* **33,** 629–634.

Flaks J.G. and Cohen S.S. (1959). Virus induced acquisition of metabolic function. I. Enzymatic formation of 5-hydroxymethyl-deoxycytidylate. *J. Biol. Chem.* **234,** 1501–1506.

Folkmanis A., Takeda Y., Simuth J., Gussin G. and Echols H. (1976). Purification and properties of a DNA binding protein with characteristics expected for the Cro protein of phage lambda, a repressor essential for lytic growth. *Proc. Natl. Acad. Sci. USA* **73,** 2249–2253.

Forrest G.L. and Cummings D.J. (1971). Head proteins from T-even phages. II. Physical and chemical characterization. *J. Virol.* **8,** 41–55.

Forsheit A.B. and Ray D.S. (1971). Replication of phage M13. VI. Attachment of M13 DNA to a fast sedimenting host cell componnent. *Virology* **43,** 647–664.

Forsheit A.B., Ray D.S. and Lica L. (1971). Replication of phage M13. V. Single strand synthesis during M13 infection. *J. Mol. Biol.* **57,** 117–127.

Foster T.J., Howe T.G.B. and Richmond K.M.V. (1975). Translocation of the tetracycline resistance determinant from R100-1 to E.coli chromosome. *J. Bacteriol.* **124,** 1153–1158.

Fox M.S. (1960). Fate of transforming DNA following fixation by transformable bacteria. *Nature* **187,** 1004–1006.

Fox M.S. (1966). On the mechanism of integration of transforming DNA. *J. Gen. Physiol.* **49,** 183–196.

Fox M.S. and Allen M.K. (1964). On the mechanism of DNA integration in pneumococcal transformation. *Proc. Natl. Acad. Sci.* **52,** 412–419.

Fox M.S. and Hotchkiss R.D. (1960). Fate of transforming DNA following fixation by transformable bacteria. *Nature* **187,** 1002–1003.

Frame R. and Bishop J.O. (1971). The number of sex factors per chromosome in E.coli. *Biochem. J.* **121,** 93–103.

Francke B.F. and Ray D.S. (1971). Formation of the parental replicative form DNA of phage $\phi$X174 and the initial events in its replication. *J. Mol. Biol.* **61,** 565–586.

Francke B. and Ray D.S. (1972). Cis limited action of the gene A product of phage $\phi$X174 and the essential bacterial site. *Proc. Natl. Acad. Sci. USA* **69,** 475–479.

Francke B. and Hofschneider P.H. (1966). Formation of a biologically intact replicative form in RNA phage M12 infected cells. *J. Mol. Biol.* **16,** 544–552.

Frankel F.R. (1966a). Studies on the nature of replicating DNA in T4-infected E.coli. *J. Mol. Biol.* **18,** 127–143.

Frankel F.R. (1966b). Studies on the nature of the replicating DNA in E.coli infected with certain amber mutants of T4. *J. Mol. Biol.* **18,** 144–155.

Frankel F.R. (1968a). DNA replication after T4 infection. *Cold Spring Harbor Symp. Quant. Biol.* **33,** 485–494.

Frankel F.R. (1968b). Evidence for long DNA strands in the replicating pool after T4 infection. *Proc. Natl. Acad. Sci. USA* **59,** 131–138.

Frankel F.R., Batcheler M.L. and Clark C.K. (1971). The role of gene *49* in DNA replication and head morphogenesis in phage T4. *J. Mol. Biol.* **62,** 439–463.

Franklin N.C. (1961). Serological study of tail structure and function in phages T2 and T4. *Virology* **14,** 417–429.

Franklin N.C. (1971a). The N operon of lambda: extent and regulation as observed in fusions to the tryptophan operon of E.coli. In The Bacteriophage Lambda, ed A.D. Hershey. (*Cold Spring Harbor Laboratory: New York*), pp. 621–638.

Franklin N.C. (1971b). Illegitimate recombination. In The Bacteriophage Lambda, ed. A.D. Hershey. (*Cold Spring Harbor Laboratory, New York*), pp. 175–194.

Franklin N.C. (1974). Altered reading of genetic signals fused to the N operon of phage lambda: genetic evidence for modification of polymerase by the protein product of the N gene. *J. Mol. Biol.* **89,** 33–48.

Franklin N.C., Dove W.F. and Yanofsky C. (1965). The linear insertion of a prophage into the chromosome of E.coli shown by deletion mapping. *Biochem. Biophys. Res. Commun.* **18,** 910–923.

Fredericq P. (1957). Colicins. *Ann. Rev. Microbiol.* **11,** 7–22.

Freifelder D. (1968a). Studies on E.coli sex factors. III. Covalently closed F′ lac DNA molecules. *J. Mol. Biol.* **34,** 31–38.

Freifelder D. (1968b). Studies on E.coli sex factors. IV. Molecular weights of the DNA of several F′ elements. *J. Mol. Biol.* **35,** 95–102.

Freifelder D.R. and Friefelder D. (1968a). Studies on E.coli sex factor. I. Specific labeling of F′ lac DNA. *J. Mol. Biol.* **32,** 15–23.

Freifelder D.R. and Freifelder D. (1968b). Studies on E.coli sex factors. II. Some physical properties of F′lac and F DNA. *J. Mol. Biol.* **32,** 25–35.

Freifelder D., Folkmanis A. and Kirschner I. (1971). Studies on E.coli sex factors: evidence that covalent circles exist within cells and the general problem of isolation of covalent circles. *J. Bacteriol.* **105,** 722–731.

Freifelder D., Kleinschmidt A.K. and Sinsheimer R.L. (1964). Electron microscopy of single stranded DNA: circularity of DNA of phage $\phi$X174. *Science* **146,** 254–255.

Friedberg E.C. and King J.J. (1969). Endonucleolytic cleavage of UV-irradiated DNA controlled by the $v^+$ gene in phage T4. *Biochem. Biophys. Res. Commun.* **37,** 646–651.

Friedberg E.C. and King J.J. (1971). Dark repair of UV-irradiated DNA by phage T4: purification and characterization of a dimer-specific phage induced endonuclease. *J. Bacteriol.* **106,** 500–507.

Friedman D.I. and Ponce-Campos R. (1975). Differential effect of phage regulator functions on transcription from various promotors. Evidence that the P22 gene *24* and the λ gene *N* products distinguish three classes of promotors. *J. Mol. Biol.* **98,** 537–550.

Friedman D.I., Jolly C.A., Mural R.J., Ponce-Campos R. and Baumann M.F. (1976). Growth of lambda variants with added or altered promotors in N-limiting bacterial mutants: evidence that an N recognition site lies in the $P_R$ promotor. *Virology* **71,** 61–73.

Frydman A., Cooke M., Meynell E. and Meynell G.G. (1970). Repressor insensitive mutants of the F sex factor. *J. Mol. Biol.* **48,** 177–179.

Fukada A. and Sinsheimer R.L. (1976a). Process of infection with phage $\phi$X174, XXXVIII. Replication of $\phi$X174 replicative form in vivo. *J. Virol.* **17,** 776–787.

Fukuda A. and Sinsheimer R.L. (1976b). Process of infection with phage $\phi$X174, XL. Viral DNA replication of $\phi$X 174 mutants blocked in progeny single stranded DNA synthesis. *J. Virol.* **18,** 218–226.

Fulton C. (1965). Continuous chromosome transfer in E.coli. *Genetics* **52,** 55–74.

Funk F. and Sinsheimer R.L. (1970). Process of infection with phage $\phi$X174. XXXV. Cistron VIII. *J. Virol.* **6,** 12–19.

Gariglio P. and Green M.H. (1973). Characterization of polycistronic late lambda mRNA. *Virology* **53,** 392–404.

Gasson M.J. and Willetts N.S. (1975). Five control systems preventing transfer of E.coli K12 sex factor F. *J. Bacteriol.* **122,** 518–525.

Gefter M.L., Becker A. and Hurwitz J. (1967). The enzymatic repair of DNA. I. Formation of circular λ DNA. *Proc. Natl. Acad. Sci. USA* **58,** 240–247.

Gefter M.L., Hirota Y., Kornberg T., Wechsler J.A. and Barnoux C. (1971). Analysis of DNA polymerases II and III in mutuants of E.coli thermosensitive for DNA synthesis. *Proc. Nat. Acad. Sci. USA* **68,** 3150–3153.

Geider K. and Kornberg A. (1974). Conversion of M13 viral strand to the double stranded replicative forms by purified proteins. *J. Biol. Chem.* **249,** 3999–4005.

Geiduschek E.P. and Sklar J. (1969). Continual requirement for a host RNA polymerase component in a phage development. *Nature* **221,** 833–836.

Gellert M. (1967). Formation of covalent circles of lambda DNA by E.coli extracts. *Proc. Natl. Acad. Sci. USA* **57,** 148–155.

Georgopoulos C.P. (1971). Bacterial mutants in which the gene N function of phage lambda is blocked have an altered RNA polymerase. *Proc. Natl. Acad. Sci. USA* **68,** 2977–2981.

Georgopoulos C.P. and Herskowitz I. (1971). E.coli mutants blocked in lambda DNA synthesis. In The Bacteriophage Lambda, ed. A.D. Hershey. (*Cold Spring Harbor Laboratory, New York*). pp. 553–564.

Georgopoulos C.P., Hendrix R.W., Casjens S.R. and Kaiser A.D. (1973). Host participation in phage lambda head assembly. *J. Mol. Biol.* **76,** 45–60.

Gerard G.F., Johnson J.C. and Boezi J.A. (1972). Release of the sigma subunit of Pseudomonas putida DNA dependent RNA polymerase. *Biochemistry* **11,** 989–997.

Ghei O.K. and Lacks S.A. (1967). Recovery of donor DNA marker activity from eclipse in pneumococcal transformation. *J. Bacteriol.* **93,** 816–829.

Gilbert J. and Kaempfer R. (1975b). Translational repression of a viral mRNA by a host protein. *J. Biol. Chem.* **250,** 5749–5755.

Gilbert W. and Dressler D. (1968). DNA replication: the rolling circle model. *Cold Spring Harbor Symp. Quant. Biol.* **33,** 473–484.

Gillin F.D. and Bode V.C. (1971). The arrangement of DNA in λ phage heads. II. λ DNA after exposure to Micrococcal nuclease at the site of head-tail joining. *J. Mol. Biol.* **62,** 503–511.

Gingery R. and Echols H. (1967). Mutants of bacteriophage λ unable to integrate into the host chromosome. *Proc. Natl. Acad. Sci. USA* **58,** 1507–1514.

Gingery R. and Echols H. (1968). Integration, excision and transducing particle genesis by bacteriophage λ. *Cold Spring Harbor Symp. Quant. Biol.* **33,** 721–728.

Godson G.N. (1971). Characterization and synthesis of φX174 proteins in UV irradiated and unirradiated cells. *J. Mol. Biol.* **57,** 541–553.

Godson G.N. (1973). DNA heteroduplex analysis of the relation between phages φX174 and S13. *J. Mol. Biol.* **77,** 467–477.

Godson G.N. (1974a). Origin and direction of φX174 double and single stranded DNA synthesis. *J. Mol. Biol.* **90,** 127–142.

Godson C.N. (1974b). Evolution of φX174. Isolation of four new φX-like phages and comparison with φX174. *Virology* **58,** 272–289.

Goebel W. and Helinski D.R. (1968). Generation of higher multiple circular DNA forms in bacteria. *Proc. Natl. Acad. Sci. USA* **61,** 1406–1413.

Goff C.G. (1974). Chemical structure of a modification of the E.coli RNA polymerase α polypeptides induced by phage T4 infection. *J. Biol. Chem.* **249,** 6181–6190.

Goff C.G. and Weber K. (1970). A T4 induced RNA polymerase α subunit modification. *Cold Spring Harbor Symp. Quant. Biol.* **35,** 101–108.

Goldman E. and Lodish H.F. (1972). Specificity of protein synthesis by bacterial ribosomes and initiation factors: absence of change after phage T4 infection. *J. Mol. Biol.* **67,** 35–47.

Goldmark P.J. and Linn S. (1972). Purification and properties of the *recB* DNAase of E.coli K12. *J. Biol. Chem.* **247,** 1849–1860.

Goldsmith M.R., Havas S.W., Ma R.I. and Kallenbach N.R. (1970). Intracellular effects on development of competence in B.subtilis. *J. Bacteriol.* **102,** 774–783.

Golomb M. and Chamberlin M (1974a). Characterization of T7 specific RNA polymerase. IV. Resolution of the major in vitro transcripts by gel electrophoresis. *J. Biol. Chem.* **249,** 2858–2863.

Golomb M. and Chamberlin M. (1974b). A preliminary map of the major transcription units read by T7 RNA polymerase on the T7 and T3 phage chromosomes. *Proc. Natl. Acad. Sci. USA* **71,** 760–764.

Goodgal S.H. (1961). Studies on transformation of H.influenzae. IV. Linked and unlinked transformations. *J. Gen. Physiol.* **45,** 205–228.

Goodgal S.H. and Herriot R.M. (1961). Studies on transformation of Hemophilus influenzae. I. Competence. *J. Gen. Physiol.* **44,** 1201–1227.

Goodgal S.H. and Notani N. (1968). Evidence that either strand of DNA can transform. *J. Mol. Biol.* **35,** 449–454.

Goodman H.M., Billeter M.A., Hindley J. and Weissmann C. (1970). The nucleotide sequence at the 5′ terminus of the Q$\beta$ RNA minus strand. *Proc. Natl. Acad. Sci. USA* **67,** 921–928.

Gottesman M.E. and Weisberg R.A. (1971). Prophage insertion and excision. In The Bacteriophage Lambda, ed. A.D. Hershey. (*Cold Spring Harbor Laboratory, New York*). pp. 113–138.

Gottesman M.E. and Yarmolinsky M.B. (1968a). Integration-negative mutants of bacteriophage $\lambda$. *J. Mol. Biol.* **31,** 487–505.

Gottesman M.E. and Yarmolinsky M.B. (1968b). The integration and excision of the bacteriophage lambda genome. *Cold Spring Harbor Symp. Quant. Biol.* **33,** 735–748.

Gottesman M.M. and Rosner J.L. (1975). Acquisition of a determinant for chloramphenicol resistance by phage lambda. *Proc. Natl. Acad. Sci. USA* **72,** 5041–5045.

Gottesman M.M., Hicks M.L. and Gellert M. (1973). Genetics and function of DNA ligase in E.coli. *J. Mol. Biol.* **77,** 531–549.

Gottesman S. and Gottesman M.E. (1975a). Elements involved in site specific recombination in phage lambda. *J. Mol. Biol.* **91,** 489–500.

Gottesman S. and Gottesman M. (1975b). Excision of prophage $\lambda$ in a cell free system. *Proc. Natl. Acad. Sci. USA* **72,** 2188–2192.

Goulian M. and Kornberg A. (1967). Enzymatic synthesis of DNA. XXIII. Synthesis of circular replicative form of phage $\phi$X 174DNA. *Proc. Nat. Acad. Sci. USA* **58,** 1723–1730.

Goulian M., Lucas Z.J. and Kornberg A. (1968). Enzymatic synthesis of DNA. XXV. Purification and properties of DNA polymerase induced by infection with phage T4$^{+}$. *J. Biol. Chem.* **243,** 627–638.

Gralla J., Steitz J.A. and Crothers D.M. (1974). Direct physical evidence for secondary structure in an isolated fragment of R17 phage mRNA. *Nature* **248,** 204–208.

Granboulan N. and Franklin R.M. (1966). Electron microscopy of viral RNA, replicative form, and replicative intermediate of the phage R17. *J. Mol. Biol.* **22,** 173–177.

Grasso R.J. and Buchanan J.M. (1969). Synthesis of early RNA in bacteriophage T4-infected E.coli B. *Nature* **224,** 882–886.

Gray T.C. and Ephrussi-Taylor H. (1967). Genetic recombination in DNA-induced transformation of pneumococcus. V. The absence of interference and evidence for the selective elimination of certain donor sites from the final recombinants. *Genetics* **57,** 125–153.

Greenstein M. and Skalka A. (1975). Replication of phage lambda DNA: in vivo studies of the interaction between the viral gamma protein and the host recBC nuclease. *J. Mol. Biol.* **97,** 543–560.

Griffith F. (1928). Significance of pneumococcal types. *J. Hygiene* **27,** 113–159.

Grindley J.N. and Anderson E.S. (1971). I-like resistance factors with the *fi*$^{+}$ character. *Genet. Res.* **17,** 267–271.

Grindley N.D.P., Humphreys G.O. and Anderson E.S. (1973). Molecular studies of R factor compatibility groups. *J. Bacteriol.* **115,** 387–398.

Grippo P. and Richardson C.C. (1971). DNA polymerase of phage T7. *J. Biol. Chem.* **246,** 6867–6873.

Grodzicker T., Arditti R.R. and Eisen H. (1972). Establishment of repression in lambdoid phage in catabolite activator protein and adenylate cyclase mutants of E.coli. *Proc. Nat. Acad. Sci. USA* **69,** 366–370.

Groner Y., Pollack Y., Berissi H. and Revel M. (1972a). Cistron specific translation control protein in E.coli. *Nature New Biol.* **239,** 16–19.

Groner Y., Scheps R., Kamen R., Kolakofsky D. and Revel M. (1972b). Host subunit of Qβ replicase in translation control factor i. *Nature New Biol.* **239,** 19–20.

Gross J.D. and Caro L. (1966). DNA transfer in bacterial conjugation. *J. Mol. Biol.* **16,** 269–284.

Gross J., Grunstein J. and Witkin E.M. (1971). Inviability of *recA*⁻ derivatives of the DNA polymerase mutant of de Lucia and Cairns. *J. Mol. Biol.* **58,** 631–634.

Grunstein M. and Hogness D.S. (1975). Colony hybridization: a method for the isolation of cloned DNAs that contain a specific gene. *Proc. Natl. Acad. Sci. USA* **72,** 3961–3965.

Guarneros G. and Echols H. (1970). New mutants of bacteriophage λ with a specific defect in excision from the host chromosome. *J. Mol. Biol.* **47,** 565–574.

Guarneros G. and Echols H. (1973). Thermal asymmetry of site specific recombination by phage λ. *Virology* **52,** 30–38.

Guerrini F. (1969). On the asymmetry of λ integration sites. *J. Mol. Biol.* **46,** 523–542.

Guerrini F. and Fox M.S. (1968a). Genetic heterozygosity in pneumococcal transformation. *Proc. Nat. Acad. Sci. USA* **59,** 429–436.

Guerrini F. and Fox M.S. (1968b). Effects of DNA repair in transformation in heterozygotes of pneumococcus. *Proc. Natl. Acad. Sci. USA* **59,** 1116–1123.

Guerry P. and Falkow S. (1971). Polynucleotide sequence relationships among some bacterial plasmids. *J. Bacteriol.* **107,** 372–374.

Guerry P., Falkow S. and Datta N. (1974). R62, a naturally occurring hybrid R plasmid. *J. Bacteriol.* **119,** 144–151.

Guha A. and Szybalski W. (1968). Fractionation of the complementary strands of phage T4 based on the asymmetric distribution of the poly(U) and poly(U,G) binding sites. *Virology* **34,** 608–616.

Guha A., Szybalski W., Salser W., Bolle A. and Geiduschek E.P. (1971). Controls and polarity of transcription during phage T4 development. *J. Mol. Biol.* **59,** 329–350.

Guiney D.G. and Helinski D.R. (1975). Relaxation complexes of plasmid DNA and protein. III. Association of protein with the 5′ terminus of the broken DNA strand in the relaxed complex of plasmid Col E1. *J. Biol. Chem.* **250,** 8796–8803.

Gupta S.L., Chen J., Schaefer L., Lengyel P. and Weissman S.M. (1970). Nucleotide sequence of a ribosome attachment site of phage f2 RNA. *Biochem. Biophys. Res. Commun.* **39,** 883–888.

Gurney T.J. and Fox M.S. (1968). Physical and genetic hybrids formed in bacterial transformation. *J. Mol. Biol.* **32,** 83–100.

Gussin G.N. (1966). Three complementation groups in R17. *J. Mol. Biol.* **21,** 435-453.

Guthrie C. (1975). The nucleotide sequence of the dimeric precursor to glutamine and leucine tRNAs coded by phage T4. *J. Mol. Biol.* **95,** 529–538.

Guthrie C., Seidman J.G., Altman S., Barrell B.G., Smith J.D. and McClain W.H. (1973). Identification of tRNA precursor molecules made by phage T4. *Nature New Biol.* **246,** 6–11.

Guyer M.S. and Clark A.J. (1976). Cis dominant, transfer deficient mutants of the E.coli K12 sex factor. *J. Bacteriol.* **125,** 233–247.

Hall B.D., Nygaard A.P. and Green M.H. (1964). Control of T2 specific RNA synthesis. *J. Mol. Biol.* **9,** 143–153.

Hall B.D., Green M., Nygaard A.P. and Boezi J. (1963). The copying of DNA in T2 infected E.coli. *Cold Spring Harbor Symp. Quant. Biol.* **28,** 201–203.

Hall C.E., Maclean E.C. and Tessman I. (1959). Structure and dimensions of $\phi$X174 from electron microscopy. *J. Mol. Biol.* **1,** 192–194.

Hall Z.W. and Lehman I.R. (1968). An in vitro transversion by a mutationally altered T4 induced DNA polymerase. *J. Mol. Biol.* **36,** 321–334.

Hamer D.H. and Thomas C.A. jr. (1976). Molecular cloning of DNA fragments produced by restriction endonucleases Sal I and Bam I. *Proc. Natl. Acad. Sci. USA* **73,** 1537–1541.

Hamilton D.L. and Luftig R.B. (1972). Phage T4 head morphogenesis. III. Some novel properties of gene 13 defective heads. *J. Virol.* **9,** 1047–1056.

Hamilton D.L. and Luftig R.B. (1976). Phage T4 head morphogenesis. VII. Terminal stages of head maturation. *J. Virol.* **17,** 550–567.

Hamlett N.V. and Berger H. (1975). Mutations altering genetic recombination and repair of DNA in phage T4. *Virology* **63,** 539–567.

Harada K., Suzuki M., Kerneda M. and Mitsuhasi S. (1960). On the drug resistance of enteric bacteria. 2. Transmission of the drug resistance among Enterobacteriaceae. *Jap. J. Exp. Med.* **30,** 289–299.

Harada K., Kameda M., Suzuki M. and Mitsuhashi S. (1963). Drug resistance of enteric bacteria. II. Transduction of transmissible drug resistance (R) factors with phage epsilon. *J. Bacteriol.* **86,** 1332–1338.

Harada K., Kameda M., Suzuki M., and Mitsuhashi S. (1964). Drug resistance of enteric bacteria. III. Acquisition of transferability of nontransmissible R(TC) factor in cooperation with F factor and formation of FR(TC). *J. Bacteriol.* **88,** 1257–1265.

Hardy K.G. (1975). Colicinogeny and related phenomena. *Bacteriol. Rev.* **39,** 464–515.

Hardy K.G., Meynell G.G. and Dowman J.E. (1973). Two major groups of colicin factors: their evolutionary significance. *Mol. Gen. Genet.* **125,** 217–230.

Harm W. (1963). Mutants of phage T4 with increased sensitivity to ultraviolet. *Virology* **19,** 66–71.

Harris A.W., Mount D.W.A., Fuerst C.R. and Siminovitch L. (1967). Mutations in phage λ affecting host cell lysis. *Virology* **32,** 553–569.

Harrison D.P., Brown D.T. and Bode V.C. (1973). The lambda head-tail joining reaction: purification, properties and structure of biologically active heads and tails. *J. Mol. Biol.* **79,** 437–450.

Haselkorn R., Vogel M. and Brown R.D. (1969). Conservation of the rifamycin sensitivity of transcription during T4 development. *Nature* **221,** 836–838.

Hashimoto H. and Rownd R.H. (1975). Transition of the R factor NR1 in Proteus mirabilis: level of drug resistance of nontransitioned and transitioned cells. *J. Bacteriol.* **123,** 56–68.

Hausmann R. and Gomez B. (1967). Amber mutants of phages T3 and T7 defective in phage directed DNA synthesis. *J. Virol.* **1,** 779–792.

Hausmann R. and La Rue K. (1969). Variations in sedimentation patterns among DNAs synthesized after infection of E.coli by different amber mutants of phage T7. *J. Virol.* **3,** 278–281.

Hayashi M.N., Hayashi M. and Spiegelman S. (1963). Restriction of in vivo genetic transcription to one of the complementary strands of DNA. *Proc. Natl. Acad. Sci. USA* **50,** 664–672.

Hayashi Y. and Hayashi M. (1970). Fractionation of $\phi$X174 specific mRNA. *Cold Spring Harbor Symp. Quant. Biol.* **35,** 171–177.

Hayes S. and Szybalski W. (1973). Control of short leftward transcripts from the immunity of *ori* regions in induced phage lambda. *Molec. Gen. Genet.* **126,** 275–290.

Hayes W. (1952a). Recombination in Bact. coli K12: unidirectional transfer of genetic material. *Nature* **169,** 118–119.

Hayes W. (1952b). Genetic recombination in Bact. coli K12: analysis of the stimulating effect of ultraviolet light. *Nature* **169,** 1017–1018.

Hayes W. (1953a). The mechanism of genetic recombination in E.coli. *Cold Spring Harbor Symp. Quant. Biol.* **18,** 75–93.

Hayes W. (1953b). Observations on a transmissible agent determining sexual differentiation in Bacterium coli. *J. Gen. Microbiol.* **8,** 72–88.

Hayes W. (1957). The kinetics of the mating process in E.coli. *J. Gen. Microbiol.* **16,** 97–119.

Hayes W. (1968). The genetics of bacteria and their viruses. (*Blackwell Scientific Publications: Oxford*). Second edition.

Hedges R.W. and Datta N. (1971). $fi^-$ R factors giving chloramphenicol resistance. *Nature* **234,** 220–221.

Hedges R.W. and Datta N. (1972). R124, a $fi^{\infty}$ R factor of a new compatibility class. *J. Gen. Microbiol.* **71,** 403–405.

Hedges R.W. and Datta N. (1973). Plasmids determining I pili constitute a compatibility complex. *J. Gen. Microbiol.* **77,** 19–25.

Hedges R.W. and Jacob A. (1974). Transposition of ampicillin resistance from RP4 to other replicons. *Molec. Gen. Genet.* **132,** 31–40.

Hedgpeth J., Goodman, H.M. and Boyer H.W. (1972). DNA nucleotide sequence restricted by the RI endonuclease. *Proc. Natl. Acad. Sci. USA* **69,** 3448–3452.

Heffron F., Rubens C. and Falkow S. (1975). Translocation of a plasmid DNA sequence which mediates ampicillin resistance: molecular nature and specificity of insertion. *Proc. Natl. Acad. Sci. USA* **72,** 3623–3627.

Heffron F., Sublett R., Hedges R.W., Jacob A. and Falkow S. (1975). Origin of the TEM-beta-lactamase gene found on plasmids. *J. Bacteriol.* **122,** 250–256.

Heinemann S.F. and Spiegelman W.F. (1970). Role of the gene N product in phage lambda. *Cold Spring Harbor Symp. Quant. Biol.* **35,** 315–318.

Helinski D.R. (1973). Plasmid determined resistance to antibiotics: molecular properties of R factor. *Ann. Rev. Microbiol.* **27,** 437–470.

Helmstetter C.E., Cooper S., Pierucci O. and Revelas E. (1968). On the bacterial life sequence. *Cold Spring Harbor Symp. Quant. Biol.* **33,** 809–822.

Helmuth R., and Achtman M. (1975). Operon structure of DNA transfer cistrons on the F sex factor. *Nature* **257,** 652–656.

Henderson D. and Weil J. (1975a). Recombination-deficient deletions in phage lambda and their interaction with chi mutations. *Genetics* **79,** 143–174.

Henderson D. and Weil J. (1975b). The nature and origin of a class of essential gene substitutions in phage $\lambda$. *Virology* **67,** 124–135.

Hendrix R.W. (1971). Identification of proteins coded in phage lambda. In The Bacteriophage Lambda, ed. A.D. Hershey. (*Cold Spring Harbor Laboratory, New York*). pp. 355–370.

Hendrix R.W. and Casjens S.R. (1974a). Protein fusion: a novel reaction in phage λ head assembly. *Proc. Natl. Acad. Sci. USA* **71,** 1451–1455.

Hendrix R.W. and Casjens S.R. (1974b). Protein cleavage in phage λ tail assembly. *Virology* **61,** 156–159.

Hendrix R.W. and Casjens S.R. (1975). Assembly of phage lambda heads: protein processing and its genetic control in petit λ assembly. *J. Mol. Biol.* **91,** 187–200.

Henry T.J. and Knippers R. (1974). Isolation and function of the gene A initiator of phage φX 174, a highly specific DNA endonuclease. *Proc. Natl. Acad. Sci. USA* **71,** 1549–1553.

Henry T.J. and Pratt D. (1969). The proteins of phage M13. *Proc. Natl. Acad. Sci. USA* **62,** 800–807.

Hercules K., Jovanovich S. and Sauerbier W. (1976). Early gene expression in phage T7. I. In vivo synthesis, inactivation, and translational utilization of early mRNAs. *J. Virol.* **17,** 642–658.

Hercules K., Schweiger M. and Sauerbier W. (1974). Cleavage by RNAase III converts T3 and T7 early precursor RNA into translatable message. *Proc. Natl. Acad. Sci. USA* **71,** 840–844.

Herman R.K. (1965). Reciprocal recombination of chromosome and F merogenote in E.coli. *J. Bacteriol.* **90,** 1664–1667.

Herschman H.R. and Helinski D.R. (1967). Purification and characterization of colicin E2 and E3. *J. Biol. Chem.* **242,** 5360–5368.

Hershey A.D. (1946). Spontaneous mutations in bacterial viruses. *Cold Spring Harbor Symp. Quant. Biol.* **11,** 67–77.

Hershey A.D. (1957). Some minor components of phage T2 particles. *Virology* **4,** 237–264.

Hershey A.D. and Burgi E. (1965). Complementary structure of interacting sites at the ends of lambda DNA molecules. *Proc. Natl. Acad. Sci. USA* **53,** 325–328.

Hershey A.D. and Chase M. (1951). Genetic recombination and heterozygosis in phage. *Cold Spring Harbor Symp. Quant. Biol.* **16,** 471–479.

Hershey A.D. and Chase M. (1952). Independent functions of viral protein and nucleic acid in growth of bacteriophage. *J. Gen. Physiol.* **26,** 36–56.

Hershey A.D. and Rotman R. (1949). Genetic recombination between host range and plaque type mutants of phage in single bacterial cells. *Genetics* **34,** 44–71.

Hershey A.D., Burgi E. and Ingraham L. (1963). Cohesion of DNA molecules isolated from phage lambda. *Proc. Natl. Acad. Sci. USA* **49,** 748–755.

Hershey A.D., Dixon J. and Chase M. (1953). Nucleic acid economy in bacteria infected with phage T2. I. Purine and pyrimidine composition. *J. Gen. Physiol.* **36,** 777–785.

Hershfield M.S. (1973). On the role of DNA polymerase in determining mutation rates. Characterization of the defect in T4 DNA polymerase caused by the ts L88 mutation. *J. Biol. Chem.* **248,** 1417–1423.

Hershfield M.S. and Nossal N.G. (1972). Hydrolysis of template and newly synthesized DNA by the 3′ to 5′ exonuclease activity of the T4 DNA polymerase. *J. Biol. Chem.* **247,** 3393–3404.

Hershfield V., Boyer H.W., Yanofsky C., Lovett M.A. and Helinski D.R. (1974). Plasmid

Col El as a molecular vehicle for cloning and amplification of DNA. *Proc. Natl. Acad. Sci. USA* **71,** 3455–3459.

Hershfield V., Boyer H.W., Chow L. and Helinski D.R. (1976). Characterization of a mini-Col E1 plasmid. *J. Bacteriol.* **126,** 447–453.

Herskowitz I. and Signer E.R. (1970a). A site essential for expression of all late genes in bacteriophage λ. *J. Mol. Biol.* **47,** 545–556.

Herskowitz I. and Signer E.R. (1970b). Control of transcription from the *r* strand of phage lambda. *Cold Spring Harbor Symp. Quant. Biol.* **35,** 355–368.

Herskowitz I. and Signer E.R. (1974). Substitution mutations in phage λ with new specificity for late gene expression. *Virology* **61,** 112–119.

Hill C. and Holland I.B. (1967). Genetic basis of colicin E susceptibility in E.coli. I. Isolation and properties of refractory mutants and the preliminary mapping of their mutations. *J. Bacteriol.* **94,** 677–686.

Hinckle D.C. and Chamberlin M.J. (1972a). Studies of the binding of E.coli RNA polymerase to DNA. I. The role of sigma subunit in site selection. *J. Mol. Biol.* **70,** 187–196.

Hinckle D.C. and Chamberlin M.J. (1972b). Studies of the binding of RNA polymerase to DNA. II. The kinetics of the binding reaction. *J. Mol. Biol.* **70,** 187–196.

Hinckle D.C. and Richardson C.C. (1974). Phage T7 DNA replication in vitro. Requirements for DNA synthesis and characterization of the product. *J. Biol. Chem.* **249,** 2974–2985.

Hinckle D.C. and Richardson C.C. (1975). Phage T7 DNA replication in vitro. Purification and properties of the gene 4 protein of phage T7. *J. Biol. Chem.* **250,** 5523–5529.

Hinckle D.C., Mangel W.F. and Chamberlin M.J. (1972). Studies of the binding of E.coli RNA polymerase to DNA. IV. The effect of rifampicin on binding and on RNA chain initiation. *J. Mol. Biol.* **70,** 209–220.

Hindley J. and Staples D.H. (1969). Sequence of a ribosome binding site in phage Qβ RNA. *Nature* **224,** 964–967.

Hindley J., Staples D.H., Billeter M.A. and Weissmann C. (1970). Location of the coat cistron on the RNA of phage Qβ. *Proc. Natl. Acad. Sci. USA* **67,** 1180–1187.

Hirota Y. (1960). The effect of acridine dyes on mating type factors in E.coli. *Proc. Natl. Acad. Sci. USA* **46,** 57–64.

Hirota Y., Gefter M. and Mindich L. (1972). A mutant of E.coli defective in DNA polymerase II activity. *Proc. Natl. Acad. Sci. USA* **69,** 3238–3242.

Hirota Y., Fujii T. and Nishimura Y. (1966). Loss and repair of conjugal fertility and infectivity of the resistance factor and sex factor in E.coli. *J. Bacteriol.* **91,** 1298–1304.

Hirota Y., Nishimura Y., Ørskov F. and Ørskov I. (1964). Effect of drug resistance factor R on the F properties of E.coli. *J. Bacteriol.* **87,** 341–351.

Hirsch H.J., Saedler H. and Starlinger P. (1972). Insertion mutations in the control region of the *gal* operon of E.coli. II. Physical characterization of the mutations. *Molec. Gen. Genet.* **115,** 266–276.

Hirsch H.J., Starlinger P. and Brachet P. (1972). Two kinds of insertions in bacterial genes. *Molec. Gen. Genet.* **119,** 191–206.

Hoffmann-Berling H. and Mazé R. (1964). Release of male specific phages from surviving host bacteria. *Virology* **22,** 305–313.

Hofschneider P.H. (1963). Untersuchungen uber "kleine" E.coli K12 bacteriophagen. I. Mitt. Die isolierung und einige eigenschaften der "kleinen" bacteriophagen M12, M13 und M20. *Z. Naturforsch* **18b,** 203–205.

Hofschneider P.H. and Preuss A. (1963). M13 bacteriophage liberation from intact bacteria as revealed by electron microscopy. *J. Mol. Biol.* **7,** 450–451.

Hohn B. (1975). DNA as substrate for packaging into phage lambda in vitro. *J. Mol. Biol.* **98,** 93–106.

Hohn B. and Hohn T. (1974). Activity of empty, headlike particles for packaging of DNA of phage lambda in vitro. *Proc. Natl. Acad. Sci. USA* **71,** 2372–2376.

Hohn B. and Korn D. (1969). Cosegregation of a sex factor with the E.coli chromosome during curing by acridine orange. *J. Mol. Biol.* **45,** 385–395.

Hohn B. Lechner H. and Marvin D.A. (1971). Filamentous bacterial viruses. I. DNA synthesis during the early stages of infection with fd. *J. Mol. Biol.* **56,** 143–154.

Hohn B., von Schutz H. and Marvin D.A. (1971). Filamentous bacterial viruses. II. Killing of bacteria by abortive infection with fd. *J. Mol. Biol.* **56,** 155–165.

Hohn T. and Hohn B. (1973). A minor pathway leading to plaque forming particles in phage lambda. Studies on the function of gene D. *J. Mol. Biol.* **79,** 649–662.

Hohn T., Flick H. and Hohn B. (1975). Petit λ, a family of particles from phage lambda infected cells. *J. Mol. Biol.* **98,** 107–120.

Holliday R. (1964). A mechanism for gene conversion in fungi. *Genet. Res.* **5,** 282–304.

Homyk T. jr. and Weil J. (1974). Deletion analysis of two nonessential regions of the T4 genome. *Virology* **61,** 505–523.

Hong J., Smith G.R. and Ames B.N. (1971). Cyclic AMP concentration in the bacterial host regulates the viral decision between lysogeny and lysis. *Proc. Natl. Acad. Sci. USA* **68,** 2258–2262.

Honigman A., Oppenheim A., Oppenheim A.B. and Stevens W.F. (1975). A pleiotropic regulatory mutation in λ phage. *Molec. Gen. Genet.* **138,** 85–112.

Honigman A., Hu S.-L., Chase R. and Szybalski W. (1976). 4S *oop* RNA is a leader sequence for the immunity-establishment transcription in phage lambda. *Nature* **262,** 112–116.

Hopkins N. (1970). Bypassing a positive regulator: isolation of a λ mutant that does not require N product to grow. *Virology* **40,** 223–229.

Hori K., Eoyang L., Bannerjee A.K. and August J.T. (1967). Replication of RNA viruses. V. Template activity of synthetic ribopolymers in the Qβ RNA polymerase reaction. *Proc. Natl. Acad. Sci. USA* **57,** 1790–1797.

Horii Z.I. and Clark A.J. (1973). Genetic analysis of the recF pathway to genetic recombination in E.coli K12: isolation and characterization of mutants. *J. Mol. Biol.* **80,** 327–344.

Horiuchi K. (1975). Genetic studies of RNA phages. In RNA Phages, ed. N.D. Zinder. (*Cold Spring Harbor Laboratory: New York*). pp. 29–50.

Horiuchi K. and Matsuhashi S. (1970). Three cistrons in phage Qβ. *Virology* **42,** 49–60.

Horiuchi K., Lodish H.F. and Zinder N.D. (1966). Mutants of the phage f2. VI. Homology of temperature sensitive and host dependent mutants. *Virology* **28,** 438–447.

Horiuchi K., Webster R.E., Matsuhashi S. (1971). Gene products of phage Qβ. *Virology* **45,** 429–439.

Horiuchi K., Vovis G.F., Enea V. and Zinder N.D. (1975). Cleavage map of phage fl: location of the E.coli B-specific modification sites. *J. Mol. Biol.* **95,** 147–165.

Horvitz H.R. (1973). Polypeptide bound to the host RNA polymerase is specified by T4 control gene 33. *Nature New Biol.* **244,** 137–140.

Horvitz H. (1974a). Control by bacteriophage T4 of two sequential phosphorylations of the alpha subunit of E.coli RNA polymerase. *J. Mol. Biol.* **90,** 727–738.

Horvitz H. (1974b). Phage T4 mutants deficient in alteration and modification of the E.coli RNA polymerase. *J. Mol. Biol.* **90,** 739–750.

Hosoda J. and Cone R. (1970). Analysis of T4 phage proteins. I. Conversion of precursor proteins into lower molecular weight peptides during normal capsid formation. *Proc. Natl. Acad. Sci. USA* **66,** 1275–1281.

Hosoda J. and Levinthal C. (1968). Protein synthesis by E.coli infected with phage T4D. *Virology* **34,** 709–727.

Hosoda J. and Matthews E. (1968). DNA replication in vivo by a temperature sensitive polynucleotide ligase mutant of T4. *Proc. Natl. Acad. Sci. USA* **61,** 997–1004.

Hosoda J. and Matthews E. (1971). DNA replication in vivo by polynucleotide ligase defective mutants of T4. II. Effect of chloramphenicol and mutations in other genes. *J. Mol. Biol.* **55,** 155–179.

Hotchkiss R.D. (1951). Transfer of penicillin resistance in pneumococci by the desoxyribonucleate derived from resistant cultures. *Cold Spring Harbor Symp. Quant. Biol.* **16,** 457–461.

Hotchkiss R.D. (1954). Cyclical behaviour in pneumococcal groups and transformability occasioned by environmental changes. *Proc. Natl. Acad. Sci. USA* **40,** 49–55.

Hotchkiss R.D. (1957). Criteria for quantitative genetic transformation of bacteria. In The Chemical Basis of Heredity, eds. W.D. McElroy and B. Glass. (*Johns Hopkins Press: Baltimore*), pp. 321–335.

Hotchkiss R.D. (1966). Gene, transforming principle, and DNA. In Phage and the Origins of Molecular Biology, eds. J. Cairns, G.S. Stent and J.D. Watson. (*Cold Spring Harbor Laboratory: New York*). pp. 180–200.

Hotchkiss R.D. and Gabor M. (1970). Bacterial transformation, with special reference to recombination process. *Ann. Rev. Genetics.* **4,** 193–224.

Howatson A.F. and Kemp C.L. (1975). The structure of tubular head forms of phage λ; relationship to the capsid structure of petit λ and normal λ heads. *Virology* **67,** 80–84.

Howe C.C., Buckley P.J., Carlson K.M. and Kozinski A.W. (1973). Multiple and specific initiation of T4 DNA replication. *J. Virol.* **12,** 130–148.

Hradecna Z, and Szybalski W. (1967). Fractionation of the complementary strands of phage λ DNA based on the asymmetric distribution of the poly(I,G) binding sites. *Virology* **32,** 633–643.

Hsu M.T. (1974). Electron microscope analysis of partial denaturation of F factor DNA. *J. Bacteriol.* **118,** 425–433.

Hsu T. and Weiss B. (1969). Selective translation of T4 template RNA ribosomes from T4 infected E.coli. *Proc. Natl. Acad. Sci. USA* **64,** 345–351.

Hu S., Ohtsubo E. and Davidson N. (1975). Electron microscope heteroduplex studies of sequence relations among plasmids of E.coli: structure of F13 and related F-primes. *J. Bacteriol.* **122,** 749–763.

Hu S., Ohtsubo E., Davidson N. and Saedler H. (1975a). Electron microscope heteroduplex studies of sequence relations among bacterial plasmids: identification and mapping of the insertion sequences IS1 and IS2 in F and R plasmids. *J. Bacteriol.* **122,** 764–775.

Hu S., Ptashne K., Cohen S.N. and Davidson N. (1975b). $\alpha\beta$ sequence of F is IS3. *J. Bacteriol.* **123,** 687–692.

Huang W.M. and Lehman I.R. (1972). On the exonuclease activity of phage T4 DNA polymerase. *J. Biol. Chem.* **247,** 3139–3146.

Huberman J.A., Kornberg A. and Alberts B.M. (1971). Stimulation of T4 bacteriophage DNA polymerase by the protein product of T4 gene 32. *J. Mol. Biol.* **62,** 39–52.

Hurwitz J. and Wickner S. (1974). Involvement of two protein factors and ATP in in vitro DNA synthesis catalyzed by DNA polymerase III of E.coli. *Proc. Natl. Acad. Sci. USA* **71,** 6–10.

Hutchison C.A. and Sinsheimer R.L. (1966). The process of infection with $\phi$X174. X. Mutations in a $\phi$X lysis gene. *J. Mol. Biol.* **18,** 429–447.

Hyman R.W. (1971). Physical mapping of T7 mRNA. *J. Mol. Biol.* **61,** 369–376.

Hyman R.W. and Summers W.C. (1972). Isolation and physical mapping of T7 gene 1 mRNA. *J. Mol. Biol.* **71,** 573–582.

Hyman R.W., Brunovskis I. and Summers W.C. (1973). Base sequence homology between coliphages T7 and $\phi$II and between T3 and $\phi$II as determined by heteroduplex mapping in the electron microscope. *J. Mol. Biol.* **77,** 189–196.

Hyman R.W., Brunovskis I. and Summers W.C. (1974). A biochemical comparison of the related phages T7, $\phi$1, $\phi$11, W31, H and T3. *Virology* **57,** 189–206.

Ingram L.C. (1973). DNA-DNA hybridization of R factors. *J. Bacteriol.* **115,** 1130–1134.

Inman R.B. and Schnos M. (1971). Structure of branch points in replicating DNA: presence of single stranded connections in λ DNA branch points. *J. Mol. Biol.* **56,** 319–326.

Inman R.B. and Schnos M. (1974). In Principles and Techniques of Electron Microscopy, volume 4. Ed. M.A. Hayat. (*Van Nostrand, Reinhold; New York*).

Inoko H. and Imai M. (1976). Isolation and genetic characterization of the *nitA* mutants of E.coli affecting the termination factor rho. *Molec. Gen. Genet.* **143,** 211–221.

Inokuchi H., Dove W.F. and Freifelder D. (1973). Physical studies of RNA involvement in phage λ DNA replication and prophage excision. *J. Mol. Biol.* **73,** 721–728.

Inouye H., Pollack Y. and Petre J. (1974). Physical and functional homology between ribosomal protein S1 and interference factor i. *Eur. J. Biochem.* **45,** 109–117.

Inselburg J. (1971). R factor DNA in chromosomeless progeny of E.coli. *J. Bacteriol.* **105,** 620–628.

Inselburg J.W. (1973a). Formation of catenated colicin factor E1 DNA molecules in a recombination deficient strain of E.coli. *J. Bacteriol.* **113,** 1084–1086.

Inselburg J. (1973b). Colicin factor DNA: a single nonhomologous region in Col E2-E3 heteroduplex molecules. *Nature New Biol.* **241,** 234–237.

Inselburg J. (1974a). Replication of colicin E1 plasmid DNA in minicells from a unique replication initiation site. *Proc. Natl. Acad. Sci. USA* **71,** 2256–2259.

Inselburg J. (1974b). Incompatibility exhibited by colicin plasmids E1, E2 and E3 in E.coli. *J. Bacteriol.* **119,** 478–483.

Inselburg J. and Fuke M. (1971). Isolation of catenated and replicating DNA molecules of colicin factor E1 from minicells. *Proc. Natl. Acad. Sci. USA* **68,** 2839–2842.

Ippen-Ihler K., Achtman M. and Willetts N. (1972). Deletion map of the E.coli sex factor F: the order of eleven transfer cistrons. *J. Bacteriol.* **110,** 857–863.

Isaacs L.N., Echols H. and Sly W.S. (1965). Control of lambda mRNA by the cI immunity region. *J. Mol. Biol.* **13,** 963–967.

Ishibashi M. (1967). F pilus as $f^+$ antigen. *J. Bacteriol.* **93,** 379–389.

Ishii T. and Yanagida M. (1975). Molecular organization of the shell of the T-even phage head. *J. Mol. Biol.* **97,** 655–660.

Isobe T., Black L.W. and Tsugita A. (1976). Primary structure of phage T4 internal protein II and characterization of the cleavage upon phage maturation. *J. Mol. Biol.* **102,** 349–366.

Iwakura Y., Ishihama A. and Yura T. (1973). RNA polymerase mutants of E.coli. II. Streptolydigin resistance and its relation to rifampicin resistance. *Molec. Gen. Genet.* **121,** 181–196.

Iwaya M. and Denhardt D.T. (1971). The mechanism of replication of $\phi$X174 single stranded DNA II. The role of viral proteins. *J. Mol. Biol.* **57,** 159–175.

Iwaya M. and Denhardt D.T. (1973a). Mechanism of replication of $\phi$X single stranded DNA. IV. The parental viral strand is not conserved in the replicating DNA structure. *J. Mol. Biol.* **73,** 279–290.

Iwaya M. and Denhardt D.T. (1973b). Mechanism of replication of $\phi$X single stranded DNA. V. Dispersive and conservative transfer of parental DNA into progeny DNA. *J. Mol. Biol.* **73,** 291–305.

Iwaya M., Eisenberg S., Bartok K. and Denhardt D.T. (1973). Mechanism of replication of single stranded $\phi$X174 DNA. VII. Circularization of the progeny viral strand. *J. Virol.* **12,** 808–818.

Jackson D.A., Symons R.H. and Berg P. (1972). Biochemical method for inserting new genetic information into DNA of SV40: circular SV40 DNA molecules containing lambda phage genomes and the galactose operon of E.coli. *Proc. Natl. Acad. Sci. USA* **69,** 2904–2909.

Jacob F. (1955). Transduction of lysogeny in E.coli. *Virology* **1,** 207–220.

Jacob F. and Campbell A. (1959). Sur le systeme de repression assurant l'immunite chex les bacteries lysogenes. *CR Acad. Sci. Paris* **248,** 3219–3221.

Jacob F. and Fuerst C.R. (1958). The mechanism of lysis by phage studied with defective lysogenic bacteria. *J. Gen. Microbiol.* **18,** 518–526.

Jacob F. and Monod J. (1961). Genetic regulatory mechanisms in the synthesis of proteins. *J. Mol. Biol.* **3,** 318–356.

Jacob F. and Wollman E.L. (1953). Induction of phage development in lysogenic bacteria. *Cold Spring Harbor Symp. Quant. Biol.* **18,** 101–121.

Jacob F. and Wollman E.L. (1954). Etude genetique d'un bacteriophage tempere d'E.coli. I. Le systeme genetique du bacteriophage λ. *Ann. Inst. Pasteur* **87,** 653–673.

Jacob F. and Wollman E.L. (1955). Etude genetique d'un phage tempere d'E.coli. III. Effet du rayonnement ultraviolet sur la recombinaison genetique. *Ann. Inst. Pasteur* **88,** 724–749.

Jacob F. and Wollman E.L. (1956a). Sur les processus de conjugaison et de recombinaison genetique chez E.coli. I. L'induction par conjugaison ou induction zygotique. *Ann. Inst. Pasteur* **91,** 486–510.

Jacob F. and Wollman E. (1956b). Recombinaison genetique et mutants de fertilite chez E.coli. *Comptes Rendues Acad. Sci. Paris* **242,** 303–306.

Jacob F. and Wollman E.L. (1957). Analyse des groups de liason génétique de differentes souches donatrices. *Compt. Rend. Acad. Sci. Paris* **245,** 1840–1843.

Jacob F. and Wollman E.L. (1958a). Genetic and physical determinations of chromosomal segments in E.coli. *Symp. Soc. Exp. Biol.* **12,** 75–92.

Jacob F. and Wollman E.L. (1961). Sexuality and the genetics of bacteria. (*Academic Press, New York*).

Jacob F., Brenner S. and Cuzin F. (1963). On the regulation of DNA replication in bacteria. *Cold Spring Harbor Symp. Quant. Biol.* **28,** 329–348.

Jacob F., Fuerst C.R. and Wollman E.L. (1957). Recherches sur les bacteries lysogenes defectives. II. Les types physiologiques lies aux mutations du prophage. *Ann. Inst. Pasteur* **93,** 724–745.

Jacob F., Ryter A. and Cuzin F. (1966). On the association between DNA and the membrane in bacteria. *Proc. Roy. Soc. B.* **164,** 267–278.

Jacob F., Schaeffer P. and Wollman E.L. (1960). Episomic elements in bacteria. *Symp. Soc. Gen. Microbiol.* **10,** 67–91.

Jacob F., Sussman R. and Monod J. (1962). Sur la nature du represseur assurant l'immunite des bacteries lysogenes. *Compt. Rend. Acad. Sci. Paris* **254,** 4214–4216.

Jacobson A. (1972). Role of F pili in the penetration of phage fl. *J. Virol.* **10,** 835–843.

Jaenisch R., Hofschneider P.H. and Preuss A. (1966). Uber infeckiose substrukturen aus bacteriophagen. VII. On the tertiary structure and biological properties of $\phi$X174 replicative form. *J. Mol. Biol.* **21,** 501–516.

Jakes K., Zinder N.D. and Boon T. (1974). Purification and properties of colicin E3 immunity protein. *J. Biol. Chem.* **249,** 438–444.

Jaskunas S.R., Burgess R.R. and Nomura M. (1975). Identification of a gene for the $\alpha$ subunit of RNA polymerase at the *str-spc* region of the E.coli chromosome. *Proc. Natl. Acad. Sci. USA* **72,** 5036–5040.

Javor G.T. and Tomasz A. (1968). An autoradiographic study of genetic transformation. *Proc. Natl. Acad. Sci. USA* **68,** 1216–1222.

Jayaraman R. (1972). Transcription of the phage T4 DNA by E.coli RNA polymerase in vitro: identification of some immediate early and delayed early genes. *J. Mol. Biol.* **70,** 253–264.

Jazwinski S.M. and Kornberg A. (1975). DNA replication in vitro starting with an intact $\phi$X 174 phage. *Proc. Natl. Acad. Sci. USA* **72,** 3863–3867.

Jazwinski S.M., Lindberg A.A. and Kornberg A. (1975a). The lipopolysaccharide receptor for phages $\phi$X174 and S13. *Virology* **66,** 268–282.

Jazwinski S.M., Lindberg A.A. and Kornberg A. (1975b). The gene H spike protein of phages $\phi$X174 and S13. I. functions in phage receptor recognition and in transfection. *Virology* **66,** 283–293.

Jazwinski S.M., Marco R. and Kornberg A. (1973). A coat protein of the phage M13 virion participates in membrane-oriented synthesis of DNA. *Proc. Natl. Acad. Sci. USA* **70,** 205–209.

Jazwinski S.M., Marco R. and Kornberg A. (1975). The gene H spike protein of phages $\phi$X174 and S13. II. Relation to synthesis of the parental replicative form. *Virology* **66,** 294–305.

Jeng Y., Gelfand D., Hayashi M., Shlesser R. and Tessman E.S. (1970). The eight genes of phages $\phi$X174 and S13 and comparison of the phage specified proteins. *J. Mol. Biol.* **49,** 521–526.

Jeppesen P.G.N., Argetsinger-Steitz J., Gesteland R.R. and Spahr P.F. (1970). Gene order in the bacteriophage R17 RNA. *Nature* **226,** 230–237.

Joenje H. and Venema G. (1975). Different nuclease activities in competent and noncompetent B.subtilis. *J. Bacteriol.* **122,** 25–33.

Joenje H., Konings W.N. and Venema G. (1975). Interactions between exogenous DNA and membrane vesicles isolated from competent and noncompetent B.subtilis. *J. Bacteriol.* **121,** 771–776.

Johnson E.M., Falkow S. and Baron L.S. (1964). Recipient ability of S.typhosa in genetic crosses with E.coli. *J. Bacteriol.* **87,** 54–60.

Johnson P.H. and Sinsheimer R.L. (1974). Structure of an intermediate in the replication of phage $\phi$X174 DNA: the initiation site for DNA replication. *J. Mol. Biol.* **83,** 35–46.

Jordan E. (1964). The location of the b2 deletion of phage λ. *J. Mol. Biol.* **10,** 341–344.

Jordan E., Saedler H. and Starlinger P. (1968). 0° and strong polar mutations in the *gal* operon are insertions. *Molec. Gen. Genet.* **102,** 353–363.

Joyner A., Isaacs L.N., Echols H. and Sly W.S. (1966). DNA replication and mRNA production after induction of wild type λ phage and λ mutants. *J. Mol. Biol.* **19,** 174–186.

Kacian D.L., Mills D.R., Kramer F.R. and Spiegelman S. (1972). A replicating RNA molecule suitable for detailed analysis of extracellular evolution and replication. *Proc. Natl. Acad. Sci. USA* **69,** 3038–3042.

Kahn P.L. (1968). Isolation of high frequency recombining strains from E.coli containing the V colicinogenic factor. *J. Bacteriol.* **96,** 205–214.

Kahn P. and Helinski D.R. (1964). Relationship between colicinogenic factors E1 and V and an F factor in E.coli. *J. Bacteriol.* **88,** 1573–1579.

Kaiser A.D. and Jacob F. (1957). Recombination between related temperate phages and the genetic control of immunity and prophage localization. *Virology* **4,** 509–521.

Kaiser A.D. and Masuda T. (1970). Evidence for a prophage excision gene in λ. *J. Mol. Biol.* **47,** 557–564.

Kaiser D. (1971). Lambda DNA replication. In The Bacteriophage Lambda, ed A.D. Hershey. (*Cold Spring Harbor Laboratory: New York*). pp. 195–210.

Kaiser D., Syvanen M. and Masuda T. (1975). DNA packaging steps in phage lambda head assembly. *J. Mol. Biol.* **91,** 175–186.

Kamen R.I. (1970). Characterization of the subunits of Qβ replicase. *Nature* **228,** 527–533.

Kamen R.I. (1975). Structure and function of the Qβ RNA replicase. In RNA Phages, ed. N.D. Zinder. (*Cold Spring Harbor Laboratory: New York*). pp. 203–234.

Kamen R.I., Monstein H.J. and Weissmann C. (1974). The host factor requirement of Qβ RNA replicase. *Biochim. Biophys. Acta* **366,** 292–299.

Kamen R.I., Kondo M., Romer W. and Weissbach C. (1972). Reconstitution of the Qβ replicase lacking subunit α with protein interference factor I. *Eur. J. Biochem.* **31,** 44–51.

Karam J.D. and Bowles M.G. (1974). Mutation to overproduction of phage T4 gene products. *J. Virol.* **13,** 428–438.

Karu A.E., Mackay V., Goldmark P.J. and Linn S. (1973). The *recBC* DNAase of E.coli K12. Substrate specificity and reaction intermediates. *J. Biol. Chem.* **248,** 4874–4884.

Karu A.E., Sakaki Y., Echols H. and Linn S. (1975). The gamma protein specified by phage λ. Structure and inhibitory activity for the *recBC* enzyme of E.coli. *J. Biol. Chem.* **250,** 7377–7387.

Kasai T. and Bautz E.K.F. (1969). Regulation of gene specific RNA synthesis in phage T4. *J. Mol. Biol.* **41,** 401–417.

Katsura I. and Kuhl P.W. (1975). Morphogenesis of the tail of phage λ. III. Morphogenetic pathway. *J. Mol. Biol.* **91,** 257–274.

Katz E.R. and Brenner S. (1975). Genetic map of the beginning of the *rIIB* cistron of T4. *Molec. Gen. Genet.* **143,** 101–104.

Kedes L.H., Cohn R.H., Lowry J.C., Chang A.C.Y. and Cohen S.N. (1975). The organization of sea urchin histone genes. *Cell* **6,** 359–370.

Keil T.V. and Hofschneider P.H. (1973). Secondary structure of RNA phage M12 replicative intermediates in vivo. *Biochim. Biophys. Acta* **312,** 294–312.

Kellenberger E. (1968). Studies on the morphopoiesis of the head of phage T-even. V. The components of the T4 capsid and of other, capsid-related structures. *Virology* **34,** 549–561.

Kellenberger E. and Boy de la Tour E. (1965). Studies on the morphopoiesis of the head of phage T-even. II. Observations on the fine structure of polyheads. *J. Ult. Res.* **13,** 343–358.

Kellenberger E. and Edgar R.S. (1971). Structure and assembly of phage particles. In The Bacteriophage Lambda, ed. A.D. Hershey. (*Cold Spring Harbor Laboratory, New York*). pp. 271–295.

Kellenberger E., Eiserling F.A. and Boy de la Tour E. (1968). Studies on the morphopoiesis of the head of the phage T-even. III. The cores of head-related structures. *J. Ult. Res.* **21,** 335–360.

Kellenberger E., Sechaud J. and Ryter A. (1959). Electron microscopical studies of phage multiplication. IV. The establishment of the DNA pool of vegetative phage and the maturation of phage particles. *Virology* **8,** 478–498.

Kellenberger E., Bolle A., Boy de la Tour E., Epstein R.H., Franklin N.C., Jerne N.K., Reale-Scafati A., Sechaud J., Bendet I., Goldstein D. and Lauffer N.R. (1965). Functions and properties related to the tail fibres of phage T4. *Virology* **26,** 419–440.

Kellenberger G.M., Zichichi M.L. and Epstein H.T. (1962). Heterozygosis and recombination of phage λ. *Virology* **17,** 44–55.

Kellenberger G., Zichichi M.L. and Weigle J. (1961a). Exchange of DNA in the recombination of bacteriophage lambda. *Proc. Natl. Acad. Sci. USA* **47,** 869–878.

Kellenberger G., Zichichi M.L. and Weigle J. (1961b). A mutation affecting the DNA content of bacteriophage lambda and its lysogenising abilities. *J. Mol. Biol.* **3,** 399–408.

Kellenberger G. and Weisberg R.A. (1971). Recombination in phage λ. I. Exchange of DNA promoted by phage and bacterial recombination mechanisms. In The Bacteriophage Lambda, ed. A.D. Hershey. (*Cold Spring Harbor Laboratory, New York*). pp. 407–416.

Kelley T.J. jr. and Thomas C.A. jr. (1969). An intermediate in the replication of phage T7 DNA molecules. *J. Mol. Biol.* **44,** 459–475.

Kells S.S. and Haselkorn R. (1974). Phage T4 short tail fibres are the product of gene 12. *J. Mol. Biol.* **83,** 473–486.

Kemp C.L., Howatson A.F. and Siminovitch L. (1968). Electron microscope studies of mutants of lambda phage. I. General description and quantitation of viral products. *Virology* **36,** 490–502.

Kemper B. and Brown D.T. (1976). Function of gene *49* of phage T4. II. Analysis of intracellular development and the structure of very fast sedimenting DNA. *J. Virol.* **18,** 1000–1015.

Kemper B. and Janz E. (1976). Function of gene *49* of phage T4. I. Isolation and biochemical characterization of very fast sedimenting DNA. *J. Virol.* **18,** 992–999.

Kent J.L. and Hotchkiss R.D. (1964). Kinetic analysis of multiple linked recombinations in pneumococcal transformation. *J. Mol. Biol.* **9,** 308–322.

Kerr C. and Sadowsky P.D. (1972). Gene *6* exonuclease of phage T7. *J. Biol. Chem.* **247,** 305–310.

Kerr C. and Sadowski P.D. (1974). Packaging and maturation of DNA of phage T7 in vitro. *Proc. Natl. Acad. Sci. USA* **71,** 3545–3549.

Kiger J.A. and Sinsheimer R.L. (1969). Vegetative lambda DNA. IV. Fractionation of replicating lambda DNA on benzoylated-naphthoylated DEAE cellulose. *J. Mol. Biol.* **40,** 467–490.

Kikuchi Y. and King J. (1975a). Genetic control of phage T4 baseplate morphogenesis. I. Sequential assembly of the major precursor, in vivo and in vitro. *J. Mol. Biol.* **99,** 645–672.

Kikuchi Y. and King J. (1975b). Genetic control of phage T4 baseplate morphogenesis. II. Mutants unable to form the central part of the baseplate. *J. Mol. Biol.* **99,** 673–694.

Kikuchi Y. and King J. (1975c). Genetic control of phage T4 baseplate morphogenesis. III. Formation of the central plug and overall assembly pathway. *J. Mol. Biol.* **99,** 695–716.

Kim J.S. and Davidson N. (1974). Electron microscope heteroduplex study of sequence relations of T2, T4 and T6 phage DNAs. *Virology* **57,** 93–111.

Kim J.-S., Sharp P.A. and Davidson N. (1972). Electron microscope studies of heteroduplex DNA from a deletion mutant of phage ϕX174. *Proc. Natl. Acad. Sci. USA* **69,** 1948–1952.

King J. (1968). Assembly of the tail of phage T4. *J. Mol. Biol.* **32,** 231–262.

King J. (1971). Phage T4 tail assembly: four steps in core formation. *J. Mol. Biol.* **58,** 693–709.

King J. and Laemmli U.K. (1971). Polypeptides of the tail fibers of T4. *J. Mol. Biol.* **62,** 465–477.

King J. and Laemmli U.K. (1973). Phage T4 tail assembly: structural proteins and their genetic identification. *J. Mol. Biol.* **75,** 315–338.

King J. and Mykolajewycz N. (1973). Phage T4 tail assembly: proteins of the sheath, core and baseplate. *J. Mol. Biol.* **75,** 339–358.

King J. and Wood W.B. (1969). Assembly of phage T4 tail fibers: the sequence of gene product interaction. *J. Mol. Biol.* **39,** 583–601.

Kingsbury D.T. and Helinski D.R. (1973). Temperature sensitive mutants for the replication of plasmids in E.coli: requirement for DNA polymerase I in the replication of the plasmid col E1. *J. Bacteriol.* **114,** 1116–1124.

Kleckner N., Chan R.K., Tye B.-K. and Botstein D. (1975). Mutagenesis by insertion of a drug resistance element carrying an inverted repetition. *J. Mol. Biol.* **97,** 561–576.

Kleinschmidt A.K., Lang D., Jacherts D. and Zahn R.K. (1962). Darstellung und Langenmessungen des gesamten Desoxyribonucleinsäureinhaltes von T2-Bakteriophagen. *Biochim. Biophys. Acta* **61,** 857–864.

Klem E.B., Hsu W.-T. and Weiss S.B. (1970). The selective inhibition of protein initiation by T4 phage-induced factors. *Proc. Natl. Acad. Sci. USA* **67,** 696–701.

Kline B.C. and Miller J.R. (1975). Detection of nonintegrated plasmid DNA in the folded chromosome of E.coli: physicochemical approach to studying the unit of segregation *J. Bacteriol.* **121,** 165–172.

Klug A. and Berger J.E. (1964). An optical method for the analysis of periodicities in electron micrographs, and some observations on the mechanism of negative staining. *J. Mol. Biol.* **10,** 565–569.

Knippers R. and Sinsheimer R.L. (1968). The process of infection with phage ϕX174. XX. Attachment of the parental DNA of phage ϕX174 to a fast sedimenting cell component. *J. Mol. Biol.* **34,** 17–29.

Knippers R., Komano T. and Sinsheimer R.L. (1968). The process of infection with phage ϕX174. XXI. Replication and the fate of the replicative form. *Proc. Natl. Acad. Sci. USA* **59,** 572–581.

Knippers R., Whalley J.M. and Sinsheimer R.L. (1969). The process of infection with phage

$\phi$X174. XXX. Replication of double stranded $\phi$X DNA. *Proc. Natl. Acad. Sci. USA* **64,** 275–282.

Knippers R., Salivar W.O., Newbold J.E. and Sinsheimer R.L. (1969a). The process of infection with $\phi$X174. XXVI. Transfer of the parental DNA of phage $\phi$X174 into progeny phage particles. *J. Mol. Biol.* **39,** 641–654.

Knippers R., Razin A., Davis R. and Sinsheimer R.L. (1969). The process of infection with phage $\phi$X174. XXIX. In vivo studies on the synthesis of the single stranded DNA of progeny $\phi$X174 phage. *J. Mol. Biol.* **45,** 237–263.

Koch G. and Hershey A.D. (1959). Synthesis of phage-precursor protein in bacteria infected with T2. *J. Mol. Biol.* **1,** 260–279.

Kolakofsky D. and Weissmann C. (1971a). Possible mechanism for transition of viral RNA from polysome to replication complex. *Nature New Biol.* **231,** 42–45.

Kolakofsky D. and Weissmann C. (1971b). Q$\beta$ replicase as repressor of Q$\beta$ RNA-directed protein synthesis. *Biochim. Biophys. Acta* **246,** 596–599.

Komano T., Knippers R. and Sinsheimer R.L. (1968). The process of infection with phage $\phi$X174. XXII. Synthesis of progeny single stranded DNA. *Proc. Natl. Acad. Sci. USA* **59,** 911–916.

Kondo M., Gallerani R. and Weissmann C. (1970). Subunit structure of Q$\beta$ replicase. *Nature* **228,** 525–527.

Konings R.N.H., Ward R., Francke B. and Hofschneider P.H. (1970). Gene order of RNA bacteriophage M12. *Nature* **226,** 604–606.

Konings R.H., Jansen J., Cuypers C. and Shoenmakers J.G. (1973). Synthesis of phage M13 specific proteins in a DNA dependent cell free system. II. In vitro synthesis of biologically active gene 5 protein. *J. Virol.* **12,** 1466–1472.

Konings R.N.H., Hulsebos T. and Van Den Hondel C.A. (1975). Identification and characterization of the in vitro synthesized gene products of phage M13. *J. Virol.* **15,** 570–584.

Konisky J. (1972). Characterization of colicin Ia and Ib. Chemical studies of protein structure. *J. Biol. Chem.* **247,** 3750–3755.

Konisky J. and Cowell B.S. (1972). Interaction of colicin Ia with bacterial cells. Direct measurement of Ia-receptor interaction. *J. Biol. Chem.* **247,** 6524–6529.

Konisky J. and Liu C.-T. (1974). Solubilization and partial characterization of the colicin I receptor of E.coli. *J. Biol. Chem.* **249,** 835–840.

Konrad E.B. and Lehman I.R. (1975). Novel mutants of E.coli that accumulate very small DNA replicative intermediates. *Proc. Natl. Acad. Sci. USA* **72,** 2150–2154.

Konrad E.B., Modrich P. and Lehman I.R. (1973). Genetics and enzymatic characterization of conditional lethal mutant of E.coli with a temperature sensitive DNA ligase. *J. Mol. Biol.* **77,** 519–530.

Kontomichalou P., Mitani M. and Clowes R.C. (1970). Circular R factor molecules controlling penicillinase synthesis, replicating in E.coli under either relaxed or stringent control. *J. Bacteriol.* **104,** 34–44.

Kooistra J. and Venema G. (1970). Fate of donor DNA in some poorly transformable strains of H.influenzae. *Mutat. Res.* **9,** 245–253.

Kooistra J. and Venema G. (1974). Fate of donor DNA in a highly transformation-deficient strain of H.influenzae. *J. Bacteriol.* **119,** 705–717.

Kopecko D.J. and Cohen S.N. (1975). Site specific *recA*-independent recombination between bacterial plasmids: involvement of palindromes at the recombinational loci. *Proc. Natl. Acad. Sci.* **72,** 1373–1377.

Korn L.J. and Yanofsky C. (1976). Polarity suppressors increase expression of the wild type tryptophan operon of E.coli. *J. Mol. Biol.* **103,** 395–410.

Kosturko L.D. and Kozinski A.W. (1976). Late events in T4 phage production. I. Late DNA replication is primarily exponential. *J. Virol.* **17,** 794–800.

Kourilsky P., Bourguignon M.F. and Gros F. (1971). Kinetics of viral transcription after induction of prophage. In The Bacteriophage Lambda, ed. A.D. Hershey. (*Cold Spring Harbor Laboratory: New York*). pp. 647–666.

Kourilsky P., Marcaud L., Sheldrick P., Luzzati D. and Gros F. (1968). Studies on the mRNA of phage λ. I. Various species synthesized early after induction of the prophage. *Proc. Natl. Acad. Sci. USA* **61,** 1013–1020.

Kourilsky P., Bourguignon M.F., Bouquet M. and Gros F. (1970). Early transcription controls after induction of prophage lambda. *Cold Spring Harbor Symp. Quant. Biol.* **35,** 305–314.

Kozak M. and Nathans D. (1971). Fate of maturation protein during infection by phage MS2. *Nature New Biol.* **234,** 209–211.

Kozinski A.W. (1961). Uniform sensitivity to $^{32}P$ decay among progeny of $^{32}P$ free phage ϕX174 grown on $^{32}P$ labeled bacteria. *Virology* **13,** 377–378.

Kozinski A.W. (1968). Molecular recombination in the ligase-negative T4 amber mutant. *Cold Spring Harbor Symp. Quant. Biol.* **33,** 375–92.

Kozinski A.W. and Doermann A.H. (1975). Repetitive DNA replication of the incomplete genomes of phage T4 petite particles. *Proc. Natl. Acad. Sci. USA* **72,** 1734–1738.

Kozinski A.W. and Kozinski P.B. (1969). Covalent repair of molecular recombinants in the ligase-negative amber mutant of T4 bacteriophage. *J. Virol.* **3,** 85–88.

Kozinski A.W. and Kosturko L.D. (1976). Late events in T4 phage production. II. Giant phages contain concatemers generated by recombination. *J. Virol.* **17,** 801–804.

Kozinski A.W., Doermann A.H. and Kozinski P.B. (1976). Absence of interparental recombination in multiplicity reconstitution from incomplete phage T4 genomes. *J. Virol.* **18,** 873–884.

Kozinski A.W., Kozinski P.B. and James R. (1967). Molecular recombination in T4 bacteriophage DNA. I. Tertiary structure of early replicative and recombining DNA. *J. Virol.* **1,** 758–770.

Kozloff L.M. and Lute M. (1957). Viral invasion. II. The role of zinc in phage invasion. *J. Biol. Chem.* **228,** 529–536.

Kozloff L.M., Crosby L.K. and Lute M. (1975c). Phage T4 baseplate components. III. Location and properties of the phage structural thymidilate synthetase. *J. Virol.* **16,** 1409–1419.

Kozloff L.M., Lute M. and Crosby L.K. (1975a). Phage T4 baseplate components. I. Binding and location of the folic acid. *J. Virol.* **16,** 1391–1400.

Kozloff L.M., Lute M., Crosby L.K., Rao N., Chapman V.A. and DeLong S.S. (1970a). Phage tail components. I. Pteroyl polyglutamates in T-even phages. *J. Virol.* **5,** 726–739.

Kozloff L.M., Verses C., Lute M. and Crosby L.K. (1970b). Phage tail components. II. Dihydrofolate reductase in T4D phage. *J. Virol.* **5,** 740–753.

Kozloff L.M., Crosby L.K., Lute M. and Hall D.H. (1975b). Phage T4 baseplate components. II. Binding and location of phage induced dihydrofolate reductase. *J. Virol.* **16,** 1401–1408.

Krahn P.M. and Paranchych W. (1971). Heterogeneous distribution of A protein in R17 phage preparations. *Virology* **43,** 533–535.

Krahn P.M., O'Callaghan R.J., and Paranchych W. (1972). Stages in phage R17 infection. VI. Injection of A protein and RNA into the host cell. *Virology* **47,** 628–637.

Kramer R.A., Cameron J.R. and Davis R.W. (1976). Isolation of phage lambda containing yeast rRNA genes: screening by in situ RNA hybridization to plaques. *Cell* **8,** 227–232.

Kramer R.A., Rosenberg M. and Steitz J.A. (1974). Nucleotide sequences of the 5′ and 3′ termini of phage T7 early messenger RNAs synthesized in vivo: evidence for sequence specificity in RNA processing. *J. Mol. Biol.* **89,** 767–776.

Kranias E.C. and Dumas L.B. (1974). Replication of phage $\phi$X174 DNA in a temperature sensitive *dnaC* mutant of E.coli C. *J. Virol.* **13,** 146–154.

Krisch H.M., Bolle A. and Epstein R.H. (1974). Regulation of the synthesis of phage T4 gene 32 protein. *J. Mol. Biol.* **88,** 89–104.

Kuhl P.W. and Katsura I. (1975). Morphogenesis of the tail of phage $\lambda$. In vitro intratail complementation. *Virology* **63,** 221–237.

Kumar S., Calef E. and Szybalski W. (1970). Regulation of the transcription of E.coli phage lambda by its early genes N and tof. *Cold Spring Harbor Symp. Quant. Biol.* **35,** 331–340.

Kuo C.-H., Eoyang L. and August J.T. (1975). Protein factors required for the replication of phage Q$\beta$ in vitro. In RNA Phages, ed. N.D. Zinder. (*Cold Spring Harbor Laboratory: New York*). pp. 259–278.

Kushner S.R., Nagaishi H., and Clark A.J. (1971). Indirect suppression of *recB* or *recC* mutations by exonuclease deficiency. *Proc. Natl. Acad. Sci. USA* **68,** 1366–1370.

Kushner S.R., Nagaishi H., Templin A. and Clark A.J. (1972). Genetic recombination in E.coli: the role of exonuclease. *Proc. Natl. Acad. Sci. USA* **69,** 824–827.

Kutter E.M. and Wiberg J.S. (1968). Degradation of cytosine-containing bacterial and phage DNA after infection of E.coli B with phages T4D wild type and with mutants defective in genes *46, 47, 56*. *J. Mol. Biol.* **38,** 395–411.

Kutter E., Beug A., Sluss R., Jensen L. and Bradley D. (1975). The production of undegraded cytosine-containing DNA by phage T4 in the absence of dCTPase and endonucleases II and IV, and its effects on T4 directed protein synthesis. *J. Mol. Biol.* **99,** 591–608.

Labaw L.W. (1953). The origin of phosphorus in the T1, T5, T6 and T7 phages of E.coli. *J. Bacteriol.* **66,** 429–436.

Lacks S. (1962). Molecular fate of DNA in genetic transformation of Pneumococcus. *J. Mol. Biol.* **5,** 119–131.

Lacks S. (1966). Integration efficiency and genetic recombination in pneumococcal transformation. *Genetics* **53,** 207–235.

Lacks S. (1970). Mutants of Diplococcus pneumoniae that lack DNAase and other activities possibly pertinent to genetic transformation. *J. Bacteriol.* **101,** 373–383.

Lacks S. and Greenberg B. (1973). Competence for DNA uptake and DNAase action external to cells in the genetic transformation of Diplococcus pneumoniae. *J. Bacteriol.* **114,** 152–163.

Lacks S. and Greenberg B. (1976). Single strand breakage on binding of DNA to cells in the genetic transformation of Diplococcus pneumonia. *J. Mol. Biol.* **101,** 255–276.

Lacks S. and Hotchkiss R.D. (1960a). A study of the genetic material determining an enzyme activity in Pneumococcus. *Biochim. Biophys. Acta* **39,** 508–517.

Lacks S. and Hotchkiss R.D. (1960b). Formation of amylomaltase after genetic transformation of Pneumococcus. *Biochim. Biophys. Acta* **45,** 155–163.

Lacks S. and Neuberger M. (1975). Membrane location of a DNAase implicated in the genetic transformation of D.pneumoniae. *J. Bacteriol.* **124,** 1321–1329.

Lacks S., Greenberg B. and Carlson K. (1967). Fate of donor DNA in pneumococcal transformation. *J. Mol. Biol.* **29,** 327–347.

Lacks S., Greenberg B. and Neuberger M. (1974). Role of a DNAase in the genetic transformation of Diplococcus pneumoniae. *Proc. Natl. Acad. Sci. USA* **71,** 2305–2309.

Lacks S., Greenberg B. and Neuberger M. (1975). Identification of a DNAase implicated in genetic transformation of D.pneumoniae. *J. Bacteriol.* **123,** 222–232.

Laemmli U.K. (1970). Cleavage of structural proteins during the assembly of the head of phage T4. *Nature* **227,** 680–685.

Laemmli U.K. and Favre M. (1973). Maturation of the head of phage T4. I. DNA packaging events. *J. Mol. Biol.* **80,** 575–600.

Laemmli U.K. and Johnson R.A. (1973). Maturation of the head of phage T4. II. Head related aberrant $\tau$-particles. *J. Mol. Biol.* **80,** 601–611.

Laemmli U.K. and Quittner S.F. (1974). Maturation of the head of phage T4. IV. The proteins of the core of the tubular polyheads and in vitro cleavage of the head proteins. *Virology* **62,** 483–499.

Laemmli U.K., Amos L.A. and Klug A. (1976). Correlation between structural transformation and cleavage of the major head protein of T4 phage. *Cell* **7,** 191–204.

Laemmli U.K., Beguin F. and Gujer-Kellenberger E. (1970a). A factor preventing the major head protein of phage T4 from random aggregation. *J. Mol. Biol.* **47,** 69–85.

Laemmli U.K., Teaff N. and D'Ambrosia J. (1974). Maturation of the head of phage T4. III. DNA packaging into preformed heads. *J. Mol. Biol.* **88,** 749–766.

Laemmli U.K., Molbert E., Showe M. and Kellenberger E. (1970b). Form determining function of the genes required for the assembly of the head of phage T4. *J. Mol. Biol.* **49,** 99–113.

Lam S.T., Stahl M.M., McMilin K.D. and Stahl F.W. (1974). Rec-mediated recombinational hot spot activity in phage lambda. II. A mutation which causes hot spot activity. *Genetics* **77,** 425–433.

Landers T.A., Blumenthal T. and Weber K. (1974). Function and structure in RNA phage Q$\beta$ RNA replicase. The roles of the different subunits in transcription of synthetic templates. *J. Biol. Chem.* **249,** 5801–5808.

Lane H.E.D. and Denhardt D.T. (1974). The *rep* mutation. III. Altered structure of the replicating E.coli chromosome. *J. Bacteriol.* **120,** 805–814.

Lane H.E.D. and Denhardt D.T. (1975). The *rep* mutation. IV. Slower movement of replication forks in E.coli *rep* strains. *J. Mol. Biol.* **97,** 99–112.

Lanni F. and Lanni Y. (1953). Antigenic structure of bacteriophage. *Cold Spring Harbor Symp. Quant. Biol.* **18,** 159–168.

Lark K.G. and Wechsler J.A. (1975). DNA replication in *dnaB* mutants of E.coli: gene product interaction and synthesis of 4S pieces. *J. Mol. Biol.* **92,** 145–163.

Lawn A.M., Meynell E., Meynell G.G. and Datta N. (1967). Sex pili and the classification of sex factors in the Enterobacteriaceae. *Nature* **216,** 343–346.

Lawson J.W. and Gooder H. (1970). Growth and development of competence in the group H streptococci. *J. Bacteriol.* **102,** 820–825.

LeBlanc D.J. and Falkow S. (1973). Studies on superinfection immunity among transmissible plasmids in E.coli. *J. Mol. Biol.* **74,** 689–701.

LeClerc J.E. and Setlow J.K. (1975). Single strand regions in the DNA of competent H.influenzae. *J. Bacteriol.* **122,** 1091–1102.

Lecocq J.-P. and Dambly C. (1976). A bacterial RNA polymerase mutation that renders lambda growth independent of the *N* and *cro* functions at 42°C. *Molec. Gen. Genet.* **145,** 53–62.

Lederberg E.M. and Lederberg J. (1953). Genetic studies of lysogenicity in interstrain crosses in E.coli. *Genetics* **38,** 51–64.

Lederberg J. (1947). Gene recombination and linked segregations in E.coli. *Genetics* **32,** 505–525.

Lederberg J. (1949). Aberrant heterozygotes in E.coli. *Proc. Natl. Acad. Sci, USA* **35,** 178–184.

Lederberg J. (1952). Cell genetics and hereditary symbiosis. *Physiol. Rev.* **32,** 403–428.

Lederberg J. and Lederberg E.M. (1952). Replica plating and indirect selection of bacterial mutants. *J. Bacteriol.* **63,** 399–406.

Lederberg J. and Tatum E.L. (1946a). Novel genotypes in mixed cultures of biochemical mutants of bacteria. *Cold Spring Harbor Symp. Quant. Biol.* **11,** 113–114.

Lederberg J. and Tatum E.L. (1946b). Gene recombination in E.coli. *Nature* **158,** 558.

Lederberg J., Cavalli L.L. and Lederberg E.M. (1952). Sex compatibility in E.coli. *Genetics* **37,** 720–730.

Lederberg J., Lederberg E., Zinder N.D. and Lively E.R. (1951). Recombinational analysis of bacterial heredity. *Cold Spring Harbor Symp. Quant. Biol.* **16,** 413–440.

Lee A.S. and Sinsheimer R.L. (1974). A cleavage map of phage $\phi$X174 genome. *Proc. Natl. Acad. Sci. USA* **71,** 2882–2886.

Lee C.S., Davis R.W. and Davidson N. (1970). A physical study by electron microscopy of the terminally repetitious, circularly permuted DNA from the phage particles of E.coli 15. *J. Mol. Biol.* **48,** 1–22.

Lee H.J., Ohtsubo E., Deonier R.C. and Davidson N. (1974). Electron microscope heteroduplex studies of sequence relations among plasmids of E.coli. V. ilv$^+$ deletion mutants of F14. *J. Mol. Biol.* **89,** 585–588.

Lee-Huang S. and Ochoa S. (1971). Messenger discriminating species of initiation factor f3. *Nature New Biol.* **234,** 236–239.

Lee-Huang S. and Ochoa S. (1973). Purification and properties of two messenger discriminating species of E.coli initiation factor 3. *Arch. Biochem. Biophys.* **156,** 84–96.

Lembach K.J., Kuninaka A. and Buchanan J.M. (1969). The relationship of DNA replication to the control of protein synthesis in protoplasts of T4 infected E.coli B. *Proc. Natl. Acad. Sci. USA* **62,** 446–453.

Leonard C.G. (1973). Early events in development of Streptoccocal competence. *J. Bacteriol.* **114,** 1198–1205.

Leonard C.G. and Cole R.M. (1972). Purification and properties of Streptococcal competence factor isolated from chemically defined medium. *J. Bacteriol.* **110,** 273–280.

Lerman L.S. and Tolmach L.J. (1957). Genetic transformation. I. Cellular incorporation of DNA accompanying transformation in Pneumococcus. *Biochim. Biophys. Acta* **26,** 68–82.

Levine A.J. and Sinsheimer R.L. (1969a). The process of infection with phage $\phi$X174. XXV. Studies with $\phi$X174 mutants blocked in progeny replicative form DNA synthesis. *J. Mol. Biol.* **39,** 619–639.

Levine A.J. and Sinsheimer R.L. (1969b). The process of infection with $\phi$X174. XXVII. Synthesis of a viral specific chloramphenicol resistant protein in $\phi$X174 infected cells. *J. Mol. Biol.* **39,** 655–668.

Leung D., Behme M.T. and Ebisuzaki K. (1975). Effect of DNA delay mutations of phage T4 on genetic recombination. *J. Virol.* **16,** 203–213.

Levine J.S. and Strauss N. (1965). Lag period characterizing the entry of transforming DNA into B.subtilis. *J. Bacteriol.* **89,** 281–287.

Levinthal C. and Visconti N. (1953). Growth and recombination in bacterial viruses. *Genetics* **38,** 500–511.

Levisohn R., Konisky J. and Nomura M. (1967). Interaction of colicins with bacterial cells. IV. Immunity breakdown studied with colicin Ia and Ib. *J. Bacteriol.* **96,** 811–821.

Lewin B. (1974a). Gene Expression, **1,** Bacterial Genomes. (*John Wiley and Sons, New York and London*).

Lewin B. (1974b). Interaction of regulator proteins with recognition sequences of DNA. *Cell* **2,** 1–8.

Lickfield K.G., Menge B., Hohn B. and Hohn T. (1976). Morphogenesis of phage λ: electron microscopy of thin sections. *J. Mol. Biol.* **103,** 299–318.

Lieb M. (1966). Studies of heat inducible lambda phage. I. Order of genetic sites and properties of mutant prophages. *J. Mol. Biol.* **16,** 149–163.

Lin N.S.C. and Pratt D. (1972). Role of phage M13 gene 2 in viral DNA replication. *J. Mol. Biol.* **72,** 37–49.

Lin N.S.-C. and Pratt D. (1974). Phage M13 gene 2 protein: increasing its yield in infected cells and identification and localization. *Virology* **61,** 334–342.

Lin S.-Y. and Riggs A.D. (1975). The general affinity of *lac* repressor for E.coli DNA: implications for gene regulation in procaryotes and eucaryotes. *Cell* **4,** 107–112.

Lindahl G., Sironi G., Bialy H. and Calendar R. (1970). Phage λ: abortive infection of bacteria lysogenic for phage P2. *Proc. Natl. Acad. Sci. USA* **66,** 587–594.

Lindqvist B.H. and Sinsheimer R.L. (1967a). The process of infection with phage $\phi$X174. XIV. Studies on macromolecular synthesis during infection with a lysis-defective mutant. *J. Mol. Biol.* **28,** 87–94.

Lindqvist B.H. and Sinsheimer R.L. (1967b). The process of infection with phage $\phi$X174. XV. Phage DNA synthesis in abortive infections with a set of conditional lethal mutations. *J. Mol. Biol.* **30,** 69–80.

Lindqvist B.H. and Sinsheimer R.L. (1968). The process of infection with phage $\phi$X174. XVI. Synthesis of the replicative form and its relationship to viral single stranded DNA synthesis. *J. Mol. Biol.* **32,** 285–302.

Linney E. and Hayashi M. (1973). Two proteins of gene A of $\phi$X174. *Nature New Biol.* **245,** 6–8.

Linney E., Hayashi M.N. and Hayashi M. (1972). Gene A of $\phi$X174. I. Isolation and identification of its products. *Virology* **50,** 381–387.

Linial M. and Malamy M.H. (1970). The effect of F factors on RNA synthesis and degradation after infection of E.coli with $\phi$II. *Cold Spring Harbor Symp. Quant. Biol.* **35,** 263–268.

Litt M., Marmur J., Ephrussi-Taylor H.E. and Doty P. (1958). The dependence of pneumococcal transformation on the molecular weight of deoxyribose nucleic acid. *Proc. Natl. Acad. Sci. USA* **44,** 144–152.

Little J.W. and Gottesman M. (1971). Defective lambda particles whose DNA carries only a single cohesive end. In The Bacteriophage Lambda, ed. A.D. Hershey. (*Cold Spring Harbor Laboratory, New York*). pp. 371–394.

Livingston D.M., Hinckle D.C. and Richardson C.C. (1975). DNA polymerase III of E.coli. Purification and properties. *J. Biol. Chem.* **250,** 461–469.

Lobban P.E. and Kaiser A.D. (1973). Enzymatic end to end joining of DNA molecules. *J. Mol. Biol.* **78,** 453–471.

Lodish H.F. (1968a). Polar effects of an amber mutation in f2 phage. *J. Mol. Biol.* **32,** 47–58.

Lodish H.F. (1968b). Independent translation of the genes of phage f2 RNA. *J. Mol. Biol.* **32,** 681–686.

Lodish H.F. (1968c). Phage f2 RNA: control of translation and gene order. *Nature* **220,** 345–350.

Lodish H.F. (1970a). Specificity in bacterial protein synthesis: role of initiation factors and ribosomal subunits. *Nature* **226,** 705–708.

Lodish H.F. (1970b). Secondary structure of phage f2 RNA and the initiation of in vitro protein synthesis. *J. Mol. Biol.* **50,** 689–702.

Lodish H.F. (1971). Thermal melting of phage f2 RNA and initiation of synthesis of the maturation protein. *J. Mol. Biol.* **56,** 627–632.

Lodish H.F. (1975). Regulation of in vitro protein synthesis by phage RNA by RNA tertiary structure. In RNA Phages, ed. N.D. Zinder. (*Cold Spring Harbor Laboratory: New York*). pp. 301–318.

Lodish H.F. and Robertson H.D. (1969). Cell free synthesis of phage f2 maturation protein. *J. Mol. Biol.* **45,** 9–22.

Lodish H.F. and Zinder N.D. (1966a). Replication of the RNA of bacteriophage f2. *Science* **152,** 372–378.

Lodish H.F. and Zinder N.D. (1966b). Mutants of the bacteriophage f2. VIII. Control mechanisms for phage-specific syntheses. *J. Mol. Biol.* **19,** 333–348.

Lodish H.F., Cooper S. and Zinder N.D. (1964). Host-dependent mutants of the bacteriophage f2. IV. On the biosynthesis of a viral RNA polymerase. *Virology* **24,** 60–70.

Lodish H.F., Horiuchi K. and Zinder N.D. (1965). Mutants of phage f2. On the production of noninfectious phage particles. *Virology* **27,** 139–155.

Loeb T. (1960). Isolation of a bacteriophage for the $F^+$ and Hfr mating types of E.coli K12. *Science* **131,** 932–933.

Loeb T. and Zinder N.D. (1961). A bacteriophage containing RNA. *Proc. Natl. Acad. Sci. USA* **47,** 282–289.

Lovett M.A. and Helinski D.R. (1975). Relaxation complexes of plasmid DNA and protein. II. Characterization of the proteins associated with the unrelaxed and relaxed complexes of plasmid ColE1. *J. Biol. Chem.* **250,** 8790–8795.

Lovett M.A., Guiney D.G. and Helinski D.R. (1974). Relaxation complexes of plasmids colE1 and colE2: unique site of the nick in the open circular DNA of the relaxed complexes. *Proc. Natl. Acad. Sci. USA* **71,** 3854–3857.

Lovett M.A., Sparks R.B. and Helinski D.R. (1975). Bidirectional replication of plasmid R6K DNA in E.coli: correspondence between origin of replication and position of single strand break in relaxed complex. *Proc. Natl. Acad. Sci.* **72,** 2905–2909.

Low B. (1966). Low recombination frequency for markers very near the origin in conjugation in E.coli. *Genet. Res.* **6,** 469–473.

Low B. (1968). Formation of merodiploids in matings with a class of $rec^-$ recipient strains of E.coli K12. *Proc. Natl. Acad. Sci.* **60,** 160–167.

Low K.B. (1972). E.coli K12 F prime factors, old and new. *Bacteriol. Rev.* **36,** 587–607.

Lozeron H.A., Dahlberg J.E. and Szybalski W. (1976). Processing of the major leftward mRNA of coliphage lambda. *Virology* **71,** 262–277.

Ludwig R.A. and Summers W.C. (1975). A restriction fragment analysis of the T7 left early region. *Virology* **68,** 360–373.

Ludwig R.A. and Summers W.C. (1976). Localization of RNA polymerase binding sites on T7 DNA. *Virology* **71,** 278–290.

Luftig R.B. and Ganz C. (1972). Phage T4 head morphogenesis. IV. Comparison of gene 16-, 17- and 49- defective head structures. *J. Virol.* **10,** 545–554.

Luckfield K.G., Menge B., Hohn B. and Hohn T. (1976). Morphogenesis of phage lambda: electron microscopy of thin sections. *J. Mol. Biol.* **103,** 299–318.

Luftig R.B. and Lundh N.P. (1973). Phage T4 head morphogenesis. V. The role of DNA in maturation of an intermediate in head assembly. *Virology* **51,** 432–442.

Luftig R.B., Wood W.B. and Okinara R. (1971). Phage T4 head morphogenesis. On the nature of gene 49 defective heads and their role as intermediates. *J. Mol. Biol.* **57,** 555–574.

Luria S. (1947). Reactivation of irradiated phage by transfer of self reproducing units. *Proc. Natl. Acad. Sci. USA* **33,** 253–264.

Luria S.E. (1951). The frequency distribution of spontaneous phage mutants as evidence for the exponential rate of phage reproduction. *Cold Spring Harbor Symp. Quant. Biol.* **16,** 463–470.

Luria S.E. and Delbruck M. (1943). Mutations of bacteria from virus sensitivity to virus resistance. *Genetics* **28,** 491–511.

Luzzati D. (1970). Regulation of λ exonuclease synthesis: role of the N gene product and λ repressor. *J. Mol. Biol.* **49,** 515–519.

Lwoff A. and Gutmann A. (1950). Recherches sur un Bacillus megatherium lysogene. *Ann. Inst. Pasteur* **78,** 1–29.

Lyons L.B. and Zinder N.D. (1972). The genetic map of the filamentous phage fl. *Virology* **49,** 45–60.

Maas R. (1963). Exclusion of an F lac episome by an Hfr gene. *Proc. Natl. Acad. Sci. USA* **50,** 1051–1055.

Maas W.K. and Goldschmidt A.D. (1969). A mutant of E.coli permitting replication of two F factors. *Proc. Natl. Acad. Sci. USA* **62,** 873–880.

Maas R. and Maas W.K. (1962). Introduction of a gene from E.coli B into Hfr and F⁻ strains of E.coli K12. *Proc. Natl. Acad. Sci. USA* **48,** 1887–1893.

MacFarren A.C. and Clowes R.C. (1967). A comparative study of two F-like colicin factors, ColV2 and colV3, in E.coli K12. *J. Bacteriol.* **94,** 365–377.

Mackay D.J. and Bode V.C. (1976a). Events in lambda injection between phage adsorption and DNA entry. *Virology* **72,** 154–166.

Mackay D.J. and Bode V.C. (1976b). Binding to isolated phage receptors and lambda DNA release in vitro. *Virology* **72,** 167–181.

Mackay V. and Linn S. (1974). The mechanism of degradation of duplex DNA by the recBC enzyme of E.coli K12. *J. Biol. Chem.* **249,** 4286–4294.

MacHattie L.A., Ritchie D.A., Thomas C.A. jr and Richardson C.C. (1967). Terminal repetition in permuted T2 phage DNA molecules. *J. Mol. Biol.* **23,** 355–364.

Mailhammer R., Yang H.-L., Reiness G. and Zubay G. (1975). Effects of phage T4 induced modification of E. coli RNA polymerase on gene expression in vitro. *Proc. Natl. Acad. Sci. USA* **72,** 4928–4932.

Malamy M.H. (1970). Some properties of insertion mutations in the *lac* operon. In The

Lactose Operon, ed. J.R. Beckwith and D. Zipser. (*Cold Spring Harbor Laboratory: New York*). pp. 359–373.

Malamy M.H., Fiandt M. and Szybalski W. (1972). Electron microscopy of polar insertions in the *lac* operon of E.coli. *Molec. Gen. Genet.* **119,** 207–222.

Mandel M. and Higa A. (1970). Calcium dependent phage DNA infection. *J. Mol. Biol.* **53,** 159–162.

Mangel W.F. and Chamberlin M.J. (1974a). Studies of RNA chain initiation by E.coli RNA polymerase bound to T7 DNA. I. An assay for the rate and extent of RNA chain initiation. *J. Biol. Chem.* **249,** 2995–3001.

Mangel W.F. and Chamberlin M.J. (1974b). Studies of RNA chain initiation by E.coli RNA polymerase bound to T7 DNA. II. The effect of alterations in ionic strength on chain initiation and on the conformation of binary complexes. *J. Biol. Chem.* **249,** 3002–3006.

Mangel W.F. and Chamberlin M.J. (1974c). Studies of RNA chain initiation by E.coli RNA polymerase bound to T7 DNA. III. The effect of temperature on RNA chain initiation and on the conformation of binary complexes. *J. Biol. Chem.* **249,** 3007–3013.

Maniatis T. and Ptashne M. (1973a). Multiple repressor binding at the operators in phage λ. *Proc. Natl. Acad. Sci. USA* **70,** 1531–1535.

Maniatis T. and Ptashne M. (1973b). Structure of the λ operators. *Nature* **246,** 133–136.

Maniatis T., Jeffrey A. and Kleid D.G. (1975). Nucleotide sequence of the rightward operator of phage lambda. *Proc. Natl. Acad. Sci. USA* **72,** 1184–1188.

Maniatis T., Ptashne M. and Maurer R. (1973). Control elements in the DNA of phage lambda. *Cold Spring Harbor Symp. Quant. Biol.* **38,** 857–868.

Maniatis T., Ptashne M., Barrell B.G. and Donelson J. (1974). Sequence of a repressor-binding site in the DNA of phage λ. *Nature* **240,** 394–397.

Maniatis T., Ptashne M., Backman K., Kleid D., Flashman S., Jeffrey A. and Maurer R. (1975). Recognition sequences of repressor and polymerase in the operators of phage lambda. *Cell* **5,** 109–113.

Maniatis T., Kee S.G., Efstratiadis A. and Kafatos F.C. (1976). Amplification and characterization of a $\beta$ globin gene synthesized in vitro. *Cell* **8,** 163–181.

Manly K.F., Signer E.R. and Radding C.M. (1969). Non-essential functions of bacteriophage λ. *Virology* **37,** 177–188.

Marchin G.L., Comer M.M. and Neidhardt F.C. (1971). Viral modification of the valyl-tRNA synthetase of E.coli. *J. Biol. Chem.* **247,** 5132–5145.

Mark D.F. and Richardson C.C. (1976). E.coli thioredoxin: a subunit of phage T7 DNA polymerase. *Proc. Natl. Acad. Sci. USA* **73,** 780–784.

Marmur J. and Lane D. (1960). Strand separation and specific recombination in DNAs: biological studies. *Proc. Natl. Acad. Sci. USA* **46,** 453–461.

Marsh R.C., Breschkin A.M. and Mosig G. (1971). Origin and direction of phage T4 DNA replication. II. A gradient of marker frequencies in partially replicated T4 DNA as assayed by transformation. *J. Mol. Biol.* **60,** 213–234.

Marrs B.L. and Yanofsky C. (1971). Host and phage specific mRNA degradation in T7 infected E.coli. *Nature New Biol.* **234,** 168–170.

Marvin D.A. and Hoffmann-Berling H. (1963). Physical and chemical properties of two new small phages. *Nature* **197,** 517–518.

Marvin D. and Hohn B. (1969). Filamentous bacterial viruses. *Bacteriol. Rev.* **33,** 172–209.

Marvin D.A. and Schaller H. (1966). The topology of DNA from the small filamentous phage fd. *J. Mol. Biol.* **15,** 1–7.

Marvin D.A. and Wachtel E.J. (1975). Structure and assembly of filamentous bacterial viruses. *Nature* **253,** 19–23.

Marvin D.A., Pigram W.J., Wiseman R.L., Wachtel E.J. and Marvin F.J. (1974). Filamentous bacterial viruses. XII. Molecular architecture of the class 1 (fd, 1f, 1ke) virion. *J. Mol. Biol.* **88,** 581–600.

Masamune Y., Frenkel G.D. and Richardson C.C. (1971). A mutant of phage T7 deficient in polynucleotide ligase. *J. Biol. Chem.* **246,** 6874–6879.

Masker W.E. and Richardson C.C. (1976a). Phage T7 DNA replication in vitro. V. Synthesis of intact chromosomes of phage T7. *J. Mol. Biol.* **100,** 543–556.

Masker W.E. and Richardson C.C. (1976b). Phage T7 DNA replication in vitro. VI. Synthesis of biologically active T7 DNA. *J. Mol. Biol.* **100,** 557–568.

Mason W.S. and Haselkorn R. (1972). Product of T4 gene 12. *J. Mol. Biol.* **66,** 445–470.

Matsubara K. (1972). Interference in phage growth by a resident plasmid λdv. I. The mode of interference. *Virology* **50,** 713–726.

Matsubara K. (1974a). Preparation of plasmid λdv from phage λ: role of promotor-operator in the plasmid replicon. *J. Virol.* **13,** 596–602.

Matsubara K. (1974b). Interference in phage growth by a resident plasmid λdv. II. Role of the promotor-operator. *J. Virol.* **13,** 603–612.

Matsubara K. (1976). Genetic structure and regulation of a replicon of plasmid λdv. *J. Mol. Biol.* **102,** 427–440.

Matsubara K. and Kaiser A.D. (1968). λ dv: an autonomously replicating DNA fragment. *Cold Spring Harbor Symp. Quant. Biol.* **33,** 769–775.

Matsubara K. and Takeda Y. (1976). Role of *tof* gene in production and perpetuation of lambda-dv plasmid. *Molec. Gen. Genet.* **142,** 225–230.

Maurer R., Maniatis T. and Ptashne M. (1974). Promotors are in the operators in phage lambda. *Nature* **249,** 221–223.

Mazur B.J. and Model P. (1973). Regulation of phage fl single stranded DNA synthesis by a DNA binding protein. *J. Mol. Biol.* **78,** 285–300.

Mazur B.J. and Zinder N.D. (1975a). Evidence that gene VII is not the distal portion of gene V of phage fl. *Virology* **68,** 284–285.

Mazur B.J. and Zinder N.D. (1975b). The role of gene V protein in fl single strand synthesis. *Virology* **68,** 490–502.

McCarthy C. and Nester E.W. (1967). Macromolecular synthesis in newly transformed cells of B.subtilis. *J. Bacteriol.* **94,** 131–140.

McCarty M. and Avery O.T. (1964a). Studies on the chemical nature of the substance inducing transformation of pneumococcal types. II. Effect of desoxyribonuclease on the biological activity of the transforming substance. *J. Exp. Med.* **83,** 89–96.

McCarty M. and Avery O.T. (1946b). Studies on the chemical nature of the substance inducing transformation of pneumococcal types. III. An improved method for the isolation of the transforming substance and its application to pneumococcal types II, III and VI. *J. Exp. Med.* **83,** 97–103.

McCarty M., Taylor H.E. and Avery O.T. (1946). Biochemical studies of environmental factors essential in transformation of pneumococcal types. *Cold Spring Harbor Symp. Quant. Biol.* **11,** 177–183.

McClain W.H. and Champe S.P. (1970). Genetic alterations of the rIIB cistron polypeptides of phage T4. *Genetics* **66,** 11–21.

McClain W.H., Guthrie C. and Barrell B.G. (1972). Eight transfer RNAs induced by infection of E.coli with phage T4. *Proc. Natl. Acad. Sci. USA* **69,** 3703–3707.

McClain W.H., Marchin G.L., Neidhardt F.C., Chace K.V., Rementer M.L. and Hall D.W. (1975). A gene of phage T4 controlling the modification of host valyl-tRNA synthetase. *Virology* **67,** 385–394.

McClure S.C.C. and Gold M. (1973). Intermediates in the maturation of phage λ DNA. *Virology* **54,** 19–27.

McClure S.C.C., MacHattie L. and Gold M. (1973). A sedimentation analysis of DNA found in E.coli infected with phage λ mutants. *Virology* **54,** 1–18.

McDermit M., Pierce M., Stacey D., Shimaji M., Shaw R. and Wulff D.L. (1976). Mutations masking the lambda *cin-1* mutation. *Genetics* **82,** 417–422.

McFadden G. and Denhardt D.T. (1974). Mechanism of replication of φX174 single stranded DNA. IX. Requirement of the E.coli dnaG protein. *J. Virol.* **14,** 1070–1075.

McFadden G. and Denhardt D.T. (1975). The mechanism of replication of φX174 DNA. XIII. Discontinuous synthesis of the complementary strand in an E.coli host with a temperature sensitive polynucleotide ligase. *J. Mol. Biol.* **99,** 125–142.

McMacken R., Kessler S. and Boyce R. (1975). Strand breakage of phage λ supercoils in infected lysogens. I. Genetic and biochemical evidence for two types of nicking processes. *Virology* **66,** 356–371.

McMacken R., Mantei N., Butler B., Joyner A. and Echols H. (1970). Effect of mutations in the cII and cIII genes of phage λ on macromolecular synthesis in infected cells. *J. Mol. Biol.* **49,** 639–655.

McMilin K.D. and Russo V.E.A. (1972). Maturation and recombination of phage λ DNA molecules in the absence of DNA duplication. *J. Mol. Biol.* **68,** 49–56.

McMilin K.D., Stahl M.M. and Stahl F.W. (1974). Rec-mediated recombinational hot spot activity in phage lambda. I. Hot spot activity associated with spi$^-$ deletions and bio substitutions. *Genetics* **77,** 409–423.

Meezan E. and Wood W.B. (1971). The sequence of gene product interaction in T4 tail core assembly. *J. Mol. Biol.* **58,** 685–692.

Meselson M. and Weigle J. (1961). Chromosome breakage accompanying genetic recombination in bacteriophage. *Proc. Natl. Acad. Sci, USA* **47,** 857–868.

Meselson M. (1964). On the mechanism of genetic recombination between DNA molecules. *J. Mol. Biol.* **9,** 734–745.

Meselson M. (1967). Reciprocal recombination in prophage lambda. *J. Cell. Physiol.* **70,** suppl. **1,** 113–118.

Meyer B.J., Kleid D.G. and Ptashne M. (1975). λ repressor turns off transcription of its own gene. *Proc. Natl. Acad. Sci. USA* **72,** 4785–4789.

Meynell E. and Cooke M. (1969). Repressor minus and operator constitutive derepressed mutants of F-like R factors: their effect on chromosomal transfer by HfrC. *Genet. Res.* **14,** 309–313.

Meynell E. and Datta N. (1966). The relation of resistance transfer factors to the F factor of E.coli. *Genet. Res.* **7,** 134–140.

Meynell E. and Datta N. (1967). Mutant drug resistance factors of high transmissibility. *Nature* **214,** 885–887.

Meynell G.G. and Lawn A.M. (1968). Filamentous phages specific for the I sex factor. *Nature* **217,** 1184–1186.

Meynell E., Meynell G.G. and Datta N. (1968). Phylogenetic relationships of drug resistance factors and other transmissible bacterial plasmids. *Bacteriol. Rev.* **32,** 55–83.

Miao R. and Guild W.R. (1970). Competent Diplococcus pneumoniae accept both single and double stranded DNA. *J. Bacteriol.* **101,** 361–372.

Milanesi G., Brody E.N. and Geiduschek E.P. (1969). Sequence of the in vitro transcription of T4 DNA. *Nature* **221,** 1014–1016.

Milanesi G., Brody E.N., Grau O. and Geiduschek E.P. (1970). Transcription of the phage T4 template in vitro: separation of delayed early from immediate early transcription. *Proc. Natl. Acad. Sci. USA* **66,** 181–188.

Miller L.K. and Sinsheimer R.L. (1974). Nature of $\phi$X174 linear DNA from a DNA-ligase-defective host. *J. Virol.* **14,** 1503–1514.

Miller M.J. and Wahba A.J. (1974). Inhibition of synthetic and natural messenger translation. II. Specificity and mechanism of action of a protein isolated from E.coli MRE 600 ribosomes. *J. Biol. Chem.* **249,** 3808–3813.

Miller M.J., Niveleau A. and Wahba A.J. (1974). Inhibition of synthetic and natural messenger translation. I. Purification and properties of a protein isolated from E.coli MRE 600 ribosomes. *J. Biol. Chem.* **249,** 3803–3807.

Miller R.C. jr. (1972). Association of replicative T4 DNA and bacterial membranes. *J. Virol.* **10,** 920–924.

Miller R.C. jr. (1975a). Replication and molecular recombination of T-phage. *Ann. Rev. Microbiol.* **29,** 355–376.

Miller R.C. jr. (1975b). T4 DNA polymerase (gene 43) is required in vivo for repair of gaps in recombinants. *J. Virol.* **15,** 316–321.

Miller R.C. jr., Lee M., Scraba D.G. and Paetkau V. (1976). The role of phage T7 exonuclease (gene 6) in genetic recombination and production of concatemers. *J. Mol. Biol.* **101,** 223–234.

Millette R.L., Trotter C.D., Herrlich P. and Schweiger M. (1970). In vitro synthesis, termination and release of active mRNA. *Cold Spring Harbor Symp. Quant. Biol.* **35,** 135–142.

Milliken C.E. and Clowes R.C. (1973). Molecular structure of an R factor, its component drug resistance determinants and transfer factor. *J. Bacteriol.* **113,** 1026–1033.

Mills D.R., Kramer F.R. and Spiegelman S. (1973). Complete nucleotide sequence of a replicating RNA molecule. *Science* **180,** 916–937.

Mills D.R., Kramer F.R., Dobkin C., Nishihara T. and Spiegelman S. (1975). Nucleotide sequence of microvariant RNA: another small replicating molecule. *Proc. Natl. Acad. Sci. USA* **72,** 4252–4256.

Min Jou W., Haegeman G., Ysebaert M. and Fiers W. (1972). Nucleotide sequence of the gene coding for the bacteriophage MS2 coat protein. *Nature* **237,** 82–88.

Minkley E.G. jr. (1974a). Transcription of the early region of phage T7: characterization of the in vivo transcripts. *J. Mol. Biol.* **83,** 289–304.

Minkley E. G. jr. (1974b). Transcription of the early region of phage T7: specificity and selectivity in the in vitro initiation of RNA synthesis. *J. Mol. Biol.* **83,** 305–332.

Minkley E.G. and Pribnow D. (1973). Transcription of the early region of phage T7. Selective initiation with dinucleotides. *J. Mol. Biol.* **77,** 255–277.

Minton K., Durphy M., Taylor R., and Friedberg E.C. (1975). The ultraviolet endonuclease of phage T4. Further characterization. *J. Biol. Chem.* **250,** 2823–2829.

Mitra S., Reichard P., Inman R.B., Bertsch L.L. and Kornberg A. (1967). Enzymic synthesis of DNA. XXII. Replication of a circular single stranded DNA template by DNA polymerase of E.coli. *J. Mol. Biol.* **24,** 429–447.

Mitsuhashi S., Harada K. and Hashimoto H. (1960a). Multiple resistance of enteric bacteria and transmission of drug resistance to other strains by mixed cultivation. *Jap. J. Exp. Med.* **30,** 179–184.

Mitsuhashi S., Harada K., Karneda M., Suzuki M. and Egawa R. (1960b). On the drug resistance of enteric bacteria. 3. Transmission of the drug resistance from Shigella to $F^-$ or Hfr strains of E.coli K12. *Jap. J. Exp. Med.* **30,** 301–306.

Mitsuhashi S., Harada K., Hashimoto H., Kameda M. and Suzuki M. (1962). Combination of two types of transmissible drug resistance factors in a host bacterium. *J. Bacteriol.* **84,** 9–16.

Model P. and Zinder N.D. (1974). In vitro synthesis of phage fl proteins. *J. Mol. Biol.* **83,** 231–251.

Modrich P. and Richardson C.C. (1975a). Phage T7 DNA replication in vitro. A protein of E.coli required for phage T7 DNA polymerase activity. *J. Biol. Chem.* **250,** 5508–5514.

Modrich P. and Richardson C.C. (1975b). Phage T7 DNA replication in vitro. Phage T7 DNA polymerase: an enzyme composed of phage and host specified subunits. *J. Biol. Chem.* **250,** 5515–5522.

Molineux I.J. and Gefter M.L. (1975). Properties of the E.coli DNA binding (unwinding) protein interaction with nucleolytic enzymes and DNA. *J. Mol. Biol.* **98,** 811–826.

Møller J.K., Bak A.L., Christiansen C., Christiansen G. and Stenderup A. (1976). Extrachromosomal DNA in R factor harboring Enterobacteriaceae. *J. Bacteriol.* **125,** 398–403.

Monk M. and Clowes R.C. (1964a). Transfer of the colicin I factor in E.coli K12 and its interaction with the F fertility factor. *J. Gen. Microbiol.* **36,** 365–384.

Monk M. and Clowes R.C. (1964b). The regulation of colicin synthesis and colicin factor transfer of E.coli K12. *J. Gen. Microbiol.* **36,** 385–392.

Monk M. and Kinross J. (1972). Conditional lethality of *recA* and *recB* derivatives of a strain of E.coli K12 with a temperature sensitive DNA polymerase I. *J. Bacteriol.* **109,** 971–978.

Moody E.E.M. and Hayes W. (1972). Chromosome transfer by autonomous transmissible plasmids: the role of the bacterial recombination (rec) system. *J. Bacteriol.* **111,** 80–85.

Moody M.F. (1965). The shape of the T even phage head. *Virology* **26,** 567–576.

Moody M.F. (1967a). Structure of the sheath of phage T4. I. Structure of the contracted sheath and polysheath. *J. Mol. Biol.* **25,** 167–200.

Moody M.F. (1967b). Structure of the sheath of phage T4. II. Rearrangement of the sheath subunits during contraction. *J. Mol. Biol.* **25,** 201–208.

Moody M.F. (1973). Sheath of phage T4. III. Contraction mechanism deduced from partially contracted sheaths. *J. Mol. Biol.* **80,** 613–636.

Moore C.H., Farron F., Bohnert D. and Weissmann C. (1971). Possible origin of a minor virus specific protein (A1) in Q$\beta$ particles. *Nature New Biol.* **234,** 204–206.

Morris C.F., Hershberger C.L. and Rownd R. (1973). Strand specific nick in open circular R factor DNA: attachment of the linear strand to a proteinaceous cellular component. *J. Bacteriol.* **114,** 300–308.

Morris C.F., Sinha N.K. and Alberts B.M. (1975). Reconstruction of phage T4 DNA replication apparatus from purified components: rolling circle replication following de novo chain initiation on a single stranded circular DNA template. *Proc. Natl. Acad. Sci. USA* **72,** 4800–4804.

Morris C.F., Hashimoto H., Mickel S. and Rownd R. (1974). Round of replication mutant of a drug resistance factor. *J. Bacteriol.* **118,** 855–866.

Morrison D.A. (1971). Early intermediate state of transforming DNA during uptake by B.subtilis. *J. Bacteriol.* **108,** 38–44.

Morrison D.A. and Guild W.R. (1972a). Activity of DNA fragments of defined size in B. subtilis transformation. *J. Bacteriol.* **112,** 220–223.

Morrison D.A. and Guild W.R. (1972b). Transformation and DNA size: extent of degradation on entry varies with size of donor. *J. Bacteriol.* **112,** 1157–1168.

Morrison D.A. and Guild W.R. (1973). Structure of DNA on the cell surface during uptake by Pneumococcus. *J. Bacteriol.* **115,** 1055–1062.

Morrison T.G. and Malamy M.H. (1971). T7 translational control mechanisms and their inhibition by F factors. *Nature* **231,** 37–31.

Morrison T.G., Blumberg B. and Malamy M. (1974). T7 protein synthesis in F′ episome containing cells: assignment of specific proteins to three translational groups. *J. Virol.* **13,** 386–393.

Morrow J.F., Cohen S.N., Chang A.C.Y., Boyer H.W., Goodman H.M. and Helling R.B. (1974). Replication and transcription of eucaryotic DNA in E.coli. *Proc. Natl. Acad. Sci. USA* **71,** 1743–1747.

Morse M.L., Lederberg E.M. and Lederberg J. (1956a). Transduction in E.coli K12. *Genetics* **41,** 142–156.

Morse M.L., Lederberg E.M. and Lederberg J. (1956b). Transductional heterogenotes in E.coli. *Genetics* **41,** 758–779.

Mortelmans K. and Friedberg E.C. (1972). DNA repair in phage T4: observations on the roles of the *x* and *v* genes and of host factors. *J. Virol.* **10,** 730–736.

Mosig G.A. (1970). Preferred origin and direction of phage T4 DNA replication. I. A gradient of allele frequencies in crosses between normal and small T4 particles. *J. Mol. Biol.* **53,** 503–514.

Mount D.W.A., Harris A.W., Fuerst C.R. and Siminovitch L. (1968). Mutations in bacteriophage λ affecting particle morphogenesis. *Virology* **35,** 134–149.

Mousset S. and Thomas R. (1968). Dilysogenic excision: an accessory expression of the termination function? *Cold Spring Harbor Symp. Quant. Biol.* **33,** 749–754.

Mousset S. and Thomas R. (1969). Ter, a function which generates the ends of the mature λ chromosome. *Nature* **221,** 242–244.

Murialdo H. and Siminovitch L. (1971). The morphogenesis of phage λ III. Identification of the genes specifying morphogenetic proteins. In The Bacteriophage Lambda, ed. A.D. Hershey. (*Cold Spring Harbor Laboratory, New York*). pp. 711–724.

Murialdo H. and Siminovitch L. (1972a). The morphogenesis of phage λ. IV. Identification of gene products and control of the expression of morphogenetic information. *Virology* **48,** 785–823.

Murialdo H. and Siminovitch L. (1972b). The morphogenesis of phage λ. V. Form determining function of the genes required for assembly of the head. *Virology* **48,** 824–835.

Murray N.E. and Murray K. (1974). Manipulation of restriction targets in phage lambda to form receptor chromosomes for DNA fragments. *Nature* **251,** 476–481.

Nagel de Zwaig R. and Luria S.E. (1967). Genetics and physiology of colicin-tolerant mutants of E.coli. *J. Bacteriol.* **94,** 1112–1123.

Nagel de Zwaig R. and Puig J. (1964). The genetic behaviour of colicinogenic factor E1. *J. Gen. Microbiol.* **36,** 311–321.

Nagel de Zwaig R., Anton D.N. and Puig J. (1962). The genetic control of colicinogenic factors E2, I and V. *J. Gen. Microbiol.* **29,** 473–484.

Nakashima Y. and Konigsberg W. (1974). Reinvestigation of a region of the fd coat protein sequence. *J. Mol. Biol.* **88,** 598–599.

Nakashima Y., Dunker A.K., Marvin D.A. and Konigsberg W. (1974). The amino acid sequence of a DNA binding protein, the gene 5 product of fd filamentous phage. *Febs Lett.* **40,** 290–292.

Nash H.A. (1975a). Integrative recombination in phage lambda: analysis of recombinant DNA *J. Mol. Biol.* **91,** 501–514.

Nash H.A. (1975b). Integrative recombination of phage lambda DNA in vitro. *Proc. Natl. Acad. Sci. USA* **72,** 1072–1076.

Nathans D., Oeschger M.P., Eggen K. and Shimura Y. (1966). Phage specific proteins in E.coli infected with an RNA phage. *Proc. Natl. Acad. Sci. USA* **56,** 1844–1851.

Nathans D., Oeschger M.P., Polmar S.K. and Eggen K. (1969). Regulation of protein synthesis directed by phage MS2 RNA. I. Phage protein and RNA synthesis in cells infected with suppressible mutants. *J. Mol. Biol.* **39,** 279–292.

Nester E.W. (1964). Penicillin resistance of competent cells in DNA transformation of B.subtilis. *J. Bacteriol.* **87,** 867–875.

Nester E.W. and Stocker B.A.D. (1963). Biosynthetic latency in early stages of deoxyribonucleic acid transformation in Bacillus subtilis. *J. Bacteriol.* **86,** 785–796.

Newbold J.E. and Sinsheimer R.L. (1970a). The process of infection with phage $\phi$X174. XXXII. Early steps in the attachment process: attachment, eclipse and DNA penetration. *J. Mol. Biol.* **49,** 49–66.

Newbold J.E. and Sinsheimer R.L. (1970b). The process of infection with phage $\phi$X174. XXXIV. Kinetic of the attachment and eclipse steps of the infection. *J. Virol.* **5,** 427–431.

Newcombe H.B. (1949). Origin of bacterial variants. *Nature* **164,** 160.

Newman J. and Hanawalt P. (1968). Intermediates in T4 DNA replication in a T4 ligase deficient strain. *Cold Spring Harbor Symp. Quant. Biol.* **33,** 145–150.

Nichols J.L. (1970). Nucleotide sequence from the polypeptide chain termination region of the coat protein cistron in phage R17 RNA. *Nature* **225,** 147–152.

Nijkamp H.J., Bøvre K. and Szybalski W. (1970). Controls of rightward transcription in phage λ. *J. Mol. Biol.* **54,** 599–604.

Nikolaev N., Silengo L. and Schlessinger D. (1973). Synthesis of a large precursor to rRNA in a mutant of E.coli. *Proc. Natl. Acad. Sci. USA* **70,** 3361–3365.

Niles E.G. and Condit R.C. (1975). Translational mapping of phage T7 RNAs synthesized in vitro by purified T7 RNA polymerase. *J. Mol. Biol.* **98,** 57–68.

Niles E.G., Conlon S.W. and Summers W.C. (1974). Purification and physical characterization of T7 RNA polymerase from T7 infected E.coli B. *Biochemistry* **13,** 3904–3911.

Nishimura A., Nishimura Y. and Caro L. (1973). Isolation of Hfr strains from $R^+$ and ColV2$^+$ strains of E.coli and derivation of an R'*lac* factor by transduction. *J. Bacteriol.* **116,** 1107–1112.

Nishimura Y., Ishibashi M., Meynell E. and Hirota Y. (1967). Specific pilation directed by a fertility factor and a resistance factor of E.coli. *J. Gen. Microbiol.* **49,** 89–98.

Nishimura Y., Caro L., Berg C.M. and Hirota Y. (1971). Chromosome replication in E.coli. IV. Control of chromosome replication and cell division by an integrated episome. *J. Mol. Biol.* **55,** 441–456.

Nisioka T., Mitani M. and Clowes R. (1969). Composite circular forms of R factor DNA molecules. *J. Bacteriol.* **97,** 376–385.

Nisioka T., Mitani M. and Clowes R.C. (1970). Molecular recombination between R factor DNA molecules in E.coli host cells. *J. Bacteriol.* **103,** 166–177.

Nomura M. (1963). Mode of action of colicins. *Cold Spring Harbor Symp. Quant. Biol.* **28,** 315–324.

Nomura M. (1967a). Colicins and related bacteriocins. *Ann. Rev. Microbiol.* **21,** 257–284.

Nomura M. (1967b). Mechanism of action of colicins. *Proc. Natl. Acad. Sci. USA* **52,** 1514–1521.

Nomura M. and Benzer S. (1961). The nature of the deletion mutants in the rII region of phage T4. *J. Mol. Biol.* **3,** 684–691.

Nomura M. and Witten C. (1967). Interaction of colicins with bacterial cells. III. Colicin tolerant mutations in E.coli. *J. Bacteriol.* **94,** 1093–1011.

Nordstrom K., Ingram L.C. and Lundback A. (1972). Mutation in R factors of E.coli causing an increased number of R factor copies per chromosome. *J. Bacteriol.* **110,** 562–569.

North T.W., Stafford M.E. and Mathews C.K. (1976). Biochemistry of DNA-defective mutants of phage T4. VI. Biological functions of gene 42. *J. Virol.* **17,** 973–982.

Nossal N.G. (1974). DNA synthesis on a double stranded DNA template by the T4 phage DNA polymerase and the T4 gene 32 DNA unwinding protein. *J. Biol. Chem.* **249,** 5668–5676.

Nossal N.G. and Hershfield M.S. (1971). Nuclease activity in a fragment of phage T4 DNA polymerase induced by the amber mutant amB22. *J. Biol. Chem.* **246,** 5415–5426.

Notani G.W. (1973). Regulation of phage T4 gene expression. *J. Mol. Biol.* **73,** 231–250.

Notani G.W., Engelhardt D.L., Konigsberg W. and Zinder N.D. (1965). Suppression of a coat protein mutant of the phage f2. *J. Mol. Biol.* **12,** 439–447.

Notani N. and Goodgal S.H. (1966). On the nature of recombinants formed during transformation in H.influenzae. *J. Gen. Physiol.* **49,** suppl., 197–209.

Notani N.K. and Setlow J.K. (1974). Mechanism of bacterial transformation and transfection. *Prog. Nucleic Acid Res. Mol. Biol.* **14,** 39–100.

Notani N.K., Setlow J.K., Joshi V.R. and Allison D.P. (1972). Molecular basis for the transformation defects in mutants of H.influenzae. *J. Bacteriol.* **110,** 1171–1180.

Novick A. and Szilard L. (1951). Virus strains of identical phenotype but different genotype. *Science* **113,** 34–35.

Novick R.P. (1969). Extrachromosomal inheritance in bacteria. *Bacteriol. Rev.* **33,** 210–263.

Ochiai K., Yamanaka T., Kimura K. and Sawada O. (1959). Studies on inheritance of drug resistance between Shigella strains and E.coli strains. (*in Japanese*). *Nihon Iji Shimpo* **1861,** 34–46.

Oda K., Sakakibara Y. and Tomizawa J. (1969). Regulation of transcription of the λ phage genome. *Virology* **39,** 901–918.

Oey J.L. and Knippers R. (1972). Properties of the isolated gene 5 protein of phage fd. *J. Mol. Biol.* **68,** 125–138.

O'Farrell P.Z. and Gold L. M. (1973a). Phage T4 gene expression. Evidence for two classes of prereplicative cistrons. *J. Biol. Chem.* **248,** 5502–5511.

O'Farrell P.Z. and Gold L.M. (1973b). Transcription and translation of prereplicative phage T4 genes in vitro. *J. Biol. Chem.* **248,** 5512–5519.

Ogawa H. and Tomizawa J. (1967). Phage lambda DNA with different structures formed in infected cells. *J. Mol. Biol.* **23,** 265–276.

Ohki M. and Tomizawa J. (1968). Asymmetric transfer of DNA strands in bacterial conjugation. *Cold Spring Harbor Symp. Quant. Biol.* **33,** 651–657.

Ohtsubo E. (1970). Transfer defective mutants of sex factors in E.coli. II. Deletion mutants of an F prime and deletion mapping of cistrons involved in genetic transfer. *Genetics* **64,** 189–197.

Ohtsubo E., Nishimura Y. and Hirota Y. (1970). Transfer-defective mutants of sex factors in E.coli. I. Defective mutants and complementation analysis. *Genetics* **64,** 173–188.

Ohtsubo E., Deonier R.C., Lee H.J. and Davidson N. (1974a). Electron microscope heteroduplex studies of sequence relations among plasmids of E.coli. IV. The F sequences in F14. *J. Mol. Biol.* **89,** 565–584.

Ohtsubo E., Lee H.J., Deonier R.C. and Davidson N. (1974b). Electron microscope heteroduplex studies of sequence relations among plasmids of E.coli. VI. Mapping of F14 sequences homologous to $\phi$80dmetBJF and $\phi$80dargECBH phages. *J. Mol. Biol.* **89,** 599–618.

Ohtsubo H. and Ohtsubo E. (1976). Isolation of inverted repeat sequences, including IS1, IS2 and IS3, in E.coli plasmids. *Proc. Natl. Acad. Sci. USA* **73,** 2316–2320.

Oka A. and Inselburg J. (1975). Replicative intermediates of colicin E1 plasmid DNA in minicells. *Proc. Natl. Acad. Sci. USA* **72,** 829–833.

Okamoto T., Sugimoto K., Sugisaki H. and Takanami M. (1975). Studies on phage fd DNA. II. Localization of RNA initiation sites on the cleavage map of the fd genome. *J. Mol. Biol.* **95,** 33–44.

Okazaki R., Okazaki T., Sakabe K., Sugimoto K., Kainuma R., Sugino A. and Iwatsuki N. (1968). In vivo mechanism of DNA chain growth. *Cold Spring Harbor Symp. Quant. Biol.* **33,** 129–144.

Onorata L. and Showe M.K. (1975). Gene *21* dependent proteolysis in vitro of purified gene *22* product of phage T4. *J. Mol. Biol.* **92,** 395–412.

Oppenheim A.B. and Riley M. (1966). Molecular recombination following conjugation in E.coli. *J. Mol. Biol.* **20,** 331–357.

Oppenheim A.B. and Riley M. (1967). Covalent union of parental DNAs following conjugation in E.coli. *J. Mol. Biol.* **28,** 503–511.

Ordal G.W. (1973). Mutations in the right operator of phage lambda: physiological effects. *J. Mol. Biol.* **79,** 723–729.

Ordal G.W. and Kaiser A.D. (1973). Mutations in the right operator of phage lambda: evidence for operator-promotor interpenetration. *J. Mol. Biol.* **79,** 709–722.

Ørskov I. and Ørskov F. (1960). An antigen termed $f^+$ occurring in $F^+$ E.coli strains. *Acta Pathol. Microbiol. Scand.* **48,** 37–46.

Ottolenghi E. and Hotchkiss R.D. (1962). Release of genetic transforming agent from pneumococcal cultures during growth and disintegration. *J. Exp. Med.* **116,** 491–519.

Ou J.T. and Anderson T.F. (1970). Role of pili in bacterial conjugation. *J. Bacteriol.* **102,** 648–654.

Overby L.R., Barlow G.H., Doi R.H., Jacob M. and Spiegelman S. (1966a). Comparison of two serologically distinct RNA phages. *J. Bacteriol.* **91,** 442–448.

Overby L.R., Barlow G.H., Doi R.H., Jacob M. and Spiegelman S. (1966b). Comparison of two serologically distinct RNA phages. II. Properties of the nucleic acids and coat proteins. *J. Bacteriol.* **92,** 739–745.

Ozeki H. and Howarth S. (1961). Colicin factors as fertility factors in bacteria: S.typhimurium LT2. *Nature* **190,** 986–988.

Ozeki H., Stocker B.A.D. and De Margerie H. (1959). Production of colicin by single bacteria. *Nature* **184,** 337–339.

Ozeki H., Stocker B.A.D. and Smith S.M. (1962). Transmission of colicinogeny between strains of S.typhimurium grown together. *J. Gen. Microbiol.* **28,** 671–687.

Pacumbaba R.R. and Center M.S. (1973). Association of replicating phage T7 DNA with bacterial membranes. *J. Virol.* **12,** 855–861.

Padmanabhan R., Wu R. and Bode V.C. (1972). Arrangement of DNA in λ phage heads. III. Location and number of nucleotides cleaved from λ DNA by microccoal nuclease attack on heads. *J. Mol. Biol.* **69,** 201–208.

Pakula R. and Walczak W. (1963). On the nature of competence of transformable streptococci. *J. Gen. Microbiol.* **31,** 125–133.

Palchoudhury S.R. and Iyer V.N. (1971). Compatibility between two F factors in an E.coli strain bearing a chromosomal mutation affecting DNA synthesis. *J. Mol. Biol.* **57,** 319–334.

Paranchych W. (1975). Attachment, ejection and penetration stages of the RNA phage infectious process. In RNA Phages, ed. N.D. Zinder. (*Cold Spring Harbor Laboratory: New York*). pp. 85–112.

Paranchych W., Krahn P.M. and Bradley R.D. (1970). Stages in R17 infection. IV. Phage heterogeneity and host-phage interactions. *Virology* **41,** 465–473.

Parkinson J.S. (1968). Genetics of the left arm of the chromosome of bacteriophage lambda. *Genetics* **59,** 311–325.

Parkinson J.S. (1971). Deletion mutants of phage λ. II. Genetic properties of att-defective mutants. *J. Mol. Biol.* **56,** 385–401.

Parkinson J. and Huskey R. (1971). Deletion mutants of bacteriophage lambda. I. Isolation and initial characterization. *J. Mol. Biol.* **56,** 369–384.

Parma D.H., Ingraham L.J. and Snyder M. (1972). Tandem duplications of the rII region of phage T4D. *Genetics* **71,** 319–335.

Paulson J.R., Lazaroff S. and Laemmli U.K. (1976). Head length determination in phage T4: the role of the core protein P22. *J. Mol. Biol.* **103,** 155–174.

Perlman D. and Rownd R.H. (1975). Transition of R factor NR1 in Proteus mirabilis: molecular structure and replication of NR1 DNA. *J. Bacteriol.* **123,** 1013–1034.

Perlman D. and Rownd R.H. (1976). Two origins of replication in composite R plasmid DNA. *Nature* **259,** 281–285.

Perlman D., Twose T.M., Holland M.J. and Rownd R.H. (1975). Denaturation mapping of R factor DNA. *J. Bacteriol.* **123,** 1035–1042.

Pero J. (1970). Location of the phage λ gene responsible for turning off λ exonuclease synthesis. *Virology* **40,** 65–71.

Pero J. (1971). Deletion mapping of the site of action of the tof gene product. In The Bacteriophage Lambda, ed. A.D. Hershey. (*Cold Spring Harbor Laboratory: New York*). pp. 599–608.

Peters G.G. and Heywood R.S. (1974). Transcriptional terminal in vitro: the 3′ terminal sequence of phage T7 early mRNA. *Biochem. Biophys. Res. Commun.* **61,** 809–816.

Peterson R.B., Kievitt K.D. and Ennis H.L. (1972). Membrane protein synthesis after infection of E.coli B with phage T4: the rIIB protein. *Virology* **50,** 520–527.

Pfister P., DeVries J.K. and Maas W.K. (1976). Expression of a mutation affecting F incompatibility in the integrated but not the autonomous state of F. *J. Bacteriol.* **127,** 348–353.

Piechowska M. and Fox M.S. (1971). Fate of transforming DNA in B.subtilis. *J. Bacteriol.* **108,** 680–689.

Piechowska M., Soltyk A. and Shugar D. (1975). Fate of heterologous DNA in B.subtilis. *J. Bacteriol.* **122,** 610–622.

Pierucci O. (1969). Regulation of cell division in E.coli. *Biophys. J.* **9,** 90–112.

Pilarski L.M. and Egar J.B. (1973). Role of DNA topology in transcription of phage lambda in vitro. II. DNA topology protects the template from exonuclease attack. *J. Mol. Biol.* **76,** 257–266.

Pirrotta V. (1973). Isolation of the operators of phage lambda. *Nature New Biol.* **244,** 13–15.

Pirrotta V. (1975). Sequence of the $O_R$ operator of phage lambda. *Nature* **254,** 114–117.

Pirrotta V. and Ptashne M. (1969). Isolation of the 434 phage repressor. *Nature* **222,** 541–544.

Pittard J. and Adelberg E.A. (1964). Gene transfer by F′ strains of E.coli K12. III. An analysis of the recombination events occurring in the F′ male and in the zygotes. *Genetics* **49,** 995–1007.

Pittard J. and Ramakrishnan T. (1964). Gene transfer by F′ stains of E.coli K12. IV. Effect of a chromosomal deletion on chromosome transfer. *J. Bacteriol.* **88,** 367–373.

Pittard J. and Walker E.M. (1967). Conjugation in E.coli: recombination events in terminal regions of transferred DNA. *J. Bacteriol.* **94,** 1656–1663.

Postel E.H. and Goodgal S.H. (1966). Uptake of single stranded DNA in H.influenzae and its ability to transform. *J. Mol. Biol.* **16,** 317–327.

Postel E.H. and Goodgal S.H. (1967). Further studies on transformation with single stranded DNA of H.influenzae. *J. Mol. Biol.* **28,** 247–259.

Postel E.H. and Goodgal S.H. (1972a). Competence mutants. II. Physical and biological fate of donors transforming DNA. *J. Bacteriol.* **109,** 292–297.

Postel E.H. and Goodgal S.H. (1972b). Competence mutants. III. Responses to radiations. *J. Bacteriol.* **109,** 298–306.

Potter H. and Dressler D. (1976). On the mechanism of genetic recombination: electron microscopic observation of recombination intermediates. *Proc. Natl. Acad. Sci. USA* **73,** 3000–3004.

Prashad N. and Hosoda J. (1972). Role of genes 46 and 47 in phage T4 reproduction. II. Formation of gaps in parental DNA of polynucleotide ligase defective mutants. *J. Mol. Biol.* **70,** 617–636.

Pratt D. and Erdahl W.S. (1968). Genetic control of phage M13 DNA synthesis. *J. Mol. Biol.* **37,** 181–200.

Pratt D., Laws P. and Griffith J. (1974). Complex of phage M13 single stranded DNA and gene 5 protein. *J. Mol. Biol.* **82,** 425–440.

Pratt D., Tzagoloff H. and Beaudoin J. (1969). Conditional lethal mutants of the small filamentous phage M13. II. Two genes for coat proteins. *Virology* **39,** 42–53.

Pratt D., Tzagoloff H. and Erdahl W.S. (1966). Conditional lethal mutants of the small filamentous phage M13. I. Isolation, complementation, cell killing, time of cistron action. *Virology* **30,** 397–410.

Pretorius H.T., Klein M. and Day L.A. (1975). Gene V protein of fd phage. Dimer formation and the role of tyrosyl groups in DNA binding. *J. Biol. Chem.* **250,** 9262–9269.

Pribnow D. (1975a). Nucleotide sequence of an RNA polymerase binding site at an early T7 promotor. *Proc. Natl. Acad. Sci. USA* **72,** 784–788.

Pribnow D. (1975b). Phage T7 early promotors: nucleotide sequences of two RNA polymerase binding sites. *J. Mol. Biol.* **99,** 419–444.

Pritchard R.H., Barth P.T. and Collins J. (1969). Control of DNA synthesis in bacteria. *Symp. Soc. Gen. Microbiol.* **19,** 293–298.

Prives C.L. and Silverman P.M. (1972). Replication of RNA viruses: structure of a 6S RNA synthesized by the Q$\beta$ RNA polymerase. *J. Mol. Biol.* **71,** 657–670.

Protass J.J. and Korn D. (1966). Function of the N cistron of phage $\lambda$. *Proc. Natl. Acad. Sci. USA* **55,** 1089–1095.

Ptashne K. and Cohen S.N. (1975). Occurrence of insertion sequence (IS) regions on plasmid DNA as direct and inverted nucleotide sequence duplications. *J. Bacteriol.* **122,** 776–781.

Ptashne M. (1967a). Isolation of the lambda repressor. *Proc. Natl. Acad. Sci. USA* **57,** 306–312.

Ptashne M. (1967b). Specific binding of the $\lambda$ phage repressor to $\lambda$ DNA. *Nature* **214,** 232–234.

Ptashne M. (1971). Repressor and its action. In The Bacteriophage Lambda, ed. A.D. Hershey. (*Cold Spring Harbor Laboratory: New York*). pp. 221–238.

Ptashne M. and Hopkins N. (1968). The operators controlled by the $\lambda$ phage repressor. *Proc. Natl. Acad. Sci. USA* **60,** 1282–1287.

Ptashne M., Backman K., Humayun M.Z., Jeffrey A., Maurer R., Meyer B. and Saucer R.T. (1976). Autoregulation and function of a repressor in phage $\lambda$. *Science* **194,** 156–161.

Pulitzer J.F. and Geiduschek E.P. (1970). Function of T4 gene 55. II. RNA synthesis by temperature sensitive gene 55 mutants. *J. Mol. Biol.* **49,** 489–508.

Punch J.D. and Kopecko D.J. (1972). Positive and negative control of R factor replication in P.mirabilis. *J. Bacteriol.* **109,** 336–349.

Putnam F.W., Miller D., Palm L. and Evans E.A. jr. (1952). Biochemical studies of virus reproduction. X. Precursors of phage T7. *J. Biol. Chem.* **199,** 177–191.

Rabussay D. and Geiduschek E.P. (1976). Regulation of gene action in lytic phages. In Comprehensive Virology, Vol. 10, eds. H. Fraenkel-Conrat and R. Wagner. (*Plenum Press, New York*).

Radding C.M. (1970). The role of exonuclease and $\beta$ protein of phage $\lambda$ in genetic recombination. I. Effect of red mutants on protein structure. *J. Mol. Biol.* **52,** 491–500.

Radding C.M. (1973). Molecular mechanisms in recombination. *Ann Rev. Genet.* **7,** 87–111.

Radding C.M. and Echols H. (1968). The role of the N gene of phage $\lambda$ in the synthesis of two phage-specified proteins. *Proc. Natl. Acad. Sci. USA* **60,** 707–712.

Radding C.M. and Shreffler D.C. (1966). Regulation of $\lambda$ exonuclease. II. Joint regulation of exonuclease and a new $\lambda$ antigen. *J. Mol. Biol.* **18,** 251–261.

Radding C.M., Szpirer J. and Thomas R. (1967). The structural gene for $\lambda$ exonuclease. *Proc. Natl. Acad. Sci. USA* **57,** 277–283.

Rahmsdorf H.J., Pai S.H., Ponta H., Herrlich P., Roskoski R. jr., Schweiger M. and Studier

F.W. (1974). Protein kinase induction in E.coli by phage T7. *Proc. Natl. Acad. Sci. USA* **71,** 586–589.

Rambach A. and Tiollais P. (1974). Phage lambda having Eco RI endonuclease sites only in the nonessential region of the genome. *Proc. Natl. Acad. Sci.* **71,** 3927–3930.

Randall-Hazelbauer L. and Schwartz M. (1973). Isolation of the phage lambda receptor from E.coli. *J. Bacteriol.* **116,** 1436–1446.

Ranhand J.M. and Lichstein H.C. (1969). Effect of selected antibiotics and other inhibitors on competence development in H.influenzae. *J. Gen. Microbiol.* **55,** 37–43.

Ratner D. (1974). T4 transcriptional control gene 55 codes for a protein bound to E.coli RNA polymerase. *J. Mol. Biol.* **89,** 803–809.

Ray D.S. (1969). Replication of phage M13. II. The role of replicative forms in single strand synthesis. *J. Mol. Biol.* **43,** 631–643.

Ray D.S. (1970). Replication of phage M13. Synthesis of M13 specific DNA in the presence of chloramphenicol. *J. Mol. Biol.* **53,** 239–250.

Ray D.S., Preuss A. and Hofschneider P.H. (1966). Replication of the single stranded DNA of the male specific phage M13. Circular forms of the replicative DNA. *J. Mol. Biol.* **21,** 485–491.

Ray P. and Murialdo H. (1975). The role of gene nu3 in phage lambda head morphogenesis. *Virology* **64,** 247–263.

Ray P., Sinha N.K., Warner H.R. and Snustad D.P. (1972). Genetic location of a mutant of phage T4 deficient in the ability to induce endonuclease II. *J. Virol.* **9,** 184–186.

Ray U. and Skalka A. (1976). Lysogenization of E.coli by phage lambda: complementary activity of the host's DNA polymerase I and ligase and phage replication proteins O and P. *J. Virol.* **18,** 511–517.

Reader R.W. and Siminovitch L. (1971a). Lysis defective mutants of phage lambda: genetics and physiology of *S* cistron mutants. *Virology* **43,** 607–622.

Reader R.W. and Siminovitch L. (1971b). Lysis defective mutants of phage lambda: on the role of the *S* function in lysis. *Virology* **43,** 623–637.

Reichardt L.F. (1975a). Control of phage lambda repressor synthesis after phage infection: the role of the *N, cII, cIII* and *cro* products. *J. Mol. Biol.* **93,** 267–288.

Reichardt L.F. (1975b). Control of phage lambda repressor synthesis: regulation of the maintenance pathway by the *cro* and *cI* products. *J. Mol. Biol.* **93,** 289–310.

Reichardt L. and Kaiser A.D. (1971). Control of λ repressor synthesis. *Proc. Natl. Acad. Sci. USA* **68,** 2185–2189.

Remaut E. and Fiers W. (1972). Studies on phage MS2 RNA. XVI. The termination signal of the A protein cistron. *J. Mol. Biol.* **71,** 243–262.

Reeves P.R. (1965). The bacteriocins. *Bacteriol. Rev.* **29,** 25–45.

Reeves P. and Willetts N. (1974). Plasmid specificity of the origin of transfer of sex factor F. *J. Bacteriol.* **120,** 125–130.

Reuben R.C. and Gefter M.L. (1973). A DNA binding protein induced by phage T7. *Proc. Natl. Acad. Sci. USA* **70,** 1846–1850.

Reuben R.C. and Gefter M.L. (1974). A DNA binding protein induced by phage T7. Purification and properties of the protein. *J. Biol. Chem.* **249,** 3843–3850.

Revel H. (1965). Synthesis of β-galactosidase after F-duction of $lac^+$ genes into E.coli. *J. Mol. Biol.* **11,** 23–34.

Revel H.R. and Luria S.E. (1970). DNA glucosylation in T-even phage: genetic determination and role in phage-host interactions. *Ann. Rev. Genetics* **4,** 177–192.

Reyes O., Gottesman M. and Adyha S. (1976). Suppression of polarity of insertion mutations in the *gal* operon and *N* mutations in phage lambda. *J. Bacteriol.* **126,** 1108–1112.

Ribolini A. and Baylor M. (1975). Novel multi nonsense suppressor in phage T4D. *J. Mol. Biol.* **98,** 615–630.

Richardson C.C., Masamune Y., Live T.R., Jacquemin-Sablon A., Weiss B. and Fareed G. (1968). Studies on the joining of DNA by polynucleotide ligase of phage T4. *Cold Spring Harbor Symp. Quant. Biol.* **33,** 647–650.

Richardson J.P. (1970). Rho factor function in T4 transcription. *Cold Spring Harbor Symp. Quant. Biol.* **35,** 127–134.

Richardson J.P., Grimley C. and Lowery C. (1975). Transcription termination factor rho activity is altered in E.coli with *suA* gene mutations. *Proc. Natl. Acad. Sci. USA* **72,** 1725–1728.

Richter A. (1961). Attachment of wild type F factor to a specific chromosomal region in a variant strain of E.coli K12: the phenomenon of episomic alternation. *Genet. Res.* **2,** 333–345.

Ringrose P. (1970). Sedimentation analysis of DNA degradation products resulting from the action of colicin E2 in E.coli. *Biochim. Biophys. Acta* **213,** 320–334.

Ritchie D.A., Thomas C.A. jr., MacHattie L.A. and Wensink P.C. (1967). Terminal repetition in non permuted T3 and T7 phage DNA molecules. *J. Mol. Biol.* **23,** 365–376.

Riva S., Cascino A., and Geiduschek E.P. (1970a). Coupling of late transcription to viral replication in phage T4 development. *J. Mol. Biol.* **54,** 85–89.

Riva S., Cascino A. and Geiduschek E.P. (1970b). Uncoupling of late transcription from DNA replication in phage T4 development. *J. Mol. Biol.* **54,** 103–120.

Roberts J.W. (1969a). Promotor mutation in vitro. *Nature* **223,** 480–482.

Roberts J.W. (1969b). Termination factor for RNA synthesis. *Nature* **224,** 1168–1175.

Roberts J.W. (1970). The rho factor: termination and anti-termination in λ. *Cold Spring Harbor Symp. Quant. Biol.* **35,** 121–126.

Roberts J.W. (1975). Transcription termination and late control in phage λ. *Proc. Natl. Acad. Sci. USA* **72,** 3300–3304.

Roberts J.W. and Gussin G.N. (1967). Polarity in an amber mutant of phage R17. *J. Mol. Biol.* **30,** 565–570.

Robertson H.D. (1975). Functions of replicating RNA in cells infected by RNA phages. In RNA Phages, ed. N.D. Zinder. (*Cold Spring Harbor Laboratory: New York*). pp. 113–146.

Robertson H., Webster R.E. and Zinder N.D. (1968). Phage coat protein as repressor. *Nature* **218,** 533–536.

Roa M. and Scandella D. (1976). Multiple steps during the interaction between phage lambda and its receptor protein in vitro. *Virology* **72,** 182–194.

Robertson H.D., Webster R.E. and Zinder N.D. (1968). Purification and properties of RNAase III from E.coli. *J. Biol. Chem.* **243,** 82–91.

Romero E. and Meynell E. (1969). Covert fi$^-$ R factors in fi$^+$ R$^+$ strains of bacteria. *J. Bacteriol.* **97,** 780–786.

Rosenberg M., Kramer R.A. and Steitz J.A. (1974). T7 early messenger RNAs are the direct products of RNAase III cleavage. *J. Mol. Biol.* **89,** 777–782.

Roth T.F. and Hayashi M. (1966). Allomorphic forms of phage $\phi$X174 replicative DNA. *Science* **154,** 658–660.

Rownd R. (1969). Replication of a bacterial episome under relaxed control. *J. Mol. Biol.* **44,** 387–402.

Rownd R. and Mickel S. (1971). Dissociation and reassociation of RTF and r-determinants of the R factor NR1 in Proteus mirabolis. *Nature New Biol.* **234,** 40–43.

Rownd R., Nakaya R. and Nakamura A. (1966). Molecular nature of the drug resistance factors of the Enterobacteriaceae. *J. Mol. Biol.* **17,** 376–393.

Rubenstein I., Thomas C.A. jr. and Hershey A.D. (1961). The molecular weights of T2 phage DNA and its first and second breakage products. *Proc. Natl. Acad. Sci. USA* **47,** 1113–1122.

Rupp W.D. and Ihler G. (1968). Strand selection during bacterial mating. *Cold Spring Harbor. Symp. Quant. Biol.* **33,** 647–650.

Rush M.G. and Warner R.C. (1968). Molecular recombination in a circular genome: ϕX174 and S13. *Cold Spring Harbor Symp. Quant. Biol.* **33,** 459–466.

Russel M. (1973). Control of phage T4 DNA polymerase synthesis. *J. Mol. Biol.* **79,** 83–94.

Russell R.L. (1974). Comparative genetics of the T-even phages. *Genetics* **78,** 967–988.

Russell R.L. and Huskey R.J. (1974). Partial exclusion between T-even phages: an incipient genetic isolation mechanism. *Genetics* **78,** 989–1014.

Russo V.E.A. (1973). On the physical structure of lambda recombinant DNA. *Molec. Gen. Genet.* **122,** 353–366.

Rust P. and Sinsheimer R.L. (1967). The process of infection with ϕX174. XI. Infectivity of the complementary strand of the replicative form. *J. Mol. Biol.* **23,** 545–552.

Sabet S.F. and Schnaitman C.A. (1971). Localization and solubilization of colicin receptors. *J. Bacteriol.* **108,** 422–430.

Sadowski P.D. and Bakyta I. (1972). T4 endonuclease IV: improved purification procedure and resolution from T4 endonuclease III. *J. Biol. Chem.* **247,** 405–412.

Sadowski P.D. and Hurwitz J. (1969). Enzymatic breakage of DNA. I. Purification and properties of endonuclease II from T4 phage-infected E.coli. *J. Biol. Chem.* **244,** 6182–6191.

Sadowski P.D. and Kerr C. (1970). Degradation of E.coli B DNA after infection with DNA-defective amber mutants of phage T7. *J. Virol.* **6,** 149–155.

Sadowski P.D. and Vetter D. (1973). Control of T4 endonuclease IV by the D2a region of phage T4. *Virology* **54,** 544–546.

Sadowski P.D. and Vetter D. (1976). Genetic recombination of phage T7 DNA in vitro. *Proc. Natl. Acad. Sci. USA* **73,** 692–696.

Sadowski P.D., Warner H.R.R., Hercules K., Munro J.L., Mendelsohn S. and Wiberg J.S. (1971). Mutants of phage T4 defective in the induction of T4 endonuclease II. *J. Biol. Chem.* **246,** 3431–3433.

Saedler H. and Heiss B. (1973). Multiple copies of the insertion DNA sequences IS1 and IS2 in the chromosome of E.coli K12. *Molec. Gen. Genet.* **122,** 267–277.

Saedler H., Besemer J., Kemper B., Rosenwith B. and Starlinger P. (1972). Insertion mutations in the control region of the *gal* operon of E.coli. I. Biological characterization of the mutations. *Molec. Gen. Genet.* **115,** 258–265.

Saigo K. and Uchida H. (1974). Connection of the right hand terminus of DNA to the proximal end of the tail in phage lambda. *Virology* **61,** 524–536.

Sakai H., Hashimoto S. and Komano T. (1974). Replication of DNA in E.coli C mutants temperature sensitive in the initiation of chromosome replication. *J. Bacteriol.* **119,** 811–820.

Saitoh T. and Hiraga S. (1975). F DNA superinfected into phenocopies of donor strains. *J. Bacteriol.* **121,** 1007–1013.

Sakaki Y., Karu A.E., Linn S. and Echols H. (1973). Purification and properties of the

$\gamma$-protein specified by phage lambda: an inhibitor of the host recBC recombination enzyme. *Proc. Natl. Acad. Sci. USA* **70,** 2215–2219.

Sakakibara Y. and Tomizawa J.-I. (1971). Regulation of transcription of lambda phage operator mutants. *Virology* **44,** 463–472.

Sakakibara Y. and Tomizawa J.-I. (1974a). Termination point of replication of colicin E1 plasmid DNA in cell extracts. *Proc. Natl. Acad. Sci. USA* **71,** 4935–4939.

Sakakibara Y. and Tomizawa J.-I. (1974b). Replication of colicin E1 plasmid DNA in cell extracts: II. Selective synthesis of early replicative intermediates. *Proc. Natl. Acad. Sci. USA* **71,** 1403–1407.

Salstrom J.S. and Pratt D. (1971). Role of phage M13 gene 5 in single stranded DNA production. *J. Mol. Biol.* **61,** 489–501.

Salivar W.O., Henry T. and Pratt D. (1967). Purification and properties of diploid particles of phage M13. *Virology* **32,** 41–51.

Salivar W.O., Tzagoloff H. and Pratt D. (1964). Some physical and biological properties of the rod shaped phage M13. *Virology* **24,** 359–371.

Salser W., Bolle A. and Epstein R. (1970). Transcription during phage T4 development: a demonstration that distinct subclasses of the early RNA appear at different times and that some are turned off at late times. *J. Mol. Biol.* **49,** 271–296.

Sanderson K.E., Ross H., Ziegler L. and Makela P.H. (1972). $F^+$, Hfr and F′ strains of S.typhimurium and S.abony. *Bacteriol. Rev.* **36,** 608–637.

Sato K. and Campbell A. (1970). Specialised transduction by lambda phage from a deletion lysogen. *Virology* **41,** 474–487.

Sato K. and Sekiguchi M. (1976). Studies on temperature-dependent ultraviolet sensitive mutants of phage T4: the structural gene of T4 endonuclease. *J. Mol. Biol.* **102,** 15–26.

Sauerbier W. and Hercules K. (1973). Control of gene function in phage T4. IV. Post-transcriptional shutoff of expression of early genes. *J. Virol.* **12,** 538–547.

Saxe L.S. (1975a). The action of colicin E2 on supercoiled λ DNA. I. Experiments in vivo. *Biochemistry* **14,** 2051–2057.

Saxe L.S. (1975b). The action of colicin E2 on supercoiled λ DNA. II. Experiments in vitro. *Biochemistry* **14,** 2058–2063.

Scaife J. (1966). F prime factor formation in E.coli K12. *Genet. Res.* **8,** 189–196.

Scaife J. (1967). Episomes. *Ann. Rev. Microbiol.* **21,** 601–638.

Scaife J. and Gross J.D. (1962). Inhibition of multiplication of an F-lac factor in Hfr cells of E.coli K12. *Biochem. Biophys. Res. Commun.* **7,** 403–407.

Scaife J. and Gross J.D. (1963). The mechanism of chromosome mobilization by an F′ factor in E.coli K12. *Genet. Res.* **4,** 328–331.

Scaife J. and Pekhov A.P. (1964). Deletion of chromosome markers in association with F-prime factor formation in E.coli K12. *Genet. Res.* **5,** 495–498.

Scandella D. and Arber W. (1974). An E.coli mutant which inhibits the injection of phage λ DNA. *Virology* **58,** 504–513.

Scandella D. and Arber W. (1976). Phage λ DNA injection into E.coli *pel*⁻ mutants is restored by mutations in phage genes *V* or *H*. *Virology* **69,** 206–215.

Schachner M. and Zillig W. (1971). Fingerprint maps of tryptic peptides from subunits of E.coli and T4 modified DNA dependent RNA polymerase. *Eur. J. Biochem.* **22.** 513–519.

Schaeffer P., Edgar R.S. and Rolfe R. (1960). Sur l'inhibition de la transformation bacterienne par des DNAs de compositions variees. *Comptes Rendues Soc. Biol.* **154,** 1978–1983.

Schaller H., Gray C. and Herrmann K. (1975). Nucleotide sequence of an RNA polymerase binding site from the DNA of phage fd. *Proc. Natl. Acad. Sci. USA* **72,** 737–741.

Schaller H., Uhlmann A. and Geider K. (1976). A DNA fragment from the origin of single strand to double strand replication of phage fd. *Proc. Natl. Acad. Sci. USA* **73,** 49–53.

Schaller H., Voss H. and Gucker S. (1969). Structure of the DNA of phage fd. II. Isolation and characterization of a DNA fraction with double strand-like properties. *J. Mol. Biol.* **44,** 445–458.

Schekman R. and Ray D.S. (1971). Polynucleotide ligase and $\phi$X174 single strand synthesis. *Nature New Biol.* **231,** 170–173.

Schekman R., Weiner A. and Kornberg A. (1974). Multienzyme systems of DNA replication. *Science* **186,** 987–993.

Schekman R., Iwaya M., Bromstrup K. and Denhardt D.T. (1971). The mechanism of replication of $\phi$X174 single stranded DNA. III. An enzymatic study of the structure of the replicative form II DNA. *J. Mol. Biol.* **57,** 177–199.

Schekman R., Wickner W., Westergaard O., Brutlag D., Geider K., Bertsch L.L. and Kornberg A. (1972). Initiation of DNA synthesis: synthesis of $\phi$X174 replicative form requires RNA synthesis resistant to rifampicin. *Proc. Natl. Acad. Sci. USA* **69,** 2691–2695.

Schekman R., Weiner J.H., Weiner A. and Kornberg A. (1975). Ten proteins required for conversion of $\phi$X174 single stranded DNA to duplex form in vitro. Resolution and reconstitution. *J. Biol. Chem.* **250,** 5859–5865.

Scherberg N.H. and Weiss S.B. (1970). Detection of phage T4 and T5 coded tRNAs. *Proc. Natl. Acad. Sci. USA* **67,** 1164–1171.

Schiff N., Miller M.J. and Wahba A.J. (1974). Purification and properties of the initiation factor 3 from T4 infected and uninfected E.coli MRE 600. Stimulation of translation of synthetic and natural messengers. *J. Biol. Chem.* **249,** 3797–3802.

Schlegel R.A. and Thomas C.A. jr. (1972). Some special structural features of intracellular phage T7 concatemers. *J. Mol. Biol.* **68,** 319–346.

Schmidt D.A., Mazaitis A.J., Kasai T. and Bautz E.K.F. (1970). Involvement of a phage T4 $\sigma$ factor and an anti-terminator protein in the transcription of early T4 genes in vivo. *Nature* **225,** 1012–1016.

Schmidt F., Besemer J. and Starlinger P. (1976). The isolation of IS1 and IS2 DNA. *Molec. Gen. Genet.* **145,** 145–154.

Schnos M. and Inman R.B. (1970). Position of branch points in replicating lambda DNA. *J. Mol. Biol.* **51,** 61–74.

Schroder C.H. and Kaerner H.C. (1972). Replication of phage $\phi$X 174 replicative form DNA in vivo. *J. Mol. Biol.* **71,** 351–362.

Schwartz M. (1975). Reversible interaction between phage lambda and its receptor protein. *J. Mol. Biol.* **99,** 185–202.

Schwartz S.A. and Helinski D.R. (1971). Purification and characterization of colicin E1. *J. Biol. Chem.* **246,** 6318–6327.

Scocca J.J., Poland R.L. and Zoon K.C. (1974). Specificity in DNA uptake by transformable H.influenzae. *J. Bacteriol.* **118,** 369–373.

Sechaud J., Streisinger G., Emrich J., Newton J., Lanford H., Reinhold H. and Stahl M.M. (1965). Chromosome structure in phage T4. II. Terminal redundancy and heterozygosis. *Proc. Natl. Acad. Sci. USA* **54,** 1333–1339.

Sedat J.W. and Sinsheimer R.L. (1970). The in vivo $\phi$X mRNA. *Cold Spring Harbor Symp. Quant. Biol.* **35,** 163–170.

Sedgwick B. and Setlow J.K. (1976). Single stranded regions in transforming DNA after uptake by competent H.influenzae. *J. Bacteriol.* **125,** 588–594.

Seeburg P.H. and Schaller H. (1975). Mapping and characterization of promoters in phages fd, fl and M13. *J. Mol. Biol.* **92,** 261–279.

Seidman J.G., Comer M.M. and McClain W.H. (1974). Nucleotide alterations in the phage T4 glutamine tRNA that affects ochre suppressor activity. *J. Mol. Biol.* **90,** 677–690.

Sekiya T. and Khorana H.G. (1974). Nucleotide sequence in the promotor region of the E.coli tyrosine tRNA gene. *Proc. Natl. Acad. Sci. USA* **72,** 2978–2982.

Senear A.W. and Steitz J.A. (1976). Site specific interaction of Q$\beta$ host factor and ribosomal protein S1 and Q$\beta$ and R17 phage RNAs. *J. Biol. Chem.* **251,** 1902–1912.

Senior B.W. and Holland I.B. (1971). Effect of colicin E3 upon the 30S ribosomal subunit of E.coli. *Proc. Natl. Acad. Sci. USA* **68,** 959–963.

Sertic V. and Boulagakov N. (1935). Classification et identification des typhi phages. *Comp. Rend. Soc. Biol.* **119,** 1270–1272.

Serwer P. (1974). Fast sedimenting phage T7 DNA from T7 infected E.coli. *Virology* **59,** 70–88.

Seto H. and Tomasz A. (1974). Early stages in DNA binding and uptake during genetic transformation of Pneumococci. *Proc. Natl. Acad. Sci. USA* **71,** 1493–1498.

Seto H. and Tomasz A. (1975). Protoplast formation and the leakage of intramembrane cell components: induction by the competence activator of pneumococci. *J. Bacteriol.* **121,** 344–353.

Seto H., Lopez R., Garrigan O. and Tomasz A. (1975). Nucleolytic degradation of homologous and heterologous DNA molecules at the surface of competent pneumococci. *J. Bacteriol.* **122,** 676–685.

Shah D.B. (1976). Replication and recombination of gene 59 mutant of phage T4D. *J. Virol.* **17,** 175–182.

Shah D.B. and Berger H. (1971). Replication of gene 46–47 amber mutants of phage T4D, *J. Mol. Biol.* **57,** 17–34.

Shalitin C. and Naot Y. (1971). Role of gene 46 in phage T4 DNA synthesis. *J. Virol.* **8,** 142–153.

Shapiro B.M., Siccardi A.G., Hirota Y. and Jacob F. (1970). Membrane protein alterations associated with mutations affecting the initiation of DNA synthesis. *J. Mol. Biol.* **52,** 75–89.

Shapiro L. and Bendis I. (1975). RNA phages of bacteria other than E.coli. In RNA Phages, ed. N.D. Zinder. (*Cold Spring Harbor Laboratory: New York*). pp. 397–410.

Sharp P.A., Cohen S.N. and Davidson N. (1973). Electron microscope studies of sequence relations among plasmids of E.coli. II. Structure of drug resistance (R) factors and F factors. *J. Mol. Biol.* **75,** 235–255.

Sharp P.A., Hsu M.-T. and Davidson N. (1972). Note on the structure of prophage $\lambda$. *J. Mol. Biol.* **71,** 499–502.

Sharp P.A., Hsu M.-T., Ohtsubo E. and Davidson N. (1972). Electron microscope heteroduplex studies of sequence relations among plasmids of E.coli. I. Structure of F′ factors. *J. Mol. Biol.* **71,** 471–497.

Sheehy R.J., Orr C. and Curtis R. III. (1972). Molecular studies on entry exclusion in E.coli minicells. *J. Bacteriol.* **112,** 861–869.

Sheehy R.J., Perry A., Allison D.P. and Curtis R. III. (1973). Molecular nature of R factor DNA isolated from S.typhimurium minicells. *J. Bacteriol.* **114,** 1328–1335.

Shimada K. and Campbell A. (1974a). Int constitutive mutants of phage λ. *Proc. Natl. Acad. Sci. USA* **71,** 237–241.

Shimada K. and Campbell A. (1974b). Lysogenization and curing by int-constitutive mutants of phage λ. *Virology* **60,** 157–165.

Shimada K., Weisberg R.A. and Gottesman M.E. (1972). Prophage λ at unusual chromosomal locations. I. Location of the secondary attachment sites and properties of the lysogens. *J. Mol. Biol.* **63,** 483–503.

Shimada K., Weisberg R.A. and Gottesman M.E. (1973). Prophage lambda at unusual chromosomal locations. II. Mutations induced by phage λ in E.coli K12. *J. Mol. Biol.* **80,** 297–314.

Shimada K., Weisberg R.A. and Gottesman M.E. (1975). Prophage lambda at unusual chromosomal locations. III. The components of the secondary attachment sites. *J. Mol. Biol.* **93,** 415–430.

Shimizu K. and Sekiguchi M. (1976). 5′ - 3′ exonucleases of phage T4. *J. Biol. Chem.* **251,** 2613–2619.

Shine J. and Dalgarno L. (1974). The 3′ terminal sequence of E.coli 16S rRNA: complementarity to nonsense triplets and ribosome binding sites. *Proc. Natl. Acad. Sci. USA* **71,** 1342–1346.

Shoemaker N.B. and Guild W.R. (1972). Kinetics of integration of transforming DNA in pneumococcus. *Proc. Natl. Acad. Sci. USA* **69,** 3331–3335.

Showe M.K. and Black L.W. (1973). The assembly core of phage T4, an intermediate in head formation. *Nature New Biol.* **242,** 70–75.

Shull F.W., Fralick J.A., Stratton L.P. and Fisher W.D. (1971). Membrane association of conjugally transferred DNA in E.coli. *J. Bacteriol.* **106,** 626–632.

Shulman M. and Gottesman M. (1971). Lambda *att*$^2$: a transducing phage capable of intramolecular *int-xis* promoted recombination. In The Bacteriophage Lambda, ed. A.D. Hershey. (*Cold Spring Harbor Laboratory, New York*). pp. 477–488.

Shulman M. and Gottesman M. (1974). Attachment site mutants of phage lambda. *J. Mol. Biol.* **81,** 461–482.

Shulman M.J., Mizuuchi K. and Gottesman M.M. (1976). New *att* mutants of phage lambda. *Virology* **72,** 13–22.

Shulman M.J., Hallick M.L., Echols H. and Signer E.R. (1970). Properties of recombination deficient mutants of phage λ. *J. Mol. Biol.* **52,** 501–520.

Sicard A.M. and Ephrussi-Taylor H. (1965). Genetic recombination in DNA-induced transformation of Pneumococcus. II. Mapping the amiA region. *Genetics* **52,** 1207–1227.

Siccardi A.G., Shapiro B.M., Hirota Y. and Jacob F. (1971). On the process of cellular division in E.coli. IV. Altered protein composition and turnover of the membranes of the thermosensitive mutants defective in chromosome replication. *J. Mol. Biol.* **56,** 475–490.

Sidikaro J. and Nomura M. (1974). E3 immunity substance. A protein from E3 and colicinogenic cells that accounts for their immunity to colicin E3. *J. Biol. Chem.* **249,** 445–453.

Siddiqui O. and Fox M.S. (1973). Integration of donor DNA in bacterial conjugation. *J. Mol. Biol.* **77,** 101–124.

Siden E.J. and Hayashi M. (1974). Role of the gene B product in phage φX174 development. *J. Mol. Biol.* **89,** 1–16.

Sidikaro J. and Nomura M. (1974). E3 immunity substance. A protein from E3 and colicinogenic cells that accounts for their immunity to colicin E3. *J. Biol. Chem.* **249,** 445–453.

Siegel P.J. and Schaechter M. (1973). Phage T4 head maturation: release of progeny DNA from the host cell membrane. *J. Virol.* **11,** 359–367.

Siegel R.B. and Summers W.C. (1970). The process of infection with phage T7. III. Control of phage specific RNA synthesis in vivo by an early phage gene. *J. Mol. Biol.* **49,** 115–123.

Sigal N., Delius H., Kornberg T., Gefter M.L. and Alberts B.M. (1972). A DNA unwinding protein isolated from E.coli: its interaction with DNA and with DNA polymerases. *Proc. Natl. Acad. Sci. USA* **69,** 3537–3541.

Signer E.R. (1966). Interaction of prophages at the $att_{80}$ site with the chromosome of E.coli. *J. Mol. Biol.* **15,** 243–255.

Signer E.R. (1968). Lysogeny: the integration problem. *Ann. Rev. Microbiol.* **22,** 451–488.

Signer E.R. (1971). General recombination. In The Bacteriophage Lambda, ed. A.D. Hershey. (*Cold Spring Harbor Laboratory, New York*). pp. 139–174.

Signer E.R. and Beckwith J. (1966). Transposition of the lac region of E. coli. III. The mechanism of attachment of coliphage $\phi$80 to the bacterial chromosome. *J. Mol. Biol.* **22,** 33–51.

Signer E.R. and Weil J. (1968a). Recombination in phage $\lambda$. I. Mutants deficient in general recombination. *J. Mol. Biol.* **34,** 261–271.

Signer E.R. and Weil J. (1968b). Site specific recombination of phage $\lambda$. *Cold Spring Harbor Symp. Quant. Biol.* **33,** 715–720.

Signer E.R., Weil J. and Kimball P.C. (1969). Recombination in bacteriophage $\lambda$. III. Studies on the nature of the prophage attachment region. *J. Mol. Biol.* **46,** 543–563.

Signer E.R., Echols H., Weil J., Radding C., Shulman M., Moore L. and Manly K. (1968). The general recombination system of phage $\lambda$. *Cold Spring Harbor Symp. Quant. Biol.* **33,** 711–714.

Silberstein S., Inouye M. and Studier F.W. (1975). Studies on the role of phage T7 lysozyme during phage infection. *J. Mol. Biol.* **96, 1**–12.

Silver R.P. and Cohen S.N. (1972). Nonchromosomal antibiotic resistance in bacteria. V. Isolation and characterization of R factor mutants exhibiting temperature sensitive repression of fertility. *J. Bacteriol.* **110,** 1082–1088.

Silver R.P. and Falkow S. (1970a). Specific labeling and physical characterization of R factor DNA in E.coli. *J. Bacteriol.* **104,** 331–339.

Silver R.P. and Falkow S. (1970b). Studies on resistance transfer factor DNA in E.coli. *J. Bacteriol.* **104,** 340–344.

Silver S.D. (1963). Transfer of material during mating in Escherichia coli. Transfer of DNA and upper limits on the transfer of RNA and protein. *J. Mol. Biol.* **6,** 349–360.

Silverman P.M. and Valentine R.C. (1969). The RNA injection step of phage f2 infection. *J. Gen. Virol.* **4,** 111–124.

Simon L.D. (1972). Infection of E.coli by T2 and T4 phages as seen in the electron microscope: T4 head morphogenesis. *Proc. Natl. Acad. Sci. USA* **69,** 907–911.

Simon L.D. and Anderson T.F. (1967a). The infection of E.coli by T2 and T4 phages as seen in the electron microscope. I. Attachment and penetration. *Virology* **32,** 279–297.

Simon L.D. and Anderson T.F. (1967b). The infection of E.coli by T2 and T4 phages as seen in the electron microscope. II. Structure and function of the baseplate. *Virology* **32,** 298–305.

Simon L.D., Snover D. and Doermann A.H. (1974). Bacterial mutation affecting T4 phage DNA synthesis and tail production. *Nature* **252,** 451–455.

Simon L.D., McLaughlin J.M., Snover D., Ou J., Grisham C. and Loeb M. (1975). E.coli membrane lipid alteration affecting T4 capsid morphogenesis. *Nature* **256,** 379–383.

Simon M.N. and Studier F.W. (1973). Physical mapping of the early region of phage T7. *J. Mol. Biol.* **79,** 249–266.

Simon M.N., Davis R.W. and Davidson N. (1971). Heteroduplexes of DNA molecules of lambdoid phages: physical mapping of their base sequence relationships by electron microscopy. In The Bacteriophage Lambda, ed. A.D. Hershey. (*Cold Spring Harbor Laboratory: New York*). pp. 313–328.

Simon T.J., Smith C.A. and Kaplan N.O. (1975). Action of phage T4 UV endonuclease on duplex DNA containing one ultraviolet irradiated strand. *J. Biol. Chem.* **250,** 8748–8752.

Sinha N.K. and Snustad D.P. (1971). DNA synthesis in phage T4 infected E.coli: evidence supporting a stoichiometric role for gene 32 product. *J. Mol. Biol.* **62,** 267–271.

Sinsheimer R.L. (1959a). Purification and properties of bacteriophage $\phi$X174. *J. Mol. Biol.* **1,** 37–42.

Sinsheimer R.L. (1959b). A single-stranded DNA from bacteriophage $\phi$X174. *J. Mol. Biol.* **1,** 43–53.

Sinsheimer R.L. (1968). Bacteriophage $\phi$X174 and related viruses. *Prog. Nuc. Acid Res. Mol. Biol.* **8,** 115–170.

Sinsheimer R.L., Knippers R. and Komano T. (1968). Stages in the replication of phage $\phi$X174 DNA in vivo. *Cold Spring Harbor Symp. Quant. Biol.* **33,** 443–447.

Sinsheimer R.L., Starman B., Nagler C. and Guthrie S. (1962). The process of infection with bacteriophage $\phi$X174. I. Evidence for a 'replicative form'. *J. Mol. Biol.* **4,** 142–160.

Sironi G., Bialy H., Lozeron H.A. and Calendar R. (1971). Phage P2: interaction with phage $\lambda$ and with recombination deficient bacteria. *Virology* **46,** 387–396.

Skalka A. (1966). Regional and temporal control of genetic transcription in phage $\lambda$. *Proc. Natl. Acad. Sci. USA* **55,** 1190–1195.

Skalka A. (1971). Origin of DNA concatemers during growth. In The Bacteriophage Lambda, ed. A.D. Hershey. (*Cold Spring Harbor Laboratory: New York*). pp. 535–548.

Skalka A., Poonian M. and Bartl P. (1972). Concatemers in DNA replication: electron microscopic studies of partically denatured intracellular $\lambda$ DNA. *J. Mol. Biol.* **64,** 541–550.

Skare J., Niles E.G. and Summers W.C. (1974). Localization of the leftmost initiation site for T7 late transcription, in vivo and in vitro. *Biochemistry* **13,** 3912–3916.

Skurray R.A., Hancock R.E.W. and Reeves P. (1974). Con$^-$ mutants: class of mutants in E.coli K12 lacking a major cell wall protein and defective in conjugation and adsorption of a bacteriophage. *J. Bacteriol.* **119,** 726–735.

Skurray R.A., Nagaishi H. and Clark A.J. (1976). Molecular cloning of DNA from F sex factors of E.coli K12. *Proc. Natl. Acad. Sci. USA* **73,** 64–68.

Sly W.S., Rabideau K. and Kolber A. (1971). The mechanism of $\lambda$ virulence. II. Regulatory mutations in classical virulence. In The Bacteriophage Lambda, ed. A.D. Hershey. (*Cold Spring Harbor Laboratory: New York*). pp. 575–588.

Smilowitz H. (1974). Phage fl infection: fate of the parental major coat protein. *J. Virol.* **13,** 94–99.

Smith G.R., Eisen H., Reichardt L. and Hedgpeth J. (1976). Deletions of lambda phage locating a *prm* mutation within the rightward operator. *Proc. Natl. Acad. Sci. USA* **73,** 712–716.

Smith M. and Skalka A. (1966). Some properties of DNA from phage infected bacteria. *J. Gen. Physiol.* **49,** suppl. **2,** 127–142.

Smith L.H. and Sinsheimer R.L. (1976a). The in vitro transcription units of phage φX174. I. Characterization of synthetic parameters and measurement of transcript molecular weights. *J. Mol. Biol.* **103,** 681–698.

Smith L.H. and Sinsheimer R.L. (1976b). The in vitro transcription units of phage φX174. II. In vitro initiation sites of φX174 transcription. *J. Mol. Biol.* **103,** 699–710.

Smith L.H. and Sinsheimer R.L. (1976c). The in vitro transcription units of phage φX174. III. Initiation with specific 5′ end oligonucleotides of in vitro φX174 RNA. *J. Mol. Biol.* **103,** 711–736.

Smith S.M., Ozeki H. and Stocker B.A.D. (1963). Transfer of colE1 and colE2 during high frequency transmission of colI in S.typhimurium. *J. Gen. Microbiol.* **33,** 231–242.

Snell D.T. and Offord R.E. (1972). The amino acid sequence of the B protein of phage ZJ-2. *Biochem. J.* **127,** 167–178.

Snustad D.P., Warner H.R., Parson K.A. and Anderson D.L. (1972). Nuclear disruption after infection of E.coli with a phage T4 mutant unable to induce exonuclease II. *J. Virol.* **10,** 124–133.

Snustad D., Bursch C.J.H., Parson K.A. and Hefeneider H. (1976a). Mutants of phage T4 deficient in the ability to induce nuclear disruption: shutoff of host DNA and protein synthesis, gene dosage experiments, identification of a restrictive host, and possible biological significance. *J. Virol.* **18,** 268–288.

Snustad D.P., Tigges M.A., Parson K.A., Bursch C.J.H., Caron F.M., Koerner J.F. and Tutas D.J. (1976b). Identification and preliminary characterization of a mutant defective in the phage T4-induced unfolding of the E.coli nucleoid. *J. Virol.* **17,** 622–641.

Snyder R.L. and Montgomery D.L. (1974). Inhibition of T4 growth by an RNA polymerase mutation of E.coli: physiological and genetic analysis of the effects during phage development. *Virology* **62,** 184–196.

Sobell H.M. (1973). Symmetry in protein-nucleic acid interactions and its genetic implications. *Adv. Genet.* **17,** 411–490.

Sogo J.M., Greenstein M. and Skalka A. (1976). The circle mode of replication of phage lambda: the role of covalently closed templates and the formation of mixed catenated dimers. *J. Mol. Biol.* **103,** 537–562.

Soltyk A., Shugar S. and Piechowska M. (1975). Heterologous DNA uptake and complexing with cellular constituents in competent B.subtilis. *J. Bacteriol.* **124,** 1429–1438.

Souther A., Bruner R. and Elliot J. (1972). Degradation of E.coli chromosome after infection by phage T4: role of phage gene D2a. *J. Virol.* **10,** 979–984.

Spahr P.F. and Gesteland R.F. (1968). Specific cleavage of phage R17 RNA by an endonuclease isolated from E.coli MRE 600. *Proc. Natl. Acad. Sci. USA* **59,** 876–883.

Spencer H.T. and Heriot R.M. (1965). Development of competence of Hemophilus influenzae. *J. Bacteriol.* **90.** 911–920.

Speyer J.F. (1965). Mutagenic DNA polymerase. *Biochem. Biophys. Res. Commun.* **21,** 6–8.

Spiegelman S., Haruna I., Holland I.B., Beaudreau G. and Mills D. (1965). The synthesis of a self-propagating and infectious nucleic acid with a purified enzyme. *Proc. Natl. Acad. Sci. USA* **54,** 919–927.

Spiegelman S., Pace N.R., Mills D.R., Levisohn R., Eikhom T.S., Taylor M.M., Peterson R.L. and Bishop D.H.L. (1968). The mechanism of RNA replication. *Cold Spring Harbor Symp. Quant. Biol.* **33,** 101–124.

Spiegelmann W.G., Heinemann S.F., Brachet P., da Silva L.P. and Eisen H. (1970). Regulation of the synthesis of phage lambda repressor. *Cold Spring Harbor Symp. Quant. Biol.* **35,** 325–330.

Spiegelman W.G., Reichardt L.F., Yaniv M., Heinemann S.F., Kaiser A.D. and Eisen H. (1972). Bidirectional transcription and the regulation of phage lambda repressor synthesis. *Proc. Natl. Acad. Sci. USA* **69,** 3156–3160.

Spizizen J., Reilly B.E. and Evans A.H. (1966). Microbial transformation and transfection. *Ann. Rev. Microbiol.* **20,** 371–400.

Sriprakash K.S., Lundh N., Moo-On M.H. and Radding C.M. (1975). The specificity of λ exonuclease. Interactions with single stranded DNA. *J. Biol. Chem.* **250,** 5438–5446.

Stahl F.W. and Stahl M.M. (1976). On recombination between close and distant markers in phage lambda. *Genetics* **82,** 577–593.

Stahl F.W. and Steinberg C.M. (1964). The theory of formal phage genetics for circular maps. *Genetics* **50,** 531–538.

Stahl F.W., Crasemann J.N. and Stahl M.M. (1975). Rec-mediated recombinational hot spot activity in phage lambda. III. Chi mutations are site-mutations stimulating rec-mediated recombination. *J. Mol. Biol.* **94,** 203–212.

Stahl F.W., Edgar R.S. and Steinberg J. (1964). The linkage map of phage T4. *Genetics* **50,** 539–552.

Stahl F.W., McMilin K.D., Stahl M.M., Malone R.E., Nozu Y. and Russo V.E.A. (1972a). A role for recombination in the production of "free loader" λ phage particles. *J. Mol. Biol.* **68,** 57–68.

Stahl F.W., McMilin K.D., Stahl M., and Nozu Y. (1972b). An enhancing role for DNA synthesis in formation of phage λ recombinants. *Proc. Natl. Acad. Sci. USA* **69,** 3598–3601.

Stahl F.W., McMilin K.D., Stahl M.M., Crasemann J.M., and Lam S. (1974). The distribution of crossovers along unreplicated lambda phage chromosome. *Genetics* **77,** 395–408.

Staples D.H. and Hindley J. (1971). Ribosome binding site of Qβ RNA polymerase cistron. *Nature New Biology* **234,** 211–212.

Staples D.H., Hindley J., Billeter M.A. and Weissman C. (1971). Localization of Qβ maturation cistron binding site. *Nature New Biology* **234,** 202–204.

Stallions D.R. and Curtis R. III. (1971). Chromosome transfer and recombinant formation with DNA temperature sensitive strains of E.coli. *J. Bacteriol.* **105,** 886–895.

Steinberg C.M. and Edgar R.S. (1961). On the absence of high negative interference in triparental crosses. *Virology* **15,** 511–513.

Steinberg C.M. and Edgar R.S. (1962). A critical test of a current theory of genetic recombination in bacteriophage. *Genetics* **47,** 187–208.

Steinberg R.A. and Ptashne M. (1971). In vitro repression of RNA synthesis by purified lambda phage repressor. *Nature New Biology* **230,** 76–80.

Steinhart W.L. and Herriot R.M. (1968). Fate of recipient DNA during transformation in H.influenzae. *J. Bacteriol.* **96,** 1718–1724.

Steitz J.A. (1968a). Identification of the A protein as a structural component of phage R17. *J. Mol. Biol.* **33,** 923–936.

Steitz J.A. (1968b). Isolation of the A protein from phage R17. *J. Mol. Biol.* **33,** 937–946.

Steitz J.A. (1969). Polypeptide chain initiation: nucleotide sequences of the three ribosomal binding sites in phage R17 RNA. *Nature* **224,** 957–963.

Steitz J.A. (1972). Oligonucleotide sequence of replicase initiation site in Qβ RNA. *Nature New Biology* **236,** 71–75.

Steitz J.A. (1973). Discriminatory ribosome rebinding of isolated regions of protein synthesis initiation from the RNA of R17. *Proc. Natl. Acad. Sci. USA* **70,** 2605–2609.

Steitz J.A. (1975). Ribosome recognition of initiator regions in the RNA phage genome. In RNA Phages, ed. N.D. Zinder. (*Cold Spring Harbor Laboratory: New York*). pp. 319–352.

Steitz J.A., Dube S.K. and Rudland P.S. (1970). Control of translation by T4 phage: altered ribosome binding at R17 sites. *Nature* **226,** 824–827.

Sternberg N. (1973a). Properties of a mutant of E.coli defective in phage lambda head formation (groE). I. Initial characterization. *J. Mol. Biol.* **76,** 1–24.

Sternberg N. (1973b). Properties of a mutant of E.coli defective in phage lambda head formation (groE). II. The propagation of phage lambda. *J. Mol. Biol.* **76,** 25–44.

Sternberg N. and Weisberg R.A. (1975). Packaging of prophage and host DNA by phage lambda. *Nature* **256,** 97–103.

Steven A.C., Aebi U. and Showe M.K. (1976). Folding and capsomere morphology of the P23 surface shell of phage T4 polyheads from mutants in five different head genes. *J. Mol. Biol.* **102,** 373–399.

Steven A.C., Couture E., Aebi U. and Showe M.K. (1976). The structure of T4 polyheads. II. A pathway of polyhead transformations as model for T4 capsid maturation. *J. Mol. Biol.* **106,** 187–222.

Stevens A. (1972). New small polypeptides associated with DNA dependent RNA polymerase of E.coli after infection with phage T4. *Proc. Natl. Acad. Sci. USA* **69,** 603–607.

Stevens A. (1974). DNA dependent RNA polymerases from two T4 phage infected systems. *Biochemistry* **13,** 493–502.

Stevens W.F., Adhya S. and Szybalski W. (1971). Origin and bidirectional orientation of DNA replication in phage lambda. In The Bacteriophage Lambda, ed. A.D. Hershey. (*Cold Spring Harbor Laboratory: New York*). pp. 515–534.

Stocker B.A.D. (1963). Transformation of Bacillus subtilis to motility and prototrophy: micro-manipulative isolation of bacteria of transformed phenotype. *J. Bacteriol.* **86,** 797–804.

Stocker B.A.D., Smith S.M. and Ozeki H. (1963). High infectivity of S.typhimurium newly infected by the ColI factor. *J. Gen. Microbiol.* **30,** 201–221.

Stone K.R. and Cummings D.J. (1970). Isolation and characterization of two basic internal proteins from the T-even phages. *J. Virol.* **6,** 445–454.

Stone K.R. and Cummings D.J. (1972). Comparison of the internal proteins of the T-even phages. *J. Mol. Biol.* **64,** 651–670.

Stouthamer A.H., De Haan P.G. and Bulten E.J. (1963). Kinetics of F curing by acridine orange in relation to the number of F particles in E.coli. *Genet. Res.* **4,** 305–317.

Strack H.B. and Kaiser A.D. (1965). On the structure of the ends of lambda DNA. *J. Mol. Biol.* **12,** 36–49.

Strathern A. and Herskowitz I. (1975). Defective prophage in E.coli K12 strains. *Virology* **67,** 136–143.

Stratling W. and Knippers R. (1973). Function and purification of gene *4* protein of phage T7. *Nature* **245,** 195–197.

Stratling W., Krause E. and Knippers R. (1973). Fast sedimenting DNA in phage T7-infected cells. *Virology* **51,** 109–119.

Stratling W., Ferdinand F.J., Krause E. and Knippers R. (1973). Phage T7 DNA replication in vitro: an experimental system. *Eur. J. Biochem.* **38,** 160–169.

Strauss N. (1970). Early energy dependent step in entry of transforming DNA. *J. Bacteriol.* **101,** 35–37.

Streips U.N. and Welker N.E. (1971a). Competitive inducing factor of B.stearothermophilus. *J. Bacteriol.* **106,** 955–959.

Streisinger G. (1956a). The genetic control of host range and serological specificity in phages T2 and T4. *Virology* **2,** 377–387.

Streisinger G. (1956b). Phenotypic mixing of host range and serological specificities in phages T2 and T4. *Virology* **2,** 388–398.

Streisinger G. and Bruce V. (1960). Linkage of genetic markers in phages T2 and T4. *Genetics* **45,** 1289–1296.

Streisinger G. and Franklin N.C. (1956). Mutation and recombination at the host range genetic region of phage T2. *Cold Spring Harbor Symp. Quant. Biol.* **21,** 103–111.

Streisinger G. and Weigle J. (1956). Properties of phages T2 and T4 with unusual inheritance. *Proc. Natl. Acad. Sci. USA* **42,** 504–510.

Streisinger G., Edgar R.S. and Denhardt G.H. (1964). Chromosome structure in phage T4. I. Circularity of the linkage map. *Proc. Natl. Acad. Sci. USA* **51,** 775–779.

Streisinger G., Emrich J. and Stahl M.M. (1967). Chromosome structure in phage T4. III. Terminal redundancy and length determination. *Proc. Natl. Acad. Sci. USA* **57,** 292–295.

Strike P. and Emmerson P.T. (1972). Coexistence of polA and recB mutations of E.coli in the presence of *sbc,* a mutation which indirectly suppresses *recB. Molec. Gen. Genetics* **116,** 177–180.

Studier F.W. (1969). The genetics and physiology of phage T7. *Virology* **39,** 562–574.

Studier F.W. (1972). Bacteriophage T7. *Science* **176,** 367–376.

Studier F.W. (1973a). Genetic analysis of non essential phage T7 genes. *J. Mol. Biol.* **79,** 227–236.

Studier F.W. (1973b). Analysis of phage T7 early RNAs and proteins on slab gels. *J. Mol. Biol.* **79,** 237–248.

Studier F.W. (1975). Gene *0.3* of phage T7 acts to overcome the DNA restriction system of the host. *J. Mol. Biol.* **94,** 283–295.

Studier F.W. and Maizel J.V. jr. (1969). T7 directed protein synthesis. *Virology* **39,** 575–586.

Studier F.W. and Movva N.R. (1976). SAMase gene of T3 is responsible for overcoming host restriction. *J. Virol.* **19,** 136–145.

Sugimoto K., Okamoto T., Sugisaki H. and Takanami M. (1975). The nucleotide sequence of an RNA polymerase binding site on phage fd DNA. *Nature* **253,** 410–414.

Sugino Y. and Hirota Y. (1962). Conjugal fertility associated with resistance factor R in E.coli. *J. Bacteriol.* **84,** 902–910.

Sugiura M., Okamoto T. and Takanami M. (1970). RNA polymerase sigma factor and the selection of initiation sites. *Nature* **225,** 598–600.

Sugiyama T. and Nakada D. (1968). Translational control of phage MS2 RNA cistrons by MS2 coat protein; gel electrophoretic analysis of proteins synthesized in vitro. *J. Mol. Biol.* **31,** 431–440.

Sugiyama T., Hebert R.R. and Hartman K.A. (1967). Ribonucleoprotein complexes formed between phage MS2 RNA and MS2 coat protein in vitro. *J. Mol. Biol.* **25,** 455–463.

Sumida-Yasumoto C., Yudelovich A. and Hurwitz J. (1976). DNA synthesis in vitro dependent upon $\phi$X174 form I DNA. *Proc. Natl. Acad. Sci. USA* **73,** 1887–1891.

Summers W.C. and Siegel R.B. (1969). Control of template specificity of E.coli RNA polymerase by a phage coded protein. *Nature* **223,** 1111–1113.

Summers W.C. and Siegel R.B. (1970). Transcription of late phage RNA by T7 RNA polymerase. *Nature* **228,** 1160–1162.

Summers W.C. and Szybalski W. (1968). Totally asymmetric transcription of phage T7 in vivo: correlation with poly(G) binding sites. *Virology* **34,** 9–16.

Summers W.C., Brunovskis I. and Hyman R.W. (1973). The process of infection with phage T7. VII. Characterization and mapping of the major in vivo transcription products of the early region. *J. Mol. Biol.* **74,** 291–300.

Susman M. (1970). General bacterial genetics. *Ann. Rev. Genetics* **4,** 135–176.

Sussman R. and Jacob F. (1962). Sur un systeme de repression thermosensible chez le bacteriophage $\lambda$ d'un E.coli. *CR Acad. Sci. Paris* **254,** 1517–1519.

Syvanen M. (1974). In vitro genetic recombination of phage lambda. *Proc. Natl. Acad. Sci. USA* **71,** 2496–2499.

Syvanen M. (1975). Processing of bacteriophage lambda DNA during its assembly into heads. *J. Mol. Biol.* **91,** 165–174.

Szpirer J. and Brachet P. (1970). Relations physiologiques entre les phages temperes $\lambda$ et $\phi$80. *Molec. Gen. Genet.* **108,** 78–92.

Szybalski W., Bøvre K., Fiandt M., Guha A., Hradneca Z., Kumar S., Lozeran H.A., Maher V.M., Nijkamp H.J.J., Summer W.C. and Taylor K. (1969). Transcriptional controls in developing phages. *J. Cell. Physiol.* **74,** suppl. **1,** 33–70.

Szybalski W., Bøvre K., Fiandt M., Hayes S., Hradneca Z., Kumar S., Lozeron H.A., Nijkamp H.J.J. and Stevens W.F. (1970). Transcriptional units and their controls in E.coli phage lambda: operons and scriptons. *Cold Spring Harbor Symp. Quant. Biol.* **35,** 341–354.

Tabak H.F., Griffith J., Geider K., Schaller H. and Kornberg A. (1974). Initiation of DNA synthesis. VII. A unique location of the gap in the M13 replicative duplex synthesized in vitro. *J. Biol. Chem.* **249,** 3049–3054.

Takahashi S. (1974). The rolling circle replicative structure of a phage lambda DNA. *Biochem. Biophys. Res. Commun.* **61,** 657–663.

Takahashi S. (1975). Structure of the oligomeric circular molecules of a phage $\lambda$ DNA. *Virology* **64,** 319–329.

Takahashi S. (1976). Starting point and direction of rolling circle intermediate of phage lambda DNA. *Molec. Gen. Genet.* **142,** 137–154.

Takahashi H., Coppo A., Manzi A., Martire G. and Pulitzer J.F. (1975). Design of a system of conditional lethal mutations (*tab/k/com*) affecting protein-protein interactions in phage T4 infected E.coli. *J. Mol. Biol.* **96,** 563–578.

Takanami M., Okamoto T. and Sugiura M. (1970). The starting nucleotide sequence and size of mRNA transcribed in vitro on phage DNA templates. *Cold Spring Harbor Symp. Quant. Biol.* **35,** 179–188.

Takanami M., Okamoto T. and Sugiura M. (1971). Termination of RNA transcription on the replicative form of phage fd. *J. Mol. Biol.* **62,** 81–88.

Takeda Y., Matsubara K. and Ogata K. (1975a). Regulation of early gene expression in phage lambda: effect of *tof* mutation on strand specific restrictions. *Virology* **65,** 374–384.

Takeda Y., Ogata K. and Matsubara K. (1975b). Genetic analysis of the *tof* gene in the phage λ genome. *Virology* **65,** 385–391.

Taketo A. (1973). Sensitivity of E.coli to viral nucleic acid. VI. Capacity of *dna* mutants and DNA polymerase-less mutants for multiplication of φA and φX174. *Molec. Gen. Genet.* **122,** 15–22.

Taketo A. (1975). Replication of φA and φX174 in E.coli mutants thermosensitive in DNA synthesis. *Molec. Gen. Genet.* **139,** 285–291.

Tate W.P. and Petersen G.B. (1974). Structure of the filamentous bacteriophages: orientation of the DNA molecule within the phage particle. *Virology* **62,** 16–25.

Tatum E.L. (1945). X-ray induced mutant strains of E.coli. *Proc. Natl. Acad. Sci. USA* **31,** 215–219.

Tatum E.L. and Lederberg J. (1947). Gene recombination in the bacterium E.coli. *J. Bacteriol.* **53,** 673–684.

Taylor A. (1971). Endopeptidase activity of phage lambda endolysin. *Nature New Biol.* **234,** 144–145.

Taylor A.L. and Adelberg E.A. (1961). Evidence for a closed linkage group in Hfr males of Escherichia coli K12. *Biochem. Biophys. Res. Comm.* **5,** 400–404.

Taylor A.L. and Thoman M.S. (1964). The genetic map of E.coli K12. *Genetics* **50,** 659–677.

Taylor A.L. and Trotter C.D. (1972). Linkage map of E.coli. *Bacteriol. Rev.* **36,** 504–524.

Taylor K., Hradecna Z. and Szybalski W. (1967). Asymmetric distribution of the transcribing regions on the complementary strands of phage λ DNA. *Proc. Natl. Acad. Sci. USA* **57,** 1618–1625.

Tessmam E.S. (1965). Complementation groups in phage S13. *Virology* **25,** 303–321.

Tessman E.S. (1966a). Mutants of phage S13 blocked in infectious DNA synthesis. *J. Mol. Biol.* **17,** 218–236.

Tessman E.S. and Shleser R. (1963). Genetic recombination between phages S13 and φX174. *Virology* **19,** 239–240.

Tessman E.S. and Tessman I. (1959). Genetic recombination in phage S13. *Virology* **7,** 465–467.

Tessman I. (1959). Mutagenesis in phages φX174 and T4 and properties of the genetic material. *Virology* **9,** 375–385.

Tessman I. (1966b). Genetic recombination of phage S13 in a recombination-deficient mutant of E.coli K12. *Biochem. Biophys. Res. Commun.* **22,** 169–174.

Tessman I. (1968). Selective stimulation of one of the mechanisms for genetic recombination of phage S13. *Science* **161,** 481–482.

Tessman I., Ishiwa H., Kumar S. and Baker R. (1967). Phage S13: a seventh gene. *Science* **156,** 824–825.

Tevethia M.J. and Caudill C.P. (1971). Relationship between competence for transformation of B.subtilis with native and single stranded DNA. *J. Bacteriol.* **106,** 808–811.

Tevethia M.J. and Mandel M. (1971). Effects of pH on transformation of B.subtilis with single stranded DNA. *J. Bacteriol.* **106,** 802–807.

Thach S.S. and Thach R.E. (1973). Mechanism of viral replication. I. Structure of replication complexes of R17 phage. *J. Mol. Biol.* **81,** 367–380.

Thirion J.P. and Hofnung M. (1972). On some genetic aspects of phage λ resistance in E.coli K12. *Genetics* **71,** 207–216.

Thomas CA. jr. (1963). The arrangement of nucleotide sequences in T2 and T5 DNA molecules. *Cold Spring Harbor Symp. Quant. Biol.* **28,** 395–396.

Thomas C.A. jr. and MacHattie L.A. (1964). Circular T2 DNA molecules. *Proc. Natl. Acad. Sci. USA* **52,** 1297–1301.

Thomas C.A. jr. and MacHattie L.A. (1967). The anatomy of viral DNA molecules. *Ann. Rev. Biochem.* **36,** 485–518.

Thomas J.O. (1974). Chemical linkage of the tail to the right hand end of phage λ DNA. *J. Mol. Biol.* **87,** 1–9.

Thomas M., Cameron J.R. and Davis R.W. (1974). Viable molecular hybrids of phage lambda and eucaryotic DNA. *Proc. Natl. Acad. Sci. USA* **71,** 4579–4583.

Thomas R. (1955). Recherches sur la cinetique des transformations bacteriennes. *Biochim. Biophys. Acta.* **18,** 467–481.

Thomas R. (1966). Control of development of temperate phages. I. Induction of prophage genes following heteroimmune superinfection. *J. Mol. Biol.* **22,** 79–95.

Thomas R. (1970). Control of development in temperate phages. III. Which prophage genes are and which are not trans-activable in the presence of immunity? *J. Mol. Biol.* **49,** 393–404.

Thompson B.J., Camien M.N. and Warner R.C. (1976). Kinetics of branch migration in double stranded DNA. *Proc. Natl. Acad. Sci. USA* **73,** 2299–2303.

Thompson B.J., Escarmis C., Parker B., Slater W.C., Doniger J., Tessman I. and Warner R.C. (1975). Figure-8 configuration of dimers S13 and $\phi$X 174 replicative form DNA. *J. Mol. Biol.* **91,** 409–420.

Tillack T.W. and Smith D.W.E. (1968). The effect of phage T2 infection on the synthesis of tRNA in E.coli. *Virology* **36,** 212–222.

Timmis K. and Marvin D.A. (1974a). Filamentous bacterial viruses. XV. DNA replication and abortive infection. *Virology* **59,** 281–292.

Timmis K. and Marvin D.A. (1974b). Filamentous bacterial viruses. XVI. Inherent temperature sensitivty of gene 5 protein and its involvement in abortive infection. *Virology* **59,** 293–300.

Timmis R., Cabello F. and Cohen S.N. (1974). Utilization of two distinct modes of replication by a hybrid plasmid constructed from separate replicons. *Proc. Natl. Acad. Sci. USA* **71,** 4556–4560.

Timmis K., Cabello F. and Cohen S.N. (1975). Cloning, isolation and characterization of replication regions of complex plasmid genomes. *Proc. Natl. Acad. Sci. USA* **72,** 2242–2246.

Tiraby J.G. and Fox M.S. (1973). Marker discrimination in transformation and mutation of pneumococcus. *Proc. Natl. Acad. Sci. USA* **70,** 3541–3545.

Tiraby J.G. and Fox M.S. (1974). Marker discrimination and mutagen induced alterations in pneumococcal transformation. *Genetics* **77,** 449–458.

Tiraby G. and Sicard M.A. (1973a). Integration efficiency in DNA-induced transformation of Pneumococcus. II. Genetic studies of mutant integrating all the markers with high efficiency. *Genetics,* **75,** 35–48.

Tiraby G. and Sicard A.M. (1973b). Integration efficiencies of spontaneous mutant alleles of amiA locus in pneumococcal transformation. *J. Bacteriol.* **116,** 1130–1135.

Tiraby G., Fox M.S. and Bernheimer H. (1975). Marker discrimination in DNA-mediated transformation of various Pneumococcus strains. *J. Bacteriol.* **121,** 608–618.

Tomasz A. (1965). Control of the competent state in Pneumococcus by a hormone-like cell product: an example of a new type of regulatory mechanism in bacteria. *Nature* **208,** 155–159.

Tomasz A. (1969). Some aspects of the competent state in genetic transformation. *Ann. Rev. Genetics* **3,** 217–232.

Tomasz A. (1970). Cellular metabolism in genetic transformation of pneumococci: requirement for protein synthesis during induction of competence. *J. Bacteriol.* **101,** 860–871.

Tomasz A. and Beiser S.M. (1965). Relationship between the competence antigen and the competence activator substance in pneumococci. *J. Bacteriol.* **90,** 1226–1232.

Tomasz A. and Hotchkiss R.D. (1964). Regulation of the transformability of pneumococcal cultures by macromolecular cell products. *Proc. Natl. Acad. Sci. USA* **51,** 480–487.

Tomasz A. and Mosser J.L. (1966). On the nature of the pneumococcal activator substance. *Proc. Natl. Acad. Sci. USA* **55,** 58–66.

Tomasz A. and Zanati E. (1971). Appearance of a protein ''agglutinin'' on the spheroplast membrane of pneumococci during induction of competence. *J. Bacteriol.* **105,** 1213–1215.

Tomasz A., Zanati E. and Ziegler R. (1971). DNA uptake during genetic transformation and the growing zone of the cell envelope. *Proc. Natl. Acad. Sci. USA* **68,** 1848–1852.

Tomizawa J.I. (1971). Functional cooperation of genes *O* and *P*. In The Bacteriophage Lambda, ed. A.D. Hershey. (*Cold Spring Harbor Laboratory: New York*). pp. 549–552.

Tomizawa J.I. and Anraku N. (1965). Molecular mechanisms of genetic recombination of phage. IV. Absence of polynucleotide interruption in DNA of T4 and lambda phage particles with special reference to heterozygosis. *J. Mol. Biol.* **11,** 509–527.

Tomizawa J. and Ogawa T. (1968). Replication of phage lambda DNA. *Cold Spring Harbor Symp. Quant. Biol.* **33,** 533–551.

Tomizawa J.I., Sakakibara Y. and Kakefuda T. (1974). Replication of colicin E1 plasmid DNA in cell extracts. Origin and direction of replication. *Proc. Natl. Acad. Sci. USA* **71,** 2260–2264.

Tonegawa S. and Hayashi M. (1970). Intermediates in the assembly of $\phi$X174. *J. Mol. Biol.* **48,** 219–242.

Tooze J. and Weber K. (1967). Isolation and characterization of amber mutants of phage R17. *J. Mol. Biol.* **28,** 311–330.

Travers A. (1969). Phage sigma factor for RNA polymerase. *Nature* **223,** 1107–1110.

Travers A. (1970a). RNA polymerase and T4 development. *Cold Spring Harbor Symp. Quant. Biol.* **35,** 241–252.

Travers A. (1970b). Postivie control of transcription by a phage sigma factor. *Nature* **225,** 1009–1012.

Travers A. and Burgess R.R. (1969). Cyclic reuse of the RNA polymerase sigma factor. *Nature* **222,** 537–540.

Trenkner E., Bonhoeffer F. and Gierer A. (1967). The fate of the protein component of phage fd during infection. *Biochem. Biophys. Res. Commun.* **28,** 932–939.

Tresguerres E.F., Nandadasa H.G. and Pritchard R.H. (1975). Suppression of initiation-negative strains of E.coli by integration of sex factor F. *J. Bacteriol.* **121,** 554–561.

Tromans W.J. and Horne R. W. (1961). The structure of bacteriophage $\phi$X174. *Virology* **15,** 1–7.

Truitt C.L. and Walker J.R. (1974). Growth of phages lambda, $\phi$X174 and M13 requires the *dnaZ* (previously *dnaH*) gene product of E.coli. *Biochem. Biophys. Res. Commun.* **61,** 1036–1042.

Tseng B.Y. and Marvin B.Y. (1972a). Filamentous bacterial viruses. V. Asymmetric replication of fd duplex DNA. *J. Virol.* **10,** 371–383.

Tseng B.Y. and Marvin D.A. (1972b). Filamentous bacterial viruses. VI. Role of fd gene 2 in DNA replication. *J. Virol.* **10,** 384–391.

Tutas D.J., Wehner J.M. and Koerner J.F. (1974). Unfolding of the host genome after infection of E.coli with phage T4. *J. Virol.* **13,** 548–550.

Uhlenhopp E.L., Zimm B.H. and Cummings D.J. (1974). Structural aberrations in the T-even phage. VI. Molecular weight of DNA from giant heads. *J. Mol. Biol.* **89,** 689–702.

Uhlin B.E. and Nordstrom K. (1975). Plasmid incompatibility and control of replication: copy mutants of the R factor R1 in E.coli K12. *J. Bacteriol.* **124,** 641–649.

Unger R.C. and Clark A.J. (1972). Interaction of the recombination pathways of phage λ and its host E.coli K12: effects on exonuclease V activity. *J. Mol. Biol.* **70,** 539–548.

Unger R.C., Echols H. and Clark A.J. (1972). Interaction of the recombination pathways of phage lambda and host E.coli: effects on λ recombination. *J. Mol. Biol.* **70,** 531–538.

Valentine R.C. and Wedel H. (1965). The extracellular stages of RNA phage infection. *Biochem. Biophys. Res. Commun.* **21,** 106–112.

Valentine R.C., Engelhardt D.L. and Zinder N.D. (1964). Host dependent mutants of the phage f2. II. Rescue and complementation of mutants. *Virology* **23,** 159–163.

Valenzuela M.S. and Inman R.B. (1975). Visualization of a novel junction in phage λ DNA. *Proc. Natl. Acad. Sci. USA* **72,** 3024–3028.

Valenzuela M.S., Freifelder D. and Inman R.B. (1976). Lack of a unique termination site for the first round of phage lambda DNA replication. *J. Mol. Biol.* **102,** 569–582.

Van de Vate C. and Symonds N. (1974). A stable duplication as an intermediate in the selection of deletion mutants of phage T4. *Genet Res.* **23,** 87–105.

Vandenberghe A., Min Jou W. and Fiers W. (1975). 3′ terminal nucleotide sequence (n = 361) of phage MS2 RNA. *Proc. Natl. Acad. Sci. USA* **72,** 2559–2562.

Van den Hondel C.A. and Schoenmakers J.G.G. (1975). Studies on phage M13 DNA. I. A cleavage map of the M13 genome. *Eur. J. Biochem.* **53,** 547–558.

Van den Hondel C.A., Konings R.N.H. and Schoenmakers J.G.G. (1975a). Regulation of gene activity in phage M13 DNA: coupled transcription and translation of purified genes and gene fragments. *Virology* **67,** 487–497.

Van den Hondel C.A., Weijers A., Konings R.N.H. and Schoenmakers J.G.G. (1975b). Studies on phage M13. II. The gene order of the M13 genome. *Eur. J. Biochem.* **53,** 559–567.

Van Dieijen G., Van der Laken C.J., Van Knippemberg P.H. and Van Duin J. (1975). Function of E.coli ribosomal protein S1 in translation of natural and synthetic mRNA. *J. Mol. Biol.* **93,** 351–366.

Vapnek D. and Rupp W.D. (1970). Asymmetric segregation of the complementary sex factor DNA strands during conjugation in E.coli. *J. Mol. Biol.* **53,** 287–303.

Vapnek D. and Spingler E. (1974). Asymmetry and extent of in vivo transcription of R plasmid DNA in E.coli. *J. Bacteriol.* **120,** 1274–1278.

Vapnek D., Lipman M.B. and Rupp W.D. (1971). Physical properties and mechanism of transfer of R factors in E.coli. *J. Bacteriol.* **108,** 508–514.

Venema G., Pritchard R.H. and Venema-Schröder T. (1965a). Fate of transforming deoxyribonucleic acid in Bacillus subtilis. *J. Bacteriol.* **89,** 1250–1255.

Venema G., Pritchard R.H. and Venema-Schröder T. (1965b). Properties of the newly introduced transforming deoxyribonucleic acid in Bacillus subtilis. *J. Bacteriol.* **90,** 343–346.

Verhoef C. and De Haan P.G. (1966). Genetic recombination in E.coli. I. Relation between linkage of unselected markers and map distance. *Mutat. Res.* **3,** 101–110.

Vermeulen C.A. and Venema G. (1974a). Electron microscope and autoradiographic study of ultrastructural aspects of competence and DNA absorption in B.subtilis: ultrastructure of competent and noncompetent cells and cellular changes during development of competence. *J. Bacteriol.* **118,** 334–341.

Vermeulan C.A. and Venema G. (1974b). Electron microscope and autoradiographic study of ultrastructural aspects of competence and DNA absorption in B.subtilis: localization of uptake and of transport of transforming DNA in competent cells. *J. Bacteriol.* **118,** 342–350.

Vidaver G.A. and Kozloff L.M. (1957). The rate of synthesis of DNA in E.coli B infected with T2$r^+$ phage. *J. Biol. Chem.* **225,** 335–347.

Vielmetter W., Bonhoeffer F. and Schutte A. (1968). Genetic evidence for transfer of a single DNA strand during bacterial conjugation. *J. Mol. Biol.* **37,** 81–86.

Vinuela E., Algranati I.D. and Ochoa S. (1967). Synthesis of virus specific proteins in E.coli infected with the RNA phage MS2. *Eur. J. Biochem.* **1,** 3–11.

Vinuela E., Algranati I.D., Feix G., Garwes D., Weissmann C. and Ochoa S. (1968). Virus specific proteins in E.coli infected with some amber mutants of phage MS2. *Biochim. Biophys. Acta* **155,** 558–565.

Visconti N. and Delbruck M. (1953). The mechanism of genetic recombination in phage. *Genetics* **38,** 5–33.

Voll M.J. and Goodgal S.H. (1961). Recombination during transformation in Hemophilus influenzae. *Proc. Natl. Acad. Sci. USA* **47,** 505–512.

Voll M.J. and Goodgal S.H. (1965). Loss of activity of transforming DNA after uptake by H.influenzae. *J. Bacteriol.* **90,** 873–883.

Vollenweider H.J., Koller T., Weber H. and Weissmann C. (1976). Physical mapping of Q$\beta$ replicase binding sites on Q$\beta$ RNA. *J. Mol. Biol.* **101,** 367–378.

Vovis G.F., Horiuchi K. and Zinder N.D. (1975). Endonuclease R.EcoRII restriction of phage fl DNA in vitro ordering of genes V and VII, location of an RNA promotor for gene VIII. *J. Virol.* **16,** 674–684.

Wada C. and Yura T. (1974). Phenethyl alcohol resistance in E.coli. III. A temperature sensitive mutation (*dnaP*) affecting DNA replication. *Genetics* **77,** 199–220.

Wahba A.J., Miller M.J., Niveleau A., Landers T.A., Carmichael G.C., Weber K., Hawley D.A. and Slobin L.I. (1974). Subunit I of Q$\beta$ replicase and 30S ribosomal proteins S1 of E.coli. *J. Biol. Chem.* **249,** 3314–3316.

Wake R.G., Kaiser A.D. and Inman R.B. (1972). Isolation and structure of phage $\lambda$ head mutant DNA. *J. Mol. Biol.* **64,** 519–540.

Walz A. and Pirrotta V. (1975). Sequence of the $P_R$ promotor of phage $\lambda$. *Nature* **254,** 119–121.

Walz A., Pirrotta V. and Ineichen K. (1976). Lambda repressor regulates the switch between $P_R$ and $P_{rm}$ promotors. *Nature* **262,** 665–669.

Wang J.C. and Kaiser A.D. (1973). Evidence that the cohesive ends of mature $\lambda$ DNA are generated by the gene A product. *Nature New Biology* **241,** 16–17.

Ward S. and Dickson R.C. (1971). Assembly of phage T4 tail fibers. III. Genetic control of the major tail fiber polypeptides. *J. Mol. Biol.* **62,** 479–492.

Ward S., Luftig R.B., Wilson J.H., Edleman H., Lyle H. and Wood W.B. (1970). Assembly of T4 tail fibers. II. Isolation and characterization of tail fiber precursors. *J. Mol. Biol.* **54,** 15–31.

Warner H.R., Snustad D.P., Jorgensen S.E. and Koerner J.F. (1970). Isolation of phage T4 mutants defective in the ability to degrade host DNA. *J. Virol.* **5,** 700–708.

Watanabe I. (1964). Persistent infection with an RNA bacteriophage. *Nihon Rinsho* **22,** 243–251.

Watanabe T. (1963a). Infective heredity of multiple drug resistance in bacteria. *Bacteriol. Rev.* **27,** 87–115.

Watanabe T. (1963b). Episome mediated transfer of drug resistance in Enterobacteriaceae. VI. High frequency resistance transfer system in E.coli. *J. Bacteriol.* **85,** 788–794.

Watanabe T. (1967). Evolutionary relationships of R factors with other episomes and plasmids. *Fed. Proc.* **26,** 23–28.

Watanabe T. and Fukasawa T. (1961a). Episome mediated transfer of drug resistance in Enterobacteriaceae. I. Transfer of resistance factors by conjugation. *J. Bacteriol.* **81,** 669–678.

Watanabe T. and Fukasawa T. (1961b). Episome mediated transfer of drug resistance in Enterobacteriaceae. II. Elimination of resistance factors with acridine dyes. *J. Bacteriol.* **81,** 679–683.

Watanabe T. and Fukasawa T. (1961c). Episome mediated transfer of drug resistance in Enterobacteriaceae. III. Transduction of resistance factors. *J. Bacteriol.* **82,** 202–209.

Watanabe T. and Fukasawa T. (1962). Episome mediated transfer of drug resistance in Enterobacteriaceae. IV. Interactions between resistance transfer factor and F factor in E.coli K12. *J. Bacteriol.* **83,** 727–735.

Watanabe T., Fukasawa T. and Takano T. (1962). Conversion of male bacteria of E.coli K12 to resistance to F phages by infection with the episome 'resistance transfer factor'. *Virology* **17,** 218–219.

Watanabe T., Sakaizumi S. and Furuse C. (1968). Superinfection with R factors by transduction in E.coli and S.typhimurium. *J. Bacteriol.* **96,** 1796–1802.

Watanabe T., Nishida H., Ogata C., Arai T. and Sato S. (1964). Episome-mediated transfer of drug resistance in Enterobacteriaceae. VII. Two types of naturally occurring R factors. *J. Bacteriol.* **88,** 716–726.

Watanabe T., Takano T., Arai T., Nishida H. and Sato S. (1966). Episome-mediated transfer of drug resistance in Enterobacteriaceae. X. Restriction and modification of phages by fi$^-$ R factors. *J. Bacteriol.* **92,** 477–486.

Watson J.D. (1972). Origin of concatemeric T7 DNA. *Nature New Biol.* **239,** 197–201.

Weber K. and Konigsberg W. (1975). Proteins of the RNA phages. In RNA Phages, ed. N.D. Zinder. (*Cold Spring Harbor Laboratory: New York*). pp. 51–84.

Weber H., Billeter M.A., Kahane S., Weissmann C., Hindley J. and Porter A. (1972). Molecular basis for repressor activity of Q$\beta$ replicase. *Nature New Biol.* **237,** 166–169.

Webster R.E., Engelhardt D.L., Zinder N.D. and Konigsberg W. (1967). Amber mutants and chain termination in vitro. *J. Mol. Biol.* **29,** 27–43.

Wechsler J.A. (1975). Genetic and phenotypic characterization of *dnaC* mutations. *J. Bacteriol.* **121,** 594–599.

Weidel W. and Primosigh J. (1958). Biochemical parallels between lysis by virulent phage and lysis by penicillin. *J. Gen. Microbiol.* **18,** 513–517.

Weigle J. (1966). Assembly of phage lambda in vitro. *Proc. Natl. Acad. Sci. USA* **55,** 1462–1466.

Weigle J., Meselson M. and Paigen K. (1959). Density alterations associated with transducing ability in the phage lambda. *J. Mol. Biol.* **1,** 379–386.

Weil J. (1969). Reciprocal and non-reciprocal recombination in phage λ. *J. Mol. Biol.* **43,** 351–355.

Weil J. and Signer E.R. (1968). Recombination in phage λ. II. Site specific recombination in phage λ. *J. Mol. Biol.* **34,** 273–279.

Weiner A.M. and Weber K. (1971). Natural readthrough at the UGA termination signal of Qβ coat protein cistron. *Nature New Biol.* **234,** 206–208.

Weiner A.M. and Weber K. (1973). A single UGA codon functions as a natural termination signal in the phage Qβ coat protein. *J. Mol. Biol.* **80,** 887–889.

Weiner J.H., Bertsch L. and Kornberg A. (1975). The DNA unwinding protein of E.coli. Properties and functions in replication. *J. Biol. Chem.* **250,** 1972–1980.

Weiner J.H., McMacken R. and Kornberg A. (1976). Isolation of an intermediate which precedes *dnaG* RNA polymerase participation in enzymatic replication of phage φX174 DNA. *Proc. Natl. Acad. Sci. USA* **73,** 752–756.

Weintraub S.B. and Frankel F.R. (1972). Identification of the T4 rIIB gene product as a membrane protein. *J. Mol. Biol.* **70,** 589–616.

Weisbeek P.J. and Van Arkel G.A. (1976). On the origin of φX174 replicative form DNA replication. *Virology* **72,** 72–79.

Weisbeek P.J. and Sinsheimer R.L. (1974). A DNA-protein complex involved in phage φX174 particle formation. *Proc. Natl. Acad. Sci. USA* **71,** 3054–3058.

Weisbeek P.J., Vereijken J.M., Baas P.D., Jansz H.S. and Van Arkel G.A. (1976). The genetic map of phage φX174 constructed with restriction enzyme fragments. *Virology* **72,** 61–71.

Weisberg R.A. and Gottesman M. (1971). The stability of int and xis functions. In The Bacteriophage Lambda, ed. A.D. Hershey. (*Cold Spring Harbor Laboratory, New York*). pp. 489–500.

Weiss S.B., Hsu W.-T., Fift J.W. and Scherberg N.H. (1968). Transfer RNA coded by the T4 phage genome. *Proc. Natl. Acad. Sci. USA* **61,** 114–121.

Weissbach A., Bartl P. and Salzman L.A. (1968). The structure of replicative lambda DNA—electron microscope studies. *Cold Spring Harbor Symp. Quant. Biol.* **33,** 525–532.

Weltzien H.U. and Jesaitis M.A. (1971). The nature of the colicin K receptor of E.coli Cullen. *J. Exp. Med.* **133,** 534–553.

Weissmann C., Colthart L. and Libonati M. (1968). Determination of viral plus and minus RNA strands by an isotope dilution assay. *Biochemistry* **7,** 865–874.

Weissmann C., Feix G. and Slor H. (1968). In vitro synthesis of phage RNA: the nature of the intermediates. *Cold Spring Harbor Symp. Quant. Biol.* **33,** 83–100.

Weissmann C., Borst T., Burdon R.H., Billeter M.A. and Ochoa S. (1964a). Replication of viral RNA. III. Double-stranded replicative form of $MS_2$ phage RNA. *Proc. Natl. Acad. Sci. USA* **51,** 682–690.

Weissmann C., Borst T., Burdon R.H., Billeter M.A. and Ochoa S. (1964b). Replication of viral RNA. IV. Properties of RNA synthetase and enzymatic synthesis of $MS_2$ phage RNA. *Proc. Natl. Acad. Sci. USA* **51,** 890–897.

Wendt L.W., Ippen K.A. and Valentine R.C. (1966). General properties of F pili. *Biochem. Biophys. Res. Commun.* **23,** 375–380.

Wensink P.C., Finnegan D.J., Donelson J.E. and Hogness D.S. (1974). A system for mapping DNA sequences in the chromosomes of D.melanogaster. *Cell* **3,** 315–326.

Westergaard O., Brutlag D. and Kornberg A. (1973). Initiation of DNA synthesis. IV. Incorporation of the RNA primer into the phage replicative form. *J. Biol. Chem.* **248,** 1361–1364.

Whitaker P.A., Yamada Y. and Nakada D. (1975). F factor mediated restriction of phage T7: synthesis of RNA and protein in T7 infected E.coli $F^-$ and $F^+$ cells. *J. Virol.* **16,** 1380–1390.

White R.L. and Fox M.S. (1974). On the molecular basis of high negative interference. *Proc. Natl. Acad. Sci. USA* **71,** 1544–1548.

White R.L. and Fox M.S. (1975a). Genetic heterozygosity in unreplicated phage λ recombinants. *Genetics* **81,** 33–50.

White R.L. and Fox M.S. (1975b). Genetic consequences of transfection with heteroduplex phage lambda DNA. *Molec. Gen. Genet.* **141,** 163–172.

Whitney E.N. (1971). The *tolC* locus in E.coli K12. *Genetics* **67,** 39–53.

Wiberg J.S. (1966). Mutants of phage T4 unable to cause breakdown of host DNA. *Proc. Natl. Acad. Sci. USA* **55,** 614–621.

Wiberg J.S., Dirksen M.L., Epstein R.H., Luria S.E. and Buchanan J.M. (1962). Early enzyme synthesis and its control in E.coli infected with some amber mutants of phage T4. *Proc. Natl. Acad. Sci. USA* **48,** 293–302.

Wiberg J.S., Mendelsohn S., Warner V., Hercules K., Aldrich C. and Munro J.L. (1973). SP62, a viable mutant of phage T4D defective in regulation of phage enzyme synthesis. *J. Virol.* **12,** 775–792.

Wickner R.B., Wright M., Wickner S. and Hurwitz J. (1972b). Conversion of φX174 and fd single stranded DNA to replicative forms in extracts of E.coli. *Proc. Natl. Acad. Sci. USA* **69,** 3233–3237.

Wickner S. and Hurwitz J. (1974). Conversion of φX174 viral DNA to double stranded form by purified E.coli proteins. *Proc. Natl. Acad. Sci. USA* **71,** 4120–4124.

Wickner S. and Hurwitz J. (1975a). Interaction of E.coli *dnaB* and *dnaC(D)* gene products in vitro. *Proc. Natl. Acad. Sci. USA* **72,** 921–925.

Wickner S. and Hurwitz J. (1975b). Association of φX 174 DNA-dependent ATPase activity with an E.coli protein, replication factor Y, required for in vitro synthesis of φX 174 DNA. *Proc. Natl. Acad. Sci. USA* **72,** 3342–3346.

Wickner S. and Hurwitz J. (1976). Involvement of E.coli *dnaZ* gene product in DNA elongation in vitro. *Proc. Nat. Acad. Sci. USA* **73,** 1053–1057.

Wickner S., Wright M. and Hurwitz J. (1973). Studies on in vitro DNA synthesis. Purification of the *dnaG* gene product from E.coli. *Proc. Natl. Acad. Sci. USA* **70,** 1613–1618.

Wickner S., Wright M. and Hurwitz J. (1974). Association of DNA-dependent and independent ribonucleoside triphosphatase activities with *dnaB* gene product of E.coli. *Proc. Natl. Acad. Sci. USA* **71,** 783–787.

Wickner S., Berkower I., Wright M. and Hurwitz J. (1973b). Studies on in vitro DNA synthesis: purification of the *dnaC* gene product containing *dnaD* activity from E.coli. *Proc. Natl. Acad. Sci. USA* **70,** 2369–2373.

Wickner W. and Kornberg A. (1973). DNA polymerase III star requires ATP to start synthesis on a primed DNA. *Proc. Natl. Acad. Sci.* **70,** 3679–3683.

Wickner W. and Kornberg A. (1974a). A novel form of RNA polymerase from E. coli. *Proc. Natl. Acad. Sci. USA* **71,** 4425–4428.

Wickner W. and Kornberg A. (1974b). A holoenzyme form of DNA polymerase III. Isolation and properties. *J. Biol. Chem.* **249,** 6244–6249.

Wickner W., Brutlag D., Schekman R. and Kornberg A. (1972a). RNA synthesis initiates in vitro conversion of M13 DNA to its replicative form. *Proc. Natl. Acad. Sci. USA* **69,** 965–969.

Wickner W., Schekman R., Geider K. and Kornberg A. (1973c). A new form of DNA polymerase III and a copolymerase replicate a long, single stranded primer-template. *Proc. Natl. Acad. Sci. USA* **70,** 1764–1767.

Wilkins B.M. (1969). Chromosome transfer from $Flac^+$ strains of E.coli K12 mutant at recA, recB or recC. *J. Bacteriol.* **98,** 599–604.

Willetts N.S. (1970). The interaction of an I-like R factor and transfer-deficient mutants of F*lac* in E.coli K12. *Molec. Gen. Genetics* **108,** 365–373.

Willetts N.S. (1971). Plasmid specificity of two proteins required for conjugation in E.coli K12. *Nature New Biol.* **230,** 183–185.

Willetts N.S. (1972a). The genetics of transmissible plasmids. *Ann. Rev. Genetics* **6,** 257–268.

Willetts N.S. (1972b). Location of the origin of transfer of the sex factor F. *J. Bacteriol.* **112,** 773–778.

Willetts N.S. (1974). Mapping loci for surface exclusion and incompatibility on the F factor of E.coli K12. *J. Bacteriol.* **118,** 778–782.

Willetts N.S. (1975). Recombination and the E.coli K12 sex factor F. *J. Bacteriol.* **121,** 36–43.

Willetts N.S. and Achtman M. (1972). Genetic analysis of transfer by the E.coli sex factor F, using P1 transductional complementation. *J. Bacteriol.* **110,** 843–851.

Willetts N.S. and Maule J. (1974). Interactions between the surface exclusion systems of some F-like plasmids. *Genet. Res.* **24,** 81–90.

Willetts N.S. and Paranchych W. (1974). Inhibition of Flac transfer by the $Fin^+$ I-like plasmid R62. *J. Bacteriol.* **120,** 101–105.

Willetts N.S., Maule J. and McIntire S. (1976). The genetic locations of *traO, finP* and *tra-4* on the E.coli K12 sex factor F. *Genet. Res.* **26,** 255–264.

Williams R.C. and Richards K.E. (1974). Capsid structure of phage lambda. *J. Mol. Biol.* **88,** 547–550.

Wilson J.H. and Abelson J.N. (1972). Phage T4 tRNA. II. Isolation and characterization of two phage coded nonsense suppressors. *J. Mol. Biol.* **69,** 57–74.

Wilson J.H. and Kells S. (1972). Phage T4 tRNA. I. Isolation and characterization of two phage coded nonsense suppressors. *J. Mol. Biol.* **69,** 39–56.

Wilson J.H., Luftig R.B. and Wood W.B. (1970). Interaction of phage T4 tail fiber components with a lipopolysaccharide fraction from E.coli. *J. Mol. Biol.* **51,** 423–434.

Wilson J.H., Kim J.S. and Abelson J.N. (1972). Phage T4 tRNA. III. Clustering of the genes for T4 tRNA. *J. Mol. Biol.* **71,** 547–556.

Wolfson J. and Dressler D. (1972). Regions of single stranded DNA in the growing points of replicating phage T7 chromosomes. *Proc. Natl. Acad. Sci. USA* **69,** 2682–2686.

Wolfson J., Dressler D. and Magazin M. (1971). Phage T7 DNA replication: a linear replicating intermediate. *Proc. Natl. Acad. Sci. USA* **69,** 499–504.

Wollman E.L. (1953). Sur le déterminisme génétique de la lysogénie. *Ann. Inst. Pasteur* **84,** 281–284.

Wollman E.L. and Jacob F. (1954). Etude genetique d'un bacteriophage tempere d'E.coli. II. Mecanisme de la recombinaison genetique. *Ann. Inst. Pasteur* **87,** 674–690.

Wollman E.L. and Jacob F. (1955). Sur le mecanisme du transfert de materiel genetique au cours de la recombinaison chez E.coli K12. *CR Acad. Sci. Paris* **240,** 2449–2451.

Wollman E.L. and Jacob F. (1957). Sur les processus de conjugaison et de recombinaison chez E.coli. II. La localisation chromosomique du prophage λ et les conséquences génétiques de l'induction zygotique. *Ann. Inst. Pasteur* **93,** 323–329.

Wollman E.L. and Jacob F. (1958). Sur les processus de conjugaison et de recombinaison chez E.coli. V. Le mecanisme du transfert de materiel genetique. *Ann. Inst. Pasteur* **95,** 641–666.

Wollman E.L., Jacob F. and Hayes W. (1956). Conjugation and genetic recombination in E.coli. *Cold Spring Harbor Symp. Quant. Biol.* **21,** 141–162.

Wood T.H. (1968). Effects of temperature, agitation and donor strain on chromosome transfer in E.coli K12. *J. Bacteriol.* **96,** 2077–2084.

Wood W.B. (1974). Bacteriophage T4. In Handbook of Genetics, ed. R.C. King. (*Plenum: New York*). pp. 327–331.

Wood W.B. and Henninger M. (1969). Attachment of tail fibers in phage T4 assembly: some properties of the reaction in vitro and its genetic control. *J. Mol. Biol.* **39,** 603–618.

Wovcha M.G., Tomich P.K., Chiu C.-S. and Greenberg G.R. (1973). Direct participation of dCMP hydroxymethylase in synthesis of phage T4 DNA. *Proc. Natl. Acad. Sci. USA* **70,** 2196–2220.

Wright M., Wickner S. and Hurwitz J. (1973). Studies on in vitro DNA synthesis. Isolation of dnaB gene product from E.coli. *Proc. Natl. Acad. Sci. USA* **70,** 3120–3124.

Wu R. (1970). Nucleotide sequence analysis of DNA. I. Partial sequence of the cohesive ends of bacteriophage λ and 186 DNA. *J. Mol. Biol.* **51,** 501–521.

Wu R. and Geiduschek E.P. (1975). The role of replication proteins in the regulation of phage T4 transcription. I. Gene 45 and hydroxymethyl-C-containing DNA. *J. Mol. Biol.* **96,** 513–538.

Wu R. and Kaiser A.D. (1967). Mapping the 5′ terminal nucleotides of the DNA of phage lambda and related phages. *Proc. Natl. Acad. Sci. USA* **57,** 170–177.

Wu R. and Kaiser A.D. (1968). Structure and base sequence in the cohesive ends of phage lambda DNA. *J. Mol. Biol.* **35,** 523–537.

Wu R. and Taylor E. (1971). Nucleotide sequence analysis of DNA. II. Complete nucleotide sequence of the cohesive ends of bacteriophage λ DNA. *J. Mol. Biol.* **57,** 491–511.

Wu R. and Yeh Y.-C. (1974). DNA arrested mutants of gene 59 of phage T4. II. Replicative intermediates. *Virology* **59,** 108–122.

Wu R., Geiduschek E.P. and Cascino A. (1975). The role of replication proteins in the regulation of phage T4 transcription. II. Gene 45 and late transcription uncoupled from replication. *J. Mol. Biol.* **96,** 539–562.

Wulff D.L. (1976). Lambda *cin-1* a new mutation which enhances lysogenization by phage lambda, and the genetic structure of the lambda *cy* region. *Genetics* **82,** 401–416.

Wurtz M., Kistler J. and Hohn T. (1976). Surface structure of in vitro assembled phage lambda polyheads. *J. Mol. Biol.* **101,** 39–56.

Yamada Y. and Nakada D. (1976). Early to late switch in phage T7 development; no translational discrimination between T7 early mRNA and late mRNA. *J. Mol. Biol.* **100,** 35–46.

Yamada Y., Whitaker P.A. and Nakada D. (1974). Early to late switch in phage T7 development. Functional decay of T7 early mRNA. *J. Mol. Biol.* **89,** 293–304.

Yamada Y., Whitaker P.A. and Nakada D. (1975). Chemical stability of phage T7 early mRNA. *J. Virol.* **16,** 1683–1687.

Yamamoto M. and Uchida H. (1973). Organization and function of phage T4 tail. I. Isolation of heat sensitive T4 tail mutants. *Virology* **52,** 234–245.

Yamamoto M. and Uchida H. (1975). Organization and function of the tail of phage T4. II. Structural control of the tail contraction. *J. Mol. Biol.* **92,** 207–224.

Yanagida M., DeRosier D.J. and Klug A. (1972). Structure of the tubular variants of the head of phage T4 (polyheads). II. Structural transition from a hexamer to a 6+1 morphological unit. *J. Mol. Biol.* **65,** 489–499.

Yanagida M., Boy de la Tour E., Alff-Steinberg C. and Kellenberger E. (1970). Studies on the morphopoiesis of phage T-even. VIII. Multilayered polyheads. *J. Mol. Biol.* **50,** 35–58.

Yarmolinsky M.B. (1971). Making and joining DNA ends. In The Bacteriophage Lambda, ed. A.D. Hershey. (*Cold Spring Harbor Laboratory, New York*). pp. 97–112.

Yarus M.J. and Sinsheimer R.L. (1967). The process of infection with $\phi$X174. XIII. Evidence for an essential bacterial site. *J. Virol.* **1,** 135–144.

Yen K.M. and Gussin G.N. (1973). Genetic characterization of a *prm*$^-$ mutant of phage $\lambda$. *Virology* **56,** 300–312.

Yoshida M. and Rudland P.S. (1972). Ribosomal binding of phage RNA with different components of initiation factor f3. *J. Mol. Biol.* **68,** 465–482.

Yoshimori R., Roulland-Dussoix D. and Boyer H.W. (1972). R factor controlled restriction and modification of DNA: restriction mutants. *J. Bacteriol.* **112,** 1275–1279.

Young E. and Sinsheimer R. (1964). Novel intracellular forms of lambda DNA. *J. Mol. Biol.* **10,** 562–564.

Young E. and Sinsheimer R. (1967). Vegetative phage lambda DNA. II. Physical characterization and replication. *J. Mol. Biol.* **30,** 165–200.

Young E. and Sinsheimer R. (1968). Vegetative lambda DNA. III. Pulse labeled components. *J. Mol. Biol.* **33,** 49–59.

Young R.A. and Blumenthal T. (1975). Phage Q$\beta$ RNA replicase. Subunit relationships determined by intramolecular crosslinking. *J. Biol. Chem.* **250,** 1829–1832.

Zachary A., Simon L.D. and Litwin S. (1976). Lambda head morphogenesis as seen in the electron microscope. *Virology* **72,** 429–442.

Zahler S.A. (1958). Some biological properties of phages S13 and $\phi$X174. *J. Bacteriol.* **75,** 310–315.

Zechel K., Bouché J.-P. and Kornberg A. (1975). Replication of phage G4. A novel and simple system for the initiation of DNA synthesis. *J. Biol. Chem.* **250,** 4684–4689.

Zelle M.R. and Lederberg J. (1951). Single cell isolations of diploid heterozygous E.coli. *J. Bacteriol.* **61,** 351–355.

Zichichi M.L. and Kellenberger G. (1963). Two distinct functions in the lysogenization process: the repression of phage multiplication and the incorporation of the prophages in the bacterial genome. *Virology* **19,** 450–460.

Zillig W., Fujiki H., Blum W., Janekovic D., Schweiger M., Rahmsdorf H.-J., Ponta H. and Hirsch-Kaufmann M. (1975). In vivo and in vitro phosphorylation of DNA dependent RNA polymerase of E.coli by phage T7 induced protein kinase. *Proc. Natl. Acad. Sci. USA* **72,** 2506–2510.

Zinder N.D. and Cooper S. (1964). Host dependent mutants of the phage f2. I. Isolation and primary classification. *Virology* **23,** 152–158.

Zinder N.D. and Lederberg J. (1952). Genetic exchange in Salmonella. *J. Bacteriol.* **64,** 679–699.

Zinger N.D., Valentine R.C., Roger M. and Stoeckenius W. (1963). fl: a rod shaped male specific phage that contains DNA. *Virology* **20,** 638–640.

Zissler J., Signer E. and Schaefer F. (1971a). The role of recombination in growth of phage λ. I. The gamma gene. In The Bacteriophage Lambda, ed. A.D. Hershey. (*Cold Spring Harbor Laboratory, New York*). pp. 455–468.

Zissler J., Signer E. and Schaefer F. (1971b). The role of recombination in growth of phage λ. II. Inhibition of growth by prophage P22. In The Bacteriophage Lambda, ed. A.D. Hershey. (*Cold Spring Harbor Laboratory, New York*). pp. 469–476.

Zuccarelli A.J., Benbow R.M. and Sinsheimer R.L. (1972). Deletion mutants of phage ϕX174. *Proc. Natl. Acad. Sci. USA* **69,** 1905–1910.

# INDEX